Methods in Enzymology

Volume 109

HORMONE ACTION

Part I

Peptide Hormones

METHODS IN ENZYMOLOGY

EDITORS-IN-CHIEF

Sidney P. Colowick   Nathan O. Kaplan

Methods in Enzymology

Volume 109

Hormone Action

Part I

Peptide Hormones

EDITED BY

Lutz Birnbaumer

DEPARTMENT OF CELL BIOLOGY
BAYLOR COLLEGE OF MEDICINE
HOUSTON, TEXAS

Bert W. O'Malley

DEPARTMENT OF CELL BIOLOGY
BAYLOR COLLEGE OF MEDICINE
HOUSTON, TEXAS

1985

ACADEMIC PRESS, INC.

(*Harcourt Brace Jovanovich, Publishers*)

Orlando San Diego New York London
Toronto Montreal Sydney Tokyo

ACADEMIC PRESS, INC.
Orlando, Florida 32887

United Kingdom Edition published by
ACADEMIC PRESS INC. (LONDON) LTD.
24–28 Oval Road, London NW1 7DX

LIBRARY OF CONGRESS CATALOG CARD NUMBER: 54-9110

ISBN 0-12-182009-2

PRINTED IN THE UNITED STATES OF AMERICA

85 86 87 88 9 8 7 6 5 4 3 2 1

Table of Contents

Section I. Receptor Assays

Section II. Identification of Receptor Proteins

Section III. Methods for Identification of Internalized Hormones and Hormone Receptors

Section IV. Preparation of Hormonally Responsive Cells and Cell Hybrids

Section V. Purification of Membrane Receptors and Related Techniques

Section VI. Assays of Hormonal Effects and Related Functions

Section VII. Antibodies in Hormone Action

Section VIII. General Methods

Contributors to Volume 109

Article numbers are in parentheses following the names of contributors.
Affiliations listed are current.

GRETI AGUILERA (10), *Endocrinology and Reproduction Research Branch, National Institute of Child Health and Human Development, National Institutes of Health, Bethesda, Maryland 20205*

G. AILHAUD (33), *Centre de Biochimie du CNRS, Faculte des Sciences, Parc Valrose, 06034 Nice Cédex, France*

CLAUDE D. ARNAUD (4, 5), *Department of Medicine, Veterans Administration Medical Center, and Departments of Medicine and Physiology, University of California, San Francisco, California 94121*

GILBERT ASHWELL (19), *National Institute of Arthritis, Diabetes, and Digestive and Kidney Diseases, National Institutes of Health, Bethesda, Maryland 20205*

ALBERT BAUKAL (10), *Endocrinology and Reproduction Research Branch, National Institute of Child Health and Human Development, National Institutes of Health, Bethesda, Maryland 20205*

ERNST BAYER (26), *Physiologisch-Chemisches Institut der Universität, D-8700 Würzburg, Federal Republic of Germany*

SUZANNE K. BECKNER (30, 31), *Laboratory of Cellular and Developmental Biology, National Institute of Arthritis, Diabetes, Digestive and Kidney Diseases, National Institutes of Health, Bethesda, Maryland 20205*

HAROLD R. BEHRMAN (25), *Reproductive Biology Section, Department of Obstetrics and Gynecology and Pharmacology, Yale University School of Medicine, New Haven, Connecticut 06510*

TIMOTHY P. BENDER (55), *National Cancer Institute/Navy Medical Oncology Branch, Naval Hospital, Bethesda, Maryland 20814*

JOHN J. M. BERGERON (18), *Department of Anatomy, McGill University Medical School, Montreal, Quebec H3A 2B2, Canada*

L. BIRNBAUMER (1, 38, 45), *Department of Cell Biology, Baylor College of Medicine, Houston, Texas 77030*

P. F. BLACKMORE (43), *Howard Hughes Medical Institute, and the Department of Physiology, Vanderbilt University School of Medicine, Nashville, Tennessee 37232*

DANIEL F. BOWEN-POPE (8), *Department of Pathology, School of Medicine, University of Washington, Seattle, Washington 98195*

BARRY L. BROWN (24), *Department of Human Metabolism and Clinical Biochemistry, University of Sheffield Medical School, Sheffield S10 2RX, England*

MICHAEL S. BROWN (36), *Department of Molecular Genetics, University of Texas Health Science Center at Dallas, Dallas, Texas 75235*

SUSAN A. BUHROW (62), *Department of Biological Chemistry, The John Hopkins University School of Medicine, Baltimore, Maryland 21205*

ROBERT E. CANFIELD (50), *Department of Medicine, Columbia University College of Physicians and Surgeons, New York, New York 10032*

GRAHAM CARPENTER (9), *Department of Biochemistry and Division of Dermatology, Vanderbilt University School of Medicine, Nashville, Tennessee 37232*

KEVIN J. CATT (10), *Endocrinology and Reproduction Research Branch, National Institute of Child Health and Human Development, National Institutes of Health, Bethesda, Maryland 20205*

JANICE YANG CHOU (34), *Human Genetics Branch, National Institute of Child Health and Human Development, National Institutes of Health, Bethesda, Maryland 20205*

JUAN CODINA (38, 45), *Department of Cell Biology, Baylor College of Medicine, Houston, Texas 77030*

PEDRO CUATRECASAS (35), *Department of Molecular Biology, The Wellcome Research Laboratories, Research Triangle Park, North Carolina 27709*

MICHAEL P. CZECH (14), *Department of Biochemistry, University of Massachusetts Medical School, Worcester, Massachusetts 01605*

FREDERICK J. DARFLER (30, 31), *Laboratory of Cellular and Developmental Biology, National Institute of Arthritis, Diabetes, Digestive and Kidney Diseases, National Institutes of Health, Bethesda, Maryland 20205*

WILLIAM H. DAUGHADAY (59), *Metabolism Division, Department of Medicine, Washington University School of Medicine, St. Louis, Missouri 63110*

PIERRE N. E. DE GRAAN (11), *Division of Neurobiology, Rudolf Magnus Institute for Pharmacology, and Institute of Molecular Biology, State University Utrecht, NL-3508 TB Utrecht, The Netherlands*

ROBERT B. DICKSON (21), *Medical Breast Cancer Section, Medicine Branch, Division of Cancer Treatment, National Cancer Institute, National Institutes of Health, Bethesda, Maryland 20205*

JEAN DJIANE (12, 52), *Laboratoire de Physiologie de la Lactation, Institut National de la Recherche Agronomique, CNZR, 78350 Jouy-en-Josas, France*

PAULINE R. M. DOBSON (24, 63), *Department of Human Metabolism and Clinical Biochemistry, University of Sheffield Medical School, Sheffield S10 2RX, England*

ISABELLE DUSANTER-FOURT (52), *Laboratoire de Physiologie de la Lactation, Institut National de la Recherche Agronomique, CNRZ, 78350 Jouy-en-Josas, France*

ALEX N. EBERLE (11), *Laboratory of Endocrinology, Department of Research, University Hospital and University Children's Hospital, CH-4031 Basel, Switzerland*

PAUL H. EHRLICH (50), *Sandoz Research Institute, Sandoz Inc., East Hanover, New Jersey 07936*

JOSEPH EICHBERG (41), *Department of Biochemical and Biophysical Sciences, University of Houston, Houston, Texas 77004*

W. H. EVANS (20), *National Institute for Medical Research, Mill Hill, London NW7 1AA, England*

J. H. EXTON (43), *Howard Hughes Medical Institute, and the Department of Physiology, Vanderbilt University School of Medicine, Nashville, Tennessee 37232*

J. N. FAIN (39), *Section of Biochemistry, Division of Biology and Medicine, Brown University, Providence, Rhode Island 02192*

DAVID R. FERRY (42), *Rudolf Buchheim-Institut für Pharmakologie, Justus Liebig Universität, Giessen, D-63 Giessen, Federal Republic of Germany*

JAMES B. FIELD (45), *Division of Endocrinology, Department of Medicine, Baylor College of Medicine, Houston, Texas 77030*

FRANCES M. FINN (37), *Protein Research Laboratory, University of Pittsburgh, School of Medicine, Pittsburgh, Pennsylvania 15261*

C. FOREST (33), *Centre de Biochimie du CNRS, Faculte des Sciences, Parc Valrose, 06034 Nice Cédex, France*

H. G. FRIESEN (54), *Department of Physiology, Faculty of Medicine, University of Manitoba, Winnipeg, Manitoba R3E OW3, Canada*

JERRY D. GARDNER (7, 23), *Digestive Diseases Branch, National Institute of Arthritis, Diabetes, and Digestive and Kidney Diseases, National Institutes of Health, Bethesda, Maryland 20205*

THOMAS GLASER (26), *Troponwerke, Neurobiology Department, D-5000 Köln, Federal Republic of Germany*

HARTMUT GLOSSMANN (10, 42), *Institut für Biochemische Pharmakologie, A-6020 Innsbruck, Austria*

JOSEPH L. GOLDSTEIN (36), *Department of Molecular Genetics, University of Texas Health Science Center at Dallas, Dallas, Texas 75235*

P. GRIMALDI (33), *Centre de Biochimie du CNRS, Faculte des Sciences, Parc Valrose, 06034 Nice Cédex, France*

BERND HAMPRECHT (26, 27), *Physiologisch-Chemisches Institut der Universität, D-8700 Würzburg, Federal Republic of Germany*

JOHN A. HANOVER (21), *Laboratory of Molecular Biology, National Cancer Institute, National Institutes of Health, Bethesda, Maryland 20205*

JOE HARFORD (19), *National Institute of Arthritis, Diabetes, and Digestive and Kidney Diseases, National Institutes of Health, Bethesda, Maryland 20205*

CHARLES A. HARRINGTON (41), *Analytical Neurochemistry Laboratory, Texas Research Institute of Mental Sciences, Houston, Texas 77030*

JOSE A. HEDO (47), *Diabetes Branch, National Institute of Arthritis, Diabetes, Digestive and Kidney Diseases, National Institutes of Health, Bethesda, Maryland 20205*

JOHN T. HERBERG (2, 17), *Department of Cell Biology, Baylor College of Medicine, Houston, Texas 77030*

JOHN D. HILDEBRANDT (38, 45), *Worcester Foundation for Experimental Biology, Schrewsbury, Massachusetts 01545*

KLAUS H. HOFMANN (37), *Protein Research Laboratory, University of Pittsburgh, School of Medicine, Pittsburgh, Pennsylvania 15261*

LOUIS-MARIE HOUDEBINE (52), *Laboratoire de Physiologie de la Lactation, Institut National de la Recherche Agronomique, CNRZ, 78350 Jouy-en-Josas, France*

RENÉ E. HUMBEL (60), *Biochemisches Institut, University of Zürich, CH-8057 Zürich, Switzerland*

RAVI IYENGAR (2, 17), *Department of Cell Biology, Baylor College of Medicine, Houston, Texas 77030*

STEVEN JACOBS (35), *Department of Molecular Biology, The Wellcome Research Laboratories, Research Triangle Park, North Carolina 27709*

LEONARD JARETT (15), *Department of Pathology and Laboratory Medicine, University of Pennsylvania, Philadelphia, Pennsylvania 19104*

CAROLE L. JELSEMA (40), *Laboratory of Cell Biology, National Institutes of Mental Health, National Institutes of Health, Bethesda, Maryland 20205*

ROBERT T. JENSEN (7, 23), *Digestive Diseases Branch, National Institute of Arthritis, Diabetes, and Digestive and Kidney Diseases, National Institutes of Health, Bethesda, Maryland 20205*

INHAE JI (16), *Department of Biochemistry, University of Wyoming, Laramie, Wyoming 82071*

TAE H. JI (16), *Department of Biochemistry, University of Wyoming, Laramie, Wyoming 82071*

LEONARD R. JOHNSON (6), *Department of Physiology and Cell Biology, University of Texas Medical School, Houston, Texas 77025*

MARK L. JOHNSON (46), *Department of Biochemistry, St. Jude Children's Research Hospital, Memphis, Tennessee 38101*

C. RONALD KAHN (47, 48), *Research Division, Joslin Diabetes Center, and Department of Medicine, Brigham and Women's Hospital, and Harvard Medical School, Boston, Massachusetts 02215*

MASATO KASUGA (48), *Third Department of Internal Medicine, School of Medicine, University of Tokyo, Tokyo, Japan*

MASAO KATOH (12, 52), *Laboratory of Molecular Endocrinology, Royal Victoria Hospital, Montreal H3A 1A1 Quebec, Canada*

PAUL A. KELLY (12, 52), *Laboratory of Molecular Endocrinology, Royal Victoria Hospital, Montreal H3A 1A1 Quebec, Canada*

MASOOD N. KHAN (18), *Department of Medicine, McGill University Medical School, Montreal, Quebec H3A 1A1, Canada*

C. SHIRLEY LIN (49), *Department of Immunology and Inflammation Research, Merck Sharp and Dohme Research Laboratories, Rahway, New Jersey 07065*

MICHAEL C. LIN (30, 31), *Laboratory of Cellular and Developmental Biology, National Institute of Arthritis, Diabetes, Digestive and Kidney Diseases, National Institutes of Health, Bethesda, Maryland 20205*

FRIDOLIN LÖFFLER (27), *Physiologisch-Chemisches Institut der Universität, D-8700 Würzburg, Federal Republic of Germany*

JUDITH L. LUBORSKY (25), *Reproductive Biology Section, Department of Obstetrics and Gynecology, Yale University School of Medicine, New Haven, Connecticut 06510*

VINCENT C. MANGANIELLO (40), *Laboratory of Cellular Metabolism, National Heart, Lung, and Blood Institute, National Institutes of Health, Bethesda, Maryland 20205*

P. MANJUNATH (56), *Reproduction Research Laboratory, Clinical Research Institute of Montreal, Montreal, Quebec H2W 1R7, Canada*

BERNICE MARCUS-SAMUELS (51), *Diabetes Branch, National Institute of Arthritis, Diabetes, Digestive and Kidney Diseases, National Institutes of Health, Bethesda, Maryland 20205*

IDA K. MARIZ (59), *Metabolism Division, Department of Medicine, Washington University School of Medicine, St. Louis, Missouri 63110*

JOAN MASSAGUÉ (14), *Department of Biochemistry, University of Massachusetts Medical School, Worcester, Massachusetts 01605*

RAFAEL MATTERA (45), *Department of Cell Biology, Baylor College of Medicine, Houston, Texas 77030*

J. MAXWELL MCKENZIE (53), *Department of Medicine, University of Miami School of Medicine, Miami, Florida 33101*

R. MOLENAAR (22), *Department of Biochemistry II, Medical Faculty, Erasmus University, 3000 DR Rotterdam, The Netherlands*

JOEL MOSS (40), *Laboratory of Cellular Metabolism, National Heart, Lung, and Blood Institute, National Institutes of Health, Bethesda, Maryland 20205*

WILLIAM R. MOYLE (50), *Department of Obstetrics and Gynecology, University of Medicine and Dentistry of New Jersey, Piscataway, New Jersey 08854*

R. NEGREL (33), *Centre de Biochimie du CNRS, Faculte des Sciences, Parc Valrose, 06034 Nice Cédex, France*

ROBERT A. NISSENSON (4, 5), *Department of Medicine, Veterans Administration Medical Center, and University of California, San Francisco, California 94121*

SAMUEL R. NUSSBAUM (49), *Endocrine Unit, Massachusetts General Hospital, and Harvard Medical School, Boston, Massachusetts 02114*

THOMAS F. PARSONS (57), *International Genetic Engineering, Inc., Santa Monica, California 90404*

IRA PASTAN (21), *Laboratory of Molecular Biology, National Cancer Institute, National Institutes of Health, Bethesda, Maryland 20205*

JOHN G. PIERCE (57), *Department of Biological Chemistry, UCLA School of Medicine, Los Angeles, California 90024*

BARRY I. POSNER (18), *Department of Medicine, McGill University Medical School, Montreal, Quebec H3A 1A1, Canada*

JOHN T. POTTS, JR. (49), *Massachusetts General Hospital, and Harvard Medical School and Medical Services, Boston, Massachusetts 02114*

FRIEDRICH PROPST (26), *Frederick Cancer Research Facility, Frederick, Maryland 21701*

ELAINE W. RAINES (58), *Department of Pathology, University of Washington, Seattle, Washington 98195*

GEORG REISER (26), *Physiologisch-Chemisches Institut der Universität, D-7400 Tübingen, Federal Republic of Germany*

FERNANDO A. P. RIBEIRO-NETO (45), *Division of Endocrinology, Department of Medicine, Baylor College of Medicine, Houston, Texas 77030*

JOHN R. RODGERS (46), *Howard Hughes Medical Institute, Baylor College of Medicine, Houston, Texas 77030*

FRANCISCO J. ROJAS (1), *Department of Obstetrics and Gynecology, University of Texas Health Sciences Center at San Antonio, San Antonio, Texas 78284*

F. F. G. ROMMERTS (22), *Department of Biochemistry II, Medical Faculty, Erasmus University, 3000 DR Rotterdam, The Netherlands*

JEFFREY M. ROSEN (46), *Department of Cell Biology, Baylor College of Medicine, Houston, Texas 77030*

MICHAEL ROSENBLATT (49), *Biological Research, Merck Sharp and Dohme Research Laboratories, West Point, Pennsylvania 19486*

ALAN S. ROSENTHAL (49), *Department of Immunology and Inflammation Research, Merck Sharp and Dohme Research Laboratories, Rahway, New Jersey 07065*

WALTER ROSENTHAL (38), *Department of Cell Biology, Baylor College of Medicine, Houston, Texas 77030*

RUSSELL ROSS (8, 58), *Departments of Pathology and Biochemistry, School of Medicine, University of Washington, Seattle, Washington 98195*

M. R. SAIRAM (56), *Reproduction Research Laboratory, Clinical Research Institute of Montreal, Montreal, Quebec H2W 1R7, Canada*

SENKITI SAKAI (12), *Department of Animal Breeding, Faculty of Agriculture, University of Tokyo, Tokyo 113, Japan*

D. M. SALMON (32), *Hormones Division, National Institute for Biological Standards and Control, Holly Hill, Hampstead, London NW3 6RB, England*

BERNARD P. SCHIMMER (29), *Banting and Best Department of Medical Research, University of Toronto, Toronto, Ontario M5G 1L6, Canada*

WOLFGANG J. SCHNEIDER (36), *Department of Molecular Genetics, University of Texas Health Science Center at Dallas, Dallas, Texas 75235*

MICHAEL SCHRAMM (28), *Department of Biological Chemistry, The Hebrew University of Jerusalem, 91904 Jerusalem, Israel*

JOYCE A. SCHROER (55), *Laboratory of Immunogenetics, National Institute of Allergy and Infectious Diseases, National Institutes of Health, Bethesda, Maryland 20205*

D. SCHULSTER (32), *Hormones Division, National Institute for Biological Standards and Control, Holly Hill, Hampstead, London NW3 6RB, England*

RONALD D. SEKURA (38, 44, 45), *Laboratory of Developmental and Molecular Immunity, National Institute of Child Health and Human Development, National Institutes of Health, Bethesda, Maryland 20205*

ERIC M. SHOOTER (3), *Department of Neurobiology, Stanford University School of Medicine, Stanford, California 94305*

J. S. A. SIMPSON (54), *Diagnostic Assays Department, Hazleton Biotechnologies Corporation, Vienna, Virginia 22180*

ROBERT M. SMITH (15), *Department of Pathology and Laboratory Medicine, University of Pennsylvania, Philadelphia, Pennsylvania 19104*

JAMES V. STAROS (62), *Department of Biochemistry, Vanderbilt University, School of Medicine, Nashville, Tennessee 37232*

SONIA STEINER (28), *Department of Biological Chemistry, The Hebrew University of Jerusalem, 91904 Jerusalem, Israel*

PHILIP G. STRANGE (63), *Department of Biochemistry, The Medical School, Queen's Medical Center, Nottingham NG7 2UH, England*

THOMAS W. STRICKLAND (57), *AMGen, Thousand Oaks, California 91320*

MARJORIE E. SVOBODA (61), *Department of Pediatrics, Division of Pediatric Endocrinology, University of North Carolina School of Medicine, Chapel Hill, North Carolina 27514*

SIMEON I. TAYLOR (51), *Diabetes Branch, National Institute of Arthritis, Diabetes, Digestive and Kidney Diseases, National Institutes of Health, Bethesda, Maryland 20205*

ANNE P. TEITELBAUM (4, 5), *Department of Medicine, University of California, San Francisco, California 94121*

LISA H. UNDERHILL (51), *Diabetes Branch, National Institutes of Arthritis, Diabetes, Digestive and Kidney Diseases, National Institutes of Health, Bethesda, Maryland 20205*

RONALD D. VALE (3), *Department of Neurobiology, Stanford University School of Medicine, Stanford, California 94305*

H. J. VAN DER MOLEN (22), *Department of Biochemistry II, Medical Faculty, Erasmus University, 3000 DR Rotterdam, The Netherlands*

JUDSON J. VAN WYK (61), *Department of Pediatrics, Division of Pediatric Endocrinology, University of North Carolina School of Medicine, Chapel Hill, North Carolina 27514*

M. A. WALLACE (39), *Section of Biochemistry, Division of Biology and Medicine, Brown University, Providence, Rhode Island 02912*

MORRIS F. WHITE (48), *Research Division, Joslin Diabetes Center, Boston, Massachusetts 02215*

MARK C. WILLINGHAM (21), *Laboratory of Molecular Biology, National Cancer Institute, National Institutes of Health, Bethesda, Maryland 20205*

CLEMENT W. T. YEUNG (13), *Playfair Neuroscience Unit and Department of Biochemistry, University of Toronto, Toronto, Ontario M5T 2S8, Canada*

CECIL C. YIP (13), *Banting and Best Department of Medical Research, University of Toronto, Toronto, Ontario M5G 1L6, Canada*

MARGARITA ZAKARIJA (53), *Department of Medicine, University of Miami School of Medicine, Miami, Florida 33101*

PETER P. ZUMSTEIN (60), *Dana-Farber Cancer Institute, Boston, Massachusetts 02115*

Preface

The field of hormone action is undoubtedly one of the fastest growing areas of biological science. A rough assessment of the rate of growth of this field, as determined from an evaluation of journal articles and programs of national meetings, leads us to the surprising conclusion that an approximate tenfold expansion has occurred over the last decade. Research in hormone action not only has grown into a dominant effort in endocrinology and reproductive biology, but has also captured a large share of the more general disciplines of biochemistry, cell biology, and molecular biology. This development has occurred because of the dynamic aspects of the field and the increasing interest inherent in the new discipline of regulatory biology. None of these advances could have occurred without a widespread concurrent development of new techniques or adaptation of relevant techniques from other disciplines for studies on hormones and the mechanisms involved in hormone action.

In this volume of *Methods in Enzymology* a series of techniques and methods of study as they relate to research on peptide hormones and their mechanisms of action have been compiled. It has been subdivided into sections on receptor assays, identification of receptor proteins on cell surfaces, methods for the identification of internalized hormones and hormone receptors, preparation of hormonally responsive cells and cell hybrids, purification of membrane receptors, assays for hormonal effects and related functions, the use of antibodies in the study of hormone action, and, finally, into a section on general methods which includes a variety of methods for the modification of protein hormones, purification, some of the newer growth factors, as well as other methods of interest.

As always, the techniques gathered are not all-inclusive, some of the research areas are presented in a fragmentary way, and, undoubtedly, important methods have escaped our attention. We hope, however, that the approaches and methods that we have collected in this volume will be, as they have already been, of very general applicability and an aid both to researchers and to the more rapid advancement of the field of hormones and hormone action.

LUTZ BIRNBAUMER
BERT W. O'MALLEY

METHODS IN ENZYMOLOGY

EDITED BY

Sidney P. Colowick and Nathan O. Kaplan

VANDERBILT UNIVERSITY
SCHOOL OF MEDICINE
NASHVILLE, TENNESSEE

DEPARTMENT OF CHEMISTRY
UNIVERSITY OF CALIFORNIA
AT SAN DIEGO
LA JOLLA, CALIFORNIA

METHODS IN ENZYMOLOGY

EDITORS-IN-CHIEF

Sidney P. Colowick and Nathan O. Kaplan

VOLUME VIII. Complex Carbohydrates
Edited by ELIZABETH F. NEUFELD AND VICTOR GINSBURG

VOLUME IX. Carbohydrate Metabolism
Edited by WILLIS A. WOOD

VOLUME X. Oxidation and Phosphorylation
Edited by RONALD W. ESTABROOK AND MAYNARD E. PULLMAN

VOLUME XI. Enzyme Structure
Edited by C. H. W. HIRS

VOLUME XII. Nucleic Acids (Parts A and B)
Edited by LAWRENCE GROSSMAN AND KIVIE MOLDAVE

VOLUME XIII. Citric Acid Cycle
Edited by J. M. LOWENSTEIN

VOLUME XIV. Lipids
Edited by J. M. LOWENSTEIN

VOLUME XV. Steroids and Terpenoids
Edited by RAYMOND B. CLAYTON

VOLUME XVI. Fast Reactions
Edited by KENNETH KUSTIN

VOLUME XVII. Metabolism of Amino Acids and Amines (Parts A and B)
Edited by HERBERT TABOR AND CELIA WHITE TABOR

VOLUME XVIII. Vitamins and Coenzymes (Parts A, B, and C)
Edited by DONALD B. MCCORMICK AND LEMUEL D. WRIGHT

VOLUME XIX. Proteolytic Enzymes
Edited by GERTRUDE E. PERLMANN AND LASZLO LORAND

VOLUME XXXIV. Affinity Techniques (Enzyme Purification: Part B)
Edited by WILLIAM B. JAKOBY AND MEIR WILCHEK

VOLUME XXXV. Lipids (Part B)
Edited by JOHN M. LOWENSTEIN

VOLUME XXXVI. Hormone Action (Part A: Steroid Hormones)
Edited by BERT W. O'MALLEY AND JOEL G. HARDMAN

VOLUME XXXVII. Hormone Action (Part B: Peptide Hormones)
Edited by BERT W. O'MALLEY AND JOEL G. HARDMAN

VOLUME XXXVIII. Hormone Action (Part C: Cyclic Nucleotides)
Edited by JOEL G. HARDMAN AND BERT W. O'MALLEY

VOLUME XXXIX. Hormone Action (Part D: Isolated Cells, Tissues, and Organ Systems)
Edited by JOEL G. HARDMAN AND BERT W. O'MALLEY

VOLUME XL. Hormone Action (Part E: Nuclear Structure and Function)
Edited by BERT W. O'MALLEY AND JOEL G. HARDMAN

VOLUME XLI. Carbohydrate Metabolism (Part B)
Edited by W. A. WOOD

VOLUME XLII. Carbohydrate Metabolism (Part C)
Edited by W. A. WOOD

VOLUME XLIII. Antibiotics
Edited by JOHN H. HASH

VOLUME XLIV. Immobilized Enzymes
Edited by KLAUS MOSBACH

VOLUME XLV. Proteolytic Enzymes (Part B)
Edited by LASZLO LORAND

VOLUME XLVI. Affinity Labeling
Edited by WILLIAM B. JAKOBY AND MEIR WILCHEK

VOLUME XLVII. Enzyme Structure (Part E)
Edited by C. H. W. HIRS AND SERGE N. TIMASHEFF

VOLUME XLVIII. Enzyme Structure (Part F)
Edited by C. H. W. HIRS AND SERGE N. TIMASHEFF

VOLUME XLIX. Enzyme Structure (Part G)
Edited by C. H. W. HIRS AND SERGE N. TIMASHEFF

VOLUME L. Complex Carbohydrates (Part C)
Edited by VICTOR GINSBURG

VOLUME LI. Purine and Pyrimidine Nucleotide Metabolism
Edited by PATRICIA A. HOFFEE AND MARY ELLEN JONES

VOLUME LII. Biomembranes (Part C: Biological Oxidations)
Edited by SIDNEY FLEISCHER AND LESTER PACKER

VOLUME LIII. Biomembranes (Part D: Biological Oxidations)
Edited by SIDNEY FLEISCHER AND LESTER PACKER

VOLUME LIV. Biomembranes (Part E: Biological Oxidations)
Edited by SIDNEY FLEISCHER AND LESTER PACKER

VOLUME LV. Biomembranes (Part F: Bioenergetics)
Edited by SIDNEY FLEISCHER AND LESTER PACKER

VOLUME LVI. Biomembranes (Part G: Bioenergetics)
Edited by SIDNEY FLEISCHER AND LESTER PACKER

VOLUME LVII. Bioluminescence and Chemiluminescence
Edited by MARLENE A. DELUCA

VOLUME LVIII. Cell Culture
Edited by WILLIAM B. JAKOBY AND IRA PASTAN

VOLUME LIX. Nucleic Acids and Protein Synthesis (Part G)
Edited by KIVIE MOLDAVE AND LAWRENCE GROSSMAN

VOLUME LX. Nucleic Acids and Protein Synthesis (Part H)
Edited by KIVIE MOLDAVE AND LAWRENCE GROSSMAN

VOLUME 61. Enzyme Structure (Part H)
Edited by C. H. W. HIRS AND SERGE N. TIMASHEFF

VOLUME 62. Vitamins and Coenzymes (Part D)
Edited by DONALD B. MCCORMICK AND LEMUEL D. WRIGHT

VOLUME 63. Enzyme Kinetics and Mechanism (Part A: Initial Rate and Inhibitor Methods)
Edited by DANIEL L. PURICH

VOLUME 64. Enzyme Kinetics and Mechanism (Part B: Isotopic Probes and Complex Enzyme Systems)
Edited by DANIEL L. PURICH

VOLUME 65. Nucleic Acids (Part I)
Edited by LAWRENCE GROSSMAN AND KIVIE MOLDAVE

VOLUME 66. Vitamins and Coenzymes (Part E)
Edited by DONALD B. MCCORMICK AND LEMUEL D. WRIGHT

VOLUME 67. Vitamins and Coenzymes (Part F)
Edited by DONALD B. MCCORMICK AND LEMUEL D. WRIGHT

VOLUME 68. Recombinant DNA
Edited by RAY WU

VOLUME 69. Photosynthesis and Nitrogen Fixation (Part C)
Edited by ANTHONY SAN PIETRO

VOLUME 70. Immunochemical Techniques (Part A)
Edited by HELEN VAN VUNAKIS AND JOHN J. LANGONE

VOLUME 71. Lipids (Part C)
Edited by JOHN M. LOWENSTEIN

VOLUME 72. Lipids (Part D)
Edited by JOHN M. LOWENSTEIN

VOLUME 73. Immunochemical Techniques (Part B)
Edited by JOHN J. LANGONE AND HELEN VAN VUNAKIS

VOLUME 74. Immunochemical Techniques (Part C)
Edited by JOHN J. LANGONE AND HELEN VAN VUNAKIS

VOLUME 88. Biomembranes (Part I: Visual Pigments and Purple Membranes, II)
Edited by LESTER PACKER

VOLUME 89. Carbohydrate Metabolism (Part D)
Edited by WILLIS A. WOOD

VOLUME 90. Carbohydrate Metabolism (Part E)
Edited by Willis A. Wood

VOLUME 91. Enzyme Structure (Part I)
Edited by C. H. W. HIRS AND SERGE N. TIMASHEFF

VOLUME 92. Immunochemical Techniques (Part E: Monoclonal Antibodies and General Immunoassay Methods)
Edited by JOHN J. LANGONE AND HELEN VAN VUNAKIS

VOLUME 93. Immunochemical Techniques (Part F: Conventional Antibodies, Fc Receptors, and Cytotoxicity)
Edited by JOHN J. LANGONE AND HELEN VAN VUNAKIS

VOLUME 94. Polyamines
Edited by HERBERT TABOR AND CELIA WHITE TABOR

VOLUME 95. Cumulative Subject Index Volumes 61–74 and 76–80
Edited by EDWARD A. DENNIS AND MARTHA G. DENNIS

VOLUME 96. Biomembranes [Part J: Membrane Biogenesis: Assembly and Targeting (General Methods; Eukaryotes)]
Edited by SIDNEY FLEISCHER AND BECCA FLEISCHER

VOLUME 97. Biomembranes [Part K: Membrane Biogenesis: Assembly and Targeting (Prokaryotes, Mitochondria, and Chloroplasts)]
Edited by SIDNEY FLEISCHER AND BECCA FLEISCHER

VOLUME 98. Biomembranes [Part L: Membrane Biogenesis (Processing and Recycling)]
Edited by SIDNEY FLEISCHER AND BECCA FLEISCHER

VOLUME 99. Hormone Action (Part F: Protein Kinases)
Edited by JACKIE D. CORBIN AND JOEL G. HARDMAN

VOLUME 100. Recombinant DNA (Part B)
Edited by RAY WU, LAWRENCE GROSSMAN, AND KIVIE MOLDAVE

VOLUME 101. Recombinant DNA (Part C)
Edited by RAY WU, LAWRENCE GROSSMAN, AND KIVIE MOLDAVE

VOLUME 102. Hormone Action (Part G: Calmodulin and Calcium-Binding Proteins)
Edited by ANTHONY R. MEANS AND BERT W. O'MALLEY

VOLUME 103. Hormone Action (Part H: Neuroendocrine Peptides)
Edited by P. MICHAEL CONN

VOLUME 104. Enzyme Purification and Related Techniques (Part C)
Edited by WILLIAM B. JAKOBY

VOLUME 105. Oxygen Radicals in Biological Systems
Edited by LESTER PACKER

VOLUME 106. Posttranslational Modifications (Part A)
Edited by FINN WOLD AND KIVIE MOLDAVE

VOLUME 107. Posttranslational Modifications (Part B)
Edited by FINN WOLD AND KIVIE MOLDAVE

VOLUME 108. Immunochemical Techniques (Part G: Separation and Characterization of Lymphoid Cells)
Edited by GIOVANNI DI SABATO, JOHN J. LANGONE, AND HELEN VAN VUNAKIS

VOLUME 109. Hormone Action (Part I: Peptide Hormones)
Edited by LUTZ BIRNBAUMER AND BERT W. O'MALLEY

VOLUME 110. Steroids and Isoprenoids (Part A) (in preparation)
Edited by JOHN H. LAW AND HANS C. RILLING

VOLUME 111. Steroids and Isoprenoids (Part B) (in preparation)
Edited by JOHN H. LAW AND HANS C. RILLING

VOLUME 112. Drug and Enzyme Targeting (in preparation)
Edited by KENNETH J. WIDDER AND RALPH GREEN

VOLUME 113. Glutamate, Glutamine, Glutathione, and Related Compounds (in preparation)
Edited by ALTON MEISTER

Section I

Receptor Assays

[1] Assay for the Glucagon Receptor

By FRANCISCO J. ROJAS and L. BIRNBAUMER

The direct *in vitro* measurement of glucagon receptor binding using ^{125}I-labeled glucagon as a tracer has been a commonly employed method dating from the work of Rodbell *et al.*[1] more than a decade ago. Preparation of "iodoglucagon" was originally accomplished by using either I_2 as the oxidizing agent, or mixing glucagon with carrier-free ^{125}I and chloramine T, followed by preparative polyacrylamide gel electrophoresis for its purification. In either case, it was assumed that a monoiodinated species of glucagon had been prepared and that the biological activity of glucagon was not altered by iodination. As a receptor probe, this "iodoglucagon" was found to bind with high affinity to binding sites in liver parenchymal cell plasma membranes[1] and this binding was discovered to be influenced by GTP.[2] This eventually led to the establishment that hormonal stimulation of adenylyl cyclases is absolutely dependent on guanine nucleotide[3] and that adenylyl cyclases are two-component systems.[4,5] However, it was also found that both association and dissociation rates of "iodoglucagon" to and from the membranes were too slow to account for the rates at which adenylyl cyclase activity was stimulated.[6] The discrepancy between kinetics of binding and adenylyl cyclase stimulation was considered to be indicative of the existence of a large proportion of the total glucagon-specific sites being permanently uncoupled from the adenylyl cyclase system.[6] This interpretation assumed that a chemically homogeneous receptor probe had been utilized.

Subsequent studies performed by Bromer *et al.*[7] and Desbuquois[8] demonstrated that iodinated species of glucagon are between 4 and 10

[1] M. Rodbell, H. M. J. Krans, S. L. Pohl, and L. Birnbaumer, *J. Biol. Chem.* **246,** 1861 (1971).

[2] M. Rodbell, H. M. J. Krans, S. L. Pohl, and L. Birnbaumer, *J. Biol. Chem.* **246,** 1872 (1971).

[3] M. Rodbell, L. Birnbaumer, S. L. Pohl, and H. M. J. Krans, *J. Biol. Chem.* **246,** 1877 (1971).

[4] E. M. Ross and A. G. Gilman, *J. Biol. Chem.* **252,** 6966 (1977).

[5] J. K. Northup, P. C. Sternweis, M. D. Smigel, L. S. Schleifer, E. M. Ross, and A. G. Gilman, *Proc. Natl. Acad. Sci. U.S.A.* **77,** 6516 (1980).

[6] L. Birnbaumer and S. L. Pohl, *J. Biol. Chem.* **248,** 2056 (1973).

[7] W. W. Bromer, M. E. Boucher, and J. M. Patterson, *Biochem. Biophys. Res. Commun.* **53,** 134 (1973).

[8] B. Desbuquois, *Eur. J. Biochem.* **53,** 569 (1975).

ISBN 0-12-182009-2

times more potent in adenylyl cyclase assays than native glucagon, the potency depending on the degree of iodination. Also, the glucagon molecule was demonstrated to be extremely sensitive to oxidative breakdown and that under chloramine T- and lactoperoxidase/H_2O_2-mediated iodinations, much of the material obtained is severely damaged[8,9] and probably represented a heterogeneous mixture of iodinated forms of undefined composition. It became therefore apparent that "iodoglucagon" isolated from gel electrophoresis[1] may be composed of glucagon derivatives of which only a small fraction are biologically active iodoglucagon molecules with the remainder being one or more species of glucagon that do not activate adenylyl cyclase, reducing the potency of the total mixture, but that bind to liver membranes with anomalous kinetics, creating the impression of existence of heterogeneous sets of sites. It is also clear from the above that the resolution capacity of the preparative polyacrylamide gel electrophoresis method used was insufficient to separate an active monoiodoglucagon species usable as probe in receptor studies.

Here, we describe a new iodination procedure for glucagon using 1,3,4,6-tetrachloro-3α,6α-diphenylglycouril (Iodogen) as the oxidizing agent, and the subsequent separation in pure form of [^{125}I-Tyr^{10}]monoiodoglucagon by reverse-phase high-pressure liquid chromatography (HPLC) over C_{18}-μ Bondapak columns. The newly synthesized [^{125}I]monoiodoglucagon is shown to be a suitable probe for studying structural and functional properties of glucagon receptors.

Reagents

Carrier-free ^{125}I was purchased from Isotex Diagnostics, Friendswood, Texas. Inorganic carrier-free ^{32}P used for synthesis of [α-^{32}P]ATP was purchased from Union Carbide, Tuxedo, New York. Glucagon (recrystallized, insulin-free) was a generous gift from Dr. W. W. Bromer, Eli Lilly and Co., Indianapolis, Indiana. Iodogen (1,3,4,6-tetrachloro-3α,6α-diphenylglycouril) was purchased from Pierce Chemical Co., Rockford, Illinois, and cetyltrimethylammonium chloride from Eastman Kodak Co., Rochester, New York. Oxoid cellulose-acetate filters were obtained from Oxoid Ltd., Basingstoke, Hants, England and soaked 2–10 hr in 10% bovine serum albumin (BSA) before use. [α-^{32}P]ATP was synthesized by the method of Walseth and Johnson[10] and purified by ion exchange chromatography over DEAE-Sephadex A-25 as described elsewhere.[11] GTP was from Sigma Chemical Co. (St. Louis, Missouri) and

[9] L. Birnbaumer, T. L. Swartz, F. J. Rojas, and A. J. Garber, *J. Recept. Res.* **3,** 3 (1982).

[10] T. F. Walseth and R. A. Johnson, *Biochim. Biophys. Acta* **562,** 11 (1979).

[11] L. Birnbaumer, H. N. Torres, M. M. Flawia, and R. B. Fricke, *Anal. Biochem.* **93,** 124 (1979).

GMP-P(NH)P was from Boehringer Mannheim (Elkhart, Indiana). All nucleotides were purified by ion exchange chromatography before used.[11] Creatine phosphokinase, myokinase, and creatine phosphate were purchased from Sigma Chemical Co. and purified from guanine nucleotide-like contaminants as described.[12] C_{18} reverse-phase columns were μBondapak-C_{18} 0.4 × 30 cm from Waters Association, Waltham, Massachusetts. Methanol and acetonitrile used for HPLC were from Burdick and Jackson Laboratories, Inc., Muskegon, Michigan. BSA was Cohn fraction V from Armour Pharmaceutical Co., Phoenix, Arizona. All BSA-containing solutions were filtered through 0.45-μm Millipore HA type filters before use. All other materials and chemicals were of the highest purity commercially available.

Methods

Iodination of Glucagon

Solutions

GLUCAGON. A stock suspension of 10 mg/ml of glucagon is made in 10 m*M* cetyltrimethylammonium chloride. Ten-microliter aliquot is then diluted by adding 230 μl of distilled water. For iodination, 90 μl of this final solution will provide 9 nmol of glucagon.

PHOSPHATE BUFFER. Potassium phosphate buffer (0.4 *M*), pH 7.2, is made from a 1 *M* stock solution.

SODIUM METABISULFITE. A freshly prepared solution of 22 mg/ml sodium metabisulfite is made in distilled water. Ten microliters of this is then diluted by adding 560 μl of distilled water. The concentration of the resulting dilution is 2 m*M*.

TYROSINE. Two millimolar of tyrosine in water is made from a 10 m*M* stock solution. With the exception of phosphate buffer, all stock solutions are kept at −20° and thawed just before iodination.

Preparation of the Reaction Vessel

Eight milligrams of Iodogen is first dissolved in 2 ml of chloroform and diluted 1:10 times with the same solvent. Ten microliters of this dilution containing 9 nmol of Iodogen is then applied onto the inner bottom of a 10 × 75 glass test tube and thoroughly dried under a gentle stream of nitrogen gas at room temperature. Tubes with plated Iodogen can be stored in the dark and at −20° for up to 8 months.

[12] R. Iyengar, P. W. Mintz, T. L. Swartz, and L. Birnbaumer, *J. Biol. Chem.* **255,** 11875 (1980).

Radioiodination Procedure

Radioiodination of glucagon is carried out by using as an iodide-oxidizing agent the water-insoluble chloramide Iodogen at a molar ratio of glucagon: ^{125}I: Iodogen of 1:1:1. To the tube containing the plated Iodogen (9 nmol), the following solutions are added sequentially: 90 μl (containing 9 nmol) of the diluted solution of glucagon prepared as described under "Solutions," 90 μl of 0.4 *M* potassium phosphate, pH 7.2, and 60 μl of 20 mCi carrier-free ^{125}I in 0.1 *N* sodium hydroxide (~9 nmol). These additions are readily completed in less than 10 sec by using a set of pipets (e.g., Pipetman) preloaded with the different solutions. Following additions of carrier-free ^{125}I, the tube is vortexed repeatedly for 1.0 min at room temperature and the reaction stopped by adding 10 μl of 2 m*M* sodium metabisulfite and 10 μl of 2 m*M* tyrosine. The soluble mixture is then removed and injected immediately onto the column for chromatography.

Reverse-Phase High-Pressure Liquid Chromatography (HPLC) and Separation of [^{125}I-Tyr^{10}]Monoiodoglucagon

Reverse-phase HPLC is performed on μBondapak-C_{18} columns (0.4 × 30 cm) at a constant flow rate of 1.0 ml/min using the following solvents: a 40:60 mixture of methanol and 10 m*M* H_3PO_4 in H_2O, adjusted to pH 3.0 after mixing with freshly redistilled triethylamine (Solvent A), and a 50:50 mixture of acetonitrile and 0.1 *M* Tris–HCl, pH 9.0 in H_20 (Solvent B). These solvents are previously filtered through 0.45-μm Millipore filters and deaereated before use. The column receives the solvents from one or two pumps (Waters Model 6000A, or similar) directed by gradient programmer (Model 660, or similar). Back pressure is about 1200 psi. Injections are made without interruption of pressure of flow (Waters Model U6K injector, or similar) located between the solvent pumps and the column. Column eluates are monitored for adsorption at 254 nm with the aid of a UV absorbance detector (Waters Model 440, or similar) fitted with an 8 μl flow cell and then fractionated with the aid of a fraction collector. Chromatography, UV monitoring, and fraction collection are carried out at room temperature (22–25°). Prior to use, the column is washed with methanol (30 min), water (30 min), and Solvent A (30 min). Next, the iodinated reaction mixture to be chromatographed is injected onto the head of the column and after washing with Solvent A for 8 min, the chromatographic separation is effected by subjecting the column to a nonlinear gradient between Solvents A and B formed of two consecutive linear gradients. The first one is a linear gradient from 0 to 60% Solvent B from t_1 = 8 min to t_2 = 40 min, and the second one is a linear gradient from

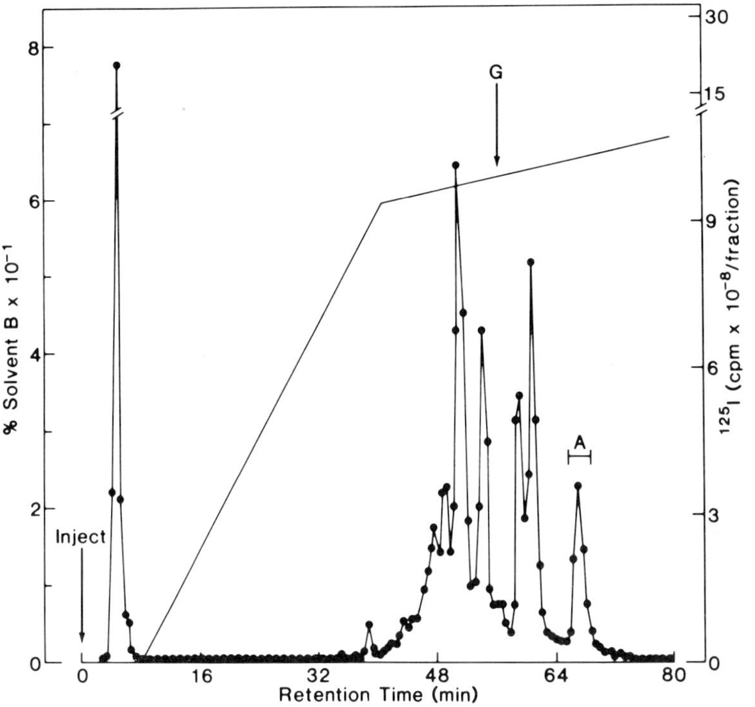

FIG. 1. Reverse-phase high-pressure liquid chromatography of glucagon iodination products using Iodogen as the oxidizing agent. Elution pattern of ^{125}I-labeled materials obtained after iodinating glucagon at molar glucagon : ^{125}I : Iodogen ratios of 1 : 1 : 1 using 20 mCi of ^{125}I. The reaction was allowed to proceed for 1 min at room temperature after which the reaction was stopped, the soluble mixture was injected onto the μBondapak-C_{18} column, and subjected to the HPLC procedure as described in the text. Fractions (450 μl) were collected and the radioactivity of each fraction quantitated. Arrow indicates the elution position of native glucagon. The last peak eluting to the right of native glucagon is shown as peak A.

60 to 100% Solvent B from t_2 = 40 min to t_3 = 260 min. Fractions (450 μl) are collected and the radioactivity of each fraction is quantitated. A typical elution pattern of ^{125}I-labeled materials is shown in Fig. 1. Arrow indicates the elution position of native glucagon corresponding to approximately 54 min of retention time. The last peak eluting to the right of native glucagon (peak A) corresponds to [^{125}I-Tyr^{10}]monoiodoglucagon as demonstrated recently in a detailed report.[13] Basically, analysis of the composition of peak A indicated that (1) radioactivity of the peak was precipitable by trichloroacetic acid; (2) after extreme digestion of a

[13] F. J. Rojas, T. L. Swartz, R. Iyengar, A. Garber, and L. Birnbaumer, *Endocrinology* **113**, 711 (1983).

sample of Peak A with pronase followed by ion exchange chromatography, the radioactivity comigrated with monoiodotyrosine and not with diiodotyrosine or iodohistidine; and (3) digestion with trypsin followed by separation of peptides by gel filtration, the radioactivity comigrated with the peptide fragment containing the tyrosine-10 and not with the smallest peptide fragment containing the tyrosine-13.[13] Also, the [^{125}-Tyr^{10}]monoiodoglucagon was demonstrated to be biologically active and 5- to 8-fold more potent than native glucagon[13] which agrees with the data by Bromer *et al.*[7] showing a similar increase in potency for iodoglucagon preparations synthesized with nonradioactive I_2.

After gradient elution, the column was washed with Solvent B (30 min), with water (30 min), and methanol (30 min) in which it was finally stored.

Storage of Iodinated Glucagon Purified by HPLC

Fractions of HPLC purified [^{125}I-Tyr^{10}]monoiodoglucagon are pooled, diluted with neutralized BSA, and stored in small aliquots at −70°. In general, we store aliquots of 50 μl containing 10–15 × 10^6 cpm (2400 cpm/fmol), in 0.1% BSA and carry out fresh iodinations every 2 to 3 weeks. The yield of [^{125}I-Tyr^{10}]monoiodoglucagon is in the order of 0.8 to 1.0 mCi per iodination. The molarity of the iodinated glucagon solution is calculated on the basis of ^{125}I content according to the composition of peak A.

Binding of Iodinated Glucagon to Liver Plasma Membranes and Separation of Free and Bound Hormone

Rat liver plasma membranes prepared by the abbreviated[14] method of Neville[15] are incubated at a final concentration of 0.02–0.08 mg protein/ml in a final reaction volume of 100 μl containing 0.1% BSA, 1.0 m*M* EDTA, 20 m*M* Tris–HCl, pH 7.5 (Buffer I), [^{125}I-Tyr^{10}]Monoiodoglucagon, guanine nucleotides, Mg ion, and other additions. The mixture is incubated in a shaking bath for 20 min at 32.5°, or at 22° in the case that effects of guanine nucleotides upon binding are being studied. The lower incubation temperature was chosen because we have previously noted[13] that nucleotide addition reduces affinity of receptors in a temperature-dependent manner, and that at 32.5°, this reduction in affinity is too large to allow for a reliable measurement of guanine nucleotide-affected [^{125}I]-monoiodoglucagon binding. After incubation, the reaction is stopped by adding 5 ml of ice-cold Buffer I to the test tubes. The mixtures are vor-

[14] S. L. Pohl, L. Birnbaumer, and M. Rodbell, *J. Biol. Chem.* **246,** 1849 (1971).
[15] D. M. Neville, Jr., *Biochim. Biophys. Acta* **154,** 540 (1968).

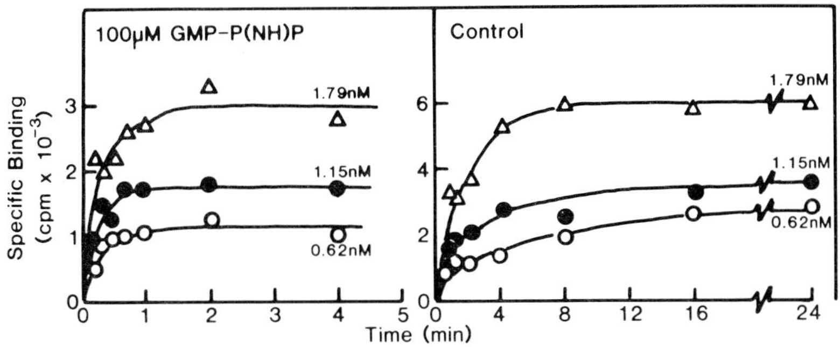

FIG. 2. Time course of specific binding of [^{125}I]monoiodoglucagon to rat liver plasma membranes. Rat liver membranes (3.6 μg of protein) were incubated in triplicate in a final volume of 200 μl of 20 m*M* Tris–HCl, pH 7.5, containing 1.0 m*M* EDTA and 0.1% BSA (Buffer I) in the presence of the indicated concentrations of [^{125}I]monoidoglucagon (2400 cpm/fmol). Incubations were carried out at 22° for the indicated time periods in the presence of 100 μM GMP-P(NH)P (left panel) as well as in the absence of added guanine nucleotide (right panel). Parallel assays were carried out in duplicate in the presence of 1 μM unlabeled glucagon. Assays were stopped by dilution with ice-cold Buffer I and immediately filtered through Oxoid filters as described in the text. Radioactivity retained on the filters was quantitated by gamma spectroscopy. Specific binding was calculated by substracting nonspecific binding from total binding.

texed and immediately filtered through 0.45-μm Oxoid cellulose-acetate filters that have presoaked overnight in 10% BSA and washed with ice-cold Buffer I immediately before receiving the stopped and diluted reaction mixture. The tubes are rinsed once with 5 ml of ice-cold Buffer I and the rinses filtered through the Oxoid filters which are then immediately washed with a final 5-ml aliquot of ice-cold Buffer I. The whole procedure takes no more than 20 sec. The amount of radioactivity remaining on the Oxoid filters is quantitated by scintillation counting in a Searle-Analytic model 1195 autogammaspectrometer. Under the above conditions and using 0.5 n*M* [^{125}I-Tyr^{10}]monoiodoglucagon prepared as described above, the nonspecific binding measured in the presence of membrane protein and an excess of unlabeled glucagon (10^{-6} *M*) is about 2% of the total counts added and about 10% of the total counts bound to membranes. In general, the nonspecific binding value is nearly identical to the value obtained when binding to Oxoid filters alone is measured in the absence of membranes.

Binding of [^{125}I]monoiodoglucagon proceeds rather rapidly at temperatures above 30° reaching equilibrium within 1 min at 0.5–0.6 n*M* of [^{125}I]-monoiodoglucagon added. The reaction rates can be reduced somewhat by decreasing the temperature of incubation and are affected by guanine nucleotides. Figure 2 illustrates associated rates of varying concentra-

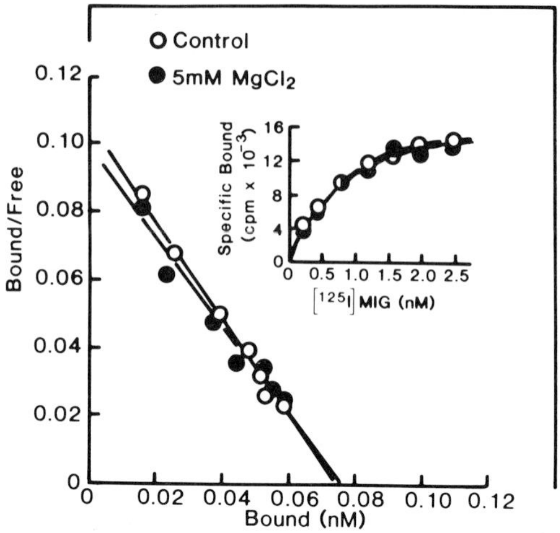

FIG. 3. Equilibrium binding analysis in the absence and presence of Mg ion. Liver membranes (3.6 μM protein/100 μl assay) were incubated in triplicate with increasing concentrations of [^{125}I]monoidoglucagon in medium containing 0.1% BSA, 50 mM Tris–HCl, pH 7.6, and either 5.0 mM EDTA (control) or 5 mM $MgCl_2$. Parallel assays were carried out in duplicate in the presence of 1 μM unlabeled glucagon. Incubations (20 min, 32.5°) were stopped and [^{125}I]monoiodoglucagon bound was quantitated by the Oxoid filtration assay described in the text. Nonspecific binding was 2300 ± 200 cpm (mean ± SD of quadruplicates). Specific binding was calculated and analyzed according to Scatchard giving a K_D of 0.75 nM and a B_{max} of 2.1 pmol/mg protein. The main figure shows the Scatchard plot of the data depicted in the inset.

tions of [^{125}I]monoiodoglucagon to liver plasma membranes as obtained at 22° in the presence and absence of 100 μM GMP-P(NH)P. It can be seen that equilibrium binding is reached considerably faster in the presence of guanine nucleotide than in its absence and that binding at equilibrium is less in the presence than in the absence of GMP-P(NH)P (note different scales on the y axis on left and right panels of Fig. 2). Under both conditions, binding is unchanged for up to 30 min.

Equilibrium binding experiments show that liver membranes bind [^{125}I]monoiodoglucagon in a saturable manner. This is illustrated in Fig. 3 which presents Scatchard plots of liver membranes incubated with increasing concentrations of [^{125}I]monoiodoglucagon. Specific binding is obtained by determining the difference between radioactivity bound when [^{125}I]monoiodoglucagon is added alone and that bound when 10^{-6} M unlabled glucagon is added to determine nonspecific binding. It can be seen that essentially linear Scatchard plots are obtained with K_D values of 0.75 nM and a B_{max} of 2.1 pmol/mg protein. Figure 3 also shows that Mg ion has no effect on the affinity of the glucagon receptor for its ligand.

Adenylyl Cyclase Assay and Rate of Activation of Liver Membrane Adenylyl Cyclase by [^{125}I]Monoiodoglucagon

Iodinated glucagon obtained by HPLC is diluted serially in ice-cold Buffer I to yield final concentrations between 10^{-11} and 10^{-7} *M*. Molarity of the initial solution of glucagon dissolved in distilled water is calculated from its absorbance at 280 nm using a molar adsorption coefficient of 8050. Molarity of HPLC purified radioactive glucagon is assumed to equal that of ^{125}I. Aliquots of the serially diluted sample (10 μl) are then tested for their capacity to stimulate adenylyl cyclase activity in 2–5 μg rat liver plasma membranes as assayed in 10 min at 32.5° in a final volume of 50 μl containing (final concentrations) 0.1 *M* [α-^{32}P]ATP (~2000 cpm/pmol), 5 m*M* $MgCl_2$, 10 μM GTP, 1.0 m*M* EDTA, 0.1% BSA, 1.0 m*M* [^{3}H]cAMP (~10,000 cpm), a nucleoside triphosphate regenerating system composed of 20 m*M* creatine phosphate, 0.2 mg/ml creatine phosphokinase (200 U/mg) and 0.02 mg/ml myokinase (200 U/mg), and 25 m*M* Tris–HCl, pH 7.5. These assays are stopped by the addition of 100 μl stopping solution consisting of 10 m*M* cAMP, 40 m*M* ATP, and 1% sodium dodecyl sulfate. The [^{32}P]cAMP formed and the [^{3}H]cAMP added to monitor recovery are

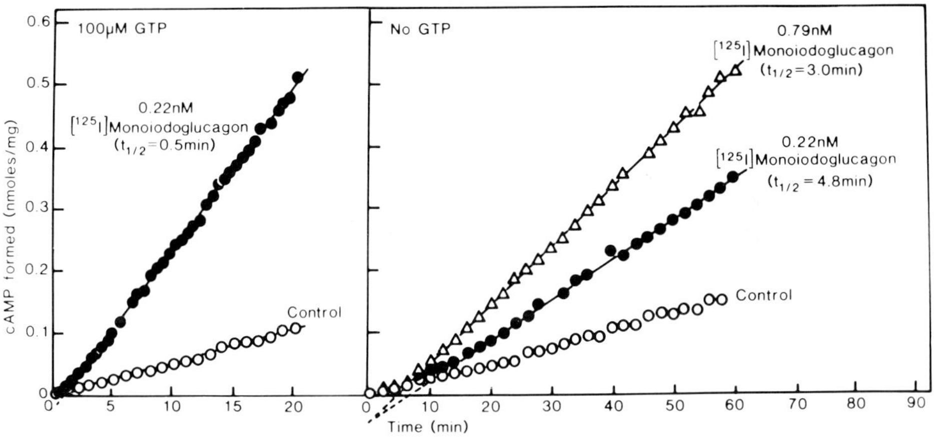

FIG. 4. Time courses of activation of liver membrane adenylyl cyclase by [^{125}]monoidoglucagon in the presence (left panel) and the absence (right panel) of guanine nucleotide. Reactions were carried out under standard adenylyl cyclase assay conditions as the indicated concentrations of GMP-P(NH)P and [^{125}I]monoidoglucagon. Each curve is derived from a single incubation from which 50-μl aliquots containing the cAMP formed by a 3.6 μg of liver membrane protein were withdrawn at the indicated times and quantitated as described in the text. The $t_{1/2}$ values were calculated from the apparent activation rate constants k according to $t_{1/2} = \ln 2/k$; k values were calculated from the data shown according to $k = v_{\text{steady state}}/y$ intercept.

isolated according to the method of Salomon *et al.*[16] using Dowex and alumina chromatography, as modified by Bockaert *et al.*[17]

As illustrated in Fig. 4, the rates of activation of liver membrane adenylyl cyclase are affected by both the concentration of hormone added and the presence of guanine nucleotide. In the absence of guanine nucleotide, the times required to obtain steady-state stimulation of adenylyl cyclase with 0.22 and 0.79 nm [^{125}I]monoiodoglucagon are approximately 15 and 9 min (calculated $t_{1/2}$ values were 4.8 and 3.0 min, respectively). In the presence of 100 μM GTP and using 0.22 nM [^{125}I]monoiodoglucagon, the steady-state activity is obtained much faster (calculated $t_{1/2}$ = 0.5 min). These time courses of stimulation of adenylyl cyclase by [^{125}I]monoiodoglucagon (Fig. 4) correlate well with the time courses of [^{125}I]monoiodoglucagon binding to the membranes (Fig. 3). We conclude from these studies that in contrast to the previous findings with ill-defined "iodoglucagon"[1,2,6], the kinetics of binding of [^{125}I]monoiodoglucagon and those of adenylyl cyclase stimulation correlate in such manner as to justify a cause-effect relationship between the binding site occupancy and adenylyl cyclase stimulation.

We feel that the availability of the newlv synthesized [^{125}I-Tyr^{10}]-monoiodoglucagon as described here will provide new insight regarding the elucidation of the mechanism by which the glucagon receptor activates the system(s) it couples to, and represents an effective tool for studies including variations in receptor levels and properties in disease states.

Acknowledgment

Supported in part by NIH Grants AM-19318 and AM-27682.

[16] Y. Salomon, C. Londos, and M. Rodbell, *Anal. Biochem.* **58,** 541 (1974).
[17] J. Bockaert, M. Hunzicker-Dunn, and L. Birnbaumer, *J. Biol. Chem.* **251,** 2653 (1976).

[2] Wheat Germ Lectin-Sepharose Affinity Adsorption Assay for the Soluble Glucagon Receptor

By RAVI IYENGAR and JOHN T. HERBERG

Introduction

The hepatic glucagon-sensitive adenylyl cyclase has been widely used as a model system to study the signal transduction process in adenylyl cyclase systems.[1-3] However, purification of the receptor has not been possible up to now. This has, in part, been due to lack of availability of a reliable assay for the solubilized receptor. Methods commonly used for other soluble receptors such as gel filtration and polyethylene glycol precipitation have not proven useful for assaying the glucagon receptor due to unacceptably high backgrounds. Consequently, we attempted to develop a simple assay that would be based on a unique property of the receptor.

Rationale

In the past few years, it has been found that many hormone receptors are glycoproteins. Hence, we tested if the glucagon receptor was also a glycoprotein. For this, we used the receptor that had [^{125}I-Tyr^{10}]monoiodoglucagon ([^{125}I]MIG) covalently attached to it by use of the heterobifunctional cross-linker, hydroxysuccinimidyl-*p*-azidobenzoate.[4-5a] The covalently labeled receptor was solubilized with Lubrol-PX. The solubilized receptor was exposed to wheat germ lectin-Sepharose (WGL-Sepharose) in the presence and absence of various sugars. The supernatants after the exposure to the Sepharose-gels were analyzed by SDS-gel electrophoresis. It was found that the glucagon receptor was adsorbed onto WGL-Sepharose but not Sepharose. Further adsorption to WGL-Sepharose was blocked by *N*-acetylglucosamine (GlcNAc) but not glucose or galactose (Fig. 1). The sugar specificity for the adsorption of the

[1] M. Rodbell, L. Birnbaumer, S. L. Pohl, and H. M. J. Krans, *J. Biol. Chem.* **246,** 1877 (1971).

[2] R. Iyengar, *J. Biol. Chem.* **256,** 11042 (1981).

[3] R. Iyengar and L. Birnbaumer, *Proc. Natl. Acad. Sci. U.S.A.* **79,** 5179 (1982).

[4] G. L. Johnson, V. I. MacAndrew, and P. F. Pilch, *Proc. Natl. Acad. Sci. U.S.A.* **78,** 875 (1981).

[5] R. Iyengar and J. T. Herberg, *J. Biol. Chem.* **259,** 5222 (1984).

[5a] J. T. Herberg, J. Codina, K. A. Rich, F. J. Rojas, and R. Iyengar, *J. Biol. Chem.* **259,** 9285 (1984).

ISBN 0-12-182009-2

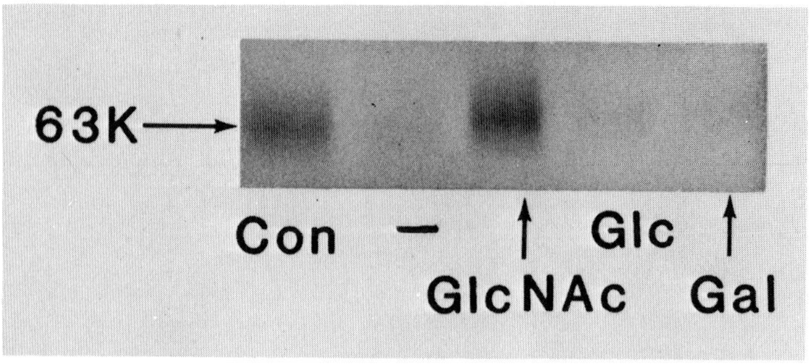

FIG. 1. Sugar specific adsorption of the M_r = 63,000 glucagon receptor to WGL-Sepharose. Liver membranes that had [^{125}I]MIG covalently attached were extracted with 1% Lubrol-PX. The 100,000 *g* supernatant was then exposed to Sepharose or WGL-Sepharose. Of the Lubrol extract 100 μl was exposed to 70 μl of packed Sepharose or WGL-Sepharose in a final volume of 600 μl. Exposure to WGL-Sepharose was carried out without any further additions (−) or in the presence of 0.1 *M* GlcNAc, 0.1 *M* glucose (Glc), or 0.1 *M* galactose (Gal). The supernatants were removed and 70-μl aliquots were electrophoresed on SDS-polyacrylamide gels. The gel was then dried and subjected to autoradiography. An autoradiogram (48 hr exposure) of M_r = 63,000 region is shown.

glucagon receptor is in accordance with the previously published sugar specificity for wheat germ lectin.[6] This observation indicates that the receptor is a glycoprotein. The primary structure of glucagon is known and it has been shown that it has no sugars covalently attached to it. Thus, it appeared that difference in covalently attached sugars between the hormone and the receptor could be used to develop an assay for the solubilized receptor. We reasoned that the hormone–receptor complex could be specifically adsorbed onto the lectin-Sepharose while the free hormone would remain in solution. The solid phase containing the hormone–receptor complex could be separated from the free hormone by low-speed centrifugation and removal of the supernatant. The hormone–receptor complex formed could then be estimated by counting the lectin-Sepharose in a gamma counter.

Materials

[^{125}I-Tyr[10]]Monoiodoglucagon is synthesized and purified according to the procedure of Rojas *et al.*[7]

[6] L. L. Adair and S. Kornfield, *J. Biol. Chem.* **249,** 4696 (1974).

[7] F. J. Rojas, T. L. Swartz, R. Iyengar, A. J. Garber, and L. Birnbaumer, *Endocrinology* **113,** 711 (1983).

WGL-Sepharose is synthesized according to the procedure of Adair and Kornfield[6] with minor modifications. One hundred milligrams of wheat germ lectin (Sigma Chemical Co., Catalog L-1005) is reacted with 30 ml of packed CNBr-activated Sepharose 6B-CL in the presence of 10 m*M* $NaHCO_3$, 100 m*M* NaCl, and 200 m*M* GlcNAc overnight at 4°. The gel is then washed with 2 liters of ice-cold 10 m*M* $NaHCO_3$, 100 m*M* NaCl, and then reacted with 1.5 g of glycine in 10 m*M* $NaHCO_3$, 100 m*M* NaCl overnight at 4°. After the coupling reactions, the gel is washed again with 2 liters of 10 m*M* $NaHCO_3$ and 100 ml NaCl and stored at 4° as a 1:2 suspension in the same solution. Sepharose 6B-CL was activated according to the procedure of Parikh *et al.*[8] with 200 mg of CNBr/ml of packed gel in the presence of 2 *M* Na_2CO_3. The CNBr was dissolved in acetonitrile. The WGL-Sepharose retains its binding characteristics for 4–6 weeks and shows no significant leaching of the wheat germ lectin.

Methods

Preparation of CHAPS Extract

Liver membranes are suspended in 20 m*M* 3-[(3-cholamidopropyl)dimethylammonio]-1-propane sulfate (CHAPS), 0.5 *M* NaCl, 0.5 m*M* EDTA, 1.0 m*M* $MgCl_2$, and 25 m*M* NaHepes, pH 8.0, at a final concentration of 6–8 mg protein/ml and stirred for 30 min at 0–4°. The detergent suspension was centrifuged at 100,000 *g* for 60 min. The clear supernatant is carefully removed and made 16% (w/v) with respect to sucrose using 40% sucrose. Aliquots are quick frozen and stored at −70° until use. This procedure results in 25–30% solubilization of the glucagon receptor.

Binding Assay for the Solubilized Receptor

The CHAPS extract is incubated in a final volume of 300 μl in the presence of 0.1–1 n*M* [^{125}I]MIG, 2 m*M* $MgCl_2$, 1 m*M* EDTA, 25 m*M* NaHepes, pH 8.0, and 12% (w/v) sucrose. [^{125}I]MIG was added in 30 μl of 0.1% BSA. CHAPS extract protein was added in 60 μl so as to achieve a final CHAPS concentration of 3 m*M* or less. Incubations are carried out in 16 × 100 mm plastic centrifuge tubes (Sarstedt 25.468) for 90–180 min on ice. At the end of the incubation, 300 μl of 1:4 (v/v) suspension of WGL-Sepharose in 25 m*M* NaHepes, 8% (w/v) sucrose, 0.5% Lubrol-PX, 50 m*M* KCl, 1 m*M* EDTA, and 2 m*M* $MgCl_2$ (wash buffer) is added to each tube. Prior to use, the WGL-Sepharose is washed with 10 volumes of

[8] I. Parikh, S. March, and P. Cuatrecasas, this series, Vol. 34, p. 77.

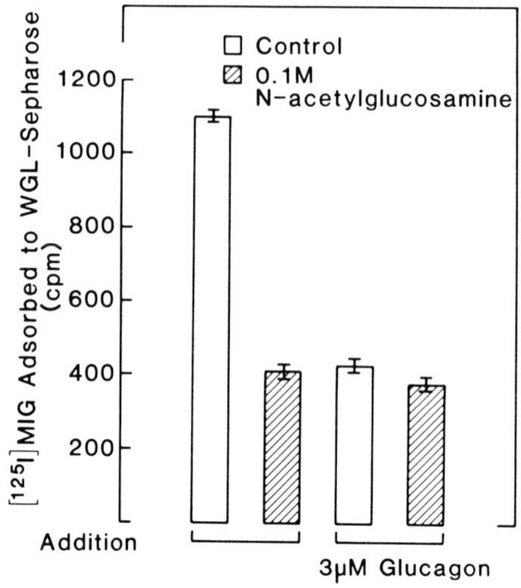

FIG. 2. [^{125}I]MIG binding to CHAPS extract of rat liver membranes. CHAPS extract protein was incubated ith 0.2 n*M* [^{125}I]MIG in the absence and presence of 3 μM glucagon for 90 min as described under Methods. After the binding reaction, the suspension of WGL-Sepharose with or without 0.1 *M* GlcNAc was added. The samples were further incubated for 2 hr with moderate shaking and then processed as described under Methods. Values are mean ± SD of triplicate determinations.

the wash buffer. The samples are then shaken for 2 hr in the cold at a moderate speed so as not to impede the gel from settling. At the end of the 2-hr period, the samples are diluted to 5 ml with ice-cold wash buffer and centrifuged at 1500 *g* for 15 min in a refrigerated IEC centrifuge. The supernatant is aspirated and the gel is resuspended again in 5 ml of wash buffer. After a second centrifugation, the supernatant is removed and the gel is counted in a Tracor gamma counter. When [^{125}I]MIG was used in the absence of CHAPS extract protein, 0.08 to 0.3% (600–2500 cpm out of 720,000 cpm) of the total counts adsorbed to the gel in a manner unaffected by the absence or presence of 0.1 *M* GlcNAc.

Analysis of the Data

A typical binding experiment using the solubilized receptor is shown in Fig. 2. Two independent methods of estimating the nonspecific binding are used. The first method utilizes the presence of excess unlabled glucagon during the binding reaction. The second method uses the presence of GlcNAc during exposure of the hormone receptor complex to WGL-Sepharose to block the adsorption of the [^{125}I]MIG–receptor complex

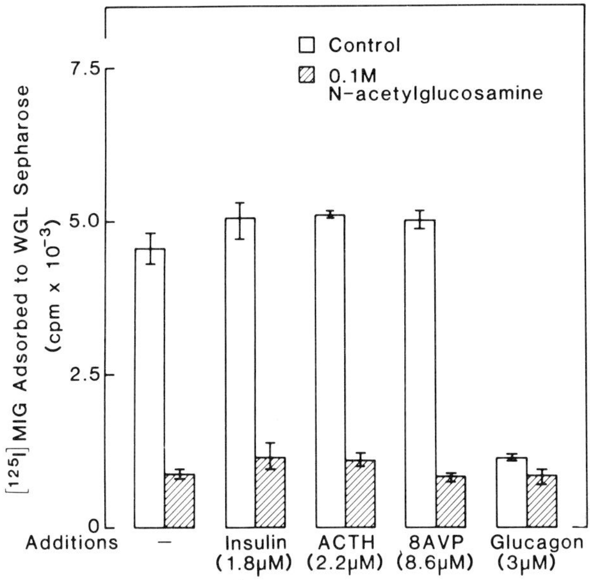

FIG. 3. Effect of various peptide hormones on [^{125}I]MIG binding to the solubilized receptor and the effect of GlcNAc during the adsorption to WGL-Sepharose. The binding reaction was performed in a final volume of 300 μl at 1 n*M* [^{125}I]MIG for 90 min at 0°. During the binding reaction, 210 μg of CHAPS extract protein was present. The concentration of CHAPS was 2.4 *M*. When present, the concentrations of various hormones were as follows: insulin, 1.8 μM; ACTH, 2.2 μM; 8AVP, 8.6 μM; and glucagon, 3.0 μM. After the binding reaction, the suspension of WGL-Sepharose with or without GlcNAc was added. The final concentration of GlcNAc was 0.1 *M*. The samples were further incubated for 2 hr with moderate shaking, then processed as described under Methods. Values are mean ± SD of triplicate determination.

onto the lectin-Sepharose. It can be readily seen that the amount of [^{125}I]MIG adsorbed to WGL-Sepharose is the same in the presence of excess unlabeled glucagon during the binding reaction, or GlcNAc during exposure to WGL-Sepharose. Further addition of excess unlabeled glucagon during the binding reaction followed by addition of GlcNAc during exposure to WGL-Sepharose does not alter the amount of [^{125}I]MIG bound. These observations indicate the nonspecific background in this assay results from direct interaction of [^{125}I]MIG with the WGL-Sepharose rather than by adsorption to a glycoprotein, which, in turn, interacts with the WGL-Sepharose. In Fig. 2, the difference in counts between no additions and the other three conditions is taken as the specific counts bound to the glucagon receptor. The assumption here is that this difference in counts represent [^{125}I]MIG binding to WGL-Sepharose

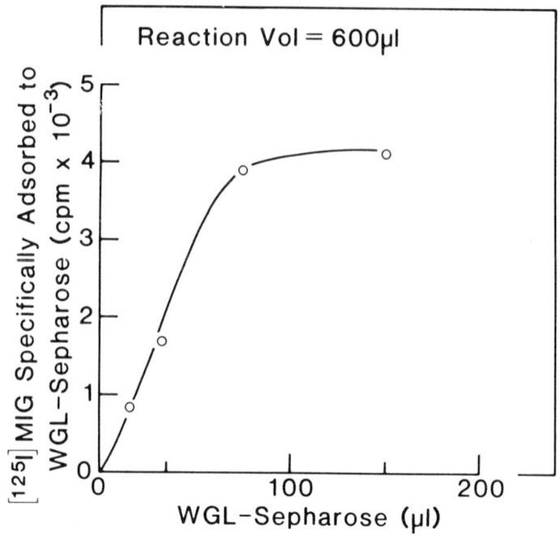

FIG. 4. [^{125}I]MIG specifically adsorbed to varying amounts of WGL-Sepharose. CHAPS extract protein was incubated with 1 n*M* [^{125}I]MIG and other additives for 90 min at 0° in a final volume of 300 μl. Then, indicated amounts of WGL-Sepharose was added as a slurry (total vol = 300 μl). The samples were then shaken gently for 2 hr in the cold. The gels were washed and counted. Nonspecific binding was measured in the presence of 1 μM glucagon. Nonspecific binding was 18 μl = 920 cpm, 38 μl = 1680, 75 μl = 2784, and 150 μl = 4168 cpm. All values are means of triplicate determinations. Coefficient of variance was less than 10% in all cases.

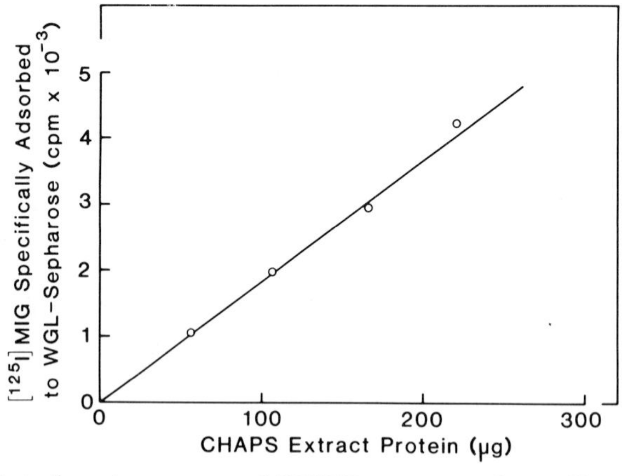

FIG. 5. Effect of varying amounts of CHAPS extract protein on [^{125}I]MIG specifically bound. The indicated amounts of CHAPS extract protein were incubated with [^{125}I]MIG as described under Methods. In all cases during the binding reaction, the concentration of CHAPS was 3 m*M*. Nonspecific binding measured in the presence of 3 μM glucagon was as follows: 55 μg protein, 1046 cpm; 110 μg protein, 1237 cpm; 165 μg protein, 1552 cpm; and 220 μg, 1562 cpm. The nonspecific counts were subtracted from the total to obtain specific binding. All values are mean of triplicate determinations. Coefficient of variance was less than 7% in all cases.

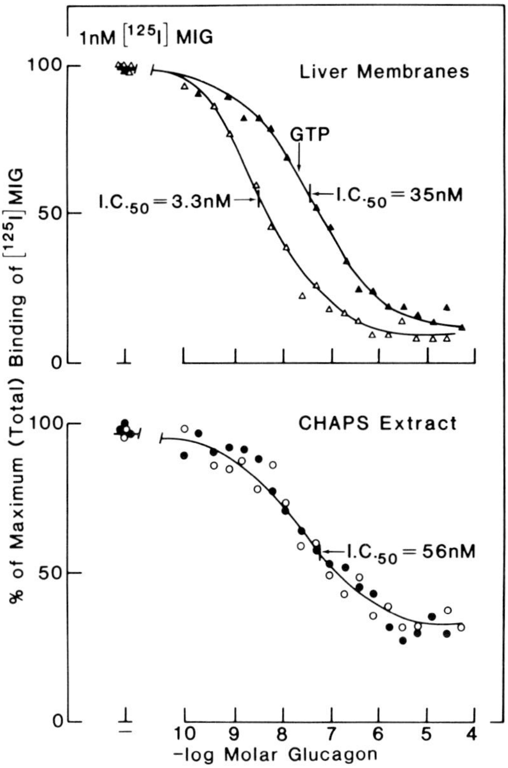

FIG. 6. Effect of various concentrations of unlabeled glucagon on [^{125}I]MIG binding to liver membranes (upper panel) and CHAPS extract protein (lower panel). [^{125}I]MIG binding to membranes and CHAPS extract protein was carried out in the presence of 25 m*M* NaHepes, 1 m*M* EDTA, 2 m*M* $MgCl_2$, 100 μM GTP, a nucleoside triphosphate regenerating system consisting of 20 m*M* creatine phosphate, 0.2 mg/ml creatine phosphokinase and 0.02 mg/ml myokinase, and 1 n*M* [^{125}I]MIG. The assay for [^{125}I]MIG binding to membranes also contained 0.1% BSA and was carried out in a final volume of 100 μl containing 75 μg protein. After a 20 min incubation at 32°, the samples were filtered through cellulose acetate filters and counted. Binding of [^{125}I]MIG to CHAPS extract protein was carried out in a final volume of 300 μl containing 85 μl of protein per assay tube. Total binding (cpm) was liver membranes without GTP, 13,600 ± 513; with GTP, 3623 ± 127; CHAPS extract without GTP, 2378 ± 113; with GTP, 2133 ± 30 cpm.

via the glucagon receptor which is a glycoprotein. To demonstrate that indeed the glycoprotein specifically recognizes glucagon, we compared the effect of addition of various peptide hormones during the binding reaction. One such experiment is shown in Fig. 3. Here it can be seen that only glucagon blocks [^{125}I]MIG binding while insulin, ACTH, and vasopressin are without effect. This experiment establishes that the glycoprotein specifically recognizes glucagon, indicating that it is the glucagon receptor.

The assay is performed at saturating levels of WGL-Sepharose, such that increasing the amount of WGL-Sepharose does not result in increased [^{125}I]MIG–receptor complex bound to the gel (Fig. 4). In contrast, the extent of [^{125}I]MIG specifically bound to WGL-Sepharose is proportional to the amount of CHAPS extract protein present during the binding incubation as shown in Fig. 5. These two parameters need to be cross calibrated so that for each batch of WGL-Sepharose used, the specific binding is proportional to the extract protein added.

The concentration range in which the glucagon competes with [^{125}I]MIG for binding to the receptor in detergent solution is similar to that observed in the membranes, provided that guanine nucleotides are present during the membrane binding assay (Fig. 6). Taken together, all these data indicate that the binding activity we measure in CHAPS solution represent the solubilized glucagon receptor.

Comments

A number of precautions taken during the performance of the assay are useful in obtaining optimal results. We have found that removal of bovine serum albumin from the binding reaction mixture greatly reduces the nonspecific binding of [^{125}I]MIG to the WGL-Sepharose. Consequently, our assay never contain more than 0.01% bovine serum albumin. Plasticware is used throughout, since it appears plastic adsorbs much less [^{125}I]MIG than glass. During the coupling of WGL to the CNBr-activated Sepharose, a top stirrer should be used for mixing. Vigorous shaking should be avoided since it generally results in significant decreases in the binding capacity of the gels. The WGL-Sepharose is not used for more than 4–5 weeks after it is synthesized. There appears to be significant time-dependent decrease in the binding capacity of the wheat germ lectin.

Acknowledgments

This research was supported by NIH Grants AM-26905, CA-29808, AM-27685 (DERC), and an Established Investigator Award to RI from the American Heart Association.

[3] Assaying Binding of Nerve Growth Factor to Cell Surface Receptors

By RONALD D. VALE and ERIC M. SHOOTER

Introduction

In the early 1950s, Levi-Montalcini and Hamburger discovered a factor produced by a mouse sarcoma which induced neurite outgrowth from sympathetic and dorsal root ganglion cells. On the basis of its activity, this substance was named nerve growth factor (NGF). NGF was shown to be a polypeptide, and much has been learned of its protein structure and physiological actions in the 30 years which have elapsed since its discovery.[1,2] It is acknowledged that NGF plays an essential role in the development of the sympathetic and sensory nervous systems, both in terms of promoting neuronal survival and possibly in chemotactically directing axons towards appropriate target tissues. NGF (referring to the β subunit of the 7 S NGF complex) can be prepared in milligram quantities from the submaxillary glands of adult male mice by a variety of previously reported methods.[3,4]

An important breakthrough in attempting to understand the mechanism of action of NGF has been the identification of specific cell surface receptors for this hormone. In the early 1970s, several investigators prepared high specific activity, radioiodinated derivatives of NGF which retained biological activity and were able to show specific binding of [^{125}I]NGF to cells which respond to the hormone. This chapter describes methods both for the iodination of NGF and for assaying NGF receptors by reversible binding techniques. The procedures presented are ones currently in use in this laboratory. Alternative methods employed by other investigators will be mentioned but will not be presented in detail. For a complete discussion of NGF receptors, the reader is referred to some recent reviews on this subject.[5,6]

[1] H. Thoenen and Y.-A. Barde, *Physiol. Rev.* **60,** 1284 (1980).
[2] B. A. Yankner and E. M. Shooter, *Annu. Rev. Biochem.* **51,** 845 (1982).
[3] L. E. Burton, W. Wilson, and E. M. Shooter, *J. Biol. Chem.* **253,** 7807 (1978).
[4] W. C. Mobley, A. Schenker, and E. M. Shooter, *Biochemistry* **15,** 5543 (1976).
[5] R. A. Bradshaw, N. V. Costrini, C. J. Morgan, J. F. Tait, and S. A. Weinman, *Prog. Clin. Biol. Res.* **79,** 271 (1982).
[6] R. D. Vale, C. E. Chandler, A. Sutter, and E. M. Shooter, *Recept. Recognition, Ser. B* (in press).

ISBN 0-12-182009-2

Iodination

In 1973, Herrup and Shooter[7] prepared an iodinated derivative of NGF using a lactoperoxidase labeling technique. As many as 4 mol of iodine could be successfully incorporated per mole of NGF without a loss of biological activity. The lactoperoxidase labeling procedure for NGF was subsequently modified by Sutter *et al.*[8] and is the preparation currently being used in this laboratory. This iodination results in the incorporation of 0.5 mol of iodine per mole of NGF, and the $[^{125}I]$NGF product binds with high specificity to cell surface receptors on a number of cell types.

The reaction is carried out at room temperature in an iodination hood in a 10 × 75-mm glass tube. Reagents are added in the following order:

1. 0.1 *M* potassium phosphate solution, pH 7.4 (40.5 ml of 0.1 *M* K_2HPO_4 and 9.5 ml of 0.1 *M* KH_2PO_4)
2. 3 mCi $Na^{125}I$ (10 mCi/ml)
3. 50 μg NGF (generally 50 μl from a 1 mg/ml NGF solution in 0.02% acetic acid)
4. 150 ng lactoperoxidase (Sigma) (15 μl from a 30 μg/ml stock solution)
5. 15 μl of a 0.003% H_2O_2 solution (prepared by a 10,000 fold dilution from a 30% stock solution just prior to use)

The total reaction volume is 200 μl, and the volume of 0.1 *M* phosphate buffer can be adjusted to accommodate changes in volumes of the other reactants. Generally, 30 μl of $Na^{125}I$ (corresponding to 3 mCi) and 50 μl of a 1 mg/ml NGF solution are added. In this case, the total volume is increased to 200 μl by adding 90 μl of phosphate buffer.

After each addition, reactants are gently mixed. When the reaction mixture is complete, the tube is then capped and left at room temperature for 30 min. At this point, a fresh 10,000-fold dilution of a 30% H_2O_2 stock is made and an additional 15 μl of this 0.003% solution is added.

After a further 30 min incubation (60 min reaction time in total), the reaction is stopped by the addition of 200 μl of 0.4% acetic acid followed by 600 μl of 0.01 *M* sodium acetate, pH 4 (90 ml of 0.01 *M* sodium acetate and 410 ml of 0.01 *M* acetic acid) containing 0.5 *M* NaCl and 1 mg/ml bovine serum albumin (iodination buffer). Triplicate 10-μl aliquots are removed from the iodinated mixture and added to 1 ml samples of iodin-

[7] K. Herrup and E. M. Shooter, *Proc. Natl. Acad. Sci. U.S.A.* **70,** 3884 (1973).

[8] A. Sutter, R. J. Riopelle, R. M. Harris-Warrick, and E. M. Shooter, *J. Biol. Chem.* **254,** 5972 (1979).

ation buffer for a later determination of trichloroacetic acid (TCA) precipitability.

In order to remove some of the unincorporated ^{125}I, the reaction mixture is dialyzed overnight. Dialysis membrane (Spectropore, Scientific Products, molecular weight cutoff--3500), presoaked in iodination buffer, is cut so that it is one layer thick. It is placed on top of the iodination tube and secured in place either with a tight-fitting plastic cap which had its top cut away with a razor blade or with a section of tight-fitting tygone tubing. The tube is inverted, and the dialysis membrane is immersed in 200 ml of 0.01 *M* acetate buffer, pH 4, 0.5 *M* NaCl (without bovine serum albumin). The solution in changed after 3 hr, and dialysis is continued overnight. In order to determine recovery of NGF after dialysis, triplicate 10 μl samples are transferred to 1 ml of iodination buffer for subsequent determinations of TCA precipitability.

In order to minimize levels of nonspecific [^{125}I]NGF binding during the receptor binding assay, the dialyzed [^{125}I]NGF is filtered through Centriflo CF 50A filters (Amicon, 2100 CF 50A) to remove aggregated molecules of [^{125}I]NGF. This step reduces nonspecific binding by as much as a factor of 10. Excess adsorption of [^{125}I]NGF to the filter can be avoided by presoaking the Centriflo filters for 24 hr at 4° in iodination buffer containing 1 mg/ml protamine sulfate. The Centriflo filter is placed in a plastic support (Amicon, CSIA) which fits on top of a standard 50-ml plastic tissue culture centrifuge tube and is dried by centrifugation at 2000 rpm for 10 min in a clinical centrifuge. The filter and support are then placed over another 50 ml centrifuge tube, the dialyzed [^{125}I]NGF is added to the bottom of the filter and the solution is passed through the filter by centrifugation as described above. The solution is collected from the bottom of the tube, the volume measured, and triplicate 10 μl samples of the [^{125}I]NGF solution are aliquoted into 1 ml samples of iodination buffer for TCA precipitability determinations.

TCA precipitabiity from samples of [^{125}I]NGF removed (1) prior to dialysis, (2) after dialysis, and (3) after centrifugation through the Centriflo filter is determined at this stage. Initially, as mentioned above, 10-μl aliquots are added to 1 ml of iodination buffer. These samples are then further diluted by adding 10 μl of the above solution to 490 μl of iodination buffer, followed by the addition of 500 μl of 20% TCA. Samples are vortexed, incubated on ice for 20 min, and centrifuged for 10 min at 2500 rpm in a clinical centrifuge. One-half volume of the supernatants (500 μl) is removed to a new tube. The pellet plus the remaining one-half volume supernatant and the one-half volume supernatant samples are counted separately in a gamma counter and triplicate samples are averaged. The

TABLE I
IODINATION OF NERVE GROWTH FACTOR[a]

	cpm in pellet	cpm in supernatant	Total cpm	TCA precipitable (%)
Predialysis	295300	57400	352700	83.7
Postdialysis	285300	8000	293400	97.3
Filtered	117200	4600	121900	96.2

[a] Nerve growth factor was iodinated according to the method described in the text. This table shows the recovery of [^{125}I]NGF at various stages in the preparation. The specific activity of this particular preparation was 59 cpm/pg and the final concentration of [^{125}I]NGF was 19.9 μg/ml.

one-half volume supernatant counts are subtracted from pellet plus one-half volume supernatant samples, and the one-half volume supernatant counts are doubled to yield final radioactivity in the pellets and supernatants, respectively.

Using these values of TCA precipitability, it is possible to calculate the specific activity and recovery of the [^{125}I]NGF preparation (see Table I). Specific activity is determined by multiplying the TCA precipitable radioactive counts in the predialysis sample by 10^4 (dilution correction) and dividing by 5×10^7 pg NGF (the starting material added). The final concentration of [^{125}I]NGF can be determined by comparing the difference in TCA precipitable counts in the predialysis sample versus the filtered [^{125}I]NGF by the following ratio:

$$\text{Final } [^{125}\text{I}]\text{NGF Conc. } (\mu\text{g/ml}) = \frac{\text{TCA ppt cpm after filtration}}{\text{TCA ppt cpm before dialysis}} \times 50\ \mu\text{g/ml}$$

Table I summarizes the results of an [^{125}I]NGF preparation. Incorporation of ^{125}I into TCA precipitable NGF generally is between 75 and 90%. The majority of the [^{125}I]NGF (90%) is recovered from dialysis; however, approximately 50% of [^{125}I]NGF is lost during centrifugation through the Centriflo filter, the step included to reduce nonspecific binding of [^{125}I]NGF during the receptor binding assay. Recoveries of [^{125}I]NGF off the Centriflo filter can be improved by washing the filter, after the initial centrifugation step, with 0.5 ml of iodination buffer containing 0.1% Triton X-100 and centrifuging as before (P. Grob and M. Bothwell, unpublished observations). This step is reported to increase the recovery of [^{125}I]NGF to 90%, and the Triton X-100 is diluted to such an extent in a typical [^{125}I]NGF binding assay that it does not interfere with this assay.

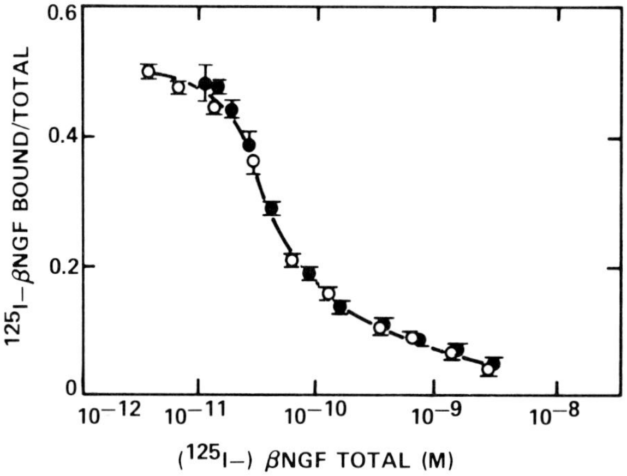

FIG. 1. Comparison of the binding affinities of [^{125}I]NGF and unlabled NGF. Dorsal root ganglia cells (2.8×10^6/ml) were incubated at 37° for 45 min with various concentrations of [^{125}I]NGF (○) (specific activity 58 cpm/pg), or with 0.3 ng/ml of [^{125}I]NGF and increasing concentrations of unlabeled NGF (●). Triplicate 100-μl aliquots were assayed for [^{125}I]NGF binding as described in the text. Bars indicate the standard derivation of the mean. Reproduced with permission from Sutter *et al.*[8]

The final [^{125}I]NGF preparation has TCA precipitability greater than 90%, final concentrations in the range of 20–30 μg/ml and specific activities between 50 and 80 cpm/pg. As shown in Fig. 1, the binding properties of unlabeled and ^{125}I-labeled NGF to cell surface receptors are very similar. [^{125}I]NGF also exhibits biological activity which is identical to the unlabeled hormone, as determined by assaying NGF-induced neurite outgrowth in dorsal root ganglion cells. [^{125}I]NGF is stored at 4° and shows optimum specific binding activity within a week after iodination. For best results, it is recommended that [^{125}I]NGF be used within 2 weeks of preparation.

A variety of other preparations of iodinated NGF have also been described. These include labeling procedures involving chloramine T,[9] a Bolton and Hunter succinimide ester reagent,[10] and more recently, a solid phase iodination procedure using Enzymo beads.[11] Tait *et al.*[12] have prepared [^{125}I]NGF with very high specific activity (4–6 mol iodine/mol of

[9] W. A. Frazier, L. F. Boyd, and R. A. Bradshaw, *J. Biol. Chem.* **249,** 5513 (1974).

[10] S. P. Banerjee, P. Cuatrecasas, and S. H. Snyder, *J. Biol. Chem.* **251,** 5680 (1976).

[11] S. E. Buxser, D. J. Kelleher, L. Watson, P. Puma, and G. L. Johnson, *J. Biol. Chem.* **258,** 3741 (1983).

[12] J. F. Tait, S. A. Weinman, and R. A. Bradshaw. *J. Biol. Chem.* **256,** 11086 (1981).

NGF dimer) using a lactoperoxidase catalyzed reaction. This highly radioactive preparation enabled these investigators to examine dissociation of NGF from its receptor at very low ligand concentrations under dilution conditions. The preparation of $[^{125}I]$NGF described here allows one to measure reliabily binding of $[^{125}I]$NGF to a number of cell types at concentrations as low as 10 p*M*.

Determining NGF Receptor Binding

In order to successfully assay NGF receptors on cell surfaces, one must (1) prepare NGF labeled with ^{125}I to high specific activity in order that receptors with high affinity (K_D on the order of 50 p*M*) and low site numbers (a few thousand per cell or less) can be detected, (2) have a reasonably pure cell population with which to study so that heterogeneous binding cannot be attributed to the presence of a variety of cell types, and (3) have an assay system which is rapid and reliable and will yield accurate values for the affinities and numbers of receptors in the preparation being studied. The preparation of a $[^{125}I]$NGF derivative has already been discussed. The following section will describe methods for preparing cells for binding assays and determining the levels of specific $[^{125}I]$NGF binding to these cells. In addition, procedures are discussed which distinguish two populations of NGF receptors on the basis of their steady-state binding, dissociation kinetics, trypsin sensitivity, and Triton X-100 solubility.

Cell Preparation

Sympathetic and sensory neurons are primary target tissues for NGF. The age of chick embryonic dorsal root ganglia used for binding studies is important, since ganglia obtained from older embryos lose their responsiveness to NGF, a change which occurs concomitantly with a decrease in their levels of NGF receptors.[13] Unlike sensory ganglia, no significant difference in NGF receptor numbers or affinities was detected in chick lumbar sympathetic ganglia between embryonic days 6.5 to 20.[14]

Dorsal root ganglia are prepared from 8-day-old chick embryos by the following procedure.[8] Dorsal root ganglia are dissected and placed in Ca^{2+},Mg^{2+}-free phosphate buffered saline (CMF-PBS: 137 m*M* NaCl, 2.68 m*M* KCl, 1.5 m*M* KH_2HPO_4, 0.65 m*M* Na_2HPO_4,pH 7.4). In order to facilitate dissociation, ganglia (200–1000) are incubated in 5 ml of CMF-PBS containing 0.012% trypsin (Worthington) and 0.0012% DNase I (Worthington) for 10 min at 37°. Soybean trypsin inhibitor (Worthington)

[13] K. Herrup and E. M. Shooter, *J. Cell Biol* **67,** 118 (1975).

[14] E. W. Godfrey and E. M. Shooter, unpublished observations (1980).

and fetal calf serum are then added to final concentrations of 150 μg/ml and 10%, respectively. Dissociation is achieved by gentle trituration through a 5-ml pipet. The cell suspension is filtered then through a Swinnex 13 filter holder loaded with three layer of 40 μm nylon mesh (Tetko Inc. #HC3-48). An enriched neuronal population (80–90%) can be obtained by preplating the cell suspension on 35 cm^2 tissue culture plates and incubating at 37° in an incubator with an atmosphere of 88% air/12% CO_2 for 60 min. Nonneuronal cells such as glia and fibroblasts adhere to the dish, and medium containing the less adherent neurons is removed. Nonneuronal cells have receptors for NGF but do not display the high affinity receptor found on sensory neurons.[15] Next, neurons are centrifuged (500 *g*, 5 min) and resuspended in PBS containing Ca^{2+} and Mg^{2+} (0.9 m*M* $CaCl_2$, 0.5 m*M* $MgCl_2$ added to the previously mentioned CMF-PBS solution) along with 1 mg/ml each of glucose and bovine serum albumin. This PBS solution containing glucose and bovine serum albumin shall be referred to as "binding buffer." This procedure yields 40,000 cells per dorsal root ganglion while between 30 and 40 ganglia can be successfully dissected from each egg.

Godfrey and Shooter[14] modified the above procedure for obtaining a cell dissociate from chick lumbar sympathetic ganglia. After dissection, ganglia are kept in ice cold CMF-PBS containing 1 mg/ml glucose, then cleaned of adhering connective tissue with forceps. Ganglia are then exposed to 0.01% trypsin (0.1% for embryos older than 11 days) and 0.001% DNase I in CMF-PBS for 10 min at 37°. Fetal calf serum (5%) and soybean trypsin inhibitor (80 μg/ml) are added, and the ganglia are triturated (30–40 times) with a 9-in. pasteur pipet. The cell suspension is then filtered through nylon mesh as described above. The yield of cells increases from 10,000 per ganglion on day 8 to 40,000 per ganglion at day 14, then falls to 14,000 per ganglion at day 20. Cell viability is about 95%.

Because dissecting tissue from animals is time consuming and material is scarce, a number of laboratories have begun to study the characteristics of NGF receptors on cell lines which can be grown continuously in culture. The PC12 cell line, derived from a rat pheochromocytoma,[16] the A875 cell line, derived from a human melanoma,[17] and some cell lines derived from neuroblastomas,[18] all contain NGF receptors. The affinities

[15] A. Sutter, R. J. Riopelle, R. M. Harris-Warrick, and E. M. Shooter, *in* "Transmembrane Signalling" (M. Bilensky, R. J. Collier, D. F. Steiner, and C. F. Fox, eds.), p. 659. Alan R. Liss, Inc., New York, 1979.

[16] A. S. Tischler and L. A. Greene, *Nature (London)* **258,** 341 (1975).

[17] R. N. Fabricant, J. E. De Larco, and G. J. Todaro, *Proc. Natl. Acad. Sci. U.S.A.* **74,** 565 (1977).

[18] K. H. Sonnenfeld and D. N. Ishii, *J. Neurosci. Res.* **8,** 375 (1982).

TABLE II
NGF RECEPTORS ON VARIOUS CELL TYPES

Cell type	Presence of two receptor species: Steady-state analysis	Presence of two receptor species: Kinetic analysis	Equilibrium dissociation constant (*M*)	Receptor site per cell
Chick embryonic dorsal root ganglia (day 8)[a,b]	Yes	Yes	Site I: 2.3×10^{-11}	Site I: 3,000
			Site II: 1.7×10^{-9}	Site II: 45,000
	Yes	Yes	Site I: 3.3×10^{-11}	Site I: 4,000
	Yes	Yes	Site II: 1.7×10^{-9}	Site II: 47,000
Chick embryonic sympathetic ganglia (day 9)[c]	Yes	Yes	Site I: 3.0×10^{-11}	Site I: 4,000
			Site II: 1.8×10^{-9}	Site II: 38,000
PC12 pheo-chromoytoma cell line[d,e]	Yes	Yes	Site I: 5.0×10^{-11}	Site I: 2,500
			Site II: 1.0×10^{-9}	Site II: 50,000
	No	Yes	Site I: 2×10^{-10}	Site I: 15,000
			Site II: 2×10^{-10}	Site II: 45,000
A875 melanoma cell line[f]	No	No	Site II 1×10^{-9}	Site II: 500,000–700,000
Neuroblastoma cell lines[g]				
SH-SY5Y	No	No	Site I: 4.9×10^{-11}	Site I: 700
MC-IXC	Yes	Yes	Site I: not determined	Site I: not determined
			Site II: 1.7×10^{9}	Site II: 250,000

[a] A. Sutter, R. J. Riopelle, R. M. Harris-Warrick, and E. M. Shooter, *J. Biol. Chem.* **254,** 5972 (1979).
[b] E. J. Olender, B. J. Wagner, and R. W. Stack, *J. Neurochem.* **37,** 436 (1981).
[c] E. W. Godfrey and E. M. Shooter, unpublished results (1980).
[d] G. E. Landreth, D. A. Estell, and E. M. Shooter, unpublished results (1981).
[e] A. L. Schechter and M. A. Bothwell, *Cell* **24,** 867 (1981).
[f] F. N. Fabricant, J. E. De Larco, and G. J. Todaro, *Proc. Natl. Acad. Sci. U.S.A.* **74,** 565 (1977).
[g] K. H. Sonnenfeld and D. N. Ishii, *J. Neurosci. Res.* **8,** 375 (1982).

and numbers of NGF receptors on various cell types are listed in Table II. All of these continous cell lines are derived from tumors of tissues of neural crest origin. The PC12 and neuroblastoma cell lines respond to NGF by producing neurites. The A875 cell line does not exhibit this classic response to NGF, but NGF does cause an increase in cell number, possibly implicating NGF as a survival or proliferative factor for these melanoma cells. One potentially useful attribute of the A875 cell line is the large number of receptors which are present as shown in Table II.

Our laboratory has primarily studied receptors on the PC12 and A875 cell lines. PC12 cells are grown in 100-cm^2 Falcon dishes in Dulbecco's modified Eagle's medium (DMEM) containing 10% fetal calf serum and 5% horse serum (both obtained from Grand Island Biological Company) at 37° in an atmosphere of 88% air/12% CO_2. A875 melanoma cells are grown in 75-cm^2 Falcon flasks in DMEM containing 10% fetal calf serum at 37° in an atmosphere of 88% air/12% CO_2. These cells are removed for passage at confluency with an incubation in 5 ml of CMF-PBS containing 1 m*M* ethylenediamine tetraacetic acid (EDTA) for 5 min at 37°.

Binding of [^{125}I]NGF to PC12 or A875 cells is easily performed in cell suspensions. PC12 cells are first washed twice on the dish with 10 ml of PBS containing 1 mg/ml each of glucose and bovine serum albumin (binding buffer). Cells are then removed by trituration with 10 ml of binding buffer, counted in a hemocytometer, and used immediately. A875 cells are removed by incubation with 5 ml of CMF-PBS containing 1 m*M* EDTA for 5 min at 37°. Cells are centrifuged (500 *g*, 5 min) and resuspended in 10 ml of binding buffer and washed twice more in this manner before being counted in a hemocytometer and used for experimentation. Since the A875 cell line is derived from a melanoma of human origin, certain precautions are used when handling them. Gloves are always used, and material which comes in contact with these cells is treated with detergent and autoclaved.

It should be mentioned that many dynamic properties of the receptor may be different for cells in suspension as opposed to cells attached to a substratum. For example, down regulation of the receptor in PC12 cells, presumably a reflection in part of the internalization and degradation of the receptor, occurs more rapidly in attached than suspended cells. The inclusion of fetal calf serum as opposed to bovine serum albumin in the medium may similarly affect regulation of NGF receptors.

Assaying NGF Binding

In order to assay [^{125}I]NGF binding, a cell suspension is prepared in an appropriate physiological buffer, either the phosphate-buffered saline previously described or Krebs–Ringer solution (137.3 m*M* NaCl, 4.74 m*M* KCl, 2.56 m*M* $CaCl_2$, 1.18 m*M* KH_2PO_4, 1.18 m*M* $MgSO_4$) buffered with 10 m*M* 4-(-2-hydroxylethyl)-1-piperazineethanesulfonic acid (Hepes) to pH 7.4. Bovine serum albumin (1 mg/ml) is added to the buffer to minimize nonspecific adsorption of [^{125}I]NGF, and plastic as opposed to glass tubes are used for all incubations for the same reasons. Glucose (1 mg/ml) is present to maintain cell viability. [^{125}I]NGF binding is sensitive to the pH of the medium (Fig. 2). Maximal binding to PC12 cells is achieved at

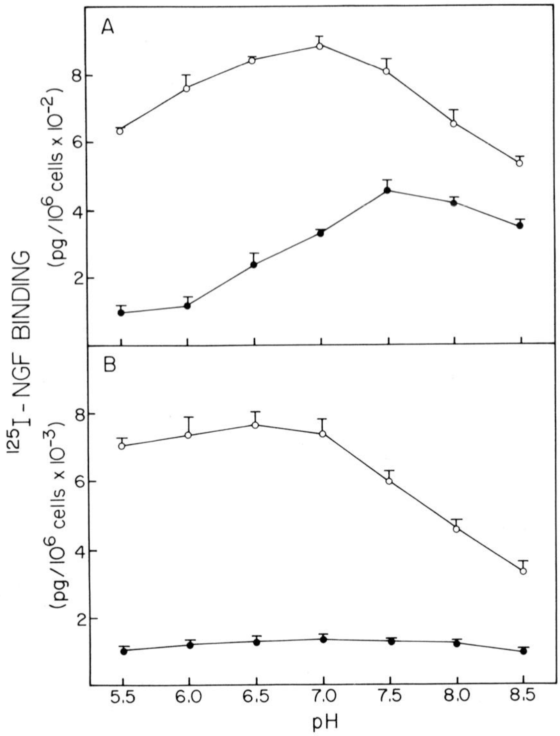

FIG. 2. Effect on pH on the binding of [^{125}I]NGF to PC12 cells. PC12 cells (1.2 × 10^6/ml) were incubated at the indicated pH in Krebs–Ringer solution containing 1 mg/ml each of glucose and bovine serum albumin and 10 m*M* Hepes. [^{125}I]NGF was then added at a final concentration of 50 p*M* (A) or 500 p*M* (B) and after 45 min at 23°, total, nonspecific and slowly dissociating binding were assayed as described in the text. Values for specific total (○) and slowly dissociating binding (●) are the means of triplicate determinations. Bars indicate standard deviations.

about 6.5 and falls to about 40% of the maximal level at pH 8.5. Our assays are performed at the physiological pH of 7.4 A cell concentration of 1 × 10^6/ml is generally adequate to assay [^{125}I]NGF binding with reasonable confidence.

To measure [^{125}I]NGF binding, [^{125}I]NGF bound to cells must be separated from that which is free in solution. It is important that this separation be efficient and rapid so that [^{125}I]NGF bound to sites with rapidly dissociating kinetics can be accurately measured. Such a separation can be achieved by centrifuging a suspension of cells containing [^{125}I]NGF through a solution of 0.15 *M* sucrose in binding buffer. After a given incubation period, 100-μl aliquots of the cell suspension (in triplicate) are layered gently over 200 μl of the 0.15 *M* sucrose solution in 400-μl microfuge tubes which had their caps previously removed. Thereafter, the tubes are centrifuged for 30 sec at 10,000 *g* in a model B Beckman microfuge. Cells require only about 2 sec to pellet to the bottom of the tube, thus enabling [^{125}I]NGF receptor complexes with short half-lifes to be detected. The microfuge tubes are then frozen in a dry ice/ethanol bath and placed in a block with a horizontal slit which allows one to cut the plastic tubes with a razor blade just above the cell pellet. The bottom of the tubes, containing cell-bound [^{125}I]NGF, and the tops of the tubes, containing free [^{125}I]NGF, are counted separately in a gamma counter. To measure nonspecific binding, cells are incubated with [^{125}I]NGF in the presence of an excess amount of unlabeled NGF (generally 10 μg/ml). The concentration of unlabeled NGF should be at least 100- to 200-fold above the K_d of the lower affinity NGF receptor. Nonspecific binding generally accounts for less than 10% of the total binding. This assay also works well for measuring binding of [^{125}I]NGF to membrane preparations.[19] A longer centrigugation time is required, however (minimum of 1 min), in order to completely pellet the membranes, and this time may depend upon the membrane preparation which is employed. Assay procedures for detecting solubilized receptors have also been described for superior cervical ganglia cells[20] and more recently for PC12 and A875 cells.[11]

In interpreting data, it is important to consider the conditions under which one assays NGF receptor binding. After initial binding, NGF–receptor complexes could become internalized, relocalized on the cell surface, phosphorylated or covalent crosslinked, as have been observed for other hormone–receptor complexes. Such events might alter the binding properties of the receptor. Furthermore, internalization of the NGF–receptor complex, which is known to occur, places the complex in an environment in which bound [^{125}I]NGF is no longer in equilibrium with free [^{125}I]NGF in the medium. One must carefully consider such a situation if one wishes to examine binding data by analysis which presumes the ligand and receptor to be in a reversible equilibrium. In order to exclude the influence of internalization, experiments can be performed at 4° a temperature at which endocytosis does not occur. Furthermore, internalization may affect steady-state levels of [^{125}I]NGF binding to a greater extent with longer incubations with the ligand. If binding of [^{125}I]NGF to PC12 cells is followed at 37°, maximum binding is achieved after 30 min

[19] R. J. Riopelle, M. Klearman, and A. Sutter, *Brain Res.* **199,** 63 (1980).
[20] N. V. Costrini and R. A. Bradshaw, *Proc. Natl. Acad. Sci. U.S.A.* **76,** 3242 (1979).

and remains at a steady state for the next 60 min. Thereafter, binding begins to decrease such that by 10 hr, its level drops to 25% of the previous maximum.[21] This decrease in binding is the result of a loss of cell surface receptors due to internalization and is accompanied by the degradation of [^{125}I]NGF in lysosomes and subsequent release of [^{125}I]monoiodotyrosine into the medium. Lysosomal degradation, but not internalization, can be blocked by the addition of primary amines such as methylamine or chloroquine which are capable of raising the intralysosomal pH. The decrease in receptor number after incubation with hormone has been termed "down regulation." This process has been observed for a number of hormone–receptor systems. However, even before a decrease in binding is observed, a portion of the cell-associated NGF is most likely internalized, and therefore the total binding measured is a composite of both membrane bound and intracellular [^{125}I]NGF.

Detection of Two Discrete Receptor Species

Analogous to many hormone–receptor systems, NGF binds to two receptor species which can be distinguished on the basis of (1) equilibrium binding, (2) dissociation kinetics, (3) proteolytic sensitivity, (4) pH sensitivity, and (5) Triton X-100 solubility. Understanding the roles and possible interrelationships of the two NGF receptors subtypes is a topic of great interest. In this section, methods are described which allow one to distinguish [^{125}I]NGF binding to these two receptor populations.

Steady-State Binding Analysis. There are a variety of techniques which are useful for evaluating reversible binding interactions between a ligand and its receptor. A general assumption is that this reaction obeys second-order chemical kinetics, so that equilibrium binding can be described on the basis of the association and dissociation kinetics of the ligand and receptor. A common procedure to characterize receptor binding is to perform saturation or competition experiments over a range of ligand concentrations and to determine the amount of bound and free ligand at each concentration of ligand used. It is important that the separation of bound versus free ligand be made accurately and that the ligand and receptor indeed be at equilibrium at the time of the assay in order to avoid erroneous interpretations.

The relationships between bound versus free ligand can be analyzed by a number of linear transformations,[22] the most popular one being Scatchard analysis.[23] By plotting the bound/free ratio on the ordinate versus

[21] P. G. Layer and E. M. Shooter, *J. Biol. Chem.* **258,** 3012 (1983).
[22] J. M. Boeynaems and J. E. Dumont, *J. Cyclic Nucleotide Res.* **1,** 123 (1975).
[23] G. Schatchard, *Ann. N.Y. Acad. Sci.* **51,** 660 (1949).

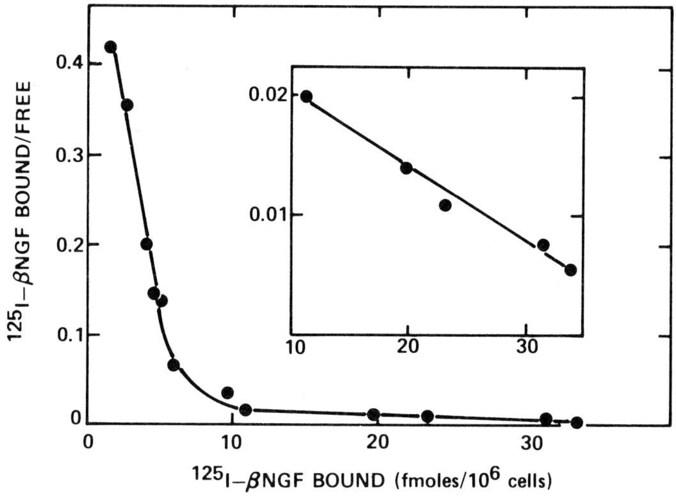

FIG. 3. Scatchard analysis of equilibrium of [^{125}I]NGF to 8-day-old chick embryo sensory ganglion cells. Cells (0.6 × 10^6/ml) were incubated at 37° for 45 min with various concentrations of [^{125}I]NGF (3 p*M* to 3.7 n*M*). Triplicate determinations of binding were made at each point as described in the text, nonspecific binding was subtracted, and the data were transformed into a Scatchard plot. Binding data for the low affinity region are expanded in the insert. Reproduced with permission from Sutter *et al.*[8]

the concentration of bound ligand on the abscissa, one can derive information on the affinity of the receptor as well as the number of receptors present in the preparation. One should be wary of even this simple form of data analysis, as there are a number of artifacts which can arise and lead to improper interpretations,[24,25] especially if Scatchard plots are not linear as discussed below.

Scatchard analysis for several ligand-receptor interactions, including NGF receptors (see Fig. 3), reveals curvilinear rather than linear plots and can therefore not be explained by a ligand interacting with a single population of noncooperative receptors. A number of different models have been evoked to explain nonlinear Scatchard plots including multiple receptor species, receptor cooperativity, interaction of ligand dimers (which could pertain to NGF since it is a dimer of two identical 13,000 molecular weight subunits) with a homogeneous class of receptors, and receptor interactions with a membrane bound effector molecule which can modulate receptor affinity.[22,26,27] Nonlinear curve-fitting computer

[24] J. G. Norby, P. Ottolenghi, and J. Jensen, *Anal. Biochem.* **102,** 318 (1980).

[25] I. M. Klotz, *Science* **217,** 1247 (1982).

[26] D. Rodbard and H. A. Feldman, this series, Vol. 36, p. 3.

[27] C. De Lisi and R. Chabay, *Cell Biophys.* **1,** 117 (1979).

programs have been used to model complex ligand–receptor interactions, thereby allowing one to compare the fit predicted by a particular model with the binding data. A discussion of these methods is beyond the scope of this article, but detailed information on this topic is available in a number of articles and reviews.[28–31] It is often difficult, if not impossible, to distinguish between models based upon equilibrium binding alone; however, the combination of equilibrium binding analysis with kinetic studies can provide valuable information of the mechanisms involved in ligand–receptor interactions.

Sutter *et al.*[8] observed curvilinear Scatchard plots for [^{125}I]NGF binding to dorsal root ganglion cells (Fig. 3). Kinetic analysis revealed that this complex binding was due to the presence of two binding species rather than to one receptor interacting in a negatively cooperative manner. One binding site was a high affinity, low capacity receptor (K_d = 23 p*M*; sites per cell = 3000), and the other a low affinity, high capacity receptor (K_d = 1700 p*M*; sites per cell = 45,000). The two receptor species were also observed at 2° as well as 37°, indicating that such a result was not an artifact of internalization. As shown in Table II, two classes of NGF receptors have been distinguished on other cell types as well.

Kinetic Analysis. The two receptor subtypes can also be distinguished on the basis of their dissociation kinetics. If [^{125}I]NGF is bound to cells and an excess (500- to 1000-fold) of unlabeled NGF is added, a portion of the [^{125}I]NGF dissociates within seconds while the remainder of the [^{125}I]NGF dissociates much more slowly ($t_{1/2}$ of dissociation = 10 and 30 min for chick dorsal root ganglion and PC12 cells, respectively) (Fig. 4). The rapidly and slowly dissociating populations correspond to NGF bound to the low and high affinity receptor species, respectively. The precise ratio of rapidly to slowly dissociating cell-bound [^{125}I]NGF depends upon the concentration of [^{125}I]NGF used. Low concentrations of ligand will bind primarily to high affinity sites which exhibit slow dissociation kinetics. With increasing concentrations of [^{125}I]NGF, more rapidly dissociating binding will be observed as lower affinity receptors become occupied.

Determining the amount of [^{125}I]NGF bound to rapidly and slowly dissociating receptors can be facilitated by performing the dissociation experiment with excess unlabeled NGF at 4° instead of 37°. As shown in

[28] R. D. Vale, A. DeLean, R. J. Lefkowitz, and J. M. Stadel, *Mol. Pharmacol.* **22,** 619 (1982).

[29] A. De Lean and D. Rodbard, *in* "The Receptors: A Comprehensive Treatise" (R. D. O'Brien, ed.), Vol. 1, p. 143. Plenum, New York, 1979.

[30] D. Hunston, *Anal. Biochem.* **63,** 99 (1975).

[31] H. A. Feldman, *Anal. Biochem.* **48,** 317 (1972).

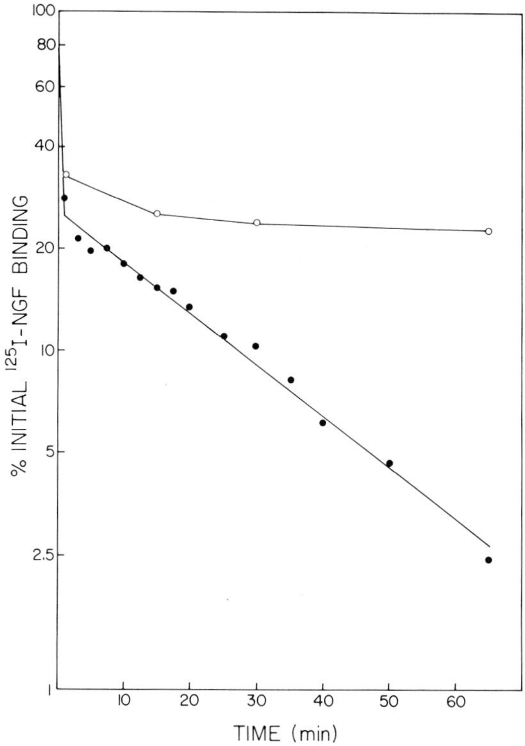

FIG. 4. Dissociation kinetics of [^{125}I]NGF bound to PC12 cells. PC12 cells (1×10^6/ml) were incubated with 615 p*M* [^{125}I]NGF for 45 min at 37°. Binding was assayed at this time as described in the text. Dissociation of cell-bound [^{125}I]NGF was then followed after the addition of 500 n*M* unlabeled NGF at 37° (●) or 4° (○). Triplicate determinations were made at each time point, and results are expressed as the percentage of the initial specific binding prior to dissociation. The line through the data points of the 37° dissociation is the best fit provided by linear regression.

Fig. 4, at 4°, [^{125}I]NGF -dissociates completely from rapidly dissociating sites, whereas in contrast to the experiment performed at 37°, [^{125}I]NGF remains stably bound to slowly dissociating receptors for a period of at least 30 min. The intercept of the line of slowly dissociating [^{125}I]NGF binding at 37° to the ordinate is approximately the same value as the binding remaining after a 30 min dissociation with unlabeled ligand at 4°. Thus, by adding a mixture of cells and [^{125}I]NGF to a 500- to 1000-fold excess of unlabeled NGF (generally 10 μg/ml) for 30 min at 4°, it is possible to selectively dissociate [^{125}I]NGF from rapidly dissociating receptors and thereby obtain an estimate of [^{125}I]NGF bound to slowly dissociating sites. By subtracting slowly dissociating binding from total binding, both corrected for nonspecific binding, one can calculate the amount of binding to the rapidly dissociating receptor subtype. As shown in Fig. 2, the proportion of slowly and rapidly dissociating binding depends upon the pH of the medium, with the ratio of slowly to rapidly dissociating binding being greater at more alkaline pH.

The assay for slowly dissociating binding is standardly performed as follows. After a given incubation of cells with [^{125}I]NGF, a 400-μl aliquot of the cell suspension is removed and added to 10 μl of a 0.5 mg/ml solution of unlabeled NGF in the bottom of a plastic 12 × 75-mm plastic tube on ice. After a 30 min incubation at 4°, three 100-μl aliquots are layered over a 0.15 *M* sucrose solution and centrifuged as described previously. Nonspecific binding is subtracted from the radioactivity to yield specific slowly dissociating [^{125}I]NGF binding. The simplicity of this assay allows one to examine ratios of rapidly and slowly dissociating [^{125}I]NGF binding for a large number of samples.

If the initial incubation of [^{125}I]NGF with cells is performed at 37°, some of slowly dissociating binding is in fact due to internalization of [^{125}I]NGF. It is not known or agreed upon how much internalized NGF contributes to the slowly -dissociating binding component; however, it is clear that slowly dissociating binding cannot be entirely explained by internalization since this component is also observed if the initial [^{125}I]NGF incubation is performed at 4° or if metabolic inhibitors are included in the medium. Furthermore, after a 30 min incubation at 37°, the majority of [^{125}I]NGF released from slowly dissociating sites is TCA precipitable, indicating that it has not been degraded in lysosomes. It is thus reasonable to conclude that slowly dissociating binding does reflect a property of a cell-surface NGF receptor.

Trypsin Sensitivity. The two classes of NGF receptors on PC12 cells can also be distinguished from one another by their resistance to trypsin at shown in Fig. 5.[32] If trypsin is incubated with PC12 cells prior to the addition of [^{125}I]NGF, 90–100% of the receptors binding activity is abolished (not shown). However, if [^{125}I]NGF is added first to the cells and trypsin is then added, one observes that some cell-bound [^{125}I]NGF is resistant to trypsin degradation. Trypsin releases the majority of bound [^{125}I]NGF within 5 min, after which time very little of the remaining cell-associated ligand can be removed. The trypsin-resistant [^{125}I]NGF corresponds to ligand associated with the slowly dissociating receptor. If the dissociation with excess unlabeled NGF is performed after trypsin treatment, no further [^{125}I]NGF is released. Similarly, if trypsinization is performed after dissociating [^{125}I]NGF bound to rapidly dissociating recep-

[32] G. E. Landreth and E. M. Shooter, *Proc. Natl. Acad. Sci. U.S.A.* **77,** 4751 (1980).

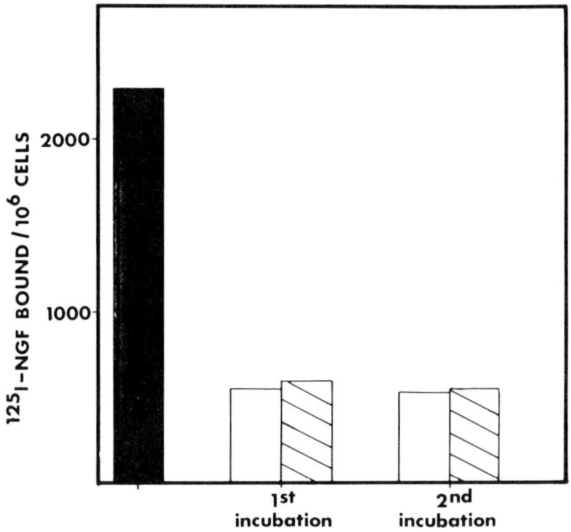

FIG. 5. Comparison of high affinity and trypsin-resistant [^{125}I]NGF binding. PC12 cells were preincubated for 30 min at 37° with 170 p*M* [^{125}I]NGF, the suspension was cooled to 0.5°, and total specific binding was determined (solid bar). Half of the sample received 180 p*M* unlabeled NGF prior to incubation for 30 min at 0.5°. The amount of remaining bound [^{125}I]NGF was determined (open bar, 1 st incubation). Trypsin was then added to this sample at a final concentration of 0.5 mg/ml and the incubation was continued for an additional 30 min at 0.5° before the remaining bound [^{125}I]NGF was measured (open bar, 2nd incubation). With the second sample (striped bars), the order of addition was reversed: the cells first recieived trypsin as described above and at the end of 30 min, the amount of bound [^{125}I]NGF was determined. Soybean trypsin inhibitor was added to a final concentration of 0.5 mg/ml followed by unlabeled NGF; after 30 min at 0.5;dg;, bound [^{125}I]NGF was measured. [^{125}I]NGF binding is expressed as pg bound per 10^6 cells. Reproduced with permission from Landreth and Shooter.[32]

tors, the remaining slowly dissociating [^{125}I]NGF can not be degraded. Although some of the trypsin-resistant [^{125}I]NGF may be internalized, it also represents a property of the NGF–slowly dissociating receptor complex at the cell surface, since trypsin-resistant NGF binding is observed if the initial [^{125}I]NGF incubation is performed at 4°.

It is not clear how ligand binding converts NGF receptors from a trypsin-sensitive to a resistant state. NGF itself is relatively refractory to inactivation by trypsin, so the trypsin sensitivity of the [^{125}I]NGF–receptor complex appears to reside with the receptor. It is possible that NGF induces a conformational change in the receptor which protects it from proteolysis or that some NGF–receptor complexes are localized in the membrane in such a fashion (possibly as receptor aggregates) so as to protect them from attack by trypsin. Another possibility is that the receptor contains only one protease susceptible site at or near the ligand binding site and that NGF binding competes with trypsin for access to this site. Thus, NGF bound to rapidly dissociating receptor will dissociate after a short period of time and will not protect the receptor from proteolytic attack, while NGF which is more stably associated with the slowly dissociating receptor will sterically block access of proteolytic enzymes to the protease-sensitive site on the receptor.

Trypsin-resistant [^{125}I]NGF binding can be routinely assayed in the following manner. [^{125}I]NGF is first incubated with cells for a given period of time. Then, a 400-μl aliquot of the cell suspension is added to 20 μl of a 10 mg/ml solution of trypsin (final concentration = 0.5 mg/ml) in the bottom of a 12 × 75-mm plastic tube and incubated for 30 min on ice. If there are broken cells in the preparation, trypsin can release a tangle of DNA from nuclei, but this DNA can be degraded by adding, along with the trypsin, 2 μl of a 10 mg/ml solution of DNase I (final concentration = 50 μg/ml). After the trypsin treatment, residual [^{125}I]NGF binding is assayed as described in detail previously. Nonspecific binding is subtracted from these radioactive counts. It is convenient to conduct parallel determination of slowly dissociating and trypsin-resistant binding since both are incubated for the same amount of time prior to being assayed.

Triton X-100 Solubility. Slowly and rapidly dissociating NGF receptors are also solubilized to a different extent by the detergent Triton X-100, as was first observed by Schechter and Bothwell.[33] Rapidly dissociating receptor binding can be completely solubilized by 0.5% Triton X-100 in 0.3 *M* sucrose, 3 m*M* $MgCl_2$, 20 m*M* Tris–HCl, pH 7.4, while between 40 and 100% of [^{125}I]NGF bound to slowly dissociating receptors is insoluble in this Triton X-100 solution. The Triton X-100 insoluble material of PC12 cells consists of nuclei and an array of cytoskeletal elements as judged by electron microscopy. Histones, actin, myosin, and tubulin are prominent proteins seen in electrophoretic analysis of Triton X-100 extracted cells. Since cytoskeletal elements are prominent consituents of Triton X-100 insoluble material, it has been suggested that [^{125}I]NGF binding which is insoluble in Triton X-100 may reflect an association of the receptor with the cytoskeleton. Definitive evidence of a linkage of the receptor to cytoskeletal proteins nonetheless remains to be demonstrated.

Triton X-100 insoluble [^{125}I]NGF binding can be assayed in one of two ways. The first method is to centrifuge PC12 cells and [^{125}I]NGF (500 *g*, 5 min), and to resuspend the cells in an equal volume of the Triton X-100/sucrose solution described above with 1 m*M* phenylmethylsulfonyl fluoride (PMSF) followed by an incubation on ice for 5 min. Then, triplicate

[33] A. L. Schechter and M. A. Bothwell, *Cell* **24**, 867 (1981).

100-μl samples are laid over the sucrose solution and centrifuged as described for the standard binding assay. Alternatively, 100-μl aliquots of cells and [^{125}I]NGF can be directly layered over 200 μl of the 0.5% Triton X-100, 0.3 *M* sucrose, 3 m*M* $MgCl_2$, 20 m*M* Tris–HCl, pH 7.4, solution instead of the 0.15 *M* sucrose solution and centrifuged in an identical fashion. As cells pass into the Triton X-100 solution, they are rapidly solubilized and material which is insoluble in this detergent and is sufficiently large to pellet to the bottom of the tube is considered Triton X-100 insoluble. Nonspecific binding samples are also centrifuged through the Triton X-100 solution and generally about 60–75% of the nonspecifically bound [^{125}I]NGF is insoluble after the centrifugation through the detergent solution. The two methods for determining Triton X-100 insoluble NGF binding provide similar results, although the rapid procedure yields somewhat higher levels of Triton X-100 insoluble binding. The rapid procedure is preferred as it limits problems such as proteolysis of cytoskeletons or receptors after detergent extraction as well as rebinding of [^{125}I]NGF to low affinity sites on the exposed cytoskeleton.

Conclusions

The preparation of a radiolabeled derivative of NGF along with a rapid method for determining the amount of cell-bound ligand have allowed the detection of NGF receptors on a number of cell types. Binding experiments have provided a good deal of information on these receptors. For example, the receptor population in many types of cells is heterogeneous, and various methods have been devised to distinguish the two receptor subtypes. Other techniques such as covalent crosslinking of [^{125}I]NGF to its receptor have recently provided new information on the receptor(s)[34,35] and are useful approaches that complement studies which employ reversible binding techniques. Furthermore, purification of the receptor, which has recently been described,[36] will provide more detailed knowledge of the biochemistry of the receptor which will hopefully assist in elucidating the mechanism whereby these receptors generate intracellular signals.

Acknowledgments

We would like to thank Drs. Tony Young, Rivka Sherman-Gold, Stuart Feinstein, Earl Godfrey, Gary Landreth, and Mark Bothwell for their kind help in reviewing this manuscript.

[34] J. Massague, B. J. Guillette, M. P. Czech, C. J. Morgan, and R. A. Bradshaw, *J. Biol. Chem.* **256,** 9419 (1981).

[35] M. Hosang and E. M. Shooter, *J. Biol. Chem.* in press (1984).

[36] P. Puma, S. E. Buxser, L. Watson, D. J. Kelleher, and G. L. Johnson, *J. Biol. Chem.* **258,** 3370 (1983).

[4] Assays for Calcitonin Receptors

By ANNE P. TEITELBAUM, ROBERT A. NISSENSON, and CLAUDE D. ARNAUD

Twenty years after the discovery of calcitonin,[1] its role in mineral homeostasis is still unclear. Calcitonin, which is secreted from the C cells of the thyroid in response to elevated serum calcium,[2] inhibits bone resorption and stimulates renal calcium excretion,[3] thereby reducing the calcium concentration in blood. These actions are counterregulatory to those of parathyroid hormone (PTH), which is secreted by the parathyroid gland in response to hypocalcemia. PTH increases the calcium concentration in blood by stimulating bone resorption and decreasing renal calcium excretion. However, it is generally acknowledged that the relative contribution of calcitonin to this regulatory system may be small.

The major target organs for calcitonin are bone and kidney. These tissues have specific, high-affinity receptors for both calcitonin and PTH that are coupled to activation of adenylate cyclase. Evidence indicates that these hormones act on different cell types in bone and kidney. Calcitonin affects osteoclast-like cells (prepared by the sequential digestion method of Wong and Cohn[4]), whereas PTH affects osteoblast-like cells. The calcitonin receptors demonstrated in intact bone[5] and the calvarial membrane[6] have not yet been assigned to a particular cell type.

The physiologic importance of calcitonin binding sites in other tissues is unclear. Cultured cell lines derived from human breast[7,8] and lung carcinomas[9] exhibit high-affinity receptors for calcitonin that are coupled to cyclic AMP formation. Specific binding of calcitonin has also been ob-

[1] P. F. Hirsch, E. F. Voelkel, and P. L. Munson, *Science* **146,** 412 (1964).

[2] D. H. Copp, *Recent Prog. Horm. Res.* **20,** 59 (1964).

[3] A. E. Broadus, *in* "Endocrinology and Metabolism" (P. Felig, J. D. Baxter, A. E. Broadus, and L. A. Frohman, eds.), p. 982. McGraw-Hill, New York, 1982.

[4] G. Wong and D. V. Cohn, *Nature (London)* **252,** 713 (1974).

[5] A. H. Tashjian, Jr., D. R. Wright, J. L. Ivey, and A. Pont, *Recent Prog. Horm. Res.* **34,** 285 (1978).

[6] S. J. Marx, C. J. Woodard, and G. D. Aurbach, *Science* **178,** 999 (1972).

[7] T. J. Martin, D. M. Findlay, I. MacIntyre, J. A. Eisman, V. P. Michelangeli, J. M. Moseley, and N. C. Partridge, *Biochem. Biophys. Res. Commun.* **96,** 150 (1980).

[8] S. J. Lamp, D. M. Findlay, J. M. Moseley, and T. J. Martin, *J. Biol. Chem.* **256,** 12269 (1981).

[9] D. M. Findlay, M. deLuise, V. P. Michelangeli, M. Ellison, and T. J. Martin, *Cancer Res.* **40,** 1311, (1980).

METHODS IN ENZYMOLOGY, VOL. 109

ISBN 0-12-182009-2

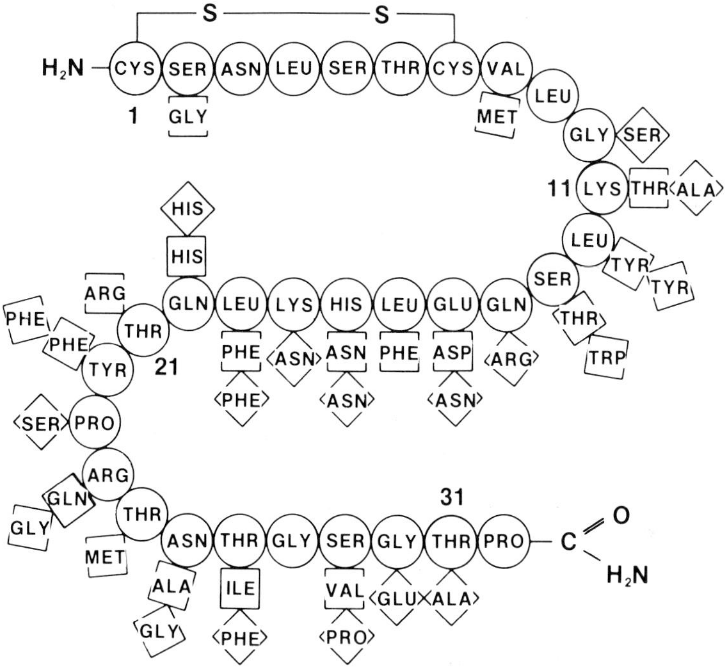

FIG. 1. Structure of calcitonin. ○, Human; □, salmon; ◇, porcine. Adapted from H. Rasmussen, *in* "Textbook of Endocrinology" (R. H. Williams, ed.), p. 696. Saunders, Philadelphia, Pennsylvania, 1974.

served in cultured human lymphocytes,[10] brain,[11] lung,[12] and testes.[13] The description of assays in this chapter will focus on their use in the study of the well-established target organs for calcitonin, bone and kidney.

Assay Methods

Calcitonin is a 32-amino acid peptide, active forms of which contain an interchain disulfide bridge between cysteine residues at positions 1 and 7 and a proline amide at position 32 on the carboxyl terminus (Fig. 1). Binding sites for calcitonin exhibit the same rank order of selectivity for

[10] S. J. Marx, G. D. Aurbach, J. R. Gavin, III, and D. W. Buell, *J. Biol. Chem.* **249,** 6812 (1974).

[11] A. J. Rizzo and D. Goltzman, *Endocrinology* **108,** 1672 (1981).

[12] M. Fouchereau-Peron, M. S. Moukhtar, A. A. Benson, and G. Milhaud, *Proc. Natl. Acad. Sci. U.S.A.* **78,** 3973 (1981).

[13] A. Chausmer, C. Stuart, and M. Stevens, *J. Lab. Clin. Med.* **96,** 933 (1980).

various synthetic analogs of the hormone, regardless of the tissue source.[14] Similarly, the order of potency for the calcitonin analogs found to stimulate adenylate cyclase is the same as that required for inhibition of radiolabeled calcitonin binding.[14] The potency of salmon calcitonin is one to two orders of magnitude greater than that of the native sequences of human, bovine, and porcine calcitonin.[7,10,14] These native peptides are generally available for research. Substitute analogs of human calcitonin, which can be used in studies of the relation between structure and function, can be obtained from Ciba-Geigy (Basel, Switzerland).

Radiolabeling of Calcitonin

The radioligand used in virtually all calcitonin binding studies is ^{125}I-labeled salmon calcitonin. The lack of methionine residues in this peptide permits the use of the strong oxidant chloramine-T (Eastman Chemical, Rochester, NY) in the iodination reaction. Under these conditions, biologic activity of the hormone is retained. The high affinity of calcitonin receptors for the salmon sequence requires the use of a radioligand of high specific activity.

The following procedure for the radiolabeling of calcitonin is a modification of the Hunter and Greenwood[15] method for peptide radioiodination.

1. Add to a 12 × 75 mm disposable glass test tube:
   - 25 μl of 0.3 *M* sodium phosphate, pH 7.5
   - 1 mCi Na^{125}I (1–10 μl)
   - 1 μg salmon calcitonin (1–5 μl in 10 m*M* acetic acid)
   - 10 μl of 1–2 m*M* chloramine-T in 0.3 *M* phosphate buffer, pH 7.5 (prepared as closely as possible to initiation of the reaction).
2. Shake gently for 10 sec.
3. Stop reaction by adding a fresh preparation of 50 μl of 25 m*M* sodium metabisulfite in 0.3 *M* phosphate buffer, pH 7.5. β-Mercaptoethylamine–HCl (7.0 m*M*) can be used in place of sodium metabisulfite as a reducing agent. Separation of iodinated hormone from free ^{125}I is generally achieved either by adsorption to and subsequent elution from microfine silica (QUSO G32; Philadelphia Quartz Co., Philadelphia) or by column chromatography.

[14] D. Goltzman, *Endocrinology* **106,** 510 (1980).

[15] W. M. Hunter and F. C. Greenwood, *Nature (London)* **194,** 495 (1962).

QUSO Purification

1. Add 10–60 mg QUSO to 1 ml of [^{125}I]calcitonin iodination mixture (addition of bovine serum albumin (≃2%) is optional; it reduces adsorption to glass test tubes).
2. Vortex and centrifuge at 2000 *g* for 3 min at 4°. Discard supernatant.
3. Wash pellet with 2 ml of 2 m*M* β-mercaptoethanol. Centrifuge at 2000 *g* for 3 min at 4°. Discard supernatant.
4. Elute radioactivity with 1 ml of 20% acetone with 1% glacial acetic acid. [^{125}I]Calcitonin in the supernatant retains binding activity for 4 weeks when stored at −20°.

BioGel P-10 Purification

1. Add urea to the iodinated mixture to a final concentration of 8 *M*.
2. Apply mixture to BioGel P-10 column (100–200 mesh, 0.9 × 60 cm; BioRad Laboratories, Richmond, CA) equilibrated at 4° with 0.1 *M* ammonium acetate, pH 5.0 with 2% bovine albumin.
3. [^{125}I]Calcitonin is eluted with above column buffer at ≃50 ml and free ^{125}I at ≃120 ml. When stored at 4°, the radiolabeled hormone is stable for 2 weeks.

Sephadex G-25 Purification[16]

1. Place iodination mixture on 0.7 × 45 cm Sephadex G-25 column equilibrated with 10 m*M* acetic acid.
2. Elute [^{125}I]calcitonin in same column buffer. (Storage and stability data were not included in this report.)

Binding Assays

Intact Bone[5]

1. Incubate individual halves of calvariae from 5-day-old rats in 400 μl of Oxman's modification of Eagle's minimal essential medium (MEM) containing 5 mg/ml bovine serum albumin at 37° in 5% CO_2, 95% air for 30 min.
2. Add 100 μl of the above medium containing radiolabeled calcitonin and test substances. Include 2 μg/ml calcitonin in the incubation for estimation of adsorption to glassware and binding to nonspecific binding sites.

[16] S. J. Marx, C. J. Woodard, G. D. Aurbach, H. Glossman, and H. T. Keutmann, *J. Biol. Chem.* **248,** 4797 (1973).

3. Separation of bound and free hormone is accomplished by removing bones and washing them by swirling in iced Gey's solution. Determine radioactivity in bones and incubation medium.

Skeletal Plasma Membranes[17]

Tissue Preparation

1. Dissect calvariae of 18-day-old fetal rats (Sprague–Dawley) or 24-day-old fetal rabbits (New Zealand albino) from membranous tissue. Place in iced 0.25 *M* sucrose.
2. Freeze frontal bone (≃500 mg) in liquid nitrogen. Pulverize in stainless-steel mortar.
3. Add 6–8 vol of 0.25 *M* sucrose, 50 m*M* Tris–HCl, 20 m*M* EDTA, pH 7.5. Homogenize in glass for 30 sec using a motor-driven Teflon pestle or homogenize 3 times (20 sec each time) in a Virtis homogenizer at half-maximal speed.
4. Filter through glass wool or two layers of cheesecloth. Centrifuge filtrate at 2200 *g* for 15 min at 4°.
5. Resuspend pellet in 6–8 vol of 0.25 *M* sucrose, 10 m*M* Tris–HCl, 1 m*M* EDTA, pH 7.5. Centrifuge at 2200 *g* for 15 min at 4°.
6. Resuspend pellet in 50 m*M* Tris–HCl, pH 7.4, for assay.

Assay Procedure

1. Incubate membranes, radioiodinated calcitonin, and unlabeled peptide in sufficient volume of 50 m*M* Tris–HCl, pH 7.5, containing 2% heat-inactivated albumin to obtain replicate aliquots for separation of bound and free hormone.
2. Layer 100-μl aliquots on 200 μl of iced assay buffer in a plastic microfuge tube and centrifuge for 5 min at 5°.
3. Aspirate supernatant and wash pellet with 0.3 ml of 10% sucrose; aspirate wash buffer.
4. Cut tip of microfuge tube just above membrane pellet. Determine radioactivity in pellet.

Renal Plasma Membranes

Tissue Preparation[18]

All steps are performed in buffers at 4°.

[17] L. R. Chase, S. A. Fedak, and G. D. Aurbach, *Endocrinology* **84,** 761 (1969).

[18] D. F. Fitzpatrick, G. R. Davenport, L. Forte, and E. J. Landon, *J. Biol. Chem.* **244,** 3561 (1969).

1. Dissect renal capsule from Sprague–Dawley rats (150–200 g), dice with fine scissors, and wash in iced 1 m*M* EDTA, 5 m*M* Tris–HCl buffer, pH 7.5, containing 9.0 g/liter NaCl.
2. Exchange medium for 3 vol (relative to wet weight of tissue) of 0.25 *M* sucrose, 1 m*M* EDTA, 5 m*M* Tris–HCl, pH 7.5 (SET buffer).
3. Disrupt tissue with 2–3 short (5 sec) bursts of a Polytron homogenizer and further homogenize using 10 strokes of a motor-driven Teflon pestle.
4. Centrifuge (1475 *g*, 10 min) and discard supernatant. Resuspend pellets in ≈3 vol of SET buffer and centrifuge at 1475 *g* for 10 min.
5. Resuspend pellet in 1 vol of 2 *M* sucrose, 1 m*M* EDTA, 5 m*M* Tris–HCl, pH 7.5. Centrifuge at 13,000 *g* for 10 min.
6. Dilute supernatant 8-fold with 1 m*M* EDTA, 5 m*M* Tris–HCl, pH 7.5, and centrifuge at 20,000 *g* for 15 min.
7. Remove the lighter-colored, upper layer of the pellet by gentle washing with SET buffer. Repeat centrifugation at 20,000 *g* twice more and homogenize white, fluffy plasma membranes using a Dounce homogenizer with an A pestle.

Receptor Assay[16]

1. Incubate membranes with radioiodinated calcitonin in 50 m*M* Tris–HCl, pH 7.5, with 2% bovine serum albumin, 0.5 m*M* dithiothreitol (to inhibit oxidation) and 125 μg/ml ACTH(1–24) (to inhibit degradation of tracer).
2. Separate bound calcitonin from free by centrifugation (see assay for bone plasma membranes).

Primary Kidney Cell Cultures[19]

Cell Preparation

1. Remove whole kidneys from 1- to 3-day-old Sprague–Dawley rats using aseptic technique.
2. Wash with Eagle's minimal essential medium (MEM) and mince.
3. Transfer tissue to Ca- and Mg-free Hanks' balanced salt solution containing 0.5 mg/ml collagenase (CLS II; Worthington Biochemical Corp., Freehold, NJ), 1 mg/ml hyaluronidase, 2.5 m*M* $CaCl_2$, 2 m*M* citrate, and 1 mg/ml glucose.
4. Incubate in a 95% O_2 and 5% CO_2 atmosphere at 37° for 60 min with agitation. Disperse cells and tubules by gently pipetting with a

[19] W-T. H. Chao and L. R. Forte, *Endocrinology* **112,** 745 (1983).

10 ml serologic pipet every 15 min, adding deoxyribonuclease (2 U/μl) during the last 10 min to prevent cell aggregation.

5. Filter through one layer of nylon (250 μm pores). Centrifuge filtrate at 100 *g* for 2 min.
6. Wash cells 3 times with MEM and resuspend in growth medium (Coon's modification of Ham's F-12 medium containing 10% calf serum, 1.2 m*M* HEPES, and antibiotics).
7. Plate in 24-well culture plates at a density sufficient for confluence in 24–48 hr. Culture cells at 37° in 95% air, 5% CO_2, changing growth medium after 24 hr.

Binding Assay

1. Add ^{125}I-labeled calcitonin to monolayer cultures in growth medium above.
2. After incubation, remove medium and rinse cells twice with iced isotonic sodium phosphate buffer, pH 7.5.
3. Dissolve cells with 2 successive 1-ml aliquots of 0.3 *N* NaOH. Determine radioactivity bound to cells.

Calcitonin Metabolism

Studies in rats in vivo[20] have shown that biologically active, ^{125}I-labeled calcitonin localizes primarily in kidney (17% of dose after 30 min) and to a smaller extent in bone (8%) after iv injection. Both tissues metabolized calcitonin but metabolites accumulated more rapidly in kidney than in bone. Chromatography of tissue extracts on BioGel P-6 columns (1.5 × 75 cm) revealed formation of degradation products of small molecular weight but did not permit conclusions about the site of cleavage within the calcitonin molecule.

Other studies in the rat[14,18] and rabbit[14] have shown that ^{125}I-labeled calcitonin is degraded by the plasma membrane fractions derived from the kidney. Several criteria were used to assess calcitonin metabolism. Radioactivity that failed to precipitate on addition of 20% trichloroacetic acid (1%, with albumin as carrier) was considered to be degraded hormone. These degradation products did not bind to receptors. Furthermore, radioactivity eluted from membranes (with 100 m*M* acetic acid) previously incubated with [^{125}I]calcitonin associated even more readily with fresh membranes.[14] Renal membranes from rats metabolized the hormone more rapidly than did those from rabbits.[14] Calcitonin degradation by renal plasma membranes in rats was not affected by calcitonin, but was partially inhibited by EDTA, trypsin inhibitor, and ACTH.[18]

[20] P. J. Scarpace, J. G. Parthemore, and L. J. Deftos, *Endocrinology* **103,** 128 (1978).

Renal calcitonin receptors in rat tissue were degraded during a 2-hr incubation of membranes at room temperature.[21] Inclusion of 5 m*M* EDTA or 5000 Kallikrein inhibition units of trypsin inhibitor per milliliter reduced the receptor loss that ordinarily occurs during the 2 hr of incubation from 18 to 11%.

Metabolism of calcitonin by primary cultures of rat kidney cells was evaluated by chromatography on BioGel P-10 columns (0.9 × 60 cm, 0.1 *M* ammonium acetate, pH 5.0 with 2% bovine albumin). Fragments of calcitonin of small molecular weight accumulated in the medium that was bathing the cells incubated with ^{125}I-labeled hormone.[19]

Calcitonin bound to receptors in human breast cancer cells (T47D) becomes progressively inaccessible to elution by isotonic acid washing (0.15 *M* NaCl, 50 m*M* glycine, pH 2.5). Metabolic inhibitors, such as azide, dinitrophenol, and cyanide, and low temperatures inhibit the apparent cell uptake of ^{125}I-labeled calcitonin, suggesting that hormone–receptor complexes can be internalized by these cells. Furthermore, ammonium chloride (10 m*M*), an established inhibitor of lysosomal function, both enhances cell uptake of calcitonin and decreases formation of hormone degradation products. Together, these data provide strong evidence that calcitonin is internalized by a receptor-mediated endocytotic process in these cells.[22]

The calcitonin receptor in T47D cells has been covalently labeled with a photoreactive derivative of calcitonin.[23] The photoreactive moiety, *N*-(β-aminoethyl)-4-azide-2-nitroaniline, which was linked to the hormone using transglutaminase, did not affect the ability of calcitonin to bind to receptors or activate adenylate cyclase. Sodium dodecyl sulfate–polyacrylamide gel electrophoresis (SDS–PAGE) (10% polyacrylamide, 0.25% bisacrylamide) of photolyzed (366 n*M*) monolayer cultures revealed specific binding of the hormone derivative to a single component (molecular weight = 85,000).

Regulation of Calcitonin Receptors

When intact calcitonin-responsive cells are exposed to calcitonin, the cyclic AMP response of the cells becomes desensitized to further stimulation by the hormone.[8,24,25] Refractoriness to the hormone is also accompa-

[21] L. R. Chase, *Endocrinology* **96,** 70 (1975).

[22] D. M. Findlay, K. W. Ng, M. Niall, and T. J. Martin, *Biochem. J.* **206,** 343 (1982).

[23] J. M. Moseley, D. M. Findlay, T. J. Martin, and J. J. Gorman, *J. Biol. Chem.* **257,** 5846 (1982).

[24] W-T. H. Chao and L. R. Forte, *Endocrinology* **111,** 252 (1982).

[25] D. M. Findlay, M. deLuise, V. P. Michelangeli, and T. J. Martin, *J. Endocrinol.* **88,** 271 (1981).

nied by a loss of calcitonin binding sites, which is due, at least in part, to persistent occupancy of receptors.[24] Restoration of the responsiveness of breast cancer cells requires 24 hr—an extended period in comparison to many other hormone–receptor systems. This prolonged refractoriness may be caused by a slow rate of dissociation of the calcitonin–receptor complex, similar to that observed in renal plasma membranes.[26] Another factor that may contribute to the slow recovery of calcitonin responsiveness is the relative insensitivity of calcitonin receptors to guanyl nucleotides, which are known to affect the regulation of hormonal responsiveness in intact cells and cell membranes.[26]

Recent evidence indicates that the protective effect of calcitonin on bone may be important in the etiology of postmenopausal osteoporosis. Serum levels of immunoreactive calcitonin are lower in women of all ages[27] and may be further decreased after menopause.[28] In addition, the chronic calcitonin deprivation caused by thyroidectomy might be responsible for the bone loss observed in patients who undergo this procedure.[29] Knowledge of this hormone's mechanism of action in the skeleton would be enhanced by the availability of an intact bone cell system for the study of calcitonin receptors.

[26] N. Loreau, C. LaJotte, F. Wahbe, and R. Ardaillou, *J. Endocrinol.* **76,** 533 (1978).
[27] C. J. Hillyard, J. C. Stevenson, and I. MacIntyre, *Lancet* **1,** 961 (1978).
[28] J. C. Stevenson, C. J. Hillyard, G. Abeyasekera, K. G. Phang, I. MacIntyre, S. Campbell, O. Young, P. T. Townsend, and M. I. Whitehead, *Lancet* **1,** 693 (1981).
[29] M. T. McDermott, G. S. Kidd, P. Blue, V. Ghaed, and F. D. Hofeldt, *J. Clin. Endocrinol. Metab.* **56,** 936 (1983).

[5] Assay for Parathyroid Hormone Receptors

By ROBERT A. NISSENSON, ANNE P. TEITELBAUM, and CLAUDE D. ARNAUD

Parathyroid hormone (PTH) is a single chain, 84-amino acid polypeptide that maintains calcium and phosphorus homeostasis through direct actions on bone and kidney. The cellular effects of PTH are thought to be initiated by receptor-mediated activation of adenylate cyclase in the plasma membrane and the consequent increase in cyclic AMP levels. Interest in these membrane effects of PTH has arisen with the discoveries of a genetic disease of the PTH receptor-adenylate cyclase system

ISBN 0-12-182009-2

(pseudohypoparathyroidism)[1,2] and of abnormalities in target cell responses to PTH in other pathologic states (e.g., chronic renal failure).[3] In what follows, we will focus on methods that we have used to identify and quantitate PTH receptors in kidney and bone.

As with other radioligand binding assays, the PTH radioreceptor assay presents three major methodological hurdles: the preparation of a biologically active, labeled ligand, the establishment of a suitable receptor-containing preparation, and a rapid and convenient means for separating bound from free radioligand.

Preparation of Labeled PTH

Parathyroid hormone contains methionine residues at positions 8 and 18, which are sensitive to oxidation. Because these amino acids reside in the bioactive region of the PTH molecule, their oxidation results in a biologically inactive product. Radioiodination procedures employing strong oxidants, such as the chloramine-T technique,[4] have thus proved useless for developing probes of the PTH receptor using native-sequence hormone. Our laboratory has overcome this problem by radioiodinating PTH under conditions of mild oxidation controlled by the use of an electrolysis assembly (Fig. 1).[5] Our procedure is based upon that previously described by Sammon and colleagues.[6,7]

An aliquot of 6 mCi $Na^{125}I$ is incubated with 25 μg synthetic, purified bovine (b) PTH(1–34) (Bachem, Inc., Torrance, CA) in 200 μl of 0.5 *M* potassium phosphate (pH 7.5) plus 0.1 *M* NaCl. Incubation is carried out with continuous stirring in an ice-cold platinum crucible, which serves as the anode compartment of the electrolysis assembly (Fig. 1). The incubation is initiated by the application of an initial current (10–20 μA) sufficient to achieve a potential difference of 680 mV. The applied current is continually decreased to maintain a constant potential difference of 680 mV. After 20 min, the reaction is terminated by the addition of β-mercap-

[1] Z. Farfel, A. S. Brickman, H. R. Kaslow, V. M. Brothers, and H. R. Bourne, *N. Engl. J. Med.* **303,** 237 (1980).

[2] M. A. Levine, R. W. Downs, Jr., M. Singer, S. J. Marx, G. D. Aurbach, and A. M. Spiegel, *Biochem. Biophys. Res. Commun.* **94,** 1319 (1980).

[3] S. Tomlinson, G. N. Hendy, D. M. Pemberton, and J. L. H. O'Riordan, *Clin. Sci. Mol. Med.* **51,** 59 (1976).

[4] W. M. Hunter and F. C. Greenwood, *Nature (London)* **194,** 495 (1962).

[5] R. A. Nissenson and C. D. Arnaud, *J. Biol. Chem.* **254,** 1469 (1979).

[6] P. J. Sammon, J. S. Brand, W. F. Neuman, and L. G. Raisz, *Endocrinology* **92,** 1596 (1973).

[7] W. F. Neuman, M. W. Neuman, P. J. Sammon, and K. Lane, *Calcif. Tissue Res.* **18,** 241 (1975).

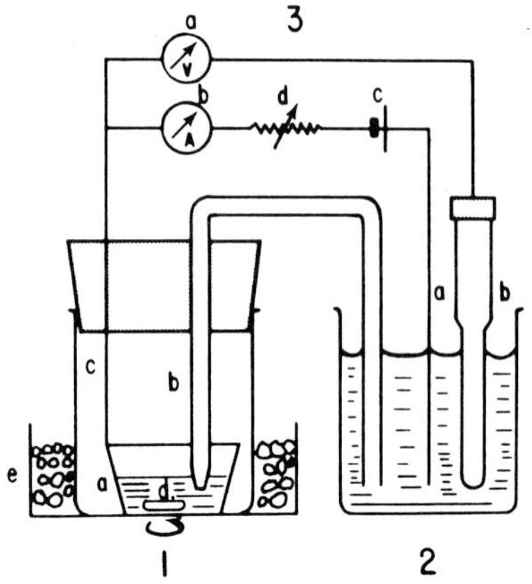

FIG. 1. Electrolytic iodination apparatus. (1) Reaction vessel; (a) platinum crucible (1-cm diameter) anode; (b) agar/salt bridge [preparation: gradually warm 30 mg agar in 10 ml 90% saturated KCl (commercially available), with stirring; keep covered; fill U-shaped glass tube by suction; avoid bubbles; seal ends with agar; store in saturated KCl]; (c) 10-ml beaker and close-fitting rubber stopper with holes for platinum wire and salt bridge; (d) Magnetic stirring bar; (e) ice bath. (2a) Cathode, and (b) calomel reference electrode, both in saturated KCl. (3a) Potentiometer; (b) microammeter; (c, d) variable potential source. Reproduced from A. P. Teitelbaum, *in* "Assay of Calcium-regulating Hormones" (D. D. Bikle, ed.), p. 191. Springer-Verlag, New York, 1983, with the permission of the publisher.

toethanol (1.0 m*M*). A small aliquot is removed for determination of specific activity by trichloroacetic acid precipitation (typical results ≅100 μCi/μg), and to the remainder is added 300 μl of 10 m*M* HCl, 1.0 m*M* β-mercaptoethanol, and 0.25% bovine serum albumin (BSA). This mixture is applied to a small column of Dowex 1-4X (preequilibrated in the same buffer) to remove unreacted ^{125}I. [^{125}I]bPTH(1–34) is eluted with successive washes with the acidic eluant, and the preparation is immediately desalted on a 1 × 50-cm column of BioGel P-2 using 0.1 *M* acetic acid/0.1% BSA as eluant. Typically, 1 mCi of [^{125}I]bPTH(1–34) is recovered at this stage.

Purification of Labeled PTH

When tested in a highly sensitive, PTH-responsive renal adenylate cyclase assay,[8] this [^{125}I]bPTH(1–34) preparation has ≤20% of the biologic activity of unlabeled bPTH(1–34). This reduced activity is attribut-

[8] R. A. Nissenson, S. R. Abbott, A. P. Teitelbaum, O. H. Clark, and C. D. Arnaud, *J. Clin. Endocrinol. Metab.* **52,** 840 (1981).

able not to the introduction into the peptide of iodine atoms per se, but to residual oxidation of methionine despite the controlled labeling conditions. To isolate the unoxidized, biologically active portion of the [^{125}I]bPTH(1–34) preparation, we take advantage of the inability of oxidized hormone to bind to target cell receptors. Electrolytically labeled [^{125}I]bPTH(1–34) is routinely purified by affinity adsorption to crude chicken renal membranes. For membrane isolation, 10 female chickens (8–12 weeks old) are decapitated and their kidneys excised and placed in ice-cold 0.9% NaCl, 1 m*M* EDTA, and 10 m*M* Tris–HCl, pH 7.5. Fat and connective tissue are removed and kidneys are washed several times with ice-cold buffer. All subsequent steps are performed at 4°. Kidneys are homogenized in 3 volumes (relative to wet weight) of ice-cold 0.25 *M* sucrose, 1 m*M* EDTA, and 10 m*M* Tris–HCl, pH 7.5 (SET buffer), using 10 strokes of a motor-driven, loose Teflon pestle. After centrifugation at 1475 *g* for 10 min, the pellets are washed once with SET buffer, then resuspended in 1 volume (relative to wet weight) of 2.0 *M* sucrose, 1 m*M* EDTA, and 10 m*M* Tris–HCl, pH 7.5. After centrifugation at 13,300 *g* for 10 min, the supernatant is saved and the pellet resuspended in 2.0 *M* sucrose, 1 m*M* EDTA, and 10 m*M* Tris–HCl, pH 7.5, and recentrifuged at 13,300 *g* for 10 min. The supernatant, containing most of the particulate material, is combined with the previous supernatant and diluted 8-fold with 1 m*M* EDTA and 10 m*M* Tris–HCl, pH 7.5. This suspension is then centrifuged at 30,000 *g* for 15 min. The fluffy layer of the pellet is resuspended in SET buffer and recentrifuged at 30,000 *g* for 15 min. The fluffy layer of the pellet is then resuspended in 40 to 50 ml of SET buffer to yield a protein concentration of 8–10 mg/ml. This suspension is stored in 0.5-ml aliquots at −80°.

Affinity purification is performed by incubating 15–25 μCi of electrolytically labeled [^{125}I]bPTH(1–34) with 0.5 ml of the crude chicken renal membrane suspension in a 2.0-ml final volume containing 25 m*M* Tris–HCl (pH 7.5), 2 m*M* $MgCl_2$, and 0.1% BSA. After 2 hr of incubation at 4°, the tubes are centrifuged at 2000 *g* for 5 min. The supernatants are discarded, and the pellets washed three times by resuspension in 5 ml of ice-cold 25 m*M* Tris–HCl (pH 7.5) and 2.0 m*M* $MgCl_2$, followed by centrifugation at 2000 *g* for 5 min. Bound [^{125}I]bPTH(1–34) is then eluted from the pellets by resuspension in 3 ml of ice-cold 0.1 *M* acetic acid. After centrifugation at 2000 *g* for 5 min, supernatants containing purified [^{125}I]bPTH(1–34) are frozen and lyophilized overnight. The residues after lyophilization are resuspended in 10 m*M* acetic acid and centrifuged at 2000 *g* for 15 min to remove residual particulate material. Supernatants are pooled and stored at a concentration of 1 to 2 × 10^6 cpm/ml at 4°. No detectable loss of binding activity is observed during 4 weeks of storage.

A typical yield of receptor-purified [^{125}I]bPTH(1–34) is 5% of the original electrolytically labeled hormone that was added.

The identity of the eluted radioactivity is verified by its comigration with electrolytically labeled [^{125}I]bPTH(1–34) and with purified unlabeled bPTH(1–34) on cellulose thin-layer chromatography (TLC) (solvent: butanol, pyridine, water, and acetic acid in a ratio of 15:10:12:3), carboxymethylcellulose TLC (solvent: 0.1 *M* ammonium acetate, pH 5.5), and chromatoelectrophoresis. The radiohomogeneity of the receptor-purified material should be greater than 95% when evaluated by each of these procedures.

Comments

The efficacy of this affinity purification procedure is illustrated in Fig. 2. The specific binding activity of [^{125}I]bPTH(1–34) in chicken renal plasma membranes is increased several-fold by the receptor purification procedure, whereas nonspecific binding (in the presence of a large excess of unlabeled PTH) is negligible. Adenylate cyclase bioassay[8] of affinity-purified [^{125}I]bPTH(1–34) can be used to verify the biopotency of each labeled hormone preparation.

The labeling and purification methods described above are quite reliable and reproducible, but the need for affinity purification lengthens the procedure while diminishing substantially the yield of ligand. An alternative approach is the use of a methionine-free PTH analog [^{8}Nle, ^{18}Nle, ^{34}Tyr], bPTH(1–34)amide, which is available commercially and can be labeled to high specific activity with conventional iodination techniques without loss of biologic activity. This ligand has been used successfully to label renal PTH receptors.[9–11] However, in our experience the binding activity of the analog varies substantially from one iodination to the next, perhaps due to its marked propensity to adsorb to glass and plastic surfaces. We thus continue to employ almost exclusively electrolytically labeled, affinity-purified [^{125}I]bPTH(1–34) for our studies of PTH receptors.

PTH Binding Assays

Renal plasma membranes and cultured cells from embryonic chick calvariae are the model systems that we use for the study of PTH receptors.

[9] G. V. Segré, M. Rosenblatt, B. L. Reiner, J. E. Mahaffey, and J. T. Potts, Jr., *J. Biol. Chem.* **254,** 6980 (1979).

[10] E. Bellorin-Font and K. J. Martin, *Am. J. Physiol.* **241,** F364 (1981).

[11] L. R. Forte, S. G. Langeluttig, R. E. Poelling, and M. L. Thomas, *Am. J. Physiol.* **242,** E154 (1982).

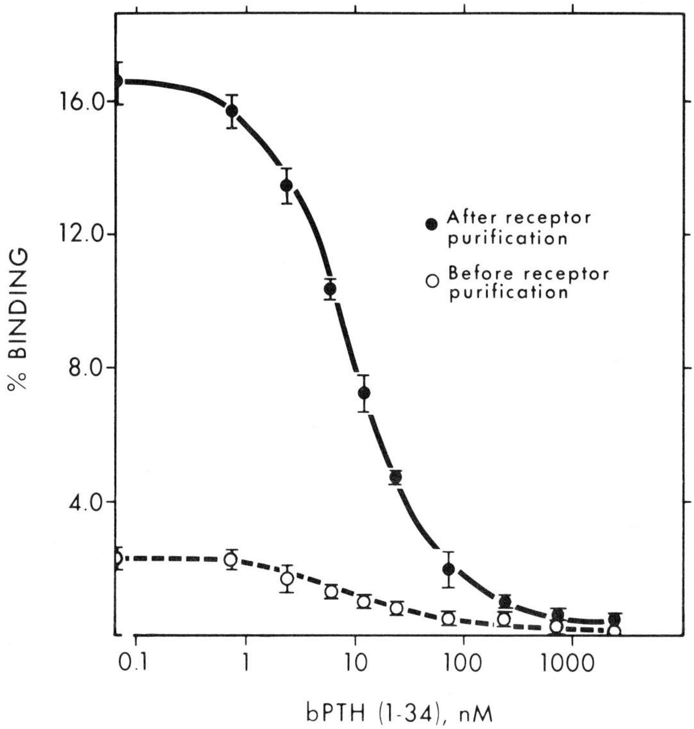

FIG. 2. [^{125}I]bPTH(1–34) binding to chicken renal plasma membranes before and after receptor purification of the labeled hormone. Binding was measured in the presence of 7500 cpm (about 50 pg) of [^{125}I]bPTH(1–34). The percentage binding of [^{125}I]bPTH(1–34) is plotted as a function of the concentration of unlabeled bPTH(1–34) in the medium. The final membrane concentration was 0.25 mg/ml. Each point is the mean ± SE of quadruplicate determinations. From Nissenson and Arnaud.[5]

Isolation of Canine Renal Cortical Plasma Membranes

Required Equipment

1. Standard surgical and dissection instruments
2. Polytron homogenizer
3. Potter-Elvehjem homogenizer with loose-fitting Teflon pestle
4. High-speed refrigerated centrifuge
5. Ultra-cold (−80°) freezer

Required Solutions

1. 0.9% NaCl, 1.0 m*M* EDTA, and 5 m*M* Tris–HCl, pH 7.5 (solution A)

2. 0.25 *M* sucrose, 1.0 m*M* EDTA, and 5 m*M* Tris–HCl, pH 7.5 (solution B)
3. 2.0 *M* sucrose, 1.0 m*M* EDTA, and 5 m*M* Tris–HCl, pH 7.5 (solution C)
4. 1.0 m*M* EDTA and 5 m*M* Tris–HCl, pH 7.5 (solution D)

Protocol

Canine renal cortical plasma membranes are isolated by a modification of the procedure of Fitzpatrick *et al.*[12]

1. Kidneys from adult mongrel dogs are placed into ice-cold solution A. All further steps are performed at 3–4°.
2. Whole kidneys are sliced longitudinally, and cortical tissue is dissected away from the medulla and corticomedullary regions.
3. Cortical sections are stripped of capsular tissue and fat, and dispersed in 3 volumes (relative to wet weight) of solution B with three 5-sec bursts of a Polytron homogenizer (setting 4).
4. Complete homogenization is effected with 10 strokes of a motor-driven, loose Teflon pestle.
5. The homogenate is centrifuged at 1475 *g* (10 min), and the pellets are resuspended in solution C with 3 strokes of a motor-driven, loose Teflon pestle and recentrifuged at 13,300 *g* for 10 min.
6. The supernatant recentrifuged at 13,300 *g* is diluted 8-fold with solution D and centrifuged at 20,000 *g* for 15 min. This sediments plasma membranes (top "fluffy" layer) and mitochondria (bottom "compact" layer).
7. The plasma membranes are resuspended by swirling gently in solution B, followed by recentrifugation at 20,000 *g* for 15 min. This cycle of resuspension and centrifugation is repeated at least 3 times.
8. Final resuspension of membranes is in solution B at about 3 mg protein/ml. Plasma membranes are frozen in dry ice and acetone in multiple aliquots and stored at −80°. In our experience, PTH binding activity is stable for at least 6 months under these storage conditions. The yield of purified plasma membranes varies, averaging about 30 mg protein/10 g (wet weight) kidney. It is recommended that each preparation be tested for enrichment (relative to homogenate) in ouabain-sensitive Na^+,K^+-ATPase, a marker for renal

[12] D. F. Fitzpatrick, G. R. Davenport, L. Forte, and E. J. Landon, *J. Biol. Chem.* **244,** 3561 (1969).

basolateral membranes. The specific activity of this enzyme shouid be enriched at least 6- to 8-fold in the membrane preparation.

Measurement of [^{125}I]bPTH(1–34) Binding: Canine Renal Membranes

Routine binding assays are done in a final volume of 0.1 ml containing 50 m*M* Tris–HCl, 50 m*M* Hepes (pH 7.5), 2.0 m*M* $MgCl_2$, 0.1% BSA, approximately 7000 cpm (25–65 pg) of receptor-purified [^{125}I]bPTH(1–34), and, where appropriate, unlabeled bPTH(1–34) or other peptides.[13] Binding is initiated by adding 25 to 50 μg of highly purified renal plasma membranes. After 60 min of incubation at 30°, the sample is diluted with 2.0 ml of ice-cold washing buffer containing 25 m*M* Tris–HCl (pH 7.5), 2.0 m*M* $MgCl_2$, 0.1% BSA, and 125 m*M* NaCl. Samples are immediately filtered under vacuum through 0.2-μm cellulose acetate Millipore filters at 4°. Each filter is washed three times with 10 ml of ice-cold washing buffer. Filtration rates vary from 5 to 10 sec/ml. Filters are then analyzed in a γ-scintillation counter for ^{125}I radioactivity. Blank binding to the filters in the absence of plasma membranes is less than 1.0% of the added [^{125}I]bPTH(1–34). Although extremely low, blank values are routinely subtracted from all binding values.

Measurement of [^{125}I]bPTH(1–34) Binding: Bone Cells

Primary cultures of embryonic chick bone cells are obtained from collegenase-dispersed calvariae as previously described.[14,15] Cells reach confluence and are used for experiments after 6–7 days of growth in culture. At this time, culture medium is replaced with 0.5 ml of medium 199 containing 20 m*M* Hepes and 0.1% BSA at pH 7.4. [^{125}I]bPTH(1–34) and an appropriate amount of unlabeled hormone or diluent (10 m*M* acetic acid and 0.1% BSA) are added, each in a volume of 10 μl. Incubations are carried out at room temperature (24°). The monolayers are washed four times with 0.9 ml of Hanks' solution at 4°. Then 0.1 ml of Hanks' solution is added, and the cells are scraped from the dishes with a rubber policeman and transferred to test tubes with a Pasteur pipet. Cell-bound and medium radioactivity are measured in a well-type γ-counter. Dishes containing no cells are incubated and washed in the same manner, and the binding values obtained (~0.5%) are subtracted from the total binding per

[13] A. P. Teitelbaum, R. A. Nissenson, and C. D. Arnaud, *Endocrinology* **111,** 1524 (1982).

[14] P. J. Nijweide, A. van der Plas, and J. P. Scherft, *Calcif. Tissue Int.* **33,** 529 (1981).

[15] N. B. Pliam, K. O. Nyiredy, and C. D. Arnaud, *Proc. Natl. Acad. Sci. U.S.A.* **79,** 2061 (1982).

culture. Binding is expressed as a percentage of the total amount of [^{125}I]bPTH(1–34) added to each culture dish.

Comments

A major difficulty in handling submicrogram quantities of PTH is the propensity of the hormone to adsorb to glass and plastic surfaces. We find that this problem can be largely avoided by diluting stock solutions of labeled or unlabeled PTH in 10 m*M* acetic acid containing at least 0.1% BSA. Dilutions should be made in borosilicate glass test tubes. Coating such tubes with silanols, such as an aqueous solution of octadecyltrialkoxysilane (AquaSil, Pierce Chemical Company, Rockford, IL), followed by coating with BSA may further reduce adsorption. Alternatively, polypropylene tubes coated with cetyl alcohol may be suitable.[16]

Although our PTH receptor studies have been carried out exclusively with [^{125}I]bPTH(1–34) or its methionine-free analog, there are two recent reports of the successful use of [^{125}I]bPTH(1–84) to identify PTH receptors in canine[17] and rabbit[18] kidney. In these studies, nonspecific binding of labeled PTH was substantial (30–50% of bound hormone), whereas nonspecific binding of affinity-purified [^{125}I]bPTH(1–34) is routinely 5% of bound hormone.[5,13] Thus, the advantages of native-sequence ligand for identifying PTH receptors remain to be established.

Acknowledgments

We appreciate the editorial assistance of Mary Jean Moore. This work was supported by NIH Grants AM 27755 and AM 21614 and by funds from the Research Service of the Veterans Administration.

[16] P. Q. Barrett and W. F. Neuman, *Biochim. Biophys. Acta* **541,** 223 (1978).

[17] R. E. Rizzoli, T. M. Murray, S. J. Marx, and G. D. Aurbach, *Endocrinology* **112,** 1303 (1983).

[18] R. Kremer, H. P. J. Bennett, J. Mitchell, and D. Goltzman, *J. Biol. Chem.* **257,** 14048 (1982).

[6] Gastrin Receptor Assay

By LEONARD R. JOHNSON

Acting as an endocrine substance gastrin has two physiological actions on the oxyntic gland mucosa of the stomach. These are the stimulation of acid secretion from the parietal cells and the stimulation of cell division

ISBN 0-12-182009-2

and mucosal growth, an effect elicited from the mucous neck cell which is the stem or progenitor cell for most of the oxyntic gland mucosa.[1]

Several laboratories have reported the development of gastrin receptor assays. Lewin *et al.*[2,3] used tritiated gastrin having a specific activity of 60 Ci/mmol to demonstrate reversible binding to both partially purified plasma membranes and to intact gastric mucosal cells which had been separated by pronase digestion. This binding was time and temperature dependent and proportional to the amount of membranes or the number of parietal cells present. Maximum binding was 50 fmol/mg protein or 12.5 fmol gastrin per million cells. The equilibrium K_D was 0.9×10^{-8} *M*, and Scatchard analysis suggested a single class of binding sites. Unfortunately, the binding in these studies was not shown to be specific, and the binding interaction did not have an affinity in the physiological range of hormone concentration.

Baur and Bacon[4] have described an assay using ^{125}I synthetic human gastrin I to measure gastrin receptors present in membranes of canine antral smooth muscle cells. Maximal binding was about 50 fmol/mg protein and specific binding amounted to 53% of the total counts bound after binding to the filters was subtracted. Specific binding was equal to 0.12% of the 60,000 cpm of labeled gastrin added to the membranes. In other words, only about 72 counts were specifically bound out of a total of 60,000.

Using ^{125}I-labeled gastrin I, Brown and Gallagher[5] demonstrated a specific gastrin receptor in the 250–20,000 *g* fraction of rat oxyntic gland membranes. Binding was reversible, temperature and pH dependent, and was inhibited by tryptic digestion of the membrane preparation. Cholecystokinin, gastrin and its analogs, and secretin inhibited binding, while cimetidine did not. Observed maximal binding of gastrin occurred at a concentration of label equal to 2.0×10^{-10} *M*, and there was no binding to membranes from antral mucosa. Using radioiodinated 15-Leu gastrin 17-I with a specific activity of 2000 cpm/fmol we developed a gastrin binding assay[6,7] that appeared to be quite similar to that of Brown and Gallagher.

[1] L. R. Johnson, *Annu. Rev. Physiol.* **39,** 135 (1977).
[2] M. Lewin, A. Soumarmon, J. P. Bali, S. Bonfils, J. P. Girman, J. L. Morgat, and P. Fromageot, *FEBS Lett.* **66,** 168 (1976)
[3] A. Soumarmon, A. M. Cheret, and M. J. M. Lewin, *Gastroenterology* **73,** 900 (1977).
[4] S. Baur and V. C. Bacon, *Biochem. Biophys. Res. Commun.* **73,** 928 (1976).
[5] J. Brown and N. D. Gallagher, *Biochim. Biophys. Acta.* **538,** 42 (1978).
[6] K. Takeuchi, G. R. Speir, and L. R. Johnson, *Am. J. Physiol.* **237,** E284 (1979).
[7] K. Takeuchi, G. R. Speir, and L. R. Johnson, *Am. J. Physiol.* **237,** E295 (1979).

Iodination Procedure

Synthetic 15-Leu G-17-I was obtained from Research Plus, Denville, NJ. $Na^{125}I$ (350 mCi/ml) was obtained from Isotex, Houston, TX. The high specific activity $[^{125}I]$G-17-I is prepared as described by Dockray *et al.*[8] Briefly, 25 μl of 250 m*M* phosphate buffer (pH 7.4) is added to 2.8 nmol (6 μg) G-17-I dissolved in 6 μl of 50 m*M* NH_4HCO_3. To this mixture, we add 10 μl of $Na^{125}I$ (350 mCi/ml) and 9 nmol of chloramine-T dissolved in 5 μl of 250 m*M* phosphate buffer (pH 7.4). The reaction proceeds for 10 sec and is stopped by the addition of 52 nmol (10 μg) of $Na_2S_2O_5$ in 20 μl of phosphate buffer (pH 7.4). The reaction mixture is applied to Sephadex G-10; further purification is performed using ion-exchange chromatography. The specific activity of the monoiodinated gastrin collected in peak II was determined to be 2000 cpm/fmol. This peak is used for the determination of biological activity and binding analysis.

The specific activity of the gastrin was calculated by counting an aliquot of the peak and determining the concentration by radioimmunoassay (RIA). RIA-inhibition curves were prepared. In the first curve the concentration of labeled gastrin remains constant and doubling amounts of unlabeled gastrin are added. In the second curve the unlabeled gastrin is held constant and doubling amounts of labeled gastrin are used. When plotted as bound-to-free (B/F) ratio against the logarithm of unlabeled gastrin concentration, the two curves should superimpose. The number of counts corresponding to a given concentration of gastrin can be determined by comparing the two curves at a specific B/F ratio. The specific activity of the labeled gastrin was verified by using the method developed by Akera and Cheng[9] to determine the K_D and B_{max} of the interaction of gastrin with its receptor.[7]

Use of 15-Leu G-17 ensures the preparation of a biologically active labeled gastrin.[6] Natural gastrin contains a methionyl residue in position 15, which is oxidized during the chloramine-T procedure. This destroys the biological activity of the hormone. The leucine substituted compound is fully active and is not affected by the oxidation step.

Tissue Preparation

Male Sprague–Dawley rats (200 g) are etherized, and the stomachs are removed, placed in ice-cold 0.9% NaCl, and rinsed free of debris and blood. After removal of the antrum, the remaining glandular mucosal

[8] G. J. Dockray, J. H. Walsh, and M. I. Grossman, *Biochem. Biophys. Res. Commun.* **69,** 339 (1976).

[9] T. Akera and V.-J. K. Cheng, *Biochim. Biophys. Acta* **470,** 412 (1977).

portion of the stomach is scraped. All steps are performed at 4°. The scrapings are placed in ice-cold buffer A (120 nM NaCl, 4.8 mM KCl, 25.3 mM $NaHCO_3$, 1.2 mM KH_2PO_4, 2.47 mM $MgSO_4$, 1.29 mM $CaCl_2$, 11.15 mM $C_6H_{12}O_6$, and 1 mg/liter penicillin (Sigma, St. Louis, MO). The buffer volume is 2.5 ml per stomach.

This buffer is centrifuged at 200 g at 4° for 10 min. The mucous and buffy layers are aspirated and discarded; the pellet is retained. This centrifugation step is essential for the implementation and reproducibility of the final filtration step in the binding assay. The pellet is resuspended in buffer B (250 mM sucrose, 25 mM KCl, 10 mM $MgCl_2$, 50 mM Tris, pH 7.4), at a concentration of 4 ml per stomach and then homogenized in a Dounce glass–glass homogenizer (Kontes K-88530) using pestle A (10 strokes), then pestle B (10 strokes). The homogenate is filtered through nylon mesh and centrifuged at 270 g at 4° for 10 min. The pellet is discarded and the supernatant is centrifuged at 30,000 g at 4° for 45 min. The final pellet is resuspended in buffer C (10 mM Tris, pH 7.4). The protein concentration of this suspension was determined by the Lowry technique[10] and the Bradford technique.[11] The protein concentration is adjusted to 7.5–10 mg/ml (150–200 μg/assay tube) for use in the binding assay.

If this membrane fraction is further fractionated through a 0–40% sucrose gradient, specific gastrin binding is found in the same fractions containing specific ouabain binding (Fig. 1). Since ouabain binds to Na^+,K^+-ATPase which is located on the basolateral cell membranes, this indicates that the gastrin receptors present in this preparation share the same location.

Binding Assay

Two parallel sets of tubes are used to evaluate the binding of gastrin. Series A contains 40 μl buffer D (50 mM N-2-hydroxyethylpiperazine-N'-2-ethanesulfonic acid (HEPES), 10 mM $MgCl_2$, and 1.5% bovine serum albumin, pH 7.4), 20 μl of membrane preparation and 10 μl buffer E (1% bovine serum albumin, 10 mM HEPES). Series B contains 40 μl buffer D, 20 μl of membrane preparation, and 10 μl of 4.8×10^{-4} M 15-Leu G-17-I dissolved in buffer E. The method of O'Keefe *et al.*[12] is modified as follows: the tubes are preincubated at 30° for 20 min, followed by the

[10] D. H. Lowry, N. J. Rosebrough, A. L. Farr, and R. J. Randall, *J. Biol. Chem.* **193,** 265 (1951).

[11] M. A. Bradford, *Anal. Biochem.* **72,** 248 (1976).

[12] E. O'Keefe, M. D. Hollenberg, and P. Cuatrecasas, *Arch. Biochem. Biophys.* **164,** 518 (1974).

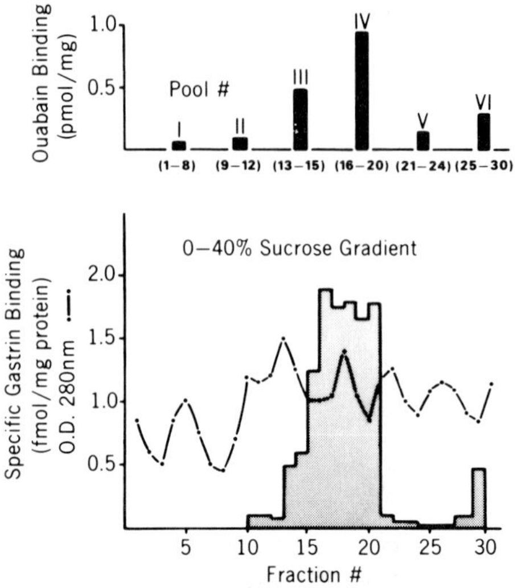

FIG. 1. Comparison of specific gastrin binding and specific ouabain binding of various sucrose gradient fractions of membrane pellet from 270–30,000 *g* oxyntic gland mucosal preparation. Ouabain binding (top panel) and gastrin binding (bottom panel) are present in same fractions, indicating that the gastrin binding in 270–30,000 *g* preparation is to basolateral membranes.

addition of 10 μl monoiodo-15-Leu G-17-I in buffer E at 1-min intervals. The tubes are incubated at 30° for 30 min. Single-point assays are performed with 200–300 p*M* labeled gastrin ± 1000-fold molar excess of unlabeled gastrin. The incubation is terminated by the addition of 720 μl of buffer E and suction filtration[13] through a Gelman GA-6 filter. The filter paper, which had been soaked overnight in 10% bovine serum, is rinsed 3 times with 2 ml of buffer E. All filtration is carried out at 0° within 30 sec. The filter paper was counted in a Searle analytic gamma counter with a counting efficient of 80%.

The counts obtained with series B tubes (nonspecific binding) are then subtracted from identical series A tubes (total binding) to obtain specific binding. The amount of membrane protein present is the assay tube has slight effects on both K_D and B_{max}.[6] The important point is to keep this constant within an assay and between any experiments which are to be compared. Optimal specific binding occurs when 180 to 260 μg of membrane protein are present.

Specific gastrin binding is complete after 30 min incubation and remains constant for up to 2 hr.[6] In subsequent studies we have used an

[13] I. D. Goldfine, J. Roth, and L. Birnbaumer, *J. Biol. Chem.* **247**, 1211 (1972).

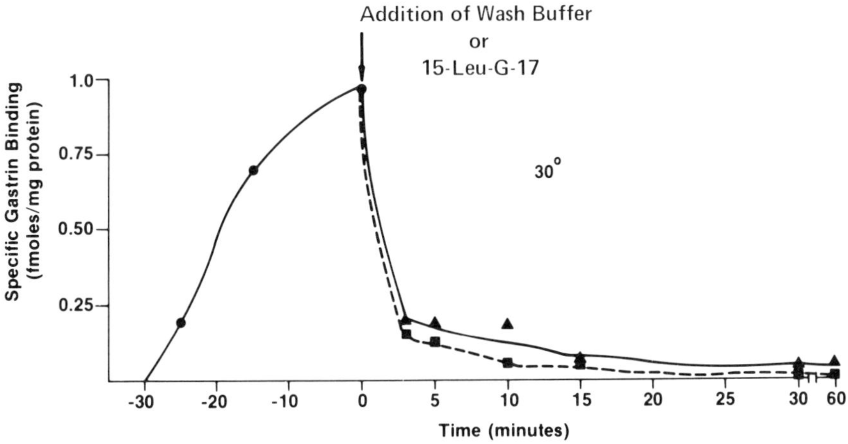

FIG. 2. Specific binding of [^{125}I]15-Leu G-17 to rat gastric mucosal membrane preparations. After a 30 min incubation (●), either 10 m*M* HEPES, pH 7.4 wash buffer (■) equaling a 10-fold dilution or 5×10^{-7} *M* unlabeled gastrin (▲) was added. Second incubation proceeded at 30° for 60 min. For other conditions see text. From Takeuchi *et al.*[7]

incubation period of 30 min at a temperature of 30°, which also proved to be optimum.[6]

If at the end of 30 min, wash buffer or a 1000-fold excess of 15-Leu G-17 is added to the incubation, bound gastrin rapidly dissociates.[7] Dissociation is 90% complete within 5 min and virtually complete within 30 min (Fig. 2).

Optimal binding of gastrin to its receptor occurs at pH 7.0. Since binding at this point was not significantly greater than at pH 7.4, we have used the more physiological 7.4 in subsequent studies.[6]

Using the techniques and assay conditions described above, we can expect to obtain specific gastrin binding equal to significantly more than 50% of total binding. A typical binding assay is illustrated in Fig. 3. Scatchard[14] plot analysis of this curve is shown in Fig. 4 and yields the following binding constants: $K_D = 4.17 \times 10^{-10}$, $K_A = 0.24 \times 10^{10}\ M^{-1}$, $B_{max} = 4.14$ fmol/mg protein. The K_D of this interaction correlates with circulating concentrations of gastrin in the fed, male rat.

Tissue Specificity

Specific binding of gastrin has been demonstrated in oxyntic gland and duodenal mucosa, but not in mucosa from the antral region of the stomach.[6] This distribution fits with the known actions of gastrin, since it

[14] G. Scatchard, *Ann. N.Y. Acad. Sci.* **51,** 660 (1949).

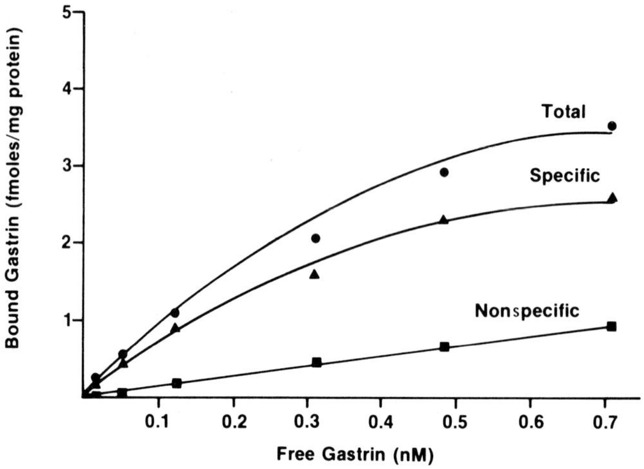

FIG. 3. Total, specific, and nonspecific binding in gastric mucosal membrane preparation. Binding studies were performed as described in the text. Amount of protein in each incubation tube was 160 μg. Upper curve (●) is from incubations with increasing concentrations of ^{125}I-labeled gastrin. Lower curve (■) is from incubations containing same concentrations of ^{125}I-labeled gastrin in addition to a 1000-fold molar excess of unlabeled gastrin; it represents nonspecific binding in the incubate. Middle curve (▲) is difference between nonspecific and total binding and represents specific binding. From Takeuchi *et al.*[6]

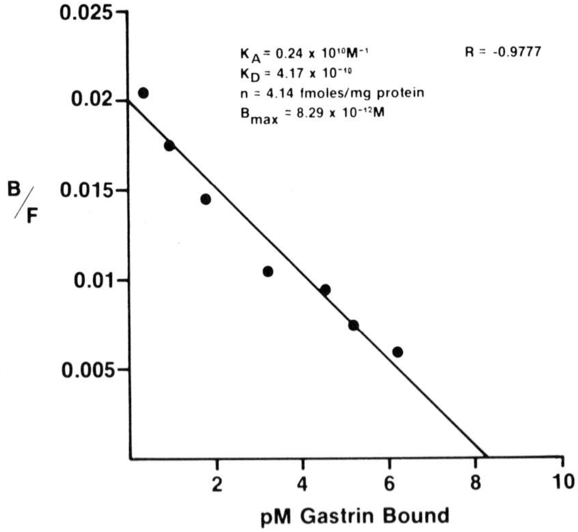

FIG. 4. Representative Scatchard plot analysis of ^{125}I-labeled gastrin specific binding in gastric mucosal membranes. A least-squares linear regression analysis resulted in a correlation coefficient of −0.9777 and a single binding site. Binding constants were calculated as shown. In this transformation the slope of the line equals $-1/K_D$ or $-K_A$, and the value of the x intercept equals the binding capacity (B_{max}). From Takeuchi *et al.*[6]

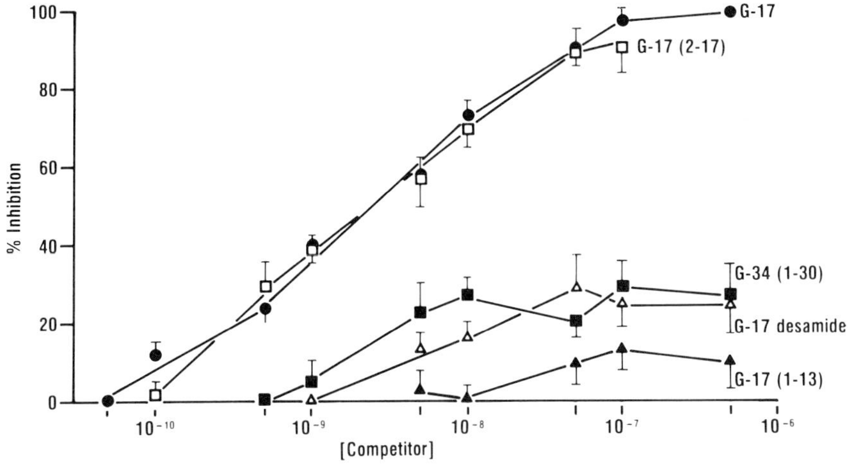

FIG. 5. Dose–response curves of competitors showing specific binding of 200 pM [^{125}I]15-Leu G-17 (●). Data are expressed as percentages of inhibition of specific gastrin binding against concentration of competitor. From Takeuchi *et al.*[15]

stimulates acid secretion and growth of oxyntic gland mucosa and growth of duodenal mucosa. Gastrin has no known effects on antral mucosa, which is the main site of production of the hormone. We have not been able to detect gastrin receptors in tissues outside of the gastrointestinal tract. These include liver, spleen, kidney, and skeletal muscle.

Molecular Specificity

Gastrin and its analogs inhibit specific binding in a competitive manner.[6] In order of decreasing potency 15-Leu G-17 > CCK > caerulein > pentagastrin. Secretin, which is a noncompetetive inhibitor of the biological effects of gastrin in the rat, is a noncompetetive inhibitor of binding as well. Apparently, in binding to its own receptor, secretin produces a conformational change in the gastrin receptor.[6] Analogs of G-17 having as many as 9 amino acids removed from the N-terminal end of the hormone compete for the receptor almost as well as G-17 itself.[15] Removal of the C-terminal active tetrapeptide eliminates binding to the receptor (Fig. 5). In fact, if one substitutes phenylalanine for the C-terminal phenylalamide residue all binding is eliminated as is biological activity.[15]

Correlation of Binding and Biological Activity

In the newborn rat, sensitivity to gastrin as determined by both acid secretion and the trophic response develops at the time of weaning. The

[15] K. Takeuchi, G. R. Speir, and L. R. Johnson, *Am. J. Physiol.* **239,** G395 (1980).

gastrin receptor is also absent from the mucosa of rats until this time.[16] Injection of corticosterone on day 7 of life cause a premature appearance of both gastrin receptors and gastrin response on day 10. Using identically labeled gastrin, Soll[17] has recently shown an excellent correlation between gastrin specific binding and aminopyrine uptake in isolated canine parietal cells.

[16] K. Takeuchi, W. Peitsch, and L. R. Johnson, *Am. J. Physiol.* **240,** G163 (1981).
[17] A. H. Soll, D. A. Amirian, L. P. Thomas, T. J. Reedy, and J. D. Elashoff, *J. Clin. Invest.* **73,** 1434 (1984).

[7] Assay for the Cholecystokinin Receptor on Pancreatic Acini

By ROBERT T. JENSEN and JERRY D. GARDNER

Introduction

Receptors for cholecystokinin have been identified in brain and pancreas using radioiodinated cholecystokinin.[1–5] In general, these receptors have a relatively high affinity for cholecystokinin and caerulein (an analog of the C-terminal portion of cholecystokinin isolated from the skin of the frog *Hyla caerulea*) and a relatively low affinity for gastrin.[1–5] Removing the sulfate from the tyrosine residue in cholecystokinin causes a significant decrease in the apparent affinity of the peptide for the receptors, whereas removing the sulfate from the tyrosine residue in gastrin causes little or no change in the apparent affinity of gastrin for the receptors. In pancreas, butyryl derivatives of cyclic GMP, proglumide, a derivative of glutaramic acid, and benzotript, a derivative of tryptophan, function as specific cholecystokinin receptor antagonists in that these compounds can occupy cholecystokinin receptors but are incapable of eliciting a biologi-

[1] R. T. Jensen, G. F. Lemp, and J. D. Gardner, *Proc. Natl. Acad. Sci. U.S.A.* **77,** 2079 (1980).
[2] H. Sankaran, I. D. Goldfine, C. W. Deveney, K.-Y. Wong, and J. A. Williams, *J. Biol. Chem.* **255,** 1849 (1980).
[3] S. E. Hays, M. C. Beinfeld, R. T. Jensen, F. K. Goodwin, and S. M. Paul, *Neuropeptides* **1,** 53 (1980).
[4] A. Saito, I. D. Goldfine, and J. A. Williams, *J. Neurochem.* **37,** 483 (1981).
[5] R. D. Innes and S. H. Snyder, *Proc. Natl. Acad. Sci. U.S.A.* **77,** 6917 (1980).

METHODS IN ENZYMOLOGY, VOL. 109

ISBN 0-12-182009-2

cal response.[6-8] In brain, butyryl derivatives of cyclic GMP inhibit binding of radiolabeled cholecystokinin; however, in brain, it cannot be determined whether agents that interact with the cholecystokinin receptor possess agonist or antagonist activity.[4]

Preparation of Radiolabeled Cholecystokinin

For binding studies with cholecystokinin receptors, radiolabeled hormone of moderately high specific activity must be prepared by a method that does not cause loss of biologic activity of the labeled peptide. Because cholecystokinin has two methionines in the biologically active COOH-terminus as well as a sulfated tyrosine, which is essential for high affinity binding, preparation of radiolabeled cholecystokinin by the use of oxidizing agents, such as chloramine T, has not resulted in a biologically active radiolabeled peptide. A satisfactory radiolabeled peptide can be obtained by labeling pure porcine cholecystokinin with the iodinated Bolton–Hunter reagent.

Iodination of Porcine Cholecystokinin for Binding Studies

One millicurie of ^{125}I-Bolton–Hunter reagent is dried under a gentle stream of nitrogen in an iodination vial. Five micrograms of natural porcine cholecystokinin (>99% pure, Gastrointestinal Hormone Research Laboratories, Karolinska Institute, Stockholm, Sweden) is dissolved in 5 μl of 2.5 m*M* acetic acid, and added to the iodination vial. Twenty microliters of 50 m*M* sodium borate (pH 10) is then added, the iodination vial is vortexed vigorously, and the mixture is incubated at 0° for 30 min, after which time 250 μl of 0.2 *M* glycine in 50 m*M* sodium borate (pH 10) is added. The reaction is continued at 0° for 5 min and then 250 μl of 6 *M* guanidine hydrochloride is added and the vial is vortexed vigorously.

Purification of Iodinated Cholecystokinin

A column of Sephadex-G50 superfine (1.5 × 50 cm) that has been preequilibrated with 50 m*M* acetic acid containing 0.1% gelatin is prepared and kept at 4°. The iodination mixture is applied to the column. The reaction vial is rinsed once with an additional 250 μl of 6 *M* guanidine hydrochloride with vigorous vortexing and then once with 250 μl of

[6] S. R. Peiken, A. J. Rottman, S. Batzri, and J. D. Gardner, *Am. J. Phys.* **235,** E743 (1978).
[7] N. Barlas, R. T. Jensen, M. C. Beinfeld, and J. D. Gardner, *Am. J. Phys.* **242,** G161 (1982).
[8] W. F. Hahne, R. T. Jensen, G. F. Lemp, and J. D. Gardner, *Proc. Natl. Acad. Sci. U.S.A.* **78,** 6304 (1981).

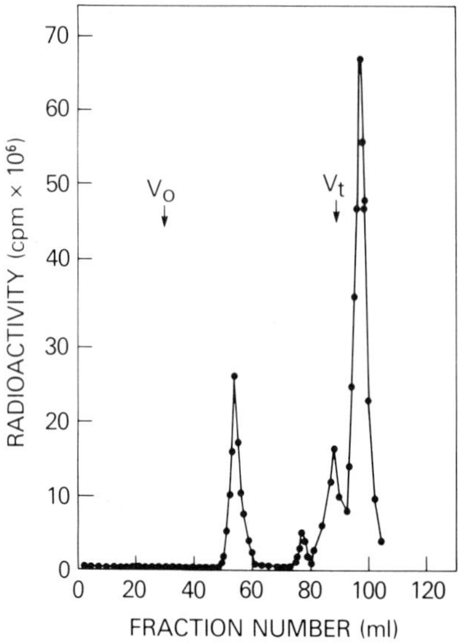

FIG. 1. Elution profile of ^{125}I radioactivity after application of radioiodination mixture following conjugation of ^{125}I-Bolton–Hunter reagent to cholecystokinin. Radioactivity is plotted as a function of the fraction number (1 ml). V_0 = void volume and V_t = total volume of the column.

50 m*M* acetic acid with 0.1% (w/v) gelatin. The column is eluted with 50 m*M* acetic acid containing 0.1% (w/v) gelatin at 4° and 1.0-ml fractions are collected. An example of a typical elution profile is shown in Fig. 1. [^{125}I]CCK appears in fractions 50–61 and the tracer with optimal binding characteristics is in fractions 55–57.

Characteristics of Iodinated Cholecystokinin

The [^{125}I]CCK from fractions 55–57 is more than 95% acid precipitable by 10% (v/v) trichloroacetic acid and can be stored for 4–6 weeks at −20° without loss of activity. If fractions are repeatedly thawed and refrozen, accelerated loss of activity will occur; therefore, eluates should be aliquoted in small fractions. [^{125}I]CCK has a specific activity of 900–1300 Ci/mmol and is fully biologically active.

Binding Studies with Dispersed Pancreatic Acini

Assay for Cholecystokinin Receptor Binding

Assay Solution. The assay incubation solution contains 25.5 m*M* HEPES (pH 7.4), 100 m*M* NaCl, 6.0 m*M* KCl, 2 m*M* KH_2PO_4, 1.2 m*M*

$MgCl_2$, 11.5 m*M* glucose, 5 m*M* Na-fumurate, 5 m*M* Na-glutamate, 5 m*M* Na-pyruvate, 0.5 m*M* $CaCl_2$, 2 m*M* glutamine, 1% (w/v) albumin, 0.008% trypsin inhibitor, 1% (v/v) amino acid mixture, and 1% (v/v) essential vitamin mixture. This is the same solution that is used to measure the ability of CCK to stimulate amylase secretion from pancreatic acini.

Dispersed Acini from Guinea Pig Pancreas. Dispersed acini are prepared as outlined in the previous section.[9] For a typical experiment, dispersed acini from 3 guinea pigs are suspended in 30 ml of assay solution.

Incubation. The incubation is performed in 12 × 75-mm siliconized borosilicate glass tubes to which are added unlabeled peptide or sample to be assayed, 0–50 μl; assay solution, 0–50 μl; dispersed pancreatic acini, 1000 μl; and [^{125}I]CCK (100,000 cpm), 5–30 μl.

The dispersed pancreatic acini must be well suspended and all clumps should be removed or dispersed. The reaction is started by adding [^{125}I]CCK.

Incubation Conditions. Tubes are gassed with 100% O_2, covered with parafilm, and shaken continuously in a constant temperature bath at 37°. After 30 min, duplicate 400-μl samples are transferred to individual microfuge tubes (total capacity, 1.8 ml) containing 700 μl of chilled (4°) wash solution [assay solution containing 4% (w/v) albumin and 0.1% (w/v) bacitracin]. Before removing each sample, the incubation tube should be vortexed to be certain that the acini are dispersed homogenously.

Separation of Bound and Free [^{125}I]CCK. The microfuge tubes are centrifuged for 10 sec in a Beckman Model B Microfuge. The supernatant is aspirated by vacuum and discarded. The tubes are agitated by vortexing and placed so that ice covers the tips of the tubes. Iced wash solution (700 μl) is added and the tubes are centrifuged again. The supernatant is aspirated and the acini are washed a third time with 700 μl of iced wash solution. After the third wash, the supernatant is aspirated, the microfuge tube is inserted into a 16 × 100-mm borosilicate glass tube, and radioactivity is measured by counting the glass tube in a crystal scintillation counter.

Results with [^{125}I]CCK Binding to Dispersed Pancreatic Acini

When dispersed acini are incubated with [^{125}I]CCK at 37° binding of the labeled peptide reaches a steady-state by 45 min and remains constant from 45–90 min at which time 3–6% of the added [^{125}I]CCK is bound (total binding). In the presence of unlabeled CCK (1 μ*M*) or CCK-8 (0.3 μ*M*)

[9] J. D. Gardner and R. T. Jensen, this volume [23].

[10] R. T. Jensen, G. F. Lemp, and J. D. Gardner, *J. Biol. Chem.* **257,** 5554 (1982).

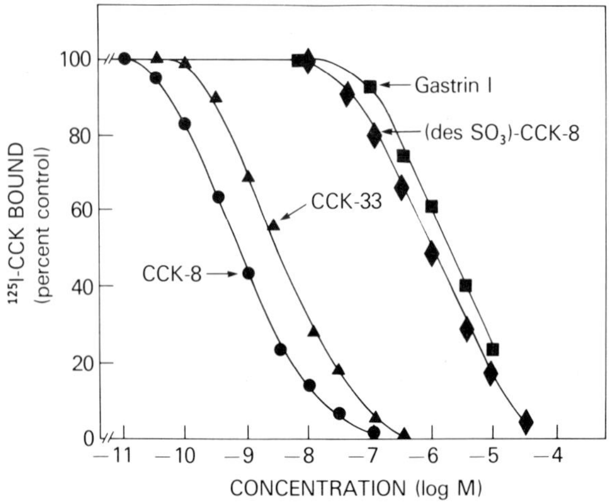

FIG. 2. Ability of various peptides to inhibit binding of [^{125}I]CCK to dispersed pancreatic acini. Binding is expressed as the percentage of radioactivity that was saturably bound in the absence of added nonradioactive peptides. In each experiment each value was determined in triplicate and results given are means for eight separate experiments. CCK-33, Cholecystokinin; CCK-8, octapeptide of cholecystokinin; (des SO_3)-CCK-8, desulfated octapeptide of cholecystokinin.

0.2–0.5% of the added radioactivity is bound (nonsaturable binding). Saturable binding is defined as total binding minus nonsaturable binding. Under the conditions of the assay nonsaturable binding should always be less than 20% of the total binding and is usually less than 10%. To determine nonsaturable binding, synthetic CCK-8 (0.3 μM) is generally used because CCK is difficult to obtain and CCK-8 is three times more potent in inhibiting binding of [^{125}I]CCK than is CCK (Fig. 2).

The amount of saturable binding seen with different concentrations of unlabeled peptide can be determined and the results expressed as bound/free at a particular peptide concentration or as the percentage of the saturable binding obtained with no added peptide (Fig. 2).

Bound [^{125}I]CCK dissociates slowly (47% dissociated by 150 min at 37°). The rate of dissociation is temperature dependent (6% dissociation by 150 min at 4°), and is increased by 50% by adding unlabeled CCK (65% dissociated by 150 min at 37°).

Analysis by Scatchard plots of equilibrium binding data with unlabeled CCK reveals a curvilinear plot with 9000 binding sites per acinar cell. CCK is 3-fold less potent than CCK-8 (Fig. 2), but 1000-fold more potent than gastrin I. Removing of the sulfate from the tyrosine residue in CCK-8 results in a 1000-fold decrease in the affinity of the peptide for the CCK receptor.[1]

Comparison of the ability of CCK-related peptides to occupy the CCK-receptor and to stimulate amylase secretion indicates that maximal enzyme secretion occurs when 40% of the receptors are occupied.[1,9]

[8] Methods for Studying the Platelet-Derived Growth Factor Receptor

By Daniel F. Bowen-Pope and Russell Ross

Introduction

Platelet-derived growth factory (PDGF) is a basic (p$I \simeq 10$) 30,000-dalton protein (see this volume [58] for a detailed biochemical description). It circulates in normal blood sequestered within the platelet alpha granule. It is likely that PDGF is released from platelets at sites of vascular damage and that it contributes toward the cell proliferation and connective tissue formation seen in healing wounds and in arteriosclerotic lesions.[1] It is active at low concentrations ($ED_{50} \simeq 10^{-11}$ *M*) in stimulating the metabolism and multiplication of cultured connective tissue cell types (e.g., see reference 2 for a partial list). Radioiodinated PDGF shows saturable (e.g., 60,000–120,000 receptors per diploid human fibroblast) high affinity binding to cultured PDGF-responsive cells. The apparent dissociation constant reported for this binding interaction by different groups has ranged from 10^{-11} *M*[3] to 10^{-10} *M*[4] to 10^{-9} *M*.[5,6] We believe that these differences are due to technical difficulties, including slow approach to equilibrium and local depletion of ligand concentration. This report will focus on factors which affect [^{125}I]PDGF binding, or the evaluation of binding, on the evidence that the observed binding is to a functional receptor, and on the development of a radioreceptor assay for PDGF.

1 R. Ross and J. A. Glomset, *N. Engl. J. Med.* **295,** 420 (1976).

2 D. F. Bowen-Pope, P. E. DiCorleto, and R. Ross, *J. Cell Biol.* **96,** 679 (1983).

3 D. F. Bowen-Pope and R. Ross, *J. Biol. Chem.* **257,** 5161 (1982).

4 L. T. Williams, P. Tremble, and H. N. Antoniades, *Proc. Natl. Acad. Sci. U.S.A.* **79,** 5867 (1982).

5 C.-H. Heldin, B. Westermark, and åA. Wasteson, *Proc. Natl. Acad. Sci. U.S.A.* **78,** 3664 (1981).

6 J. S. Huang, S. S. Huang, B. Kennedy, and T. F. Deuel, *J. Biol. Chem.* **257,** 8130 (1982).

ISBN 0-12-182009-2

Preparation and Characterization of [^{125}I]PDGF

Radioiodination Protocol

PDGF of the highest purity available is used. These preparations are eluted from phenyl-Sepharose using 50% (v/v) ethylene glycol in 0.1 *N* ammonium acetate pH 7.0[7] (also see this volume [58]) and are judged to be greater than 90% pure by SDS–PAGE. For a typical iodination (essentially as described in reference 3), 20 μg of PDGF in 1–2 ml of ethylene glycol solution is dispensed into a Surfasil (Pierce)-treated 13 × 100-mm glass tube and placed in an ice bath. Carrier-free Na^{125}I (10 mCi) (Amersham) is added in 100 μl of dilute NaOH, mixed well, and followed by 50 μl of 0.1 ng/ml iodine monochloride prepared immediately before use by dilution of a 3.2 g/ml stock ICl (a liquid at room temperature. Available from Eastman Kodak) into 0.1 *N* HCl. All solutions containing ICl must be handled using glass pipets, as the ICl will react with plastic. The reaction mixture is rapidly mixed by swirling and incubated on ice for 20 min. The reaction is quenched by adding 0.2 ml of 5 mg/ml L-tyrosine in 0.05 *N* NaOH. To prevent loss of PDGF by adsorption during dialysis, 5–10 ml of 0.25% gelatin in 10 m*M* acetic acid is added and the solution transferred to dialysis tubing with a nominal molecular weight exclusion of 6000–8000 (Spectrapor 1). Five daily 2-liter changes of dialysis buffer (10 m*M* acetic acid, 150 m*M* NaCl) are needed to achieve greater than 99.5% precipitability. Use of BSA in place of gelatin greatly hinders removal of TCA-soluble [^{125}I]PDGF by dialysis. Specific activities routinely obtained range from 25,000 to 35,000 cpm/ng (1050–1350 cpm/fmol). The iodinated preparation retains biological activity during storage at 4 or −20° for at least 2 months.

We have used the ICl radioiodination procedure in preference to using Iodogen,[4,6] or chloramine T[5,8] for several reasons. The buffer in which we obtain pure PDGF greatly reduces the efficiency of iodination by these techniques. The ICl procedure is also one of the mildest of the oxidation techniques. It permits relatively extensive iodination and thus allows accurate assessment of the effect of iodination on biological activity.

Biological Activity of [^{125}I]PDGF and Characterization of Individual [^{125}I]PDGF Preparations

The iodine monochloride method uses unlabeled ^{127}ICl as an oxidizing agent ($^{127}I^+ + {}^{125}I^- \leftrightarrow {}^{127}I^- + {}^{125}I^+$) so that the specific activity of the ^{125}I is

[7] E. W. Raines and R. Ross, *J. Biol. Chem.* **257,** 5154 (1982).

[8] J. C. Smith, J. P. Singh, J. S. Lillquist, D. S. Goon, and C. D. Stiles, *Nature (London)* **296,** 154 (1982).

necessarily reduced. As a consequence, although the radioiodinated preparation contains an average of 0.25 molecules of ^{125}I per molecule of PDGF, the total average iodination is 1.7 molecules of ^{125}I + ^{127}I per PDGF[3] and very little PDGF would escape some extent of iodination. To systematically investigate the possibility that iodination by this method may affect the biological activity of PDGF, we prepared a series of [^{125}I]PDGF derivatives in which the average extent of iodination varied from 0 to 2.6 I/PDGF. There was no consistent effect of extent of iodination on either mitogenic activity or on binding affinity.[3]

After iodination, each preparation of [^{125}I]PDGF is evaluated as follows.

TCA Precipitation. Precipitability by 10% TCA in the presence of 0.25% gelatin carrier protein must be greater than 99% or the preparation is redialyzed.

SDS–Polyacrylamide Gel Electrophoresis. A sample (10^4 cpm) of each iodinated preparation is analyzed by SDS–PAGE on a 15% separating gel followed by autoradiography.[7] Each of the 27,000–31,000-dalton forms of PDGF which can be resolved by silver staining gels of unlabeled PDGF is labeled with ^{125}I in proportion to the quantity present, suggesting that all are equally good substrates for iodination. When [^{125}I]PDGF is analyzed after reduction, the 14,400–17,500 subunits are all found to be labeled, suggesting that there is at least one possible iodination site on each of the two subunits proposed to constitute native PDGF. Although not routinely checked, we have found[7] that SDS–PAGE of material obtained by solubilization of cells incubated with [^{125}I]PDGF under conditions in which greater than 95% of the radioactivity is specifically bound shows that all radioactivity is of the molecular weight ascribed to PDGF.

Functional Purity. The fraction of TCA-precipitable radioactivity capable of high-affinity binding to cultured PDGF-responsive cells is determined by comparing reduction in TCA-precipitable radioactivity with reduction of high-affinity cell binding during sequential incubations with test cultures. Figure 1 shows the results of such a determination made using two different preparations of [^{125}I]PDGF. By extrapolation, 18 and 42% of the TCA-insoluble radioactivity remains in solution after all radioactivity capable of high-affinity binding to the test cultures has been removed, suggesting that, respectively, 18 and 42% of the radioactivity was not associated with biologically active PDGF. The nature of the inactive TCA-insoluble radioactivity is not certain. In the example shown in Fig. 1, the preparation showing the greatest percentage of innactive material was known to contain a 38K contaminating protein which could be visualized on silver-stained SDS–PAGE. This protein was radioiodinated, as demonstrated by SDS–PAGE autoradiographs of the iodinated prepara-

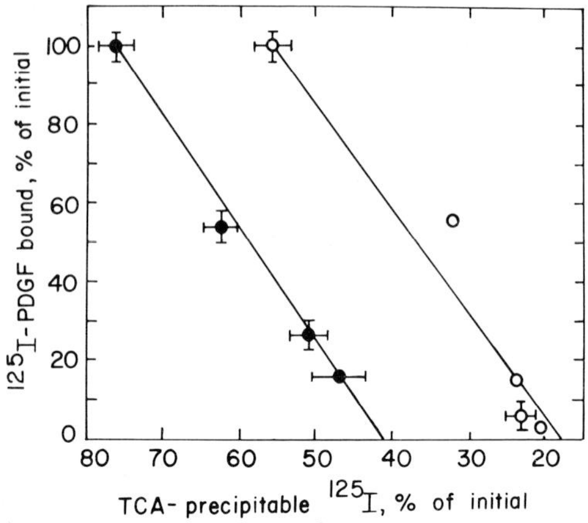

FIG. 1. Percentage of iodinated preparation which cannot bind specifically to 3T3 cells. Confluent, quiescent cultures of Swiss 3T3 cells were incubated for 2 hr at 4° in 0.5 ml of binding medium containing 1 ng of [^{125}I]PDGF prepared from a stock containing no contaminants demonstrable by SDS–PAGE (○) or from a stock containing a 38 kilodalton contaminant (●). The binding medium was then collected and cell-bound ^{125}I was determined by solubilizing the monolayer with solubilization buffer. Triplicate aliquots of the collected medium were used to determine TCA-precipitable ^{125}I in the used binding medium. The remaining medium was reincubated at 0.5 ml per culture on at least three fresh cultures. The procedure was performed a total of 4 times. Nonspecific binding was determined for each incubation by including 200 μg partially purified (CMS III) PDGF in one well. These media were not collected and reused. The results are plotted as specific binding of [^{125}I]PDGF (percentage of initial value) versus TCA-precipitable (percentage of initial) material remaining in the binding medium after binding. The means ± SEM of at least triplicate determinations are plotted. This experimental protocol can be modified to make it less time consuming by using multiple binding periods of 30 min at 37° rather than 2 hr at 4°. Binding is more rapid at 37 than at 4° and measurable degradation does not occur during a 30-min incubation, at least as indicated by the lack of appearance of TCA-soluble radioactivity in the medium. From Bowen-Pope and R. Ross.[3]

tion, but was not seen in extracts of cultures incubated with the radioiodinated preparation. This supports the possibility that the inactive material represents iodinated contaminants. The possibility that some PDGF is damaged during iodination will be discussed below.

Concentration of Active PDGF. The concentration of active PDGF in the preparation is determined by two independent methods:

1. To determine mitogenic potency, confluent quiescent cultures of mouse 3T3 cells or diploid human fibroblasts are incubated for 20–22 hr with increasing concentrations of [^{125}I]PDGF or with unlabeled PDGF

standards. The test medium is then aspirated and replaced with growth medium containing 10 μCi/ml [^{3}H]thymidine. After 2 hr this medium is aspirated and the cultures rinsed 3 times with TCA and solubilized with 0.8 ml/well 0.25 *N* NaOH. For cultures incubated with [^{125}I]PDGF, this extract is transferred to a plastic tube to determine ^{125}I content by gamma counting. Of each extract 0.6 ml is then added to 5 ml of Aquasol (New England Nuclear) for liquid scintillation counting. The liquid scintillation values are corrected for contributions from ^{125}I by multiplying the gamma determinations by relative counting efficiency of ^{125}I in the ^{3}H channel. The correction is relatively small (less than 3%) when using 10 μCi/ml [^{3}H]thymidine for 2 hr, in part because [^{3}H]thymidine incorporation is large, and in part because much of the [^{125}I]PDGF which was cell associated at the beginning of the labeling period is degraded before [^{3}H]thymidine incorporation is determined. By comparing the ED_{50} of the test preparation with that of our reference PDGF standards (see this volume [58]), the concentration of biologically active PDGF in the preparation is calculated. This assayed value for active PDGF concentration is then used for all calculations involving this preparation.

2. To determine the concentration of [^{125}I]PDGF capable of high-affinity binding to test cultures, increasing concentrations of the iodinated preparation are used to construct a saturation binding curve. Parallel cultures are incubated with increasing concentrations of unlabeled reference PDGF standards for 3 hr at 4°, then with a single concentration of [^{125}I]PDGF for 1 hr at 4° as will be described for the radioreceptor assay (RRA). By comparing the $K_{D(app)}$ for the saturation binding experiment, which reflects the concentration of PDGF in the iodinated test preparation, with the ED_{50} for the RRA, which reflects the concentration of reference PDGF standards, the concentration of PDGF in the iodinated preparation can be calculated. We have found that the mitogenesis and binding methods for determining the PDGF content of the iodinated preparation give comparable results. The values range from 75 to 90% of the values predicted if no PDGF were damaged during iodination or lost during handling and dialysis. Since redialysis of [^{125}I]PDGF gives recoveries of about 85–90%, we believe that much of the observed loss of PDGF results from handling rather than from iodination damage.

Calculating Specific Activity of [^{125}I]PDGF. To calculate the specific activity of biologically active [^{125}I]PDGF present in the preparation, the value for cpm capable of high-affinity binding to test cultures (see section on Functional Purity, above) is divided by the value for mitogenically active PDGF present (see section on Concentration of Active PDGF, above).

Saturable Binding of [^{125}I]PDGF

In order to define in one place the basic solutions and procedures used to determine specific [^{125}I]PDGF binding, we will describe here our procedure for determining high-affinity saturable [^{125}I]PDGF binding. The rationale for this procedure will then be discussed in following sections.

Cultures of test cells in 2 cm^2 24-well cluster trays are rinsed once with 0.5 ml cold "binding rinse" (phosphate-buffered saline, pH 7.4, with 1 m*M* $CaCl_2$ and 0.1% BSA). The binding rinse is aspirated and replaced with the desired concentration of [^{125}I]PDGF in 1.0 ml "binding medium" [Ham's medium F12 (Gibco) buffered at pH 7.4 with 25 m*M* HEPES [4-(2-hydroxyethyl)-1-piperazineethanesulfonic acid] and with either 2% calf serum chromatographed to remove PDGF (CMS-I[9]) or with 0.25% BSA]. Nonspecific binding is determined using 0.1 or 0.2 ng/ml [^{125}I]PDGF incubated together with 20 ng/ml pure PDGF or 50 μg/ml partially purified PDGF (side fractions from Sephacryl S-200 molecular seiving, see this volume [58]). Nonspecific [^{125}I]PDGF binding has been shown[3] to be linear with [^{125}I]PDGF concentration in the range of the dissociation constant so that the nonspecific binding values appropriate for the remaining concentrations can be determined by calculation. To determine cell number, parallel culture wells are incubated in binding medium without [^{125}I]PDGF. The trays containing the above additions are incubated at 4° on a platform oscillating at 2.5 rotations per second. After at least 3 hr the trays are rinsed 4 times with 0.5 ml/well binding rinse. The wells for cell counts receive 100 μl 0.25% trypsin in PBS and are covered with tape to prevent accidental contamination by solubilization buffer. The remaining wells are extracted by adding 1 ml/well solubilization buffer (1% Triton X-100 with 0.1% BSA in H_2O), pipetting up and down a few times, and transferring to a plastic tube for gamma counting. Cell counting can be accomplished when convenient after trypsinization is complete (usually by 10 min) using an electronic particle counter.

An example of saturable binding to several cell types is shown in Fig. 12A and will be discussed further in the section identifying the [^{125}I]PDGF binding site as a mitogen receptor. Note: to convert PDGF concentrations from ng/ml to molarity, multiply by 3.3×10^{-11}.

Determination of Cell Associated [^{125}I]PDGF

Test Cultures

Binding experiments are performed using monolayer cultures in 24-well cluster trays. The trays marketed by Costar are convenient in that

[9] A. Vogel, E. Raines, B. Kariya, M. J. Rivest, and R. Ross, *Proc. Natl. Acad. Sci. U.S.A.* **75,** 2810 (1978).

the relatively deep wells prevent spillage when swirling 1.0 ml/well. In an attempt to standardize conditions when measuring receptor number and affinity on different cell types, and in order to be able to measure mitogenic response to PDGF and binding of [^{125}I]PDGF in parallel cultures, we prepare test cultures in the following fashion. Cells are seeded at 1–2 × 10^4 cells per 2 cm^2 culture well in growth medium (DMEM containing 5–10% calf serum or 10% fetal calf serum, depending on the cell type). When the cultures have grown almost to confluence (3–10 days, depending on cell type), the medium is replaced with DMEM containing 1% human plasma-derived serum or 2% calf serum which has been chromatographed on carboxymethyl Sephadex to remove PDGF (CMS-I[9]). Two days later, most nontransformed PDGF-responsive cell types are confluent and quiescent and can be used to test for [^{3}H]thymidine incorporation in response to added PDGF or can be used to measure [^{125}I]PDGF binding. Unfortunately, not all cell types can be brought into this standard state. Most transformed cells are difficult to bring to quiescence. This makes them unsuitable for measurement of stimulation of [^{3}H]thymidine incorporation and adds the variable of cell cycle stage to the comparison between [^{125}I]PDGF binding by transformed cells and by their normal counterparts. In addition, many transformed cell types, e.g., Kirsten-transformed NIH-3T3 cells, are not well attached to the culture surface and tend to detach during the incubation and rinse procedures, especially when performed at 4°. Nevertheless, most cultured cells can be brought to a reasonable approximation of the state described and this state will be assumed in this report unless noted otherwise.

Distinguishing [^{125}I]PDGF Binding to Cells from [^{125}I]PDGF Binding to the Culture Substrate

PDGF is an extremely "sticky" molecule, exhibiting both hydrophobic properties and a large net positive charge in the physiological pH range (for details see this volume [58]). Binding of [^{125}I]PDGF to laboratory glassware can be minimized by coating with Surfasil (Pierce) and by adding 0.1% or more BSA as a carrier. Tissue culture surfaces cannot be treated in this fashion, however, since normal cell attachment and growth require a charged culture surface. If low concentrations of [^{125}I]PDGF in medium containing 0.25% BSA are incubated for several hours in fresh cell-free culture wells, about 11% of the total input radioactivity binds to the culture surface and can be extracted by 0.25 *N* NaOH. About 50% of this binding is competed by high concentrations (50 μg/ml) of partially purified PDGF (less than 1% pure). If the cell-free culture wells are preincubated with growth medium for 3 days followed by mitogen-deficient medium for an additional 2 days, as though they contained cells being prepared for a binding assay, NaOH-extractable dish-bound [^{125}I]PDGF is reduced about 5-fold, presumably due to blockage of binding sites on the culture surface by constituents of the culture medium. This source of nonreceptor-mediated binding can be much further reduced by selectively solubilizing the cell monolayer using "solubilization buffer" (1% Triton X-100 in H_2O with 0.1% BSA added as a carrier). One milliliter per 2 cm^2 well is added, squirted up and down several times over the culture surface with a pasteur pipet, and transferred to a plastic tube for gamma counting. It is important not to allow the solubilization buffer to stand in the wells longer than necessary since [^{125}I]PDGF bound to be a bare culture well is slowly eluted by solubilization buffer ($t_{1/2}$ = 3 hr). It is usually convenient to add solubilization buffer to 4–12 wells at a time. Rinsed cultures can be stored before solubilization by freezing at −20° and thawed when convenient.

In addition to dish-bound [^{125}I]PDGF which can be extracted with Triton or with 0.25 *N* NaOH, some radioactivity binds in some fashion not extractable even with 1 *M* guanidine[10] or 6 *M* urea.[8] This component can be quantitated either by subtracting total recovered cpm from input cpm or by cutting up and gamma counting the plastic culture surface after extraction. The importance of [^{125}I]PDGF which is not extractable by solubiliztion buffer is difficult to evaluate. It is possible that this adsorbed [^{125}I]PDGF can be recognized by cell surface receptors. PDGF-responsive cells which are plated onto culture dishes which have been preincubated with PDGF, then rinsed to remove nonbound PDGF, grow considerably more rapidly than do cells plated into untreated dishes.[8,10] This might result from recognition by the cells of the surface-adsorbed PDGF. However, we have found that [^{125}I]PDGF adsorbed to the tissue culture surface does appear in the culture medium in TCA-soluble and TCA-insoluble forms during the growth of cultured cells on the treated surface. It is possible that the adsorbed PDGF is not effective *in situ,* but must first dissociate from the plastic surface before binding productively the cell-surface receptors.

Nonspecific Binding

To determine "nonspecific binding" (nsb) we have routinely used either 50 μg/ml of partially purified PDGF (0.2–2% pure) or 20 ng/ml pure PDGF. The former gives values for nsb under our binding conditions ranging from about 1% of total binding (Triton-extractable cpm) for confluent Swiss 3T3 cells at K_D concentrations of [^{125}I]PDGF to about 15% of total binding for sparse cultures of diploid human fibroblasts. The major

[10] A. Faggiotto, D. F. Bowen-Pope, and R. Ross, unpublished observations (1982).

variable determining % nsb is the amount of specific binding—the absolute nsb as percentage of input cpm is relatively constant at about 0.7%. Where practical, we prefer to use 20 ng/ml pure PDGF to determine nsb. This is at least 100 times the K_D observed under nondepleting conditions. Use of high concentrations of impure material, although necessary when sufficient pure material is not available, can cause underestimation of nonreceptor-mediated binding. As discussed above, PDGF shows significant binding to tissue culture plastic. Although much of this binding is either blocked by culture medium components during the growth of the test cultures, or is not extracted by the Triton solubilization procedure, tissue culture plastic does show some residual Triton-extractable [^{125}I]PDGF binding and some of this binding is competed by very high concentrations of the impure, very cationic preparations of partially purified PDGF. This is not a problem in cases where specific binding is high (e.g., confluent 3T3 cells), but can significantly underestimate nsb for cultures showing very few receptors. As a compromise, therefore, we currently use 20 ng/ml pure PDGF to determine nsb for cultures which express few receptors and 50 μg/ml partially purified PDGF for cultures which express many receptors and which are therefore not sensitive to small underestimations of nsb, but which are sensitive to depletion of added PDGF with attendant overestimation of nsb (see below).

Binding Solutions

Binding Medium

For initial characterization of the PDGF receptor, we measured [^{125}I]PDGF binding in medium able to support a mitogenic response to PDGF. For this purpose we used Ham's medium F-12 buffered at pH 7.4 with HEPES and supplemented with 2% calf serum from which the PDGF had been removed by incubation with carboxymethyl-Sephadex (CMS-I[9]). This allows the dependence of [^{125}I]PDGF binding on concentration and cell type to be compared as closely as possible to the dependence of mitogenic stimulation on these variables. As will be discussed below, these comparisons support the conclusion that the [^{125}I]PDGF is binding to the PDGF mitogen receptor and not to unrelated binding sites. In order to simplify preparation of the binding medium, we have now replaced the chromatographed serum with 0.25% BSA with no change in the results obtained from saturation binding assays or radioreceptor assays (Fig. 2). Attempts to replace Ham's F-12 medium with a simple salt solution (Ringer's saline supplemented with 1 m*M* $MgSO_4$ and buffered at pH 7.4 with 25 m*M* HEPES) resulted in a reduction in the apparent binding affinity

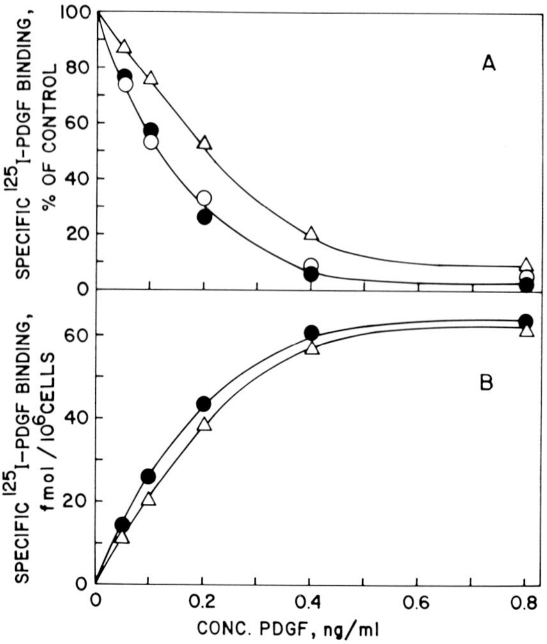

FIG. 2. Effect of different binding media on saturation binding and on the radioreceptor assay. (A) Radioreceptor assay. Subconfluent cultures (2.64×10^4 cells/culture) of diploid human fibroblasts in 24-well cluster trays were incubated with the indicated concentrations of pure PDGF in 1 ml of binding medium at 4° for 3 hr with gentle shaking. The binding media used were Ham's medium F-12 with 2% chromatographed calf serum (●), Ham's medium F-12 with 0.25% BSA (○), or Ringer's saline with 1 m*M* $MgSO_4$ and 0.25% BSA (△). All binding media were buffered at pH 7.4 with 25 m*M* HEPES. The cultures were then rinsed once with binding rinse and incubated in 1 ml of the same binding medium containing 0.5 ng/ml [^{125}I]PDGF. Specific [^{125}I]PDGF binding was determined as described in the text. Mean of triplicates. (B) Saturation binding. Parallel cultures were incubated in the binding media used for A containing the indicated concentration of [^{125}I]PDGF. After 3 hr at 4° with gentle shaking, specific [^{125}I]PDGF binding was determined. Symbols are as used for A except that the values for Ham's F-12 with chromatographed serum or with BSA were not significantly different, and are plotted as (●). Mean of triplicates.

which is particularly pronounced in the RRA procedure (Fig. 2A). For this reason, we currently measure [^{125}I]PDGF binding in Ham's medium F-12 buffered at pH 7.4 with 25 m*M* HEPES and supplemented with 0.25% crude BSA (Cohn Fraction V, Sigma).

Dependence of Binding on pH and Cation Concentration

To determine the dependence of [^{125}I]PDGF binding on specific cations while minimizing the exposure of the test cells to suboptimal ionic conditions, we reduced the binding incubation period to 1 hour. The table shows that in this experiment [^{125}I]PDGF binding was not affected by omission of Ca^{2+} and Mg^{2+} even in the presence of up to 0.1 m*M* EDTA. At higher concentrations, EDTA causes significant changes in cell morphology and cell detachment, so that the decrease in [^{125}I]PDGF binding could reflect indirect effects on cell properties rather than a requirement for low concentrations of divalent cations for binding per se. [^{125}I]PDGF binding was reduced to less than 20% of control if both monovalent and divalent salts were replaced with isotonic sucrose. Binding was only partially restored by addition of the divalent salts, but was brought to 90% of control by adding 50% of physiological levels of monovalent salts.

[^{125}I]PDGF binding is independent of pH within a broad range near

EFFECT OF CATION CONCENTRATION ON [^{125}I]PDGF BINDING

Monovalent salts (% of value in Hanks' saline[a]	Ca^{2+} conc. (m*M*)	Mg^{2+} conc. (m*M*)	EDTA conc. (m*M*)	[^{125}I]PDGF binding (% value in complete binding medium[b])
100	1	1	0	100
	5	0	0	98 ± 2
	1	0	0	111 ± 3
	0.1	0	0	106 ± 5
	0	0	0	104 ± 7
	0	0	0.05	106 ± 5
	0	0	0.1	99 ± 5
	0	0	0.2	67 ± 4
	0	0	0.5	56 ± 1
0	0	0	0	17 ± 0
	1	1	0	37 ± 2
	1	0	0	30 ± 2
	0	1	0	30 ± 1
3.13	0	0	0	19 ± 5
6.25	0	0	0	20 ± 2
12.5	0	0	0	24 ± 4
25	0	0	0	45 ± 3
50	0	0	0	90 ± 1

[a] All binding media contained 2% dialyzed calf CMS-I and 25 m*M* HEPES, pH 7.4. In addition, they contained the indicated concentrations of added Ca^{2+} and Mg^{2+} and either the monovalent salts present in Hanks' saline, or isotonic sucrose (250 m*M*), or isotonic sucrose partially replaced with Hanks' salts. The results are expressed as [^{125}I]PDGF bound in test medium as a percentage of [^{125}I]PDGF bound in ionically complete binding medium.

[b] Confluent, quiescent cultures of Swiss 3T3 cells were rinsed twice with ice-cold isotonic sucrose buffered at pH 7.4 with 25 m*M* HEPES. 1.0 ml of ice-cold binding medium containing 0.5 ng [^{125}I]PDGF/ml was added, and the cultures were incubated for 1 hr at 4° with gentle agitation. Specific [^{125}I]PDGF binding was determined as described in the text. From Bowen-Pope and Ross.[3]

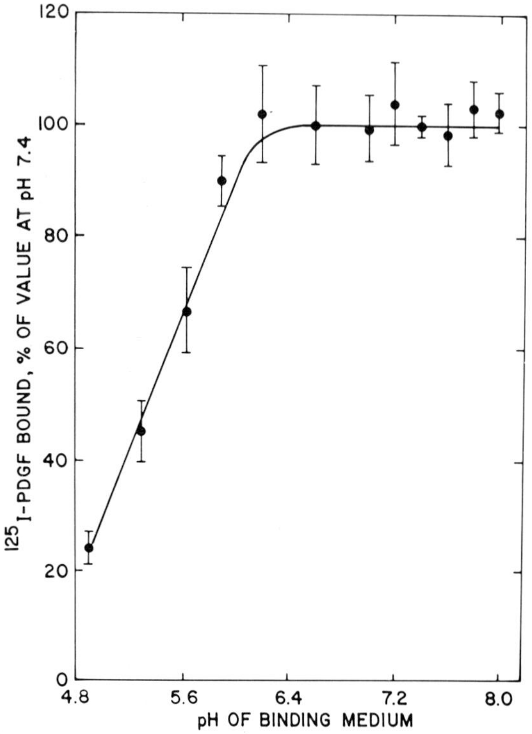

FIG. 3. Dependence of [^{125}I]PDGF binding on pH. Binding medium was adjusted to higher or lower pH by addition of NaOH or HCl. Confluent, quiescent cultures of Swiss 3T3 were incubated for 2 hr at 4° in these media with 1.2 ng [^{125}I]PDGF (28,800 cpm). Washing of the cultures, Triton extraction of cell-bound [^{125}I]PDGF, and correction for nonspecific binding were as described in the text. The results are plotted as the means ± standard error of the mean of triplicate determinations. Parallel cultures were incubated in media at pH 5.3, 6.0, 7.0, 7.4, and 8.0 without added [^{125}I]PDGF. After rinsing as in the binding assay, these cultures were trypsinized and cell number per culture determined by Coulter counting. There was no significant difference in number of cells per well after incubation at any of these pH values. From Bowen-Pope and Ross.[3]

physiological pH (Fig. 3). Below pH 6, binding diminishes precipitously. As will be discussed below, prebound [^{125}I]PDGF can also be dissociated by incubation with acidified saline.

Binding Conditions

General Problems

The extremely high affinity of the PDGF receptor ($K_{D(app)} \leq 10^{-11}$ *M*) combined with the relatively large numbers of PDGF receptors on some commonly used cultured cell types (e.g., about 120,000 per cell on Swiss mouse 3T3 cells) causes some problems not encountered in studying binding of other ligands to cells. The crux of the problem is that the number of PDGF receptors present in the assay can be comparable to, or can exceed, the concentration of added [^{125}I]PDGF. For example, a 2 cm^2 culture well containing 4 × 10^4 Swiss 3T3 cells (approximately 120,000 PDGF receptors per cell) will contain 4.8 × 10^9 PDGF receptors. One milliliter of binding medium containing [^{125}I]PDGF at 10^{-11} *M*, i.e., near the K_D, will contain only 6 × 10^9 molecules of [^{125}I]PDGF. Clearly, the problem of depletion of ligand will be more severe at concentrations of [^{125}I]PDGF less than the K_D. This situation does not arise when studying, for example, EGF binding to Swiss 3T3 cells, since the ligand concentration range of interest is 10- to 100-fold higher ($K_{D(app)} = 10^{-9}$ *M*) and the number of EGF receptors per cell is lower (approximately 60,000 per cell). When studying [^{125}I]PDGF binding, this situation must be circumvented experimentally, or corrected for mathematically, or the true K_D will be overestimated.

Volume Dependence

Ideally, the amount of [^{125}I]PDGF bound would be a function of the initial [^{125}I]PDGF concentration and would be independent of the volume of the labeling solution. Ideal behavior will be seen, however, only if the receptor concentration of the assay is substantially less than the ligand concentration, and in practice is not achieved under many circumstances. Figure 4 shows the volume dependence of [^{125}I]PDGF binding to monolayer cultures of Swiss 3T3 cells (part A) and to diploid human fibroblasts (part B). The Swiss 3T3 cells have more PDGF receptors per cell (approximately 120,000) than do human fibroblasts (approximately 60,000) and were present at a 3-fold higher cell density to accentuate the difference in number of receptors per assay well. It is clear from Fig. 4 that binding of low concentrations of [^{125}I]PDGF is very volume dependent for the 3T3 cell cultures and only slightly volume dependent for the HF cell cultures. As would be predicted if the volume dependence reflected depletion, binding of the highest concentration of [^{125}I]PDGF shows little volume dependence even for 3T3 cell cultures. To minimize depletion, we routinely use 1 ml/2 cm^2 culture well. This is the largest volume that can be used without risk of spillage during the mixing described below and is a larger ratio of volume to culture surface than can be used with larger (e.g., 60 mm dishes) culture dishes without spillage problems.

Kinetics of [^{125}I]PDGF Binding at 4°

A second inconvenient manifestation of the depletion problem is the slow approach to equilibrium seen with low concentrations of [^{125}I]PDGF.

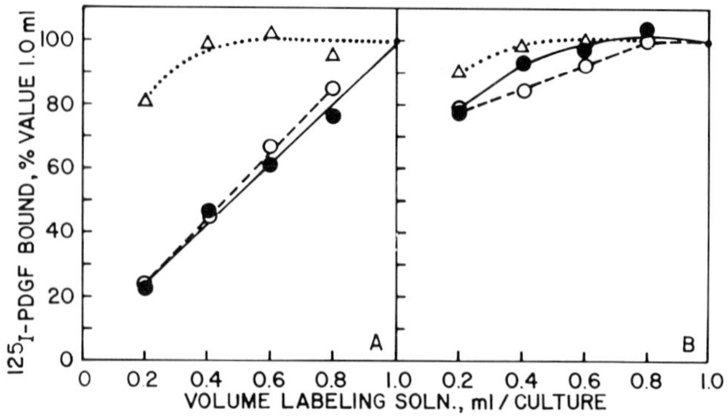

FIG. 4. Dependence of [^{125}I]PDGF binding on volume of labeling solution. Cultures of Swiss 3T3 cells (A: 4.0×10^4 cells) or diploid human fibroblasts (B: 1.4×10^4 cells) in Costar 2 cm^2 24-well culture trays were incubated in the indicated volume of labeling solution at 4° for 4 hr with gentle shaking. The labeling solutions contained 0.05 (○), 0.5 (●), or 5.0 (△) ng/ml [^{125}I]PDGF (14,916 bindable cpm/ng). Results (mean of triplicate determinations) are expressed as specific [^{125}I]PDGF bound as percentage of [^{125}I]PDGF bound using 1.0 ml labeling solution/culture. These values were, for Swiss 3T3 cells: 440 cpm at 0.05 ng/ml; 3201 cpm at 0.5 ng/ml; 8962 cpm at 5 ng/ml; for human fibroblasts: 100 cpm at 0.05 ng/ml; 414 cpm at 0.5 ng/ml; 668 cpm at 5 ng/ml.

Figure 5A shows that equilibrium binding of low concentrations of [^{125}I]PDGF is not achieved even by 8 hr at 4°. The rate of binding is increased when the cultures are incubated on a shaking platform oscillating at 2.5 cycles per second, but still does not reach a constant value by 6–8 hr. The effect of shaking is greater for binding to 3T3 monolayers (high receptor density) than for binding to subconfluent fibroblast cultures (low receptor density). When high concentrations of [^{125}I]PDGF are used (Fig. 5B) binding to both high and low receptor density cultures does reach a constant value within 6–8 hr. We interpret these kinetic data in terms of the depletion phenomenon discussed above. The layer of medium immediately in contact with the monolayer, termed the "unstirred layer" or "diffusion boundary layer,"[11] is not instantaneously and completely mixed with the bulk labeling medium. When using low concentrations of [^{125}I]PDGF, this layer is rapidly depleted and the rate of [^{125}I]PDGF binding becomes limited by the rate of access of [^{125}I]PDGF from the bulk medium. As a result, the depletion problems in the medium to which the cells are actually exposed is much more severe than the calculations of depletion of total [^{125}I]PDGF in the assay would suggest. The problem can be alleviated somewhat by mixing the bulk medium as well as possible on an oscillating table (compare binding to stationary and shaking cultures in Fig. 5) and by keeping the receptor density in the monolayer as low as

[11] N. G. Maroudas, *Cell* **3,** 217 (1974).

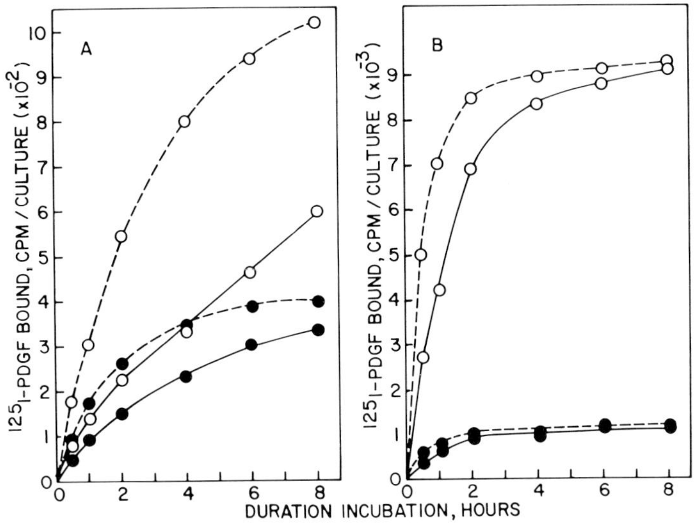

FIG. 5. Kinetics of [^{125}I]PDGF binding at 4° with or without continuous mixing. Diploid human fibroblasts (2.3 × 10^4 cells/2 cm^2 well; ●) or Swiss 3T3 cells (8.9 × 10^4 cells/2 cm^2 well; ○) were incubated at 4° in 1 ml/well [^{125}I]PDGF at 0.1 ng/ml (A) or 10 ng/ml (B). One set of cultures was incubated on a stationary surface (solid lines) while a second set was incubated on a platform oscillating at 2.5 cycles per second to impart a definite rotary motion to the labeling medium (broken lines). Cell-associated cpm were determined at the times indicated by Triton extraction. Means of triplicate determinations are plotted.

possible, e.g., by plating at a lower cell density or by using cells with fewer receptors per cell (compare binding to fibroblasts and to 3T3 cell cultures in Fig. 5). Nevertheless, it is clear from Fig. 5A that steady-state binding of low concentrations of [^{125}I]PDGF may not be completely achieved during a binding period of 3–4 hr at 4° even if cultures at relatively low receptor densities are mixed in this fashion.

Kinetics of [^{125}I]PDGF Binding at 37°

When [^{125}I]PDGF binding is measured at 37° (Fig. 6), total cell-associated [^{125}I]PDGF increases to a maximum by about 1 hr and then declines. TCA-soluble radioactivity appears in the medium during the decline phase, suggesting that the decline may represent degradation of the cell-associated [^{125}I]PDGF–receptor complexes. Depletion of [^{125}I]PDGF present in the medium accounts for some, but not all, of the reduction in binding at 9 hr, since replacing the medium with fresh [^{125}I]PDGF solution accomplished only a small increase in binding. We and others[3,4,6,12] have interpreted this phenomenon in terms of the model developed to explain a comparable phenomenon with [^{125}I]EGF binding as representing internal-

[12] C.-H. Heldin, Å. Wasteson, and B. Westermark, *J. Biol. Chem.* **257,** 4216 (1982).

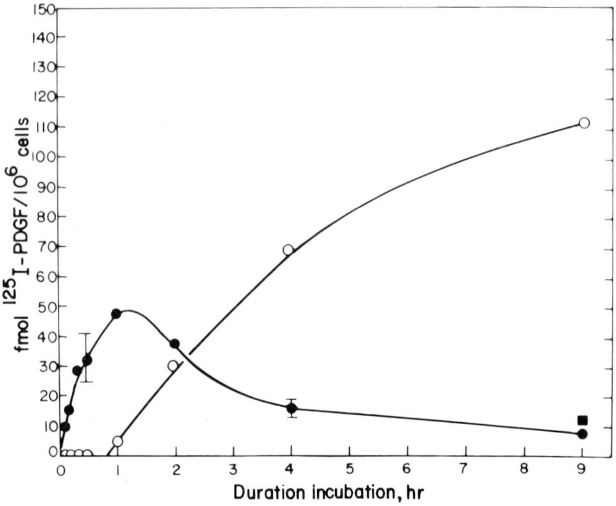

FIG. 6. Kinetics of [^{125}I]PDGF binding and degradation at 37°. Confluent (5.2×10^4 cells/culture), quiescent cultures of Swiss 3T3 cells were incubated at 37° in 0.5 ml of binding medium containing 40 p*M* [^{125}I]PDGF. At the times indicated on the abscissa, triplicate cultures were sacrificed for determination of specifically bound [^{125}I]PDGF and for determination of trichloroacetic acid-soluble [^{125}I]PDGF present in the incubation medium. The latter values were corrected for trichloroacetic acid-soluble [^{125}I]PDGF present in unused binding medium (1.5% of total ^{125}I). Binding medium incubated at 37° without cells for 9 hr was still 98% trichloroacetic acid precipitable. The mean ± SE of triplicate determinations is plotted: ●, cell-associated ^{125}I; ○, TCA-soluble ^{125}I in the medium. Additional triplicate cultures were incubated 8.5 hr with [^{125}I]PDGF, then for 0.5 hr in fresh [^{125}I]PDGF-containing medium before cell-associated ^{125}I (■) was measured. Adapted from Bowen-Pope and Ross.[3]

ization and subsequent intracellular degradation of the bound [^{125}I]PDGF and possibly of the receptor.[13] This interpretation is supported by our observations[14] that [^{125}I]PDGF bound to the cell surface (acetic acid dissociable) at 4° is internalized (acetic acid resistant) with a $t_{1/2}$ = 20 min after warming, while the appearance of TCA-soluble degradation products shows a $t_{1/2}$ = 60–90 min.

Radioreceptor Assay for PDGF

We will first describe our standard radioreceptor assay (RRA) protocol, then discuss some of the parameters which affect the sensitivity and specificity of the assay.

[13] G. Carpenter, K. J. Lembach, M. M. Morrison, and S. Cohen, *J. Biol. Chem.* **250,** 4297 (1975).

[14] M. E. Rosenfeld, D. F. Bowen-Pope, and R. Ross, *J. Cell. Physiol.*, in press (1984).

Protocol

Subconfluent monolayers of diploid human fibroblasts are prepared by plating 1.5×10^4 cells per 2 cm^2 culture well in Costar 24-well cluster trays in DMEM supplemented with 1% human plasma-derived serum (PDS). Cultures are usually used between the first and seventh day after plating. Cultures are set on an ice tray and rinsed once with ice-cold binding rinse to help chill the trays and remove residual plating medium. One milliliter/well of test substance in binding medium is added and the cultures incubated in a refrigerated room on an oscillating platform (2.5 cycles per second) for 3–4 hr. The trays are then placed on ice, aspirated, rinsed once with cold binding rinse, and incubated for 1 hr as above with 1 ml/well binding medium containing 0.5 ng/ml [^{125}I]PDGF. Labeling is terminated with 4 rinses with binding rinse and cell-associated [^{125}I]PDGF determined by extraction with solubilization buffer, as described previously. Standard curves are run using 0, 0.05, 0.1, 0.2, 0.4, 0.8, and 20 ng/ml purified PDGF, the last concentration being used to determine nonspecific binding. When the test substance is known or suspected to contain a component which might interfere indirectly with PDGF binding, e.g., a soluble PDGF-binding protein, an internal standard curve is run by adding the above concentrations of PDGF in the presence of test substance. The concentration of test substance must then be chosen such that the PDGF content of the sample itself is not sufficient to compete for more than 30% of the specific [^{125}I]PDGF binding. This ensures that competition by the added PDGF standards can still be accurately determined. Depending on the density of the test monolayer, the amount of PDGF needed to reduce subsequent [^{125}I]PDGF binding by 50% is 0.1–0.2 ng/ml (Figs. 2, 7B, and 8).

We have observed some variation from tray to tray in the amount of ^{125}I binding in the absence of added PDGF or test substance. This variation is seen even using trays plated and used in parallel. Since the variation can be as much as 20% from tray to tray (usually within 5%), we routinely include a group of controls in each tray and calculate percentage competition separately for each tray.

Effect of Sequence of Incubations

The RRA as described above is not a standard simultaneous competition assay, but rather depends on the ability of PDGF–receptor complexes formed during the period of preincubation with test substance to remain blocked during postincubation with [^{125}I]PDGF. Experiments specifically demonstrating the stability of the complex will be shown in a later section. The rationales for this protocol are illustrated by the experiments shown in Fig. 7 in which the two RRA protocols are used to assay pure

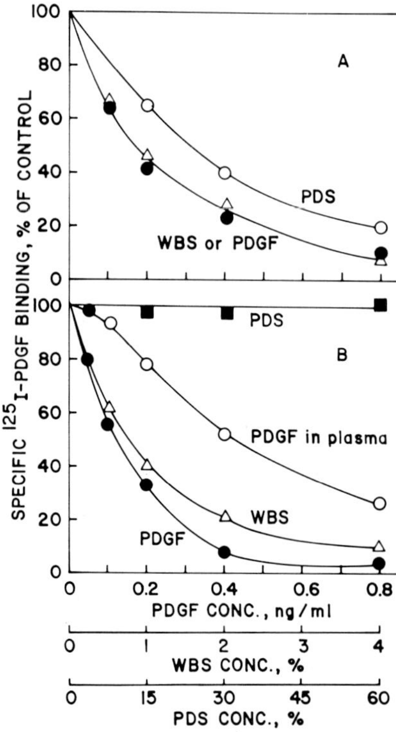

FIG. 7. Radioreceptor assays for PDGF. Confluent (4.7×10^4 cells/2.4 cm^2 culture well) cultures of human foreskin fibroblasts were incubated for 3 hr at 4° with gentle shaking with 1.0 ml/culture of the indicated concentration of test substance: (○), human PDS; (△), human WBS; (●), PDGF; (■), PDGF in 60% human PDS. For (A) 0.12 ng/ml [^{125}I]PDGF was present with each test substance. The incubation was terminated and bound [^{125}I]PDGF determined as described in the text. For (B) the 3-hr incubation with test substance was terminated with one rinse with "binding rinse" and followed by a 1-hr incubation at 4° with gentle shaking in medium without test substance containing 0.25 ng/ml [^{125}I]PDGF. The incubation was terminated and bound [^{125}I]PDGF determined as described in the text. All data points represent the mean of determinations from 3 cultures. The standard deviations from the mean were usually ≤10% and always ≤20%. From Bowen-Pope *et al.*[15]

PDGF, human whole blood serum, and human plasma-derived serum. When test substance and [^{125}I]PDGF are incubated with test cultures simultaneously, the assay appears to demonstrate the presence of significant PDGF in plasma-derived serum (PDS), as well as in whole blood serum (Fig. 7A). This result is *not* obtained using the sequential incubation protocol (Fig. 7B)—no PDGF is detected in PDS. We interpret this difference as follows: Plasma contains substances able to bind [^{125}I]PDGF such that it is no longer capable of binding to the cell-surface PDGF

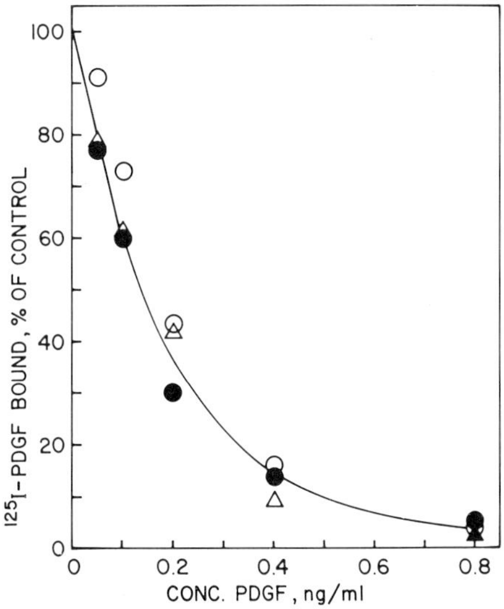

FIG. 8. Effect of [^{125}I]PDGF concentration on the sensitivity of the radioreceptor assay. Confluent cultures of diploid human fibroblasts in Costar 24-well cluster trays were incubated with gentle shaking for 3 hr at 4° in 1.0 ml binding medium containing the concentration of unlabeled PDGF shown on the abscissa. The cultures were then rinsed once with cold binding rinse and incubated with gentle shaking for 1 hr at 4° in 1.0 ml binding medium containing 0.1 ng/ml (○), 0.5 ng/ml (●), or 5.0 ng/ml (△) of [^{125}I]PDGF. Specific binding (mean of triplicate determinations) is expressed as percentage of specific binding to control cultures incubated in the absence of unlabeled PDGF. Specific binding to control cultures was 101, 379, and 550 cpm/culture, respectively.

receptor.[15,16] In the simultaneous incubation protocol, this results in a decreased [^{125}I]PDGF binding to the monolayer which is not distinguished from reduced [^{125}I]PDGF binding resulting from competition by PDGF in the sample. Hence the spurious "detection" of PDGF in PDS. In the sequential incubation protocol, the [^{125}I]PDGF is never exposed directly to the test substance and cannot be inactivated by complex formation. The presence of PDGF-binding proteins in plasma can be readily confirmed by molecular sieving of [^{125}I]PDGF incubated with or without plasma. In the presence of plasma, significant [^{125}I]PDGF elutes with a greatly increased apparent size[16,16a] consistent with the formation of a complex between the [^{125}I]PDGF and a plasma binding protein(s). The presence of this binding protein(s) in plasma shows up in the sequential

[15] D. F. Bowen-Pope, T. W. Malpass, D. M. Foster, and R. Ross, *Blood* **64,** 458 (1984).

[16] E. W. Raines, D. F. Bowen-Pope, and R. Ross, *Proc. Natl. Acad. Sci. U.S.A.* **81,** 3424 (1984).

[16a] S. J. Huang, S. S. Huang, and T. F. Deuel, *J. Cell Biol.* **97,** 383 (1983).

format of the RRA as a shift in the standard curve: more PDGF must be added to block cell-surface PDGF receptors since some of the PDGF is inactivated by binding to plasma proteins. The same situation applies to PDGF which might be present in the plasma test sample; hence the use of internal standards. Figure 7B shows that the presence of 60% plasma decreases the sensitivity of the assay by about 4-fold. Even the use of internal standards does not permit evaluation of the amount of PDGF present as an irreversible complex with a binding protein nor of PDGF bound to a low-capacity binding component, such that there is not sufficient excess binding capacity in the test sample to bind a significant fraction of the internal standards. Ideally, the PDGF content of such samples would be assayed only after being purified away from the binding components.

The second reason for using the sequential incubation protocol is that it gives a greater sensitivity than the simultaneous incubation protocol. Robard *et al.*[17] can be consulted for a discussion of the mathematical basis for differences in the sensitivity of various assay protocols. The final reason for using the sequential incubation protocol is that achieving maximum sensitivity with the simultaneous incubation protocol necessitates use of [^{125}I]PDGF concentrations much less than the K_D. With a $K_{D(app)}$ of 0.1–0.2 ng/ml (0.3–0.6 $\times$ 10^{-11} *M*), this severely limits the cpm/culture obtained using 2 cm^2 wells of subconfluent fibroblasts. The sequential incubation protocol can be used with any concentration of [^{125}I]PDGF with identical results (Fig. 8). This is because the labeling period determines the relative number of unoccupied PDGF receptors and subsequent binding of ^{125}I-ligand is proportional to receptor number at all ^{125}I-ligand concentrations. We have chosen 0.5 ng/ml [^{125}I]PDGF because it gives close to maximum specific [^{125}I]PDGF binding (about 500 cpm) during the short (1 hr) labeling period without showing high values of nonspecific binding (7.5–15% of total binding).

Factors Which Affect Assay Sensitivity

Since the sequential incubation protocol is based on the occupancy of a measurable fraction of the PDGF receptors on the monolayer, the sensitivity of the assay will be reduced if the number of receptors is large. In practice we have found that the subconfluent cultures of human fibroblasts described above give optimal sensitivity (ED_{50} 0.1–0.2 ng/ml) with low but acceptable cpm (approximately 500 cpm in control wells). The cells are plated and maintained in 1% human PDS, since the cells are

[17] D. Rodbard, H. J. Ruder, J. Vaitukaitis, and H. S. Jacobs, *J. Clin. Endocrinol. Metao.* **33,** 343 (1970).

healthy in this medium but multiply only slowly (low concentrations of calf serum could probably be substituted). As a consequence, the cultures can be maintained at a desirable cell density (confluent and below) for long periods of time (2 weeks with a feeding at 1 week) until needed. Confluent cultures of 3T3 cells (approximately 10^5 cells/well) can be used, but the sensitivity is greatly reduced (ED_{50} = approximately 1 ng/ml), presumably as a result of higher receptor density. Sparsely plated Swiss 3T3 cells show improved sensitivity ($ED_{50} \geq 0.15$ ng/ml, depending on cell density) but have a greater tendency than diploid human fibroblasts to show morphological changes (retraction) and some detachment during prolonged incubation at 4°.

Characteristics of the test sample which can affect assay sensitivity include pH and the presence of PDGF-binding components. As shown in Fig. 3, the pH of the test solution must be above pH 6 to permit maximum binding and, as discussed above for Fig. 7, the presence of PDGF-binding components can be at least partially corrected for by using internal standards. We have found that the 4° RRA is not sensitive to the presence of levels of toxic buffer constituents or contaminants which interfere with assays of mitogenic activity (see this volume [58]).

Dissociation of Bound PDGF

Kinetics of Dissociation

Figure 9 shows that [^{125}I]PDGF bound to human fibroblasts at 4° does not dissociate significantly during a 6-hr postincubation in fresh binding medium without PDGF. Dissociation is not appreciable even if the postincubation medium contains 10 ng/ml unlabeled PDGF (sufficient to inhibit virtually all rebinding of dissociated [^{125}I]PDGF under these conditions), 100 μg/ml goat antihuman-PDGF IgG (sufficient to inhibit rebinding by greater than 95%), or 60% human plasma-derived serum (containing enough PDGF binding protein to reduce rebinding by about 70%). When the concentration of unlabeled PDGF present in the postbinding medium is varied from 0.1 to 100 ng/ml, no significant dissociation is observed (data not shown), either induced by the unlabeled PDGF or revealed by inhibition of rebinding of dissociated [^{125}I]PDGF.

Dissociation by Low pH

Just as the initial binding of [^{125}I]PDGF is reduced when the pH of the binding medium is decreased below pH 6, it is possible to dissociate prebound [^{125}I]PDGF by briefly incubating at 4° in saline at a pH less than

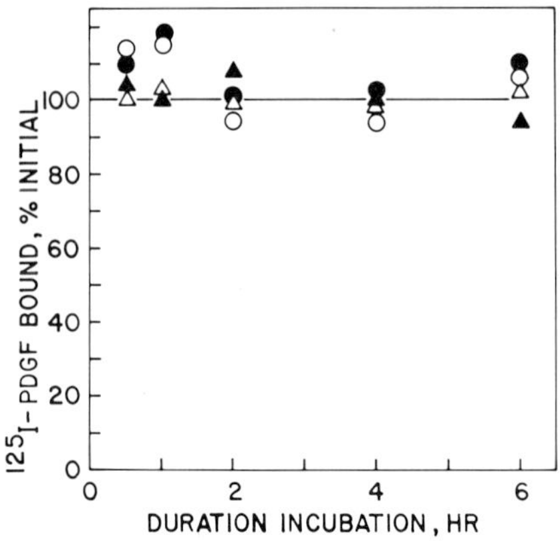

FIG. 9. Dissociation of cell-surface bound [^{125}I]PDGF at 4°. Confluent cultures of diploid human fibroblasts in 24-well cluster trays were incubated with 0.5 ng/ml [^{125}I]PDGF for 3 hr at 4° and rinsed 5 times with 4° binding rinse to remove nonbound [^{125}I]PDGF. The cultures were then incubated at 4° with gentle shaking with binding medium (○) or with binding medium supplemented with 10 ng/ml pure PDGF (●), 60% human plasma-derived serum (△), or with 100 μg/ml goat antihuman PDGF IgG (▲). At the times indicated, the incubation was terminated by aspiration and cell-associated [^{125}I]PDGF determined as described. Non-specific binding (mean = 18 cpm/culture; no systemic variation during postincubation) was determined from parallel cultures incubated with 0.5 ng/ml [^{125}I]PDGF plus 20 μg/ml partially purified PDGF and has been subtracted. The mean of triplicate determinations is plotted as percentage of value (670 cpm/culture) obtained with no postincubation.

6 (Fig. 10B), essentially as described for prebound [^{125}I]EGF.[18] At the unadjusted pH of 20 m*M* acetic acid in saline with 0.25% BSA (pH 3.7), prebound [^{125}I]PDGF is rapidly dissociated (Fig. 10A) with 90% of the prebound counts dissociated by 1 min. This brief incubation does not damage the culture nor irreversibly decrease [^{125}I]PDGF binding as measured by three criteria[2]: (1) binding of [^{125}I]PDGF to acid-treated cultures was not significantly different from binding to untreated cultures (Fig. 10B). (2) Less than 2% of the cells stained with trypan blue. (3) Cultures of 3T3 cells treated with acidified saline were able to respond normally to subsequent addition of PDGF as assayed by [^{3}H]thymidine incorporation measured 20 hr later. Although saline at pH 3.7 does not extract or irreversibly damage the PDGF receptor, rinsing with saline at lower pH does reduce subsequent [^{125}I]PDGF binding (Fig. 10B). For this reason, we use pH 3.7, even though extraction of prebound [^{125}I]PDGF is not as complete as at lower pH.

We have used sensitivity to rapid dissociation by acidified saline as an

[18] H. T. Haigler, F. R. Maxfield, M. C. Willingham, and I. Pastan, *J. Biol. Chem.* **255,** 1239 (1980).

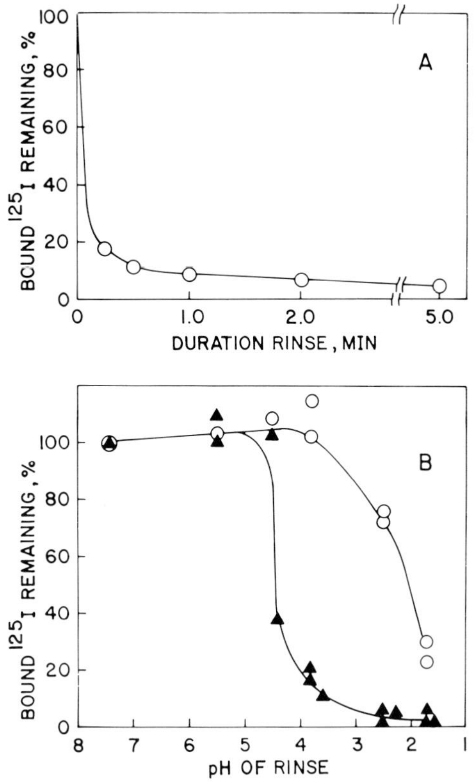

FIG. 10. Dissociation of bound [^{125}I]PDGF by acidified saline and effect of acidified saline rinsing on subsequent binding. Confluent cultures of Swiss 3T3 cells in 24-well cluster trays were incubated with 0.5 ng/ml [^{125}I]PDGF at 4° and rinsed 5 times to remove nonbound [^{125}I]PDGF. (A) Labeled cultures were incubated with 1.0 ml of 20 m*M* acetic acid (pH 3.6) in 150 m*M* NaCl and 0.25% BSA. The acidified saline was gently swirled on an ice bath for the time indicated, then removed by aspiration and cell-associated [^{125}I]PDGF determined by Triton extraction. Means of triplicate determinations corrected for nonspecific binding are plotted as a percentage of [^{125}I]PDGF bound before acid rinsing. (B) Labeled cultures were incubated for 1 min at 4° with 1 ml cold acidified saline adjusted to the pH indicated with NaOH or HCl. Results from 3 separate experiments are expressed as for (A) (▲). Parallel unlabeled cultures were treated with the same solutions of acidified saline, then incubated with 0.5 ng/ml [^{125}I]PDGF at 4° and specifically bound [^{125}I]PDGF determined. Results are expressed as percentage of specific binding obtained when cultures were rinsed with binding rinse at pH 7.4. Results from two separate experiments are shown (○).

assay for [^{125}I]PDGF bound to the cell surface. The inability of this procedure to remove internalized [^{125}I]PDGF has been confirmed by experiments in which cultures were preincubated at 4° with [^{125}I]PDGF to permit binding to cell-surface receptors and then we rinsed and incubated at 37°.

The cell-associated radioactivity becomes resistant to dissociation by acetic acid with a half time of 20 min after warming.[14] This is consistent with the kinetics of internalization of [^{125}I]PDGF as visualized by electron microscope autoradiography and by electron microscopy of PDGF adsorbed to electron-dense colloidal gold particles.[14,18a]

Scatchard Analysis

Figure 11 shows [^{125}I]PDGF binding to three different cell types plotted according to the method of Scatchard.[19] The data cannot be fit well to a single straight line, but instead shows a downward "hook" at low values of bound [^{125}I]PDGF. It is possible that the hook results from positive cooperative behavior in the binding of [^{125}I]PDGF to its receptor. It is also possible that the hook reflects some artifact in the measurement of binding. Very low concentrations of [^{125}I]PDGF could be especially vulnerable to the degradation, inactivation, or binding to the plastic surface. A second source of error in the hook region is incomplete equilibrium between bound and free [^{125}I]PDGF. As shown in Fig. 5, very low concentrations of [^{125}I]PDGF approach equilibrium binding very slowly. This could account for some of the low ratios of bound/free seen at the lowest concentrations of [^{125}I]PDGF. Since very long incubation of some cell types (e.g., Swiss 3T3) at 4° causes some morphological, and thus possibly functional, changes in the test cells, we have not attempted to achieve complete equilibration by prolonging the incubation period, preferring to make these measurements when a cell-free PDGF receptor system is established, with the realization that existing methods tend to overestimate K_D.

Given the possible problems in trying to derive a true dissociation constant from Scatchard analysis of [^{125}I]PDGF binding data, we currently determine the apparent dissociation constant $K_{D(app)}$ as the concentration of [^{125}I]PDGF giving half-maximal specific [^{125}I]PDGF binding determined directly from a plot of bound vs final free [^{125}I]PDGF concentration. These values agree relatively well with $K_{D(app)}$ determined from the slope of the Scatchard plot at a half-maximal value of bound/free (for the 3 cell types shown in Fig. 11 the values are 7.5, 6.5, and 10.5 p*M* from the saturation plot and 12.2, 8.6, and 17 p*M* from the Scatchard plot).

The molecular domains on PDGF and on the PDGF receptor which are involved in the binding interaction seem to have been highly con-

[18a] J. Nilsson, J. Thyberg, C.-H. Heldin, A. Wasteson, and B. Westermark, *Proc. Natl. Acad. Sci. U.S.A.* **80,** 5592 (1983).

[19] G. Scatchard, *Ann. N.Y. Acad. Sci.* **51,** 660 (1949).

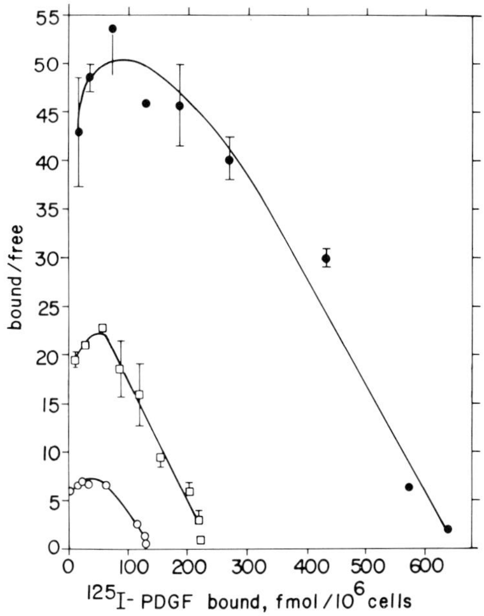

FIG. 11. Scatchard analysis of [^{125}I]PDGF binding. The binding data from 3 cells types used in the experiment of Fig. 12B are plotted according to the method of Scatchard: ●, high-binding strain of Swiss 3T3; □, Swiss 3T3; ○, monkey smooth muscle cells. Adapted from Bowen-Pope and Ross.[3]

served during evolution since serum from all tested higher chordates contains material which blocks binding to mouse 3T3 cells[20] and since cells from many species have been found to bind human [^{125}I]PDGF with comparable affinity (e.g., Fig. 12A), including a cell line (BF-2) derived from bluegill fish ($K_{D(app)}$ = 13 pM) and chicken embryo cells ($K_{D(app)}$ = 15 pM).[21] The major experimental difficulty in comparing the $K_{D(app)}$ for different cell types has been the effect of depletion of [^{125}I]PDGF in the binding medium. When cell types, like Swiss 3T3, which have many receptors per cell, are assayed at a confluent cell density, the $K_{D(app)}$ is high (greater than 1 ng/ml; 33 pM) compared to a value of 0.1–0.2 ng/ml (3–6 pM) obtained using subconfluent diploid human fibroblasts. If sparse cultures of Swiss 3T3 cells are used, the $K_{D(app)}$ is as low as 0.2 ng/ml, depending on cell density. Based on the effects of depletion, and especially of local depletion, on the binding of different concentrations of [^{125}I]PDGF (see earlier sections), we believe that the high values of $K_{D(app)}$

[20] J. P. Singh, M. A. Chaikin, and C. D. Stiles, *J. Cell Biol.* **95,** 667 (1982).

[21] D. F. Bowen-Pope, R. A. Seifert, and R. Ross, *in* "Cell Proliferation: Recent Advances" (A. L. Boynton and H. L. Leffert, eds.). Academic Press, New York, in press.

obtained with cultures with a high receptor density represent artifactual overestimations of the true K_D rather than actual increases in K_D with increasing cell density. In view of this problem, it is necessary when comparing the $K_{D(app)}$ for [^{125}I]PDGF binding to different cell types to keep the receptor density low, even if this requires using cultures which are not in the "standard state" (confluent, quiescent monolayer) defined earlier. Artifacts resulting from depletion and the attendant problem of slow approach to equilibrium at low concentrations probably account for most of the differences between the published values for $K_{D(app)}$ reported by different laboratories which range from 10^{-11} *M* or lower[3,22] to 10^{-10} *M*[4] to 10^{-9} *M*.[5,6]

The PDGF Binding Site Is a Mitogen Receptor

Several lines of evidence support the conclusion that the cell surface binding of [^{125}I]PDGF is to a specific functional mitogen receptor. (1) All cell types which respond mitogenically to PDGF bind [^{125}I]PDGF. Most cell types which do not respond mitogenically to PDGF do not bind [^{125}I]PDGF. The cells which bind but do not respond are all abnormal in one of two ways—either, like transformed cells, they show a maximum growth rate in the absence of PDGF and could not, therefore, be stimulated further by a functional PDGF receptor, or, like diploid cells cultured to "senescence," they are incapable of responding mitogenically to any growth stimulus. (2) Swiss 3T3 cells which were selected by [^{3}H]thymidine suicide against mitogenic response to PDGF showed a 20-fold reduction in [^{125}I]PDGF binding. It seems likely that this reduced binding was the defect through which responsiveness was reduced. (3) For all cell types tested, the concentration of PDGF stimulating half-maximal [^{3}H]thymidine incorporation is comparable to the concentration of [^{125}I]PDGF producing half-maximal [^{125}I]PDGF binding (Fig. 12), suggesting that the observed binding may initiate the stimulation. A rigorous quantitative comparison between the concentration dependences of binding and mitogenesis such as would be needed to determine whether a maximal mitogenic response can be achieved before all PDGF receptors are occupied is difficult to make: The ^{125}I-binding assay is performed at 4° while the mitogenesis assays are performed at 37°. Binding can be measured at 37° under non-steady-state conditions (e.g., see Fig. 6) using short (30–45 min) incubation periods to avoid degradation of bound [^{125}I]PDGF. Under these conditions the $K_{D(app)}$ is elevated several fold. Whether this is due to the effects of temperature per se or to the non-steady-state binding conditions is not yet known. The second barrier to

[22] P. E. DiCorleto and D. F. Bowen-Pope, *Proc. Natl. Acad. Sci. U.S.A.* **80,** 1919 (1983).

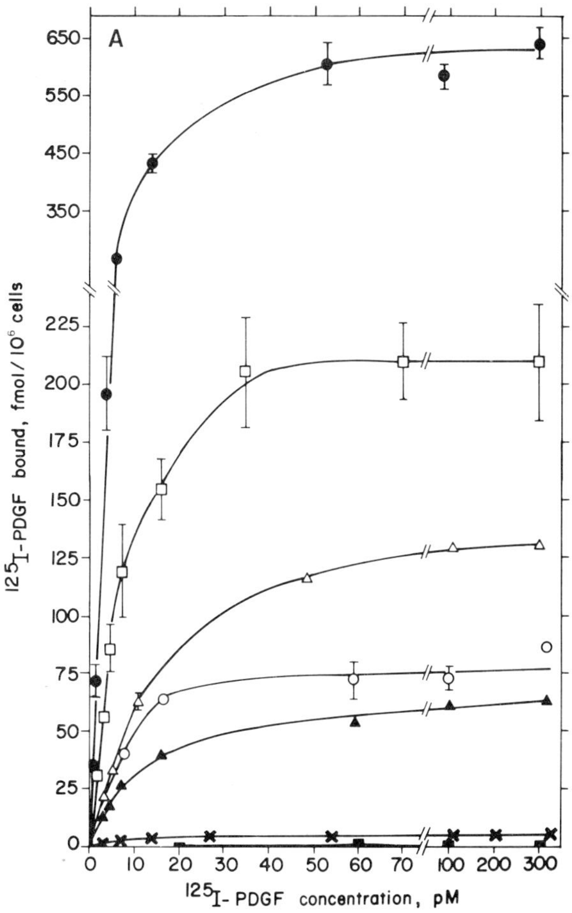

FIG. 12. [^{125}I]PDGF binding and stimulation of [^{3}H]thymidine incorporation in different cell types. Each cell type was plated in Costar 24-well trays and grown to confluence in a 1 : 1 mixture of Dulbecco–Vogt modified Eagle's medium and Ham's F-12 medium *f* 5% calf serum. The growth medium was then replaced with a 1 : 1 mixture of Dulbecco–Vogt modified Eagle's medium and Ham's F-12 medium *f* 2% calf CMS-I for 48 hr. (A) [^{125}I]PDGF binding was determined using triplicate cultures for each concentration by incubating with 1 ml of binding medium containing increasing amounts of [^{125}I]PDGF. The cultures were incubated for 5.5 hr at 4° with gentle agitation before measuring cell-bound [^{125}I]PDGF by Triton extraction. Nonspecific binding was determined as described in the text, and has been subtracted. The concentrations plotted on the abscissa represent the concentrations of intact [^{125}I]PDGF at the end of the binding period. The cell types used were ●, high-binding Swiss 3T3; □, Swiss 3T3; △, monkey smooth muscle; ○, human foreskin fibroblasts; ▲, human smooth muscle cells; ×, Swiss 3T3 variant clone PF2; □, A431 carcinoma. (B) Parallel cultures were incubated for 18 hr with increasing concentrations of unlabeled PDGF before incorporation of [^{3}H]thymidine into trichloracetic acid-insoluble material was measured. The results are plotted as fold stimulation (ratio of [^{3}H]thymidine incorporation in the presence of added PDGF/[^{3}H]thymidine incorporation in the absence of added PDGF) versus the initial concentration of PDGF. [^{3}H]Thymidine incorporation was not determined for human smooth muscle cells or A431 cells. From Bowen-Pope and Ross.[3]

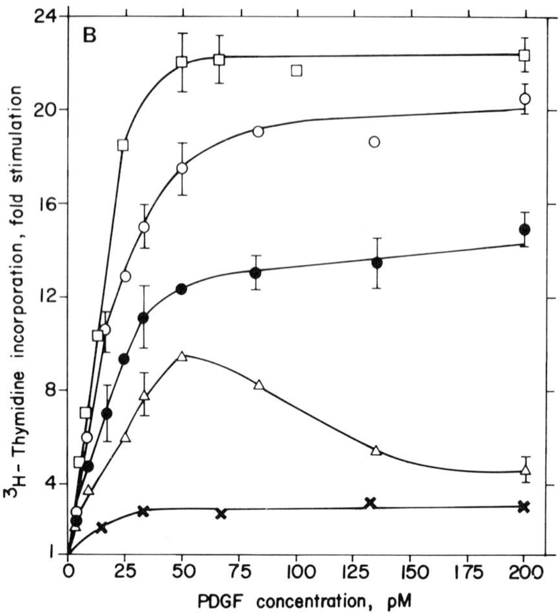

FIG. 12B. See legend on p. 95.

directly comparing the $K_{D(app)}$ with the ED_{50} for mitogenesis is that, during the 18–22 hr 37° incubation used in mitogenesis studies, the concentration of intact PDGF in the medium continuously decreases as it is bound and degraded. It is not clear how to quantitatively express the relationship between mitogenesis and PDGF concentration in a system in which the concentration of mitogen is rapidly decreasing and in which the importance of the mitogen to the cell may be changing (e.g., see reference 23). (4) Binding is very specific for PDGF. Other purified human alpha-granule proteins (thrombospondin and platelet factor 4) at 5×10^{-5} *M*, whole plasma (including a very wide spectrum of proteins) at 60% v/v, bovine insulin at 10^{-3} *M*, partially purified fibroblast growth factor at 10^{-5} *M*, and mouse EGF at 2.6×10^{-6} *M*, all fail to compete for [^{125}I]PDGF binding.[3,21] Since PDGF has a very high isoelectric point (near pH 10), polycations become likely candidates for competition. We have found that polylysine reduces [^{125}I]PDGF significantly only at concentrations (ED_{50} − 10 μg/ml) slightly below those producing cell damage visible at the light and electron microscope levels. Huang *et al.*[6] have reported that protamine sulfate at

[23] W. J. Pledger, C. D. Stiles, H. N. Antoniades, and C. D. Scher, *Proc. Natl. Acad. Sci. U.S.A.* **74,** 4481 (1977).

high concentrations (ED_{50} = 5 μg/ml) can compete for [^{125}I]PDGF binding to 3T3 cells and can cause release of prebound [^{125}I]PDGF without being itself active as a mitogen. It is likely that this competition reflects only the charge of the polycations and is not of physiological significance.

Molecular Properties of the PDGF Receptor

Size

We have obtained approximate values for the size of the PDGF receptor by using covalent cross-linking reagents.[24] [^{125}I]PDGF was incubated with test culture at 4° to permit binding to cell-surface receptors and nonbound [^{125}I]PDGF was removed by rinsing with Krebs–Ringer phosphate buffer. The cross-linking reagent, EGS (ethylene glycol bis-succinimidyl succinate: Pierce) or DSS (disuccinimidyl succinate: Pierce), was dissolved in dimethyl sulfoxide and added, with mixing, at a 1:50 or 1:100 dilution into the KR buffer for the indicated final concentration of cross-linking agent. Upon mixing, this solution was immediately added to the KR buffer-rinsed cell cultures and allowed to incubate at 15° for 15 min. The cultures were then rinsed once with KR buffer and 0.1 ml of 2 × SDS sample buffer [0.2 *M* Tris–HCl, pH 6.8, 10% (v/v) glycerol, 6% (w/v) SDS, and 0.005% (w/v) bromphenol blue] was added to quench the cross-linking reaction and solubilize the cells. The solubilized cell monolayer was mixed with the SDS sample buffer with a Teflon spatula and collected into a tube. The dish was rinsed with an additional 50 μl of the 2 × SDS sample buffer using the Teflon spatula and this rinse was pooled with the original extract and heated at 100° for 3–5 min. Samples were analyzed by SDS–polyacrylamide gel electrophoresis (4.5% stacking gel, 5 or 5–15% linear gradient separating gels) followed by autoradiography (Kodak X-Omat-R film with Chronex enhancing screen). Exposure of the cultures to 1–3 m*M* EGS before extraction resulted in the appearance of a high-molecular-weight radioiodinated complex which was not seen when specific binding of [^{125}I]PDGF to the cell surface receptor had been blocked by excess unlabeled PDGF, down-regulated by prior exposure to [^{125}I]PDGF at 37°, or when cell types which do not express PDGF receptors were used. The apparent size of this complex was 193K when estimated using EGS and 188K when estimated using DSS. Subtracting 30K to correct for the contribution of the [^{125}I]PDGF to the complex yields estimates of 159–164K for the PDGF receptor component. This size was not significantly reduced when the complex was reduced with 2-mercap-

[24] K. C. Glenn, D. F. Bowen-Pope, and R. Ross, *J. Biol. Chem.* **257,** 5172 (1982).

toethanol before electrophoresis, suggesting that it was not composed of disulfide bonded subunits. Since SDS disrupts noncovalent interactions between proteins, we cannot determine whether the receptor exists in the cell membrane as a complex of 159–164K protein with other identical or nonidentical components, or whether it exists alone. The labeled component is at least partially proteinaceous since trypsin treatment of test cultures results in reduced labeling with a concomitant appearance of a 94K-dalton complex—presumably a product of partial proteolysis. This affinity-labeling procedure was used to determine and compare the molecular weight of the PDGF receptor on PDGF-responsive cells other than 3T3 cells. Figure 13 shows that EGS cross-linked [^{125}I]PDGF into a 193 K-dalton complex on 3T3 cells, human fibroblasts (low passage WI-38), and monkey arterial smooth muscle cells. This conservation of molecular size among different species is consistent with the conservation of the binding affinity discussed earlier.

In addition to the 159–164K complex, the sample contains [^{125}I]PDGF migrating at its native molecular weight of 30K and very large material seen at the top of the separating and stacking gels. We interpret this latter material as complexes of [^{125}I]PDGF with its receptor plus additional membrane components, perhaps other components of the native receptor complex. The observation that partial proteolysis resulted in a single new labeled fragment suggests that the [^{125}I]PDGF itself is linked to unique protein. These findings have been confirmed[24a] using a photoactivatable cross-linking reagent.

Phosphotyrosine Protein Kinase Activity

We have studied PDGF-stimulated protein phosphorylation using partially purified membranes prepared in the folloiwng fashion.[25] Cells are scraped from 150 mm dishes in the presence of 20 ml of 5 m*M* HEPES, pH 7.4, 2 m*M* $MgCl_2$, and 5 m*M* 2-mercaptoethanol. The cells are then homogenized with 20 strokes in a Dounce homogenizer using a tight-fitting pestle. The homogenate is centrifuged at 18,000 rpm for 30 min and the pellet resuspended for assay by homogenization in 40 m*M* HEPES pH 7.4, 200 m*M* NaCl, 0.4% Triton X-100. Membrane phosphorylation assays are carried out in a total volume of 30 μl containing (in final concentrations) 20 m*M* HEPES pH 7.4, 100 m*M* NaCl, 0.2% Triton X-100, 10 m*M* $MnCl_2$, 40 μM App(NH)p, 5 m*M* *p*-nitrophenyl phosphate, 10 μM [γ-^{32}P]ATP (10,000–20,000 cpm/pmol), and approximately 30–40 μg mem-

[24a] C.-H. Heldin, B. Ek, and L. Ronnstrand, *J. Biol. Chem.* **257,** 421b (1983).
[25] L. J. Pike, D. F. Bowen-Pope, R. Ross, and E. G. Krebs, *J. Biol. Chem.* **258,** 9383 (1983).

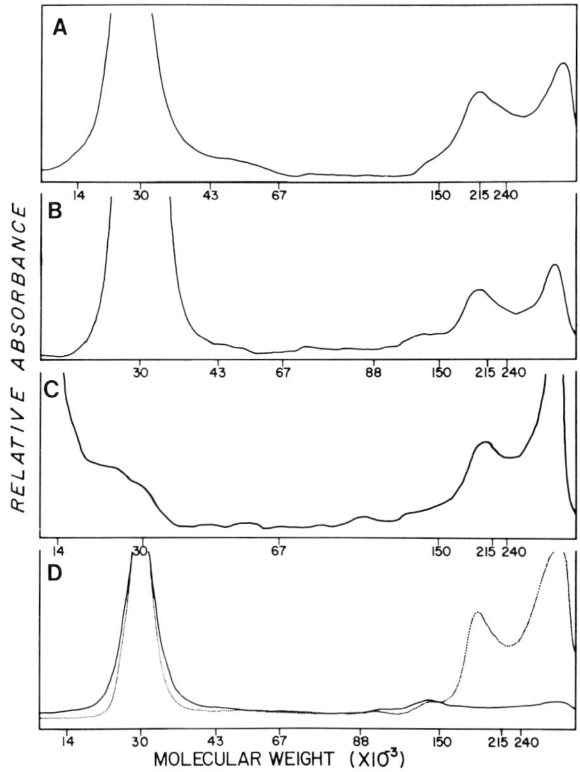

FIG. 13. Comparison of [^{125}I]PDGF affinity-labeled samples of several cell types and the role of EGS in the affinity-labeling procedure. Swiss 3T3 cells (A), human fibroblasts (WI-38) (B), and monkey smooth muscle cells (C) were incubated with 8 ng/ml of [^{125}I]PDGF for 4 hr at 4° with agitation. These cells were then rinsed, solubilized in SDS sample buffer, and electrophoresed on 5–15% polyacrylamide linear gradients gels as described in the test, with the exception that 2% (v/v) 2-mercaptoethanol was added to the monkey smooth muscle cell sample to limit the formation of large aggregates. Densitometric scans of 3-, 4-, or 28-day (A, B, and C, respectively) autoradiographic exposures of the stained and dried gels (without the tacking gels) are shown with the positions of Coomassie-stained molecular weight markers run on adjacent lanes marked on the abscissa. D shows densitometric scans of a 7-day autoradiogram of a [^{125}I]PDGF affinity-labeled 3T3 sample incubated with or without 1 *M* hydroxylamine, pH 8.5, for 6 hr at 37°. From Glenn *et al.*[24]

brane protein. Based on whole cell binding studies, this amount of NR6-3T3 cell membrane protein contains 90–135 fmol of PDGF binding activity. App(NH)p is included to reduce nonspecific hydrolysis of ATP; *p*-nitrophenyl phosphate is added to inhibit protein phosphatases. Routine assays contain 5 μg BSA with or without 70 n*M* PDGF or 540 n*M* EGF.

Assays are started by addition of the membrane suspension and incubated at 30° for 3 min. The reactions are terminated by addition of 30 μl sodium dodecyl sulfate (SDS) buffer (2.5% SDS, 100 m*M* dithiothreitol, 125 m*M* Tris pH 6.7, 2.5 m*M* EDTA, 50% glycerol) and subsequently boiled for 3 min prior to application to a 5 to 10% polyacrylamide gel (30). Gels are stained to locate standard proteins, dried, and autoradiographed for 8 to 24 hr.

Using the above or similar protocols, it has been shown[25–28] that PDGF stimulates the *in vitro* incorporation of ^{32}P from [γ-^{32}P]ATP onto tyrosine residues on a protein of approximately the same size (170K) as that determined for the PDGF receptor by affinity cross-linking studies. By analogy to the EGF receptor, which has been shown to have autokinase activity, this supports the conclusion that the 164K membrane component observed in affinity cross-linking experiments is, or is closely associated with, the PDGF receptor. As for the 164K dalton component, the approximately 170K phosphoprotein has been demonstrated on several PDGF responsive cell types from different species, is not seen on cells which do not bind [^{125}I]PDGF, and can be greatly reduced by preexposure of cells to PDGF at 37°. It is not yet certain that the plasma membrane PDGF-stimulated protein kinase activity is a property of the PDGF receptor itself, nor what role this activity may play in the function of the PDGF receptor. Since many cytoplasmic proteins are rapidly phosphorylated on tyrosine in response to PDGF[29] and since phosphorylation of many proteins has been shown to affect their enzymatic activity,[30] it is possible that activation of the PDGF receptor-associated kinase is involved in the response of the cells to PDGF binding.

Acknowledgments

We thank R. Seifert and E. Raines for critical reading of the manuscript and M. Hillman for tireless typing.

[26] B. Ek, B. Westermark, åA. Wasteson, and C.-H. Heldin, *Nature (London)* **295,** 419 (1982).
[27] B. Ek and C.-H. Heldin, *J. Biol. Chem.* **257,** 10486 (1982).
[28] J. Nishimura, J. S. Huang, and T. F. Deuel, *Proc. Natl. Acad. Sci. U.S.A.* **79,** 4303 (1982).
[29] J. A. Cooper, D. F. Bowen-Pope, E. W. Raines, R. Ross, and T. Hunter, *Cell* **31,** 263 (1982).
[30] P. Cohen, *Nature (London)* **296,** 613 (1982).

[9] Binding Assays for Epidermal Growth Factor

By GRAHAM CARPENTER

The binding of epidermal growth factor (EGF) to its receptor has been intensely studied to elucidate the mechanisms by which this small (M_r = 6045) polypeptide regulates the proliferation of mammalian cells (see references 1–4 for recent reviews). Due to the relative ease in obtaining this growth factor in purified form, its chemical stability, and the diversity of responsive cell types, EGF has become a model mitogen for studies of growth control, particularly in cell culture systems. The interactions of radiolabeled EGF with intact cells or subcellular preparations have revealed a great deal about the metabolic processing of EGF–receptor complexes *in vivo* and the biochemical nature of the membrane receptor for EGF. The EGF receptor is a glycoprotein (M_r = 170,000) which is widely distributed among mammalian cells; most cells having an average of approximately 50,000 to 100,000 receptors per cell. The predominance of published data indicates that the EGF receptor has no subunit structure and contains a binding site for EGF and an EGF-activated tyrosine kinase activity located in separate domains of a single polypeptide chain.[5,6]

In this chapter assays will be described for measuring the interaction of radiolabeled EGF with its receptor on intact cells, in crude membrane preparations, and in detergent-solubilized receptor preparations. Other contributions in this series have described procedures for the isolation of EGF from mouse submaxillary glands,[7] methodology for the purification of the EGF receptor-kinase from A-431 cells and the assay of tyrosine kinase activity,[8] and methods to study the receptor mediated endocytosis of EGF.[9] Another chapter in this volume describes methodology for affinity labeling of the ATP site on the EGF receptor-kinase.[10]

[1] G. Carpenter and S. Cohen, *Annu. Rev. Biochem.* **48,** 193 (1979).
[2] G. Carpenter, *Handbook Exp. Pharmacol.* **57,** 89 (1981).
[3] M. Das, *Int. Rev. Cytol.* **78,** 233 (1982).
[4] H. Haigler, *in* "Growth and Maturation Factors" (G. Guroff, ed.), Vol. 1, p. 117. Wiley, New York, 1983.
[5] S. Cohen, H. Ushiro, C. Stoscheck, and M. Chinkers, *J. Biol. Chem.* **257,** 1523 (1982).
[6] S. A. Buhrow, S. Cohen, and J. Staros, *J. Biol. Chem.* **257,** 4019 (1982).
[7] S. Cohen and C. R. Savage, Jr., this series, Vol. 37, p. 424.
[8] S. Cohen, this series, Vol. 99, p. 378.
[9] H. Haigler, this series, Vol. 98, p. 283.
[10] S. A. Buhrow and J. Staros, this volume [62].

METHODS IN ENZYMOLOGY, VOL. 109

ISBN 0-12-182009-2

Binding of [^{125}I]EGF to Intact Cells

The interactions of [^{125}I]EGF with intact cells can be used to derive a great deal of information about receptor activity under various physiological circumstances and, in particular, to study the process of hormone–receptor complex internalization and degradation. Assays for the specific binding of labeled EGF to intact cells are relatively easy to perform, require simple materials and technical maneuvers, and can be performed quickly. Importantly, there are several ways to manipulate the basic assay to obtain additional information about the internalization and degradation processes. The binding assay as described below is that which the author routinely employs as a basic procedure.

Materials

[^{125}I]EGF: The iodination of EGF is performed as described by Carpenter and Cohen.[11] For each experiment the stock solution of [^{125}I]EGF is mixed with unlabeled EGF to obtain a desired concentration and specific activity. This solution is prepared in the binding media.

Unlabeled EGF: This is used to determine the level of nonspecific binding of [^{125}I]EGF. Its preparation from mouse submaxillary glands is as described by Cohen and Savage.[7]

Hanks' balanced salt solution: This or another balanced salt solution is prewarmed (37°).

Binding medium: Any balanced salt solution or growth medium can be employed and is prewarmed (37°). Generally, we have employed Dulbecco's modified Eagle's medium supplemented with 20 m*M* HEPES, pH 7.5, 2 g/liter sodium bicarbonate, 50 ng/ml Garamycin, and 0.1% bovine serum albumin (BSA).

Hanks' BSA: Hanks' balanced salt solution containing 0.1% BSA is prechilled in an ice bucket and used to wash cell monolayers at the end of the incubation period to remove unbound [^{125}I]EGF.

Cells: The binding assay is designed for cells which are "attached" to a supporting surface such as a culture dish. If the cells are easily detached by mechanical procedures (such as washing) it may be necessary to carefully monitor any cell loss during the assay. We generally use fibroblastic or epithelial cells grown to confluence in 60 mm culture dishes. These cells have, on the average, approximately 100,000 receptors for EGF per cell. By using higher specific activity [^{125}I]EGF in the binding assay it is possible to reduce the number of cells.

[11] G. Carpenter and S. Cohen, *J. Cell. Biol.* **71,** 159 (1976).

1 *N* sodium hydroxide: After the removal of free [^{125}I]EGF, this solution or alternatively an SDS solution is used to solubilize cell-associated radioactivity, for counting.

Basic Procedure

1. The cell monolayers are gently washed twice with warm Hanks' balanced salt solution. Usually 3 ml of Hanks is used per wash for cell monolayers in 60 mm dishes.

2. An appropriate volume of binding medium (prewarmed) is added, 1.5 ml for monolayers in 60 mm dishes.

3. Unlabled EGF is added to those culture dishes in which the nonspecific binding of radioactivity is to be measured. For each data point a measurement is made of the total [^{125}I]EGF binding and the nonspecific binding of radioactivity (defined as that which is not displaced by an excess of non-radioactive EGF). Specific binding is then calculated and defined as the difference between the total binding and the nonspecific binding. It should be recognized that this is essentially an operational definition and by these criteria specific binding of labeled polypeptides to substances such as glass and talc have been reported.[12] The amount of unlabeled EGF added to determine nonspecific binding should be at least 20-fold more than the amount of [^{125}I]EGF that will be added. Ideally one would like to add the labeled and unlabeled EGF simultaneously, but in practice the unlabeled polypeptide is usually added up to 1 min before the labeled growth factor. If the unlabeled growth factor is added in a small volume (usually 1–10 μl) it is important to mix this into the binding media before proceeding with the addition of [^{125}I]EGF. In most assays of [^{125}I]EGF binding to intact cells the nonspecific binding amounts to approximately 10% or less of the total binding.

4. Add the [^{125}I]EGF to each culture dish, mix by gently swirling the media, and place the cells in a 37° incubator. For cells which have a usual receptor number and affinity, such as human fibroblasts or 3T3 cells, a final [^{125}I]EGF concentration of 10 ng/ml should be saturating. Generally about 1–5% of the total cpm put into the assay can be expected to become cell-bound. Therefore, if one would like to measure cell-bound radioactivity on the order of 10,000 cpm, it will be necessary to add approximately 1×10^6 cpm to the assay.

5. After incubation for the desired amount of time (with most cell types binding is maximal after a 1 hr incubation at 37°), the cells are removed from the incubator and washed, at least four times, with ice-cold Hanks containing 0.1% BSA to remove unbound radioactivity. With cells

[12] P. Cuatrecasas and M. D. Hollenberg, *Biochem. Biophys. Res. Commun.* **62,** 31 (1975).

in 60 mm culture dishes a wash volume of 1.5 ml is sufficient. Depending on the cell type, it may be important to perform these washes gently. For example, pipetting directly onto the cell monolayer may dislodge cells; we pipet the washing solution against the side of the culture dishes. It is also important to perform these washes as rapidly as possible and to keep the washing solution cold.

6. After the last wash has been aspirated from the culture dish, the cells are solubilized by adding 1 *N* sodium hydroxide (1.5 ml to a 60 mm culture dish). The dishes are incubated at 37° for aproximately 1 hr. The solubilized cell-associated radioactivity is then transferred to a counting tube, and the dish rinsed with 0.5 ml of sodium hydroxide.

7. After determining the level of radioactivity in the different tubes, the level of specific binding is calculated by subtracting the amount of nonspecific cell associated radioactivity from the amount of total cell-associated radioactivity. An aliquot of the [^{125}I]EGF solution should be counted to permit a specific activity value (cpm per ng EGF) to be calculated. Frequently it is useful to determine cell numbers on parallel cultures so the data can be expressed as molecules of [^{125}I]EGF bound per cell.

The results obtained with the basic protocol outlined above yield a specific binding value which represents the total amount of cell-associated radioactivity. When the conditions described above are followed, most importantly incubation at 37°, a small percentage of the total cell-associated radioactivity will be intact [^{125}I]EGF that is bound to the cell surface and a smaller amount will be degraded [^{125}I]EGF that has not been released from the cell. The vast majority of the radioactivity (>80%) will be [^{125}I]EGF that has been internalized, but not extensively degraded. It is important in the design and interpretation of binding experiments that the investigator understands the nature and cellular localization of the radioactivity that is being measured. The binding procedures outlined above can be altered to permit measurements of either [^{125}I]EGF bound strictly at the cell surface or [^{125}I]EGF that has been internalized or both.

If the binding reaction is carried out at a low temperature (usually 5°), with an extended incubation time to allow binding equilibrium to be reached, then all the cell-associated radioactivity will represent [^{125}I]EGF bound at the surface. When the assays are performed at 5° with adherent cells, it will be necessary to monitor possible detachment of the cell from the culture dish during and after the low temperature incubation. At low temperature incubation conditions, the binding reaction can be considered reversible; whereas, at 37° the binding is largely irreversible due to the rapid metabolism of cell-bound [^{125}I]EGF. The application of Scatchard analysis[13] to ligand binding data for the determination of K_D and B_{max} assumes that a reversible equilibrium is attained. This condition is not usually satisifed by binding assays carried out on viable cells at 37°. Although the practical consequences of these essentially irreversible binding conditions at 37° may not be critical in most instances, it does complicate detailed, e.g., thermodynamic, interpretation of the binding data.

When the binding assays are performed at 37°, it is possible by a relatively simple maneuver to determine the amount of radioactivity bound on the cell surface, the amount located within the cell, and the total cell-associated radioactivity. The dissociation of receptor–EGF complexes is facilitated by lowering the pH to 4.5 or below.[14] Therefore, [^{125}I]EGF bound at the cell surface can be dissociated and removed by washing cell monolayers with a low pH buffer. Internalized radioactivity will not be extracted by these washes. To remove surface-bound [^{125}I]EGF, Haigler *et al.*[14] employed a buffer of 0.2 *M* sodium acetate and 0.5 *M* sodium chloride at pH 4.5 or below. Ascoli[15] modified this technique to dissociate surface-bound [^{125}I]hCG by employing a buffer of 50 m*M* glycine hydrochloride and 100 m*M* sodium chloride adjusted to pH 3. This laboratory employs the Ascoli buffer adjusted to pH 4.0. The low pH washes are carried out at 0° over a period of approximately 10 min. The radioactivity that remains associated with the cells following this type of wash represents internalized growth factor. If a parallel set of dishes is washed with the same buffer adjusted to pH 7.5, then one derives a measurement of the total cell-associated radioactivity. Subtraction of the cell-associated radioactivity present after the low pH wash from that present after the wash at neutral pH gives an estimate of the surface-bound [^{125}I]EGF. The efficiency of this procedure can be judged by carrying out the binding at 5° and washing with the two buffers. Under this condition the low pH buffer should remove nearly all (>90%) of the cell-bound radioactivity. Gentle trypsinization also can be employed to distinguish [^{125}I]EGF bound at the cell surface from internalized [^{125}I]EGF.[11]

Inhibitors can be employed with caution to provide useful information concerning [^{125}I]EGF binding to intact, viable cells at 37°. A wide variety of compounds, nearly all amines, can be used to inhibit the lysosomal degradation of [^{125}I]EGF. The most widely used are methylamine (or ammonium chloride) or chloroquine at concentrations of approximately 10 or 0.1 m*M*, respectively.[11] The inhibitors are preincubated with the cells for 15–30 min prior to the addition of [^{125}I]EGF. Dansylcadaverine[14]

[13] G. Scatchard, *Ann. N.Y. Acad. Sci.* **51,** 660 (1949).

[14] H. T. Haigler, F. R. Maxfield, M. C. Willingham, and I. Pastan, *J. Biol. Chem.* **255,** 1239 (1980).

[15] M. Ascoli, *J. Biol. Chem.* **257,** 13,306 (1982).

or phenylarsine oxide[16] have been employed to inhibit the internalization of cell-bound [^{125}I]EGF. Both of these reagents, however, are quite toxic to cultured cells and their usefulness for physiological experiments is limited.

Often the intact cell receptor assay is employed to assay the concentration of EGF in biological fluids or extracts or to detect other molecules, such as phorbol esters, antibodies, hormones, etc., which influence [^{125}I]EGF binding. There are two strategies to use for these types of studies. The most commonly used is a direct competition assay in which the sample substance and [^{125}I]EGF are added simultaneously to the binding assay. When this approach is used, it is important to appreciate the fact that reduced binding of the labeled ligand can result from events unrelated to binding competition. This is particularly true when complex fluids or extracts are being analyzed. For example, protease activity in the sample can result in an alteration of the labeled ligand concentration or affinity of the receptor. The competitive binding assays can be performed in a sequential manner, however, to avoid some of these potential artifacts. In this assay the cells are incubated first with the test substance, washed to remove any unbound material, and then [^{125}I]EGF is added. This sequential assay eliminates the possibility of a false positive due to effects of the test material on the ligand.

The assay conditions described above will apply to most all cell types—normal or transformed. Although EGF receptors are rather ubiquitous,[1,2] cells of the circulatory system, chick embryo fibroblasts, and Chinese hamster ovary (CHO) cells do not have significant [^{125}I]EGF binding capacity. Also cells transformed by certain tumor viruses, predominately RNA tumor viruses, have greatly decreased [^{125}I]EGF binding capacity.[17] A 3T3 cell variant designated NR-6 has been described and is useful as it does not bind [^{125}I]EGF nor respond biologically to EGF as 3T3 cells do.[18] Recently, significant use has been made in EGF receptor studies of a cell line designated A-431. These cells were derived from a human epidermoid carcinoma of the vulva and display a very large number of receptors for [^{125}I]EGF: approximately 2.5×10^6 receptors per cell.[19] Use of this or similar cell lines in the binding assay described above requires modification of the amount of [^{125}I]EGF necessary to achieve saturation—about 100–200 ng [^{125}I]EGF per ml. The binding

[16] D. L. Lowe, J. B. Baker, W. C. Koonce, and D. D. Cunningham, *Proc. Natl. Acad. Sci. U.S.A.* **78,** 2340 (1981).

[17] G. J. Todaro, J. E. DeLarco, and S. Cohen, *Nature (London)* **264,** 26 (1976).

[18] R. M. Pruss and H. R. Herschman, *Proc. Natl. Acad. Sci. U.S.A.* **74,** 2918 (1977).

[19] H. Haigler, J. F. Ash, S. J. Singer, and S. Cohen, *Proc. Natl. Acad. Sci. U.S.A.* **75,** 3317 (1978).

characteristics and other properties of the A-431 cell line have been reviewed.[20]

Binding of [^{125}I]EGF to Membrane Fractions

The specific binding of [^{125}I]EGF to membranes, prepared by a wide variety of procedures, can be measured rather easily. The situation is less complex than binding to intact cells, since there is no metabolism of the receptor–hormone complexes. Although metabolism of hormone–receptor complexes is not a problem in these assays, it is necessary to bear in mind that both the labeled ligand and/or the receptor may be sensitive to proteases present in the membrane preparation. With subcellular preparations the binding reaction is reversible and kinetic analyses can be more meaningfully applied.

Materials

HEPES buffer, pH 7.5
BSA
Membrane preparation
[^{125}I]EGF
Unlabeled EGF
Cellulose acetate filters (0.2 or 0.5 μm pore size)
20 m*M* HEPES buffer, pH 7.5, containing 0.1% BSA

Procedure

The binding assays are carried in a 100 or 200 μl total volume in small polypropylene tubes. Aliquots of concentrated HEPES buffer and BSA solution are added so that the final concentrations are, respectively, 20 m*M* and 0.1%. The precise pH of the binding assay is not critical as the profile shows a broad optimum between 6.5 and 8.0. An aliquot of the membrane preparation is added; 20 μg or greater depending on the concentration of receptors in the original cells or tissue. An excess of unlabeled EGF (20-fold or greater relative to the amount of labeled EGF) is added to those tubes in which nonspecific binding is to be measured, and the binding reaction is initiated by adding [^{125}I]EGF to each tube. The final concentration of labeled EGF will depend on the characteristics of the particular membranes being assayed. (Alternatively the assay can be initiated by the addition of membranes to reaction tubes containing labeled growth factor and other components.) The binding assay can be con-

[20] C. M. Stoscheck and G. Carpenter, *J. Cell. Biochem.* **23,** 191 (1983).

ducted at 0°, room temperature, or 37° depending on the objectives of the study and, perhaps, characteristics of the membrane preparation.

To measure receptor-bound [^{125}I]EGF it is necessary to separate the bound and free ligand. We routinely use vacuum filtration; however, other rapid methods such as centrifugation of the particulate membranes have been employed. Cellulose acetate filters (25 mm, 0.2 to 0.5 μm pore size) are placed on a multisample filtration manifold, and wetted with a BSA solution. The top of the manifold is put in place to secure a seal around the surface of the filters, and the vacuum source, usually a small pump, is applied. The reaction mixtures are filtered by diluting to a larger volume (1 ml) and rapidly transferring them to the filters with a Pasteur pipet. Unless there is an excess of membrane protein, the reaction mixture should pass quickly through the filters. Each filter is then washed 4 times with 20 m*M* HEPES, pH 7.5, containing 0.1% BSA. The filters are removed from the manifold, placed in vials, and counted in a gamma spectrometer.

Binding of [^{125}I]EGF to Solubilized Receptor

Purification of the EGF receptor and certain measurements of receptor activity require the separation of the receptor from other membrane components. Solubilization of the EGF receptor is not particularly difficult; however, retention of binding activity after detergent treatment can be quite difficult. For example, solubilization of placental membranes, which have a relatively high concentration of EGF receptors, in Triton X-100 does not yield active receptor, while solubilization of membranes from A-431 cells produces quite active receptor preparations.[21] The reason for these different sensitivities to detergent is not known.

"Solubilized" receptor is meant to apply to receptor that is not pelleted after centrifugation of at least 100,000 *g* for 1 hr. The usual procedure for solubilizing EGF receptor from membrane preparations is to mix the membranes with a buffer (20 m*M* HEPES, pH 7.5) containing Triton X-100 and 10% glycerol. The amount of Triton used is dicated by the amount of membrane protein to be solubilized. Generally, one uses a ratio of 5–10 mg of the detergent for each mg of membrane protein. At very low protein concentrations the Triton concentration should not be less than 0.1%. The inclusion of glycerol is primarily to stabilize the protein kinase activity of the EGF receptor. The mixture of membranes and detergent is allowed to sit at either room temperature or 0° for approximately 20 min and then is centrifuged at 100,000 *g* for 90 min. The solubilized receptor

[21] G. Carpenter, *Life Sci.* **24,** 1691 (1979).

can be assayed immediately or stored at −20°, after quick freezing in dry ice-methanol.

Materials

HEPES, pH 7.5
BSA
20 m*M* HEPES plus Triton and glycerol (adjusted to concentrations identical to those in the solubilized membrane preparations)
Solubilized membranes
[^{125}I]EGF
Unlabeled EGF
Polyethylene glycol 8000 (or 6000) 20.4% (w/v) in water
0.1% γ-globulin in 0.1 *M* phosphate buffer, pH 7.4
8.5% (w/v) polyethylene glycol in 0.1 *M* phosphate buffer, 7.4
0.4% BSA
Cellulose acetate filters (0.2 or 0.5 μm pore size)

Procedure

The final volume of the binding assay is usually 100–200 ml. Stock solutions of HEPES buffer, pH 7.5, and BSA are added separately or as one addition to final concentrations of 20 m*M* and 0.1%, respectively. Solubilized membranes or a solution of HEPES buffered Triton-glycerol is added. The "minus membranes" control, adjusted to the same final concentrations of Triton and glycerol, is important to include. With each membrane preparation the optimal Triton concentration for binding may vary. Our assays are normally carried out in the presence of 0.2% Triton X-100.[21] If the Triton concentration is too high [^{125}I]EGF binding may be inhibited and if it is reduced below 0.1% aggregation of the receptor may occur. The final glycerol concentration is not critical. The reaction is started by adding [^{125}I]EGF (total binding) or excess unlabeled EGF plus [^{125}I]EGF (nonspecific binding). The binding assays can be carried out at 0°, room temperature, or 37°. The binding at 37°, however, is approximately 2-fold greater than that obtained at 5°.

The separation of soluble [^{125}I]EGF–receptor complexes from free [^{125}I]EGF is performed by selective precipitation of the complexes with polyethylene glycol, essentially as first described by Cuatrecasas[22] for the analysis of solubilized ^{125}I-labeled insulin–receptor complexes. One-half milliliter of the 0.1% γ-globin solution (in 0.1 *M* phosphate buffer, pH 7.4) is added to the tube containing the binding assay and immediately 0.5 ml

[22] P. Cuatrecasas, *Proc. Natl. Acad. Sci. U.S.A.* **69,** 318 (1972).

of the 20.4% polyethylene glycol solution is added and mixed by gentle vortex. This is done at room temperature. The mixtures are then quickly filtered on cellulose acetate filters which have been presoaked in a 0.4% BSA solution. The filters are subsequently washed four times with a solution of 8.5% polyethylene glycol in 0.1 *M* phosphate buffer, pH 7.4, and counted in a gamma spectrometer.

This assay is technically perhaps the most difficult, least precise, and least accurate of the three binding assays. With crude extracts the nonspecific binding can amount to nearly 50% of the total binding. The same assay performed with purified EGF receptor produces low nonspecific values more typical of the other assays (10% or less). In addition, the affinity of the receptor for [^{125}I]EGF in solubilized membranes or vesicles prepared from A-431 cells is changed. Binding of [^{125}I]EGF to intact A-431 cells or to membrane preparations is saturated at approximately 100–200 ng [^{125}I]EGF/ml. However, when the membranes (or vesicles) are solubilized, saturation of binding is not achieved until the [^{125}I]EGF concentration reaches 800–1000 ng/ml. This is not strictly a detergent effect nor is it due to the polyethylene glycol assay. Solubilized purified receptor assayed by the same procedure is saturated in the presence of 100–200 ng [^{125}I]EGF/ml. An additional difficulty with the solubilized receptor assay is that one cannot be sure that polyethylene glycol has precipitated all the solubilized [^{125}I]EGF–receptor complexes.

Acknowledgments

The author acknowledges support of an Established Investigator Award from the American Heart Association and research Grants BC-294 from the American Cancer Society and CA24071 from the National Cancer Institute.

[10] Radioligand Assay for Angiotensin II Receptors

By HARTMUT GLOSSMANN, ALBERT BAUKAL, GRETI AGUILERA, and KEVIN J. CATT

Introduction

The receptor sites for angiotensin II, like those for other peptide hormones and neurotransmitters, are located on the plasma membranes of target cells that are specifically responsive to the endogenous peptide and

ISBN 0-12-182009-2

its agonist analogs. The actions of angiotensin II are best defined in the adrenal gland, vascular smooth muscle, and kidney, and specific receptors for the octapeptide have been identified in each of these tissues. Several other tissues that are known to be responsive to exogenous angiotensin II (uterine muscle, urinary bladder, liver, and brain) have also been shown to contain specific angiotensin binding sites, with properties similar to those of the receptors in the well-defined target tissues that are regulated by the endogenously formed peptide.[1] Both tritiated angiotensin II and monoiodinated ^{125}I-labeled derivatives of angiotensin II or its agonist and antagonist analogs, have been employed to demonstrate and characterize the receptors in isolated cells and tissue homogenates or membrane-rich fractions.[1,2]

The angiotensin II receptors of the adrenal cortex have been extensively characterized with respect to their binding properties, and the structural requirements of the AII molecule for hormone–receptor interaction have been well defined. In the adrenal glomerulosa cell, there is a close correlation between the dissociation constants (measured by binding-inhibition studies with membrane-rich adrenal particles) and the activation constants (for steroid secretion) of agonist analogs and COOH-terminal and NH_2-terminal fragments of angiotensin II.[3]

The structural characteristics of the angiotensin II receptor appear to be similar in different tissues, as indicated by the similar molecular weights of detergent-solubilized adrenal and uterine smooth muscle receptor proteins.[4] However, the *in vitro* binding properties of the adrenal angiotensin II receptor differ from those in smooth muscle, pituitary, and brain in terms of the cation dependence and effects of SH-reducing agents upon ligand binding. Such differences should be taken into consideration in the design of procedures for measurement of angiotensin receptors in different tissues.

In most of the established target tissues for angiotensin II, binding of the octapeptide to its receptors has been measured with either tritiated or monoiodinated angiotensin II. Although tritiated angiotensin II has the advantage of high biological activity,[5] its low specific activity (50 to 60 Ci/mmol) is a drawback for the identification and assay of high-affinity receptor sites. Monoiodo[^{125}I]angiotensin II has several advantages due to its

[1] A. M. Capponi, G. Aguilera, J. L. Fakunding, and K. J. Catt, *in* "Biochemical Regulation of Blood Pressure" (R. L. Soffer, ed.), p. 205. Wiley, New York, 1981.

[2] M.-A. Devynck and P. Meyer, *Biochem. Pharmacol.* **27,** 1 (1978).

[3] A. M. Capponi and K. J. Catt, *J. Biol. Chem.* **254,** 5120 (1979).

[4] A. M. Capponi and K. J. Catt, *J. Biol. Chem.* **255,** 12081 (1980).

[5] M.-A. Devynck and P. Meyer, *Am. J. Med.* **61,** 758 (1976).

high specific activity (up to 2200 Ci/mmol) and the nature of the radiation emitted. However, diiodoangiotensin II of low biological activity can also be formed during the radioiodination procedure, and should be completely separated from the monoiodinated species to obtain angiotensin II tracer with optimal binding activity.

Both ^{125}I-angiotensin II and [^{3}H]angiotensin II have been employed as tracers to label the receptors in various organs and tissues (see Table 1). In radioiodinated angiotensin II the ^{125}I atom is predominately located on the Tyr^4 residue, though labeling of the His^6 residue may also occur. The introduction of iodide at the Tyr^4 residue lowers the pK of the aromatic hydroxyl group and introduces a bulky hydrophobic moiety. It has been suggested that such structural alterations in the radioiodinated peptide could impair its interaction with the receptor by comparison with the tritium-labeled hormone.[5] However, when the binding properties of both ligands were studied in the bovine adrenal cortex[6] and the rat mesenteric artery,[7] the receptor concentrations and binding constants for each radioligand were in good agreement. Also, the close correlation noted above between the K_d values for receptor binding of angiotensin fragments and analogs and their K_{act} values for stimulation of aldosterone secretion by isolated glomerulosa cells,[3] as shown in Table 2, further indicates the validity of receptor characterization with the monoiodinated peptide.

The assay procedure for measurement of angiotensin II receptors with monoiodinated ^{125}I-angiotensin II in partially purified adrenal membranes and isolated adrenal cells is described below. The same general principles can be applied to the assay of receptors with [^{3}H]angiotensin II and radiolabeled agonist and antagonist analogs of the octapeptide. The Asp^1 or Asn^1, and Val^5 or Ile^5 forms of AII, can be used for radioiodination to prepare tracer suitable for receptor binding studies. For some purposes, the Sar^1 derivative of AII is preferable for labeling, since it has higher affinity for AII receptors and may provide higher specific binding of tracer when small numbers of receptor sites are to be analyzed.

Preparation and Storage of Monoiodo-Angiotensin II

Monoiodinated ^{125}I-angiotensin II can be readily prepared by the chloramine T, lactoperoxidase, or iodogen method, and is obtainable from commercial sources with specific activities of up to 2200 Ci/mmol.

[6] H. Glossmann, A. Baukal, and K. J. Catt, *J. Biol. Chem.* **249,** 825 (1974).

[7] S. Gunther, M. A. Gimbrone, and R. W. Alexander, *Circ. Res.* **47,** 278 (1980).

TABLE I
PROPERTIES OF ANGIOTENSIN II RECEPTORS LABELED WITH [^{125}I]MONOIODO ANGIOTENSIN II

Tissue/ cells	Species	Binding constants	Assay conditions	Temperature	Separation method	Comments	Ligand stability	Receptor stability	References[a]
Adrenal cortex (partially purified plasma membranes)	Bovine	One low affinity binding site in absence of NaCl K_D = 25.6 nM, N = 3400 fmol Two binding sites in the presence of 140 M NaCl: K_{D1} = 0.47 nM, N_1 = 105 fmol, K_{D2} = 23.2 nM, N_2 = 3400 fmol	50 mM Tris–HCl, pH 7.4, 0.2% BSA, 0.3 mg/ml glucagon, 5 mM DTT	22°	Filtration. Nitrocellulose filters HAWP (0.45 μm)	Nucleotides inhibit in the presence of 140–150 mM NaCl. With a nucleotide regenerating system added: GTP ≈ Gpp(NH)p > ITP > UTP. CTP and ATP are ineffective. Minor effects of nucleotides seen with low (<25 mM) NaCl. K_D and N values for the high-affinity binding component are quite variable (K_D = 0.4–2 nM, N = 100–500 fmol/mg of protein)	Extensive degradation, partially prevented by glucagon and dithiothreitol as evaluated by TLC and rebinding	Large losses when frozen in liquid nitrogen. Loss of high-affinity binding when stored at 2° for >4 hr	a,b,c,d
Adrenal cortex (purified capsular layer membranes)	Rat	One high affinity binding site: K_D = 0.5 nM, N = 2300 fmol	50 mM Tris–HCl, pH 7.4, 120 mM NaCl, 5 mM DTT, 0.2% BSA	22°	Filtration. Nitrocellulose filters HAWP (0.45 μm)	Tracer binding enhanced by NaCl, DTT, or EDTA	n.d.	Apparently stable when sucrose gradient purified plasma membranes are stored in liquid nitrogen	e

(*continued*)

TABLE I (*continued*)

Tissue/cells	Species	Binding constants	Assay conditions	Temperature	Separation method	Comments	Ligand stability	Receptor stability	References[a]
Adrenal cortex (glomerulosa cells)	Rat	K_D = 0.37 nM, N = 25.5 fmol per 10^5 cells	Medium 199 supplemented with 0.1% or 0.2% BSA	37°	Filtration. Nitrocellulose filters HAWP (0.45 μm)	Low affinity (K_D = 41 nM) binding site may represent 67–75% of the total receptor population	Extensive degradation. After 40 min 50% of the ligand is degraded as measured by RIA, TLC, and rebinding	n.d.	f,g
	Dog	K_D = 0.90 nM, N = 37 fmol per 10^5 cells							
Liver (purified plasma membranes)	Rat	One low affinity binding site in the absence of free Me^{2+}, K_D = 13.4 nM, N = 580 fmol Two binding sites in the presence of 10 mM Mg^{2+}, K_{D1} = 0.21 nM, N_1 = 230 fmol, K_{D2} = 2.9 nM, N_2 = 1820 fmol	20 mM Tris–HCl, pH 7.4, 100 mM NaCl, 0.2% BSA	12°	Filtration. Glassfiber filters presoaked with 0.2% BSA	Nucleotides inhibit in the presence of Mg^{2+}: Gpp(NH)p ≈ GTPγS ≈ GTP > ITP > ATP. Na^+ (100 mM) can partially substitute for Mg^{2+}	Excellent as evaluated by TLC and high-voltage electrophoresis	Apparently stable when stored in liquid nitrogen in the presence of 5 mM EDTA. Rapid loss of binding capacity which was prevented by Mg^{2+} when preincubated at 2 to 4°	h,i
Brain (cerebellar cortex)	Calf	One binding site: K_D = 0.2 nM, N = 8 fmol In some experiments: K_{D1} = 0.08 nM, N_1 = 6 fmol, K_{D2} = 0.46 nM, N_2 = 10 fmol	0.1 M sodium phosphate buffer, pH 7.0, 5 mM Na_2 EDTA, 10 M glucagon, 0.5% BSA, 0.1 mM PMSF	37°	Filtration. Glassfiber filters Whatman GFB-2	Membranes are preincubated for 30 min at 37° with 5 mM EDTA, 0.1 M sodium phosphate buffer, and 0.1 mM phenylmethylsulfonylfluoride. High-affinity binding requires sodium ion. Tris, K^+, Li^+, as well as other cations also stimulate	Excellent as evaluated by TLC and rebinding	n.d.	j,k

Mesenteric artery (crude membranes)	Rat	$K_D = 0.91$ nM, $N =$ 54 fmol	50 mM Tris–HCl, pH 7.2, 5 mM $MgCl_2$, 0.25% BSA	22° or 25°	Filtration. Glassfiber filters Whatman GF/C presoaked with 0.25% BSA	Binding stimulated by Mn^{2+}, Mg^{2+}, Ca^{2+}, and Na^+. GTP and Gpp(NH)p inhibit in the presence of free Me^{2+}. EDTA and EGTA (1 mM) inhibit ~25% and DTT (5 mM) inhibits 45%	Excellent as evaluated by TLC	n.d.	l,m
Kidney (epithelial membranes)	Rat	$K_D = 0.62$ nM, $N =$ 299 fmol	20 mM Tris–HCl, pH 7.4, 5 mM Na_2 EDTA, 120 mM NaCl, 0.2% BSA	22°	Filtration. Glassfiber filters Whatman GF/B	Binding stimulated by Na^+, K^+, Rb^+, and Li^+. DTT and guanyl nucleotides inhibit. Low affinity, high capacity site is also found	Extensive degradation. After 5 min incubation 50% of the ligand is degraded as measured by TLC	n.d.	n

[a] References: (a) H. Glossmann, A. Baukal, and K. J. Catt, *J. Biol. Chem.* **249,** 825 (1974). (b) H. Glossmann, A. Baukal, and K. J. Catt, *J. Biol. Chem.* **249,** 664 (1974). (c) H. Glossmann, A. Baukal, and K. J. Catt, *Science* **185,** 281 (1974). (d) K. J. Catt, A. Baukal, J.-M. Ketelslegers, J. Douglas, S. Saltman, P. Fredlund, and H. Glossmann, *Acta Physiol. Latinoam.* **24,** 515 (1974). (e) J. Douglas, G. Aguilera, T. Kondo, and K. J. Catt, *Endocrinology* **102,** 685 (1978). (f) G. Aguilera, A. Capponi, A. Baukal, K. Fujita, R. Hauger, and K. J. Catt, *Endocrinology* **104,** 1279 (1979). (g) J. Douglas, S. Saltman, P. Fredlund, T. Kondo, and K. J. Catt, *Circ. Res.* **38,** II-108 (1976). (h) C. P. Campanile, J. K. Crane, M. J. Peach, and J. C. Garrison, *J. Biol. Chem.* **257,** 4951 (1982). (i) J. K. Crane, C. P. Campanile, and J. C. Garrison, *J. Biol. Chem.* **257,** 4959 (1982). (j) J. P. Bennett and S. H. Snyder, *J. Biol. Chem.* **251,** 7423 (1976). (k) J. P. Bennett and S. H. Snyder, *Eur. J. Pharmacol.* **67,** 1 (1980). (l) S. Gunther, M. A. Gimbrone, Jr., and R. W. Alexander, *Circ. Res.* **47,** 278 (1980). (m) G. B. Wright, R. W. Alexander, L. S. Ekstein, and M. A. Gimbrone, Jr., *Circ. Res.* **50,** 462 (1982). (n) H. M. Cox, K. A. Munday, and J. A. Poat, *Br. J. Pharmacol.* **79,** 63 (1983). (o) H. M. Cox, J. A. Poat, and K. A. Munday, *Biochem. Pharmacol.* (in press).

TABLE II
VALUES FOR DISSOCIATION CONSTANTS[a,b]

Peptide	Symbol	Dissociation constant (M) Adrenal	Dissociation constant (M) Uterus	Activation constant (M)
1. [Ile^5]AII	AII	8.8×10^{-10}	5.4×10^{-10}	4.7×10^{-11}
2. [Val^5]AII		1.6×10^{-9}	1.1×10^{-9}	3.7×10^{-11}
3. [Sar^1]AII		4.0×10^{-10}	2.2×10^{-10}	1.4×10^{-11}
4. des-Asp^1-[Ile^5]AII	C7	5.3×10^{-10}	5.1×10^{-10}	1.2×10^{-10}
5. [Ile^5]AI	AI	8.5×10^{-8}	3.2×10^{-8}	7.3×10^{-9}[c]
6. des-Phe^8-[Val^5]AII	N7	4.1×10^{-7}	1.0×10^{-7}	
7. des-Asp^1, des-Arg^2-[Val^5]AII	C6	3.5×10^{-8}	1.1×10^{-8}	2.0×10^{-8}
8. des-Asp^1, des-Arg^2, des-Val^3-AII	C5	5.5×10^{-7}	4.1×10^{-7}	1.5×10^{-7}
9. Val-His-Pro-Phe	C4	4.6×10^{-5}	2.1×10^{-5}	
10. His-Pro-Phe	C3	5.4×10^{-4}	1.2×10^{-4}	
11. [Sar^1, Ala^8]AII		1.2×10^{-9}	9.3×10^{-10}	
12. [Sar^1, Ile^8]AII		9.7×10^{-10}	7.4×10^{-10}	2.5×10^{-10}
13. des-Asp^1-[Ile^8]AII		1.1×10^{-9}	1.1×10^{-9}	
14. LHRH		1.5×10^{-6}	1.3×10^{-6}	
15. des-pGlu^1-LHRH		3.3×10^{-6}	4.0×10^{-6}	2.5×10^{-6}

[a] Values of dissociation constants for binding of angiotensin I and II, angiotensin fragments, and analogs as well as luteinizing hormone-releasing hormone to adrenal membranes, to uterus membranes, and the respective activation constants for aldosterone production.

[b] K_D values were calculated from the relationship $K_D = K_D(\text{AII})[C/C_{\text{AII}}(1 + B/F) - B/F]$, where $K_D(\text{AII})$ = dissociation constant for angiotensin II; C and C_{AII} = concentrations giving 50% displacement of the tracer, respectively, for the compound tested and angiotensin II; B/F = bound/free ratio for [^{125}I]angiotensin II at the midpoint of the displacement curve. The activation constants are determined as the peptide concentrations giving half-maximal stimulation of aldosterone production in dispersed dog adrenal cells. LHRH, Luteinizing hormone-releasing hormone; SQ 20,881, a Squibb converting enzyme inhibitor.

[c] Measured in the presence of 10 μM SQ 20,881.

Radioiodination Procedure

1. Dissolve synthetic [Asp^1, Ile^5]AII (1 mg) in 1 ml 0.01 M acetic acid at room temperature to prepare a stock solution of about 1 mM. The actual concentration of the peptide can be checked from the optical density at 275 nm, using the molar extinction coefficient of tyrosine (1340).

2. Transfer 10 μl (10 μg AII) into a 5 × 75-mm glass tube, followed by 10 μl of 0.5 M phosphate buffer, pH 7.4, and 1 mCi (10 μl) of carrier-free

^{125}I (17 Ci/μg) in 0.1 *M* NaOH. Add 10 μl of a freshly-prepared solution of chloramine T (1 mg/ml), and after gentle shaking for 30 sec stop the reaction by adding 25 μl of sodium metabisulfite (1 mg/ml) in 0.05 *M* phosphate buffer.

Purification and Storage of Tracer

1a. Column fractionation: Transfer the contents of the tube to a DEAE Sephadex[8] or G-10 column[9] for separation of the radiolabeled monoiodoangiotensin II from free iodide and unlabeled peptide.

1b. HPLC purification: To prepare labeled angiotensin II of higher specific activity and maximum bindability, HPLC more effectively resolves the unlabeled peptide and diiodo-angiotensin II from monoiodo[^{125}I]angiotensin II. This can be achieved by several separation procedures, of which fractionation by reverse phase chromatography on a C-18 silica column with elution by 20 to 30% acetonitrile in 0.1 *M* ammonium acetate buffer, pH 7.8, gives satisfactory resolution of the monoiodo peptide.

2. Storage: After either method of purification, the labeled peptide is diluted in 0.1% bovine serum albumin in 50 m*M* phosphate buffer pH 7.4. For storage, the freshly prepared tracer, or the lyophilized iodinated derivative dissolved in 1% bovine serum albumin in distilled water, is kept at −20° after flash freezing as 50-μl aliquots each of 2 μCi. Each frozen aliquot is used only once after thawing. Under these conditions, monoiodoangiotensin II tracer usually preserves its original binding activity for at least 2 months.

Determination of Specific Activity and Maximum Binding Activity

The specific activity of ^{125}I-angiotensin II tracer is determined by radioligand-receptor assay, using the self-displacement method originally described for tracer analysis by radioimmunoassay.[10] The maximum binding activity of the tracer is determined by equilibrating a constant amount of labeled peptide with increasing amounts of particulate adrenal receptors. This value reflects the proportion of tracer that is biologically active (i.e., capable of specific binding to excess receptors) and usually ranges from 60 to 80% of the added radioiodinated hormone. During storage,

[8] M. D. Nielsen, M. Jorgensen, and J. Giese, *Acta Endocrinol.* (*Copenhagen*) **67,** 104 (1971).

[9] K. J. Catt, H. Glossmann, and A. Baukal, *in* "Mechanisms of Hypertension" (M. P. Sambhi, ed.), p. 200. Excerpta Medica, Amsterdam, 1973.

[10] S. A. Berson, R. S. Yalow, S. M. Glick, and J. Roth, *Metab., Clin. Exp.* **3,** 1135 (1964).

^{125}I-angiotensin II appears to undergo "decay catastrophe,"[9] in which decay of the ^{125}I atom is accompanied by commensurate destruction of the radioiodinated peptide, resulting in the simultaneous disappearance of both the biologically active peptide and the attached radionucleotide. For this reason, the specific activity of monoiodoangiotensin II tracer remains fairly constant during storage for up to 2–3 months.

Binding Studies with [^{125}I]Angiotensin II

Principles of the ^{125}I-Angiotensin II Binding Assay

The angiotensin II receptor assay is performed by incubation of tracer (in the absence or presence of unlabeled hormone or hormone analogs) with broken cell preparations or isolated cells. After a given time, the receptor-bound and free tracer hormone are separated by filtration or centrifugation. Since ^{125}I-angiotensin II may adhere to plastic and glass surfaces, it is advisable to count pipet tips as well as other dispensing devices for retained radioactivity. Nonspecific adsorption can be minimized by the use of plastic or siliconized glassware, and by keeping the ligand in albumin-containing solutions. Since the tracer is susceptible to degradation by various peptidases (especially in broken cell preparations), its integrity should be checked in a rebinding assay[6] or other methods, as described below.

Similar to other peptide hormones, receptor-bound ^{125}I-angiotensin II usually exhibits retention or enhancement of receptor binding activity after elution from its specific binding sites. In contrast, the free ligand may undergo extensive degradation, leading to a decrease in the concentration of free hormone, and consequently in the rate and extent of tracer binding. Such degradative changes can be detected by subjecting aliquots of the incubation medium to rebinding assay with fresh adrenal particles, or by less discriminating procedures including RIA, TLC,[11] HPLC, or high voltage electrophoresis.[12] The rebinding assay is a highly sensitive index of tracer degradation, though it does not discriminate between undegraded angiotensin II and biologically active metabolites such as des-Asp^1-AII, which also binds to the adrenal receptor.[13] For the rebinding assay, the initial incubation with particles or cells is performed in a centri-

[11] J. Douglas, P. Bartley, T. Kondo, and K. J. Catt, *Endocrinology* **102,** 1921 (1978).

[12] C. P. Campanile, J. K. Crane, M. J. Peach, and J. C. Garrison, *J. Biol. Chem.* **257,** 4951 (1982).

[13] M.-A. Devynck, M. G. Pernollet, P. G. Matthews, M. C. Khosla, F. M. Bumpus, and P. Meyer, *Proc. Natl. Acad. Sci. U.S.A.* **74,** 4029 (1977).

fuge tube, and after a given time the medium is separated from cells or particles by sedimentation and kept on ice until analyzed by reincubation with fresh adrenal particles.

The receptor-bound tracer can be released from particles or cells with 0.05 *M* acetic acid, followed by neutralization with NaOH or drying down from acetone.[14] The recovery is monitored, and a control tracer is incubated without particles (or cells) and carried throughout the entire procedure including the acidification–neutralization and/or lyophilization steps. The free tracer and that eluted from particulate receptors are then compared with the control tracer for their abilities to bind to fresh adrenal particles or cells.

Preparation of Bovine Adrenal Particulate Fraction

Bovine adrenals obtained within 10 min after death are placed on crushed ice and dissected free of fat and connective tissue.[6] The adrenals may be immediately sliced into 0.3–0.5 cm sections and kept in ice-cold Krebs–Ringer buffer (pH 7.4) containing 0.2% glucose and 1% bovine serum albumin for transport, or kept on crushed ice for dissection within 1 hr. Working in Krebs–Ringer buffer in Petri dishes kept on ice, the medulla is removed with a sharp scalpel and the cortex or the zona glomerulosa layer is dissected from the capsule. The tissue is then minced with fine scissors into small pieces of about 0.5 mm diameter, which are allowed to settle in a 200 ml beaker filled with Krebs–Ringer buffer.

The tissue fragments are then washed once in buffer and dispersed with a large Dounce homogenizer in ice-cold 20 m*M* $NaHCO_3$ at a wet weight to volume ratio of 1 : 20, employing 10 strokes of the loose pestle. Alternatively, a motor-driven Teflon-glass homogenizer can be used. The homogenates are combined in a large beaker and stirred for 15–20 min at 2 to 4°. The completeness of cell breakage is monitored by phase contrast microscopy.

The stirred homogenate is filtered through course fiberglass screen and nylon gauze or (subsequently) through two and four layers of cheese cloth. The filtered homogenate is spun at 1500 *g* for 10 min, and the supernatant centrifuged for 30 min at 20,000 *g* at 2–4°. The resulting pellet is resuspended in 20 m*M* $NaHCO_3$ (4 ml/1 g wet weight of starting material) by two or three strokes of the Dounce homogenizer and spun as above. The pellet is either resuspended in ice-cold 20 m*M* Tris–HCl

[14] G. Aguilera, A. Capponi, A. Baukal, K. Fujita, R. Hauger, and K. J. Catt, *Endocrinology* **104,** 1279 (1979).

pH 7.4 and immediately used for assay, or further purified as described below. The yield is 10–12% of the filtered homogenate protein.

Further Purification of Bovine Adrenal Cortex Membranes

The 20,000 *g* pellet is resuspended by gentle dispersion in a Dounce homogenizer (loose pestle) in 0.25 *M* sucrose containing 10 m*M* Tris–HCl, pH 7.2 and 1 m*M* Na_2EDTA, at a ratio of 1 ml medium per g (wet weight) of the original material. The resuspended pellet is layered on a discontinuous sucrose gradient (31.5, 38.5, and 42.5% w/w sucrose in the above medium) and spun for 120 min in a Beckman SW 27 rotor at 2–4°. The particulate material collecting at the upper interface (between the overlay and 31.5% sucrose) is removed and diluted 20-fold with medium, then sedimented at 40,000 *g* for 40 min at 2–4°. The yield of this procedure is 2–3% of the filtered homogenate protein.[6] An alternative method for obtaining highly purified plasma membranes by zonal centrifugation has been described.[15]

Both crude and purified membranes should be used within 4–6 hr after preparation. Considerable losses of angiotensin-receptor binding occur when the preparation is frozen in liquid nitrogen and rethawed, reflecting the lability of the free receptor sites.

Purification of Plasma Membrane-Enriched Rat Adrenal Glomerulosa Particles

Capsular layers from 20–30 adrenal glands are prepared as described below and finely minced with scissors, then homogenized in 30 ml ice-cold 20 m*M* $NaHCO_3$ with a Dounce homogenizer, using 5 strokes with the loose pestle and 5 strokes with the tight pestle.[16] Homogenates are filtered twice through nylon gauze and spun for 20 min at 600 *g*. The supernatant is centrifuged for 30 min at 20,000 *g* and the pellet is resuspended in 0.25 *M* sucrose in 20 m*M* Tris–HCl, pH 7.4. The suspension is layered on a discontinuous sucrose gradient (50, 40, 36, and 30% w/w in the above buffer) and spun in the Beckman SW 40 rotor for 90 min at 66,000 *g* at 2–4°. The particles aspirated from the interface between the 30 and 36% sucrose layers are diluted 10-fold with 20 m*M* Tris–HCl buffer, pH 7.4, and spun for 60 min at 20,000 *g*. The pellets are then resuspended

[15] W. Schlegel and R. Schwyzer, *Eur. J. Biochem.* **72,** 415 (1977).

[16] J. Douglas, G. Aguilera, T. Kondo, and K. J. Catt, *Endocrinology* **102,** 685 (1978).

in 50 m*M* Tris–HCl pH 7.4 prior to their use in the receptor binding assay. By this method, a 7- to 10-fold enrichment over the 20,000 *g* pellet can be obtained for plasma membrane markers including AII receptors, adenylate cyclase activity, 5′-nucleotidase, and Na^+,K^+-ATPase activity.

Preparation of Rat Adrenal Glomerulosa Cells

Male Sprague–Dawley rats are killed by decapitation or exposure to carbon dioxide, and the adrenal glands are removed and dissected free of fat. The translucent capsular layer, consisting predominantly of zona glomerulosa cells, is separated from the underlaying tissue by manual compression within a cellulose tissue to extrude the medulla and the fasciculata-reticularis zones. The capsules are finely minced with scissors and washed extensively with potassium-free medium 199. The washed tissue fragments are suspended in a plastic beaker with 10 ml medium 199 containing 2 mg/ml bovine serum albumin, 2 mg/ml collagenase Type II (Worthington), and 3.5 m*M* KCl, gassed with 95% O_2–5% CO_2, and incubated with shaking at 37° for 30 min. After the fragments are allowed to settle, the supernatant is discarded and the cells are dispersed by aspiration through Tygon tubing into a plastic syringe[17] in medium 199 containing 2 mg/ml bovine serum albumin, 0.1 mg/ml soyabean trypsin inhibitor, 5 μg/ml DNase, and 3.5 m*M* KCl.

The aspiration procedure is repeated until the dispersion medium is clear, and residual fragments are subjected to a second digestion followed by dispersion as above. Dispersed cells are collected by centrifugation at 200 *g* for 10 min at 22° and suspended in medium 199 supplemented with 2 mg/ml bovine serum albumin and 3.5 m*M* KCl. The yield is 8–10 $\times 10^6$ cells per 100 adrenals. A similar procedure can be applied to the preparatio of bovine glomerulosa cells,[18] but with less satisfactory viability when the time between tissue collection and cell dispersion becomes prolonged. In the rat glomerulosa cell preparation, contaminating fasciculata cells comprise about 5% of the cell number, and can be removed by filtration over a 1 cm column of Sephadex G-15[19] or by density gradient sedimentation.[20]

[17] K. Fujita, G. Aguilera, and K. J. Catt, *J. Biol. Chem.* **254,** 8567 (1979).

[18] A. Peytremann, R. D. Brown, W. E. Nicholson, D. R. Island, G. W. Liddle, and J. G. Hardman, *Steroids* **24,** 451 (1974).

[19] B. C. Williams, J. G. McDougall, P. J. Hyatt, J. G. B. Bell, S. A. S. Tait, and J. F. Tait, *J. Endocrinol.* **81,** 109p (1979).

[20] J. F. Tait, S. A. S. Tait, R. P. Gould, and M. R. J. Mee, *Proc. R. Soc. London, Ser. B* **185,** 375 (1974).

Practical Aspects of Angiotensin Receptor Binding Assays

Nonspecific Binding. For measurement of nonspecific binding, particulate receptors and ^{125}I-angiotensin II tracer are incubated with a saturating concentration of unlabeled angiotensin II, usually 0.1–1 μM. It is also advisable to determine whether nonspecific adsorption occurs to assay tubes and filters, and to include tubes which do not contain membranes or cells in each assay. With crude adrenal cortex membranes (0.1–0.4 mg of protein) the filter-bound radioactivity (nonspecific binding) in the presence of 1 μM angiotensin II is identical to that retained by HAWP filters when the tracer is filtered without particles. Usually, only 0.8–1% of the total radioactivity present in the assay tube is nonspecifically adsorbed to particles and filters.

Incubation Conditions. In adrenal particles, angiotensin II receptors are usually assayed at temperatures from 12 to 22° for 45 to 120 min. The binding reaction is much slower at 4°, but when performed overnight gives tracer binding equivalent to that attained during incubation at room temperature for 45–60 min. Although incubation temperatures of 30 and 37° have also been employed (see Table 1), degradation of both tracer peptide and receptor sites is more extensive at elevated temperatures despite the presence of peptidase inhibitors. Incubations are performed in total volumes of 100 to 250 μl in 12 × 75-mm glass tubes for adrenal particles, or at 37° under 95% O_2–5% CO_2 in plastic vials or 12 × 75-mm plastic tubes for isolated adrenal cells.

Buffers and Additions for Angiotensin Assay with Particles. Buffers containing 20 or 50 mM Tris–HCl (pH 7.4) supplemented with 0.2% bovine serum albumin (heat-inactivated for 30 min at 56°) are used for the incubation and dilution of tracer and unlabeled peptides. Several additives including EDTA (1–5 mM), dithiothreitol (0.5–5 mM), bacitracin (100 μM), glucagon (0.3 mg/ml), $ACTH_{1-24}$ (45 μM), and phenylmethylsufonylfluoride (100 μM) have been used alone or in combination to inhibit tracer degradation. In particular, agents such as dithiothreitol (DTT) and EDTA are often employed in angiotensin II binding assays for this purpose. In adrenal particles, DTT concentrations of up to 5 mM improve binding by inhibiting degradation of the free ligand. However, in smooth muscle and pituitary particles, DTT concentrations above 1 mM can markedly decrease binding by reduction of essential receptor disulfide bonds (Fig. 1). Although EDTA is an effective inhibitor of tracer degradation, under some conditions this chelator can also reduce receptor binding of tracer angiotensin II. This can be avoided by the use of EGTA, which is equally effective in preventing ligand degradation but does not interfere with ligand binding. In some tissues, notably the adrenal zona glomerulosa, sodium ions stabilize a high affinity conformation of the angiotensin

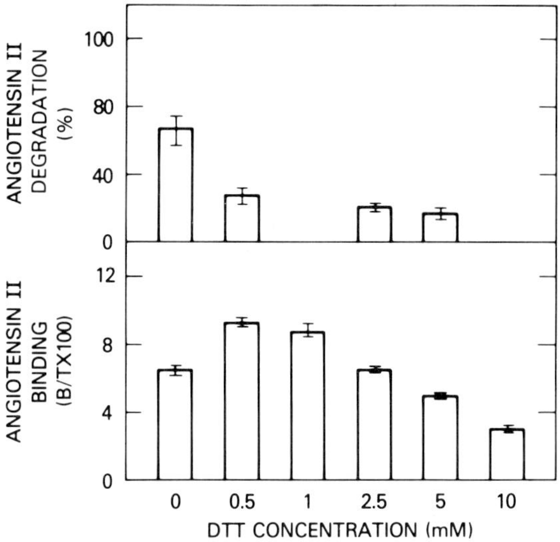

FIG. 1. Effects of DDT upon ^{125}I-angiotensin II tracer degradation (above) and receptor binding (below) in rat anterior pituitary particles.

II receptor, with half-maximal stimulation between 30 and 40 m*M* and maximum effect at 140–200 m*M*.[21] An example of the effect of sodium ions upon angiotensin II binding in bovine adrenal cortex particles is shown in Fig. 2. However, sodium ions inhibit binding in smooth muscle and pituitary gland, whereas divalent cations often increase binding. Optimal binding conditions for muscle and pituitary receptors are provided by incubation in 50 m*M* Tris–HCl buffer pH 7.4 containing 5 m*M* $MgCl_2$, 2 m*M* EGTA, 0.5% BSA, and 0.5–1 m*M* DTT.

Typical Assay for Adrenal Cortex Particles. A standard protocol for AII receptor assays is to add the following reagents in sequence to 12 × 75-mm glass tubes standing on ice: 50 μl ^{125}I-angiotensin II (0.1–0.3 pmol) in 50 m*M* Tris–HCl buffer pH 7.4 containing 0.2% BSA; 50 μl unlabeled peptide solution (final conc. 10^{-11}–10^{-8} *M* for binding-inhibition standards; 10^{-7} *M* for nonspecific; buffer alone for total binding); 50 μl of selected additives to optimize binding and to decrease tracer degradation (e.g., dithiothreitol, EDTA, NaCl, or $MgCl_2$) and 100 μl adrenal cortex particles in the above buffer (50–200 μg of protein per tube). After 45 to 60 min of incubation at 22°, followed by addition of 4 ml of ice-cold phosphate-buffered saline or 20 m*M* Tris–HCl buffer (pH 7.4) to each tube, the contents are passed through an 0.1% BSA-saturated Millipore HAWP (0.45 μm) filter or Whatman glass fiber filter under slight vacuum. The

[21] H. Glossmann, A. Baukal, and K. J. Catt, *Science* **185,** 281 (1974).

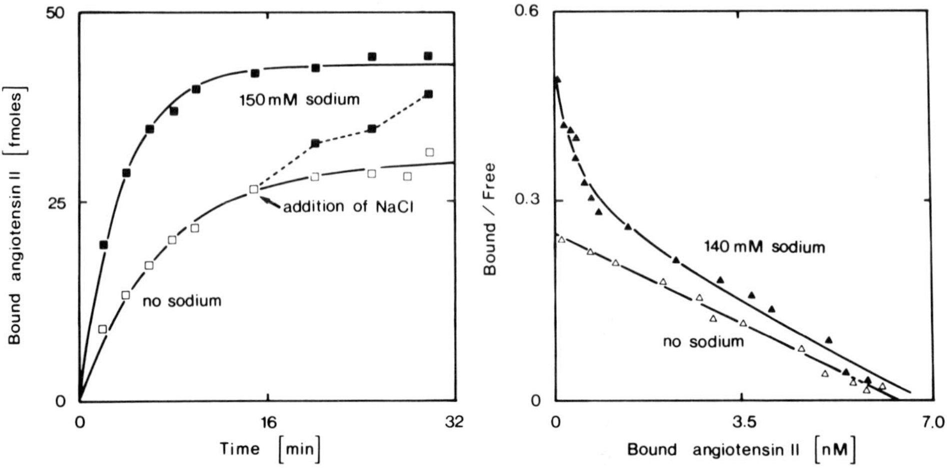

FIG. 2. Effects of sodium ion on the kinetics and equilibrium binding of ^{125}I-angiotensin II to bovine adrenal cortex receptors. Left: Rate of angiotensin II binding at 22° in the absence and presence of 150 m*M* NaCl. Right: Scatchard analysis of equilibrium binding data derived in the absence or presence of 140 m*M* NaCl.

tubes and filters are washed twice with the same volume of cold buffer, and the dried filters are analyzed for receptor-bound radioactivity in an automatic gamma spectrometer.

Properties of Angiotensin II Receptors

Equilibrium Binding Data

The equilibrium binding of ^{125}I-angiotensin II to bovine adrenal particles in the absence or presence of 150 m*M* NaCl is shown in Fig. 1. Saturation of the receptors was achieved by reducing the specific activity of the ligand with unlabeled angiotensin II (e.g., by adding 0.1–100 n*M* angiotensin II). The time course of formation of the ^{125}I-angiotensin–receptor complex is also illustrated. The equilibrium dissociation constants are in the range of 0.2 to 0.5 n*M* for the high-affinity binding site of the adrenal cortex. However, biphasic saturation isotherms are sometimes obtained, and the measured affinity and density of the binding sites appear to be critically dependent on ionic and other experimental conditions (see Table 1).

Pharmacological Profile of the Binding Sites

The binding and activation properties of angiotensin II receptors in canine adrenal gland and smooth muscle are shown in Table 2. Further references to the properties of receptors present in the individual target tissues for angiotensin II are given in reference 1.

Cation Dependence

High-affinity angiotensin binding to particulate receptor fractions of certain target tissues is markedly influenced by the cation composition of the incubation medium.[21–23] In the adrenal, several cations apparently induce or stabilize a receptor conformation that has altered kinetic properties and increased susceptibility to the inhibitory effect of guanyl nucleotides. In this respect, Na^+ and divalent cations including Ca^{2+}, Mg^{2+}, and Mn^{2+} may substitute for one another (see Table I). Adrenal binding of AII is significantly enhanced by increasing sodium concentrations up to 150 m*M* (Fig. 2) and also by increasing potassium concentrations. This effect is largely due to increased receptor affinity, and is most prominent in the adrenal glomerulosa zone and renal glomeruli. In other tissues such as smooth muscle and pituitary gland, sodium ions have little effect or inhibit angiotensin binding to receptors, whereas divalent cations (Mg^{2+}, Mn^{2+}, and Ca^{2+}) increase binding. This effect of divalent cations is due to an increase in the number of available AII receptors in smooth muscle and renal glomeruli, and to a combined effect on receptor affinity and number in the adrenal zona glomerulosa. The concentration-dependent effects of Mg^{2+} and Ca^{2+} upon angiotensin II binding, and the inhibitory actions of EGTA on binding in the presence of the divalent cations, are shown in Fig. 3.

Effects of Guanyl Nucleotides

High-affinity angiotensin receptor binding in the presence of sodium is sensitive to inhibition by guanyl nucleotides. In the absence of a nucleotide regenerating system, Gpp(NH)p and $GTP_{\gamma}S$ are the most potent inhibitors ($IC_{50} < 1\ \mu M$), whereas GTP is of comparable potency only in the presence of a regenerating system.[24] In the adrenal gland, the decrease in AII receptor binding affinity in the presence of guanyl nucleotides is largely attributable to an increase in the dissociation rate constant.[24] However, in rat kidney cortex epithelial membranes the rate of dissociation of bound AII is reported to be reduced by guanyl nucleotides.[25] The inhibitory effects of guanyl nucleotides on AII binding are now known to reflect the interaction between the AII receptors and the inhibitory guanyl nucle-

[22] J. K. Crane, C. P. Campanile, and J. C. Garrison, *J. Biol. Chem.* **257,** 4959 (1982).

[23] G. B. Wright, R. W. Alexander, L. S. Ekstein, and M. A. Gimbrone, *Circ. Res.* **50,** 462 (1982).

[24] H. Glossmann, A. Baukal, and K. J. Catt, *J. Biol. Chem.* **249,** 664 (1974).

[25] H. M. Cox, J. A. Poat, and K. A. Munday, *Biochem. Pharmacol.* **32,** 3601 (1983).

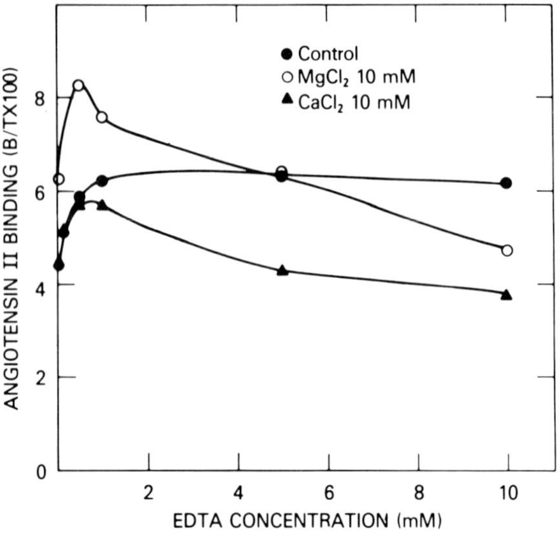

FIG. 3. Effects of divalent cations and EDTA on binding of tracer angiotensin II to rat adrenal glomerulosa receptors.

otide regulatory protein of the adenylate cyclase system, and AII has been shown to inhibit cyclase activity in liver, kidney, and adrenal glomerulosa zona.

Effects of Reducing Agents

Dithiothreitol has been reported to inhibit angiotensin II receptor binding in particulate and/or membrane-rich fractions from the rat adrenal cortex,[26] mesenteric artery,[7] anterior pituitary,[27] and kidney epithelium.[28] In both kidney and adrenal cortex, preincubation with angiotensin II protects against the inhibitory effects of dithiothreitol in particulate receptor fractions. As noted above, the inclusion of low DTT concentrations in AII receptor assays often enhances binding by reducing degradation of the tracer peptide, but at higher DTT concentrations the inhibitory effects are predominant. For this reason, it is essential to examine the effects of DTT upon angiotensin binding in any individual receptor system before employing it as an assay reagent to reduce tracer degradation.

[26] R. S. L. Chang, V. L. Lotti, and M. E. Keegan, *Biochem. Pharmacol.* **31,** 1903 (1982).

[27] R. L. Hauger, G. Aguilera, A. J. Baukal, and K. J. Catt, *Mol. Cell. Endocrinol.* **25,** 203 (1982).

[28] H. M. Cox, K. A. Munday, and J. A. Poat, *Br. J. Pharmacol.* **79,** 63 (1983).

Section II

Identification of Receptor Proteins

[11] General Principles for Photoaffinity Labeling of Peptide Hormone Receptors

By ALEX N. EBERLE and PIERRE N. E. DE GRAAN

Introduction

Photoaffinity labeling of peptide hormone receptors has become an invaluable tool for receptor localization and isolation and for mechanistic studies of hormone–receptor interaction. It represents a special category of affinity labeling in which the ligand is activated at the receptor site to a highly reactive intermediate by UV light. Photogenerated peptide labels have the advantage over chemical (electrophilic) labels in that they are stable in aqueous solution until activation and do not require nucleophilic centers for insertion into the ligand binding site. For this reason, photoaffinity labels have been successfully applied to peptides in a variety of systems whereas chemical affinity labels have hardly ever led to covalent insertion into peptide hormone receptors.[1,2]

Photoaffinity experiments were initiated by Westheimer and his group in 1962 for the active site labeling of chymotrypsin using α-ketocarbenes as reactive species generated from α-diazocarbonyl precursors.[3,4] α-Ketocarbenes were successfully used for labeling of several enzymes, antibodies and receptors with their respective ligands.[4] However, reports on the application of this type of label to peptides remained scarce,[2,5] mainly because of the inherent chemical reactivity of the diazo group and because of Wolff rearrangement of α-ketocarbenes both of which may militate against specific labeling. The chemically more stable aryl azides, introduced by Knowles and colleagues in 1969 as precursors for reactive aryl nitrenes,[6] proved to be very suitable photolabels for peptides in a variety of experiments (see Table III). However, like all other photolabels available to date, arylnitrenes have also certain drawbacks. The selection of the photoreactive group therefore needs careful consideration.

[1] R. Walter, I. L. Schwartz, O. Hechter, T. Dousa, and P. L. Hoffman, *Endocrinology* **91,** 39 (1972).

[2] A. N. Eberle, W. Hübscher, and R. Schwyzer, *Helv. Chim. Acta* **60,** 2895 (1977).

[3] A. Singh, E. R. Thornton, and F. H. Westheimer, *J. Biol. Chem.* **237,** PC3006 (1962).

[4] V. Chowdhry and F. H. Westheimer, *Annu. Rev. Biochem.* **48,** 293 (1979).

[5] D. B. P. Goodman, J. M. Stadel, and H. Rasmussen, *J. Membr. Biol.* **40** (Spec. Issue), 323 (1978).

[6] G. W. J. Fleet, R. R. Porter, and J. R. Knowles, *Nature (London)* **244,** 511 (1969); G. W. J. Flett, J. R. Knowles, and R. R. Porter, *Biochem. J.* **128,** 499 (1972).

ISBN 0-12-182009-2

The general scheme for photoaffinity labeling of a peptide hormone receptor consists of three steps: (1) binding of the peptide ligand (L) containing the photolabel (X) to the receptor (R); (2) photoactivation of X to a highly reactive species (X′), and (3) formation of a chemical bond between peptide and receptor through X′:

$$\mathrm{R + L{-}X} \underset{k_2}{\overset{k_1}{\rightleftharpoons}} \mathrm{R \cdot L{-}X} \xrightarrow[h\nu]{k_3} \mathrm{R{-}X'{-}L}$$

Apart from the covalent insertion reaction at the receptor (k_3), which at low ligand concentration is a first-order process, activated hormone (L–X′) may attach to nonspecific sites (NS) in a second-order reaction (k_4):

$$\mathrm{NS + L{-}X} \xrightarrow[h\nu]{k_4} \mathrm{NS{-}X'{-}L}$$

Consequently the hormone concentration should be kept as low as possible so as to minimize the side-reaction k_4. Well-designed photoreactive ligands L–X should be employed whose affinity for the receptor is not substantially reduced compared to that of the natural ligand L.

The procedure for UV irradiation (wavelength, intensity, and duration) is determined by the light sensitivity of the biological material, on the one hand, and the photolability of the photolabel, on the other. Ideally quantitative photolysis of the ligand should not lead to any measurable destruction of the target (receptor, membrane, or functional cell). Although specific characteristics of the UV beam, such as intensities at the site of photolysis, have rarely been reported in the literature, the equipment plays a dominant role and should allow enough flexibility for changing any of the above mentioned parameters. Equally important factors are the identification of photolabeled products and the control of specificity of the labeling.

The aim of this chapter is (1) to summarize those aspects of photoaffinity labeling which are relevant for the application to peptide hormones, such as the choice of a suitable photolabel, the design of photoreactive peptides, the equipment and procedures for photolysis, the control and analysis of the photoreaction, and (2) to present an updated (though incomplete) list with examples. More general information can be found in several excellent review articles[4,7–9] and in particular in a comprehensive

[7] J. R. Knowles, *Acc. Chem. Res.* **5,** 155 (1972); D. Creed, *Photochem. Photobiol.* **19,** 459 (1974); F. J. Darfler and A. M. Tometsko, *Chem. Biochem. Amino Acids, Pept., Proteins* **5,** 31 (1978); R. J. Guillory, *Curr. Top. Bioenerg.* **9,** 267 (1979); J. V. Staros, *Trends Biochem. Sci.* **5,** 320 (1980); P. S. Linsley, M. Das, and C. F. Fox, *Recept. Recognition, Ser. B* **11,** 87 (1981).

treatise by Bayley and Knowles[10] in this series and by a recent monograph by Bayley.[10a]

Choice of Photolabel

The main criterion for the photogenerated intermediate is its nonselective reaction, i.e., its capability of attacking even carbon–hydrogen bonds. Other important factors are the chemical stability and ease of photolysis of the precursor. The commonly used carbenes and nitrenes, which are generated by photoelimination of a nitrogen molecule from α-diazocarbonyl or aryl diazirine and aryl azide groups, respectively, react with neighboring chemical functions by insertion into single bonds, hydrogen abstraction (→ radical formation), (cyclo-)addition to multiple bonds, and coordination to nucleophilic centers:

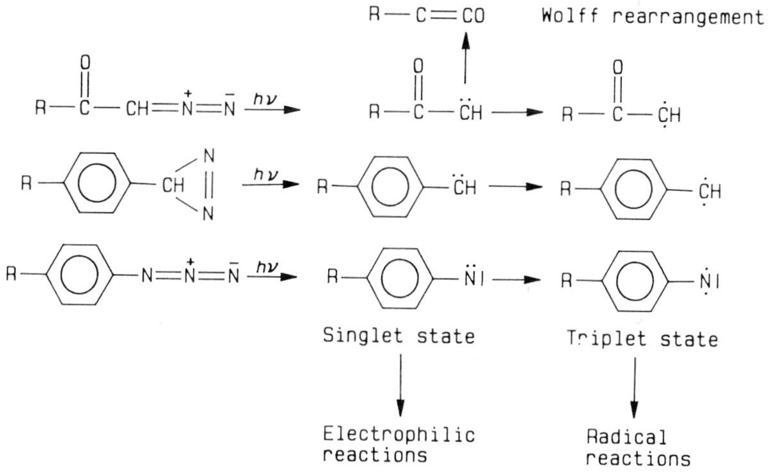

The type of reaction is determined by the spin state of the photogenerated species: in singlet carbenes/nitrenes, electrophilic reactions (in-

[8] T. H. Ji, *Biochim. Biophys. Acta* **559,** 36 (1979); this series, Vol. 91, p. 580; *Biol. Reprod.* **28,** 50 (1983).

[9] N. J. Turro, *Ann. N.Y. Acad. Sci.* **346,** 1 (1980).

[10] H. Bayley and J. R. Knowles, this series, Vol. 46, p. 69.

[10a] H. Bayley, "Photogenerated Reagents in Biochemistry and Molecular Biology." Elsevier, Amsterdam, 1983.

sertion, addition, nucleophiles) tend to dominate whereas radical reactions are typical for the triplet state leading to abstraction of a hydrogen atom.[9] The resulting triplet radical pair can produce the insertion product only after spin intersystem crossing. In low-affinity hormone–receptor systems this process may be slower than the diffusion of the radical pair. However in high-affinity systems with low k_2, triplet states will lead to insertion, thus increasing the labeling yields considerably. This is particularly important for nitrenes which are less reactive and indiscriminate for insertion reactions than carbenes. Because of the longer half-life of nitrenes the singlet state tends to rearrange into the lower energy triplet state which can produce insertion into C—H bonds.[11] In general however, according to their distinct electrophilic character, (singlet) nitrenes prefer an O—H or N—H over a C—H bond and are therefore less suitable candidates for the labeling of hydrocarbon-rich areas (membrane lipids, hydrophobic pockets) than carbenes.[12]

Aryl azides are stable to acid, base, and most procedures used in peptide synthesis, except reducing conditions, such as hydrogenation and thiols. Their absorption maximum is around 260 nm (unsubstituted rings), but they can be photolyzed at wavelengths up to 365 nm[13] although shorter wavelengths have occasionally been reported to be essential. Introduction of electron-withdrawing groups, e.g., a nitro group in the meta position of the azide, facilitates photolysis of the azide at longer wavelengths and increases the reactivity of the nitrene. At the same time the nitrene becomes more electrophilic and hence more discriminate in its reaction (→ higher reactivity with solvent water). A notable side-reaction of aryl nitrenes is the formation of an azirine that reacts with nucleophiles to give the seven-membered azepine ring-structure (for details, see ref. 10).

α-Diazocarbonyl derivatives are chemically much more reactive than aryl azides: the diazogroup of unsubstituted α-diazoketones is hydrolyzed at low pH, and it is a good leaving group in nucleophilic displacement reactions (reaction with —SH, —NH_2 or water). In addition, as mentioned above, carbenes generated from α-diazoketones may undergo Wolff rearrangement whereby a high proportion of generally much less reactive ketene is formed (up to 60%[4]) and hence the labeling efficiency may be low. Wolff rearrangement is reduced by replacing $^{\alpha}$C—H with

[11] I. M. McRobbie, O. Meth-Cohn, and H. Suschitzky, *Tetrahedron Lett.* pp. 925 and 929 (1976).
[12] H. Bayley and J. R. Knowles, *Biochemistry* **17,** 2414 (1978).
[13] E. Escher and R. Schwyzer, *FEBS Lett.* **46,** 347 (1974).

$^{\alpha}$C—COOEt[14] and more effectively with $^{\alpha}$C—CF_3[15] or $^{\alpha}$C—SO_2—(C_6H_4)—CH_3 (*p*-toluenesulfonyl).[16] The two latter substitutions stabilize the diazo compound (good heat and acid stability) and enhance its photolysis at ~350 nm, but also increase its electrophilic nature (→ reaction with water). A further side-reaction of some N-acetylated diazo compounds is the formation of the corresponding triazole above pH ~7 (Dimroth rearrangement).[10,17] α-Diazoketones (and aryl diazirines) have absorption maxima at around 250 nm (high ε) and 350 nm (low ε); both wavelengths can be used for photolysis though much longer irradiation times are usually necessary at 350 nm.

The drawbacks of α-ketocarbenes are overcome with aryl carbenes generated from aryl diazirines.[18] They combine the advantage of good chemical stability with the high reactivity of carbenes. Since photolysis of unsubstituted diazirines may generate long-lived linear diazo compounds, it is advisable to substitute the C—H by C—CF_3: the (3-trifluoromethyldiazirino)phenyl group,[19] originally used for the labeling of membrane components,[20] offers a promising alternative to aryl azides. It is equally stable to acid, base, and other peptide chemical manipulations, and it can be photolyzed smoothly at 350 nm. However, the more laborious preparation of the compound and the lack of commercial reagents have precluded its wide-spread application for peptide receptor labeling up to now. Experience will have to show in which type of experiment diazirines should be given preference over aryl azides or whether the high reactivity of carbenes may even be a disadvantage in some situations because of possible nonspecific insertion into membrane lipids.

Two other groups of photolabels have been described for peptides: the *p*-nitrophenyl group and α,β-unsaturated ketones. The first label, used in the form of *p*-nitrophenylalanine, was shown to be as suitable for the labeling of the chymotrypsin active site[13] or of angiotensin II[21] and bradykinin receptors[22] as *p*-azidophenylalanine when activated at the same wavelength (365 nm). In contrast, MSH receptors could not be labeled

[14] R. J. Vaughan and F. H. Westheimer, *J. Am. Chem. Soc.* **91,** 217 (1969).

[15] V. Chowdhry, R. Vaughan, and F. H. Westheimer, *Proc. Natl. Acad. Sci. U.S.A.* **73,** 1406 (1976).

[16] V. Chowdhry and F. H. Westheimer, *Bioorg. Chem.* **7,** 189 (1978).

[17] D. J. Brunswick and B. S. Cooperman, *Biochemistry* **12,** 4074 (1973).

[18] H. Bayley and J. R. Knowles, *Biochemistry* **17,** 2420 (1978).

[19] J. Brunner, H. Senn, and F. M. Richards, *J. Biol. Chem.* **255,** 3313 (1980).

[20] J. Brunner and F. M. Richards, *J. Biol. Chem.* **255,** 3319 (1980).

[21] E. Escher and G. Guillemette, *J. Med. Chem.* **22,** 1047 (1978).

[22] E. Escher, E. Laczko, G. Guillemette, and D. Regoli, *J. Med. Chem.* **24,** 1409 (1981).

with *p*-nitrophenyl-containing α-MSH whereas the *p*-azidophenyl group—at the same position of the molecule—gave good insertion.[23]

The photochemistry of the nitro group is poorly understood, and the reactions are not well predictable. Most probably they proceed via a radical mechanism.[21] UV irradiation of α,β-unsaturated ketones results in excitation of the n-π^* band producing a diradical triplet state that selectively abstracts hydrogen from C—H bonds in preference to the stronger O—H bonds of solvent water.[24] As an example, 4-acetylbenzoylpentagastrin could be covalently attached to albumin by irradiation at 320 nm.[25] The disadvantage of this type of photolabel resides in its high lipophilicity which may considerably alter the characteristics of the hormone, particularly of short peptides.

Aryl diazonium compounds are potential photolabels if used in acidic milieu: they are light sensitive and produce radicals upon UV irradiation.[26] At physiological pH, however, diazonium ions are chemically very reactive, forming azo links to Tyr and His residues, which would counteract photolabeling. A study with the model compound diazobenzene sulfonate revealed the potential of this probe,[27] but attempts to label angiotensin II[22] and bradykinin receptors[23] with *p*-diazoniophenylalanine-containing hormone derivatives were not successful. Further investigations to exploit the interesting properties of aryl diazonium ions as dual photoaffinity/chemical affinity probe for hormone receptors are therefore necessary.

Design of Photoreactive Peptides

Specific introduction of a photolabel into a distinct site of a peptide hormone is a prerequisite for reproducibility and correct interpretation of photolabeling experiments. Structure–activity studies usually form a valuable basis for the choice of suitable sites for modification. The ideal photoreactive peptide should have physicochemical and biological properties identical to those of the natural ligand when tested without covalent labeling (i.e., in the dark). This means that the photolabel should not disturb peptide-receptor binding and receptor activation. Nevertheless the photoactivated peptide should react covalently with the receptor.

An elegant method for the introduction of photolabels into peptides is "isosteric" replacement of a Phe or Tyr residue by a structurally related

[23] A. N. Eberle, *J. Receptor Res.* **4** (1984).

[24] N. J. Turro, "Molecular Photochemistry." Benjamin, New York, 1965.

[25] R. E. Galardy, L. C. Craig, J. D. Jamieson, and M. P. Printz, *J. Biol. Chem.* **249,** 3510 (1974).

[26] S. de Jonge, *Recl. Trav. Chim. Pays-Bas* **72,** 846 (1952).

[27] A. M. Tometsko, J. Turula, and J. Comstock, *Int. J. Pept. Protein Res.* **12,** 143 (1978).

amino acid, such as *p*-nitrophenylalanine (Pnp) (**1**), *p*-azidophenylalanine (Pap) (**2**), *p*-azido-*o*-nitrophenylalanine (Panp) (**3**), or *p*-(3-trifluoromethyl)diazirinophenylalanine (PdP) (**4**):

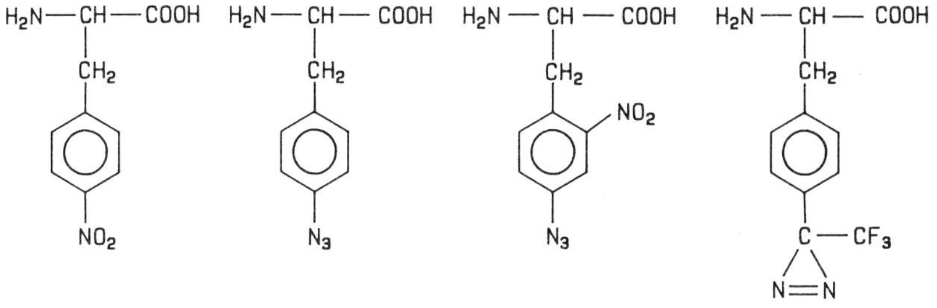

1 (Pnp) 2 (Pap) 3 (Panp) 4 (Pdp)

Pap (**2**) is prepared from Pnp[28] (**1**) via *p*-aminophenylalanine and *p*-diazoniophenylalanine in a three-step synthesis devised by Schwyzer and collaborators.[29,30] The synthesis of Panp (**3**) requires an additional step for the introduction of the nitro group.[30] Pap was successfully applied to the preparation of photoreactive α-MSH,[31,32] angiotensin II,[21] bradykinin,[22] substance P,[33] and to a neurophysin ligand.[34] Occasionally, *m*-azidophenylalanine[30] (**2a**) was used instead of Pap, e.g., for the synthesis of a photolabile enkephalin derivative.[35] Labeling of hormone receptors with Pdp[36] (**4**) has not yet been reported. The advantage of replacing an aromatic residue of a peptide by one of these photoreactive amino acids resides in the ensuing minimal structural alteration of the hormone, an important factor particularly for small peptides. On the other hand, this approach always affords a *de novo* synthesis of the peptide.

Alternatively, photolabels can be introduced specifically into amino, sulfhydryl, and carboxyl groups, as well as into the side-chains of Arg, His, Tyr, and Trp with one of the heterobifunctional reagents listed in

[28] F. Bergel and J. A. Stock, *J. Chem. Soc.* p. 90 (1959).

[29] R. Schwyzer and M. Caviezel, *Helv. Chim. Acta* **54,** 1395 (1971).

[30] E. Escher and R. Schwyzer, *Helv. Chim. Acta* **58,** 1465 (1975).

[31] A. N. Eberle and R. Schwyzer, *Helv. Chim. Acta* **59,** 2421 (1976).

[32] A. N. Eberle, P. N. E. de Graan, and W. Hübscher, *Helv. Chim. Acta* **64,** 2645 (1981).

[33] E. Escher, R. Couture, G. Champagne, J. Mizrahi, and D. Regoli, *J. Med. Chem.* **25,** 470 (1982).

[34] Y. S. Klausner, W. M. McCormick, and I. M. Chaiken, *Int. J. Pept. Protein Res.* **11,** 82 (1978).

[35] M. Smolarksy and D. E. Koshland, Jr., *J. Biol. Chem.* **255,** 7244 (1980).

[36] H. Bayley, personal communication.

Table I.[37–59] Specific modification of a carboxyl group is possible only if the peptide is partially protected at all amino, thiol, and other carboxyl groups. The same is true for the modification of other functional groups when more than one such site is present. Amino groups of peptides are quantitatively modified with the acid chlorides or bromides and active esters of *p*-azidobenzoic or *p*-azidophenylacetic acid (compound **5–8**) and their respective *o*-nitro-derivatives (e.g., **9**, **10**). Other reagents, such as imidates (**11**, **12**) and fluorides (**13**) of aryl azides or *p*-azidophenyliso-thiocyanate[45] (**14**) are often preferred for incorporation into large peptides or for labeling of membrane proteins. Imidates offer the advantage that they form positively charged amidines with amino groups at alkaline pH, thus retaining the protonated character of the group. Compounds

[37] S. H. Hixson and S. S. Hixson, *Biochemistry* **14,** 114 (1975).

[38] S. H. Hixson and S. S. Hixson, *Biochemistry* **14,** 4251 (1975).

[39] C. C. Yip, C. W. T. Yeung, and M. L. Moule, *Biochemistry* **19,** 70 (1980).

[40] A. N. Eberle, P. N. E. de Graan, and F. C. G. van de Veerdonk, *in* "Neuroendocrinology of Vasopressin, Corticoliberin and Opiomelanocortins" (A. J. Baertschi and J. J. Dreifuss, eds.), p. 231. Academic Press, New York, 1982.

[41] A. N. Eberle, unpublished results.

[42] P. Thamm, D. Saunders, and D. Brandenburg, *in* "Insulin; Chemistry, Structure and Function of Insulin and Related Hormones" (D. Brandenburg and A. Wollmer, eds.), p. 309. de Gruyter, Berlin, 1980.

[43] R. V. Lewis, M. F. Roberts, E. A. Dennis, and W. S. Allison, *Biochemistry* **16,** 5650 (1977).

[44] T. H. Ji, *J. Biol. Chem.* **252,** 1566 (1977).

[44a] D. Levy, *Biochim. Biophys. Acta* **322,** 329 (1973).

[44b] R. A. G. Smith and J. R. Knowles, *Biochem. J.* **141,** 51 (1974).

[45] H. Sigrist and P. Zahler, *FEBS Lett.* **113,** 307 (1980).

[46] M. Yagub and P. Guire, *J. Biomed. Mater. Res.* **8,** 291 (1974).

[47] P. Guire, D. Fliger, and J. Hodgson, *Pharmacol. Res. Commun.* **9,** 131 (1977).

[48] A. J. Lomant and G. Fairbanks, *J. Mol. Biol.* **104,** 243 (1976).

[49] W. Burgermeister, M. Nassal, T. Wieland, and E. J. M. Helmreich, *Biochim. Biophys. Acta* **729,** 219 (1983).

[50] M. Nassal, *Liebigs Ann. Chem.* 1510 (1983).

[51] E. Escher, H. Robert, and G. Guillemette, *Helv. Chim. Acta* **62,** 1217 (1979).

[52] M. Das and C. F. Fox, *Proc. Natl. Acad. Sci. U.S.A.* **75,** 2644 (1978).

[53] E. F. Vanin, S. J. Burkhard, and I. I. Kaiser, *FEBS Lett.* **124,** 89 (1981); T. T. Ngo, C. F. Yam, H. M. Lenhoff, and J. Ivy, *J. Biol. Chem.* **256,** 11313 (1981).

[54] I. Schwartz and J. Offengand, *Proc. Natl. Acad. Sci. U.S.A.* **71,** 3951 (1974).

[55] F. Seela and H. Rosemeyer, *Hoppe-Seiler's Z. Physiol. Chem.* **358,** 129 (1977).

[56] V. G. Budker, D. G. Knorre, V. V. Kravchenko, O. I. Lavrik, G. A. Nevinsky, and N. M. Teplova, *FEBS Lett.* **49,** 159 (1974).

[57] W. E. Trommer, H. Kolkenbrock, and G. Pfleiderer, *Hoppe-Seyler's Z. Physiol. Chem.* **356,** 1455 (1975).

[58] K. Muramoto and J. Ramachandran, *Biochemistry* **19,** 3280 (1980).

[59] C. D. Demoliou and R. M. Epand, *Biochemistry* **19,** 4539 (1980).

TABLE I
HETEROBIFUNCTIONAL REAGENTS FOR SPECIFIC INTRODUCTION OF PHOTOLABELS INTO PEPTIDES

Reagent	Structure	Reference[a]
Amino groups[b]		
5	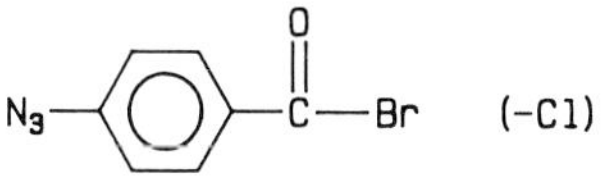	37, 38
6	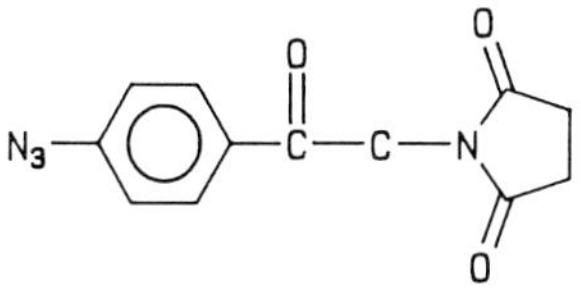	39[c]
7	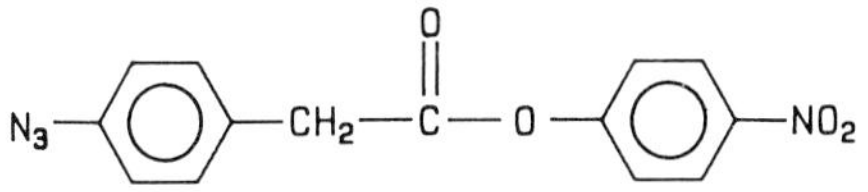	40
8	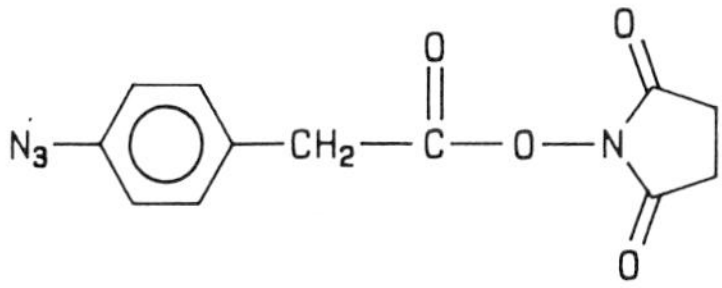	41
9	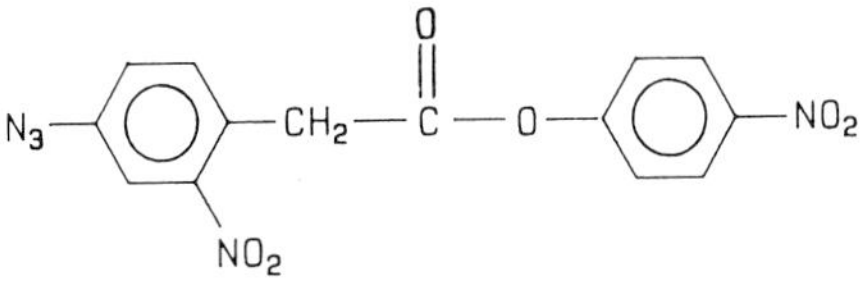	
10	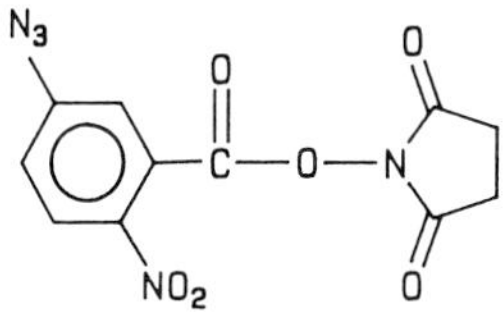	43[c]
11	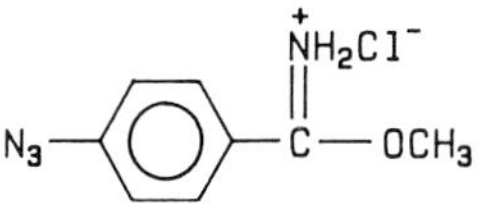	44[c]
12	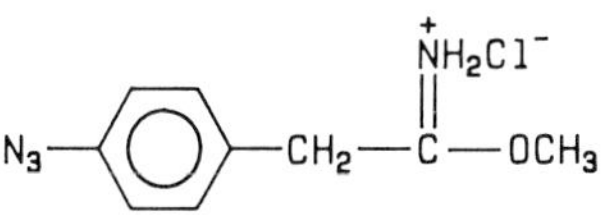	44

(*continued*)

TABLE I (*continued*)

Reagent	Structure	Reference[a]
13	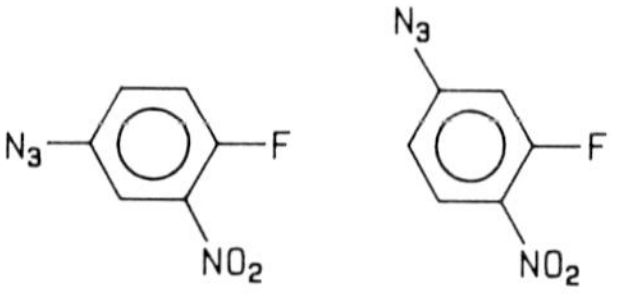	6, 44a, 44b[c,d]
14	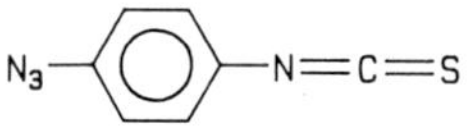	45
15	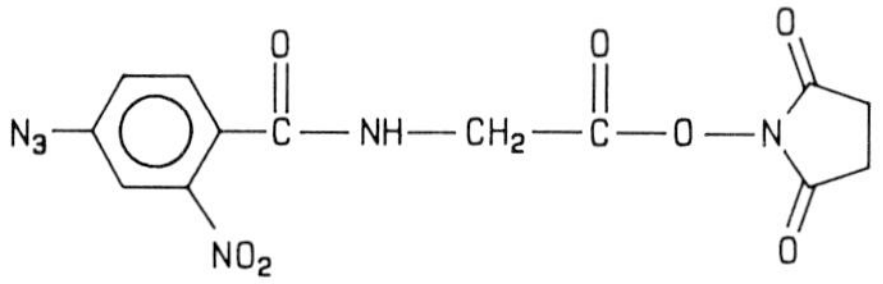	39
16	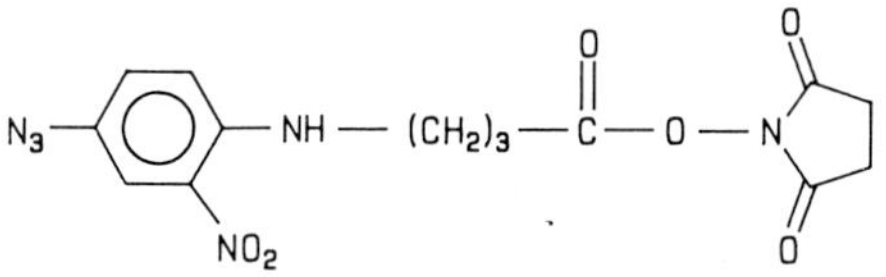	46, 47
17	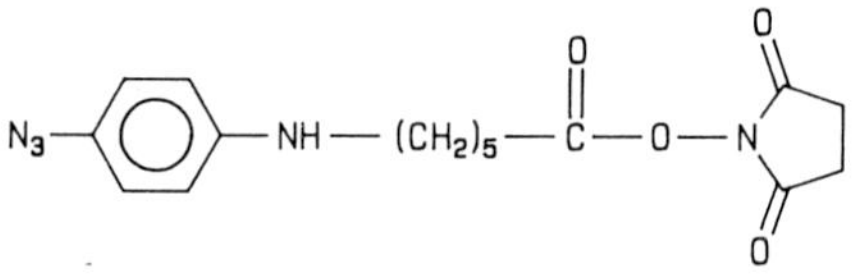	48[c]
18	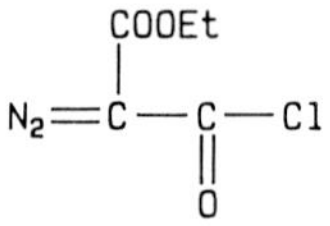	14
19	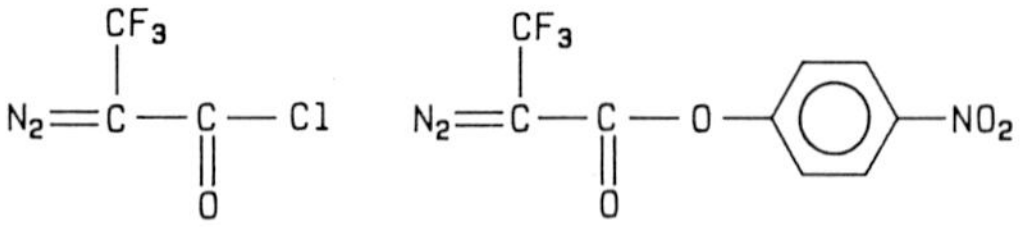	15[c]
20	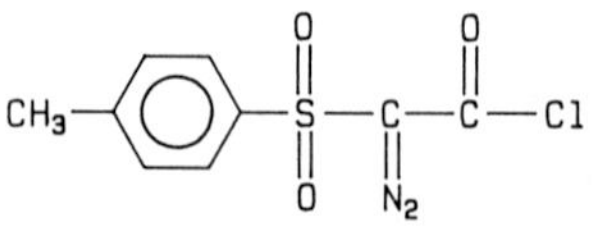	16

TABLE I (*continued*)

Reagent	Structure	Reference[a]
21	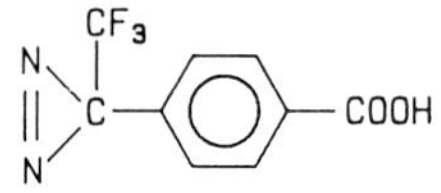	49, 50
Carboxyl groups		
22	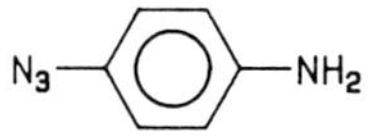	51
23	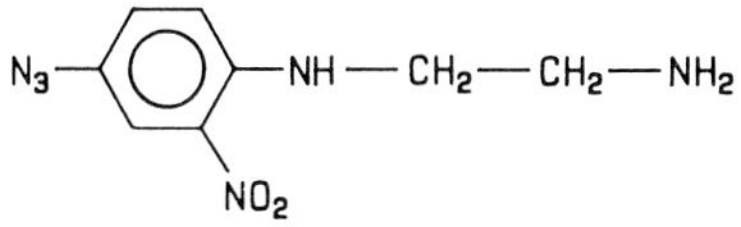	52
24	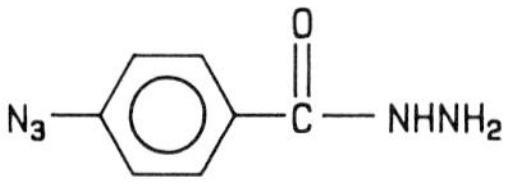	41
Arginine		
25	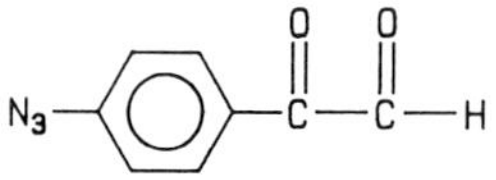	53[c]
Cysteine		
26	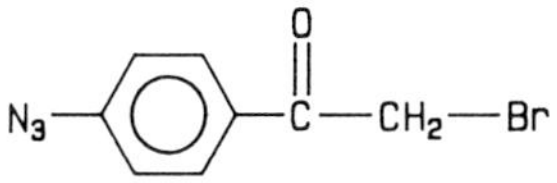	38, 54[c]
27	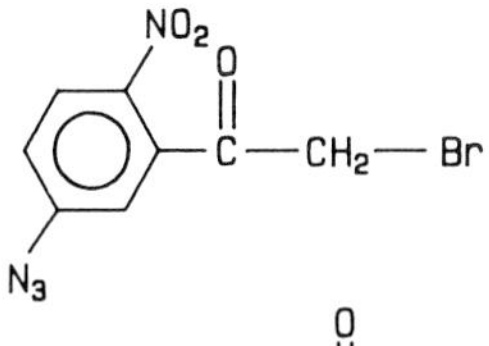	55
28	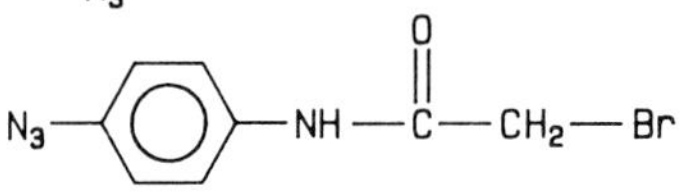	56
29	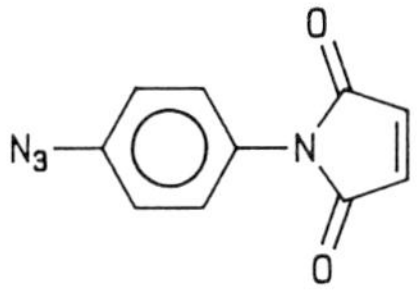	57

(*continued*)

TABLE I (*continued*)

Reagent	Structure	Reference[a]
Tryptophan		
30	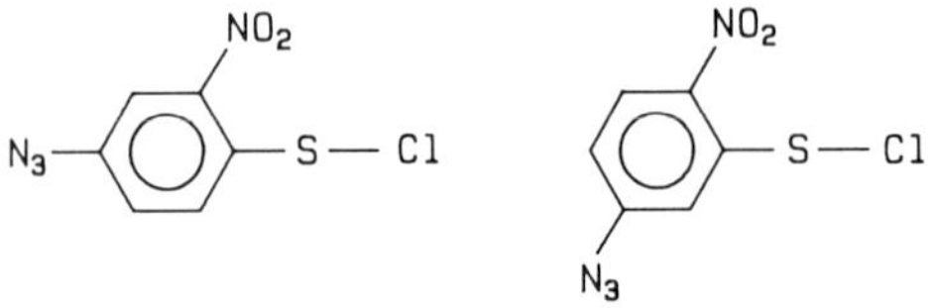	58, 59
Tyrosine/histidine		
31	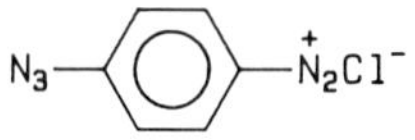	51

[a] Numbers refer to text footnotes.
[b] Amino-specific reagents can also react with free thiols.
[c] Commerically available through Pierce Chemical Company, Rockford, Ill.
[d] The 3-fluoro-2,4-dinitrophenylazide is ~1000 times more reactive with amino groups than the mono-nitro reagent, but it may form furoxan on photolysis.

with a spacer (**15–17**) may be useful when the photolabel cannot be inserted into the binding area of the peptide. The carbene-generating reagent 2-diazo-3,3,3-trifluoropropionate[15] is easily introduced into amino groups as acid chloride (**19**) or as *p*-nitrophenyl ester (**19a**), and it is stable even in 1 *M* HCl (in contrast to the unreactive *p*-nitrophenyl 2-diazoacetate[60]). The same is true for *p*-toluenesulfonyldiazoacetyl chloride[16] (**20**) and 4-(3-trifluoromethyldiazirino)benzoic acid [4-(1-azi-2,2,2,-trifluoroethyl)benzoic acid][49,50] (**21**). The guanidino-specific *p*-azidophenylglyoxal[53] (**25**) reacts extensively with arginine at pH 7.0–7.5, but the reaction product slowly decomposes at alkaline pH, regenerating the original guanidino group. The Trp-specific 2-nitro-4-azido-phenylsulfenyl chloride[58,59] (NAPS-Cl; **30**) also reacts with thiols and, at pH >6, with amino groups. Because most peptide hormones lack free Cys, NAPS-Cl can be incorporated specifically into the 2-position of the Trp indole ring at acidic pH. Two similar reagents, 2,4-dinitro-5-azidophenylsulfenyl chloride[61] (DNAPS-Cl) and 2-nitro-5-azido-phenylsulfenyl chloride[58] (2,5-NAPS-Cl; **30a**), were investigated with ACTH, but proved to be much less stable (DNAPS-Cl) or slightly less reactive (2,5-NAPS-Cl) than NAPS-Cl.

[60] J. Frank and R. Schwyzer, *Experientia* **26,** 1207 (1970).
[61] E. Canova-Davis and J. Ramachandran, *Biochemistry* **19,** 3275 (1980).

Cleavable Photolabels

Photolabels attached to the hormone via a cleavable group are particularly suitable for receptor isolation: the ligand remains covalently bound to the receptor during the entire purification process and is then removed by a cleaving agent before the identification and analysis of the receptor. Several cleavable groups have been incorporated into photolabels for this purpose, including disulfide (—S—S—), glycol (—CHOH—CHOH—), and azo (—N=N—) bridges.

Disulfide bridges are most commonly used for reversible attachment of the photolabel because they can be split under relatively mild conditions, such as incubation with 10–100 m*M* 2-mercaptoethanol, dithiothreitol, or dithioerythritol at pH 7–9. Several photolabels have been described (Table II)[61a–66] which either contain a disulfide bridge or form it upon reaction with a free —SH group introduced into the peptide prior to the reaction. An example of the latter approach is the formation of 2-mercaptotryptophan via reduction of 2-(2,4-dinitrophenylsulfenyl)-Trp.[61] 2-Mercaptotryptophan reacts readily and reversibly with NAPS-Cl (**30**)[67,68] as demonstrated with (Trp(NAPSS)^9)-ACTH whose photolabel is released quantitatively with 2% 2-mercaptoethanol.[67] The advantage of this procedure resides in the minimal structural change of the hormone after cleavage which may be important, e.g., for removal of the peptide with antibodies. There are, however, two major disadvantages in the use of disulfide containing photolabels: (1) reducing agents during labeling and isolation of receptors must be excluded, and (2) asymmetrical disulfides are susceptible to exchange at acidic as well as slightly alkaline pH.

Reagents containing a glycol[8,69,70] or azo bridge[51,71] or an amidine-forming imidate[44,71a] offer an alternative approach to cleavage. Glycols are stable under reducing conditions and can be cleaved with periodate.

[61a] E. F. Vanin and T. H. Ji, *Biochemistry* **20,** 6754 (1981).

[62] M. Das, T. Miyakawa, C. F. Fox, R. M. Druss, A. Aharonov, and H. R. Hershman, *Proc. Natl. Acad. Sci. U.S.A.* **74,** 2790 (1977).

[63] D. J. Kiehm and T. H. Ji, *J. Biol. Chem.* **252,** 8524 (1977).

[64] R. B. Moreland, P. K. Smith, E. K. Fujimoto, and M. E. Dockter, *Anal. Biochem.* **121,** 321 (1982).

[65] C. K. Huang and F. M. Richards, *J. Biol. Chem.* **252,** 5514 (1977).

[66] J. Henkin, *J. Biol. Chem.* **252,** 4293 (1977).

[67] K. Muramoto and J. Ramachandran, *Biochemistry* **20,** 3376 and 3380 (1981).

[68] A. N. Eberle, *J. Recept. Res.* **3,** 313 (1983).

[69] L. C. Lutter, F. Ortanderl, and H. Fasold, *FEBS Lett.* **48,** 288 (1974).

[70] J. R. Coggins, E. A. Hopper, and R. N. Perham, *Biochemistry* **15,** 2527 (1976).

[71] R. Uy and F. Wold, *in* "Biomedical Application of Immobilized Enzymes and Proteins" (T. M. S. Chang, ed.), Vol. 1, p. 15. Plenum, New York, 1977.

[71a] L. C. Packman and R. N. Perham, *Biochemistry* **21,** 5171 (1982).

TABLE II
CLEAVABLE HETEROBIFUNCTIONAL PHOTOLABELS FOR MODIFICATION OF AMINO AND THIOL GROUPS

Reagent	Structure	Reference[a]
Amino groups		
32	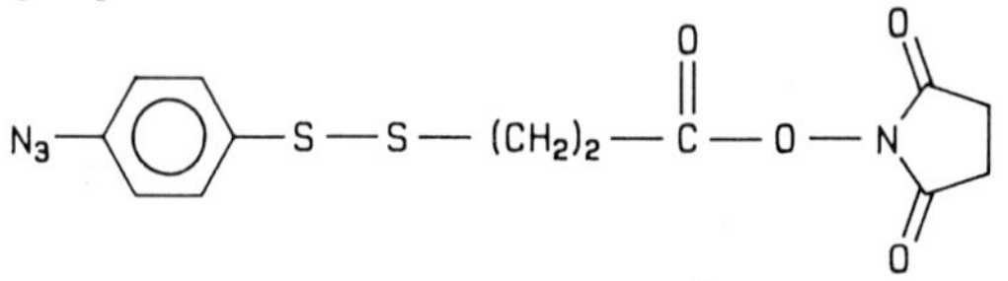	8, 61a[b]
33	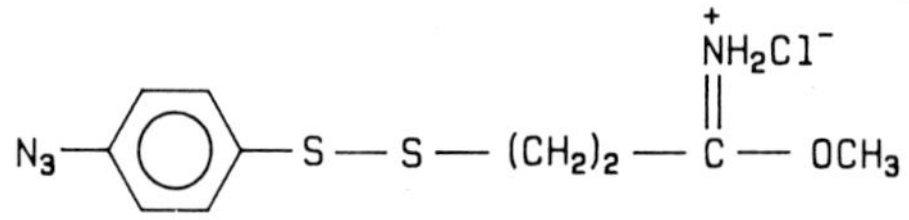	8, 61a, 62
Thiol groups		
34	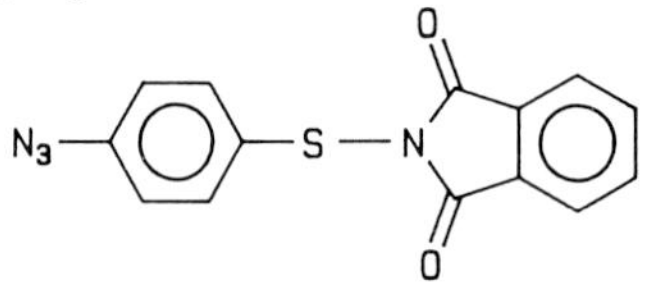	61a, 63, 64
35	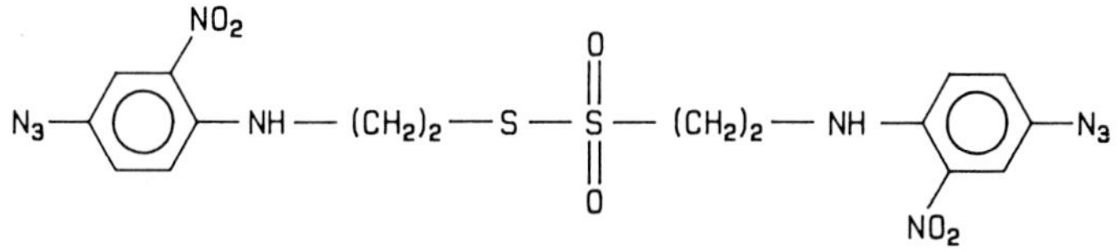	65
36	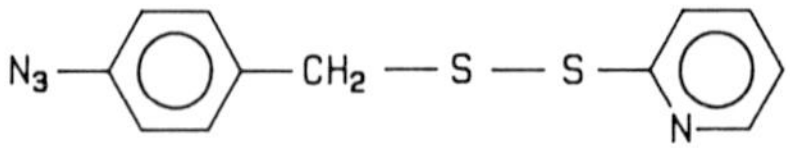	66

[a] Numbers refer to text footnotes.
[b] Commercially available through Pierce Chemical Company, Rockford, Ill.

This is mainly useful for cleavage on polyacrylamide gels by soaking them in 15 m*M* sodium periodate for 4–10 hr.[69,70] In a similar way, incubation of gels with amidine-containing photoproducts in 2 *M* methylamine–HCl, pH 11.5, in 75% acetonitril (at 37°) produces quantitative cleavage within 3 hr.[71a] Azo bridges can be introduced into a peptide either by using a

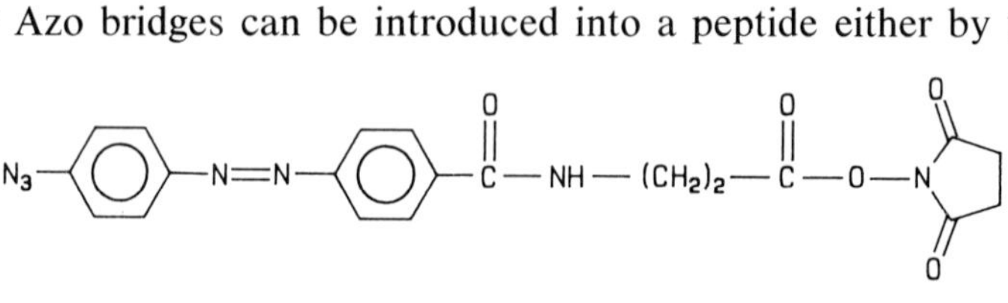

37

photolabel containing the azo bridge, such as *N*-[4-(*p*-azidophenylazo)-benzoyl]3-aminopropyl-*N'*-oxysuccinimide ester[71] (**37**), or by condensing the Tyr/His-specific *p*-azidophenyldiazonium reagent[51] (**31**) to either of the two residues. Azo compounds can be split with sodium dithionite,[51] but it has not yet been demonstrated whether the method is applicable to biological systems.

Radioactive and Fluorescent Photolabels

In order to facilitate the identification of photolabeled products, it is desirable to label the peptide hormone with radioisotopes, such as ^{125}I or tritium, or to attach a fluorescent group. Iodination of the peptide can be performed either before or after the introduction of the photolabel into the hormone. The latter approach is preferred because it is technically simpler. Although tyrosine residues are more readily iodinated than aryl azides, a partial incorporation of iodine into the photolabel may still occur causing a change of its characteristics. Therefore some authors prefer to introduce the photolabel into the iodinated hormone (see Table III). Iodinated photolabels should be assessed for biological activity and binding properties prior to their application. If a peptide hormone is inactivated by the iodination procedure or the presence of an iodine atom, or if it is devoid of any suitable residue for iodination, a radioactive photolabel may be used instead. Ji and Ji[72] have devised some heterobifunctional reagents that can be iodinated in acetone (to prevent hydrolysis of the active ester) either within the photolabile moiety (**38**) or at an adjacent tyrosine residue (**39**):

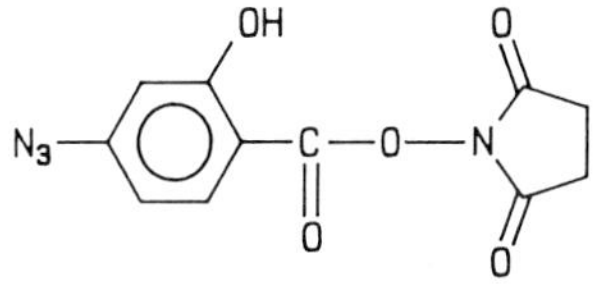

38

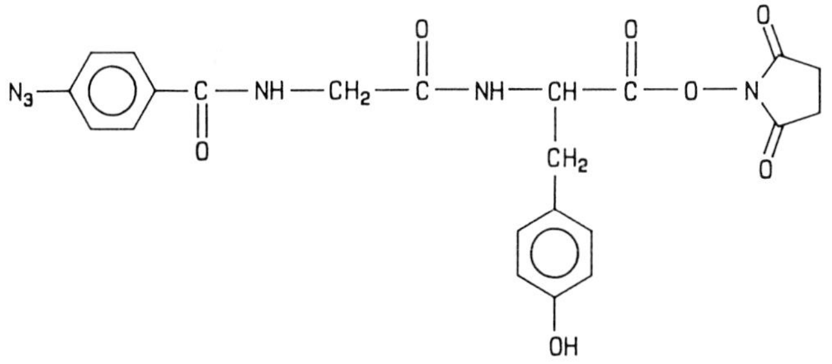

39

[72] T. H. Ji and I. Ji, *Anal. Biochem.* **121,** 286 (1982).

Photolabels containing electron-withdrawing groups, such as a nitro group, cannot be iodinated by standard techniques. A two-step procedure involving electrophilic thallation of the ring and subsequent nucleophilic substitution of the thallium by $^{125}I^-$ can be applied,[73] but the drastic thallation reaction precludes its use for sensitive molecules.

Peptides containing tritium as radioactive marker are frequently preferred over those labeled with ^{125}I. Although tritiated compounds exhibit lower specific radioactivity and hence require longer exposition in autoradiography than those labeled with ^{125}I, tritium labeling also offers some advantages: tritiated peptides can be prepared in isosteric form to the natural hormone and can be used for a relatively long period of time as they are generally much more stable than iodinated ones. It is however recommended to attach the photolabel to tritiated peptides just before use because β-irradiation can damage the photolabile group, particularly when the compound is stored in dry form. Direct insertion of tritium into the photolabel is preferred when photolabeled products are to be structurally analyzed; the radioactivity remains firmly attached to the site of photo-insertion and is not released by, e.g., enzymatic hydrolysis. However, this approach affords a four-step synthesis, as originally devised for α-MSH[31]:

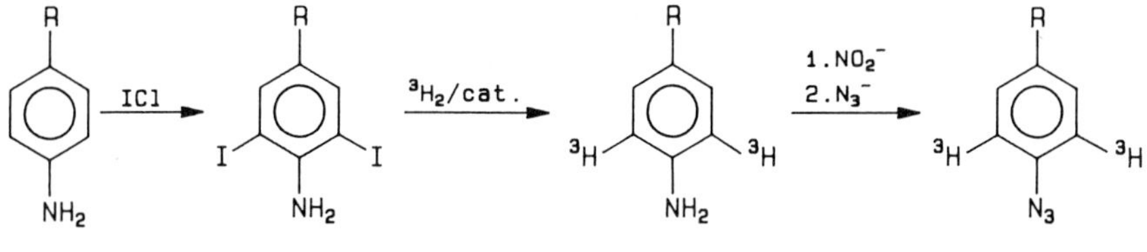

40 41 42 43

Alternatively tritium can be introduced into a phenylazide (or a phenyldiazirine) containing a formyl group by reduction of the aldehyde with [^3H]$NaBH_4$[73a]:

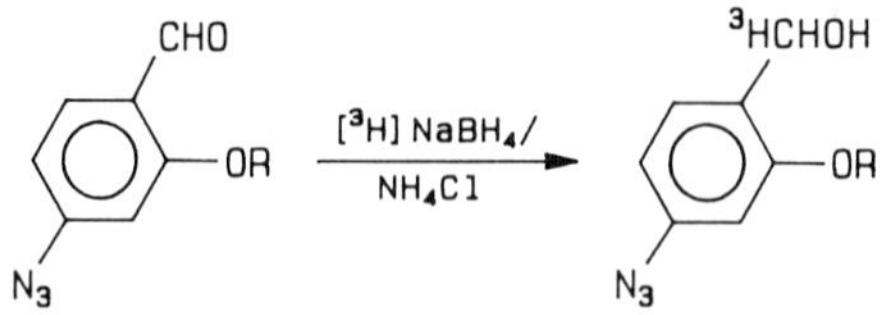

44 45

[73] F. Tejedor and J. P. G. Ballesta, *Anal. Biochem.* **128,** 115 (1983).
[73a] J. A. Maassen, *Biochemistry* **18,** 1288 (1979).

The fluorescent reagent (dansyldiazomethyl)phosphinate (**46**) introduced by Stackhouse and Westheimer[74] has several interesting properties:

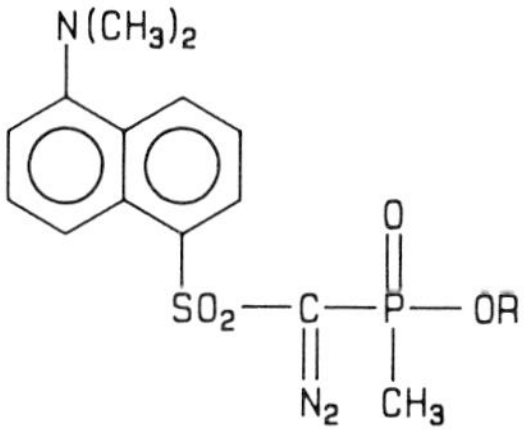

46

The reagent exhibits a high extinction coefficient at 350 nm, yields highly fluorescent products upon photolysis and can also be prepared in tritiated form from commercially available [^{3}H]dansyl chloride. The reagent—though not yet applied to peptides—may prove useful for the identification of photolabeled products during isolation or even *in situ* in biological targets.

Equipment and Procedures for Photolysis

For efficient photolysis of the photolabile peptide hormone a strong and stable light source is required emitting a broad spectrum in the ultraviolet, near ultraviolet, or visible range. High radiant fluxes are necessary when narrow bands of the spectrum have to be selected by filtration (see below) or when large surface areas (e.g., tissue culture dishes) have to be irradiated. Xenon lamps (250 to 1000 W) (→ continuous spectrum) or medium-pressure mercury and high-pressure mercury/xenon lamps (→ discontinuous spectrum with high intensities at the emission lines) are the most suitable light sources and are commercially available through several companies.[75] Such equipment is relatively expensive, and for small sample irradiation, a less expensive lamp may be sufficient (examples for such equipment can be found in ref. 10 or 10a). Flash photolysis represents an alternative to continuous irradiation. Repetitive flashes from a strong flash source may be sufficient for complete photolysis. In this way the biological system is exposed to the UV irradiation only for milliseconds although the light intensity is much higher. Details and procedures are described elsewhere.[8]

[74] J. Stackhouse and F. H. Westheimer, *J. Org. Chem.* **46,** 1891 (1981).

[75] E.g., Oriel Corp., Stamford, Connecticut; Spindler & Hoyer KG, Göttingen, FRG; Illumination Industries, Sunnyvale, California.

The light source should be equipped with a suitable system to select the proper wavelength (filters or monochromator; see below) and/or with a cuvette containing a cooling device and a filter solution. Cooling is required when powerful lamps are used in order to maintain the UV beam at low and constant temperature. Filter solutions are particularly useful for eliminating infrared components and cutting off short wavelengths: acetic acid limits the spectrum to >245 nm, cupric sulfate or potassium phthalate to >310 nm, and cobaltous nitrate from approximately 340 to 380 nm. The concentration of the filtering component depends on the pathlength of the cell (1% to saturated solution). Alternatively, cut-off filters used in photography (Kodak, Zeiss, Balzers, Schott) or Pyrex glass (for $\lambda > 280$ nm) eliminate undesirable short wavelengths.[76] For most purposes a broad spectral region is perfectly suitable. However, irradiation of UV-sensitive cells, such as nerve cells or cells in tissue culture, may require a defined wavelength. Narrow bands of the spectrum of a Xenon lamp can be selected using interference filters (e.g., from Balzers) or a monochromator. The direction of the UV beam can be altered almost without loss in intensity using planar UV mirrors (e.g., from Balzers, Schott, Zeiss).

Although photolysis at 254 nm can be very fast and produce high labeling efficiencies,[10] most authors prefer wavelengths of >310 nm for the irradiation of peptide hormone receptor systems in membranes or on intact cells because photodestruction is thereby considerably reduced. The irradiation conditions (wavelength, intensity, time) should be carefully chosen by an empirical approach at the onset of an experiment, by measuring the decomposition of the photolabel (by UV or IR spectroscopy) and simultaneously following the photostability of the biological system (with a binding or bioassay) and of the ligand (by amino acid analysis, thin-layer chromatography, HPLC, or bioassay). Once the optimal irradiation conditions are selected, it is advisable to determine the light sensitivity at the site of irradiation, and to report irradiation conditions in detail.[77]

Specific covalent attachment can be increased by repeating the photolysis cycle: the hormone–receptor complex is irradiated for a short period of time, noncovalently bound ligand is removed by washing, dialysis, or gel filtration. Subsequently the system is reequilibrated with fresh photolabile ligand and again irradiated. This procedure can be repeated several

[76] J. G. Calvert and J. N. Pitts, Jr., "Photochemistry." Wiley, New York, 1966.

[77] Light intensity measurements are reported in Ref. 22 (18 mW/cm^2 at 365 nm; measured with a spectroline radiometer), ref. 111 (1 mW/cm^2 at 338 nm; measured with a Schottky pin-10 UV-enhanced photodiode calibrated with a thermopile), Ref. 23 (180 mW/cm^2), and Ref. 72 (0.46 mW/cm^2). For experimental details see Ref. 10a.

times and is less damaging to living cells than continuous exposure to UV light. The labeling efficiency can also be affected by the pH of the medium, the equilibration time prior to photolysis, and by the temperature during irradiation. At low temperature receptor turnover and dissociation of the hormone–receptor complex are diminished, resulting in a higher degree of labeling. An interesting observation was made during photolabeling of an enzyme–substrate system in the frozen state: the rate of incorporation was enhanced by a factor of about 10 without affecting the rate of photodecomposition.[78] Whenever possible, photolysis should be carried out under nitrogen or argon to prevent chromophore-sensitized photooxidation of the biological system; solutions should be purged with gas prior to and during irradiation. Reducing agents, such as thiols (in particular dithiols), should not be present in the incubation medium before and during photolysis of labels containing azido or diazo groups because they may destroy these groups.[79,80]

Specificity, Control, and Analysis of Receptor Labeling

General considerations concerning specificity and analysis of photoproducts in photoaffinity labeling can be found in the review by Bayley and Knowles.[10] Only those aspects particularly relevant to peptide photolabels are emphasized here. Extent and specificity of receptor labeling can be determined either by bioassay, binding assay, or with radioactive photolabels. Bioassays are very valuable for following covalent insertion of peptide ligands into functional receptors when the biological system alters its activity upon photoaffinity labeling (e.g., irreversible stimulation or inhibition; see Table III). Binding assays are performed with radioactive peptides after the labeling with nonradioactive photolabel. If intact cells are employed the time allowed to elapse between photoreaction and binding studies should be short because covalently labeled receptors may be processed rapidly (degradation, internalization). Bioassays and binding assays need not necessarily give identical results because a small percentage of labeled receptors, barely detectable with the typical binding assay, may be sufficient to produce a change in the biological response (see also ref. 10). It is better therefore to combine bioassays with the use of radioactive photoaffinity labels, provided that nonspecific insertion and noncovalent binding can be kept to a minimum. Nonspecific insertion due to

[78] J. J. Ferguson, Jr., *Photochem. Photobiol.* **32,** 137 (1980).

[79] J. V. Staros, H. Bayley, D. N. Standring, and J. R. Knowles, *Biochem. Biophys. Res. Commun.* **80,** 568 (1978).

[80] Y. Takagaki, C. M. Gupta, and H. G. Khorana, *Biochem. Biophys. Res. Commun.* **95,** 589 (1980).

diffusion from the receptor after photoactivation may be reduced by the addition of scavangers, such as *p*-aminobenzoic acid[47] or *p*-aminophenylalanine,[13] to the medium during photolysis. Noncovalent "specific" and unspecific binding to sites other than receptors and subsequent insertion after photolysis[81] can be reduced or avoided by using low ligand concentrations or by adding soluble proteins (such as BSA which however itself may become labeled) or, if possible, mild detergents to the assay medium. Furthermore, it should be noted that iodoaryl compounds are sensitive to UV irradiation and can be cleaved to iodine and aryl radicals which may insert into receptors.[24] Such insertions cause problems when ^{125}I-labeled peptides not containing a photolabel are irradiated as controls.

The use of radioactive photoaffinity labels is essential for further analysis of the insertion reaction by, e.g., autoradiography of whole cells or by isolation of the photoproducts. The simplest method of characterizing radioactive ligand–receptor complexes are chromatography or SDS-gel electrophoresis with subsequent autoradiography of the gels. Other methods, such as gradient centrifugation, isoelectrofocussing, or two-dimensional electrophoresis, may also be applicable. From studies with enzymes it became evident that the insertion reaction hardly ever occurs at a unique site, but that a population of very similar photoproducts is formed instead,[10] probably with slightly different net charges. Localization of the binding site within a peptide hormone receptor has not yet been reported.

A number of control experiments should always be included into photoaffinity experiments: (1) A non-UV-irradiated control (darkness or nonphotolyzing irradiation) is especially important if the biological response is used as criterion for the extent of labeling. (2) UV irradiation of the biological system alone should not decrease its response to the native hormone or reduce subsequent photolabeling. (3) Prephotolyzed photolabel should neither be damaged by UV irradiation, exhibit altered binding characteristics in the non-UV-irradiated system, nor insert into receptors. (4) Photolysis should be quantitative under the conditions used. (5) An excess of nonphotolysable agonist or antagonist, or prephotolyzed label should protect the receptor from covalent labeling. Optimal conditions for protection are usually a low photolabel concentration, a high concentration of the protecting peptide, and a short photolysis time (flash photolysis). (6) Insertion should be saturable. (7) Unrelated peptides carrying the same photolabel should be without effect.

[81] In addition to the generally occurring unspecific binding of peptides to the cell surface, a specific component (displacable by natural hormone) has become particularly relevant in photoaffinity labeling in view of the finding that peptide hormones may strongly and specifically interact with membrane lipids; B. Gysin and R. Schwyzer, *Arch. Biochem. Biophys.* **225,** 467 (1983); *FEBS Lett.* **158,** 12 (1983).

Applications to Peptide Hormone–Receptor Systems

Photoaffinity labeling of peptide hormone receptors requires a suitable biological target: (1) intact tissues or isolated cells, (2) isolated membranes, or (3) isolated receptors. Each of these biological systems has its specific advantages and limitations. Detailed information concerning each system can be found in the literature cited in Table III[82–123] which summa-

[82] J. Ramachandran, J. Hagman, and K. Muramoto, *J. Biol. Chem.* **256,** 11424 (1981).
[83] J. Ramachandran, K. Muramoto, M. Kerez-Keri, G. Keri, and D. I. Buckley, *Proc. Natl. Acad. Sci. U.S.A.* **77,** 3967 (1980).
[84] Y. C. Kwok and G. J. Moore, *Mol. Pharmacol.* **18,** 210 (1980).
[85] A. M. Capponi and K. J. Catt, *J. Biol. Chem.* **255,** 12081 (1980).
[86] J. M. Moseley, D. M. Findlay, T. J. Martin, and J. J. Gorman, *J. Biol. Chem.* **257,** 5846 (1982).
[87] R. E. Galardy, B. E. Hull, and J. D. Jamieson, *J. Biol. Chem.* **255,** 3148 (1980).
[88] M. Smolarsky and D. E. Koshland, Jr., *J. Biol. Chem.* **255,** 7244 (1980).
[89] T. T. Lee, R. E. Williams, and C. F. Fox, *J. Biol. Chem.* **254,** 11787 (1979).
[90] U. Nagai, Y. Kudo, K. Sato, N. Taki, and S. Shibata, *Biomed. Res.* **2,** 229 (1981).
[91] R. A. Hock, E. Nexø, and M. D. Hollenberg, *Nature (London)* **277,** 403 (1979).
[92] C. Demoliou-Mason and R. M. Epand, *Biochemistry* **21,** 1989 and 1996 (1982).
[93] G. L. Johnson, V. I. MacAndrew, Jr., and P. F. Pilch, *Proc. Natl. Acad. Sci. U.S.A.* **78,** 875 (1981).
[94] M. D. Bregman and D. Levy, *Biochem. Biophys. Res. Commun.* **78,** 584 (1977).
[95] E. Hazum, *Endocrinology* **109,** 1281 (1981).
[96] E. Hazum, *FEBS Lett.* **128,** 111 (1981).
[97] E. Hazum and A. Nimod, *Proc. Natl. Acad. Sci. U.S.A.* **79,** 1747 (1982).
[98] E. Hazum and D. Keinan, *Biochem. Biophys. Res. Commun.* **107,** 695 (1982).
[98a] F. Fahrenholz, G. Tóth, P. Crause, P. Eggena, and I. L. Schwartz, *J. Biol. Chem.* **258,** 14861 (1983).
[99] I. Ji, B. Y. Yoo, C. Kaltenbach, and T. H. Ji, *J. Biol. Chem.* **256,** 10853 (1981).
[100] I. Ji and T H. Ji, *Proc. Natl. Acad. Sci. U.S.A.* **77,** 7167 (1980).
[101] I. Ji and T H. Ji, *Proc. Natl. Acad. Sci. U.S.A.* **78,** 5465 (1981).
[102] C. Hofmann, T. H. Ji, B. Miller, and D. F. Steiner, *J. Supramol. Struct. Cell. Biochem.* **15,** 1 (1981).
[103] C. C. Yip, C. W. T. Yeung, and M. L. Moule, *J. Biol. Chem.* **253,** 1743 (1978).
[104] S. Jacobs, E. Hazum, Y. Shechter, and P. Cuatrecasas, *Proc. Natl. Acad. Sci. U.S.A.* **76,** 4918 (1979).
[105] M. H. Wisher, M. D. Baron, R. H. Jones, and P. H. Sönksen, *Biochem. Biophys. Res. Commun.* **92,** 492 (1980).
[106] C. C. Yip, M. L. Moule, and C. W. T. Yeung, *Biochemistry* **21,** 2940 (1982).
[107] C. W. T. Yeung, M. L. Moule, and C. C. Yip, *Biochemistry* **19,** 2196 (1980).
[108] D. Brandenburg, C. Diaconescu, D. Saunders, and P. Thamm, *Nature (London)* **286,** 821 (1980).
[109] P. Berhanu, J. M. Olefsky, P. Tsai, P. Thamm, D. Saunders, and D. Brandenburg, *Proc. Natl. Acad. Sci. U.S.A.* **79,** 4069 (1982).
[110] L. Kuehn, H. Meyer, M. Rutschmann, and P. Thamm, *FEBS Lett.* **113,** 189 (1980).
[111] P. N. E. de Graan and A. N. Eberle, *FEBS Lett.* **116,** 111 (1980).
[112] P. N. E. de Graan, A. N. Eberle, and F. C. G. van de Veerdonk, *FEBS Lett.* **129,** 113 (1981).

rizes photoaffinity labeling experiments with a variety of peptide hormones. In addition, model receptors, such as albumin or γ-globulin, are frequently used for exploiting the potential of new photoreactive peptides; the ensuing information is however limited, and therefore only few examples of model receptors are included in Table III.

Intact isolated cells that retain their ability to respond to the hormonal stimulus have the broadest range of application in photoaffinity labeling. Data on binding and extent of labeling can be correlated with a cellular response. Furthermore, in some systems covalent hormone–receptor complexes produce a longlasting biological response or an irreversible inhibition (see Table III). Such permanently activated receptors provide unique information on the type of hormone–receptor interaction (agonist- or antagonist-like), on receptor turnover, and on factors affecting signal transduction through the membrane.[23,68,111–113] Moreover, intact cells are suitable for light and electron microscopic localization of receptors with fluorescent, radiolabeled, or electron-dense hormone derivatives, and for following the "fate" of the receptor (processing, clustering, internalization, degradation). The disadvantage of isolated cells in photoaffinity labeling is their UV sensitivity and their relatively high UV absorbance which limits the amount of material that can be irradiated at one time.

As membrane fractions can be irradiated in greater quantities, at shorter wavelengths and at higher intensities than cells, they are generally preferred as the biological target for photolabeling when receptors are to

[113] P. N. E. de Graan, A. N. Eberle, and F. C. G. van de Veerdonk, *Mol. Cell. Endocrinol.* **26,** 327 (1982).

[114] J. Messague, B. J. Guillette, M. P. Czeck, C. J. Morgan, and R. A. Bradshaw, *J. Biol. Chem.* **256,** 9419 (1981).

[115] Y. S. Klausner, W. M. McCormick, and I. M. Chaiken, *Int. J. Pept. Protein Res.* **11,** 82 (1978).

[116] D. M. Abercrombie, W. M. McCormick, and I. M. Chaiken, *J. Biol. Chem.* **257,** 2274 (1982).

[117] J. M. Stadel, D. B. P. Goodman, R. E. Galardy, and H. Rasmussen, *Biochemistry* **17,** 1403 (1978).

[118] M. D. Coltrera, J. T. Potts, and M. Rosenblatt, *J. Biol. Chem.* **256,** 10555 (1981); S. R. Goldring, G. A. Tyler, S. M. Krane, J. T. Potts, and M. Rosenblatt, *Biochemistry* **23,** 498 (1984).

[119] M. W. Draper, R. A. Nissenson, J. Winer, J. Ramachandran, and C. D. Arnaud, *J. Biol. Chem.* **257,** 3714 (1982).

[120] D. W. Borst and M. Sayase, *Biochem. Biophys. Res. Commun.* **105,** 194 (1982).

[121] B. Bhanmick, R. M. Bala, and M. D. Hollenberg, *Proc. Natl. Acad. Sci. U.S.A.* **78,** 4279 (1981).

[122] P. R. Buckland, C. R. Richards, R. D. Howells, E. D. Jones, and B. Rees Smith, *FEBS Lett.* **145,** 245 (1982).

[123] K. Nikolics, I. Teplan, and J. Ramachandran, *Int. J. Pept. Protein Res.* **24** (1984).

TABLE III

PHOTOAFFINITY LABELING OF PEPTIDE HORMONE RECEPTORS ON INTACT CELLS AND ISOLATED MEMBRANES

Photolabel/site of attachment/ radiolabel	Target	Result of photolabeling	Comments	Reference[a]
Adrenocorticotropin				
30a/Trp[9] ^{3}H	Rat adrencortical cells (freshly isolated and in tissue culture)	Persistent activation of corticosterone and cAMP production; covalent binding to a 100 kD protein	2,5-NAPS-ACTH is a partial agonist of ACTH; analysis by SDS–PAGE	82, 83
and **30**/Trp[9]				61
30/2-thio-Trp[9]	Model studies in solution and with BSA	Dimer/polymer-ACTH formation; covalent attachment to BSA	Peptide ligand is cleaved from BSA by 2% mercaptoethanol	67
Angiotensin II				
1 or **2**/pos. 4 or pos. 8	Rabbit aorta strips, rat aorta strips, rat stomach strips	Irreversible inhibition of angiotensin II induced smooth muscle contraction	Structure-activity studies; flash and conventional UV irradiation	21
5 or **17**/ N-terminal	Rat uterine smooth muscle	Irreversible blockage of the angiotensin receptor	(Nle[8])-angiotensin(2–8); evidence for spare receptors	84
10/N-terminal ^{125}I	Rat/dog adrenal cortex and uterus membrane fraction	Covalent binding to proteins of about 126 kD and 64 kD	Photolabel introduced after iodination; analysis by gel filtration, density centrifugation, SDS–PAGE	85
Bradykinin				
31/Tyr[5]	Rabbit aorta strips; rabbit jugular vein strips	Partial inhibition of bradykinin-induced smooth muscle contraction	Des-Arg[9]-bradykinin, for labeling of B1 receptors (aorta) and B2 receptors (jugular vein)	51
and **1** or **2**/pos. 5 or pos. 8				22
Calcitonin				
23/N-terminal ^{125}I	T47D human breast cancer cells	Covalent binding to a 85 kD protein	Transglutaminase-mediated attachment of the photolabel to calcitonin	86

(*continued*)

TABLE III (*continued*)

Photolabel/site of attachment/ radiolabel	Target	Result of photolabeling	Comments	Reference[a]
Cholecystokinin				
10/N-terminal	Dispersed guinea pig pancreatic acini	Longlasting secretory protein discharge	CCK-8 as photolabile ligand; repetitive photolysis (6×)	87
Enkephalin				
2a/pos. 4	Bovine caudate nucleus membranes	Inhibition of [^{3}H]etorphine binding to opiate receptors		88
13/C-terminal	NG-108 cell membranes	Irreversible inactivation of enkephalin receptor	(D-Ala^2,Met^5)-enkephalin, substituted with Tyr^6-NH—$(CH_2)_2$—NH photolabel	89
6/C-terminal	Guinea pig ileum	Irreversible receptor activation (inhibition of contraction response to electrical stimulation)	(Leu^5)-enkephalin, substituted with —NH—$(CH_2)_2$—NH photolabel or with –Lys(dansyl)—NH—(CH_2)—NH photolabel	90
Epidermal growth factor				
33/N-terminal ^{125}I or **23**/n.s.[b]	Mouse 3T3 cells	Covalent binding to a 190 kD membrane protein	Photolabel introduced after iodination; analysis by SDS–PAGE	62 52
16/N-terminal ^{125}I	Human placental membranes	Covalent binding to a 180 kD membrane protein	Photolabel introduced after iodination; analysis by SDS–PAGE	91
Glucagon				
28/Trp^{25} ^{125}I	Rat liver plasma membranes	Irreversible stimulation of adenylate cyclase; covalent binding to several membrane proteins	Labeling decreased by GTP	92
6	Rat liver plasma membranes	Covalent binding to a 53 kD protein	Photolabel introduced into [^{125}I]glucagon after binding to the receptor; analysis by SDS–PAGE	93

13/Lys^{12} ^{125}I	Rat hepatocyte plasma membranes	Covalent binding to a 23–25 kD protein	Inactive in adenylate cyclase stimulation	94
Gonadoliberin				
6/D-Lys^{6} ^{125}I	Rat granulosa cell membranes and rat pituitary cell membranes	Covalent binding to 60 and 48 kD proteins	(D-Lys^{6})-GnRH; analysis by SDS–PAGE; identical results with the antagonist (D-pGlu^{1},D-Phe^{2},D-Trp^{3},D-Lys^{6})–GnRH	95 96 97 98
30/D-Lys^{6}	Rat pituitary cells	Prolonged activation of LH secretion	[D-Lys(NAPS)^{6}]-GnRH(1–9)-ethylamide; desensitization of LH release mechanism by covalent attachment to receptor	123
Human choriogonadotropin				
6/n.s. ^{125}I	Porcine granulosa cells	Covalent binding to three proteins (96, 76, 73 kD)	Photolabel coupled as -Gly-Gly-*N*-hydroxysuccinimide ester	99 100
15/n.s.	Porcine granulosa cells	Covalent binding of both α- and β-subunits to membrane proteins	Photolabel coupled to α- and β-subunit; analysis by SDS–PAGE	101
Insulin				
6/n.s. ^{125}I	Cultured human hepatoma cells	Covalent binding to membrane proteins of 130, 95, and 40 kD	Analysis of solubilized membranes by SDS–PAGE	102
	Rat adipocyte membranes	Covalent binding to a 130 kD protein	Photolabel introduced after iodination	103
13/n.s. ^{125}I	Rat liver membranes; human placental membranes	Covalent binding to a 135 kD protein	Analysis by SDS–PAGE	104
7/Al N-term. **2**/pos. B1 **6**/Lys^{B29}	Liver plasma membranes of rat, mouse, guinea pig	Covalent binding to one (130 kD) or two (130 and 90 kD) proteins	Analysis by SDS–PAGE	105 106 106
6/Lys^{B29} and B1 N-terminal ^{125}I			Two photolabels simultaneously present	39

(*continued*)

TABLE III (*continued*)

Photolabel/site of attachment/ radiolabel	Target	Result of photolabeling	Comments	Reference[a]
6/B1 N-term.	Isolated rat adipocytes	Covalent binding to two or three membrane proteins	Inhibition of labeling by anti-receptor antibodies	107
6/Lys^{B29}				106
^{125}I				
9/B2 N-term.	Isolated rat adipocytes	Longlasting increase in lipogenesis	Des-Phe^{B1}-insulin; internalization of photolabeled receptor	108
^{125}I	Purified receptor from porcine liver membranes	Covalent binding to two proteins (125 and 90 kD)		109
			Nondenaturing gels: 600 and 300 kD proteins	110
α-Melanotropin				
2/pos. 13 or pos. 2	Melanophores of skin of *Xenopus* larvae or of *Anolis carolinensis*	Specific irreversible stimulation of pigment dispersion	Estimation of receptor turnover; dual role of Ca^{2+} in signal transduction; structure–activity studies; comparison of photolabels	111
				112
7 or **21**/N-terminal				113
				31, 32,
30/Trp^9				23, 40
36/N-terminal	Skin of *Anolis carolinensis*	Longlasting pigment dispersion made reversible by cleavage of ligand	Peptide ligand is split from receptor by 1% mercaptoethanol	36
Nerve growth factor				
6	Membranes from rabbit superior cervical ganglia	Covalent binding to 143 and 112 kD proteins	Photolabel introduced into [^{125}I]NGF after binding to receptor	114
Neurophysin binding peptide (Met-Tyr-Phe-amide)				
2/pos. 3	Neurophysin I and II	Specific irreversible binding	Specific attachment to Tyr^{49} of neurophysin	115
				116

Oxytocin				
10/N-terminal	Toad urinary bladder	Permanent inhibition of oxytocin-induced stimulation of urea permeability	Photolabel attached via Gly	5 117
Parathormone				
13 or **16**/n.s. ^{125}I	Canine renal cortical plasma membranes	Persistent activation of adenylate cyclase; covalent binding to a 60 (or to 60 and 70) kD protein	(Nle^{8,18},Tyr^{34})-bPTH(1–34); photolabel introduced after iodination	118
28/Trp^{23} ^{125}I				119
Prolactin				
6	Rat liver microsomal membranes	Covalent binding to a 36 kD protein	Photolabel introduced into [^{125}I]prolactin after binding to the receptor	120
Somatomedin				
16/n.s. ^{125}I	Human placental cell membranes	Covalent binding to a 140 kD protein	Photolabel introduced after iodination	121
Substance P				
1 or **2**/pos. 7 or pos. 8	Guinea pig trachea	No irreversible effects	Modification of pos. 7 reduces activity of SP	33
Thyrotropin				
6/n.s.	Affinity purified TSH receptors from porcine and human thyroid	Covalent binding to the receptor of 45/50 kD	Analysis by SDS–PAGE	122
Vasopressin				
2/pos. 3	Toad urinary bladder	Irreversible stimulation of hydroosmotic pressure	Ligand: [Asu^{1,6},Pap^{3}]-AVP; repetitive irradiation cycles[c]	98a

[a] Numbers refer to text footnotes.
[b] n.s., Nonspecific.
[c] Asu, α-aminosuberic acid.

be isolated and characterized. It is however important to prove the presence of functional receptors with a suitable bioassay, e.g., adenylate cyclase activation,[94,118,119] and to assess specific binding by using radiolabeled photoreactive peptide in the presence and absence of an excess of natural ligand.

Reports on the photoaffinity labeling of soluble peptide hormone receptors are scarce[110,122] because of the difficulty in obtaining isolated functional receptors to peptides. However, this approach could provide precise information on the characteristics of hormone–receptor interactions and prove the identity of soluble receptors with those labeled in membranes or on cells. In addition, covalently labeled functional receptors could be an interesting tool in recombination experiments.

Acknowledgment

Parts of this study were supported by the Swiss National Science Foundation.

[12] Photoaffinity Labeling of Prolactin Receptors

By Paul A. Kelly, Masao Katoh, Jean Djiane, and Senkiti Sakai

Prolactin (PRL), the pituitary hormone responsible for mammary development and lactation, has specific receptors distributed in a large number of tissues in addition to the mammary gland.[1] Prolactin receptors have been purified[2,3] and antisera prepared against the purified receptor have been shown to mimic the action of prolactin.[4,5]

Recent studies on the mechanism of prolactin action have demonstrated the existence of a low-molecular-weight soluble factor (second messenger) which transfers the hormonal information from the receptor

[1] P. A. Kelly, A. De Léan, C. Auclair, L. Kusan, G. S. Kledzik, and F. Labrie, *in* "Clinical Neuroendocrinology: A Pathophysiological Approach" (G. Tolis, F. Labrie, J. B. Martin, and F. Naftolin, eds.), p. 183. Raven Press, New York, 1962.

[2] R. P. C. Shiu and H. G. Friesen, *J. Biol. Chem.* **249,** 7902 (1974).

[3] M. T. Haeuptle, M. L. Aubert, J. Djiane, and J. P. Kraehenbuhl, *J. Biol. Chem.* **258,** 305 (1983).

[4] J. Djiane, L. M. Houdebine, and P. A. Kelly, *Proc. Natl. Acad. Sci. U.S.A.* **78,** 7445 (1981).

[5] A. A. M. Rosa, J. Djiane, L. H. Houdebine, and P. A. Kelly, *Biochem. Biophys. Res. Commun.* **106,** 243 (1982).

ISBN 0-12-182009-2

to the nuclei.[6,7] It is possible that a conformational change of prolactin receptors following prolactin binding is responsible for the generation of the intracellular prolactin relay. It is therefore important to elucidate the molecular characteristics of prolactin receptors which are, for the most part, much less well defined than those of insulin receptors. To that end, we have utilized a direct electrophoretic analysis of the M_r of purified PRL receptors and an electrophoretic analysis of photoaffinity-labeled receptors.

Receptor Preparation from Rabbit Mammary Gland and Other Tissues

New Zealand White rabbits of various lactation stages were pretreated with CB-154 (2α-bromocryptine, Sandoz) for 36 hr to increase the content of prolactin receptors.[8] Mammary glands were homogenized in 0.25 *M* sucrose and crude microsome fractions were prepared.[9] Other tissues, such as ovary, liver, kidney, and adrenal from the same rabbits and mammary glands from a lactating pig were treated in a similar manner.

Binding Assays. [^{125}I]oPRL or ^{125}I-labeled human growth hormone (hGH)[9a] was used as labeled hormone. A modified method of Hunter and Greenwood[10] using a 5 μg hormone and 500 ng chloramine-T with incubation of 4 min at room temperature was employed for iodination. Iodinated hormone was purified on a Sephadex G-75 column (0.9 $\times$ 60 cm). Specific activity, as calculated by isotope recovery, ranged from 48 to 98 μCi/μg. For the binding assay of particulate receptors, each tube, contained approximately 100,000 cpm of ^{125}I-labeled hormone. Binding was performed in the presence or absence of excess (1 μg) unlabeled oPRL. The total volume was adjusted to 0.5 ml with 25 m*M* Tris buffer, pH 7.4, 10 m*M* $MgCl_2$, 0.1% BSA (assay buffer). Tubes were incubated for 15–17 hr at room temperature and incubation was terminated by adding 6 volumes of cold assay buffer followed by centrifugation at 2300 *g* for 15 min. Tubes were counted in a LKB gamma spectrometer with a counting efficiency of 50%.

[6] B. Teyssot, L. M. Houdebine, and J. Djiane, *Proc. Natl. Acad. Sci. U.S.A.* **78,** 6729 (1981).

[7] B. Teyssot, J. Djiane, P. A. Kelly, and L. M. Houdebine, *Biol. Cell.* **43,** 81 (1982).

[8] J. Djiane, P. Durand, and P. A. Kelly, *Endocrinology* **100,** 1348 (1977).

[9] P. A. Kelly, G. Leblanc, and J. Djiane, *Endocrinology* **104,** 1631 (1979).

[9a] Abbreviations used: CHAPS, 3-[3-cholaminodopropyl)dimethylammonio]-1-propane sulfonate; hGH, human growth hormone; oPRL, ovine prolactin; DTT, dithiothreitol; HSAB, *N*-hydroxysuccinimidyl-4-azidobenzoate; DSS, disuccinimidyl suberate; SDS, sodium dodecyl sulfate; BSA, bovine serum albumin; M_r, molecular weight, determined by electrophoretic analysis.

[10] W. M. Hunter and F. C. Greenwood, *Nature (London)* **194,** 495 (1962).

For the soluble receptors, the tubes were incubated in the same manner as for the particulate receptors except that the final CHAPS concentration was maintained between 1 and 5 m*M*. Hormone–receptor complexes were precipitated by adding 0.5 ml of cold 0.1% bovine γ-globulin in 0.1 *M* phosphate buffer, pH 7.4, and 1 ml of 32% (w/v) solution of polyethylene glycol (PEG-8000). Tubes were vortexed and centrifuged at 2300 *g* for 15 min. Specific binding was calculated as the difference between the counts per minute bound in the absence and presence of excess unlabeled oPRL.

In order to follow protein concentration throughout the various purification steps, the method of Lowry[11] was routinely used. The modified method of Bensadoun and Weinstein[12] was used for the partially purified fractions.

Solubilization and Purification of Prolactin Receptor from Rabbit Mammary Gland. Crude microsomes, suspended in 25 m*M* Tris–HCl buffer, pH 7.4, 10 m*M* $MgCl_2$ at a final protein concentration of 5–8 mg/ml, were solubilized with CHAPS at several concentrations. After stirring with CHAPS for 30 min, at room temperature, the suspension was centrifuged at 105,000 *g* for 60 min at 4°. The clear supernatant was removed and used at once or stored frozen at −20° to avoid the formation of precipitates. For purification, microsomes were pretreated with 1 m*M* CHAPS and the supernatants were discarded since this fraction did not contain the PRL binding activity. The resulting pellets were then suspended in the original buffer and solubilized with 5 m*M* CHAPS.

Purification of mammary gland prolactin receptors was carried out using affinity chromatography. Briefly, oPRL was coupled to Affi-Gel 10 according to the method of Shiu and Friesen.[2] Ten milligrams of PRL was used per 25 ml Affi-Gel 10 with a coupling efficiency of 73–98%. The CHAPS extracts, containing 1 m*M* phenylmethylsulfonylfluoride, were applied to 25 ml oPRL Affi-Gel 10 packed in a 3 × 30-cm column at a flow rate of 1 bed volume per hr at room temperature. The column was washed with 10 to 20 volumes of 0.1 *M* borate buffer, pH 7.4, containing 1 m*M* CHAPS (column buffer) followed by 1 volume of 4 *M* urea dissolved in column buffer and 4 volumes of column buffer. The receptor was eluted with 1 volume of 5 *M* $MgCl_2$ in column buffer. Ten or twenty milliliter fractions were collected from the beginning of the elution and active fractions (50–70 ml) were combined and applied to Sephadex G-100 columns (180–300 ml), previously equilibrated with column buffer, to remove the $MgCl_2$. The fractions eluted in the void volume were com-

[11] O. H. Lowry, N. Rosebrough, A. Farr, and R. Randall, *J. Biol. Chem.* **193,** 265 (1951).
[12] A. Bensadoun and D. Weinstein, *Anal. Biochem.* **70,** 241 (1975).

bined, concentrated 20 to 30 times by lyophilization after dialysis against 0.01 *M* borate buffer, pH 7.4, 0.1 m*M* CHAPS and stored at −20°. Repeated lyophilization of partially purified receptor resulted in a loss of approximately 6% of binding activity after each lyophilization.

Generally, Triton X-100 has been used for solubilization of PRL receptor.[2,3] Although this detergent has the merit that it can solubilize receptors efficiently and in a nondenaturing form, aggregation of the PRL molecule is induced[2] resulting in a decrease of affinity for its receptor. Therefore, initially CHAPS, a newly developed zwitterionic detergent, was tested in this study. This detergent has already been reported not to induce aggregation of the oPRL molecule[13] and to be effective in solubilizing membrane receptors[14] or protein[15] in a nondenaturing state.

Crude microsomes of rabbit mammary tissue were solubilized with CHAPS for 30 min and compared to the solubilization with Triton X-100. No significant change was observed with the solubilization efficiency of CHAPS by changing the duration of solubilization from 15 to 2 hr (data not shown). Solubilization efficiency plateaued at 5–7.5 m*M* and with approximately 32–37% protein and 28–33% PRL binding sites being solubilized by 5–20 m*M* CHAPS. These values were much lower than those obtained for Triton X-100 at the concentration of 1% (v/v), which solubilized 45% protein and 64% binding sites from microsomes. After solubilization, affinity constants, which were calculated by Scatchard analysis[16] were increased 8 times for 5 m*M* CHAPS and 4 times for 1% Triton X-100 (data not shown).

CHAPS, at the concentration between 0.8 and 6 m*M* in the incubation mixture with [^{125}I]oPRL did not affect the binding activity of PRL receptor, suggesting that PRL receptors from mammary tissue do not require an optimum concentration of CHAPS as was observed for prolactin receptors from mouse liver[13] and opiate receptors[15] which were solubilized with the same detergent. The optimum concentration of polyethylene glycol to precipitate hormone receptor complexes was 16% (w/v), slightly higher than that for Triton X-100 solubilized mammary PRL receptor.[2]

In particulate fractions, lactogenic receptors have the same affinity for hGH and oPRL.[17,18] A similar observation was made for the CHAPS solubilized PRL receptor, in the presence of 1 or 5 m*M* CHAPS. Similar

[13] D. S. Liscia, T. Alhadi, and B. K. Vonderhaar, *J. Biol. Chem.* **257,** 9401 (1982).

[14] L. M. Hjelmeland, *Proc. Natl. Acad. Sci. U.S.A.* **77,** 6368 (1980).

[15] W. F. Simonds, G. Koski, R. A. Streaty, L. M. Hjelmeland, and W. A. Klee, *Proc. Natl. Acad. Sci. U.S.A.* **77,** 4623 (1980).

[16] G. Scatchard, *Ann. N.Y. Acad. Sci.* **51,** 660 (1949).

[17] R. P. C. Shiu, P. A. Kelly, and H. G. Friesen, *Science* **180,** 968 (1973).

[18] B. I. Posner, P. A. Kelly, R. P. C. Shiu, and H. G. Friesen, *Endocrinology* **95,** 521 (1973).

inhibition curves of the binding of [^{125}I]oPRL to the soluble receptors were obtained for both oPRL (30.5 IU/mg) and hGH (2.2 IU/mg). Affinity constants calculated from Scatchard plots (not shown) were 11.6×10^{-9} M^{-1} (1 mM CHAPS), 12.2×10^{-9} M^{-1} (5 mM) for hGH, and 16.0×10^{-9} M^{-1} (1 mM) and 14.7×10^{-9} M^{-1} (5 mM) for oPRL, respectively. Complete inhibition was achieved by as little as 10^{-8} M (approximately 200 ng/ml) of unlabeled hormone.

Since it has become clear that oPRL has a similar binding affinity with hGH to CHAPS solubilized receptors in the presence of up to 5 mM CHAPS as shown above, this hormone can be used as a ligand for affinity chromatography. PRL binding activity was not eluted in the 4 M urea fraction but rather in the 5 M $MgCl_2$ fraction from oPRL Affi-Gel 10 columns. Active $MgCl_2$ fractions were collected, separated from $MgCl_2$ on a Sephadex G-100 column, and concentrated by lyophilization followed by determination of protein concentration and PRL binding capacity. Table I shows the summary of purification of PRL receptor from 3 different preparations of mammary tissue. Only 0.014% protein was recovered in the $MgCl_2$ fraction and an average 650-fold purification from microsome was achieved, resulting in a recovery of total binding capacity at 8.9%. These values are comparable to those we previously obtained using Triton X-100 as a detergent and hGH as a ligand for affinity chromatography (0.11% protein recovery, 836-fold purification and 9% recovery of binding sites).

Electrophoresis Analysis of Partially Purified Receptor. Partially purified receptor was dialyzed against 2 ml Tris–HCl, pH 7.4, 0.1% SDS, and lyophilized before electrophoresis. The powders were dissolved in electrophoresis sample buffer and boiled for 2 min in the absence or presence of 10 mM DTT. Electrophoresis was performed on a 5–15 or 9–15% gradient gel slab of 1.5 mm thick according to the method of Laemli.[19] Gels were stained by 0.2% Coomassie Brilliant Blue R-250 or by silver as described by Wray *et al.*[20] Affinity labeled or radiolabeled receptors were electrophoresed similarly.

Three techniques were employed to detect the proteins separated by SDS–polyacrylamide electrophoresis, that is, Coomassie Brilliant Blue staining, silver staining, and autoradiography of the gel containing radioiodinated purified receptor on Kodak X-Ray film XAR-5 for 2–4 days at −70°. One faint band at $M_r = 32{,}000$ was occasionally detected by Coomassie Brilliant Blue staining. However, silver staining, which is much more sensitive, detected at least 7 major bands of $M_r = 30{,}000$,

[19] U. K. Laemli, *Nature (London)* **227,** 680 (1970).

[20] W. Wray, T. Boulikas, V. P. Wray, and R. Hancock, *Anal. Biochem.* **118,** 197 (1981).

TABLE I
SUMMARY OF PURIFICATION OF PRL RECEPTOR FROM RABBIT MAMMARY GLAND[a]

	Protein recovery (%)	PRL binding capacity (pmol/mg)	Purification (fold)	Binding sites recovered (%)	K_a (nM^{-1})
Crude microsome	100	0.37 ± 0.15	(1)	100	3.8 ± 1.7
CHAPS extract	22.6 ± 3.4	0.72 ± 0.33	(1.87)	43.0 ± 9.5	27.6 ± 9.5
Affinity purified	0.014 ± 0.002	258 ± 126	(655)	8.9 ± 1.7	16.2 ± 7.2

[a] Crude microsomes from 3 different preparations of lactating rabbit mammary gland (680, 1600, and 3300 mg protein, respectively) were solubilized and purified. PRL binding capacity and affinity constants were determined by Scatchard analysis. The values represent the mean ± SEM.

32,000, 39,000, 46,000, 51,000, 64,000, and 68,000. In some preparations, a larger M_r = 114,000 band was observed. This indicates that further purification is necessary to obtain a more homogeneous preparation.

As shown in Fig. 1, when this affinity-purified material was iodinated and analyzed by SDS–polyacrylamide gel electrophoresis and autoradiography, a M_r = 32,000 band was the most intense. In addition, faint M_r = 25,000, 40,000, and 53,000 bands could be observed under reducing condition. The largest M_r = 53,000 band migrated somewhat faster under nonreducing condition.

Affinity Labeling of Microsomal PRL Receptor. The effect of increasing the concentration of photoactive cross-linker, HSAB is shown in Fig. 2. As can be seen, a single band at M_r = 32,000 is seen. The only effect of increasing the HSAB concentration is to increase nonspecific background. The molecular weight of the binding component is calculated as the difference in the molecular weight of this prolactin–receptor complex and that of prolactin. The nonphotoactive agent, DSS, also resulted in good affinity labeling of the receptor although the bands were less intense than with HSAB.

To elucidate the significance of disulfide linkages among the components of PRL receptor, the effect of increasing concentration of DTT, which was used to reduce S—S bonds were added when sample was boiled with the electrophoresis sample buffer (Fig. 3). In the absence of DTT, the oPRL, which was freed from complexes during solubilization and electrophoresis, appeared at M_r = 23,000 and hormone–receptor complexes appeared at M_r = 57,000. No other binding components were observed. When more than 2 mM DTT was present, oPRL molecule

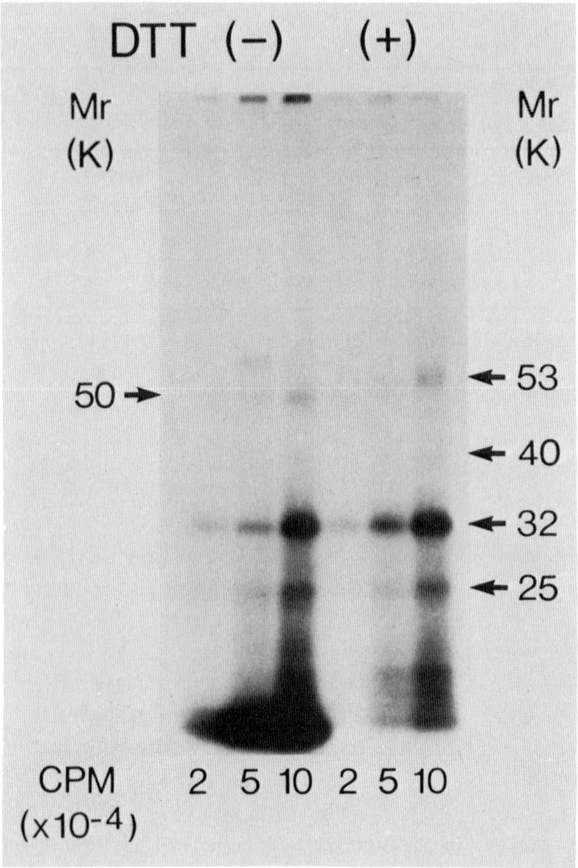

FIG. 1. SDS–polyacrylamide gel electrophoresis of ^{125}I-labeled partially purified receptor. Affinity purified receptor was iodinated and increasing concentrations of ^{125}I-labeled receptor were applied to the gel. Samples were pretreated with electrophoresis sample buffer in the presence or absence of 10 m*M* DTT at 100° for 2 min. After electrophoresis, the gel was dried and an autoradiograph was taken.

migrated more slowly indicating its M_r at 26,000. Hormone–receptor complex also migrated at a higher M_r at 58,000.

The average M_r of oPRL and the PRL binding component, observed in several experiments, is summarized in Table II. Similar to the results described in Fig. 3, the oPRL molecule migrated somewhat more slowly after reduction by 10 m*M* DTT. As a result, larger M_r values were obtained for the reduced hormone-binding component complex. Accordingly, the M_r of PRL binding component was calculated at 31,300 and 32,900 under reducing and nonreducing conditions, respectively.

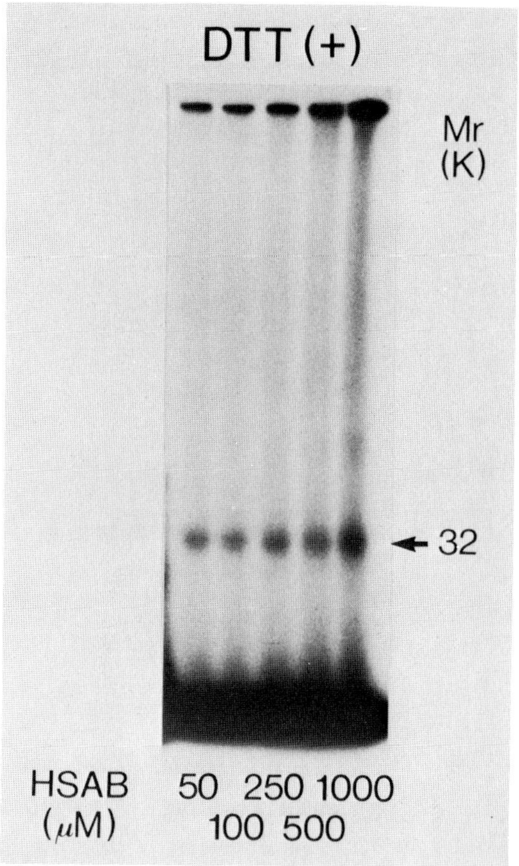

FIG. 2. Autoradiogram of the effect of increasing concentrations of HSAB on the affinity labeling of rabbit mammary gland microsomal proteins with [^{125}I]oPRL. Microsomes were photoaffinity labeled and electrophoresed on a 7.5% gel.

TABLE II
MOLECULAR WEIGHT OF THE MICROSOMAL PRL BINDING COMPONENT FROM RABBIT MAMMARY GLAND DETERMINED BY AFFINITY LABELING

	DTT (+) (n = 9)	DTT (−) (n = 6)
Binding site–PRL complex	58,000 ± 300[a]	56,000 ± 700
oPRL	27,200 ± 200	23,600 ± 200
Binding site	31,300 ± 400	32,900 ± 700

[a] Mean ± SEM.

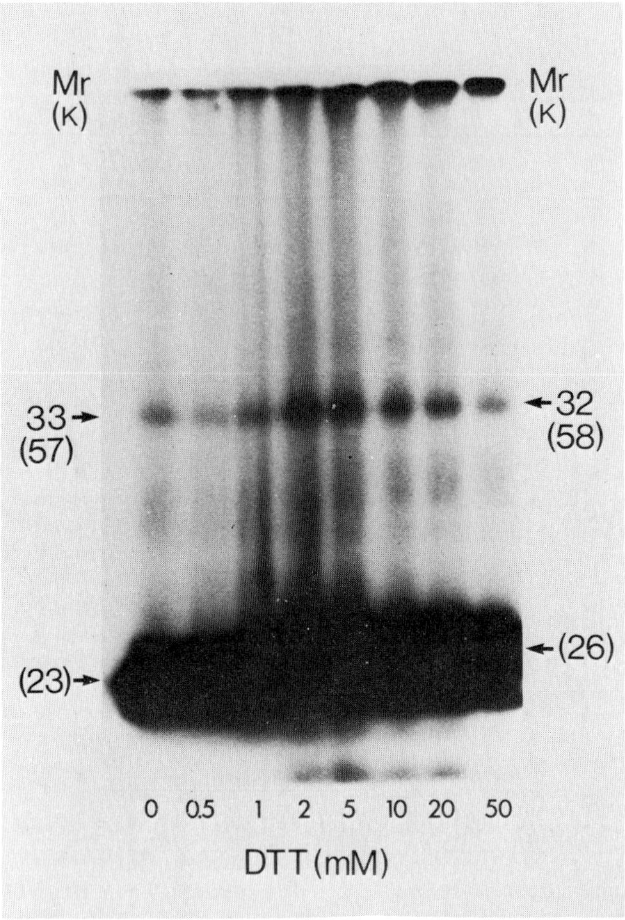

FIG. 3. Autoradiogram of the effect of increasing DTT concentrations on the migration of affinity labeled PRL binding components with [^{125}I]oPRL on a 9–15% gradient SDS-electrophoresis gel.

Microsomes from liver, kidney, adrenal, and ovary of lactating rabbits, which have been shown to contain PRL receptors,[18,21] as well as mammary gland of a lactating pig were also photoaffinity labeled (Fig. 4). Similar M_r of the major binding component to that of rabbit mammary gland (M_r = 31,000) was observed for all tissues examined with the minor exceptions that liver revealed 2 weak bands (M_r = 29,000 and 40,000) and that pig mammary gland showed a minor M_r = 65,000 binding component.

[21] R. P. C. Shiu and H. G. Friesen, *Biochem. J.* **140,** 301 (1974).

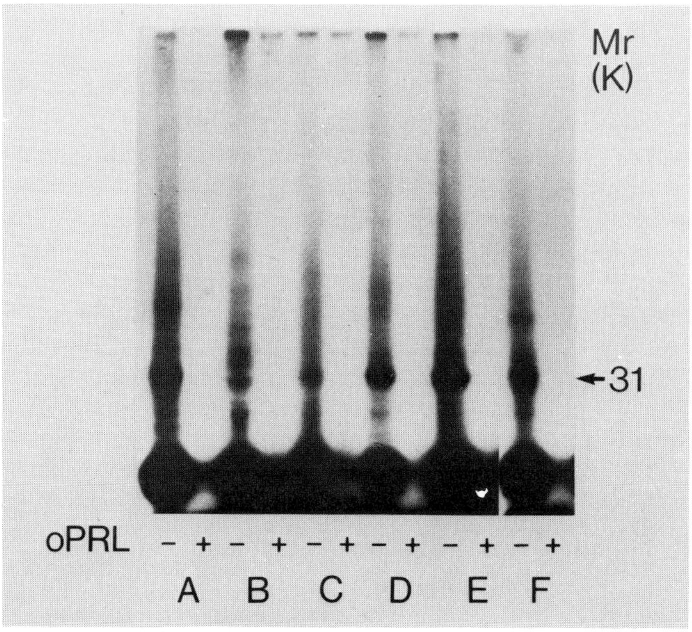

FIG. 4. Autoradiogram of the M_r of PRL binding components in various rabbit tissues and mammary gland from a lactating pig. Microsomal proteins from rabbit mammary gland (A), liver (B), kidney (C), ovary (D), adrenal (E), and mammary gland from a lactating pig (F) were affinity labeled with [^{125}I]oPRL using 500 μM HSAB, electrophoresed and autoradiographed.

Female rat liver also has been shown to contain high concentrations of PRL receptor. Figure 5 depicts the binding component from rat liver microsome (lane B) which M_r was 35,000–36,000, a little larger than that of rabbit tissues (lane A). PRL binding component of purified plasma membrane (lane C, Fig. 5) and Golgi fraction (autoradiogram not shown) also showed the same M_r.

Photoaffinity Labeling of Affinity-Purified PRL Receptor. Partially purified PRL receptors were also affinity labeled. After the incubation with [^{125}I]oPRL as described above, hormone–receptor complexes were precipitated with PEG-8000 at a final concentration of 16% and the tubes centrifuged. Pellets were labeled using HSAB as described above. After photolysis, 4 volumes of cold (−20°) ethanol was added resulting in a final ethanol concentration of 80% and the tubes centrifuged at 2300 *g* for 15 min. Pellets were treated in a similar manner as described for particulate receptors.

One major M_r = 33,000 binding component can be seen in Fig. 6, whereas 2 minor components, with a M_r of 63,000 and 80,000 assuming

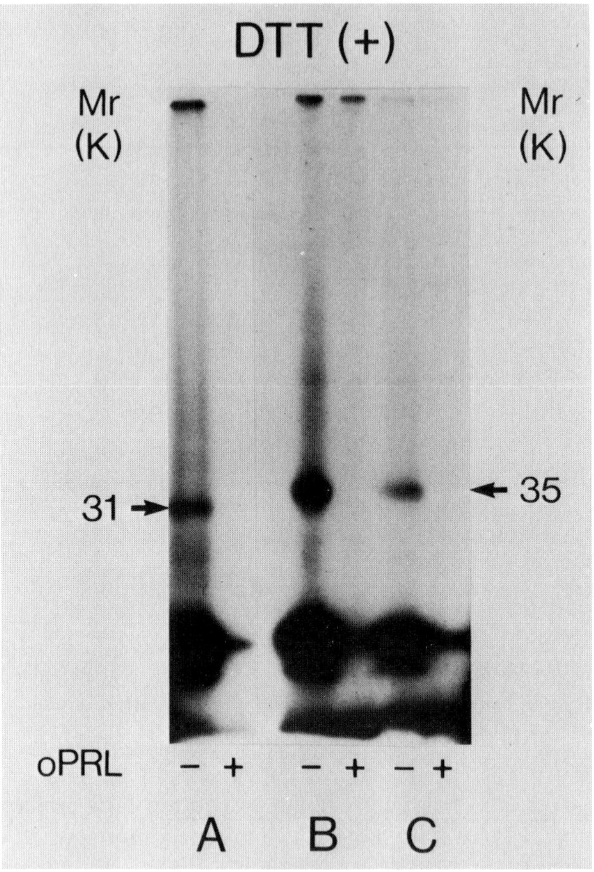

FIG. 5. Comparison of M_r of PRL binding components from rabbit mammary gland with that from rat liver. Microsomes from rabbit mammary gland (A) and rat liver (B) and purified plasma membranes from rat liver (C) were prepared. Each receptor preparation was affinity labeled using 500 μM HSAB and analyzed on a 5–15% gradient gel.

1 : 1 cross-linking between hormone and binding component, are also visible under reducing conditions. A single M_r = 65,000 was observed under nonreducing conditions.

Relevance of These Observations. The present data indicate that the major binding component in the membrane of rabbit mammary tissue has a relative molecular mass of 31,000–32,000 when affinity labeling techniques combined with SDS–polyacrylamide gel electrophoresis were employed. We have mainly used HSAB, a photoreactive reagent as a cross-linker. This reagent was more effective than DSS, an amino group specific

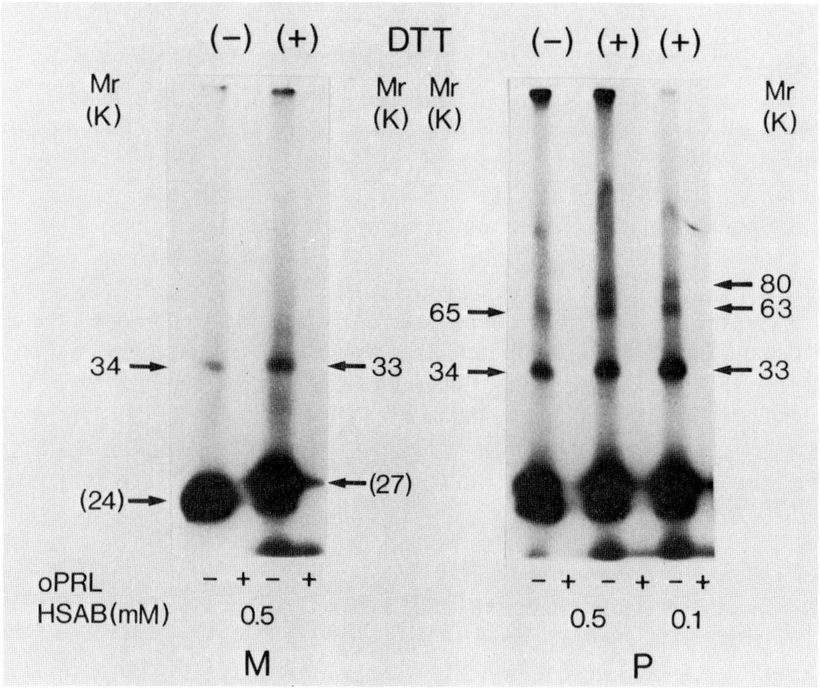

FIG. 6. Autoradiogram comparing binding components from microsomal (M) and affinity purified (P) PRL receptors. Affinity purified receptor (1.2 μg per incubation), characterized in Table I, was photoaffinity labeled with [^{125}I]oPRL using 100 or 500 μ*M* HSAB and analyzed on a 5–15% gradient gel under reducing and nonreducing conditions. Microsomal receptors were affinity labeled and electrophoresed on the same gel and indicated in the left portion of the figure (M).

homo-bifunctional cross-linker, for affinity labeling of rabbit microsomal PRL receptors with [^{125}I]oPRL and was previously employed for direct cross-linking of hormone–receptor complexes without producing photoreactive derivatives of [^{125}I]oPRL previously.[22,23]

As shown in Table II, using electrophoresis analysis under reducing conditions, we obtained from rabbit mammary gland microsomes a single M_r = 58,500 binding component–[^{125}I]oPRL complex, a similar value to that reported previously for rabbit mammary gland solubilized microsomes bound to [^{125}I]hGH (M_r = 57,000) and for rat liver microsomes

[22] G. L. Johnson, V. I. MacAndrew, and P. F. Pilch, *Proc. Natl. Acad. Sci. U.S.A.* **78,** 875 (1981).

[23] J. Massagué, B. J. Guillette, M. P. Czech, C. J. Morgan, and R. A. Bradshaw, *J. Biol. Chem.* **256,** 9419 (1981).

bound to [^{125}I]oPRL (M_r = 60,000). A similar binding component–[^{125}I]oPRL complex was observed for affinity-purified receptor as a major band (Fig. 6). This band did not disappear nor change its migration rate even when an excess amount of HSAB was used or when HSAB was replaced by other cross-linkers. The band disappeared only when excess unlabeled oPRL was incubated with receptors. However, under the conditions employed by us, [^{125}I]oPRL, which was freed from complexes during solubilization or electrophoresis, migrated at the position of M_r = 27,000 under reducing conditions resulting in the M_r of binding components at 31,100. Intact unlabeled oPRL also showed a slightly faster migration rate under the same conditions. The molecular weight of PRL can be calculated from its amino acid sequence[24] and sedimentation experiments[25] as 22,500 and 23,300, respectively, and a similar value was obtained by our nonreducing electrophoresis (Table II, Fig. 3). This discrepancy between the actual M_r which appeared on SDS-gels under reducing conditions and the theoretical value is considered to result from an effect of DTT on all or some of the 3 S—S bonds within the oPRL molecule. Nonreduced oPRL migrates faster as was shown for BSA,[26] or some of the binding components for insulin,[27] insulin-like growth factor II,[28] or for multiplication stimulating activity.[29] We therefore employed the apparent M_r of oPRL that was observed on a same gel.

Similar M_r binding components were also observed for other tissues from rabbits and pig mammary glands and slightly higher M_r components from rat liver. These observations suggest the similarity of PRL binding component in various tissues and in various animal species, as was confirmed using antiserum raised against affinity purified PRL receptor from rabbit mammary tissue.[29a]

This M_r = 31,000 binding component does not seem to bind to each other or to other binding components or membrane proteins through disulfide linkages, since no other binding components were observed and the M_r = 31,000 band did not disappear, nor was altered in the absence of DTT. Although the amount of proteins bound with [^{125}I]oPRL (not shown) or of radioactivity of labeled purified receptor (Fig. 5) which did not

[24] C. H. Li, J. S. Dixon, T. B. Los, K. D. Schmidt, and Y. A. Pankov, *Arch. Biochem. Biophys.* **141,** 704 (1970).

[25] P. G. Squire, B. Starman, and C. H. Li, *J. Biol. Chem.* **238,** 1389 (1963).

[26] G. Fairbanks, T. L. Steck, and D. F. H. Wallach, *Biochemistry* **10,** 2606 (1971).

[27] M. Kasuga, J. A. Hedo, K. M. Yemada, and C. R. Kahn, *J. Biol. Chem.* **257,** 10392 (1982).

[28] J. Massagué and M. P. Czech, *J. Biol. Chem.* **257,** 5038 (1982).

[29] J. Massagué, B. J. Guillette, and M. P. Czech, *J. Biol. Chem.* **256,** 2122 (1981).

[29a] M. Katoh, J. Djiane, G. Leblanc, and P. A. Kelly, *Mol. Cell. Endocrinol.* **34,** 191 (1984).

penetrate the gel was increased by reducing the concentration of DTT and this phenomenon was also mentioned by Haeuptle *et al.*,[3] the high M_r region of each gel lane was completely clear. Consequently, we do not believe that the PRL receptor exists as high molecular weight entities that are constituted from small $M_r = 31,000$ binding components through S—S bonds, as has been observed for insulin receptors[30–32] suggesting a similar structure to that of insulin-like growth factor II[28] and of multiplication stimulating activity[29] as well as PRL receptor, in rat liver[33] which have been considered to exist as a single binding component.

Contrary to the results obtained from the electrophoretic analysis of denatured and dissociated solubilized PRL receptor, larger molecular weights have been reported using gel-filtration chromatography (M_r = 220,000[2] and 200,000[7] from rabbit mammary gland and M_r = 300,000 from rabbit liver),[3] Furgason plot analysis of polyacrylamide gel electrophoresis data in the presence of Triton X-100 (M_r of hormone–receptor complex = 103,000 to 170,000 for rat liver, tadpole liver, and tailfin),[34] and sedimentation experiments (M_r of hormone–receptor complex = 99,800).[35] More recently, we have obtained M_r 133,000 peak of rabbit mammary receptors from high-performance liquid chromatography (HPLC) equipped by gel exclusion columns in the presence of Triton X-100. Only one group has observed the small M_r binding site under nondissociating condition with a M_r = 37,000 in mouse liver which was solubilized with CHAPS.[13,36] It could be possible that the higher molecular weight forms of the receptor contain aggregated forms of PRL receptors, two or more M_r = 31,000 binding components, although it is not clear whether other components, which do not contribute to the hormone binding, exist.

Another method to obtain information on molecular characteristics of PRL binding components is to analyze the purified receptor protein directly. The purification of this receptor has been attempted over the past decade. In all cases, affinity chromatography using hGH or oPRL as a ligand was utilized as a first step after solubilization[6,9] resulting in a 60- to 300-fold improvement of specific binding capacity from microsomes. Our purification attained better purity, 650-fold purification from microsomes

[30] P. F. Pilch and M. P. Czech, *J. Biol. Chem.* **255,** 1722 (1979).

[31] S. Jacobs, E. Hazum, Y. Shechter, and P. Cuatrecasas, *Proc. Natl. Acad. Sci. U.S.A.* **76,** 4918 (1979).

[32] J. Massagué, P. F. Pilch, and M. P. Czech, *Proc. Natl. Acad. Sci. U.S.A.* **77,** 7137 (1980).

[33] D. W. Borst and M. Sayare, *Biochem. Biophys. Res. Commun.* **105,** 194 (1982).

[34] F. E. Carr and R. C. Jaffe, *Endocrinology* **109,** 943 (1981).

[35] R. C. Jaffe, *Biochemistry* **21,** 2936 (1982).

[36] D. S. Liscia and B. K. Vonderhaar, *Proc. Natl. Acad. Sci. U.S.A.* **79,** 5930 (1982).

and a specific activity of 250 pmol/mg protein; however, the theoretical purity was only 0.8% assuming the molecular mass of PRL binding component as 31,000 and 1 : 1 coupling between hormones and binding components. Consequently, electrophoresis analysis of affinity-purified receptor revealed at least 7 major bands stained by the silver technique (Fig. 4). Although a M_r = 31,000 band had a strong binding activity, two other bands, M_r = 65,000 and 80,000, were also observed having faint binding activity (Fig. 6). It is yet to be determined whether these minor components are different from the major component and tend to be difficult to affinity label with [^{125}I]oPRL when they exist in membrane fragments, as was shown for the β-subunit of insulin receptor.[31,37,38]

Recently, Haeuptle *et al.*[3] purified PRL receptors from rabbit mammary gland by affinity chromatography utilizing the high affinity between biotimylated hGH–receptor complexes and insolubilized streptavidine and iodinated purified receptor showed a single M_r = 35,000 band by SDS electrophoresis, indicating a theoretical purification factor of 28,000. However, according to our data, as can be seen in Figs. 4 and 5, iodination by the chloramine-T method did not label all the proteins which were detected by silver staining. Consequently, we conclude that further purification is necessary for this affinity purified protein and that care should be taken to determine the purity of the preparation since this receptor might be a glycoprotein and is difficult to stain by some of the classical staining techniques.

Acknowledgments

The authors wish to express their gratitude to the National Hormone and Pituitary Program, National Institute of Arthritis, Metabolism and Digestive Diseases for providing human growth hormone and ovine prolactin. These studies were supported in part by grants from the National Cancer Institute (Canada) and the Medical Research Council of Canada.

[37] J. Massagué and M. P. Czech, *Diabetes* **29,** 945.
[38] C. C. Yip, C. W. T. Yeung, and M. L. Moule, *Biochemistry* **19,** 70 (1980).

[13] Photoaffinity Labeling of the Insulin Receptor

By CECIL C. YIP and CLEMENT W. T. YEUNG

Reactive aryl nitrenes generated from the photolysis of aryl azides are able to form intermolecular covalent bonds by insertion into carbon–hydrogen bonds, addition to unsaturated carbon–carbon double bonds, or

ISBN 0-12-182009-2

by the extraction of hydrogen. Photoreactive aryl azide derivatives of insulin have been used successfully to label the insulin receptor and to identify its disulfide-linked subunits.[1] When the light-sensitive derivatives are used, the various procedures described below are carried out in semidarkness or in red safety light. Where darkroom facility is not available, the chromatographic columns must be protected from light by being wrapped in aluminum foil. As well, the fractions collected have to be protected from light.

Preparation and Purification of $N^{\alpha B1}$-Monoazidobenzoyl Insulin, B1-MABI

$N^{\alpha A1},N^{\varepsilon B29}$-di(Boc)insulin (120 mg, 0.02 mmol), prepared according to the method of Geiger *et al.*,[2] and purified,[3] and *p*-azidobenzoyl *N*-hydroxysuccinimide ester (82 mg, 0.32 mmol), synthesized by the method of Galardy *et al.*,[4] are reacted for 24 hr at 24° in 5 ml of redistilled dimethylformamide (DMF) containing 5 μl of redistilled triethylamine. The crude reaction product is precipitated by the addition of 2.5 volume of ice-cold diethyl ether and recovered by centrifugation. The precipitate is washed twice with ether, once with fresh acetone, and air dried. The precipitate, dissolved in 5 ml of 0.05 *M* NH_4HCO_3, is gel filtered on a column (2.5 × 196 cm) of Sephadex G-50 in the same buffer. Fractions are collected at 8-min intervals. Protein content in each fraction is measured at 280 nm. Fractions containing the monomeric insulin derivative of M_r 6000 are pooled and lyophilized (yield: 84 mg, 70% of the starting material). The lyophilized material (70 mg) is treated with 5 ml of anhydrous trifluoroacetic acid for 1 hr at 24° to remove the protecting Boc groups. The acid is removed by rotary evaporation. The residue is dissolved in a minimal volume of 10% acetic acid, desalted on a column (2.5 × 30 cm) of Sephadex G-25 in 10% acetic acid, and lyophilized (yield: 66 mg, 93% of the monomer material used). The crude material (50 mg) is purified by ion-exchange chromatography on a column (2.5 × 40 cm) of CM-cellulose in 0.9 *M* acetic acid containing 7 *M* urea and 0.015 *M* NaCl. Fractions of about 5.3 ml each are collected at 10-min intervals and the fractions containing the product (Fig. 1) are pooled and desalted on a column (5 × 45 cm) of Sephadex G-25 in 10% acetic acid. The product, B1-MABI,

[1] C. C. Yip, M. L. Moule, and C. W. T. Yeung, *Biochemistry* **21,** 2940 (1982).
[2] R. Geiger, H. H. Schone, and W. Pfaff, *Hoppe-Seyler's Z. Physiol. Chem.* **352,** 1487 (1971).
[3] G. Krail, D. Brandenburg, and H. Zahn, *Makromol. Chem., Suppl.* **1,** 7 (1975).
[4] R. E. Galardy, L. C. Craig, J. D. Jamieson, and M. P. Printz, *J. Biol. Chem.* **249,** 3510 (1974).

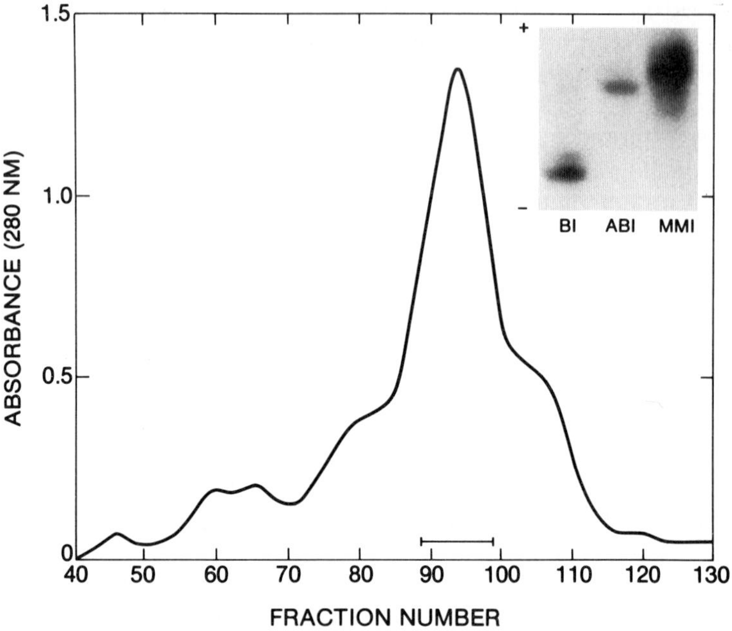

FIG. 1. Elution profile of crude B1-MABI on CM-cellulose. Fractions containing the insulin derivative as indicated by the bar were pooled to recover the product. The insert shows the purity and electrophoretic property of the derivative in polyacrylamide gel containing urea and acetic acid: BI, bovine insulin; ABI, B1-MABI; MMI, $N^{\alpha B1}$-(*N*-mesyl-L-methionyl)insulin. Reprinted with permission from Yeung *et al.*[7] Copyright 1980 American Chemical Society.

recovered by lyophilization (yield: 20 mg, 40% of the crude material used), is stored in the dark at −20°. The overall yield of the purified product is about 27% of the di(Boc)insulin. The purity of the product is assessed by polyacrylamide gel electrophoresis as described by Poole *et al.*,[5] modified to obtain 15% gel containing 8 *M* urea and 0.9 *M* acetic acid at pH 3.0.

Preparation and Purification of $N^{\varepsilon B29}$-Monoazidobenzoyl Insulin, B29-MABI

A suspension of crystalline bovine Zn-insulin (96 mg, 0.016 mmol) in 4 ml of redistilled DMF and 80 μl of redistilled triethylamine is reacted with 64 mg (0.25 mmol) of *p*-azidobenzoyl-*N*-hydroxysuccinimide ester under

[5] T. Poole, B. S. Leach, and W. W. Fish, *Anal. Biochem.* **60,** 596 (1974).

stirring for 60 min at room temperature. The reaction is stopped by the addition of 6 ml (1.5 volume) of cold absolute ethanol and 10 ml (2.5 volume) of cold diethyl ether to precipitate the product. The mixture is kept in an ice bath for 60 min and the precipitate is recovered by centrifugation and dried under nitrogen. The precipitate dissolved in 4 ml of 1.5 *M* acetic acid containing 7 *M* urea and 0.02 *M* NaCl is applied to a column (2.5 × 45 cm) of SP-Sephadex C-25 equilibrated in the same buffer. The column is developed using a gradient generated by mixing 500 ml of the starting buffer with 500 ml of this buffer containing 0.4 *M* NaCl. Fractions of 5 ml each are collected by time and measured at 280 nm for protein. The desired peak fractions (Fig. 2) eluted at a conductivity of 7–8 mmho are pooled, dialyzed against water, and then lyophilized. The lyophilized product (31 mg, 30% of the insulin used) is further purified on a smaller column (1.5 × 28 cm) of SP-Sephadex C-25 using a gradient generated by mixing 150 ml each of the two buffers. Fractions of 2 ml each are collected. The desired fractions are pooled, desalted on a column (5 × 41 cm) of Sephadex G-25 in 10% acetic acid, and recovered by lyophilization.

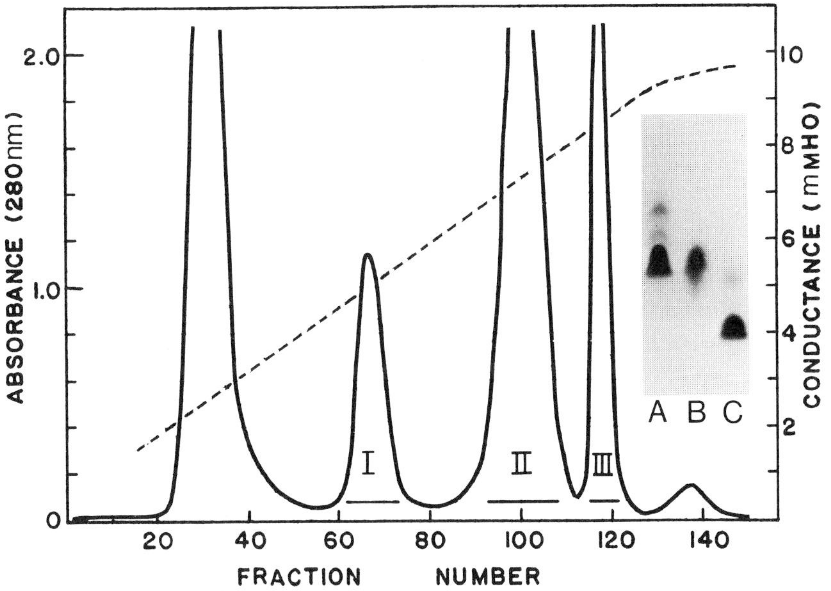

FIG. 2. Elution profile of crude (azidobenzoyl)insulins on SP-Sephadex. Pooled fraction I, II, and III contained, respectively, triazidobenzoyl insulin, diazidobenzoyl insulin, and the desired product B29-MABI. The inset shows the electrophoretic purity of the product; A, pooled fraction III of B29-MABI; B, $N^{\alpha B1}$-(*N*-mesyl-L-methionyl)insulin; C, bovine insulin. Adapted and reprinted with permission from Yip *et al.*[6] Copyright 1980 American Chemical Society.

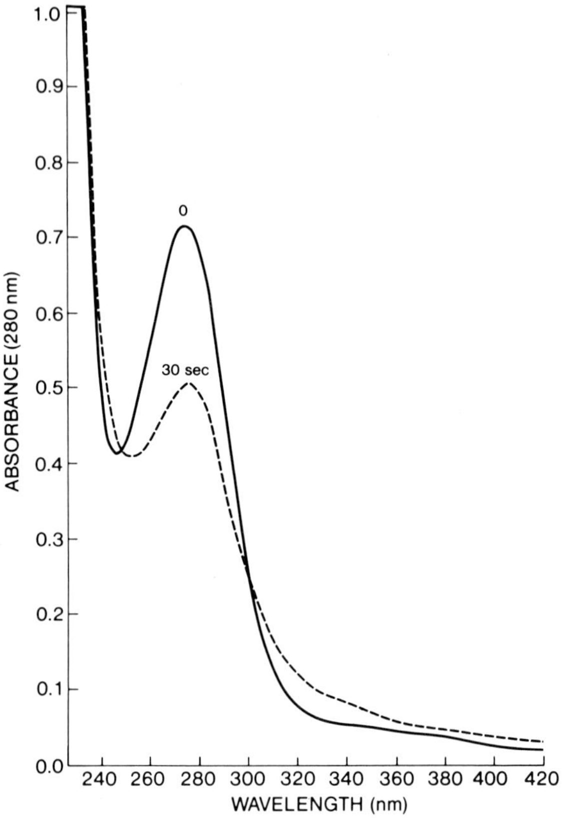

FIG. 3. Light sensitivity of B29-MABI as demonstrated by its rapid spectral changes after exposure to light. A solution of the insulin derivative (0.15 mg/ml) in 0.05 *M* ammonium bicarbonate was exposed to a focused beam of light from a 100-W high-pressure mercury lamp for 30 sec as indicated.

The final yield is about 20 mg. The purity of the derivative is determined as described above for B1-MABI.

Both insulin derivatives are subject to amino acid analysis and amino terminal determination to establish the sites of derivatization.[6,7] The light sensitivity of these insulins is demonstrated by their spectral changes upon exposure to light (Fig. 3).

[6] C. C. Yip, C. W. T. Yeung, and M. L. Moule, *Biochemistry* **19,** 70 (1980).

[7] C. W. T. Yeung, M. L. Moule, and C. C. Yip, *Biochemistry* **19,** 2196 (1980).

Purification of Azidobenzoyl Insulins by High-Performance Liquid Chromatography (HPLC)

As an alternative to ion-exchange chromatography described above, high-performance liquid chromatography provides an efficient method for the purification of smaller amounts of the photoreactive insulins. In the case of B29-MABI, we react 1 mg of bovine Zn-insulin with 1 mg of the azidobenzoyl ester in 100 μl of DMF containing triethylamine (20 μl of triethylamine in 1 ml of DMF) at room temperature for 60 min. The crude product is precipitated overnight at $-20°$ by the addition of 150 μl of cold absolute ethanol and 250 μl of cold diethyl ether. The precipitate is recovered by centrifugation and dried with nitrogen.

Reverse-phase HPLC is performed at room temperature on a column (4×250 mm) of Lichrosorb RP-8, 10 μm. We use a Waters HPLC system consisting of one Model 6000A and one Model M-45 pumps, a Model 680 Automated Gradient Controller, a Model 441 Absorbance Detector with changeable wavelength, and a Model U6K manual sample injector. The precipitate of the crude product is dissolved in 300 μl of 0.006 *N* HCl and applied to the column in one injection. As much as 1 mg can be processed in a single run. The column is developed under the following conditions using Solution A and Solution B in a linear gradient:

Solution A: water/acetonitrile: 80/20 containing 0.05 *M* triethylamine, adjusted to pH 3.0 with phosphoric acid.

Solution B: water/acetonitrile: 50/50 containing triethylamine and phosphoric acid, pH 3.0.

Conditions: Initial—75% A and 25% B
Final—40% A and 60% B
Time—60 min
Detector—280 or 214 nm
Flowrate—1.0 ml/min

The peak (Fig. 4A) containing the desired derivative, B29-MABI, is collected, neutralized immediately with 2 *N* NaOH, and dialyzed against water. The product is recovered by lyophilization. The yield is about 10–20% of the insulin used. The purity of the product is established by either acid-urea PAGE[5] or HPLC (Fig. 4B).

Photolabeling of Insulin Receptor on Isolated Adipocytes and Other Cell Types

The procedure described below for adipocytes is applicable to other isolated cell types, such as hepatocytes or lymphocytes. Rat adipocytes

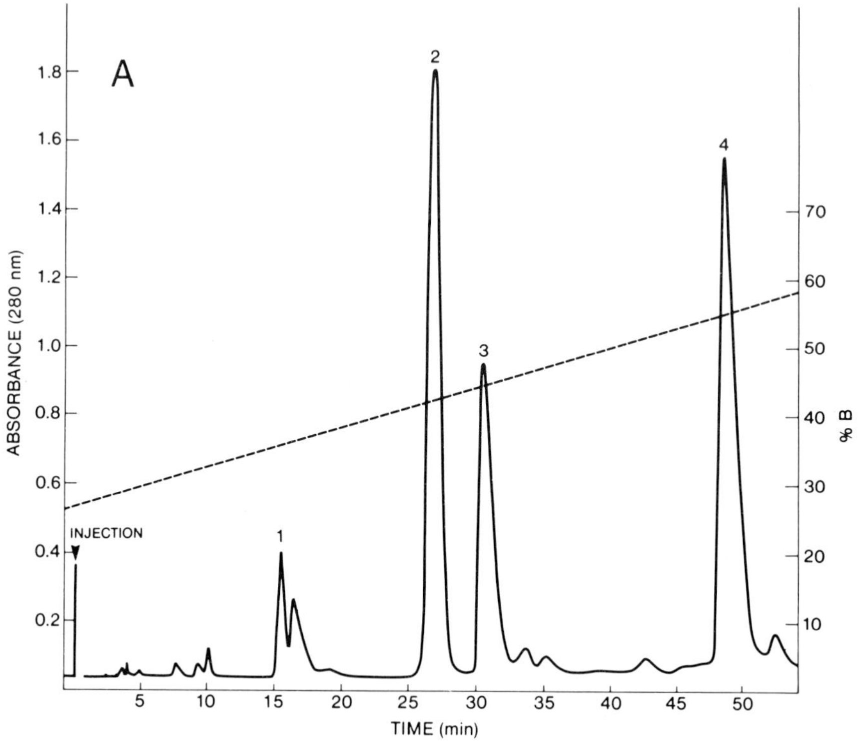

FIG. 4. (A) HPLC separation of crude reaction products of (azidobenzoyl)insulins: Peak 2, hydrolyzed ester; Peak 3, B29-MABI; Peak 4, diazidobenzoyl insulin. Peak 3 was recovered as the desired product.

are isolated from epididymal fat pads by collagenase digestion as described by Rodbell.[8] Typically, approximately 7×10^6 cells in 4 ml of Krebs–Ringer bicarbonate buffer containing 1% human serum albumin and glucose (0.1 mg/ml) are incubated at 18° for 40 min in the dark with approximately 10 n*M* of either B1-MABI or B29-MABI which has been iodinated with [^{125}I]iodine.[6] Proteolytic inhibitors such as bacitracin (0.8 mg/ml), benzamidine (1 m*M*), chloroquine (0.25 m*M*), *N*-ethylmaleimide (1 m*M*), phenylmethanesulfonic acid (1 m*M*), *p*-chloromercuriphenylsulfonic acid (0.1 m*M*), and Trasylol (1000 units/ml) can be added to the medium if required. The incubation is carried out in 50-ml round-bottom polycarbonate centrifuge tubes and the cells are kept in suspension by gentle stirring with a small magnetic stirring bar. After incubation the cell suspension is exposed for 30 sec to a focused vertical light beam of a 100-

[8] M. Rodbell, *J. Biol. Chem.* **239,** 375 (1964).

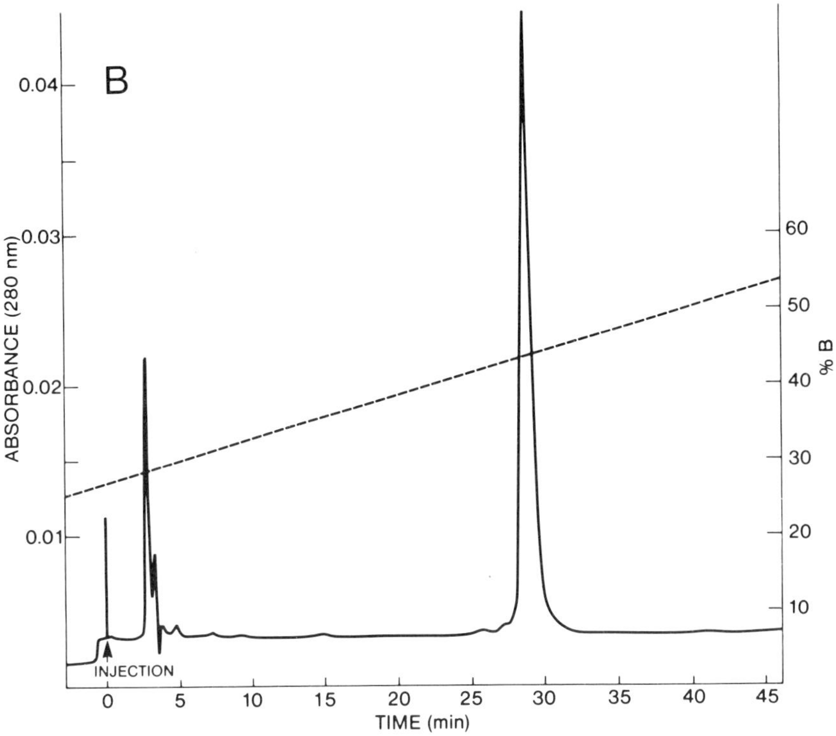

FIG. 4B. See legend on p. 176.

W high-pressure mercury lamp (PRA Photochemical Research Associate Inc., London, Ontario, Canada). The light beam is cooled by being passed through a reservoir (the reservoir is a 15 cm length of Lucite tube of 5 cm diameter sealed at both ends with plain quartz lens) of circulating cold tap water and is aimed directly down the polycarbonate tube at the cell suspension. An ultraviolet transmission filter (BG-3, Leitz) is used when viability of the cells is required for further experimental manipulation. The cells are maintained in suspension during this brief period of photolysis. After photolysis the adipocytes are washed twice with the Krebs–Ringer bicarbonate albumin buffer and twice with sucrose–Tris buffer (0.25 *M* sucrose in 5 m*M* Tris, pH 7.5). Crude plasma membrane fraction is prepared from the adipocytes by the method of McKeel and Jarett.[9] In the case of other cell types, the washed cells are pelleted down by centrifugation and solubilized directly in the sample buffer used for electrophoresis.

[9] D. W. McKeel and L. Jarett, *J. Cell Biol.* **44,** 417 (1970).

Polyacrylamide Gel Electrophoresis and Autoradiography

The membrane preparation or the cell pellet is solubilized by boiling for 10 min in 100–200 μl of 2% sodium dodecyl sulfate (SDS) in 125 mM Tris–HCl, pH 6.8. If reduction of the membrane proteins is required, dithiothreitol (0.1 M) is added to this solubilization buffer. The solubilized sample, 50–100 μl, is analyzed by slab gradient SDS–polyacrylamide gel electrophoresis. The slab gel is prepared by mixing an equal volume of a 3 and a 10% acrylamide solution in a gradient maker. These two acrylamide solutions are prepared by dilution of a water solution of 30% acrylamide–0.8% bis(acrylamide) with buffer to contain the appropriate concentration of acrylamide, 0.1% SDS, 375 mM Tris–HCl, pH 8.8, ammonium persulfate (1.25 mg/10 ml), and N,N,N',N'-tetramethylethylenediamine (10 μl/30 ml). The 3 and 10% acrylamide solutions contain also 6.25 and 16.25% sucrose, respectively. A stacking gel layer of 3% acrylamide in

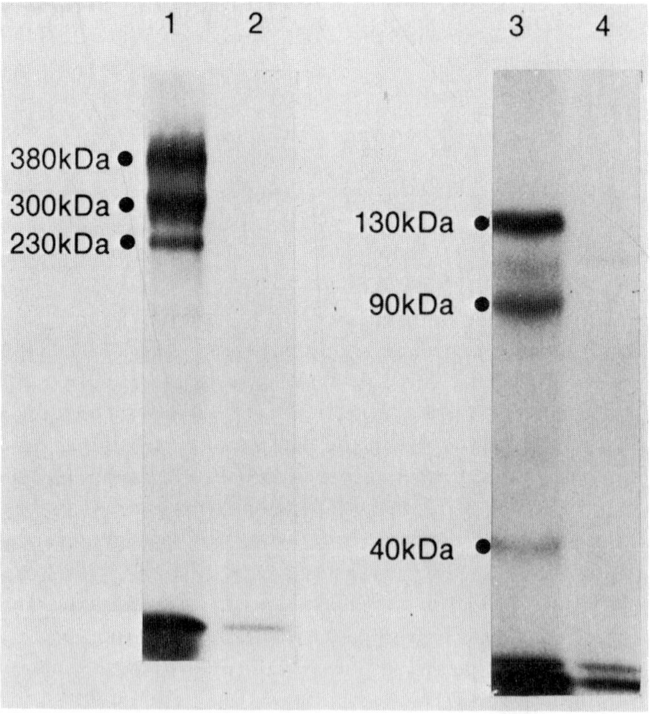

FIG. 5. Autoradiogram showing the specific photoaffinity labeling of the three redox forms of insulin receptor of M_r 380,000, 300,000 and 230,000 (lane 1) and their subunits of M_r 130,000, 90,000, and 40,000 after reduction (lane 3). Lane 2 and lane 4 were obtained when photolabeling was carried out in the presence of excess insulin.

0.05% sucrose–125 mM Tris–HCl, pH 6.8 is used. Electrophoresis is carried out at a constant voltage of 200 V. The electrophoresis buffer contains 0.1% SDS, 0.025 M Tris, and 0.192 M glycine. Under these conditions, a linear relationship between electrophoretic mobility and log molecular weight is obtained for protein standards covering the molecular weight range of 660,000 to 67,000. The gel slab is stained in Coomassie Blue (0.05% of the dye in 25% isopropanol–10% acetic acid) overnight and destained in 10% acetic acid–10% isopropanol. The destained gel slab is soaked in 10% glycerol before drying using a slab-gel dryer. To prevent cracking in storage, the dry gel is laminated in plastic using a sheet laminator. Autoradiography is carried out at −76° using Kodak X-Omat AR film with a Du Pont Cronex Lightning-Plus intensifying screen. A typical autoradiogram demonstrating the specific photolabeling of insulin receptor and its subunits in adipocytes is shown in Fig. 5.

Acknowledgment

Works described in this communication were supported in part by the Medical Research Council, Canada and the Charles H. Best Foundation.

[14] Affinity Cross-Linking of Receptors for Insulin and the Insulin-Like Growth Factors I and II

By JOAN MASSAGUÉ and MICHAEL P. CZECH

Photoaffinity labeling of a peptide hormone receptor is a technique based on the attachment of a reactive group to the corresponding high-affinity ligand. After binding to a specific receptor, the reactive group in the derivatized ligand is activated by exposure to light, generally of short wavelength, generating a highly reactive intermediate that can form covalent bonds with a variety of chemical groups in the occupied binding site. While photoaffinity labeling has proven to be an excellent approach to identify receptor structures for various peptide hormones, including insulin and insulin-like growth factor I, the limitations of this technique are the need to prepare and purify a derivatized ligand, the risk to damage the biological properties of the native ligand during chemical derivatization, and the possible alteration of the biological system under study due to ultraviolet light exposure. These problems can be circumvented by the use of receptor affinity cross-linking, an alternative methodology based on the generation of covalent bonds between the receptor and unmodified

ISBN 0-12-182009-2

ligand by addition of a bifunctional agent to ligand–receptor complexes. Affinity cross-linking was first developed to affinity label insulin receptors,[1,2] and has been subsequently used to identify and characterize receptors for insulin-like growth factors I[3,4] and II,[3–5] platelet-derived growth factor,[6,7] epidermal growth factor,[8] transforming growth factors,[8,9] growth hormone,[10] human interferon α2,[11] glucagon,[12] and nerve growth factor.[13,14] This chapter will describe procedures for affinity cross-linking receptors for insulin and the insulin-like growth factors I and II.

Cross-Linking Agents

Pilch and Czech[1] introduced the most generally used cross-linking agent for peptide hormone receptor studies, disuccinimidyl suberate (DSS). DSS and its analogs (Fig. 1) are homobifunctional agents that exhibit two succinimidyl groups. These succinimidyl moieties can be displaced by primary aliphatic amino groups such as those present in the amino terminus or in lysine residues in a polypeptide chain, resulting in covalent acylation of the reacting amino group. When one molecule of DSS reacts with one amino group in the receptor and one amino group in the receptor-bound ligand, a specifically tagged receptor complex is obtained. The presence of an easily detectable label, generally ^{125}I, in the ligand allows the visualization of the cross-linked complex when it is subjected to electrophoresis, chromatography, or other analytical techniques.

[1] P. F. Pilch and M. P. Czech, *J. Biol. Chem.* **254,** 3375 (1979).

[2] P. F. Pilch and M. P. Czech, *J. Biol. Chem.* **255,** 1722 (1980).

[3] J. Massagué and M. P. Czech, *J. Biol. Chem.* **257,** 5038 (1982).

[4] M. Kasuga, E. Van Obberghen, S. P. Nissley, and M. M. Rechler, *J. Biol. Chem.* **256,** 5305 (1981).

[5] J. Massagué, B. J. Guillette, and M. P. Czech, *J. Biol. Chem.* **256,** 2122 (1981).

[6] K. Glenn, D. F. Bowen-Pope, and R. Ross, *J. Biol. Chem.* **257,** 5172 (1982).

[7] C.-H. Heldin, B. Ek, A. Wasteson, and B. Westermark, *J. Cell Biochem. Suppl.* **6,** 146 (1982).

[8] J. Massagué, M. P. Czech, K. Iwata, J. E. DeLarco, and G. J. Todaro, *Proc. Natl. Acad. Sci. U.S.A.* **79,** 6822 (1982).

[9] J. Massagué, *J. Biol. Chem.* **258,** 13614 (1983).

[10] D. B. Donner, *J. Biol. Chem.* **258,** 2736 (1983).

[11] C. R. Faltynek, A. A. Branca, S. McCandless, and C. Baglioni, *Proc. Natl. Acad. Sci. U.S.A.* **80,** 3269 (1983).

[12] G. L. Johnson, V. I. McAndrew, and P. F. Pilch, *Proc. Natl. Acad. Sci. U.S.A.* **78,** 875 (1981).

[13] J. Massagué, B. J. Guillette, M. P. Czech, C. J. Morgan, and R. A. Bradshaw, *J. Biol. Chem.* **256,** 9419 (1981).

[14] P. Puma, S. E. Buxser, L. Watson, D. J. Kelleher, and G. L. Johnson, *J. Biol. Chem.* **258,** 3370 (1983).

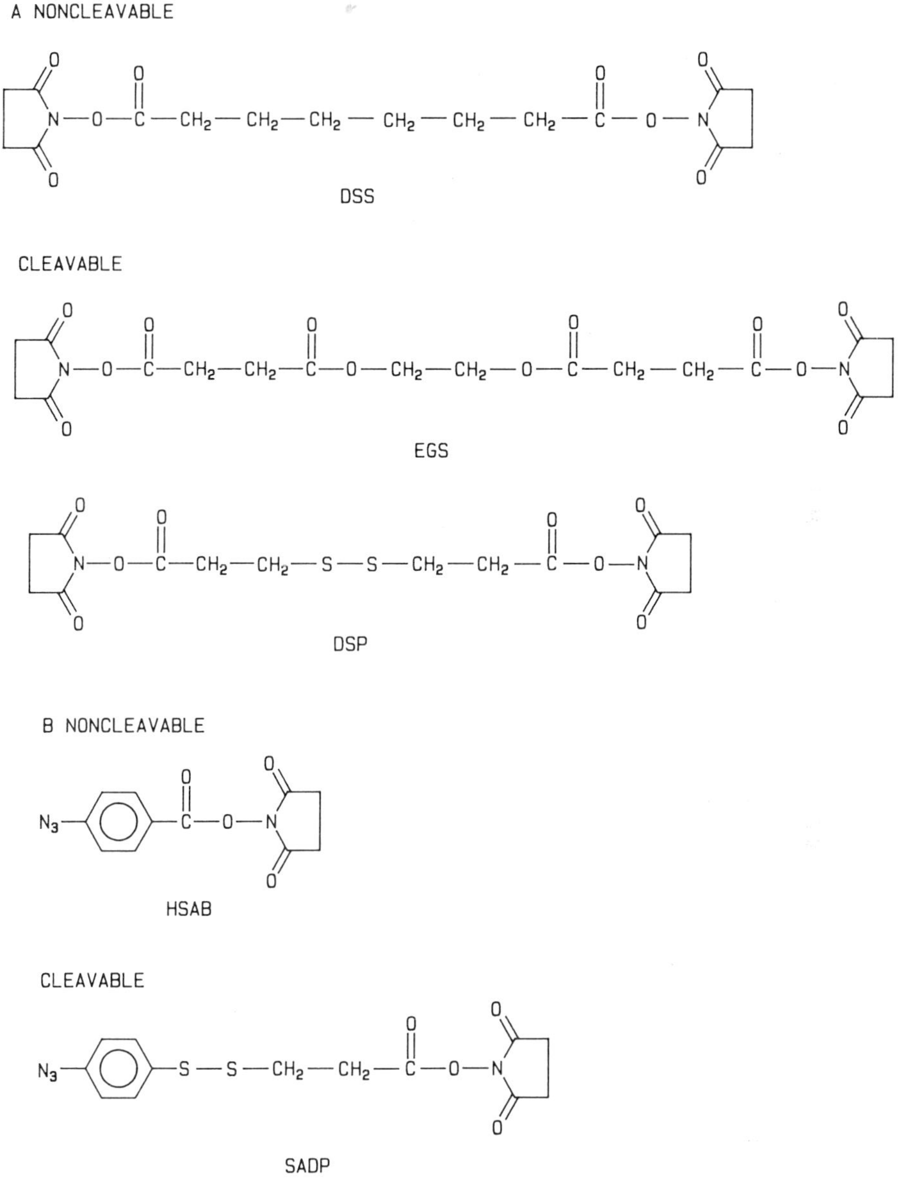

FIG. 1. Structures of cross-linking agents most commonly used in receptor affinity-cross-linking studies. (A) Homobifunctional; (B) heterobifunctional. DSS, Disuccinimidyl suberate; EGS, ethylene glycol bis(succinimidyl succinate); DSP, dithiobis(succinimidyl propionate); HSAB, *N*-hydroxysuccinimidyl-4-azidobenzoate; SADP, *N*-succinimidyl-(4-azidophenyldithio)propionate.

The cross-linked complexes generated by DSS are irreversible. However, a number of cleavable analogs of DSS are available for experimental situations in which controlled reversion of crosslinks is desired. Cleavable crosslinkers that can be used for affinity labeling of receptors for insulin, IGF-I and IGF-II include dithiobis(succinimidyl propionate) that can be cleaved by reduction with dithiothreitol or 2-mercaptoethanol, and ethylene glycol bis(succinimidyl succinate) that is cleaved by incubation at 37° for 6 hr with hydroxylamine at pH 8.5 (Fig. 1).

A limitation of the affinity cross-linking technique using homobifunctional agents of the bissuccinimidyl family is the requirement for appropriately spaced amino groups in the receptor and ligand molecules. In systems in which the chemical requirements for homobifunctional cross-linking are not met, such as the glucagon[13] and nerve growth factor[14] receptor systems, this limitation can be circumvented by the use of heterobifunctional agents such as hydroxysuccinimidyl azidobenzoate (HSAB) (Fig. 1). HSAB exhibits a succinimidyl group that can be displaced by primary amines and an azido group that generates a highly reactive nitrene upon activation with ultraviolet radiation. Addition of HSAB to ligand–receptor complexes followed by exposure of the mixture to ultraviolet light can result in the derivatization of amino groups in the receptor and/or ligand *in situ,* and the subsequent formation of cross-links through the activated nitrene group.

All cross-linking agents mentioned above are commercially available.[15]

Affinity Cross-Linking Procedures

The protocols for affinity cross-linking receptors for insulin and the insulin-like growth factors I and II (IGFs I and II) involve four basic steps: (1) formation of ligand–receptor complexes under equilibrium binding conditions, (2) addition of the suitable cross-linking agent to preformed ligand–receptor complexes, (3) termination of the cross-linking reaction by quenching of excess unreacted cross-linking agent, and (4) preparation of the affinity labeled sample for subsequent analysis. Affinity cross-linking is usually performed on receptor-containing samples in which binding of the corresponding ^{125}I-labeled ligand has reached equilibrium. To maximize the intensity and specificity of affinity labeling, binding conditions allowing a high level of receptor occupancy with a relatively low level of nonspecific binding at equilibrium are desirable. Optimal binding conditions may vary for each particular hormone and receptor system under study, but they generally involve incubation for

[15] Pierce Chemical Co., Rockford, Illinois.

45–60 min at 22° or for 75–90 min at 10° with a concentration of ligand that effects and approximately half-maximal receptor occupancy at equilibrium.

The cross-linking agent of choice for irreversible labeling of insulin, IGF-I, and IGF-II receptors is DSS. Affinity labeling of these receptors using DSS is characterized by a relatively high efficiency. Routinely, 10–15% of the total receptor-bound ^{125}I-labeled ligand becomes covalently attached to the receptor[2,3] by DSS. In addition, high specificity of labeling is frequently obtained with DSS, the labeled receptor bands being usually the only major labeled products obtained in the cross-linking reaction.[3,16] The cleavable agent ethylene glycol bis(succinimidyl succinate) exhibits an efficiency and specificity similar to DSS.[5]

To affinity cross-link receptors in isolated membrane fractions to receptor-bound, ^{125}I-labeled insulin IGF-I or IGF-II the ligand binding reaction is carried out in the presence of a balanced salt solution (for example, 125 m*M* NaCl, 5 m*M* KCl, 1.2 m*M* $MgSO_4$, 1.2 m*M* $CaCl_2$) containing 2–10 mg/ml of bovine serum albumin, and buffered at the optimal pH for binding with 10 m*M* sodium phosphates or 10–50 m*M* 4-(2-hydroxyethyl)-1-piperazineethanesulfonate (HEPES). Incubation with the ^{125}I-labeled ligand at 10° or at lower temperature is recommended to minimize ligand degradation. A higher level of nonspecifically labeled products obtained after affinity cross-linking is commonly observed when the binding is performed at higher (22 or 37°) temperature.[16,17] Before addition of the cross-linking agent membranes are separated from the bulk of free ligand by dilution with 5–10 volumes of ice-cold binding solution followed by centrifugation to sediment the particulate material. This centrifugation can be done at 30,000 *g* for 20 min for maximal recoveries, but it can usually be substituted by a 3 min spin at 12,000 *g* in a microcentrifuge with little loss of membrane material in the supernatant. The resulting pellet is resuspended with ice-cold, albumin-free binding medium to a final 0.5 to 1.0 mg/ml organelle protein concentration and placed on an ice bath. DSS freshly dissolved in dimethyl sulfoxide is added at a 1/100 dilution to the resuspended sample with constant agitation. Incubation of the reacting mixture proceeds at 0–4° to minimize ligand dissociation and to decrease lateral mobility of membrane proteins that would favor cross-links among membrane components. Final DSS concentrations in the 0.1–0.3 m*M* range are the optimal under these conditions.

The cross-linking reaction is terminated by dilution with a solution containing an excess of primary amino groups that will quench the un-

[16] J. Massagué and M. P. Czech, *J. Biol. Chem.* **257,** 6729 (1982).

[17] J. Massagué, P. F. Pilch, and M. P. Czech, *J. Biol. Chem.* **256,** 3182 (1981).

reacted DSS still remaining in solution. Ten millimolar Tris–HCl, 1 m*M* EDTA, pH 7.0 is a commonly used quenching solution. The affinity-labeled membranes are immediately sedimented by centrifugation, and the resulting pellet solubilized with medium containing 1% Triton X-100 or 1% sodium dodecyl sulfate, and Tris–HCl. This step is important to ensure a complete quenching of DSS that has partition into the polar domains of the membranous material. If the affinity labeled sample has to be analyzed by sodium dodecyl sulfate polyacrylamide gel electrophoresis (SDS–PAGE), direct solubilization in Tris-containing electrophoresis sample buffer is recommended.

Cell surface insulin and IGF receptors in isolated rat adipocytes and in a variety of cultured cell lines can be affinity cross-linked using this protocol with minor modifications.[3,16–18] Whenever possible, affinity cross-linking of cell surface receptors in cultured cells is performed in suspensions rather than in monolayers of these cells. Cultured cell lines can be detached from culture vessels by a brief (10 min at 37°) incubation in the presence of a solution containing 125 m*M* NaCl, 5 m*M* KCl, 50 m*M* Hepes, 5 m*M* glucose, 1 m*M* EDTA, pH 7.4. The suspended cells are washed and resuspended with the appropriate binding medium. After binding to the corresponding ^{125}I-labeled ligand, the excess unbound ligand may be removed by centrifugation at 1000 *g* for 5–10 min, and resuspension of the cell pellet with fresh medium. However, this step can be omitted if it would adversely affect cellular viability or the stability of ligand–receptor complexes. DSS (30–50 m*M* in dimethyl sulfoxide) is added at a final concentration 0.3–0.5 m*M* to the cell suspension (0.5–2.0 $\times 10^6$ cells/ml) at 10–15°, and the cross-linking reaction is allowed to proceed for 15 min at this temperature. The reaction is terminated by dilution, centrifugation, and resuspension of the cell pellet with 250 m*M* sucrose, 10 m*M* Tris–HCl, 1 m*M* EDTA, pH 7.0. For further analysis of the affinity-labeled cellular material a crude membrane fraction may be obtained from the labeled cells by homogenization in the isotonic sucrose buffer, and centrifugation at 30,000 *g* for 30 min after removing nuclei and other organelles by a 10 min spin at 5000 *g*. Alternatively, the intact cell pellet obtained after affinity cross-linking can be directly solubilized in detergent-containing buffer, clarified by centrifugation, and analyzed directly by SDS–PAGE or chromatography.

Analysis of Affinity Cross-Linked Receptor Complexes

The receptors for IGF-II identified in variety of tissues and cultured cell lines by affinity cross-linking[3–5] or purified by affinity chromatogra-

[18] J. Massagué, L. A. Blinderman, and M. P. Czech, *J. Biol. Chem.* **257,** 13958 (1982).

phy[19,20] consists of one single polypeptide chain of M_r = 250,000–270,000 not disulfide-linked to any other membrane component. This receptor species has higher affinity for IGF-II than for IGF-I, and no affinity for insulin.[3] It can be affinity crosslinked with [^{125}I]IGF-II or [^{125}I]IGF-I, but IGF-II is more effective in competing with the labeled tracer than IGF-I. Identification of affinity-crosslinked IGF-II receptors in one-dimension dodecyl sulfate-polyacrylamide gels is therefore straightforward. In contrast, identification of receptors for insulin and IGF-I is complicated by the fact that these two receptor types are highly homologous disulfide-linked complexes consisting of two M_r = 115,000–135,000 subunits (α subunits) and two M_r = 90,000–95,000 subunits (β subunits).[3,21] In addition, the disulfides linking these receptor subunits in the (β-S-S-α) S-S (α-S-S-β) (M_r = 350,000–400,000) oligomeric receptor complex exhibit a distinctive susceptibility to reduction by dithiothreitol or 2-mercaptoethanol.[16,22] Thus, the disulfides linking one (α-S-S-β) receptor half to the other (class I disulfides) are very sensitive to reduction even under non-denaturing conditions. Treatment of intact cells or isolated membranes with dithiothreitol either before or after affinity cross-linking with ^{125}I-labeled insulin or [^{125}I]IGF-I yields labeled (α-S-S-β) receptor fragments that migrate as a M_r = 210,000 band in SDS–PAGE. In contrast, addition of reductant after denaturation with SDS is required to reduce the class II disulfides yielding free α and β receptor subunits.

Another complication of the insulin and IGF-I receptor systems is the presence of M_r = 320,000 and M_r = 290,000 receptor complexes in addition to the native M_r = 350,000 receptor complexes in most types of membrane preparations. These two lower M_r receptor complexes originate from the native M_r = 350,000 receptor form by limited proteolysis of the β subunits at a site near the middle of their amino acid sequence that is very susceptible to attack by elastase and lysosomal proteases.[3,17] The M_r = 45,000–50,000 fragment of the β subunit that remains disulfide-linked to the receptor core after cleavage at this site has been termed the β_1 receptor subunit and is found in the (β-S-S-α)S-S(α-S-S-β_1) (M_r = 320,000) and (β_1-S-S-α)S-S(α-S-S-β_1) (M_r = 290,000) receptor forms. The three oligomeric receptor forms can be easily separated by SDS–PAGE using highly porous gels (5% polyacrylamide, 100 : 1 acrylamide : bis-acrylamide ratio).

[19] C. L. Oppenheimer and M. P. Czech, *J. Biol. Chem.* **258,** 8539 (1983).

[20] G. P. August, S. P. Nissley, M. Kasuga, L. Lee, L. Greenstein, and M. M. Rechler, *J. Biol. Chem.* **258,** 9033 (1983).

[21] J. Massagué, P. F. Pilch, and M. P. Czech, *Proc. Natl. Acad. Sci. U.S.A.* **77,** 7137 (1980).

[22] J. Massagué and M. P. Czech, *in* "Frontiers in Biochemical and Biophysical Studies of Proteins and Membranes" (T.-Y. Liu, S. Sakikabara, A. Schechter, K. Yagi, H. Yajima, and K. T. Yasunobu, eds.), p. 453.

Typically the α subunit in receptors for insulin and IGF-I is intensely labeled by affinity cross-linking whereas the β subunit is very poorly labeled using this technique. After SDS–PAGE under reducing conditions and autoradiography of affinity-labeled insulin and IGF-I receptors, the labeled β subunit is often masked by the background and by minor nonspecifically labeled bands. Rigorous identification of β receptor subunits requires bidimensional SDS–PAGE analysis, the first dimension being run in the absence of reductant and the second dimension in the presence of 50 m*M* dithiothreitol. The first dimension gel, usually a standard 3-mm-diameter tube gel, should be a discontinuous,[23] highly porous (5% polyacrylamide, 100 : 1 acrylamide–bisacrylamide ratio) gel to allow resolution of the high M_r unreduced receptor complexes. After electrophoresis, this gel is soaked in a solution containing 50 m*M* dithiothreitol, 100 m*M* Tris–HCl, pH 6.8, 1% SDS, for 30 min at 37° to cause the dissociation of disulfide-linked complexes. This gel is then layered on top of a slab gel[23] containing 6% polyacrylamide (37.5 : 1 acrylamide : bisacrylamide ratio). The space between the first dimension gel and the second gel is filled with a connecting gel that contains 1% agarose, 50 m*M* dithiothreitol, 100 m*M* Tris–HCl, pH 6.8, 1% SDS, and 10 μg/ml bromphenol blue. Alternatively, the receptor complexes resolved by electrophoresis in the first gel can be localized by autoradiography and excised from this gel. The gel fragments are frozen at −70°, finely minced on a chilled mortar, suspended in SDS–PAGE sample buffer containing 50 m*M* dithiothreitol, and electrophoresed on the second gel. After second dimension electrophoresis the gels can be fixed and stained for protein,[24] dried, and subjected to autoradiography using XAR-5 Kodak film and Du Pont Cronex Lightning Plus enhancing screens.

The displacement of the label by an excess unlabeled ligand present during incubation of membranes or cells with ^{125}I-labeled insulin or [^{125}I]IGF-I is normally an acceptable criterium for affinity-labeling specificity. However, exceptions to this rule can occasionally happen. For example, bovine serum albumin present in the incubation buffers can become labeled by ^{125}I-labeled insulin fragments during affinity cross-linking of samples in which extensive degradation of the ^{125}I-labeled insulin tracer has occurred.[17] An excess unlabeled insulin added to parallel samples can protect the ^{125}I-labeled insulin from proteolysis preventing the labeling of albumin and giving therefore a pseudospecific displacement of ligand from albumin binding sites.

Another source of abnormal affinity-labeled species is the high sensi-

[23] U. K. Laemmli, *Nature (London)* **227,** 680 (1970).

[24] G. Fairbanks, T. L. Steck, and D. F. H. Wallach, *Biochemistry* **10,** 2606 (1971).

tivity of insulin and IGF-I receptors to proteolysis. In addition to an elastase-like lysosomal enzyme that converts the β subunit into a M_r = 45,000–50,000 fragment, a second type of protease, probably a thiol-proteinase, has been observed in dithiothreitol-treated rat adipocyte membranes that will degrade the α insulin receptor subunit into a free M_r = 80,000 fragment.[16] Affinity-labeled receptor degradation products can be easily identified by comparative peptide mapping[25] of affinity-labeled species excised from an SDS–PAGE gel.[16,17]

Sincc insulin, IGF-I, and IGF-II exhibit varying degrees of affinity for each other's receptors, affinity cross-linking of heterologous receptor species by these ligands can occur, particularly in cell lines or tissues that contain high numbers of the heterologous receptors. Thus, affinity-labeling of IGF-I and IGF-II receptors by crosslinking to [^{125}I]IGF-II and [^{125}I]IGF-I, respectively, is usually observed.[3,4] Similarly, [^{125}I]IGF-I and ^{125}I-labeled insulin can affinity-label each other's receptor. To discriminate between the three receptor types, incubation of cells or membranes with a given ^{125}I-labeled ligands is done in the presence of various concentrations of unlabeled insulin, IGF-I and/or IGF-II. The relative ability of the native ligands to inhibit the labeling will identify what receptor typc is being affinity-cross-linked. This distinction is especially necessary for the rigorous identification of receptors for insulin versus IGF-I because of the high degree of structural homology between these two receptor types.

[25] D. W. Cleveland, S. G. Fisher, M. W. Kirschner, and U. K. Laemmli, *J. Biol. Chem.* **252,** 1102 (1977).

[15] The Preparation of Biologically Active Monomeric Ferritin–Insulin and Its Use as a High Resolution Electron Microscopic Marker of Occupied Insulin Receptors

By ROBERT M. SMITH and LEONARD JARETT

In recent years a number of laboratories have performed combined ultrastructural and biochemical analyses of biological processes recognizing that both types of analyses are required in order to achieve a complete understanding of the events which regulate cellular function. These combined studies have been particularly prevalent in areas such as endocrinology and pharmacology where the effects of hormones, drugs and other agents on cellular structure may be related to their biochemical functions. Our laboratory has studied the interaction of insulin, a major anabolic

ISBN 0-12-182009-2

hormone in mammalian organisms, with its targe tissues. We were interested in determining the ultrastructural nature of the events involved in the normal binding, degradation, internalization, and processing of the hormone and its receptor. In addition we wanted to determine the effects of various agents which modify insulin action, binding or processing. Biochemical studies by several investigators[1] have been the basis for important speculations about the organization and the cellular processing of the hormone receptor complex. The development of reliable ultrastructural methodologies should complement these studies and provide a more complete understanding of the biological processes involved.

Morphological analyses have used different techniques from the biochemical studies. These techniques have included light and electron microscopic autoradiography using ^{125}I-labeled insulin,[2–5] light microscopy using fluorescently labeled insulin,[6] and high-resolution electron microscopy.[7–16] Performance of high-resolution electron microscopic analyses of hormone receptor sites requires that a ligand, either the native hormone, an agonist, or a monovalent antibody to the receptor, be chemically modified to make it electron dense. This process is accomplished either by conjugating to it an electron-dense molecule such as ferritin, hemocyanin or gold, or by linking it to a product, such as peroxidase, which subsequently can be chemically treated to produce an electron-dense product. We elected to use ferritin as the electron-dense molecule because of its

[1] J. M. Olefsky, S. Marshall, P. Berhanu, M. Saekow, K. Heidenreich, and A. Green, *Metab., Clin. Exp.* **31,** 670 (1982).

[2] J. J. M. Bergeron, G. Levine, R. Sikstrom, D. O'Shaughnessy, B. Kopriwa, N. J. Nadler, and B. I. Posner, *Proc. Natl. Acad. Sci. U.S.A.* **74,** 5051 (1977).

[3] J.-L. Carpentier, P. Gorden, M. Amherdt, E. Van Obherghen, C. R. Kahn, and L. Orci, *J. Clin. Invest.* **61,** 1051 (1978).

[4] M. Fehlmann, J.-L. Carpentier, A. LeCam, P. Thamm, D. Saunders, D. Bradenburg, L. Orci, and P. Freychet, *J. Cell Biol.* **93,** 82 (1982).

[5] I. D. Goldfine, A. L. Jones, G. T. Hradek, K. Y. Wong, and J. S. Mooney, *Science* **202,** 760 (1978).

[6] J. Schlessinger, Y. Schechter, M. C. Willingham, and I. Pastan, *Proc. Natl. Acad. Sci. U.S.A.* **75,** 2659 (1978).

[7] L. Jarett and R. M. Smith, *J. Biol. Chem.* **249,** 7024 (1974).

[8] L. Jarett and R. M. Smith, *Proc. Natl. Acad. Sci. U.S.A.* **72,** 3526 (1975).

[9] L. Jarett and R. M. Smith, *J. Supramol. Struct.* **6,** 45 (1977).

[10] D. M. Nelson, R. M. Smith, and L. Jarett, *Diabetes* **27,** 530 (1978).

[11] L. Jarett and R. M. Smith, *J. Clin. Invest.* **63,** 571 (1979).

[12] L. Jarett, J. B. Schweitzer, and R. M. Smith, *Science* **210,** 1127 (1980).

[13] R. M. Smith and L. Jarett, *J. Histochem. Cytochem.* **30,** 650 (1982).

[14] R. M. Smith and L. Jarett, *Proc. Natl. Acad. Sci. U.S.A.* **79,** 7302 (1982).

[15] R. M. Smith and L. Jarett, *J. Cell. Physiol.* **115,** 199 (1983).

[16] L. Jarett and R. M. Smith, *Proc. Natl. Acad. Sci. U.S.A.* **80,** 1023 (1983).

availability, size, and chemical characteristics. Several hormones, including epidermal growth factor,[17] luteinizing hormone,[18] luteinizing hormone releasing factor,[19] and melanocyte stimulating hormone[20] have also been linked to ferritin and used to describe the distribution of occupied receptor sites on various target tissues.

We describe a rapid, reproducible method for preparing a monomeric ferritin–insulin conjugate. The methodology has been previously described[13]; much of the present data comes from a recent routine preparation chosen for illustrative purposes. In order for the conjugate to be a valid marker for the insulin receptor and allow certain analyses to be made, a set of strict criteria must be met. Monomeric ferritin–insulin fulfills these criteria which include (1) the conjugate contains immunologically active insulin which is indistinguishable from native insulin; (2) the biologically active insulin is equivalent in potency to the immunologic concentration of insulin and is indistinguishable from native insulin; (3) the specificity and affinity of the conjugate for the insulin receptor is equal to native insulin; (4) the conjugate is stable during storage and incubation with tissues; (5) the conjugate is monovalent and monomeric in terms of both insulin and ferritin; and (6) the final conjugate contains no free, unconjugated insulin.

This monomeric ferritin insulin conjugate has provided a method for studying the organization of the insulin receptor on cell surfaces, insulin uptake, and recycling as well as documenting the effects of a variety of agents on the distribution of insulin receptors on cell surfaces, and on insulin uptake and distribution in intracellular structures.[7–12,14–16]

Preparation and Characterization of Monomeric Ferritin-Labeled Insulin

Materials

Porcine insulin (lot PJ5682, 23.1 U/mg) was a gift from Dr. R. Chance, Eli Lilly Co., Indianapolis, IN. Horse spleen ferritin (6× crystallized, cadmium free) was purchased from Miles Laboratories, Inc., Elkhart, IN, and was used without prior purification. Glutaraldehyde (25%, EM grade, stored in glass ampoules under inert gas) was purchased from Electron Microscopy Sciences, Fort Washington, PA. Previous studies used glu-

[17] J. A. McKanna, H. T. Haigler, and S. Cohen, *Proc. Natl. Acad. Sci. U.S.A.* **76,** 5689 (1979).

[18] J. L. Lubarsky and H. R. Behrman, *Mol. Cell. Endocrinol.* **15,** 61 (1979).

[19] C. R. Hopkins and H. Gregory, *J. Cell Biol.* **75,** 528 (1977).

[20] A. DiPasquale, J. M. Varga, G. Moellmann, and J. McGuire, *Anal. Biochem.* **84,** 37 (1978).

taradehyde from various sources and no differences were found between similar grades of glutaraldehyde.[13] BioGel A1.5M was a product of Bio-Rad Laboratories, Richmond, CA. Bovine serum albumin and collagenase were purchased from Sigma Chemical Co., St. Louis, MO and ^{125}I was purchased from Amersham, Arlington Heights, IL. Other reagents and supplies were purchased from standard sources or as indicated in the text.

Preparation of Monomeric Ferritin-Labeled Insulin (Fm-I)

Table I summarizes the standard protocol used to prepare Fm-I. Ferritin was superactivated with glutaraldehyde by using a modification of a previously described method.[21] Ferritin (15 to 20 mg) was diluted in 4 ml of 0.05 *M* sodium phosphate buffer, pH 8.0 (phosphate buffer). The ferritin concentration was determined by measuring the optical absorption at 440 nm and using the extinction coefficient of OD_{440} 1.000 = 0.65 mg ferritin.[22] A freshly opened vial of glutaraldehyde was diluted to 2.5% with phosphate buffer and 0.20 to 0.26 ml of the cross-linker was added to the ferritin suspension with constant stirring. The final molar ratio in the suspension was >500 : 1 (glutaraldehyde to ferritin). After mixing for 1 hr at room temperature the ferritin suspension was applied to a 3 × 30 cm Sephadex G-200 column and eluted with phosphate buffer to separate the superactivated ferritin from the free glutaraldehyde. The void volume material, which contained the superactivated ferritin, was collected and diluted if necessary to about 20 ml with phosphate buffer. The excess glutaraldehyde in the original reaction resulted in virtual saturation of the reactive amino groups in the ferritin by the bifunctional glutaraldehyde providing more than 100 activated amino groups. The reaction modestly increased the amount of ferritin oligomers. The ferritin as purchased contained 5–7% oligomeric forms which was increased to 12–15% after the superactivation with glutaraldehyde.

Porcine insulin was dissolved in 0.1 *M* HCl and diluted with phosphate buffer to a concentration of 7 mg/ml. One milliliter of the insulin solution was slowly added to the activated ferritin with constant vigorous initial mixing and then slowly mixed at room temperature for 6 hr. The molar ratio was about 35 : 1 (insulin to ferritin). The exact nature of the reaction between the aldehyde groups on the activated ferritin molecule and the insulin molecule is not known. The pH of the reaction would favor the cross-linking occurring through the epsilon amino groups of B-29 lysine versus the N-terminal amino groups. Insulin probably existed in the dimeric or hexameric form due to the high concentration of the hormone at

[21] Y. Kishida, B. R. Olsen, R. A. Berg, and D. J. Prockop, *J. Cell Biol.* **64,** 3331 (1975).
[22] A. F. Schick and S. J. Singer, *J. Biol. Chem.* **236,** 2477 (1961).

TABLE I
SUMMARY OF THE PREPARATION OF MONOMERIC FERRITIN–INSULIN

1. Superactivation of ferritin: 15–20 mg ferritin in 4 ml of 0.05 *M* sodium phosphate buffer, pH 8.0, is mixed for 1 hr with 0.2–0.26 ml 2.5% glutaraldehyde. The activated ferritin is separated from the excess glutaraldehyde by collecting the ferritin fraction at the void volume of a G-200 Sephadex column eluted with phosphate buffer
2. Conjugation to insulin: The superactivated ferritin is mixed for 6 hr with 7 mg porcine insulin which was dissolved in 0.1 *M* HCl and diluted in phosphate buffer
3. Neutralization of unreacted aldehyde groups: 2 ml of 2.25 *M* lysine–HCl is added to the reaction mixture and stirred for 16 hr at room temperature
4. Concentration of ferritin, and ferritin–insulin: The reaction mixture is ultracentrifuged at 40,000 rpm for 90 min. The supernatants are discarded and the pellets resuspended in less than 5 ml of phosphate buffer
5. Purification of monomeric ferritin–insulin: The resuspended material from step 4 is applied to a BioGel 1.5M agarose column and eluted with phosphate buffer. The ferritin oligomer peak is discarded and the ferritin monomer peak is collected
6. Futher purification: Steps 4 and 5 are repeated
7. Concentration, storage, and characterization of monomeric ferritin–insulin: The monomeric ferritin–insulin is centrifuged at 40,000 rpm for 90 min and the pellets resuspended in 1–2 ml phosphate-buffered saline, pH 7.4. The conjugate is stored at 4°. Characterization was performed by RIA, radioreceptor assay, and bioassay as described in the text

pH 8.0.[23] If an insulin dimer or hexamer were linked to the ferritin via a single amino group in the hormone, upon dilution of the conjugate one should find high concentrations of free insulin when the dimers and hexamers dissociate into monomers. This was not found and it appears likely that the cross-linking occurred between aldehyde groups and the small amount of monomeric insulin present. The dimer and hexamer of insulin would hide the probable reactive site and therefore would be unlikely to react. As the monomeric insulin was coupled to ferritin, the equilibrium between monomers, dimers, and hexamers would replenish the monomeric insulin, assuring a constant concentration of monomeric insulin.

Unreacted aldehyde groups in the ferritin were blocked by adding 2 ml of 2.25 *M* lysine–HCl in phosphate buffer and mixing for another 16 hr at room temperature. Blocking the unreacted aldehyde groups was essential to prevent nonspecific binding of the Fm-I to cell membranes and to prevent oligomer formation during subsequent centrifiguations of the conjugate. The mixture was centrifuged at 40,000 rpm for 90 min to concentrate the ferritin–insulin and free ferritin. (Since the Fm-I preparation contained both labeled and unlabeled ferritin, as indicated below, all further reference to Fm-I will assume the presence of the unlabeled material.) The tubes were immediately removed from the centrifuge and the

[23] B. H. Frank, A. H. Pekar, and A. J. Veros, *Diabetes* **21,** Suppl. 2, 486 (1972).

supernatants, containing unconjugated insulin, were removed as completely as possible. The pellets were resuspended in a minimal amount of phosphate buffer (<5 ml) and was applied to a 3 × 45-cm BioGel A1.5M agarose column and eluted with phosphate buffer. The ferritin migrated as two distinct bands. The first band contained ferritin oligomers and was eluted at the void volume and discarded. The second band contained 0.95% ferritin monomers and was collected and pooled. The monomer fraction was concentrated by ultracentrifugation as described above. The pellets were resuspended in phosphate buffer and rechromatographed on the BioGel column. There was virtually no oligomer band associated with the rechromatographed ferritin. The monomer fraction was collected and the Fm-I was concentrated by ultracentrifugation as before. The final pellets were resuspended in 1 to 2 ml of 0.9% NaCl in 0.05 *M* sodium phosphate, pH 7.4. The ferritin concentration was determined as above and the conjugate was stored at 4°. Under the conditions described, the final solution contained about 7 μg of insulin or 0.1% of the initial insulin and about 41% of the initial ferritin. The average ferritin to insulin ratio was 15 : 1.

Determination of Free, Unlabeled Insulin

The amount of free insulin, i.e., insulin not attached to ferritin molecules, must be insignificant in the final preparation. This was determined by adding 2.25×10^8 cpm of ^{125}I-labeled insulin to a preparation after the

TABLE II
SEPARATION OF UNREACTED ^{125}I-LABELED INSULIN FROM MONOMERIC FERRITIN–INSULIN DURING PURIFICATION OF THE CONJUGATE

Purification procedure	Total ^{125}I-labeled insulin recovered (%)
Neutralization of unreacted aldehyde groups; ^{125}I-labeled insulin added[a]	100.0
Concentrated ferritin following first ultracentrifugation	8.5
Monomer ferritin fraction from first BioGel column	0.4
Concentrated ferritin following second ultracentrifugation	0.04
Monomer fraction from second BioGel column	0.001
Concentrated monomeric ferritin–insulin	0.0001

[a] Following the addition of lysine to block unreacted aldehyde groups on the activated ferritin, 2.25×10^8 cpm of ^{125}I-labeled insulin was added to the reaction mixture. The ^{125}I-labeled insulin did not cross-link to the ferritin therefore the separation of ^{125}I-labeled insulin from the monomeric ferritin insulin should parallel the removal of free, unreacted porcine insulin.

activated aldehyde groups on the ferritin had been blocked. The Fm-I was purified as described and the results of purification are shown in Table II. Less than 0.0001% of the ^{125}I-labeled insulin was recovered in the final Fm-I fraction demonstrating that the centrifugation steps and chromatography columns effectively removed the free insulin. In this preparation the free insulin would be less than 0.1% of the total insulin present in the purified conjugate.

Determination of Insulin Concentration and Activity

The insulin concentration in the purified Fm-I was determined using a routine radioimmunoassay with native porcine insulin as standards. Samples were diluted to an appropriate concentration range and triplicate determinations were made. Results of a typical radioimmunoassay are shown in Fig. 1. The Fm-I competed with ^{125}I-labeled insulin for binding

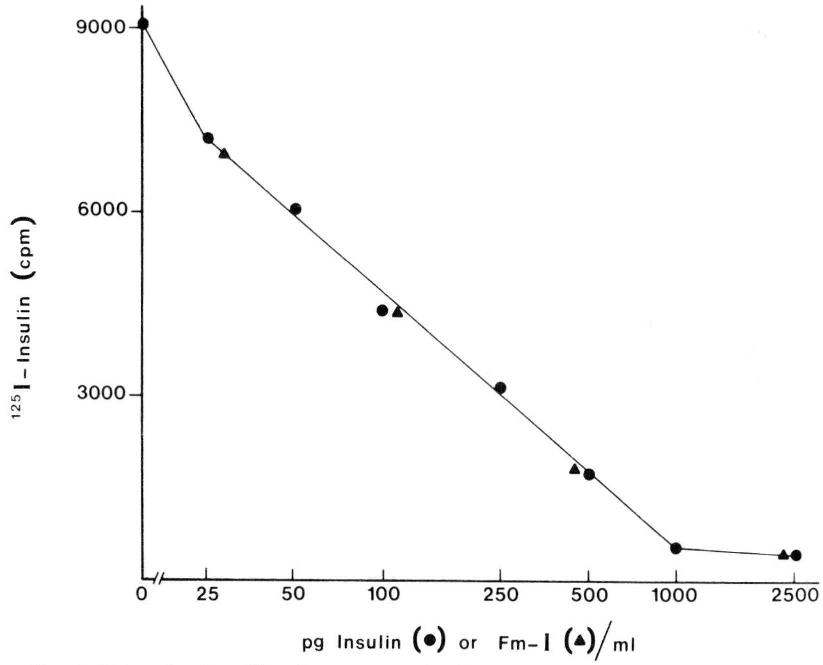

FIG. 1. Determination of insulin concentration in monomeric ferritin–insulin by radioimmunoassay. The purified monomeric ferritin insulin was diluted 1/500 and then serially diluted 1/4 with RIA buffer. Aliquots of 200 μl were incubated in a final volume of 1 ml with 10^4 cpm ^{125}I-labeled insulin in a standard insulin RIA. Porcine insulin was used as standards. The data showed that monomeric ferritin–insulin (▲) reacted in a dose-dependent manner identically to native porcine insulin (●). These data when corrected for dilutions showed that the insulin concentration in the monomeric ferritin–insulin was approximately 4.5 μg/ml.

to the antiinsulin antibody in a dose-dependent manner identically to native porcine insulin. From these data it was possible to establish the concentration of immunologically active insulin in the conjugate and calculate the apparent molar ratio of ferritin to insulin, assuming the molecular weights of 450,000 and 6000 for ferritin and insulin, respectively. The biological activity of the insulin conjugated to the ferritin was determined by comparing the ability of Fm-I and native porcine insulin to stimulate glucose oxidation in adipocytes. Adipocytes were prepared from 120 g male Sprague–Dawley rats by previously published methods.[7] The cells were diluted in Krebs–Ringer phosphate buffer, pH 7.4, with 3% bovine serum albumin and added to vials containing 0.55 m*M* D-[^{14}C]glucose and 0.2 to 4 ng porcine insulin/ml or similar concentration of Fm-I as determined in the radioimmunoassay. The cells were incubated at 37° for 1 hr and the $^{14}CO_2$ trapped on filter papers with hyamine hydroxide as described.[24] Figure 2 shows the results of a bioassay performed on the same Fm-I preparation as in Fig. 1. Fm-I stimulated the conversion of glucose to CO_2 in a dose-dependent manner similar to that found for native insulin, and the biological activity of the Fm-I was very nearly identical to the concentration of insulin determined in the radioimmunoassay. This would suggest that the immunologically active insulin was accessible to the insulin receptor and that any damage to or inactivation of the insulin molecule as a result of its linkage to the ferritin molecule had not affected its biologically active region. This finding was similar to results found with other types of modified insulin molecules[25] where insulin has been modified only at the B-29 lysine.

The apparent molar ratio of ferritin to insulin (15 : 1) and the equal immunological and biological concentrations of insulin suggest that few if any ferritin molecules contain more than one insulin molecule. The statistical probability of having more than one insulin linked to a ferritin is low when only 8% of the ferritin molecules contain insulin; such conjugates would be hard to detect by current assays. If multiple insulin molecules were coupled to a single ferritin molecule, then the biological and immunological activity would probably not be equal. The insulin molecules would be readily accessible to antiinsulin antibody in the immunoassay but only one insulin on the ferritin would bind to a receptor and cause a biological effect. The other insulin molecules on the ferritin would be unlikely to have access to a receptor due to steric inhibition by the ferritin molecule. We have shown this to be the case.[13] Incubation of superactivated ferritin with insulin for 18 hr increased the amount of insulin coupled

[24] J. Gliemann, *Diabetologia* **3**, 382 (1967).
[25] M. J. Ellis, S. C. Darby, R. H. Jones, and P. H. Sonksen, *Diabetologia* **15**, 403 (1978).

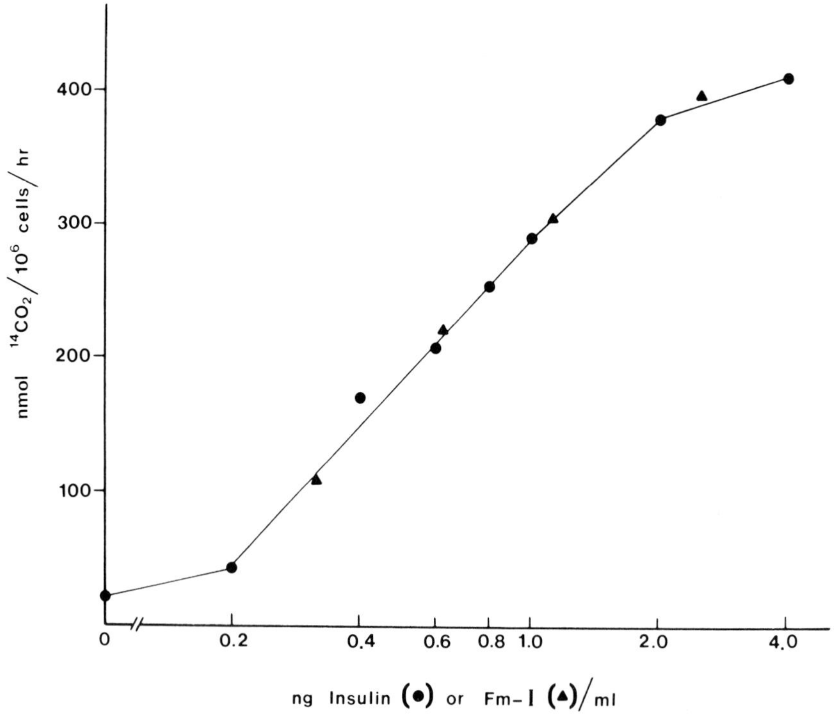

FIG. 2. Determination of the biological activity of monomeric ferritin–insulin in an adipocyte glucose oxidation assay. The monomeric ferritin–insulin preparation assayed in Fig. 1 was diluted on the basis of the RIA results to appropriate concentrations and added to the reaction vessels. Porcine insulin standards were prepared and assayed simultaneously. The conversion of [^{14}C]glucose to $^{14}CO_2$ was determined by published methods.[7] The results of this assay showed that monomeric ferritin–insulin (▲) stimulated glucose oxidation in adipocytes in a dose-dependent manner identically to native porcine insulin (●) and that the biological activity of the conjugated hormone was identical to its immunological concentration.

to ferritin and the immunologically detectable insulin was four times the biologically measurable insulin.

Specificity of Fm-I Binding to Insulin Receptors

The binding of Fm-I to the insulin receptor was directly assessed by comparing the extent of competition of Fm-I or native insulin with ^{125}I-labeled insulin for binding to receptors. These assays were performed by incubating adipocytes, IM9 lymphocytes, H4(IIEC3) cultured hepatoma cells, adipocyte plasma membranes, and liver plasma membranes with 0.4

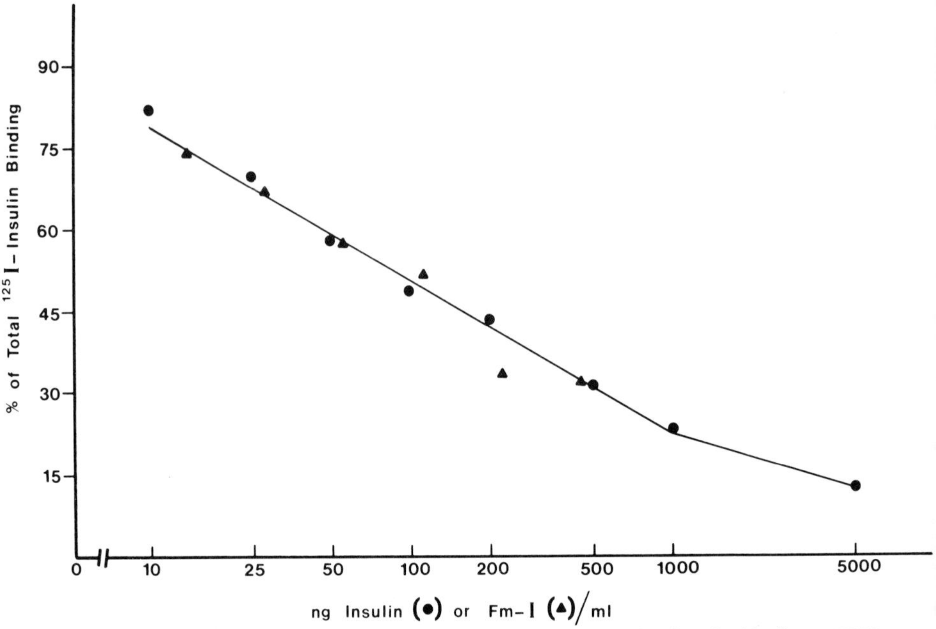

FIG. 3. Comparison of monomeric ferritin–insulin and porcine insulin binding to IM9-lymphocytes. IM9-lymphocytes were incubated for 60 min at 15° with 0.4 ng/ml ^{125}I-labeled insulin and increasing concentrations of monomeric ferritin–insulin (▲) or porcine insulin (●). A 75-μl aliquot of the incubation mixture was layered over 300 μl 0.25 *M* sucrose in 10 m*M* sodium phosphate buffer and the cells pelleted in a microfuge. The pellets were counted to determine cell-associated ^{125}I-labeled insulin. Nonspecific binding was determined in the presence of 25 μg/ml porcine insulin. The values reported represent specifically bound ^{125}I-labeled insulin. These data show that monomeric ferritin–insulin competes with ^{125}I-labeled insulin for binding to the IM9-lymphocyte insulin receptor almost identically to native porcine insulin.

ng/ml ^{125}I-labeled insulin, prepared as described,[26] and increasing concentrations of native insulin or Fm-I. Incubation conditions are given in detail in the legends to the figures. Receptor bound insulin was determined by microfuge gradient centrifugation through 0.25 *M* sucrose in 0.05 *M* sodium phosphate buffer, pH 7.4 (IM9s, H4(IIEC3)s, adipocyte, and liver plasma membranes) or by microfuge gradient centrifugation through dinonyl phthalate (adipocytes) as previously described.[27] The results of typical assays with the cells are shown in Figs. 3 to 5. As might have been

[26] J. B. Schweitzer, R. M. Smith, and L. Jarett, *Proc. Natl. Acad. Sci. U.S.A.* **77,** 4692 (1980).

[27] G. T. Hammons, R. M. Smith, and L. Jarett, *J. Biol. Chem.* **257,** 11563 (1982).

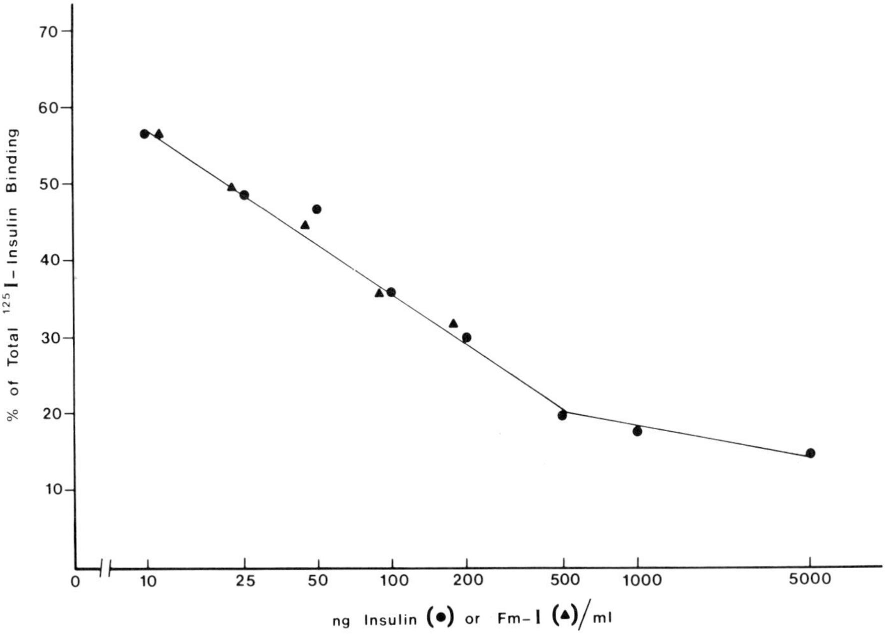

FIG. 4. Comparison of monomeric ferritin–insulin and porcine insulin binding to H4(IIEC3) cultured hepatoma cells. H4(IIEC3) hepatoma cells were incubated for 60 min at 24° with 0.4 ng/ml ^{125}I-labeled insulin and increasing concentrations of monomeric ferritin–insulin (▲) or porcine insulin (●). Total and specific binding of ^{125}I-labeled insulin was determined as described in Fig. 3. The results indicate that monomeric ferritin–insulin is indistinguishable from porcine insulin in its ability to compete with ^{125}I-labeled insulin for binding to the IIEC3 hepatoma insulin receptor.

expected from the biological activity assay, Fm-I competed with ^{125}I-labeled insulin for binding to the adipocyte insulin receptor in a dose-dependent manner identically to native insulin of similar concentration. We had previously shown similar data using isolated adipocyte plasma membranes.[13] We have shown that the ultrastructural distribution pattern of insulin receptors was not the same on all cell types.[28] Receptor sites on adipocytes are predominantly grouped while the receptors on liver membranes are mostly single sites. These differences might have some effect on the binding of Fm-I to different tissues if the ferritin molecule interferes even slightly in the binding process. However binding data indicated Fm-I binding in all of the tissues studied was identical to native insulin.

[28] R. M. Smith and L. Jarett, *in* "Insulin, Its Receptor and Diabetes" (M. D. Hollenberg, ed.). Dekker, New York (in press).

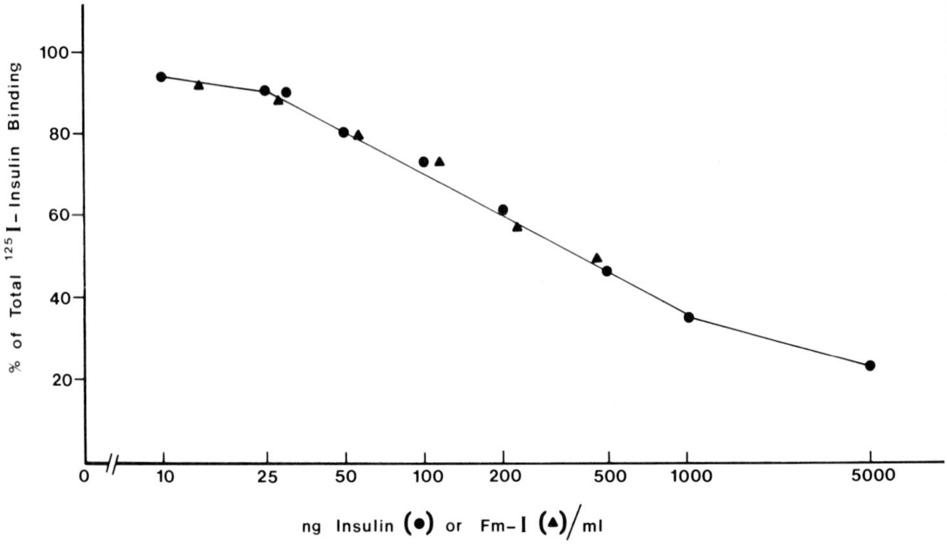

FIG. 5. Comparison of monomeric ferritin–insulin and porcine insulin binding to rat adipocytes. Intact rat adipocytes were incubated for 30 min at 24° with 0.4 ng/ml ^{125}I-labeled insulin and increasing concentrations of monomeric ferritin–insulin (▲) or porcine insulin (●). A 300-μl aliquot of the incubation mixture was placed in a microfuge tube and then over layered with 100 μl of dinonyl pthalate. The tubes were microfuged for 1 min and the cell layer cut from the microfuge tube for counting. Nonspecific binding in the presence of 25 μg/ml porcine insulin was subtracted from total binding. The results show that monomeric ferritin insulin is indistinguishable from porcine insulin in its ability to compete with ^{125}I-labeled insulin for binding to rat adipocytes.

Stability of the Conjugate and Reproducibility of the Method

In addition to quantitating the insulin concentration, experiments have been performed to test the stability of the Fm-I. Preparations have been stored at 4° for up to 6 months with no loss of insulin activity or dissociation of the conjugate (Table III). This was determined by ultracentrifugation of aliquots of a preparation to sediment the Fm-I. The supernatants were saved and the pellets resuspended to their original volumes. The insulin concentration of both was determined by radioimmunoassay. The procedure was repeated at various intervals. If the insulin was dissociating the concentration of insulin in the supernatant would have increased. The small amount of insulin present in the supernatants, which remained virtually constant over 6 months, was attributed to a small amount of ferritin which failed to sediment. Similar protocols were performed after incubating cells or membrane fractions with Fm-I and the data in Table IV indicate that under incubation conditions insulin did not dissociate from the conjugate.

TABLE III
STABILITY OF MONOMERIC–FERRITIN INSULIN DURING STORAGE AT 4°[a]

Elapsed time from preparation date (days)	Pellet (ng recovered)	Supernatant (ng recovered)
3	915.4 ± 26.6	4.2 ± 0.6
14	927.2 ± 17.0	2.7 ± 1.1
35	896.3 ± 16.5	3.2 ± 0.7
64	905.1 ± 12.2	4.6 ± 1.2
182	918.7 ± 18.8	2.9 ± 1.6

[a] An aliquot of monomeric ferritin–insulin was ultracentrifuged at 40,000 rpm for 90 min after being stored for the indicated time period at 4°. The supernatant was removed and the pellet resuspended to its original volume. Insulin concentrations of both the pellet and supernatant were determined by RIA, and the results expressed in terms of the total insulin recovered. The small amount of insulin activity found in the supernatants was probably due to incomplete sedimentation of the ferritin.

TABLE IV
STABILITY OF MONOMERIC FERRITIN–INSULIN DURING INCUBATION WITH ADIPOCYTE PLASMA MEMBRANES[a]

	− Membranes (ng insulin)	+ Membranes (ng insulin)
Supernatant	0.2 ± 0.01	0.2 ± 0.01
Ferritin pellet	41.6 ± 0.9	38.2 ± 1.2

[a] Purified adipocyte plasma membranes were incubated with 40 ng monomeric ferritin insulin in 1 ml of Krebs–Ringer phosphate buffer, pH 7.4, with 3% bovine serum albumin, for 30 min at 24°. The membranes were pelleted by microfuge centrifugation and the membrane-free supernatant was ultracentrifuged at 40,000 rpm for 90 min. The insulin concentration was determined in both the supernatant and the resuspended pellets by RIA. The results indicate that incubation with membranes did not result in dissociation of the insulin from the conjugate.

TABLE V
REPRODUCIBILITY OF THE PREPARATION OF MONOMERIC FERRITIN–INSULIN[a]

Analysis	Mean	Range
Ferritin : insulin (molar ratios)	15.2 : 1	10.2–17.1 : 1
Immunologic : biologic (concentrations)	1 : 1	0.9–1.1 : 1
Recovery of insulin (%)	0.10	0.094–0.112
Total insulin recovered (μg)	7.0	6.6–7.8

[a] The above analyses were made on 25 separate preparations of monomeric ferritin–insulin during the past 3 years. The results indicate that the current method has been highly reproducible.

In the past two years at least 25 preparations of Fm-I have been made in our laboratory by various individuals. When the described procedure was followed, the results have been very consistent as demonstrated in Table V. Modifications, such as increasing the time of incubation of the insulin and activated ferritin, have resulted in predictable differences.[13]

Electron Microscopic Studies

Estimation of Oligomer Contamination of Fm-I

In order to assess the relative purity of the Fm-I and the amount of contamination with oligomeric forms, Fm-I was diluted in deionized water with 0.1% bacitracin (added as a wetting agent) and applied to a carbon-coated Formvar-coated 300 mesh copper grid. The samples were negatively stained with 2.5% ammonium molybdate in 2.0% ammonium acetate. The specimen was observed in an JEOL 100CX electron microscope and random areas of the specimen were photographed. The total number of ferritin particles and the number of closely aligned and probably oligomeric ferritin molecules were counted. These results were compared to that found for a similar dilution of cationic ferritin (Miles Laboratories Inc., Elkhart, IN) which is known to be entirely monomeric. Electron micrographs of Fm-I in Fig. 6 and cationic ferritin (not shown) showed that the Fm-I was virtually identical in appearance to cationic ferritin and quantitation of the Fm-I showed that the concentration of oligomeric ferritin was less than 3% of the total (data not shown).

Use of Fm-I to Localized Occupied Insulin Receptors

Incubation of various tissues with ferritin insulin has been described in detail elsewhere[7–12,14–16,18] and is outlined in the legends to subsequent figures. After incubation the cells were washed by centrifugation through

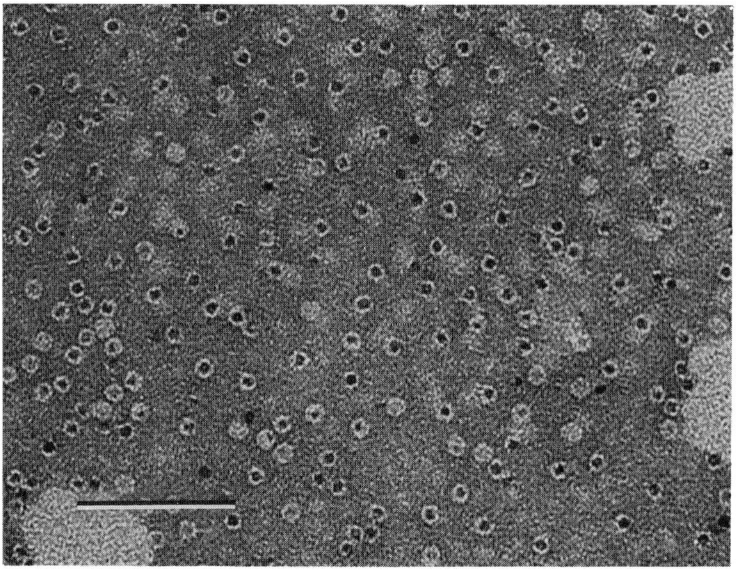

FIG. 6. Electron micrograph of negatively stained monomeric ferritin–insulin. The lack of closely approximated ferritin particles, which would be observed with ferritin oligomers, indicates that the conjugate is virtually pure monomeric ferritin. Bar = 0.1 μm; magnification, ×200,000.

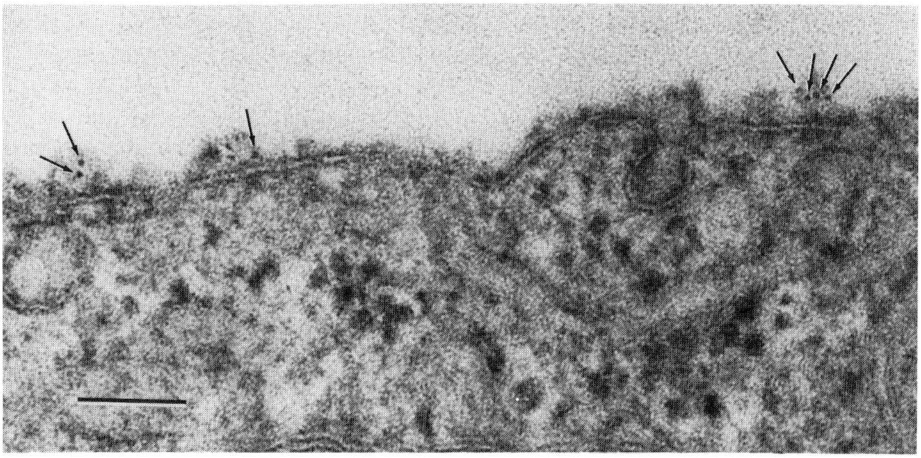

FIG. 7. Electron micrograph of Fm-I receptor sites on intact rat adipocytes. Isolated adipocytes were prepared and incubated with 40 ng/ml Fm-I for 30 min at 37° in Krebs–Ringer phosphate buffer, pH 7.4. The cells were washed and prepared for electron microscopy by previously published methods.[8] The arrows indicate Fm-I molecules bound to the insulin receptor in the glycocalyx of the adipocyte plasma membrane. Insulin receptors were found predominantly in small groups on the rat adipocyte and these groups did not aggregate during prolonged incubation.[9] Bar = 0.1 μm; magnification, ×175,000.

buffer and fixed with 2% glutaraldehyde in 0.1 *M* sodium cacodylate (membrane preparations)—or in Krebs–Ringer phosphate buffer (intact cell preparations). The tissues were routinely postfixed in 2% OsO_4 in 0.1 *M* sodium cacodylate buffer, stained in block with uranyl acetate, and embedded in Spurr resin. Thin sections were stained with saturated alcoholic uranyl acetate and examined in the electron microscope. Fm-I receptor quantitation has been described in detail elsewhere.[11]

Monomeric ferritin insulin has been used to qualititatively and quantitatively determine the original distribution, microaggregation, and endocytosis of insulin receptors on a variety of cell types. A comprehensive review of those studies has been prepared.[28] It is sufficient to say that our analyses have shown that there are signficant differences in the native distribution of insulin receptors, both in terms of the location of the receptor sites relative to specialized structures such as microvilli, and the organization of the receptor sites into groups of multiple receptors. Two cell

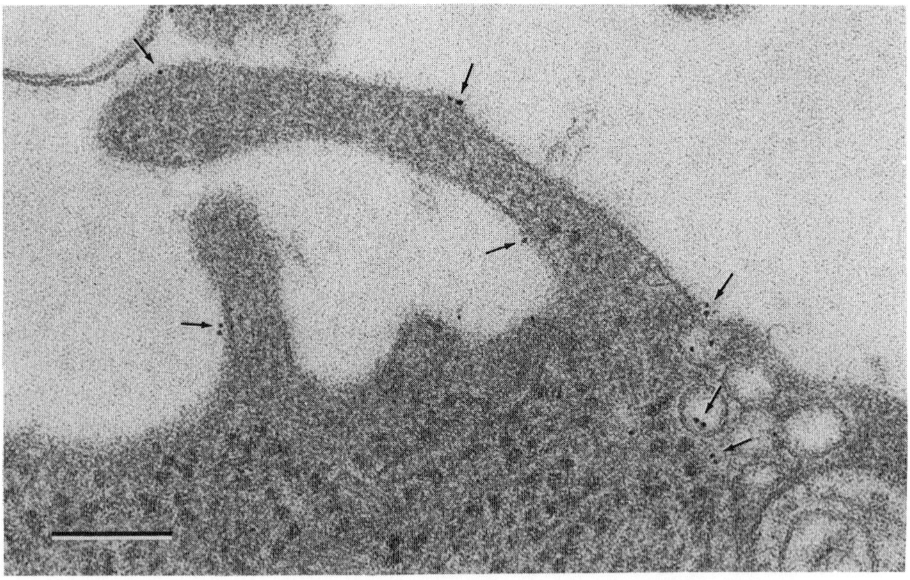

FIG. 8. Electron micrograph of Fm-I receptor sites on 3T3-L1 adipocytes. Cultured 3T3-L1 adipocytes were suspended in Krebs–Ringer phosphate buffer, pH 7.4, and incubated with 40 ng/ml Fm-I for 30 min at 37°. The cells were washed and prepared for electron microscopy by routine methods. The arrows indicate Fm-I molecules found to insulin receptors on the microvilli and pinocytotic vesicles. Quantitative analysis of the receptor distribution indicated that most of the receptors are initially located on the microvilli as single or paired receptor sites. The receptors microaggregate into small clusters and move to the intervillous cell surface and are endocytosed in pinocytotic invaginations.[28] Bar = 0.1 μm; magnification, $\times$175,000.

types which demonstrate some of these differences are shown in Figs. 7 and 8. A number of studies have proven that insulin receptors on the rat adipocyte are normally grouped together and the groups are randomly distributed over the entire cell surface.[9,11,12] The receptor sites do not aggregate and are randomly endocytosed in smooth pinocytotic invaginations.[15] Figure 7 demonstrates Fm-I binding to rat adipocytes.

In contrast to the rat adipocyte, the insulin receptors on the 3T3-L1 adipocyte are initially located on the microvillous projections of the cell and are found primarily as single or paired receptor sites. During incubation the insulin receptors appear to microaggregate and move to the cell surface where they concentrate in smooth pinocytotic invaginations and coated pits. The binding of Fm-I to the 3T3-L1 adipocyte is shown in Fig. 8. The observation of these differences in the ultrastructure of the insulin receptor is only part of a growing body of evidence that the insulin receptor may not be identical on all cell types. Monomeric ferritin–insulin, as well as other similar conjugates, provides a technique which should facilitate detailed investigation of insulin receptor dynamics on a variety of insulin responsive and resistant cell types.

[16] Cross-Linking of hCG to Luteal Receptors

By TAE H. JI and INHAE JI

A general approach to labeling receptors with photoactivable derivatives of polypeptide ligands has been discussed.[1-3] Most of polypeptide hormones are molecules of a single polypeptide or have covalently linked subunits. As a result, cross-linking of these hormones to receptors are straightforward. Lutropin (LH) and human choriogonadotropin (hCG) consist of two noncovalently associated, dissimilar subunits, designated α and β.[4] The cross-linking results are more complex and, therefore, we have prepared a variety of hCG derivatives. The results suggest that both the α and β subunits can photoaffinity label the receptor and the receptor consists of heterologous subunits.[5-7]

[1] T. H. Ji, *in* "Membranes and Neoplasia: New Approaches and Strategies" (V. T. Marchesi, ed.), p. 171. Liss, New York, 1976.

[2] T. H. Ji, *Biochim. Biophys. Acta* **559,** 39 (1979).

[3] T. H. Ji, this series, Vol. 91, p. 580.

[4] J. G. Pierce and T. F. Parsons, *Annu. Rev. Biochem.* **50,** 469 (1981).

[5] T. H. Ji, D. J. Kiehm, and C. R. Middaugh, *J. Biol. Chem.* **255,** 2990 (1980).

[6] I. Ji and T. H. Ji, *Proc. Natl. Acad. Sci. U.S.A.* **77,** 7167 (1980).

[7] I. Ji, B. Y. Yoo, C. Kaltenbach, and T. H. Ji, *J. Biol. Chem.* **258,** 10853 (1981).

ISBN 0-12-182009-2

Preparation of Reagents

The *N*-hydroxysuccinimide esters (NHS) of 4-azidobenzoic acid (ABA), 4-azidobenzoylglycine (ABG), and 4-azidobenzoylglcylglycine (ABGG) were synthesized as follows.[5] *N*-Hydroxysuccinimide (41 mmol in 50 ml of dry acetone) and dicyclohexylcarbodiimide (42 mmol in 20 ml of dry acetone) were introduced sequentially into 40 mmol of 4-azidobenzoic acid suspended in 100 ml of dry acetone. This mixture was stirred for 5 hr and the precipitate was recovered by filtration. The filtrate was evaporated to dryness and the resulting powder (NHS-ABA) was washed with boiling petroleum ether. Azidobenzoylglycine (ABG) and azidobenzoylglycyclglycine (ABGG) were prepared[6] by reacting 10 mmol of NHS-ABA in 60 ml of dioxane with 10 mmol of glycine or glycylglycine, respectively, in 20 ml of 1 *M* sodium bicarbonate for 5 hr at 25°. The reaction mixture was filtered, the filtrate concentrated to remove dioxane, and acidified with hydrochloric acid. The precipitate was collected and recrystallized in ethyl alcohol. The resulting ABG or ABGG was esterified with *N*-hydroxysuccinimide as before to make NHS-ABG or NHS-ABGG.

Derivatization and Radioiodination

The reagents were freshly dissolved in dimethyl sulfoxide to a concentration of 100 m*M* and diluted 1 : 500 in phosphate-buffered saline (pH 7.4). Fifty microliters was added to 10 μg of hCG in 50 μl of 0.1 *M* Na_2HPO_4 (pH 7.5) and incubated for 30 min at 25°. To the solution were added 7 μg of chloramine-T in 7 μl PBS and 1 mCi of sodium [^{125}I]iodide in 10 μl of 0.1 *M* NaOH; 20 sec later, radioiodination was stopped by addition of 10 μg of sodium metabisulfite in 10 μl PBS and 50 μl of 16% sucrose in PBS. The derivatized and radioiodinated hCG was immediately fractionated on a Sephadex G-75 (superfine) column (0.6 × 15 cm) which was preequilibrated and eluted with 0.1% gelatin (Knox) in PBS at 5°.

For subunit photoaffinity labeling experiments, isolated subunits were derivatized, radioiodinated, and purified. The treated subunit, ABG-^{125}I-α or ABG-^{125}I,β was recombined with the other untreated subunit in 0.1 *M* Na_2HPO_4 (pH 7.5) and incubated at 37° for 16 hr. The reconstituted hormones ABG-^{125}I-α,β or α,ABG-^{125}I-β were purified on a Sephadex G-75 (superfine) column (0.7 × 100 cm) which was preequilibrated and eluted with 10 m*M* Na_2HPO_4 (pH 7.5), 0.9% NaCl, and 0.1% gelatin (Knox).

Biological Activities of hCG Derivatives

Protein hormones are susceptible to structural perturbation during chemical modification, radioiodination, and column chromatography.

Modified hCG had a higher binding activity when it was radioiodinated by lactoperoxidase catalysis than by the chloramine-T method. The enzyme, however, complicates the interpretation of photoaffinity-labeling results since radioiodination not only of the hormone but also of the enzyme occurs. Other impurities are also present in enzyme preparations. These radiolabeled contaminants appear on gels along with photoaffinity-labeled complexes.

The sequence of derivatization and radioiodination has some effect on the binding activity of AB-[^{125}I]hCG.[7] When the hormone was derivatized first and then radioiodinated, the binding activity was higher than when the hormone was treated in the reverse order. We also found that the hormone could be radioiodinated immediately after derivatization, without purification on a column. This elimination of the column chromatographic step enhanced the binding activity of AB-[^{125}I]hCG. The sequence of hormone treatment in subsequent experiments was derivatization at 100 μM of the reagent, radioiodination by the chloramine-T method, and final purification on a Sephadex G-75 (superfine) column. Under these conditions approximately half of the hormone is derivatized.[6]

The hormone derivatives have been tested for the binding affinity, number of binding sites, competitive binding in the presence of other hormones and nonbinding proteins, binding specificity toward cells possessing and lacking the receptor and finally progesterone synthesis.

The equilibrium association constants of [^{125}I]hCG and ABGG-[^{125}I]hCG differ slightly, but not significantly, being 5.1×10^9 and $2.98 \times 10^9\ M^{-1}$, respectively. The number of binding sites on granulosa cells was equal for both (6520 ± 430/granulosa cell, $n = 4$). Nonspecific binding (the binding in the presence of 1000-fold higher concentration of untreated hCG) was less than 15% of the total binding above 100 pM and less than 2% below 100 pM. The specific binding of ABGG-[^{125}I]hCG to granulosa cells was completely blocked to basal level by excess untreated hCG or LH but not by FSH, prolactin, insulin, or bovine serum albumin.[7]

To determine which hormone derivative is most potent, the binding activity of three different [^{125}I]hCG derivatives, AB-[^{125}I]hCG, ABG-[^{125}I]hCG, and ABGG-[^{125}I]hCG, was evaluated by competitive binding in the presence of increasing concentrations of untreated hCG. The bindings of [^{125}I]hCG and ABG-[^{125}I]hCG were similar and their nonspecific binding less than 5%. AB-[^{125}I]hCG and ABGG-[^{125}I]hCG, however, did not bind as well as ABG-[^{125}I]hCG; ABGG-[^{125}I]hCG had the lowest specific and highest nonspecific binding.[7] Therefore, ABG-[^{125}I]hCG was chosen for subsequent experiments.

ABG-[^{125}I]hCG binding to granulosa cells appeared to be relatively specific since the hormone derivative did not bind to erythrocytes, peripheral lymphocyes, or H4 hepatoma cells.[7] Furthermore, granulosa cells cultured for 2 weeks failed to bind the hormone derivative. It is known that the binding capacity of granulosa cells declines as they are cultured in the absence of insulin; we observed 10% binding following 6 days of culture and only background levels after 2 weeks.

Finally, the physiological response of the hormone derivatives was examined. Both [^{125}I]hCG and ABG-[^{125}I]hCG were similarly effective in stimulating progesterone production, whereas ABG-^{125}I-α lacks biological activity. In conclusion, ABG-[^{125}I]hCG is able to bind specifically, with a high affinity, to receptors on granulosa cells and to induce progesterone synthesis.

Photoaffinity Labeling

Granulosa cells were incubated with ABG-[^{125}I]hCG for 30 min at 37° and hormone binding was terminated by rinsing the cells twice. After photolysis, the cells were solubilized and electrophoresed. The photolyzed samples revealed four new bands with slower electrophoretic mobility in addition to the α, β, and $\alpha\beta$ bands. Of the four bands, three were conspicuous, but the fourth band was difficult to detect on print, although it is distinctly visible on autoradiographs. In the nonphotolyzed sample, they were absent. They were also absent when the sample was irradiated with a tungsten lamp or when the cells were incubated with ABG-[^{125}I]hCG in the presence of 50 ng/ml of untreated hCG and photolyzed. The positions of the three conspicuous bands on the autoradiograph correspond to molecular weights of 96,000, 76,000, and 73,000 and the fourth faint band corresponds to that of 120,000. Photoaffinity labeling by ABG-^{125}I-α,β produced 73K, 76K, and 96K bands. In contrast, α, ABG-^{125}I-β produced the 106K, 88K, and 83K molecular weight bands. Because both the reagent and [^{125}I]iodine were confined to only one of the subunits, the results demonstrate that the ^{125}I-β derivative was cross-linked directly to membrane components in the 106K, 88K, and 83K bands, and the ^{125}I-α derivative to those in the 96K, 76K, and 73K bands.

Lutropin (LH) is one of the three known glycoprotein hormones of pituitary origin, the other two being follitropin (FSH) and thyrotropin (TSH). Human choriogonadotropin (hCG) of placental origin resembles pituitary LH. These hormones consist of two noncovalently associated, dissimilar subunits, designated α and β. Within the same mammalian species, the α subunits of these hormones have virtually identical amino acid sequences. In contrast, the β subunits have distinct sequences and account for hormone specificity. Binding and the subsequent biological response are produced only by the $\alpha\beta$ dimer and not from either of the dissociated subunits. Results of studies with modified hormones, photoaffinity labeling,[6–8] and monoclonal anti-hCG[9] are consistent with the view that both subunits of LH/hCG interact with the hormone receptor. The same conclusion was drawn from a conventional hormone–receptor cross-linking study.[10]

In conclusion, our photoaffinity labeling of the LH/hCG receptor system on porcine granulosa cells[6–8] has demonstrated that both the α and β subunits of hCG directly photoaffinity label the hormone receptor. Three new bands appear on SDS–PAGE as a consequence of photoaffinity labeling by *each* subunit: the molecular weights of the three bands (106K, 88K, and 83K) produced by the subunit are larger by approximately 10K than those of the three bands (96K, 76K, and 73K) labeled by the α subunit. Although it could be a coincidence that the molecular weight of the β subunit is approximately 10K larger than that of the α subunit, the similarity in these differences suggests the possibility that both the α and β subunits have labeled the same polypeptides.

[8] I. Ji and T. H. Ji, *Proc. Natl. Acad. Sci. U.S.A.* **78,** 5465 (1981).
[9] W. R. Moyle, P. H. Ehrlich, and R. E. Canfield, *Proc. Natl. Acad. Sci. U.S.A.* **79,** 2245 (1982).
[10] R. V. Rebois, *J. Cell Biol.* **94,** 70 (1982).

[17] Covalent Labeling of the Hepatic Glucagon Receptor

By JOHN T. HERBERG and RAVI IYENGAR

Principle

This procedure for covalently labeling the hepatic glucagon receptor utilizes the light-sensitive heterobifunctional cross-linker hydroxysuccinimidyl-*p*-azidobenzoate (HSAB) to link the bound [^{125}I-Tyr^{10}]monoiodoglucagon ([^{125}I]MIG) to the receptor protein. The procedure was first described by Johnson *et al.*[1] to covalently attach ^{125}I-labeled glucagon to its liver membrane receptor. We have subsequently used the procedure with the well-defined probe, [^{125}I]MIG.[2,3] The method involves first the

[1] G. L. Johnson, V. I. MacAndrew, and P. F. Pilch, *Proc. Natl. Acad. Sci. U.S.A.* **78,** 875 (1981).
[2] F. J. Rojas, T. L. Swartz, R. Iyengar, A. J. Garber, and L. Birnbaumer, *Endocrinology* **113,** 711 (1983).
[3] R. Iyengar and J. T. Herberg, *J. Biol. Chem.* **259,** 5222 (1984).

ISBN 0-12-182009-2

binding of the labeled hormone to its receptor and the removal of the excess unbound label. This is then followed by incubation with the cross-linker, in the dark and then under ultraviolet illumination to covalently couple the bound [^{125}I]MIG. HSAB contains an amino reactive group as well as an aryl azide which, upon light activation, is converted to an aryl nitrene that reacts in a chemically unspecific manner.

Materials

Rat liver membranes are prepared according to the method of Neville[4] as described by Pohl *et al.*[5]

[^{125}I]MIG is synthesized and purified on a C18-μBondapak column using a high-pressure liquid chromatography system according to the procedure of Rojas *et al.*[2]

HSAB is purchased from Pierce Chemical Co. and dissolved in dimethyl sulfoxide. The HSAB solution is prepared fresh for each experiment.

Methods

[^{125}I]MIG Binding to Liver Membranes. Liver membranes (0.2–1.0 mg protein/ml) are incubated in a mixture containing 20 m*M* phosphate buffer, pH 7.5, 1 m*M* EDTA, 1% (w/v) BSA, 1 n*M* [^{125}I]MIG, and appropriate additives. When used, final concentration of glucagon is 1–3 μM and that of GTP 0.1 m*M*, or as specified. Incubations are carried out at 32° for 15 min. Incubation volumes vary from 0.1 for routine filtration assays to 10 ml for cross-linking and further analysis of the receptor.

When [^{125}I]MIG binding to membranes are analyzed by filtration, the 100 μl samples are diluted to 1 ml with ice-cold 20 m*M* phosphate, 1% (w/v) BSA, pH 7.5. The diluted sample is filtered through 0.45-μm cellulose acetate filters (Oxoid) which had been previously soaked in 10% BSA. The tube containing the sample is washed once with 1 ml of 1% BSA. The filters are then washed with 10 ml of ice-cold 20 m*M* phosphate buffer, 0.1% BSA (wash buffer). The filters are then counted in a Searle gamma counter to determine bound radioactivity.

Large volume (2–10 ml) binding assays were carried out in 30 ml ultracentrifuge tubes. After the incubation, the samples (1-ml aliquots) are diluted to 25 ml with ice-cold 20 m*M* wash buffer. All further handling of the membranes was at 0–4° unless specified otherwise. The membranes were sedimented by centrifugation at 100,000 *g* for 20 min. The superna-

[4] D. M. Neville, Jr., *Biochim. Biophys. Acta* **154,** 540 (1968).

[5] S. L. Pohl, L. Birnbaumer, and M. Rodbell, *J. Biol. Chem.* **246,** 1849 (1971).

COMPARISON OF THE FILTRATION AND CENTRIFUGATION METHODS FOR THE REMOVAL OF UNBOUND [^{125}I]MIG AFTER EXPOSURE OF LIVER MEMBRANES TO [^{125}I]MIG

Additions to assay[a]	Method of separation: Filtration[b]	Centrifugation[c]
	[^{125}I]MIG bound (fmol/mg)	
None	227 ± 6	236 ± 7
100 μM GTP	55 ± 3	51 ± 5
1 μM glucagon	13 ± 1	12 ± 5

[a] The assay was carried out in a final volume of 1 ml at 0.6 mg protein/ml.
[b] 100-μl aliquots were withdrawn at the end of the incubation and filtered as described in the text.
[c] 200 μl sample was diluted to 2.0 ml and centrifuged as described in detail in the text.

tant was removed by aspiration and the pellet was resuspended in 1 ml of 20 m*M* phosphate buffer, pH 7.5, diluted to 25 ml with phosphate buffer and sedimented by centrifugation. The final pellet is resuspended in 20 m*M* phosphate buffer, pH 7.5 at a concentration of 5–10 mg protein/ml. This two-step washing procedure results in removal of all unbound [^{125}I]MIG, such that the pmol of [^{125}I]MIG bound per mg membrane protein was essentially identical to that seen by filtration (see the table).

Cross-Linking of Bound [^{125}I]MIG to Liver Membranes. HSAB is weighed in the dark and dissolved in dimethyl sulfoxide in a brown bottle to give a concentration of 20 m*M*. The membranes containing [^{125}I]MIG-occupied receptors are mixed with the HSAB solution such that the final concentration is 200 μM. The samples are kept in the dark for 3 min. Subsequently, they are irradiated in a final volume of 1.0 ml for 9 min under a mercury arc lamp (Gates Inc. A. H. Thomas #6281-H10) at a distance of 16 cm. The samples are held on ice during the UV irradiation. After the irradiation, the samples are diluted to 10 ml with 25 m*M* Tris–HCl, pH 7.5, 100 μM GTP, and 1 μM glucagon. This mixture is incubated for 15 min at 32°. The membranes are then sedimented by centrifugation at a 100,000 *g* for 20 min.

SDS-Gel Electrophoretic Analysis of Covalently Labeled Membranes. SDS–PAGE are run according to the procedure of Laemmli.[6] Routinely, samples are analyzed on a 10% gel with a 3% stacking gel. Prior to appli-

[6] V. Laemmli, *Nature* (*London*) **227,** 680 (1970).

cation onto the gel, the membranes are incubated for 1 hr at 32° in the presence of 1% SDS and 2.5% 2-mercaptoethanol. Heating at higher temperatures is avoided since this lead to aggregation and subsequent retention of the samples in the stacking gel during electrophoresis. Twenty-five to fifty micrograms of membrane protein in a volume of 50 μl per sample is generally applied to the gel. After electrophoresis, the gels are stained with Coomassie blue, destained, and dried. The dry gels are then exposed to Kodak X-Omat film in cassettes equipped with intensifying screens (Cronex Xtra-Lite). Generally, exposure is for 48 hr at −70°. Subsequently, the autoradiograms are developed and scanned with a Kontes Fiberoptics scanner equipped with a Hewlett-Packard integrator. In most instances, 48 hr exposure allows for scanning within the linear ranges of the instrument.

Analysis of the Data

When liver membranes that had [^{125}I]MIG bound are treated with HSAB and then analyzed by SDS-gel electrophoresis, it is observed that label is associated with one macromolecular entity of 63,000 Da. Inclusion of GTP during the binding reaction results in a significant decrease in the label associated with 63,000 Da protein, while addition of glucagon abolishes labeling of the 63,000 Da protein (Fig. 1). It is also noteworthy in Fig. 1 that inclusion of GTP and unlabeled glucagon affect not only the label at M_r = 63,000 region but also the label that runs with the dye front, which is presumably "free" [^{125}I]MIG. This is predictable since only the bound [^{125}I]MIG is present during the cross-linking reaction and since GTP and glucagon reduce total binding and do not affect the nonspecific component is relatively small (2–5% of total), the label at the dye front represents the fraction of the receptor bound [^{125}I]MIG not cross-linked by HSAB treatment. In all the experiments that we have performed, additives that affect specific binding also affect the amount of label that travels with the dye front.

The specificity with which the 63,000 Da protein recognizes [^{125}I]MIG is shown in Fig. 2. In this experiment, the binding reaction was performed in the absence or presence of various peptide hormones. Of the four hormones tested, only addition of glucagon results in abolishment of labeling of the 63,000 Da protein. Addition of insulin, ACTH, or 8-arginine vasopressin had no effect on the labeling of the 63,000 Da protein. This experiment indicates that the 63,000 Da protein specifically recognizes glucagon. The concentration range in which [^{125}I]MIG affects the labeling of the 63,000 Da peptide was further analyzed. One such experiment is shown in Fig. 3. Here, varying concentrations of glucagon were included

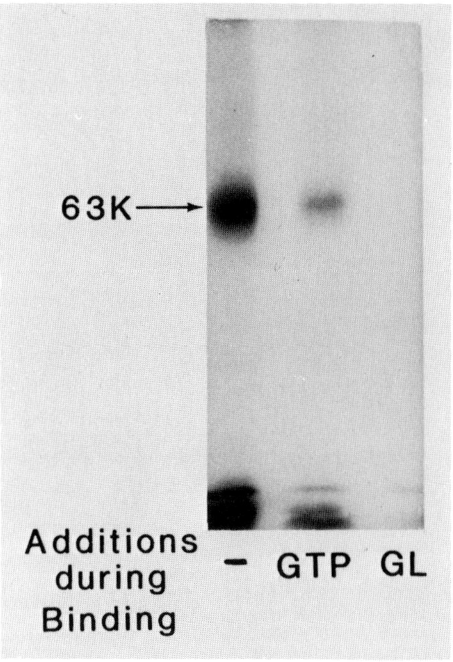

FIG. 1. Attachment of rat liver plasma membrane bound [^{125}I]MIG to a macromolecular moiety after treatment with the heterobifunctional cross-linker HSAB. Liver membranes (0.27 mg protein) were exposed to 1 nM [^{125}I]MIG in a final volume of 1.0 ml. The binding reaction was carried out in the presence of 100 μM GTP or 3 μM glucagon or without these ligands. At the end of the incubation period, the membranes were diluted 20-fold and washed twice by centrifugation. Membranes were suspended in phosphate buffer, pH 7.5, to a final volume of 1.0 ml and incubated with 0.2 mM HSAB in the dark and under UV illumination. After irradiation, the samples were diluted to 10 ml with 25 mM Tris–HCl, pH 7.5, and washed by centrifugation. The pellets were resuspended in 0.5 ml of 20 mM phosphate buffer, pH 7.5. Proteins were estimated in the final pellets. Of the appropriate membrane proteins 35 μg was applied onto the gel. The samples were electrophoresed, the gel was dried, and subjected to autoradiography. A 48 hr autoradiogram is shown.

during the binding reaction. The bound [^{125}I]MIG was covalently attached to the membranes and equivalent amounts of membrane protein were analyzed by SDS-gel electrophoresis. It is observed that the concentration range in which glucagon blocks labeling of the 63,000 Da peptide agrees well with the range in which it interacts with the glucagon receptor and stimulates adenylyl cyclase.[7]

[7] R. Iyengar, J. Abramowitz, M. E. Riser, and L. Birnbaumer, *J. Biol. Chem.* **255,** 3558 (1980).

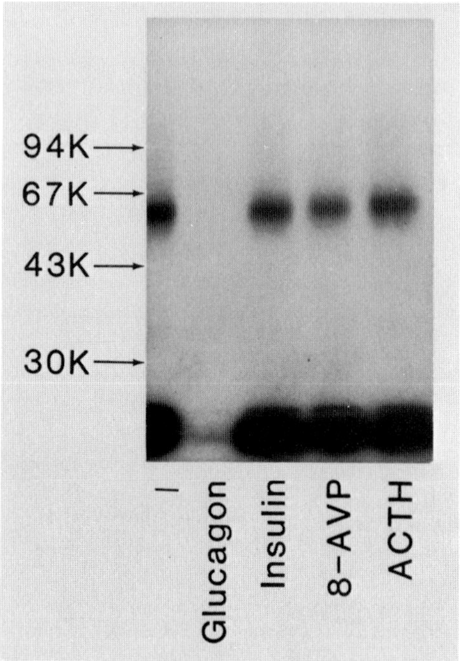

FIG. 2. Effect of various peptide hormones on the binding of [^{125}I]MIG to liver membranes and the HSAB-mediated covalent attachment of [^{125}I]MIG to a liver membrane protein. Liver membranes (0.32 mg/ml) were incubated with 1 n*M* [^{125}I]MIG without and with the other specified peptide hormones. The hormones used were insulin (1.8 μM), ACTH (2.2 μM), 8-AVP (8.6 μM), and glucagon (3 μM). After the incubation, the membranes were washed free of unbound label and incubated with HSAB in the dark and under UV illumination. The membranes were then washed and proteins and bound counts in the final pellets were estimated. It was found that incubation in the absence of any unlabeled hormone or in the presence of insulin, ACTH, or 8-AVP resulted in a 29,000–32,000 cpm bound per 40 μg membrane protein. When glucagon was present, 1200 cpm were bound to 40 μg membrane protein. Forty micrograms of protein for each sample was loaded on the gel. The samples were then subjected to electrophoresis followed by autoradiography. An autoradiogram (48 hr exposure) is shown.

To further establish that the 63,000 Da receptor labeled is capable of interacting with the stimulatory regulator of adenylyl cyclase, we quantitatively compared the effect of varying concentrations of GTP on the amount of [^{125}I]MIG bound to receptors in the membrane and on the amount of labeling obtained after crosslinking. One such experiment is shown in Fig. 4. In this experiment, [^{125}I]MIG binding was performed in

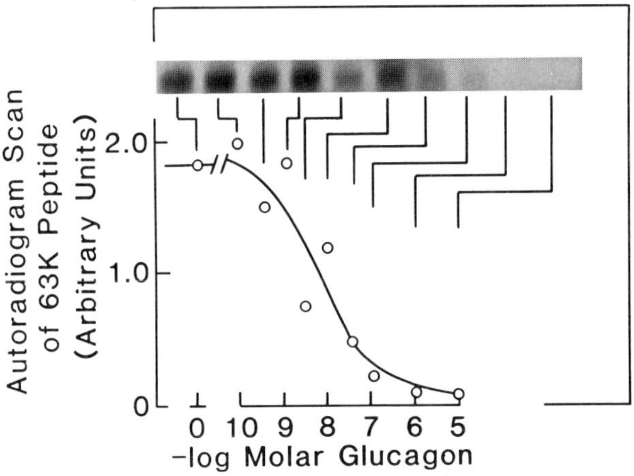

FIG. 3. Effect of varying concentrations of glucagon on the extent of HSAB-mediated [125I]MIG cross-linked to the M_r = 63,000 peptide. Liver membranes were incubated with 1 n*M* [125I]MIG and indicated concentrations of unlabeled glucagon. After the binding reaction, the membranes were washed free of unbound label. The bound [125I]MIG was cross-linked by incubation with HSAB. The membranes were washed free of cross-linker and the proteins estimated in the final pellets. Equivalent amounts of proteins (35 μg) were applied to the gel for electrophoresis. After electrophoresis, the gels were dried and subjected to autoradiography (48 hr exposure). The autoradiogram was scanned using a Kontes Fiber Optics Scanner equipped with a Hewlett-Packard Integrator. The areas under the individual peaks for the M_r = 63,000 peptide is shown as a function of the glucagon concentration used. The region of autoradiogram used for the scanning is shown as the inset.

the presence of varying concentrations of GTP. Aliquots were withdrawn and the amount of [125I]MIG bound per mg protein was determined. Another aliquot of membranes were incubated with HSAB in the dark first and under UV light afterwards to covalently attach the bound [125I]MIG. The peptide(s) labeled by this treatment were then analyzed by SDS–PAGE and autoradiography. As illustrated, increasing concentrations of GTP lead to a progressive decrease in binding and in the labeling of the 63,000 Da peptide. Further, the concentration range in which GTP affected on the labeling of the 63,000 Da peptide (upper panel) correlated well with the range in which GTP affected [125I]MIG binding to the receptor (lower panel).

Since GTP effects on hormone binding occur due to receptor-regulatory component interactions, this experiment allows us to conclude that the 63,000 Da protein functionally interacts with the stimulatory regulator. The experiments shown in Figs. 1–4 indicate that the 63,000 Da protein that is labeled with [125I]MIG using HSAB specifically recognizes

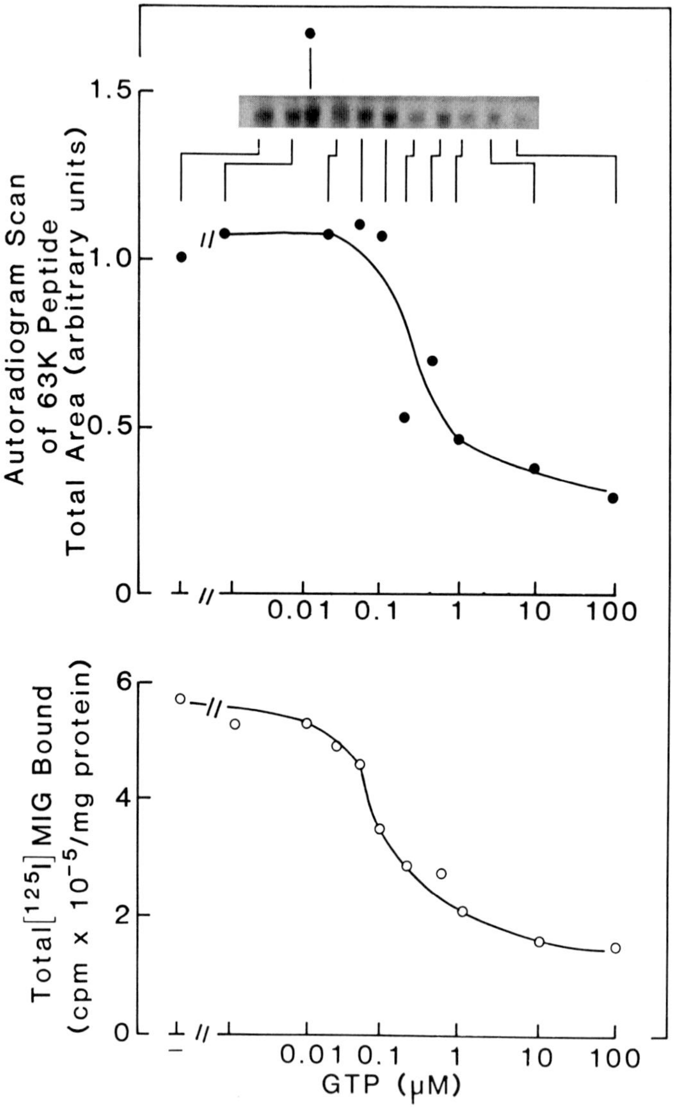

FIG. 4. Effect of varying concentrations of GTP on the extent of [^{125}I]MIG covalently attached to the M_r = 63,000 peptide (upper panel) and binding of [^{125}I]MIG to liver membranes (lower panel). Liver membranes (0.25 ml/mg) were incubated with 1 nM [^{125}I]MIG, 1 mM ATP, a nucleoside triphosphate regenerating system consisting of 20 mM creatine phosphate, 0.2 mg/ml creatine phosphokinase, and 0.02 mg/ml myokinase and indicated concentrations of GTP in a final volume of 3 ml. After 15 min at 32°, three 100-μl aliquots were withdrawn and filtered to estimate hormone binding to membranes (lower panel). The remaining membranes were washed free of unbound [^{125}I]MIG and incubated with HSAB in

glucagon and interacts with the stimulatory regulator of adenylyl cyclase, thus indicating that it is the structural correlate of the binding and adenylyl cyclase stimulating activity.

Acknowledgments

This research was supported by NIH Grants AM-26905, CA-29808, AM-27685 (DERC), and an Established Investigator Award to R.I. from the American Heart Association.

the dark and under UV illumination. The membranes were then washed to remove the cross-linker and proteins were estimated in the final pellet. Thirty-one micrograms of protein for each sample was applied to the gel. The samples were then subjected to electrophoresis. The gel dried was then exposed to X-ray film. The autoradiogram (48 hr exposure) was scanned using a Kontes Fiber Optics Scanner equipped with a Hewlett-Packard Integrator. The areas under individual peaks for the M_r = 63,000 peptide is shown as a function of the GTP concentration (upper panel). The region of autoradiogram used for the scanning is shown as inset.

Section III

Methods for Identification of Internalized Hormones and Hormone Receptors

[18] Peptide Hormone Receptors in Intracellular Structures from Rat Liver

By BARRY I. POSNER, MASOOD N. KHAN, and JOHN J. M. BERGERON

This chapter presents methods which have been used in our laboratories for the preparation and analysis of peptide hormone receptor-rich structures derived from within the rat hepatocyte. A new entity—the intermediate or "unique" vesicle—has been observed to contribute to the heterogeneity of Golgi fractions.[1] Though we only present a method for resolving intermediate vesicles from Golgi elements we are confident that in the near future a more general quantitative scheme for preparing receptor-rich intracellular structures will be available.

Sherman or Sprague–Dawley rats 150 to 250 g each have been used in these studies. The animals are fed *ad libitum* to the day before use and are fasted overnight before the experiment. They are killed by decapitation and exsanguinated before the livers are removed. All sucrose solutions (w/w solution) are prepared with reagent grade material (Fisher Scientific Limited). In earlier studies the rats received ethanol (0.6 g/100 g body wt) in a 50% (w/v) solution by stomach tube 90 min prior to sacrifice.[2–10] In more recent studies the use of ethanol has been discontinued.[1,11]

Golgi Subfractions by a Differential Centrifugation Procedure[1,2,11]

Livers are collected individually or in batches in ice-cold 0.25 *M* sucrose (5 ml/g liver), weighed, minced with scissors, and drained. All sub-

[1] M. N. Khan, B. I. Posner, R. J. Khan, and J. J. M. Bergeron, *J. Biol. Chem.* **257,** 5969 (1982).
[2] J. H. Ehrenreich, J. J. M. Bergeron, P. Siekevitz, and G. E. Palade, *J. Cell Biol.* **59,** 45 (1973).
[3] J. J. M. Bergeron, B. I. Posner, R. Sikstrom, and Z. Josefsberg, *J. Biol. Chem.* **253,** 4058 (1978).
[4] B. I. Posner, J. J. M. Bergeron, and Z. Josefsberg, *J. Biol. Chem.* **253,** 4067 (1978).
[5] B. I. Posner, Z. Josefsberg, D. Raquidan, and J. J. M. Bergeron, *Proc. Natl. Acad. Sci. U.S.A.* **75,** 3302 (1978).
[6] B. I. Posner, Z. Josefsberg, and J. J. M. Bergeron, *J. Biol. Chem.* **254,** 12494 (1979).
[7] B. I. Posner, R. M. Gonzalez, and H. J. Guyda, *Can. J. Biochem.* **58,** 1075 (1980).
[8] Z. Josefsberg, B. I. Posner, B. Patel, and J. J. M. Bergeron, *J. Biol. Chem.* **254,** 209 (1979).
[9] B. I. Posner, B. Patel, A. K. Verma, and J. J. M. Bergeron, *J. Biol. Chem.* **255,** 735 (1980).
[10] B. I. Posner, A. K. Verma, B. A. Patel, and J. J. M. Bergeron, *J. Cell Biol.* **93,** 560 (1982).
[11] B. I. Posner, B. Patel, M. N. Khan, and J. J. M. Bergeron, *J. Biol. Chem.* **257,** 5789 (1982).

ISBN 0-12-182009-2

sequent procedures are carried out in a cold room at 4°. The same volume of ice-cold 0.25 *M* sucrose is added to the minced tissue which is transferred to a Potter-Elvehjem glass-Teflon homogenizer. The mixture is homogenized with 6 full strokes of the motor-driven Teflon pestle at 1000 rpm. The homogenate is centrifuged at 4° at 8800 g_{max} for 10 min (8000 rpm in a Beckmann J2-21 using a JA-17 rotor). The supernatant is decanted into chilled 25 ml polycarbonate tubes and the pellet is discarded. The supernatant is centrifuged at 200,000 g_{av} for 30 min (52,000 rpm in a Spinco L2-65 or L2-65B using a 60 Ti rotor). The supernatant is decanted and discarded and the pellet gently vortexed in a small vol of 0.25 *M* sucrose and then decanted into a Dounce homogenizer. The following final volumes of sucrose solutions are added in order to the pellet based on the wet weight of liver from which the pellet was derived: 0.25 *M* sucrose, 0.33 ml/g liver; 2.0 *M* sucrose, 0.85 ml/g liver; and distilled water, 0.33 ml/g liver. The pellet, in 1.5 ml/g liver of 1.15 *M* sucrose, is gently homogenized with a loose-fitting pestle to yield an even suspension.

While the above centrifugation steps are proceeding the following sucrose gradient (see tabulation below) is being prepared in polyallomer tubes (for SW27 rotor), or cellulose nitrate tubes (for SW40 rotor). The sucrose solutions are prepared the day before and kept ice cold. The centrifuge tubes and sucrose solutions are embedded in crushed ice. Each succeeding sucrose layer is placed beneath the others using a syringe with a long needle.

	Volume (ml)/tube	
Sucrose solution (*M*)	SW27 rotor tubes	SW40 rotor tubes
0.25	7.0	2.4
0.60	7.0	2.4
0.86	7.0	2.4
1.00	7.0	2.4
Microsome suspension	~10.0	~4.0

The microsome suspension (see above) is slowly injected beneath the four sucrose layers using a syringe and long needle so as to fill the tube. The approximate homogenate volumes required are noted above. Centrifugation is carried out at: 100,000 g_{av} for 190 min (25,000 rpm, SW27 rotor), or 200,000 g_{av} for 95 min (40,000 rpm, SW40 rotor). Each fraction is removed from the gradient (Fig. 1) with a 90° angle needle and syringe:

Golgi light (Gl) at 0.25/0.60 *M* sucrose interface
Golgi intermediate (Gi) at 0.60/0.86 *M* sucrose interface

Golgi heavy (Gh) at 0.86/1.00 *M* sucrose interface
Small vesicles (SV) at 1.00/1.15 *M* sucrose interface

If it is desired that each fraction is maintained in isotonic conditions then each is placed in a separate cellulose nitrate tube and diluted to 0.25 *M* sucrose as follows: Gl, fill with 0.25 *M* sucrose; Gi, add equal volume of distilled water then fill with 0.25 *M* sucrose; Gh, add 2 volumes of distilled water and fill with 0.25 *M* sucrose; and SV, add 2.5 volumes of distilled water and fill with 0.25 *M* sucrose. The samples are then centrifuged at 200,000 g_{av} for 30 min (53,000 rpm, 60 Ti rotor). Corresponding pellets may be combined. For binding studies, all pellets are resuspended in 25 m*M* Tris–10 m*M* $MgCl_2$, pH 7.4.

Characterization

Protein Recovery. The recovery of protein is depicted in Fig. 1.

Morphology. The subfractions generated as depicted in the scheme of Fig. 1 have distinctive morphological features. Gl consists largely of vesicles containing lipoprotein particles. Gi is similar, but is somewhat more heterogeneous. The vesicles in this fraction show a wider range of size and there are some tubular and flattened saccular elements as well. The Gh fraction is quite heterogeneous. Lipoprotein containing vesicles are minority elements. There is an abundance of flattened saccules and tubular elements and a significant proportion of small vesicles. SV was formerly part of our Gh fraction prepared without a 1.0 *M* sucrose layer. It consists largely of small vesicles. It is not clear to what extent these represent vesiculated plasmalemma or endocytotic vesicles.

Enzyme Composition. It has been recently appreciated that various endoplasmic reticulum marker enzymes (viz. NADPH cytochrome *c* reductase and glucose-6-phosphatase) are present in bona fide Golgi compo-

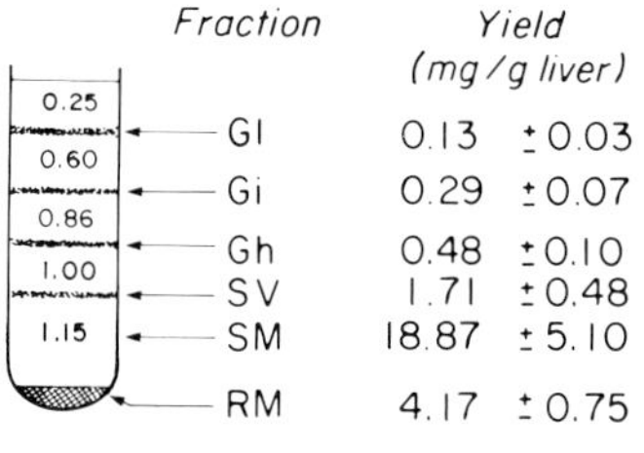

FIG. 1. Schematic depiction of fractions generated from microsomes by centrifugation in the discontinuous sucrose gradient. Sucrose concentrations and the yield of each subfraction are noted. Gl, Gi, and Gh refer to Golgi light, intermediate and heavy fractions respectively. SM refers to those smooth microsomal elements constituting the residual load zone. RM is the residual pellet containing predominantly rough microsomes. Each determination is the mean ± standard deviation of 36 fractionations.

nents[12] in addition to those regarded as plasma membrane markers.[13] However, progress in approaching the Golgi biochemically followed the recognition that the enzymes which characterize the Golgi and are especially enriched therein are various terminal glycosyltransferases.[14]

We assay galactosyltransferase (EC 2.4.1.38) by a modification of the method of Bretz and Staubli.[15]

Reagents

1. Sodium cacodylate buffer, 0.35 *M*
   $MnCl_2$, 0.32 *M*
   2-Mercaptoethenol, 0.31 *M*
   (pH adjusted to 6.6)
2. UDP-[^{3}H]Galactose, 10 μCi/ml
   (evaporate original material from New England Nuclear and redissolve in H_2O)
3. UDP-Galactose, 10 m*M*
   Adenosine triphosphate, 50 m*M*
4. Ovomucoid, 175 mg/ml
5. Triton X-100 in H_2O, 2% (w/v)
6. Phosphotungstic acid in 0.5 *N* HCl, 1.0% (w/v)

(Note: Solutions 1, 2, 4, and 5 can be stored frozen (−20°) for several months; solution 3 must be freshly prepared; and solution 6 can be stored at 4° for an indefinite time).

Procedure

The assay solution is prepared by mixing solutions 1, 2, 3, 4, and 5 in the ratio 5 : 1 : 4 : 4 : 4 (vol). Samples, containing 2 to 50 μg protein in a volume up to 220 μl, are mixed with 180 μl of assay solution and assayed in triplicate. Duplicate controls in which sample is added after terminating the reaction are prepared in parallel. After incubating at 37° for 30 min the reaction is stopped by adding 2 ml of ice-cold phosphotungstic acid–HCl. The mixture is vortexed and the tubes are left on ice for 30 min before centrifuging at 1600 g_{av} for 10 min. The pellet is washed 2 or 3 times with 2-ml aliquots of phosphotungstic acid and then once with 2 ml of ether : ethanol, 1 : 1 (v/v). The pellet is left to dry overnight at room temperature and dissolved in 0.5 ml Protosol (New England Nuclear) or 1 ml of 2 *M*

[12] K. E. Howell, A. Ito, and G. E. Palade, *J. Cell Biol.* **79,** 581 (1978).
[13] M. G. Farquhar, J. J. M. Bergeron, and G. E. Palade, *J. Cell Biol.* **60,** 8 (1974).
[14] R. Bretz, H. Bretz, and G. E. Palade, *J. Cell Biol.* **84,** 87 (1980).
[15] R. Bretz and W. Staubli, *Eur. J. Biochem.* **77,** 181 (1977).

NH_4OH. The solution is neutralized with glacial acetic acid and counted in 12 ml of Aquasol-2 (New England Nuclear). Enzyme activity is calculated as pmol galactose incorporated into protein/30 min/mg protein.

Sialyltransferase (EC 2.4.99.1) is assayed by a modification of the method of Bretz *et al.*[14]

Reagents

1. Sodium cacodylate buffer, 0.4 *M*
   2-Mercaptoethanol, 0.16 *M*
2. CMP-[^{3}H]sialic acid, 10 μCi/ml
   (evaporate original material from New England Nuclear and redissolve in H_2O)
3. CMP-*N*-acetylneuraminic acid, 10 m*M*
   (ammonium salt)
4. Asialofetuin, 175 mg/ml
   [Fetuin (Sigma Chem. MO) is acid hydrolyzed in 0.05 *N* H_2SO_4 by the method of Spiro,[16] the lyophilized protein is dissolved in H_2O]
5. Triton X-100 in H_2O, 2% (w/v)
6. Phosphotungstic acid in 0.5 *N* HCl, 1% (w/v)
7. Trichloroacetic acid in H_2O (TCA), 10% (w/v)
   All solutions are stored frozen at −20° except for solutions 5 and 6 which can be stored at 4°

Procedure

The assay solution is prepared by mixing solutions 1 to 5 in the ratios 25 : 10 : 10 : 10 : 10 (vol). Samples containing 1 to 8 μg protein in a volume up to 35 μl are mixed with 65 μl of the assay solution. Duplicate controls in which sample is added after terminating the reaction are prepared in parallel. After incubating at 37° for 30 min the reaction is stopped by adding 1 ml of phosphotungstic acid. The mixture is vortexed and left on ice for 30 min prior to centrifuging at 1600 g_{av} for 10 min. The pellet is washed 2 or 3 times with 2 ml ice-cold TCA and once with 1 ml ether : ethanol, 1 : 1 (v/v). The pellet is left overnight at room temperature to dry and is dissolved in 0.5 ml Protosol or 1 ml 2 *M* NH_4OH before neutralizing with glacial acetic acid and counting in Aquasol-2. Enzyme activity is calculated as nmol of CMP-neuraminic acid incorporated into protein/30 min/mg protein.

In distinguishing between various vesicle populations it is often useful

[16] R. G. Spiro, *J. Biol. Chem.* **235,** 2860 (1960).

to measure acid phosphatase (EC 3.1.3.2). Our procedure is based on the method of Gianetto and de Duve[17] with minor modifications.

Reagents

1. β-Glycerophosphate, disodium salt, 0.5 *M*
   Sodium acetate buffer,pH 5.0, 0.6 *M*
2. Triton X-100 in H_2O, 1% (w/v)
3. Trichloroacetic acid (TCA), 16% (w/v)

Procedure

The reaction mixture is prepared in phosphorus-free, clean glass tubes by mixing 50 μl of solution 1 and 50 μl of solution 2 with sample containing 2 to 25 μg of protein. The volume is adjusted to 500 μl with distilled water. Control tubes are those without added sample. After 60 min at 37° the reaction is stopped by adding 500 μl of 16% TCA. The clear supernatant is collected after centrifugation at 2400 g_{av} for 15 min and inorganic phosphorus is determined by the method of Chen *et al.*[18] The results are corrected for readings of the control incubations. A unit of enzyme is μmol of P_i released/hr/mg protein.

Hormone Binding. The Golgi fractions are enriched in binding sites for insulin, insulin-like growth factors, lactogen, and growth hormone.[19] In order to maximize binding to these fractions they must be frozen and thawed 4 times or incubated in the presence of 0.1% Triton X-100 to permeabilize the vesicles. This latency is explained as due to the presence of receptors on the inner aspect of the vesicles. Hormone binding is measured essentially as previously described.[4–6]

Reagents

1. 125 m*M* Tris–HCl, 50 m*M* $MgCl_2$, 0.5% bovine serum albumin (BSA), pH 7.4
2. ^{125}I-labeled hormone
   (prepared by chloramine-T procedure as described elsewhere[20])
3. Unlabeled hormone prepared in solution 1
   (insulin, oPRL, and bGH, 50 μg/ml; hGH, 20 μg/ml)
4. Glass fiber filters (2.4 cm, Reeve Angel or Whatman 934-AH) pre-soaked for 24 hr in 25 m*M* sodium acetate buffer, pH 5.0, containing 0.1% BSA

[17] R. Gianetto and C. de Duve, *Biochem. J.* **59,** 432 (1955).

[18] P. S. Chen, Jr., T. Y. Toribara, and H. Warner, *Anal. Chem.* **28,** 1756 (1956).

[19] B. I. Posner, J. J. M. Bergeron, Z. Josefsberg, M. N. Khan, R. J. Khan, B. A. Patel, R. A. Sikstrom, and A. K. Verma, *Recent Prog. Horm. Res.* **37,** 539 (1981).

[20] B. I. Posner, *Diabetes* **23,** 209 (1974).

Procedure

Solution 2 (100 μl) is incubated in triplicate with 800 μl of sample (20 to 100 μg protein) and 100 μl of solution 1 (final vol of 1.0 ml) for 24 to 30 hr at 4° with constant shaking. Duplicate control tubes for determining nonspecific binding are run under identical conditions except that 100 μl of solution 1 is replaced by 100 μl of solution 3. The incubations are terminated by adding 3 ml of an ice cold 1 : 5 dilution (with H_2O) of solution 1. Bound and free radioactivity are separated by filtration on presoaked glass fiber filters using a Millipore manifold assembly. After filtration the filters are washed with 20 ml of a 1 : 5 dilution of solution 1. Radioactivity retained on the filters is counted in a gamma-scintillation counter as described elsewhere. Specific binding is the difference between radioactivity bound in the absence (total binding) and in the presence (nonspecific binding) of excess unlabeled hormone and is expressed as a percentage of total radioactivity in the incubation. In circumstances where the sample is suspended in 25 m*M* Tris–HCl–10 m*M* $MgCl_2$–0.1% BSA a 1 : 5 dilution of solution 1 rather than solution 1 itself is added to the incubation.

Intact Golgi by a Single Step Gradient Procedure[21]

The following solutions are prepared in advance: 5 m*M* $MgCl_2$, 25 m*M* KCl, 50 m*M* Tris, pH 7.4 (TKM); 0.25 *M* sucrose TKM (0.25 M STKM); and 2.3 *M* sucrose TKM (2.3 M STKM). Excised livers are placed in ice-cold 0.25 *M* STKM, weighed, and adjusted to 6 ml buffer per g liver. The tissue is minced, rinsed, and homogenized as above, in a Potter-Elvehjem homogenizer with 4 full strokes of the motor-driven Teflon pestle. Twenty-five milliliters of homogenate is mixed with 13 ml of 2.3 *M* STKM to give a final concentration of 1.02 *M* STKM and 6 ml is pipetted into each of six 14 ml cellulose nitrate tubes (for SW40 rotor). In preparation for gradient formation rinse the gradient maker with 0.25 *M* STKM and allow this to fill the valve space before emptying both chambers. Then add 4 ml of 0.25 *M* STKM to the mixing chamber. An equal volume of 1.02 *M* STKM solution, made by mixing 8 volumes of 2.3 *M* STKM with 10 volumes of TKM, is added to the reservoir chamber of the gradient maker. With the stirrer moving the valve is opened and the outlet tube set against the wall of the SW40 tube just above the homogenate. On completing the overlying gradient the tubes are centrifuged at 200,000 g_{av} for 90 min (40,000 rpm, SW40 rotor). The appearance of the centrifuge tube

[21] J. J. M. Bergeron, R. A. Rachubinski, R. A. Sikstrom, B. I. Posner, and J. Paiement, *J. Cell Biol.* **92,** 139 (1982).

contents after centrifugation is depicted in Fig. 2. The material from the Golgi region is removed with a 90° angle needle and syringe. The suspension is diluted in 0.25 *M* STKM if isotonicity must be preserved as for morphologic studies or in TKM or 25 m*M* Tris–10 m*M* $MgCl_2$, pH 7.4 for binding studies.

Characterization

Protein Recovery. This is depicted in Fig. 2. The recovery is similar to that of the Golgi intermediate fraction (Fig. 1).

Morphology. The intact Golgi preparation is morphologically quite different from the Golgi fractions prepared above. In the intact preparation one sees oriented stacks of fenestrated saccules with associated vesicles containing lipoprotein particles. These vesicles are often seen to have continuity with the fenestrated tubules.[21] Presumably, in the course of preparing Gl, Gi, and Gh the individual components (vesicles and saccules) become separated from one another.

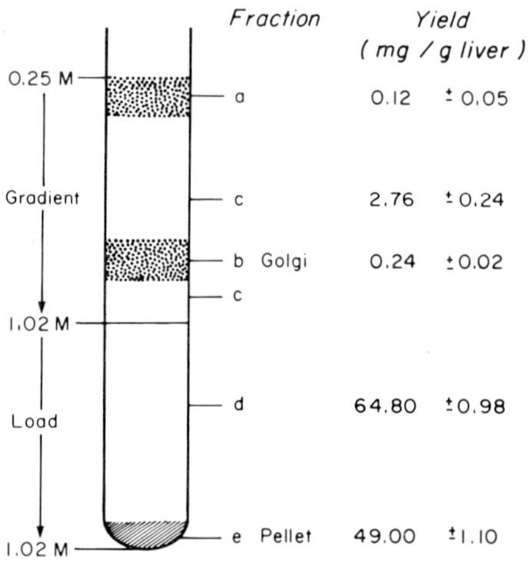

FIG. 2. Diagrammatic representation of continuous gradient used to isolate intact Golgi fraction. The homogenate is made to 1.02 *M* STKM and a linear gradient generated above the load zone (vol of gradient : vol of load zone; 1 : 1). After 1.5 hr of centrifugation at 190,000 g_{av} a prominent band (b) is observed 1/3 of the distance up the density gradient overlay with another band noted at the air/0.25 *M* STKM interface (fraction a). Five fractions are removed from the gradient with a bent blunt needle (#16, Becton Dickenson and Co., Rutherford, N.J.) and syringe. Fraction a, air : 0.25 *M* STKM interface; fraction b, intact Golgi fraction; fraction c, remainder of gradient after removal of fraction b; fraction d, residual load zone; fraction e, pellet. On the right is indicated the mean ± SEM of protein determinations carried out on 12 fractionations.

Enzyme Composition. There are few noteworthy differences between the intact Golgi preparation and the other Golgi fractions. The intact Golgi fraction contains a high concentration of galactosyltransferase and detectable, though much lower, levels of 5′-nucleotidase, acid phosphatase, and various endoplasmic reticulum enzymes (viz. glucose-6-phosphatase).

Hormone Binding. The pattern and extent of ^{125}I-labeled hormone binding closely resembles that seen in the other Golgi fractions.

Comment

The intact Golgi fraction is morphologically different but biochemically similar to the other Golgi fractions. It is more easily prepared in a one-step procedure. Of particular interest is the demonstration that it is more potent than the other Golgi fractions in effecting endogenous glycosylation.[21]

Isolation of Nonlysosomal, Non-Golgi Vesicles (i.e., Intermediate or "Unique" Vesicles) by Percoll Density Gradient Centrifugation

Nonlysosomal, non-Golgi vesicles, highly enriched in peptide hormone receptors, are isolated by further fractionation of Golgi intermediate or heavy fractions on an isoosmotic, continuous, self-generating Percoll gradient.[1,22]

Solutions

1. 0.25 *M* sucrose solution (8.55 g sucrose/100 ml distilled water)
2. 62% sucrose solution (62 g sucrose/100 ml distilled water)
3. 2.5 *M* sucrose solution.
4. Percoll (Pharmacia Fine Chemicals, Sweden), 100% solution in H_2O
5. Density marker beads (Pharmacia Fine Chemicals, Sweden)
6. 25 m*M* Tris–HCl, 10 m*M* $MgCl_2$, pH 7.4

Percoll Gradient Fractionation

The density of a freshly prepared Golgi intermediate (or heavy) fraction is determined by weighing 100-μl aliquots in triplicate. The sucrose concentration in the sample is computed from tabulated specific gravities of sucrose solutions.[23] The Golgi fractions (5–10 ml; 1 mg/ml protein) is

[22] M. N. Khan, B. I. Posner, A. K. Verma, R. J. Khan, and J. J. M. Bergeron, *Proc. Natl. Acad. Sci. U.S.A.* **78,** 4980 (1981).

[23] Scientific Tables, *in* "Documents Geigy" (K. Diem, ed.), p. 320.

diluted to 24 ml (the volume of the centrifuge tube) with a 100% Percoll solution (4.13 ml) and water or 2.5 *M* sucrose as required to yield a final solution of 0.25 *M* sucrose–17.2% (v/v) Percoll at a final density of 1.05 g/ml.

A continuous gradient is generated *in situ* by centrifuging at 60,000 g_{av} for 30 min at 4° (Spinco Ti 60, fixed angle rotor). Twelve 2 ml fractions are collected sequentially from the top of the centrifuge tube by pumping a 62% sucrose solution into the bottom of the centrifuge tube. The density of individual fractions is determined by weighing 100-μl aliquots in triplicate, or by running an additional tube containing identical Percoll and sucrose concentrations and density marker beads as described by Mickelson *et al.*[24]

Most of the assays for marker enzymes, protein, and hormone binding can be done in off-the-gradient fractions. Where it is desirable to remove Percoll the fraction is diluted 1:3 with 25 m*M* Tris–HCl–10 m*M* $MgCl_2$ (pH 7.4) and centrifuged at 82,000 g_{av} for 120 min at 4° (SW27 rotor). The supernatant is removed with a Pasteur pipet and the membranous material resting on the Percoll pellet is resuspended in a small volume (~2 ml) of Tris–HCl, $MgCl_2$ buffer, and either assayed immediately or frozen at −20°.

Characterization

With Percoll gradient centrifugation two populations of receptor-rich structures can be resolved from the Golgi intermediate and heavy fractions. Those entities which have a density greater than 1.055 are vesicles containing lipoprotein particles which are devoid of both galactosyl- and sialyltransferases. These structures morphologically resemble the lipoprotein-containing lighter structures which cosediment with glycosyltransferases. In addition the heavier vesicles contain modest levels of acid phosphatase and other lysosomal enzymes (approx. 1/4 to 1/3 the concentration in lysosomes). Because these structures appear to share features of both Golgi vesicles and lysosomes we have referred to them as intermediate or "unique" vesicles to emphasize their apparent distinctiveness.[1,22]

The analysis of fractions off the Percoll gradients is complicated by the interference of Percoll with the Lowry method of protein estimation. We have devised a modified Lowry procedure for protein estimation in which absorbance at 750 and 420 nm of unknowns, bovine serum albumin standards in 0.25 *M* sucrose and Percoll solutions is determined.[25] In this procedure Percoll-containing samples develop a thick white precipitate on

[24] J. R. Mickelson, M. L. Breaser, and B. B. Marsh, *Anal. Biochem.* **109,** 255 (1980).
[25] M. N. Khan, R. J. Khan, and B. I. Posner, *Anal. Biochem.* **117,** 108 (1981).

the addition of 1.2 *N* NaOH which disappears after 20 min in a boiling water bath. Since the relation between the concentration of protein or Percoll and the absorbance values at the above two wavelengths is linear one can devise a simple equation for computing protein concentrations. The method is suitable for samples containing as much as 30% Percoll.[25]

Methods for Quantitative Autoradiography of Subcellular Fractions

Subcellular fractions are routinely processed for random sampling by the filtration protocol outlined by Baudhuin *et al.*[26] as modified by Wibo *et al.*[27] Briefly, after fixation in 2.5% glutaraldehyde in 0.05 *M* sodium cacodylate, buffer pH 7.4, Golgi fractions are recovered onto Millipore membranes (Type HA, 0.45 μm pore size), using the automated filtration apparatus of Baudhuin *et al.*[26] then postfixed in 1% OsO_4 and block-stained in uranyl acetate. Membrane pellicles are then dehydrated, embedded in Epon 812, and thin sections cut with block face orientated in order to assure a cross-section through the depth of the filtered pellicle. These thin sections can then be viewed directly in the electron microscope following grid staining in uranyl acetate and lead citrate. This method which has been standardized by Dr. B. Kopriwa[28] is as follows.

Reagents

Celloidon (Randolph Finishing Products, Carlstadt, N.J.)
Isoamylacetate (Tousimis, New Jersey)
Ilford L-4 nuclear emulsion (Ilford Canada, Markham, Ont.)
Gold chloride (Canlab, Montreal, Canada)
Potassium thiocyanate (Fisher Scientific Limited, Canada)
Potassium bromide (Fisher Scientific Limited, Canada)
Elon (*p*-methylaminophenol sulfate) (Kodak, New Jersey)
Anhydrous sodium sulfite (Fisher Scientific Limited, Canada)
Sodium thiosulfate (Fixer) (Fisher Scientific Limited, Canada)

Equipment

Semiautomatic coating apparatus (Averlaid, Toronto).

Procedures

Precleaned microscope slides are further cleaned by wiping with lens tissue. The slides are dipped into a solution of 1% celloidon in isoamylacetate and dried vertically in a dust-free cabinet overnight. Thin sections

[26] P. Baudhuin, P. Evrard, and J. Berthet, *J. Cell Biol.* **32,** 181 (1967).
[27] M. Wibo, A. Amar-Costesec, J. Berthet, and H. Beaufay, *J. Cell Biol.* **51,** 52 (1971).
[28] B. M. Kopriwa, *Histochemie* **37,** 1 (1973).

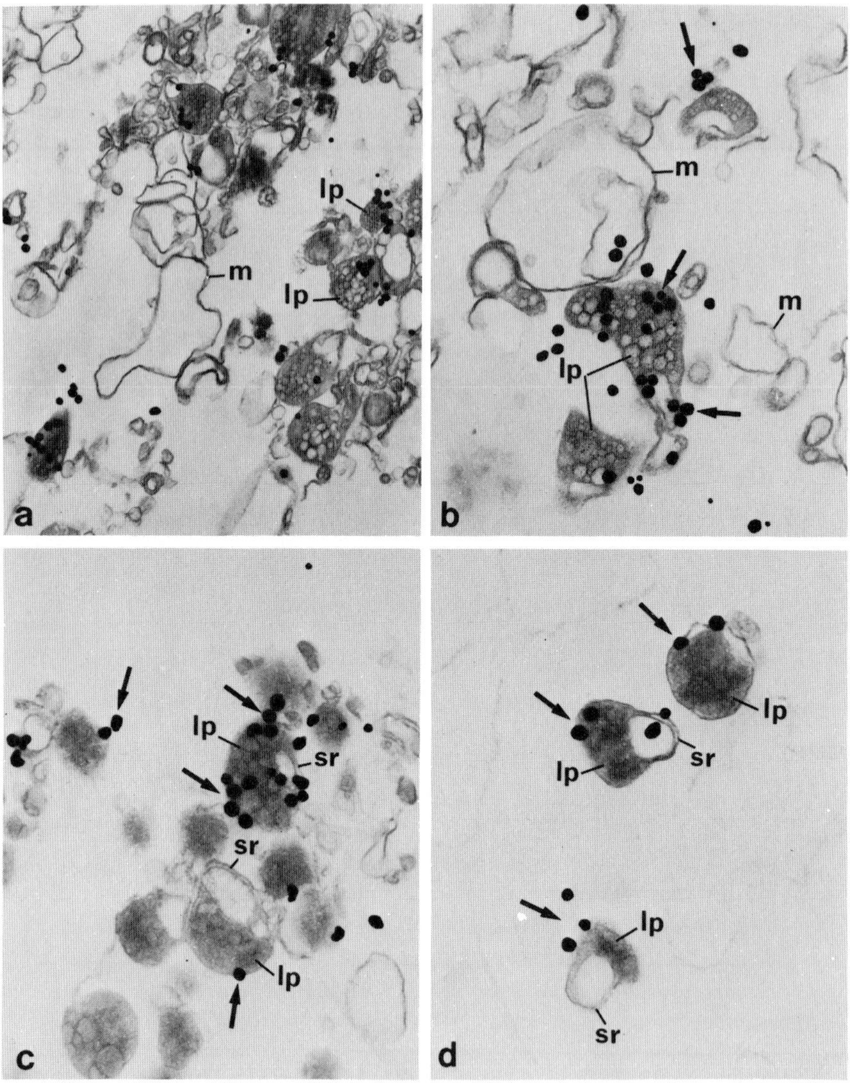

FIG. 3. Electron microscope autoradiography of ^{125}I-labeled insulin in Golgi preparations isolated 5 min after ^{125}I-labeled insulin injection *in vivo*. The starting material was a Golgi intermediate fraction which was further subfractionated on Percoll gradients into subfractions of light (a, b), intermediate (c), or heavy density (d). The autoradiographic "fine" silver grains overlie lipoprotein-containing vesicular structures in all regions of the Percoll gradient. Taken from Khan *et al.*[1] a, ×23,000; b, c, d, ×34,000.

(~50 nm in thickness) are placed on the celloidon-covered glass slides. These sections are further covered with 5 nm of carbon using an Edwards carbon evaporator especially modified to assure a uniform application of carbon. The slides are subsequently dipped in Ilford L-4 nuclear emulsion (diluted ~1 : 2 : 5 in distilled water) at 40° using the semiautomatic dipping device to ensure covering by a monolayer of emulsion. The slides are dried vertically, then stored in lightproof boxes at 8° with added Drierite in each box. After suitable exposure (~60 days) the emulsion is developed for "fine grain" development by the "solution-physical" procedure after prior treatment with gold in order to stabilize the exposed latent images. Thus slides are dipped vertically into distilled water (1 min) followed by a freshly prepared gold thiocyanate solution (0.4 mg of gold chloride in 100 ml of 0.5% potassium thiocyanate, 0.6% potassium bromide), for 1 min followed by 7 min development in freshly prepared Agfa-Gevaert developer (0.75 g Elon, 0.5 g anhydrous sodium sulfite, 0.2 g potassium thiocyanate in 100 ml distilled water).[28] Slides are subsequently placed in distilled water for 30 sec followed by fixer (2 min in 24% sodium thiosulfate) and washed in five changes of distilled water. Sections are finally transferred to EM grids by flotation of the cut celloidin-emulsion couplet onto the surface of distilled water.

The key feature of the above technique involves the use of gold deposition about the exposed latent images. This treatment ensures the symmetrical deposition of silver about the exposed latent images by the Agfa-Gavaert developer. The final visualized micrograph is therefore one of exposed latent images and not, as for all other autoradiographic methods[29] a micrograph showing developed silver crystals where the latent images have been randomly developed. An example for the autoradiographic analysis of ^{125}I-labeled insulin distribution of Golgi fractions is illustrated in Fig. 3.

[29] B. M. Kopriwa, *Histochemistry* **44,** 201 (1975).

[19] Assessment of Receptor Recycling in Mammalian Hepatocytes: Perspectives Based on Current Techniques

By JOE HARFORD and GILBERT ASHWELL

Introduction

Endocytosis is the process by which cells engulf extracellular materials within infoldings of their plasma membrane that pinch off to form cytoplasmic vesicles.[1] This general process is utilized by cells in the uptake of particulates (phagocytosis) as well as for the nonselective internalization of extracellular fluid (fluid phase pinocytosis). Additionally, when coupled with high affinity receptor molecules on the cell surface, endocytosis provides a selective and highly efficient mechanism for the internalization of substances from the extracellular space. Within the past several years, the list of receptors mediating endocytosis of specific ligands has grown rapidly and has been the subject of several excellent reviews.[2-5] From these studies, it has become increasingly clear that the mechanism of uptake and receptor reutilization follows, in many cells, a common pattern with only minor and subtle alterations designed to meet specific metabolic requirements.

One such system, the hepatic receptor for asialoglycoproteins, has been extensively examined in a number of laboratories[6,7] and is the subject matter of the present chapter. As is not uncommon in a rapidly developing discipline, differences in experimental design or conditions have led to differences in interpretation of data. Clearly, it is not within the purview of this chapter to resolve such differences. Rather, the intent is to emphasize the various methodologies employed and to indicate the conclusions drawn.

To facilitate assessment of the diverse experimental approaches employed in the study of receptor reutilization, a series of fundamental questions have been posed and treated separately in the text. These include (1) What are the survival half-lives of the receptor and ligand? (2) What is the

[1] S. C. Silverstein, R. M. Steinman, and Z. A. Cohn, *Annu. Rev. Biochem.* **46,** 669 (1977).
[2] J. L. Goldstein, R. G. W. Anderson, and M. S. Brown, *Nature* (*London*) **279,** 679 (1979).
[3] I. H. Pastan and M. C. Willingham, *Annu. Rev. Physiol.* **42,** 239 (1981).
[4] R. M. Steinman, I. S. Mellman, W. A. Muller, and Z. A. Cohn, *J. Cell Biol.* **96,** 1 (1983).
[5] M. S. Brown, R. G. W. Anderson, and J. L. Goldstein, *Cell* **32,** 663 (1983).
[6] J. Harford and G. Ashwell, *in* "The Glycoconjugates" (M. Horowitz, ed.), Vol. 4, Part B, p. 531. Academic Press, New York, 1982.
[7] G. Ashwell and J. Harford, *Annu. Rev. Biochem.* **51,** 531 (1982).

ISBN 0-12-182009-2

number of cell surface receptors? (3) Do cell surface receptors move into the cell in a complex with ligand? (4) Is there an additional pool of intracellular receptors and to what extent does this pool participate in the process? (5) Do receptors cycle to and from the cell surface in the absence of ligand? and (6) What events subsequent to receptor internalization account for sparing of receptors and their return to the cell surface?

It should be evident from the above that a critical evaluation of the techniques involved in receptor reutilization does not lend itself readily to the more conventional format for describing well-established methodologies. Instead, the experimental approaches and the evidence bearing on each of the above questions are reviewed, in turn, in an attempt to provide a realistic picture of the currently available technology. Omissions or misrepresentation of the extensive literature are unintentional and reflect the limitations of the authors.

The Survival Half-Lives of Asialoglycoproteins and Their Receptors

Following the initial observation that desialylation dramatically reduced the residency times of glycoproteins in serum,[8] evidence was provided implicating hepatic lysosomes as the major site of asialoglycoprotein catabolism.[9] The transfer of these ligands from serum to lysosomes is rapid occurring between 5 and 13 min after injection.[10] Based on the large amounts of asialoglycoproteins removed by the liver and the inability to detect ligand-induced receptor synthesis, Regoeczi *et al.*[11] concluded that receptors were reutilized. This conclusion was also reached by Tanabe *et al.*[12] who determined a survival half-time of 88 hr for the receptor protein in rat liver. This was in marked contrast to the half-life of injected ligands which is on a time scale of minutes. No change in the calculated receptor half-life resulted from introduction of large amounts of asialoglycoprotein into the circulation of the animals. These data strongly suggested that the metabolic pathways for receptor and ligand diverge allowing the sparing of receptor during ligand degradation.

Subsequently, corroboration was provided with isolated hepatocytes

[8] A. G. Morell, R. A. Irvine, I. Sternlieb, I. H. Scheinberg, and G. Ashwell, *J. Biol. Chem.* **243,** 155 (1968).

[9] G. Gregoriadis, A. G. Morell, I. Sternlieb, and I. H. Scheinberg, *J. Biol. Chem.* **245,** 5833 (1970).

[10] J. H. LaBadie, K. P. Chapman, and N. N. Aronson, Jr., *Biochem. J.* **152,** 271 (1975).

[11] E. Regoeczi, M. T. Debanne, M. W. C. Hatton, and A. Koj, *Biochim. Biophys. Acta* **541,** 372 (1978).

[12] T. Tanabe, W. E. Pricer, Jr., and G. Ashwell, *J. Biol. Chem.* **254,** 1038 (1979).

blocked in protein synthesis.[13–15] In these studies, receptor reutilization was inferred from uptake of ligand in excess of the total receptor content of the cells. Calculation of the number of ligands handled by a receptor molecule is dependent on knowledge of the number of receptors participating in ligand uptake and catabolism. Estimates, based on the assumption that only cell surface receptors participate, indicate that 5–20 times the surface receptor complement are internalized per hour even in the absence of protein synthesis,[13–15] a finding consistent with some form of receptor reutilization. Warren and Doyle[16] examined the turnover of the asialoglycoprotein receptor in monolayer cultures of rat hepatocytes by assessing the turnover of receptor labeled either via lactoperoxidase-catalyzed iodination or metabolically with L-[^{35}S]methionine. Using either method of labeling, a half-life of approximately 20 hr was calculated. Based on this estimate of the turnover time of receptor and the measured rate of ligand catabolism, the authors concluded that each receptor molecule was capable of delivering about 1000 molecules of ligand to lysosomes for degradation. The discrepancy in half-life values in cultured hepatocytes[16] and *in vivo*[12] presumably reflects the different systems studied and may be due to a loss of polarity and/or nitrogen imbalance in the cultured cells. Nonetheless, receptor turnover in the isolated cells resembled turnover *in vivo* in that both were unaffected by maximal uptake of ligand and in both cases receptor reutilization was indicated.

Recently, the asialoglycoprotein receptor has been characterized in the continuous human hepatoma line HepG2.[17] The value of ligand uptake/functional receptor/minute was found to be quite similar to that in freshly isolated hepatocytes. Given the relatively low rate of receptor synthesis in HepG2 cells,[18] it was calculated that each receptor must handle one ligand every 16 min.[19]

The Number of Cell Surface Receptors

Striking disparity exists in the literature with regard to the number of receptors for asialoglycoproteins on the surface of isolated hepatocytes. Generally, cell surface receptors are measured as the amount of ligand-

[13] C. J. Steer and G. Ashwell, *J. Biol. Chem.* **255,** 3008 (1980).

[14] J. U. Baenziger and D. Fiete, *Cell* **22,** 611 (1980).

[15] H. Tolleshaug, *Int. J. Biochem.* **13,** 45 (1981).

[16] R. Warren and D. Doyle, *J. Biol. Chem.* **256,** 1346 (1981).

[17] A. L. Schwartz, S. E. Fridovich, B. B. Knowles, and H. F. Lodish, *J. Biol. Chem.* **256,** 8878 (1981).

[18] A. L. Schwartz and D. Rup, *J. Biol. Chem.* **258,** 11249 (1983).

[19] A. L. Schwartz, S. E. Fridovich, and H. F. Lodish, *J. Biol. Chem.* **257,** 4230 (1982).

specific binding at 4° although surface binding to the hepatoma HepG2 cells has been measured at 37° after exposure to 10–12 m*M* azide to inhibit internalization.[19] One source of disparity may arise from the perfusion with collagenase used to isolate cells. Collagenase preparations are known to contain protease activity that might inactivate a significant number of receptors. Indeed, when isolated cells were intentionally reexposed to collagenase at 37°, cell surface binding was reduced 50–90%.[20]

It has been shown that isolated hepatocytes increase their number of cell surface receptors by as much as 2- to 3-fold upon incubation at 37°.[13,20–23] In two instances, this increase was reported to be unaffected by inhibitors of protein synthesis.[20,22] Enigmatically, the number of surface binding sites for a galactose-terminated fetuin glycopeptide appear to increase significantly more than the number of sites for asialoorosomucoid.[23] Moreover, the apparent affinity constants for the fetuin glycopeptide were altered by incubation at 37° whereas those for asialoorosomucoid were virtually unchanged. The significance of this observation remains uncertain.

In the most complete study on the effect of temperature on cell surface receptor number, Weigel and Oka[22] demonstrated that surface binding is reversibly modulated by temperature changes. By cycling isolated hepatocytes between 25 and 37° the surface binding capacity could be modulated up and down. In this range, the number of surface receptors/cell was shown to depend on the final temperature but to be independent of the pathway to that temperature. The authors favored an explanation of this phenomenon which involved recycling of the receptor even in the absence of added ligand with the externalization phase of the cycle being more temperature sensitive (see section on receptor recycling below).

Washing freshly isolated cells with the calcium chelator, EDTA, also resulted in an increase (~30%) in cell surface receptors.[21] This was interpreted as being due to removal of endogenous ligands by the chelator. This concept is consistent with the findings of Schwartz *et al.*[24] who found cell surface receptor numbers ranging from 40,000 to 560,000/cell. Presumably the variability in the measurable binding sites reflects the presence of inhibitory substances (ligands) in the sera or albumin preparations included in the incubation and wash medium.

Given that the number of measurable surface receptors may vary due to proteolytic inactivation, temperature-induced receptor redistribution

[20] P. L. Zeitlin and A. L. Hubbard, *J. Biol. Chem.* **92,** 634 (1982).
[21] P. H. Weigel, *J. Biol. Chem.* **255,** 6111 (1980).
[22] P. H. Weigel and J. A. Oka, *J. Biol. Chem.* **258,** 5089 (1983).
[23] D. Fiete, M. D. Brownell, and J. U. Baenziger, *J. Biol. Chem.* **258,** 817 (1983).
[24] A. L. Schwartz, D. Rup, and H. F. Lodish, *J. Biol. Chem.* **255,** 9033 (1980).

or occupation/inhibition, caution is warranted in the interpretation of receptor number on isolated hepatocytes. Differences in each of these parameters may contribute to the variation in cell surface receptor number reported by different laboratories. However, it is reassuring that preparations with a relatively low number (<100,000/cell) of receptors[13,21] or with a relatively high number (100,000–500,000/cell) of receptors[14,16,24,25] appear to be strikingly similar in the pathway of internalization and catabolism of asialoglycoproteins (see also Ref. 20). Moreover, it has been a consistent finding that ligand in excess of the surface receptor population is internalized at 37°, thus supporting the contention that receptors are reutilized.

Internalization of Receptor with Ligand

Entry of receptor into the cell during endocytosis was indicated by the elegant morphological studies of Wall *et al.*[26] Lactosaminated ferritin was shown to be recognized by the asialoglycoprotein receptors and internalized *in vivo*. This ligand was seen in close apposition to the membrane of both coated vesicles and coated vesicular profiles in the peripheral cytoplasm. Moreover, close apposition was not apparent in larger irregular structures wherein ferritin appeared at later times. This observation was interpreted to indicate that internalized ligands remain bound to the membrane (and presumably the receptor) within the endocytic vesicles. That receptors enter the cell is also supported by the kinetic studies of Weigel.[27] At 37°, occupied receptors were depleted from the cell surface at 5 times the rate of appearance of unoccupied receptors. At 18°, this disparity was even greater with receptors appearing at only 6% the rate of endocytosis. The time lag required to replenish the initial content of receptors suggests that receptor enters the cell with the ligand. It must be noted however that other laboratories using isolated hepatocytes[28] or perfused liver[29] have not observed a lag in receptor replenishment at short times (5 and 3.5 min, respectively) of endocytosis (see below). The reason for this apparent discrepancy remains unclear.

Similar kinetic studies have been performed in human hepatoma (HepG2) cells.[19,30] After prebinding asialoorosomucoid to the cell surface,

[25] H. Tolleshaug and T. Berg, *Hoppe-Seyler's Z. Physiol. Chem.* **361,** 1155 (1980).

[26] D. A. Wall, G. Wilson, and A. L. Hubbard, *Cell* **21,** 79 (1980).

[27] P. Weigel, *Biochem. Biophys. Res. Commun.* **101,** 1419 (1981).

[28] K. Bridges, J. Harford, G. Ashwell, and R. D. Klausner, *Proc. Natl. Acad. Sci. U.S.A.* **79,** 350 (1982).

[29] D. A. Wall and A. L. Hubbard, *J. Cell Biol.* **90,** 687 (1981).

[30] A. J. Ciechanover, A. L. Schwartz, and H. F. Lodish, *Cell* **32,** 267 (1983).

cells were warmed to 37°. At the start of the 37° incubation, only 13% of the total population of receptor was internal yet 45–55% of the cell surface receptors disappeared then reappeared. At the end of the study, the starting distribution (13% internal) was restored. It was calculated[30] that at least 34% of the surface receptors had been internalized and recycled by the end of the experiment. Similarly, during steady state internalization of ligand, a 17% decrease in surface receptors was noted.[19] Both results support receptor internalization during ligand endocytosis.

Intracellular receptor–ligand complex has been detected and quantitated during endocytosis of asialoorosomucoid by freshly isolated rat hepatocytes in suspension[28] or short-term (<2 days) monolayer cultures of these cells.[31] In both cases, surface receptors were occupied at 4° and after washing free of unbound ligand, the cells were warmed to 37°. At various times, the cells were rechilled, stripped of residual surface ligand by an EGTA wash, and subjected to a solubilization/precipitation assay capable of distinguishing receptor–ligand complex from free ligand.[28,31,32] Although the kinetics of appearance and disappearance of intracellular receptor–ligand complex differed somewhat between the suspension and monolayer cells, in both instances, computer-assisted compartmental modeling indicated the intracellular complex to be an obligate intermediate in ligand catabolism.

Intracellular Receptors

Identification of the asialoglycoprotein receptor as a component of isolated plasma membranes[33] was followed by the demonstration of binding activity associated with other subcellular fractions.[34] Indeed the majority of receptors appeared to be intracellular. This conclusion was subsequently confirmed in studies employing subcellular fractionation of rat liver and receptor-specific antibody,[12] by sonic disruption of cells,[14] or permeabilization/solubilization of cells by nonionic detergents.[13,14,16,35–37] Clearly, the considerations outlined above regarding variations in measurable surface receptor will, by definition, influence estimates of the ratio of

[31] J. Harford, K. Bridges, G. Ashwell, and R. D. Klausner, *J. Biol. Chem.* **258,** 3191 (1983).

[32] R. L. Hudgin, W. E. Pricer, Jr., G. Ashwell, R. J. Stockert, and A. G. Morell, *J. Biol. Chem.* **249,** 5536 (1974).

[33] W. E. Pricer, Jr. and G. Ashwell, *J. Biol. Chem.* **246,** 4825 (1971).

[34] W. E. Pricer, Jr. and G. Ashwell, *J. Biol. Chem.* **251,** 7539 (1976).

[35] R. J. Stockert, U. Gartner, A. G. Morell, and A. W. Wolkoff, *J. Biol. Chem.* **255,** 2830 (1980).

[36] P. H. Weigel and J. A. Oka, *J. Biol. Chem.* **257,** 1201 (1982).

[37] P. H. Weigel and J. A. Oka, *J. Biol. Chem.* **258,** 5095 (1983).

extracellular to intracellular receptors. This is especially true of the binding activity that becomes measurable after incubation at 37° since this appears to reflect receptor redistribution. In contrast, Geuze *et al.*[38] reported that indirect immunofluorescence and immunocytochemistry detected receptor only on the surface of hepatocytes in frozen liver sections. However, using improved methods and antireceptor antibody conjugated to colloidal gold these authors have more recently found 65% of the receptors intracellularly.[39] Moreover, in the latter study, the distributions of the receptor as determined by quantitative immunocytochemistry on liver sections from untreated rats and rats infused for 60 min with asialofetuin were indistinguishable.

The intracellular asialoglycoprotein receptor appears to be indistinguishable from that on the cell surface. Isolation of the receptor was accomplished by ligand affinity chromatography of individual preparations enriched in plasma membranes, Golgi complex, lysosomes, or smooth microsomes. In each case, the receptor preparation exhibited the same binding properties as well as similar subunit structure and immunological reactivity.[33] Dissociation constants determined from equilibrium binding studies in the presence or absence of a permeabilizing concentration of digitonin, to assess total or surface receptors, have been reported recently to be identical.[37] In addition to the major subcellular components shown to possess receptor activity,[12,33] a receptor-rich vesicle population has been isolated whose properties do not fall directly within the precinct of heretofore characterized organelles.[40] The precise origin and role of these vesicles remain to be determined.

What then is the purpose of this pool of intracellular receptors? Hypotheses generally fall into four categories: (1) Intracellular receptors reflect molecules in the biosynthetic pathway on the way to a function related exclusively to the surface population of receptors; (2) intracellular receptors serve as a reservoir for replacement of internalized surface receptors; (3) intracellular receptors serve as a routing mechanism for internalized ligand; and (4) intracellular receptors serve a function within the cell unrelated to the endocytosis of extracellular ligand. Of course, these postulates are not mutually exclusive.

The contention that at least a portion of the intracellular receptor represents newly synthesized molecules is supported by the results of

[38] H. J. Geuze, J. W. Slot, G. J. A. M. Strous, H. F. Lodish, and A. L. Schwartz, *J. Cell Biol.* **92,** 865 (1982).

[39] H. J. Geuze, J. W. Slot, G. J. A. M. Strous, H. F. Lodish, and A. L. Schwartz, *Cell* **32,** 277 (1983).

[40] M. T. Debanne, W. H. Evans, N. Flint, and E. Regoeczi, *Nature (London)* **298,** 398 (1982).

Nakada *et al.*[41] who reported that newly synthesized receptor was associated with rough microsomes. Schwartz and Rup[18] have examined biosynthesis of the receptor in human HepG2 cells. Using antireceptor antibodies, these authors estimated a transit time of ~1 hr for newly synthesized receptor to reach the plasma membrane. If a similar transit time exists in rat liver or isolated rat hepatocyes, the relatively long half lives of 88 hr[12] and 20 hr,[16] respectively, would suggest that biosynthetic transit would not account for the bulk of the intracellular receptor.

Several attempts have been made to assess the functional equivalence of the intracellular and extracellular pools of asialoglycoprotein receptor. Stockert *et al.*[35] blocked uptake of asialoorosomucoid by a perfused liver with antireceptor antibodies. Although uptake remained blocked for at least 90 min, the capacity of a liver homogenate was reduced less than 15%. These findings were interpreted as indicating that functional receptors were not added to the cell surface under these conditions. The same laboratory also addressed this question by exposing hepatocytes to neuraminidase.[42] This treatment had previously been demonstrated to abolish asialoorosomucoid binding but with retention of the ability to bind and catabolize desialylated ovine submaxillary mucin, a ligand of higher affinity.[43] Although the intracellular receptor pool was unaffected by neuraminidase treatment, the enzyme-treated cells maintained their altered specificity, i.e., continued catabolism of desialylated mucin of up to 20 times surface binding capacity and inability to bind or degrade asialoorosomucoid. As with the antibody experiment, these results were interpreted as indicating a segregation of extracellular receptor from the relatively larger pool of intracellular receptor. Clearly, a caveat should be considered regarding the possibility that antibody bound to surface receptors or neuraminidase treatment may disrupt normal cycling and/or affect the ability to detect it.

Several lines of evidence suggest that the intracellular pool of receptors participates in the replacement of internalized surface receptors. Shortly after internalization of [^{3}H]asialoorosomucoid by freshly isolated hepatocytes, at a time when almost all of the internalized ligand remained receptor bound (see above), the ability of the cells to bind [^{125}I]asialoorosomucoid was assessed.[28] This was done with and without removal of occupied surface receptors by an EGTA wash. The results indicated that unoccupied receptors had replaced internalized receptors such that a constant number of surface receptors (occupied plus unoccupied) was main-

[41] H. Nakada, T. Sawamura, and Y. Tashiro, *J. Biochem.* (*Tokyo*) **89,** 135 (1981).

[42] R. J. Stockert, D. J. Howard, A. G. Morell, and I. H. Scheinberg, *J. Biol. Chem.* **255,** 9028 (1980).

[43] R. J. Stockert, A. G. Morell, and I. H. Scheinberg, *Science* **197,** 667 (1977).

tained despite internalization of half of the surface receptor population. The rate constants calculated in this study for internalization and intracellular dissociation of the receptor–ligand complex suggest that a relatively large intracellular pool must participate in receptor replacement if constancy of surface receptor number is to be maintained during endocytosis. Weigel[27] has also presented evidence for the coincident appearance of unoccupied receptors during asialoorosomucoid endocytosis in isolated hepatocytes. In this case, functional receptor appearance on the cell surface was reported to be slower than the internalization of ligand (see above). A related experiment in perfused liver was performed by Wall and Hubbard.[29] After one round of binding and internalization of ligand, surface receptors were cleared of ligand by EGTA. It was found that the number of receptors present on the cell surface appeared to be constant even when the hepatocytes were actively engaged in ligand internalization. The participation of intracellular receptors is also supported by the finding that within 1 to 2 hr in the presence of excess asialoorosomucoid, a steady state was attained in which approximately 70% of the intracellular receptors were occupied with ligand.[37] The transient depletion in cell surface receptors upon ligand internalization by HepG2 cells[30] (see above) may reflect the relatively small pool of intracellular receptors (~13% of the total receptor) in these cells.[19]

Although the experiments outlined above support the belief that intracellular receptors participate in receptor recycling in rat liver and isolated hepatocytes, some experiments indicate that they are not absolutely required. As was just mentioned, HepG2 cells have relatively few intracellular receptors.[19] Additionally, the asialoglycoprotein receptor was introduced artificially into mouse cells (L-929) devoid of this receptor. Rat hepatocyte membrane vesicles[44] or reconstituted vesicles containing the rat liver receptor[45] were fused with the L cells to yield cells capable of endocytosis and degradation of asialoglycoproteins. The introduced receptor did not appear to redistribute to produce an internal pool analogous to hepatocytes. Nonetheless, the chimeric cells appeared capable of at least two rounds of binding and catabolism of ligand thereby suggesting a meshing of the exogenous receptor with the degradative machinery of the L cell and receptor reutilization. However, in this case, binding and degradation of ligand required a period of 24–48 hr compared to a time frame of minutes in isolated hepatocytes. Thus, intracellular receptors may facilitate the rapidity with which hepatocytes handle asialoglycoproteins.

[44] D. Doyle, E. Hou, and R. Warren, *J. Biol. Chem.* **254,** 6853 (1979).
[45] H. Baumann, E. Hou, and D. Doyle, *J. Biol. Chem.* **255,** 10001 (1980).

Receptor Recycling in the Absence of Ligand

As detailed above, several lines of evidence support the transient internalization of ligand–receptor complexes. Evidence for movement of receptor in the absence of ligand is more difficult to obtain. This is due primarily to the fact that ligands are easily tagged with radioactivity or conjugated to electron-dense particles or enzymes for cytochemistry. Covalent labeling of receptors *in situ* is technically more difficult. Photoactivatable glycopeptide reagents for site-specific labeling of membrane receptors[46] may prove useful for labeling the asialoglycoprotein receptor. However, receptors labeled this way resemble more closely occupied as opposed to unoccupied receptors rendering even this approach suspect in discerning the traffic in unoccupied receptor.

Several agents and treatments of hepatocytes appear to decrease the number of surface receptors in the absence of ligand. This observation has been interpreted to indicate that a constitutive cycle of receptor internalization and externalization is responsible for maintenance of the steady-state level of surface receptors. If this were so, selective inhibition of receptor externalization would be expected to lead to ligand-independent redistribution resulting in less surface receptors. Kolset *et al.*[47] reported that treatment of hepatocytes with 100 μM colchicine reduced the binding capacity to 75% of control levels. A more dramatic effect was observed when cells were treated with 1 mM chloroquine.[48] Here binding capacity was reduced by 85%, as determined by Scatchard plots. However, under these conditions, chloroquine treatment was not reversible. In contrast, alterations in surface receptors mediated by temperature changes appear to be fully reversible.[21,22] The most favored interpretation of these results was a selective temperature effect on the externalization phase of ligand-independent receptor recycling. Thus, reduction of temperature reduced the proportion of active receptors on the surface whereas raising the temperature restored these receptors. Redistribution of receptors between the intracellular and extracellular pool is also altered during liver regeneration.[49,50] The ability of regenerating livers to internalize asialoglycoproteins was reduced[49] without a decrease in the total number of receptors[50] but with an apparently selective loss of cell surface receptor. Normal distribution was restored as regeneration was completed. Here, too,

[46] J. U. Baenziger and D. Fiete, *J. Biol. Chem.* **257,** 4421 (1982).

[47] S. O. Kolset, H. Tolleshaug, and T. Berg, *Exp. Cell Res.* **122,** 159 (1979).

[48] H. Tolleshaug and T. Berg, *Biochem. Pharmacol.* **28,** 2919 (1979).

[49] U. Garnter, R. J. Stockert, A. G. Morell, and A. W. Wolkoff, *Hepatology* **1,** 99 (1981).

[50] D. J. Howard, R. J. Stockert, and A. G. Morell, *J. Biol. Chem.* **257,** 2856 (1982).

disruption of the receptor recycling mechanism was suggested as a possible explanation for selective depletion of surface receptors.

The hypothesis that ligand binding triggers endocytosis has also been suggested. Baenziger and Fiete[14] reported that not all glycopeptides recognized by the isolated asialoglycoprotein receptor were endocytosed by hepatocytes. A model was suggested in which "correct" spacing of terminal galactose or *N*-acetylgalactosamine residues was seen as inducing the receptor to enter a new state capable of interaction with other cellular components and resulting in endocytosis. The findings of Ciechanover *et al.*[30] also support the ligand triggering of endocytosis. Upon warming of chilled cells to 37° a transient decrease in surface receptors was found to be dependent upon occupation of surface receptors with ligand. No equivalent decrease in unrelated receptors, such as insulin or transferrin, was observed to result from asialoglycoprotein endocytosis. That externalization of constantly recycling receptors may be slowed by occupation with ligand to yield similar results was not excluded in this study. Receptor cross-linking/aggregation has been implicated as the trigger in a number of endocytic systems.[4] If a signal for asialoglycoprotein endocytosis does exist, it appears unlikely that receptor cross-linking is involved since the extent of endocytosis was unaffected by a 10^6-fold range in the average number of ligands bound per cell.[36] Also indirectly arguing against cross-linking as a trigger for endocytosis, is the inability of hepatocytes to endocytose dimers and trimers of glycopeptides that they do not take up as a monomer.[14]

Interpretation of studies of this type have been complicated by the provocative finding that treatment of cells with the carboxylic acid ionophores, nigericin or monensin, reduces the number of cell surface binding sites for asialoglycoprotein and glycopeptides in the absence of added ligand.[23] These results are reminiscent of the earlier results with chloroquine[48] and might be similarly viewed as supportive of ligand-independent receptor recycling and an inhibition of receptor externalization. However, Fiete *et al.*[23] also reported that the amount of receptor on the cell surface detected by competitive radioimmunoassay and by immunoprecipitation of surface labeled receptor was not significantly altered by the ionophore treatment. The authors postulate that a relatively low degree of alkalinization of the hepatocyte cytosol mediated by the ionophores results in a transmembrane modulation of the ligand binding properties of cell surface receptors.

Postinternalization Events in Asialoglycoprotein Endocytosis

Subsequent to binding and internalization of the asialoglycoprotein receptor–ligand complex, several critical steps in the pathway of ligand catabolism occur. The complex must dissociate at some point and its components segregate if ligand is to be destroyed and receptor reutilized. Ligands must then make their way to lysosomes, where catabolism occurs, and receptors must return to the plasma membrane. This overall process has been examined morphologically in a number of laboratories.[20,26,29,38,39,51–56] Collectively, these studies support the pathway of endocytosis via coated pits, i.e., specialized regions of the plasma membrane that have been implicated in adsorptive endocytosis of several ligands in a variety of cell types.[2,57]

Such morphological studies are relevant to the processes of dissociation of ligand from receptor, segregation of the receptor from ligand, and delivery of ligand to lysosomes. As mentioned earlier, coated vesicular profiles in the peripheral cytoplasm shortly after uptake of lactosaminated ferritin reveal ferritin to be in close apposition to the inner vesicular membrane.[26] Somewhat later, larger and more irregular vesicles were seen containing more ferritin per vesicle and with the ferritin no longer in apposition to the membrane. It was suggested that the loss of proximity to the membrane was indicative of dissociation of ligand from receptor. Geuze *et al.*[39] used double-label immunoelectron microscopy on ultrathin cryosections of rat liver for the simultaneous localization of ligand and receptor. Both were found associated with the membrane of coated vesicles near the cell surface. Other uncoated vesicles revealed ligand apparently free in the lumen and receptor concentrated in tubular extensions largely free of ligand. It was proposed that these tubular extensions are intermediates in the recycling of receptors to the cell surface. Dunn *et al.*[53] have presented evidence that the fusion of intermediate endocytic vesicles with lysosomes ceases at temperatures at or below 17°. At this temperature, [^{125}I]asialofetuin was located in electron-lucent vesicles different from the structures involved in the initial internalization of ligand.

In addition to morphological examination, attempts have been made to analyze biochemically the events of endocytosis subsequent to internalization. Utilizing assays for cell surface-bound ligand, intracellular receptor-bound ligand, intracellular unbound ligand, and degraded ligand, rate coefficients for each of the transitions in the pathway have been calculated.[28,31] Certain differences were apparent when suspensions of freshly

[51] A. L. Hubbard, G. Wilson, G. Ashwell, and H. Stukenbrok, *J. Cell Biol.* **83,** 47 (1979).

[52] A. L. Hubbard and H. Stukenbrok, *J. Cell Biol.* **83,** 65 (1979).

[53] W. A. Dunn, A. L. Hubbard, and N. N. Aronson, Jr., *J. Biol. Chem.* **255,** 5971 (1980).

[54] M. Horisberger and M. von Lanthan, *J. Histochem. Cytochem.* **26,** 960 (1978).

[55] R. J. Stockert, H. B. Haimes, A. G. Morell, P. M. Novikoff, A. B. Novikoff, N. Quintana, and I. Sternlieb, *Lab. Invest.* **43,** 556 (1980).

[56] V. Kolb-Bachofen, *Biochim. Biophys. Acta* **645,** 293 (1981).

[57] C. J. Steer and R. D. Klausner, *Hepatology* **3,** 437 (1983).

isolated hepatocytes[28] were compared with monolayer cultures of hepatocytes.[31] Most notably, in monolayers, dissociation of the intracellular receptor-ligand complex was best fit by two rather than one rate coefficient. Nonetheless, the endocytic pathway in both systems involve the obligate intermediate states of intracellular receptor-bound and intracellular unbound ligand. The observation of two rate coefficients for dissociation of internalized receptor–ligand complex has recently been confirmed by Weigel and Oka.[58] In this study digitonin was used to permeabilize freshly isolated hepatocytes and release unbound ligand from the cells. The experimental design also differed from that of Bridges *et al.*[28] in that cells in the study of Weigel and Oka[58] were preincubated for 45 min at 37° prior to exposure to ligand. The primary monolayer cultures of Harford *et al.*[31] wherein biphasic dissociation kinetics were observed were incubated for 16 hr at 37° before exposure to ligand. The mechanism by which these treatments might yield biphasic dissociation kinetics remains obscure. Interestingly, the slower of the two dissociation processes appeared to be preferentially inhibited at temperatures below 18°.[58] Wolkoff *et al.*[60] also found that reduced temperature (18°) inhibited intracellular ligand dissociation. Kinetic parameters of the pathway have also been calculated for asialoglycoprotein endocytosis and catabolism in HepG2 cells.[19]

A variety of metabolic inhibitors have proved to be useful in elucidating sequential steps in the endocytic pathway. The transition between internal receptor-bound and internal unbound ligand was inhibited by ammonium chloride,[28] monensin or chloroquine,[9] or by reducing the incubation temperature to 18°C.[60] More significantly, ammonium chloride[28] and monensin[59] were shown to lead to accumulation of ligand in a subcellular fraction of lower density than either plasma membranes or lysosomes. These results were interpreted to indicate a pH-mediated dissociation of the receptor–ligand complex within acidic, nonlysosomal endocytic vesicles. The weak bases, ammonium chloride and chloroquine, as well as the proton ionophore, monensin, would be expected to neutralize such acidic organelles. Thus, intracellular dissociation of the receptor–ligand complex appears to be obligatory for delivery of ligand to lysosomes. Leupeptin which inhibits lysosomal degradation but not intracellular dissociation of the receptor–ligand complex, does not impede transfer of ligand to lysosomes.[28]

Access of endocytosed asialoglycoprotein to lysosomes is also impeded by colchicine and cytochalasin B[47,60] as is removal of sodium from the incubation medium.[60,61] Using an assay based on the ability of monen-

[58] P. Weigel and J. Oka, *J. Biol. Chem.* **258,** 10253 (1983).
[59] J. Harford, A. W. Wolkoff, G. Ashwell, and R. D. Klausner, *J. Cell Biol.* **96,** 1824 (1983).
[60] A. W. Wolkoff, R. D. Klausner, G. Ashwell, and J. Harford, *J. Cell Biol.* **98,** 375 (1984).
[61] J. U. Baenziger and D. Fiete, *J. Biol. Chem.* **257,** 6007 (1982).

sin to mediate reassociation of receptor and ligand, the combination of colchicine and cytochalasin B was identified as an inhibitor of the segregation of receptor and ligand subsequent to their dissociation. The assay is based on the presupposition that receptor and ligand in the same vesicle are able to reassociate under appropriate conditions, whereas, after segregation into separate structures, reassociation is not possible. Monensin, by neutralizing the acidic environment responsible for receptor–ligand dissociation, produces conditions condusive to their reassociation. Upon prolonged incubation, the ability to reassociate was lost, a finding consistent with segregation. Colchicine plus cytochalasin B markedly slows the rate of dissociation suggesting that microtubules and microfilaments participate in the segregation of receptor and ligand. Colchicine and cytochalasin B have also been found to enhance the toxicity of diphtheria toxin fragment A-asialoorosomucoid hybrid toxins by reducing their degradation.[62] Data in this study were interpreted as supporting the hypothesis that two receptor-mediated pathways are involved in ligand uptake and transport to lysosomes of which only one is colchicine sensitive.

Baenziger and Fiete[61] reported that replacement of sodium by potassium in incubation media results in marked inhibition of asialoglycoprotein catabolism by isolated hepatocytes. They further demonstrated that, in this "high K^+" buffer, ligand was not delivered to lysosomes but that receptors recycled. These findings indicated that delivery of ligand to lysosomes is not a prerequisite for receptor reutilization. Using the assay for monensin-mediated reassociation described above, Wolkoff *et al.*[60] have shown that in media where potassium is the only monovalent cation, intracellular dissociation and segregation of receptor and ligand proceed unimpeded. These results are fully consistent with the ability of these cells to reutilize receptor. It was further shown in this study that the absence of sodium, rather than the presence of potassium was responsible for the "high K^+" effect. Isoosmotic replacement of sodium with sucrose yielded results indistinguishable from those obtained in sodium-free, potassium medium. The exact nature of the sodium requirement in ligand delivery to lysosomes remains unclear.

Endocytosis does not obligitorily result in ligand degradation in lysosomes. For example, a small portion of injected asialoglycoproteins is removed from the circulation via the asialoglycoprotein receptor but escapes degradation and appears in the bile.[63,64] This may be related to the findings of Tolleshaug *et al.*[65] who showed that, subsequent to endocyto-

[62] T.-M. Chang and D. W. Kullberg, *J. Biol. Chem.* **257,** 12563 (1982).

[63] R. L. Burger, R. J. Schneider, C. S. Mehlman, and C. S. Allen, *J. Biol. Chem.* **250,** 7707 (1975).

[64] P. Thomas and J. W. Summers, *Biochem. Biophys. Res. Commun.* **80,** 335 (1978).

[65] H. Tolleshaug, P. A. Chindemi, and E. Regoeczi, *J. Biol. Chem.* **256,** 6526 (1981).

sis of low amounts of asialotransferrin, isolated hepatocytes exocytose the preponderance of the ligand. However, a more recent paper[66] supports the contention that this asialotransferrin that returns undegraded to the cell exterior is internalized by interaction with the transferrin receptor of hepatocytes rather than the asialoglycoprotein receptor. Nonetheless, return to the cell exterior of ligands unrelated to transferrin has also been observed by other investigators.[67–70] The significance of these observations remains to be determined. The pathway of ligand movement is doubtless more complex than we currently envision.

[66] S. P. Young, A. Bomford, and R. Williams, *J. Biol. Chem.* **258,** 4972 (1983).
[67] D. T. Connolly, R. R. Townsend, K. Kawaguchi, W. R. Bell, and Y. C. Lee, *J. Biol. Chem.* **257,** 939 (1982).
[68] D. A. Wall, R. R. Townsend, Y. C. Lee, and A. L. Hubbard, *J. Cell Biol.* **95,** 425a (1982).
[69] T.-M. Chang, *Fed. Proc., Fed. Am. Soc. Exp. Biol.* **42,** 1824 (1983).
[70] J. Harford and A. W. Wolkoff, unpublished observation (1983).

[20] Preparation of Low-Density "Endosome" and "Endosome"-Depleted Golgi Fractions from Rat Liver

By W. H. Evans

Located on the surfaces of animal cells are a variety of distinctive binding sites or receptors where circulating ligands, e.g., hormones, metabolites, antibodies, viruses, toxins etc. bind in a highly specific manner. The so-formed ligand–receptor complexes are rapidly transferred from the plasma membrane toward the lysosomes or the Golgi apparatus via intermediary compartments now mapped out by morphological and immunocytochemical methods.[1–3] The new intermediary membrane systems are being studied by subcellular fractionation techniques, and the membrane fractions where the internalized ligands are concentrated are

[1] M. G. Farquhar and G. Palade, *J. Cell Biol.* **91,** 77s (1981).
[2] I. H. Pastan and M. C. Willingham, *Science* **214,** 504 (1981).
[3] H. J. Geuze, J. W. Slot, G. J. A. M. Strouse, H. F. Lodish, and A. L. Schwartz, *Cell* **32,** 277 (1983).

ISBN 0-12-182009-2

termed receptosomes,[4] endosomes,[5] endocytic vesicles,[6] ligandosomes,[7] diacytosomes,[8] intermediate vesicles,[9] pinosomes etc.

Most ligands, including polypeptide hormones, after concentration in coated pits at the cell surface, appear to follow similar receptor-mediated internalization routes. The present article describes a method for preparing, from rat liver homogenates, of "endosome" fractions where insulin, prolactin, and various asialoglycoproteins are concentrated 25- to 60-fold relative to tissue homogenates. Furthermore, the hormones and metabolites are recovered in an undegraded state in these fractions. An extension of the method permits "endosome"-depleted Golgi apparatus subfractions to be collected from the same tissue homogenate. Methods for subfractionating "endosomes" are briefly outlined and their biochemical properties are reviewed. In addition to the presence of intact ligand, a proton-transporting ATPase is emerging as a functional enzymic marker for this "endosome" fraction.

Preparation of Liver "Endosome" Fractions

Low density "endosome" fractions were first isolated by following the subcellular distribution of various iodinated ligands, but they can also be prepared without resort to the use of these as markers.[8] Radioiodinated polypeptide hormones and asialoglycoproteins, approximately 2 min after intraportal administration to livers maintained at 37°, concentrate in subcellular components equilibrating at low densities in gradients of sucrose or Percoll. The equilibration of the ligand-transporting components at low density positions in sucrose gradients was first observed by analytical centrifugation of the liver homogenates,[10] and these observations have provided the basis for the development of a procedure for the isolation and characterization of "endosome" subcellular fractions.

Experimental Procedure

Tracer amounts (<1 μg) of porcine insulin, human prolactin, asialoglycoproteins (asialotransferrin Type 3, asialo-alkaline phosphatase) are io-

[4] R. B. Dickson, L. Beguinot, J. A. Hanover, N. D. Richert, M. C. Willingham, and I. Pastan, *Proc. Natl. Acad. Sci. U.S.A.* **80,** 5335 (1983).

[5] C. R. Hopkins, *Nature (London)* **304,** 684 (1983).

[6] A. L. Hubbard, D. A. Wall, and W. A. Dunn, *in* "Structural Carbohydrates in the Liver" (H. Popper, W. Reutter, F. Gudat, and E. Kottgen, eds.), pp. 265–386. MTP Press, Lancaster, 1983.

[7] G. D. Smith and T. J. Peters, *Biochim. Biophys. Acta* **716,** 24 (1982).

[8] M. T. Debanne, W. H. Evans, N. Flint, and E. Regoeczi, *Nature (London)* **298,** 398 (1982).

[9] B. I. Posner, M. N. Khan, and J. J. M. Bergeron, *Endocrine Rev.* **3,** 280 (1982).

[10] G. D. Smith, W. H. Evans, and T. J. Peters, *FEBS Lett.* **120,** 104 (1980).

dinated to high specific activity by the iodogen procedure[11] (although other iodination methods that also minimize damage to the ligands may also be employed). After separation of the iodinated ligand from free iodine by gel filtration on Sephadex G-25 columns equilibrated in a phosphate-buffered saline, pH 7.4, 0.1% bovine serum albumin, portions (not more than 0.5 ml per rat) are injected (plastic syringe equipped with a needle 0.4 mm diameter) into the hepatic portal or caudal veins. The liver is removed approximately 10–15 min later and transferred into ice-cold 0.25 *M* sucrose. The following procedure applies to two livers removed from 100–250 g weight rats of either sex. The livers, after washing and blotting, are weighed and cut into small pieces using scissors. They are homogenised in 0.25 *M* sucrose (3 vol/g wet weight) by using 10 strokes of a loose-fitting (L) 35 ml Dounce homogenizer (clearance 0.119 mm) (Blaessig Glass Co., Rochester, New York]. The homogenate is filtered through nylon gauze, 50–100 mesh (Swiss Silk Bolting Cloth Manufacturing Co. Ltd., Zurich, Switzerland) and rehomogenized using 6 strokes of a tight-fitting (T) pestle of the Dounce homogenizer (clearance 0.072 mm). The filtered homogenate is centrifuged using polycarbonate tubes, in a 8 × 50 ml angle rotor (Sorvall SS-34; Beckman JA20 or equivalent) at 1000 *g* for 10 min; the pellet is washed twice by resuspension (1.5 vol/g liver wet weight) and centrifugation in 0.25 *M* sucrose, and the combined supernatants (vol ~ 100 ml) are centrifuged at 33,000 g_{av} for 8 min at speed in a Beckman Type 30 angle rotor.[12] The supernatants are collected, taking meticulous care to ensure that loose material resting on the multilaminate pellet is not disturbed. The combined supernatants (vol 100 ml) are layered onto 6 of the following sucrose (w/v) gradients made in 38-ml cellulose nitrate tubes: 1 ml 70% sucrose, 5 ml 43%, a continous gradient by mixing 7.5 ml 40% and 7.5 ml 15% sucrose. The SW27/28 (or equivalent) rotor is centrifuged for 3.5 or 4 hr, respectively, at maximum speed. The sucrose gradients are unloaded from the bottom of the tubes by using a hollow needle attached to a peristaltic pump and approximately 40 × 0.7 ml fractions collected. The density of the sucrose is determined using a refractometer and the distribution of radioligand is determined using a gamma scintillation counter.

The general scheme is shown in Fig. 1. The distribution across the sucrose gradient of [^{125}I]prolactin 30 sec and 10 min after injection into the portal vein is shown in Fig. 2. Similar time-dependent movements of radioiodinated insulin and asialotransferrin Type 3 to a low-density posi-

[11] P. J. Fraker and J. C. Speck, *Biochem. Biophys. Res. Commun.* **80,** 849 (1978).

[12] Programmable centrifuges (Beckman L8 or equivalent) can be set for w^2t 2.61 × 10^9 rad^2/sec. Beckman L5 acceleration dial is set to half maximum.

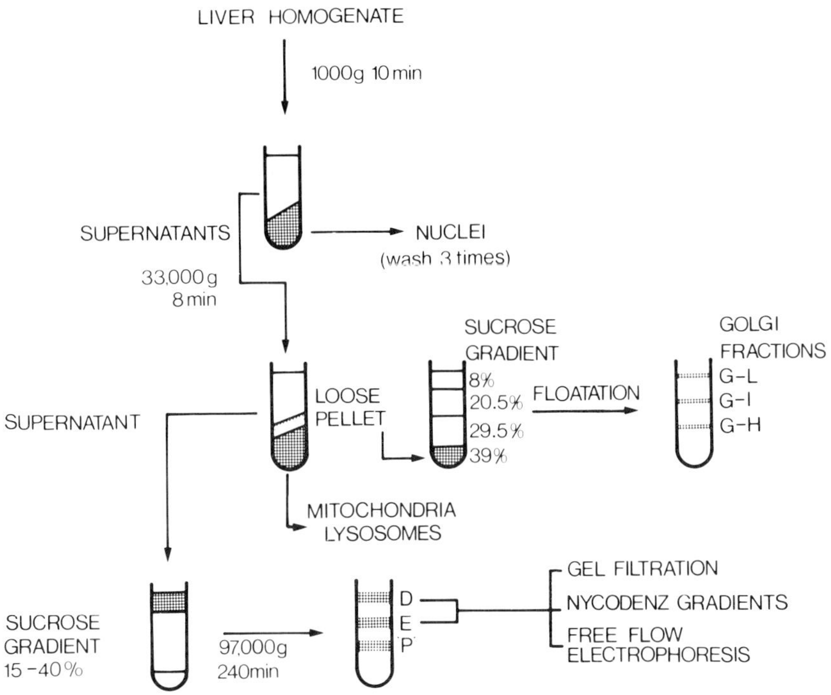

FIG. 1. Preparation of low-density "endosome" fractions D and E, and "endosome"-depleted Golgi apparatus G-L, G-I, and G-H subfractions from rat liver homogenates. Fraction P (density 1.145–1.165 g/cm^3) is of unknown origin. Approximately 0.25 and 0.15% of homogenate protein is recovered (after gel filtration step) in fractions D and E, respectively.

tion also occur.[13] Fraction E (Fig. 1) corresponds to components of sucrose density 1.120–1.140 g/cm^3, where the ligand peak is located 30 sec to 1 min, and fraction D corresponds to components containing ligands that equilibrate at density 1.095–1.117 g/cm^3 at periods 5–30 min after administration of isotopes at 37°. In Percoll gradients, fractions E and D (Fig. 1) correspond to components located at densities 1.050 and 1.043 g/cm^3, respectively. Fractions in these two density ranges are pooled.

The distribution of radioiodinated ligands and subcellular marker enzymes in this procedure is shown in Fig. 3. Although there is an increase in specific activities of galactosyl- and sialyltransferase activities in the ligand-containing fraction D, a high recovery of these Golgi apparatus

[13] W. H. Evans, N. Flint, M. Debanne, and E. Regoeczi, *in* "Structural Carbohydrates in the Liver" (H. Popper, W. Reutter, F. Gudat, and E. Kottgen, eds.), pp. 301–311, MTP Press, Lancaster, 1983.

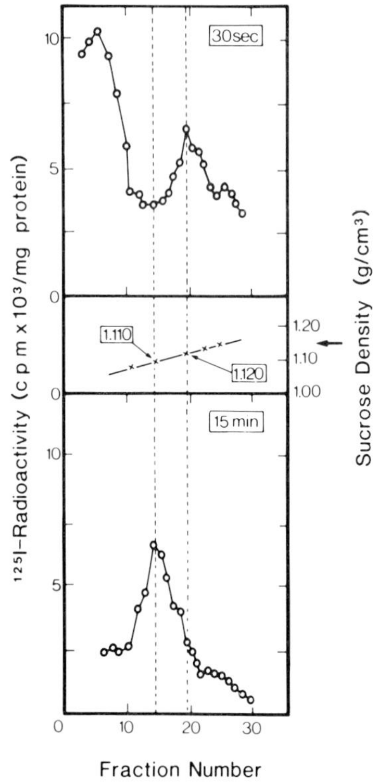

FIG. 2. Time dependence of the position of [^{125}I]prolactin in the sucrose-density gradient step of the procedure described in the text, and shown in Fig. 1. At 30 sec after injection into the portal vein of a female rat, the [^{125}I]prolactin peak locates at density 1.120 g/cm³, but rapidly moves to density 1.110 g/cm³. Fractions 1–10 contain liver-soluble components where free ligand is collected.

marker components, as well as lysosomal markers, was obtained in the mitochondrial–lysosomal pellet (see below). Note that asialoorosmucoid, an asialoglycoprotein rapidly hydrolyzed in lysosomes, is not concentrated in the "endosome" D fraction isolated 10–15 min after injected of isotopes.

Further Purification of "Endosome" Fractions D and E

The fractions pooled from the sucrose gradient have a yellow coloration, and this is removed and the specific activity of radioligands is increased substantially by the following procedures. Fractions, concentrated to 5–10 ml by vacuum dialysis, are applied to a Sepharose 2B column (80 × 1 cm) equilibrated in 0.15 *M* NaCl, 1 m*M* EDTA, and

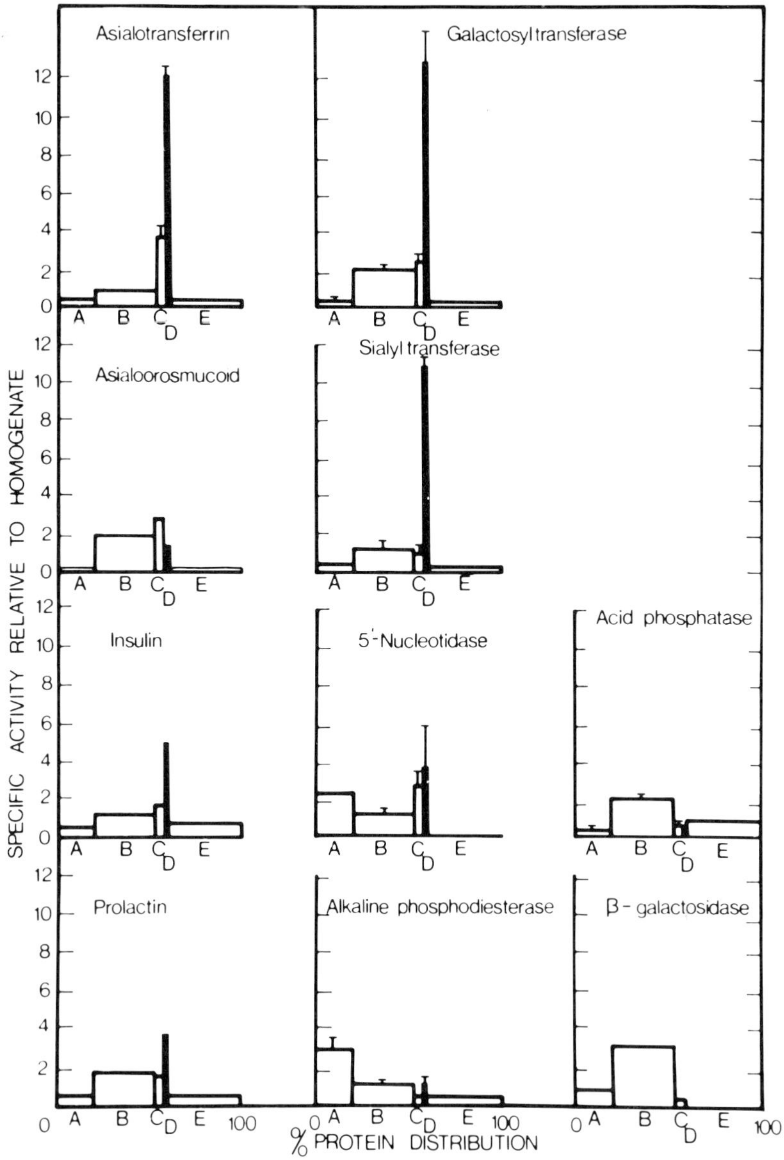

FIG. 3. Subcellular distribution of radioligands and marker enzymes in the fractionation scheme. Fraction A contains mainly nuclei, plasma membrane sheets, and cell debris, and fraction B contains mitochondria and lysosomes, with a loose overlay of Golgi apparatus components. Fractions C, D, and E are the components lighter (E) or heavier (C) than the low-density "endosome" fraction D, collected at density range 1.095–1.117 g/cm^3. Fractions P and E shown in Fig. 1 were not collected in this scheme. The three radioiodinated ligands were injected into the portal vein 10–15 min before livers were homogenized. Recoveries of protein, ligands, and marker enzymes varied from 81 to 100%.

10 m*M* Tris–HCl, pH 7.4; inclusion of 0.3 *M* sucrose in the equilibrating buffer may prove helpful. About 60–85% of the applied radioactivity is eluted in the void volume accompanied by about 20–25% of the protein. Soluble liver proteins, free ribosomes, and the yellow coloration are eluted in later fractions. An alternative method for purifying and concentrating the membrane components involves centrifugation of the fractions pooled from the sucrose gradients on 40% (w/v) sucrose cushions at 27,000 g_{av} for 2 hr in a Beckman SW27 rotor or equivalent. The "endosomes" are collected as a whitish compact band at the density interface, whereas the components with a yellow coloration migrate into the sucrose cushion.

Subfractionation of "Endosomes"

Morphological examination of low-density "endosome" fractions, after removal of the yellow coloration and ribosomes as described above, shows the fraction to contain vesicular membrane profiles (Fig. 4). Morphometric analyses indicate that at least three major categories of structures are present—vesicles with a clear lumen (average diameter 135 nm), multivesicular bodies (average diameter 155 nm), and granular bodies (average diameter 105 nm). The relative proportion of these components can vary according to the density position on the sucrose gradient, or the subfractionation technique (e.g., free flow electrophoresis) adopted, but the granular components are usually in the majority. Analysis of fractions D and E by free flow electrophoresis showed that two populations of ligand–receptor-containing membranes could be separated.[8] Also, fractionation of fraction D for 16 hr at 100,000 *g* into 13.8–27.6% (w/v) isoosmotic Nycodenz gradients (Nyegaard & Co., Oslo, Norway) yields a pellet and three whitish components; the radioiodinated ligands are associated mainly with the membrane vesicles in the two lightest components of density 1.050 and 1.090 g/ml.[14,15] These results indicate the morphological and biochemical complexity of the low-density "endosome" fractions prepared from liver homogenates.

Biochemical Properties of Low-Density "Endosome" Fractions

The isolation of low-density "endosome" fractions was achieved by following the subcellular distribution of radioiodinated ligands. Table I shows the recoveries in the "endosome" fraction of ligands and various

[14] W. H. Evans and N. Flint, unpublished work.

[15] T. Saermark, N. Flint, and W. H. Evans, *Biochem. Soc. Trans.*, **12,** 1073 (1984).

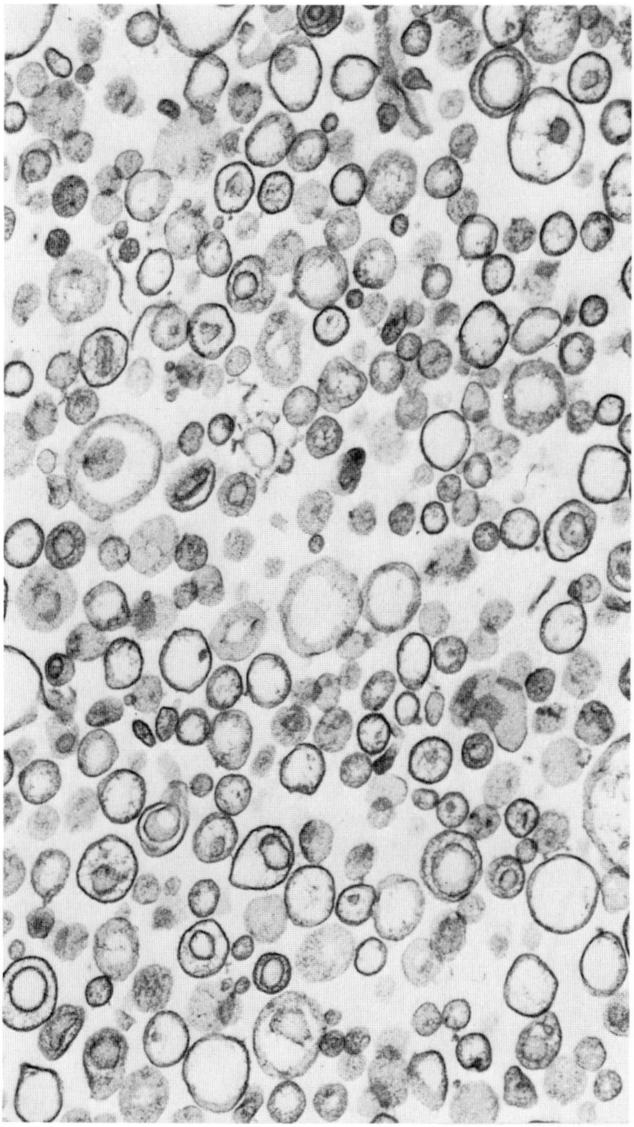

FIG. 4. Electron micrograph of the low-density "endosome" fraction after purification by gel exclusion. ×42,250.

TABLE I
RECOVERY AND RELATIVE SPECIFIC ACTIVITIES OF RADIOIODINATED LIGANDS AND VARIOUS MARKER ENZYMES IN THE LOW-DENSITY "ENDOSOME" D FRACTION PREPARED AS SHOWN IN FIG. 1

Component in fraction	Recovery from homogenate (%)	Specific activity relative to homogenate
Protein	0.25 ± 0.04	—
Asialotransferrin type 3	19.4 ± 0.9	62.0
Asialoorosomucoid	2.7	6.5
Insulin	7.9 ± 1.1	25.0
Prolactin	16.2	39.0
Galactosyltransferase	21.0	65.0[a]
Sialyltransferase	9.5	51.0
5′-Nucleotidase	6.7 ± 2.4	16.5
Alkaline phosphodiesterase	2.5 ± 0.4	6.5
Monensin-activated Mg^{2+}-ATPase[a]	—	20[a]
Adenylate cyclase (basal)	1.4	1.0
Leucine aminopeptidase	0.3	0.1
γ-Glutamyl transpeptidase	0.4	0.3
Acid phosphatase	1.0	2.2
β-Galactosidase	0.009	0.1

[a] Further fractionation of "endosome" fraction D in isoosmotic Nycodenz gradients reduces galactosyltransferase specific activities and recoveries by one-sixth of this value, and increases the monensin-activated Mg^{2+}-ATPase specific activity 3-fold. Recoveries of radioiodinated ligands calculated 10–15 min after injection into portal vein.

enzymic markers. The ligands are increased in specific activity relative to the homogenate 25- to 62-fold; the enzymic markers, with the exception of the glycosyltransferases, are not enriched to any great extent. Thus, the "endosome" fraction is deficient in a sinusoidal plasma membrane marker (adenylate cyclase), canalicular plasma membrane markers (5′-nucleotidase, leucine aminopeptidase), lysosomal markers (acid phosphatase, β-galactosidase), and an endoplasmic reticulum marker (NADP–cytochrome *c* reductase).[16] The question of to what extent galactosyl- and sialyltransferases are constitutive components of the endosome fractions, or are Golgi contaminants is now being resolved. The free-flow electrophoresis technique separates galactosyl- and sialyltransferase-containing vesicles away from a proportion of the ligand-containing components, but an appreciable amount of glycosyltransferases still coelectrophoreses with the ligand and receptor-containing vesicles.[8] However,

[16] W. H. Evans, "Preparation and Characterisation of Mammalian Plasma Membranes." North Holland/Elsevier, Amsterdam, 1978.

overnight centrifugation at 100,000 *g* in Nycodenz gradients produces a population of vesicles (~30% of the protein applied to the gradients) of density 1.050 g/cm^3 containing ligands (asialoglycoproteins, prolactin, insulin) and extremely low galactosyltransferase activity.[14] A second peak (density 1.090 g/cm^3) containing ligands coequilibrated with a peak of galactosyltransferase activity (~8–10% of homogenate activity). These subfractionation results indicate that a proportion of the ligand-transporting "endosome" components can be isolated relatively free of glycosyltransferase activities.

Polyacrylamide gel electrophoretic analysis indicates that "endosome" fractions have distinctive polypeptides, but many of the components are of similar electrophoretic mobility to those in the various plasma membrane domains.[14] Polypeptides common to "endosome" and plasma membrane fractions may correspond to the subunits of receptors for asialoglycoproteins, insulin, and prolactin that are present in these fractions.[8,13]

"Endosome" Subcellular Markers—Current Status

The present status of subcellular markers useful for the isolation of "endosome" fractions from liver and other tissues and cells may thus be summarized. "Endosome" fractions consist of smooth-surfaced membrane vesicles where a variety of ligands, transferred from plasma membrane receptors, are entrapped; the specific activities of the radioiodinated ligands are manyfold higher than in the cell/tissue homogenate. The ligands in the "endosome" fractions are undegraded as assessed by polyacrylamide gel electrophoresis (asialoglycoproteins) or gel filtration (polypeptide hormones). The fractions are generally of low density (1.11–1.13 g/cm^3) in sucrose gradients, and are deficient in enzymic markers characteristic of plasma membranes, lysosomes, and endoplasmic reticulum. "Endosome" subfractions can be isolated from liver free of Golgi markers by free-flow electrophoresis or centrifugation in Nycodenz density gradients, indicating that those parts of the endocytic compartment from where they originate can be distinguished biochemically, if not morphologically, from the Golgi apparatus. Finally, a monensin-activated Mg^{2+}-ATPase, functioning as an ATP-driven proton pump is enriched, relative to tissue homogenates, ~40- to 60-fold in "endosome" subfractions.[15]

The isolated "endosomes" acidify in the presence of ATP (shown by 9-aminoacridine uptake), a process reversed when monensin is added.[15] The occurrence in the "endosome" fractions of a proton-transporting ATPase, and the presence of undegraded ligands emerge as *combined*

TABLE II
RECOVERY (%) FROM HOMOGENATE OF [^{125}I]PROLACTIN AND [^{125}I]INSULIN[a]

Hormone	Minutes after injection	Endosome fraction D	Golgi apparatus subfractions L	I	H	Golgi total
Prolactin	0.5	2.9	0.03	0.18	0.57	0.78
	2.0	5.2	0.26	0.34	0.53	1.13
	10.0	17.0	0.79	2.4	0.77	3.96
Insulin	5.0	6.5	0.08	0.47	1.24	1.79
	10.0	7.4	0.19	0.58	1.26	2.03

[a] Recovery in low-density "endosome" D fraction (1.070–1.120 g/cm^3) and "endosome"-depleted Golgi fractions at various intervals after injection into the portal vein. The fractions were prepared as shown in Fig. 1.

functional markers foı the membranes deriving from the endocytic networks of hepatocytes[15,17] and other cells.[18] It is important to note that lysosomes also contain an ATP-driven proton pump, but the hydrolytic enzymes present rapidly hydrolyze the ligands. Finally, "endosome" fractions contain ligand-binding components, and it is likely that these comprise intracellular pools that feature in receptor recycling and regulation.

Preparation of "Endosome"-Depleted Golgi Apparatus Fractions

The fractionation procedure described for the preparation of "endosomes" can also be used to obtain three Golgi apparatus subfractions starting from the same liver homogenate (Fig. 1). The three Golgi apparatus subfractions correspond morphologically and have similar recoveries of galactosyltransferase to those prepared by standard methods.[19,20]

The loose region of the pellet obtained after centrifuging the postnuclear supernatant at 33,000 *g* for 8 min at speed is collected by swirling in 72% (w/v) sucrose. After dispersion, in a loose-fitting Dounce homogenizer, the material is adjusted to 39% (w/v) (approximate vol 50 ml) and is overlayed with equal amounts of 29.5, 20.5, and 8% (w/v) sucrose using six cellulose nitrate tubes. After centrifugation at 97,000 g_{av} for 3 hr in a

[17] M. J. Geisow and W. H. Evans, *Exp. Cell Res.* **150,** 36 (1984).

[18] C. Galloway, G. E. Dean, M. Marsh, G. Rudnick, and I. Mellman, *Proc. Natl. Acad. Sci. U.S.A.* **80,** 1300 (1983).

[19] J. H. Ehrenreich, J. J. M. Bergeron, P. Siekevitz, and G. E. Palade, *J. Cell Biol.* **59,** 45 (1973).

[20] J. J. M. Bergeron, *Biochim. Biophys. Acta* **555,** 493 (1979).

Beckman SW27 rotor, the G-L, G-I, and G-H subfractions are collected respectively at the 8–20.5, 20.5–29.5 and 29.5–39% interfaces. Table II shows that the recoveries of insulin and prolactin in these Golgi subfractions are lower than those obtained simultaneously in the "endosome" fraction, indicating that the low-density "endosome" fraction is a major repository of these polypeptide hormones at short intervals after uptake by liver.

Acknowledgments

I thank Dr. M. Debanne (Department of Pathology, McMaster Univeristy) and Mr. N. Flint (National Institute for Medical Research, Mill Hill) for help in the development of the methods described. NATO Research Grant 88.82 catalyzed the collaboration between the two laboratories.

[21] Isolation of Receptosomes (Endosomes) from Human KB Cells

By ROBERT B. DICKSON, JOHN A. HANOVER, IRA PASTAN, and MARK C. WILLINGHAM

Introduction

Receptor-mediated endocytosis allows a cell to internalize various ligands, including polypeptide hormones, growth factors, viruses, and toxins from its environment at a much greater rate than materials entering not bound to a receptor. The process of receptor-mediated endocytosis involves binding of a ligand to a specific cell membrane receptor, concentration of ligand–receptor complexes in clathrin-coated pits,[1,2] and rapid transfer of the ligand into an uncoated vesicle termed a receptosome[3] or endosome.[4] Vesicles with the morphologic characteristics of receptosomes have been found in almost all types of animal cells. A simple

[1] I. H. Pastan and M. C. Willingham, *Science* **214,** 504 (1981).
[2] J. L. Goldstein, R. G. W. Anderson, and M. S. Brown, *Nature* (*London*) **279,** 679 (1979).
[3] M. C. Willingham and I. Pastan, *Cell* **21,** 66 (1980).
[4] A. Helenius, M. Marsh, and J. White, *Trends Biochem. Sci.* **5,** 104 (1980).

ISBN 0-12-182009-2

description of a receptosome is an isolated vesicle with an acidic pH that forms from coated pits at the cell surface and carries ligands by saltatory motion[1] from the cell surface to a specialized portion of the Golgi. This entire process takes about 10–30 min, depending on cell type.[1,5] Sorting and processing appear to occur in the Golgi.[6] Following endocytosis and Golgi delivery, many ligands such as α_2-macroglobulin, epidermal growth factor (EGF), low-density lipoprotein, insulin, asialoglycoproteins, and lysosomal enzymes are transferred to lysosomes,[1,2] but transferrin behaves differently. After releasing its iron, it is returned undegraded to the cell surface.[7,8] Many viruses also enter cells via coated pits and receptosomes[9–11] and appear to escape into the cytosol from this acidic vesicle.[4,12–14] A considerable amount of cell membrane proteins and phospholipids enters the cell along with the ligand as the receptosome is formed. This process probably in large part accounts for the rapid turnover of cell membrane constituents. The current study[15] presents a method to isolate receptosomes and characterizes some of their properties.

Receptosome Isolation Protocol

Strategy for the Procedure. The strategy chosen to isolate receptosomes was to load them with ^{125}I-labeled EGF ([^{125}I]EGF) and purify a vesicle fraction enriched in this marker. To do this, a small number of KB cells in monolayer culture was labeled with tracer [^{125}I]EGF, and these cells were combined with a large number of unlabeled KB cells; such large numbers are most easily grown in suspension culture. Cells are homogenized and the homogenate is subjected sequentially to isopycnic centrifugation on colloidal silica to remove lysosomes rapidly, to gel filtration on Sephacryl S-1000 to remove soluble proteins and the bulk of the Golgi membranes, and finally to isopycnic centrifugations on sucrose to remove other contaminating membranes, including plasma membrane.

5 M. C. Willingham and I. Pastan, *J. Cell Biol.* **94,** 207 (1982).
6 S. Goldfischer, *J. Histochem. Cytochem.* **30,** 717 (1982).
7 R. B. Dickson, J. A. Hanover, M. C. Willingham, and I. Pastan, *Biochemistry* **22,** 5667 (1983).
8 J. Renswoude, K. R. Bridges, J. B. Harford, and R. D. Klausner, *Proc. Natl. Acad. Sci. U.S.A.* **79,** 6186 (1982).
9 S. Dales, *Bacteriol. Rev.* **37,** 103 (1973).
10 R. B. Dickson, M. C. Willingham, and I. Pastan, *J. Cell Biol.* **89,** 29 (1981).
11 K. S. Matlin, H. Reggio, A. Helenius, and K. Simons, *J. Cell Biol.* **91,** 601 (1981).
12 B. Tycko and F. Maxfield, *Cell* **28,** 643 (1982).
13 M. Marsh, E. Bolzau, and A. Helenius, *Cell* **32,** 931 (1983).
14 D. J. P. FitzGerald, R. Padmanabhan, I. H. Pastan, and M. C. Willingham, *Cell* **32,** 607 (1983).
15 R. B. Dickson, L. Beguinot, J. A. Hanover, N. D. Richert, M. C. Willingham, and I. Pastan, *Proc. Natl. Acad. Sci. U.S.A.* **80,** 5335 (1983).

Reagents

KB carcinoma cells (from American Type Culture Collection, Rockville, MD) grown in monolayer[5] and suspension[16]

^{125}I-labeled epidermal growth factor (from Bethesda Research Laboratories, Bethesda, MD), 360 μCi/μg labeled by chloramine-T method[17,18]

Basic buffer for isolation (TES): 10 m*M* triethanolamine (Sigma), pH 6.7, 1 m*M* EDTA, 0.25 *M* sucrose, 0.02% (w/v) sodium azide, 1% aprotinin (v/v; Sigma)

Isoosmotic Percoll: To 9 volumes of Percoll (Pharmacia Fine Chemicals, Uppsala, Sweden), add 1 volume 2.5 *M* sucrose containing 0.02% (w/v) azide

Density marker beads (Pharmacia)

2.5 *M* sucrose solution

0.63 *M* sucrose, 10 m*M* triethanolamine, pH 6.7, 1 m*M* EDTA, 1% aprotinin (v/v), 0.02% (w/v) sodium azide

1.38 *M* sucrose, 10 m*M* triethanolamine, pH 6.7, 1 m*M* EDTA, 1% aprotinin (v/v), 0.02% (w/v) sodium azide

Sephacryl S-1000 (Pharmacia) column, 50 ml bed volume

Equipment

1275 Minigamma counter (LKB Instruments)

IEC PR-J centrifuge (International Equipment Company)

Nitrogen cavitation bomb (Parr Instruments, Moline IL)

Dounce homogenizer, tight fitting (Arthur Thomas)

Sorvall RC-5B centrifuge, SS34 rotor (DuPont Instruments)

L5-65 Centrifuge, Ti-65 rotor, SW28 rotor (Beckman Instruments)

Auto densiFlow 11C pump (Buchler)

Refractometer (Zeiss)

Membrane filtration apparatus (50 ml, 43 mm size), PM 30 membranes (Amicon)

Human KB Cells: Culture and Labeling with [^{125}I]EGF. KB cells were grown in suspension in spinner flasks[16] (1–3 liter from Bellco) by using Eagle's spinner medium (Gibco) with 5% heated horse serum (Gibco; heated to 56°, 30 min). Cells were seeded at 2–4 $\times$ 10^5 cells/ml and diluted or harvested at 8–12 $\times$ 10^5 cells/ml. For each preparation 1.5–3 $\times$ 10^9 cells were centrifuged and combined with 1–3 $\times$ 10^7 [^{125}I]EGF-labeled cells grown in monolayer. Monolayer cells were grown as previously described[5] and used for [^{125}I]EGF labeling when ≈70% confluent.

[16] M. Green and W. S. M. Wald, this series, Vol. 58 [36].

[17] W. H. Hunter and F. C. Greenwood, *Nature* (*London*) **194,** 495 (1962).

[18] H. T. Haigler, F. R. Maxfield, M. C. Willingham, and I. Pastan, *J. Biol. Chem.* **225,** 1239 (1980).

TABLE I
RECEPTOSOME PURIFICATION SUMMARY[a]

Step	[^{125}I]EGF Radioactivity (cpm × 10^{-3})	[^{125}I]EGF Recovery (%)	Protein Amount (mg)	Protein Recovery (%)	Purification (fold)
Homogenate	1185	100	222	100	1.0
Supernatant	1000	84	122	54	1.6
Percoll	466	39	66	25	1.6
Sephacryl S-1000	320	28	3.6	1.6	18
Sucrose					
Gradient 1	188	16	1.5	0.7	23
Gradient 2	127	11	0.65	0.3	37

[a] Cells were incubated with [^{125}I]EGF to label receptosomes. At each step, the amount of [^{125}I]EGF and protein (Bradford method) was determined and expressed as a percentage of initial homogenate value. Fold purification of [^{125}I]EGF per mg of protein is expressed relative to homogenate.

To label endocytic vesicles with [^{125}I]EGF, two T-150 flasks (Costar, Cambridge, MA) containing 1–3 × 10^7 cells were preincubated for 1 hr at 37° in serum-free medium, cooled to 4°, and gently rocked at 4° for 1 hr with 10 n*M* [^{125}I]EGF. After washing five times with 4° serum-free medium, cells were incubated at 37° for 8 min by adding 10 ml of prewarmed (37°) serum-free medium. This allowed endocytosis of the majority of [^{125}I]EGF to occur.[5,6,18] Endocytosis was stopped by rapidly replacing the cell medium with 4°C serum-free medium and placing the flasks on ice. With KB cells, the critical time interval for receptosome labeling after warming was between 5 and 11 min. This time interval minimized labeling of clathrin-associated structures[8] and maximized EGF concentration in receptosomes. In some experiments, cells were labeled with L-[^{35}S]methionine (Amersham) prior to exposure to [^{125}I]EGF.[19]

Membrane Fractionation. KB cells grown in suspension were harvested by centrifugation (2000 rpm, 5 min in an International IEC PR-J centrifuge). The cell pellet was washed three times with 20 ml TES. The pellet of washed cells was resuspended in 15 ml TES and combined with [^{125}I]EGF-labeled cells (harvested from two flasks in 5 ml TES using a rubber scraper). Cells were placed in a conical tube and pressurized to 80 psi (1 psi = 6.89 kPa) for 30 min in a nitrogen cavitation bomb. After disruption by slow release of pressure, cells were homogenized in a tight-

[19] N. D. Richert, M. C. Willingham, and I. Pastan, *J. Biol. Chem.* **258,** 8902 (1983).

fitting Dounce homogenizer by using 30–100 strokes. Cell disruption was monitored by using trypan blue staining (Gibco) and continued until >90% stained, intact nuclei were observed. Receptosomes appear to be disrupted by excessive homogenization, osmotic shock, or pelleting by centrifugation because these treatments were observed to release substantial amounts of [^{125}I]EGF.

Summaries of the separation of [^{125}I]EGF and protein are presented in Table I and Fig. 1. Summaries of marker enzyme data during fractionation are presented in Tables II and III (see also Marker Enzyme Composition, below). The cell homogenate (20 ml) was centrifuged (DuPont Sorvall RC-5B) to remove unbroken cells and nuclei for 10 min at 500 g. The supernatant was collected, and the pellet resuspended in 20 ml TES and recentrifuged as above. Combined supernatants were added to isoosmotic Percoll stock and TES to yield 18% Percoll in a total volume of 60 ml. An isoosmotic Percoll stock was prepared by a 1 : 9 dilution of 2.5 M sucrose : Percoll (Pharmacia Fine Chemicals, Uppsala, Sweden) and designated as "100%." The sample (60 ml of Percoll-containing supernatant) was lay-

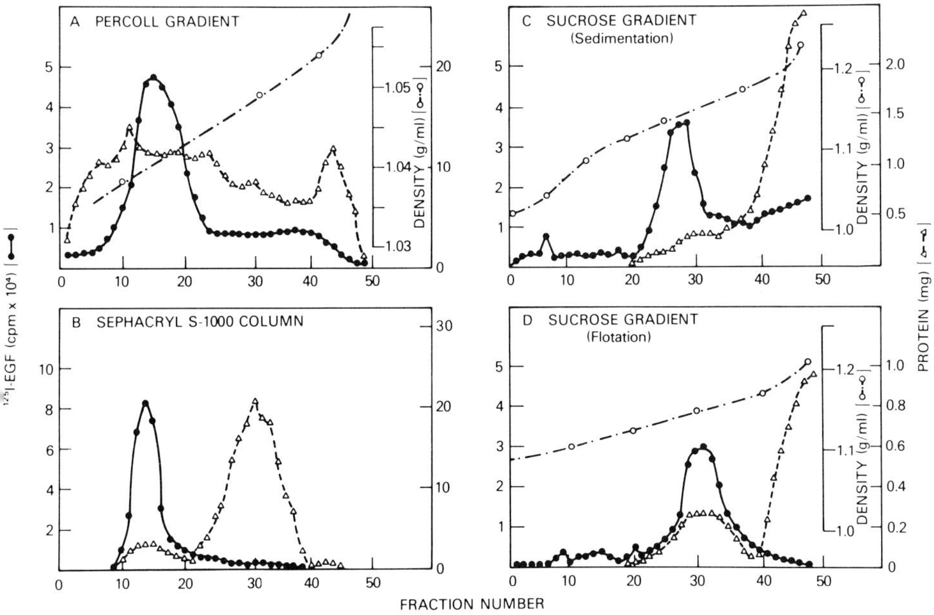

FIG. 1. Summary of purification steps in membrane fractionation. Endocytic vesicles of KB cells were labeled with [^{125}I]EGF and the cells homogenized. Membranes were fractionated and [^{125}I]EGF and protein (Bradford method) determined. In each purification step the peak of [^{125}I]EGF was pooled. In B, V_o of the column was at fraction 12, whereas V_i was at fraction 31.

TABLE II
ACTIVITY OF MARKER ENZYMES DURING MEMBRANE FRACTIONATION

Step	5′-Nucleotidase		Galactosyltransferase (ovalbumin substrate)		β-Hexosaminidase	
	Specific activity (pg/min/μg)	Total activity (pg/min)	Specific activity (fmol/min/μg)	Total activity (fmol/min)	Specific activity (nmol/min/μg)	Total activity (nmol/min)
Homogenate	73	320×10^5 (100%)	0.27	200×10^3 (100%)	50	200×10^6 (100%)
Percoll	116	130×10^5 (40%)	0.34	84×10^3 (42%)	8.0	8.9×10^6[a] (4%)
Sephacryl S-1000	89	65×10^5 (20%)	0.61	1.8×10^3[b] (0.9%)	61	4.4×10^6 (2%)
Sucrose gradient 1	68	3.2×10^5[c] (1%)	0.58	1.0×10^3 (0.5%)	7.4	0.22×10^6 (0.1%)
Sucrose gradient 2	54	0.7×10^5 (0.22%)	0.43	0.9×10^3 (0.44%)	5.6	0.071×10^6 (0.03%)

[a] Eighty percent of the β-hexosaminidase activity present on the Percoll gradient was in a denser class of membranes than [^{125}I]EGF and found between 1.05 and 1.12 g/ml.

[b] Galactosyltransferase interacted with the Sepharcyl S-1000 column. Its activity was smeared (at a low level) through all collected fractions, and membranes with high activity (4.3 fmol/min/μg) were recovered from the top of the top of the Sephacryl column at the end of the separation.

[c] Of the activity lost in this step of purification, 70% was recovered in a peak at 1.17 g/ml sucrose and 20% was deeper in the gradient, peaking near the interface with the sucrose cusion (1.23 g/ml sucrose).

TABLE III
SPECIFIC ACTIVITY MARKER ENZYMES IN MEMBRANE FRACTIONS[a]

Membrane fraction	$[^{125}I]$EGF (cpm/μg)	Na^+,K^+-ATPase (mmol/min/μg)	Adenylate cyclase (fmol/min/μg)	5′-Nucleotidase (pg/min/μg)	Galactosyltransferase: Ovalbumin substrate (fmol/min/μg)	Galactosyltransferase: GlcNAc substrate (fmol/min/μg)	β-Hexosaminidase (nmol/min/μg)
Homogenate	2.0	170	115	73	0.28	7.6	50
Lysosome	2.0	1.7	N.D.	46	0.028	1.0	574
Plasma membrane	N.D.[b]	280	319	297	0.49	1.9	5.6
Receptosome	76	99	34	54	0.53	16	5.6

[a] Enzyme markers in cell fractions. Several marker enzyme activities and $[^{125}I]$EGF were determined in homogenate, lysosome, plasma membrane, and receptosome containing fractions. Data are expressed per μg of protein (Bradford method).

[b] N.D., Not done.

ered over 2 ml of 2.5 *M* sucrose and centrifuged in a Ti-65 rotor (Beckman L5-65 ultracentrifuge) for 1 hr at 16,500 rpm. Fractions (1 ml) were collected from the top using a Buchler Auto densiFlow IIC pump, setting #2. The peak of [^{125}I]EGF, monitored using a 1275 minigamma counter (LKB Instruments) (1.038–1.043 g/ml, density range, determined with marker beads from Pharmacia), was pooled. These two steps removed most of the lysosomes (96% of β-hexosaminidase was in the dense fractions of the Percoll gradient) and some plasma membranes (60% of the 5′ nucleotidase was in the lighter fractions) (see Marker Enzyme Composition, below). The positions of a variety of marker enzymes on Percoll gradients have been determined.[7]

The pooled Percoll peak was concentrated to 3–5 ml by Amicon filtration (70 psi by using a TES-soaked PM 30 filter) and chromatographed on a Sephacryl S-1000 (Pharmacia) column (bed volume, 50 ml) by using TES buffer. The column was routinely pretreated with Percoll-fractionated cell membranes (from 2×10^9 cells) to decrease nonspecific adsorption of receptosomes to the column. The column pretreatment consisted of chromatographing 5 ml of Percoll-fractionated unlabeled membrane concentrate, followed by washing with 500 ml TES buffer. It is not necessary to repeat this pretreatment for the life of the column. For chromatography of the [^{125}I]EGF labeled experimental sample, column fractions (1 ml) were collected and the turbid peak of membranes near V_0 containing the peak of [^{125}I]EGF was saved. This step removed soluble proteins, Percoll, and the vast majority of galactosyltransferase (a marker enzyme for Golgi stacked cisternae). Only 2.1% of the galactosyltransferase activity applied to the column emerged with receptosomes. After chromatography, the column was washed with 1 liter of TES; adsorbed galactosyltransferase gradually was released by washing.

The vesicles (5–8 ml) were next layered on a 30-ml linear sucrose gradient (0.63–1.38 *M* sucrose in TES made with a Buchler Auto densiFlow IIC pump, setting #2) over 2 ml of a 2.5 *M* sucrose cushion. Centrifugation was carried out for 18 hr at 24,000 rpm in an SW28 rotor. Fractions (0.5 ml) were collected from the top (as with the Percoll gradient, above) and the peak of [^{125}I]EGF was pooled (1.12–1.16 g/ml, density range, monitored by refractive index). This step removed more markers for plasma membrane and lysosomes, which were in the denser fractions. The vesicles were recentrifuged on a linear 0.63–1.38 *M* sucrose gradient by using a flotation protocol. To accomplish this, pooled fractions from the first gradient were adjusted to 1.42 *M* sucrose by dropwise addition of 2.5 *M* sucrose; the sample (5–8 ml) was layered over a 2-ml cushion of 2.5 *M* sucrose, and a 30 ml linear 0.63–1.38 *M* sucrose gradient was formed (as above), but on top of the sample. Centrifugation, sample collection, and pooling were as for the first sucrose gradient. At the end of the purification, 0.22% of 5′ nucleotidase, 0.44% of galactosyltransferase, and 0.03% of β-hexosamindiase activities remained with the [^{125}I]EGF-enriched receptosome fraction. The final material was processed for electron microscopy, used in other tests of purity, or frozen at −70° for subsequent studies of lipid, enzyme, and polypeptide composition. A typical experiment (Table I) resulted in a 37-fold enrichment of [^{125}I]EGF-containing vesicles with a yield of 11%. In other experiments, up to 54-fold enrichment was achieved.

Receptosome Characterization

Tests of Purity. The final receptosome preparation was subjected to two other separation techniques to evaluate purity. Nonequilibrium sucrose density gradient centrifugation was carried out in an SW28 rotor on a linear 0.63–1.38 *M* sucrose gradient as with equilibrium studies (above), but only for 2.5 hr (Fig. 2). Under these conditions, in contrast to homogenate, the receptosome fraction demonstrated a single peak of [^{125}I]EGF and protein. While a minor peak of more densely sedimenting membranes was observed, no additional purification of [^{125}I]EGF/protein was obtained.

Free-flow electrophoresis (Fig. 3) was carried out by using a Dasaga FF-48 (Brinkman) apparatus (800 V, 45 mA, loading setting 8.0, collected setting 2.5). While homogenate protein was extremely heterogeneous by this technique, the purified receptosome fraction demonstrated only a single band of protein and [^{125}I]EGF. No further purification was obtained.

Lipid Composition. One milliliter of total homogenate or of the final receptosome fraction (0.5–1.5 mg of protein, respectively) was added to 3.7 ml of methanol/chloroform, 2 : 1 (v/v), to extract lipids.[20] Lipids were then dissolved in chloroform (0.5 ml) for determination of phospholipid phosphorous[21] or total cholesterol.[22]

The data in Table IV show the cholesterol and phospholipid content of the receptosome fraction compared with whole homogenate, Receptosomes, like plasma membrane and Golgi,[23,24] have a relatively high cholesterol content.

[20] M. Kates, "Techniques of Lipidology," p. 351. Elsevier/North-Holland, New York, 1972.

[21] G. R. Bartlett, *J. Biol. Chem.* **234,** 466 (1959).

[22] A. Zlatkis, B. Zak, and A. J. Boyle, *J. Lab. Clin. Med.* **41,** 486 (1963).

[23] T. W. Keenan and D. J. Moore, *Biochemistry* **9,** 19 (1970).

[24] W. H. Evans, "Preparation and Characterization of Mammalian Plasma Membranes," p. 126. Elsevier/North-Holland, New York, 1978.

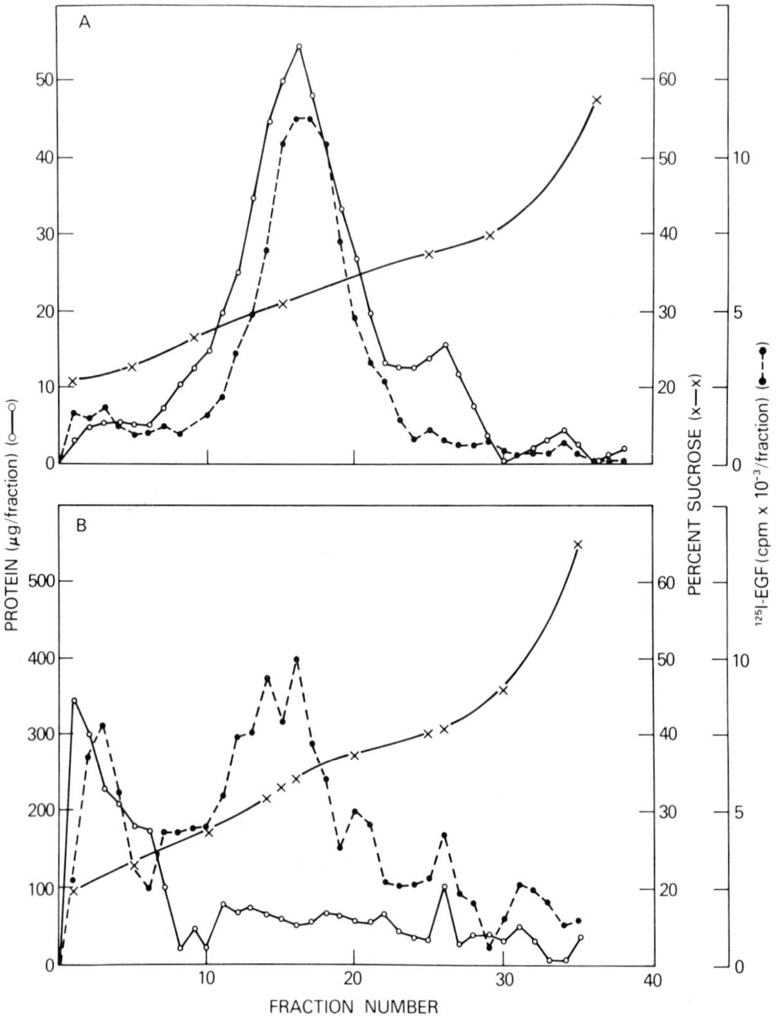

FIG. 2. Nonequilibrium density gradient centrifugation. Homogenate (B) or a purified [^{125}I]EGF-labeled receptosome fraction (A) was layered over a continuous linear 0.63–1.38 *M* sucrose density gradient and centrifuged at 29,000 rpm (SW28 rotor, Beckman) for 2 hr. Protein and [^{125}I]EGF were determined. Two lysosomal marker enzymes (β-hexosaminidase and β-galactosidase) were also determined and found concentrated in fractions 1–6 (loading zone). While lysosomes appear to have a lower S value than receptosomes, lysosomal marker enzymes sedimented deeper than [^{125}I]EGF labeled receptosomes after equilibrium (18 hr) centrifugation on 0.63–1.38 *M* linear sucrose gradients (indicating a higher density for lysosomes) (data not shown).

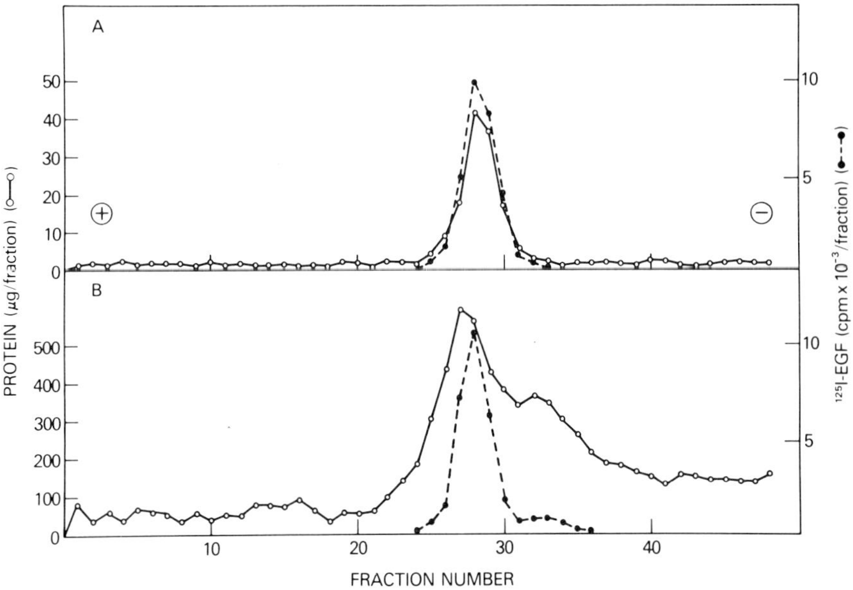

FIG. 3. Free flow electrophoresis. Homognate (B) or purified [^{125}I]EGF-labeled receptosome fraction (A) applied to a Desaga FF4 free flow electrophoresis apparatus. The electric field was 800 V, dozing rate was 9, collecting rate was 2.5, and temperature was 4°. Protein and radioactivity were determined for each fraction.

Marker Enzyme Composition. A variety of enzymes characteristic of different cellular organelles was measured at each stage of purification. Protein was determined by the method of Lowry *et al.*[25] (bovine serum albumin standard) or Bradford[26] (bovine IgG standard). Na^+,K^+-ATPase,[27] 5′ nucleotidase,[28] adenylate cyclase,[29] β-galactosidase,[30] β-hexosaminidase,[31] and galactosyltransferase with either ovalbumin[30] or *N*-acetylglucosamine[32] as substrates were determined as described. Table III shows that, relative to whole homogenate, receptosomes were depleted of the standard enzyme markers for plasma membrane (Na^+,K^+-ATPase, adenylate cyclase, or 5′ nucleotidase) or lysosomes (β-hex-

[25] O. H. Lowry, N. J. Rosebrough, A. L. Varr, and R. J. Randall, *J. Biol. Chem.* **193,** 265 (1951).

[26] M. M. Bradford, *Anal. Biochem.* **72,** 248 (1976).

[27] Sigma Chemical Company, "Sigma Technical Bulletin No. 366-UV." St. Louis, Missouri.

[28] C. H. Fiske and Y. SubbaRow, *J. Biol. Chem.* **66,** 375 (1925).

[29] Y. Solomon, *Adv. Cyclic Nucleotide Res.* **10,** 35 (1979).

[30] L. H. Rome, A. J. Garvin, M. M. Allietta, and E. F. Neufeld, *Cell* **17,** 143 (1979).

[31] Y.-T. Li and S.-T. Li, this series, Vol. 28, p. 702.

[32] B. Fleischer, S. Fleischer, and H. Ozawa, *J. Cell Biol.* **43,** 59 (1969).

TABLE IV
LIPID COMPOSITION OF RECEPTOSOME FRACTION[a]

Fraction	Phospholipids (μg/mg of protein[b])	Cholesterol (μg/mg of protein[b])	Cholesterol/ phospholipid[c]
Homogenate	200	16.7	0.072;0.088
Receptosomes	186	71	0.33;0.39

[a] Lipids were extracted and total phospholipid and cholesterol were determined.
[b] Results are the average of three determinations.
[c] Average of three determinations each with two separate preparations.

osaminidase). A marker for Golgi stacks,[6] galactosyltransferase, was slightly enriched in the receptosome fraction, but the enrichment was one-fifteenth of that of [^{125}I]EGF. Membranes highly enriched in galactosyltransferase were mainly found adsorbed to the Sephacryl S-1000 column.

For comparison of enzyme and polypeptide composition (see below), a plasma membrane-enriched fraction from KB cells was prepared by using sucrose gradient centrifugation.[33] A lysosome-enriched fraction from KB cells was prepared for the same purposes by using Percoll gradient centrifugation.[34]

Polypeptide Composition: SDS Gel Electrophoresis. Samples of various cell fractions were analyzed by using SDS gel electrophoresis on 5–17.5% gradient gels.[35] Gels were fixed and stained in 50% trichloroacetic acid with 0.25% Coomassie blue and destained. All of the major protein bands were also observed by [^{35}S]methionine labeling and autoradiography, indicating that the major proteins were cellular and not serum in origin. Figure 4 shows that the major bands have M_r of 120,000, 92,000, 70,000–75,000, 68,000, 55,000, 43,000 (actin), and 36,000. Actin was found to a variable extent in different preparations; its presence may be due to its very high concentration in contaminating microvilli (see Electron Microscopy, below). The polypeptide composition of the receptosome fraction was different from that of a plasma membrane-containing or lysosome-containing fraction.

Electron Microscopy. For *in situ* examination of receptosomes, cells were warmed to 37°C for 8 min after incubation with ferritin-labeled EGF at 4° as described.[36] For examination of vesicle fractions, 1 ml of a suspen-

[33] T. D. Butters and R. C. Hughes, *Biochem. J.* **140,** 469 (1974).
[34] E. Harms, J. Kartenbeck, G. Barai, and J. Schneider, *Exp. Cell Res.* **131,** 251 (1981).
[35] U. K. Laemmli, *Nature (London)* **227,** 680 (1970).
[36] M. C. Willingham, H. T. Haigler, D. J. P. FitzGerald, M. Gallo, A. V. Rutherford, and I. Pastan, *Exp. Cell Res.* **146,** 163 (1983).

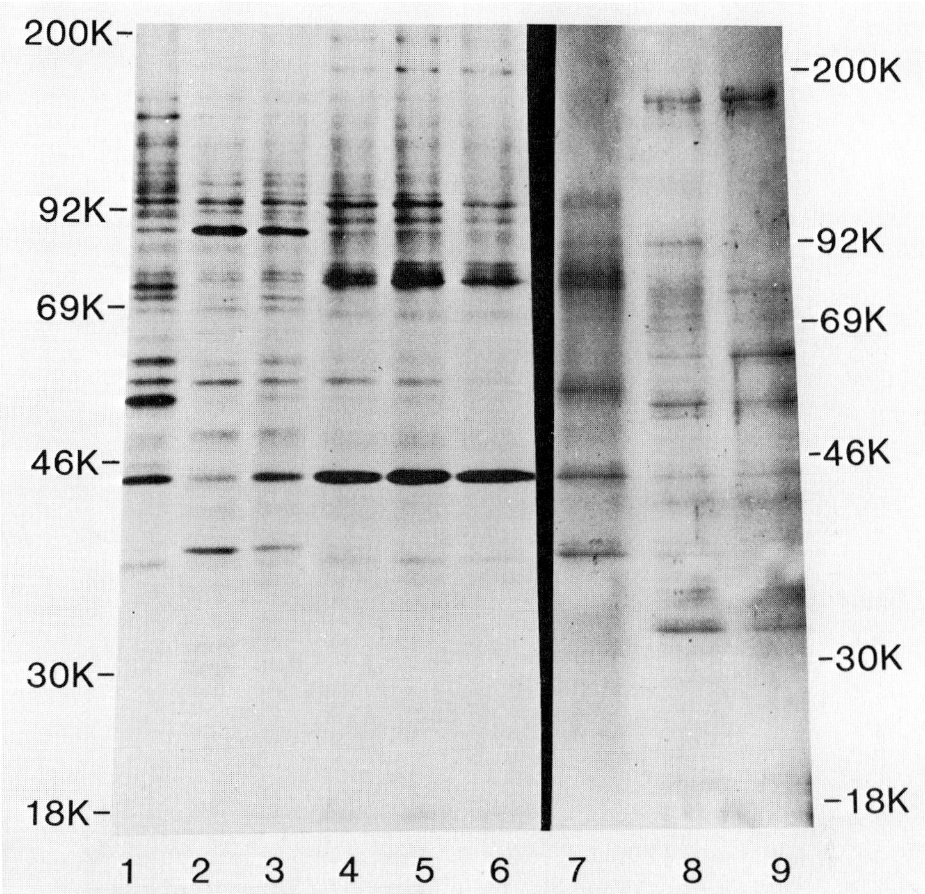

FIG. 4. SDS gel electrophoresis of membrane fractions at various stages of purification and comparison with plasma membrane- and lysosome-containing fractions. Membranes were fractionated and SDS gel electrophoresis was carried out on a 5–17.5% gradient gel. Lane 1, homogenate; lane 2, low-speed supernatant; lane 3, Percoll pool; lane 4, Sephacryl S-1000 pool; lane 5, sucrose gradient 1 pool; lane 6, sucrose gradient 2 (flotation) pool (lanes 1–6, [^{35}S]methionine labeling and autoradiography); lane 7, sucrose gradient 2 pool; lane 8, plasma membrane fraction; lane 9, lysosome fraction (lanes 7–9, Coomassie blue staining).

sion of vesicles (≈300 μg of protein) was fixed in 2% glutaraldehyde at 23° for 15 min, then diluted to 10 ml with Dulbecco's phosphate-buffered saline (Gibco), centrifuged at 105,000 *g* (Ti-65 rotor in a Beckman L5-65 ultracentrifuge) for 30 min to form a vesicle pellet, and processed as described.[15]

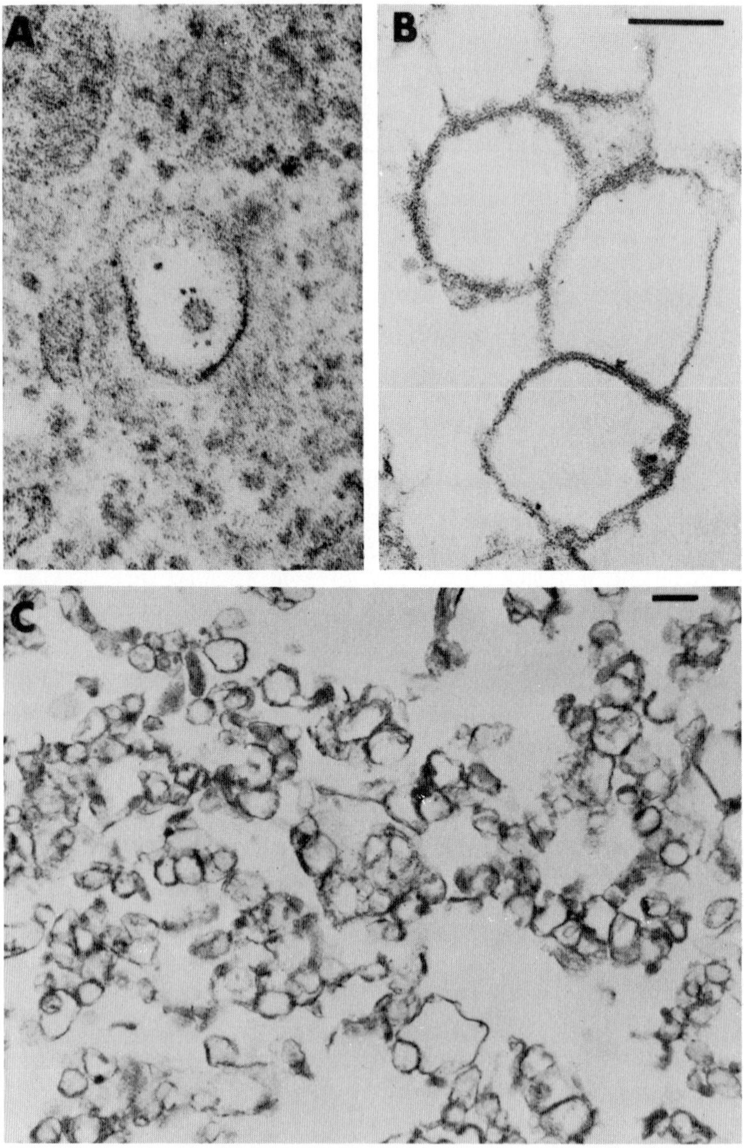

FIG. 5. Electron microscopy. Intact cells labeled with ferritin EGF (A) show a typical receptosome structure. (B and C) Appearance of membranes in the most highly purified receptosome-containing fraction after isolation. A and B, ×107,900 (bar = 0.1 μm); C, ×24,000 (bar = 0.2 μm).

An electron micrograph of the most highly purified preparation of receptosomes (Fig. 5B and C) demonstrates a population of vesicles 2000–4000 Å in diameter, closely resembling the size of the structures found in intact cells (Fig. 5A). However, the frilly material characteristically found associated with a segment of the membrane of some receptosomes[3] was absent. A small amount of contamination with microvilli was noted. Organelles having characteristics of mitochondria, coated vesicles, or rough endoplasmic reticulum were not noted in the purified material, but were present in less purified fractions. Since there is not yet available a histochemical or morphologic marker to unambiguously label receptosomes, the exact degree of purity of this membrane fraction remains unknown.

Future Prospects

Studies in intact cells have shown that receptosomes (endosomes) are acidic vesicles[8,12] that contain various ligands and receptors. These vesicles appear to fuse with the trans reticular Golgi and thereby transfer their contents into the Golgi system.[5] Receptosomes may have a role in receptor down-regulation and recycling. They also are involved in virus infection and toxin entry into cells.[1,4,9] The current preparation should be of use in studying the biochemical basis of these processes.

Acknowledgments

The authors thank J. Rose (National Institutes of Health) for KB cells, N. D. Richert and L. Beguinot for assistance with gel electrophoresis, D. Werth for helpful discussion, R. Alderson for assays of adenylate cyclase, E. Lovelace and A. Harris for assistance with cell culture, R. Steinberg for photography, A. Rutherford for assistance with electron microscopy, and R. M. Coggin for typing the manuscript. R.B.D. was supported by National Institutes of Arthritis, Metabolism, and Digestive Diseases National Research Service Award F32 AMO6318-Os; J.A.H. is supported by a grant from the Jane Coffin Childs Fund for Medical Research.

Section IV

Preparation of Hormonally Responsive Cells and Cell Hybrids

[22] Preparation of Isolated Leydig Cells

By F. F. G. ROMMERTS, R. MOLENAAR, and H. J. VAN DER MOLEN

Introduction

Different *in vitro* preparations of Leydig cells have been used for studies on the biochemical mechanisms involved in regulation of steroidogenesis. Initially, many attempts were made to study Leydig cell function *in vitro* in intact testicular tissue.[1,2] However, complete testis tissue contains mainly seminiferous tubules and only a small fraction of Leydig cells. Moreover, under *in vitro* conditions cells inside the tissue may lack oxygen and nutrients from the medium. Teasing of testis tissue or dissection of interstitial tissue may overcome these problems, but causes cell damage resulting in a decreased steroidogenic response to LH.[3,4] These difficulties have stimulated the development of techniques for isolation of Leydig cells using collagenase treatment and mechanical agitation. Since the description of the latter technique (in 1975 by Dufau and Catt in Methods of Enzymology, Volume XXXIX, p. 252) for rat testes, several alternative methods for cell isolation and characterization have been developed and the techniques have been used to isolate Leydig cells also from porcine and mouse testes, and from Leydig cell tumors.

A careful investigation and standardization of the methods appears necessary, because for Leydig cells isolated from adult rats a large variation in the capacity for steroid production has been observed. A variation in LH-stimulated testosterone production ranging from 3 to 100 ng/10^6 Leydig cells/hr and a stimulation factor ranging from 4- to 12-fold has been reported.[5–12] Moreover, the recovery of Leydig cells after the isola-

[1] M. L. Dufau, K. J. Catt, and T. Tsuruhara, *Endocrinology* **90,** 1032 (1972).
[2] F. F. G. Rommerts, B. A. Cooke, J. W. C. M. van der Kemp, and H. J. van der Molen, *FEBS Lett.* **24,** 251 (1972).
[3] F. F. G. Rommerts, B. A. Cooke, J. W. C. M. van der Kemp, and H. J. van der Molen, *FEBS Lett.* **33,** 114 (1973).
[4] L. F. Aldred and B. A. Cooke, *Int. J. Androl.* **5,** 191 (1982).
[5] F. H. A. Janszen, B. A. Cooke, M. J. A. van Driel, and H. J. van der Molen, *J. Endocrinol.* **70,** 345 (1976).
[6] J. Y. Browning, R. d'Agata, and H. E. Grotjan, *Endocrinology* **109,** 667 (1981).
[7] S. B. Cigorraga, S. Sorrell, J. Bator, K. J. Catt, and M. L. Dufau, *J. Clin. Invest.* **65,** 699 (1980).
[8] A. H. Payne, J. R. Downing, and K. L. Wong, *Endocrinology* **106,** 1424 (1980).
[9] K. Purvis, O. P. F. Clausen, and V. Hansson, *J. Reprod. Fertil.* **60,** 77 (1980).
[10] G. C. C. Chen, T. Lin, E. Murono, J. Osterman, B. T. Cole, and H. Nankin, *Steroids* **37,** 63 (1981).

ISBN 0-12-182009-2

tion procedure is low (less than 10%) and there are indications that extensive cell damage can occur during tissue dispersion.[4,12,13] When isolated cells are used for investigations on the biochemical mechanisms involved in regulation of steroidogenesis, it is essential to work with homogeneous preparations of viable cells and to prevent various degrees of functional heterogeneity and other artifacts which may be caused by the isolation procedure. The present chapter concerns relevant observations on the preparation, purification, and characterization of isolated Leydig cells.

Isolation and Purification of Leydig Cells

Choice of Tissue

Leydig cells can be isolated from Leydig cell tumor tissue or testicular tissue. Leydig tumor tissues have been grown in mice[14] and in rats[15] and pure Leydig cells which respond to LH with increased steroidogenesis can easily be obtained. Properties of these cells, however, may be different from normal Leydig cells. When Leydig cells are isolated from adult rat testes, the testis tissue should not be teased to prevent cells becoming inactive[4]; on the other hand, many active Leydig cells can be isolated from mouse testes by pure mechanical disruption of the tissue.[16]

The recovery of isolated Leydig cells from testes of mature rats is only 10%, even when teasing is omitted. Many purification steps are required before satisfactory cell preparations can be obtained. Testis tissues containing relatively low numbers of germinal cells usually give a much higher recovery and a better quality of Leydig cells. For this purpose testes from 3- to 4-week-old piglets[17] or rats[18] have been used frequently. Relatively high numbers of active Leydig cells can be isolated also from testes of mature rats with largely reduced numbers of germinal cells due to hypophysectomy, prenatal irradiation, or experimental cryptorchidism.[13]

[11] B. A. Cooke, R. Magee-Brown, M. Golding, and C. J. Dix, *Int. J. Androl.* **4,** 355 (1981).
[12] R. M. Sharpe and I. Cooper, *J. Reprod. Fertil.* **65,** 475 (1982).
[13] R. Molenaar, F. F. G. Rommerts, and H. J. van der Molen, *Int. J. Andro.* **6,** 261 (1983).
[14] W. B. Neaves, *J. Natl. Cancer Inst.* (*U.S.*) **55,** 623 (1975).
[15] B. A. Cooke, L. M. Lindh, F. H. A. Janszen, M. J. A. van Driel, C. P. Bakker, M. P. I. van der Plank, and H. J. van der Molen, *Biochim. Biophys. Acta* **583,** 320 (1979).
[16] M. Schumacher, G. Schàfer, A. F. Holstein, and H. Hilz, *FEBS Lett.* **91,** 333 (1978).
[17] J. P. Mather, J. M. Saez, and F. Haour, *Steroids* **38,** 35 (1981).
[18] F. F. G. Rommerts, M. J. A. van Roemburg, L. M. Lindh, J. A. J. Hegge, and H. J. van der Molen, *J. Reprod. Fertil.* **65,** 289 (1982).

Dispersion of Tissue

Tissues are generally dispersed by mechanical forces frequently in combination with collagenase action. These treatments, however, can damage the cells. Collagenase affects mainly the properties of the cell membrane, whereas mechanical damage causes rather aspecific artifacts. The best method for dispersion of the tissue may depend therefore on the aim of the investigation. Mechanical dispersion may be the best choice for investigations of membrane properties, whereas enzyme digestion may be optimal when intracellular processes are investigated.

Active Leydig cells from mouse testes can be isolated by mechanical forces alone.[16] For mouse testis tissue this can be accomplished in a reproducible way by flushing the testes 10 times up and down in a siliconized glass tube of 4 mm internal diameter with constrictions of 2 mm internal diameter. The first strokes should be gentle and the last few strokes fast.

Rat Leydig cells from testis tissue of mature animals are sensitive to mechanical forces. However, application of these damaging forces is essential, because cells will not be dispersed using collagenase (10 mg/ml) treatment only, not even after perfusion of the testis with collagenase (10 mg/ml) followed by rupture of the capillary blood vessels and incubation at 37° for 30 min.[13]

Leydig cells from rat testis tissue have been dispersed using different combinations of enzyme concentrations and incubation conditions (with varying shaking frequencies, amplitudes, duration of incubation, temperature). Standardization of mechanical forces is difficult, because the agitation of the tissue at a given shaking frequency and amplitude of the incubation vessels depends very much on the geometry of the vessel, the volume of the incubation fluid, and the amount of tissue. Similar problems are encountered with the choice of the collagenase. Different sources and batches of collagenase have been used for cell dispersion and the results show that the purest collagenase is not the best choice. Most of the collagenase preparations used (sometimes in combination with anti-trypsin) have been impure and unknown impurities are probably more important than the collagenase alone. Hence, it appears that specification of the amount of collagenase used, is less important than the source, the type, and possibly the batch number of the enzyme.

In our experience, the best procedure for release of Leydig cells from testis tissue of mature rats is as follows: Rats are killed by CO_2 inhalation and testes are removed immediately and are decapsulated without teasing or cutting of testis tissue. Two decapsulated testes are added to 7 ml culture medium or Krebs Ringer bicarbonate containing 1 mg/ml collagenase (Sigma, type 1; 135 U/mg). The mixture is shaken for 20 min at 37° in capped 50 ml plastic conical tubes (Falcon) longitudinally at approx. 80 cpm and 2 cm amplitude. If a lower temperature or a lower collagenase concentration (not less than 0.1 mg/ml) is used, longer incubation periods or increased shaking frequencies are required for release of the same amount of cells. However, at increased shaking frequencies, cell damage is increased.

Purification of Specific Cell Fractions from Dispersed Cells

After dispersion of the tissue, a mixture of tissue fragments, isolated cells, and cell debris remains. Tissue fragments can be removed by sedimentation at 1 *g* for a few minutes or by filtration through gauze. Cells can be separated from cell debris by centrifugation at 150 *g* for a few minutes. Different subfractions can be isolated according to density or size. Methods using density gradient centrifugation for purification of Leydig cells have mainly used metrizamide,[19] Ficoll,[5] or Percoll.[16] In this way Leydig cells can be separated from germinal cells and erythrocytes.

Recently, functionally different subpopulations of cells have been isolated.[5,13,19] Cell preparations after tissue dispersion contain also damaged cells and for many investigations they can be considered as contaminating cells. It is not clear how cell damage affects the sedimentation properties.

Isolated intact interstitial cells can attach to a plastic surface and will spread out within an hour. Nonviable cells do not spread out and these cells can be removed by washing. In this way the quality of the cell preparation can be improved by selection of viable, attached cells. Slight membrane damage can be repaired during the initial phases of the incubation period.

Isolation of Leydig Cells from Mature Rat Testes

Leydig cells from mature rats can be isolated from the collagenase-dispersed tissue as follows: 10 ml suspension of cells and tissue fragments (7 ml medium and 3 ml testis tissue) is mixed with 15 ml 0.9% NaCl solution at room temperature in a 50 ml conical plastic tube (Falcon) by inversion of the tube 5 times. Large tissue fragments are allowed to sediment when the tube remains vertical for approximately 1–2 min. The supernatant is filtered through nylon gauze of 60 μm and the filtrate is collected. The percentage of Leydig cells in these preparations can be increased by density centrifugation through Ficoll. For this purpose a 26% Ficoll solution is prepared as follows[5]:

[19] A. H. Payne, K. L. Wong, and M. M. Vega, *J. Biol. Chem.* **255,** 7118 (1982).

Ficoll 400 (26 g) is dissolved in 25 ml distilled water and 50 ml Krebs Ringer bicarbonate (KRB) solution without glucose. After the Ficoll is dissolved, the volume is adjusted to 100 ml with KRB and 50 ml portions are sterilized by heating at 120° for 20 min. Immediately before use 0.2 g bovine albumine is added to 50 ml solution and the pH is adjusted to 6.5 with CO_2. An equal volume of 26% Ficoll is added to the 60 μm filtrate of the collagenase-dispersed cells and the resulting cell suspension is mixed by inverting the tube several times. It is important that the resulting suspension is homogeneous and that there is no unmixed Ficoll. The cell suspension (in 13% Ficoll) is subsequently centrifuged in plastic tubes for 10 min at 1500 *g*. The top layer containing fat and broken cells is removed with filter paper and the supernatant containing germinal cells and damaged cells is discarded. The sedimented cells are resuspended in medium by swirling the tube with a few drops of medium. Several steps of the cell preparation method are summarized in Fig. 1. Leydig cells can also be separated in different subpopulations by density *gradient* centrifugation in Percoll. For this, nine parts of Percoll (Pharmacia) are mixed with 1 part 10-fold concentrated Earle's balanced salt solution containing 10% fetal calf serum or 0.1% serum albumine to give a final osmolarity of 300

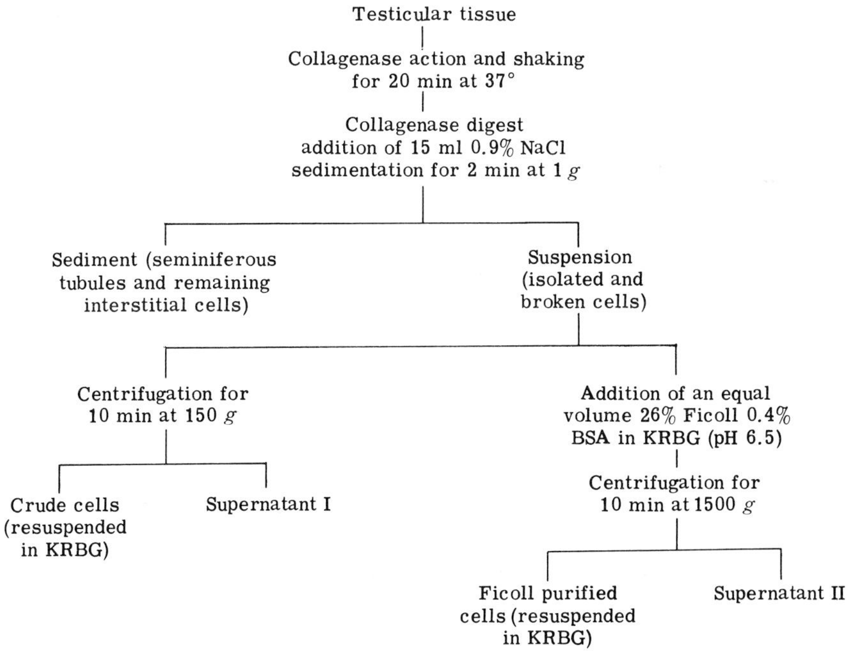

FIG. 1. Fractionation of Leydig cells.

mOsmol/kg. Linear density gradients (0–90%) can be prepared by dilution of the 90% Percoll solution with Earle's salt buffer containing 1% fetal calf serum or 0.1% serum albumin.[11,16] After centrifugation of 10^8 cells in 3 ml medium for 20 min at 800 *g* at room temperature, four visible bands are obtained. After isolation of the different bands the cells can be isolated after dilution of the Percoll with 3 volumes of medium and sedimentation at 150 *g* for 15 min at room temperature.

Highly purified Leydig cells can be recovered from the third band from the top, corresponding to a Percoll concentration of 40–50% and a density of 1.07 g/ml. Leydig cells can be recovered also from other bands with different densities. It is not clear to which extent the Leydig cells from these bands are physically damaged or whether they represent cells with intrinsic functional differences which are apparently reflected in the cell density. Part of the variation in density may be caused by phagocytosis of Percoll particles by Leydig cells.[20] Percoll may also affect the membrane properties.[20]

Isolation of Leydig Cells from Tumor Tissue, Mouse Testes, or Immature Rat Testes

Leydig cells isolated from tumor tissue or testis tissue from immature rats and mice can be recovered after centrifugation of the dispersed cells for 10 min at 150 *g* at room temperature and are generally not further purified with centrifugation techniques. After removal of the supernatant, cells can be resuspended by mixing the pellet with a few drops of medium. Contaminating erythrocytes, which are abundantly present in cell preparations from tumor tissue, can be removed by suspending and discarding the top layer of the cell pellet.

Selection of Viable Leydig Cells

Suspensions of Leydig cells usually contain damaged Leydig cells, germinal cells, and erythrocytes. The quality and purity of the cell preparation can be further improved by selection of the viable cells which can attach to plastic. For isolation of Leydig cells cell suspensions containing 10^5–10^6 cells are seeded in plastic dishes (diameter 35 mm), containing 2 ml Krebs Ringer bicarbonate or culture medium, prepared from modified Eagle's medium with Earle's liquid and nonessential amino acids containing extra glutamin (0.6 mg/ml) and antibiotics (100 U penicillin/ml, 100 μg streptomycin/ml, and 0.6 μg Fungizone/ml). Most of the incubation fluids used contain 1% fetal calf serum (Gibco). Cells are allowed to attach

[20] J. S. Wakefield, J. S. Gale, M. V. Berridge, T. W. Jordan, and H. C. Ford, *Biochem. J.* **202,** 795 (1982).

during incubation periods lasting from 60 min (for rat Leydig cells) to 24 hr (for mouse Leydig cells). During this period most of the viable somatic cells will attach to the plastic surface, while the germinal cells and erythrocytes remain floating. Floating cells can be removed by washing. Attachment of cells is less selective when cell suspensions are incubated without serum. All cell types will aspecifically stick to plastic or glass by centrifugation of cell suspensions in medium without serum for a few minutes at 1000 *g*.

Characterization of Cell Preparations

Yield of Isolated Cells

Collagenase dispersion of testis tissue from mature rats resulted in isolated cells and tissue fragments, but also in broken cells. The distribution of interstitial cells in the various fractions (see Fig. 1) has been investigated using nonspecific esterase activity, which is predominantly localized in the endoplasmic reticulum of the interstitial tissue.[21,22] Enzyme solutions were prepared by sonication of 0.5-ml aliquots of the cell suspensions for 15 sec at 20 kHz with an amplitude of 5 μm at 0°. The solutions could be stored at −20° for several weeks without loss of enzyme activity. The enzyme solution was added to 0.1 *M* Tris–HCl buffer (pH 8), containing about 1 m*M* *p*-nitrophenyl acetate and the rate of hydrolysis of *p*-nitrophenyl acetate at room temperature was measured at 400 nm. The high specific esterase activity in the cells sedimented through Ficoll indicates that interstitial cells have been purified, albeit with a low recovery (Table I).[13] Significant amounts of esterase activity were also present in the supernatant after centrifugation of the cells. The presence of this intracellular membrane-bound enzyme in the extracellular space indicates that many interstitial cells had been ruptured. 3β-Hydroxysteroid dehydrogenase activity, which in intact cells is also present in the endoplasmic reticulum, could also be detected in the extracellular space.

Approximately 50% of the enzyme activity remains in the residual clumps of seminiferous tubules (Table I). Thus, tissue dispersion is incomplete, on the one hand, and results in cell damage, on the other hand. Indications for cell damage during isolation of Leydig cells have also been observed by others.[4,12,23] Nevertheless, in our experience, the described

[21] F. F. G. Rommerts, L. G. van Doorn, H. Galjaard, B. A. Cooke, and H. J. van der Molen, *J. Histochem. Cytochem.* **21,** 572 (1973).

[22] G. J. van der Vusse, M. L. Kalkman, and H. J. van der Molen, *Biochim. Biophys. Acta* **348,** 404 (1974).

[23] L. F. Aldred and B. A. Cooke, *J. Steroid Biochem.* **19,** 359 (1983).

TABLE I
PHENYL ESTERASE ACTIVITY IN VARIOUS PREPARATIONS OF RAT TESTES AFTER TISSUE DISPERSION

Fraction	Phenyl esterase activity: Percentage of interstitial activity	Phenyl esterase activity: Specific activity (μmol/min/mg protein)
Total testis	100	0.24 ± 0.01 (4)
(Tubuli and) remaining interstitial cells	52.0 ± 11.9[a] (3)	0.16 ± 0.04 (4)
Isolated and broken cells	46.9 ± 14.5 (3)	0.27 ± 0.07 (3)
Crude cells	23.3 ± 3.3 (3)	1.15 ± 0.46 (5)
Supernatant I	13.8 ± 2.9 (3)	—
Ficoll purified cells	10.1 ± 3.6 (3)	2.07 ± 0.69 (2)

[a] Of total testis activity 90% is localized in interstitial tissue.[21] Activity of the remaining interstitial cells was calculated from the total activity after correction for the tubular activity. Means ± SD; number of cell preparations in parentheses.

procedure for isolating Leydig cells from normal mature rats appears optimal, although cell damage cannot be prevented. The recovery of intact Leydig cells increases when cells are isolated from testes depleted of germinal cells (i.e., from immature rats, irradiated rats, hypophysectomized rats, etc.).

Quality of Isolated Cells

Because of the damage which may be inflicted by the isolation procedure, the viability of isolated cells must be characterized extensively and, where possible, damaged cells should be removed or allowed to recover. The quality of Leydig cell preparations can be assessed by measuring the effects on steroid production of LH (contribution of intact cells) and NADPH (contribution of cells with damaged cell membrane).

Steroid production by collagenase-dispersed cells prepared from mature rat testes with a 10-fold increased shaking frequency (crude preparation method) could not be stimulated by LH, but in the presence of NADPH testosterone production was stimulated 7-fold (Table II). Better cells were prepared with the standard collagenase dispersion technique. The content of viable cells was increased after attachment to plastic and preincubation (Table II), although cells with damaged cell membranes were still present. In contrast, cells isolated from tumor tissue or from

TABLE II
EFFECT OF LH AND NADPH ON STEROID PRODUCTION BY DIFFERENT LEYDIG CELL PREPARATIONS[a]

Cell preparation	Tissue	Stimulation factor[b] LH	NADPH
Crude method	Mature testis	1.0 ± 0.0 (3)	7.0 ± 0.5 (3)
Standard method			
Freshly isolated	Mature testis	8.6 ± 3.5 (4)	2.6 ± 0.4 (4)
Preincubated and attached to plastic surface	Mature testis	15.5 ± 4.0 (5)	2.1 ± 1.1 (5)
Preincubated and attached to plastic surface	Immature testis	23.8 ± 2.8 (4)	1.1 ± 0.6 (3)
Preincubated and attached to plastic surface	Tumor tissue	5.1 ± 0.9 (3)	1.2 ± 0.4 (4)

[a] Steroidogenesis in cells from mature rats was measured via testosterone estimations. For tumor cells and Leydig cells from immature rats pregnenolone formation in the presence of inhibitors of 3β-hydroxysteroid dehydrogenase and 17α-hydroxylase was measured.[18]

[b] Steroid production with LH (1000 ng/ml for tumor tissue and 100 ng/ml for testis tissue) or NADPH (generating system containing 10 mM glucose 6-phosphate, 1 mM $NADP^+$, and 0.3 unit glucose 6-phosphate per ml) divided by the basal steroid production.

testes of immature rats do not contain a detectable amount of damaged cells which respond to NADPH (Table II).

The combination of the effects of LH and NADPH on steroid production is characteristic for the quality of cell preparations and can be expressed as a "quality factor" which is defined as the stimulation obtained after addition of LH, divided by the stimulation obtained after addition of NADPH.

The quality of cell preparations can be evaluated also with histochemical procedures. Mitochondrial diaphorase has been used for this purpose. When the activity of this enzyme is stimulated by NADH or NADPH, this is an indication for leaking cell membranes. Diaphorase histochemistry can be performed by incubating cells in Krebs Ringer bicarbonate buffer with nitro blue tetrazolium (NBT) 0.2 mg/ml and NADH (0.4 mg/ml) for 1 hr at 37°. The results of this histochemical test are complementary to the effect of NADPH on steroid production (Table III). Cells attached to plastic have the highest quality factor and contain the lowest percentage

TABLE III
EFFECT OF CELL ATTACHMENT TO PLASTIC AND PREINCUBATION ON THE QUALITY FACTOR OF CELL PREPARATIONS AND THE PERCENTAGE OF DIAPHORASE-POSITIVE CELLS

Leydig cell preparation	Quality factor	Percentage diaphorase-positive cells
Mature rats		
Ficoll-purified	4.0 ± 2.5 (3)	27.8 ± 3.2 (3)
Mature rats		
Ficoll-purified		
attached to plastic	8.7 ± 3.6[a] (4)	2.0 ± 2.3[b] (4)
Immature rats		
collagenase dispersed		
attached to plastic	21.6 ± 4 (3)	<0.1
Tumor tissue		
collagenase dispersed		
attached to plastic	4.3 ± 0.7 (3)	<0.1

[a] Significant increase ($p < 0.05$) after attachment.
[b] Significant decrease ($p < 0.005$) after attachment.

diaphorase-positive cells. In our experience only preparations of tumor Leydig cells and Leydig cells from immature rats did not contain cells permeable to NAD(P)H. All cells, however, show diaphorase activity after freeze-damaging the membranes. The ability of the cells to exclude pyridine nucleotides was found to be a more sensitive index of cell viability than trypan blue exclusion[23] (determined after exposure of the cells for 10 min to 0.2% trypan blue). However, with an even more sensitive test, using intracellular retention of accumulated fluorescein, which is endogeneously formed via hydrolysis of fluorescein diacetate which can pass intact cell membranes,[24] variations in the fluorescence could be shown in cell preparations which did not react in the diaphorase test.

Identification of Leydig Cells

Testis tissue from normal rats contains many different cell types and only a few percent Leydig cells. Leydig cell preparations are therefore frequently contaminated with other cell types. Germinal cells and Sertoli cells can be recognized using light microscopy without marker enzymes. The presence of other cell types, such as peritubular cells (myoid cells), fibroblastic cell types, and macrophages, cannot be assessed without

[24] B. Rotman and B. W. Papermaster, *Proc. Natl. Acad. Sci. U.S.A.* **55,** 134 (1966).

marker enzymes. For identification of Leydig cells histochemical 3β-hydroxysteroid dehydrogenase and esterase activity have been used most frequently.[5,11,13] In addition, immunohistochemical or autoradiographic detection methods for LH receptors have been used.[25] Under certain conditions, Leydig cells may have undetectable 3β-hydroxysteroid dehydrogenase activity, either due to intrinsically low enzyme activities or to technical problems such as enzyme denaturation or poor permeability of cell membranes for NAD^+ and NBT.

Nonspecific esterase can be used also to identify Leydig cells[13] in testes of mature rats, mice or rat tumor Leydig cells, but not in cell preparations from immature rat testes. Staining of cells for esterase can be carried out without rupturing the cell membranes, because the substrate for esterase can pass the cell membrane. In many cases macrophages may constitute the major contaminating cell type in Leydig cell preparations.[26] Recent results[27] have shown that marker functions of macrophages and Leydig cells, such as phagocytic activity and 3β-hydroxysteroid dehydrogenase or esterase activity, respectively, may partly overlap, depending on the age of the animal. The purity of cell suspensions must therefore be assessed using different parameters.

Histochemical methods used for 3β-hydroxysteroid dehydrogenase and esterase are as follows.

3β-Hydroxysteroid Dehydrogenase

Washed intact cells attached to plastic are covered with the staining solution (Krebs Ringer bicarbonate buffer, containing nitroblue tetrazolium (0.2 mg/ml), NAD^+ (0.4 mg/ml), 5α-androstane-3β-ol-17-on (30 μg/ml) as substrate, and 3% Ficoll. Cells and solution are frozen in liquid nitrogen and are thawed at 37° in a humid atmosphere to permeabilize the membranes. The mixture is subsequently incubated at 37° for 1 hr. Incubation is stopped by addition of 70% ethanol. Cells can be kept under glycerol or embedded in gelatine. The medium of crude cell preparations may contain 3β-hydroxysteroid dehydrogenase, originating from broken cells. This extracellular enzyme may cause the production of a blue precipitate of reduced nitroblue tetrazolium outside the cell membrane. The precipitate may subsequently deposit on Leydig cells as well as on con-

[25] K. Purvis, O. P. F. Clausen, N. M. Ulrik, and V. Hansson, *in* "Testicular Development, Structure and Function" (A. Steinberger and E. Steinberger, eds.), p. 211. Raven Press, New York, 1980.

[26] A. K. Christensen, *in* "Handbook of Physiology" (R. O. Greep and E. B. Astwood, eds.), Sect. 7, Vol. 5, p. 57. Am. Physiol. Soc., Washington, D.C., 1975.

[27] R. Molenaar, F. F. G. Rommerts, and H. J. van der Molen, "Hormone Action and Testicular Function." Proceedings 8th Testis Workshop, Bethesda, 1983.

taminating cells, causing an overestimation of the number of Leydig cells. On the other hand, the number of Leydig cells may be underestimated if freeze-damaging of the cell membranes is not complete. The effectiveness of the membrane permeabilization method can be checked with the diaphorase test.

Naphthyl Esterase

Intact cells attached to plastic are incubated in the staining solution, which is prepared by mixing successively 200 μl $NaNO_2$ 4% with 200 μl pararosaniline 4% in 2 *M* HCl, 125 μl 1% α-naphthyl acetate in 50% acetone, and finally 5 ml 0.15 *M* Na_2HPO_4. The staining solution must be prepared immediately before use and must be kept in the dark. Incubation is carried out for approximately 20–60 sec at room temperature and is stopped by addition of 70% ethanol. A red precipitate is formed around intact Leydig cells, whereas permeable cells show the precipitate inside the cells. Histochemical 3β-hydroxysteroid dehydrogenase and esterase are commonly used to determine the percentage of Leydig cells in cell suspensions. From the percentage of Leydig cells and the total cell count, the number of Leydig cells per ml can be calculated. This procedure cannot be applied to cells attached to plastic, because of large variations in the composition and number of cells in different areas of the dish. Moreover, the percentage of the Leydig cells attaching to plastic is also variable. Leydig cell attachment is often low (20–30%) when cell preparations from normal rats are used.

The total number of Leydig cells present can be estimated also by a quantitative measurement of the marker enzyme activity. The relationship between the number of Leydig cells and quantitative marker enzyme activity can be established with cell suspensions and, with this relationship, the number of attached cells can be calculated from the enzyme activity in the attached cells. For this purpose, esterase activity is preferentially used as a marker enzyme, because it is 1000-fold more active than 3β-hydroxysteroid dehydrogenase[22] and can be measured with simple methods.

The substrate solution is prepared by mixing 100 μl 1% α-naphthyl acetate in 50% acetone with 10 ml 0.1 *M* phosphate buffer pH 7.4, containing 1% albumin. Leydig cells are incubated under continuous shaking for 2–10 min at room temperature in 1 ml of the substrate solution. If necessary, the incubation solution can be separated from the attached cells after the incubation period. The reaction is completely terminated by addition of 1 ml 2% sodium dodecyl sulfate to the recovered incubation fluid, either with or without cells. After the incubation with substrate

TABLE IV
STEROIDOGENIC ACTIVITIES OF VARIOUS LEYDIG CELL PREPARATIONS AFTER THE STANDARD ISOLATION PROCEDURE AND ATTACHMENT TO PLASTIC[a]

Leydig cells from	Steroid	ng P or T[b] ($h \times 10^6$ Leydig cells)	
		Without additions	With LH
Mature rats	T	67 ± 13 (3)	568 ± 55 (3)
Mature rats	P	82 ± 7 (3)	828 ± 13 (3)
Immature rats	P	7 ± 2 (4)	153 ± 19 (4)
Mice	T	69 ± 11 (4)	750 ± 75 (4)
Rat tumor tissue	T	41 ± 9 (5)	250 ± 31 (5)

[a] Pregnenolone production was estimated when cells were incubated in the presence of SU-10603 (2×10^{-5} *M*) and cyanoketone (5×10^{-6} *M*) to inhibit pregnenolone metabolism. Leydig cells from rat or mice testes were incubated with 100 ng LH/ml. Tumor Leydig cells were incubated with 1000 ng LH/ml. Mean results ± SD and the number of cell preparations are given.

[b] P, pregnenolone; T, testosterone.

solution the cells are still viable and can be used for further experiments. After addition of 2 ml freshly prepared 2 m*M* Fast-blue BB to the reaction mixture in SDS, the color is developed during 1 hr in the dark. The naphthol concentration is determined by measuring the extinction at 540 nm. Using this technique the naphthyl esterase activity of three different dilutions of Ficoll purified Leydig cell suspensions (10^3–10^5 Leydig cells/ml; estimated by cell counting and histochemistry) showed an almost linear relationship between the amount of cells and esterase activity.[13] From this relationship the number of attached cells in a plastic dish could be calculated after measuring the naphthyl esterase activity. The relationship between the number of cells and esterase activity must be determined for each individual cell preparation since a variation in enzyme activity per 10^6 Leydig cells was found for different preparations of Leydig cells.

Functional Properties of Isolated Cells

Freshly isolated cells, even after application of extensive purification methods, still contain cells with damaged membranes. A recovery period for cell repair is therefore required. Incubation of cells may change functional properties of Leydig cells, such as the number of LH receptors[28] and

[28] M. G. Hunter, R. Magee-Brown, C. J. Dix, and B. A. Cooke, *Mol. Cell. Endocrinol.* **25,** 35 (1982).

the response to actinomycin D and LH.[29] However, it is known that almost all cells in primary culture will loose some functional properties. For isolated Leydig cells changes in steroidogenic activities have been observed after 24 hr incubation.[18] Changes may occur even within a few hours after cell isolation, for instance 17α-hydroxylase in tumor Leydig cell is reduced to 11% of the original activity within 4 hr after isolation of the cells[18] and Leydig cells from immature rats become sensitive to adenosine within the first 4 hr after isolation.[30] These observations indicate that after cell isolation, functional properties of the cells may undergo continuous changes. It is difficult therefore to indicate the optimal period for investigation of isolated Leydig cells. However, we have routinely obtained reproducible results when short-term experiments were performed between 1 and 5 hr after cell isolation. Steroidogenic activities of various cell preparations are given in Table IV. Future developments on cell isolation techniques should certainly include investigations on the required microenvironment for maintaining the functional characteristics of Leydig cells *in vitro*.

[29] B. A. Cooke, F. H. A. Janszen, M. J. A. van Driel, and H. J. van der Molen, *Mol. Cell. Endocrinol.* **14,** 181 (1979).

[30] F. F. G. Rommerts, R. Molenaar, J. W. Hoogerbrugge, and H. J. van der Molen, *Biol. Reprod.* **30,** 842 (1984).

[23] Preparation of Dispersed Pancreatic Acinar Cells and Dispersed Pancreatic Acini

By JERRY D. GARDNER and ROBERT T. JENSEN

Before the development of techniques for preparing dispersed cells, studies of pancreatic function *in vitro* used slices or fragments of the gland. The major drawback to slices or fragments is that they do not allow one to take multiple identical samples of tissue during the course of an incubation. This drawback was overcome with the development of techniques for preparing a homogeneous suspension of dispersed pancreatic cells. Dispersed cell suspensions not only permit one to take multiple identical samples during the course of an incubation, but also enable one to fractionate the various cell types and isolate and purify the cells of interest. Because dispersed pancreatic cells maintain their viability and retain their responsiveness to secretagogues, they have enabled investiga-

ISBN 0-12-182009-2

tors to explore in detail the various biochemical steps that constitute the mechanisms of action of different pancreatic secretagogues.

Dispersed Pancreatic Acinar Cells

The procedure for preparing dispersed pancreatic acinar cells (Fig. 1, left) is a minor modification of the technique developed by Amsterdam and Jamieson.[1-3] This preparation contains at least 96% acinar cells and these cells retain their responsiveness to pancreatic secretagogues for up to 5 hr *in vitro*. These cells have been used to measure binding of secretagogues to their cells surface receptors, changes in cellular cyclic nucleotides such as cyclic AMP and cyclic GMP, changes in the transport of cations such as calcium, sodium, and potassium, and stimulation of secretion of various pancreatic enzymes. Dispersed acinar cells have been prepared using pancreas from mice, guinea pigs or rats.

Stock Solution. 100 m*M* NaCl, 6 m*M* KCl, 11.5 m*M* glucose, 2 m*M* KH_2PO_4, 25.5 m*M* HEPES (pH 7.4), 5 m*M* Na pyruvate, 5 m*M* Na glutamate, 5 m*M* Na fumarate, 2 m*M* glutamine, 0.008% (w/v) soybean trypsin inhibitor, 1% (v/v) Eagle's basal amino acid mixture (100 times concentrated), and 1% (v/v) essential vitamin mixture (100 times concentrated, Microbiological Associates, Bethesda, MD).

Digestion Solution A. Stock solution plus 0.1 m*M* $CaCl_2$, 1.2 m*M* $MgCl_2$, 0.75 mg/ml Type I collagenase (Sigma Chemical Co., St. Louis, MO), and 1.5 mg/ml Type I-S hyaluronidase (Sigma Chemical Co., St. Louis, MO).

Digestion Solution B. Stock solution plus 2 m*M* EGTA.

Digestion Solution C. Same as digestion solution A except use 1.25 mg/ml collagenase and 2 mg/ml hyaluronidase.

Wash Solution A. Stock solution plus 0.1 m*M* $CaCl_2$, 1.2 m*M* $MgCl_2$.

Wash Solution B. Stock solution plus 1 m*M* $CaCl_2$, 1.2 m*M* $MgCl_2$ and 4% (w/v) albumin.

Incubation Solution. Stock solution plus 0.5 m*M* $CaCl_2$, 1.2 m*M* $MgCl_2$ and 1% (w/v) albumin.

The animal is killed by a blow to the head and the pancreas is removed, trimmed of fat and mesentery, and pinned to a wax tray. Five milliliters of digestion solution A is injected into the tissue using a syringe and a 25-gauge needle. The gland and the digestion solution are transferred to a siliconized 25-ml Erlenmeyer flask, gassed for 30 sec with 100%

[1] A. Amsterdam and J. D. Jamieson, *Proc. Natl. Acad. Sci. U.S.A.* **69,** 3028 (1972).

[2] A. Amsterdam and J. D. Jamieson, *J. Cell Biol.* **63,** 1037 (1974).

[3] A. Amsterdam and J. D. Jamieson, *J. Cell Biol.* **63,** 1057 (1974).

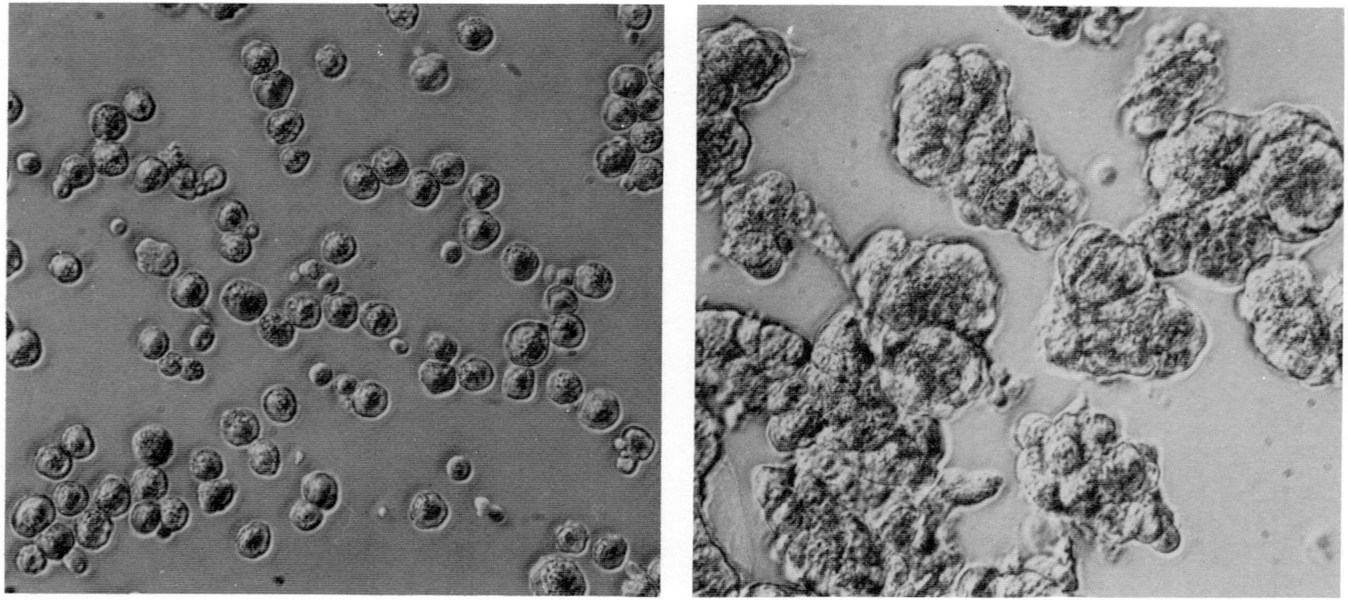

FIG. 1. Phase contrast microscopy of dispersed acinar cells (left) and dispersed acini (right) prepared from guinea pig pancreas. ×320.

O_2, capped, and incubated in a Dubnoff metabolic shaking incubator (160 oscillations per minute) at 37° for 15 min. After 15 min, the digestion solution is decanted and replaced with 8 ml of digestion solution B. The flask is gassed with 100% O_2, capped, and returned to the incubator for 5 min at 37°. After 5 min the digestion solution is decanted and replaced with 8 ml of fresh digestion solution B. The flask is gassed with 100% O_2, capped, and returned to the incubator for 5 min at 37°. At the end of the second 5-min incubation the digestion solution is decanted and the pancreas is washed twice with 10 ml of wash solution A. The washes are discarded. Five milliliter of digestion solution C is added to the flask. The flask is gassed with 100% O_2, capped, and returned to the incubator at 37° for 20 min. After 20 min the pancreas is dispersed by pipetting up and down 5 times through a Pasteur pipet and then passing it in and out of a syringe bearing a 19-gauge needle 5 times. The resulting cell suspension is passed through nylon mesh (Spectramesh 70 μm, Spectrum Medical Industries, Los Angeles, CA) to remove large clumps of nondispersed tissue. To separate the acinar cells from the digestion solution and cell debris, the cell suspension is gently layered into two 15-ml conical centrifuge tubes each containing 6 ml of wash solution B. The tubes are centrifuged at 50 *g* for 5 min and the supernatant is discarded. The cell pellets are combined and washed twice with 6 ml of wash solution B by centrifugation at 50 *g* for 5 min and resuspension. After the second wash, the supernatant is removed and discarded and the acinar cells are suspended in an appropriate volume of Incubation Solution.

Dispersed Pancreatic Acini

The procedure for preparing dispersed pancreatic acini (Fig. 1, right) is a minor modification of the technique reported by Peikin *et al.*[4] Pancreatic acini can be used for at least 5 hr *in vitro* and have been used to measure the same functions that have been measured using dispersed acinar cells. The major advantage of pancreatic acini is that they secrete much more enzyme in response to pancreatic secretagogues than do dispersed acinar cells (Fig. 2). Dispersed acini have been prepared using pancreas from mice, guinea pigs or rats.

Stock Solution. 100 m*M* NaCl, 6 m*M* KCl, 11.5 m*M* glucose, 2 m*M* KH_2PO_4, 25.5 m*M* HEPES (pH 7.4), 5 m*M* Na pyruvate, 5 m*M* Na glutamate, 5 m*M* Na fumarate, 2 m*M* glutamine, 0.008% (w/v) soybean trypsin inhibitor, 1% (v/v) Eagle's basal amino acid mixture (100 times concentrated), and 1% (v/v) essential vitamin mixture (100 times concentrated, Microbiological Associates, Bethesda, MD).

[4] S. R. Peikin, A. J. Rottman, S. Batzri, and J. D. Gardner, *Am. J. Physiol.* **234,** E743 (1978).

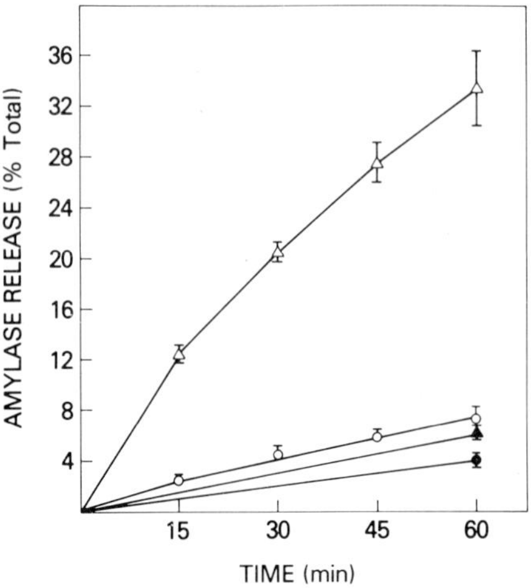

FIG. 2. Comparison of amylase release from dispersed acinar cells with that from dispersed acini prepared from guinea pig pancreas. Cells and acini were suspended in incubation solution and incubated at 37°. The percentage of the amylase activity in the tissue at the beginning of the incubation that was released into the incubation medium during the incubation was determined at the times indicated. Closed symbols represent results obtained with dispersed acinar cells; open symbols represent results obtained with dispersed acini. Circles represent control incubations; triangles represent results with 1 n*M* C-terminal octapeptide of cholecystokinin. Results shown are means from 3 separate experiments and vertical bars represent ±1 SD. In each experiment, each value was determined in duplicate.

Digestion Solution. Stock solution plus 0.05 mg/ml purified collagenase (462 units/mg; type CLSPA, Worthington Biochemical Corp., Freehold, NJ), 0.2% (w/v) albumin, and 2 m*M* $CaCl_2$.

Wash Solution. Stock solution plus 2 m*M* $CaCl_2$ and 4% (w/v) albumin.

Incubation Solution. Stock solution plus 0.5 m*M* $CaCl_2$ and 1% (w/v) albumin.

The animal is killed by a blow to the head and the pancreas is removed, trimmed of fat and mesentery, and pinned to a wax tray. Five milliliters of digestion solution is injected into the tissue using a syringe and a 25-gauge needle. The gland and the digestion solution are transferred to a siliconized 25-ml Erlenmeyer flask, gassed for 30 sec with 100% O_2, capped, and incubated in a Dubnoff metabolic shaking incubator (160 oscillations per minute) at 37° for 10 min. After 10 min of incubation, the digestion solution is decanted and replaced with 5 ml of fresh digestion solution. The flask is gassed, capped, and placed in the incubator for 10 min at 37°. After the second 10-min incubation, the digestion solution is again replaced with 5 ml of fresh digestion solution and the flask is gassed, capped, and placed in the incubator for 10 min at 37°. After the third 10-min incubation, the digestion solution is decanted and replaced with 5 ml of fresh digestion solution. The flask is gassed and capped, and the tissue is disrupted by shaking the flask vigorously by hand for 15 sec. The tissue is then dispersed by pipetting it 5 times through each of two siliconized glass pipets of decreasing internal diameter (3 and 1 mm). After discarding the duct system and large tissue fragments, the suspension of acini is gently layered into two 15-ml conical centrifuge tubes each containing 6 ml of wash solution. The tubes are centrifuged at 200 *g* for 5 sec and the supernatant is discarded. The cell pellets are combined and any large clumps of tissue are removed with a Pasteur pipet. The acini are washed twice with wash solution by centrifugation (200 *g* for 5 sec) and resuspended. After the last wash, the supernatant is discarded and the acini are suspended in an appropriate volume of incubation solution.

[24] The Preparation, Culture, and Incubation of Rat Anterior Pituitary Cells for Static and Dynamic Studies of Secretion

By PAULINE R. M. DOBSON and BARRY L. BROWN

The preparation of isolated viable cells from tissues has greatly facilitated studies on both cell morphology and function. There has been in recent years an increased awareness of the advantages associated with the use of isolated cells compared to tissue pieces including increased diffusion of solutes, greater precision of measurement due to sample randomization, increased sensitivity to regulation, and elimination of the problems of tissue atrophy. Several tissue dissociation techniques have been developed for use with various tissues. The success of these much depends upon the type and stage of development of the particular tissue. The importance of selecting a particular technique with respect to the studies for which the isolated cells are intended has been emphasized.[1]

In this presentation, we will concentrate on the special requirements for the preparation of suspensions of anterior pituitary cells (see also Ref. 2). In our studies trypsin is used as the dissociating enzyme; a choice which has been dependent on a number of considerations, e.g., it has specific inhibitors, and it generates single cell suspensions which are responsive to modifiers after a suitable recovery period.

[1] C. R. Kahn, *J. Cell Biol.* **70,** 261 (1976).

[2] C. R. Hopkins and M. G. Farquhar, *J. Cell Biol.* **59,** 276 (1973).

ISBN 0-12-182009-2

Although collagenase has been used successfully for the dissociation of anterior pituitary cells either alone or in conjunction with other enzymes, e.g., hyaluronidase,[3] our experience suggests that either complete dissociation does not occur or that the cells rapidly reaggregate. Also, collagenase has been shown to remove sialic acid from the cell membrane.[4] As sialic acid has been associated with cell adhesion, and also binds calcium, its removal may be an adverse property if the aim is to study the involvement of calcium in the control of pituitary function. However, it has been reported that pure preparations of collagenase do not possess this trait.[5] Neuraminidase also attacks sialic acid.[2,6–9] For this reason, the method routinely employed by this group, which is based on the method of Hopkins and Farquhar,[2] excludes the neuraminidase stage.

The aim of this report is to present our current strategies for both cell preparation and the subsequent maintenance and use of those cells. We emphasize the use of the cells to study secretion, the ultimate biological response, however, minor modifications will permit the study of other parameters such as changes in cyclic nucleotides[10,11] and phosphatidylinositol breakdown.[12,13]

Materials

Trypsin was obtained from Difco Laboratories, Detroit, Michigan, and Soyabean trypsin inhibitor (SBTI) was from Sigma Chemicals, Poole, Dorset, England. Cytodex beads were purchased from Pharmacia, Uppsala, Sweden. BioGel P-2 200–400 mesh was purchased from Bio-Rad Laboratories. Flow Laboratories, Irvine, Ayrshire, Scotland, supplied the Linbro 24-well plates, and Dulbecco's modification of Eagles medium (DMEM) with 20 m*M* HEPES. DMEM (bicarbonate buffered), fetal calf serum, "Fungizone," and penicillin and streptomycin were from Gibco Laboratories, Paisley, Scotland.

[3] W. Vale, G. Grant, M. Amoss, R. Blackwell, and R. Guillemin, *Endocrinology* **91,** 562 (1972).

[4] J. Rosenthal and J. Fain, *J. Biol. Chem.* **245,** 5888 (1971).

[5] H. Nakano, C. Fawcett, and S. McCann, *Endocrinology* **98,** 278 (1976).

[6] A. S. G. Curtis, *Prog. Biophys. Mol. Biol.* **27,** 317 (1973).

[7] J. N. Mehrishi, *Prog. Biophys. Mol. Biol.* **25,** 1 (1972).

[8] J. J. Deman, E. A. Bruyneel, and M. M. Mareel, *J. Cell Biol.* **60,** 641 (1974).

[9] G. A. Langer, J. S. Franks, L. M. Nudd, and K. Saraydarian, *Science* **193,** 1013 (1976).

[10] G. D. Barnes, B. L. Brown, T. G. Gard, D. Atkinson, and R. P. Ekins, *Mol. Cell. Endocrinol.* **12,** 273 (1978).

[11] M. P. Schrey, B. L. Brown, and R. P. Ekins, *Mol. Cell. Endocrinol.* **11,** 249 (1978).

[12] J. G. Baird, P. R. M. Dobson, R. J. H. Wojcikiewicz, J. E. Merritt, and B. L. Brown, *Biosci. Rep.* **3,** 1091 (1983).

[13] J. G. Baird, R. J. H. Wojcikiewicz, P. R. M. Dobson, and B. L. Brown, *Biochem. Soc. Trans.* **12,** 327 (1984).

Dissection of Rat Anterior Pituitaries

Sprague–Dawley rats are used in our studies. The selected rats are 8 to 10 weeks old (180–200 g). Animals are killed by a blow to the back of the neck, followed by stretching of the spinal column. After decapitation, the skull is cut away, and the brain is folded back to reveal the pituitary gland lodged in the sella turcica. The posterior pituitary appear as a pink-colored circle of tissue overlaying the larger anterior pituitary. The posterior pituitary can be "plucked-out" with forceps, and the remaining anterior pituitary gland transferred immediately to HEPES-buffered DMEM containing penicillin (100 units/ml) and streptomycin (100 μg/ml), and also Fungizone (2.5 μg/ml).

Batch Preparation of Cell Dissociation Reagents

The following stock solutions are prepared in advance and stored at −20°.

NaCl, 9.0% (1540 m*M*, 10× isotonic)
KCl, 5.75% (770 m*M*, 5× isotonic)
$CaCl_2$, 6.10% (550 m*M*, 5× isotonic)
($CaCl_2 \cdot 6H_2O$, 12.0%)
KH_2PO_4, 10.55% (770 m*M*, 5× isotonic)
$MgSO_4 \cdot 7H_2O$, 19.00% (770 m*M*, 5× isotonic)
$NaHCO_3$, 6.5% (770 m*M*, 5× isotonic)

Dilute the $NaHCO_3$ stock solution 1/5 (8–40 ml) with water. Bubble this diluted solution for 10 min with 5% CO_2.

Prepare low Ca^{2+}/Mg^{2+} KRB from the stock solutions:

NaCl (×10 stock), 10 ml
KCl (×5 stock), 0.8 ml
$CaCl_2$ (×5 stock), 23.6 μl
KH_2PO_4 (×5 stock), 0.2 ml
$MgSO_4$ (×5 stock), 16.9 μl
$NaHCO_3$ (diluted and bubbled as above), 21.0 ml

Make up the volume to 130 ml with water, controlling the pH to 7.4–7.6. Add 390 mg BSA, 130 mg glucose, 6.5 mg DNase, and 130 μl Phenol red stock solution (12 mg/ml stock).

In low Ca^{2+}/Mg^{2+} KRB prepared as above, make up *separate* solutions (40 ml) of 0.25% trypsin and 0.25% SBTI (Soyabean trypsin inhibitor).

Prepare Ca^{2+}- and Mg^{2+}-free KRB from the stock solutions:

NaCl (×10 stock), 5.0 ml
KCl (×5 stock), 0.4 ml
KH_2PO_4 (×5 stock), 0.1 ml
$NaHCO_3$ (diluted and bubbled as above), 10.5 ml

Make up the volume to 65 ml with water. Add 195 mg BSA, 65 mg glucose, 3.25 mg DNase, and 65 μl Phenol red stock solution (12 mg/ml stock).

In a laminar flow cabinet filter each of the 4 solutions through a sterile Millipore filter (0.22 or 0.45 μm) in the following order into sterile containers: Ca^{2+}- and Mg^{2+}-free KRB, KRB, trypsin, deionized H_2O (to wash filter—discard) and SBTI.

In a laminar flow cabinet aliquot the 4 sterile solutions into sterile bijou bottles: Ca^{2+}- and Mg^{2+}-free KRB, trypsin, and SBTI (each in aliquots of 2–3 ml) and KRB (aliquots of 5–6 ml). Cap, label, and freeze the aliquots immediately.

Cell Preparation

Plastic or siliconized glass vessels are used to avoid the adhesion of cells to surfaces. As far as possible, aseptic conditions are employed.

Before use, equilibrate the thawed solutions prepared as above by pouring each into a 30 mm petri dish, and transferring these to a CO_2 incubator for a few minutes.

Wash the anterior pituitary glands in KRB solution three times.

Chop the tissue into pieces approximately 0.5 mm^3.

Wash the pieces once in KRB solution, and incubate these with trypsin solution for 35–40 min at 37° in a 5% CO_2 incubator.

Wash the pieces once with KRB solution, then transfer to SBTI solution for 4 min at 37° under CO_2/air.

Wash the tissue pieces in Ca^{2+}- and Mg^{2+}-free KRB solution. Disperse the pieces by gently drawing them through a syringe fitted with a wide gauge (19-gauge) needle (NB cells are fragile in Ca^{2+}- and Mg^{2+}-free conditions).

Cells are transferred to medium (DMEM) containing 0.5% BSA and counted using a hemocytometer.

Static Incubations

Suspension Cultures. If experiments require suspension cultures, prepared cells can be maintained overnight as acutely dispersed cells are generally unresponsive due to receptor damage.

Freshly dissociated cells are suspended in serum-free culture medium (DMEM) containing 0.5% BSA in a screw-capped container in which is placed a magnetic stirring bar. Air is displaced from the container by a stream of 5% CO_2, 95% air. The culture is stirred slowly overnight. (As much as 25% of the cell population can be lost in this overnight incubation.)

On the following day, the cells are transferred to a conical container and centrifuged at 200 *g* for 5 min to pellet the cells. The cells are resuspended in fresh medium (HEPES-buffered DMEM containing 0.5% BSA), recounted, and diluted in medium to the appropriate density. For secretion studies the suspension is diluted to 1–3 × 10^5 cells/ml, and aliquots of 0.8 ml are dispensed into plastic tubes (60 × 10 mm) to which a further 0.2 ml medium ± modifier is added for the appropriate duration. On completion, the incubates are centrifuged 200 *g* for 5 min to pellet the cells, and a portion of the supernatant is drawn off and stored at −20° for hormone assay.

Monolayer Cultures. Freshly prepared cells are resuspended in DMEM containing 10% fetal calf serum, penicillin (100 units/ml), and streptomycin (100 μg/ml) to the appropriate dilution (normally 1–3 × 10^5 cells/ml). The suspension is dispensed into the wells of 24-well multiwell plates (1 ml/well) and maintained at 37° in a 5% CO_2 incubator for 3 days. On the third day, the monolayers are washed with HEPES-buffered DMEM containing 0.5% BSA. This medium ± modifier is added to the wells and the incubation period is commenced in a 37° waterbath. On completion, the supernatant is drawn off and a portion is retained at −20° for hormone assay.

For studies of cyclic nucleotides, 65% ethanol is added directly to the monolayers. The plates are stored overnight at −20° to aid protein precipitation. The precipitates are scraped into the medium. These are transferred to test tubes and centrifuged (100 *g*/15 min). The supernatant is dried either under vacuum or in a stream of compressed air.

Dynamic Incubations

Perifusion and superfusion methods provide the advantage that a population of pituitary cells can be subjected to a continual flow of material *in vitro,* thereby both supplying fresh modifier and removing the products of the response as is likely to occur *in vivo*. Also, as the outflow is collected continuously into fractions which are then analyzed, this is a useful means of monitoring the dynamics of the system. The methods employed by this group are either that of Schrey *et al.,*[14] using BioGel as the supporting matrix, or substitution of BioGel by Cytodex beads.[15,16] The BioGel cell column method routinely uses approximately 10^7 cells admixed with 0.5 g of BioGel and packed into a column made from a plastic syringe as de-

[14] M. P. Schrey, B. L. Brown, and R. P. Ekins, *Mol. Cell. Endocrinol.* **8,** 271 (1977).

[15] M. A. Smith and W. W. Vale, *Endocrinology* **107,** 1425 (1980).

[16] R. J. H. Wojcikiewicz, P. R. M. Dobson, and B. L. Brown, *Biochim. Biophys. Acta* (in press).

scribed by Schrey *et al.*[14] The cell column is placed in a water bath and medium (usually KRB with 0.5% BSA and 0.02% glucose) is perfused at a rate of 0.4–0.6 ml/min.

When using a Cytodex cell column, cells are seeded directly after dissociation, onto Cytodex beads (contained in bacteriological petri dishes) at a density of 1–2 × 10^5 cells/mg. Cytodex in Ham's F10 containing 10% fetal calf serum, penicillin, and streptomycin. On the third day, cells are washed with normal KRB solution (i.e., containing 2 m*M* Ca^{2+} and 2 m*M* Mg^{2+}) containing 0.5% BSA and 0.02% glucose. The Cytodex beads are then packed into the perifusion apparatus. Usually a smaller column is used (i.e., approximately 20 mg Cytodex) and is perfused at a rate of 0.4–0.5 ml/min.

[25] Isolation and Functional Aspects of Free Luteal Cells

By JUDITH L. LUBORSKY and HAROLD R. BEHRMAN

Introduction

During each ovarian cycle, changes in the endocrine function of the ovary are dependent on differentiation of follicular cells and luteal cells. Differentiation of ovarian cells is expressed as changes in morphology, cellular metabolism, and their responsiveness to gonadotropins. Ovarian cell function has been studied *in vitro* with intact follicles and copora lutea, tissue slices, or more recently with isolated granulosa or luteal cells. Luteal cell function in particular has been studied in granulosa cells that have been allowed to luteinize in culture[1–3] or by direct isolation of luteal cells from lutenized tissue.[4–7]

Since the function of the corpus luteum must be terminated for a subsequent ovulation to occur or extended for a pregnancy to be maintained, the functional state of corpus luteum is delicately balanced by both

[1] G. L. Kumari and C. P. Channing, *J. Steroid Biochem.* **11,** 781 (1979).
[2] K. M. Henderson and K. P. McNatty, *J. Endocrinol.* **73,** 71 (1977).
[3] R. W. Tureck and J. F. Strauss, *J. Clin. Endocrinol. Metab.* **54,** 367 (1982).
[4] T-C. Liu and J. Gorski, *Endocrinology* **88,** 419 (1971).
[5] D. Gospodarowicz and F. Gospodarowicz, *Endocrinology* **90,** 1427 (1972).
[6] D. H. Wu, W. G. Wiest, and A. C. Enders, *Endocrinology* **98,** 1378 (1976).
[7] K. R. Simmons, J. L. Caffrey, J. L. Phillips, J. H. Abel, and G. D. Niswender, *Proc. Soc. Exp. Biol. Med.* **152,** 366 (1976).

ISBN 0-12-182009-2

tropic and lytic endocrine regulators. Isolated luteal cell preparations were developed in order to examine cell regulation independent of variables present *in vivo,* such as changes in blood flow and endogenous hormone concentrations. Isolated cells can be uniformly and rapidly treated with drugs and hormones. Their use does not have the problems of sample variation and nonuniform exposure of cells that are inherent in isolated corpora lutea and luteal tissue slices. Luteal cells isolated from sheep,[7–9] cow,[5,10] pig,[11] rat,[6,12–14] guinea pig,[15] rabbit,[4] monkey,[16] and human corpora lutea[17–20] have been used to study aspects of their cellular morphology, biochemical responses and hormone receptor function.

General Methods of Luteal Cell Isolation

In general, methods of luteal cell isolation employ enzymatic treatment of minced or sliced luteal tissue with collagenase (to break down connective tissue) and deoxyribonuclease (DNase) (to break down released DNA that may cause cells to clump). Additional enzymes such as hyaluronidase or Pronase are also used in some laboratories.[5,6,10,12] Enzyme treatment is usually coupled with mechanical treatment to further disperse cells.

A systematic study of conditions for rat luteal cell isolation examined duration of incubation, and combinations of variable concentrations of collagenase, trypsin, and hyaluronidase, as well as bovine serum albumin (BSA) and Ca^{2+} concentrations necessary to maintain maximum viability, morphology, and steroidogenesis. A final combination of all three enzymes (2940 U trypsin, 2320 U collagenase, and 16,900 U hyaluronidase/10 ml of medium) plus 0.5% BSA and 3.3 mM Ca^{2+} was chosen.[6]

[8] R. J. Rodgers and J. D. O'Shea, *Aust. J. Biol. Sci.* **35,** 441 (1982).

[9] C. E. Ahmed, H. R. Sawyer, and G. D. Niswender, *Endocrinology* **109,** 1380 (1981).

[10] J. Ursely and P. Leymarie, *J. Endocrinol.* **83,** 303 (1979).

[11] M. Lemon and M. Loir, *J. Endocrinol.* **72,** 351 (1977).

[12] R. F. Wilkinson, E. Anderson, and J. Aalberg, *J. Ultrastruct. Res.* **57,** 168 (1976).

[13] H. R. Behrman, S. L. Preston, and A. K. Hall, *Endocrinology* **107,** 656 (1980).

[14] J. L. Luborsky and H. R. Behrman, *Mol. Cell. Endocrinol.* **15,** 61 (1979).

[15] M. C. Richardson and M. J. Peddie, *J. Reprod. Fertil.* **66,** 117 (1982).

[16] R. L. Stouffer, W. E. Nixon, B. J. Gulyas, D. K. Johnson, and G. D. Hodgen, *Steroids* **27,** 543 (1976).

[17] M. T. Williams, M. S. Roth, J. M. Marsh, and W. J. LeMaire, *J. Clin. Endocrinol. Metab.* **48,** 427 (1979).

[18] M. C. Richardson and G. M. Masson, *J. Endocrinol.* **87,** 247 (1980).

[19] L. T. Goldsmith, M. Essig, P. Sarosi, P. Beck, and G. Weiss, *J. Clin. Endocrinol. Metab.* **53,** 8980 (1981).

[20] M. L. Polan, A. H. DeCherney, F. P. Haseltine, H. C. Mezer, and H. R. Behrman, *J. Clin. Endocrinol. Metab.* **56,** 288 (1983).

Luteinizing hormone (LH) receptor binding and function varies with different preparations of collagenase[21,22]; these changes are thought to be due to variations in trypsin activity. However, trypsin improved separation of endothelial and large and small luteal cells of ovine corpus luteum, and the luteal cells retained their ability to respond to LH.[8] A systematic study of tryspin effects on primate luteal cells showed that a brief exposure (10 min) did not change human chorionic gonadotropin (hCG)-induced progesterone secretion, while a prolonged exposure (3 hr) significantly reduced both basal and hCG-stimulated progesterone secretion.[23] Variations in cell isolation with different collagenase preparations may be due to contamination with clostripain; collagenase free of clostripain was reported to produce uniform and consistent preparations of pancreatic cells.[24]

After cell dispersion with enzyme(s), intercellular junctions are not completely separated. Frequently, some cells are well separated while the majority are still in small clumps or have attached debris from broken cells. Recently, additional treatment with trypsin and/or a divalent cation chelator was introduced to facilitate the separation of intercellular junctions.[8,14] Trypsin in low Ca^{2+} medium containing ethyleneglycol bis(β-aminoethylether) (EGTA) separates intercellular junctions.[8] A brief exposure to low Ca^{2+} medium containing ethylenediaminetetraacetic acid (EDTA) also improves separation of luteal cells.[14] Prolonged exposure (greater than 3–5 min) or high concentrations of EDTA (>2 mM) are deleterious to cell function.[14] A sequential treatment with EDTA and hypotonic sucrose has been reported to enhance separation of gap junctions between granulosa cells.[25]

Cell suspensions have been further enriched by separation on gradients of Ficoll,[8,10] Percoll,[13,26] sucrose,[9] or BSA.[11,12] These gradients were used simply to remove membrane debris and dead cells[13,26] or to obtain cells of different sizes.[8,10–12] Removal of membrane debris and broken cells is particularly important for morphological and receptor binding studies but does not appear to be essential for studies of cellular metabolism. Final yields of cells are about $1–10 \times 10^7$ cells/g of luteal tissue.[4,5,7,8,10,16,26]

[21] M. L. Dufau and K. J. Catt, this series, Vol. 37, p. 252.

[22] J. L. Luborsky and H. R. Behrman, unpublished observations (1982).

[23] B. J. Gulyas, L. C. Yuan, and G. D. Hodgen, *Steroids* **35,** 43 (1980).

[24] G. R. Gunther, G. S. Schultz, B. E. Hull, H. A. Alicea, and J. D. Jamieson, *J. Cell Biol.* **75,** 368 (1979).

[25] K. L. Campbell and A. R. Midgley, *J. Cell Biol.* **75,** 246 (1977).

[26] B. C. McNamara, C. E. G. Cranna, R. Booth, and D. A. Stansfield, *Biochem. J.* **192,** 559 (1980).

Morphological Aspects of Isolated Luteal Cells

Isolated luteal cells retain the morphological characteristics of steroid secreting cells after isolation. They contain abundant mitochondria, variable amounts of lipid droplets, and an extensive smooth endoplasmic reticulum, similar to cells *in vivo*.[12,27,28] Two size ranges of isolated cells, of 15–25 and 30–50 μm in diameter, from monkey, rat, pig, sheep, and cow corpus luteum have been reported.[8,10–12,29] A comparison of cell types *in vivo* and *in vitro* suggests that the larger cells are extremely fragile and that there is a lower recovery after isolation than for the smaller cells.[8,10,12] Other cells, such as endothelial and blood cells, are found in isolated luteal cell preparations, and these can be reduced by further purification.[8] It was suggested by several authors[8,11] that the small and large cell types originate from thecal and granulosa cells, respectively; it has *also* been suggested[12] that these cells are part of a continuum of differentiating luteal cells. The actual source of the different size cells has not been established but it is clear that they differ morphologically and metabolically. The large cells contain more progesterone, are less responsive to LH, have more surface projections, and contain more mitochondria.[8,10–12,29]

Isolated luteal cells have been maintained in culture for extended periods of time. Histochemical analysis of bovine luteal cells cultured for seven days showed lipid material that stained with Sudan Black and Oil Red, as well as 3β-hydroxysteroid dehydrogenase and glucose-6-phosphate dehydrogenase, characteristic of luteal cells.[30] Ultrastructural analysis indicates that cells retain their typical steroidogenic features,[12,29] but small cells tend to lose their lipid droplets by extrusion during culture.[12]

LH receptors have been localized on surfaces of rat[14] and monkey[31] luteal cells with a ferritin–LH and a horseradish peroxidase–hCG conjugate, respectively. In both cases, receptors are located at intervals over the entire cell surface. Rate luteal cells in suspension do not appear to maintain their surface polarity to the same degree as found *in vivo,* with a basal smooth surface and an apical microvillus surface containing LH receptors.[28] Cultured monolayers of rat luteal cells [12] may be more similar to cells *in vivo* but this has not been documented with respect to receptor localization. Primate luteal cells have been reported to lack morphological

[27] A. K. Christensen and S. W. Gillim, *in* "The Gonads" (K. W. McKerns, ed.), p. 415. North-Holland Publ., Amsterdam, 1969.

[28] W. Anderson, Y.-H. Kang, M. E. Perotti, T. A. Bramley, and R. J. Ryan, *Biol. Reprod.* **20,** 362 (1979).

[29] B. J. Gulyas, R. L. Stouffer, and G. D. Hodgen, *Biol. Reprod.* **20,** 779 (1979).

[30] D. Gospodarowicz and F. Gospodarowicz, *Exp. Cell Res.* **75,** 353 (1972).

[31] B. J. Gulyas, S. Matsuura, H.-C. Chen, L. C. Yuan, and G. D. Hodgen, *Biol. Reprod.* **25,** 609 (1981).

polarity both *in vivo*[32] and *in vitro*[31]; this has not been correlated with studies of receptor localization.

Responses of isolated luteal cells to hormones and growth factors involve a morphological component or shape change. In general, cells appear more spherical and polygonal when treated with LH (hCG) or dibutyryladenosine 3′ : 5′-cyclic monophosphate [$(Bu)_2$cAMP] and during enhanced steroidogenic activity.[3,33,34] Cell shape is probably determined by microtubules and microfilaments since shape changes are inhibited by colcemid, vinblastin, or cytochalasin B.[33] In the absence of hormone or after its removal, more flattened, oblong cells are observed.[3] It is possible that LH may control and maintain morphological differentiation. Glucocorticoids and testosterone inhibit cell growth but progesterone does not.[33] In the presence of fibroblast growth factor (FGF) and epidermal growth factor (EGF), cell growth is stimulated and cells become more flattened.[35] Cells in suspension culture have not been reported to undergo shape changes.

Biochemical/Endocrine Function of Isolated Luteal Cells

Isolated luteal cells have been used in numerous studies to examine the regulation of steroidogenesis by LH. The functional capacity of freshly isolated luteal cells has been reported to reflect their functional capacity *in vivo* with respect to progesterone production in response to LH.[16,26,36] Effects of LH on human luteal cells during the cycle[34] and primate luteal cells during pregnancy[36] parallel that seen *in vivo*.[36] LH stimulated progesterone production and increases in adenosine 3′ : 5″-cyclic monophosphate (cAMP) levels and cAMP binding to protein kinase[37,38] are directly correlated. Aminoglutethimide inhibits cholesterol side chain cleavage and progesterone production from cholesterol esters[26,39] and compactin blocks cholesterol biosynthesis *in vitro*.[26] In the presence of aminoglutethimide but not compactin, rapid (short-term) responses of cells to LH were blocked, suggesting stored cholesterol but not cholesterol biosynthesis is required for acute responses to LH.[26] Also,

[32] B. J. Gulyas, *Am. J. Anat.* **139,** 95 (1974).

[33] D. Gospodarowicz and F. Gospodarowicz, *Endocrinology* **75,** 458 (1975).

[34] K. Kamei, *Nippon Sanka Fujinka Gakki Zasshi* **34,** 261 (1982).

[35] D. Gospodarowicz, C. R. Ill, and C. R. Birdwell, *Endocrinology* **100,** 1121 (1977).

[36] R. L. Stouffer, W. E. Nixon, B. J. Gulyas, and G. D. Hodgen, *Endocrincology* **100,** 596 (1977).

[37] G. B. Sala, M. L. Dufau, and K. J. Catt, *J. Biol. Chem.* **254,** 2077 (1979).

[38] W. Y. Ling, M. T. Williams, and J. M. Marsh, *J. Endocrinol.* **86,** 45 (1980).

[39] M. Lemon and P. Mauleon, *J. Reprod. Fertil.* **64,** 315 (1982).

inhibition of cholesterol ester cleavage by cannabinoids reduced progesterone production.[40] Ca^{2+} has been shown to augment LH-dependent steroidiogenesis.[41,42] In isolated cells LH action is blocked by 2-bromopalmitate indicating involvement of fatty acid oxidation.[43] LH selectively stimulates incorporation of ^{32}P into phosphatidic acid, phosphatidylinositol, and polyphosphoinositides.[44] A direct action of biogenic amines acting at separate receptors, on LH stimulated progesterone production was shown.[45–47] Biogenic amine action on cAMP accumulation in isolated luteal cells was shown to decrease with increasing corpus luteum age.[47] Danazol was reported to inhibit LH, cholera toxin, and $(Bu)_2$cAMP-stimulated progesterone production but not corresponding increases in cAMP, suggesting it acts distal to the hormone receptor site.[48]

Responsiveness of the two cell types to LH appears to differ. Large cells produce higher basal levels of progesterone but are less responsive to LH[8,10,11] and there are less LH receptors on large cells.[31] In addition, in a continuous perfusion system, experiments with large and small cells "in series" showed that small cells can increase the amount of progesterone produced by large cells but not the reverse.[11]

During maintenance of primate and other luteal cells in long-term culture, progesterone and estrogen production gradually decline[19,29] and this is not prevented by addition of thyroxine, insulin, cortisol, or cholesterol.[49] Also, bovine serum was reported to decrease progesterone production in bovine luteal cells.[50] Lipoproteins increase progesterone production in cultured rat,[51] bovine,[50] and human[3] luteal cells. Luteal macrophages were reported to maintain progesterone production of mouse luteal cells.[52]

[40] S. Burstein, S. A. Hunter, and T. S. Shoupe, *Res. Commun. Chem. Pathol. Pharmacol.* **24,** 413 (1979).
[41] N. Kamiya, *Nippon Naibunpi Gakkai Zasshi* **58,** 833 (1982).
[42] L. J. Dorflinger, P. J. Albert, A. T. Williams, and H. R. Behrman, *Endocrinology* **114,** 1208 (1984).
[43] C. H. Tan and J. Robinson, *Life Sci.* **30,** 1205 (1982).
[44] J. S. Davis, R. V. Farese, and J. M. Marsh, *Endocrinology* **109,** 469 (1981).
[45] R. C. Rhodes and R. D. Randel, *Comp. Biochem. Physiol.* **72,** 113 (1982).
[46] A. W. Jordan, J. L. Caffrey, and G. D. Niswender, *Endocrinology* **103,** 385 (1978).
[47] E. Norjavaara, G. Selstam, and K. Ahren, *Acta Endocrinol. (Copenhagen)* **100,** 613 (1982).
[48] M. Menon, S. Azhar, and K. M. Menon, *Am. J. Obstet. Gynecol.* **136,** 524 (1980).
[49] B. J. Gulyas, L. C. Yuan, and G. D. Hodgen, *Biol. Reprod.* **23,** 21 (1980).
[50] J. L. Pate and W. A. Condon, *Mol. Cell. Endocrinol.* **28,** 551 (1982).
[51] S. Azhar and K. M. Menon, *J. Steroid Biochem.* **16,** 175 (1982).
[52] T. M. Kirsch, A. C. Friedman, R. L. Vogel, and G. L. Flickinger, *Biol. Reprod.* **25,** 629 (1981).

Prolactin (PRL) maintains LH receptor and steroidogenic function in rat luteal cells *in vivo*.[53] *In vitro*, PRL maintains basal steroidogenesis in rat luteal cells without a specific stimulation of its own.[5,54] PRL also suppresses 20α-steroid dehydrogenase in the rat.[6,55] Its role in human luteal function is not as clear. In humans, PRL was reported to have a dual, age-dependent, luteolytic and luteotropic action, without an effect on responses to hCG.[56] Only a minimal effect of PRL was seen on primate luteal cells.[57]

Prostaglandin-$F_{2\alpha}$ ($PGF_{2\alpha}$) has been shown to be luteolytic in a variety of animals. In some studies the newly formed corpus luteum is refractory to $PGF_{2\alpha}$,[2,15,58] but later, a marked inhibition of LH stimulated cAMP and progesterone production by $PGF_{2\alpha}$ in isolated luteal cells[59–62] occurs. $PGF_{2\alpha}$ appears to have no effect on bovine luteal cells[63] but is luteolytic *in vivo*.[64] In early luteal cells of humans, in the presence of no LH, or low levels of LH, $PGF_{2\alpha}$ stimulated progesterone production.[18] From these studies, $PGF_{2\alpha}$ acts directly and acutely at the LH receptor, as well as at other sites such as on steroidogenesis[60,62] and *in vivo* on delivery of hormone from blood.[64]

Estrogen is thought to be involved in luteolysis of human, primate, and bovine corpus luteum. Direct inhibition by estrogen of LH and $(Bu)_2$cAMP stimulated progesterone production but not LH-stimulated cAMP production suggests that estrogen does reduce bovine,[65] human,[66,67] and primate[68] luteal cell function, but at a point beyond LH

[53] H. R. Behrman, D. L. Grinwich, M. Hichens, and G. D. Macdonald, *Endocrinology* **103,** 349 (1978).

[54] K. Shiota and W. G. Wiest, *Adv. Exp. Med. Biol.* **112,** 169 (1979).

[55] M. Lahav, S. A. Lamprecht, A. Amsterdam, and H. R. Lindner, *Mol. Cell. Endocrinol.* **6,** 293 (1977).

[56] T. Sawada, *Nippon Sanka Fujinka Gakki Zasshi* **34,** 2212 (1982).

[57] R. L. Stouffer, J. L. Coensgen, and G. D. Hodgen, *Steroids* **35,** 523 (1980).

[58] R. L. Stouffer, W. E. Nixon, and G. D. Hodgen, *Biol. Reprod.* **20,** 897 (1979).

[59] J. P. Thomas, L. J. Dorflinger, and H. R. Behrman, *Proc. Natl. Acad. Sci. US.A.* **75,** 1334 (1978).

[60] A. W. Jordan, *Biol. Reprod.* **25,** 327 (1981).

[61] A. K. Hall and J. Robinson, *J. Endocrinol.* **81,** 157 (1979).

[62] H. R. Behrman, J. L. Luborsky, C. Y. Pang, K. Wright, and L. J. Dorflinger, *Adv. Exp. Biol. Med.* **112,** 557 (1979).

[63] J. E. Hixon and W. Hansel, *Adv. Exp. Med. Biol.* **112,** 613 (1979).

[64] H. R. Behrman, *Annu. Rev. Physiol.* **41,** 685 (1979).

[65] M. T. Williams and J. M. Marsh, *Endocrinology* **103,** 1611 (1978).

[66] M. Thibier, N. el-Hassan, M. R. Clark, W. J. LeMaire, and J. M. Marsh, *J. Clin. Endocrinol. Matab.* **50,** 590 (1980).

[67] M. Richardson and G. M. Masson, *J. Endocrinol.* **91,** 197 (1981).

[68] R. L. Stouffer, L. A. Bennett, and G. D. Hodgen, *Endocrinology* **106,** 519 (1980).

induced increases in cAMP.[65,66] Isolated primate luteal cells also synthesize and secrete estrogen and respond to LH with an increase in estrogen production; neither estrogen production nor the estrogen/progesterone ratio increases with the age of the corpus luteum.[68] Testosterone has been reported to inhibit LH action in porcine[69] but not human[67] luteal cells.

Recently, a direct inhibitory action of luteinizing hormone releasing hormone (LHRH) on rat luteal cells was shown. LHRH binds to specific receptors on luteal cells and inhibits LH-induced increases in cAMP accumulation and progesterone secretion.[70–73] LHRH stimulates incorporation of ^{32}P into phospholipid.[74] The antigonadotropic action of LHRH is similar to that for $PGF_{2\alpha}$ and these agents appear to increase intracellular Ca^{2+} levels.[42] At this time, the physiological significance of LHRH receptors on ovarian cells is not established. LHRH or an LHRH-like peptide may be one of several intraovarian factors that regulate ovarian function.[75] In this regard follicular fluid injected into monkeys reduced progesterone production and responsiveness to LH in subsequently isolated luteal cells.[76]

The corpus luteum has been identified as the source of the peptide hormone, relaxin. Long-term cultures of isolated rat[77] and human[19] luteal cells have been used to examine the factors regulating relaxin secretion. Its release in rats is stimulated by various combinations, but not singly, of LH, PRL, progesterone, epinephrine, and estrogen; $(Bu)_2cAMP$ did not stimulate relaxin secretion.

The role of microfilaments and microtubules in steroidogenesis has been investigated in isolated luteal cells. Cytochalasin B and D decreased progesterone secretion in response to LH, $(Bu)_2cAMP$, cholera toxin and 8-bromo-cAMP in rat luteal cells without a reduction of cAMP levels.[78] Cytochalasin B also decreased LH and $(Bu)_2cAMP$ stimulated progesterone secretion in bovine luteal cells.[79] Microtubule inhibitors had little effect on steroidogenesis when applied directly to luteal cells[78] but if injected *in vivo,* progesterone production in subsequently isolated cells

[69] M. G. Hunter, *J. Reprod. Fertil.* **63,** 471 (1981).
[70] A. K. Hall and H. R. Behrman, *J. Endocrinol.* **88,** 27 (1981).
[71] H. R. Behrman, S. L. Preston, and A. K. Hall, *Endocrinology* **107,** 656 (1980).
[72] J. Massicotte, J. P. Borgus, R. Lachance, and F. Labrie, *J. Steroid Biochem.* **14,** 239 (1981).
[73] R. N. Clayton, J. P. Harwood, and K. J. Catt, *Nature (London)* **282,** 90 (1979).
[74] C. K. Leung, V. Raymond, and F. Labrie, *Endocrinology* **112,** 1138 (1983).
[75] A. T. Williams and H. R. Behrman, *Semin. Reprod. Endocrinol* **1,** 269 (1983).
[76] R. L. Stouffer and G. D. Hodgen, *J. Clin. Endocrinol. Metab.* **51,** 669 (1980).
[77] L. T. Goldsmith, H. S. Grob, and G. Weiss, *Ann. N.Y. Acad. Sci.* **380,** 60 (1982).
[78] S. Azhar and K. M. Menon, *Biochem. J.* **194,** 19 (1981).
[79] M. T. Williams and J. M. Marsh, *Adv. Exp. Biol. Med.* **112,** 549 (1979).

was decreased[80] without morphological evidence of microtubule disruption. Thus, cellular microfilaments were suggested to be involved in hormone regulated steroidogenesis.

Hormone Receptor Function of Isolated Luteal Cells

LH receptor binding studies were employed to quantitate specific properties of hormone-receptor interaction in relation to cellular function. Binding of [^{125}I]hCG to bovine luteal cells and membranes was compared and it was concluded that the enzymatic treatment used to isolate cells did not change LH receptor binding kinetics or affinity. Binding to bovine cells was saturable, rapid and temperature dependent with an estimated binding capacity of 5×10^4 receptors/cell and an equilibrium binding constant (K_D) of 5.3×10^{-10} *M*.[81] Prostaglandin E_1 (PGE_1) binding to bovine luteal cells was rapid, saturable, reversible, and specific with a K_D = 2.4 n*M* and 1.8×10^5 receptors/cell, with most properties similar to receptors in isolated membranes.[82]

Binding of [^{125}I]hCG to cells from rats desensitized with hCG was compared to cells from untreated rats. The desensitized or refractory state was accompanied by a significant loss of hCG binding concomitant with decreased cAMP and progesterone production in response to LH.[83] This reduced steroid production was also due to an additional lesion in steroid biosynthesis, rather than to a simple loss of LH receptors. In studying recovery from desensitization, it was found that a full complement of LH receptors was required for maximum sensitivity (i.e., an LH response is obtained, but more hormone is needed to obtain a maximum response). Similarly, cells isolated from primate corpus luteum after hCG exposure were refractory to hCG.[84]

Rat luteal cells increased their sensitivity to LH (cAMP) without a change in LH receptor binding during prolonged culture.[70] This increased sensitivity was impaired by early exposure to LH, $PGF_{2\alpha}$, LHRH, or an LHRH analog. Since there was no change in LH receptor binding, it was suggested that receptor desensitization may precede receptor loss and may be due to rapid uncoupling of the receptor and adenylate cyclase. A similar inhibition of LH action by $PGF_{2\alpha}$ was seen in *in vivo* studies with $PGF_{2\alpha}$,[53,62] but this may be due to additional action of $PGF_{2\alpha}$ *in vivo*.

[80] S. Azhar and E. Reaven, *Am. J. Physiol.* **243,** E380 (1982).

[81] S. Papaionannou and D. Gospodarowicz, *Endocrinology* **97,** 114 (1975).

[82] M. T. Lin and C. V. Rao, *Mol. Cell. Endocrinol.* **9,** 311 (1978).

[83] J. P. Harwood, M. Conti, P. M. Conn, M. L. Dufau, and K. J. Catt, *Mol. Cell. Endocrinol.* **11,** 121 (1978).

[84] R. L. Stouffer, W. E. Nixon, and G. D Hodgen, *Biol. Reprod.* **18,** 858 (1978).

Mechanisms for changes in the LH receptor content of cells have been studied. Loss of receptors from the ovine luteal cell surface occurred with a $t_{1/2}$ of 9.6 hr and about 85% was lost within 24 hr. Cell-associated radioactivity increased up to 4 hr, was maintained, and then decreased during 12–24 hr. This time course is similar to that seen *in vivo* during down-regulation.[62] Receptor loss was three times faster than membrane protein turnover. A majority of bound [^{125}I]hCG was internalized and degraded.[85] Inhibitors of transglutaminase (inhibits receptor clustering) did not inhibit internalization of [^{125}I]hCG[86] but did produce an intracellular accumulation of [^{125}I]hCG and an inhibition of hCG degradation. HCG binding may result in an increase in LH receptor aggregation, but LH receptors are not necessarily internalized as large clusters in coated pits.[14] Furthermore, inhibition of protein synthesis did not decrease the total number of LH receptors per cell, although it did decrease LH receptor degradation and the number of receptors on the cell surface. Thus, synthesis of new receptors is not required for continued binding and internalization, and it was suggested that receptors are recycled.[85]

The effects of EGF and FGF and their binding to cells is similar to that *in vivo;* bovine luteal cells lose their proliferative response to EGF but retain their responsiveness to FGF.[87] Analysis of EGF binding showed that both granulosa cells and luteal cells bind EGF, but in luteal cells there is no mitogenic respnse. In fact, EGF receptors increased with luteinization (0.2 to 1 $\times$ 10^5). Down-regulation, or loss of EGF receptors in response to EGF, occurred in both granulosa and luteal cells. Low concentrations of EGF produced a greater loss of receptors than those that were occupied.[87]

Preparation of Free Luteal Cells

Animals. Ovaries from gonadotropin primed rats, given 50 IU pregnant mares serum (Gestyl, Organon) and 25 IU hCG (A.P.L., Ayerst) 64 hr later, were removed 4–8 days after hCG injection.

Enzymatic Treatment. Ovaries are rapidly trimmed in medium A (GIBCO No. 320-1380 containing 0.1% BSA), weighed, and minced with a clean, sharp razor. The fragments are washed to remove excess blood cells. The fragments are added to a 25-ml Erlenmeyer flask containing 2000 IU collagenase (Worthington Biochemical Corporation) and 3000 IU DNase (Worthington Biochemical Corporation) per gram of tissue in 5 ml of medium A. An additional 5 ml of medium A is added and the solution is

[85] D. E. Suter and G. D. Niswender, *Endocrinology* **112,** 838 (1983).

[86] C. E. Ahmed and G. D. Niswender, *Endocrinology* **109,** 1388 (1981).

[87] I. Vlodavsky, K. D. Brown, and G. Gospodarowicz, *J. Biol. Chem.* **253,** 3744 (1978).

gently mixed and aerated with 95% O_2/5% CO_2 for 1 min. The tissue is incubated for 1 hr at 37° with agitation (Dubnoff incubator, 100 cycles/min). During the incubation the fragments are pipetted five or six times with a large bore, Pasteur pipet with a flame polished tip. At the end of incubation the tissue is centrifuged (100 *g*, 5 min) in a 15-ml round bottom centrifuge tube and the supernatant discarded.

Low Ca^{2+} Treatment. After enzyme treatment, the majority of cells are in small clumps. Luteal cells are more difficult to separate than granulosa cells since they develop extensive intercellular junctions during luteinization.[12] To improve separation of these junctions, cells are resuspended in 10 ml of medium A containing 1 m*M* EDTA. The solution is pipetted gently about six times to resuspend the pellet, with a Pasteur pipet with a flame polished tip, incubated for 2 min (22°), centrifuged (100 *g*, 5 min), and the supernatant discarded. The cells are extremely fragile in low Ca^{2+} conditions; additional time or increased concentrations of EDTA will damage cells.

Mechanical Dispersion. The pellet is resuspended in 3 ml of medium A. Cells are mechanically dispersed by repeated pipetting with a flame polished Pasteur pipet, allowing clumps to settle (about 0.5 min) and removing the supernatant containing isolated cells. Cells are filtered through nylon mesh (Nyten, Tetko) supported by a disposable plastic funnel into a clean 15-ml polypropylene centrifuge tube. The pellet is resuspended in 1 ml of medium A and the procedure repeated until only a few small clumps of cells and connective tissue remain. Pooled cells are centrifuged (100 *g*, 5 min). This process minimizes mechanical manipulation of isolated cells. Cells may be used without further processing at this point; cells are resuspended in medium B (GIBCO No. 380-2360 containing 0.1% BSA) and preincubated 60 min before use. Alternatively, the cells are resuspended in 2 ml of medium A and further enriched on a Percoll column to remove membrane debris, some red blood cells, and nonfunctional luteal cells.

Enrichment of Luteal Cell Preparation. A discontinuous Percoll gradient is prepared by layering 3 ml each of 40, 30, 20, and 10% (bottom to top) Percoll solutions in a 15 ml conical polypropylene centrifuge tube. A Percoll stock solution is prepared from 9 ml of Percoll plus 1 ml of medium C (GIBCO No. 199, 10×) and is diluted with medium A. The cell suspension is layered on top and centrifuged through the gradient (100 *g*, 30 min). The excess medium is removed and cells at the 10–20% and 20–30% interfaces are combined. Cells are rinsed with medium B (100 *g*, 10 min) and resuspended to about 2 to 4 × 10^6 cells/ml in medium B. One column will separate cells from 14 ovaries.

Cell Yield and Viability. The cells are counted in a hemocytometer. Generally this method yields about 1–2 × 10^6 cells/ovary. Before Percoll enrichment the preparation contains about 50–60% luteal cells and after Percoll enrichment it contains about 70–80% luteal cells. Cell viability is about 90% when tested for the ability of cells to exclude trypan blue.[88] Similar but more quantitative results are obtained when the LDH content (LDH kit, Statzyme, Worthington) of medium is tested before and after 90 min of incubation at 37°. Total protein synthesis, as measured by incorporation of [^{14}C]amino acids into perchloric acid precipitable proteins,[89] remains constant over 24 hr.

Methods of Incubation and Preparation of Samples for Analysis. For short-term experiments luteal cells are incubated in 12 × 75 glass tubes in medium B at 10^5 to 10^6 cells/ml/tube. The incubation is terminated by immersion of the tubes in a boiling water bath for 10 min. Samples are then frozen until analysis by radioimmunoassay for cAMP[90] or progesterone.[91] For hormone binding experiments 2 ml of ice cold medium B is added, the cells are centrifuged (1000 *g*, 15 min), the supernatant aspirated, and the radioactivity in the cell pellet counted.

For longer term experiments, cells are cultured for up to 48 hr in disposable plastic tissue culture trays (CoStar) with 10^6 cells/well/0.5 ml in medium B containing 50 IU/ml myocostatin, 100 IU/ml penicillin–streptomycin, and 10% fetal calf serum (GIBCO). The cells are cultured on individual nitrocellulose filters (Millipore, HAWG 01300) soaked in medium 2 for 16 hr at 4°.[70] The presoaked filters are held in place by narrow plastic rings. For measurement of secreted cAMP, the medium is transferred to test tubes with 1 m*M* theophylline and frozen until they are analyzed. To measure intracellular cAMP, the filter with attached cells is placed in 1 ml of medium B containing 1 m*M* theophylline and heated in a boiling water bath for 10 min. For measurement of [^{125}I]hCG binding, [^{125}I]hCG of high specific activity[92] is incubated with cells on filters and the medium removed, the filters rinsed, and placed in test tubes and the radioactivity counted.

[88] H. J. Phillips, *in* "Tissue Culture: Methods and Applications" (P. F. Kruse and M. K. Patterson, eds.), p. 406. Academic Press, New York, 1973.

[89] R. J. Mans and E. D. Novelli, *Biochem. Biophys. Res. Commun.* **3,** 540 (1960).

[90] A. L. Steiner, *in* "Methods of Hormone Radioimmunoassay" (B. M. Jaffee and H. R. Behrman, eds.), p. 3. Academic Press, New York, 1979.

[91] G. P. Orczyk, M. Hichens, G. Arth, and H. R. Behrman, *in* "Methods of Hormone Radioimmunoassay" (B. M. Jaffee and H. R. Behrman, eds.), p. 701. Academic Press, New York, 1979.

[92] M. Hichens, D. L. Grinwich, and H. R. Behrman, *Prostaglandins* **7,** 449 (1979).

In experiments where adenosine triphosphate (ATP) production was measured, incubations are carried out in 0.5 ml of medium B with 1–3 × 10^5 cells/tube. Reactions are stopped by the addition of 0.1 ml of 50% trichloroacetic acid (TCA) and samples are stored at 4° until they are analyzed. The TCA treated cells are extracted four times with 3 ml of diethyl ether, the pH adjusted to 7 with 0.02 *N* NaOH, and ATP assayed by the luciferin-luciferase assay (Sigma, L0663).[93]

For ultrastructural analysis[14] the cells are rinsed gently with medium B and after centrifugation (100 *g*, 5 min), 1 ml of 2% glutaraldehyde, in 0.067 *M* cacodylate buffer at pH 7.4 containing 0.5% tannic acid (350 mOs*M*) is added to the pellet for 20–30 min at 23°. With a large bore plastic pipet the cell pellet is transferred to a 1.5-ml microfuge tube, spun briefly (>3 sec) in a Beckman microfuge B and rinsed with 0.1 *M* cacodylate buffer containing sucrose (310 mOs*M*). One-half milliliter of 2% OsO_4 is added for 1 hr at 22°. Buffer containing sucrose is then added and cells are centrifuged (<3 sec). The pellet is rinsed again. During this procedure the cells are not handled except to "lift" the pellet with a small flat wooden spatula. The tips of each microfuge tube (5 mm length) are cut with a sharp clean razor and the pellet scooped onto filter paper tied over the end of a polyethylene test tube (1.0–1.5 cm opening) open at both ends. The pellet is covered with a second piece of filter paper, which is tied with surgical thread. The tubes are carried through routine dehydration (5 min each in an ethanol series, 2 hr in 50 : 50 propylene oxide : resin, overnight with two changes in 100% resin) and embedded without handling the cells. After infiltration, the upper layer of filter paper is removed and the cell pellet lifted (scooped) into a drop of resin in the bottom of an embedding capsule which is then filled with resin. For Epon 812 (Fullam) or Embed-812 (Polysciences), capsules were polymerized overnight at 60°. Sections were examined in a Hitachi 12 electron microscope.

Applications of Dispersed Luteal Cells

The dispersed rat luteal cell preparation contains greater than 80% luteal cells with the remainder endothelial and blood cells. Based on light microscopic (Fig. 1) and electron microscopic (Fig. 2) observation there are two sizes of luteal cells. Their average diameters are about 15 and 25 μm. Both types of cells contain abundant smooth endoplasmic reticulum, mitochondria, and lipid droplets similar to cells *in vivo*.[27] The larger cells appear to contain less lipid droplets, more smooth endoplasmic reticulum, and relatively little smooth surface area (Figs. 4 and 5).

[93] T. J. Brennan, R. Ohkawa, S. D. Gore, and H. R. Behrman, *Endocrinology* **112,** 449 (1983).

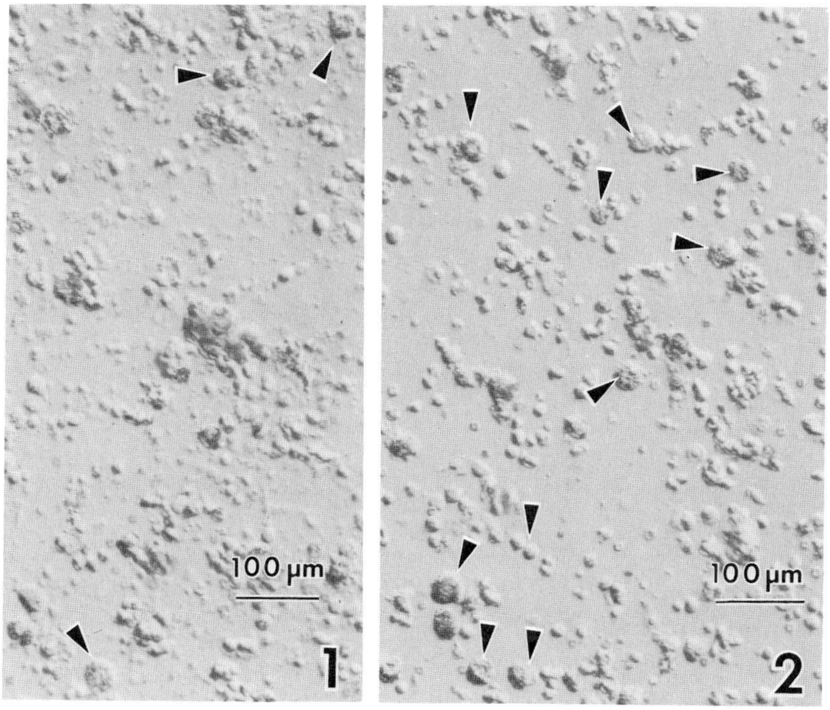

FIG. 1. Isolated rat luteal cells after dissociation of corpora lutea. This preparation contains about 60% luteal cells (arrows) and debris which is evident as particulate matter. Nomarski optics. ×120.

FIG. 2. The same preparation of isolated rat luteal cells after enrichment with a discontinuous Percoll density gradient. Substantial membrane debris has been removed as well as large clumps of cells and some blood cells. This preparation contains about 80% luteal cells (arrows). Nomarski optics. ×120.

Rat luteal cells bind [^{125}I]hCG similar to isolated membranes[92] with a K_D of $0.5 \times 10^{10}\ M^{-1}$.[94] They respond to LH with a dose-dependent increase in cAMP and progesterone production.[13,59] Responsiveness to LH is inversely related to the age of the corpus luteum.[20,95]

LH receptors, localized with a ferritin–LH conjugate, are distributed at irregular intervals along the luteal cell surface. Both specific internalization of bound ferritin–LH by small vesicles, and nonspecific bulk uptake were observed and are rapid processes.[14] Binding and microaggregation of receptors appear to be related to ferritin–LH concentration.[96] LH

[94] J. L. Luborsky, L. J. Dorflinger, K. Wright, and H. R. Behrman, *Endocrinology* **115,** in press (1984).

[95] H. R. Behrman, A. K. Hall, S. L. Preston, and S. D. Gore, *Endocrinology* **110,** 38 (1982).

[96] J. L. Luborsky and H. R. Behrman, *Endocrinology* **115,** in press (1984).

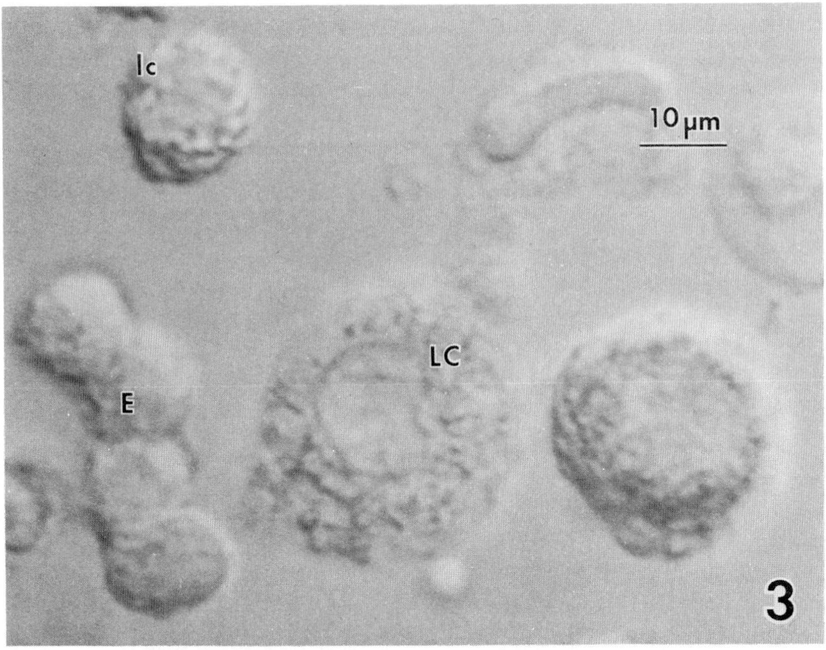

FIG. 3. A higher magnification view of the luteal cells shown in Fig. 2, to show larger luteal cells (LC), smaller luteal cells (lc), and probable endothelial cells (E) which are also present. Luteal cells typically have a more granular cytoplasm than endothelial cells. Nomarski optics. ×1200.

receptor aggregation decreased in the presence of $PGF_{2\alpha}$ and increased in the presence of $(Bu)_2cAMP$[96] suggesting that LH receptor aggregation is also related to cell function.

With this preparation of luteal cells we showed a direct action of $PGF_{2\alpha}$ on LH stimulated cAMP and progesterone production. $PGF_{2\alpha}$ rapidly inhibits the acute action of LH by inhibition of cAMP production.[59,62] Luteal cells have specific receptors for $PGF_{2\alpha}$.[97,98] LHRH and D-Trp^6 LHRH also inhibit LH stimulated cAMP and progesterone production.[70] The rapid inhibition by $PGF_{2\alpha}$ and LHRH was not due to a decrease in LH binding. However, following the initial rapid inhibition, $PGF_{2\alpha}$ was shown to reduce the subsequent equilibrium binding level of $[^{125}I]$hCG (i.e., after 2 to 3 hr) by about 10%. This decrease was not due to a change in the initial association or dissociation rate but resulted in a decreased binding capacity for LH.[94] Upon initial binding, LH increases the number of

[97] J. L. Luborsky, K. Wright, and H. R. Behrman, *Adv. Exp. Biol. Med.* **112,** 633 (1979).
[98] K. Wright, J. L. Luborsky, and H. R. Behrman, *Mol. Cell. Endocrinol.* **13,** 25 (1979).

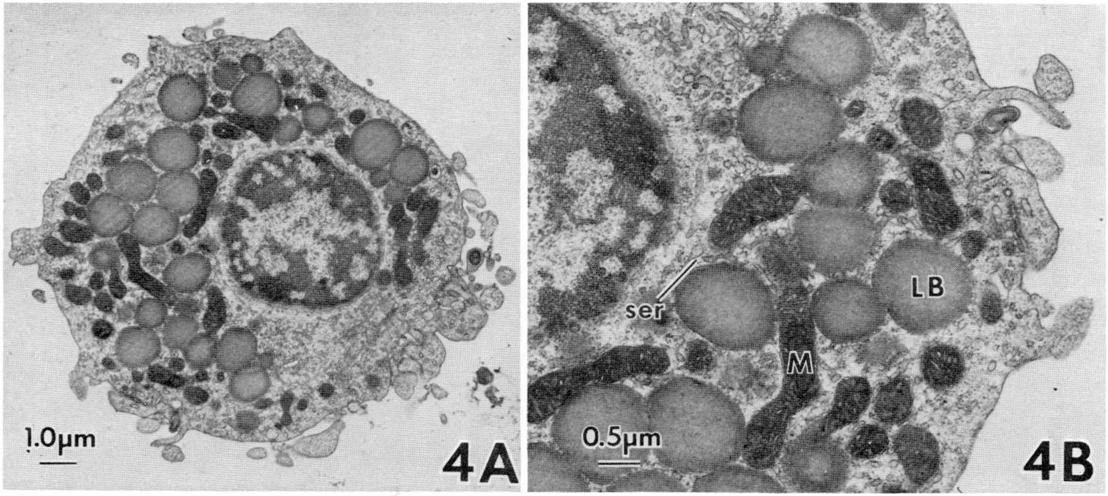

FIG. 4. Electron micrograph of a small isolated rat luteal cell. The membrane surface is usually smoother than that of larger cells (A, ×6000). The cytoplasm of smaller (B, ×13,000) and larger luteal cells is similar but smaller cells contain less mitochondria (M) and smooth endoplasmic reticulum (ser) and variable quantities of lipid bodies (LB).

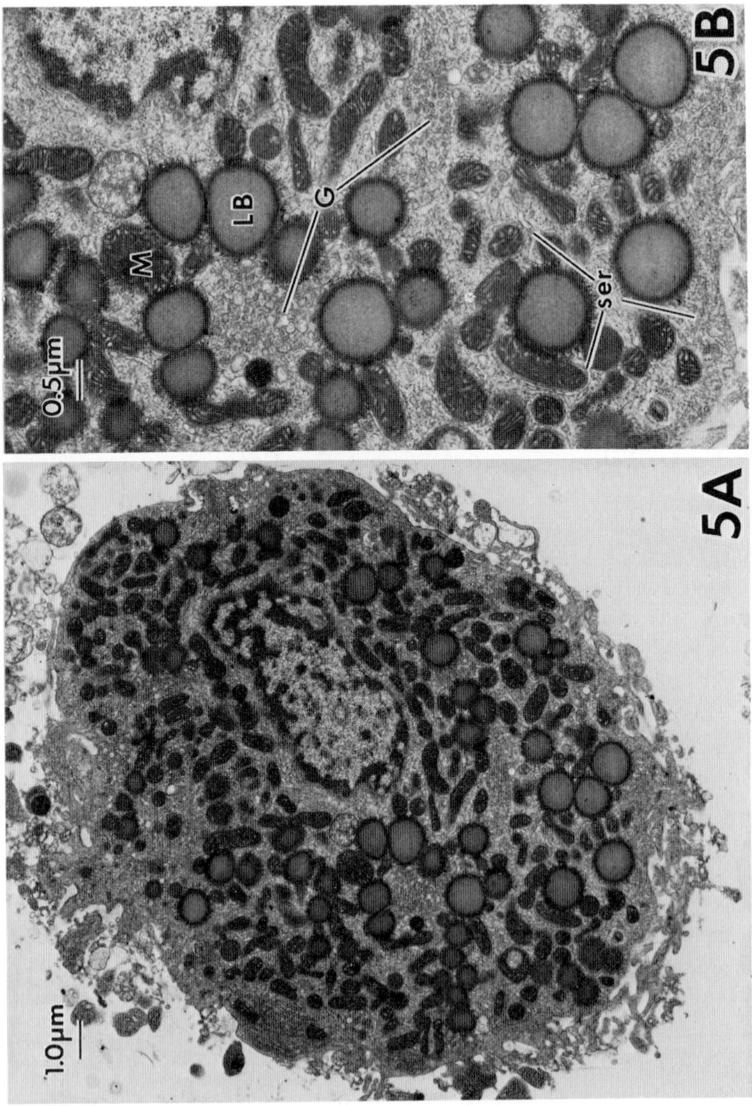

FIG. 5. Electron micrograph of a large isolated rat luteal cell (A, ×6000). The cytoplasm contains abundant smooth endoplasmic reticulum (ser) and mitochondria (M), scattered Golgi elements (G), and lipid bodies (LB) (B, ×12,500).

available receptors by a similar small increment (about 10%) and this is blocked by $PGF_{2\alpha}$.[94] Thus, inhibition of LH action by $PGF_{2\alpha}$ occurs at the level of the coupling mechanism between the LH receptor and adenylate cyclase. In long-term cultures, LH-dependent increases in cAMP and progesterone production are inhibited by a prior pulse (24 hr earlier) of LH, $PGF_{2\alpha}$, LHRH, or D-Trp^6 LHRH. An associated loss of LH receptors, similar to that seen *in vivo*, was not observed.[70] In this same study, untreated cells exhibited an increased (100×) sensitivity to LH at 24 hr.

We also reported with luteal cells the first evidence that adenosine amplified the action of a polypeptide hormone.[20,99] In rat luteal cells, adenosine amplifies LH action partly by an intracellular action (about 80%) and partly by interaction with a cell surface catalytic site.[99] Leydig cells do not respond to adenosine with an amplification of LH.[99] Adenosine also amplifies FSH action in human and rat granulosa cells and LH action in human luteal cells.[20] In addition, adenosine attenuates the inhibition of LH by $PGF_{2\alpha}$, but this action is not associated with a change in LH receptor binding or prostaglandin synthesis.[95] Adenosine is less active in rat and human luteal cells from older corpora lutea.[20,95]

The basis for this purine amplification of LH action in luteal cells appears to be due to a rapid increase in ATP levels which follows a similar time course to the increase in cAMP production in response to LH.[93] Neither LH nor $PGF_{2\alpha}$, however, changes ATP levels. Thus, the ability of the cell to produce cAMP appears to be dependent on ATP levels and adenosine increases this capacity.

The mechanism of action of antigonadotropins such as $PGF_{2\alpha}$ and LHRH was further investigated with respect to their possible action on transmembrane ion flux cations. Ouabain and monensin, inhibitors of Na^+ flux, inhibited LH-dependent increases in cAMP and progesterone production in intact cells but did not inhibit cAMP production in isolated membranes.[100] Prior treatment with LH or removal of extracellular Ca^{2+} prevented subsequent inhibition by ouabain and monensin. Thus, Ca^{2+} appeared to regulate inhibition of LH action. This was confirmed in another study with $PGF_{2\alpha}$ and LHRH.[42] LH-stimulated cAMP production is enhanced in the absence of extracellular Ca^{2+}, the calcium ionophore A23187 inhibited LH stimulated cAMP, and LH-stimulated adenylate cyclase is directly inhibited by Ca^{2+} in isolated membranes. Verapamil, which blocks Ca^{2+} channels, did not prevent inhibition of LH action by $PGF_{2\alpha}$ or LHRH. Thus acute increases in intracellular Ca^{2+}, rather than an influx of extracellular Ca^{2+}, appear to inhibit activation of adenylate

[99] A. K. Hall, S. L. Preston, and H. R. Behrman, *J. Biol. Chem.* **256,** 10390 (1981).
[100] S. D. Gore and H. R. Behrman, *Endocrinology* **114,** 2020 (1984).

cyclase. This suggests that an increase in intracellular Ca^{2+}, independent of an extracellular source, mediates the action of $PGF_{2\alpha}$ and LHRH.

In conclusion, we have used isolated luteal cells to show a relationship between LH receptor topography and cell function, $PGF_{2\alpha}$ and LHRH inhibition of LH action, activation of adenylate cyclase by LH is regulated by the intracellular Ca^{2+} concentration, and that adenosine amplifies LH and FSH action by an increase in ATP levels.

Acknowledgments

Supported by NIH Grants HD 10718 (HRB) and HD 14098 (JLL).

[26] Culture and Characteristics of Hormone-Responsive Neuroblastoma × Glioma Hybrid Cells

By BERND HAMPRECHT, THOMAS GLASER, GEORG REISER, ERNST BAYER, and FRIEDRICH PROPST

Due to the complexity of the mammalian nervous system, results from biochemical or pharmacological experiments with pieces or homogenates of nervous tissue are difficult to interpret. Especially it is hard to assign a certain effect observed to a certain cell type. The most desirable situation would be to have at one's disposal the numerous cell types as homogenous cell populations for studying their individual differentiated functions and their mechanisms of intercellular communication. Only recently a few cases have been reported of apparently relatively homogeneous populations of certain cell types from nervous tissue. Therefore, the hope to be able to study molecular mechanisms of nervous tissue functions has been resting completely on model systems derived from tumors of the nervous system.

Initially, Sato's laboratory developed clonal rat glioma[1] and mouse neuroblastoma[2] cell lines that still expressed differentiated functions known to occur in glial cells and neurons, respectively. Other groups

[1] P. Benda, J. Lightbody, G. Sato, L. Levine, and W. Sweet, *Science* **161,** 370 (1968).
[2] G. Augusti-Tocco and G. Sato, *Proc. Natl. Acad. Sci. U.S.A.* **64,** 311 (1969).

ISBN 0-12-182009-2

expanded the list of neural tumor cell lines considerably.[3–5] The methodology for handling tumor cell lines of neuronal character has been described.[6] By somatic cell hybridization several hybrid cell lines have been generated[7–9] from mouse neuroblastoma and rat glioma cells some of which[8–11] display a long list of neuronal properties and in which these differentiated functions are well expressed. Therefore, these cells are especially suited for studying such functions among which are susceptibilities to peptide hormones.

Generation and Characteristics of Hybrid Lines

Neuroblastoma × Glioma Hybrid Cell Lines 108CC5 and 108CC15

Generation. Neuroblastoma × glioma hybrid cells were generated[8–10] by cell fusion of the 6-thioguanine-resistant clonal mouse neuroblastoma cells N18TG2 (ref. 12) and the bromodeoxyuridine-resistant rat glioma cells C6-BU-1 (ref. 7), selection and cloning. For this purpose, 10^4 N18TG2 and 2×10^6 C6-BU-1 cells per plastic Petri dish 60 mm in diameter were cocultured for 1 day before they were fused[7] with the aid of inactivated Sendai virus. Beginning 3 days after fusion, cells were grown in selective HAT medium[13] for 18 days before hybrid cell colonies were picked with the aid of steel cloning cylinders. The surviving hybrid cell clones 108CC5 and 108CC15 originated from different culture dishes. Unfortunately, these clonal cell lines have been renamed.[14] Thus, for clones 108CC5 and 108CC15 also the designations NG108-5 and NG108-15, respectively, are being used in the literature.

Production of Hybrid Cell Tumors. The hybrid cells grow in groups and overgrow easily into multiple cell layers or clusters long before confluency has been reached. This is not surprising, since the hybrids are derived from tumor cells and can themselves form tumors in athymic

[3] T. Amano, E. Richelson, and M. Nirenberg, *Proc. Natl. Acad. Sci. U.S.A.* **69,** 258 (1972).
[4] D. Schubert, S. Heinemann, W. Carlisle, H. Tarikas, B. Kimes, J. Patrick, J. H. Steinbach, W. Culp, and B. L. Brandt, *Nature (London)* **249,** 224 (1974).
[5] L. A. Greene and A. S. Tischler, *Proc. Natl. Acad. Sci. U.S.A.* **73,** 2424 (1976).
[6] D. Schubert and W. Carlisle, this series, Vol. 58, p. 584.
[7] T. Amano, B. Hamprecht, and W. Kemper, *Exp. Cell Res.* **85,** 399 (1974).
[8] B. Hamprecht, *Colloq. Ges. Biol. Chem.* **25,** 391 (1974).
[9] B. Hamprecht, *Angew. Chem., Int. Ed. Engl.* **15,** 194 (1976).
[10] B. Hamprecht, *Int. Rev. Cytol.* **49,** 99 (1977).
[11] R. Heumann, M. Öcalan, V. Kachel, and B. Hamprecht, *Proc. Natl. Acad. Sci. U.S.A.* **76,** 4674 (1979).
[12] J. Minna, D. Glazer, and M. Nirenberg, *Nature (London) New Biol.* **235,** 225 (1972).
[13] J. W. Littlefield, *Science* **145,** 709 (1964).
[14] W. A. Klee and M. Nirenberg, *Proc. Natl. Acad. Sci. U.S.A.* **71,** 3474 (1974).

nude mice.[15] For the production of hybrid cell tumors, hybrid cells, for the last 3 days grown in culture medium consisting of 90% Dulbecco's modified Eagle's medium (DMEM) and 10% fetal calf serum (FCS) and supplemented with 0.1 m*M* hypoxanthine and 16 μM thymidine, are detached from their culture vessel by mild trypsinization (see below). A suspension of 1 million viable cells in 0.2 ml DMEM is injected subcutaneously on the left and right side of 3- to 4-week-old nude BALB/c mice. With a take of 100%, solid tumors weighing 1–4 g develop in the course of 2 to 4 weeks. In an isotonic medium tumors can easily be disintegrated mechanically to form cell suspensions. The tumor cells reintroduced into culture display all the differentiated functions tested that are known to occur in the hybrid cells used for inoculation.[15] Thus, growth in nude mice is a relatively simple and cheap means of mass culturing hybrid cells.

Changes of Properties of Hybrid Cell Line 108CC5 during Continuous Culture. The parental and hybrid cell lines are highly aneuploid.[10] While the modal chromosome numbers are 81 and 49 for the parental lines neuroblastoma N18TG2 and glioma C6-BU-1, respectively, they are 254 and 160 for the hybrids 108CC5 and 108CC15, respectively.[10] With continuous culture, measured by the number of subcultivations (passage number), the hybrid cells lose chromosomes. Correspondingly their cellular content of DNA and protein and their volume (7000 and 5000 μm^3 at passage numbers 14 and 53, respectively) decrease in a parallel fashion.[16] At a passage number of about 60 an apparent plateau with probably no further change of these gross cell parameters is reached. All the differentiated functions tested were still present at high passage numbers (60), although the cells became more flattened and thus less refractile than at low passage numbers (e.g., 20). Nevertheless, to be sure to work under reproducible conditions, it is recommended to strictly count the passage numbers, create a large stock of frozen cells of low passage numbers, and use the hybrid cells in a narrow bracket of passage numbers. In our laboratory, for example, we may start a culture at passage number 12 and discard the cells at passage number 20.

Neuronal Properties of the Hybrid Cell Lines 108CC5 and 108CC15. Every characteristic generally ascribed to neurons has been observed with the hybrid cells (Table I).[17–28] Therefore, the hybrid cells have been used for studying these properties in detail (for reviews, see refs. 10 and 29). They have proved most useful in the elucidation of the mode of action of various neurohormones, especially the opioids (Table II).[30–47]

[15] R. Heumann, D. Stavrou, G. Reiser, M. Öcalan, and B. Hamprecht, *Eur. J. Cancer* **13,** 1417 (1977).

[16] R. Heumann, G. Valet, D. Maison, J. Kemper, G. Reiser, and B. Hamprecht, *J. Cell Sci.* **27,** 141 (1977).

[17] B. Hamprecht, J. Traber, and F. Lamprecht, *FEBS Lett.* **42,** 221 (1974).

TABLE I
NEURONAL PROPERTIES OF NEUROBLASTOMA × GLIOMA HYBRID CELLS 108CC5 AND/OR 108CC15

Neuronal property	References
Extension of long processes	8–10, 17, 18
Clear and dense core vesicles	18
Excitable membranes (inward current of action potentials carried by Na^+ or Ca^{2+})	8–10, 19
Formation of functional synpases	20
Neurotransmitter enzymes	
Choline acetyltransferase (EC 2.3.1.6)	8–10
Dopamine-β-hydroxylase (EC 1.14.2.1)	17
Synthesis of neurohormones	
Acetylcholine	21, 22
Leu- and Met-enkephalin	23
Dynorphin-(1-8),α-neoendorphin	24
β-Endorphin	24
Vasoactive intestinal peptide	25
Angiotensin	26
Hydra head activator-like activity	27
Uptake systems for	
Catecholamines	22
Taurine	28
Depolarization-induced Ca^{2+}-dependent release of acetylcholine	21
Receptors for neurohormones	See Table II

[18] M. Daniels and B. Hamprecht, *J. Cell Biol.* **63,** 691 (1974).

[19] G. Reiser, R. Heumann, W. Kemper, E. Lautenschlager, and B. Hamprecht, *Brain Res.* **130,** 495 (1977).

[20] P. Nelson, C. Christian, and M. Nirenberg, *Proc. Natl. Acad. Sci. U.S.A.* **73,** 123 (1976).

[21] R. McGee, P. Simpson, C. Christian, M. Mata, P. Nelson, and M. Nirenberg, *Proc. Natl. Acad. Sci. U.S.A.* **75,** 1314 (1978).

[22] K. Kürzinger, Ph.D. Thesis, University of Munich (1978).

[23] T. Glaser, K. Hübner, and B. Hamprecht, *J. Neurochem.* **39,** 59 (1982).

[24] B. R. Seizinger, T. Glaser, V. Höllt, and B. Hamprecht, unpublished observations (1983).

[25] T. Glaser, J. Besson, G. Rosselin, and B. Hamprecht, *Eur. J. Pharmacol.* **90,** 121 (1983).

[26] M. C. Fishman, E. A. Zimmerman, and E. E. Slater, *Science* **214,** 921 (1981).

[27] C. H. Schaller, R. Heumann, and B. Hamprecht, unpublished observations (1977).

[28] K. Kürzinger and B. Hamprecht, *J. Neurochem.* **37,** 956 (1981).

[29] Y. Kimhi, *in* "Excitable Cells in Tissue Culture" (P. G. Nelson and M. Lieberman, eds.), p. 173. Plenum, New York, 1981.

[30] B. Hamprecht and J. Schultz, *Hoppe Seyler's Z. Physiol. Chem.* **354,** 1633 (1973).

[31] S. K. Sharma, W. A. Klee, and M. Nirenberg, *Proc. Natl. Acad. Sci. U.S.A.* **72,** 3092 (1975).

[32] J. Traber, Ph.D. Thesis, University of Munich (1976).

[33] F. Propst, L. Moroder, E. Wünsch, and B. Hamprecht, *J. Neurochem.* **32,** 1495 (1979).

TABLE II
HORMONAL RESPONSES OF NEUROBLASTOMA × GLIOMA HYBRID CELLS 108CC15

Nature of response	Hormone or analog	References
Stimulation of formation of cyclic AMP	Prostaglandin E_1	30
	Adenosine	31, 32
	Glucagon	33
	Secretin	33
Inhibition of formation of cyclic AMP	Acetylcholine (muscarinic receptor)	34
	Noradrenaline (α-adrenergic receptor)	35
	Opioids	
	Morphine	36, 37
	Enkephalins	38, 39
	Endorphins	40
	Dermorphins	41
	Somatostatin	42
Stimulation of formation of cyclic GMP	Bradykinin	43
	Angiotensin II	43
Depolarization of plasma membrane	Acetylcholine	8
	Catecholamines	35, 44
Inhibition of dopamine-induced depolarization	Morphine	45
Hyperpolarization of plasma membrane	Bradykinin	46
	Angiotensin II	46a
Increase of sodium permeability of plasma membrane	Substance P	47

[34] J. Traber, K. Fischer, C. Buchen, and B. Hamprecht, *Nature* (*London*) **255,** 558 (1975).

[35] J. Traber, G. Reiser, K. Fischer, and B. Hamprecht, *FEBS Lett.* **52,** 327 (1975).

[36] J. Traber, K. Fischer, S. Latzin, and B. Hamprecht, *Nature* (*London*) **253,** 120 (1975).

[37] S. K. Sharma, M. Nirenberg, and W. A. Klee, *Proc. Natl. Acad. Sci. U.S.A.* **72,** 590 (1975).

[38] M. Brandt, R. J. Gullis, K. Fischer, C. Buchen, B. Hamprecht, L. Moroder, and E. Wünsch, *Nature* (*London*) **262,** 311 (1976).

[39] W. A. Klee and M. Nirenberg, *Nature* (*London*) **263,** 609 (1976).

[40] M. Brandt, C. Buchen, and B. Hamprecht, *FEBS Lett.* **80,** 251 (1977).

[41] T. Glaser, K. Hübner, R. de Castiglione, and B. Hamprecht, *J. Neurochem.* **37,** 1613 (1981).

[42] J. Traber, T. Glaser, M. Brandt, W. Klebensberger, and B. Hamprecht, *FEBS Lett.* **81,** 351 (1977).

[43] G. Reiser, U. Walter, and B. Hamprecht, *Brain Res.* **290** (1984).

[44] P. R. Myers and D. R. Livengood, *Nature* (*London*) **255,** 235 (1975).

[45] P. R. Myers, D. R. Livengood, and W. Shain, *Nature* (*London*) **257,** 238 (1975).

[46] G. Reiser and B. Hamprecht, *Brain Res.* **239,** 191 (1982).

[46a] H. Höpp, G. Reuter, G. Reiser, and B. Hamprecht, unpublished observations (1984).

[47] G. Reiser, K. Folkers, and B. Hamprecht, *Regul. Pept.* **5,** 85 (1982).

(Neuroblastoma × Glioma) × Neuroblastoma Hybrid–Hybrid Cell Line NH15-CA2

Generation. This hybrid–hybrid was obtained by fusion of hybrid line 108CC15 and the bromodeoxyuridine-resistant mouse neuroblastoma line N115-BU-8.[11] Besides being twice as voluminous as the cholinergic hybrid line 108CC15, NH15-CA2 displays as its most characteristic property the production of both acetylcholine and noradrenaline,[11] i.e., it is simultaneously cholinergic and adrenergic. Otherwise these cells appear to share most neuronal properties with hybrid lines 108CC5 and 108CC15: They can extend long neurite-like processes, they are even more excitable than the parental hybrid, and may spontaneously fire trains of action potentials. Another important feature of the hybrid–hybrid cells is the strong increase in specific activity of choline acetyltransferase elicited by prolonged exposure to media conditioned by glioma cells or by primary cultures of heart cells or glial cells.[48] Therefore, NH15-CA2 cells constitute a valuable indicator system for the purification of the factors contained in the conditioned media or elsewhere.

Culture and Handling of Hybrid and Hybrid–Hybrid Cells

The cell lines 108CC5, 108CC15, and NH15-CA2 are treated identically.

Culture Medium

a. Composition
   1. Dulbecco's modified Eagle's medium (DMEM),[49–51] 900 ml
   2. Fetal calf serum, 100 ml
   3. 10 m*M* hypoxanthine in approx. 0.1 m*M* NaOH, 10 ml
   4. 0.5 m*M* aminopterin in water, 2 ml
   5. 16 m*M* thymidine in water, 1 ml

   Solutions 3 to 5 are added together as a premixed stock solution that had been sterilized by filtration before having been stored frozen at −18°.
b. Glucose ($C_6H_{12}O_6$) concentration: 4.5 g/liter
c. Osmolarity: 325–340 mosmol/liter at 22°
d. pH: 7.3 at 37°, if equilibrated with the growth atmosphere (90% air, 10% CO_2, ~100% relative humidity)

[48] R. Heumann, M. Öcalan, and B. Hamprecht, *FEBS Lett.* **107,** 37 (1979).
[49] R. Dulbecco and G. Freeman, *Virology* **8,** 396 (1959).
[50] H. J. Morton, *In Vitro* **6,** 89 (1970).
[51] L. P. Prutzky and R. W. Pumper, *In Vitro* **9,** 486 (1974).

e. Tests for bacterial[52] and mycoplasmal (see below) contamination are carried out.

Start of Cell Culture

1. Thaw cell suspension (1 ml) stored frozen (see below) in a water bath (37°) as rapidly as possible.

2. Transfer cell suspension into a polystyrene culture flask (~75 cm^2 growth area) containing 20 ml of growth medium.

3. After attachment of most of the cells (4–12 hr later) replace medium by fresh culture medium (10–20 ml, according to cell density).

Subcultivation

Solutions

1. Trypsin stock solution 0.5% (w/v): Dissolve 500 mg (~35,000 units) of trypsin (crystallized twice; Boehringer, Mannheim) in a mixture of 5 ml 20× Puck's medium D1 (160 g NaCl, 8 g KCl, 0.6 g KH_2PO_4, 0.6 g $Na_2HPO_4 \cdot 2H_2O$ dissolved in water to 1000 ml) and 95 ml of water.

2. Trypsin solution[53] 0.005% (w/v). Mix trypsin stock solution, 10 ml, Puck's medium D1 20× (see above), 50 ml, glucose/sucrose solution 20× (20 g glucose · H_2O, 400 g sucrose dissolved in water to 1000 ml), 50 ml, and water to 1000 ml. Osmolarity ~330 mosmol/liter. Filter sterilly and store frozen in portions of 50 ml. Test samples for sterility (bacteria[52] and mycoplasma, see below).

Protocol

1. Suck off medium from the cell layer in the 75-cm^2 flask.

2. Wash cell lawn with 5 ml of an ice-cold isotonic solution of 0.005% trypsin.

3. Detach cells by exposure to 3–5 ml of the trypsin solution in an incubator of 37° for 2–5 min and subsequent rocking of the flask.

4. Stop action of trypsin by the addition of 10 ml of growth medium.

5. Transfer cell suspension to 50 ml centrifuge tube and spin down in a refrigerated low speed centrifuge at approximately 500 *g* for several minutes.

6. Resuspend cells in fresh growth medium of known volume.

7. Count cells in a hemocytometer and determine viability by assessing exclusion of nigrosin.[54,55]

[52] G. D. McGarrity, this series, Vol. 58, p. 18.

[53] A. J. Blume, F. Gilbert, S. Wilson, J. Farber, R. Rosenberg, and M. Nirenberg, *Proc. Natl. Acad. Sci. U.S.A.* **67,** 786 (1970).

[54] H. J. Phillips and J. F. Terryberry, *Exp. Cell Res.* **13,** 341 (1957).

[55] J. P. Kaltenbach, M. H. Kaltenbach, and W. B. Lyons, *Exp. Cell Res.* **15,** 112 (1958).

8. With growth medium dilute cell suspension to the cell concentration required for inoculation of new culture vessels.

9. Evenly distribute cell suspension into new culture vessels (flasks, dishes or wells in multiple-well trays).

10. Culture cells in a cell incubator at 37°, in an atmosphere of 90% air, 10% CO_2 and approximately 100% relative humidity.

11. *Important note:* For energy production these cells rely to a large extent on anaerobic glycolysis. They rapidly produce lactic acid from glucose. In cultures acidified this way, cells detach in clumps and some form multinuclear giant cells. On their surfaces protrusions and large vacuoles may appear and the viability drops rapidly. Therefore, it is mandatory that the *medium never becomes acidic* enough to generate a yellowish color of the phenol red used as a pH indicator in the medium. This requires frequent inspection of the cultures because the rate of pH change progresses rapidly. It is of limited use to increase the volume of medium with the intention to prolong the period the cells can be left without attention. With the increased fluid layer the cells have less access to oxygen, more strongly depend on anaerobic glycolysis, and thus more quickly acidify the medium. Cultures that have become acidic for a prolonged period should be discarded, since it takes many passages for them to recover from the maltreatment.

Storage and Shipment

Hybrid cells harvested as for subculturing are resuspended in freezing medium (92.5 vol DMEM and 7.5 vol dimethyl sulfoxide) that was sterilized by filtration through a 0.22-μm bacterial filter. Cells are frozen in 1-ml portions in polypropylene screw cap tubes in a temperature-controlled fashion.[56] Cells are stored in the vapor phase of a liquid nitrogen storage tank.

Shipping cells in culture flasks completely filled with culture medium has consistently met with little success. Therefore, cells have to be shipped frozen in dry ice.

Screening for Mycoplasma

The procedure is a modification of the method described by Chen[57] using the Hoechst 33258 Bisbenzimidazol dye.

Solution 1: Phosphate-buffered saline (pH 7.2; 340 mosmol/liter)
Potassium phosphate, 10 m*M*
NaCl, 150 m*M*

[56] L. L. Coriell, this series, Vol. 58, p. 29.
[57] T. R. Chen, *Exp. Cell Res.* **104,** 255 (1977).

Solution 2: Fixative: glacial acetic acid, 1 vol, and methanol 3 vol Prepare freshly, keep in ice

Solution 3: Solution of fluorescence dye: 1 μM Hoechst 33258; 2-[2-(4-hydroxyphenyl)-6-benzimidazolyl]-6-(1-methyl-4-piperazyl) benzimidazole, Calbiochem catalog No. 382061, in water

Protocol

1. Fix with the aid of sterile silicon greese to the bottom of plastic Petric dishes (diameter 50 mm) 3 coverslips (18 × 18 mm).
2. Seed into the Petri dishes 10,000 C6 glioma cells[1] suspended in 4.5 ml of a medium composed of 90% DMEM and 10% fetal calf serum. (a) Add 0.5 ml of culture medium exposed to cells or 0.5 ml of lysate of cells to be tested for contamination by mycoplasma (e.g., hybrid cells). (b) Alternatively, for negative control, add 0.5 ml of culture medium not previously exposed to cells. (c) Alternatively, for positive control, add 0.5 ml of culture medium previously exposed to a cell culture known to be contaminated by mycoplasma.
3. Culture the glioma cells under the conditions described above for the hybrid cells for 6 days.
4. Suck off culture medium.
5. Wash cells once with 5 ml solution 1.
6. Slowly dip cover slip into ice-cold fixative (solution 2) and keep it there for 10 min.
7. Airdry coverslip.
8. Incubate coverslip at room temperature in solution 3 for 10 min.
9. Wash coverslip twice with phosphate-buffered saline for 5 min.
10. Once briefly wash coverslip with deionized water.
11. Airdry.
12. Mount coverslip, cells down, onto microscope slide by using a saturated solution of sucrose.
13. Inspect preparations by using a fluorescence microscope. In the case of the Leitz Dialux the exciter filter BP349-380 nm, the beam splitter RKP400, the barrier filter LP430, an oil immersion objective 40×, and an ocular of 12.5× were used. In the case of a Zeiss fluorescence microscope, the filter unit I487702 containing the exciter filters UG1 and UG5 (360 nm), the beam splitter FT429, the barrier filter LP418, an oil immersion objective 40×, and an ocular of 10× were used.
14. Mycoplasma are recognized as extranuclear (often even extracellular) fluorescent particles that mostly occur in clusters.

Morphological Differentiation of Hybrid Cells

During prolonged exposure to $N^6,O^{2'}$-dibutyryl adenosine 3′ : 5′-cyclic monophosphate (dibutyryl cyclic AMP) hybrid cells[8–10] and hybrid–hybrid cells extend long neurite-like processes. After treatment with the cyclic AMP analog for 10–12 days, a network of fibers develops in the culture with the fibers emerging from and ending in islets formed by a few to many hybrid cells. Along with the morphological differentiation visible under the phase contrast microscope a more subtle but dramatic differentiation can be observed with the electron microscope.[18] In addition the hybrids and hybrid–hybrids become highly excitable.[8–10,19] In the hybrids,[8–10] but not the hybrid–hybrids,[11] the specific activity of choline acetyltransferase increases strongly by treatment with dibutyryl cyclic AMP.[8–10] The optimal concentration of the analog of cyclic AMP for eliciting morphological differentiation is 1 m*M*. At 0.5 m*M* it is almost without effect. Of the many ways and agents tried[10] as inducers of morphological differentiation in hybrid cells only one other analog of cyclic AMP has proven successful: 8-*p*-chlorophenylthio-cyclic AMP. However, in contrast to dibutyryl cyclic AMP, it does not promote electrical maturation of the hybrid cells.[58]

Preparations of dibutyryl cyclic AMP often contain butyric acid as a contaminant, which may have adverse effects on the differentiation process of these cells.[59] Therefore, it should be removed from the commercial powder by extraction with diethylether. For this purpose several grams of the powder are suspended in ~30–40 ml of ether in a 50 ml polypropylene screw cap centrifuge tube (Falcon Plastics; as used for sterile centrifugation of cultured cells). After centrifugation and discarding of the ether phase the process of extraction is repeated several times, until, after evaporation of the ether the smell of butyric acid has almost vanished.

The quality of serum in the culture medium may be of utmost influence on the differentiation process. In recent years we have found it increasingly difficult to generate morphological differentiation by dibutyryl cyclic AMP if 10% fetal calf serum is present in the culture medium. Previously this had never been a problem at all.[8–10] We find[58] excellent morphological and electrical differentiation in hybrid cells cultured in a medium containing only 1% (v/v) fetal calf serum, but the 5 special ingredients of the N2 medium of Bottenstein and Sato.[60] Below are given the two procedures used in our laboratory for morphological and electrical differentiation of

[58] G. Reiser and B. Hamprecht, *Exp. Cell Res.* **141,** 498 (1982).

[59] B. Hamprecht, unpublished observation.

[60] J. E. Bottenstein and G. H. Sato, *Proc. Natl. Acad. Sci. U.S.A.* **76,** 514 (1979).

the hybrid cells. Cultures prepared this way are well suited for electrophysiological investigations by using microelectrodes inserted into the cells.

The cell density in the culture dish at the time of application of the inducing medium is of the most critical importance for the success of morphological differentiation. If it is too low, most cells will die; if it is too high, morphological differentiation is inhibited.

Protocol for Morphological and "Electrical" Differentiation of Hybrid Cells

Solution 1: 4 m*M* solution of dibutyryl cyclic AMP in DMEM. Dissolve 1.88 g of the nucleotide in 1 liter of DMEM. Filter sterilely and store at +4°.

Solution 2: 10× stock solution of the 5 special ingredients of N2 medium[60]: Dissolve in 1 liter of DMEM: 50 mg insulin, 1 g transferrin, 161 mg putrescine dihydrochloride, 1 ml of 0.3 m*M* aqueous Na_2SeO_3 (dissolve 5.2 mg in 100 ml of water), 1 ml 0.2 μM progesterone in ethanol (dissolve 6.3 mg of progesterone in 100 ml of 96% ethanol). Filter sterilely and store at +4°.

Differentiation Medium A

For 1 liter of this medium mix DMEM, 637 ml, fetal calf serum, 100 ml, solution 1, 250 ml, and HAT premix (see Culture and handling of hybrid and hybrid–hybrid cells), 13 ml.

Differentiation Medium B

For 1 liter of this medium mix DMEM, 627 ml, fetal calf serum, 10 ml, solution 1, 250 ml, solution 2, 100 ml, and HAT premix (see Culture and handling of hybrid and hybrid–hybrid cells), 13 ml.

1. Seed 40,000 viable hybrid cells suspended in 4 ml of culture medium (see Culture and handling of hybrid and hybrid–hybrid cells) into plastic Petri dishes 50 mm in diameter.
2. Incubate in cell incubator (37°; 100% relative humidity; 90% air, 10% CO_2) for 1–2 days.
3. Replace medium by 3 ml of differentiation medium A or B.
4. Three days later replace medium by 5 ml of the same medium, prewarmed to 37°.
5. Subsequently renew medium as soon as it starts getting a yellow touch. If in doubt the decision should always be for renewal of medium. If

the medium ever turns yellowish the odds are that the experiment will fail. Thus, frequent and persistent observation of the cultures is mandatory for the success of the experiment.

6. After a total exposure to dibutyryl cyclic AMP of 8–12 days the cells are optimal for electrophysiological measurements. After shorter exposure the cells are still markedly less excitable; after longer exposure the cultures become too crowded and their excitability tends to decrease again.

Procedures for Testing the Hormonal Regulation of Intracellular Levels of Cyclic AMP in Hybrid Cells

If the hormone the action of which is to be investigated is at hand in sufficient quantities, incubations can be carried out with cells grown in large culture Petri dishes (85 or 50 mm in diameter; procedures A1[61] and A2.[61,62] If a hormonal preparation is available only in very limited amounts the exposure of cell layers to the hormone should be carried out in small wells of multiwell trays (24 wells, diameter of well 16 mm; procedures B1 and B2[61]). The procedures described are given in a paradigmatic way for an exposure of hybrid cells to prostaglandin E_1 (PGE_1) and Leu-enkephalin in plates 50 mm in diameter. All incubations are carried out in triplicate.

Procedures A (for Petri Dishes)

Materials. Before the experimental incubation the following items have to be prepared.

Cell Cultures

1. Three days before the incubation, initiate at least as many replica cultures as needed for the exposure to the hormones and for the various controls, for cell counting, for determination of cellular protein, and for reserves.
2. To this end transfer into each culture dish (50 mm in diameter) a suspension of 2–3 × 10^5 viable hybrid cells in 6 ml of growth medium (see above).
3. Three days later, at the time of the experiment, each culture dish will contain ~1 million viable cells (corresponding to 1 mg of cellular protein), viability 90–96%.

[61] T. Glaser, Ph.D. Thesis, University of Munich (1981).
[62] E. Bayer and B. Hamprecht, unpublished results.

Incubation Medium A. This essentially consists of DMEM buffered with *N*-2-hydroxyethylpiperazine-*N'*-2-ethanesulfonate (HEPES) instead of $NaHCO_3$.

1. Dissolve in 20 liters of water DMEM powder, 2 packages (Gibco), 200 g, D-glucose · H_2O (final concentration 19.4 m*M*), 77 g, HEPES (final concentration 25 m*M*), 119 g, and NaCl, 24 g.
2. Adjust to pH 7.4 (22°) by the addition of ~30–35 ml of 5 *M* NaOH.
3. Measure osmolarity. It should be around 330 mosmol/liter.
4. Store frozen at −20° in 1-liter plastic bottles in quantities of 800 ml.
5. Before experimental incubation place the bottle with incubation medium in a water bath of 37° and keep it there during the incubation procedure.

Anion Exchange Columns

1. Place columns (Pasteur pipets, total length 23 cm; diameter of the wide upper part 6 mm) in a rack. The cylindrical column should be straight without the usual narrowing just below its upper orifice.
2. Attach to the upper orifice of columns conical plastic funnels adapted from the upper parts of Chromaflex plastic columns (Kontes Glass Co., Vineland, N.J.).
3. Plug each column with glass wool.
4. Add a suspension of BioRad AG1 × 8 (200–400 mesh, Cl^- form) to generate columns of 4 cm height.
5. Before use pass 10 ml of 5 *M* HCl through the columns. For columns that have already been used this serves to regenerate the resin.
6. Wash with 20 ml of water.

Water bath. Enough water is filled into water bath (38°) that its surface stands a few millimeters above a metal grid that is placed horizontally in the water bath. The grid serves as a support to the culture dishes which are prevented from tilting and floating during incubation. The water bath stands next to the cell incubator.

Stop watches. Sufficient number of stop watches is needed to monitor the time of incubation of each culture dish.

Incubation with Hormones, Procedure A1

This procedure is used if a lyophilizer is available.

1. Take dish from cell incubator and aspirate growth medium.
2. Wash cell layer once with 3 ml of prewarmed (37°) incubation medium A.
3. Add 3 ml of prewarmed incubation medium A.
4. Add 30 μl of a 10 μM solution of Leu-enkephalin in water (final concentration 0.1 μM).

5. Immediately thereafter add 9 μl of a 0.1 mM solution of prostaglandin E_1 in 96% aqueous ethanol (final concentration of PGE_1 0.3 μM, of ethanol 0.3% v/v). The corresponding control dishes receive the solvents only.

6. Swirl around to evenly distribute the additions in the dish and start stop watch.

7. Place dish on the water bath. Cover it with its lid in an inverted position and place a small weight on it to press the dish down onto the steel grid support.

8. Incubate at 37° for 10 min.

9. Completely remove incubation fluid from the dish by aspiration if the cyclic AMP released by the cells is not determined. During the brief incubation period the release of cyclic AMP into the medium is constantly around 10% of total cellular cyclic AMP and therefore neglected. However, if the cyclic AMP released by the cells is also of interest, the culture medium is removed with a pipettor and transferred directly onto an anion exchange column (see below).

10. Add 2 ml of 96% ethanol and 50 μl of an aqueous solution of [^{3}H]cyclic AMP (3000–4000 cpm, specific radioactivity ~30 Ci/mmol).

11. Transfer liquid (cell extract) from dish onto the anion exchange column.

12. Wash with 10 ml of water. Discard the washing fluid emerging from the column.

13. Elute cyclic AMP by 3 ml of 0.1 M aqueous HCl into plastic scintillation minivials (capacity 5 ml).

14. Lyophilize content of scintillation minivials.

15. Dissolve residue in scintillation minivials in 500 to 1000 μl of H_2O.

16. For determination of recovery of cyclic AMP, transfer 250 to 400 μl of this solution onto Whatman GF/B glass fiber filters (2.5 cm in diameter) that are placed horizontally in a well of a polypropylene board that contains multiple cylindrical wells. The upper part of a well is 4 mm in height and has a diameter of 26 mm; the lower part of a well is concentrical to the upper one, is 6 mm in height, and has a diameter of 22 mm.

17. Dry the filter in an oven at 100° for 45–90 min, dependent on the volume of fluid applied.

18. By using forceps immerse dry filter into glass scintillation vial (capacity 25 ml) filled with 5 to 15 ml of scintillation fluid nonmiscible with water and devoid of Triton X-100 (e.g., Rotiszint 11, K. Roth Co., Karlsruhe, F.R.G.) and determine radioactivity in a liquid scintillation spectrometer. The radioactive cyclic AMP does not dissolve in this fluid.

19. With the aid of forceps carefully recover the filter from the scintillation vial and discard it. The scintillation vial is now ready for the next filter to be analyzed for radioactivity. Occasionally the scintillation fluid

in the scintillation vial has to be replenished. This way the same vial can be used up to 50 times or more. From time to time it is necessary to determine that after removal of the filter no residual radioactivity remains in the vial. Should pieces of the filter carrying radioactivity remain in the vial, the scintillation fluid can be reused only after filtration. Normally, after removal of the filter the radioactivity of the vial does not exceed background level. This procedure effectively helps in cutting down the volume of radioactive waste.

Incubation with Hormones, Procedure A2

This procedure is used if a vacuum centrifuge for rapid evaporation of fluids is at hand (e.g., Speed Vac concentrator). Cyclic AMP is assayed without prior purification by anion exchange chromatography.

1. Follow steps 1 to 9 of procedure A1.
2. Add 1 ml of ~100% ethanol.
3. Transfer ethanolic cell extract into polypropylene tube (75 × 12 mm).
4. Wash culture dish with 1 ml of ~100% ethanol and combine washing fluid with cell extract.
5. Evaporate ethanol from the polypropylene tube in the Speed Vac concentrator under moderate heating.
6. Dissolve residue in 500 μl of water and, if necessary, freeze at −20° until the assays of cyclic AMP are carried out.

Assay for Cyclic AMP (subsequent to procedures A1 or A2). The procedure is a modification of the method of Gilman.[63,64] Protein kinase inhibitor is not used in the assay.

Materials. Eppendorf snap cap vials.

Solutions

Solution 1: Solution of [^{3}H]cyclic AMP. Dissolve [^{3}H]cyclic AMP (specific radioactivity ~30 Ci/mmol) in 1 *M* sodium acetate buffer, pH 4.0 at 22° containing 0.1% (w/v) bovine serum albumin, to a concentration of 2 × 10^6 cpm/ml.

Solution 2: Solution of unlabeled cyclic AMP: 500 n*M* in aqueous solution of 0.1% (w/v) bovine serum albumin.

Solution 3: Cyclic AMP binding protein is prepared as described,[64] with the exception that the solution contains 0.1% (w/v) bovine serum albumin.

Solution 4: Potassium phosphate buffer, 20 m*M*, pH 6.0.

[63] A. G. Gilman, *Proc. Natl. Acad. Sci. U.S.A.* **67,** 305 (1970).

[64] A. G. Gilman and F. Murad, this series, Vol. 38, p. 49.

Assay Protocol

1. Pipet 10 μl of solution 1 in all incubation vials.

2. To the vials used for establishing the calibration curve are added 0, 4, 10, 20, 40, 60, and 80 μl of solution 2 in duplicate.

3. To the vials used for analyzing the samples of unknown cyclic AMP content the cyclic AMP extracted from the cells is added in a volume up to 150 μl. The final concentration of cyclic AMP should ideally be such that the value measured will correspond to the medium part of the calibration curve (see below, item 14). Otherwise relatively large errors ensue. Thus, repeating the assay by using properly diluted cell extracts may be necessary in some instances.

4. To all vials enough water is added to make at this stage the total volume 160 μl.

5. Add 40 μl of solution 3 to all vials except that for producing the blank value; in the latter case 40 μl of water is added.

6. Mix gently with Vortex mixer and incubate for at least 2 hr at 4° (e.g., overnight).

7. Dilute reaction mixture by the addition of 800 μl of solution 4.

8. Pass by suction content of plastic vial through a nitrocellulose filter (2.5 cm in diameter, 0.45 μm pore width, Schleicher and Schüll, Dassel, F.R.G.) that has been wetted by solution 4.

9. Wash filter with 10 ml of solution 4.

10. Roll filter and insert it into a well (diameter 15 mm, depth 15 mm) of a multiwell polypropylene plastic tray.

11. Dry filters in an oven of 100° (~30 min).

12. By using forceps gently place brittle filter roll into glass scintillation vial (capacity 25 ml) filled with 10–15 ml of scintillation fluid nonmiscible with water and devoid of Triton X-100 (Rotiszint 11, K. Roth Co., Karlsruhe, F.R.G.) and determine radioactivity in liquid scintillation spectrometer.

13. Continue as described in item 19 of method A1 (see above).

14. The standard curve for evaluating the cyclic AMP contents of cell extracts is obtained by plotting I_0/I_x vs the amount (picomol) of cyclic AMP in the assay. I_0, I_x, mean values of radioactivities found in the absence and presence of unlabeled cyclic AMP, respectively. Routinely, those data are evaluated by computer.

Procedures B (for Multiwell Trays)

All incubations are carried out in triplicate.

Materials

1. Multiwell trays (Linbro trays) 24 wells per tray, diameter of well 16 mm.

2. Cell culture: 1.5 ml of culture medium containing 1.0 to 1.5×10^4 viable hybrid cells (see above Subcultivation) is placed in each well of a multiwell tray. The cells are cultured in a cell incubator (see above Subcultivation) for 4 days.

3. Solutions of hormones. All hormones and drugs are dissolved in incubation medium A (see procedure A1) at 30 times the final concentrations to be used in the incubations. In the case of PGE_1 this means that a 0.1 mM stock solution in ethanol is diluted to about 3 μM with incubation medium A. The solutions are kept in ice, unless precipitates form at low temperature. In that case they are kept at room temperature.

4. Anion exchange columns (see procedure A1).

Incubation with Hormones, Procedure B1

This method is used if a lyophilizer is available.

1. Remove multiwell tray from cell incubator and aspirate the growth medium with an appropriate manifold from 6 wells at a time.

2. Wash cell layer in each well by ~0.5 ml of ice-cold incubation medium A (see procedure A).

3. Place tray onto a perforated metal grid that is located slightly submerged in a plastic trough filled with a slurry of ice and water.

4. Pipet into each well 270 μl of incubation medium. Each incubation is carried out in duplicate or triplicate. In each tray a separate set of controls is included.

5. Add a total of 30 μl of solutions of hormones and/or drugs and/or vehicle. Final volume 300 μl. In order to carry out the additions as fast as possible (within a fraction of a minute) use a repetitive syringe.

6. Rotate the tray to mix the contents of the wells.

7. Place tray in water bath (as described for plates in procedures A) of 39°, with the lid of the tray lying inversed on the bottom part of the multiwell tray. This prevents evaporation of water from the wells. A temperature of 39° is required in order to reach a final temperature of 37° in the wells.

8. Incubate for 10 min.

9. Suck off medium as in 1. During the brief incubating period the release of cyclic AMP into the medium is constantly around 10% of total cellular cyclic AMP and therefore neglected.

10. To each well rapidly add 1 ml of 96% ethanol.

11. Add 10 μl of an aqueous solution of [^{3}H]cyclic AMP (3000 cpm: specific radioactivity approximately 30 Ci/mmol) for determination of recovery.

12. Transfer ethanolic cell extract to anion exchange column.

13. Follow steps 12 to 14 of procedure A1.

14. Dissolve residue in scintillation minivial in 500 μl of H_2O.

15. Use 250 μl of this solution for determination of recovery of cyclic AMP by following steps 16–19 of procedure A1.

Incubation with Hormones, Procedure B2

Use of this procedure as described with procedure A2.

1. Follow steps 1 to 9 of procedure B1.
2. To each well add from a 5 ml pipet about 0.5 ml of ~100% ethanol.
3. Transfer ethanolic cell extract into polypropylene tube (75 × 12 mm).
4. Wash each well with ~0.5 ml of ~100% ethanol and transfer the wash fluid into a polypropylene tube.
5. Evaporate ethanol in Speed Vac concentrator (step 5, procedure A2).
6. Dissolve residue in 500 μl of water, etc. (step 6, procedure A2).

Assay for Cyclic AMP (subsequent to procedures B1 and B2). This procedure is identical to that used subsequent to procedures A1 or A2 (described above) with the exception that the following concentrations of solutions and volumes added are different.

1. Solution 1 (solution of [^{3}H]cyclic AMP): contains 10^6 cpm/ml.
2. Solution 2 (solution of unlabeled cyclic AMP): 100 n*M*.
3. Solution 3 (cyclic AMP binding protein): contains binding protein at half the concentration[64] used in the solution of the cyclic AMP assay subsequent to procedures A1 and A2. In addition it contains 0.1% (w/v) bovine serum albumin.
4. Step 2 of assay procedure (subsequent to procedures A1 or A2): to the duplicate vials used for establishing the calibration curve are added 0, 2, 5, 10, 50, 75, and 100 μl of solution 2 (see above).

Hormonal Regulation of Adenylate Cyclase in Homogenates of Hybrid Cells[65,66]

The homogenates of hybrid cells employed in these assays can be stored frozen above liquid nitrogen without loss of enzyme activity. Therefore, it is worthwhile to grow hybrid cells in large batches and prepare homogenates sufficient for a large number of assays. This strategy avoids the variability in activity and hormonal sensitivity from one cell preparation to another.

[65] F. Propst, Ph.D. Thesis, University of Innsbruck (1981).

[66] F. Propst and B. Hamprecht, *J. Neurochem.* **36,** 580 (1981).

Cell Culture

1. Inoculate into plastic Petri dishes (150 mm in diameter) 0.2–1.0 × 10^6 viable hybrid cells suspended in 70 ml of growth medium (see above Subcultivation).

2. After a considerable degree of confluency is reached (usually 5–6 days later), remove growth medium. If, before homogenization, the cells are to be incubated with special agents (e.g., with choleratoxin), this incubation is carried out at this stage.[65]

3. Wash cells twice with 5 ml of ice-cold buffer A (buffer A: 50 m*M* HEPES, 250 m*M* sucrose, pH 7.7 at 22°).

4. Scrape off cells with a rubber scraper and transfer cells, by using a total of 2 ml of buffer A, to a centrifuge tube.

5. Centrifuge cells down (400 *g*/10 min/0°). Discard supernatant fluid.

Preparation of Cell Homogenate (Method 1)

1. Resuspend cells by trituration in 1 ml of buffer B per dish. This provides an approximate protein concentration of 10 mg/ml. Buffer B: 0.5 m*M* ATP (GTP-free, Sigma catalog No. 2383), 5 m*M* $Mg(CH_3COO)_2$, 20 m*M* creatine phosphate, 25 U/ml creatine kinase, dissolved in buffer A.

2. In order to guarantee rapid freezing, transfer the suspension into a plastic tube such that the height exceeds the diameter of the liquid column.

3. An aliquot portion is used for determination of protein.[67] Remember that about 12% of that protein found is due to the presence of creatine kinase.

4. Add 1–2 mg of deoxyribonuclease to avoid formation of clumps in the homogenate to be prepared.

5. Immerse tube for 5–10 min into liquid nitrogen. Thaw contents by swirling tube in a water bath of 37°. This cycle of freezing and thawing is repeated at least once until, after mixing of 1 droplet of suspension with 1 droplet of a solution of nigrosin,[54,55] under the microscope cells are no longer seen to exclude nigrosin.

6. The homogenate is frozen in portions of 0.5–1.0 ml and stored over liquid nitrogen.

Preparation of Cell Homogenate (Method 2)

1–4. Steps are identical to those of method 1.

5. At 4° place centrifuge tube containing cell suspension into nitrogen

[67] O. H. Lowry, N. J. Rosebrough, A. L. Farr, and R. J. Randall, *J. Biol. Chem.* **193,** 265 (1951).

cavatation bomb (Parr Instrument, Moline, Illinois) and apply a pressure of 6.9×10^8 Pa (1000 psi) for 10 min. Release pressure and collect homogenate in a plastic tube.

6. Incubate tube at 37° for 5 min to allow degradation of DNA. This cell homogenate can be used for the preparation of hybrid cell membranes (see below).

7. Freeze as in step 6 of method 1.

Comment. Of the five homogenization procedures compared, two methods were eliminated due to serious drawbacks. Homogenization by rotating knives (Ultra-Turrax, Janke & Kunkel K. G., Staufen im Breisgau, F.R.G.) yielded homogenates in which adenylate cyclase could no longer be inhibited by α_2-adrenergic action of noradrenaline. In homogenates prepared by ultrasound irradiation almost no adenylate cyclase activity was left. The remaining three methods yielded comparable results. However, one of them, cell destruction in a Potter-Elvehjem type of homogenizer, turned out to be too cumbersome. For obtaining a high yield of broken cells the gap between the rotating Teflon pestle and glass cylinder should be as narrow as possible, otherwise ~100 up-and-down strokes in a hypoosmotic medium (100 mosmol/liter) are required. During the homogenization process, the degree of homogenization has to be monitored by the nigrosin test (see above).

Thus, two procedures, freezing and thawing (method 1) and nitrogen cavatation (method 2) remained. The former method appears to be the most gentle one, since the highest activities of adenylate cyclase and the strongest effects of hormones were observed in homogenates prepared this way.

In isotonic HEPES buffers the results are independent of the pH (in the range pH 7.6 to 8.2) during the homogenization procedure. Homogenization in isotonic buffers is preferable. The presence of 0.2 m*M* EGTA, 0.1–0.5% (w/v) bovine serum albumin, 0.1 or 0.5 m*M* ascorbic acid ± 0.1% (w/v) bovine serum albumin, 1 or 5 m*M* dithiothreitol or mercaptoethanol was without ameliorating effect on the quality of homogenates prepared according to method 1. ATP and Mg^{2+} in the homogenization buffer appear to be beneficial for maintaining enzyme activity during incubation of the homogenate (warm-up period before the assay). Adenylate cyclase activity and hormonal responsiveness of the homogenates are stable at 37° for at least 30 min.

Preparation of Hybrid Cell Membranes[65]

1. The hybrid cell homogenate prepared according to method 2 (steps 1 through 6; see above) is centrifuged at 31,000 *g* for 15 min.

2. Discard supernatant. Resuspend pellet in the same volume as the homogenate of buffer B (see above Preparation of cell homogenate, method 1, step 1) lacking creatine phosphate and creatine kinase.

3. Repeat centrifugation.

4. Repeat steps 2 and 3.

5. Resuspend pellet in buffer B; determine protein concentration.[67]

6. Freeze suspension in portions of 0.5–1.0 ml and store over liquid nitrogen.

Adenylate Cyclase Assay

The assay by Salomon *et al.*[68,69] was modified.[65,66]

Solutions

Solution 1: 250 m*M* sucrose, 50 m*M* HEPES, pH 8.0 at 22°. Store frozen.

Solution 2: 35 m*M* $Mg(CH_3COO)_2$, 200 m*M* creatine phosphate, 250 U/ml creatine kinase, 5 m*M* isobutylmethylxanthine in solution 1. Solution 2, prepared as a large stock for many series of experiments and stored frozen can be thawed and refrozen without loss of activity. However, one should always make sure that the slowly dissolving isobutylmethylxanthine has dissolved completely.

Solution 3: [α-^{32}P]ATP (30–90 cpm/pmol ATP), 10 m*M* ATP in solution 1. Store frozen. The specific radioactivity is chosen such that the lowest values measured in a sample assayed for adenylate cyclase activity would not fall below 200 cpm.

Solution 4: Hormones and other additions to the incubation mixture for assaying adenylate cyclase activity were dissolved in solution 1 at 10-fold the final concentration to be used in the assay. Store frozen.

Solution 5: 2% (w/v) sodium dodecyl sulfate, 40 m*M* ATP, 1.4 m*M* cyclic AMP in solution 1. Store frozen.

Solution 6: [^{3}H]Cyclic AMP (specific radioactivity 20–30 Ci/mmol), 10^5 cpm/ml, in water. Store frozen.

Assay Protocol

1. Pipet 10 μl of solution 2 into Eppendorf snap cap plastic vials placed in ice.

[68] Y. Salomon, C. Londos, and M. Rodbell, *Anal. Biochem.* **58,** 541 (1974).

[69] Y. Salomon, *Adv. Cyclic Nucleotide Res.* **10,** 35 (1979).

2. Wherever required add 10 μl of the solution 4.
3. Fill up volume to 60 μl by the addition of solution 1.
4. Add 10 μl of solution 3.
5. Approximately 5 min before starting the incubation place reaction vessels into water bath of 37°.
6. Briefly before use, the frozen cell homogenate or cell membrane preparation is thawed in a water bath of 37° (note that the thawed preparations are not to be refrozen for a second use), diluted by homogenization buffer to the protein concentration desired, and immediately before use prewarmed at 37° for 3 min.
7. To start the reaction, add 30 μl of cell homogenate to the vial and mix by Vortex.
8. At the end of the 10-min incubation period add 100 μl of solution 5 and mix by Vortex.
9. Add 50 μl of solution 6 as a tracer for determining the recovery of cyclic AMP during the subsequent procedure of purification of the nucleotide.
10. Add 0.8 ml of water.
11. The purification of cyclic AMP and the determination of radioactivity of ^{3}H and ^{32}P are carried out as described.[69]

Procedure for Testing Hormonal Regulation of Intracellular Levels of Cyclic GMP in Hybrid Cells[43]

So far two hormones, bradykinin[43] and angiotensin II,[43] have been found to elevate the level of cyclic GMP in the hybrid cells. The procedure of hormone exposure of the cells follows largely that for hormonal stimulation of formation of cyclic AMP (see above): All incubations are carried out in triplicate.

Materials

Cell cultures: As described above for the hormonal regulation of cyclic AMP (procedures A1 and A2).

Incubation medium B: NaCl, 145 m*M*, KCl, 5.4 m*M*, $CaCl_2$, 1.8 m*M*, $MgCl_2$, 1.0 m*M*, Na_2HPO_4, 2.0 m*M*, glucose, 20 m*M*, and HEPES, 20 m*M*.

The solution is adjusted to pH 7.4 (22°) by adding a concentrated solution of tris(hydroxymethyl) aminomethane. Osmolarity approximately 330 mosmol/liter. Filter sterilly and store at +4°. Other details are as for the incubation medium A used for studying the hormonal regulation of cyclic AMP levels (see above).

Water bath: See above, Materials used for studying the hormonal regulation of cyclic AMP levels.

Incubation with Hormones

1. Take dish (50 mm in diameter) from cell incubator and aspirate growth medium.
2. Wash cell layer twice with 2 ml of incubation medium B (37°).
3. Add 2 ml of incubation medium B (37°) containing the hormone and possibly other ingredients in the final concentration. Control dishes receive incubation medium B only.
4. Place dish on water bath of 37°. Start stop watch.
5. Incubate for the usually short periods of 15–60 sec.
6. Quickly aspirate medium, add 1 ml of approximately 100% (v/v) ethanol, and cover dishes until the next step is carried out.
7. Transfer cell extract to a polypropylene tube (75 × 12 mm).
8. Wash dish with 1 ml of approximately 100% (v/v) ethanol and transfer washing fluid into the polypropylene tube.
9. Evaporate ethanol in a Speed Vac concentrator.
10. Redissolve residue in an appropriate volume of water (e.g., 1 ml) by using a Vortex mixer. Store samples at −20° until cyclic GMP assays are carried out.

Assay for Cyclic GMP

1. Centrifuge in a refrigerated cell centrifuge for 5 min at the highest speed possible to firmly pack the cells that had been transferred with the alcohol. Use supernatant for determination of cyclic GMP.
2. Cyclic GMP is acetylated and determined by radioimmunoassay essentially as described[70] by using [^{125}I]succinylcyclic GMP tyrosine methylester as radioactively labeled ligand. The only modifications are that (a) during the acetylation procedure all volumes are scaled down to 40% of the values recommended[71] and (b) the concentration of bovine serum albumin in the acetate buffer containing the antiserum is 30 instead of 5 mg/ml.

Procedure for Testing Hormonal Regulation of Uptake of Guanidinium Ions into Hybrid Cells[71]

Noradrenaline[35] and acetylcholine[8,10] depolarize the hybrid cells. Such electrophysiological studies are tedious and do not allow the establish-

[70] G. Brooker, J. F. Harper, W. L. Terasaki, and R. D. Moyland, *Adv. Cyclic Nucleotide Res.* **10,** 1 (1979).

[71] G. Reiser, A. Günther, and B. Hamprecht, *J. Neurochem.* **40,** 493 (1982).

ment of clear-cut dose–response relationships. Therefore, a method for measuring receptor-regulated ion transport is desirable that circumvents these drawbacks. Such studies are possible only if the receptor involved in the depolarization phenomenon of the hybrid cells does not desensitize very quickly in the presence of the hormone. In the case of Na^+ ions, transport is generally studied by using the dangerous radioisotope ^{22}Na. Fortunately, in many cases Na^+ can be replaced by other monovalent cations such as Li^+ and guanidinium. An example is the inward current of the action potential of the hybrid cells.[71,72] Recently it was observed that the undecapeptide substance P can cause a strong increase in the rate of influx of guanidinium into hybrid cells.[47] The procedure for studying such hormonally regulated guanidinium fluxes is described below.

Material

Cell culture. See above, Hormonal regulation of intracellular levels of cyclic AMP in hybrid cells.

Choline medium. This medium is identical with incubation medium B (see above, Procedure for testing hormonal regulation of intracellular levels of cyclic GMP in hybrid cells) except that 145 m*M* NaCl is replaced by 145 m*M* choline chloride.

Incubation medium C: Essentially consists of a solution of 1 or 10 m*M* guanidinium chloride and 50–100 nCi [^{14}C]guanidinium (Amersham) per ml in choline medium. If 10 m*M* guanidinium is used, the concentration of choline chloride is reduced to 135 m*M*. The incubation medium added to each plate also contains the individual additions, e.g., substance P, in their final concentrations.

Incubation Procedure

All incubations are carried out in triplicate.

1. Aspirate growth medium from culture plate (50 mm in diameter).
2. Wash cell lawn twice with 3 ml of choline medium (37°).
3. Add incubation medium (37°) and incubate on water bath (see above, Procedure for testing the hormonal regulation of intracellular levels of cyclic AMP in hybrid cells) for 10 min.
4. Aspirate incubation medium and immediately wash culture dish 4 times with 3 ml ice-cold choline medium. Suck off washing fluid as completely as possible.
5. Let the open plate dry at the air for 1 hr.
6. Add 1.0 ml 0.4 *M* NaOH and place dish in a shaker overnight to dissolve the cells.

[72] G. Reiser, F. Scholz, and B. Hamprecht, *J. Neurochem.* **39,** 228 (1982).

7. Transfer alkaline solution into scintillation vial.

8. Wash culture dish with 0.4 ml of 1.25 *M* HCl. Transfer washing fluid into the scintillation vial.

9. Add 15 ml of scintillation fluid (Rotiszint 22, C. Roth, Karlsruhe, F.R.G.) containing Triton X-100 and determine radioactivity.

10. Express rates of guanidinium uptake (means of triplicate ± SD) by referring them to cellular protein,[67] in the dimension nmol/min/mg protein.

Determination of Protein

1. Of each batch of cultured cells 3 dishes are used for determination of cellular protein.

2. Aspirate culture medium.

3. Wash plates twice with 5 ml of incubation medium kept in a water bath of 37°. It is irrelevant whether the one for studying hormonal action on cyclic AMP levels is used or the one for studying cyclic GMP levels. The only points that matter are osmolarity (330 mosmol/liter), pH 7.4 and the presence of approximately 2 m*M* Ca^{2+}.

4. Add a measured volume of 2 ml of 0.1–0.4 *M* NaOH and place dish in a shaker overnight to dissolve the cells.

5. Transfer alkaline solution to plastic tubes and store in freezer until determination of protein.

6. Determine protein by the method of Lowry *et al.*[67]

Comment

In neuroblastoma × glioma hybrid cells quite a number of hormones have been shown to regulate, via receptors in the plasma membrane, several effector systems such as adenylate cyclase, guanylate cyclase, and sodium channels. It is likely that future work on these cells will demonstrate additional effector systems to be controlled by hormone receptors, e.g., other membrane-bound enzymes, ion channels, and transport systems. Probably the hybrids would constitute a useful system for studying the molecular mechanisms underlying such hormone–receptor–effector systems. In addition, if, of the hormones known, one finds those that regulate a given effector system in hybrid cells, that effector system can serve as an indicator for finding new hormones or hormone antagonists in extracts of animals and plants, respectively. At present, 8 out of about 50 hormonal agents have been shown to regulate the formation of cyclic AMP in the hybrids.[33] Therefore, this effector system could already be used for the purpose indicated. Or the hybrids could be used for finding new toxins that open or close ion channels. Of course, this principle of detecting and quantitizing (new) highly active agents applies to any differentiated function that is expressed by any cell type in culture. Examples are the hybrid cells as indicator of the opioids produced by the same hybrid cells[23] or the hybrid–hybrid cells as indicators of glial and heart derived factors that regulate choline acetyltransferase.[48]

[27] Primary Glial Cultures as a Model for Studying Hormone Action

By BERND HAMPRECHT and FRIDOLIN LÖFFLER

Primary cultures of perinatal mouse or rat brain consist of many cell types[1] such as astroblasts,[2] oligodendroblasts,[3] ependymal cells,[4,5] capillary endothelial cells,[6] phagocytic cells[6–8] (macrophages and/or microglia), and mesenchymal cells,[9–11] but are devoid of neurons.[7] In spite of this cellular heterogeneity one cell type, the astroblast, appears to dominate quantitatively.[11] Most of these cells have been identified immunocytochemically by using antibodies against cellular markers, in most cases cell-type specific proteins.[1]

Several hormones including peptides regulate the intracellular concentration of cyclic AMP in these cultures[12,13] (see the table). Many other

[1] M. C. Raff, K. L. Fields, S.-I. Hakamori, R. Mirsky, R. M. Pruss, and J. Winter, *Brain Res.* **174,** 283 (1979).

[2] E. Bock, M. Møller, C. Nissen, and M. Sensenbrenner, *FEBS Lett.* **83,** 207 (1977).

[3] M. C. Raff, R. Mirsky, K. L. Fields, R. P. Lisak, S. H. Dorfman, D. H. Silberberg, N. A. Gregson, S. Leibovitz, and M. C. Kennedy, *Nature (London)* **274,** 813 (1978).

[4] A. G. Gilman and B. K. Schrier, *Mol. Pharmacol.* **8,** 400 (1972).

[5] D. van Calker, M. Müller, and B. Hamprecht, *in* "Neural Growth and Differentiation" (E. Meisami and M. A. B. Brazier, eds.), p. 11. Raven Press, New York, 1979.

[6] D. Hansson, Å. Sellström, L. I. Persson, and L. Rönnbäck, *Brain Res.* **188,** 233 (1980).

[7] F. Löffler, Ph.D. Thesis, University of Munich (1983).

[8] F. Löffler, and B. Hamprecht, in preparation (1984).

[9] M. Schachner, G. Schoonmaker, and R. O. Hynes, *Brain Res.* **158,** 149 (1978).

[10] A. Pateau, K. Mellström, B. Westermark, D. Dahl, M. Haltia, and A. Vaheri, *Exp. Cell Res.* **329,** 337 (1980).

[11] P. E. Stieg, H. K. Kimelberg, J. E. Mazurkiewicz, and G. A. Bauker, *Brain Res.* **199,** 493 (1980).

[12] B. Hamprecht, M. Brandt, F. Propst, D. van Calker, and F. Löffler, *Adv. Cyclic Nucleotide Res.* **14,** 637 (1981).

[13] F. Löffler, D. van Calker, and B. Hamprecht, *EMBO J.* **1,** 297 (1982).

ISBN 0-12-182009-2

HORMONES THAT REGULATE THE INTRACELLULAR CONCENTRATIONS OF CYCLIC AMP IN PRIMARY GLIAL-RICH CULTURES OF PERINATAL MURINE BRAIN[a]

Hormonal effect on level of cyclic AMP	
Increase	Inhibition of increase
Noradrenaline (β-receptor)[1,2]	Noradrenaline (α-receptor)[1,2]
Secretin[3]	Somatostatin[3]
Vasoactive intestinal peptide (VIP)[3]	Adenosine (A_1 receptor)[6,7]
Adrenocorticotropin, melanotropins[4]	
Parathyrin[5]	
Calcitonin[5]	
Adenosine (A_2 receptor)[6,7]	
Histamine (H_2 receptor)[8]	
Prostaglandins E_1 and E_2[8–11]	

[a] References:

[1] D. van Calker, M. Müller, and B. Hamprecht, *J. Neurochem.* **30,** 713 (1978).

[2] K. D. McCarthy and J. de Vellis, *J. Cyclic Nucleotide Res.* **4,** 15 (1978).

[3] D. van Calker, M. Müller, and B. Hamprecht, *Proc. Natl. Acad. Sci. U.S.A.* **77,** 6907 (1980).

[4] D. van Calker, F. Löffler, and B. Hamprecht, *J. Neurochem.* **40,** 418 (1983).

[5] F. Löffler, D. van Calker, and B. Hamprecht, *EMBO J.* **1,** 297 (1982).

[6] D. van Calker, M. Müller, and B. Hamprecht, *Nature* (*London*) **276,** 839 (1978).

[7] D. van Calker, M. Müller, and B. Hamprecht, *J. Neurochem.* **33,** 999 (1979).

[8] D. van Calker, M. Müller, and B. Hamprecht, *in* "Neural Growth and Differentiation" (E. Meisami and M. A. B. Brazier, eds.), p. 11. Raven Press, New York, 1979.

[9] A. G. Gilman and B. K Schrier, *Mol. Pharmacol.* **8,** 410 (1972).

[10] F. Löffler, Ph.D. Thesis, University of Munich (1983).

[11] F. Löffler and B. Hamprecht, in preparation.

hormones and candidate neurohormones do not affect noticeably the adenylate cyclase system of the cultured cells.[14] From the size of most of the hormonal stimulations it has been concluded that the action must be on the most abundant cells, the astroblasts.[5,15] The fact that a hormone listed as inhibitory (see the table) interferes with the action of a stimulating hormone (see the table) has been interpreted as evidence for the

[14] D. van Calker, F. Löffler, and B. Hamprecht, *J. Neurochem.* **40,** 418 (1983).

[15] D. van Calker, M. Müller, and B. Hamprecht, *J. Neurochem.* **30,** 713 (1978).

copresence of the corresponding receptors in the same cell surface.[12,15] Thus, the interference of one group of hormones with the action of another group of hormones has been used to allocate a set of receptors to one cell type within a heterogeneous population of cells. This set of receptors could serve as an identity label during the process of isolation of that cell type from the mixture.[5,12,15] As suggested in the case of neuroblastoma × glioma hybrid cells[15a] the primary cultures of perinatal murine brain might be used for discovering new hormonal agents regulating adenylate cyclase.

Procedure for Establishing Primary Cultures of Perinatal Murine Brain

The procedure follows that of Schrier[16,17] with some modifications.[18]

Material

Nylon bag filters. From Scrynel nylon mesh cloth, 211 and 135 μm pore diameter (NY211HC and NY135HC, Züricher Beuteltuchfabrik AG, Rüschlikon, Switzerland) form enlongated bags (2 × 5 and 1 × 10 cm, respectively) by placing two welding sutures with a plastic foil welding apparatus.

Solutions

Solution 1: Penicillin–streptomycin stock solution: Dissolve 2 million units of penicillin G (Na^+ or K^+ salt) and 2 g of streptomycin sulfate in 100 ml of water. Filter sterilly. Keep frozen in plastic tubes in portions of 4 ml.

Solution 2: 70% (v/v) ethanol.

Solution 3: Medium D[19] supplemented per liter with 1 g of glucose, 20 g of sucrose, and 10 ml of penicillin–streptomycin stock solution (approximately 330 mosmol/liter). Filter sterilly.

Solution 4: Culture medium. Mix: sterile Dulbecco's modified Eagle's medium[20–22] (DMEM), 900 ml, fetal calf serum, 100 ml, and penicillin–streptomycin stock solution, 1 ml.

[15a] B. Hamprecht, T. Glaser, G. Reiser, E. Bayer, and F. Propst, this volume [26].
[16] B. K. Schrier, *J. Neurobiol.* **4,** 117 (1973).
[17] B. K. Schrier, S. H. Wilson, and M. Nirenberg, this series, Vol. 32, p. 765.
[18] D. van Calker, Ph.D. Thesis, University of Munich (1977).
[19] R. G. Ham and T. T. Puck, this series, Vol. 5, p. 90.
[20] R. Dulbecco and G. Freeman, *Virology* **8,** 396 (1959).
[21] H. J. Morton, *In Vitro* **6,** 89 (1970).
[22] L. P. Prutzky and R. W. Pumper, *In Vitro* **9,** 468 (1974).

Protocol for the preparation of primary glial-rich culture from newborn mouse or rat brain.

1. From 10 neonatal mouse or 6 neonatal rat brains one can generate enough viable cells to inoculate approximately 50 plates 50 mm in diameter.
2. Briefly immerse newborn animal in solution 2.
3. Wash the animal in solution 3.
4. Decapitate and remove total brain as described.[16]
5. Collect brains in a plastic Petri dish (85 mm in diameter) filled with 10 ml solution 3 and placed on ice.
6. Fill bag (211 μm pore width) with 10–20 mouse or 6–12 rat brains. Pass the brain cells through the net into 20 ml of solution 3 by gentle massage of the bag using bent forceps with blunt tips.
7. Filter the suspension through the bag of 135 μm pore width that stands in a sterile 50 ml plastic centrifuge tube (Falcon Plastics).
8. Centrifuge (5 min, 400 *g*) in a refrigerated cell centrifuge.
9. Resuspend cells in 15 ml or 20 ml solution 4 (37°) per mouse or rat brain, respectively, to yield a suspension of at least 10^6 viable cells per ml.
10. Count cells in a hemocytometer and determine viability by the exclusion of nigrosin.[23,24] Viabilities generally observed are at least 50% for mouse and 40% for rat brain cells.
11. Dilute with solution 4 down to 6×10^5 viable cells/ml.
12. Place 5 ml of cell suspension into culture dish of 50 mm in diameter.
13. Incubate cells in cell incubator (37°; 90% air, 10% CO_2; approximately 100% relative humidity).
14. Six to 7 days later renew culture medium (solution 4).
15. Continue cell culture without further renewal of culture medium until use of the cultures for studies of hormonal influences (2–4 weeks after the initiation of the culture).

Procedure for Studying the Hormonal Regulation of the Level of Cyclic AMP in Primary Glial-Rich Cultures

Solutions

Solution 1: Incubation medium: Is identical with DMEM except that it contains 24.6 g instead of 37 g of $NaHCO_3$ per liter and additional NaCl to yield an osmolarity of 330 mosm/liter.

[23] H. J. Phillips and J. F. Terryberry, *Exp. Cell Res.* **13,** 341 (1957).
[24] J. P. Kaltenbach, M. H. Kaltenbach, and W. B. Lyons, *Exp. Cell Res.* **15,** 112 (1958).

Starting before the experimental incubation of cells with hormones, the solution is equilibrated by passing through it a mixture of 90% air and 10% CO_2.

Solution 2: Solutions in solution 1 of cyclic nucleotide phosphodiesterase inhibitors 3-isobutyl-1-methylxanthine[25] (0.5 m*M*) or 4-(3-butoxy-4-methoxybenzyl)-2-imidazolidinone (Ro 20-1724, Hoffmann-La Roche, Basel, Switzerland)[26] (0.5 m*M*). The solution is equilibrated with CO_2 as described above for solution 1.

Solutions 3: Solutions of the peptide hormones secretin, vasoactive intestinal peptide, adrenocorticotropin, melanotropins, calcitonin, or somatostatin are prepared in water at 100× the final concentration to be used in the incubation with cells (maximally 1 m*M*). Analogously, parathyrin is dissolved in 10 m*M* acetic acid containing 0.5% (w/v) bovine serum albumin.

Protocol

1. Aspirate culture medium.
2. Wash with 2 ml of solution 1 (37°).
3. Add 1.5 ml of solutions 1 or 2.
4. Add 100× concentrated solutions of hormones. If both inhibitory and stimulating hormones are used together, the former is added before the latter.
5. Swirl around to mix all ingredients well, start stop watch, and place dish uncovered into cell incubator (37°; 10% CO_2 in air) for 10 min (standard incubation time) or any other time desired.
6. Follow procedures A1 or A2 for testing hormonal regulation of cyclic AMP levels in neuroblastoma × glioma hybrid cells [26].[15a]

Comment

If small numbers of cells are available only, e.g., cells grown in wells of multiwell trays (24 wells, 16 mm in diameter) methods B1 or B2 described for neuroblastoma × glioma hybrid cells [26][15a] should be used.

[25] J. Schultz and B. Hamprecht, *Naunyn-Schmiedeberg's Arch. Pharmacol.* **278,** 215 (1973).
[26] H. Sheppard and B. Wiggan, *Mol. Pharmacol.* **7,** 111 (1970).

[28] Study of Receptor Function by Membrane Fusion: The Glucagon Receptor in Liver Membranes Fused to a Foreign Adenylate Cyclase

By SONIA STEINER and MICHAEL SCHRAMM

Hormone receptors which are linked to adenylate cyclase may not express their maximal activity in the system in which they are being studied. This might be due to genetic or experimentally induced changes in the components of the system or in the cell membrane in which it resides. Also, purification of the receptor would obviously remove the other components of the adenylate cyclase. To permit the assay of receptor function under such conditions, the preparation is fused with an excess of cell membranes containing a healthy adenylate cyclase system. When hormone is subsequently added, the transferred receptor will activate the foreign adenylate cyclase. Prior to fusion the adenylate cyclase in the preparation contributing the receptor is inactivated by *N*-ethylmaleimide to ascertain that all the activity which will be assayed will be due to the transferred receptor activating the heterologous enzyme.[1,2]

Preparations

Friend Erythroleukemia Cells. These are grown as a quiescent suspension on F-12 medium with 10% heat-inactivated calf serum.[3] Two–three day cultures in Roux bottles, containing 120 ml medium, reaching a density of 4×10^5 cells/ml, are used. On the day of the experiment, cells are counted, sedimented by centrifugation 5 min at 150 *g* at 25°, and washed once with half the original volume of solution A (see reagents). Cells are counted and the stock suspension is diluted to 10^7 cells/ml. The suspension is kept at 25° until use, within 2 hr.

Rat Liver Membrane Pellets. All operations are performed at 0°. Membranes prepared as described[4] and stored under liquid N_2 are brought

[1] M. Schramm, *in* "Membranes and Transport" (A. N. Martonosi, ed.), Vol. 2, p. 555. Plenum, New York, 1982.

[2] The procedure outlined below is based on the earlier study: M. Schramm, *Proc. Natl. Acad. Sci. U.S.A.* **76,** 1174 (1979).

[3] While fetal calf serum can help to start a culture from frozen stock, it should not be used for maintenance or large scale culture. It sometimes induces the *de novo* synthesis of various hormone receptors which activate adenylate cyclase, and thus the cells cannot be used to test the transfer of these receptors from other membranes.

[4] D. R. Neville, *Biochim. Biophys. Acta* **154,** 540 (1968); S. L. Phol, L. Birnbaumer, and M. Rodbell, *J. Biol. Chem.* **246,** 1849 (1971).

ISBN 0-12-182009-2

to 0° by addition of 10 volumes of 10 m*M* 4-morpholinepropanesulfonate buffer, pH 7.5. After sedimentation at 18,000 *g* for 10 min, the pellet is resuspended at 2 mg/ml in the above buffer. One-tenth volume of 50 m*M* *N*-ethylmaleimide is added and the suspension is kept for 25 min. Five volumes of the above buffer, containing 3 m*M* mercaptoethanol, are added and after 5 min the membranes are sedimented as above. The pellet is resuspended at 0.9 mg/ml in the buffer containing 1 m*M* mercaptoethanol. Phospholipid dispersion is then added in the amount of 25 μl/ml suspension of membranes, followed after 10 min by 10 μl/ml of 1 *M* $MgCl_2$. The suspension becomes more turbid and, after 10 min, is centrifuged at 18,000 *g* for 10 min in a 12 ml polycarbonate tube which will serve subsequently for the fusion reaction. After removal of the supernatant, the pellets are frozen in liquid N_2 and stored at −70° until use.

Reagents

1. Solution A: Tris, 20 m*M*, pH 7.4; $MgCl_2$, 0.8 m*M*; KCl, 5 m*M*; NaCl, 135 m*M*.
2. Polyethylene glycol medium: Polyethylene glycol 6000 (Merck), 52% w/w in solution A containing also ATP, 2 m*M*; EDTA, 0.1 m*M*; glucose, 6 m*M*; $MgCl_2$, 4 m*M*; and NaOH, 2.5 m*M*. The Mg-ATP preserves adenylate cyclase activity during fusion, while the NaOH serves to neutralize some acidity in the polyethylene glycol preparation.
3. Dilution medium: Same as (2) but without polyethylene glycol and NaOH.
4. Hypotonic buffer: Tris, 10 m*M*, pH 7.5; $MgCl_2$, 2 m*M*; and EGTA, 0.1 m*M*.
5. 4-Morpholinepropanesulfonate, pH 7.5, 10 m*M*.
6. $MgCl_2$, 1 *M*.
7. Phospholipid dispersion: A sonicated clear solution of soybean phospholipids (Sigma crude lecithin), 10 mg/ml in 10 m*M* Tris, pH 7.5, containing 0.1 m*M* EDTA.
8. Adenylate cyclase reaction mixture: Final m*M* concentrations were 4-morpholinepropanesulfonate, 50; $MgCl_2$, 1.5; [α-^{32}P]ATP, 0.6; EGTA, 0.2; creatine phosphate, 12; 3,5-cyclic AMP, 1; mercaptoethanol, 4; theophylline, 0.2; and 19 U/ml of creatine kinase.

Mg^{2+} concentrations higher than shown and addition of GTP above 0.1 μM increase basal adenylate cyclase activity and thus decrease the relative stimulation by glucagon.

Fusion Procedure

Duplicate pellets of 150 and 300 μg liver membranes are used. Each pellet at 0° receives 1 ml of the Friend cell suspension. Two additional tubes receive cells but no membranes; one serves to assess the effect of fusion per se on adenylate cyclase activity, while the other measures the adenylate cyclase without the fusion treatment. The suspensions are centrifuged 5 min at 7500 *g* at 4°. The supernatant is thoroughly removed by suction and the pellet of cells is mixed well with the pellet of liver membranes by Vortex, with a glass rod inside the tube. A tube with Friend cells only receives hypotonic buffer and is kept in ice. The tubes for fusion are transferred to a 30° bath and, after 2 min, receive 0.5 ml polyethylene glycol medium which has been briefly preincubated at 30°. The suspensions are thoroughly mixed by Vortex for 10 sec and returned to the bath. After 2 min, 1 ml of dilution medium is added and the suspension is briefly shaken by hand until homogeneous. After an additional 2 min at 30°, 9 ml of dilution medium is squirted into the tubes to further dilute the polyethylene glycol. The suspensions are cooled in ice and centrifuged for 10 min at 18,000 *g*. The pellets are suspended in the hypotonic medium with a glass rod and Vortex to bring the final volume to 0.5 ml.

Adenylate Cyclase Assay and Evaluation of Results

Duplicate aliquots of 50 μl from the fusion systems and from the Friend cell control lysed in hypotonic medium are assayed for activity[5] in the adenylate cyclase reaction mixture specified above, in a final volume of 120 μl, without further addition, with 1 μM glucagon, with 10 μM prostaglandin E_1, and with 10 mM NaF. Incubation is 10 min at 30°. Results are calculated as pmol cyclic AMP formed/min/mg protein. The glucagon receptor activity in different samples is compared on the basis of pmol cyclic AMP formed/min/mg protein in the systems containing glucagon, after subtracting the basal adenylate cyclase activity produced by the same sample in absence of an enzyme activator. This procedure was found to be much more reliable for comparing different receptor samples than the conventional rating by fold stimulation over basal activity. Small fluctuations in the basal activity significantly changes the fold stimulation, while it has little effect on the calculation of the net activation due to glucagon.

The table shows the results of a typical experiment of receptor transfer by fusion. The 150 and 300 μg samples of membrane show proportional glucagon receptor stimulations of adenylate cyclase activity. The prostaglandin E_1 stimulation serves to assess the effects of fusion on a receptor indigenous to the Friend cell, while the fluoride stimulation measures adenylate cyclase activity, bypassing the receptor but still requiring the

[5] Y. Salomon, C. Londos, and M. Rodbell, *Anal. Biochem.* **58,** 541 (1974).

ACTIVITY OF GLUCAGON RECEPTOR TRANSFERRED BY FUSION TO FRIEND CELL ADENYLATE CYCLASE

Fusion system[a] and amount of liver membranes (μg)	Net adenylate cyclase activity due to activator[b]		
	Glucagon	Prostaglandin E_1 (pmol/min/assay)	NaF
Fc–L_{NEM}[c] 150	1.0 (× 4.9)[d]	4.4	7.1
Fc–L_{NEM} 150	1.1 (× 4.4)	4.0	6.0
Fc–L_{NEM} 300	2.2 (× 9.9)	3.2	8.2
Fc–L_{nNEM} 300	2.5 (× 7.1)	2.7	9.2
Fc–Fc	−0.1	4.3	5.6
Native Fc	−0.2	9.2	4.9

[a] Fc, Friend cells; L_{NEM}, liver membranes treated with *N*-ethylmaleimide prior to fusion.

[b] The adenylate cyclase activity shown is produced by one-tenth of the amount of membranes in the fusion system. Basal activity, which was 0.3–0.4 pmol/min/assay, has been subtracted from the values shown.

[c] L_{NEM} by itself, 45 μg, assayed in the adenylate cyclase system in presence of fluoride produced no measurable adenylate cyclase activity.

[d] Numbers in parentheses show the adenylate cyclase activity in presence of glucagon relative to the basal activity of the same fused preparation (fold stimulation).

guanyl nucleotide binding protein.[6] Comparing the activities of native Friend cell adenylate cyclase with those of cells fused with each other, it can be seen that fusion per se causes a decrease in the prostaglandin E_1 activity and an increase in the NaF activity.

Accuracy and Reproducibility

When fusion systems are set up with two different amounts of receptor, each assayed in duplicate, the results vary ± 10% from the calculated average, provided that the amounts of receptor used for fusion are in the range producing a proportional response. Excess liver membranes in the fusion system produces inhibition of glucagon activated adenylate cyclase. Within the optimal range, results are readily reproducible with different batches of Friend cells. However, the cell culture may have to be started fresh from frozen stock after a few months, if the characteristics of the adenylate cyclase change.

[6] P. C. Sternweis, J. K. Northup, M. D. Smigel, and A. G. Gilman, *J. Biol. Chem.* **256,** 11517 (1981).

Application of the Fusion Procedure to Other Preparations and to Other Hormone Receptors

The procedure is rather versatile. In addition to membrane to cell fusion, membrane to membrane transfer also works quite satisfactorily. The glucagon receptor of liver membranes was thus transferred to adenylate cyclase in membranes prepared from rat salivary gland.[7] The procedure has been most extensively applied to the β-adrenergic receptor of turkey erythrocytes. For this latter system, receptor was readily assayed in native membranes,[8] in solubilized reconstituted,[9] and in delipidated preparations after readdition of lipids.[10] These very different preparations required only minor changes in the procedure here described. Fusion of the β-adrenergic receptor and subsequent adenylate cyclase assay were performed at 37 and not at 30°. Activated and nonactivated guanyl nucleotide binding protein can also be measured by the above described procedure, using cells deficient in this component.[8,11]

Acknowledgments

The procedures described are based on research supported by Grant AM-10451 from the National Institutes of Health and a grant from the United States–Israel Binational Science Foundation.

[7] M. Schramm, *in* "Membrane Bioenergetics" (C. P. Lee, G. Schatz, and L. Ernster, eds.), p. 349. Addison-Wesley, Reading, Massachusetts, 1979.

[8] G. Neufeld, M. Schramm, and N. Weinberg, *J. Biol. Chem.* **255,** 9268 (1980).

[9] S. Eimerl, G. Neufeld, M. Korner, and M. Schramm, *Proc. Natl. Acad. Sci. U.S.A.* **77,** 760 (1980).

[10] J. Kirilovsky and M. Schramm, *J. Biol. Chem.* **258,** 6841 (1983).

[11] Y. Citri and M. Schramm, *J. Biol. Chem.* **257,** 13257 (1982).

[29] Isolation of ACTH-Resistant Y1 Adrenal Tumor Cells

By BERNARD P. SCHIMMER

Introduction

The Y1 adrenal tumor cell line[1] is a well-characterized system, with functional properties similar to those of normal isolated adrenal cells. This cell line has contributed extensively to our understanding of basic mechanisms involved in hormonal regulation of adrenocortical functions,

[1] Y. Yasumura, V. Buonassisi, and G. Sato, *Cancer Res.* **26,** 529 (1966).

ISBN 0-12-182009-2

and has provided an important interface between the areas of cell biology and endocrinology.[2] Y1 cells have been of particular interest recently because they have been susceptible to genetic manipulation, and have afforded us the opportunity to apply somatic cell genetic techniques to studies of hormone action.[3,4] Our laboratory has isolated two families of mutants from the Y1 cell line—one with defective adenylate cyclase systems and one with impaired cAMP-dependent protein kinase activities. We have used these two families of mutants to establish the importance of cyclic nucleotides in the actions of ACTH on various adrencortical functions.[4–6] The objective of this report is to describe the methods which we have used to isolate ACTH-resistant mutants from Y1 adrenal tumor cells.

Properties of Y1 Adrenal Tumor Cells

An extensive description of the origin and properties of the Y1 adrenal cell line has been presented elsewhere.[2] Briefly, Y1 cells originate from a minimally deviated, mouse, adrenal tumor. These cells grow as flat epithelial cells in monolayer culture, synthesize and secrete Δ^4-3-keto-C_{21} steroids from cholesterol, and accumulate ascorbic acid and cholesterol from the culture medium. Y1 cells have a plating efficiency of 4 to 10%, grow with an average doubling time of 30 to 40 hr, and reach saturation densities at approximately 2.7×10^5 cells/cm^2. They have a nearly-diploid karyotype with a modal number of 39 acrocentric or telocentric chromosomes. When treated with ACTH (corticotropin) or cAMP (adenosine 3′,5′-monophosphate), these cells increase the rate of steroidogenesis 4- to 10-fold, stop dividing, assume a rounded morphology, and detach from the culture vessel. The ability of ACTH and cAMP to inhibit Y1 cell growth and cause cell rounding and detachment provides a basis for selection of Y1 mutants resistant to hormones and cyclic nucleotides.

Methods

Growth of Y1 Cells in Culture. The Y1 cell line currently is available from the American Type Culture Collection (No. CCL 79). We have

[2] B. P. Schimmer, *in* "Functionally Differentiated Cell Lines" (G. Sato, ed.), p. 61. Alan R. Liss, Inc., New York, 1981.

[3] B. P. Schimmer, J. Tsao, and M. Knapp, *Mol. Cell. Endocrinol.* **8,** 135 (1977).

[4] P. A. Rae, N. S. Gutmann, J. Tsao, and B. P. Schimmer, *Proc. Natl. Acad. Sci. U.S.A.* **76,** 1896 (1979).

[5] P. A. Rae, H. Zinman, J. Ramachandran, and B. P. Schimmer, *Mol. Cell. Endocrinol.* **17,** 171 (1980).

[6] J. E. Kudlow, P. A. Rae, N. S. Gutmann, B. P. Schimmer, and G. N. Burrow, *Proc. Natl. Acad. Sci. U.S.A.* **77,** 2767 (1980).

observed marked clonal variation in this population of Y1 cells,[7] and recommend that these cells be cloned immediately upon receipt of the culture in order to obtain a stock with a stable phenotype. Multiple aliquots of the stock culture should be frozen as soon as possible after cloning to preserve the line and limit uncontrolled phenotypic drift.

Details pertaining to the requirements and procedures for propagating Y1 cells have been presented in an earlier volume of this series.[7] All procedures are carried out using strict aseptic bacteriological procedures. Routinely, Y1 cells are grown in 82.5% nutrient mixture F10,[8] 15% heat-inactivated horse serum, and 2.5% heat-inactivated fetal bovine serum. Penicillin (200 U/ml) and streptomycin sulfate (200 μg/ml) are included as antibacterial agents. Duplicate stock cultures are initiated by seeding approximately 2×10^6 dispersed cells in each of two plastic tissue culture flasks (75 cm^2) containing 15 ml of growth medium. The inoculated flasks are capped loosely and maintained at 36.5° in a humidified atmosphere of 5% CO_2 in air. Culture medium is changed every third to fourth day. At the second medium change, the volume of growth medium is increased to 30 ml per flask. The cells attach to the surface of the culture flask within 2 to 3 hr after plating, and reach saturation densities of approximately 2×10^7 cells per flask after 10 to 12 days. Cells are dispersed for subculture by treatment with 0.1% Pancreatin 4X, N.F. (Grand Island Biological Co.) in phosphate-buffered saline[9] without Ca^{2+} or Mg^{2+} salts. Alternatively, 0.1% trypsin (Bacto-trypsin 1 : 250, Difco Laboratories)–0.02% EDTA in phosphate-buffered saline without Ca^{2+} or Mg^{2+} salts can be used. To subculture cells, the growth medium is removed, and the monolayer is rinsed twice with 2 ml of the proteolytic solution. The proteolytic solution is removed and the flask of cells is incubated for approximately 10 min at 36.5°. The incubation period is complete when cells can be dislodged from the monolayer by tapping the flask. The cells are washed from the wall of the flask with 10 ml of growth medium, and are dispersed by repeated, gentle pipeting of the mixture. One milliliter of fully dispersed cells is used to seed each new flask.

Y1 cells can be stored as frozen aliquots at −70° or in liquid nitrogen. Cells are treated with a proteolytic solution as for subculture, and are suspended at 2×10^6 cells/ml in growth medium supplemented with sterile dimethyl sulfoxide (10% v/v).[10] One-milliliter aliquots are placed in sterile ampules, sealed, packed in a cardboard container, and kept at −70° for

[7] B. P. Schimmer, this series, Vol. 58, p. 570.

[8] R. G. Ham, *Exp. Cell Res.* **29,** 515 (1963).

[9] R. Dulbecco and M. Vogt, *J. Exp. Med.* **99,** 167 (1954).

[10] We have found that dimethyl sulfoxide is much better than glycerol[7] as a cryoprotective agent, particularly for many of the Y1 mutants.

3.5 hr. Cells than can be unpacked and quickly transferred to liquid nitrogen for long-term storage. To recover cells from the frozen state, one vial is thawed rapidly in a 37° water bath (wear safety goggles), suspended by gentle pipetting, and divided between two 75 cm^2 tissue culture flasks containing growth medium. Cells are ready for subculture after 2 to 3 weeks.

Mutagenesis. Cloned populations of adrenal cells at low passage numbers can be treated with mutagens such as *N*-methyl-*N'*-nitro-*N*-nitrosoguanidine (MNNG) or ethyl methanesulfonate (EMS) to raise the frequency of specific mutations to detectable levels.[3,4] Alternatively, the original stock of Y1 cells maintained in continuous culture can be used without mutagenesis as a source of spontaneously occurring mutants.[11,12] A rationale for using mutagens on newly cloned populations of Y1 cells is to raise the frequency of specific mutations while minimizing the contributions of independently accumulated, spontaneous variations to the cells' phenotype.

Stock solutions of MNNG (1 mg/ml) are prepared by dissolving the MNNG in acetone at 10 mg/ml and then diluting this solution in sterile buffer (10 m*M* sodium acetate, pH 5.0). Aliquots of the stock solution are stored in the dark at −20°.

Y1 cells are plated in 150 mm tissue culture dishes and cultured until they reach approximately 2×10^6 cells per plate. Cells are transferred to 10 ml of fresh growth medium (nutrient mixture F10 plus sera and antibiotics) and MNNG is added to a final concentration of 1.25 μg/ml. Cells are left in contact with MNNG for 2.5 hr at 37°, the mutagenic agent is removed, and 50 ml of growth medium is added to each culture dish. In our hands, this concentration of mutagen reduced the plating efficiency of Y1 cells to approximately 30% of control levels and is considered to be within the effective range for induction of mutations without excessive cell damage.[13] Following mutagenesis, cultures are grown for 6 to 9 days with regular medium changes to permit the expression of mutations.[14,15] The concentration of mutagen required to achieve appropriate reductions in plating efficiency may require some adjustment depending on the batch of reagent or the subclone of Y1 cells used.

For mutagenesis with EMS, Y1 cells are treated with 300 μg of EMS/ml of culture medium for 4 days. Other procedures are as described above for mutagenesis with MNNG.

[11] B. P. Schimmer, *J. Cell. Physiol.* **74,** 115 (1969).
[12] B. P. Schimmer, *J. Biol. Chem.* **247,** 3134 (1972).
[13] L. H. Thompson, this series, Vol. 58, p. 308.
[14] E. H. Y. Chu and H. V. Malling, *Proc. Natl. Acad. Sci. U.S.A.* **61,** 1306 (1968).
[15] B. W. Penman and W. G. Thilly, *Somatic Cell Genet.* **2,** 325 (1976).

Selection. As noted above, the effects of ACTH on Y1 cells are complex and include the regulation of a number of end responses such as steroidogenesis, growth, and morphology. ACTH also brings about changes in a number of intracellular processes which secondarily support one or more of the end effects of the hormone. Therefore, many different mutations can render Y1 cells resistant to some aspect of ACTH action. Since cAMP was a likely mediator of many of the actions of ACTH on Y1 cells[2,16] a selection scheme was designed to isolated cAMP-resistant mutants with alterations early in the pathway of cAMP action. These mutations would be expected to have pleiotypic effects, altering many adrenal end-responses both to ACTH and cAMP. The selection scheme which was adopted is based on the observations that the potent cAMP analog, 8BrcAMP (8-bromoadenosine 3′,5′-monophosphate), arrests the growth of Y1 cells and causes them to retract and detach from the monolayer.[3] The combination of these effects of 8BrcAMP on growth and morphology markedly reduces the plating efficiency of newly cloned Y1 cells such that no survivors are obtained from 5×10^6 cells plated and grown for at least 14 days in the presence of 0.4 m*M* 8BrcAMP. Resistant mutants are isolated from mutagenized Y1 cells as populations which attach and grow in the presence of high concentrations of the cyclic nucleotide.

Seven to ten plates of mutagenized Y1 cells are transferred to 50 ml of growth medium containing 0.8 m*M* 8BrcAMP and maintained for 4 to 6 weeks with regular changes of medium. After this period, two distinct populations of survivors can be observed by phase-contrast microscopy. One population is resistant to the effects of 8BrcAMP on both growth and morphology and grows as flat colonies in the presence of 8BrcAMP; the second group of survivors grows in the presence of 8BrcAMP as colonies of round cells in a monolayer.[4] Individual surviving colonies of both phenotypes are isolated on the culture dishes with sterile, stainless-steel cylinders and are collected by treatment with a proteolytic solution.[13] Colonies are picked from different dishes to ensure that they reflect independent mutational events.

Dispersed cells from each colony are transferred to growth medium without 8BrcAMP and grown without further selection pressure in the usual manner. Aliquots at early passage are preserved in liquid nitrogen to restrict phenotypic drift. Isolates of 8BrcAMP-resistant cells should be retested for drug resistance to ensure that the phenotype is stable; these isolates then can be recloned to ensure that the population is homogeneous.

[16] B. P. Schimmer, *Adv. Cyclic Nucleotide Res.* **13,** 181 (1980).

Characteristics of 8BrcAMP-Resistant Clones

Mutations Affecting cAMP-Dependent Protein Kinase Activity. The mutants which are resistant to the growth-inhibitory and morphological effects of 8BrcAMP have altered cAMP-dependent protein kinase activities[4] and are designated Y1(Kin). Among the different Y1(Kin) clones, the apparent affinities of cAMP-dependent protein kinase for cAMP are reduced 5- to 600-fold.[4] All of the clones examined so far have defects which affect the structure and activity of the regulatory subunit of the soluble, type 1 cAMP-dependent protein kinase; the type 2 enzyme seems unaffected.[17,18] Several Y1(Kin) clones contain mRNAs which code for both mutant and wild-type forms of the regulatory subunit, suggesting that this mutation affects one of the structural genes for R^I and is codominant.[19] In Y1(Kin) clones, the ability of ACTH or cAMP to stimulate steroidogenesis is impaired; the degree of impaired response correlates with the extent of the defect in cAMP-dependent protein kinase activity.[4] The mutations affecting cAMP-dependent protein kinase activity also diminish the morphological effects of ACTH and the ability of ACTH to stimulate ornithine decarboxylase activity.[6] Thus, mutations in the type 1 regulatory subunit of cAMP-dependent protein kinase affect a large family of responses to ACTH.

Mutations Affecting Adenylate Cyclase Activity

The mutants which emerge from the selection with 8BrcAMP as rounded cells are only partially resistant to the growth-inhibitory effects of 8BrcAMP.[4] They are resistant to the actions of ACTH on morphology, steroidogenesis, and induction of ornithine decarboxylase activity, but are fully responsive to these actions of cAMP.[4,6] These mutant clones appear to be defective in the adenylate cyclase system and are designated Y1(Cyc). In Y1(Cyc) clones, adenylate cyclase activity is stimulated by general regulators of the enzyme such as fluoride and cholera toxin, and is selectively insensitive to ACTH. The precise nature of the mutation affecting adenylate cyclase activity is unknown, but may be associated with a specific, intracellular 68,000 Da protein.[20] Although the selection of Y1(Cyc) mutants using this procedure is unexpected, it may have general application since clones with altered adenylate cyclase activity also have

[17] N. S. Gutmann P. A. Rae, and B. P. Schimmer, *J. Cell. Physiol.* **97,** 451 (1978).

[18] P. J. Doherty, J. Tsao, B. P. Schimmer, M. C. Mumby, and J. A. Beavo, *J. Biol. Chem.* **257,** 5877 (1982).

[19] S. A. Williams and B. P. Schimmer, *J. Biol. Chem.* **258,** 10215 (1983).

[20] V. M. Watt and B. P. Schimmer, *J. Biol. Chem.* **256,** 11365 (1981).

been isolated as 8BrcAMP-resistant mutants from rat pituitary tumor cells.[21]

Caution

Since most chemical mutagens are carcinogenic, they should be handled with extreme care. Although experience suggests that there is no biohazard associated with the Y1 mouse adrenocortical tumor cell line, it seems prudent to adopt containment procedures similar to those recommended by the Medical Research Council of Canada[22] when handling them.

Acknowledgment

The procedures described here were developed with research funds provided by the National Cancer Institute of Canada and the Medical Research Council of Canada.

[21] T. F. J. Martin and S. A. Ronning, *J. Cell. Physiol.* **190,** 289 (1981).
[22] "Guidelines for the Handling of Recombinant DNA Molecules and Animal Viruses and Cells," Cat. No. MR21-1/1980. Minister of Supply and Services, Ottawa, Canada, 1980.

[30] Induction of Glucagon Responsiveness in Transformed MDCK Cells Unresponsive to Glucagon

By SUZANNE K. BECKNER, FREDERICK J. DARFLER, and MICHAEL C. LIN

A cloned line of MDCK (Madin–Darby canine kidney) cells which were transformed with Harvey murine sarcoma virus has been established.[1] This line is maintained in continuous culture under the same conditions as the parental MDCK cells[2] in Dulbecco's MEM, 5% fetal bovine serum, 5% CO_2/95% air, with 80% humidity. The morphology of this transformed line is more fibroblastic than that of normal cells; in addition a 21,000 Da protein (p21) coded by the virus has been identified on the inner surface of the plasma membrane[3] of this transformed line.

[1] E. M. Scolnick, D. Williams, J. Maryak, W. Vass, R. J. Goldberg, and W. P. Parks, *J. Virol.* **20,** 570 (1976).
[2] M. C. Lin, S. K. Beckner, and F. J. Darfler, this series [31].
[3] M. C. Willingham, I. Pastan, T. Y. Shih, and E. M. Scolnick, *Cell* **9,** 1005 (1980).

ISBN 0-12-182009-2

The growth characteristics of transformed MDCK cells are similar to normal cells (doubling time 24 hr); both grow equally well in serum free media.[4]

Like normal MDCK cells, the adenylate cyclase of the transformed line responds to a variety of hormones (β-adrenergic agonists, prostaglandin E_1, vasopressin). However, transformation results in a selective loss of glucagon responsiveness, due to an absence of glucagon binding sites at the cell surface.[5] Thus this cell line represents a good model system to examine factors which regulate the expression of differentiated functions.

Induction Conditions

Glucagon receptors can be induced in transformed cells by a variety of agents such as butyrate, PGE_1, and Ro 20-1724 [4-(3-butyoxy-4-methoxybenzyl)-2-imidazolidinone, a generous gift from Dr. W. E. Scott, Hoffmann-LaRoche, Inc.]. However, exact control of culture conditions is essential to observe optimal induction. To achieve maximal induction by any agent, cells are subcultured and plated into DMEM containing 5% fetal bovine serum such that the density of the culture at the time of induction will be approximately 200,000 cells/35 mm dish. Therefore, cells are generally seeded at 50,000 cells/35 mm dish, and induction initiated 3 days later. When the cells have achieved a density of 200,000 cells/35 mm dish, the cells are washed once with 3 ml of phosphate-buffered saline (PBS), and the inducer in defined medium is added. The washing is essential to remove serum components which inhibit induction (see below). The incubation of cells with inducer continues at 37°. Under these conditions, induction is maximal after 3 days in the presence of the inducer. If the density is 200,000 cells/plate at the time of the induction, after 72 hr the final density will be slightly less than 1×10^6/dish or 80% confluent.

The induction of glucagon receptors is most conveniently monitored by measurement of intracellular cyclic AMP following incubation for 3 min in the presence or absence of 2 μM glucagon.[2] Transformed MDCK cells do not produce large amounts of prostaglandins as do normal MDCK cells,[6] so the 2 hr incubation with DMEM[2] is not required. Cells are washed twice (10 min, room temperature) to remove inducers, and incu-

[4] M. Taub, L. M. Chuman, M. H. Saier, Jr., and G. Sato, *Proc. Natl. Acad. Sci. U.S.A.* **76**, 3338 (1979).

[5] M. C. Lin, S. M. Koh, D. D. Dykman, S. K. Beckner, and T. Y. Shih, *Exp. Cell Res.* **142**, 181 (1982).

[6] M. C. Lin, S. M. Wang, and S. K. Beckner, *in* "Prostaglandins and Cancer: First International Conference," p. 493. Alan R. Liss, New York, 1982.

bated at room temperature for 30 min with 20 mM HEPES buffer, pH 7.4, 20 μM Ro 20-1724, an inhibitor of cyclic AMP phosphodiesterase, in DMEM, 2 ml/35 mm dish. A small aliquot of concentrated hormone solution (20 μl of 200 μM) is added. The reaction is terminated after 3 min as described.[2] Glucagon responsiveness is expressed as the ratio of cyclic AMP formed in the presence of glucagon to that formed in its absence. Consideration of density is important, since the basal adenylate cyclase activity of transformed MDCK cells varies with density, from 1 pmol/3 min/10^6 cells at a density of 1×10^6 cells to 20 pmol/3 min/10^6 cells at a density of 300,000 cells.

Although the above procedure results in consistent induction of glucagon responsiveness, the degree of induction varies from one experiment to another. The reasons for this variability are not known. Although density is important, obviously other factors which affect induction remain to be identified.

The most effective inducer of glucagon receptors in transformed MDCK cells is sodium butyrate, an inducer of differentiation in a variety of systems[7]; the mechanism responsible for its effect has not yet been firmly established. The desired butyrate concentration is obtained by the addition of the appropriate amount of stock 1 M sodium butyrate (butyric acid, Sigma, with pH adjusted to 6.5 with NaOH) to defined media. The induction of glucagon receptors by butyrate increases with time (not shown) and butyrate concentration between 0.5 and 2 mM (Fig. 1). Following induction with 1 mM butyrate, glucagon responsiveness is evident by 8 hr and maximal by 72 hr. No induction is evident with less than 0.5 mM and greater than 3 mM butyrate is toxic to the cells. Butyrate significantly decreases the growth rate of cells within 24 hr. After 72 hr with 1 mM butyrate cell number is only 50% of that of untreated cells.

Glucagon receptors can also be induced in transformed MDCK cells by prostaglandin E_1 (PGE_1, Sigma), which stimulates adenylate cyclase,[5] and Ro 20-1724. Unlike butyrate, the degree of glucagon responsiveness induced by these agents is biphasic (Fig. 1). Induction by PGE_1 is not evident below 1 nM, maximal with 0.1 μM, and reduced at higher concentrations. Although 100 μM PGE_1 inhibits cell growth by 20% over the 72-hr induction period, there is no effect of lower concentrations on cell number. Similarly, induction by Ro 20-1724 is evident with 0.1 nM, maximal with 1 μM, but declines at higher concentrations (Fig. 1). Unlike butyrate, there is no effect of Ro 20-1724 on cell growth at concentrations between 0.1 nM and 0.1 mM. Therefore, the lack of induction by high concentrations of PGE_1 and Ro 20-1724 is not a reflection of cell toxicity.

[7] J. Kruh, *Mol. Cell Biol.* **42,** 65 (1982).

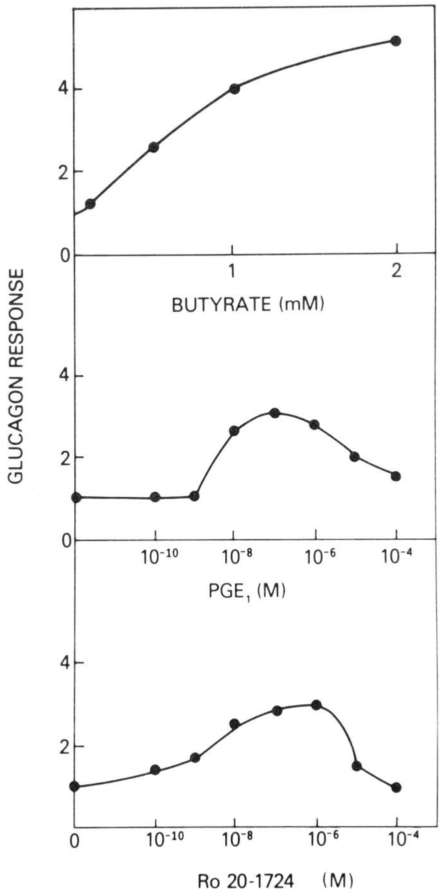

FIG. 1. Concentration dependence of induction by butyrate, PGE_1, and Ro 20-1724. Transformed MDCK cells were plated and induced as described with the indicated concentrations of butyrate, PGE_1, or Ro 20-1724 in defined medium.[2] After 72 hr, the glucagon responsiveness, expressed as the ratio of cyclic AMP formed in the presence of glucagon to that formed in its absence, was measured as described.

Inhibition of Induction

The induction of glucagon receptors in transformed MDCK cells can be prevented completely by 0.3 m*M* cycloheximide (which inhibits protein synthesis) or 1 μg/ml tunicamycin (which inhibits protein glycosylation); and partially by 2.5 μg/ml α-amanitin (which inhibits RNA polymerase II activity). An effect of these metabolic inhibitors can be best observed if the agent is added with 1 m*M* butyrate in defined media for

8 hr. This amount of time is sufficient to observe induction by butyrate but short enough to prevent toxic effects of metabolic inhibitors.

The induction of glucagon receptors by PGE_1 and Ro 20-1724, but not butyrate, occurs much better in defined media as compared to serum. In fact, if serum is added to defined media containing PGE_1, induction of glucagon receptors is inhibited in a manner dependent on the concentration of serum. In contrast, the induction by butyrate is identical in both defined and serum-containing media.

High levels of cyclic AMP also decrease the induction of glucagon responsiveness by butyrate. This observation may explain why concentrations of PGE_1 and Ro 20-1724 greater than 1 μM do not induce as extensively as do lower concentrations of these agents. Induction by any agent can be prevented by inhibitors of phosphodiesterase, cholera toxin, or cyclic AMP analogs, if any of these agents are added along with the inducer.

Applications of Inducible Hormone Receptors

Obviously, a system where hormone receptors can be manipulated provides a convenient model system to study the biogenesis of receptors, from their transcription to their packaging and insertion into the plasma membrane. Also of interest is an understanding of the mechanism of the induction process by butyrate, which causes numerous changes in many cell types,[7] as well as more physiological regulators of differentiation, such as prostaglandins and cyclic AMP. Because glucagon receptors disappear as a result of viral transformation, the understanding of this phenomenon may provide insight into the transformation process.

[31] Characterization of Hormone-Sensitive Madin–Darby Canine Kidney Cells

By MICHAEL C. LIN, SUZANNE K. BECKNER, and FREDERICK J. DARFLER

Cultured cells, especially the established lines, provide a continuous and homogeneous system for studying hormone action in intact cells. The cell line known as Madin–Darby canine kidney cells (MDCK cells),[1] derived from normal dog kidney more than 20 years ago, is ideal for this type

[1] C. R. Gaush, W. L. Hard, and T. F. Smith, *Proc. Soc. Exp. Biol. Med.* **122,** 931 (1966).

ISBN 0-12-182009-2

of research, since it retains differentiated functions in culture[2] and, in addition, responds to several hormones, including glucagon, vasopressin, β-adrenergic agonists, and prostaglandins.[3,4] The maintenance and general properties of MDCK cells were previously described by Taub and Saier.[5] This chapter will deal mainly with the optimal culture conditions for maintaining hormone responsiveness, the measurement of intracellular cyclic AMP, and the characteristics of several types of hormone sensitivity in MDCK cells.

Maintenance of Cell Cultures. Stock cells are kept in 100-mm culture dishes (Costar #3100) at 37° under 5% CO_2/95% air with 80% humidity. For subculture, medium is removed from stock cells, the monolayer culture rinsed with 10 ml phosphate-buffered saline (PBS) and 5 ml trypsin (0.05%)–EDTA (0.02%) is added. Trypsin digestion proceeds for 5 to 20 min at 37° (depending on the culture age of stock cells), until cells begin to detach from the dish when shaken. The cells are quantitatively detached by pipetting or scraping and suspended with Dulbecco's modified Eagle's medium (DMEM) containing 5% fetal bovine serum (Gibco or Hyclone) at the desired cell density for plating. The addition of medium containing 5% serum terminates the trypsin digestion. The optimal cell density for plating is about 0.5 to 2×10^5 cells/35 mm dish. Cells can be maintained in 5% serum or in defined medium as described below. The serum-free medium for MDCK cells has been reported by Taub *et al.*[6] This defined medium consists of DMEM : Ham's F-12 (1 : 1), 5 μg/ml insulin, 5 μg/ml transferrin, 50 n*M* hydrocortisone, 5 p*M* triiodothyronine, 0.1 μ*M* prostaglandin E_1(PGE_1), 10 n*M* selenium dioxide, and 10 m*M* HEPES buffer, pH 7.4. Since PGE_1 is not required for growth under our conditions, it is routinely omitted from our defined medium to maximize hormone sensitivity of cells. To maintain cells in serum-free conditions, MDCK cells are rinsed with PBS 4 hr after plating in 5% serum and defined medium is added.

Measurement of Hormone Responsiveness. The surface of MDCK cells is polarized[5]; it is believed that hormone receptors are on the serosal surface facing the culture dish. Therefore, to assure accessibility of hormone receptors, monolayer cultures which are 80% confluent are used for experiments. Normally 2 to 3 days after plating, MDCK cells in 35-mm dishes are rinsed with 3 ml PBS and incubated with 3 ml DMEM without

[2] J. Leighton, Z. Brada, L. W. Estes, and G. Justh, *Science* **163,** 472 (1969).

[3] M. J. Rindler, L. M. Chuman, L. Shaffer, and M. H. Saier, Jr., *J. Cell Biol.* **81,** 635 (1979).

[4] M. C. Lin, S. M. Koh, D. D. Dykman, S. K. Beckner, and T. Y. Shih, *Exp. Cell Res.* **142,** 181 (1982).

[5] M. Taub and M. H. Saier, Jr., this series, Vol. 58, p. 552.

[6] M. Taub, L. M. Chuman, M. H. Saier, Jr., and G. Sato, *Proc. Natl. Acad. Sci. U.S.A.* **76,** 3338 (1979).

serum or hormone supplement at 37° for 2 hr. When cultured in 5% serum, MDCK cells produce substantial quantities of PGE_2 and $PGF_{2\alpha}$, which can reach a concentration of 0.1 μM in the medium. This level of prostaglandins not only causes high basal concentrations of cyclic AMP but also leads to desensitization of hormone sensitivity, effects not seen when cells are cultured in defined medium. To minimize these effects, cells are incubated for 2 hr with DMEM prior to measurement of hormone sensitivity. After this incubation, the DMEM is replaced with 2 ml fresh DMEM containing 20 μM Ro 20-1724 [4-(3-butoxy-4-methoxy-benzyl)-2-imidazolidinone, Hoffmann-La Roche], an inhibitor of cyclic AMP phosphodiesterase, and 20 m*M* HEPES buffer, pH 7.4. After 30 min at 25° (room temperature), a small aliquot of hormone solution is added to the desired concentration to initiate the reaction. After 3 min, the medium is quickly removed and 1 to 2 ml boiling water is added which serves to terminate the reaction and to extract cyclic AMP. Dishes are scraped and each sample is brought to an identical volume (usually 2 ml) and boiled for 5 min. After centrifugation, the concentration of cyclic AMP in the supernatant is measured by a radioimmunoassay. Since the intracellular concentration of cyclic AMP is low in MDCK cells, the addition of a phosphodiesterase inhibitor simplifies detection of cyclic AMP by eliminating a concentration step.

Radioimmunoassay of Cyclic AMP. Cyclic AMP is measured by radioimmunoassay as described previously[7] with two modifications: (1) cyclic AMP is acetylated to increase the sensitivity of the assay and (2) anti-rabbit Ig antiserum is used to separate bound from free [^{125}I]succinyl cyclic AMP. Briefly the procedure is as follows: samples or standards (5 to 2000 fmol/tube) in 25 to 100 μl are made up to 200 μl with 50 m*M* acetate buffer, pH 6.2, and acetylated with 5 μl acetic anhydride : triethylamine (1 : 2). After 15 min at room temperature, [^{125}I]succinyl cyclic AMP (Meloy Labs, VA), 10,000 cpm in 100 μl, and 100 μl cyclic AMP antiserum, sufficient to bind 30–60% of the radioactive ligand, are added. After 4 hr at room temperature, carrier rabbit serum and second antiserum (anti-rabbit Ig from sheep, goat or burro, Meloy Labs) are added. After standing overnight at 4°, 2 ml of cold 10 m*M* acetate buffer, pH 6.2, is added, the tubes are centrifuged, and the radioactivity in the pellets counted. We have used the antiserum against cyclic AMP obtained from Collaborative Research and New England Nuclear with good results. Unfortunately, antiserum from Collaborative Research is no longer available. Other sources are Meloy Labs (Springfield, VA), Sigma (St. Louis, MO), Miles (Elkhart, IN), and Research Products International Corp.

[7] A. L. Steiner, this series, Vol. 38, p. 96.

(Elk Grove Village, IL). The antiserum from New England Nuclear is precoated with the second antibody and thus the assay requires only one incubation.

Hormone Responsiveness of MDCK Cells. The production of cyclic AMP in the presence of hormone is linear for about 3 min as shown in Fig. 1. During that time period, more than 90% of the cyclic AMP produced remains inside the cells. Therefore, hormone responsiveness is measured in terms of the increase in intracellular cyclic AMP produced in 3 min in the presence of hormone over the basal level. The responsiveness of MDCK cells to several hormones is shown in the table. Cells are more responsive and have lower basal concentrations of cyclic AMP when they are grown in the defined medium than in 5% fetal bovine serum. Concentration-dependent activation of cyclic AMP production is shown in Fig. 2 for three hormones. The concentrations required for half-maximal activation by glucagon, isoproterenol, and PGE_1 are 20, 80, and 100 n*M*, respectively. The binding of [^{125}I]glucagon to MDCK cells, as described in the legend to Fig. 3, is also half maximal at 20 nM (Fig. 3).

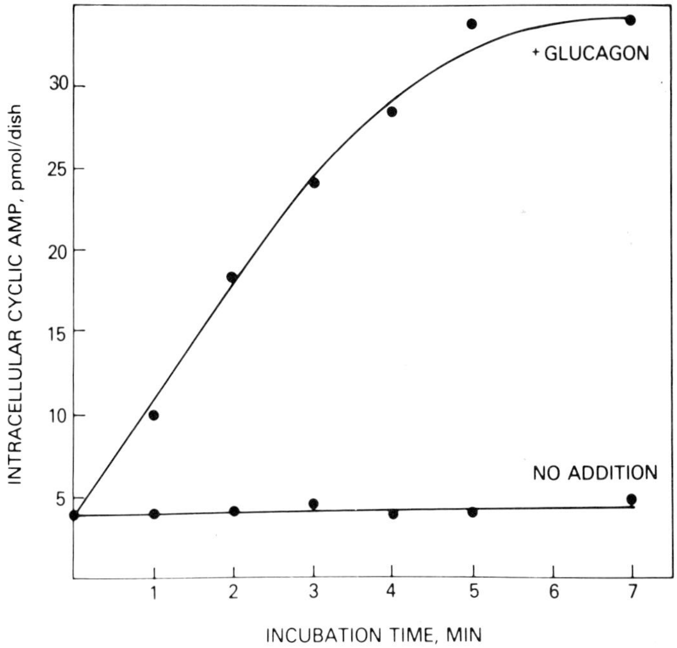

FIG. 1. Time course for cyclic AMP production in MDCK cells. Intracellular cyclic AMP produced in the absence or presence of glucagon (2 μM) was measured by a radioimmunoassay as described.

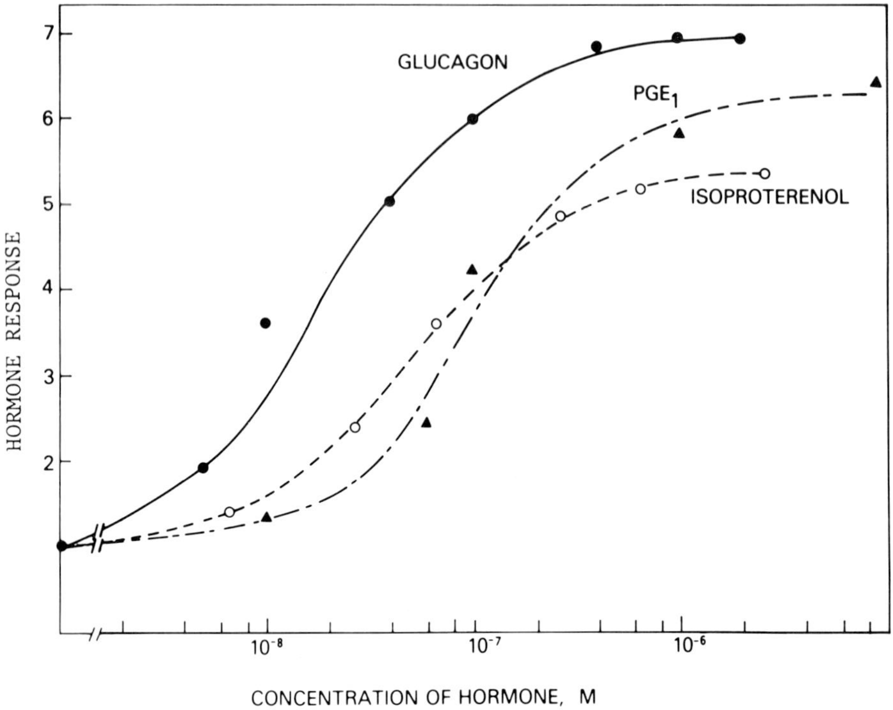

FIG. 2. Concentration-dependent activation of cyclic AMP production by glucagon (●), isoproterenol (○), and prostaglandin E_1 (▲). Intracellular cyclic AMP was measured 3 min after the addition of hormones at concentrations indicated. Hormone response is expressed as described in footnote *d* to the table.

HORMONE RESPONSIVENESS OF MDCK CELLS

Culture medium	Intracellular cyclic AMP[a] (fmol/dish)				
	No addition	+Glucagon[b]	+Vasopressin	+Isoproterenol	+PGE_1
5% fetal bovine serum	74	454 (6.2)[d]	268 (3.7)	440 (6.0)	466 (6.4)
Serum-free[c]	9	386 (43)	85 (9.4)	385 (43)	409 (45)

[a] The concentration of intracellular cyclic AMP was assayed as described in the text. Each value represents the average of three determinations which agreed within 5%.

[b] Concentrations used in the assay: 2 μM glucagon, 2 μM vasopressin, 2 μM isoproterenol, and 5 μM prostaglandin E_1.

[c] Serum-free medium contains hormone supplements as described in the text.

[d] Values in parentheses represent hormone response which is the ratio of cyclic AMP produced in the presence of hormone to that in its absence.

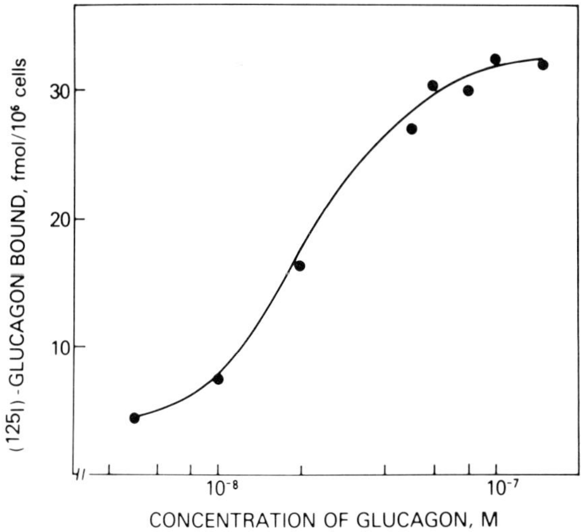

FIG. 3. Concentration-dependent binding of [^{125}I]glucagon to MDCK cells. Monolayer cultures (in 35 mm dishes) were incubated in DME medium containing 20 m*M* HEPES buffer, pH 7.4, 0.2% bovine serum albumin, and 1 m*M* bacitracin for 20 min at 37° with 1 n*M* [^{125}I]glucagon (New England Nuclear) and native glucagon at the concentrations indicated. After two washings with 2 ml of the same medium (4°), bound [^{125}I]glucagon was extracted with 1 ml 0.5 *N* NaOH and the radioactivity was counted. Nonspecific binding was estimated by the inclusion of 2 μM unlabeled glucagon.

Comments. The MDCK cell line is one of the few lines that retain sensitivity toward numerous hormones, particularly the peptide hormones, glucagon and vasopressin. In contrast, glucagon responsiveness of cell lines derived from liver is often diminished. MDCK cells have been used as a model system for studying hormone regulation of kidney functions.[6,8] The availability of a differentiated, hormone-responsive cell line facilitates studies of the cascade of biochemical events subsequent to cyclic AMP elevation and its correlation with specific kidney functions. In a companion report, we show that certain hormone responses of MDCK cells are lost following viral transformation.[9]

[8] J. S. Lever, *Proc. Natl. Acad. Sci. U.S.A.* **76,** 1323 (1979).

[9] S. K. Beckner, F. J. Darfler, and M. C. Lin, this series [30].

[32] Construction of Hormonally Responsive Intact Cell Hybrids by Cell Fusion: Transfer of β-Adrenergic Receptor and Nucleotide Regulatory Protein(s) in Normal and Desensitized Cells

By D. SCHULSTER and D. M. SALMON

Introduction

Sendai virus has been used to fuse various hormone receptors and membrane components together.[1–3] Whole cells can be similarly fused, and receptors from one cell coupled with and activate the adenylate cyclase from another cell.[4] Studies demonstrating this procedure have been performed with turkey erythrocytes (treated with *N*-ethylmaleimide to inactivate their adenylate cyclase system; E_{NEM}) and fused with either Friend mouse erythroleukemia cells (F cells) or mouse adrenal Y1 tumor cells (Y1 cells). The β-adrenergic receptors of the erythrocytes are thus transferred and can be activated to elevate cyclic AMP levels inside the F cells or Y1 cells. In this way, the functioning of receptors and other membrane components may be examined within the natural milieu of the intact cell.

Desensitized turkey erythrocytes may also be fused with normal and cholera toxin treated-human erythrocytes, in order to replenish nucleotide regulatory proteins.[5,6] Sendai virus-mediated fusion of these cells restores the β-adrenergic responsiveness of the desensitized cells.

Materials

N-Ethylmaleimide (NEM), adenosine-3′-5′-cyclic monophosphate Na^+ salt (cyclic AMP), and guanosine 5′-[β,γ-imido]triphosphate (Gpp(NH)p) are all from Sigma Chemical Co. and stored desiccated at −20°. A stock solution (20 ml) of 20 m*M* 3-isobutyl-1-methylxanthine (MIX; Alrich Chemical Co.) is made by dissolving 88.9 mg MIX in 10 ml H_2O and 0.45 ml 1 *N* NaOH, then making up to 20 ml with H_2O.

[1] J. Orly and M. Schramm, *Proc. Natl. Acad. Sci. U.S.A.* **73,** 4410 (1976).

[2] M. Schramm, J. Orly, S. Eimerl, and M. Korner, *Nature* (*London*) **268,** 310 (1977).

[3] Y. Citri and M. Schramm, *J. Biol. Chem.* **257,** 13257 (1982).

[4] D. Schulster, J. Orly, G. Seidel, and M. Schramm, *J. Biol. Chem.* **253,** 1201 (1978).

[5] D. M. Salmon and D. Schulster, *Biochem. Soc. Trans.* **10,** 493 (1982).

[6] D. Schulster and D. M. Salmon, *in* "Membrane Located Receptors for Drugs and Endogenous Agents" (E. Reid, G. M. W. Cook, and D. J. Moore, eds.), pp. 143–150. Plenum, New York, 1983.

METHODS IN ENZYMOLOGY, VOL. 109

ISBN 0-12-182009-2

Agonists are used together with 0.6 m*M* MIX and are prepared as follows.

dl-Isoproterenol hydrochloride (1.5 mg) (isoproterenol; Mann Research Co.) is dissolved in 10 ml H_2O with 30 μl of 1 *M* 2-mercaptoethanol (MCE; Merck-Schuchardt, West Germany) to give a 5×10^{-4} *M* isoproterenol working solution containing 3 m*M* MCE which is made up fresh for each experiment.

Prostaglandin E_1 (0.5 mg) (PGE_1; Upjohn Co.) is dissolved in 1.0 ml ethanol (95% v/v) and 9.0 ml of Na_2CO_3 solution (0.2 mg/ml) added. This stock PGE_1 solution is divided into 0.5-ml aliquots and stored in the freezer (−20°).

Porcine $ACTH_{1-39}$ (2.0 mg) (Armour Pharmaceutical Co., 122 IU/mg) in 1.22 ml of pH 4 saline containing 0.1% human serum albumin (w/v). This stock ACTH solution of 200 IU/ml was divided into 50-μl aliquots and stored in the freezer (−20°).

Radioactive Chemicals. [2-^{3}H]Adenine is 5.8 Ci/mmol, 1 mCi/ml (Nuclear Research Centre, Beersheba, Israel) and [γ-^{32}P]ATP is 13 Ci/mmol, 1 mCi/ml (Amersham Radiochemical Centre, UK).

Methods

Turkey Erythrocytes. Collect fresh blood in heparin from a female turkey from a vein under the wing and wash the erythrocytes twice in Na^+ salt medium (135 m*M* NaCl, 5 m*M* KCl, 0.8 m*M* $MgCl_2$; Tris/HCl buffer, pH 7.4). The buffy coat is aspirated and suspend cells finally in the Na^+ salt medium at a concentration of 20% v/v (8×10^8 cells/ml). The erythrocyte suspension is incubated for 10 min at 4° with 10 m*M* NEM to inactivate irreversibly the enzymic activity of the adenylate cyclase. The Na^+ salt medium is used throughout, otherwise the details are as previously described.[1,2]

Desensitized turkey erythrocytes are prepared by incubating cells in the Na^+ salt buffer pH 7.4 at 37° for 30 min with 50 μ*M* isoproterenol, i.e., final concentrations of 0.3 m*M* MCE and 50 μ*M* isoproterenol.

For membrane preparation, turkey erythrocytes obtained as above are resuspended at 30% (v/v) in Na^+–Tris buffer pH 7.4 containing 20 m*M* glucose. For membranes to be prepared from desensitized cells, erythrocytes are preincubated in this buffer at 37° for 30 min with a final concentration of 50 μ*M* isoproterenol and 0.3 m*M* MCE. Aliquot volumes are taken for membrane preparation.[7] The yield of membrane protein pre-

[7] L. J. Pike and R. J. Lefkowitz, *J. Biol. Chem.* **255,** 6860 (1980).

pared from 1 ml of packed cells is approximately 2 mg.[8] Membrane preparations stored at −70° do not lose activity for at least a month.

Human Erythrocytes. Fresh blood is collected in trisodium citrate from adult male volunteers and washed as described above for turkey erythrocytes.

Cholera Toxin Treatment. A 10% suspension of human erythrocytes (30 ml) in the Na^+ salt medium described above containing 20 m*M* glucose, is incubated with 6×10^{-8} *M* cholera toxin for 2 hr.[9] After the cells have been washed twice in Na^+ salt medium, they are resuspended at a concentration of 10% (v/v).

Cell Cultures. Friend T3C12 mouse erythroleukemia cells (F cells) are grown in F-12 medium (GIBCO) supplemented with 10% fetal calf serum.[1] Mouse adrenal Y1 tumor cells (Y1 cells[10] obtained from the American Type Culture Collection) are grown in F-10 medium (GIBCO) supplemented with 15% fetal calf serum. Cells were harvested[2] at room temperature by centrifuging at 100 *g* for 5 min.

Determination of Cyclic AMP. Radiolabeling with [2-^{3}H]adenine. For the assay of cyclic AMP, cells are preincubated with [2-^{3}H]adenine to produce intracellular [^{3}H]ATP.[11] F cells are harvested by centrifugation and resuspended in 2–3 ml of fresh medium (10^7 cells/ml). The suspension is preincubated with 40 μCi [2-^{3}H]adenine/10^7 cells at 37° for 2 hr. Nonlabeled turkey erythrocytes are used for most of the experiments described below, however, for some control experiments, prelabeling may be performed. In this case 2 ml erythrocytes (20% v/v) are preincubated in Na^+ salt buffer with 120 μCi [^{3}H]adenine. After labeling, cells are washed twice and resuspended in Na^+ salt medium at room temperature. Incorporation of ^{3}H into the cells is 60–80% of the label added to the medium.

After labeling, radioactive metabolites may be separated by thin layer chromatography[12] on PEI-cellulose (Polygram Cell 300 PEI from Macherey-Nagal Co., Duren, West Germany), developed with 1.5 *M* LiCl from ultraviolet absorbing areas eluted for scintillation counting by shaking for 1 hr in 1 ml of 0.7 *M* $MgCl_2$, 20 m*M* Tris–HCl, pH 7.4. In this way, it may be determined that 40% of the [^{3}H]adenine incorporated into the F cells is converted into intracellular [^{3}H]ATP, under the preincubation conditions described.

[8] O. H. Lowry, N. J. Rosebrough, A. L. Farr, and R. J. Randall, *J. Biol. Chem.* **193,** 265 (1951).

[9] S. A. Rudolph, D. E. Schafer, and P. Greengard, *J. Biol. Chem.* **252,** 7132 (1977).

[10] Y. Yasumura, V. Buonassisi, and G. Sato, *Cancer Res.* **26,** Part I, 529 (1966).

[11] J. L. Humes, M. Rounbehler, and F. Kuehl, *Anal. Biochem.* **32,** 210 (1969).

[12] K. Randerath and E. Randerath, this series, Vol. 12, Part A, p. 323.

Cyclic AMP is assessed, following sequential chromatography on columns of Dowex AG 50W-X4 and aluminum oxide,[13] and determining accumulated [^{3}H]AMP in a scintillation counter set for tritium measurement. Labeled cyclic AMP is prepared for use as a recovery indicator, using the adenylate cyclase of Y1 cell ghosts,[2] activated by NaF for the conversion of [α-^{32}P]ATP to cyclic [^{32}P]AMP. This cyclic nucleotide is purified as previously described.[13]

Binding protein assay. Cyclic AMP accumulation in intact cells is also measured using a binding protcin assay method[14] supplied by Amersham International, UK. This assay routinely measures cyclic AMP concentrations in the range 10–250 pmol/ml with a coefficient of variation of less than 7%.

Adenylate cyclase assay. Membrane protein (20–40 μg), suspended in 10 μl of a buffer (75 mM Tris and 25 mM $MgCl_2$, pH 7.4) is incubated at 37° for 5 min in a final volume of 50 μl of medium containing 0.5 mM ATP, 20 mM $MgCl_2$, 50 mM Tris, 0.1 mM EGTA, 2 mM MCE, 2 mM phosphocreatine, and 2.5 units/ml creatine kinase pH 7.4.[13] The reaction is terminated by the addition of 0.2 ml of 5% trichloroacetic acid. After centrifugation, the supernatant is back-extracted three times to remove the acid. Cyclic AMP production is measured by the binding protein assay described above.[14]

Cell Fusion Systems for Intact Cells. NEM-treated turkey erythrocytes (E_{NEM}) were fused with intact F cells (E_{NEM}–Fc) or Y1 cells (E_{NEM}–Y1).

A 50- to 100-fold excess of E_{NEM} is suspended with F cells or Y1 cells in Na^+ salt medium with 2 mM $MnCl_2$. Sendai[15] virus is added in an amount (hemagglutinating units, HAU) calculated on the basis of the number of turkey erythrocytes (see figure legends), and the cells allowed to agglutinate at 4° for 10 min. During this period aliquots of the cells are distributed into polypropylene test tubes (10 × 98 mm) and the fusion process then initiated by transfer to 37°. Agonists or other agents are added at selected times after the onset of fusion to bring the volume of each experimental system to 0.5 ml.

The agents used are 50 μM isoproterenol 10 μM PGE_1, 0.5 μM $ACTH_{1-39}$, and 10 mM sodium fluoride, all used together with 0.6 mM MIX to inhibit the phosphodiesterase.[4]

Desensitized turkey erythrocytes (tE_D) are fused with normal or cholera toxin-treated human erythrocytes (hE or hE_{CT}) as described above.

[13] Y. Salomon, C. Londos, and M. Rodbell, *Anal. Biochem.* **58,** 541 (1974).

[14] A. G. Gilman, *Proc. Natl. Acad. Sci. U.S.A.* **67,** 305 (1970).

[15] Z. Toister and A. Loyter, *Biochem. Biophys. Res. Commun.* **41,** 1523 (1970).

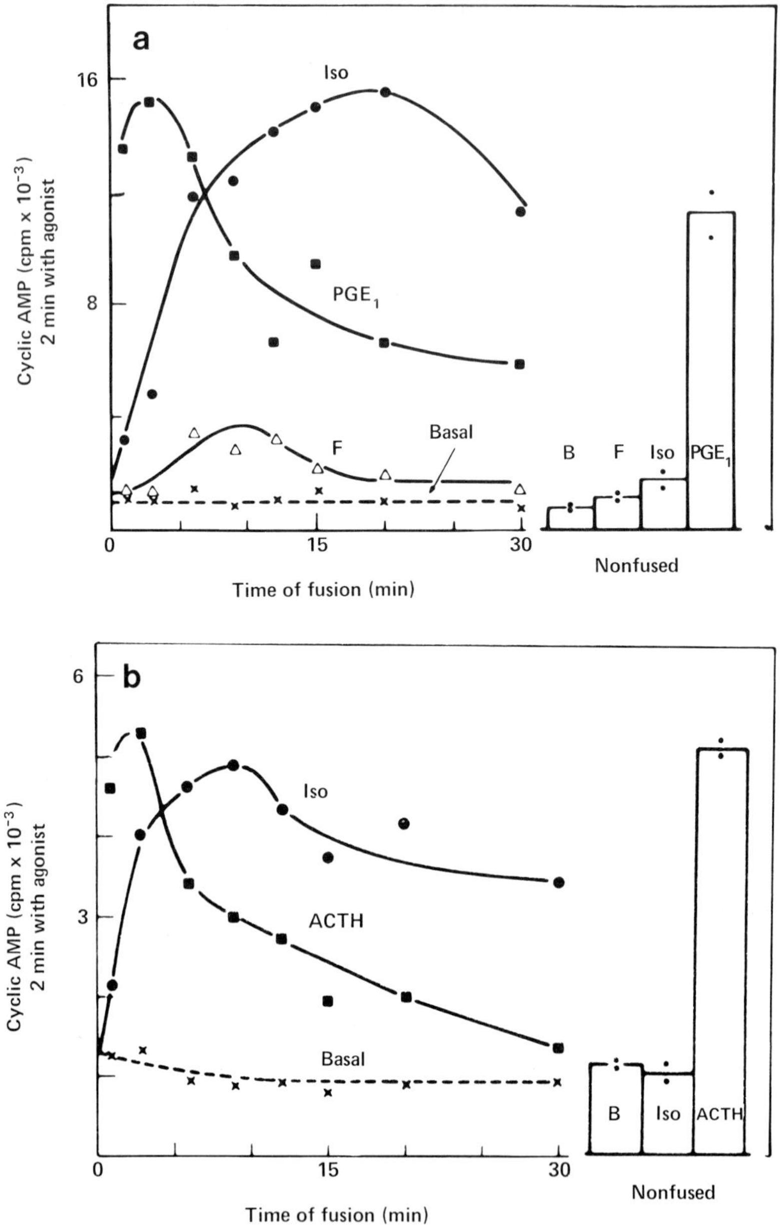

FIG. 1. (a) E_{NEM}–F cells: Changes in hormone responsiveness during the fusion process. Zero time is taken as the time of transfer of the fusion systems from 40 to 37°. Hormones or NaF, together with 0.6 m*M* 3-isobutyl-1-methylxanthine were added at different times during fusion and the reaction terminated after 2 min. Each experimental system of 0.5 ml contains 1×10^6 F cells, 7×10^7 E_{NEM}, and 200 HAU of Sendai virus. Similarly treated experimental systems, but in the absence of virus, serve as the nonfused control system

The protocol is summarized in Scheme 1. Suspensions of 1.5×10^6 tE cells are distributed with 5×10^8 hE or hE_{CT} cells per tube into Eppendorf microfuge tubes (tE/hE or tE/hE_{CT} ratio of 1 : 300). $MnCl_2$ is added to a final concentration of 2 m*M* and 10 μl (150 hemagglutinating units) of a suspension of Sendai virus then added. Agglutination is again allowed to proceed at 4° for 10 min and fusion initiated by transferring to 37° for incubation. Agonists and other agents are added at selected times to bring the final volume to 0.5 ml. Incubations are terminated by centrifugation of cells in an Eppendorf microfuge for 10 sec, removal of the supernatant, and the addition of 0.3 ml of 5% trichloroacetic acid to the cell pellet. After it has been mixed and centrifuged, the acid supernatant is back-extracted (×3) with diethylether and heated at 55° for 20 min, before the binding protein assay of cyclic AMP.

Transfer of β-Adrenergic Receptor during Fusion

The β-receptor is contributed from turkey erythrocytes in which the adenylate cyclase is irreversibly inactivated by NEM. For the fusion process, Sendai virus is added to the cells which are held at 4° for 10 min and subsequently transferred to 37° to initiate the fusion process. The cyclic AMP production in response to different stimulators, is shown in Fig. 1, at different times after the onset of fusion for both E_{NEM}–Fc (Fig. 1a) and E_{NEM}–Y1 (Fig. 1b). Incubation with the stimulator is for the brief period of 2 min since the fusion process continues during incubation with agonist and this time period is optimal for producing maximal cyclic AMP levels.[4] Isoproterenol rapidly enhances cyclic AMP production and this is visible 1 min after the onset of fusion; the maximal production of cyclic AMP levels is evident for E_{NEM}–Fc 15 min, or for E_{NEM}–Y1 10 min after fusion onset. It is of interest that the 15-fold response produced by isoproterenol after 15 min of fusion (E_{NEM}–Fc) is as high as that of the initial PGE_1 response on the native PGE_1 receptor both in the fused cell system and in the nonfused controls. Similar responses to isoproterenol are observed for E_{NEM}–Y1.

shown in the histogram. Systems are Iso, 50 μM isoproterenol (●); 10 μM PGE_1 (■); F, 10 m*M* fluoride (△); B, basal; 0.6 m*M* 3-isobutyl-1-methylxanthine only (×). (b) E_{NEM}–Y1: Changes in hormone responsiveness during cell fusion. The experimental procedure is as described in a. Each 0.5 ml experimental system contains 6×10^5 Y1 cells, 1×10^8 E_{NEM}, and 600 HAU of Sendai virus. Experimental systems as above, but without virus, serve as the nonfused controls shown in the histogram. Systems are Iso, 50 μM isoproterenol (●); ACTH, 0.5 μM $ACTH_{1-39}$ (■); basal, B, 3-isobutyl-1-methylxanthine only (×).

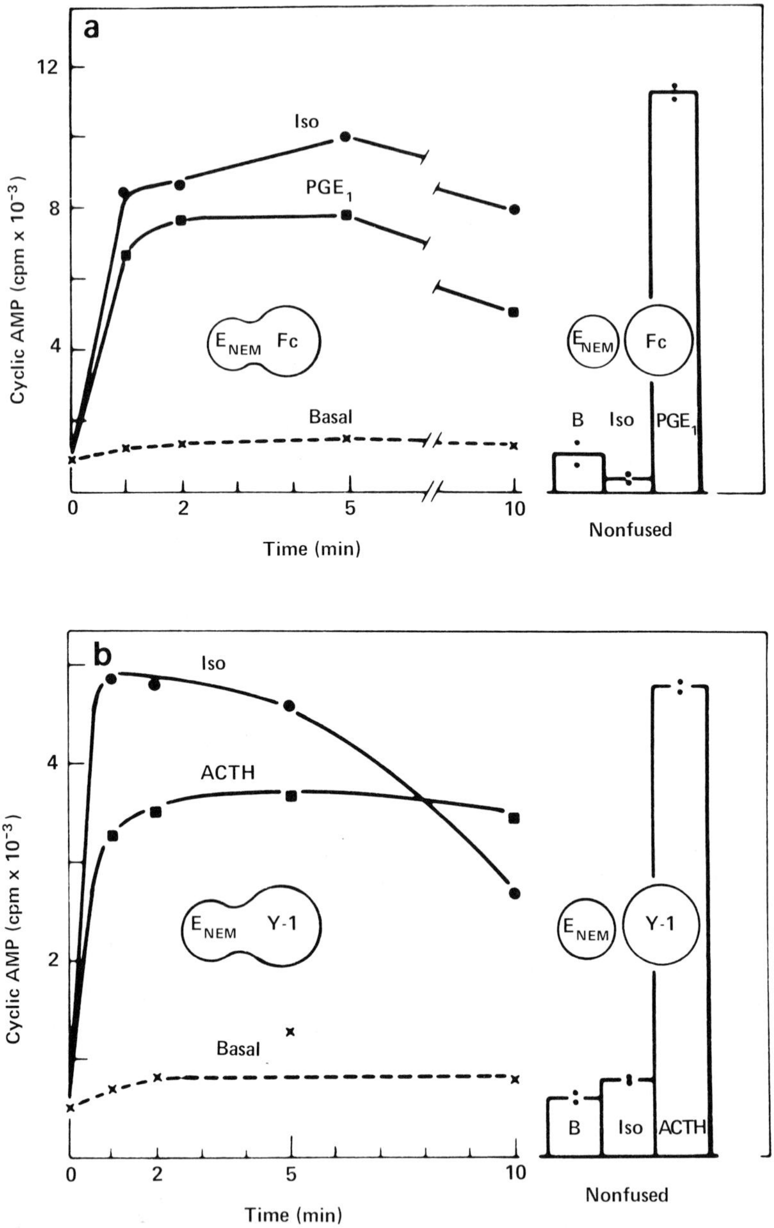

FIG. 2. (a) E_{NEM}–F cells: Kinetics of cyclic AMP production after cell fusion. Systems are symbolized by circles, joined for the fused cells and separated for the nonfused cells. Zero time is 15 min after transfer of the fusion systems from 4 to 37°. At this time isoproterenol or PGE_1, together with 0.6 m*M* 3-isobutyl-1-methylxanthine, is added. At the indicated times, the reaction is terminated. Each 0.5 ml experimental system contains 2×10^6 F cells, 1.1×10^8 E_{NEM}, and 320 HAU of Sendai virus. A mixture of cells in the absence of

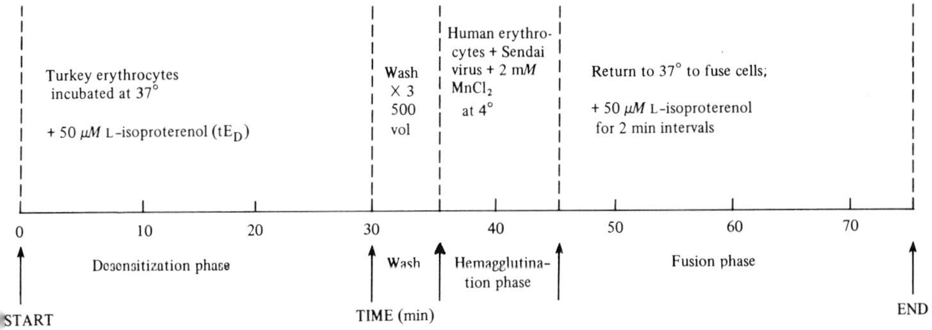

SCHEME 1. Protocol for desensitization and cell fusion.

Kinetics of Cyclic AMP Production after Cell Fusion

It is evident from Fig. 1a that 15 min after onset of fusion, the adenylate cyclase of the F cells responds maximally to the actions of the β-adrenergic receptor of the turkey erythrocyte. Therefore, the time course of cyclic AMP production in response to various stimulators is studied 15 min after onset of fusion (Fig. 2a). The response is similarly rapid for the transferred β-adrenergic receptor and for the native PGE_1 receptor. Maximal responses are observed 1 to 2 min after addition of hormone. Thereafter the responses level off, and at 10 min after hormone addition a decrease in labeled cyclic AMP is observed. These studies indicate a burst of hormone-stimulated adenylate cyclase activity with rapid attainment of a new steady-state level of cyclic AMP.

Receptor transfer is also examined in the E_{NEM}–Y1 system (Fig. 2b). The β-adrenergic response is remarkably similar to that previously shown for the E_{NEM}–Fc system, despite the difference in origin and in the native receptors of the two cultured cell types. Nine minutes after onset of fusion, isoproterenol increases the cyclic AMP level 6-fold over basal and this equals that obtained by $ACTH_{1-39}$ in the nonfused cells (Fig. 1b). It is also evident from this figure that the relative increase in cyclic AMP, that the hormone elicits, is again dependent on the time after onset of fusion. The $ACTH_{1-39}$ response of the Y1 cells fused with NEM-treated erythro-

---

virus is similarly treated and serves as the nonfused control system shown in the histogram. (b) E_{NEM}–Y1 cells: Kinetics of cyclic AMP production after cell fusion. Zero time is 12 min after the transfer of the fusion systems from 4 to 37° when 0.6 m*M* 3-isobutyl-1-methylxanthine and hormones are added. The experimental procedure is as described in a. Each 0.5 ml experimental system contains 6.6×10^5 Y1 cells, 6.4×10^7 E_{NEM}, and 400 HAU of Sendai virus. Experimental systems as above, but without virus, serve as the nonfused controls shown in the histogram. Systems are Iso, 50 μ*M* isoproterenol (●); ACTH, 0.5 μ*M* $ACTH_{1-39}$ (■); basal, B, 3-isobutyl-1-methylxanthine only (×).

cytes declines steadily from 3 min after onset of fusion, and after 30 min is less than twice the basal value. The time course of cyclic AMP production after fusion is shown in Fig. 2b and rapid responses to isoproterenol and $ACTH_{1-39}$ are observed. Maximum levels are again reached 1 to 2 min after addition of the hormones.

Five minutes after addition of hormone 80–85% of the total cyclic AMP is intracellular.[4] Thereafter the amount accumulating in the medium increases, although most of the cyclic AMP is still inside the fused cells even 20 min after the hormone is added. Thus, although the results show the response patterns for total cyclic AMP production, they are very similar to those found for intracellular cyclic AMP.

It is possible to reverse the effect of added hormone to the fused cell system, with an equimolar concentration of a β-receptor blocker such as propranolol. Although 2 μM propranolol rapidly diminished the response to 2 μM isoproterenol, this was reversed by flooding the system with 200 μM isoproterenol.[4] The transferred β-adrenergic receptor response to agonist and antagonist is therefore like that in conventional cell systems.

Desensitization in Normal and Fused Cells

The extent of desensitization is established in normal cells (Fig. 3). After 30 min exposure to isoproterenol, turkey erythrocytes accumulate less cyclic AMP when challenged for a second time with isoproterenol compared with freshly prepared cells.

The extent of desensitization is assessed by calculating the quantity of cyclic AMP produced in cells stimulated with isoproterenol and subtracting that produced in its absence (basal). In fresh erythrocytes 9.4 pmol cyclic AMP accumulated/10^6 cells/30 min, compared to 4.3 pmol in desensitized cells and the β-adrenergic response of these cells is reduced by 54%.

Desensitization to isoproterenol in turkey erythrocytes is overcome by fusing these cells with human erythrocytes which are able to contribute hormonally unaltered nucleotide regulatory protein.[16,17] The use of human erythrocytes as donors for nucleotide regulatory protein is advantageous in such studies since they possess few or no β-receptors or adenylate cyclase activity.[16,17] However, they contain as much nucleotide regulatory protein as other preparations replete in hormonally responsive adenylate cyclase components.[17] Cholera toxin treatment of human erythrocytes before fusion serves to alter nucleotide regulatory protein by

[16] H. R. Kaslow, Z. Farfel, G. L. Johnson, and H. R. Bourne, *Mol. Pharmacol.* **15,** 472 (1979).

[17] T. B. Nielson, P. M. Lad, M. S. Preston, and M. Rodbell, *Biochim. Biophys. Acta* **629,** 145 (1980).

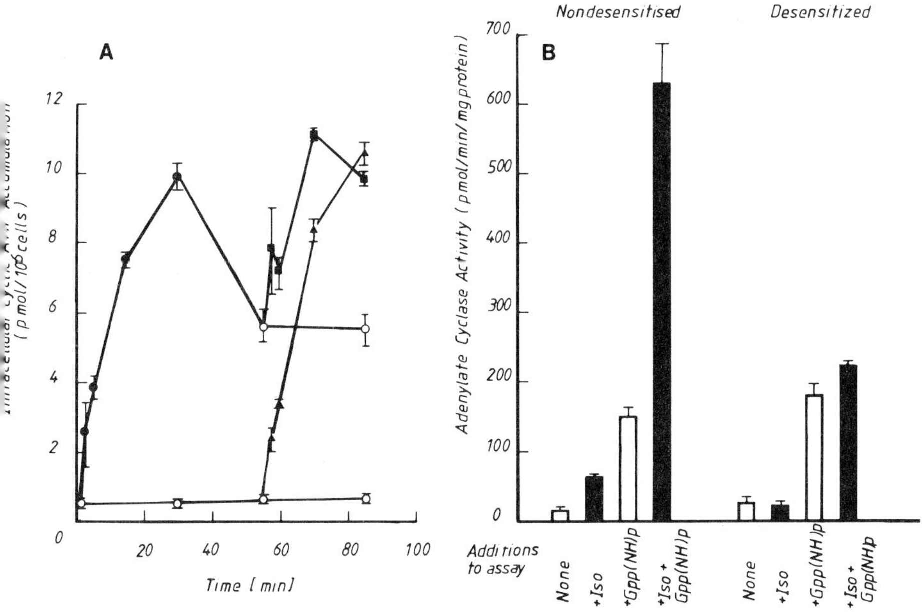

FIG. 3. Desensitization in intact cells and membranes prepared from turkey erythrocytes stimulated with isoproterenol: (A) 1.5×10^6 turkey erythrocytes are incubated at 37° for 30 min in the presence (●) or absence (○) of 50 μM isoproterenol. After washing three times with 500 vol of Tris-buffered saline they are then returned to 37° with (■, ▲) or without (○) 50 μM isoproterenol for a further 30 min. Points represent the mean ± SEM of three observations. (B) Membranes prepared from cells, as described in text, are incubated at 37° for 30 min with or without isoproterenol. The activity of adenylate cyclase in these preparations is determined with no additions (□), 50 μM isoproterenol (■), 40 μM Gpp(NH)p (□), or 50 μM isoproterenol with 40 μM Gpp(NH)p (■). Bars are the mean ± SEM of three observations.

ADP-ribosylation.[16] This results in inhibition of GTPase activity, such that it is incapable of maintaining desensitization. Nucleotide regulatory protein may thus be "activated" by cholera toxin and transferred to desensitized turkey erythrocytes, using Sendai virus cell fusion techniques. In such acute experiments, fusion is maximal over approximately a 30-min period. The most sensitive way to detect transfer of a membrane component between cells is achieved by determining rates of cyclic AMP accumulation for 2 min (Figs. 1 and 2). Cell fusion continues over 30 min.

Turkey erythrocytes are desensitized for 30 min by incubation in the presence of isoproterenol, washed, and then fused with human erythrocytes in the presence of Mn^{2+} and Sendai virus (Scheme 1). In the same experiments, desensitized turkey erythrocytes are fused with human erythrocytes that have been previously treated with cholera toxin.

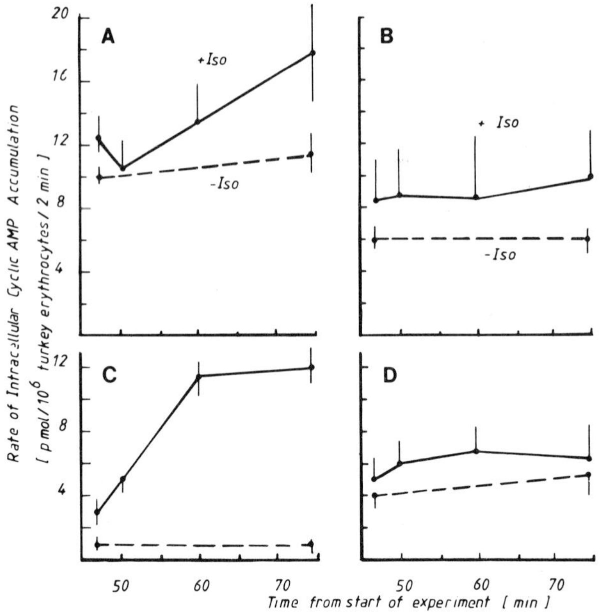

FIG. 4. β-Catecholamine stimulation of cyclic AMP in desensitized turkey erythrocytes fused with human erythrocytes. (A) Fused, normal human erythrocytes ($tE_D \times hE$): 1.5×10^6 turkey erythrocytes are desensitized (tE_D) by incubation with 50 μM isoproterenol at 37° for 30 min. After washing for 5 min (500 vol ×3) the cells were incubated at 4° for 10 min with 5×10^8 normal human erythrocytes (hE) in the presence of 150 HAU Sendai virus and 2 mM $MnCl_2$. They were then returned to 37° and incubated in presence (——) or absence (--) of 50 μM isoproterenol for 2-min periods (45–47, 50–52, 60–62, and 75–77 min). Each point is the mean ± SEM of three observations. The difference between the responses after 75 min is statistically significant by the Student's test ($p < 0.1$). (B) Nonfused controls ($tE_D + hE$): cells were treated as described in (A) except that Sendai virus was omitted. (C) Fused, cholera toxin-treated human erythrocytes ($tE_D \times hE_{CT}$): cells were incubated as described in (A) except that the human erythrocytes were pretreated for 2 hr with 6×10^8 M cholera toxin (hE_{CT}). (D) Nonfused, cholera toxin-treated human erythrocytes ($tE_D + hE_{CT}$): cells were incubated as described in (C) except that Sendai virus was omitted.

The effect of fusion on the response to isoproterenol is assessed by comparing the increment for fused and nonfused systems in the rate of intracellular cyclic AMP accumulation on the addition of isoproterenol (i.e., stimulated minus basal). When desensitized turkey erythrocytes are fused with untreated turkey erythrocytes (Fig. 4A) this increment is 6.4 pmol/10^6 turkey erythrocytes/2 min compared with 3.6 pmol in the nonfused system (Fig. 4B). These values are obtained 30 min after the initiation of fusion and are significantly different by Student's t test ($p < 0.1$).

When desensitized turkey erythrocytes are fused with cholera toxin-treated human erythrocytes (Fig. 4C), the increment in the rate of cyclic AMP accumulation caused by isoproterenol is even greater: 11.2 pmol cyclic AMP is accumulated/10^6 turkey erythrocytes/2 min in the fused system and only 1.2 pmol in the nonfused system (Fig. 4D).

In summary, fusion of normal, untreated human erythrocytes with desensitized turkey erythrocytes increases isoproterenol stimulation of cyclic AMP accumulation over basal rates. Moreover, pretreatment of the human erythrocytes with cholera toxin before they are fused with desensitized turkey erythrocytes leads to a large stimulation with isoproterenol. This is even greater and far more rapid than the response obtained if turkey erythrocytes are treated directly with cholera toxin. It is concluded that the stimulation in the fused system is due to the transfer of an ADP-ribosylated subunit of nucleotide regulatory protein.[18]

Acknowledgments

We gratefully acknowledge financial support from the European Molecular Biology Organization, and the Wellcome Trust. Professor M. Schramm and Dr. J. Orly are thanked for their contribution in the development of the work and much helpful discussion.

[18] E. Hanski and A. G. Gilman, *J. Cyclic Nucleotide Res.* **8,** 323 (1982).

[33] Establishment and Characterization of Fibroblast-Like Cell Lines Derived from Adipocytes with the Capacity to Redifferentiate into Adipocyte-Like Cells

By R. NÉGREL, P. GRIMALDI, C. FOREST, and G. AILHAUD

Over the past 10 years, fibroblast-like cells of established cell lines and cell strains which convert to adipocyte-like cells in culture have provided unique opportunities for studying the process of adipose conversion.[1–15]

[1] H. Green and O. Kehinde, *Cell* **1,** 113 (1974).
[2] H. Green and M. Meuth, *Cell* **3,** 127 (1974).
[3] H. Green and O. Kehinde, *Cell* **7,** 105 (1976).
[4] L. Diamond, T. G. O'Brien, and G. Rovera, *Nature (London)* **269,** 247 (1977).
[5] R. Négrel, P. Grimaldi, and G. Ailhaud, *Proc. Natl. Acad. Sci. U.S.A.* **75,** 6054 (1978).
[6] J. S. Greenberger, *Nature (London)* **275,** 752 (1978).
[7] A. Hiragun, M. Sato, and H. Mitsui, *In Vitro* **16,** 685 (1980).
[8] M. Darmon, G. Serrero, A. Rizzino, and G. Sato, *Exp. Cell Res.* **132,** 313 (1981).
[9] H. Kodama, Y. Amagai, H. Koyama, and S. Kasai, *J. Cell. Physiol.* **112,** 83 (1982).

METHODS IN ENZYMOLOGY, VOL. 109

ISBN 0-12-182009-2

Among the most obvious advantages of this approach are (1) the ease of conducting long-term experiments *in vitro* since the viability of adipocytes isolated from adipose tissue does not exceed a few hours, and to do so with no interference of other cell types as it is the case in explants of adipose tissue, (2) the possibility to control the cell environment and to show the direct effects of growth factors and hormones, and (3) the possibility to study morphological and biochemical events occurring during adipose conversion at both cellular and molecular levels.

In this chapter, we will describe the isolation of cell lines from the adipocyte fraction of epididymal fat pads of C57 BL/6J genetically obese (genotype ob/ob) and nonobese (genotype +/?) mice. Cells of these fibroblast-like established lines are able to differentiate into adipose-like cells having both morphological and biochemical properties characteristic of adipocytes.

Obtaining the Adipocyte Fraction

Both epididymal fat pads (2–2.5 g) of 8-week-old ob/ob or +/? mice from the same litter are removed under sterile conditions, rinsed thoroughly at room temperature with a Krebs–Ringer phosphate (KRP) solution, pH 7.4 containing 100 μg streptomycin/ml and 400 units penicillin/ml. All subsequent operations are carried out under sterile conditions with plastic material. Fat tissue is then cut into 10–30 mm^3 cubes and incubated at 37° for 45 min with KRP solution containing 2 mg/ml of collagenase (Worthington Biochemical Corporation, grade I) and 1% bovine serum albumin (Sigma Chemical Co., fraction V) (1 g tissue/3 ml KRP solution).

The cell suspension is then filtered on nylon gauze (200–250 μm pore size) and the filtrate is centrifuged after a 5-fold dilution with KRP solution (10 min at 800 *g*). The adipocyte fraction is recovered, washed gently with 15 ml of Dulbecco's modified Eagle's medium (Gibco catalogue No. 430-2100), and centrifuged 10 min at 800 *g*. This step is repeated three times.

[10] M. Lanotte, D. Scott, T. M. Dexter, and T. D. Allen, *J. Cell. Physiol.* **111,** 177 (1982).
[11] W. J. Poznanski, I. Waheed, and R. L. R. Van, *Lab. Invest.* **29,** 570 (1973).
[12] G. H. Rothblat and F. D. DeMartinis, *Biochem. Biophys. Res. Commun.* **78,** 45 (1977).
[13] R. L. R. Van and D. A. K. Roncari, *Cell Tissue Res.* **181,** 197 (1977).
[14] P. Björntorp, M. Karlsson, H. Pertoft, P. Petterson, L. Sjöström, and U. Smith, *J. Lipid Res.* **19,** 316 (1978).
[15] H. A. K. Plass and A. Cryer, *J. Dev. Biol.* **2,** 275 (1980).

Obtaining the Clonal Cell Lines

Materials

Culture dishes (Falcon, Nunc)
25-mm-diameter coverslips (Thermanox-Lux Scientific Corporation, Cat. No. 5415)
Nylon Blutex (250 μm pore size; Tripette & Renaud)
Freezing vials (volume 1.8 ml; Nunc)
Culture medium A: Dulbecco's modified Eagle's (DME) medium (Gibco Cat. No. 430-2100) supplemented with sodium bicarbonate, 3.7 g/liter, penicillin-G, 200 U/ml, streptomycin, 50 μg/ml, tetracyclin, 10 μg/ml, and 10% fetal calf serum (FCS)
Culture medium B: culture medium A without tetracycline but supplemented with 33 μM biotin and 17 μM pantothenate
Culture medium C: DME medium (Gibco Cat. No. 430-1600) supplemented with sodium bicarbonate, 1.2 g/liter, HEPES buffer, 15 mM, pH 7.4, penicillin-G, 200 U/ml, streptomycin, 10 μg/ml, biotin, 33 μM, pantothenate 17 μM, and 10% fetal calf serum
Fetal calf serum (Gibco, Seromed, Flow Laboratories)
Trypsin solution: phosphate-buffered saline pH 7.2 (PBS), Mg^{2+}- and Ca^{2+}-free, containing trypsin, 2.5 g/liter (from Gibco 1:250) and EDTA, 0.5 mM

The adipocyte fraction is recovered after the last centrifugation (approximately 5×10^4 cells in 0.5 ml). Cell number is determined with a hemacytometer. A 0.1-ml aliquot of the cell suspension is layered onto a plastic coverslip placed at the bottom of a 35-mm-diameter culture dish. The cells are "sandwiched"[11] with a second coverslip. After 15 min at room temperature, 1.5 ml of culture medium A is added. The cells are maintained at 37° in a humidified 5% CO_2 atmosphere and fed every other day for 12 days. During this period of time a significant proportion of cells (10–20%) lose all lipid droplets and become fibroblast-like (Fig. 1). Under these conditions, no multiplication of the "dedifferentiated" cells is observed during at least the first 9 days after inoculation, in contrast to cells of the stromal-vascular fraction grown in parallel which divide actively in the first 2 days after inoculation. After removal of the culture medium, coverslips are separated from each other and the cells attached to them are rinsed at 37° with KRP solution. Each coverslip is treated in a dish for 10 min at 37° with 1 ml of the trypsin solution. After cell dissociation, both coverslips are rinsed with 2 × 1 ml of culture medium A at 37° and the combined cell suspensions are centrifuged 10 min at 800 g. The cell pellet is resuspended in 5 ml of culture medium A and cells are inoculated into

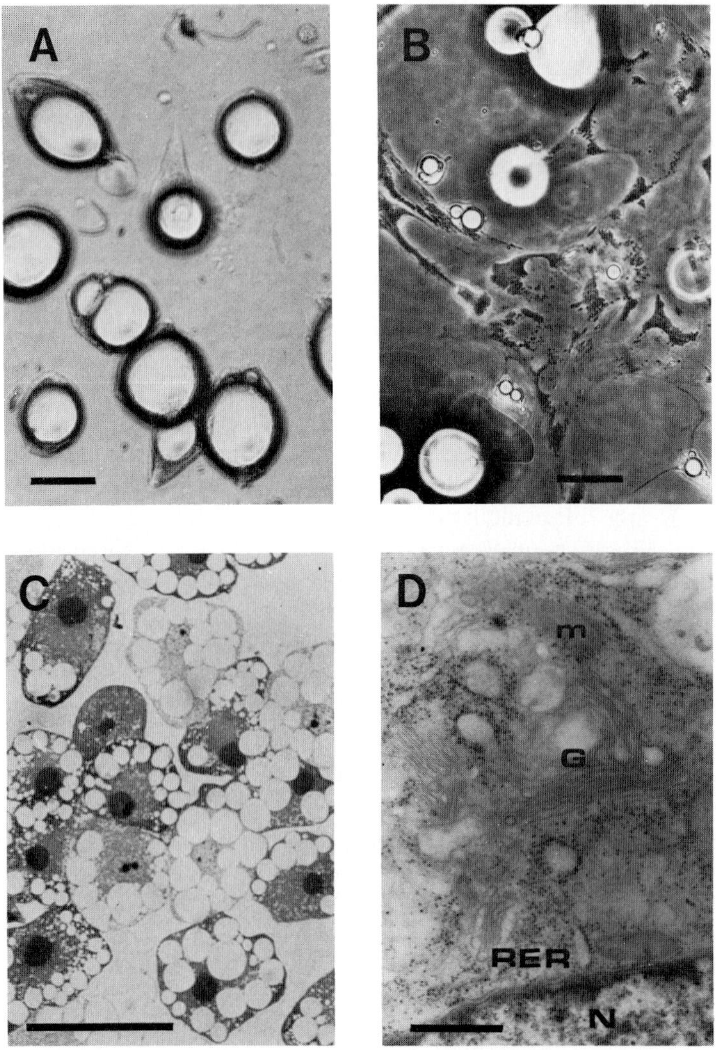

FIG. 1. Morphology of isolated adipocytes on days 2 (A) and 10 (B), and of adipocyte-like cells after establishment of ob17 clonal line (C and D). Upon culture, adipocytes ("sandwiched" between two coverslips) begin to attach to the dish (A), many of them lose triglycerides and become fibroblast-like (B). Once established, fibroblast-like cells of ob17 line susceptible to adipose conversion change their morphology after confluence, round up, accumulate triglycerides, and present a multilocular appearance. Differentiated cells are present in clusters, whereas insusceptible cells are not visible on this 1-μm-thick section (C). Differentiated cells possess a well-organized Golgi apparatus and subcellular structures: N, nucleus; G, Golgi apparatus; m, mitochondria; RER, rough endoplasmic reticulum (D). (A–C) bar = 50 μm; (D) bar = 0.5 μm. (Micrographs obtained through the courtesy of Drs. E. Baechler, C. Mourre, and C. Vannier.)

60-mm-diameter dishes (50–500 cells/cm^2). Within the next 8–10 weeks a few colonies, if any, appear (1–3 colonies per dish). The colonies are dissociated by trypsin treatment as above, and subcloned by using two different methods.

First method. This method has been used to select the ob17 cell line (see the table).[16–26] Exponentially growing cells (approximately 1/4 of the cell density attained at confluence) are treated with trypsin as described above and suspended in culture medium A. After counting with a Coulter counter, the cell suspension is further diluted to 5 cells/ml. Fifty cells are seeded in 100-mm-diameter dishes and grown for 2 weeks, with medium changes every other day, during which time cell clones develop (15–25 clones per dish). Cells of each clone are trypsinized using a cloning cylinder. An aliquot of each cell suspension is reinoculated to select clones with high density inhibition of growth and high frequency of conversion into adipose-like cells after confluence. The remaining cells are passaged twice (approximately 9 generations), trypsinized, and stored in liquid nitrogen. Reinoculation of frozen cells is defined as the first passage.

Second method. This method has been used to select the HG clone (see the table). The cells, suspended in culture medium B, are counted with a Coulter counter, further diluted to 1 cell/4 ml, and inoculated in 24-multiwell plates (16-mm diameter; 1 ml of cell suspension per well). The same medium is changed once a week. On the second week 30 n*M* insulin is added. After 2–3 weeks, clones are selected using the same criteria as above. After a number of passages in culture medium B, corresponding at least to 80 generations, cells of the HG clone were treated with 5-fluorodeoxyuridine in order to select cells showing a higher density inhibition of growth and increased frequency of adipose conversion. For that purpose, 150,000 cells are inoculated in 100-mm-diameter dishes and grown to

[16] R. Négrel and G. Ailhaud, *Biochem. Biophys. Res. Commun.* **98,** 768 (1981).

[17] P. Grimaldi, R. Négrel, J. P. Vincent, and G. Ailhaud, *J. Biol. Chem.* **254,** 6849 (1979).

[18] C. Forest, R. Négrel, and G. Ailhaud, *Biochem. Biophys. Res. Commun.* **102,** 577 (1981).

[19] J. Gharbi-Chihi, P. Grimaldi, J. Torresani, and G. Ailhaud, *J. Recept. Res.* **2,** 153 (1981).

[20] C. Forest, D. Czerucka, P. Grimaldi, C. Vannier, R. Négrel, and G. Ailhaud, *in* "The Adipocyte and Obesity: Cellular and Molecular Mechanisms" (A. Angel and D. A. K. Roncari, eds.), p. 53. Raven Press, New York, 1983.

[21] C. Vannier, H. Jansen, R. Négrel, and G. Ailhaud, *J. Biol. Chem.* **257,** 12387 (1982).

[22] C. Forest, P. Grimaldi, D. Czerucka, R. Négrel, and G. Ailhaud, *In Vitro* **19,** 344 (1983).

[23] G. Ailhaud, E. Amri, C. Cermolacce, P. Djian, C. Forest, D. Gaillard, P. Grimaldi, J. Khoo, R. Négrel, G. Serrero-Davé, and C. Vannier, *Diabete Metab.* **9,** 125 (1983).

[24] P. Grimaldi, P. Djian, R. Négrel, and G. Ailhaud, *Embo J.* **1,** 687 (1982).

[25] P. Grimaldi, P. Djian, C. Forest, P. Poli, R. Négrel, and G. Ailhaud, *Mol. Cell. Endocrinol.* **29,** 271 (1983).

[26] P. Grimaldi, C. Forest, P. Poli, R. Négrel, and G. Ailhaud, *Biochem. J.* **214,** 443 (1983).

confluence in culture medium B. Fresh medium containing 5-fluorodeoxyuridine, 25 μg/ml and uridine, 125 μg/ml, is then added and the cells maintained for 72 hr under these conditions. Cells (designated HGFu) are then rinsed with a PBS Mg^{2+}- and Ca^{2+}-free solution, trypsinized, seeded at a density of 5×10^4 cells per 100-mm-diameter dish, and grown in culture medium B. After tryptic dissociation of half-confluent cells, aliquots of the cell suspension are stored in liquid nitrogen (vide infra). Reinoculation of frozen cells is defined as the first passage.

Freezing Cells for Storage and Recovery from Storage

After removal of culture medium, cells in 100-mm-diameter dish are rinsed with PBS solution, then 1 ml of trypsin solution is added. After 5 min at 37°, trypsination is arrested by adding 20 ml of culture medium B or C. The cell suspension is centrifuged, and the cell pellet is resuspended at 4° in DME medium containing 10% dimethyl sulfoxide and 20% fetal calf serum, at a density of 8×10^5–10^6 cells/ml. The cell suspension is rapidly transferred to freezing vials (1 ml per vial). Cells are maintained 6–24 hr at −20° and vials of frozen cells are transferred to a liquid nitrogen storage tank.

To recover cells from frozen storage, one warms the bottom of the vial at 37° immediately upon removal from liquid nitrogen. As soon as the sample has melted, the vial is wiped with 95% ethanol, and the cells transferred into a 100-mm-diameter dish containing culture medium B at 37° enriched to 20% fetal calf serum. This medium is aspirated after 24 hr and replaced by 10 ml of culture medium B or C. Under these conditions 80 to 90% of thawed cells remain viable and are ready to propagate.

Propagation of Cultured Preadipocyte Cells

Cells are propagated at 37° in a 5% CO_2 humidified atmosphere in culture medium B or C. Medium is changed every 2 days. Cells are passaged when they reach a density between one-half and two-thirds of that of confluent cells, at a splitting ratio of 1 to 10 or 1 to 20. For that purpose, cells are treated with trypsin and the cell density adjusted with culture medium to 2.5–5×10^3 cells/ml. Cells are then seeded at a density of 5×10^2–10^3 cells/cm^2. Studies on adipose conversion (vide infra) are carried out with cells between the third and the thirtieth passage. The doubling times of ob17 and HGFu cells may vary with the batch of fetal calf serum used, but for a single batch growth rates are similar for both cell lines (see the table). Selection of the serum lot is very important since many batches prove to be strongly mitogenic and, in that case, continuous proliferation occurs yet no significant adipose conversion takes place.

The absence of mycoplasmal contamination has been ascertained by a fluorescence technique using the Hoechst compound 33258.[27]

Adipose Conversion

After seeding at a density of 10^3–2×10^3 cells/cm^2, cells are grown to confluence in culture medium B changed every 2 days: 40–55 $\times$ 10^3 cells/cm^2 are thus obtained within 4–5 days. Optimal concentrations of 17 n*M* insulin and 2 n*M* triiodothyronine are then added. The formation of clusters is visible within 5–7 days after confluence and maximal adipose conversion occurs within the following 10–13 days (Fig. 1). The adipose conversion is not a uniform process, not even within a given group of cells. In a given cluster, for cells susceptible to convert to adipose-like cells, the accumulation of lipids starts in central cells whereas cells at the edges are still multiplying. Adipose conversion of ob17 and HGFu cells is completely abolished by treating cells during their exponential phase of growth and after confluence with 5-bromodeoxyuridine,[2,5] or by treating confluent cells (tested on ob17 cells) with mitogens (prostaglandin $F_{2\alpha}$,[28] EDGF,[28] FGF[23]). Addition at confluence of either clofenapate or indomethacin to 5-bromodeoxyuridine-treated cells overrides this block.[29] Significant adipose conversion can occur by maintaining ob17 cells in the presence of 10% bovine plasma or 10% calf serum instead of 10% fetal calf serum, providing the addition of methylisobutylxanthine (MIX) and of bovine growth hormone (b-GH). MIX (0.1–0.5 m*M*) is added within the first 2 days after confluence only, whereas b-GH (3–12 n*M*) is present throughout the postconfluent phase. However, the best method to accelerate adipose conversion is by treating exponentially growing cells in culture medium C, between 4 and 2 days preconfluence, with optimal concentrations of phosphodiesterase inhibitors, either 0.1 m*M* MIX or 0.05 m*M* RO 20-1724 (Hoffmann-Laroche & Co.). Growth is then continued in culture medium C until confluence, followed by the addition of 2 n*M* triiodothyronine and 17 n*M* insulin. Under these conditions, cell clusters are visible within 3 days after confluence and maximum adipose conversion occurs within the following 8–10 days.

So far, the adipose conversion of cell lines and cell strains does involve a limited proportion of cells (40–90%). Cell lines will differentiate as colonies of fat cells separated from each other by cells insusceptible to adipose conversion.[30] Possible models to explain this phenomenon have

[27] W. C. Russell, C. Newman, and D. H. A. Silliamson, *Nature (London)* **253,** 461 (1975).
[28] R. Négrel, P. Grimaldi, and G. Ailhaud, *Biochim. Biophys. Acta* **666,** 15 (1981).
[29] P. Verrando, R. Négrel, P. Grimaldi, M. Murphy, and G. Ailhaud, *Biochim. Biophys. Acta* **663,** 255 (1981).
[30] P. Djian, P. Grimaldi, R. Négrel, and G. Ailhaud, *Exp. Cell Res.* **142,** 273 (1982).

FUNCTIONAL CHARACTERISTICS OF CULTURED ADIPOCYTE-LIKE CELLS

Animal strain	Clone selected	Karyotype	Colony-forming ability in soft agarose	Doubling time in FCS (hr)			Saturation density in FCS (cell number × 10^4/cm^2)			Prostaglandin production
				1%	3%	10%	1%	3%	10%	
C57 BL/6J ob/ob	ob17	Aneuploid[5] (mode 62; 15%)	Low[5]	33	21	15	0.8	2.0	4.2	PGE_2, prostacyclin[16]
C57 BL/6J +/?	HGFu	Aneuploid[22] (mode 59; 24%)	Low[22]	39	29	18	0.95	1.8	5.2	Unknown

Specific receptors for and responses to			Lipolytic and lipogenic enzymes in differentiated cells	TG[b] accumulation in differentiated cells
Insulin	β-Agonists	Triiodothyronine[a]		
Yes[17]	Yes[5,18]	Yes[19,21]	Yes[18–23]	Yes[5,20]
Yes[20,22]	Yes[22]	Yes[20]	Yes[20,22]	Yes[20,22]

[a] T_3 (triiodothyronine) is essential for adipose conversion.[19,24] Insulin modulates the synthesis of lipolytic and lipogenic enzymes.[25,26] Insulin, T_3, and catecholamines are active at physiological concentrations.[5,20–26]

[b] TG, Triglyceride.

been proposed.[3,23] During adipose cell differentiation morphological and enzymatic changes are dramatic and they are accompanied by qualitative and quantitative variations of the cell protein content. These changes include the induction of enzymes of the fatty acid and triglyceride synthesizing pathways and a subsequent triglyceride accumulation (Fig. 1). The development of hormonal responses to insulin and to β-adrenergics is also observed[18,23,25] (see the table).

The properties of adipocyte-like cells are qualitatively and quantitatively similar to those found in adipocytes isolated from white adipose tissue of rodents. The specific activities of key lipolytic and lipogenic enzymes are within the same range of magnitude in both cases.[23,31] The characteristic properties of converted cells are not shared by fibroblasts from rat skin and from bovine lung,[13,15] and by fibroblastic cell lines. Preadipocyte cells synthesize collagen and adipose conversion can be considered,[2,15,32] therefore, as the result of differentiated functions of specialized fibroblasts. Both ob17 and HGFu cells are capable of responding on a long-term basis to physiological concentrations of insulin (0.1–10 nM) and of triiodothyronine (0.15–2 nM). In addition these cells are able to respond on a short-term basis to β-agonists such as catecholamines added within a physiological range of concentrations (0.1–1 μM), with a release of unesterified fatty acid from intracellular triglyceride stores.[5,18,22]

[31] M. G. Murphy, R. Négrel, and G. Ailhaud, *Biochim. Biophys. Acta* **664**, 240 (1981).
[32] P. Verrando, R. Négrel, P. Grimaldi, and G. Ailhaud, *Biochimie* **62**, 201 (1980).

[34] Establishment of Rat Fetal Liver Lines and Characterization of Their Metabolic and Hormonal Properties: Use of Temperature-Sensitive SV40 Virus

By JANICE YANG CHOU

A major advance in the study of mechanisms of normal gene regulation and cellular differentiation has been the development of cell culture systems. Various cell culture systems have been developed; each has its advantages and disadvantages. Primary cultures of mammalian cells dedifferentiate with prolonged culture time or fail to proliferate. Although tumor cells express the tissue-specific functions and grow in culture, nevertheless, due to their malignant state, tumor cell studies may not yield

ISBN 0-12-182009-2

conclusions relevant to normal gene regulation. In order to circumvent some of these problems, we have taken a somewhat different approach. The technique we have adopted is to start with normal differentiated tissues, transform them in tissue culture with mutants of simian virus 40 (SV40) that are temperature sensitive (*ts*) in the gene required for maintenance of the transformed phenotype, and finally to select lines that are differentiated at the nonpermissive temperature and undifferentiated at the permissive temperature. Despite limitations to these temperature-sensitive cell lines, such cells may exhibit transformed and nontransformed phenotypes, thus providing an internal control. Therefore, when studies with temperature-sensitive cell line fail to corroborate results with corresponding tumor cell lines, one must question the relevance of the tumor cell studies.

The basic technique for the establishment of permanent cell lines by SV40 transformation is simple and straightforward. The success of this technique depends not only upon the homogeneity of the intended primary cell cultures, but on the number of clones screened. In general, cell lines that retain desired tissue-specific functions occur at low frequency. Therefore, a reliable, simple, and rapid screening method is one of the critical determinants in obtaining such cell lines.

Rationale

SV40 is a small DNA tumor virus which induces transformation in a variety of mammalian cells.[1] Transformation results from the integration of viral DNA into the cellular genome. It is expressed by the acquisition of certain growth characteristics not exhibited by the parental cells.[2] Transformation usually refers to acquisition of the ability to form dense foci on plastic or colonies in soft agar. The *tsA* mutants of SV40 are temperature-sensitive mutant viruses defective in the *A* gene required for maintenance of the transformed phenotype in mammalian cells.[3–6] Therefore, the *tsA* mutant-transformed cells are conditionally transformed cells that express the transformed phenotype only at the permissive temperature. At the nonpermissive temperature, these cells revert to a normal, nontransformed phenotype.

[1] J. Tooze, "The Molecular Biology of Tumour Viruses." Cold Spring Harbor Lab., Cold Spring Harbor, New York, 1973.

[2] R. G. Martin, *Adv. Cancer Res.* **34,** 1 (1981).

[3] R. G. Martin and Y. J. Chou, *J. Virol.* **15,** 599 (1975).

[4] P. Tegtmeyer, *J. Virol.* **15,** 613 (1975).

[5] J. S. Brugge and J. S. Butel, *J. Virol.* **15,** 619 (1975).

[6] M. Osborn and K. Weber, *J. Virol.* **15,** 636 (1975).

General Properties of SV40 *tsA* Mutant-Transformed Mammalian Cells

At the permissive temperature (33°), *tsA*-transformed cells exhibit the transformed phenotype characterized by decreased generation time, increased saturation density, increased efficiency of growth on plastic, increased ability to overgrow a nontransformed monolayer, and increased ability to clone in soft agar. However, these cells do not express many tissue-specific functions at the permissive temperature. At the nonpermissive temperature (40°), these cells exhibit the nontransformed, normal phenotype characterized by increased generation time, decreased saturation density, decreased efficiency of growth on plastic, decreased ability to overgrow a nontransformed monolayer, and decreased or lost ability to clone in soft agar. However, these cells usually express tissue-specific differentiated functions at the nonpermissive temperature.

The two well characterized fetal rat liver lines, *RLA255-4*[7] and *RLA209-15*,[8] transformed by the *tsA255* and *tsA209* mutants of SV40 virus, respectively, share all these properties. The method described in this chapter was used for the establishment of these two liver lines.

Establishment of SV40 *tsA* Mutant-Transformed Fetal Hepatocyte Lines

Materials

Pregnant Rat, 17–20 Days of Gestation. Rats in an earlier stage of gestation can also be used, depending upon the desired fetal liver functions.

Viruses. SV40 *tsA* mutants.[9,10] Stocks of the *tsA* mutants are prepared by infecting CV-1 monkey kidney monolayer cells at low multiplicity of infection (moi) [approximately 0.01 plaque-forming unit (PFU) per cell] at 33° and allowing the cells to proceed to lysis in modified Eagle's medium buffered with tricine[10] supplemented with 5% fetal bovine serum. The infected cultures are frozen, then thawed, and centrifuged to remove cell debris. The lysates, which contain approximately 10^7 to 10^8 PFU/ml, are used directly for transformation.

Laminar Flow Hood

Tissue Culture Incubators. Two humidified temperature-controlled incubators are required. The temperature of one is maintained at 33 to 34°

[7] S. E. Schlegel-Haueter, W. Schlegel, and J. Y. Chou, *Proc. Natl. Acad. Sci. U.S.A.* **77,** 2731 (1980).

[8] J. Y. Chou and S. E. Schlegel-Haueter, *J. Cell Biol.* **89,** 216 (1981).

[9] P. Tegtmeyer, *J. Virol.* **10,** 591 (1972).

[10] J. Y. Chou and R. G. Martin, *J. Virol.* **13,** 1101 (1974).

and the second maintained at 40°. The incubators should have a stable temperature control with less than a 1° fluctuation in temperature and rapid temperature equilibrium.

Beaucoup. A detergent-disinfectant germicidal material obtained from Huntington Laboratories, Inc., Huntington, IN. A 0.8% solution in water (v/v) is used to scrub the hoods, incubators, and to decontaminate media, dishes, flasks, and pipets that have been in contact with the viruses.

Sterile Disposable Plastic Culture Dishes, Flasks, and Pipets. These are available from commercial sources in a variety of sizes. Plastic dishes and flasks for cell culture are specifically treated to allow cell attachment and growth, but we have found that not all sterile plasticware will suffice for successful culture. In our laboratory, we routinely use the following plasticwares: Corning (Corning Glass Works, Corning, NY), Costar (Data Packaging Corporation, Cambridge, MA), or Falcon (Becton Dickinson Labware, Oxnard, CA).

Culture Media

Primary Cultures. Alpha-modified minimal essential medium (Flow Laboratories, McLean, VA) (αMEM) without arginine, supplemented with 0.4 m*M* ornithine, 100 μg/ml streptomycin, 100 U/ml penicillin, 10% fetal bovine serum, 10 μ*M* cortisol, and 5 μg/ml insulin.[11]

Transformation and Transformed Cultures. αMEM supplemented with streptomycin, penicillin, and 4% fetal bovine serum. The serum used is one of the critical determinants of success. In addition to checking each batch of serum for its efficiency in supporting cell growth, we also check its ability to maintain the expression of differentiated hepatic functions.

Preparation of Primary Fetal Hepatocyte Cultures

After decapitating the pregnant rat, the uterine horns containing the embryos are carefully dissected free and placed in a petri dish. The embryos are removed and placed in a sterile petri dish. All subsequent steps are conducted in a laminar flow hood. The fetal livers are removed from the embryos and placed in a sterile petri dish containing αMEM supplemented with penicillin and streptomycin, washed several times with this medium, then minced with sharp curved scissors or scalpels into small pieces of about 1 mm in diameter. The minced fetal livers are then digested with collagenase [Worthington Biochemical Co., Freehold, NJ, 2 mg/ml in phosphate-buffered saline (PBS)] by incubating at 37° for 20 min

[11] H. L. Leffert, T. Moran, R. Boorstein, and K. S. Koch, *Nature* (*London*) **267,** 58 (1977).

with occasional vortex mixing. The supernatant containing the dispersed cells is decanted and the remaining fetal liver pieces are again digested with collagenase. The dissociated cells are collected by pouring the digests through a double layer of sterile cheese-cloth, and centrifuging at 2000 rpm for 5 min. The pelleted cells are resuspended in the culture medium specific for primary cultures, and incubated at 37° until they attach to the plastic flasks.

Transformation

A frequent problem in establishing differentiated mammalian cell cultures is the overgrowth of these cells by fibroblasts. Therefore, in addition to using the specific culture medium for fetal hepatocytes, one should infect the primary hepatocytes with SV40 *tsA* mutants as soon as the cells are attached, preferably within 24 hr.

Primary hepatocytes are washed three times with medium to remove unplated cells and blood; then 0.5 ml of the virus stock is added per 25-cm^2 flask, giving a moi of 5–20 PFU/cell. Adsorption is allowed to proceed at 33° for 2 to 3 hr. The virus fluid is aspirated, and the cells are allowed to incubate in medium with low serum (medium for transformation) for 16 to 24 hr. The infected cultures are then treated with 0.1% trypsin–0.5 m*M* EDTA in PBS and serially diluted so individual clones can be identified. Serial dilution is required because the transformation frequency for fetal rat hepatocytes by SV40 is relatively high.

Transformation medium is replaced twice weekly and transformed clones can be identified 2 weeks postinfection. Clones are picked up using a stainless-steel cloning cylinder placed directly over the clone. The bottom of the cylinder is first coated with sterile silicone grease to form a seal isolating the clone within the cylinder from the rest of the culture. The clone within the cylinder is removed enzymatically with trypsin-EDTA.

Screening for Functional Hepatocytes

Primary rat fetal hepatocytes can be readily transformed by SV40 *tsA* mutants. However, not all transformants express the desired functions. The reasons for the loss of differentiated functions upon transformation are not clear. Presumably, functional hepatocytes dedifferentiate *in vitro* during the time period required for the establishment of stable transformation. Therefore, a major effort is generally required for the isolation, growth, and screening of large numbers of clones to obtain transformants that either resemble fetal hepatocytes *in vivo,* or at least retain many of the desired functions.

Since the synthesis and secretion of many proteins found in serum are specialized functions of mammalian liver,[12] the ability to produce the serum proteins α-fetoprotein (AFP), albumin, and transferrin can be used to screen for functional fetal hepatocytes. The following two procedures are used in our laboratory to screen for functional fetal hepatocyte lines.

Reverse Hemolytic Plaque Assay. This is a rapid, sensitive, and specific method for screening antigen-secreting cells without the necessity of prior isolation and is therefore particularly good for initial screening. The reverse hemolytic plaque assay uses erythrocytes chemically linked to immunoglobulins to serve as sensitive probes for the detection of the corresponding antigens.[13–15] Polypeptide in solution binds to the antibody at the erythrocyte surface, and the subsequent addition of facilitating antibody renders the erythrocyte susceptible to complement-mediated lysis.

The immunoglobulin fractions of rabbit antisera to rat AFP, rat albumin, and rat transferrin are collected by ammonium sulfate (2 *M*) precipitation. The precipitates are dissolved, dialyzed, and chromatographed on DEAE-Sephadex A-50 column in 0.09% NaCl. The immunoglobulin fraction eluted with 0.09% NaCl and collected in the void volume is then adjusted to 4 mg/ml and 0.9% NaCl and used directly for coupling. Erythrocytes are coupled with antibody by the $CrCl_3$ method.[16] Sheep erythrocytes (at least 2 weeks old) collected in Alsever's solution [dextrose (20.5 g/liter), NaCl (4.2 g/liter), Na-citrate (8 g/liter), and citric acid (0.55 g/liter)], are washed 3 times with 0.9% NaCl. Equal volumes of packed sheep erythrocytes, antibody (4 mg/ml), and 0.9% NaCl are gently mixed. Two to three volumes of $CrCl_3$ (0.5%, pH 5) are added dropwise with mixing. After 5 min at room temperature with occasional stirring, the reaction is terminated by adding an equal volume of PBS. Antibody-coupled erythrocytes are washed twice with PBS and stored in Alsever's solution containing penicillin (200 U/ml), streptomycin (200 μg/ml), and 3% heat-inactivated fetal bovine serum. Coupled erythrocytes are stable at 4° for at least 2 weeks.

Local hemolysis is used to detect functional colonies that secrete specific hepatic proteins. Transformed colonies grown at 33° in 25-cm^2 flasks or 60 mm petri dishes (identifiable 2 to 3 weeks postinfection of primary fetal hepatocytes with *tsA* mutants of SV40 at 33°) are shifted to 40°. After

[12] S. C. Madden and L. J. Zeidis, *in* "Liver Functions" (R. W. Brauer, ed.), p. 325. Williams & Wilkins, Baltimore, Maryland, 1958.

[13] G. A. Molinaro and S. Dray, *Nature (London)* **248,** 515 (1974).

[14] G. A. Molinaro, E. Maron, W. C. Eby, and S. Dray, *Eur. J. Immunol.* **5,** 771 (1975).

[15] M. A. Feldman and J. Y. Chou, *In Vitro* **19,** 171 (1983).

[16] J. W. Goding, *J. Immunol. Methods* **10,** 61 (1976).

an additional 1 to 2 days incubation at 40°, the flasks or dishes are washed twice with PBS and are overlayered with 1.2 ml of 5% antibody-coupled erythrocytes in αMEM containing 1% noble agar and 3% heat-inactivated fetal bovine serum equilibrated at 40 to 45° and allowed to gel at room temperature. After incubation for an additional 16 to 24 hr at 40°, 1 ml of developer containing 10% guinea pig complement, 5–10% facilitating antiserum, and 1% noble agar in Hanks' balanced salt solution equilibrated at 40 to 45° is added and allowed to gel at room temperature. A hemolytic plaque will become visible around the colony that secretes the corresponding antigen after 1 hr at 37°. Since *tsA*-transformed cells produce increased amounts of differentiated proteins at the nonpermissive temperature, the hemolytic plaque assay is usually performed with transformed colonies that have been shifted to 40°. The same plaque assay can be used to screen colonies grown at 33°. These polypeptide-secreting colonies remain viable and can be selected by placing a cloning cylinder directly over the plaque, removing the erythrocyte-agar layer which contains part of the transformed clone with a pasteur pipet, and removing the remainder of the clone with 0.1% trypsin–0.5 m*M* EDTA in PBS. Transformed colonies are propagated and cloned at 33°.

The sensitivity and specificity of the reverse hemolytic plaque assay depend upon the specificity and sensitivity of the antibody used. The sensitivities routinely obtained in our laboratory are 2 ng/spot for rat AFP, 4 ng/spot for rat transferrin, and 6 ng/spot for rat albumin. Rat transferrin, rat albumin, rabbit antisera against rat transferrin, and rat albumin are obtained from N. L. Cappel Laboratories, Inc. (Cochran, PA).

Immunoassays. Functional fetal hepatocyte lines are also identified by growing the transformed cells at 33 and 40° and using radioimmunoassays or other immunoassays to estimate the amounts of AFP, albumin, or transferrin secreted into the culture medium. This method should be used to characterize transformed cell lines isolated by the reverse hemolytic plaque assay. However, if the desired antibody is not available in sufficient quantity, immunoassays can be used directly for screening for functional clones.

Transformed colonies are picked up and grown into monolayer cultures at 33°. When confluent, they are trypsinized and two replicate sets of transformed cultures are prepared and grown at 33°. The first set of cultures is maintained at 33° throughout growth. The second set of cultures is shifted from 33 to 40° after 3 to 4 days growth at 33°. Medium is changed and collected every 24 to 48 hr. The presence of differentiated hepatic proteins in culture media is then measured. Most fetal hepatocyte lines established by transformation with SV40 *tsA* mutants retain the ability to produce one or more of these serum proteins, although few retain all hepatic functions. For example, out of the 40 randomly selected transformants, 32 (80%) produced transferrin only, 3 produced albumin plus transferrin, and only one produced AFP, albumin, and transferrin in culture.

Freezing

Cells can be frozen in αMEM supplemented with 15% fetal bovine serum and 10% dimethyl sulfoxide. The use of a liquid nitrogen freezer is the optimal method for long-term storage of these cells which survive in a −70° freezer for at least 1 month.

Characterization of the SV40 *tsA* Mutant-Transformed Hepatocyte Lines

Temperature Sensitivity for Maintenance of Transformed Phenotype

The following three criteria are used to assay for the establishment and maintenance of the transformed phenotype in these fetal liver cells.

Growth Studies. Three parallel sets of the transformed cultures are plated initially at 33°. The first set of cultures is maintained at 33° throughout growth. The second set of cultures is shifted from 33 to 40° after various periods of growth at 33° (preferably 2 to 5 days). The third set of cultures is shifted to 40° as the second set but is shifted back to 33° after 3 to 4 days at 40°. Culture medium is changed every day and cells are trypsinized and counted with a Celloscope 112 TH (Particle Data, Inc., Elmhurst, IL).

The growth curve for *RLA209-15* fetal liver cells is shown in Fig. 1. At the permissive temperature (33°), these cells grow like transformed cells; they grow rapidly to high cell densities. In general, nontransformed primary fetal hepatocytes show only limited growth *in vitro*. At the nonpermissive temperature (40°), however, these cells grow more slowly and to lower saturation densities than cells grown at 33°. This indicates that these cells are temperature sensitive for growth. The growth inhibition is reversible; when cultures grown at 40° for 4 or more days are shifted back to 33°, a marked increase in the growth rate is obtained.

Overgrowth of Nontransformed Cell Layers. Nontransformed primary human or rat fibroblasts are prepared at a density of 2×10^5 cells per ml. Each well of a 24-well plate receives 1.5 ml of the suspension so that confluence will be attained immediately after attachment. Serial (1 to 5) dilutions of the transformed cells are prepared, and equal volumes are added to each well. The first well receives 10^5 transformed cells. Medium

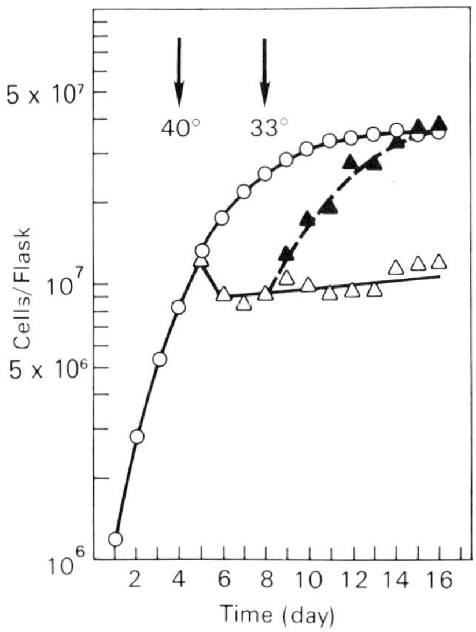

FIG. 1. Growth of *RLA209-15* cells at 33 and 40°. Cultures were plated at 33° and were shifted from 33° to 40° after 4 days growth at 33°. After incubating at 40° for 4 days, some of the cultures were shifted back to 33°. Medium was changed every day. Arrows indicate time of temperature shift. ○, 33°; △, 40°; ▲, cultures shifted from 40 to 33° after 4 days at 40°.

is replaced twice weekly at 33 and 40°. After 2 weeks, fixed cells are prepared by removing the medium, washing with PBS, rinsing with absolute methanol, fixing in methanol for 20 min at 25°, and allowing the fixed cells to dry. The cells are stained with a 0.1% solution of Evans Blue in PBS for 20 min at 25°, rinsed with water, and dried. For a *tsA*-transformed fetal liver line that is temperature sensitive for maintenance of the transformed phenotype, the efficiency of overgrowth of nontransformed cell layers at 33° is usually 10^2- to 10^3-fold higher than that at 40°.

Colony Formation in Soft Agar. Transformed cells are suspended in 1 ml of cloning agar (αMEM supplemented with 27.5% fetal bovine serum and 0.3% noble agar) at concentrations of 10^6, 10^5, 10^4, and 10^3 cells/ml. The agar-cell suspension is added to each well of the 24-well plates on top of a 0.5 ml layer of solidified 0.7% agar in αMEM supplemented with 17% fetal bovine serum. Cultures are fed with 0.5 ml of cloning agar every 7 days at 33° and every 5 days at 40°. After 2 weeks, colonies grown in soft agar are stained with 0.2 ml cloning agar containing 0.1% neutral red and counted. Not all transformed cells acquire the ability to grow in soft agar. The ability to form colonies in soft agar is a more stringent criterion for transformation than is the ability to overgrow a nontransformed cell layer. For a *tsA*-transformed fetal liver line that is temperature sensitive for

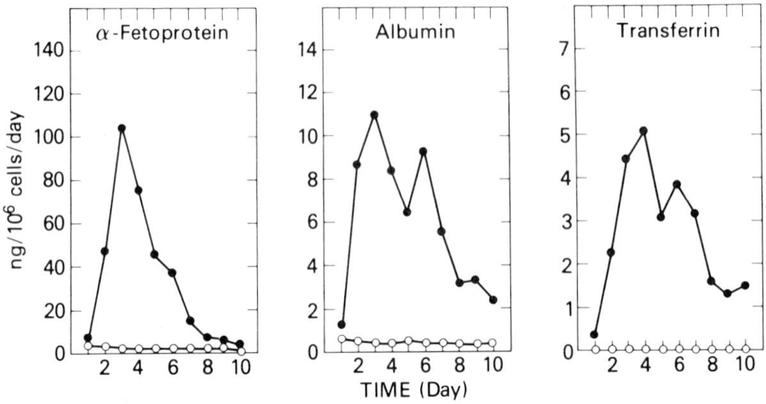

FIG. 2. Synthesis of AFP, albumin, and transferrin in *RLA209-15* cells grown at 33 and 40°. Two parallel sets of *RLA209-15* cultures were grown at 33° initially. The first set of cultures was maintained at 33° throughout growth. The second set of cultures was shifted from 33 to 40° after 4 days growth at 33° (day 0). Culture medium was changed and collected every day. AFP, albumin, and transferrin in the culture media were determined by double-antibody radioimmunoassays. ○, 33°; ●, 40°.

maintenance of the transformed phenotype, the efficiency of colony formation in soft agar at 33° is usually 10^2- to 10^3-fold higher than that at 40°.

Temperature Sensitivity for Expression of Differentiated Hepatic Functions

If the transformants retain differentiated hepatic functions, they should be able to synthesize and secrete one or more of these serum proteins into the culture medium. Since transformation of differentiated cells usually results in inhibition of the expression of tissue-specific functions, *tsA*-transformed fetal hepatocytes grown at 33° should produce low or reduced levels of AFP, albumin, and transferrin. However, a *tsA*-transformed hepatocyte line temperature sensitive for maintenance of the transformed type would produce greatly induced levels of these proteins at the nonpermissive temperature. AFP, albumin, and transferrin production by *RLA209-15* fetal hepatocyte line grown at 33 and 40° is shown in Fig. 2.

Response to the Peptide Hormone Glucagon

The presence of functional receptors for glucagon is one of the characteristics of hepatocytes.[17,18] Glucagon response can be assayed by the

[17] S. J. Pilkis, T. H. Claus, R. A. Johnson, and C. R. Park, *J. Biol. Chem.* **250,** 6328 (1975).
[18] C. Plas and J. Nunez, *J. Biol. Chem.* **250,** 5304 (1975).

increase in cAMP levels induced by this hormone.[7,8] Most, if not all, of the *tsA*-transformed hepatocyte lines that produce one or more of these serum proteins retain functional receptors for glucagon. The glucagon response occurs equally well at 33 and 40°. Further, the response to glucagon is usually enhanced by glucocorticoid at the nonpermissive temperature.[7,8]

Growth and Maintenance

The *tsA*-transformed fetal liver cell lines are maintained as monolayers at 33° in αMEM supplemented with 2% fetal bovine serum, penicillin (100 U/ml), and streptomycin (100 μg/ml), under 5% CO_2–95% air. We found that serum levels in media and the proteolytic enzyme used for subculturing affect the ability of *tsA*-transformed fetal liver cells to synthesize serum proteins. The temperature-sensitive growth characteristics and the response to glucagon are not altered by serum levels or enzyme type. AFP production in *RLA209-15* cells increases following serum removal. Continued passage of *RLA209-15* cells in high serum medium (αMEM supplemented with 10% fetal bovine serum) and subculture in trypsin-EDTA eventually decreases the production of AFP and albumin but increases the production of transferrin. For this reason, *tsA*-transformed fetal liver cell lines are maintained in αMEM containing low levels of serum (2 to 4% depending upon serum batch) and subcultured with collagenase (1 mg/ml in PBS). Cells maintained in such culture conditions retain their ability to produce these serum proteins for at least 3 months.

Conclusion

The *tsA*-transformed fetal liver cells are well suited for studying problems of development, differentiation, carcinogenesis, and the regulation of these processes by hormones and other effectors. In addition to responding to the peptide hormone glucagon, we found that the differentiated states of *RLA209-15* fetal liver cells are regulated by cAMP,[19] glucocorticoid hormones, progesterone, and retinoic acid.[20] Furthermore, these fetal cells can be induced to mature in culture as demonstrated by an increase in albumin production and a decrease in AFP production.[20] Although differentiation and maturation are biologically important phenomena, very little is known about the control mechanisms. Fetal tissues are programmed to mature under physiological conditions, and hence it is

[19] J. Y. Chou, *Oncodev. Biol. Med.* **4,** 177 (1983).
[20] J. Y. Chou and F. Ito, *Biochem. Biophys. Res. Commun.* **118,** 168 (1984).

virtually impossible to dissociate differentiation from maturation *in vivo*. The *tsA*-transformed fetal liver lines therefore provide a valuable system to study the processes of differentiation and maturation independently and in concert.

SV40 *tsA* mutants have been successfully employed in the author's laboratory to generate several types of cell lines that retain differentiated, tissue-specific functions. In addition to the fetal rat liver lines described here, we have established human placental cell lines[21,22] and adult rat liver line.[23] The use of SV40 *tsA* mutants to generate cell lines that retain differentiated functions should, in principle, be of universal applicability to any cell type that can be transformed by SV40.

[21] J. Y. Chou, *Proc. Natl. Acad. Sci. U.S.A.* **75,** 1409 (1978).
[22] J. Y. Chou, *Proc. Natl. Acad. Sci. U.S.A.* **75,** 1854 (1978).
[23] J. Y. Chou, *Mol. Cell. Biol.* **3,** 1013 (1983).

Section V

Purification of Membrane Receptors and Related Techniques

[35] Purification of Insulin Receptor from Human Placenta

By STEVEN JACOBS and PEDRO CUATRECASAS

All published schemes for the successful purification of insulin receptors have as their major purification step affinity chromatography using either antibodies to insulin receptor[1,2] or insulin[3,4] itself as the immobilized ligand. While both types of procedures produce similar results, the limited availability of antireceptor antibodies severely restricts their general usefulness.

The following is a description of the scheme we currently use for purification of insulin receptors from human placenta. It is a modification of several published methods.[4–6] Placenta membranes are solubilized with Triton X-100 and purified by sequential chromatography on concanavalin A-Sepharose, insulin-Sepharose, and wheat germ agglutinin-Sepharose. Approximately 40 μg of insulin receptor can be purified from two placentas with a final overall yield of approximately 10% (determined by recovery of insulin binding activity).

Preparation of Affinity Adsorbents

Concanavalin A-Sepharose and Wheat Germ Agglutinin-Sepharose

These are prepared as previously described.[7] Concanavalin A (200 mg) or 80 mg of wheat germ agglutinin is dissolved in 20 ml of 0.1 *M* sodium phosphate, pH 7.4, containing 0.1 *M* α-methyl mannopyranoside (concanavalin A) or 0.1 *M* *N*-acetylglucosamine (wheat germ agglutinin). Sepharose CL-4B (20 ml of packed gel) is washed with 0.1 *M* NaCl and suspended in an equal volume of water. The beads are gently stirred in a vacuum hood with a magnetic stirrer and 3 g of finely divided CNBr is added. The pH is immediately raised and maintained at 11 with 10 *N* NaOH. Ice is added as necessary to keep the mixture near room tempera-

[1] L. C. Harrison and A. Itin, *J. Biol. Chem.* **255,** 12066 (1980).

[2] R. A. Roth and D. J. Cassell, *Science* **219,** 299 (1983).

[3] P. Cuatrecasas, *Proc. Natl. Acad. Sci. U.S.A.* **69,** 1277 (1972).

[4] S. Jacobs, Y. Shechter, K. Bissell, and P. Cuatrecasas, *Biochem. Biophys. Res. Commun.* **77,** 981 (1977).

[5] P. Cuatrecasas and G. P. E. Tell, *Proc. Natl. Acad. Sci. U.S.A.* **70,** 485 (1973).

[6] T. W. Siegel, S. Ganguly, S. Jacobs, O. M. Rosen, and C. S. Rubin, *J. Biol. Chem.* **256,** 9266 (1981).

[7] P. Cuatrecasas and I. Parikh, this series, Vol. 34, p. 653.

ISBN 0-12-182009-2

ture. The reaction is complete when the solid CNBr is fully dissolved and the rate of base consumption declines. Ice is immediately added to the slurry and the contents rapidly filtered through a coarse sintered glass funnel under vacuum and rapidly washed with 1 liter of ice cold water. The moist beads are then immediately transferred to a polyethylene bottle containing the dissolved lectin and gently shaken overnight at 4°. Monoethanolamine, adjusted to pH 8.0, is added to a final concentration of 0.1 *M* and the beads are shaken for an additional 2 hr at room temperature. The beads are then washed with three liters of 0.2 *M* NaCl and stored at 4° in 50 m*M* sodium acetate, pH 6.0, containing 1 m*M* $MgCl_2$, 1 m*M* $CaCl_2$, and 20 m*M* NaN_3. Generally, more than 80% of the lectin is covalently coupled. This can be determined by measuring the OD_{280} of the coupling reaction supernatant.

Insulin Sepharose

Insulin is covalently coupled to Sepharose CL-4B via a diaminodipropylamine succinyl spacer arm.[3] Sepharose CL-4B (20 ml of packed gel) is washed, suspended in 20 ml of H_2O, and activated with 5 g CNBr as described above. The activated Sepharose is gently shaken overnight at 4° with 20 ml of 2 *M* diaminodipropylamine adjusted to pH 10 with HCl. The diaminodipropylamine agarose is washed with water and resuspended in an equal volume of water. Succinic anhydride (20 mmol) is added and the mixture is stirred at 4° while maintaining the pH at 6.0 by adding 5 *N* NaOH. After 1 hr, the reaction mixture is transferred to a polyethylene bottle and shaken overnight at 4°. The derivatized Sepharose is then washed with one liter of H_2O and suspended in 0.1 *N* NaOH for 1 hr to cleave labile succinyl esters of the agarose hydroxyl groups. The derivatized Sepharose is finally washed with 500 ml of water. The extent of coupling of diaminodipropylamine to the Sepharose and its subsequent derivatization can be checked qualitatively by the TNBS colorimetric assay.[8]

In order to couple insulin, the reactive *N*-hydroxysuccinimide ester is formed.[9] Succinyldiaminodipropylamino Sepharose (20 ml packed beads) is washed with 500 ml H_2O and then 500 ml of dioxane to obtain anhydrous conditions. The gel is suspended in 40 ml of dioxane. *N*-Hydroxysuccinimide (0.004 mol) and dicyclohexylcarbodiimide (0.004 mol) are added and the mixture stirred for 70 min at room temperature. A white precipitate which is dicyclohexyl urea should form. The gel is then washed in a coarse sintered glass funnel under vacuum with 100 ml of

[8] P. Cuatrecasas, *J. Biol. Chem.* **245,** 3059 (1970).
[9] P. Cuatrecasas and I. Parikh, *Biochemistry* **11,** 2291 (1972).

dioxane, 100 ml of methanol, and 50 ml of dioxane. The dried gel is immediately transferred to a polyethylene bottle containing 200 mg of insulin with approximately 100,000 cpm of ^{125}I-labeled insulin in 40 ml of 0.1 *M* sodium phosphate, 6 *M* urea, at pH 7.4. (The urea solution is filtered prior to the addition of insulin to remove particulate material.) The beads are then gently shaken at 4° overnight. Monethanolamine, adjusted to pH 8 with HCl, is added to a final concentration of 0.1 *M* and incubated at room temperature for 2 hr. The beads are then washed with 3 liters of NaCl, packed into a column, and washed overnight with 1 liter of 6 *N* urea, and finally washed with an additional liter of 1 *M* NaCl. The beads are stored at 4° as a 50% slurry in 20 m*M* NaN_3.

Assay of Solubilized Insulin Receptor

Solubilized insulin receptor is detected and quantitated by its ability to bind ^{125}I-labeled insulin. Polyethylene glycol is used to separate receptor-bound insulin from free insulin based on its ability to selectively precipitate insulin-receptor complexes.[10]

Solubilized insulin receptor is incubated at 22° with ^{125}I-labeled insulin (20,000 cpm) in a final volume of 0.2 ml of 50 m*M* Tris–HCl, pH 7.7, containing 0.1% bovine albumin. After 45 min, the tubes are placed in an ice bath; 0.5 ml of 0.1% bovine γ-globulin in 50 m*M* Tris–HCl, pH 7.7, and 0.5 ml of 25% polyethylene glycol (molecular weight 6000) is added. The tubes are vortexed vigorously and incubated at 4° for 15 min. They are then filtered under vacuum through 0.5-μm cellulose acetate filters (millipore EHWP). The filters are washed with 3 ml of 8.3% polyethylene glycol (molecular weight 6000) in 50 m*M* Tris–HCl, pH 7.7. Nonspecific binding is determined in tubes run in parallel in which unlabeled insulin is added to achieve a final concentration of 10 μg/ml either with or prior to the addition of ^{125}I-labeled insulin.

Insulin binding determined by this assay is severely inhibited by concentrations of Triton X-100 greater than 0.2% (v/v). Maximum values for insulin binding are obtained when the concentration of Triton X-100 in the assay is less than 0.1%.

Preparation and Solubilization of Placenta Membranes

The specific activity of insulin binding is highest in a pure plasma membrane fraction. However, methods required to produce pure plasma membranes are laborious and difficult to scale up. Furthermore, the yields

[10] P. Cuatrecasas, *Proc. Natl. Acad. Sci. U.S.A.* **69,** 318 (1972).

are prohibitively low. Therefore, a crude microsomal fraction is used as the starting material for receptor purification.

All steps are performed at 4°. The umbilical cord and membranes are stripped from two human placentas. The placentas are cut into small pieces and homogenized with a Polytron in one liter of 0.25 *M* sucrose containing 20 μg/ml PMSF. The homogenate is centrifuged at 600 *g* for 10 min and 10,500 *g* for 30 min. The supernatant is adjusted to 0.1 *M* NaCl and 0.2 m*M* $MgCl_2$, and centrifuged at 40,000 *g* for 30 min. The pellet is suspended with a Polytron in 800 ml of 50 m*M* Tris–HCl, pH 7.7, containing 20 μg/ml PMSF, and again centrifuged at 40,000 *g* for 30 min. This is repeated twice. The final pellet is resuspended in 75 ml of 50 m*M* Tris–HCl to give a protein concentration of approximately 10 mg/ml.

The placenta membranes are solubilized with Triton X-100 as follows: bacitracin and PMSF are added to achieve a final concentration of 1 mg/ml and 20 μg/ml. Sufficient Triton X-100 is added to achieve a final concentration of 2% (v/v). The membranes are then homogenized with a Polytron and shaken in the cold for 45 min. Soluble receptor is recovered in the supernatant following centrifugation at 150,000 *g* for 60 min.

In contrast to liver or fat cells, only about 40% of the insulin binding activity is solubilized from placenta with this procedure; the remainder is associated with the residual pellet. Use of higher concentrations of Triton, inclusion of EDTA or high salt, or the use of up to 0.1% SDS in addition to 2% Triton does not improve the yield. A small fraction of the remaining receptor can be extracted from the residual pellet by a second treatment with Triton. However, this is probably not worthwhile.

Affinity Chromatography

Concanavalin A-Sepharose

Concanavalin A-Sepharose can be used to purify insulin receptor either in a batch procedure or a column procedure. Since at this stage of the purification, insulin receptor is in a relatively large volume, a batch procedure is somewhat more convenient.

Solubilized insulin receptor is diluted with an equal volume of Tris–HCl, pH 7.7, containing 1 mg/ml bacitracin, 20 μg/ml PMSF, 2 m*M* $MgCl_2$, and 2 m*M* $CaCl_2$. Packed concanavalin A-Sepharose (15 ml), which has been extensively washed with this buffer (with bacitracin omitted), is added and the suspension shaken gently overnight at 4°. The slurry is then filtered through a coarse sintered glass funnel under vacuum, the breakthrough collected, and the beads washed on the funnel with one liter of 50 mM Tris–HCl, pH 7.7, containing 0.2% Triton X-100, 2 mM $MgCl_2$, and 2 mM $CaCl_2$. During these procedures, the beads are not permitted to become completely dry. The beads are then suspended in an equal volume of wash buffer and packed into a glass column. Elution buffer (10 ml) (0.5 M α-methyl mannopyranoside in 50 mM Tris–HCl, pH 7.7, containing 0.2% Triton X-100, 1 mg/ml bacitracin, and 20 μg/ml PMSF) is applied to the column, which is clamped for 45 min. Insulin receptor is then eluted with four or five column volumes of this buffer. Generally, 10 ml fractions are collected and those fractions containing insulin binding activity retained for further purification. After use, the concanavalin A-Sepharose column can be stored in sodium acetate buffer, pH 6.0, containing 1 mM $MgCl_2$, 1 mM $CaCl_2$, and 20 mM NaN_3 and reused several times.

This step results in an approximately 8-fold purification with a yield of approximately 80%. Although the extent of purification is relatively small, this step greatly enhances the performance of the following insulin-Sepharose purification. High concentrations of Triton X-100 interfere with insulin binding, but decreasing the concentration of Triton early in the purification procedure results in aggregation of membrane phospholipids and proteins making purification impossible. Concanavalin A-Sepharose purification removes the bulk of membrane phospholipids and some membrane proteins, and provides a convenient method for decreasing the concentration of Triton without excessively diluting the receptor. Concanavalin A-Sepharose purification also removes much of the inner-membrane peripheral proteins, which seem to be nonspecifically adsorbed to insulin-Sepharose.

Insulin-Sepharose

Insulin-Sepharose (5 ml) is packed into a 10-ml disposable pipet stopped with glass wool. The column is washed overnight with 4.5 M urea in 50 mM sodium acetate, pH 6.0. The column is then washed with 10 ml of 50 mM Tris–HCl, pH 7.7, containing 0.1% Triton X-100 (v/v). Insulin receptor that has been eluted from concanavalin A-Sepharose is diluted with an equal volume of 1 M NaCl in 50 mM Tris–HCl, pH 7.7, and applied to the insulin-Sepharose column at room temperature over 6 hr. The column is washed overnight at 4° with several hundred milliliters of 50 mM Tris–HCl, pH 7.7, containing 0.1% Triton and 0.5 M NaCl. The column is eluted at room temperature with 4.5 M urea in 50 mM sodium acetate, pH 6.0, containing 0.1% Triton X-100 and 20 μg/ml PMSF. This elution buffer causes time-dependent denaturation of the receptor. In order to minimize the extent of denaturation, 2.5 ml fractions are collected directly into 2.5 ml of ice cold Tris–HCl, pH 7.7, containing 0.1% Triton X-100. Fractions containing insulin-binding activity are pooled and immediately dialyzed against Tris–HCl, pH 7.7, containing 0.1% Triton X-100. Insulin-Sepharose can be stored in 20 m*M* NaN_3 and reused.

The performance of insulin-Sepharose varies from batch to batch and cannot be predicted from the amount of insulin coupled to the beads. From 50% to virtually all the insulin binding activity applied to the column is adsorbed and as much as 20% can be eluted in an active form. That fraction of receptor which is not adsorbed will be adsorbed when applied to a second column. More receptor is adsorbed if a larger column is used, but the fraction of active receptor that can be eluted decreases.

Insulin receptor eluted from insulin-Sepharose in this manner can have a specific activity for insulin binding as high as 12 μg of insulin/mg of protein. When, after reduction, it is subjected to SDS polyacrylamide gel electrophoresis, it is composed of one major band with a molecular weight of 135,000, and one minor band having a molecular weight of 90,000. A minor band of molecular weight 45,000 is variably present. These three bands together routinely make up more than 85% of the protein present. The 135,000 and 90,000 molecular weight bands are subunits of the receptor. The 45,000 molecular weight band appears to be a proteolytic fragment of the 90,000 molecular weight subunit.

Alternative Methods of Elution. The method of elution described above results in some irreversible denaturation of the receptor. This accounts for the specific activity of insulin binding being less than theoretically predicted. It also accounts for the loss of protein kinase activity that is associated with the native receptor.[11] In order to avoid these problems more gentle methods of elution have been devised. Fujita-Yamaguchi *et al.*[12] used 50 m*M* sodium acetate buffer, pH 5.0, containing 0.1% Triton X-100 and 1 *M* NaCl to elute the receptor from insulin-Sepharose. They obtained a receptor preparation with a specific activity for insulin binding of 28.5 μg/ml of protein that was fully active as a protein kinase. The yield with this procedure was also somewhat higher than with the urea method of elution. Rosen *et al.*[13] have used 3 *M* urea at pH 7.4 to elute receptor from insulin-Sepharose. The specific activity of insulin binding was only 90 pmol/mg (~0.6 μg/mg). However, this preparation retained protein kinase activity.

[11] L. M. Petruzzelli, S. Ganguly, C. J. Smith, M. H. Cobb, C. S. Rubin, and O. M. Rosen, *Proc. Natl. Acad. Sci. U.S.A.* **79,** 6792 (1982).

[12] Y. Fujita-Yamaguchi, S. Choi, Y. Sakamoto, and K. Itakura, *J. Biol. Chem.* **258,** 5045 (1983).

[13] O. M. Rosen, R. Herrera, Y. Glowe, L. M. Petruzzelli, and M. H. Cobb, *Proc. Natl. Acad. Sci. U.S.A.* **80,** 3237 (1983).

Wheat Germ Agglutinin-Sepharose

Only a small degree of further purification can be obtained by purifying the insulin-Sepharose eluate on wheat germ agglutinin-Sepharose. However, this is a very convenient method of concentrating the receptor, and exchanging the buffer or detergent. Furthermore, at this stage, the receptor can be eluted in the absence of free detergent without apparent aggregation or loss of insulin binding activity. Wheat germ agglutinin-Sepharose will also remove any free insulin which may have leaked from the insulin-Sepharose column during elution of the receptor. Because of the very small amount of glycoprotein present at this stage, insulin receptor can be quantitatively adsorbed to 0.3 ml of wheat germ agglutinin-Sepharose and eluted in comparably small volumes.

Wheat germ agglutinin-Sepharose (0.3 ml of packed gel) is loaded into a pasture pipet and washed with 100 ml of 50 m*M* Tris–HCl, pH 7.7, containing 2 m*M* $MgCl_2$ and 2 m*M* $CaCl_2$. The dialyzed eluate from the insulin-Sepharose column is applied slowly over 2 hr at room temperature. The column is washed with 20 ml of the above wash buffer. If detergent is to be eliminated, the column is washed with an additional 2 ml of the desired buffer without detergent. *N*-acetylglucosamine (0.2 ml of 0.5 *M*) in the desired buffer with or without detergent is applied to the column, which is clamped for 45 min. Insulin receptor is then eluted in the next one to two column volumes with a yield approaching 100%.

[36] Purification of the LDL Receptor

By WOLFGANG J. SCHNEIDER, JOSEPH L. GOLDSTEIN, and MICHAEL S. BROWN

Introduction

The low-density lipoprotein (LDL) receptor on the surface of mammalian cells binds plasma LDL, a cholesterol transport protein, and thereby initiates a chain of events culminating in the internalization of the LDL with concomitant delivery of cholesterol to the cell.[1] Interest in this receptor stems from several of its characteristics: (1) it plays a crucial role in regulating the level of plasma and cellular cholesterol in man and in animals[1]; (2) its absence constitutes the basic defect in the human disease

[1] J. L. Goldstein and M. S. Brown, *Annu. Rev. Biochem.* **46**, 897 (1977).

ISBN 0-12-182009-2

familial hypercholesterolemia, a common cause of premature atherosclerosis[1]; and (3) its localization in coated pits on the cell surface has called attention to the role of these specialized structures in mediating the endocytosis of a variety of receptor-bound macromolecules, such as peptide hormones, lipid-enveloped viruses, toxins, and plasma transport proteins.[2]

In mediating the endocytosis of LDL, the LDL receptor migrates continuously from one cell organelle to another.[3] After synthesis in the rough endoplasmic reticulum, the receptor travels to the Golgi complex and then to random sites on the plasma membrane. Within minutes, it binds LDL and clusters with other receptors in clathrin-coated pits. The receptor-LDL complex is internalized in coated vesicles, which rapidly shed their clathrin coat and fuse with other vesicles to form endosomes. Within the acidic environment of the endosome, the receptor and LDL separate: the receptor returns to the surface and the LDL proceeds to the lysosome where it is degraded, yielding its cholesterol for cellular metabolism. After returning to the surface, the receptor binds another LDL particle and thus initiates another cycle of endocytosis. The cholesterol released from the lysosomal degradation of LDL is used for the synthesis of membranes (most cells), steroid hormones (adrenal cortex and ovarian corpus luteum), and bile acids (hepatocytes).

The LDL receptor has an apparent molecular weight of 160,000 as determined by sodium dodecyl sulfate (SDS)–polyacrylamide gel electrophoresis[4] and density gradient centrifugation.[5] The receptor is synthesized in the rough endoplasmic reticulum as a precursor with an apparent molecular weight of 120,000.[6,7] It undergoes an apparent 40,000-Da increase in molecular weight in the Golgi apparatus prior to its insertion into the plasma membrane. The receptor has both N-linked and O-linked carbohydrate chains.[8] The posttranslational increase in size occurs concomitantly with the elongation of the O-linked chains in the Golgi apparatus.[8]

The mature receptor on the cell surface recognizes the apoprotein B component of LDL; it also recognizes apoprotein E, which is found normally in intermediate density lipoprotein (IDL) and a subfraction of high-

[2] J. L. Goldstein, R. G. W. Anderson, and M. S. Brown, *Nature (London)* **279,** 679 (1979).

[3] M. S. Brown, R. G. W. Anderson, and J. L. Goldstein, *Cell* **32,** 663 (1983).

[4] W. J. Schneider, U. Beisiegel, J. L. Goldstein, and M. S. Brown, *J. Biol. Chem.* **257,** 2664 (1982).

[5] W. J. Schneider, J. L. Goldstein, and M. S. Brown, *J. Biol. Chem.* **255,** 11442 (1980).

[6] H. Tolleshaug, J. L. Goldstein, W. J. Schneider, and M. S. Brown, *Cell* **30,** 715 (1982).

[7] H. Tolleshaug, K. K. Hobgood, M. S. Brown, and J. L. Goldstein, *Cell* **32,** 941 (1983).

[8] R. D. Cummings, S. Kornfeld, W. J. Schneider, K. K. Hobgood, H. Tolleshaug, M. S. Brown, and J. L. Goldstein, *J. Biol. Chem.* **258,** 15261 (1983).

density lipoprotein called HDL_c.[9,10] The affinity of the LDL receptor for apo E-HDL_c (which contains apo E as its sole apoprotein) is 20-fold greater than for LDL (which contains only apo B), but at saturation the receptor binds only one-fourth as many particles of apo E-HDL_c as LDL.[11] Binding of either apo B or apo-E-containing lipoproteins to the LDL receptor is absolutely dependent on Ca^{2+} and can be abolished by EDTA.

In this chapter, methods for the purification of the LDL receptor from various sources are described. These methods are designed to enable researchers to obtain chemical amounts of purified receptor from a rich source, the adrenal cortex,[12] and to obtain smaller amounts of radiolabeled LDL receptors from less abundant sources, including cultured cells.

General Reagents and Procedures

Buffers

A. 20 m*M* Tris–chloride (pH 8)
 0.15 *M* NaCl
 1 m*M* $CaCl_2$
 1 m*M* PMSF (phenylmethylsulfonyl-fluoride)
 0.1 m*M* leupeptin

B. 250 m*M* Tris–maleate (pH 6)
 2 m*M* $CaCl_2$
 1 m*M* PMSF
 0.1 m*M* leupeptin

C. 10 m*M* Tris–maleate (pH 6)
 2 m*M* $CaCl_2$
 1% (v/v) Triton X-100
 1 m*M* PMSF
 0.05 m*M* leupeptin

D. 50 m*M* Tris–maleate (pH 6)
 2 m*M* $CaCl_2$
 1% (v/v) Triton X-100
 1 m*M* PMSF
 0.05 m*M* leupeptin

[9] M. S. Brown, P. T. Kovanen, and J. L. Goldstein, *Science* **212,** 628 (1981).
[10] R. W. Mahley and T. L. Innerarity, *Biochim. Biophys. Acta* **737,** 197 (1983).
[11] T. L. Innerarity, R. E. Pitas, and R. W. Mahley, *Biochemistry* **19,** 4359 (1980).
[12] P. T. Kovanen, S. K. Basu, J. L. Goldstein, and M. S. Brown, *Endocrinology* **104,** 610 (1979).

E. 50 m*M* Tris–maleate (pH 6)
2 m*M* $CaCl_2$
40 m*M* octylglucoside
1 m*M* PMSF
0.01 m*M* leupeptin

F. 50 m*M* Tris–chloride (pH 8)
2 m*M* $CaCl_2$
50 m*M* NaCl
1 m*M* PMSF
0.05 m*M* leupeptin

G. 50 m*M* Tris–chloride (pH 7)
2 m*M* $CaCl_2$
0.1% (v/v) Nonidet P-40
1 m*M* PMSF
0.05 m*M* leupeptin

H. 25 m*M* Tris–chloride (pH 7.6)
50 m*M* NaCl
0.1 m*M* EDTA
0.02% (w/v) NaN_3

I. 10 m*M* sodium HEPES (*N*-2-hydroxyethylpiperazine-*N*′-2-ethanesulfonic acid) (pH 7.4)
150 m*M* NaCl
2 m*M* $CaCl_2$

J. 10 m*M* sodium HEPES (pH 7.4)
200 m*M* NaCl
2.5 m*M* $MgCl_2$
1 m*M* L-methionine
40 m*M* octylglucoside
0.2% (v/v) Nonidet P-40
1 m*M* PMSF
0.1 m*M* leupeptin

K. 50 m*M* Tris–chloride (pH 8)
2 m*M* $CaCl_2$
50 m*M* NaCl
20 mg/ml bovine serum albumin

L. 20 m*M* Tris–chloride (pH 8)
1 m*M* $CaCl_2$
50 m*M* NaCl

M. 200 m*M* Tris–chloride (pH 8)
2 m*M* $CaCl_2$
80 mg/ml bovine serum albumin

Preparation and Radioionation of LDL

These procedures have been described in detail elsewhere in this series.[13] The concentration of LDL is expressed in terms of its protein content, as measured by the method of Lowry *et al.*[14]

Protein Content of Membranes and Soluble Receptors

The protein content of receptor-containing samples is determined by modifications of the Lowry procedure[14] as described in Ref. 5.

Tissue Sources for Purification of LDL Receptors

Membranes from Bovine Adrenal Cortex

Immediately after steers (weighing 220–320 kg) are slaughtered, 30 adrenal glands are placed in ice-cold 0.15 *M* NaCl. We have also obtained adrenal glands from cows and bulls. No consistent differences in receptor yield between adrenals from steers, cows, and bulls have been noted. We have also not observed any great seasonal variation in receptor yield per adrenal. The adrenal medulla is separated from the cortex by sharp dissection using a 20 × 20-cm glass plate on ice. The cortex is scraped from the capsule with a razor blade and transferred into liquid nitrogen. The quick-frozen cortex can be stored at −70° without loss of LDL receptor activity for 3 months. When required, the cortices from 30 adrenals are thawed at 4° in 5 ml of buffer A per g of frozen tissue. All subsequent operations are performed at 0–4°. The tissue is homogenized with two 30-sec pulses of a Polytron homogenizer (settings No. 5 and 8) and centrifuged at 800 *g* for 10 min. The supernatant is filtered through Miracloth (Calbiochem-Behring), and the resulting filtrate (about 380 ml) is centrifuged at 100,000 *g* for 1 hr. The surface of each pellet and the walls of the centrifuge tubes are rinsed with 15 ml of buffer A to remove a lipid film. The membrane pellets are quickly frozen in liquid nitrogen and can be stored at −70° for at least 1 month without loss of LDL receptor activity.

Cultured Cells

Human fibroblasts or human A-431 epidermoid carcinoma cells are grown in monolayer and maximum expression of LDL receptors is in-

[13] J. L. Goldstein, S. K. Basu, and M. S. Brown, this series, Vol. 98, p. 241.

[14] O. H. Lowry, N. J. Rosebrough, A. L. Farr, and R. J. Randall, *J. Biol. Chem.* **193,** 265 (1951).

duced by growth in lipoprotein-deficient serum for 48 hr before harvesting the cells. Only vigorously growing cells in log-phase growth are used. Procedures for growing cells for expression of maximal LDL receptor activity have been described in detail elsewhere in this series.[13]

Purification of LDL Receptors

Chemical Amounts of Bovine Adrenal Receptors

The procedures described in this section represent a modified and updated version of an earlier purification scheme.[4,5,15]

Step 1. Solubilization. All operations are carried out at 4°. Membrane pellets prepared from 30 adrenal glands as described above are suspended in 120 ml of buffer B by sequential aspiration through a 15-gauge and a 21-gauge needle. The suspension is sonicated twice for 25 sec (Sonifier model W185, Heat Systems-Ultrasonics, Inc., Plainview, NY) with a microprobe at setting No. 6. To this suspension, 9.6 ml of 4 *M* NaCl, 62 ml H_2O, and 48 ml of 5% (v/v) Triton X-100 are added. The mixture is stirred for 10 min, and undissolved material is removed by centrifugation at 100,000 *g* for 1 hr. The resulting supernatant (230 ml containing 0.7 to 1.1 g of protein) is diluted into 690 ml of ice-cold stirred buffer C. After stirring for 10 min, the mixture is clarified by centrifugation at 5000 *g* for 10 min. At this point, make sure that the pH of the solution is 6; adjust with 2 *M* maleic acid if necessary.

Step 2. DEAE-Cellulose Chromatography. All operations are carried out at 4°. The clarified extract is applied to a column (2.6 × 8 cm) of DEAE-cellulose (DE 52, Whatman) previously equilibrated in buffer D. Ascending flow at a rate of 70 ml/hr is maintained with a peristaltic pump. The column is then washed with 60 ml of buffer D, followed by 80 ml of buffer E at a flow rate of 40 ml/hr. The column is eluted with a 100-ml linear gradient of 0 to 200 m*M* NaCl in buffer E at a flow rate of 40 ml/hr. Fractions of 4 ml each are collected, and the [^{125}I]LDL binding activity of each fraction is measured in a 200-μl aliquot as described below. Fractions containing LDL receptor activity (eluting in the second half of the gradient) are combined and quickly frozen in liquid nitrogen. This material, designated as the DEAE-cellulose fraction, can be stored at −70° for up to 2 months without loss of LDL receptor activity.

If receptor is to be purified by affinity chromatography on LDL-Sepharose (see Step 3A), the DEAE-cellulose fraction is dialyzed against 3 changes of 100 volumes each of buffer F for 24 hr. It must not be frozen

[15] W. J. Schneider, S. K. Basu, M. J. McPhaul, J. L. Goldstein, and M. S. Brown, *Proc. Natl. Acad. Sci. U.S.A.* **76,** 5577 (1979).

before further purification. If receptor is to be purified by affinity chromatography on IgG-C7-Sepharose (see Step 3B), the DEAE-cellulose fraction is mixed with 3 volumes of buffer G. At this point, it can be quickly frozen in liquid nitrogen, and stored at $-70°$ for up to 1 month without loss of receptor activity.

Step 3. Affinity Chromatography. The DEAE-cellulose fractions can be purified to apparent homogeneity by affinity chromatography on columns containing either LDL-Sepharose or a monoclonal antibody against the LDL receptor (IgG-C7) coupled to Sepharose.

3A. Chromatography on LDL-Sepharose. Freshly prepared human LDL is covalently attached to CNBr-activated Sepharose 4B (Pharmacia) according to the manufacturer's instructions, using 50 mg of LDL-protein per g of dry gel. The LDL-Sepharose is stored at 4° in buffer H and used within 4 weeks. All subsequent operations are performed at 4°.

For each milligram of protein in the DEAE-cellulose fraction, 100 μl of settled bed volume of LDL-Sepharose containing 1.2 to 1.3 mg LDL is equilibrated in buffer F in a glass column. Prior to chromatography, the DEAE-cellulose fraction is diluted with 3 volumes of buffer F and spun at 100,000 *g* for 1 hr. The resulting supernatant is applied to the column. In a typical bulk purification, 120 ml of diluted DEAE-cellulose fraction is applied to the column and recycled for a total of 90 min at a hydrostatic pressure of about 25 cm. The column is then washed with 200 column volumes of buffer F, followed by 3 volumes of H_2O (the latter can be omitted if subsequent lyophilization of the preparation is not required). Purified receptors can be eluted in either of two ways: with a solution of 5 m*M* sodium suramin (Mobay Chemical Co., FBA Pharmaceuticals, New York, NY) in 50 m*M* Tris–chloride (pH 6) or with 0.5 *M* NH_4OH. Elution with NH_4OH allows the sample to be lyophilized. In both cases, a volume of elution buffer equal to 25% of the settled bed volume of the column is added to the gel surface and the eluate discarded. An additional 1.5 column volumes of elution buffer is then applied to the column; the eluate, which contains the receptor, is quickly frozen in liquid nitrogen. If elution has been performed with NH_4OH, the frozen receptor is lyophilized and stored indefinitely at $-70°$. It can be redissolved at any time in aqueous buffers for further studies. If the receptor has been eluted with suramin, it is stored frozen and is stable indefinitely at $-70°$. The suramin can be removed by dialysis. The table summarizes the results of a typical purification using this procedure.

The LDL-Sepharose can be used three times, if the column is washed immediately after elution of receptors with 5 column volumes of 4 *M* NaCl, followed by 10 column volumes of buffer H. The bottom of the column is plugged with a stopper and the column is stored at 4° with 2 ml

PURIFICATION OF THE LDL RECEPTOR FROM BOVINE ADRENAL CORTEX[a]

Fraction	Protein (mg/fraction)		High affinity [^{125}I]LDL binding		Purification factor (fold)
	Applied	Recovered	Total (μg/fraction)	Specific activity (μg/mg protein)	
1. Intact membranes		2170 (100)[b]	673 (100)[b]	0.31	1
2. Solubilized membranes	2170	825 (38)			
3. DEAE-cellulose column	810	46 (2.1)	975 (145)	21	68
4. Affinity chromatography					
LDL-Sepharose	44	0.7 (0.032)	242 (36)	346	1116
IgG-C7 Sepharose	44	0.6 (0.028)	227 (34)	378	1219

[a] Membranes from 30 bovine adrenals were prepared and processed as described under Purification of LDL Receptors. The values for high affinity binding were determined as described under Assay Methods at a concentration of 20 μg protein/ml [^{125}I]LDL. The results shown are from one representative preparation. This procedure has been carried out 10 times using LDL-Sepharose and 95 times using IgG-C7-Sepharose. The preparations were consistently greater than 95% pure; the overall yield and binding activity of each preparation varied no more than ±20% from the values reported here.

[b] The values in parentheses represent the percentage of "protein recovered" or "total amount of high affinity [^{125}I]LDL binding" relative to Fraction 1 (intact membranes).

of buffer H layered on top of the gel. LDL-Sepharose is used within 4 weeks of its preparation.

3B. Chromatography on IgG-C7-Sepharose. The method of preparation of the monoclonal anti-LDL receptor antibody IgG-C7 has been described.[16] This antibody is of the IgG_{2b} subclass and recognizes the bovine and human LDL receptors.[16] IgG-C7 is coupled to CNBr-activated Sepharose as described above for LDL-Sepharose, except that the coupling ratio is 20 mg of IgG-C7 per g of dry gel. The IgG-C7-Sepharose is stored at 4° in buffer H and used within 4 weeks.

All subsequent operations are performed at 4°. For each milligram of protein of the DEAE-cellulose fraction, 100 μl of IgG-C7-Sepharose (settled bed volume) containing 0.5 to 0.6 mg IgG-C7 is equilibrated in buffer G in a glass column. Prior to chromatography, the DEAE-cellulose fraction (diluted with 3 volumes of buffer G as described above) is centrifuged at 100,000 *g* for 1 hr, and the supernatant is applied to the column. The sample is recycled over the column for a total of 2 hr under a hydrostatic

[16] U. Beisiegel, W. J. Schneider, J. L. Goldstein, R. G. W. Anderson, and M. S. Brown, *J. Biol. Chem.* **256,** 11923 (1981).

pressure of about 25 cm. The column is then washed with 5 column volumes of buffer G, followed by 5 column volumes of buffer G containing 0.2 *M* NaCl and 40 m*M* octylglucoside, followed by 200 column volumes of buffer G. Before elution of purified receptors, 3 column volumes of H_2O are applied to remove salts and detergent. Elution of purified receptors is achieved with 0.5 *M* NH_4OH. The elution procedure is identical to the one described for the LDL affinity column. Preparations are lyophilized immediately and stored indefinitely at −70°. The receptors can be redissolved in aqueous buffers for further studies. The table summarizes the purification procedure.

The IgG-C7-Sepharose can be used three times, if the column is washed immediately after elution of receptors with 5 column volumes of a solution of 0.2 *M* octylglucoside, followed by 5 column volumes of 4 *M* NaCl, and finally with 10 column volumes of buffer H. The columns are stored as described above for LDL-Sepharose.

The hybridoma producing IgG-C7 is available from the American Type Culture Collection (Rockville, MD).

Biosynthetically Radiolabeled Receptors from Cultured Cells

LDL receptors of cultured cells can be biosynthetically labeled with [^{35}S]methionine and then purified by immunoaffinity chromatography. The procedure described below yields a preparation of radiolabeled mature LDL receptors with a molecular weight of 160,000 as determined by SDS–polyacrylamide gel electrophoresis.[6,7]

Step 1. Biosynthetic Radiolabeling of LDL Receptors. Cells are grown in monolayer culture and switched to medium containing 10% lipoprotein-deficient serum 48 hr prior to harvest as described above. Eighteen hours prior to harvest, the cells receive medium containing 5 μM methionine, 10% lipoprotein-deficient serum, and 100–150 μCi/ml of [^{35}S]methionine. The number of culture dishes is adjusted so that a total of 5 mCi of [^{35}S]methionine is used in one experiment.

Step 2. Solubilization. After 18 hr in medium containing [^{35}S]methionine, cells are chilled to 4° for 10 min, washed twice with ice-cold buffer I, and then lysed by the addition of ice-cold buffer J (200 μl for each 60-mm dish and 500 μl for each 100-mm dish). The cells are rapidly scraped from the dishes with a rubber policeman, and insoluble material is removed by centrifugation at 4° for 30 min at 100,000 *g*. The supernatant is removed carefully and diluted with 3 volumes of buffer G. This mixture is spun at 4° for 30 min at 100,000 *g*, and the supernatant is used for affinity chromatography.

Step 3. Affinity Chromatography on IgG-C7 Sepharose. This is performed exactly as described above for the bovine adrenal receptor (see

Step 3B). For each 5 mCi of [^{35}S]methionine used to label cellular protein, 1 ml of IgG-C7-Sepharose (settled bed volume) is required. The IgG-C7-Sepharose is not reused. This procedure yields 2–6 $\times$ 10^6 dpm of radiolabeled receptor. After lyophilization the preparations can be stored for at least two half-lives of the ^{35}S-isotope without significant radiolysis.

Biosynthetically radiolabeled LDL receptors cannot be purified by affinity chromatography on LDL-Sepharose. This purification step requires prior removal of octylglucoside. When detergent is removed from the crude cellular extracts, radiolabeled proteins form mixed aggregates, making it impossible to obtain purified LDL receptors.

Assay Methods

The binding of [^{125}I]LDL to LDL receptors during purification is monitored with two assay systems: one is designed for receptors in intact membranes,[17] and the other measures LDL binding to soluble receptors.[4,5] In both systems, receptors are incubated with [^{125}I]LDL. Separation of bound and free [^{125}I]LDL is achieved by centrifugation (intact membranes) or by microporous filtration (soluble receptors). It should be pointed out that receptor activity in soluble extracts containing certain detergents (e.g., Triton X-100 or Nonidet P-40) cannot be reliably measured with any of the assays currently available. This is due to interference with these detergents in the binding, presumably through interaction of the detergents with [^{125}I]LDL or the receptor.[15] The only detergent found so far that does not cause these problems is octylglucoside.

Binding of [^{125}I]LDL to Membranes

This assay is carried out in 250-μl polyethylene microfuge tubes at room temperature as described by Basu *et al.*[17] The standard assay is conducted at pH 8 in 100 μl of buffer K containing 50–150 μg membrane protein and 20 μg protein/ml [^{125}I]LDL (200–600 cpm/mg protein) in the absence or presence of 10 m*M* EDTA. To determine the amount of membrane-bound [^{125}I]LDL, an aliquot (60 μl) is layered on top of 140 μl of fetal bovine serum in a 4.8 $\times$ 19.9 mm-cellulose nitrate airfuge tube (Beckman Instruments). The tubes are subjected to centrifugation in a Type A-100 aluminum fixed-angle rotor in a Beckman airfuge (100,000 *g* for 20 min at 4°). The supernatant is aspirated, and 100 μl of fetal bovine serum is added to the pellet without resuspension. After centrifugation at 100,000 *g* for 5 min at 4°, the supernatant is aspirated, and the bottom of the tube containing the pellet is sliced off with a razor blade and counted for

[17] S. K. Basu, J. L. Goldstein, and M. S. Brown, *J. Biol. Chem.* **253,** 3852 (1978).

radioactivity determination in a well-type gamma scintillation counter. High-affinity binding is calculated by subtracting the amount of [^{125}I]LDL bound in the presence of 10 m*M* EDTA from that bound in the absence of EDTA.

Binding of [^{125}I]LDL to Soluble LDL Receptors

Phosphatidylcholine/Acetone Precipitation of Receptors. Two hundred milligrams of egg phosphatidylcholine in ethanol (Avanti Polar-Lipids, Birmingham, AL, Cat. No. 820051) is placed in a 125-ml glass flask and the ethanol removed under a stream of nitrogen. The dry phospholipid is redissolved in diethyl ether, and a coat of the lipid covering the inner walls of the flask is formed by evaporating the ether with a stream of nitrogen. Buffer F (100 ml) is added and a suspension of liposomes is formed by hand shaking for 5 min at room temperature. The suspension is stored at 4° for up to 1 month.

Prior to precipitation with phosphatidylcholine/acetone, solutions of LDL receptors containing 40 m*M* octylglucoside must be diluted to a final concentration of 15 m*M* detergent or less; receptor solutions that contain no detergent do not require dilution. To each receptor solution, the following components are added to give the indicated final concentration: NaCl, 0.5 *M*; egg phosphatidylcholine, 0.5–1 mg/ml (prepared in buffer F as described above); and octylglucoside (if present), less than 15 m*M*. To this mixture, 0.6 volumes of acetone (precooled to −20°) is added rapidly with vigorous Vortex agitation. The precipitate is collected immediately by centrifugation at 20,000 *g* for 20 min at 4° and is designated as phospholipid/acetone precipitate.

Filtration Assay. The phospholipid/acetone precipitate is resuspended in buffer L to a final protein concentration of 20 μg/ml to 2 mg/ml by aspiration through a 22-gauge needle. Do not use a needle of any other bore. Assay mixtures (100 μl) are composed of 40 μl of resuspended phospholipid/acetone precipitate, 30 μl of buffer M, 10 μl of H_2O or 0.1 *M* EDTA, and 20 μl of 100 μg protein/ml of [^{125}I]LDL in 0.15 *M* NaCl (200–600 cpm/ng protein). The tubes are incubated for 1 hr at room temperature. Aliquots (80 μl) of each reaction mixture are then subjected to filtration exactly as described in Ref. 5 with the use of cellulose acetate membrane filters (Oxoid Ltd., Basingstoke, U.K., Cat. No. N25/45 UP). Cellulose nitrate filters are not suitable because they bind large amounts of [^{125}I]LDL. Nonspecific binding represents the amount of [^{125}I]LDL retained by the filter in incubations containing EDTA. High-affinity binding is calculated by subtracting the value for nonspecific binding from the value for total binding.

Properties of the Purified Bovine LDL Receptor

Purity and Solubility

When the LDL receptor is purified by immunoaffinity chromatography, only traces of impurities can be detected by SDS–polyacrylamide gel electrophoresis in the absence or presence of reducing agents (Fig. 1). A contaminant that is observed in some preparations has an apparent molecular weight of about 80,000. This is a proteolytic product of the receptor molecule that is retained on LDL columns as well as on monoclonal antibody IgG-C7 columns. Proteolysis can best be prevented by the addition of leupeptin and PMSF to all buffers immediately before use. The purified and lyophilized receptor is soluble in aqueous buffer of pH 6 to 11. In the absence of detergents, it is likely that the receptor exists in an aggregated, but water-soluble form. The state of aggregation has not been determined.

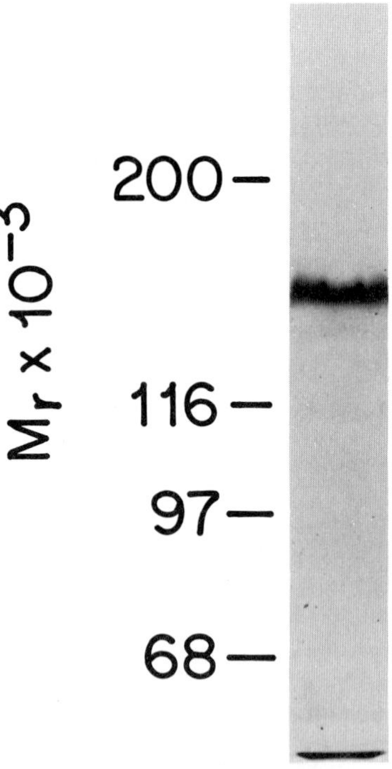

FIG. 1. SDS–polyacrylamide gel electrophoresis of purified LDL receptor from bovine adrenal cortex. The receptor was purified in the following sequence: solubilization of membranes, DEAE-cellulose chromatography, and affinity chromatography on monoclonal IgG-C7-Sepharose. The purified protein (40 μg) was subjected to SDS gel electrophoresis in the presence of reducing agent on a 5% slab gel followed by staining with Coomassie Blue. Molecular weight standards are indicated.

Binding Properties

In its purified form the receptor retains the binding properties that are characteristic of the LDL receptor in whole cells and in crude solubilized extracts.[4] The apparent K_d for [^{125}I]LDL ranges between 8 and 20 μg/ml. The purified receptor shows a 20-fold higher affinity for apo-E-HDL_c than for LDL and a binding capacity for apo E-HDL_c that is one-fourth of that for LDL. The purified receptor fails to bind methyl-LDL or typical HDL (which contains no apo E). The binding of LDL to the purified receptor requires Ca^{2+} and is inhibited by EDTA.[4]

Molecular Weight and Isoelectric Point

The receptor has an apparent molecular weight of 160,000 on SDS polyacrylamide gels in the concentration range of 4 to 8.5% when subjected to electrophoresis in the presence of 350 m*M* 2-mercaptoethanol or 100 m*M* dithiothreitol.[4] In the absence of reducing agent(s), the apparent molecular weight on SDS gels is about 130,000.[18] The isoelectric point of the receptor is ~4.6.[4,19]

Recently, a cDNA for the human LDL receptor has been isolated.[20] The nucleotide sequence revealed that the receptor is a single polypeptide chain of 839 amino acid residues with a protein molecular weight of 93,102. If the receptor contains 18 O-linked carbohydrate chains and two N-linked complex carbohydrate chains as postulated,[8] the true molecular weight of the mature glycoprotein would be about 115,000. Its migration on SDS gels as a 160,000-Da protein thus appears to be anomalous, owing to the presence of the multiple O-linked carbohydrate chains.

Carbohydrate Content

The receptor is a glycoprotein. Upon treatment with neuraminidase, the apparent M_r of the bovine adrenal LDL receptor, as determined by SDS–polyacrylamide gel electrophoresis under reducing conditions, decreases by ~9000, indicating that the receptor contains sialic acid.[4] In addition to sialic acid, the purified bovine receptor contains N-linked as well as O-linked oligosaccharide chains composed of mannose, galactose, *N*-acetylgalactosamine, *N*-acetylglucosamine, and fucose.[8]

Acknowledgments

The original research described in this article was supported by a grant (HL-20948) from the National Institutes of Health. W.J.S. is an Established Investigator of the American Heart Association.

[18] T. O. Daniel, W. J. Schneider, J. L. Goldstein, and M. S. Brown, *J. Biol. Chem.* **258,** 4606 (1983).

[19] U. Beisiegel, W. J. Schneider, M. S. Brown, and J. L. Goldstein, *J. Biol. Chem.* **257,** 13150 (1982).

[20] T. Yamamoto, C. G. Davis, M. S. Brown, M. L. Casey, J. L. Goldstein, and D. W. Russell, *Cell,* in press (1984).

[37] Synthesis of Biotinyl Derivatives of Peptide Hormones and Other Biological Materials

By FRANCES M. FINN and KLAUS H. HOFMANN

The strength of the noncovalent interaction between avidin and biotin ($K_A \sim 10^{15}\ M$)[1] makes this binding pair ideally suited for use in situations where identification or isolation of molecules present in extremely low concentrations is desirable. The biotin–avidin complex is beginning to play an increasingly important role in immunocytochemistry through the use of biotinylated immunoglobulins and avidin labeled with electron-dense or fluorescent markers.[2] Another potentially useful application for this complex is in the development of biospecific affinity chromatography columns for isolating molecules whose principal function is binding, such as, receptors, transport proteins, streptavidin, and avidin.

The use of avidin itself for construction of affinity resins has serious disadvantages owing to the high content of basic amino acid residues (9 lysines and 8 arginines per subunit) that bind to anionic surfaces. However, conversion of the positively charged ε-amino groups of the lysine side chains to half succinoates, by reacting the protein with an excess of succinic anhydride, effectively eliminates this problem without markedly changing the biotin binding properties[3] (Fig. 1). Alternatively streptavidin,[4] a biotin binding protein produced and exported by the microorganism *Streptomyces avidinii,* can be employed. This molecule binds biotin as tightly as avidin and has none of the undesirable basicity. However, it should be noted that streptavidin has no affinity for dethiobiotin,[5] a weaker binding analog of the vitamin.

In preparing biotinylated ligands for avidin-biotin affinity columns some consideration must be given to the nature of the interaction between the ligand and its binding protein or receptor. The biotin attachment site should be chosen so as not to interfere with ligand–receptor binding. Furthermore, attachment of biotin to peptides or proteins may in some instances interfere with access of the biotinylated molecule to its avidin binding site and thus alter the strength of this interaction substantially.

[1] N. M. Green, *Adv. Protein Chem.* **29,** 85 (1975).

[2] E. A. Bayer and M. Wilchek, *Methods Biochem. Anal.* **26,** 1 (1980).

[3] F. M. Finn, G. Titus, J. A. Montibeller, and K. Hofmann, *J. Biol. Chem.* **255,** 5742 (1980).

[4] K. Hofmann, S. W. Wood, C. C. Brinton, J. A. Montibeller, and F. M. Finn, *Proc. Natl. Acad. Sci. U.S.A.* **77,** 4666 (1980).

[5] H. C. Lichstein and J. Birnbaum, *Biochem. Biophys. Res. Commun.* **20,** 41 (1965).

ISBN 0-12-182009-2

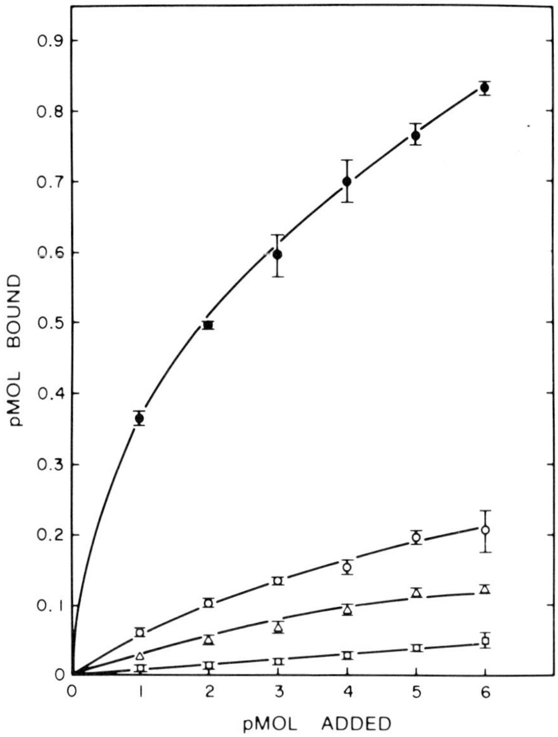

FIG. 1. Binding to rat liver membranes of [^{125}I]pHPP–avidin (●), [^{125}I]SpHPP–avidin (○), [^{125}I]SpHPP–avidin saturated with biotin (40 mol/mol) (△), and [^{125}I]SpHPP–avidin in the presence of a large excess (30 μmol) of unlabeled SpHPP–avidin (□).

A series of reagents has been developed for the introduction of biotin into polyfunctional molecules by way of carboxyl groups (biocytinamide),[6,7] amino groups (*N*-hydroxysuccinimidobiotinate),[8,9] *via* azo coupling to tyrosine and histidine side chains of proteins (*p*-aminobenzoylbiocytinamide)[6] and aldehydes produced by treating cells with periodate[10] (biotin hydrazide).[2,10,11] Recognizing that the strength of the biotin–avidin interaction may be disadvantageous in certain situations, derivatives us-

[6] K. Hofmann, F. M. Finn, and Y. Kiso, *J. Am. Chem. Soc.* **100,** 3585 (1978).
[7] H. Romovacek, F. M. Finn, and K. Hofmann, *Biochemistry* **22,** 904 (1983).
[8] J. M. Becker, M. Wilchek, and E. Katchalski, *Proc. Natl. Acad. Sci. U.S.A.* **68,** 2604 (1971).
[9] M. L. Jasiewicz, D. R. Schoenberg, and G. C. Mueller, *Exp. Cell Res.* **100,** 213 (1976).
[10] H. Heizmann and F. M. Richards, *Proc. Natl. Acad. Sci. U.S.A.* **71,** 3537 (1974).
[11] K. Hofmann, D. B. Melville, and V. du Vigneaud, *J. Biol. Chem.* **144,** 513 (1942).

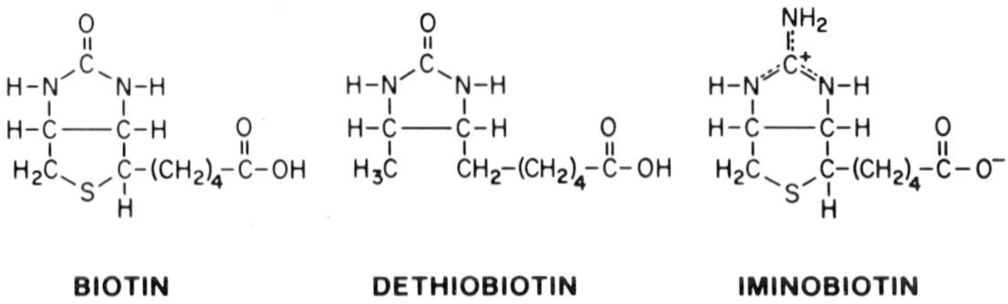

FIG. 2. Chemical structures of biotin, dethiobiotin, and iminobiotin.

ing the weaker binding (dethio-) and pH sensitive (imino-) analogs of biotin (Fig. 2) have been prepared. In order to provide space between the biotin and the ligand for those cases where steric hindrance to biotin binding results from its attachment to the ligand, as previously mentioned, "active" esters of biotin and biotin analogs containing "spacer arms" of 7 and 14 atoms have also been prepared.

The covalent attachment of a number of these derivatives to the amino terminus of the B chain of insulin has been accomplished and a general procedure is described herein. The syntheses of biotinylated corticotropins, a much more formidable task, are presented only in principle since the routes to their preparation were dictated by the fragments available and thus are not general procedures.

Finally procedures are described for modifying and labeling avidin and for assessing the rate of dissociation of biotin derivatives from avidin.

Syntheses

General Methods

Solvents are freshly distilled. DMF[11a] is kept over barium oxide lumps and is distilled *in vacuo* immediately prior to its use. DMF used for the preparation of *N*-hydroxysuccinimidobiotinate is refluxed over barium oxide and distilled *in vacuo*. Pyridine is refluxed over barium oxide and distilled. DCC (Nutritional Biochemicals) is dissolved in ether and the insoluble DCU is removed by filtration. The clear filtrate is evaporated and the residue is distilled *in vacuo*. TFA (Fisher) is distilled prior to its

[11a] Abbreviations: ACTH (corticotropin), adrenocorticotropic hormone; Boc, *tert*-butoxycarbonyl; DCHA, dicyclohexylamine; DCC, *N,N'*-dicyclohexylcarbodiimide; DCU, *N,N'*-dicyclohexylurea; DMF, dimethylformamide; DMSO, dimethyl sulfoxide; HPLC, high-pressure liquid chromatography; TCA, trichloroacetic acid; TFA, trifluoroacetic acid; THF, tetrahydrofuran; TLC, thin-layer chromatography; Z, benzyloxycarbonyl; biocytinamide, N^{ε}-biotinyl-L-lysine amide; dethiobiocytinamide, N^{ε}-dethiobiotinyl-L-lysine amide; iminobiocytinamide, N^{ε}-iminobiotinyl-L-lysine amide.

use. DMSO is distilled *in vacuo* over calcium hydride and stored over molecular sieves Type 4A, Grade 514, 8-12 mesh (Fisher). Homogeneity of the various compounds with the exception of the biotinylated insulins, ACTH derivatives, and avidins is assessed by elemental analysis, and TLC in several solvent systems. [R_f^{I} 1-butanol–glacial acetic acid–water (60 : 20 : 20); R_f^{II} chloroform–methanol–water (8 : 3 : 1); R_f^{III} 1-butanol–pyridine–glacial acetic acid–water (30 : 20 : 6 : 24)]. Compounds containing biotin, dethiobiotin, and iminobiotin are visualized on thin-layer plates by spraying with a freshly prepared 1 : 1 mixture of *p*-dimethylaminocinnamaldehyde (Sigma) (0.2% in ethanol) and 2% concentrated sulfuric acid in ethanol.[12] Compounds containing *p*-aminobenzoic acid exhibit a blue fluorescence when the plates are exposed to UV light. Spraying with dilute acetic acid, then with 1% sodium nitrite followed by 1% β-naphthol in 0.1 *N* NaOH is another method for visualizing such compounds (diazo reaction). Homogeneity of the biotinylated insulins and ACTH derivatives is assessed by HPLC. For the insulins, the solvent system is (pump A) 0.1% phosphoric acid and (pump B) acetonitrile 50% in 0.1% phosphoric acid. A linear gradient from 50 to 90% (pump B) in 20 min is employed with a pumping speed of 2 ml/min (Bondapak C_{18} column Waters). For the ACTH derivatives, the solvent system is (pump A) 0.05% TFA; (pump B) 50% acetonitrile in 0.05% TFA. A linear gradient from 20 to 80% (pump B) in 30 min with a pumping speed of 2 ml/min is employed (Bondapak C_{18} column). Acetate cycle AG 1 × 2 (Bio-Rad) is washed with two portions of 2 *N* KOH, followed by distilled water until the filtrates are neutral (pHydrion paper). The resin is then washed with two portions of 10% acetic acid and again with water until neutral. Amberlite IRA-400 (Mallinckrodt) acetate cycle is prepared as described for the AG 1 × 2. CMC (Bio-Rad) is washed with two portions of 10% acetic acid, then with water until the filtrates are neutral.

Biotin

The most widely used procedure for introducing biotin into other molecules involves acylation at pH 8.0–9.0 with *N*-hydroxysuccinimidobiotinate. Two procedures are employed to produce the latter compound. The first involves the use of *N,N'*-dicyclohexylcarbodiimide, the second utilizes *N,N'*-carbonyldiimidazole as the condensing agent. Ester prepared by the first method is contaminated with DCU which is very difficult to remove quantitatively. The disadvantaged of the *N,N'*-carbonyldiimidazole procedure is the instability of the reagent (moisture).

[12] D. B. McCormick and J. A. Roth, *Anal. Biochem.* **34,** 226 (1970).

N-Hydroxysuccinimidobiotinate

By the DCC Procedure.[8] Biotin (244 mg, 1.0 mmol) and *N*-hydroxysuccinimide (115 mg, 1.0 mmol) are dissolved in hot dimethylformamide (3 ml) and DCC (206 mg, 1.0 mmol) is added. The resultant mixture is stirred at room temperature for 2 hr, the precipitate is removed by filtration, and the filtrate is evaporated *in vacuo*. The residue is triturated with peroxide-free ether and crystallized from 2-propanol, mp 196–200°.

With N,N'-Carbonyldiimidazole.[9] *N*-Hydroxysuccinimide and biotin are dried *in vacuo* over P_2O_5. Biotin (1g, 4.1 mmol) is dissolved in DMF (20 ml) at 80° and *N,N'*-carbonyldiimidazole (Fluka) (665 mg, 4.1 mmol) is added. The solution is kept at 80° until the evolution of CO_2 ceases. The mixture is cooled at room temperature and stirring is continued for 2 hr. During this time a flocculent precipitate of biotin-imidazolid appears. At this point *N*-hydroxysuccinimide (475 mg, 4.1 mmol) in DMF (10 ml) is added and the mixture is stirred for several hours when the precipitate disappears. The solvent is removed *in vacuo,* the residue is crystallized from boiling 2-propanol; average yield 1.0 g (71%); mp 196–197°; R_f^{II} 0.62. Operation under strictly anhydrous conditions is essential for the success of this preparation. *N,N'*-carbonyldiimidazole is highly sensitive to moisture and commercially available material may be partially decomposed. It is advisable to assess the purity of commercial preparations of this reagent by measuring the CO_2 evolved when decomposing aliquots with water.

Biotinhydrazide.[2] This preparation must be performed under anhydrous conditions. Thionylchloride (1 ml) is added slowly to methanol (10 ml) cooled in an ice-salt bath. Biotin (1 g) is added and the mixture is kept at room temperature for 12 hr. The solvent is evaporated to dryness, methanol (10 ml) is added and evaporated as before. The residue (methyl biotinate) is dissolved in methanol (5 ml), hydrazine hydrate (1 ml) is added, and the mixture is kept at room temperature for 12 hr. The precipitated hydrazide is collected, washed with peroxide-free ether, and crystallized from DMF, or water; mp 227–229°.

Dethiobiotin

This compound is formed when biotin is treated with Raney Nickel.[13] The preparation of Raney Nickel is described in great detail in Organic Syntheses.[14] Biotin (1 g, 4.1 mmol) is dissolved in hot 0.5% Na_2CO_3 (100 ml) and the solution is added to approximately 40 ml of settled Raney Nickel. The suspension is heated at 75–80° with vigorous stirring for 45

[13] D. B. Melville, K. Dittmer, G. B. Brown, and V. du Vigneaud, *Science* **98,** 497 (1943).
[14] *Org. Synth.* **21,** 15 (1941).

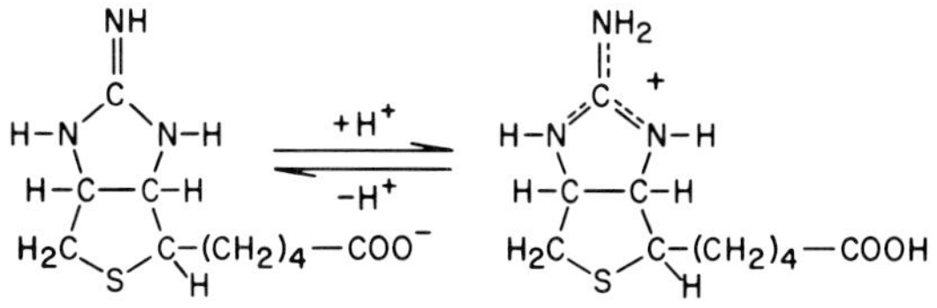

FIG. 3. Charged and uncharged forms of iminobiotin.

min. The nickel is removed by centrifugation and is washed with three 50 ml portions of 0.5% Na_2CO_3 and two 50 ml portions of water at 70° with stirring. The original solution and washings are combined, acidified with concentrated HCl to Congo red, filtered, and concentrated *in vacuo* to a small volume (approximately 30 ml). The solution is placed in a refrigerator for crystallization. The crystals are collected, washed with a small volume of ice water, and dried; yield 725 mg (83%); mp 155–156°; R_f^I 0.65; R_f^{II} 0.4; R_f^{III} 0.6.

N-Hydroxysuccinimidodethiobiotin.[15] Dethiobiotin (857 mg, 4 mmol) and *N*-hydroxysuccinimide (690 mg, 6 mmol) are dissolved in DMF (10 ml) and DCC (1.24 g, 6 mmol) in DMF (2 ml) is added at 0°. The solution is stirred at 0° for 30 min and at room temperature for 20 hr. Glacial acetic acid (0.12 ml) is added and stirring is continued for 1 hr. The suspension is cooled at 0°, the DCU is removed by filtration, and the filtrate is evaporated. The oily residue is dissolved in boiling 2-propanol (~20 ml) and the solution is cooled at 0°. The crystals are collected and recrystallized from 2-propanol; yield 980 mg (78%); mp 145–146°; R_f^I 0.5; R_f^{II} 0.8; R_f^{III} 0.6.

Iminobiotin

This compound (Fig. 3) has the unique property that its affinity for avidin and streptavidin is pH dependent. Binding is strong at pH 11.0 (uncharged guanido group) and weak at pH 4.0 (charged guanido group). This property of the molecule has been applied to the development of affinity resins for the isolation of streptavidin and avidin as discussed later.

5-(3,4-Diaminothiophan-2-yl) Pentanoic Acid Sulfate.[16] Four heavy walled glass tubes are charged each with 4 g of $Ba(OH)_2 \cdot 8\ H_2O$, biotin (200 mg) and water (15 ml). The tubes are sealed, inserted into a capped stainless-steel pipe (safety), and heated in an oil-bath at 140–145° for 20 hr. The tubes are cooled at room temperature, the contents are poured into an Erlenmeyer flask, and the suspension is heated to boiling. Carbon

[15] K. Hofmann, G. Titus, J. A. Montibeller, and F. M. Finn, *Biochemistry* **21,** 978 (1982).
[16] K. Hofmann, D. B. Melville, and V. du Vigneaud, *J. Biol. Chem.* **141,** 207 (1941).

dioxide is bubbled through the hot solution for 30 min and the heavy precipitate ($BaCO_3$) is removed on a Buchner funnel. The precipitate is washed with several portions of boiling water and filtrate and washings are pooled, heated to a boil, and acidified to Congo red with 1 *N* sulfuric acid. The suspension is filtered, through a layer of Johns Manville Hyflo Super-Cel on a Buchner funnel, the clear filtrate is concentrated to a small volume *in vacuo,* and methanol is added to the cloudpoint. The suspension is placed in a refrigerator to complete the crystallization. The crystals are collected, washed with methanol and dried; yield 662 mg (67%); $[\alpha]_D^{26}$ −13.7° (c 1% in H_2O).

Iminobiotin.[15,17] 5-(3,4-Diaminothiophan-2-yl)pentanoic acid sulfate (1.26 g, 4 mmol), is dissolved in 0.5 *N* NaOH (16 ml, 8 mmol) and cyanogen bromide (712 mg, 6.7 mmol) in acetonitrile (2 ml) is added at room temperature with stirring. A heavy precipitate soon forms and the suspension is stirred for several hours the pH being maintained at 7.0 to 7.5 by addition of 1 *N* NaOH. The precipitate is collected, washed with ice water, and dried at 100° *in vacuo*; yield 886 mg (92%), mp above 200° dec; $[\alpha]_D^{25}$ +79.7° (c 1.053 1 *N* HCl); R_f^I 0.47.

N-Hydroxysuccinimidoiminobiotinate Hydrobromide.[15,18] Carefully dried iminobiotin hydrobromide (667 mg, 2.06 mmol) [prepared from iminobiotin (1 equivalent) and aqueous HBr (1 equivalent)] and *N*-hydroxysuccinimide (260 mg, 2.26 mmol) are dissolved in DMF (6 ml) and the solution is cooled in an ice bath. DCC (466 mg, 2.26 mmol) dissolved in a small volume of DMF is added and the mixture is stirred for 1 hr at ice bath temperature and for 20 hr at room temperature. The resulting suspension is cooled in an ice bath and the DCU is removed by filtration. The filtrate is evaporated to dryness, the residue is washed with two portions of peroxide-free ether, and crystallized from 2-propanol; yield 550 mg (63%); mp 176–178°; $[\alpha]_D^{25}$ +48.5° (c 1.28, H_2O); R_f^I 0.4; R_f^{III} 0.5.

Iminobiotin AH-Sepharose 4B.[4,19] AH-Sepharose 4B (Pharmacia, 25 g) is swollen in 0.5 *M* NaCl and the gel is washed with 0.5 *M* NaCl (2 liters) and water (2 liters) on a Buchner funnel (gentle suction). A 1 : 1 slurry of the washed resin in water is prepared, iminobiotin (400 mg, 1.65 mmol) is added, and the pH is adjusted to 4.8 with 1% HBr. The solution is gently stirred until the iminobiotin is dissolved then 1-cyclohexyl-3-(2-morpholinoethyl)carbodiimide metho-*p*-toluenesulfonate (4.24 g) is slowly added while stirring, and the pH is maintained at 4.8. After 2 hr there is little change in pH. The suspension is stirred for 5 hr, the gel is

[17] K. Hofmann and A. E. Axelrod, *J. Biol. Chem.* **187,** 29 (1950).

[18] G. A. Orr, *J. Biol. Chem.* **256,** 761 (1981).

[19] G. Heney and G. A. Orr, *Anal. Biochem.* **114,** 92 (1981).

collected and washed with 1 *M* NaCl (2 liters) and water (2 liters), and a 1 : 1 slurry of the gel in water containing 0.05% sodium azide is stored in a refrigerator. Binding capacity ~10 nmol avidin/ml of settled resin.

Isolation of Streptavidin on Iminibiotinyl AH-Sepharose 4B[4]

Four liters of *Streptomyces avidinii* culture broth is centrifuged for 10 min at 10,000 *g* to remove organisms and the clear supernatant is concentrated to a volume of approximately 400 ml in an Amicon concentrator 2000 using PM 10 membranes. Ammonium sulfate is added to the concentrate at 0° to 70% saturation and the precipitate is collected by centrifugation at 10,000 *g*. The precipitate is dissolved in water (~20 ml) and the turbid solution is dialyzed against two 1 liter portions of water at 4°. The dialyzed solution is centrifuged for 20 min at 35,000 *g* and the supernatant is adjusted to pH 11.0 by addition of ammonium hydroxide. An equal volume of 50 m*M* ammonium carbonate pH 11.0 containing 0.5 *M* NaCl (pH 11.0 buffer) is added and the solution is percolated through a column of iminobiotinyl AH-Sepharose 4B (1.5 × 6 cm) equilibrated with pH 11.0 buffer. The column is washed with pH 11.0 buffer until the absorbance (280 nm) reaches blank values. Elution of the column with 50 m*M* ammonium acetate, pH 4.0 containing 0.5 *M* NaCl releases the streptavidin which is detected by absorbance measurements at 280 nm. Effluents containing streptavidin are pooled, dialyzed at 4° against distilled water (1 liter) for 12 hr, and lyophilized. The residue, dissolved in 10% acetic acid, is passed through a column of Sephadex G-50 (0.8 × 58 cm), equilibrated with 10% acetic acid, and fractions containing streptavidin are pooled and lyophilized. The streptavidin binds 4.07 ± 0.02 mol (mean ± SD of [^{14}C]biotin per mol) and exhibits a single, diffuse band on disc gel electrophoresis. The column is reequilibrated with pH 11.0 buffer and is ready to be reused. Homogeneous avidin can be readily isolated from egg white by this procedure.[19]

Preparation of Spacerarms

Figure 4 illustrates a general procedure for the preparation of biotinylated ligands in which a 6-aminohexanoic acid spacer is interposed between the biotin or biotin derivative and the insulin molecule. For example the *N*-hydroxysuccinimidoiminobiotinate (**I**) is used to acylate methyl 6-aminohexanoate (**II**) and the resulting ester (**III**) is saponified to give (**IV**). This compound is converted to the *N*-hydroxysuccinimido ester (**VI**).

Methyl 6-Aminohexanoate Hydrochloride. 6-Aminohexanoic acid (Aldrich) (5 g) is suspended in dry methanol (200 ml) and the solution is

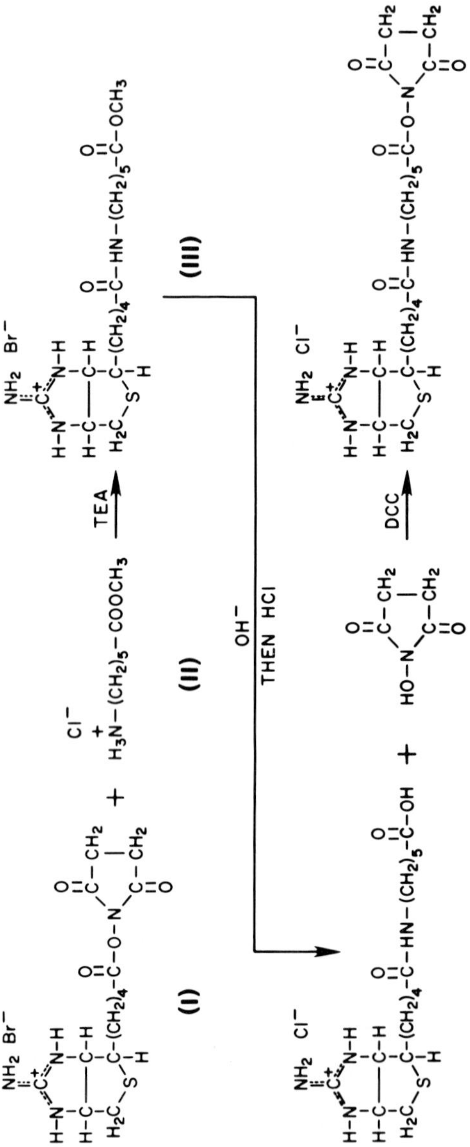

FIG. 4. Synthetic route to biotinylated spacer arms.

saturated with HCl gas while refluxing for 2 hr. The methanol is evaporated *in vacuo* (dry conditions) and the process is repeated. The residue is dissolved in methanol which is evaporated to remove excess HCl and the residue is dissolved in methanol and ether is added to the cloudpoint. The solution is placed in a refrigerator to complete the crystalization. The crystals are collected, washed with ether, dried, and recrystallized from methanol/ether; yield 5.0 g (71%) mp 119–121°. An additional amount (1.17 g) is obtained from the mother liquors.

6-(Biotinylamido)hexanoic Acid.[20,21] Triethylamine (0.153 ml, 1.1 mmol) is added to a stirred solution of *N*-hydroxysuccinimidobiotinate (341 mg, 1 mmol) and methyl 6-aminohexanoate hydrochloride (200 mg, 1.1 mmol) in DMF (5 ml) and the solution is stirred at room temperature for 20 hr. The solvent is evaporated, 1 *N* sodium hydroxide (3 ml) and enough methanol is added to give a clear solution. The mixture is stirred for 2 hr when most of the methanol is removed *in vacuo*. Water is added to the residue and the solution is acidified to Congo red with concentrated HCl. The resulting suspension is kept in a refrigerator for 2 hr, then the crystals are collected, washed with a small volume of ice water and dried; yield 327 mg (92%); mp 213–215°; R_f^{II} 0.3.

6-(6-Biotinylamidohexylamido)hexanoic Acid.[20] Triethylamine, a 10% solution in DMF (0.77 ml, 0.55 mmol), is added with stirring to a solution of *N*-hydroxysuccinimido-6-(biotinylamido)hexanoate (227 mg, 0.5 mmol) and methyl 6-aminohexanoate hydrochloride (100 mg, 0.55 mmol) in DMF (2.5 ml) and the solution is stirred at room temperature for 18 hr. The DMF is removed *in vacuo,* and 1 *N* sodium hydroxide (3 ml and enough methanol to give a clear solution) are added. The solution is stirred for 2 hr, the bulk of the methanol is removed *in vacuo,* water (approximately 5 ml) is added, and the solution is acidified to Congo red with concentrated HCl. The mixture is placed in a refrigerator, the crystals are collected and recrystallized from boiling water; yield 376 mg (85%); mp 185–190°; R_f^{I} 0.45; R_f^{II} 0.28.

6-(Dethiobiotinylamido)hexanoic Acid.[15] This compound is prepared essentially as described for the preparation of the biotin derivative from *N*-hydroxysuccinimidodethiobiotinate (311 mg, 1.00 mmol), methyl 6-aminohexanoate hydrochloride (182 mg, 1.00 mmol), and triethylamine (0.14 ml, 1.00 mmol) in DMF (5 ml). Recrystallized from boiling water; yield 300 mg (96%); mp 161–162°; R_f^{I} 0.6; R_f^{II} 0.2; R_f^{III} 0.7.

6-(6-Dethiobiotinylamidohexylamido)hexanoic Acid.[20] Triethylamine (0.153 ml, 1.1 mmol) is added with stirring to a solution of *N*-hydroxysuc-

[20] K. Hofmann, W. J. Zhang, and F. M. Finn, *Biochemistry* **23,** 2547 (1984).

[21] S. M. Costello, R. T. Felix, and R. W. Giese, *Clin. Chem.* (*Winston-Salem, N.C.*) **25,** 1572 (1979).

cinimido-6-(dethiobiotinylamido)hexanoate (425 mg, 1 mmol) and methyl 6-aminohexanoate hydrochloride (200 mg, 1.1 mmol) in DMF (5 ml) and the solution is stirred for 20 hr at room temperature. The DMF is removed *in vacuo* and 1 *N* sodium hydroxide (3 ml and enough methanol to give a clear solution) are added. The solution is stirred for 2 hr, the bulk of the methanol is removed *in vacuo,* water (approximately 3 ml) is added, and the solution is acidified to Congo red with concentrated HCl. The mixture is placed in a refrigerator. The crystals are collected and recrystallized from boiling water; yield 376 mg (85%) mp 147–150°.

6-(Iminobiotinylamido)hexanoic Acid Dihydrate.[15] This compound is prepared, essentially as described for the preparation of the corresponding biotin derivative, from *N*-hydroxysuccinimidoiminobiotinate hydrobromide (463 mg, 1.1 mmol), methyl 6-aminohexanoate hydrochloride (182 mg, 1.0 mmol), and triethylamine (0.139 ml, 1.0 mmol) in DMF (4 ml). The compound is recrystallized from water; yield 322 mg (83%); mp above 250°; $[\alpha]_D^{27}$ + 52.3° (c 1.257, 1 *N* HCl); R_f^{I} 0.3; R_f^{III} 0.5.

N-Hydroxysuccinimido 6-(Biotinylamido)hexanoate.[20,21] Initially[20] this material was prepared by the method of Sakakibara and Inukai.[22] A more convenient route is described below.

A stirred suspension of 6-(biotinylamido)hexanoate (374 mg, 1.04 mmol) and *N*-hydroxysuccinimide (180 mg, 1.57 mmol) in DMF (12 ml) is heated at 80° and DCC (322 mg, 1.56 mmol) in DMF (1.5 ml) is added. The suspension is stirred at 80° for 30 min (clear solution after 5 min) and stirring is continued at room temperature for 17 hr. The suspension is cooled in an ice bath for 2 hr, the DCU is removed by filtration, and the filtrate is evaporated to dryness *in vacuo*. The residue is crystallized from boiling isopropanol; yield 316 mg (66%); mp 160–162° dec; R_f^{II} 0.56.

N-Hydroxysuccinimido 6-(Dethiobiotinylamido)hexanoate.[15] This compound is prepared from 6-(dethiobiotinylamido)hexanoic acid (272 mg, 0.83 mmol), and *N*-hydroxysuccinimidotrifluoroacetate[22] (264 mg, 1.25 mmol) in dry pyridine (4 ml). Recrystallized from 2-propanol; yield 302 mg (86%); mp 103–105°; R_f^{II} 0.65.

N-Hydroxysuccinimido 6-(6-Dethiobiotinylamidohexylamido)hexanoate.[20] *N*-Hydroxysuccinimidotrifluoroacetate (reagent)[22] (287 mg, 1.36 mmol) is added to a suspension of 6-(6-dethiobiotinylamidohexylamido)-hexanoic acid (300 mg, 0.68 mmol) in dry pyridine (9 ml) and the mixture is stirred at room temperature for 1 hr. An additional amount of reagent (287 mg, 1.36 mmol) is added and stirring is continued for 4 hr. Insoluble material (starting acid 184 mg) is removed by filtration and the filtrate is evaporated *in vacuo*. The residue is triturated with three portions of ether

[22] S. Sakakibara and N. Inukai, *Bull. Chem. Soc. Jpn.* **38,** 1979 (1965).

and is crystallized from 2-propanol; yield 209 mg (59%) mp 130–132°; R_f^I 0.52.

N-Hydroxysuccinimido 6-(Iminobiotinylamido)hexanoate Hydrochloride.[15] 6-(Iminobiotinylamido)hexanoic acid dihydrate (169 mg, 0.43 mmol) is suspended in water, and 1 *N* hydrochloric acid (0.43 ml, 0.43 mmol) is added. The solution is evaporated and the residue is dried for 12 hr *in vacuo* over P_2O_5 and KOH pellets. To a solution of the hydrochloride and *N*-hydroxysuccinimide (50 mg, 0.43 mmol) in DMF (2 ml) is added DCC (90 mg, 0.43 mmol) in a small volume of DMF and the mixture is stirred at room temperature for 12 hr. The DCU is removed by filtration, and the filtrate is evaporated. The residue is washed with two portions of peroxide-free ether, is dissolved in a small volume of 2-propanol, and placed in a refrigerator. A small precipitate is removed by filtration and the solvent is evaporated. The residue is triturated with ethyl acetate and dried; yield 149 mg of crude product.

Biotinylation of Insulin

Random Biotinylation[23,24]

As can be expected acylation of insulin with *N*-hydroxysuccinimidobiotinate results in the formation of a mixture of seven compounds. These are: N^{α,A^1}-, N^{α,B^1}-, $N^{\varepsilon,B^{29}}$-, N^{α,A^1},N^{α,B^1}-, N^{α,A^1},$N^{\varepsilon,B^{29}}$-, $N^{\alpha,B^1}N^{\varepsilon,B^{29}}$-, and N^{α,A^1},N^{α,B^1},$N^{\varepsilon,B^{29}}$-insulins. The proportions of the various biotinylinsulins in a reaction mixture vary considerably depending on the acylation conditions used. N^{α,B^1}-biotinylinsulin is a minor component of the reaction mixture. Separation of the components can be achieved by HPLC or electrofocusing. The random approach is of little practical value as a method to prepare homogeneous monosubstituted biotinylinsulins and will not be discussed further.

Selective Biotinylation[25]

Figure 5 illustrates the general principles employed for the preparation of N^{α,B^1}-monobiotinylated insulins. Insulin (**I**) (porcine or bovine) is converted to $N^{\alpha,A^1}N^{\varepsilon,B^{29}}$-$(Boc)_2$ insulin (**II**) by treatment with Boc azide. The unprotected N-terminal amino group on the B chain is then acylated with the desired *N*-hydroxysuccinimido ester [in the example shown *N*-hy-

[23] J. M. May, R. H. Williams, and C. de Haën, *J. Biol. Chem.* **253,** 686 (1978).

[24] D. T. Pang and J. A. Shafer, *J. Biol. Chem.* **258,** 2514 (1983).

[25] K. Hofmann, F. M. Finn, H.-J. Friesen, C. Diaconescu, and H. Zahn, *Proc. Natl. Acad. Sci. U.S.A.* **74,** 2697 (1977).

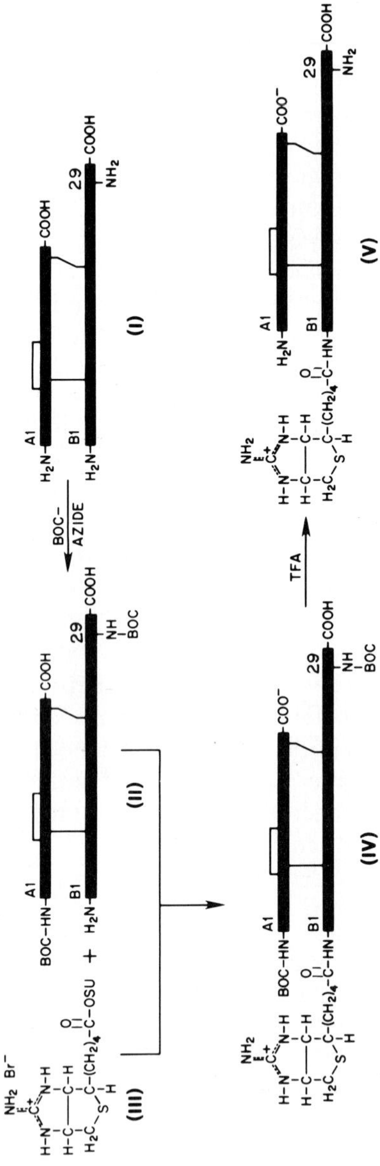

FIG. 5. Synthetic route to $N^{\alpha,B1}$-biotinylated insulins.

droxysuccinimidoiminobiotinate (**III**)] to form the protected intermediate (**IV**) which is converted to (**V**) by exposure to TFA.

$N^{\alpha,A1}$,$N^{\varepsilon,B29}$-(Boc)$_2$-Insulin.[26] Boc azide[27] (2.3 ml, 16 mmol) is added with stirring to a solution of insulin (1.2 g, 0.2 mmol) in DMSO (45 ml), water (11.5 ml), and 1 *N* sodium bicarbonate (3.3 ml) and the mixture is stirred at 35° for 5 hr. The solution is cooled at room temperature, acidified with 50% acetic acid (4 ml), and the solvents are removed *in vacuo* at a bath temperature of 35°. The residue is triturated with ether (40 ml), dried, and rubbed twice with 1% acetic acid and redried; yield 1.08 g.

$N^{\alpha,B1}$-Biotinylinsulin.[25] *N*-Hydroxysuccinimidobiotinate (200 mg, 0.59 mmol) is added at room temperature to a stirred solution of $N^{\alpha,A1}$,$N^{\varepsilon,B29}$-(Boc)$_2$ insulin (1.0 g, 0.16 mmol) and imidazole (200 mg, 2.94 mmol) in DMSO (30 ml) and the solution is stirred for 6 hr. The solution is cooled in an ice bath, and ice water (approximately 2 volumes) is added. The solution is desalted on a Sephadex G-25 column (5 × 35 cm) equilibrated with 0.05 *M* ammonium bicarbonate and fractions containing protein are pooled and lyophilized. This material is deprotected with anhydrous TFA (20 ml) (30 min at room temperature) and the TFA is removed *in vacuo* at 25°.

The residue is dissolved in 7 *M* urea/HCl, pH 3.0 (50 ml) and the pH is adjusted to 7.4 (at 4°) by addition of concentrated Tris. The solution is applied to a DE-52 (Whatman) column (5 × 60 cm), equilibrated with Tris/7 *M* urea pH 7.4 buffer. The column is developed with a NaCl gradient obtained by mixing 2 liters of starting buffer with 2 liters of the same buffer containing 0.12 *M* NaCl. Material corresponding to the major protein peak is desalted on a Sephadex G-25 column (5 × 35 cm) using 0.05 *M* ammonium bicarbonate as the eluent. The contents of tubes corresponding to the protein peak are pooled and lyophilized: yield 700 mg. Ion exchange chromatography is performed at 4°. For preparation of the Tris/urea pH 7.4 buffer free of cyanate, 7 *M* urea is adjusted to pH 3.0 to 3.5 with concentrated HCl and the solution is kept for a minimum of 3 hr at 4°, then solid Tris (6 g/liter) is added, and the pH is adjusted to 7.4. Fractions from the DE-52 column must be desalted immediately to avoid carbamylation of insulin by the cyanate formed from urea at this pH.

$N^{\alpha,B1}$-Iminobiotinylinsulin.[15] A solution of $N^{\alpha,A1}$,$N^{\varepsilon,B29}$-(Boc)$_2$ insulin (500 mg, 0.08 mmol), imidazole (100 mg, 1.47 mmol), and *N*-hydroxysuccinimidoiminobiotinate hydrobromide (100 mg, 0.24 mmol) in DMSO (15 ml) is stirred at room temperature for 6 hr. The reaction mixture is cooled

[26] R. Geiger, H. H. Schöne, and W. Pfaff, *Hoppe-Seyler's Z. Physiol. Chem.* **352,** 1487 (1971).

[27] L. A. Carpino, *J. Am. Chem. Soc.* **79,** 98 (1957).

in an ice bath and ice-cold water (~30 ml) is added. A precipitate forms which is redissolved by addition of a few drops of DMSO. The solution is desalted on a column of Sephadex G-25 (6 × 115 cm) using 0.05 *M* ammonium bicarbonate as the eluent. Fractions containing protein are pooled and lyophilized. The residue is dried *in vacuo* and is deprotected with anhydrous TFA (15 ml) for 30 min. The TFA is evaporated and the product is dried *in vacuo*. The deprotected protein is purified on a column of DE-52 (2.6 × 55 cm) in the manner described for the purification of biotinylinsulin; yield 247 mg.

Other Biotinylated Insulins. In addition to N^{α,B^1}-biotinyl (**I**), and N^{α,B^1}-iminobiotinylinsulin (**III**) the compounds listed in Fig. 6 have been prepared[15,20] using the general procedures outlined below: N^{α,A^1},$N^{\varepsilon,B^{29}}$-$(Boc)_2$-insulin (10 μmol) is acylated with the desired *N*-hydroxysuccinimido ester (40 to 80 μmol) in DMSO (4 ml) in the presence of 400 μmol of 4-methylmorpholin[28] or preferably micromoles of imidazole[25] for 15–20 hr at room temperature. The reaction mixture is diluted with 0.05 *M* ammonium bicarbonate and the solution is desalted on a column of Sephadex G-25 (3 × 54 cm) equilibrated with 0.05 *M* ammonium bicarbonate. The desalted, lyophilized product is deprotected with anhydrous TFA (1 ml) for 40 min and TFA is removed *in vacuo* at 25°. TFA ions in the product are exchanged for acetate ions on the ion exchange resin IRA-400 (Bio-Rad) and the lyophilized products are evaluated by HPLC. Impure products (HPLC) are purified by chromatography on DE-52 as described for the purification of biotinylinsulin. Derivatized insulins that contain 6-aminohexanoic acid are hydrolyzed with constant boiling HCl and the ratio of 6-aminohexanoic acid/lysine in the hydrolysate is determined with an amino acid analyzer. Dansylation of the products will show the absence of dansyl phenylalanine.

Biotinylated Corticotropins

A great deal of information is available regarding the structural features of the ACTH molecule that are essential for steroidogenic activity. $ACTH_{1-24}$ exhibits the full spectrum of biological activities of $ACTH_{1-39}$. In fact even a chain comprising only the first 20 residues of ACTH ($ACTH_{1-20}$ amide) is fully active. Thus the amino acid residues comprising positions 21–39 do not appear to be involved in the biological function of the molecule. It is also known that biological activity and binding affinity for the receptor are lost when the ε-amino groups of the lysines in positions 11, 15, and 16 are acylated. Furthermore acylation of the N-terminal

[28] G. Krail, D. Brandenburg, and H. Zahn, *Macromol. Chem., Suppl.* **1,** 7 (1975).

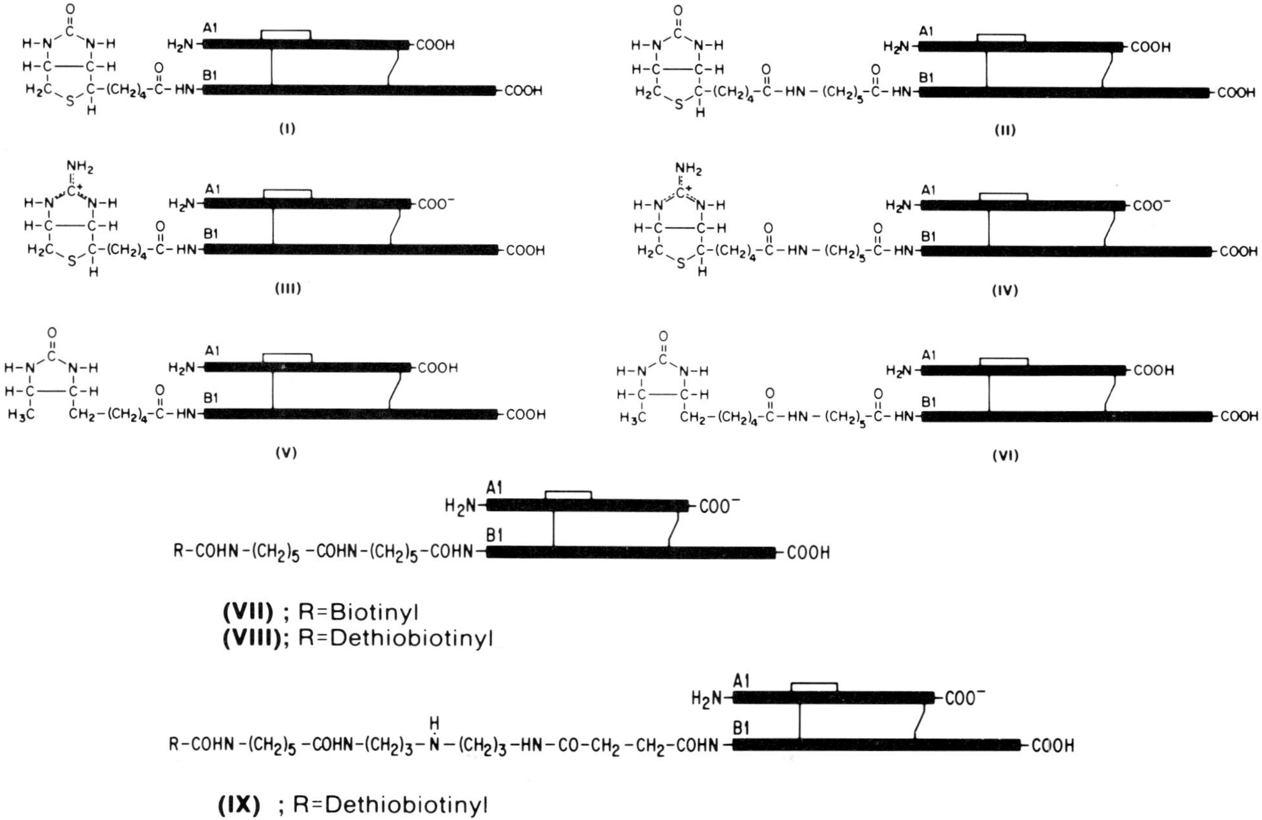

FIG. 6. Simplified structures of biotinylinsulin (**I**), 6-(biotinylamido)hexylinsulin (**II**), iminobiotinylinsulin (**III**), 6-(iminobiotinylamido)hexylinsulin (**IV**), dethiobiotinylinsulin (**V**), 6-(dethiobiotinylamido)hexylinsulin (**VI**), 6-(biotinylamidohexylamido)hexylinsulin (**VII**), 6-(dethiobiotinylamidohexylamido)hexylinsulin (**VIII**), N^{α,B^1}-[*N*-[3-[[3-[6-[6-(5 methyl-2-oxo-4-imidazolidinyl)hexanamido]hexanamido]propyl]amino]propyl]succinamoyl] insulin (**IX**).

serine residue markedly lowers biological activity.[29] Biotinylated corticotropins, capable of interacting with avidin and receptor simultaneously, cannot be prepared by acylation of $ACTH_{1-24}$ with *N*-hydroxysuccinimidobiotinate.

Since alterations at the C-terminal end of ACTH have no effect on biological activity, biotin was attached to the C-terminal carboxyl group of $ACTH_{1-24}$ in the form of biocytinamide.[6]

Although it is beyond the scope of this article to present details of the syntheses of biotinylated corticotropins, the general principles of the approach are illustrated on Fig. 7. The method, which is based on fragment condensation, involves acylation of biocytinamide (N^{ε}-biotinyl-L-lysine amide) (**I**) with a suitably protected fragment (**II**) corresponding to positions 11–24 of the ACTH molecule. The trifluoroacetyl (TFA) protecting group of the coupling product (**III**) is selectively removed to give **IV** which, in turn, is acylated with **V,** a fragment that embodies the N-terminal sequence of the ACTH molecule. The protected product (**VI**) is deprotected with TFA to give [biocytin^{25}]$ACTH_{1-25}$ amide. This same procedure is used to prepare the dethiobiocytin and iminobiocytin $ACTH_{1-25}$ amides (Fig. 8).[7] The validity of the assumptions concerning the relative unimportance of the C-terminal portion of the ACTH for biological activity is confirmed by the finding that these molecules exhibit the same steroid stimulating potency as $ACTH_{1-24}$.[6,7] With the exception of the iminobiotin analog, they also possess the ability to bind firmly to succinoylavidin.[7] Syntheses of biocytin-, dethiobiocytin-, and iminobiocytinamide acetates are described below.

Biocytins

N^{α}-Boc-N^{ε}-Z-L-Lysine Amide.[6] N^{α}-Boc-N^{ε}-Z-L-lysine DCHA salt (Fluka) (5.62 g, 10 mmol) is suspended in ice-cold ethyl acetate (250 ml) and ice-cold 0.5 *N* citric acid (180 ml) is added. The mixture is stirred for 30 min; the organic layer is separated and washed with two portions of 0.5 *N* citric acid, 10 portions of saturated sodium chloride, two portions of water, and dried over sodium sulfate. The solvent is evaporated, the residue is dried *in vacuo* over P_2O_5, redissolved in ethyl acetate (70 ml), and the solution is cooled at −10 to −15°. To this solution is added triethylamine (1.39 ml, 10 mmol) followed by isobutyl chloroformate (1.31 ml, 10 mmol) and the mixture is stirred for 10 min at −10 to −15°. To this solution, containing the mixed anhydride, is added 28% ammonium hydroxide (5 ml) and the mixture is stirred for 2 hr at 0 to −10°. The precipi-

[29] K. Hofmann, *in* "Handbook of Physiology" (E. Knobil and W. H. Sawyer, eds.), Sect. 7, Vol. 4, Part 2, p. 29. Am. Physiol. Soc., Washington, D.C., 1975.

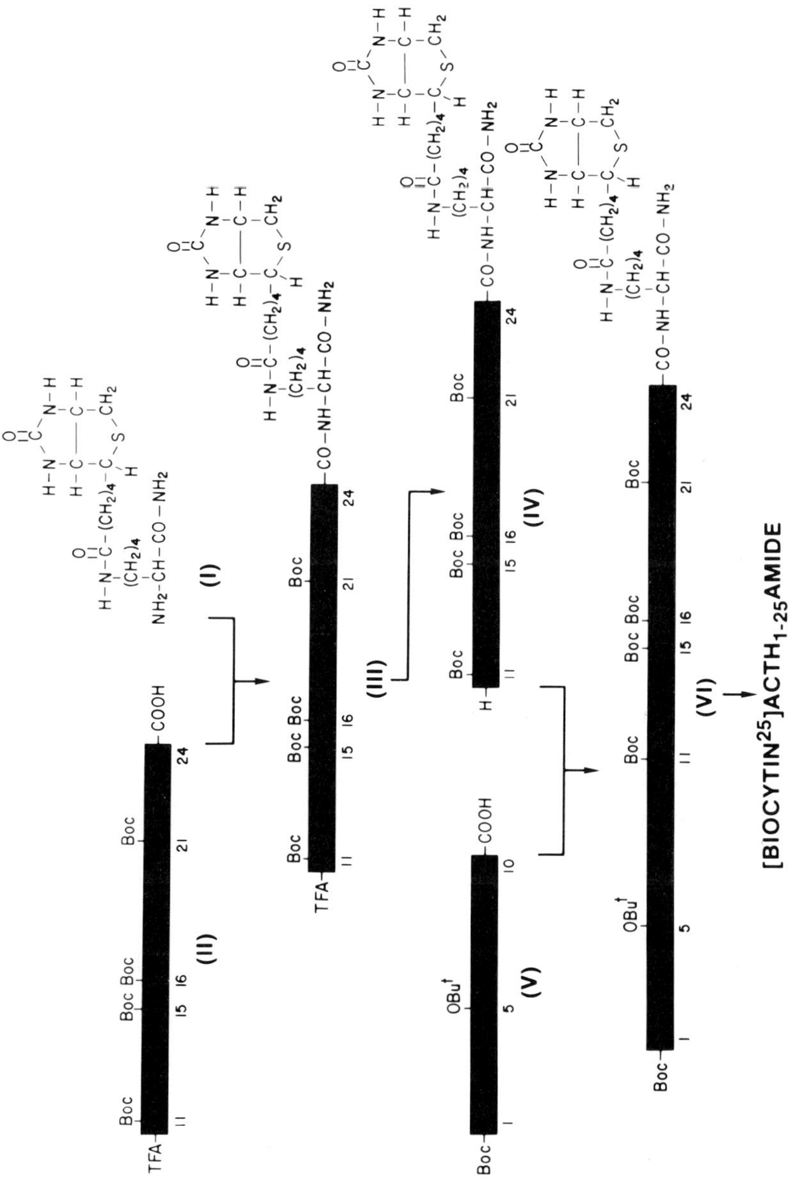

FIG. 7. Synthetic route to [biocytin]^{25}ACTH$_{1-25}$amide.

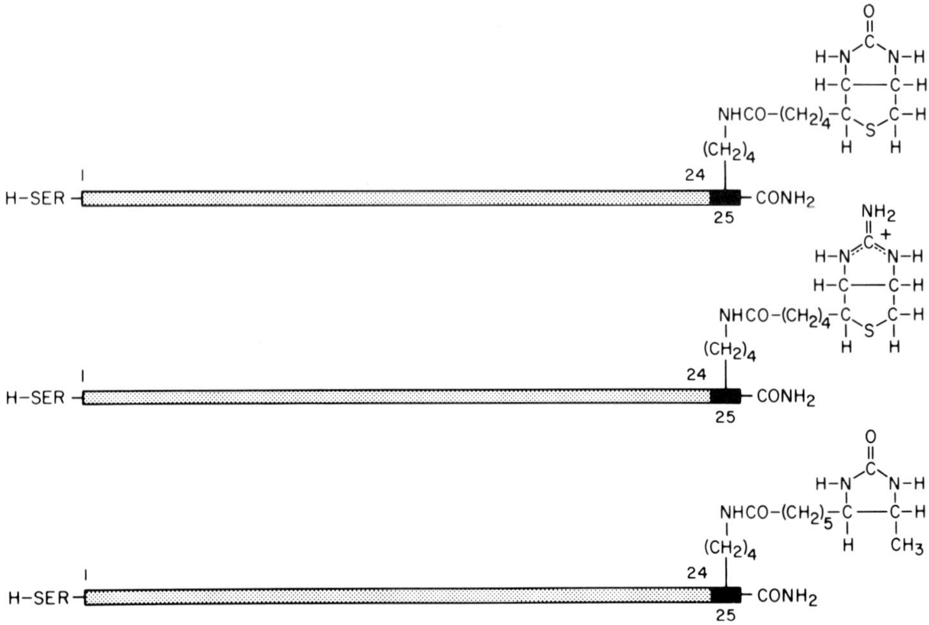

FIG. 8. Simplified structures of [biocytin]^{25}ACTH$_{1-25}$amide (top), [iminobiocytin]^{25}ACTH$_{1-25}$amide (middle), and [dethiobiotin]^{25}ACTH$_{1-25}$amide (bottom).

tate is collected and dried; yield 3.02 g. The filtrate (organic layer) is evaporated to give additional material; yield 0.82 g. Recrystallization of the combined material from ethyl acetate affords needles; yield 3.5 g (92%); mp 140–141°.

N^{α}-Boc-biocytinamide.[6] N^{α}-Boc-N^{ε}-Z-L-lysine amide (556 mg, 1.47 mmol) is hydrogenated over spongy palladium catalyst in methanol (7 ml) and 10% acetic acid (1 ml). After 5 hr the catalyst is removed by filtration through a layer of Johns Manville Hyflo Super-Cel and the filtrate is evaporated to dryness. The residue is evaporated several times with methanol and is dried over P_2O_5 and KOH *in vacuo*. This residue is dissolved in DMF (7 ml), *N*-hydroxysuccinimidobiotinate (500 mg, 1.47 mmol) is added, and the mixture is stirred at room temperature until a clear solution ensues. Triethylamine (0.2 ml, 1.44 mmol) is added and the mixture is stirred at room temperature for 44 hr. The DMF is evaporated, the residue is distributed between 0.5 *N* citric acid and 1-butanol and the butanol layers are washed with 0.5 *N* citric acid, saturated sodium chloride, 5% sodium bicarbonate and water, and the butanol is removed *in vacuo*. The residue (677 mg) is crystallized from aqueous ethanol; yield 426 mg (62%); mp 166–168°.

Biocytinamide Acetate.[6] N^{α}-Boc-biocytinamide (1.0 g) is dissolved in 90% TFA (10 ml) and the solution is kept at room temperature for 30 min. The bulk of the TFA is evaporated at room temperature and the product is precipitated with peroxide-free ether, washed with ether, and dried. This material is dissolved in 10% acetic acid (5 ml) and the solution is passed over a column of the ion exchanger Amberlite IRA-400 (Bio-Rad) in the acetate cycle. The column is eluted with 10% acetic acid and fractions containing the desired material are pooled and lyophilized. The residue is dissolved in methanol and precipitated with ethyl acetate and dried; yield 711 mg (77%) $[\alpha]_D^{27}$ + 56.2° (c 0.373, methanol); R_f^{I} 0.3; R_f^{III} 0.6.

Dethiobiocytinamide Acetate.[7] Diisopropylethylamine (0.34 ml, 2 mmol) is added to a solution of N^{α}-Boc-L-lysine amide [prepared from 760 mg, 2 mmol of N^{α}-Boc-N^{ε}-Z-L-lysine amide] (see above) and *N*-hydroxysuccinimidodethiobiotinate (716 mg, 2.3 mmol) in DMF (10 ml), and the solution is stirred at room temperature for 20 hr. The solvent is evaporated, the residue is precipitated with ethyl acetate, washed with ethyl acetate, and dried. For purification the product is precipitated from methanol with ethyl acetate/ether (1 : 1) and dried; yield 554 mg (62%) mp 93–94°; $[\alpha]_D^{25}$ + 7.2° (c 1.203, methanol); R_f^{I} 0.5; R_f^{II} 0.4; R_f^{III} 0.7. This product is deprotected with 90% TFA and the trifluoroacetate is converted to the acetate as described for the preparation of biocytinamide acetate.

Iminobiocytinamide Diacetate.[7] Triethylamine (0.139 ml, 1 mmol) is added to a solution of N^{α}-Boc-L-lysine amide [prepared from 379 mg, 1 mmol of N^{α}-Boc-N^{ε}-Z-L-lysine amide] (see above) and *N*-hydroxysuccinimidoiminobiotinate hydrobromide (463 mg, 1.1 mmol) in DMF (6 ml) and the solution is stirred at room temperature for 24 hr. The solvent is evaporated, the residue is triturated with ethyl acetate, and dried; yield 688 mg. This product is deprotected with 90% TFA (10 ml) and the trifluoroacetate salt is converted to the acetate salt in the manner described for the preparation of biocytinamide acetate; yield 509 mg. For purification the material is dissolved in water (100 ml) and one-half of the solution (50 ml) is applied to an acetate cycle CMC column (Bio-Rad) (2.3 × 14 cm) that is eluted with water (350 ml) and then in a stepwise manner with 0.015, 0.03, and 0.06% acetic acid. Fractions (20 ml) each are collected and monitored by the biotin reagent. Fractions containing the desired material (0.03 acetic acid eluates) are pooled, concentrated to a small volume *in vacuo,* and lyophilized; yield (two batches) 392 mg (80%); mp 106–108°; $[\alpha]_D^{27}$ + 45.0° (c 0.56, 5% acetic acid); R_f^{I} 0.1; R_f^{III} 0.4.

Other Biotin Reagents

p-Aminobenzoylbiocytinamide.[6] Although N^{α}-*p*-aminobenzoylbiocytinamide has thus far not been employed for introducing biotin into proteins, its synthesis is included here since it is a reagent available for coupling biotin to the side chains of tyrosine and histidine residues.

Boc-p-aminobenzoic Acid.[6] To a solution of *p*-aminobenzoic acid (4.1 g, 30 mmol) in DMF (25 ml) is added 1 *N* NaOH (30 ml) followed by di-*tert*-butyldicarbonate (Fluka) (7.6 g, 35 mmol) and the mixture is stirred at room temperature for 42 hr when the solvent is evaporated. The solid residue is distributed between water and ethyl acetate and the suspension is acidified with 20% citric acid. The organic layer is washed with five portions of 10% citric acid, two portions of saturated NaCl, and two portions of water and concentrated to a small volume. The product is precipitated by addition of petroleum ether, washed with petroleum ether, and dried. This material is dissolved in ethanol, 0.1% aqueous citric acid is added to the cloudpoint, and the mixture is kept at room temperature for 12 hr. The crystals are collected, washed with 30% aqueous ethanol, and dried; yield 3.9 g (55%); mp 197–198° dec; R_f^{II} 0.7.

N-Hydroxysuccinimido Boc-p-aminobenzoate.[6] To a solution of Boc-*p*-aminobenzoic acid (2.37 g, 10 mmol) and *N*-hydroxysuccinimide (1.15 g, 10 mmol) in THF (50 ml) is added a solution of DCC (2.06 g, 10 mmol) in THF (10 ml) and the mixture is stirred at room temperature for 20 hr. The DCU is removed by filtration, the clear filtrate is evaporated to dryness, and the residue is recrystallized from ethyl acetate/petroleum ether: yield 2.68 g (80%); mp 172–173°.

Boc-p-aminobenzoylbiocytinamide Hydrate.[6] *N*-Hydroxysuccinimido Boc-*p*-aminobenzoate (734 mg, 2.2 mmol) is added at room temperature to a solution of biocytinamide acetate (862 mg, 2.0 mmol) and triethylamine (0.278 ml, 2.0 mmol) in DMF (20 ml) and the solution is stirred at room temperature for 24 hr. The completeness of the reaction is checked by TLC and additional *N*-hydroxysuccinimido Boc-*p*-aminobenzoate is added as needed and the mixture is stirred for an additional 66 hr. Insoluble material is removed by filtration, the bulk of the DMF is evaporated, and the product is precipitated with ethyl acetate and dried. The solid residue (1.15 g) is dissolved in glacial acetic acid (5 ml) and water is added to the cloudpoint. After standing in a refrigerator for several hours the solid is collected, washed with 10% acetic acid, and dried: yield 1.07 g (88%) $[\alpha]_D^{25}$ + 40.4° (c 1.034, glacial acetic acid); R_f^{I} 0.6, R_f^{III} 0.7.

p-Aminobenzoylbiocytinamide Acetate.[6] The protected compound (800 mg) is dissolved in 90% TFA (10 ml) and the solution is kept at room temperature for 30 min. The bulk of the TFA is evaporated and the product precipitated with ether, washed with ether, and dried. Trifluoroacetate ions are exchanged for acetate ions on Amberlite IRA-400 (Bio-Rad) using 50% acetic acid as the solvent. Evaporation of the acetic acid leaves a crystalline solid which is dried and recrystallized from aqueous ethanol: yield 347 mg (48%) mp 236–238°; R_f^I 0.4, R_f^{III} 0.6.

Biotinylation of Nucleotides

Biotinyl-UTP and biotinyl-dUTP, in which biotin is linked to the C-5 position by way of an allylamine spacer arm, have been synthesized.[30] UTP or dUTP are mercurated in the C-5 position and the mercurated products are reacted with allylammonium acetate in the presence of K_2PdCl_4 to afford the corresponding 5-(3-amino)allyl nucleotides. Acylation of these compounds with *N*-hydroxysuccinimidobiotinate affords the biotinylated nucleotides.

These biotin-labeled nucleotides are efficient substrates for a variety of DNA and RNA polymerases *in vitro*. Polynucleotides containing low levels of biotin substitution (50 molecules or fewer per kilobase) have denaturation reassociation, and hybridization characteristics similar to those of unsubstituted controls. The unique features of biotinylated polynucleotides suggest that they are useful affinity probes for the detection and isolation of specific DNA and RNA sequences.

Biotinylated UTP and dUTP

Mercurated Nucleotides.[31] UTP (570 mg, 1.0 mmol) or dUTP (554 mg, 1.0 mmol) in 0.1 *M* sodium acetate, pH 6.0 (100 ml) is treated with mercuric acetate (1.59 g, 5.0 mmol). The solution is heated at 50° for 4 hr and then cooled with ice. Lithium chloride (392 mg, 9.0 mmol) is added and the solution is extracted six times with equal volumes of ethyl acetate to remove excess mercuric chloride. The efficiency of the extraction process is monitored by estimating the mercuric ion concentration in the organic layer by using 4,4′-bis(dimethylamino)-thiobenzophenone.[32] The extent of nucleotide mercuration, determined spectrophotometrically by following the iodination of an aliquot of the aqueous solution,[33] is routinely 90–100%. The nucleotide products in the aqueous layer, which often becomes cloudy during the ethyl acetate extraction, are precipitated by

[30] P. R. Langer, A. A. Waldrop, and D. C. Ward, *Proc. Natl. Acad. Sci. U.S.A.* **78,** 6633 (1981).

[31] R. M. K. Dale, E. Martin, D. C. Livingston, and D. C. Ward, *Biochemistry* **14,** 2447 (1975).

[32] A. J. Christopher, *Analyst (London)* **94,** 392 (1969).

[33] R. M. K. Dale, D. C. Ward, D. C. Livingston, and E. Martin, *Nucleic Acids Res.* **2,** 915 (1975).

addition of 3 volumes of ice-cold ethanol and collected by centrifugation. The precipitate is washed with two portions of cold absolute ethanol and once with ether and then air dried.

5-(3-Amino)allyl Nucleotides. The mercurated nucleotides are dissolved in 0.1 *M* sodium acetate, pH 5.0 and adjusted to 20 m*M* (A_{267} 200 OD units/ml). A fresh 2.0 *M* solution of allylammonium acetate is prepared by slowly adding 1.5 ml of allylamine (Aldrich) (13.3 *M*) to 8.5 ml of ice-cold 4 *M* acetic acid. Three ml (6.0 mmol) of the allylammonium acetate solution is added to 25 ml (0.5 mmol) of nucleotide solution. One nucleotide equivalent of K_2PdCl_4 (163 mg, 0.5 mmol; Alfa-Ventron, Danvers, MA) in water (4 ml) is then added to initiate the reaction; the solution gradually turns black and metal deposits appear on the walls of the reaction vessel. After standing at room temperature for 18–24 hr, the reaction mixture is passed through a 0.45-μm membrane (Nalgene) to remove most of the remaining metal precipitate. The yellow filtrate is diluted 1 : 5 with water and applied to a 100 ml column of DEAE-Sephadex A-25 (Pharmacia). The column is washed with one column volume of 0.1 *M* sodium acetate, pH 5.0, the products are eluted by using a 1-liter linear gradient (0.1–0.6 *M*) of sodium acetate, pH 8–9 or triethylammonium carbonate, pH 7.5. The desired product is present in the major UV-absorbing peak, which elutes between 0.30 and 0.35 *M* salt. Final purification is achieved by reverse-phase HPLC on columns of Partisil-ODS2 (Whatman), using either 0.5 *M* $(NH_4)_3PO_4$, pH 3.3 (analytical separations), or 0.5 *M* triethylammonium acetate, pH 4.3 (preparative separations) as eluents. 5-(3-Amino)allyl-UTP and 5-(3-amino)allyl-dUTP are the last peaks to elute from the columns and they are cleanly resolved from three, thus far unidentified, contaminants.

Biotinylated Nucleotides. *N*-Hydroxysuccinimidobiotinate (34.1 mg, 0.1 mmol) in DMF (2 ml) is added to a solution of 5-(3-amino)allyl-UTP · $4H_2O$ (70 mg, 0.1 mmol) or 5-(3-amino)allyl-dUTP · H_2O (63 mg, 0.1 mmol) in 0.1 *M* sodium borate, pH 8.5 (20 ml). The reaction mixture is kept at room temperature for 4 hr and is then applied to a 30 ml column of DEAE-Sephadex A-25 previously equilibrated with 0.1 *M* triethylammonium carbonate, pH 7.5. The column is eluted with a 400 ml linear gradient (0.1–0.9 *M* of triethylammonium carbonate). Fractions containing biotinyl-UTP or biotinyl-dUTP, which elute at 0.55–0.65 *M* triethylammonium carbonate, are desalted by rotary evaporation in the presence of methanol and then dissolved in water. Occasionally, a slightly cloudy solution is obtained. This can be clarified by filtration through a 0.45-μm filter. For long-term storage the biotinylated nucleotides are converted to the sodium salts by briefly stirring the solution in the presence of Dowex 50 (Na^+). After filtration, the biotinylated nucleotides are precipitated by

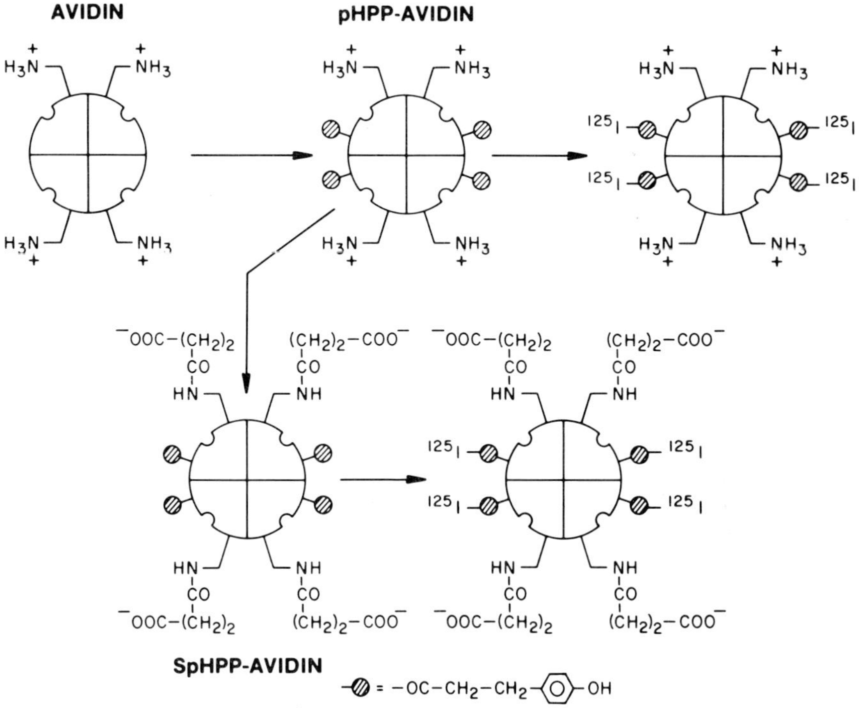

FIG. 9. Schematic presentation of avidin modifications.

addition of 3 volumes of cold ethanol, washed with peroxide-free ether, dried *in vacuo* over KOH, and stored at −20° in a desiccator.

5-(3-Amino)-allyl-UTP and 5-(3-amino)-allyl-dUTP can be reacted with the *N*-hydroxysuccinimido esters of 6-(biotinylamido)-hexanoic acid or 4-(6-biotinylamidohexyl)-aminobutyric acid to afford materials in which the biotin is separated from the nucleotides by spacer arms.[34]

Modification of Avidin and Labeling

Avidin is composed of four identical subunits each containing a single tyrosine residue that is not readily accessible to iodination. Avidin is rendered susceptible to iodination (chloramine T method) by the introduction of 3-(*p*-hydroxyphenyl)propionyl groups. ^{125}I-labeled Bolton–Hunter reagent can also be employed to label the protein.[19] The modifications of avidin are illustrated in Fig. 9.

[34] D. J. Brigati, D. Myerson, J. J. Leary, B. Spalholz, S. Z. Travis, C. K. Y. Fong, G. D. Hsiung, and D. C. Ward, *Virology* **126**, 32 (1983).

Succinoylavidin[3]

Avidin (Sigma) (51 mg, 0.75 μmol) is dissolved in 0.1 *N* HCl (1.5 ml) and 0.2 *M* pH 9.0 borate buffer (73.5 ml) is added. The solution is cooled in an ice bath and succinic anhydride (30 mg, 300 μmol) in peroxide-free dioxane (1 ml) is added with stirring. Stirring is continued for 1 hr at room temperature, then the reaction mixture is dialyzed against 4 liters of distilled water for 12 hr at 4° and lyophilized. The residue is passed over a Sephadex G-25 column (1.6 × 95 cm) in 0.05 *M* ammonium bicarbonate and fractions containing protein are lyophilized; yield 45 mg.

3-(p-Hydroxyphenyl)propionylavidin[3] *(pHPP-avidin)*

N-Hydroxysuccinimido-3-(*p*-hydroxyphenyl)propionate (Fluka) (1.05 mg, 4.0 μmol) in 2-propanol/ethyl acetate (3 : 2) (1 ml) is added to an ice-cold solution of avidin (68 mg, 1 μmol) in 0.2 *M* borate buffer pH 9.0 (20 ml), and the solution is stirred at 0° for 3 hr. The solution is lyophilized and the residue is desalted on Sephadex G-25 in 10% acetic acid. Fractions containing the desired material are pooled and lyophilized (yield essentially quantitative). The relation between the amount of active ester added and 3-(*p*-hydroxyphenyl)propionate incorporation is illustrated on Fig. 10.

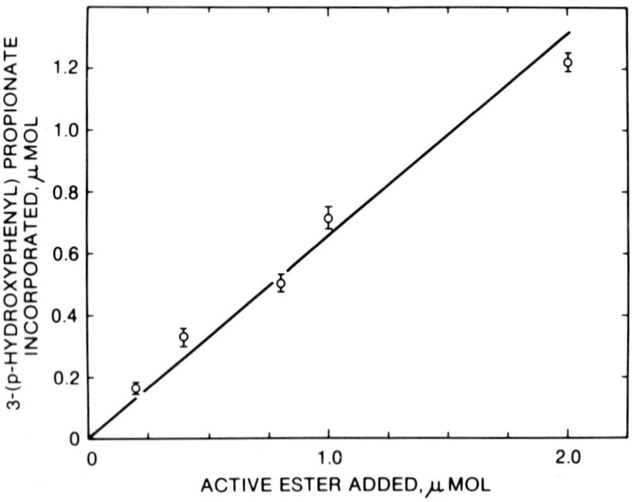

FIG. 10. Incorporation of 3-(*p*-hydroxyphenyl)-propionyl groups into avidin as a function of the concentration of *N*-hydroxysuccinimido-3-(*p*-hydroxyphenyl)propionate. For details see Finn *et al.*[3]

Succinoyl-3-(p-hydroxyphenyl)propionylavidin[3] *(SpHPP-avidin)*

3-(*p*-Hydroxyphenyl)propionylavidin (51 mg, 0.75 μmol) is dissolved in 0.1 *N* HCl (1.5 ml) and 0.2 *M* borate buffer pH 9.0 (73.5 ml) is added. The mixture is cooled in an ice bath, a solution of succinic anhydride (30 mg, 300 μmol) in peroxide-free dioxane (0.9 ml) is added, and the mixture is stirred at 0° for 1 hr and dialyzed against distilled water (2 × 1 liter) for 12 hr. The dialyzed solution is lyophilized, the residue is dissolved in 0.05 *M* ammonium bicarbonate, and the solution is desalted on Sephadex G-25 (Pharmacia) in the same buffer. Fractions containing the desired material are pooled and lyophilized; yield 42 mg. For determination of the degree of succinoylation [^{14}C]succinic anhydride is employed. An incorporation of 6 to 7 succinoyl groups per avidin subunit is realized.

Labeling 3-(p-Hydroxyphenyl)propionylavidin and Succinoyl-3-(p-Hydroxyphenyl)propionylavidin[3]

Labeling is patterned after the procedure described for insulin by De Meyts.[35] To a 1.5 ml polypropylene Beckman microfuge tube are added in the order given: 35 μl of 0.3 *M* sodium phosphate buffer, pH 7.4; 10 μl of derivatized avidin (3.4 mg/ml dissolved in 0.01 *N* HCl or 0.3 *M* sodium phosphate buffer (succinoylated derivatives); 20 μl of Na^{125}I (2 mCi); and 10 μl of chloramine-T (Sigma) (an amount sufficient to incorporate 0.25 nmol of ^{125}I into the protein, usually 0.57 μg) in phosphate buffer. After 5 min, a sample is taken for measurement of the relative proportions of TCA-precipitable and -soluble radioactivity. A second portion of chloramine-T is added when necessary, and the TCA precipitation is repeated. When approximately 50% of the radioactivity is incorporated into TCA-precipitable material, a solution of sodium metabisulfite (2.5-fold molar excess over the amount of chloramine-T added) in phosphate buffer and 100 μl of 2.5% bovine serum albumin in the same buffer are added. The iodinated avidin derivatives are isolated by gel filtration on a Sephadex G-25 column (0.6 × 35 cm) previously equilibrated with Krebs–Ringer phosphate buffer containing 10 mg/ml bovine serum albumin prepared according to De Meyts *et al.*[36] Fractions (1 ml each) corresponding to the first radioactive peak are pooled and frozen at −15°.

Displacement of Ligands from Avidin, Succinoylavidin, and Streptavidin by Biotin

Unquestionably the principal attraction of the avidin–biotin system is its uniquely strong noncovalent binding. This property theoretically as-

[35] P. De Meyts, *Methods Mol. Biol.* **9**, 301 (1976).
[36] P. De Meyts, A. R. Bianco, and J. Roth, *J. Biol. Chem.* **251**, 1877 (1976).

sures specificity when "labeled" avidins are employed to detect biotin-containing molecules in tissue or cell sections. Furthermore it simplifies the preparation of affinity resins since biotinyl ligands can be immobilized on avidin-Sepharose without the need for chemical manipulations that might alter the ligand.

In some instances, however, the strong interaction poses an obstacle to retrieval of the product. This is particularly true when the rate of dissociation of the desired product from the biotinyl ligand is very low. The use of weaker binding analogs of biotin may provide a solution to this problem. The previously mentioned application of iminobiotinyl-Sepharose columns for the isolation of streptavidin and avidin illustrates this alternative. In this case the pH dependence of iminobiotin binding is exploited to obtain highly purified material in a 2-step process.

Dethiobiotin is another analog with lower avidin affinity. Its rate of dissociation from avidin is approximately 1000-fold higher than that of biotin. Dethiobiotinyl ligands can be displaced from avidin or succinoyl-avidin by biotin.

Ligands containing biotin or its analogs may not exhibit the same affinity for avidin as the corresponding unattached molecule and such molecules may be readily displaced from avidin-Sepharose columns by biotin in a mild and highly specific manner. For example, although the rate of dissociation of the avidin–biotin complex is negligible ($t_{1/2}$, 200 days) the rate of dissociation of the avidin–biotinylinsulin complex is relatively high ($t_{1/2}$, 3 hr). For other examples see the table. Apparently attachment of insulin directly to the carboxyl group of biotin interferes with biotin's access to its avidin binding site. Thus biotin is readily able to displace biotinylinsulin from avidin. Since each biotinyl ligand may have a unique conformation, it is important to measure the rate of its dissociation from the various avidins to determine whether complex formation is possible and if so whether displacement of the biotinyl ligand by biotin is feasible.

Ligand Binding and Displacement Studies

Separation by Sephadex Chromatography.[15,37] Complexes are prepared by mixing avidin, succinoylavidin, or streptavidin (10 nmol) in 50 m*M* Tris–HCl, pH 7.6 with the desired ligand (80 nmol); a 2-fold excess) in 0.01 *N* HCl (0.6 ml). Excess [^{14}C]biotin (2 μmol 3.75 mCi/mmol) in Tris buffer is added and the solution is incubated at 25° (total volume 3.0 ml). At specified intervals aliquots (0.1 ml) are subjected to gel filtration on Sephadex G-50 columns (0.9 × 55 cm) equilibrated with the Tris buffer.

[37] R.-D. Wei and L. D. Wright, *Proc. Soc. Exp. Biol. Med.* **117,** 17 (1964).

DISSOCIATION RATES[a]

Ligand	$t_{1/2}$	k_{-1} (sec^{-1})
Complexes with succinoylavidin		
Biotin	127 days	6.3×10^{-8}
Dethiobiotin	14 hr	1.4×10^{-5}
I	3 hr	7.5×10^{-5}
II	74 days	1.1×10^{-7}
V	No complex	
VI	6 hr	3.2×10^{-5}
VII	74 days	1.1×10^{-7}
VIII	6 hr	3.2×10^{-5}
IX	14 hr	1.4×10^{-5}
Complexes with streptavidin		
I	11 days	7.6×10^{-7}
II	90 days	9.0×10^{-8}

[a] For an explanation of the roman numerals see legend to Fig. 6. The dissociation curves are biphasic.[15] Only the slower rate of dissociation has been presented. Values for $t_{1/2}$ are calculated from the rate of dissociation given in the table.

Fractions (1 ml each) are collected in polyethylene minivials, scintillation cocktail (Research Products International, 3a70B, 5 ml) is added, and the radioactivity is measured. The radioactivity in the fractions corresponding to high molecular weight material provides a measure of the rate of dissociation. Recovery of radioactivity from the column is quantitative. The total [^{14}C]biotin binding capacities of the various preparations of avidin, succinoylavidin, or streptavidin are evaluated in the same manner.

Separation by the Bentonite Procedure.[38] A reaction mixture consisting of avidin, [^{14}C]biotin, and standard or unknown in 0.2 *M* ammonium carbonate (0.6 ml) is incubated for 10 min at 23°; ammonium carbonate (1 ml) containing bentonite (Fisher Scientific) (10 mg) is added and, after an additional 5 min incubation, the suspension is filtered through cellulose nitrate filters (0.45 μm pore size, Millipore Corporation). Unbound biotin passes through the filter. The filter, containing the bentonite–avidin–[^{14}C]biotin complex, is dissolved in Bray's solution[39] (10 ml), leaving a bentonite sediment, and radioactivity is measured. A standard curve, consisting of three or more concentrations assayed in triplicate, is constructed for each assay.

[38] S. G. Korenman and B. W. O'Malley, *Biochim. Biophys. Acta* **140,** 174 (1967).
[39] G. A. Bray, *Anal. Biochem.* **1,** 279 (1960).

[38] Purification of N_s and N_i, the Coupling Proteins of Hormone-Sensitive Adenylyl Cyclases without Intervention of Activating Regulatory Ligands

By JUAN CODINA, WALTER ROSENTHAL, JOHN D. HILDEBRANDT, LUTZ BIRNBAUMER, and RONALD D. SEKURA

It is now recognized that a large number of hormones and neurotransmitters affect their target cells by modulating cAMP formation in either stimulatory or inhibitory fashion.[1,2] They do so by first interacting with their specific receptors giving rise to hormone–receptor complexes which in turn combine with so-called coupling or signal transduction proteins, also called guanine nucleotide and Mg binding regulatory components of adenylyl cyclase and abbreviated variously as G/F,[3] G,[4] or N[5,6] proteins. There are two of these coupling or N proteins: an N_s (or G_s) mediating the effects of stimulatory hormone–receptor complexes and an N_i (or G_i) mediating the effects of inhibitory or attenuating hormone–receptor complexes. The interaction of a hormone–receptor complex with an N protein then results in an increase in the proportion of the N protein in an active vs inactive conformation or state, and activated N interacts with the catalyst C of adenylyl cyclase eliciting either an increase in catalytic activity (N_s) or an inhibition of activity (N_i). For review see Ref. 7. Structurally, both N_s and N_i are $\alpha\beta\gamma$ heterotrimers.[8] The activation process of both, N_s and N_i, is dependent on a guanine nucleotide and Mg,[9,10] and seems to involve not only a conformational change but also a subunit dissociation reaction whereby the $\alpha\beta\gamma$ heterotrimers dissociate into an activated α^* subunit with guanine nucleotide bound to it (α^{*G}) and a $\beta\gamma$ complex.[11–16]

1 C. Londos, D. M. F. Cooper, and M. Rodbell, *Adv. Cyclic Nucleotide Res.* **14,** 163 (1981).

2 K. H. Jakobs, K. Aktories, and G. Schultz, *Adv. Cyclic Nucleotide Res.* **14,** 173 (1981).

3 E. M. Ross, A. C. Howlett, K. M. Ferguson, and A. G. Gilman, *J. Biol. Chem.* **253,** 6401 (1978).

4 T. Pfeuffer, *FEBS Lett.* **101,** 85 (1979).

5 M. Rodbell, *Nature (London)* **284,** 17 (1980).

6 D. M. F. Cooper, *FEBS Lett.* **138,** 157 (1982).

7 M.Hunzicker-Dunn and L. Birnbaumer, *in* "Gonadotropin Receptors and Gonadotropin Actions" (M. Ascoli, ed.). CRC Press, Miami, Florida, 1984 (in press).

8 J. D. Hildebrandt, J. Codina, R. Risinger, and L. Birnbaumer, *J. Biol. Chem.* **259,** 2039 (1984).

9 R. Iyengar and L. Birnbaumer, *J. Biol. Chem.* **256,** 11036 (1981).

10 J. D. Hildebrandt and L. Birnbaumer, *J. Biol. Chem.* **258,** 13141 (1983).

11 J. Codina, J. D. Hildebrandt, R. Iyengar, L. Birnbaumer, R. D. Sekura, and C. R. Manclark, *Proc. Natl. Acad. Sci. U.S.A.* **80,** 4276 (1983).

METHODS IN ENZYMOLOGY, VOL. 109

ISBN 0-12-182009-2

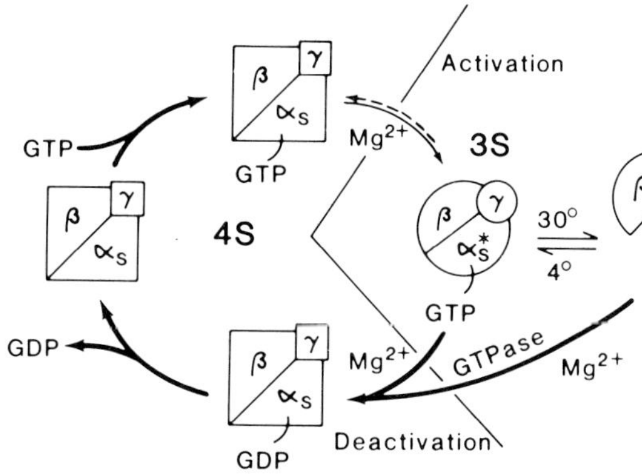

FIG. 1. Scheme of kinetic regulatory cycle of a N protein. Both N_s and N_i acquire a sedimentation behavior of about 4 S when inactive with respect to modulation of the catalytic activity of adenylyl cyclase and analyzed in 0.5% Lubrol PX-containing buffers. On interaction with guanine nucleotide *and* Mg, they isomerize, changing their sedimentation behavior to approximately 3 S and acquiring a conformation that binds GTP analogs tightly. Subsequent to this, the 3 S forms dissociate into 2 S complexes (^{*G} and $\beta\gamma$) in a temperature-dependent manner. Both the 3 S $\alpha^{*G}\beta\gamma$ and the 2 S α^{*G} complexes are thought to be capable of modulating the catalytic unit of adenylyl cyclase. Relaxation of the system can proceed via direct reversal of the above described activation pathway or, if activation had been brought about with GTP, by metabolic degradation of GTP to GDP by N yielding inactive N-GDP complexes which then release the GDP to allow for reinitiation of the cycle. Either 3 S or 2 S N proteins are assumed to be active in hydrolyzing GTP, but data to substantiate this are still missing.

In intact cells the guanine nucleotide that activates N proteins is GTP, with GDP being inactive. Both N_s and N_i are GTPases,[17-19] and activated α^{*G} does not remain as such for it hydrolyzes its GTP to GDP, and relaxes back to an unactivated form. This gives rise to a kinetic regulatory cycle such as illustrated in Fig. 1, which is applicable to both N_s and N_i. Many details of this regulatory cycle are still unknown: At which point of this

[12] A. C. Howlett and A. G. Gilman, *J. Biol. Chem.* **255,** 2861 (1980).

[13] P. C. Sternweis, J. K. Northup, M. D. Smigel, and A. G. Gilman, *J. Biol. Chem.* **256,** 11517 (1981).

[14] J. K. Northup, P. C. Sternweis, and A. G. Gilman, *J. Biol. Chem.* **258,** 11361 (1983).

[15] J. K. Northup, M. D. Smigel, P. C. Sternweis, and A. G. Gilman, *J. Biol. Chem.* **258,** 11369 (1983).

[16] J. Codina, J. D. Hildebrandt, L. Birnbaumer, and R. D. Sekura, *J. Biol. Chem.* (in press).

[17] D. R. Brandt, T. Asano, S. E. Pedersen, and E. T. Ross, *Biochemistry* **22,** 4357 (1983).

[18] T. Sunyer, J. Codina, and J. Birnbaumer, *J. Biol. Chem.* (in press).

[19] R. A. Cerione, J. Codina, J. L. Benovic, R. J. Lefkowitz, L. Birnbaumer, and M. C. Caron, *Biochemistry* (in press).

cycle does an N protein interact with the hormone–receptor complex? At which point does the hormone–receptor complex separate from N? Which form of an N protein is an active GTPase: the α^{*G} complex? The $\alpha^{*G}\beta\gamma$ complex? Both? Questions of this type as well as questions relating to possible regulation of N proteins by endocrine and pathological states, by pharmacological manipulations, and by prolonged stimulations leading to desensitized states, require an intimate knowledge of the structure of the proteins and eventually availability of antibodies to them. In this chapter we present a detailed description of a procedure for the purification of N_s and N_i from human erythrocytes. In contrast to a procedure[20] and its modification[13] as well as another procedure[21] developed by others for N_s, this procedure avoids the use of stabilizing ligands such as NaF/Mg and nonhydrolyzable GTP analogs [e.g., GMP-P(NH)P or $GTP_{\gamma}S$] which are known to affect functionally the behavior and activity of N proteins. This eliminates any possible alteration of the subunit composition of these proteins as might result from the effect of these ligands to induce their subunit dissociation. The method described below for purification of N_s is, however, longer and more tedious than that described by others using NaF/Mg.[13] On the other hand, N_i can only be purified if NaF/Mg is avoided. Whether or not the reader will chose to use this methodology for purification of N_s depends on the purposes to which he wishes to put the protein to work.

As purified, both N_s and N_i are detergent–protein complexes with the proteins being heterotrimers of $\alpha\beta\gamma$ composition and approximate M_r = 96,000. The two proteins differ in their α subunits: α_s is of M_r = 42,000 and substrate for the ADP-ribosyltransferase activity of cholera toxin[20]; α_i is of M_r = 40,000 and substrate for the ADP-ribosyltransferase activity of pertussis toxin.[11,22] The β subunits of the two N proteins appear to be the same on the basis of amino acid composition,[23] mono-[23] and two-[24] dimensional peptide mapping and immunoreactivity (Gierschick, Codina, Birnbaumer, and Spiegel, unpublished). The γ subunits of N_s and N_i comigrate on urea and polyacrylamide gradient gel electrophoresis with apparent $M_r \sim 5000$,[8] and it is not yet known whether there are distinct γ_s and γ_i or whether N_s and N_i share a common γ subunit.

[20] J. K. Northup, P. C. Sternweis, M. D. Smigel, L. S. Schleifer, E. M. Ross, and A. G. Gilman, *Proc. Natl. Acad. Sci. U.S.A.* **77,** 6516 (1980).

[21] T. Pfeuffer, B. Gaugler, and H. Metzger, *FEBS Lett.* **164,** 154 (1983).

[22] G. M. Bokoch, T. Katada, J. K. Northup, E. L. Hewlett, and A. G Gilman, *J. Biol. Chem.* **258,** 2072 (1983).

[23] D. R. Manning and A. G Gilman, *J. Biol. Chem.* **258,** 7059 (1983).

[24] J. Codina, J. D. Hildebrandt, R. D. Sekura, M. Birnbaumer, J. Bryan, C. R. Manclark, R. Iyengar, and L. Birnbaumer, *J. Biol. Chem.* **259,** 5871 (1984).

Materials

Radiochemicals and Chemicals

[α-^{32}P]ATP was synthesized using $^{32}P_i$ supplied by CentiChem according to the procedures of Walseth and Johnson,[25] purified by DEAE-Sephadex A25 chromatography[26] or purchased from ICN, Amersham or NEN. [^{32}P]NAD$^+$ was synthesized from [α-^{32}P]ATP and purified by Dowex 1 chromatography as described by Cassel and Pfeuffer[27] or purchased from NEN. Reagents for polyacrylamide gel electrophoresis were from Bio-Rad and were used without further purification. DEAE-Sephacel as well as Sephadex G-50 (coarse) and Sepharose 4B-CL were from Pharmacia; Ultrogel AcA-34 was from LKB Instruments; hydroxylapatite (HAP) was from Bio-Rad and ultrafiltration filters were from Amicon Corporation. 2-Mercaptoethanol, HEPES (Sigma Chemical Company), and ethyleneglycol (Fisher) were used without further treatments. Heptylamine-Sepharose 4B-CL was synthesized as described by Northup *et al.*[20] Bovine serum albumin (BSA) was Cohn fraction V from Sigma.

Cholate was from Sigma Chemical Company and was recrystallized six times from ethanol before used. Briefly (1) 80 g of cholic acid is dissolved in 1 liter of 95% ethanol by bringing to a boil; (2) 4 liters of cold water is added; (3) the mixture is allowed to stand at room temperature for 3 hr; (4) the clear supernatant is decanted off and discarded; (5) the remaining suspension is brought to 1 liter with 95% ethanol; (6) steps 1 through 5 are repeated five times; (7) the resulting precipitate is dried in a convection oven at 60–70° overnight.

Lubrol PX was from Sigma Chemical Company and purified as described.[13] Briefly (1) a 10% w/v solution of Lubrol PX is made; (2) a 1.0 × 18 cm column of BioRex RG 501-X8 resin (mixed bed resin of Dowex 50 and Dowex 1 with indicator dye) is made; (3) the column is subjected to a regeneration cycle by washing sequentially with 2 liters H_2O, 500 ml 2 *N* NaOH, 2 liters H_2O, 500 ml 2 *N* HCl, and 4 liters H_2O (final pH of effluent >5.0); (4) the Lubrol PX solution is passed over the BioRex column at room temperature; (5) the percolate is neutralized to pH 8.0 with NaOH and used.

All other reagents and chemicals were of the highest commercially available grade and used as obtained.

[25] T. F. Walseth and R. A. Johnson, *Biochim. Biophys. Acta* **562,** 11 (1979).

[26] L. Birnbaumer, H. N. Torres, M. M. Flawia, and R. F. Fricke, *Anal. Biochem.* **93,** 124 (1979).

[27] D. Cassel and T. Pfeuffer, *Proc. Natl. Acad. Sci. U.S.A.* **75,** 2669 (1978).

Biologicals

Blood. Outdated blood was obtained from local blood banks. The blood can be stored at 0–4° for up to 2 weeks before processing.

Cyc⁻ S49 Mouse Lymphoma Cell Membranes. Cyc⁻ S49 cells were grown in suspension cultures in Dulbecco Modified Eagle Medium (DMEM) supplemented with 10% heat inactivated (50°, 20 min) horse serum according to the procedures of Ross *et al.*[28] Cells can be obtained from the University of California at San Francisco Cell Culture Center after having obtained consent from Dr. Henry R. Bourne (Professor and Chairman, Department of Pharmacology, School of Medicine, UCSF). *Cyc⁻* membranes are prepared as described by Coffino *et al.*,[29] using Mg-free buffers throughout. Briefly for 1 liter of cell suspension (1) cells are grown at 37° to a density of 2–3 × 10^6 cells/ml; (2) cells are harvested by centrifugation for 20 min at 3000 *g* at 4° and all subsequent steps are at 4°; (3) the medium is discarded and the cell pellet is weighed (yield should be about 0.75–1.0 g wet weight); (4) cells are washed twice with 0.5 liter Puck's balanced salt solution without divalent cations; (5) cells are resuspended in 20 volumes (with respect to wet weight) of lysis medium: 25 m*M* Na—HEPES, 120 m*M* NaCl, 2 m*M* $MgCl_2$, and 1 m*M* DTT, pH 8.0; (6) cells are lysed by nitrogen cavitation/decompression using a Parr bomb after equilibration with nitrogen at 400 psi for 20 min; (7) the lysate is centrifuged at1500 *g* for 5 min and the pellet is discarded; (8) the supernatant is centrifuged at 40,000 *g* for 20 min and the supernatant is discarded; (9) the pellet is resuspended in 20 volumes (with respect to starting wet weight of cells) of wash buffer: 25 m*M* Na-HEPES and 1 m*M* DTT, pH 8.0, and recentrifuged at 40,000 *g* for 20 min; (10) this pellet is resuspended in 0.75–1.0 ml of wash buffer, (15–25 mg protein/ml), fractionated into 100-μl aliquots and stored frozen at −70° until used.

Analytical Procedures and Assays

Protein Determinations

Due to the presence of high concentrations of 2-mercaptoethanol throughout the purification procedure and because it is more sensitive than the classical Lowry procedure, proteins are determined fluorometrically by addition of fluorescamine (Fluram, Roche Diagnostics) to the samples according to instructions of the manufacturer. Buffers such as

[28] E. M. Ross, M. E. Maguire, T. W. Sturgill, R. L. Biltonen, and A. G. Gilman, *J. Biol. Chem.* **252,** 5761 (1977).

[29] P. Coffino, H. R. Bourne, and G. M. Tomkins, *J. Cell. Physiol.* **85,** 603 (1975).

Tris and other compounds containing primary amines were therefore avoided throughout.

Electrophoresis

Sample Preparation. Samples to be electrophoresed and containing up to 1% cholate, 1% Lubrol PX, 30% ethyleneglycol, and/or 20 m*M* 2-mercaptoethanol were made 50 m*M* in NaCl and then precipitated quantitatively by addition of 9 volumes of ice-cold acetone. The precipitated proteins were then resuspended in Laemmli's sample buffer[30] containing 1% sodium dodecyl sulfate and Pyronin Y as the tracking dye, and incubated at room temperature for at least 1 hr prior to electrophoresis. Boiling of samples was avoided for it led to aggregation of the α subunit of N_s and N_i. To prevent formation of insoluble potassium salts of SDS, K^+ was avoided throughout.

Sodium Dodecyl Sulfate–Polyacrylamide Gel Electrophoresis (SDS–PAGE). SDS–PAGE of proteins was carried out in 10% acrylamide slabs of 13.8 × 17.7 cm × 0.75 or 1.5 mm according to Laemmli.[30] Electrophoresis was at room temperature at a constant voltage of 100–150 V (approximately 30 mA of initial current).

Sodium Dodecyl Sulfate–Discontinuous Urea and Polyacrylamide Gradient Gel Electrophoresis (SDS–DUPAGGE). Electrophoresis was in 0.75 mm × 13.8 cm × 17.7 cm slabs using Laemmli's running buffer.[28] The gels contained a stacking gel of approximately 3 cm and a separating gel of approximately 10 cm. The stacking gel was 6.25% acrylamide/0.172% bisacrylamide made in 62.5 m*M* Tris–HCl (pH 6.9), 0.1% SDS and 0.1% TEMED, polymerized at 0.02% ammonium persulfate. The separating gel was discontinuous in composition and formed of two sections. The top half (5 ml) was 12.5% acrylamide/0.344% bisacrylamide made in 0.375 *M* Tris–HCl (pH 8.9), 0.1% SDS, and 0.01% TEMED (buffer A) with 0.02% ammonium persulfate. The bottom half of the separating gel (5 ml) was a gradient (top to bottom) from 12.5% acrylamide/0.344% bisacrylamide/4 *M* urea in buffer A with 0.02% ammonium persulfate to 25% acrylamide/0.688% bisacrylamide/8 *M* urea in buffer A with 0.01% ammonium persulfate. Gels are run at 100 V for 4 to 6 hr.

Stainings and Autoradiography. After completion of electrophoresis, gels were stained for at least 6 hr in acetic acid : methanol : water (1 : 5 : 5) containing 0.1% Coomassie Brilliant Blue and then destained overnight in 10% acetic acid. Coomassie Blue-stained gels can subsequently be stained with silver by the method of Poehling and Neuhoff[31] or Wray *et al.*[32] For

[30] U. Laemmli, *Nature (London)* **227,** 680 (1970).
[31] H. M. Poehlig and V. Neuhoff, *Electrophoresis* **2,** 141 (1981).
[32] G. W. Wray, T. Boulikas, V. P. Wray, and R. Hancock, *Anal. Biochem.* **118,** 197 (1981).

autoradiography gels were dried on Bio-Rad filter paper or between two sheets of cellophane and juxtaposed at −70° to Kodak XR-5 film for 1–3 days.

Molecular Weight Standards. Phosphorylase *b* of M_r = 97,400, BSA of M_r = 67,000, ovalbumin of M_r = 43,000, α-actin of M_r = 41,800, carbonic anhydrase of M_r = 30,000, soybean trypsin inhibitor of M_r = 20,100, and α-lactalbumin of M_r = 14,000 were obtained from Pharmacia.[24] Additional low-molecular-weight standards were prepared by cyanogen bromide cleavage of horse heart myoglobin, yielding five polypeptides with molecular weights of 16,900, 14,400, 8200, 6200 and 2500.[8] For the cleavage reaction, 100 mg myoglobin is reacted with 50 mg cyanogen bromide in 13 ml of 0.1 *N* HCl at 24° for 24 hr. The solution is diluted with 130 ml of water, lyophilized, and resuspended at a final concentration of 10 mg/ml. Standards are prepared for electrophoresis by dissolving them in Laemmli's sample buffer and heating to 100° for 3–5 min.

Adenylyl Cyclase Assays, N_s Assay, and Preactivation of N_s

Adenylyl Cyclase. This enzyme was assayed in a final volume of 50 μl containing, unless specified otherwise, 0.1 m*M* ATP, 5–8 × 10^6 cpm of [α-^{32}P]ATP (specific radioactivity > 200,000 cpm/pmol), 10 m*M* $MgCl_2$, 1.0 m*M* EDTA, 1 m*M* cAMP, 10,000–12,000 cpm [^{3}H]cAMP (specific radioactivity > 15 Ci/mmol), a nucleoside triphosphate regenerating system,[33] 25 m*M* Tris–HCl, pH 8.0, and when present, 1–10 μM GMP-P(NH)P, 10 m*M* NaF or 100 μM GTP without or with 10^{-4} *M* isoproterenol, 10 μl of *cyc*⁻ membranes (10–20 μg protein, diluted in 10 m*M* Na-HEPES, pH 8.0, and 1.0 m*M* EDTA), and 10 μl of various media containing N_s activity. Incubations were at 32.5° and [^{32}P]cAMP formed was quantitated by a modification[34] of the method of Salomon *et al.*[35] In some instances, formation of [^{32}P]cAMP can be monitored by postaddition of 10 μl containing the [α-^{32}P]ATP and [^{3}H]cAMP to 40 μl of a mixture that contains *cyc*⁻ membranes, fractions with N_s activity, and all of the above mentioned reagents and that had been subjected to a preliminary incubations at 32.5° lasting between 5 and 40 min. These incubations (e.g., 5–15 or 40–50 min assays of activity) are then stopped after 10 min and [^{32}P]cAMP formed is determined as described above.

"Reconstitution" of cyc⁻ Membrane Adenylyl Cyclase. Reconstitution of *cyc*⁻ adenylyl cyclase activity was obtained by mixing equal vol-

[33] L. Birnbaumer, P.-C. Yang, M. Hunzicker-Dunn, J. Bockaert, and J. M. Duran, *Endocrinology* **99,** 163 (1976).

[34] J. Bockaert, Hunzicker-Dunn, and L. Birnbaumer, *J. Biol. Chem.* **251,** 2653 (1976).

[35] Y. Salomon, C. Londos, and M. Rodbell, *Anal. Biochem.* **58,** 541 (1974).

umes of *cyc*$^-$ membranes (1–3 mg/ml diluted in 10 m*M* Na-HEPES, pH 8.0, and 1.0 m*M* EDTA) and N_s activity suitably diluted in 1.0% BSA, 1.0 m*M* EDTA, 20 m*M* 2-mercaptoethanol, and 10 m*M* Tris–HCl, pH 8.0, with less than 0.001% cholate, less than 0.001% Lubrol-PX, with or without 1.5 *M* KCl plus 20 m*M* 2-mercaptoethanol, letting the mixtures stand for 10–30 min on ice, and then proceeding with the assay of adenylyl cyclase activity on 20-μl aliquots of these mixtures by either of the two above described methodologies.

Treatment of Fractions Containing N_s Activity. N_s present in buffer containing 0.4–1.0% Na cholate or 0.1–1.0% Lubrol-PX, 10 m*M* Na-HEPES, pH 8.0, and up to 20 m*M* 2-mercaptoethanol, 30% ethyleneglycol, 150 m*M* NaCl, and 1 m*M* EDTA, were suitably diluted in media containing 1.0% BSA, 1.0 m*M* EDTA, and treated at 32.5° in the presence of additives such as up to 100 m*M* $MgCl_2$, 15 m*M* NaF, 100 μM GMP-P(NH)P, or $GTP_\gamma S$ for appropriate times. At the end of the treatments, the mixtures were diluted 10- to 200-fold with ice-cold media containing 1.0% BSA, 1.0–5.0 m*M* EDTA, 10 m*M* Na-HEPES, pH 8.0, with or without 1.5 *M* KCl and 20 m*M* 2-mercaptoethanol. *cyc*$^-$ reconstituting activity was then assayed as described above.

All incubations were carried out in 12 × 75-mm polypropylene test tubes (Walter Sarstedt catalog #55.526), to avoid losses of N_s protein due to adsorption to walls.

On the Linearity of Assays and Expression of Results. The conversion rates of [α-^{32}P]ATP to [^{32}P]cAMP determined depend on three parameters: time of incubation, *cyc*$^-$ membranes, and dilution of the fractions that contained *cyc*$^-$ adenylyl cyclase reconstituting activity. Conversion rates are nonlinear both with respect to the *cyc*$^-$ membrane concentration used to assay for N_s activity and with respect to the time of incubation, showing lag times that varied with the concentration of Mg in the final assay and with the history of the N_s-containing materials, i.e., preliminary treatments. On the other hand, after suitable dilution of N_s activity containing fractions, which ranged from 1 : 400 for crude cholate extracts to >1 : 10,000 after purification, the conversion rates observed at given incubation times with a given concentration of *cyc*$^-$ membranes became linear with further dilution of these fractions. Absolute "*cyc*$^-$ reconstituting" activities assayed in the different experiments will therefore vary depending on the time of incubation and *cyc*$^-$ membrane concentration in the assay and in addition, also with the batch of *cyc*$^-$ membranes used. In fact, while at a given assay condition, *cyc*$^-$ reconstituted activity does not vary by more than 20% from experiment to experiment, it may vary by as much as 3-fold upon changing from one batch of *cyc*$^-$ membranes to another, without altering the degree of dilution necessary to obtain linearity with respect to addition of a given N_s-containing fraction. The reasons for these variations in the behavior of the *cyc*$^-$ membrane batches is not clear at present, but include cell density at the moment of harvesting, growth rate of cell at moment of harvesting, and quality of the batch of heat-inactivated horse serum used as part of the growth media. Due to these characteristics of the assays, all measurements where comparisons are to be made between different N_s containing fractions should be carried out each time with a single batch of *cyc*$^-$ membrane.

Purification of N Proteins

A summary scheme of the procedures used to purify N proteins is shown in Table I. Unless stated otherwise, all procedures and manipulations are carried at 0–4°, and sodium salts of both cholic acid and EDTA are used.

Step 1: Preparation of Membranes. Red cells from 14 units of blood are washed 4 times by centrifugation (2000 *g*, 15 min) with 4-liter batches of 5 m*M* potassium phosphate buffer, pH 8.0 (P_i buffer), containing 150 m*M* NaCl. The upper layer of white cells is discarded by aspiration after each centrifugation. Washed cells (1500 ml) are lysed with 40 liters of P_i buffer, and the membranes are collected in eight 50-ml stainless-steel tubes in a Sorvall SS-34 rotor fitted with a KSR-R Szent Georgy-Blume continuous flow adaptor by centrifuging at 35,000 *g* and pumping the lysate at 110–130 ml/min with the aid of a Cole Palmer Masterflex pump. After pumping an additional 2.0 liters of P_i buffer through the system, the collected membranes are removed from the stainless-steel tubes, taking care to discard dark pellets of nonlysed cells that accumulate at the bottom of the stainless-steel tubes. The collected membranes are resuspended in 2.4 liters of P_i buffer and washed three times in 250-ml polycarbonate bottles by centrifugation at 16,500 *g* for 30 min in large six-place fixed angle rotors. The supernatants are discarded, as are the tight pellets that accumulated at the bottom of the centrifuge bottles. Typically, the light pink washed membranes from 14 units of human blood are recovered in a final volume of 500 ml at 15–20 mg protein/ml. They are made 10 m*M* in Na-HEPES, pH 8.0, and stored frozen at −70° until used.

Steps 2 and 3: Preextraction and Extraction of Washed Human Erythrocyte Membranes. Membranes (650 ml, ~12 g protein) are thawed under continuous stirring at room temperature and preextracted three times by incubating them in a final volume of 700 ml for 30–40 min at 4° with 0.1% cholate, 10 m*M* $MgCl_2$, 20 m*M* 2-mercaptoethanol, 300 m*M* NaCl, and 10 m*M* Na-HEPES, pH 8.0, followed by separation of the extracted proteins from the membranes by centrifugation at 70,000 g_{av} for 30 min in a Beckman Type 45 Ti rotor. The preextraction procedure is then com-

TABLE I. SUMMARY OF PURIFICATION SCHEME USED

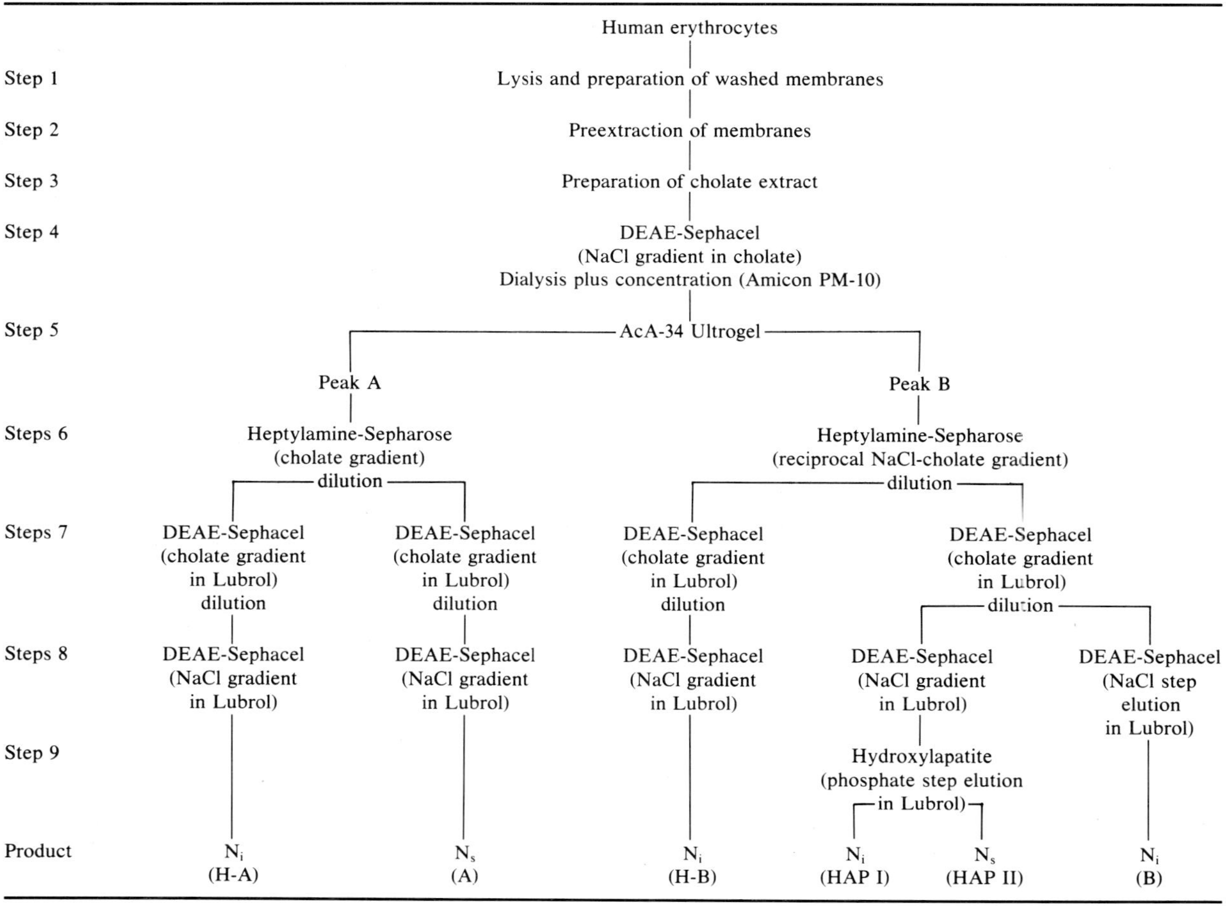

pleted by washing these membranes once with 700 ml of 1.0 mM EDTA in 10 mM Na-HEPES, pH 8.0. N proteins are then extracted from the pellet of the fourth centrifugation by resuspending in 700 of Millipore-filtered (0.45 μm porosity, type HA filters), 1% cholate, 10 mM $MgCl_2$, 20 mM 2-mercaptoethanol, and 10 mM Na-HEPES, pH 8.0, and stirring at 4° for 45 min. The extracted N proteins are separated from nonextracted proteins by centrifugation and the supernatants are collected and made 11 mM in Na-EDTA and 30% in ethyleneglycol. This mixture is referred to as "cholate extract." Typically, a 4-fold repetition of the above procedure yields from 70 units of blood ~3000 ml of cholate extract containing ~3 g of protein.

Steps 4 through 9: Chromatographies. All buffers and solutions used throughout are filtered through 0.45 μm HA Millipore filters. All procedures are carried out in Buffer A composed of 1.0 mM Na-EDTA, 20 mM 2-mercaptoethanol, 30% ethyleneglycol, and 10 mM Na-HEPES, pH 8.0. Amicon PM-10 membrane filters used to concentrate by ultrafiltration are pretreated with 1% BSA and washed extensively with distilled H_2O prior to use.

Step 4: First DEAE Ion Exchange Chromatography. Cholate extract from 70 units of blood is loaded at 1 ml/min onto a DEAE-Sephacel column (5 × 60 cm) preequilibrated with 0.9% cholate in Buffer A. This column is then washed with 1.0 liter of 0.9% cholate in Buffer A and proteins are eluted with a linear gradient (2 × 2 liters) between 0.9% cholate in Buffer A and 0.9% cholate in Buffer A plus 500 mM NaCl. As illustrated in Fig. 2, N_s activity elutes as a single peak in a volume of 650 to 725 ml at approximately 125 mM NaCl and N_i, assessed by ADP-ribosylation with pertussis toxin, elutes as an overlapping peak slightly after N_s, at approximately 150 mM NaCl. The double peak of N_s plus N_i containing 25–30% of the applied protein, and better than 90% of the N proteins, is collected. The concentration of NaCl in this eluate is reduced by dialysis against 2 × 8 liters (12 hr each) of 0.9% cholate in Buffer A. The resulting sample was then concentrated over a 2-day period to 35–40 ml by ultrafiltration over an Amicon PM-10 filter, diluted to 80 ml with 0.9% cholate in Buffer A, and again concentrated to 35–40 ml. Approximate flow rate: 20 ml/hr. Failure to reduce the concentration of NaCl prior to concentration of the sample results in an altered distribution of N proteins on subsequent gel filtration.[24]

Step 5: AcA-34 Ultrogel Exclusion Chromatography. The concentrated material, in Buffer A containing 0.9% cholate (approximately 750 mg protein), is applied at a flow rate of 1.0 ml/min to a 5 × 55 cm column of AcA-34 Ultrogel equilibrated with 0.9% cholate in Buffer A plus 100 mM NaCl. As illustrated in Fig. 3, the N_s activity elutes from this column

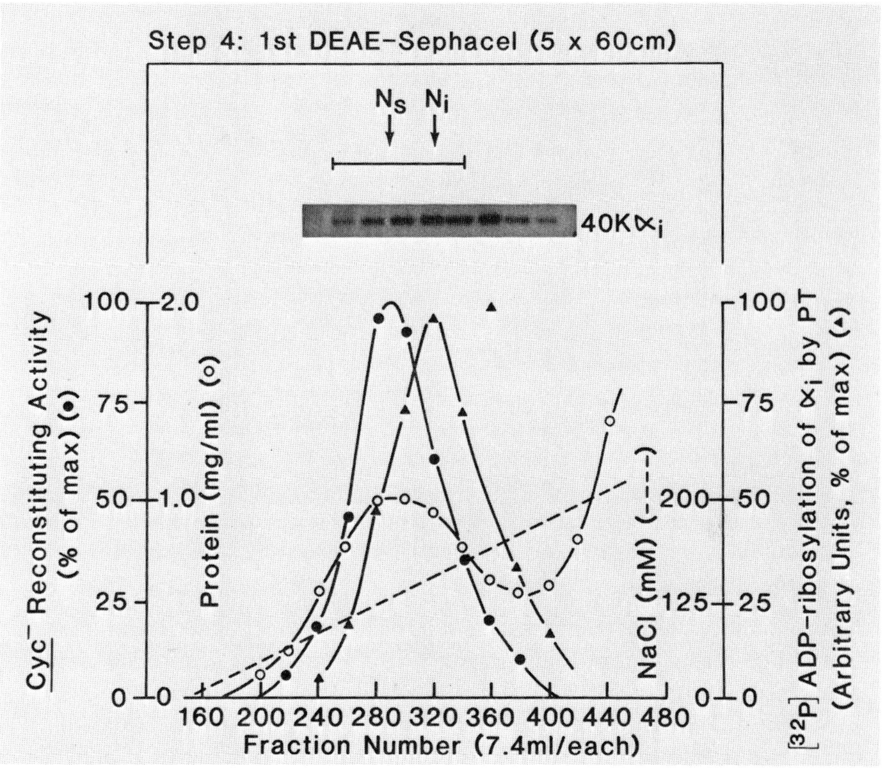

FIG. 2. Elution profile of N_s and N_i from the first DEAE-Sephacel chromatography. Routinely to obtain highly purified N_s and N_i, only fractions containing the bulk of N_s and approximately 50% of N_i are pooled and processed. N_i eluting after N_s tends to purify with contaminants of unknown functions.

in two separate peaks called A and B, which are individually processed further. N_i comigrates with N_s. Typically, peak A is in ~180 ml containing approximately 30% of the applied protein and 60% of the recovered N_s activity and peak B is in ~170 ml containing 6% of the applied protein and 25% of the recovered activity. Combination of peaks A and B leads to aggregation of materials that permit no further purification of N proteins.

Steps 6A and B: Heptylamine Sepharose 4B-CL Chromatographies. Step 6A. Peak A from the Ultrogel AcA-34 step is diluted 2.25-fold in Buffer A containing 100 m*M* NaCl so that final concentrations are 0.4% cholate and 100 m*M* NaCl in Buffer A, and is applied at a flow rate of 0.5 ml/min to a 5 × 40 cm column of heptylamine-Sepharose 4B-CL equilibrated with Buffer A plus 0.4% cholate and 100 m*M* NaCl. This column is then washed first with 1500 ml of Buffer A plus 0.4% cholate and 100 m*M*

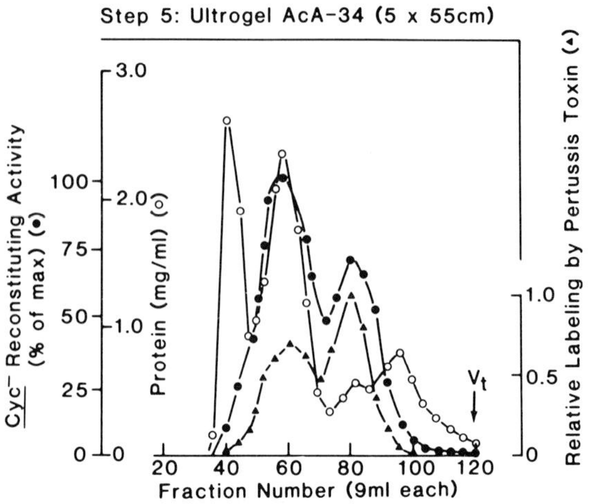

FIG. 3. Elution profiles of N_s and N_i from the Ultrogel AcA-34 column. The sample applied was that of an N_s purification containing, at this stage, also N_i. Note that both N_s and N_i subfractionate into two peaks of which the first is called A and the second B.

NaCl and then with 1000 ml of Buffer A plus 0.4% cholate and 500 m*M* NaCl. Proteins are then eluted with a linear gradient (2 × 5000 ml) of cholate from 0.4 to 4.5% in Buffer A at a flow rate of 100 ml/hr. N_s activity (25–30% of the applied activity) elutes as a single peak at ~2.25% cholate in a total volume of approximately 1200 ml. This is the source of most of the N_s. The protein recovery of the N_s part of the purification at this stage is about 40–45 mg. The bulk of N_i is eluted prior to N_s as well as partially overlapping with it. N_i is localized by ADP-ribosylation with [^{32}P]NAD$^+$ and pertussis toxin and pooled for further processing.

Step 6B. Peak B from the Ultrogel AcA-34 step is treated identically to peak A except that the size of the heptylamine-Sepharose column is smaller (2.5 × 10 cm). The column is first washed with 250 ml of Buffer A plus 0.4% cholate and 100 m*M* NaCl and then with 250 ml of Buffer A plus 0.4% cholate and 500 m*M* NaCl. Proteins are then eluted with a double reciprocal gradient (2 × 300 ml) of cholate from 0.4 to 2.0% and NaCl from 300 to 0 m*M* in Buffer A at a flow rate of 1 ml/min. As illustrated in Fig. 4, N_s activity (50–60% of the applied activity) elutes as a single peak at ~1.2% cholate and 150 m*M* NaCl in a total volume of approximately 100 ml. Polypeptides of M_r = 40,000 start to elute at approximately 0.7% cholate. However, a peak of material susceptible to ADP-ribosylation by pertussis toxin, i.e., the α subunit of N_i, elutes at 1.0% cholate. It is a part of the bulk of N_i (for further purification see below), the trailing edge of which elutes with N_s. At this point, a typical B preparation of N_s contains a total of ~5–6 mg protein.

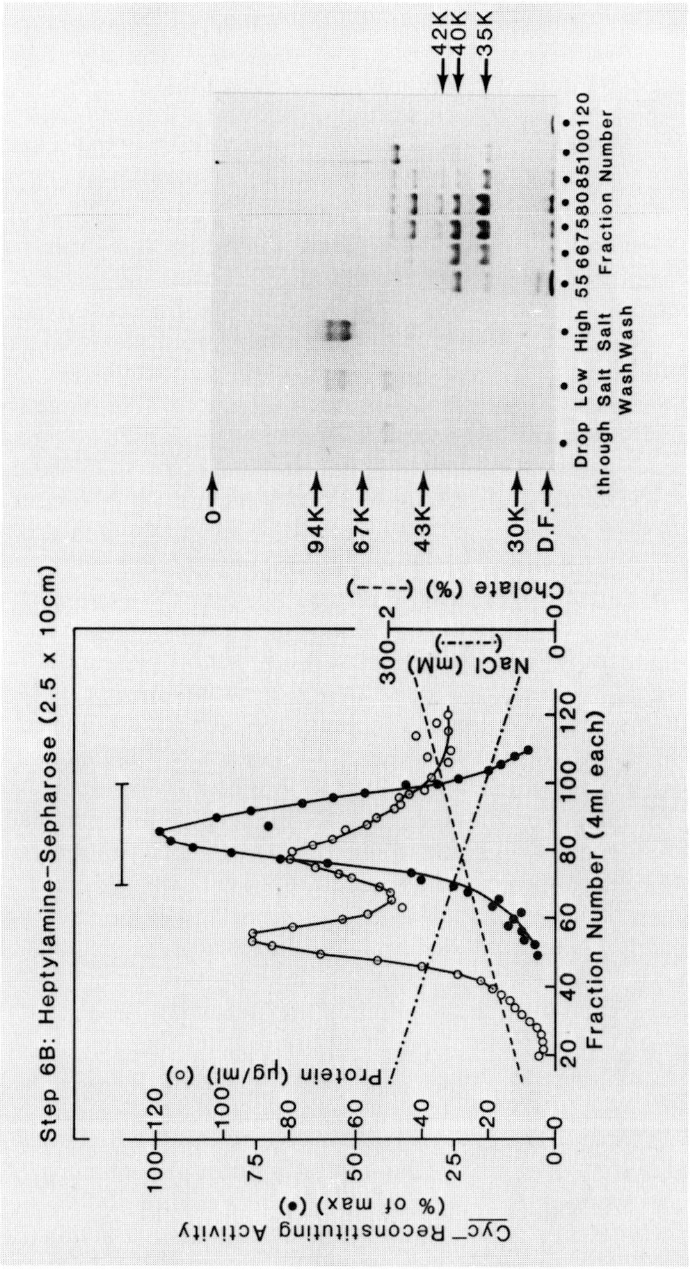

FIG. 4. Elution profile of N_s and partial SDS–PAGE protein of the fractions eluting from the heptylamine-Sepharose column. The column had received peak B of the Ultrogel AcA-34 column. Note that N_i elutes prior to N_s.

Purification of N_s from Preparations A and B

Steps 7A and 7B: Second DEAE Ion Exchange Chromatographies. The pooled fractions from preparations A or B obtained after heptylamine-Sepharose chromatography are diluted 6-fold and 4-fold, respectively, with Buffer A containing 0.5% Lubrol-PX and are applied to a 2 × 10 cm column of DEAE-Sephacel equilibrated with 0.5% Lubrol PX in Buffer A at a flow rate of 0.5 ml/min. The column is then washed with 100 ml of 0.5% Lubrol-PX in Buffer A and eluted with a linear 0–4% cholate gradient (2 × 300 ml) in Buffer A with 0.5% Lubrol-PX (flow rate 1 ml/min). In each instance, N_s activity elutes at 1.5% cholate and recovery of activity is about 75% of the applied Peak A material and essentially 100% of the applied peak B material. At this point a peak A preparation contains approximately 2 mg protein and a peak B preparation contains ~2.5 mg. In addition, SDS–PAGE analysis of the fractions eluting after N_s activity from the DEAE-Sephacel column that had received the B preparation, are found to contain material resembling purified N_i and M_r = 35,000 peptides (not shown) and constitute the B-II preparation. The pooled fractions containing N_s activity and the B preparation constitute the B-I preparation.

Steps 8A, B-I, and B-II: Third DEAE Ion Exchange Chromatography. Steps 8A and 8B-I. The A and B-I preparations obtained after the second DEAE chromatography are diluted 1 : 4 with 0.5% Lubrol-PX in Buffer A and applied to a 2 × 10 cm column of DEAE-Sephacel equilibrated with 0.5% Lubrol PX in Buffer A. The column is washed with 100 ml of equilibration buffer and then eluted at a flow rate of 0.5 ml/min with a linear 0–300 m*M* NaCl gradient (2 × 300 ml) in Buffer A containing 0.5% Lubrol PX. The elution profiles for the A and B-I preparations obtained at this stage are shown in Figs. 4 and 5. Typically, a B-I preparation yields three discrete peaks of protein: the first peak elutes at ~40 m*M* NaCl, the second minor protein peak elutes at ~75 m*M* NaCl, and the third peak elutes at ~120 m*M* NaCl. This last peak contains ~70% of the applied N_s activity and ~0.7 mg protein. In contrast, an A preparation yields only two protein peaks eluting in the positions of the first and third peaks of the B-I preparation. The second peak of the A preparation contains ~50% of the applied N_s activity, ~0.6 mg protein, and constitutes purified N_s.

Step 8 B-II: Step Elution of M_r = 35,000 Containing Protein and of N_i from DEAE-Sephacel. The pooled fractions constituting the B-II preparation obtained in step 7B were diluted 4-fold with Buffer A containing 0.5% Lubrol PX and were applied to a DEAE-Sephacel column (0.5 × 4 cm) equilibrated with Buffer A containing 0.5% Lubrol PX. A fraction *a* enriched in M_r = 35,000 peptides was eluted first by washing the column slowly with 50 m*M* NaCl in Buffer A containing 0.5% Lubrol-PX, and a fraction *b*, formed almost exclusively of about equal amounts of M_r =

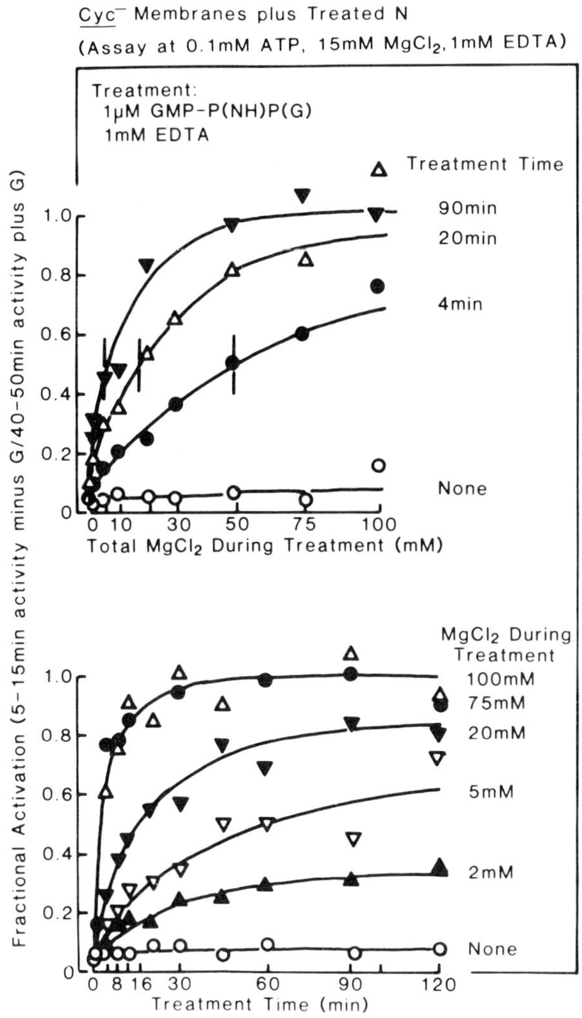

FIG. 5. Dependence of N_s activation by the GTP analog GMP-P(NH)P on time of incubation and Mg concentration. N_s was treated at 32.5° in the presence of 10 μM GMP-P(NH)P and varying concentrations of $MgCl_2$ for the times indicated, diluted as described in the text, and then assayed in the presence of *cyc*⁻ membranes for 10 min in 5–15 min incubations without and 40–50 min incubations with 10 μM GMP-P(NH)P. Ratios of activities obtained without (preactivated N_s) to activities obtained with GMP-P(NH)P (total N_s) are presented.

40,000 and M_r = 35,000 peptides was eluted next by washing the column slowly (flow rate 0.5 ml/min) with 300 mM NaCl in Buffer A containing 0.5% Lubrol PX.

Step 9: Hydroxylapatite (HAP) Chromatography. The third peak of the B-I preparation obtained in Step 8 B-I was fractionated further by chromatography over hydroxylapatite (HAP). To this end the pool of N_s activity from Step 8B-I, which by SDS–PAGE analysis of polypeptide composition contains approximately equal amounts of N_s and N_i, is applied to a 1.0 × 2.0 cm column of HAP preequilibrated with 100 m*M* NaCl and 0.1% Lubrol PX in Buffer A made with 0.1 m*M* EDTA (Buffer B). Prior to elution of the protein the column is washed with Buffer B. Proteins are then eluted with Buffer B containing 30 or 300 m*M* potassium phosphate, pH 8.0. First, on elution (flow rate 0.25 ml/min) with 30 m*M* phosphate buffer, an HAP I fraction is obtained containing in 7 ml ~40% of the recovered N_s activity and 70% of the recovered protein. This constitutes N_i with some contaminating N_s.[11,24] Upon subsequent elution (flow rate 0.25 ml/min) with 300 m*M* phosphate buffer, an HAP II fraction is obtained containing in 7 ml ~60% of the recovered N_s activity and 30% of the recovered protein. This constitutes purified N_s. HAP I and HAP II fractions are desalted by chromatography over Sephadex G-50 equilibrated with 0.1% Lubrol PX and 50 m*M* NaCl in Buffer A.

Purification of N_i Separated from N_s on Heptylamine-Sepharose Chromatography

The elution profiles of N_i from the heptylamine-Sepharose A and B columns are determined by ADP-ribosylation of 10-μl aliquots of the eluates (for details see this volume [45] on ADP-ribosylation of N_i with pertussis toxin). The fractions containing N_i, which as mentioned above elutes prior to N_s from both the A and B heptylamine-Sepharose columns, are pooled, diluted 6-fold (preparation A) or 4-fold (preparation B) with 0.5% Lubrol PX in Buffer A, and then subjected to two successive DEAE-Sephacel chromatographies developed in the same manner as the second and third DEAE-Sephacel chromatographies used to purify the A preparation of N_s. This gives rise to Steps 7 H-A and 7 H-B (second DEAE of N_i from heptylamine) and 8 H-A and 8 H-B (third DEAE of N_i from heptylamine). After each of these purification steps, the position of elution of N_i is determined either by ADP-ribosylation with pertussis toxin or by SDS–PAGE analysis of the polypeptide composition of proteins in the column eluates identifying the polypeptides by silver or Coomassie Blue staining. The final, pooled, and concentrated fractions from Steps 8 H-A and 8 H-B constitute the H-A and H-B preparations of N_i.

Yield of N_s and N_i

Table II summarizes purification, recovery, and yield of N_s. A typical procedure that starts with 60 units (pints) of outdated blood yields about

TABLE II

SUMMARY OF PURIFICATION OF N_s FROM HUMAN ERYTHROCYTES

Step	Fraction[a]	Total volume (ml)	Total protein (mg)	N_s activity[b] Total (μmol cAMP/40 min)	Specific (μmol cAMP/40 min/mg)	Recovery (%)	Purification (fold)
1	Membranes	2,600	46,480	1,594	0.033	100	1
2	Preextracted membranes	2,600	35,670	1,630	0.045	102	1.4
3	Cholate extract	2,790	2,534	1,677	0.66	105	20
4	First DEAE (dialyzed and concentrated)	45	654	1,327	2.03	83	62
Preparation A							
5	AcA 34 (peak A)	185	207	626	3.02	36	92
6A	Heptylamine A	1,300	22	333	15.1	21	458
7A	Second DEAE	86	1.8	248	138	16	4,175
8A	Third DEAE (concentrated)	5	0.73	128	175	8	5,313
Preparation B							
5	AcA 34 (peak B)	175	43.7	275	6.29	17	191
6B	Heptylamine B	125	6.5	160	24.6	10	745
7B	Second DEAE (fraction I)	50	2.35	166	70.6	10.5	2,140
8B-I	Third DEAE	6	0.66	75	114	5	3,443
9	Hydroxylapatite (fraction II, desalted)	12	0.18	30	166	2.1	5,050

[a] DEAE, DEAE-Sephacel; AcA 34, Ultrogel AcA-34; heptylamine, heptylamine-Sepharose 4B-CL. For details see text.

[b] *cyc*$^-$ reconstituting activity assayed for 0–40 min in the presence of 0.1 m*M* GMP-P(NH)P under the conditions described in Assays.

600–750 μg of N_s from Step 8 A and 150–200 μg of N_s from Step 9 II. The overall purification is about 5000-fold with respect to starting washed human erythrocyte membranes with a yield of about 10%. The isolated proteins appear to have preserved all the properties expected from studies with crude cholate extracts.[24] Figure 5 illustrates the dependence of pre-activation of N_s (Step 8 A) on Mg and time.

A typical purification procedure yields about 60–70 μg of N_i from Step 8 H-A, 300–400 μg of N_i from Step 8 H-B, 100–150 μg of N_i from Step 8 BII-b, and 200–250 μg from Step 9 I. It is possible to increase these quantitites of N_i, but not N_s, by 50–80% if a broader region of the fractions from the first DEAE-Sephacel chromatography are pooled. However, if this is done, the preparations of N_s and N_i which are obtained are

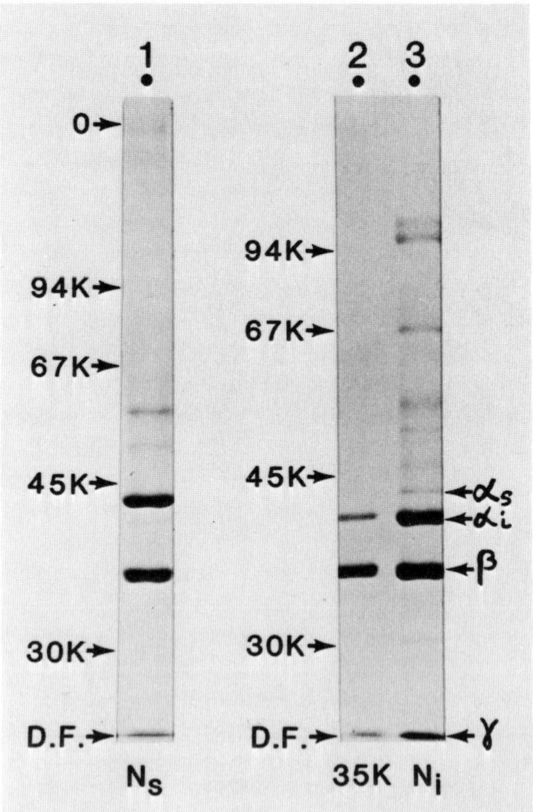

FIG. 6. SDS–PAGE analysis of various fractions of N_s and N_i purified by the described procedures.

significantly less pure. Especially, glycoproteins of M_r between 60,000 and 100,000 and a $M_r \sim 28,000$ polypeptide tend to contaminate all preparations.

Figure 6 presents an SDS–PAGE analysis of the N proteins as obtained after Step 8 third DEAE-Sephacel chromatography.

Acknowledgments

Supported in part by research grants from the U.S. Public Health Service AM-19318 and HD-09581 to L.B. and Center Grants HD-07495 and AM-27685. J.C. is a 1982–1984 trainee of U.S. Public Health Service Training Grant AM-07348.

Section VI

Assays of Hormonal Effects and Related Functions

[39] Analysis of Hormone-Induced Changes of Phosphoinositide Metabolism in Rat Liver

By M. A. WALLACE and J. N. FAIN

The relationships between hormone-stimulated phosphoinositide turnover and Ca^{2+} flux can be investigated using hepatocytes and the subcellular fractions derived from them or from whole liver. Comparison of the results obtained using intact cells to those from subcellular fractions should ultimately lead to a detailed reconstruction of the transmembrane signaling events through which hormones such as vasopressin, angiotensin II, and α_1-specific catecholamines acutely activate liver glycogenolysis. Currently it is controversial whether Ca^{2+} flux or phosphoinositide degradation is the primary event which occurs after these hormones bind to their receptors. Detailed knowledge of how Ca^{2+} interacts with the enzymes (and substrates) of phosphoinositide metabolism is essential to the understanding of how hormones influence phosphoinositide turnover. Conversely, the presence or absence of phosphoinositides, their precursors, and their degradation products may greatly influence Ca^{2+} dynamics. This section considers hormone-stimulated phosphoinositide metabolism in intact hepatocytes as well as the hepatic enzymes involved in phosphoinositide metabolism. Finally, we review the current status of studies on hormone action in broken cell preparations in liver.

Isolation of Rat Hepatocytes: Radiolabeling and Analysis of Their Phosphoinositides

Rat hepatocytes are generally isolated by a technique which involves cannulation of the liver of an anesthetized animal and subsequent perfusion with collagenase.[1,2] The cells can be washed and incubated in physiological buffer in a shaking water bath at 37°. These cells retain their sensitivity to vasopressin, angiotensin II, and catecholamines. Phospholipids can be extracted from the cells either by the method of Bligh and Dyer[3] or of Folch.[4] We commonly use the former procedure to extract cells plus media and the latter to extract phospholipids from TCA precipitates of the cells. Acidification of the $CHCl_3$: methanol (MeOH) solutions

[1] M. N. Berry and D. S. Friend, *J. Cell Biol.* **43,** 506 (1969).
[2] S. R. Wagle and W. R. Ingebretsen, Jr., this series, Vol. 35, p. 579.
[3] E. G. Bligh and W. J. Dyer, *Can. J. Biochem. Physiol.* **37,** 911 (1959).
[4] J. Folch, M. Lees, and G. H. Sloane-Stanley, *J. Biol. Chem.* **226,** 497 (1957).

ISBN 0-12-182009-2

allows better recovery of the liver polyphosphoinositides (PtdIns-4P and PtdIns-4,5P) and the use of KCl facilitates the recovery of phosphatidylinositol (PtdIns) in the organic phase.

Thus, in a typical experiment, hepatocytes are incubated in Krebs–Ringer bicarbonate buffer with 10 m*M* glucose at pH 7.4 under an atmosphere of 95% O_2–5% CO_2. Incubations are at 37° with 1–10 × 10^6 cells/ml. The incubations are stopped by addition of 80% TCA to a final concentration of 8%, and the precipitate is pelleted by brief centrifugation. After aspiration of the media, the pellet is dissolved in $CHCl_3$: MeOH : 12 *N* HCl (2 : 1 : 0.012); 2 *M* KCl is added and the mixture is thoroughly vortexed. The extractions can be performed for 30 min at room temperature, or overnight in the cold. Aqueous and organic phases are separated by centrifugation. The aqueous phase and interface are removed and the organic phase is dried *in vacuo*. The lipids can be resuspended in $CHCl_3$ or $CHCl_3$: MeOH (9 : 1) for chromatography on Mg^{2+} acetate containing Silica gel H plates made in the laboratory by mixing 3 g of magnesium-acetate with 40 g of silica gel H in 85 ml of water without Ca^{2+} binder, or on commercially available high-performance silica gel plates. Better than 90% recovery of total phospholipid phosphorus is achieved with these extractions and 70–100% of this is recovered after most chromatographic separations.

No single thin-layer chromatography system adequately resolves all the most important phospholipids. We commonly use $CHCl_3$: MeOH : 40% methylamine (126 : 70 : 20) in one-dimensional separations to look specifically at phosphatidylinositol and $CHCl_3$: MeOH : 28% NH_4OH : H_2O (90 : 90 : 19 : 10) to specifically resolve PtdIns-4P and PtdIns-4,5P in one dimension. Silica gel plates for chromatography of polyphosphoinositides should be impregnated with K^+-oxalate. Two-dimensional separations are necessary to resolve phosphatidylserine, phosphatidic acid, PtdIns, and the polyphosphoinositides simultaneously, e.g., when total phospholipid phosphorus analysis is needed. Either of the above solvents or $CHCl_3$: MeOH : 28% NH_4OH : H_2O (130 : 70 : 6 : 4) can then be used in the first dimension followed by butanol : glacial acetic acid : H_2O (6 : 1 : 1) the second dimension.[5] Phospholipids can be visualized by iodine vapor staining, but in rat liver there is usually not enough PtdIns-4P or PtdIns-4,5P for detection unless carrier standards are added. Alternatively, the polyphosphoinositides can be visualized by autoradiography if the cells or membranes were incubated with ^{32}P.

PtdIns accounts for about 7–10% of total cellular phospholipid but on a percentage basis is higher in plasma membranes. PtdIns-4P accounts for about 1% and PtdIns-4,5P about 0.5% of total cellular phospholipids. Any

[5] A. A. Abdel-Latif, S. J. Yau, and J. P. Smith, *J. Neurochem.* **22,** 383 (1974).

enrichment for these compounds in subcellular compartments of hepatocytes is undocumented. The phospholipid phosphate is determined by measuring the phosphorus in acid hydrolysates of the chromatogram scrapings using the methods of Bartlett,[6] Rouser *et al.*,[7] or Duck-Chong.[8] Harrington *et al.*[9] have described a method for the quantitation of small amounts of phosphatidylinositol after the chromatography by staining with a novel molybdate compound and then scanning densitometry. We have found this system very useful for some purposes, but the need for constructing a large number of standard curves makes the procedure cumbersome for routine work. Along the same line, analysis of phospholipids by HPLC is also rather time consuming with techniques described to date and the sensitivity of HPLC analysis does not yet surpass that of thin-layer chromatography.

1,2-Diacylglycerol and inositol phosphate(s) are the products of hormone-stimulated phosphoinositide degradation. The inositol phosphate(s) can be separated and assayed from neutralized TCA extracts of hepatocytes, or from the aqueous phase of a typical Bligh-Dyer extraction of cells plus media. The inositol phosphate(s) are sequentially eluted from Dowex-1 (X10; 200–400 mesh; formate form) columns, using 0.1 *M* formic acid with 0.2 *M* going up to 1.0 *M* ammonium formate.[10] The KCl is omitted from the initial extraction solvents when Dowex columns are to be run. Alternatively, the compounds can be separated by electrophoresis.[11] Again, these compounds are detected by radioactive labels. Improved techniques for the rapid separation and quantitation of small amounts of nonradioactive inositol phosphates are greatly needed to further research on phosphoinositide metabolism. In fact, the same can be said for assays of diglyceride, usually measured after thin-layer chromatography of lipid extracts. Changes in diglyceride are, again, routinely measured by following radioactive levels rather than small changes in mass.

Hepatocyte phosphoinositides can be labeled by incubating the cells as described above in the presence of a variety of radioactive precursors including [^{32}P]P$_i$, glycerol, glucose, acetate, various fatty acids, and inositol. Alternatively, liver phosphoinositides can be labeled by injecting the animals with 50–100 μCi of [^{3}H]inositol or [^{32}P]P$_i$ and isolating hepatocytes after 18 hr. Most research on phosphoinositides in hepatocytes has utilized [^{32}P]P$_i$ and [^{3}H]inositol. Hour-long incubations of cells with

[6] G. R. Bartlett, *J. Biol. Chem.* **234,** 466 (1959).

[7] G. Rouser, S. Fleischer, and A. Yamamoto, *Lipids* **5,** 494 (1970).

[8] C. G. Duck-Chong, *Lipids* **14,** 492 (1979).

[9] C. A. Harrington, D. C. Fenimore, and J. Eichberg, *Anal. Biochem.* **106,** 307 (1980).

[10] C. P. Downes and R. H. Michell, *Biochem. J.* **198,** 133 (1981).

[11] U. B. Seiffert and B. W. Agranoff, *Biochim. Biophys. Acta* **98,** 574 (1965).

10 μCi/ml of [^{32}P]P_i and 1–4 μCi/ml [^{3}H]inositol labels the phosphoinositides pools which are degraded in response to addition of hormones.

Both ^{32}P and [^{3}H]inositol label are rapidly lost from all phosphoinositides following hormone addition. The effects on label loss are maximal at 30–60 sec.[12–14] Surprisingly, the labels return rapidly to control values except for the ^{32}P content of phosphatidylinositol (but not the ^{32}P in polyphosphoinositides) which increases above control levels at 2.5 min and continues to increase for at least 1 hr. This is, in fact, the most often reported effect of hormones on phosphoinositide metabolism of hepatocytes—a specific increase in incorporation of [^{32}P]P_i into phosphatidylinositol.

There is no increase in incorporation of [^{3}H]inositol into phosphoinositides in hepatocytes[15] (or in brain synaptosomes[16]) under conditions in which ^{32}P uptake is dramatically increased by hormones. Since the subsequent addition of hormones to cells prelabeled with [^{3}H]inositol or [^{32}P]P_i results in an equivalent breakdown of phosphatidylinositol in response to hormones, the disparity in uptake is difficult to explain. One possibility is that [^{3}H]inositol is primarily incorporated via an exchange process which is tremendously enhanced by Mn^{2+} as noted in hepatocytes by Tolbert *et al.*[15] This process is not responsive to hormones. Also unexplained is the fact that the polyphosphoinositides will quickly label to an apparent steady state with cellular [^{32}P]ATP but they show no increased ^{32}P incorporation as does PtdIns during prolonged exposure to hormones—even though PtdIns is their direct precursor.[15,17] It could be argued that this is the result of a rapid turnover of the terminal phosphate groups so that the hormone effect could not be demonstrated at 30 min as in the studies of Tolbert *et al.*[15] However, in hepatocytes prelabeled with ^{32}P for only 5 min the uptake of label into phosphatidylinositol 4,5-bisphosphate was actually decreased during the first 2.5 min after hormone addition which was followed at 5 min by a return to control values but no stimulation of ^{32}P uptake into phosphatidylinositol 4,5-bisphosphate was seen as was true for phosphatidylinositol.[14]

[12] C. J. Kirk, J. A. Creba, C. P. Downes, and R. H. Michell, *Biochem. Soc. Trans.* **7,** 377 (1981).

[13] A. P. Thomas, J. S. Marks, K. E. Coll, and J. R. Williamson, *J. Biol. Chem.* **258,** 5716 (1983).

[14] I. Litosch, S.-H. Lin, and J. N. Fain, *J. Biol. Chem.* **258,** 13727 (1983).

[15] M. E. M. Tolbert, A. C. White, K. Aspry, J. Cutts, and J. N. Fain, *J. Biol. Chem.* **255,** 1938 (1980).

[16] S. K. Fisher and B. W. Agranoff, *J. Neurochem.* **37,** 968 (1981).

[17] J. N. Fain, S.-H. Lin, P. Randazzo, S. Robinson, and M. Wallace, *in* "Isolation, Characterization and Use of Hepatocytes" (R. A. Harris and N. W. Cornell, eds.), p. 411. Am. Elsevier, New York, 1983.

Another surprising finding is that of Litosch *et al.*[14] who have noted that while the ^{32}P and ^{3}H content of the polyphosphoinositides falls in the first few seconds after exposure to hormone, the actual content of the compounds in the cell is rising as measured by phosphorus analysis. Probably there are distinct metabolic pools of PtdIns, PtdIns-4P, and PtdIns-4,5P which undergo separate responses to hormones. An important goal of future research must be to define these pools both kinetically and as to their subcellular localization. In this latter regard, Lin and Fain[18] reported that after exposure of hepatocytes to vasopressin the initial breakdown of PtdIns was restricted to the plasma membrane. This experiment was done by injecting rats with [^{3}H]inositol, then later, isolating the hepatocytes exposing them to vasopressin *in vitro,* and quickly fractionating the cells by the method of Aronson and Touster.[19] Polyphosphoinositides were not analyzed, specifically. It might be appropriate here to note that nothing is known about hormone effects on the exchange rates of phosphoinositides between cell membranes (plasma membrane, endoplasmic reticulum, nuclear membrane, etc.), though this transfer process may be important to understanding phosphoinositide metabolism.

Of particular interest is the role that Ca^{2+} mobilization plays in the hormone induced changes of phosphoinositides. Various attempts have been made to manipulate hepatocyte Ca^{2+} content and thus mimic hormone effects on phospholipids.[13,15,20–22] Raising cell Ca^{2+} content with ionophore (A23187) does not induce rapid stimulation of phosphoinositide degradation or long-term PtdIns turnover. Depleting cell Ca^{2+} by addition of EGTA to the extracellular media (and even in the presence of ionophore to further draw out cell Ca^{2+}) does not induce phosphoinositide changes mimicking hormone action in hepatocytes, though an increase of [^{3}H]inositol incorporation into PtdIns has been found.[15,17,22] Indeed, almost irregardless of the Ca^{2+} status of intact hepatocytes, hormones can still induce changes in phosphoinositide metabolism.

This does not mean that Ca^{2+} is without effect on the enzymes of phosphoinositide metabolism, however (see below). The dose–response curves for hormone-activated phosphoinositide degradation are shifted to the right relative to response curves for other processes such as activation of glycogen phosphorylase.[23] This, plus the ability of hormones to induce

[18] S.-H. Lin and J. N. Fain, *Life Sci.* **29,** 1905 (1981).

[19] N. N. Aronson and O. Touster, this series, Vol. 31, p. 90.

[20] C. S. Kirk, T. R. Verrinder, and D. A. Hems, *Biochem. Soc. Trans.* **6,** 1031 (1978).

[21] M. M. Billah and R. H. Michell, *Biochem. J.* **182,** 661 (1979).

[22] J. N. Fain, S. Lin, I. Litosch, and M. Wallace, *Life Sci.* **32,** 2055 (1983).

[23] R. H. Michell, C. J. Kirk, C. J. Jones, C. P. Downes, and J. A. Cerba, *Philos. Trans. R. Soc. London, Ser. B* **296,** 123 (1981).

phospholipid changes in the relative absence of Ca^{2+}, suggests a close linkage between hormone–receptor occupation and phosphoinositide degradation.

Enzymes of Phosphoinositide Metabolism

The following is a compilation of some of the enzymes involved in hepatic phosphoinositide turnover including the assay conditions we normally use for these enzymes and some relevant comments concerning their sensitivity to Ca^{2+}, etc.

CDP-DG: Inositol Transferase (Enzyme I)

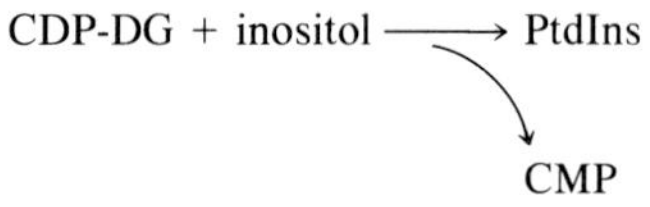

Assay Buffer and Conditions

10 m*M* HEPES, pH 8.0
1 m*M* EGTA
3 m*M* $MgCl_2$
[^{3}H]Inositol
Enzyme source: microsomes
Substrates: 1 m*M* CDP-DG
Incubations are at 37°. Reactions are stopped by addition of $CHCl_3$: MeOH (1 : 2) and extractions continued as per Bligh and Dyer. Reaction is followed by incorporation of [^{3}H]inositol into the organic phase of the extraction.

Comments. The reaction is strongly inhibited by low levels (1–10 μM) Ca^{2+}. [^{3}H]Inositol is incorporated only into PtdIns.

PA: CTP Cytidyltransferase (Enzyme II)

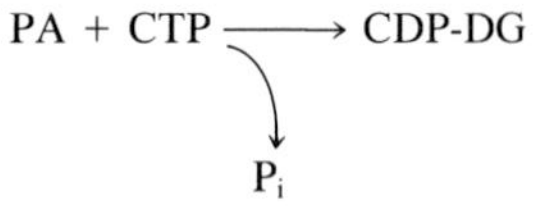

Assay Buffer and Conditions

10 m*M* HEPES, pH 8.0
1 m*M* EGTA
3 m*M* $MgCl_2$
[^{3}H]CTP + 1 m*M* CTP
Enzyme and substrate source: microsomes

Incubations are at 37° and as with the transferase above, the reaction is monitored by formation of labeled CDP-DG which extracts into the organic phase.

Comments. This reaction is also inhibited by Ca^{2+}. We find it very convenient to assay enzymes I and II in a coupled reaction using microsomes as both enzyme and substrate source. The incubation buffer is 10 m*M* HEPES, pH 8.0, 1 m*M* CTP, 1 m*M* EGTA, 3 m*M* $MgCl_2$, and [^{3}H]inositol. The reaction at 37° is followed by the formation of [^{3}H]PtdIns.

Mn^{2+} Activated PtdIns Exchange Enzyme (Enzyme III)

$$\text{PtdIns} + {}^*\text{inositol} \longrightarrow \text{Ptd}^*\text{Ins} + \text{inositol}$$

Assay Buffer and Conditions

10 m*M* HEPES, pH 8.0
1 m*M* EGTA
1 m*M* $MnCl_2$
[^{3}H]Inositol
Enzyme source and substrate: microsomes.
The reaction at 37° is followed by the formation of [^{3}H]PtdIns as above.

Comments. This reaction is only nominally inhibited by Ca^{2+}. Obviously, this enzyme contains a phospholipase D-type activity and whether this can be separated from the synthesis activity is unknown. The polyphosphoinositides do not pick up label.

Diglyceride Kinase (Enzyme IV)

$$\text{DG} \xrightarrow{\text{ATP}} \text{PA}$$

Assay Buffer and Conditions

50 m*M* Tris, pH 6.8
20 m*M* NaF
5 m*M* $MgCl_2$
2 m*M* ATP + [γ-^{32}P]ATP
0.65 m*M* diolein
0.5 m*M* deoxycholate
Enzyme source: microsomes
The reaction is run at 37°. ^{32}P-labeled PA is formed and extracted into the organic phase as above.

Comments. The enzyme can also be assayed if microsomes are preincubated for 20 min at 37° in buffer containing 50 m*M* Tris, pH 8.0, 5 m*M* $MgCl_2$, 3 m*M* CMP, and then washed. After this, the microsomes are assayed for kinase activity in 50 m*M* Tris, pH 6.8, 5 m*M* $MgCl_2$, 2 m*M* ATP + [γ-^{32}P]ATP, 5 m*M* NaF.[24] This procedure allows evaulation of divalent ion effects without the complication of using the detergent deoxycholate to activate substrate. Ca^{2+} is only slightly inhibitory.

PtdIns Kinase(s) (Enzyme V)

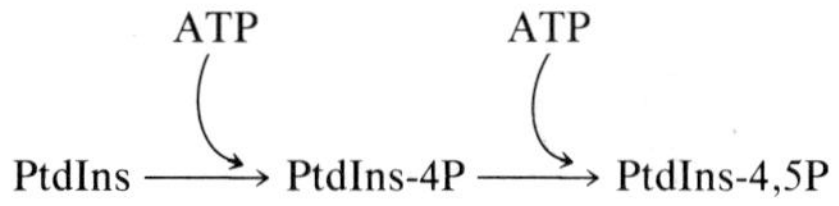

Assay Buffer and Conditions

10 m*M* HEPES, pH 7.4
10 m*M* $MgCl_2$
1 m*M* EGTA
100 m*M* KCl
1 m*M* ATP + [γ-^{32}P]ATP
Source of enzyme and substrate: plasma membrane, etc. Reaction is run at 37°. The label is incorporated into the polyphosphoinositides.

Comments. The enzyme(s) seem mildly inhibited in the presence of Ca^{2+}. They are widely distributed occurring in all subcellular fractions so far examined. Recent reports have claimed specific localization of PtdIns kinase activity in the Golgi[25] and there is probably activity in lysosomes.[26] Hawthorne and Michell[27] originally described activities at many subcellular sites.

Phosphoinositide-Specific Phospholipase C (Enzyme VI)

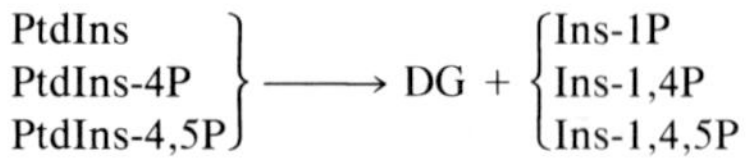

Assay Buffer and Conditions

Optimal buffers and assay conditions are unknown. Deoxycholate and/or Ca^{2+} may be necessary for activity.

[24] H. Kanoh and B. Åkesson, *Eur. J. Biochem.* **85,** 235 (1978).
[25] B. Jergil and R. Sundler, *J. Biol. Chem.* **258,** 5968 (1983).
[26] C. A. Collins and W. W. Wells, *J. Biol. Chem.* **258,** 2130 (1983).
[27] J. N. Hawthorne and R. H. Michell, *in* "Cytosols and Phosphoinositides" (H. Kindl, ed.), p. 49. Pergamon, Oxford, 1966.

Comments. While a Ca^{2+}-activated phospholipase C can be found in rat liver cytosol,[28] there is no evidence that this is a major enzyme in hepatocyte phosphoinositide metabolism. There is no evidence for Ca^{2+} control of this enzyme in intact hepatocytes. Hormone stimulated Ca^{2+} insensitive lipase C activity may be found in the plasma membrane (see below) and this could be a much more important enzyme. Recent studies with brain have indicated multiple proteins with phospholipase C activity.

Enzymes I and III have been partially purified from rat liver.[29,30] While they are "microsomal," their presence or absence in the plasma membrane needs to be definitively described. The extent to which [^{3}H]inositol incorporates into PtdIns through action of enzyme I as opposed to enzyme III in intact cells is difficult to judge. Since Ca^{2+} depletion tends to increase [^{3}H]inositol labeling of hepatocyte phospholipids, this may indicate that the Ca^{2+} manipulation has relieved an inhibitory constraint on enzyme I. The activities of enzymes I–V are not altered if hormones are present during their assay in microsomal preparations. Neither are the activities stably altered if the enzyme preparations come from hormone-treated cells or liver. All these negative findings are in accord with the current theory which predicts that the hormone-stimulated event is increased activity of a phospholipase C enzyme. Again, while a Ca^{2+} requiring phospholipase C is found in the cytosol, Ca^{2+}-mediated activation of this enzyme is not sufficient to cause the degradation of phosphoinositides such as are produced by hormone.

Hormone Stimulation of Phosphoinositide Degradation in Cell-Free Systems

In theory, the hormones considered here may cause PtdIns degradation through activation of a lipase or through activation of the substrate. If the hormone activates a soluble enzyme, then presumably an experimental system could be found to assay hormone-activated lipase using exogenous (i.e., non-plasma membrane-associated) phosphoinositides. So far, attempts to devise such an experimental system have not been successful. On the other hand, the confinement of PtdIns degradation to the hepatocyte plasma membranes as found by Lin and Fain clearly indicates either (1) hormone binding to receptor exposes PtdIns only in the plasma membrane to a lipase or (2) the important lipase is membrane associated and when activated can only react with plasma membrane phospholipids.

[28] T. Takenawa and Y. Nagai, *J. Biol. Chem.* **256,** 6769 (1981).
[29] T. Takenawa and K. Egawa, *J. Biol. Chem.* **252,** 5419 (1977).
[30] T. Takenawa, M. Saito, Y. Nagai, and K. Egawa, *Arch. Biochem. Biophys.* **182,** 244 (1977).

Thus, we have attempted to look for such events occurring in plasma membrane preparations exposed to hormones, paying particular attention to look at the Ca^{2+} requirements for these interactions.

Plasma membranes are prepared from whole liver by the procedure of Neville[31] or by the modification of that method described by Song *et al.*[32] This procedure has often been described, in brief; the preparation involves homogenization of liver in hypotonic media followed by isolation of the membranes on discontinuous sucrose gradients. By adjustment of the homogenization conditions, the procedure can be used to isolate plasma membrane from hepatocytes, also. The purity of the preparations can be assessed by the relative recoveries of marker enzymes. We commonly use 5′-nucleotidase as a marker for plasma membrane, glucose 6-phosphatase for endoplasmic reticulum, succinate dehydrogenase for mitochondria, and lactate dehydrogenase for cytoplasm.[19] In addition, it is useful to measure the cholesterol : phospholipid ratio of the membranes, which should be 0.71. Lower cholesterol : phospholipid ratios indicate, rather directly, the extent of contamination by non-plasma membrane phospholipid.

We have found hormone-activated PtdIns degradation in isolated plasma membranes.[17,33,34] This has been measured as a specific loss of phosphatidylinositol on the basis of mass[33,34] and as a loss of [^{3}H]PtdIns from membranes prepared after labeling hepatocytes or whole liver with [^{3}H]inositol.[17]

Plasma membranes from the sucrose gradients can be washed and resuspended to 1–5 mg/ml protein in an incubation buffer at pH 7.4. After equilibration, hormones are added in the presence or absence of antagonists and the incubation continued for up to 30 min. The reactions are stopped by addition of the MeOH : $CHCl_3$: HCl solution and phospholipid extractions and measurements are performed as outlined above. Using an incubation buffer containing 20 m*M* HEPES, 0.1 m*M* EGTA, and 5 m*M* $MgCl_2$ has been successful, but whether this buffer is optimal is still unknown. Added Ca^{2+} is unnecessary, but hormone effects are lost if $MgCl_2$ is absent or if the EGTA concentration is raised to 2 m*M*. The loss of the hormone effects is most probably because severe depletion of divalent cations, by itself, leads to a loss of PtdIns—the hormone being unable to spur on further degradation.[34] In these experiments the plasma membranes are both the source of substrate and enzyme. We cannot, however, rule out the possibility that the enzyme involved is a contaminate from the

[31] D. M. Neville, *Biochim. Biophys. Acta* **154,** 540 (1968).
[32] C. S. Song, W. Rubin, A. B. Rifkind, and A. Kappas, *J. Cell Biol.* **41,** 124 (1969).
[33] M. A. Wallace, P. Randazzo, S.-Y. Li, and J. N. Fain, *Endocrinology* **111,** 341 (1982).
[34] M. A. Wallace, J. Poggioli, F. Giraud, and M. Claret, *FEBS Lett.* **156,** 239 (1983).

POSSIBLE IMPORTANCE OF PHOSPHOINOSITIDE DEGRADATION STIMULATED BY HORMONES

Effect	Possible consequence
Generation of DG	Activate phospholipid/diglyceride-sensitive protein kinase, ultimately activating glycogen phosphorylase
Generation of PA	PA could act as a Ca^{2+} ionophore
Generation of inositol-1P; inositol-1,4P; inositol-1,4,5P	Might act as messengers to mobilize Ca^{2+} from intracellular stores
Release of Ca^{2+} or Mg^{2+} or other divalent ions bound to phosphoinositides	Might raise cytoplasmic Ca^{2+} and hence activate Ca^{2+}-sensitive enzymes. Would increase membrane fluidity
Loss of phosphoinositides per se	Modulate activity of enzymes specifically sensitive to PtdIns, PtdIns-4P, or PtdIns-4,5P as suggested for Ca^{2+} ATPase. Release of enzymes (to inter- and intracellular space)

cytosol, etc. In this regard, Harrington and Eichberg[35] have confirmed the findings that hormones produce PtdIns degradation in isolated plasma membranes, but they found a requirement for some cytoplasmic factors(s). The factor was not Ca^{2+}. The reason for these discrepancies is now being investigated.

The presumed products of PtdIns degradation from isolated liver plasma membranes (e.g., inositol-1P and DG) have not yet been characterized. Thus, we cannot say for certain if the reaction is catalyzed by a phospholipase C. Initial attempts to follow the fate of polyphosphoinositides in these membranes are only now underway. Optimal procedures for isolating plasma membranes which retain their normal polyphosphoinositide complement have not been established unequivocally, though the presently used methods may prove adequate.

Conclusion

Regardless of the relative importance of Ca^{2+} mobilization versus PtdIns degradation as a primary transmembrane signaling event, the consequences of phosphoinositide degradation to overall cell metabolism are of interest. The table contains some possible outcomes of hormone-stimulated phosphoinositide breakdown. Overall, it is apparent that for all the intensive research on this phenomenon in the past 30 years we are only beginning to understand how and why glycogenolytic hormones cause changes in liver phospholipids.

[35] C. A. Harrington and J. Eichberg, *J. Biol. Chem.* **258,** 1087 (1983).

[40] Effect of Bradykinin on Prostaglandin Production by Human Skin Fibroblasts in Culture

By CAROLE L. JELSEMA, JOEL MOSS, and VINCENT C. MANGANIELLO

Introduction

Prostaglandins, mediators of the action of various agents,[1,2] are synthesized primarily from arachidonic acid that is stored in cellular phospholipids; free arachidonate is generally present in only trace quantities in unstimulated cells.[3] Following exposure of cells to agonists, phospholipases are activated or permitted access to their substrates resulting in phospholipid breakdown and arachidonate release.[2] Since arachidonate is esterified primarily in the C-2 position of phosphatidylcholine and phosphatidylinositol, release can occur by direct action of phospholipase A_2[4] or the phospholipids can be degraded through other pathways before hydrolysis at the C-2 position (Fig. 1).[5] The generation of prostaglandins, prostacyclin, and thromboxane (Fig. 2) involves conversion of the released arachidonate to cyclic endoperoxides by the cyclooxygenase enzyme system[6,7]; noncyclic oxygenated metabolites, i.e., leukotrienes and HETEs, are products of the lipoxygenase system.[7,8]

In cultured human fibroblasts, bradykinin, β-adrenergic agonists, prostaglandins, and choleragen increase cAMP content.[9] Bradykinin also stimulates production of prostacyclin (PGI_2) and PGE_2 by interaction with a B-2 type receptor.[10,11] Since both of these prostaglandins are known to stimulate adenylate cyclase in these cells,[12] and inhibitors of arachidonate

[1] I. L. Bonta and M. J. Parnham, *Biochem. Pharmacol.* **27,** 1611 (1978).

[2] S. Moncoda and J. R. Vane, *Pharmacol. Rev.* **30,** 293 (1979).

[3] D. B. Wilson, S. M. Prescott, and P. W. Majerus, *J. Biol. Chem.* **257,** 3510 (1982).

[4] S. L. Hong, R. Polsky-Cynkin, and L. Levine, *J. Biol. Chem.* **251,** 776 (1976).

[5] R. L. Bell, D. A. Kennerly, N. Stanford, and P. W. Majerus, *Proc. Natl. Acad. Sci. U.S.A.* **76,** 3236 (1979).

[6] P. W. Majerus, S. M. Prescott, S. L. Hofmann, E. J. Neufield, and D. B. Wilson, *Adv. Prostaglandin Thromboxane Res.* **11,** 45 (1983).

[7] F. A. Kuehl, Jr. and R. W. Egan, *Science* **210,** 978 (1980).

[8] G. Weissman, *Adv. Inflammation Res.* **1,** 1 (1982).

[9] S. Rennard, J. Moss, B. Hom, J. Oberpriller, L. Stier, and R. C. Crystal, *Clin. Res.* **30,** 475A (1982).

[10] A. A. Roscher, V. C. Manganiello, C. L. Jelsema, and J. Moss, *J. Clin. Invest.* **72,** 626 (1983).

[11] J. Moss, V. C. Manganiello, B. E. Hom, S. Nakaya, and M. Vaughan, *Biochem. Pharmacol.* **30,** 1263 (1981).

[12] R. A. Gorman, R. D. Hamilton, and N. K. Hopkins, *J. Biol. Chem.* **254,** 1671 (1979).

METHODS IN ENZYMOLOGY, VOL. 109

ISBN 0-12-182009-2

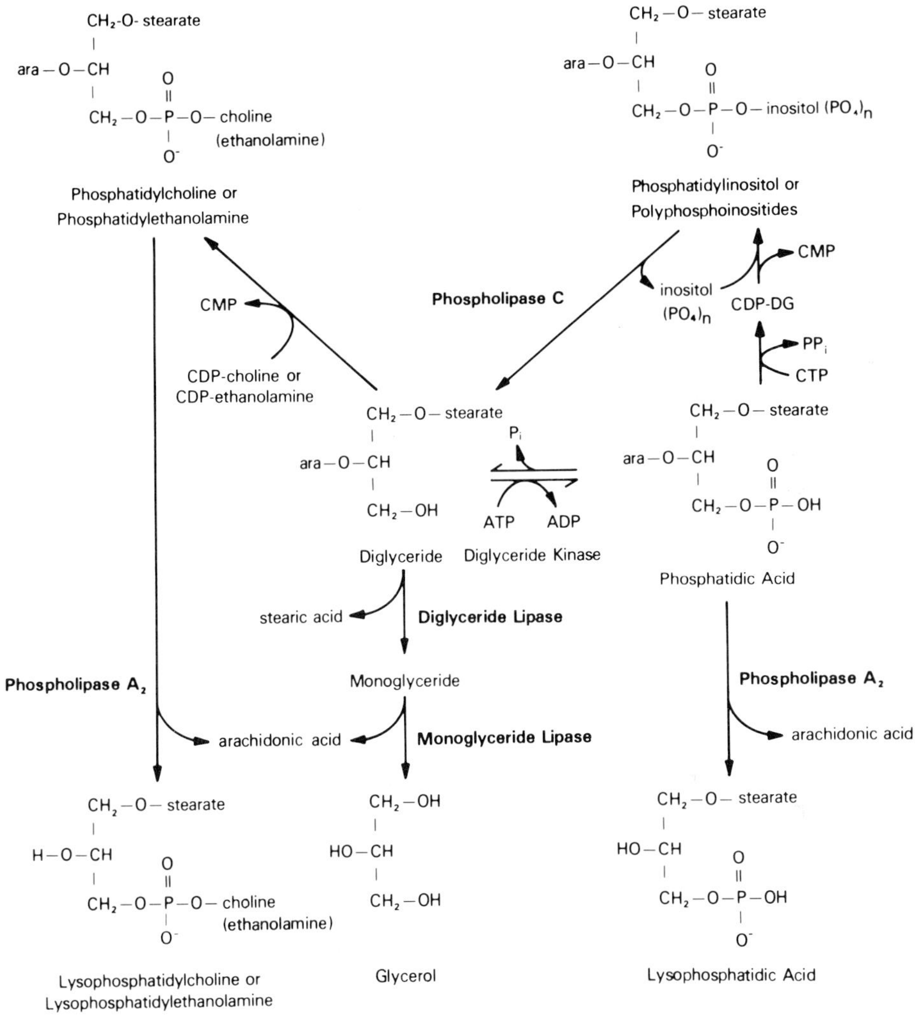

FIG. 1. Pathways for release of arachidonate from membrane phospholipids. The arachidonate in the C-2 position may be released directly by action of phospholipase A_2 or indirectly by prior action of phospholipase C. Phospholipase C activity is then followed either by resynthesis of the phospholipid and subsequent release of arachidonate via phospholipase A_2 or further degradation of the diglyceride to form free arachidonate by action of the neutral lipid lipase(s). It is not presently known whether the monoglyceride and diglyceride lipases are two separate enzymes. Ara, Arachidonate; P_i, inorganic phosphate; PP_i, pyrophosphate; myo-inositol $(PO_4)_n$, $n = 1$ for phosphatidylinositol and $n = 2$ or 3 for the polyphosphoinositides.

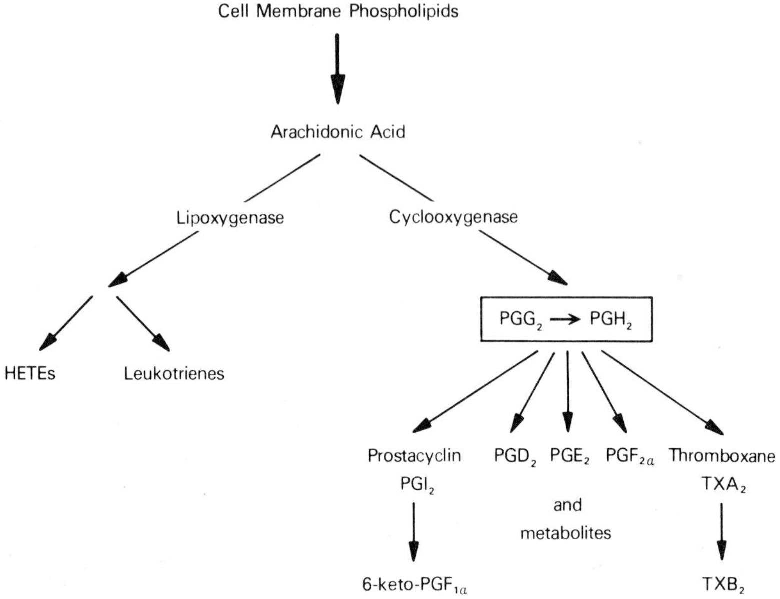

FIG. 2. Pathways for metabolism of arachidonic acid. HETEs, Hydroxyeicosatetraenoic acids.

generation or prostaglandin formation also inhibit bradykinin-induced increases in intracellular cAMP,[13] it is believed that the effect of bradykinin on cAMP is mediated through increased prostaglandin formation. Thus, the fibroblasts serve as a model system to study the mechanisms whereby effectors such as bradykinin alter intracellular cAMP content indirectly via generation of prostaglandins.

In the fibroblasts, bradykinin modulates prostaglandin formation on two different levels. First, by stimulating phospholipase activity, bradykinin increases the amount of arachidonate available for prostaglandin formation, thereby causing a general increase in all the prostaglandins (Fig. 3). Second, bradykinin specifically stimulates prostacyclin (PGI_2) formation (Table I). The importance of increased lipase or prostaglandin synthetase activity in the bradykinin-induced increases in cAMP was further assessed by analyzing the bradykinin effect on cAMP accumulation in the presence of specific lipase and synthetase inhibitors.

In studying the effect of bradykinin on prostaglandin formation in fibroblasts, two general types of assays have been employed. Radioimmu-

[13] V. C. Manganiello, J. Moss, B. E. Hom, and M. Vaughan, *Clin. Res.* **30,** 255A (1982).

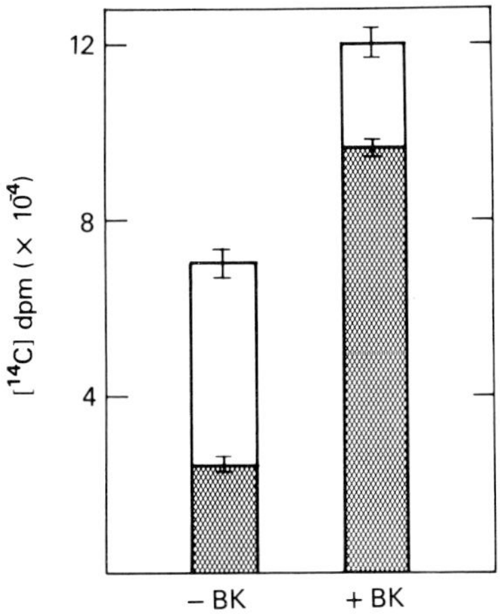

FIG. 3. Analysis of the [^{14}C]dpms released from untreated and BK-treated fibroblasts previously radiolabeled with [^{14}C]arachidonate. Fibroblasts were incubated with [^{14}C]arachidonate in Eagle's MEM supplemented with 2 m*M* glutamine. Cells were washed briefly but thoroughly with three 2 ml HBSS washes followed by incubation in the presence or absence of 0.16 μg/ml BK. After 7 min, the incubation media were removed from the cells and the total [^{14}C]dpms released into the media in control and BK-treated cells determined by counting duplicate 0.1-ml aliquots in 10 ml Biofluor. The remaining 1.8 ml were acidified and the free arachidonate (open columns) separated from prostaglandin-associated arachidonate (hatched columns) by the C_{18} column procedure.

noassays were utilized to determine specific prostaglandins. A second procedure involved initial incubation of the fibroblasts with [^{14}C]arachidonate and identification and quantification of the metabolites by chromatographic methods.

TABLE I
EFFECT OF BRADYKININ ON PROSTAGLANDIN, THROMBOXANE, AND PROSTACYCLIN PRODUCTION[a]

	ng/mg protein				
	$PGF_{2\alpha}$	PGE_2	PGB/A	PGI_2[b]	TXB_2
− Bradykinin	0.09	0.31	0.35	0.44	0.17
+ Bradykinin	0.22	0.47	0.37	6.85	0.17

[a] Measurements obtained by radioimmunoassay of incubation medium following treatment of fibroblasts for 2 min ± bradykinin in 2 ml HEPES-buffered HBSS minus Na_2CO_3 (25 m*M* HEPES, pH 7.4) at 37° as described in text.

[b] Measured as 6-keto-$PGF_{1\alpha}$.

Tissue Culture

Sterile Reagents

Eagle's minimal essential medium (MEM; GIBCO)
Fetal calf serum (FCS; GIBCO)
2 *M* glutamine (GIBCO)
Hanks' balanced salt solution (HBSS; GIBCO) brought to pH 7.4 by bubbling the salt solution with CO_2 for 5 min

Procedures

Human skin fibroblasts grown in monolayer in MEM supplemented with 2 m*M* glutamine and 10% FCS as described by Manganiello and Breslow[14] are used between passages 5 and 15 as growth rate and ligand responsiveness decreased with increasing passage.[15] Cells are detached from flasks (75 cm^2) by incubation for 2 min at 37° with 1 ml of 1.25% trypsin, then diluted 100-fold with growth medium. Subcultures, initiated with 4 ml of cell suspension per 60 × 15 mm dish or 6 ml per 100 × 20 mm dish, are incubated at 37° in an atmosphere of 95% air:5% CO_2 for 2 weeks. Medium is then replaced with 3 ml of fresh growth medium per 60 × 15 mm dish or 5 ml per 100 × 20 mm dish and cells are further incubated for 2 days. To initiate experiments, the medium is aspirated with a sterile pipette and cells are washed twice with 1 ml of HBSS (37°).

At this point, in some experiments cells are radiolabeled with [^{14}C]arachidonate using the procedures described below followed by analysis of the cellular and extracellular lipids to determine the source of the released arachidonate and to identify the arachidonate metabolites formed. In other experiments, the fibroblasts are equilibrated for 10 min in 2 ml of HBSS, pH 7.4, in the presence or absence of various drugs (e.g., indomethacin). The medium is then replaced with 2 ml of HBSS, pH 7.4, containing the indicated additions and cells are incubated at 37° on a Belco orbital shaker either in the presence or absence of bradykinin (0.16 μg/ml). After 7 min, the medium is removed and frozen at −20° prior to prostaglandin determination, and 1 ml 5% trichloroacetic acid (TCA) is then added to the cell monolayer and the cells are quick-frozen on dry ice and stored at −20° prior to determination of intracellular cAMP. In the absence of FCS or bovine serum albumin (BSA), the medium can be assayed directly for prostaglandin production. The presence of BSA or

[14] V. C. Manganiello and J. Breslow, *Biochim. Biophys. Acta* **362,** 509 (1974).
[15] Betty E. Hom, unpublished observations (1981).

FCS in the incubation medium inhibits prostaglandin formation,[16] presumably by binding the released arachidonate and thereby preventing its conversion to prostaglandin.

Assays

cAMP Determination

The cell monolayers (and TCA) are thawed and the fibroblasts scraped from the petri dish; each dish is washed twice with an additional 0.5 ml of 5% TCA, and both washes and original cell suspension are pooled and extracted with an equal volume of trichlorofluoromethane (Freon):trioctylamine (95:25) to remove the TCA.[17] In the presence of high protein levels, it may be necessary to pellet the precipitated protein by centrifugation (2000 *g*, 10 min) prior to the Freon extraction. Duplicate 0.1-ml aliquots from the upper phase are assayed for cAMP using the radioimmunoassay (RIA) kit and procedure from New England Nuclear.

Determination of Prostaglandins by Radioimmunoassay

Radioimmunoassay is a sensitive technique for the determination of prostaglandins. The usefulness of the procedure is, however, often limited by (1) the rapid metabolism of the prostaglandins and the variety of metabolites formed; (2) the availability of highly specific, well-characterized antibody preparations; and (3) the possibility of unsuspected cross-reactivity of antibodies with other prostaglandins or metabolites.

Instability of some prostaglandins requires immediate processing of the samples or brief storage at −20°. In cases where the metabolite, rather than the original prostaglandin, is to be measured, it is important to ensure complete conversion of the primary prostaglandin to its metabolite. In the fibroblasts, for example, the primary prostaglandin synthesized is PGI_2, which is measured by RIA of its 6-keto-$PGF_{1\alpha}$ metabolite. The conversion of PGI_2 to 6-keto-$PGF_{1\alpha}$ occurs extremely rapidly in standard culture medium at 37°.[18] Immediate storage of samples at −20° may, in this instance, lead to incomplete conversion of PGI_2 to 6-keto-$PGF_{1\alpha}$ and an inaccurate estimate of the prostacyclin content due to differences in

[16] K. A. Chandrabose, R. W. Bonser, and P. Cuatrecasas, *Adv. Prostaglandin Thromboxane Res.* **6,** 249 (1980).

[17] T. L. Riss, N. L. Zorich, M. D. Williams, and A. Richardson, *J. Liq. Chromatogr.* **3,** 133 (1980).

[18] J. R. Vane, *Adv. Prostaglandin, Thromboxane, Leukotriene Res.* **11,** 449 (1983).

the affinity of the antibody for PGI_2 and 6-keto-$PGF_{1\alpha}$ (data not presented). PGE_2, another prostaglandin produced by the fibroblasts, can be assayed directly.

Another problem associated with the use of a prostaglandin RIA arises when cultured cells are incubated in the presence of arachidonate. At high concentrations, arachidonate will displace the radiolabeled compound and give spuriously elevated values for prostaglandin concentration (unpublished observations). At low concentrations of arachidonate, the standard curve can be corrected by the addition of arachidonate to the RIA.

In these studies, 6-keto-$PGF_{1\alpha}$ and PGE_2 were assayed by a modification of the procedure of Jaffe *et al.*[19] Each assay contained 100 μl of incubation medium or standard (10 to 2000 pg), 50 μl of either [^{3}H]6-keto-$PGF_{1\alpha}$ or [^{3}H]PGE_2 (~7000 cpm; NEN), 50 μl of antiserum to 6-keto-$PGF_{1\alpha}$ or PGE_2 (Seragen; diluted such that ~40% of the radioligand was bound), and 0.1% gelatin in a total volume of 400 μl of 10 m*M* Tris/150 m*M* NaCl, pH 7.4. After 16 hr at 4°, 500 μl of an ice-cold 0.5% charcoal/0.05% dextran mixture is added and tubes are centrifuged (1000 *g*, 10 min, 4°). The supernatants are decanted into scintillation vials and, after addition of 10 ml Biofluor (NEN), are radioassayed for antibody-bound [^{3}H]6-keto-$PGF_{1\alpha}$ or [^{3}H]PGE_2 using an LKB scintillation counter equipped with a RIA program.

Effect of Inhibitors of Arachidonate Release and Metabolism on Bradykinin-Stimulated Prostaglandin Production

To study the role of phospholipases in the effect of bradykinin on cAMP content, we utilized two inhibitors of phospholipase A_2 and C, dexamethasone and mepacrine.[20,21] To determine the role of prostaglandin production in the bradykinin-induced increase in cAMP, we utilized two cyclooxygenase inhibitors, indomethacin and 5,8,11,14-eicosatetraenoic acid (ETYA). The interpretation of the results obtained with the inhibitors is, however, necessarily limited by the lack of absolute specificity of these compounds and their potential for multiple effects in complex biological systems.

The effect of bradykinin on prostaglandin production and cAMP content was almost completely inhibited after 24–36 hr in the presence of 1 μM dexamethasone; the effect of isoproterenol (2 μM) on cAMP accumu-

[19] B. M. Jaffe, I. W. Smith, W. T. Newton, and C. W. Parker, *Science* **171,** 494 (1971).

[20] S. L. Hong and L. Levine, *Proc. Natl. Acad. Sci. U.S.A.* **73,** 1730 (1976).

[21] S. L. Hofmann, S. M. Prescott, and P. W. Majerus, *Arch. Biochem. Biophys.* **215,** 237 (1982).

TABLE II
EFFECT OF DEXAMETHASONE (D) ON EFFECTS OF BRADYKININ, ISOPROTERENOL, AND PGE_1[a]

Additions	cAMP (pmol/mg protein)		PGI_2 (ng/mg protein)	
	− D	+ D	− D	+ D
Exp. 1				
None	35 ± 5	37 ± 5	0.6 ± 0.1	0.8 ± 0.6
Bradykinin	136 ± 14	59 ± 11	4.3 ± 1.6	1.2 ± 0.3
Exp. 2				
None	14 ± 4	13 ± 0.1		
Isoproterenol	198 ± 1	328 ± 14		
PGE_1	462 ± 12	996 ± 112		

[a] Confluent cells were incubated without or with dexamethasone (1 μM) for 72 hr (Exp. 1) or 96 hr (Exp. 2), washed, and then incubated in Hanks' medium without or with bradykinin (0.16 μg/ml) or isoproterenol (2 μM) for 7 min or PGE_1 (0.2 μg/ml) for 15 min. Data are mean values ± SEM (n = 3) in Exp. 1 or ±1/2 range of duplicates in Exp. 2.

lation was not altered. Incubation with dexamethasone (1 μM) for longer periods of time (72–96 hr), however, enhanced the effects of isoproterenol and exogenous PGE_1, but not bradykinin, on cAMP accumulation (Table II).

Confluent cells were incubated for 30 min with or without the phospholipase A_2 and C inhibitor mepacrine (10^{-6} to 10^{-4} M), the cyclooxygenase inhibitor indomethacin[22] (10^{-9} to 10^{-6} M), or an inhibitor of both cyclooxygenase and lipoxygenase, ETYA (1 μM). Indomethacin and ETYA were added in dimethyl sulfoxide (DMSO) at a final concentration of 1% which did not affect either prostaglandin or cAMP production or measurement. For inhibition of bradykinin-induced increases in cAMP content, the IC_{50} value for mepacrine was ~10 μM (Fig. 4); for indomethacin ~10 nM (Fig. 5). At ~1 μM, ETYA completely inhibited the response to bradykinin (S. Tsai, unpublished observations). These drugs also inhibited bradykinin-induced increases in release of arachidonic acid and metabolites (Figs. 4 and 5) and prostaglandin production (Table III), but did not inhibit the increase in cAMP produced by isoproterenol (data not shown).

[22] J. R. Vane, *Nature (London), New Biol.* **231,** 232 (1971).

[23] M. Hamberg, J. Svensson, and B. Samuelsson, *Proc. Natl. Acad. Sci. U.S.A.* **71,** 3824 (1974).

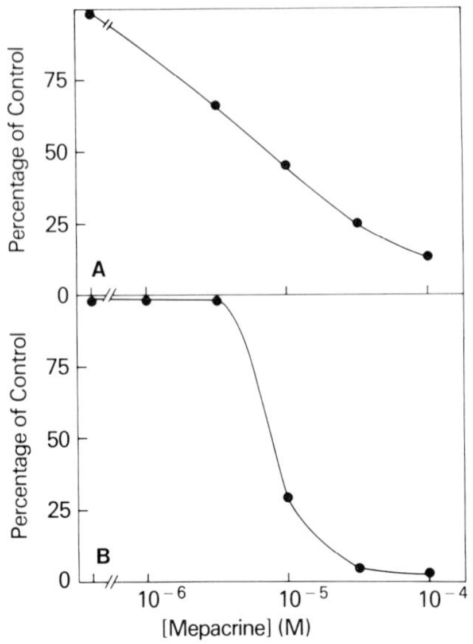

FIG. 4. Mepacrine inhibition of BK-stimulated release of arachidonic acid and metabolites and cAMP accumulation. For release of arachidonic acid and metabolites (A), confluent cells were incubated in Dulbecco's medium containing 4.5 m*M* glucose, 4 m*M* glutamine, 40 n*M* [^{3}H]arachidonic acid, and 0.5% ethanol for 30 min at 37°. Cells were washed twice with 1 ml medium without arachidonic acid and incubated with or without the indicated concentrations of mepacrine for 30 min prior to addition of BK (0.1 μg/ml). After incubation for 3 min, medium was removed to tubes on ice. Tubes were centrifuged (10 min at 27,000 *g*) and samples taken for radioassay or for chromatographic separation of arachidonic acid metabolites. For measurement of cAMP (B), confluent cells were washed and incubated at 37° in HBSS for 30 min with or without drug prior to addition of BK (0.1 μg/ml). After 5 min, medium was removed and cell cAMP content analyzed as described. Data are means of values from 2 or 3 replicate samples with the changes produced by BK in the presence of indomethacin expressed as a percent of control values. In absence of mepacrine, BK increased ^{3}H release by 7200 dpm per dish and increased cAMP content by 1000 pmol/mg protein. This experiment was performed in collaboration with D. Bureis, F. Hirata, and J. Apelrod, NIMH, NIH.

Radiolabeling of Fibroblasts with [^{14}C]Arachidonate

Identification of the Arachidonate Source

To ascertain whether free arachidonate or membrane-associated arachidonate is the primary source of bradykinin-induced prostaglandin synthesis, fibroblasts were labeled for 18 to 24 hr with [^{14}C]arachidonate, then incubated with or without bradykinin in the presence of exogenous free [^{3}H]arachidonate. [^{3}H]Arachidonate added with bradykinin would be

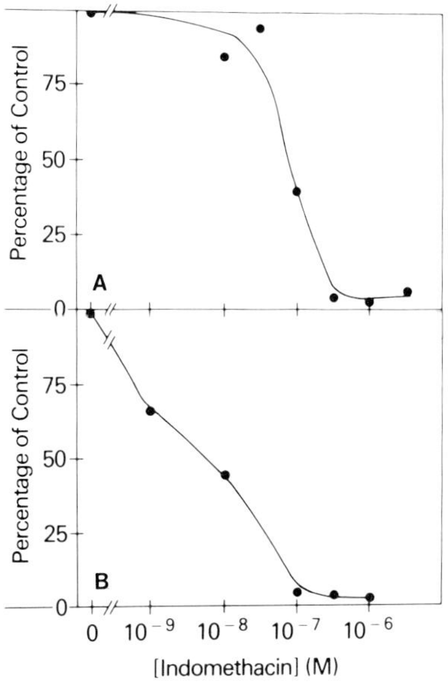

FIG. 5. Indomethacin inhibition of BK-stimulated release of arachidonic acid and metabolites (A) and cAMP accumulation (B). Confluent cells were incubated for 30 min as described in Fig. 4 with or without the indicated concentration of indomethacin prior to addition of BK (0.1 μg/ml) and analyzed for release of arachidonic acid and metabolites and cAMP accumulation. Data are presented as described in Fig. 4. In the absence of indomethacin, BK induced release of [^{3}H]arachidonic acid and metabolites by 6600 dpm/culture dish and increased cAMP content by 1400 pmol/mg protein. This experiment was performed in collaboration with D. Bareis, F. Hirata, and J. Apelrod, NIMH, NIH.

TABLE III
EFFECT OF INDOMETHACIN ON RESPONSES TO BRADYKININ[a]

Additions	cAMP (pmol/mg protein)	PGI_2 (ng/mg/protein)
None	65.2 ± 2.6	5.4 ± 1.4
Bradykinin	348.0 ± 59.7	17.4 ± 3
Indomethacin	24.4 ± 2.8	4.8 ± 1.4
Bradykinin + indomethacin	28.4 ± 8.9	5.2 ± 1.1

[a] Confluent cells were incubated for 5 min in Hanks' medium with or without indomethacin (1 μ*M*) and then for 7 min without or with bradykinin (0.16 μg/ml). Values are expressed in units/mg protein. Data are means ± SEM (n = 3).

expected to be directly available for prostaglandin synthesis; the possibility of rapid, perhaps ligand-induced, acylation of specific membrane lipids and their subsequent deacylation prior to prostaglandin synthesis, however, cannot be eliminated. Following prolonged incubation of the cells with [^{14}C]arachidonate, the radiolabel is associated primarily with membrane lipids, as verified by quantitative extraction of the cellular lipids, separation of the lipid classes and/or individual lipids by thin-layer chromatography, and quantification of the radiolabel associated with each lipid class (see Procedures for details).

Analysis of the ^{14}C and ^{3}H content of the prostaglandins formed in the presence and absence of bradykinin (as determined by separation of the free arachidonate and prostaglandin-associated radiolabel by the C18 column procedure described below) revealed that in human fibroblasts prostaglandins are derived from both prelabeled [^{14}C]arachidonate-containing lipids and free [^{3}H]arachidonate. The prostaglandins specifically formed in response to bradykinin, however, appear to derive primarily from [^{14}C]arachidonate associated with membrane lipids (data not shown).

Optimization of Incubation Conditions for Radiolabeling of Ligand-Sensitive Arachidonate Pools

Only ~5% of the [^{14}C]arachidonate incorporated into cells is usually released, and only a portion of that is released in response to ligands such as bradykinin. Incubation conditions must be designed, therefore, to optimize incorporation of radiolabeled arachidonate into the appropriate membrane lipids. Factors to be considered include the different turnover rates for various lipids, the effects of culture conditions, cell cycle, etc., on phospholipid turnover and the possible ligand selectivity for specific phospholipid-associated arachidonate pools.

In human foreskin fibroblasts, the nutritional state of the cells (fed vs starved), the composition of the incubation media (serum-free vs media containing FCS), and the duration of the radiolabeling period (acute, i.e., 0.5–3 hr vs long-term, i.e., 12–24 hr) all significantly affected both the incorporation of radiolabeled arachidonate into cells and membrane lipids as well as the subsequent release of radiolabel in response to stimuli. For most purposes, uptake of radiolabeled arachidonate serves as an adequate index of arachidonate incorporation into membrane lipids, since the amount of free arachidonate in cells is generally very low. Optimal conditions for incorporation of radiolabel, however, are not the same as those that result in maximal release of radiolabel in response to bradykinin.

Appropriate radiolabeling conditions are, therefore, best achieved by monitoring the ratio of ligand-induced release of radiolabel relative to the amount of radiolabel incorporated into cells.

Maximal effects of bradykinin on both [^{14}C]arachidonate release and [^{14}C]prostaglandin formation were observed with fibroblasts that were refed 2 days prior to radiolabeling. In such cells, optimal release of radiolabel occurred after incubation with [^{14}C]arachidonate in serum-free medium for 12–24 hr, although uptake of radiolabel was essentially complete (>90%) within 30 min (unpublished results). Radiolabeling in serum-free medium for 12–24 hr increased the bradykinin-specific response relative to cells radiolabeled in media containing FCS. The presence of FCS decreased the uptake (~80% maximum) as well as the release of radiolabel. In media containing BSA, only slightly less [^{14}C]arachidonate was incorporated into fibroblasts than with serum-free media (~90%), but basal release of radiolabel was increased, resulting in an apparent decrease in bradykinin-induced release of radiolabel (unpublished results).

Bradykinin-specific release of radiolabel could be amplified by procedures designed to enhance the radiolabeling of the bradykinin-sensitive arachidonate pool, e.g., varying the specific activity of the radiolabeled arachidonate or addition of bradykinin at the initiation of the radiolabeling period. The uptake and/or subsequent availability of arachidonate for release and conversion to prostaglandins was related to the concentration of arachidonate in the medium, with maximal uptake and bradykinin-specific release achieved with 15–60 pmole/ml arachidonate (60–80 mCi/mmol). Brief exposure of fibroblasts to bradykinin at the initiation of the radiolabeling period, while slightly depressing prostaglandin formation, led to an increase in the radiolabel released and a doubling of the bradykinin-specific response (Fig. 6).

Effect of Bradykinin on Arachidonate Release and Prostaglandin Formation

While comparison of the radiolabled arachidonate content of cellular lipids in control and bradykinin-treated cells previously labeled with [^{14}C]arachidonate allows for identification of the bradykinin-sensitive pool of arachidonate, analysis of the radiolabel released upon incubation with bradykinin allows for assessment of the specificity of the bradykinin-induced response, i.e., whether on the release and mobilization of arachidonate and/or prostaglandin synthesis and accumulation. To assess the percentage radiolabel released as prostaglandin vs free arachidonate, incubations must be performed in the absence of BSA or FCS. Both BSA

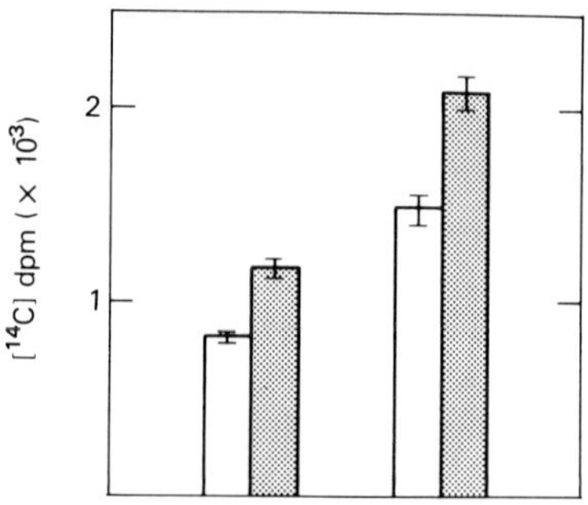

FIG. 6. Effect of BK treatment during the [^{14}C]arachidonate radiolabeling period on the subsequent release of [^{14}C]dpms in the presence or absence of BK. Fibroblasts were radiolabeled 1 hr with [^{14}C]arachidonic acid (16 μM; 60 mCi/mmol) in the presence or absence of BK as previously described using media supplemented with 10% FCS (IN1). Cells were then washed as before and incubated 7 min in the presence or absence of BK (IN2). The incubation media were collected and the total [^{14}C]dpms released determined by counting duplicate 0.1-ml aliquots. Open columns, − bradykinin; hatched columns, + bradykinin.

and FCS bind the released arachidonate, thereby preventing its immediate reutilization in either subsequent reacylation of the membrane lipids[24] or in prostaglandin formation.[25]

After radiolabeling of fibroblasts with [^{14}C]arachidonate, the bulk of the radiolabel released from cells during subsequent incubations is free arachidonate. Bradykinin increases the release of radiolabel from membrane lipids approximately 2-fold; the released radiolabel, however, is now primarily in the form of prostaglandin, not free arachidonate (Fig. 3). Of the prostaglandins formed, the predominant species is PGI_2 (Table I). Thus, in the presence of bradykinin, there appears to be a rather close coupling of phospholipase A_2 activity, arachidonate mobilization, and synthesis of a specific prostaglandin, PGI_2.

The techniques employed in the extraction, separation, and identification of free arachidonate and individual prostaglandins are described in the following section.

[24] P. C. Isakson, A. Raz, S. E. Denny, A. Wyche, and P. Needleman, *Prostaglandins* **14,** 853 (1977).

[25] S. L. Hong and L. Levine, *J. Biol. Chem.* **251,** 5814 (1976).

Procedures for Radiolabeling Cells and the Extraction, Separation, and Identification of Radiolabeled Lipid Components

Radiolabeling of Cells

Sterile Reagents

Eagle's MEM with 25 mM HEPES buffer (GIBCO)
1 M HEPES (GIBCO)
HBSS minus sodium bicarbonate (GIBCO)
Other tissue culture reagents as noted previously

Radiolabeled Compounds

[^{14}C]Arachidonic acid (60–100 mCi/mmol in ethanol) (NEN)
[^{3}H]Arachidonic acid (60–100 Ci/mmol in ethanol) (NEN)

Procedures: Preparation of Reagents. [^{14}C]Arachidonate (in ethanol) is transferred aseptically to a sterile polypropylene tube fitted with a cap through which an 18-gauge needle with 45-μM Swinnex filter has been inserted. The ethanol is then evaporated under a N_2 stream and the radiolabeled arachidonate dissolved in a small volume of sterile HBSS, pH 7.4, and sonicated 5 min in a Branson sonicating water bath. Immediately before radiolabeling the cells, the [^{14}C]arachidonate is added to an appropriate volume of media (Eagle's MEM supplemented with 2 mM glutamine and 25 mM HEPES, pH 7.4, and filtered through a 45-μM Nalgene filter) to yield a final arachidonate concentration of 16 pmol/ml (32 pmol/dish). For dual-labeled experiments, the [^{3}H]arachidonate solution is similarly dried under N_2 and the arachidonate dissolved in HBSS containing 25 mM HEPES, pH 7.4, minus Na_2CO_3 to 5 μCi/ml final concentration.

Conditions for Radiolabeling Cells with [^{14}C]Arachidonate. Following refeeding and washing of the cells as previously described (see Tissue Culture section), cells are incubated with [^{14}C]arachidonate (60–100 mCi/mmol at a concentration of 16 pmol/ml) in 2 ml HEPES-buffered Eagle's MEM supplemented with 2 mM glutamine. After a specified time at 37° in a 5% CO_2 incubator, the media are transferred from the petri dishes to scintillation vials and counted. By comparing the amount of radiolabel initially present in the media with that present after exposure to cells, the percentage of radiolabel taken up by the cells can be determined. A small percentage of the cell-associated counts (<1%) is derived from arachidonate adsorbed to the cells, rather than taken up by the cells. This does not markedly alter the determination of the percentage radiolabeled arachidonate taken up by the cells, which is generally in excess of 70% after 60 min, but could mask the bradykinin-specific release of radiolabel. Therefore, following the removal of the radiolabeling medium, cells are washed three times with 2 ml of HBSS to remove unincorporated, surface-bound [^{14}C]arachidonate.

Incubation of Arachidonate-Labeled Cells with Bradykinin. Cells are incubated with 2 ml of 25 m*M* HEPES-buffered HBSS minus Na_2CO_3, pH 7.4, in the presence or absence of 0.16 μg/ml bradykinin. For dual-labeled experiments designed to identify the specific arachidonate metabolites released in response to bradykinin, [^{3}H]arachidonate (60–100 Ci/mmol) is added to the incubation medium at a final concentration of 16 nmol/ml. The presence of HEPES and omission of Na_2CO_3 stabilize the pH of the incubation medium upon transfer of the cells from the 37°, 5% CO_2 incubator, where the cells were incubated with radiolabel, to the ambient CO_2 concentration of the 37° tissue culture hood, where the cells are incubated with bradykinin.

Cells are incubated with or without 0.16 μg/ml bradykinin on a Belco orbital shaker operated at 2 revolutions per sec. Media are removed and transferred to polypropylene RIA tubes and stored frozen at −20° prior to either RIA or chromatographic analysis of the released radiolabel. Following removal of media, 2 ml 5% TCA is added to the cell monolayer which is frozen by placing the petri dishes on dry ice. The frozen cells are then stored at −20° prior to taking 1-ml samples for cAMP analysis (as noted previously) and extraction of the cellular lipids. The ^{14}C radiolabel did not interfere with the cAMP radioimmunoassay.

Extraction and Separation of Prostaglandins and Free Arachidonate

C18 Column Extraction and Separation. This modification of the separation procedure of Skrinska and Lucas[26] allows for rapid separation of free fatty acids and prostaglandins from both nonlipid contaminants and from each other, with greater than 90% recovery for each lipid class (Table IV).

Materials and Reagents

1 ml disposable Baker octadecyl (C_{18}) columns
Baker-10 extraction apparatus with vacuum manifold
Methanol (Burdick-Jackson)
Distilled water, pH 3, with 2 *M* citrate
10% methanol prepared with pH 3 distilled water
Arachidonate (Serdary Research Laboratories)
PGI_2 (Seragen)
PGE_2 (Seragen)

[26] V. Skrinska and F. Lucas, *Prostaglandins* **22,** 365 (1981).

TABLE IV
SEPARATIONS OF ARACHIDONATE AND METABOLITES ON C_{18} COLUMNS

Eluate	[^{3}H]6-keto-PGF$_{1\alpha}$ (dpm)	[^{14}C]PGE$_2$ (dpm)	[^{3}H]TXB$_2$ (dpm)	[^{14}C]Arachidonate (dpm)
2 ml H_2O (loading volume)	3300	5	14,300	2,100
2 ml H_2O, pH 3	400	5	1,400	2,600
2 ml 10% MeOH	500	0	2,000	300
2 ml benzene	2,500	10,600	4,200	291,000
2 ml ethyl acetate[a]	118,000	184,300	301,600	400
(Applied sample)	(128,400)	(203,600)	(315,600)	(298,000)

[a] 1 ml ethyl acetate: 1 ml ethyl acetate:benzene:MeOH (40 : 60 : 30).

TXB$_2$ (Seragen)
[^{3}H]6-keto-PGF$_{1\alpha}$ (NEN)
[^{14}C]PGE$_2$ (NEN)
[^{3}H]TXB$_2$ (NEN)
[^{14}C]Arachidonate (NEN)

Procedure: Sample Preparation. Samples (1–2 ml) are acidified to pH 3 with 2 *M* citrate (100 μl/ml of sample); carrier PGI$_2$, PGE$_2$, and arachidonate are added to final concentrations of 5 μg/ml each. To prevent clogging the C_{18} columns, precipitated protein was removed by centrifugation (1000 *g*, 10 min).

Column Preparation. Extraction columns are placed in the luer fittings of the Baker-10 extraction manifold. Just prior to loading the samples, columns are washed with 2 ml methanol followed by 2 ml acidified water at a flow rate of about 1 ml/min (15–20 in. Hg). To prevent drying of the sorbent bed, the vacuum is released until the samples are loaded. The columns are then sequentially eluted with 2 ml H_2O, pH 3, 2 ml 10% methanol, 2 ml benzene, 1 ml ethyl acetate, and 1 ml benzene : ethyl acetate : methanol (60 : 40 : 30, v/v/v). In this procedure, the lipids adsorb to the C_{18} column while the nonlipid, water-soluble components are eluted with H_2O and 10% methanol. Free arachidonate is eluted with benzene and the prostaglandins with the ethyl acetate and benzene : ethyl acetate : methanol mixture. The fractions containing prostaglandins are collected either in scintillation vials for counting or in polypropylene tubes for subsequent analysis either by RIA (using an appropriate "blank" composed of the carrier lipids that were added to facilitate the column separation) and/or additional chromatographic procedures. At the same time, aliquots of the eluant were counted following addition of 10 ml Biofluor. Prior to further analysis, the samples are dried under a stream of air at 55° and dissolved in either ethanol or the RIA buffer. Reconstituted samples are mixed well and allowed to sit at room temperature for 30 min before sampling.

Prostaglandin and Fatty Acid Extraction. An alternative to the C_{18} column procedure requires initial extraction of the prostaglandins and free fatty acids by the procedure of Green *et al.*[27] followed by thin-layer chromatographic (TLC) separation. If necessary, excess protein is removed by precipitation with $-20°$ acetone (3 ml/1 ml of sample). Neutral lipids are removed from the protein-free sample by extraction at room temperature with petroleum ether (3 ml/ml of sample); after 30 min, the ether phase is removed and discarded. The sample is further extracted by addition of ethyl acetate : isopropanol : 0.2 *N* HCl (3 : 3 : 1) in a ratio of 3 ml/ml of sample, followed by vortexing twice for 15 sec and addition of 2 ml ethyl acetate and 3 ml H_2O. After mixing, the phases are separated by centrifugation and the organic phase transferred to polypropylene tubes, dried in an air stream at 55°, and reconstituted with either 1 ml RIA buffer for subsequent RIA, 200 μl ethanol for TLC separation, or 2 ml ethyl acetate for HPLC and GC–MS analysis.

Separation of Free Fatty Acids and Prostaglandins by Thin-Layer Chromatography

Materials and Reagents

Silica gel GHL Uniplates, 20 × 20 cm, 20 μM (Analtech)
50 μg/ml mixture of PGI_2, PGE_2, and arachidonate
Isooctane (trimethylpentane) (Kodak)

Procedure. Prostaglandins and free arachidonate are separated on silica gel GHL Uniplates. If carrier prostaglandins and arachidonate are not already present due to previous extraction via the C_{18} column procedure, 5 μg each of PGI_2, PGE_2, and arachidonate is added to samples in the reconstitution step (0.2 ml of 25 μg/ml lipid mixture). After application of sample (50 μl), the TLC plates are developed using the upper phase from an ethyl acetate : H_2O : acetic acid : isooctane mixture (110 : 100 : 20 : 50, v/v/v/v).[28] The mixture must be equilibrated 3 hr prior to removal of the organic phase. The lipids are visualized either by treatment with a cupric acetate spray reagent (Sigma) to specifically stain the prostaglandins or by exposure of the TLC plates to iodine vapors in a chamber to which iodine crystals have been added and heat applied to aid sublimation. After development, the TLC plates are dried for 30 min at room temperature and run

[27] K. Green, M. Hamberg, B. Samuelsson, and J. C. Frolich, *Adv. Prostaglandin Thromboxane Res.* **6,** 193 (1978).

[28] L. S. Wolfe, K. Rostworowski, and M. Manku, *in* "Prostacyclin" (J. R. Vane and S. Bergström, eds.), p. 113. Raven Press, New York, 1979.

a second time in the same development system. The radiolabel associated with the prostaglandins or the free fatty acid was determined by scraping the lipids from the TLC plates and counting the silica gel in 10 ml Biofluor (NEN) using an LKB scintillation counter.

Separation, Identification, and Quantification of Prostaglandins by Chromatographic Assays

Although often limited by the degree of sensitivity and selectivity required, chromatographic analyses allow for simultaneous separation and quantification of prostaglandins. Of the chromatographic assays available, those based on stable-isotope dilution techniques in conjunction with combined gas chromatography–mass spectrometry (GC–MS) offer maximal sensitivity and selectivity.[29] Successful use of the GC–MS, however, especially for subnanogram samples, requires preliminary purification by high-performance liquid chromatography (HPLC)[29] owing to the structural similarity of the prostaglandins and their metabolites and the fact that these lipids represent only a small fraction of the total lipids. The HPLC procedures described below are designed for separation of the prostaglandin methyl esters prior to their quantification by GC–MS. The levels of prostaglandins in most biological samples are so low that they defy detection by conventional HPLC UV and RI detectors[29]; radioactive carriers or tracers of each of the prostaglandins are added to the samples prior to processing. Blanks must, therefore, be employed in the subsequent GC–MS quantification procedures to account for these ^{3}H-labeled prostaglandin references as well as the stable isotope used as internal standards in the GC–MS procedures.

HPLC Separations

Reagents and Equipment

Liquid chromatographic system; two Model 6000 A Solvent Delivery systems, one Model U6K Injector and Model 660 Programmer (Waters and Associates)

4 mm (i.d.) × 30 cm microparticulate silicic acid column (μ-Porasil, Waters and Associates)

4 mm (i.d.) × 30 cm μBondapak C_{18} fatty acid column (Waters and Associates)

Chloroform (Burdick and Jackson, HPLC grade)

Methanol (Burdick and Jackson, HPLC grade)

[29] J. C. Frölich, *in* "The Prostaglandins" (P. W. Ramwell, ed.), Vol. 3, p. 1. Plenum, New York, 1977.

Acetonitrile (Burdick and Jackson, HPLC grade)
Benzene (Burdick and Jackson, HPLC grade)
Acetic acid (Burdick and Jackson, HPLC grade)
Diazomethane (Baker)
Benzene (Baker)
Hydroxylamine (Baker)
Pyridine (Baker)
Trimethylsilyl-imidazole (Pierce)
^{3}H-labeled prostaglandin references (NEN):
- [^{3}H]PGE_2 and PGE_1
- [^{3}H]6-keto-$PGF_{1\alpha}$
- [^{3}H]PGD_2
- [^{3}H]TXB_2
- [^{3}H]$PGF_{2\alpha}$
- [^{3}H]PGA_1
- [^{3}H]PGB_1

Procedures. Complex mixtures of prostaglandin methyl esters were readily separated by sequential use of a reversed-phase HPLC system (employing isocratic elution) and an adsorptive phase HPLC system (employing gradient elution).[30]

Sample Preparation. Prior to HPLC, the prostaglandins are separated from the bulk of the lipids by either the C_{18} column or the fatty acid and prostaglandin extraction procedure outlined above. To the ethyl acetate eluates from the C_{18} columns or the prostaglandin extracts reconstituted in ethyl acetate are added 2×10^4 cpm of each of the ^{3}H-labeled prostaglandin references. The samples together with the radiolabeled standards are then dried under N_2 and derivatized with diazomethane in ether for 5 min at room temperature. The methylated samples are treated with 0.02 ml hydroxylamine in pyridine (25 mg/ml) for 60 min at 70° in 13 × 100-cm Teflon-capped glass tubes to form the methoximes. Samples are dried under N_2 (to evaporate the pyridine), and 10 μl of trimethylsilylimidazole diluted 1 : 10 in benzene are added to convert the hydroxides to trimethylsilyl ethers and the keto groups to tetratrimethylsilyl ether, methyl ester derivatives.[31] The samples are then evaporated under N_2 and reconstituted with 200 μl water : acetonitrile : acetic acid (740 : 260 : 2).

Chromatographic Separation. The prostaglandins, prostacyclins,

[30] W. C. Hubbard, J. T. Watson, and B. J. Sweetman, *in* "Biological/Biomedical Applications of Liquid Chromatography" (G. L. Hawk, ed.), p. 31. Dekker, New York, 1981.

[31] C. N. Henby, M. Jogec, M. G. Elder, and L. Myatt, *Biomed. Mass Spectrom.* **8,** 111 (1981).

thromboxanes, and their metabolites were separated by consecutive reverse and normal phase HPLC. For reverse phase HPLC, the sample is reconstituted in 200 μl of H_2O : acetonitrile : acetic acid (740 : 260 : 2, v/v/v), injected and isocratically eluted from the reverse phase μBondapak C_{18} column using the same solvent mixture at a flow rate of 2 ml/min. Two-milliliter fractions are collected and aliquots (0.1 ml) analyzed by counting and plotting the radioactivity vs the elution volume using the LKB fraction-plot program. After each run, the column is flushed with acetonitrile for 10 min and the fractions within a given peak pooled, concentrated by lyophilization, and subsequently reconstituted in 200 μl chloroform. The samples are then injected and isocratically eluted with chloroform for 10 min on an adsorptive phase, microparticulate silicic acid column followed by nonlinear gradient elution. The gradient is obtained by increasing the concentration of chloroform : methanol (450 : 50) in chloroform from 0 to 100% over a period of 1 hr at a flow rate of 1 ml/min. After each run, the column is flushed 10 min with 10% methanol in chloroform. One-milliliter fractions are collected and the radioactivity associated with each fraction analyzed and plotted as described above. Peak fractions containing radioactivity are pooled and subsequently evaporated under N_2 at 55° for analysis by the selected ion monitoring technique (SIM) employed in conjunction with combined GC–MS.

GC–MS SIM Analysis. Prostaglandin, prostacyclin, and thromboxane metabolites are analyzed as methyl esters/methoximes/trimethylsilyl ethers following derivatization and preliminary purification by HPLC. The limits of detection by SIM analysis depends primarily on the purity of the samples, the radioactive standards employed as carriers, the sensitivity of the detection method employed following HPLC separation, as well as the possible interference from unlabeled material present in the stable isotope analog employed as the internal standard in SIM analyses. Quantification by GC–MS relies on the linearity of standard curves in the observed concentration range of the samples. Details of the GC–MS SIM analysis are described by Green *et al.*[32]

Extraction of Cellular Lipids and Separation of Major Lipid Classes

Materials and Reagents

Phospholipid mix (PL-Biochemicals)
Arachidonate (Serdary Research Laboratories)
Monoglyceride mix (Serdary Research Laboratories)

[32] K. Green, E. Granström, B. Samuelsson, and U. Axén, *Anal. Biochem.* **54**, 434 (1973).

Diglyceride mix (Serdary Research Laboratories)
Triglyceride mix (Serdary Research Laboratories)
PGI_2, PGE_2 (Seragen)
Whatman LHP—K 10 × 10 cm HPTLC silica gel plates (Pierce)
Hexane (Burdick and Jackson)
Methanol (Burdick and Jackson)

Procedure. As previously described, cells are scraped from the dish in 1 ml 5% TCA; the dishes are then washed twice with 0.5 ml 5% TCA and the sample (2.0 ml) then analyzed for cAMP and/or lipid extracted. For lipid extraction, 1 ml of sample is transferred to a glass, Teflon-capped centrifuge tube and 2.5 ml chloroform, 5 ml methanol, and 0.25 ml concentrated hydrochloric acid are added and mixed, followed by addition of 2.5 ml chloroform and 2.5 ml H_2O.[33] After vortexing, the lipids are kept capped at −20° under N_2 until the samples are extracted. With this extraction procedure, greater than 80% of the polyphosphoinositides and greater than 90% of other phospholipids and the neutral lipids are recovered (data not shown). The samples are dried under a stream of N_2 and dissolved in 400 μl chloroform. At this point, the samples can be separated into the lipid classes and the radiolabel associated with each lipid class determined. The phospholipid class of lipids can be further analyzed by two-dimensional thin-layer chromatographic separation either with or without prior separation of the polyphosphoinositides by liquid chromatography.

To separate the lipid classes, 50 μl of sample are applied in duplicate to the HPTLC silica gel plates along with 50 μl of a reference lipid mixture containing 10 mg/ml each of monoglycerides, diglycerides, triglycerides, arachidonate, PGI_2, PGE_2, TXB_2, and phospholipids. The lipids are separated using hexane : diethyl ether : acetic acid (83 : 16 : 1, v/v/v) as the mobile phase in a paper-lined chromatography tank equilibrated 15 min.[34] This procedure separates the prostaglandins, phospholipids, monoglycerides, diglycerides, triglycerides, and free fatty acids. Following development, the plates are dried at room temperature for 15 min, then the lipids visualized by exposure to iodine vapors for 2 min in a chamber containing iodine crystals which has been heated slightly to rapidly sublime the iodine. Lipids are identified by cochromatography with reference lipids and the radiolabel associated with each class determined by scraping the silica gel from the plate into scintillation vials for direct counting with 10 ml Biofluor.

[33] D. Allan and R. H. Michell, *Biochim. Biophys. Acta* **508,** 277 (1978).
[34] H. K. Mangold and D. C. Malins, *J. Am. Oil. Chem. Soc.* **37,** 383 (1960).

Separation, Identification, and Quantification of Phospholipids

Column Chromatography

Reagents and Equipment

Neomycin sulfate (Sigma)
Glycophase G/CPG-200, 200/400 mesh (Pierce)
0.7 × 5 cm glass columns

Procedures. Following lipid extraction as described above, the chloroform phase is evaporated under N_2 at 55° and reconstituted in 200 μl chloroform. If further separation of the phospholipids is required prior to TLC, an equivalent amount of methanol is added and the anionic phospholipids (PA, PS, PI, cardiolipin, 1PA, 1PS, 1PI) and polyphosphoinositides are then separated from the other phospholipids by adsorption to columns of immobilized neomycin (coupled to porous glass beads). The neomycin-glass adduct exhibits high capacity and specific affinity for the polyphosphoinositides, allows for a high flow rate with organic solvents, is stable in the presence of salts, acids, and bases, and can be used repeatedly with little loss in capacity.

Preparation of Columns. Neomycin sulfate is reductively coupled to reactive porous glass beads (Glycophase G/CPG-200, 200/400 mesh) as described by Schacht.[35] The neomycin-coated beads are packed into columns (0.7 × 2 cm), acid-stripped, converted to the required salt form, and equilibrated with the starting solvent by eluting the columns sequentially with 3 column volumes each of chloroform : methanol : 3 *N* HCl (5 : 10 : 2), chloroform : methanol : H_2O (5 : 10 : 2), 0.5 μM ammonium formate in chloroform : methanol : H_2O (5 : 10 : 2), chloroform : methanol : H_2O (5 : 10 : 2), and, finally, chloroform : methanol (1 : 1).[35]

Separation of the Phospholipids. Concentrated lipid extract is applied to the column in 2 to 3 column volumes of chloroform : methanol (1 : 1) followed by 3 column volumes of the same solvent and 3 column volumes of chloroform : methanol (1 : 2). The combined eluates containing zwitterionic and nonanionic phospholipid and their lyso derivatives are dried under N_2 at 55° before TLC. Weakly acidic phospholipids and their lyso derivatives are then eluted with 6 column volumes each of chloroform : methanol : 88% formic acid (10 : 20 : 1) and chloroform : methanol : 88% formic acid (5 : 10 : 1) followed by 2 column volumes of chloroform : methanol : H_2O (5 : 10 : 2) and, finally, 6 column volumes of 100 m*M* ammonium formate. After drying under N_2, phospholipids are

[35] J. Schacht, *J. Lipid Res.* **19,** 1063 (1978).

dissolved in 200 μl chloroform. Polyphosphoinositides are eluted in turn by increasing the salt concentration; 6 column volumes of 400 m*M* ammonium formate elute the diphosphoinositides and an equivalent volume of 1 *M* ammonium formate elutes the triphosphoinositides.[35] The polyphosphoinositide eluates are then reextracted using the lipid extraction procedure described above with the salt solution being the aqueous phase; the lower chloroform phase is then removed and the lipid dried under N_2 at 55° and dissolved in 200 μl chloroform prior to TLC.

Thin-Layer Chromatography

Reagents and Equipment

Methanol (Burdick and Jackson)
HPTLC silica gel 60 TLC plates, 10 × 20 cm (Pierce)
Reference lipid mixture containing 5 mg/ml of each phospholipid (obtained from either Serdary Research Laboratories or Sigma) in chloroform:
- Phosphatidylcholine
- Sphingomyelin
- Phosphatidylethanolamine
- Phosphatidylserine
- Phosphatidylglycerol
- Phosphatidylinositol
- Diphosphatidylglycerol (cardiolipin)
- Phosphatidic acid
- bis-Phosphatidic acid
- Diphosphatidylinositol
- Triphosphatidylinositol
- Lysophosphatidylcholine
- Lysophosphatidylethanolamine
- *N*-Monomethylphosphatidylethanolamine
- *N,N*-Dimethylphosphatidylethanolamine
- Lysophosphatidylserine
- Lysophosphatidylglycerol
- Lysophosphatidic acid
- Lysophosphatidylinositol

Procedures. Following lipid extraction and/or preliminary separation of the phospholipids on neomycin-coated glass beads, solvent evaporation, and reconstitution of the eluant or the chloroform extracts in 400 μl chloroform, duplicate 50-μl samples of the chloroform are counted directly by addition of 10 ml Biofluor after evaporation of the chloroform.

Duplicate 50-μl samples of the chloroform along with 50 μl of the reference lipid mixture are then applied with a 100-μl Hamilton syringe to the lower righthand corner of HPTLC plates approximately 1.5 cm from the bottom and 1.5 cm in from the side. The extract and reference spots are dried during application by an air stream directed to the application spot. After an additional 20–30 min at room temperature, the HPTLC plates are developed in a two-dimensional system comprised of $CHCl_3$: methanol : acetic acid : H_2O (75 : 45 : 12 : 3, v/v/v/v) in the first dimension.[36] Approximately 100 ml of the solvent system are added to the paper-lined chromatography tanks and the system is equilibrated 15 min at room temperature prior to introduction of the HPTLC plates. The plates are removed when the solvent is approximately 3 cm from the top of the plate. After 30 min at room temperature in a desiccator, the plate is rotated 90° clockwise and developed in the second dimension in a paper-lined tank equilibrated 15 min with 100 ml of the second solvent, $CHCl_3$: methanol : ammonia : H_2O (90 : 30 : 0.5 : 4, v/v/v/v). When the solvent front is ~3 cm from the edge, the plates are removed and dried 30 min in a desiccator at room temperature and exposed to iodine vapor to locate lipids. Lipid spots are scaped into vials for radioassay after addition of 10 ml of Biofluor.

Specific activities are obtained by analysis of either the phospholipid phosphate or protein content of the samples. For measurement of phospholipid phosphate, duplicate 50-μl samples are taken of either the total lipid extract or individual phospholipids that have been separated as described above and recovered by scraping the individual lipids from the TLC plates. The phospholipids are hydrolyzed and the phospholipid phosphorus content analyzed by the procedure of Rouser *et al.*[37] which involves treatment of the phospholipid extract or silica gel containing the phospholipid with 0.5 ml 70% perchloric acid at 230° for 20 min followed by analysis of the released inorganic phosphate. The silica gel is removed by low-speed centrifugation prior to analysis of the inorganic phosphate. For analysis of protein, the methanol–H_2O phase remaining after lipid extraction of the cells is evaporated and protein solubilized in 2 *N* NaOH. Duplicate aliquots are then assayed for protein using the procedure of Lowry *et al.*[38] and corrected using a linear transformation program.[39]

[36] M. M. Billah and E. G. Lapetina, *J. Biol. Chem.* **257,** 5196 (1982).

[37] G. G. Rouser, G. Kritchevsky, A. Yamamoto, G. Simon, C. Galli, and A. J. Bauman, this series, Vol. 14, p. 272.

[38] O. H. Lowry, N. J. Rosebrough, A. L. Farr, and R. J. Randall, *J. Biol. Chem.* **193,** 265 (1951).

[39] W. T. Coakley and C. J. James, *Anal. Biochem.* **85,** 90 (1978).

[41] Direct Chemical Measurement of Receptor-Mediated Changes in Phosphatidylinositol Levels in Isolated Rat Liver Plasma Membranes

By JOSEPH EICHBERG and CHARLES A. HARRINGTON

In the past few years, the mechanism and significance of the receptor-mediated stimulation by certain hormones and neurotransmitters of phosphatidylinositol metabolism, a phenomenon first described by Hokin and Hokin,[1] has become of interest to a large number of investigators. The initial step in this process in a wide variety of tissues has been generally regarded for some time as an accelerated breakdown of phosphatidylinositol by phosphodiesteratic cleavage to yield diacylglycerol and water-soluble inositol phosphates.[2,3] Quite recently, an even more rapid receptor-triggered phosphodiesteratic degradation of phosphatidylinositol 4-phosphate and phosphatidylinositol 4,5-bisphosphate has also been shown to occur by an analogous pathway in several cellular systems.[4–7] Previous efforts to demonstrate these events in cell-free preparations have been unsuccessful. However, conditions have been defined in our laboratory under which phosphatidylinositol disappearance dependent on an α_1-adrenergic receptor mediated mechanism occurs in purified rat liver plasma membranes supplemented with cytosol.[8]

Stimulated phosphatidylinositol metabolism was originally detected and its magnitude estimated by measurement of ^{32}P incorporation into the phospholipid. Since enhanced labeling of phosphatidylinositol is now known to be a secondary event in the response, it is preferable to assess the status of this compound more directly. This is usually accomplished either by following changes in radioactivity of prelabeled phospholipid or by determining the mass of phosphatidylinositol present by chemical means. In the former approach, the phospholipid is initially labeled with ^{32}P, [^{3}H]inositol, or [^{3}H]- or [^{14}C]arachidonic acid and the extent of either

[1] M. R. Hokin and L. E. Hokin, *J. Biol. Chem.* **203,** 967 (1953).
[2] R. H. Michell, *Biochim. Biophys. Acta* **415,** 81 (1975).
[3] M. J. Berridge, *Mol. Cell. Endocrinol.* **24,** 115 (1981).
[4] B. W. Agranoff, P. Murthy, and E. B. Seguin, *J. Biol. Chem.* **258,** 2076 (1983).
[5] M. J. Berridge, R. M. C. Dawson, C. P. Downes, J. R. Heslop, and R. F. Irvine, *Biochem. J.* **212,** 473 (1983).
[6] M. J. Berridge, *Biochem. J.* **212,** 849 (1983).
[7] J. A. Creba, C. R. Downes, P. T. Hawkins, G. Brewster, R. H. Michell, and C. J. Kirk, *Biochem. J.* **212,** 733 (1983).
[8] C. A. Harrington and J. Eichberg, *J. Biol. Chem.* **258,** 2087 (1983).

ISBN 0-12-182009-2

the disappearance of radioactivity from phosphatidylinositol or the appearance of labeled diacylglycerol or water-soluble inositol phosphates is determined. While this method is capable of detecting quite small changes, there is the possibility that only a limited portion of the tissue phospholipid pool is accessible to isotopic precursor and may not be identical with that fraction which undergoes metabolic stimulation in response to hormones or neurotransmitters. In addition, unless it is shown that phosphatidylinositol has been labeled to equilibrium with the appropriate immediate precursor, prior to exposure of the system to the stimulus, changes in the radioactivity cannot be taken as quantitative indicators of alterations in the amount of phosphatidylinositol. Since the required measurement of precursor specific activity as well as that of the product may be difficult or inconvenient, direct analysis of phosphatidylinositol levels constitutes a potentially attractive alternative.

The usual methods of phospholipid determination involve extraction of tissue lipids, chromatographic separation and recovery of individual phospholipids, and analysis of phosphorus by one of several procedures. This approach, although satisfactory in many instances, is limited by the completeness of the separation procedure, losses during elution, and the sensitivity and precision of the analytical method. In this article a procedure is described for the determination of changes in phosphatidylinositol concentration in plasma membrane by a combination of high-performance thin-layer chromatography followed by quantitative densitometry based either on lipid P content or by formation of a fluorescent derivative dependent on the presence of polyunsaturated fatty acyl residues.

Preparation of Hepatocyte Plasma Membranes

Reagents

0.9% NaCl
1 mM $NaHCO_3$
Surgical gauze
69.0 ± 0.5% sucrose
42.3 ± 0.1% sucrose
HEPES [4-(2-hydroxyethyl)-1-piperazine ethanesulfonic acid]
Incubation medium: 50 mM HEPES buffer, pH 7.1–10 mM $MgCl_2$–1 mM $CaCl_2$

Procedure. The method is based on those of Neville[9] and Pohl *et al.*[10] A male Sprague–Dawley rat (200–300 g) is anesthetized with diethyl ether

[9] D. M. Neville, *Biochim. Biophys. Acta* **154,** 540 (1968).
[10] S. L. Pohl, L. Birnbaumer, and M. J. Rodbell, *J. Biol. Chem.* **246,** 1949 (1971).

and is adequately ventilated to minimize tissue anoxia. The abdomen is opened and the portal vein is cannulated using a 20-gauge needle. The vena cava is then severed and the liver perfused copiously with ice-cold 0.9% NaCl until blanched. The liver is removed, trimmed free of connective tissue, and any nonblanched portions discarded. All subsequent operations are performed at 0–4°. A 10 g portion of tissue is minced and transferred to a 30-ml Dounce homogenizer (Blaessig Glass Co., Rochester, NY) fitted with a loose pestle which has been previously silylated[11] and homogenized in 25 ml of 1 m*M* $NaHCO_3$ using 8 vigorous strokes of the pestle. The suspension is poured into 425 ml of 1 m*M* $NaHCO_3$. If desired, up to 30 g of liver may be homogenized and diluted into this volume of 1 m*M* $NaHCO_3$. The mixture is stirred for 3 min and then passed first through two layers of surgical gauze and then refiltered through four layers of gauze. The filtered homogenate is spun at 150 $g(R_{max})$ for 30 min, the supernatant fluid removed by gentle aspiration, and the loose pellet transferred to the Dounce homogenizer. The pellet is resuspended with three very gentle strokes of the loose pestle and the suspension added to 62 ml of 69.0% sucrose. The mixture is diluted to approximately 110 ml with sufficient water and 69% sucrose to yield a final sucrose concentration of 44.0 ± 0.1% as determined using a refractometer with the stage maintained at 20°. Once adjusted to this concentration, the suspension is divided equally among 6 tubes and carefully overlain with sufficient 42.3% sucrose to within 3 mm of the top of each tube. Centrifugation is performed at 25,000 rpm in an SW 25.1 rotor (g_{max} = 90,300) for 150 min with the brake off. Equivalent settings may be employed if another rotor is used. Partially purified plasma membranes are located in the whitish band which migrates to the air–sucrose interface and which is cleanly separated from denser material concentrated at the interface between the two sucrose layers as well as in the pellet at the bottom of the tube. The membranes are collected using either a procelain spoonula or a syringe with the needle bent at a 90° angle, transferred to a 15-ml Corex centrifuge tube, and diluted to 10 ml with HEPES incubation medium. This suspension is centrifuged at 10,000 g_{max} for 30 min, the supernatant fluid decanted, and the pellet resuspended in sufficient HEPES incubation medium to give a final protein concentration of 1 mg/ml. If desired, the suspension is rapidly frozen in small portions in liquid nitrogen and stored at −80° until used.

[11] D. C. Fenimore, C. M. Davis, J. H. Whitford, and C. A. Harrington, *Anal. Chem.* **48,** 2289 (1976).

Preparation of Cytosol

Procedure. The liver of a male Sprague–Dawley rat is thoroughly perfused as described previously. The liver is minced and then homogenized using an ultraturax (Tekmar, Inc. Cincinnati, OH) in three volumes of HEPES incubation medium. The homogenate is centrifuged at 5000 g_{max} for 20 min, the supernatant poured off, and centrifuged again at 90,300 g_{max} for 120 min with no brake set. The clear infranatant beneath the floating fat layer is collected as completely as possible using a syringe fitted with a 20-gauge needle to which a 15 cm length of silastic tubing is attached. Cytosol so prepared may be frozen rapidly in small portions at −80°.

Incubation of Plasma Membranes

Plasma membranes (0.25 mg protein) and cytosol (1.0 mg protein) are mixed in a final volume of 0.25 ml HEPES incubation medium in the presence of norepinephrine or other additions as desired. Controls in which cytosol is omitted are always included. The reaction mixture is prepared at 0° and then transferred to a 37° water bath shaking at 80–90 oscillations per minute to start the incubation.

Extraction of Phospholipids and Preparation of Washed Lipid Extract

Reagents

$CHCl_3 : CH_3OH$ (2 : 1, v/v) containing 0.1% 12 *N* HCl
$CHCl_3 : CH_3OH : H_2O$ (75 : 25 : 2, v/v)

Procedure. Incubations are terminated by addition of 5.0 ml $CHCl_3 : CH_3OH$ (2 : 1, v/v) containing 0.1% 12 *N* HCl. The tubes are capped and shaken for 2 hr at room temperature and 1.0 ml H_2O is then added. The phases are thoroughly mixed and if necessary separated by centrifugation for a few minutes at 1000 *g*. The lower phase is removed with a silylated Pasteur pipet. To the upper phase is added 1.0 ml lower phase prepared by equilibrating solvents and water in the proportions given above and the phases mixed and separated by centrifugation. This backwash is combined with the lower phase, the mixture dried under a stream of nitrogen at 40°, and the residue redissolved in $CHCl_3 : CH_3OH : H_2O$ (75 : 25 : 2, v/v) in preparation for spotting on high-performance thin-layer plates. If capillary spotting is to be employed, 0.025 ml of solvent is added whereas if samples are to be applied by contact spotting, 0.25 ml of solvent which contains 0.5–1.0% of either *n*-octanol or *n*-dodecane (see below) is used.

Application of Samples to HPTLC Plates

Capillary Spotting. In this procedure, a Camag variable volume capillary pipet (Applied Analytical Industries, Wilmington, NC) utilizing an Evachrome like holder is utilized.

The capillary spotting pipet delivers up to 4 μl to the surface of a high-performance thin-layer chromatography plate and can be calibrated to deliver variable volumes. There are several disadvantages of this method of sample application. First, samples must necessarily be quite concentrated, and consequently viscosity of the solution may be quite high making sample application tedious and time consuming. Second, samples are spotted while dissolved in a medium in which phospholipids have reasonable chromatographic mobility and, as a result, diffusion of the applied samples leads to enlarged spots. Restriction of spot size to a maximum of 1 mm is essential to achieve the greatest resolution of the phospholipids applied to the high-performance thin-layer chromatography plate as well as the highest sensitivity in detecting phospholipids after chromatographic separation.

Contact Spotting. This technique is carried out using a contact spotter (Clarke Analytical Systems, Sierra Madre, CA) and takes advantage of the unique nonwetting properties of Teflon-like surfaces which extend even to organic solvents.[12,13] A pliable fluorinated hydrocarbon film strip which is coated with a perfluoroalkane fluid is deformed by a vacuum into a set of surface depressions located over moveable pistons on a metal surface. Into each of these depressions is placed 0.025 ml of chloroform : methanol : water (75 : 25 : 2, v/v) containing liver lipid and 0.5–1.0% of a high boiling solvent such as octanol or dodecane. The metal surface is maintained at 37° in an atmosphere of gently flowing nitrogen until the volatile solvents are evaporated. A 10 × 10-cm high-performance thin-layer chromatography plate is then placed adsorbent side down against the strip and the lipid is transferred quantitatively to the plate by switching from vacuum to positive nitrogen pressure. This essentially solventless sample application process is illustrated in Fig. 1 and is highly reproducible. For example, the spotting of 12 identical samples of [^{32}P]phosphatidylinositol yielded a relative standard deviation of less than 2.0%.

The choice of nonvolatile solvent can be important. The minute residual volume of solvent produces a small (less than 1 mm) spot on the high-performance thin-layer chromatography plate. However, the area occu-

[12] D. C. Fenimore and C. J. Meyer, *J. Chromatogr.* **186,** 555 (1979).

[13] D. C. Fenimore and C. M. Davis, *Anal. Chem.* **53,** 252A (1981).

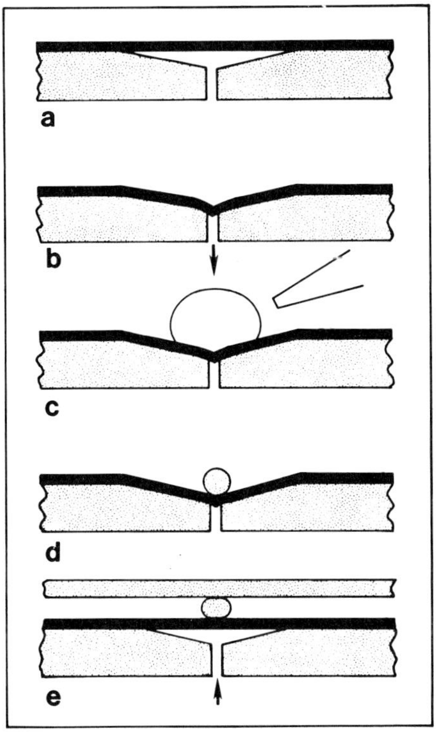

FIG. 1. Sample application by contact spotting. A specially treated fluoropolymer film is pulled into a series of depressions in a metal plate by application of vacuum (a and b). Sample solution is delivered by pipet (c), and after evaporation a residue remains (d), which is transferred to the high-performance thin-layer chromatography plate by replacing the vacuum with slight pressure (e).

pied by the transferred lipid is even smaller than this because the diffusion constants of phospholipids in these solvents are minimal. In practice, the inclusion of dodecane provides greater sensitivity of detection (about 2-fold) than if octanol is used. This is most likely due to the more polar nature of the solvent system when octanol is present which causes somewhat more spot diffusion when the moving solvent front encounters the dried sample.

A comparison of the sensitivity achieved in measurements of quantities of authentic phosphatidylinositol applied to the plate by the various methods described is shown in Fig. 2. Using the optimum spotting conditions, as little as 30 pmol of phospholipid could be reliably determined.

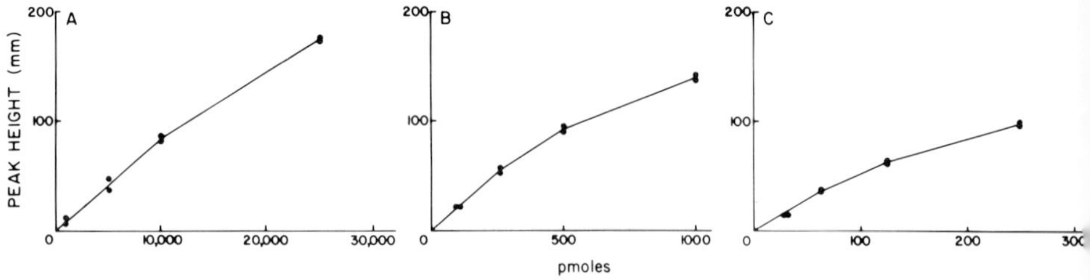

FIG. 2. Standard curves obtained with authentic phosphatidylinositol applied to high-performance thin-layer chromatography plates by different spotting procedures as described in the text. (A) Capillary spotting; (B) contact spotting using solvent containing octanol; (C) contact spotting using solvent containing dodecane. After development, plates were sprayed with molybdate-containing reagent and lipid P in each spot was determined densitometrically.

Development of the Chromatogram

Using a moveable guide, up to 15 samples containing 1–2 nmol lipid P can be spotted along one side of a 10 × 10-Silica Gel 60 high-performance thin-layer chromatography plate. In most cases, this number of samples is applied along each side of the plate.[14] Duplicate or triplicate aliquots of several different concentrations of phosphatidylinositol standard are always included. Purified beef liver phosphatidylinositol (Avanti Biochemicals, Birmingham, AL) is generally quite satisfactory. The plate is developed using a Camag linear chamber equipped with a saturation pad. Samples migrate toward the center of the plate, permitting the simultaneous development over a distance of 5 cm of up to 30 samples with 15 min. The solvent system employed is chloroform : methanol : 40% aqueous methylamine (63 : 35 : 10, v/v). Densitometric tracings of representative chromatographic separations obtained for phospholipids extracted from whole rat liver and from partially purified rat liver plasma membranes are shown in Fig. 3.

Visualization of Phosphatidylinositol

Molybdate-Containing Reagent. For determination of phosphatidylinositol based on P content, the spray reagent containing ammonium pentachlorooxymolybdate described by Kundu *et al.*[15] is used. To prepare this compound, 10 g of ammonium paramolybdate is dissolved in 75 ml of concentrated HCl. The mixture is boiled until all of the salt is dissolved and 10 ml of concentrated HI is then added to the boiling mixture with

[14] C. A. Harrington, D. C. Fenimore, and J. Eichberg, *Anal. Biochem.* **106,** 307 (1980).
[15] S. L. Kundu, S. Chakravarty, N. Bhaduri, and H. K. Saha, *J. Lipid Res.* **18,** 128 (1977).

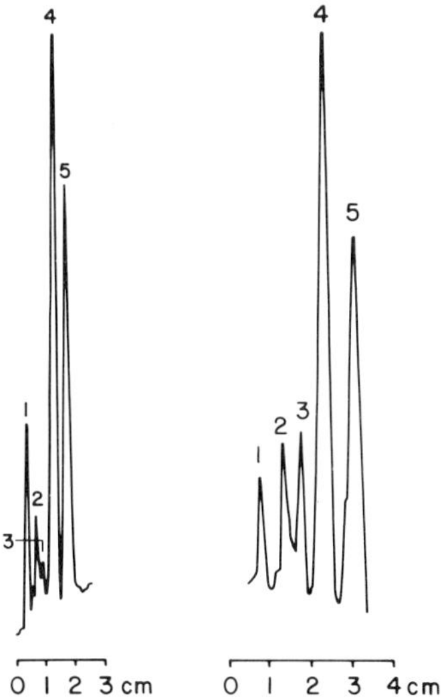

FIG. 3. Separation of phospholipids by high-performance thin-layer chromatography. Phospholipids were extracted and applied with a contact spotter using solvent containing dodecane. Densitometric tracings were obtained after visualization of lipid P using molybdate-containing reagent. Left: whole rat liver phospholipids; right: partially purified rat liver plasma membrane phospholipids.

constant stirring. After iodine vapor is no longer evolved, the solution is boiled until it becomes a pasty amorphous mass. The residue is then triturated with 75 ml of concentrated HCl and the liquid boiled off a second time. The residue is dissolved in 45 ml concentrated HCl and the solution saturated with HCl gas using a lecture bottle. The green crystals of ammonium pentachlorooxymolybdate which separate are collected by rapid filtration or centrifugation and dried over KOH *in vacuo*.

The spray reagent is prepared by dissolving 750 mg of this product in 25 ml 8–9 *N* H_2SO_4 and removing undissolved material by filtration. The solution should be stored in a container with a loose fitting top and aged for 10 to 14 days to assure maximum color development with minimal background interference. The spray reagent is stable for at least 1 year. Ammonium pentachlorooxymolybdate is stable indefinitely when stored dry and desiccated. Plates are lightly and evenly sprayed with an air-driven atomizer which produces a fine mist. Compressed gas sprayers are not suitable because they generate coarse droplets. The sprayer should be operated at 30 to 40 cm from the plate surface, a distance at which minimal movement of the sprayer itself is necessary to achieve uniform deposition of the reagent. Immediately after spraying, the plate is light yellow.

Within 10 to 20 sec, blue phospholipid spots appear and after 30 min, the background color has faded to nearly white and the plate is ready to be scanned. The intensity of color is stable for 3–4 hr after spraying.

Formation of Fluorescent Derivative of Phospholipids. The fluorogenic reagent is prepared immediately before use by combining 250 ml acetone, 6 ml concentrated HCl, and 10 ml 70–72% perchlorid acid. To induce fluorescence, the plate is dried at 40° for 15–20 min, cooled to room temperature, and dipped in the fluorogenic reagent for 1–2 min or until it becomes translucent. It is then air-dried and heated at 120° for 30–45 min. The lipid spots appear as slightly brownish-red areas. If desired, the plate may then be dipped in 10% paraffin oil in toluene until translucent and then air-dried. This seems to diminish background fluorescence and stabilizes the fluorescent spots.

Quantification of Phosphatidylinositol

Phosphatidylinositol is measured using a Shimadzu CS-910 scanning densitometer in the reflectance mode. A Camag thin-layer chromatogram scanner or equivalent instrument is equally suitable. For detection following the molybdate spray, a tungsten lamp is used at a wavelength of 600 nm. In the case of the fluorescent derivative, excitation is accomplished with a mercury lamp at 360 nm and the emitted fluorescence is determined through a 500 nm barrier filter. For each sample, peak heights are measured and the quantity of phospholipid determined from the calibration curve derived from the standards run on each plate.

Comments

1. The liver is perfused prior to fractionation primarily to remove an unidentified factor in serum which has been reported to inhibit phospholipase C activity against phosphatidylinositol.[16] In experiments in which this step was omitted, the reduction in phosphatidylinositol levels in response to hormone was not observed.

2. The isolation procedure for plasma membranes is modified from a widely used method mainly in that the final membrane pellet is suspended in HEPES containing medium. The purification of membranes achieved relative to the starting homogenate may be estimated by means of appropriate enzyme markers including adenylate cyclase, 5′-nucleotidase or Na,K-stimulated ATPase. Membranes still demonstrated responsiveness to norepinephrine in terms of phosphatidylinositol disappearance after

[16] R. M. C. Dawson, N. Hemington, and R. F. Irvine, *Eur. J. Biochem.* **112,** 33 (1980).

storage at −80° for several weeks. However, most preparations were utilized within a few days after being made.

3. The preparation of cytosol in hypotonic medium is essential since it minimizes the amount of residual lipid, probably associated with nonsedimented microsomal fragments, which is recovered in this fraction. Under the conditions employed, cytosol contained about 200 pmol phosphatidylinositol P/mg protein.

4. Overall, while sample spotting on thin-layer plates via the capillary pipet method requires the simplest apparatus, contact spotting is preferred because of the sensitivity attainable due to smaller spot size. In addition, the precision of contact spotting is such that replicate phospholipid samples generally agree within 2% upon densitometric scanning after visualization.

5. The nature of the reaction which gives rise to fluorescent derivatives of phospholipids is unknown, but clearly depends upon the presence of polyunsaturated fatty acyl groups since molecules which contain saturated and monounsaturated acyl moieties develop negligible fluorescence. It is possible that a component of the binder on the high-performance thin-layer chromatographic plates used may play a role in the fluorogenic reaction.

6. The reflectance–absorbance densitometry scans of phospholipid spots are characteristically nonlinear with concentration and for this reason peak heights rather than total area under the curves are measured. Since calibration standards are included on each plate and are subjected to identical chromatographic conditions as are the samples, procedural errors are minimized.

Acknowledgments

This work was supported in part by NIH Grant NS-12493 and a grant from the Robert A. Welch Foundation (E-675). We thank Dr. Chester M. Davis for valuable suggestions.

[42] Assay for Calcium Channels

By HARTMUT GLOSSMANN and DAVID R. FERRY

Introduction

This chapter will focus on biochemical assays for Ca^{2+}-selective channels in electrically excitable membranes which are blocked in electrophysiological and pharmacological experiments by verapamil, 1,4-dihydro-

ISBN 0-12-182009-2

pyridines, diltiazem (and various other drugs), as well as inorganic di- or trivalent cations.[1-4] The strategy employed is to use radiolabeled 1,4-dihydropyridine derivatives which block calcium channels with ED_{50} values (concentrations yielding 50% effect) in the nanomolar range. Although tritiated *d-cis*-diltiazem and verapamil can be used to label calcium channels, the 1,4-dihydropyridines offer the following advantages. The affinity for their drug receptor site within the calcium channel can be very high. Their interaction with other membrane components, including sodium channels or neurotransmitter receptors, is negligible, which makes the choice of tissue preparation and purity of the respective membrane fraction less critical. Finally a spectrum of compounds exists which regulate calcium channel function from blockade to opening.[5] The various sections cover tissue specificity of channel labeling, the complex interactions of divalent cations with the [^{3}H]nimodipine-labeled calcium channels, and the allosteric regulation of [^{3}H]nimodipine binding by the optically pure enantiomers of phenylalkylamine and benzothiazepine calcium channel blockers. A comparison of the properties of different tritiated 1,4-dihydropyridine radioligands and the iodinated channel probe [^{125}I]iodipine is given. It will be illustrated that the same methodology with minor modifications can be applied to label calcium channels with other radioligands (e.g., [^{3}H]*d-cis*-diltiazem, [^{3}H]verapamil). The autoradiographic localization of the 1,4-dihydropyridine binding sites in brain is presented and finally the procedures to solubilize and partially purify the skeletal muscle calcium channel are given.

Reagents and Chemicals

Buffers

Buffer for homogenization of tissues (buffer A): 20 m*M* $NaHCO_3$, 0.1 m*M* phenylmethylsulfonyl fluoride (PMSF).

Buffer for washing of tissue pellets and for incubation (buffer B): 50 m*M* Tris–HCl, 0.1 m*M* PMSF, pH 7.4 at 37°.

Buffer for solubilization (buffer C): 500 m*M* NaCl, 50 m*M* Tris–HCl, 0.1 m*M* PMSF with 10 m*M* 3-(3-cholamidopropyl)dimethylammonio-1-propanesulfonate (CHAPS), pH 7.4.

Buffer for sucrose density gradient centrifugation (buffer D): 500 m*M*

[1] A. Fleckenstein, *Annu. Rev. Pharmacol. Toxicol.* **17,** 149 (1977).

[2] E. Stefani and D. J. Chiriandini, *Annu. Rev. Physiol.* **44,** 352 (1982).

[3] S. Hagiwara and L. Byerly, *Annu. Rev. Neurosci.* **4,** 69 (1981).

[4] P. G. Kostyuk, *in* "Membrane Transport of Calcium" (E. Carafoli, ed.), p. 1. Academic Press, New York, 1982.

[5] M. Schramm, G. Thomas, R. Towart, and G. Franckowiak, *Nature* (*London*) **303,** 535 (1983).

NaCl, 5 m*M* $CaCl_2$, 50 m*M* Tris–HCl, pH 7.4 at 4°, 5 m*M* CHAPS in 5 and 20% (w/w) sucrose.

Buffer for filtration (buffer E): 10% (w/v) polyethyleneglycol 6000 in 10 m*M* $MgCl_2$, 20 m*M* Tris–HCl, pH 7.4 (ice-cold). Note: there are batch variations of the quality of polyethyleneglycol; we suggest that several batches be tested and then a large stock of an optimal batch stored.

Preparation of Chelex-100 treated buffer (buffer F): 500 ml 50 m*M* Tris–HCl (pH 7.4 at 37°) is pumped through 8 ml of Chelex-100 ion exchange resin (Bio-Rad) in the sodium form in a plastic column at a flow rate of 1.0 ml/min at 4°. The divalent cation depleted buffer is collected and stored in plastic.

Buffer for autoradiographic localization studies (buffer G): isoosmotic (0.170 *M*) Tris–HCl (pH 7.6 to 7.7 at 23°).

Buffer for washing of labeled tissue sections (buffer H): buffer G, supplemented with 0.25 mg/ml bovine serum albumin (essentially fatty acid free).

Radiochemicals

The following [^{3}H]1,4-dihydropyridines of ≥95% radiochemical purity are synthesized by New England Nuclear by ester exchange: [^{3}H]nimodipine (150–160 Ci/mmol), [^{3}H]nitrendipine (72–85 Ci/mmol), and [^{3}H]nifedipine (88 Ci/mmol). Ring-labeled racemic [^{3}H]PN 200 110 [isopropyl 4-(2,1,3-benzoxadiazol-4-yl)-1,4-dihydro-2,6-dimethyl-5-methoxycarbonyl-pyridine-3-carboxylate] (2.6–3.3 Ci/mmol) is synthesized by Sandoz Ltd, Basel, Switzerland and with a specific activity of 23.4 Ci/mmol in our laboratory. The optically pure (+)-[*methyl*-^{3}H]PN 200-110 (75 Ci/mmol) and the arylazide photoaffinity ligand [^{3}H]azidopine (40–50 Ci/mmol) are synthesized by Amersham. Radioligands are stored in ethanol at 100 μCi/ml for up to 3 months at −25°, and remain stable. In the presence of a large receptor excess the bindability fraction of the chiral [^{3}H]1,4-dihydopyridines remains in the range of 40–50% over 3 months. All 1,4-dihydropyridines are handled in the absence of UV light. Sodium lighting or dimmed, diffuse daylight is suitable. *d-cis* [*O-methyl*-^{3}H]Diltiazem (85 Ci/mmol) and the optically pure (−)-[^{3}H]desmethoxyverapamil (83 Ci/mmol) are synthesized by Amersham, [^{3}H]verapamil (83.9 Ci/mmol) by New England Nuclear. [^{125}I]Iodipine (~2200 Ci/mmol) [1,4-dihydro-2,6-dimethyl-4-(2-trifluoromethylphenyl)-3,5,pyridinecarboxylic acid 2-((1-(3-^{125}iodo-4-hydroxyphenyl)-3-oxopropyl)amino)ethylester] is synthesized as described[6] and can be used for up to 5 weeks after synthesis when stored in ethanol at −25°. The structures (with the exception of azidopine and (−)desmethoxyverapamil) are shown in Fig. 1.

[6] D. R. Ferry and H. Glossmann, *Naunyn-Schmiedeberg's Arch. Pharmacol.* **325,** 186 (1984).

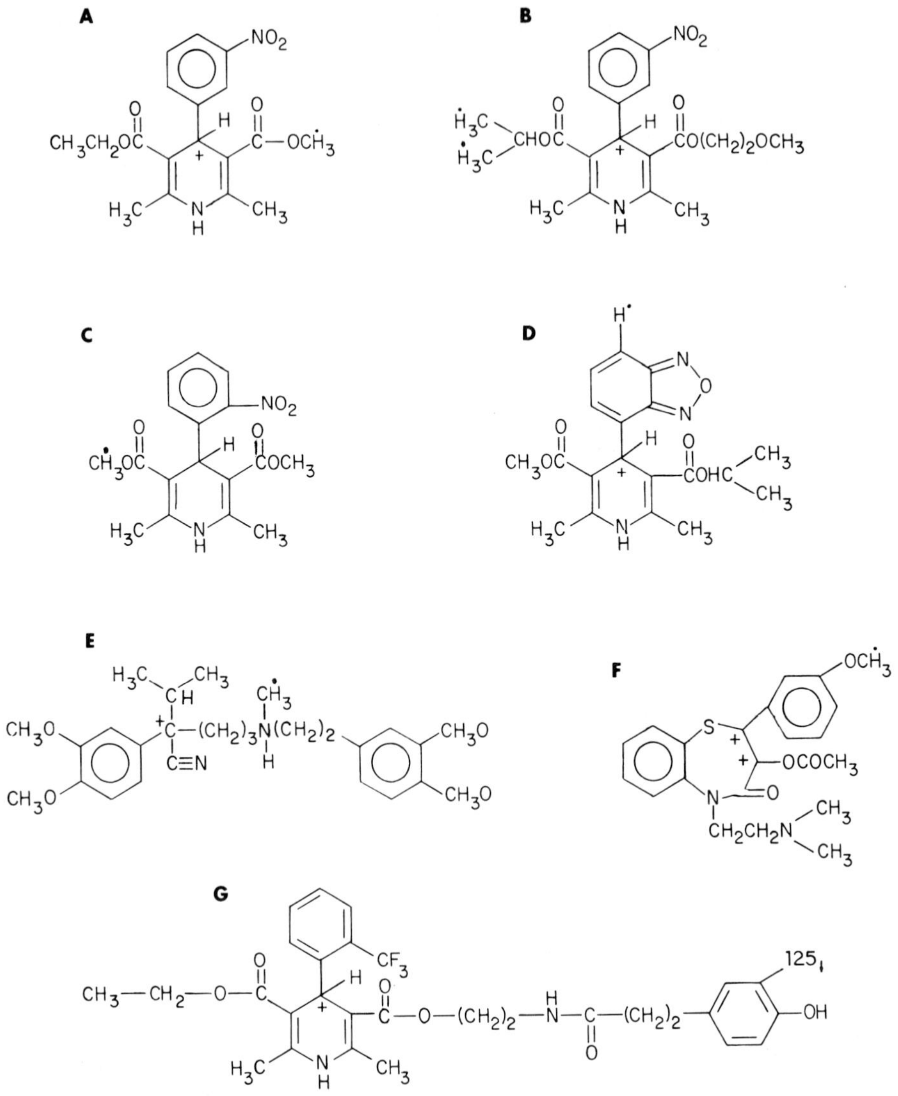

FIG. 1. Structures of radiolabeled 1,4-dihydropyridines, verapamil, diltiazem. In the structures the asterisk denotes the position of T and plus sign an asymmetric carbon atom. The drugs are (A) (±)[^{3}H]nitrendipine[1,4-dihydro-2,6-dimethyl (3-nitrophenyl)-3,5-pyridinecarboxylic acid, 3 ethyl-5-methyl ester]; (B) (±)[^{3}H]nimodipine [isopropyl (2-methoxymethyl)-1,4-dihydro-2,6-dimethyl-1-4-(3-nitrophenyl)-3,5-pyridinedicarboxylate]; (C) [^{3}H] nifedipine [dimethyl-1,4-dihydro-2,6-dimethyl-1,4-(2-nitrophenyl)-3,5-pyridine dicarboxylate]; (D) (±)[^{3}H]PN 200-110 [isopropyl 4-(2,1,3-benzoxadiazol-4-yl)-1,4-dihydro-2,6-dimethyl-5-methoxycarbonylpyridine-3-carboxylate] (the position of the tritium is indicated for the ring-labeled compound, the (+)enantiomer is labeled at the ester group); (E) (±)verapamil [5-*N*-(3,4-dimethoxyphenethyl)-*N*-methylamino-2-(3,4-dimethoxyphenyl)-2-isopropylvaleronitrile]; (F) *d,l-cis,trans*-diltiazem [3-acetoxy-2,3-dihydro-5,2-(diethylamino) ethyl-2-(*p*-methoxy phenyl)-1,5 benzothiazepine-4-(5*H*)-one hydrochloride]. The labeled compound is *d-cis*-diltiazem. (G) [^{125}I]Iodipine.

Chemicals

Unlabeled drugs can be obtained by the following drug companies: Nitrophenyl 1,4-dihydropyridine derivatives are obtained from Bayer AG, Wuppertal, F.R.G. Benzoxadiazol 1,4-dihydropyridine derivatives are from Sandoz Ltd, Basel, Switzerland. Phenylalkylamines (verapamil and D-600) are from Knoll AG, Ludwigshafen, F.R.G. Benzothiazepines (*l*- and *d-cis*-diltiazem) are from Gödecke AG, Freiburg, F.R.G. Tiapamil [*N*-(3,4-dimethoxyphenethyl)-2-(3,4-dimethoxyphenyl)]-*N*-methyl-*m*-dithiane-2-propylamine-1,1,3,3-tetraoxide) is from Hoffmann-La Roche, Basel, Switzerland. (+) and (−)Prenylamine [*N*-(2-benzhydrylethyl)-α-methylphenethylamine] are from Hoechst AG, Frankfurt, F.R.G. Optically pure enantiomers of D-600 (also known as methoxyverapamil or gallopamil) are from Grünenthal AG, Stolberg, F.R.G. or from Knoll AG, Ludwigshafen, F.R.G.

Methods

Preparation of Membrane Fragments

Cardiac Muscle and Smooth Muscle Organs. Organs (cleaned from connective tissue and fat) are finely minced with scissors and homogenized in ice-cold buffer A with a wet weight to volume ratio of 1 : 3 by an ultraturrax for 2 × 15 sec at 3/4 maximal speed and then have buffer A added to give a wet weight to volume ratio of 1 : 7. The diluted homogenate is subjected to a 15 min 1500 *g* centrifugation. The supernatants are centrifuged at 45,000 *g* for 15 min, the pellets are resuspended in buffer B, and washed twice as above.

Brain. Whole brain (minus medulla and cerebellum) is homogenized at a wet weight to volume ratio as above and then subjected to a 45,000 *g* (15 min) centrifugation. Pellets are washed twice in buffer B as above.

Skeletal Muscle. Fifteen grams of guinea pig hind limb skeletal muscles is finely minced in buffer A and homogenized at a wet weight to volume ratio of 1 : 5 with an ultraturrax as above using 2–3 × 20 sec bursts of 3/4 maximal speed. Disrupted muscle is filtered through two layers of cheese cloth and then centrifuged as described for cardiac muscle. The membrane fraction is referred to as the "microsomal fraction."[7] For preparation of t-tubule rich membranes the 45,000 *g* pellet is taken up in 0.25 *M* sucrose in buffer B (45 ml for 15 g original wet weight tissue) and is homogenized with seven complete strokes of a Dounce homogenizer (tight pestle). Seven milliliter fractions are overlayed on a discontinous sucrose gradient [8 ml of 40%, 35%, 30%, 7 ml 25% (w/w) sucrose in buffer B]. Gradients are spun in a Beckman SW27 rotor for 150 min at 2–

[7] D. R. Ferry and H. Glossmann, *FEBS Lett.* **148,** 331 (1982).

4° at 113,000 g_{max} (25,000 rpm). The material collecting at the 25% (w/w) sucrose layer/overlay interface is removed with a syringe, diluted 20-fold in buffer B, and centrifuged for 120 min at 45,000 *g*. The pellet is referred to as "t-tubule rich plasma membrane."[8]

Over a 1-year period we have observed a variation of up to 2-fold for the density of [^{3}H]1,4-dihydropyridine binding sites in skeletal muscle membranes, which may be due to improvements in the preparation of membranes or a biological variation. The membranes are kept in 2-ml aliquots in polypropylene vials and can be stored in liquid nitrogen for up to 3 months.

Solubilization of Calcium Channels

For solubilization the skeletal muscle microsomal fraction (6 mg of protein) is suspended in 5 ml ice-cold buffer B and 2.5 ml of 2 *M* NaCl is added; 2.5 ml of 40 m*M* CHAPS in buffer B is dropped into the gently shaken mixture which is kept on ice for 30 min and then centrifuged for 45,000 *g* for 60 min at 4°. The supernatant is filtered through a disposable 0.2-μm nitrocellulose filter with a syringe and is termed the "solubilized channel." At the above protein to detergent ratio, CHAPS solubilizes ~90% of the protein and 10–20% of the 1,4-dihydropyridine binding sites present in the starting membranes are recovered. Recovery and stability can be improved by inclusion of 10% (w/w) glycerol in the solubilization medium. The use of *N*-ethylmaleimide as a protease inhibitor is not recommended as it inactivates high affinity 1,4-dihydropyridine or desmethoxyverapamil binding. Digitonin may be also used to solubilize the 1,4-dihydropyridine binding sites.[9]

Basic Ligand Binding Assay

Details of such assays vary between laboratories. In this laboratory filtration manifolds (48 places), strong vacuum pumps, and 11.5 × 75-mm polystyrol incubation tubes are used. The manifold is loaded with prewetted Whatman GF/C filters. Immediately before filtration the vacuum pump is turned on, 1 ml of buffer E is poured on each filter and filters are dried by vacuum prior to filtration of the diluted contents of the assay.

Particulate Membranes

Volume of assay: 0.25 ml for [^{3}H]nimodipine, [^{3}H]nifedipine, [^{3}H]nitrendipine; 1.0 ml for [^{3}H]PN 200-110. The assay volume is higher for the

[8] H. Glossmann, D. R. Ferry, and C. B. Boschek, *Naunyn-Schmiedeberg's Arch. Pharmacol.* **323,** 1 (1983).

[9] H. Glossmann and D. R. Ferry, *Naunyn-Schmiedeberg's Arch. Pharmacol.* **323,** 279 (1983).

latter ligand when the lowest specific activity (2.6–3.3 Ci/mmol is used).

Order of additions (may be changed, depending on protocol): ^{3}H-labeled ligand, buffer B, unlabeled drugs or other additions, membranes, or solubilized channel.

Temperature of incubation: 37° (may be changed, depending on protocol).

Time of incubation: for equilibrium binding experiments: 30 min (may be changed, depending on protocol).

Nonspecific binding definition: 1 μM unlabeled nimodipine, nitrendipine, or PN 200-110.

Separation of bound and free ligand: add 3.5 ml of buffer E to the contents of the assay tube (dispensing device) and pour through the filters. Wash twice with 3.5 ml of the buffer. Dry filters and count for retained radioactivity after adding a suitable liquid scintillation cocktail.

Note: ensure that buffer E is at 0–1°.

Comment: the same methodology with minor modifications (temperature, pH of buffer, nonspecific binding definition) is employed for other ligands, e.g., [^{3}H]*d-cis*-diltiazem,[10] [^{3}H]verapamil, (−)-[^{3}H]desmethoxyverapamil, [^{125}I]iodipine (vide infra).

Solubilized Calcium Channels. The incubation is as above but the assay is performed in the presence of 0.2 mg/ml bovine serum albumin. After incubation, transfer assay tubes into melting ice and add 0.1 ml of an ice-cold mixture of bovine serum albumin/rabbit γ-globulin (5 mg of each per ml) as carrier protein, in buffer B and then 3.5 ml of buffer E. Wait 5–10 min (turbidity develops) and separate bound and free ligand as above.

Dilution of ^{3}H-Labeled Ligands for Saturation Experiments. Aliquots of the ^{3}H-labeled ligands are transferred into ice-cold buffer B with an Eppendorf pipet and a polypropylene tip. Geometric dilutions of radioactivity, required for saturation analysis, are made with a 2.5-ml Eppendorf combitip syringe. The radioligand dilution is run up and down the syringe 3–5 times and a measured aliquot is transferred into buffer. Using the same syringe, this procedure is repeated until the given number of ^{3}H-labeled ligand dilutions is obtained. The same syringe is used to pipet the ligand into the assay tubes, proceeding from the lowest to the highest concentration.

Testing the Bindability of Chiral Radiolabeled 1,4-Dihydropridines. As a routine test, to estimate bindability of the [^{3}H]1,4-dihydropyridines and of [^{125}I]iodipine, a single, low concentration of radioligand and increasing concentrations of skeletal muscle microsomal membranes are incubated. It has been shown that the bindability of [^{3}H]nimodipine is close to 50%,[7] illustrating the stability of [^{3}H]nimodipine in the presence

[10] H. Glossmann, T. Linn, M. Rombusch, and D. R. Ferry, *FEBS Lett.* **160,** 226 (1983).

of high membrane concentrations in this tissue, and that probably only the more potent 4*S*(−)enantiomer[11] binds at n*M* concentrations. Many authors seem not to pay attention to the racemic nature of a large number of the drugs active in calcium channel regulation (vide infra). Please note that the absolute configuration (*R*) and (*S*) has only been elucidated for the 1,4-dihydropyridines cited in Towart *et al.*[11]

Autoradiographic Procedures

Preparation of Tissue Sections. Unfixed tissues are used. Brains are frozen at −20° and are mounted onto microtome chucks using mounting media. They are cut with a microtome cryostate into 8-μm-thick sections. Sections are thaw-mounted onto gelatin-coated microscope slides. The slides can be stored for up to 7 days at −20° before deterioration of binding activity is observed.

1,4-Dihydropyridine Receptor Labeling. Radioactivity bound to tissue sections was measured by liquid scintillation counting after wiping off the wet tissues, with a glass fiber filter paper, from the slide. The binding of labeled 1,4-dihydropyridines to guinea pig brain sections is reversible, saturable, of high affinity (dissociation constants are in the nanomolar range) and with the appropriate pharmacology. Labeled 1,4-dihydropyridines are bound to tissue sections in buffer G at 23° for 60 min. Loosely bound radioactivity is then washed away by the following procedure. A short dip in buffer G is followed by a 30 min "wash" in ice-cold buffer H, followed by a 15–30 sec dip in ice-cold distilled water to remove salt and thereby facilitate drying with a stream of cooled ($<$10°) air, which takes 30–60 sec. Failure to include albumin in the wash buffer results in unacceptable high blank values. Even in brain sections including the heavily specifically labeled hippocampus the signal-to-noise-ratio is at best very unfavorable (note: in Ref. 12 no blank is shown). In rat brain sections the high affinity binding is almost unmeasurably low relative to blank binding without the use of albumin in the wash buffer.

The inclusion of albumin in the wash buffer has no effect on the K_D or B_{max} of [^{3}H]nimodipine or [^{3}H]PN 200-110 in guinea pig brain sections. In mapping experiments the following concentrations are used: 0.1 n*M* [^{3}H]nimodipine, 0.1 n*M* [^{3}H]PN 200-110, and 0.01 n*M* [^{125}I]iodipine. Blanks are generated with sections consecutive to those used for "total binding" by adding 2 μM nimodipine to the incubation media. Before

[11] R. Towart, E. Wehinger, and H. Meyer, *Naunyn-Schmiedeberg's Arch. Pharmacol.* **317,** 183 (1981).

[12] R. Quirion, *Neurosci. Lett.* **36,** 267 (1983).

being exposed to radioligands blanks are preincubated with 2 μM nimodipine for 10–15 min.

Generation of Autoradiograms with Film. Dried, labeled tissue sections are mounted on a square of adhesive plastic foil placed on an X-ray cassette and a ^{3}H-Ultrafilm (LKB, Stockholm, Sweden) is placed against them. The cassettes are stored, at room temperature, for the exposure period. Exposure time ranges from 3 to 8 weeks for the [^{3}H]dihydropyridine-labeled tissues[12,13] or 1–2 days for [^{125}I]iodipine. Films are developed in D-19 for 5 min at 20°, rinsed in water, and fixed 5 min at 20° in Ektaflo Fixer (1 : 4 with water). After 20 min of rinsing in running water the films are dried in air. Sections can be reexposed or counted for radioactivity.

Characterization of Calcium Channels with Direct Ligand Binding Studies

Tissue Distribution of [^{3}H]Nimodipine-Labeled Calcium Channels

Membrane fragments are prepared as described in methods. The tissue fractions are then incubated with [^{3}H]nimodipine. Table I shows the mean dpm collected/filter in the presence and absence of 1 μM unlabeled nimodipine, the signal-to-noise ratios, and the corresponding specifically bound [^{3}H]nimodipine in fmol/mg of protein. Clearly, guinea pig skeletal muscle, heart, and cerebral cortex offer excellent signal-to-noise ratios, whereas duodenum, kidney, and liver are somewhat problematic to work with. Table II shows the B_{max} and K_D values of [^{3}H]nimodipine calcium channel binding in crude heart membranes of various species. Although the B_{max} varies over a 17-fold range between species, the K_D of [^{3}H]nimodipine is essentially identical in all species so far examined, emphasizing that the calcium channel has, in evolutionary terms, highly conserved domains.

Association Kinetics

Association kinetics of [^{3}H]nimodipine with guinea pig brain membranes is followed by starting the reaction with radioligand. The addition times of radioligand and filtration times of the given assay tubes are listed in Table III, and an example of the association kinetics is shown in Fig. 2.

Dissociation Kinetics

A steady-state population of [^{3}H]nimodipine–calcium channel complexes which has been incubated for 30 min at 37° has 10 μl of 25 μM

[13] K. M. M. Murphy, R. J. Gould, and S. H. Snyder, *Eur. J. Pharmacol.* **81,** 517 (1982).

TABLE I
TISSUE DISTRIBUTION OF [^{3}H]NIMODIPINE BINDING[a]

Tissue	Bound radioligand (dpm/filter) Total	Blank	Signal-to-noise ratio	μg protein/assay tube	Specific binding (fmol/mg of protein)
Skeletal muscle	33,062	2,710	11.2	48.2	1770
Heart	15,291	3,287	3.65	91.2	371
Adrenal gland	2,890	1,303	1.22	21.1	212
Hippocampus	9,716	3,224	2.01	91.3	200
Olfactory tubercle	4,666	1,200	2.89	60.0	163
Cerebral cortex	13,518	2,582	4.24	200.0	153
Uterus	3,241	1,276	1.54	36.4	152
Duodenum	5,006	2,331	1.15	60.5	124
Kidney	19,179	15,880	0.21	300.5	31
Liver	39,698	30,926	0.28	530.0	48
Fat cell membranes	434	585	0	3.6	—
Blood platelet membranes	1,082	1,023	0	35.2	—

[a] Tissues were prepared from the guinea pig and binding experiments with [^{3}H]nimodipine (0.98 n*M*) were performed as described under Methods. Fat cell membranes (which were from hamsters) and human blood platelets did not bind [^{3}H]nimodipine specifically under the conditions employed. Signal-to-noise ratio is defined as (total dpm − blank dpm)/(blank dpm).

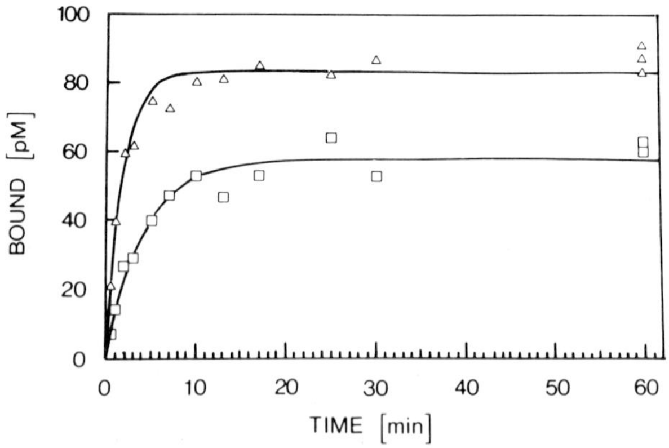

FIG. 2. Association kinetics of [^{3}H]nimodipine with guinea pig brain membranes. Guinea pig brain membranes were preincubated in buffer A for 30 min at 30° in a volume of 0.2 ml at a protein concentration of 0.46 mg/ml. Association reactions were initiated by adding 50 μl of [^{3}H]nimodipine which was prewarmed to 30°. The association reactions were performed

TABLE II
EQUILIBRIUM BINDING PARAMETERS OF [^{3}H]NIMODIPINE TO HEART MEMBRANES FROM DIFFERENT SPECIES[a]

Species	B_{max} (fmol/mg of protein)	K_D (nM)	Reference[b]
Guinea pig	350	0.26	a
Rat	400	0.24	b
Bovine	350–980	0.25	c
Frog	1700	0.29	d
Human	100	0.28	e

[a] B_{max} and K_D values for [^{3}H]nimodipine binding to heart membranes from various species. Saturation isotherms were constructed with 0.05–2 nM ^{3}H-labeled ligand (without taking account of the racemic nature of [^{3}H]nimodipine) and for the rat heart with 0.026–0.7 nM [^{3}H]nimodipine. Guinea pig, bovine, and human heart (intraventricular wall from patients with hypertrophic cardiac myopathy) equilibrium binding experiments are performed at 37°; the frog and rat heart experiments are performed at 22°.

[b] References: (a) D. R. Ferry and H. Glossmann, *Br. J. Pharmacol.* **78,** 81 p (1983); (b) R. Janis, S. C. Maurer, J. C. Sarmiento, G. T. Bolger, and D. J. Triggle, *Eur. J. Pharmacol.* **82,** 191 (1982); (c) H. Glossmann, D. R. Ferry, F. Lübbecke, H. Mewes, and F. Hofmann, *J. Recept. Res.* **3,** 177 (1983); (d) D. R. Ferry and H. Glossman (unpublished); (e) D. R. Ferry, A. J. Kaumann, and H. Glossmann (unpublished).

unlabeled nimodipine added for a given time prior to filtration of the assay mixture. The dissociation reaction is highly temperature dependent, occurring more rapidly at higher temperatures (Fig. 3). The temperature-dependent dissociation is the rational for the use of ice-cold wash buffers, to minimize dissociation of channel-bound 1,4-dihydropyridine.

Equilibrium Saturation Isotherms of Brain Calcium Channels with [^{3}H]Nimodipine

A ^{3}H-labeled ligand stock solution is diluted as described in methods, and brain membranes incubated with the different concentrations of ra-

at ^{3}H-labeled ligand concentrations (bindable enantiomer) of 0.28 nM (□) and 1.11 nM (△). The B_{max} was estimated to be 300 fmol/mg of protein which corresponds to a receptor concentration of 138 pM. Data points are fitted directly to the differential form of the second order rate equation by the error sum of squares principle. At a [^{3}H]nimodipine concentration of 0.28 nM the best parameter estimates (± asymptotic standard deviation) were k_{+1} (association rate constant) = 0.135 ± 0.034/nM/min, k_{-1} (dissociation rate constant) = 0.12 ± 0.02/min, and at a [^{3}H]nimodipine concentration of 1.11 nM: k_{+1} = 0.279 ± 0.026/nM/min and k_{-1} = 0.22 ± 0.03/min.

TABLE III

PROTOCOL FOR THE MEASUREMENT OF THE ASSOCIATION REACTION[a]

Desired graph time (min)	Graph − filtration time (min)	Addition of ^{3}H-ligand (min)	Filtration time (min)
0.5	0	54	0.5
1	0	54	1
2	0.5	53.5	1.5
3	1	53	2
5	2.5	51.5	2.5
7	4	50	3
10	6.5	47.5	3.5
13	9	45	4
17	12.5	41.5	4.5
25	20	34	5
30	24.5	29.5	5.5
60	54	0 Start	6 Finish

[a] This method of performing association reactions allows the separation of bound and free radioactivity for the desired graph times with a filtration manifold. ^{3}H-labeled ligand is added first to the assay tubes to be filtered last. Note that (filtration time) + (graph time − filtration time) = graph time. The third column shows the reading on a clock started when the first assay tubes have ^{3}H-labeled ligand added; at the last addition (graph point = 0.5 min) the clock is set to 0, and filtration time is then read from the clock.

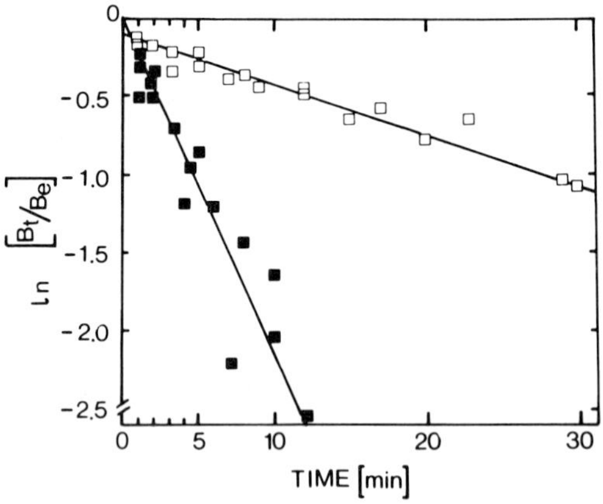

FIG. 3. Temperature dependence of [^{3}H]nimodipine dissociation from calcium channels. [^{3}H]Nimodipine was allowed to associate for 30 min at 37 or 25° with guinea pig brain membranes. Nimodipine (10 μl, 25 μM) was added for up to 30 min prior to the separation of bound and free radioactivity. The plot of the natural logarithm of the ratio of specific binding after a given time of addition of unlabeled drug (B_t) to specific binding at equilibrium (B_e) is linear. At 37° (■) the dissociation rate constant (k_{-1}), calculated from the slope of the plot, is 0.19/min. At 25° (□) the k_{-1} is 0.032/min.

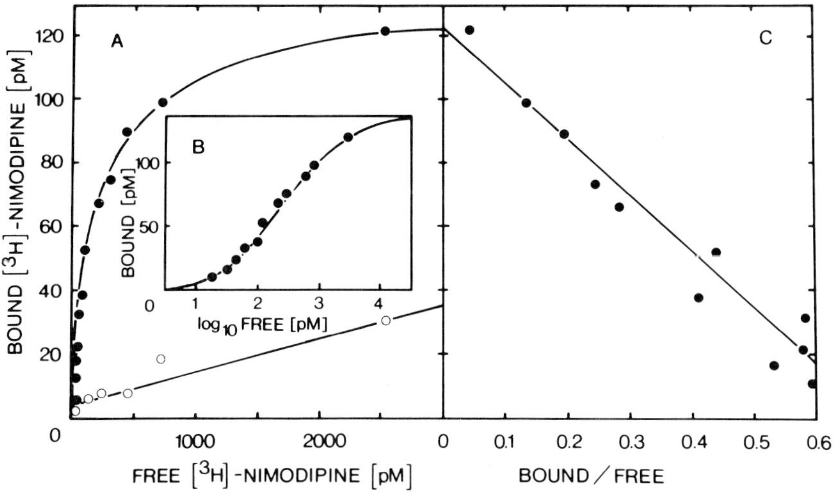

FIG. 4. Saturation isotherm of [^{3}H]nimodipine binding to guinea pig brain membranes. [^{3}H]Nimodipine at different concentrations was incubated with 0.16 mg/ml of guinea pig brain membrane protein. Blanks defined by 1 μM unlabeled nimodipine (○) have been subtracted to yield specific binding (●). A shows a plot of bound [^{3}H]nimodipine against the free [^{3}H]nimodipine (accounting only for the bindable enantiomer). B shows a semilog plot of the specific [^{3}H]nimodipine binding shown in A. Fitting these data to the general dose–response equation (inset) yielded the following parameter estimates (± asymptotic standard deviation): $K_D = 232 \pm 30$ pM, $B_{max} = 135 \pm 6$ pM, with a slope factor of 0.91 ± 0.06. In C the data are shown transformed into a Hofstee plot. Linear regression of the data gave K_D = 177 pM and B_{max} 123 pM.

dioligand. An example of such an experiment is shown in Fig. 4. Blank values defined by 1 μM unlabeled nimodipine correspond to <10% of filter-retained radioactivity at the K_D concentration of [^{3}H]nimodipine.

Divalent Cation Regulation of [^{3}H]Nimodipine-Labeled Calcium Channels

Nature of EDTA Inhibition of [^{3}H]Nimodipine Calcium Channel Labeling. When guinea pig brain membranes are incubated at 37°, EDTA and CDTA completely inhibit calcium channel binding of [^{3}H]nimodipine in a dose-dependent manner with IC_{50} values in the micromolar range. This inhibition is due to an apparent reduction in B_{max}. Similar findings with rat brain membranes and [^{3}H]nitrendipine have been reported.[14] Addition of EDTA or CDTA to [^{3}H]nimodipine-labeled calcium channels at

[14] R. J. Gould, K. M. M. Murphy, and S. H. Snyder, *Proc. Natl. Acad. Sci. U.S.A.* **79,** 3656 (1982).

equilibrium results in a monophasic decay of the complex with a k_{-1} of 0.33/min which, at 37°, is close to the k_{-1} found when unlabeled nimodipine blocks the association reaction. The dissociation reaction, induced by EDTA, is temperature dependent occurring more rapidly at higher temperatures. The channel is not irreversibly denatured by the divalent cation depletion; the high affinity 1,4-dihydropyridine binding is recovered by adding back divalent cations.[14–16]

Comment. In brain membranes EDTA and CDTA completely inhibit high affinity 1,4-dihydropyridine binding whereas the chelators inhibit only by 60% in partially purified bovine heart sarcolemma. Further, in particulate guinea pig skeletal muscle microsomes EDTA and CDTA are ineffective up to 1 m*M*; however, when the calcium channels are solubilized from skeletal muscle high affinity [^{3}H]1,4-dihydropyridine binding is as sensitive to the chelators as seen with particulate brain membranes.

Reversibility of EDTA Inhibition of [^{3}H]Nimodipine Calcium Channel Labeling. Ten milligrams of guinea pig brain membrane protein is incubated in a volume of 60 ml in 50 m*M* Tris–HCl, 5 m*M* EDTA, pH 7.4, at 37° for 30 min. The membranes are then pelleted by centrifugation at 45,000 *g* for 15 min at 4° and resuspended in 25 ml of Chelex-100 treated water, then centrifuged for 15 min at 45,000 *g* and then again resuspended in 25 ml of buffer F. Membranes are only considered "divalent cation free" if ≥95% of the starting [^{3}H]nimodipine binding is depleted. Divalent cation depletion of brain membranes requires prolonged exposure of brain membranes to EDTA at 37°. Certain published procedures for divalent cation depletion of brain membranes are performed at 0°, and then the membranes are finally suspended in 10 μM EGTA solution,[14] which means that in the "refilling" experiments the chelator is titrated by the addition of divalent cation. Assays for the refilling with divalent cations differed from the normal [^{3}H]nimodipine assay. They are conducted in a volume of 1.0 ml in buffer F to allow the use of very low protein concentrations in the range of 0.04–0.06 mg/ml, to further reduce the concentration of any residual divalent cations. Examples of refilling isotherms are shown in Fig. 5.

Pharmacological Profile of [^{3}H]Nimodipine-Labeled Calcium Channels with Reference to Tissue-Specific Variations

General Comment. Calcium antagonistic drugs are often extremely hydrophobic and can only be dissolved at concentrations of 10 m*M* in

[15] H. Glossmann, D. R. Ferry, F. Lübbecke, R. Mewes, and F. Hofmann, *Trends Pharmacol. Sci.* **3,** 431 (1982).

[16] H. Glossmann and D. R. Ferry, *Drug Dev. Eval.* **9,** 63 (1983).

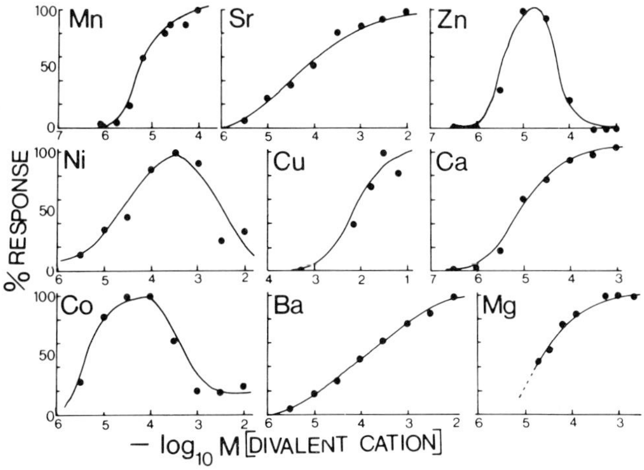

FIG. 5. Cation dependence of [^{3}H]nimodipine binding in guinea pig brain membranes. Representative example of cation refilling of EDTA-treated membranes. Incubation time was 30 min at 37° in a volume of 1.0 ml with 0.08 mg of membrane protein/ml in buffer F. The [^{3}H]nimodipine concentration was 0.79 n*M*. All divalent cations were added as chloride salts. The binding in the absence of exogenously added divalent cations was 3.57% (0.3 p*M*) of that when Ca^{2+} at 1 m*M* was added (8.4 p*M*) (96.4 0.2% reduction). Blank values were defined by 1 μM unlabeled nimodipine and were measured for every salt concentration. To calculate the "response," the specific binding (cpm) in the absence of added cations was subtracted from the specific binding in the presence of added cation and was divided by the specifically bound maximum observed and multiplied by 100. When cations caused increases in blank values at higher concentrations (e.g., Ni^{2+}, Co^{2+}, Zn^{2+}, Cu^{2+}) this was accounted for in the calculation. For cations, giving bell-shaped refilling profiles, percentage response is calculated from the maximum of the filling curve. Please note that the refilling curves are normalized to 100% with respect to the maximum response of the given ion. If refilling with Ca^{2+} (at 1 m*M*) is 100%, Mn^{2+} = 94%, Mg^{2+} = 86%, Co^{2+} = 82%, Ni^{2+} = 55%, Zn^{2+} = 36%, Cu^{2+} = 57%, Ba^{2+} = 64%, Sr^{2+} = 105%.

DMSO or ethanol. Drugs dissolved in DMSO are frozen at −20° and stored for up to 6 month without loss of activity. DMSO of up to 2% (v/v) has no effect on [^{3}H]nimodipine labeling of calcium channels. Drugs are diluted in buffer B in volumes of 1.0 ml in Eppendorf tubes transferring 0.1 ml with plastic pipet tips, changing the pipet tip for each step of the dilution. Certain 1,4-dihydropyridines [e.g., felodipine, 4-(2,3-dichlorophenyl)-1,4-dihydropyridine-2,6-dimethyl-3,5-dicarboxylic-3-ethylester,5-methylester] or compounds such as flunarizine, [*trans*-1-cinnamyl-4-(4,4′-difluorobenzhydryl)piperazine] are not soluble at concentrations of ≥10 μm in buffer.

Findings are reported below, which help to differentiate, in molecular terms, three fundamentally different classes of calcium antagonists. Also

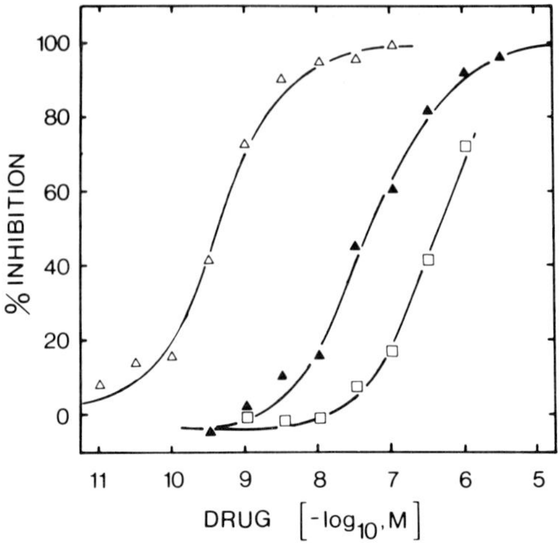

FIG. 6. Inhibition of [^{3}H]nimodipine binding to guinea pig brain membranes by class I calcium antagonists. The optically pure enantiomers of the benzoxadiazol 1,4-dihydropyridine derivative PN 200-110, (+)205-033 (△) and (−)205-034 (▲) and the pharmacologically inactive derivative isopropyl 4-[2,3,1(7-bromo)benzoxadiazol-4-yl]-1,4-dihydro-2,6-dimethyl-5-methoxycarbonyl-pyridine-3-carboxylate (Vo 2605, □) were incubated with 0.6 n*M* [^{3}H]nimodipine and guinea pig brain membranes. Blanks defined by 1 μM unlabeled nimodipine were subtracted and the percentage inhibition calculated. The IC_{50} values and Hill slopes (nH) were for (+)205-033, IC_{50} = 0.2 n*M*, nH = 1.1; (−)205-034, IC_{50} = 53 n*M*, nH = 0.90; Vo 2605, IC_{50} = 405 n*M*, nH = 1.0.

outlined is how the tissue-specific interactions of the calcium antagonists assist in the subclassification of these drugs.

Class I Calcium Antagonists. Class I calcium antagonists inhibit [^{3}H]nimodipine binding to guinea pig brain membranes in monophasic, stereoselective manner (Fig. 6). All 1,4-dihydropyridine derivatives we have investigated compete with Hill slopes of close to unity (see Table IV). The structure–activity relationship found in pharmacological experiments[17,18] is also found in binding studies. In guinea pig ileum[19] and a phaeochromocytoma cell line[20] the potencies of various 1,4-dihydropyridine calcium antagonists to occupy calcium channels labeled with [^{3}H]ni-

[17] F. Bossert, H. Horstmann, H. Meyer, and W. Vater, *Arzneim.-Forsch.* **29,** 226 (1979).

[18] R. Rodenkirchen, R. Bayer, R. Steiner, F. Bossert, H. Meyer, and E. Möller, *Naunyn-Schmiedeberg's Arch. Pharmacol.* **310,** 69 (1979).

[19] G. T. Bolger, P. J. Gengo, E. M. Luchowski, H. Siegel, D. J. Triggle, and R. Janis, *Biochem. Biophys. Res. Commun.* **104,** 1604 (1982).

[20] L. Toll, *J. Biol. Chem.* **257,** 13189 (1982).

TABLE IV

EQUILIBRIUM BINDING, KINETIC, AND BINDING-INHIBITION CONSTANTS FOR VARIOUS DRUGS OF [^{3}H]NIMODIPINE-LABELED CALCIUM CHANNELS IN BRAIN AND SKELETAL MUSCLE[a]

Brain Membranes

Equilibrium binding and kinetic parameters	Control	With 10 μM *d-cis*-diltiazem
B_{max} (fmol/mg of protein)	570 ± 99	452 ± 35
Dissociation constant (nM)	0.6 ± 0.1	0.2 ± 0.01
Association rate constant (nM^{-1} min^{-1})	0.32 ± 0.05	0.37 ± 0.10
Dissociation rate constant (min^{-1})	0.18 ± 0.01	0.05 ± 0.005

	Binding inhibition data			
	Control		With 10 μM *d-cis*-diltiazem	
Drug	IC_{50} (nM)	nH	IC_{50} (nM)	nH
Niludipine	5.7 ± 2.7	1.0 ± 0.2	0.6 ± 0.07	1.0 ± 0.07
(−)Nicardipine	28 ± 4	1.0 ± 0.06	3.2 ± 0.07	1.0 ± 0.01
(+)Nicardipine	2.3 ± 0.6	1.0 ± 0.04	0.7 ± 0.07	1.0 ± 0.01
(−)Prenylamine	590 ± 190	1.0 ± 0.05	1200 ± 300	1.1 ± 0.2
(+)Prenylamine	420 ± 45	1.0 ± 0.05	1200 ± 230	1.4 ± 0.11
	% High	% Low	% High	% Low
(−)Verapamil	30 ± 7	69 ± 7	46.3 ± 0.8	54 ± 0.8
(+)Verapamil	67 ± 5	33 ± 5	14.0 ± 6	86 ± 6
	K_D high (nM)	K_D low (nM)	K_D high (nM)	K_D low (nM)
(−)Verapamil	61 ± 30	280 ± 80	2200 ± 330	246 ± 75
(+)Verapamil	191 ± 20	70 ± 30	450 ± 210	23 ± 0.6
Diltiazem diastereoisomers	IC_{50} (μM)		EC_{50} (μM)	
d-cis-Diltiazem	—		0.38	0.07
l-cis-Diltiazem	221	60	—	

(*continued*)

TABLE IV (*continued*)

Skeletal Muscle Microsomal Membranes (particulate)

Equilibrium binding and kinetic parameters	Control	With 10 μM *d-cis*-diltiazem
B_{max} (fmol/mg protein)	2–7	10–14
Dissociation constant (nM)	1.5 ± 0.03	1.0 ± 0.15
Association rate constant (nM^{-1} min^{-1})	0.28	0.30
Dissociation rate constant (min^{-1})	1.5	0.34

Binding inhibition data

	Control		With 10 μM *d-cis*-diltiazem	
Drug	IC_{50} (nM)	nH	IC_{50} (nM)	nH
(+)205-033	2.2 ± 0.5	1.04 ± 0.08	3.5 ± 1.6	1.1 ± 0.05
(−)205-034	206 ± 42	0.9 ± 0.1	333 ± 90	1.2 ± 0.1
(±)Fendiline	540 ± 125	1.0 ± 0.10	1200 ± 205	1.2 ± 0.12
Tiapamil	290 ± 75	1.0 ± 0.15	12250 ± 4260	0.9 ± 0.09
(+)D-600	1000 ± 270	0.78 ± 0.0	5745 ± 770	0.90 ± 0.14
(−)D-600	No inhibition		1410 ± 309	0.95 ± 0.14
La^{3+}	200 × 10^3	1.3	450 × 10^3	1.2
Tetrodotoxin	No effect (1 μM)			
EDTA	No inhibition (100 μM)			

Diltiazem diastereoisomers	IC_{50} (μM)	EC_{50} (μM)
d-cis-Diltiazem	—	1.02 ± 0.130
l-cis-Diltiazem	42 ± 18	—

trendipine are similar to those required to block the responses (contraction and $^{45}Ca^{2+}$ uptake, respectively).

Class II Calcium Antagonists. Class II calcium antagonists exhibit complex, biphasic inhibition profiles in guinea-pig brain membranes. (+)Verapamil, (−)verapamil, (+)D-600, and (−)D-600 belong to this class of substances.[7,9,15,21,22] Class II calcium antagonists are not competitive inhibitors of [^{3}H]nimodipine binding, but are negative allosteric heterotropic regulators. (+)Verapamil accelerates the dissociation of [^{3}H]nimodipine–channel complexes.[16] Qualitatively similar results with racemic

[21] D. R. Ferry and H. Glossmann, *Naunyn-Schmiedeberg's Arch. Pharmacol.* **321,** 808 (1982).

[22] H. Glossmann, D. R. Ferry, F. Lübbecke, R. Mewes, and F. Hoffmann, *J. Recept. Res.* **3,** 177 (1983).

TABLE IV (*continued*)

Solubilized Skeletal Muscle Microsomes

Equilibrium binding and kinetic parameters	Control	With 10 μM *d-cis*-diltiazem
B_{max} (fmol/mg of protein)	1200 ± 70	2500 ± 140
Dissociation constant (nM)	3.6 ± 0.5	4.3
Dissociation rate constant (min^{-1})	1.5	0.38

	Binding inhibition data			
	Control		With 10 μM *d-cis*-diltiazem	
Drug	IC_{50} (nM)	nH	IC_{50} (nM)	nH
(+)205 033	9.8	1.1	7.7	1.0
(−)205 034	1200	1.0	1200	1.0
(+)D-600	480	0.9	1300	0.7
(−)D-600	65.000	0.3	1600	0.3
EDTA	17×10^3	3.5	20	2.0
CDTA	16×10^3	3.6	24	3.5
EGTA	10×10^3	3.5	24	2.1

Diltiazem diastereoisomers	EC_{50} (μM)	EC_{50} (μM)
d-cis-Diltiazem	—	3.6
l-cis-Diltiazem	59	—

[a] Equilibrium binding and kinetic parameters were calculated without taking into account the racemic nature of [^{3}H]nimodipine. The biphasic verapamil inhibition curves for brain membranes were analyzed with the LIGAND computer package. nH, Hill slope. All data were from 37° experiments. In the case of *d-cis*-diltiazem the EC_{50} value for stimulation is given. From Refs. 7, 9, and 19 with permission.

verapamil have been reported for guinea pig heart membrane calcium channels labeled by [^{3}H]nitrendipine.[23] The binding-inhibition curves of the optically pure diphenylalkylamines fit better to a two-site inhibition model than to a one-site inhibition model, possibly indicating heterogeneity of the verapamil binding sites mediating the allosteric regulation of [^{3}H]nimodipine binding.[21]

Class III Calcium Antagonists. *d-cis*-Diltiazem, bencyclane, KB 944 [diethyl-4-(benzothiazol-2-yl)benzyl phosphonate], disopyramide, and

[23] F. J. Ehlert, W. R. Roeske, E. Itoga, and H. I. Yamamura, *Life Sci.* **30,** 2191 (1982).

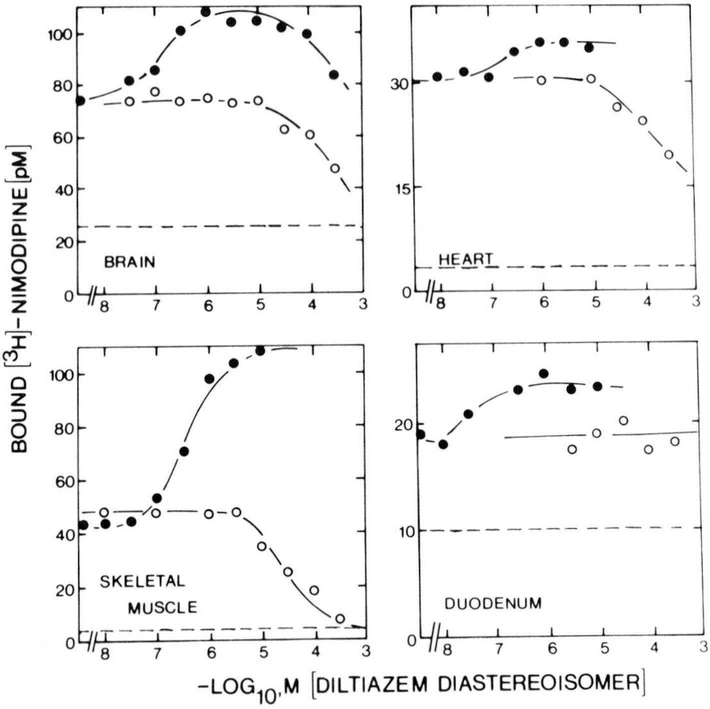

FIG. 7. Tissue-specific regulation of [^{3}H]nimodipine binding to guinea pig tissues by the diltiazem diastereoisomers. The effect of *d-cis*-diltiazem (●) and *l-cis*-diltiazem (○) on [^{3}H]nimodipine binding (0.7–1.5 nM) to guinea pig tissues at 37° is illustrated. Blanks defined by 1 μM nimodipine are indicated in the figure by dashed lines. The following protein concentrations were employed: brain 0.3–0.4 mg/ml, skeletal muscle 0.02–0.04 mg/ml, heart 0.08–0.1 mg/ml, and duodenum 0.3–0.4 mg/ml.

several other drugs belong to this class of calcium antagonists. They stimulate [^{3}H]nimodipine binding, e.g., in guinea pig brain membranes (at 37°) by acting as positive heterotropic allosteric regulators. *d-cis*-Diltiazem, at 10 μM, increases the half-life of [^{3}H]nimodipine–calcium channel complexes from 3.5 to 14 min at 37° (see Table IV).

Tissue-Specific Regulation of [^{3}H]Nimodipine Binding by the Class III Calcium Antagonist d-cis-Diltiazem. *d-cis*-Diltiazem stimulates [^{3}H]nimodipine binding to guinea pig brain, heart, skeletal muscle, and duodenual membranes, whereas *l-cis*-diltiazem does not stimulate [^{3}H]nimodipine binding[24] (Fig. 7). Figure 8 shows that the stimulation of [^{3}H]nimodipine binding in brain membranes is due to a decrease in K_D, whereas in skeletal muscle and heart membranes a B_{max} increase occurs. Duodenum mem-

[24] D. R. Ferry and H. Glossmann, *Br. J. Pharmacol.* **78,** 81p (1983).

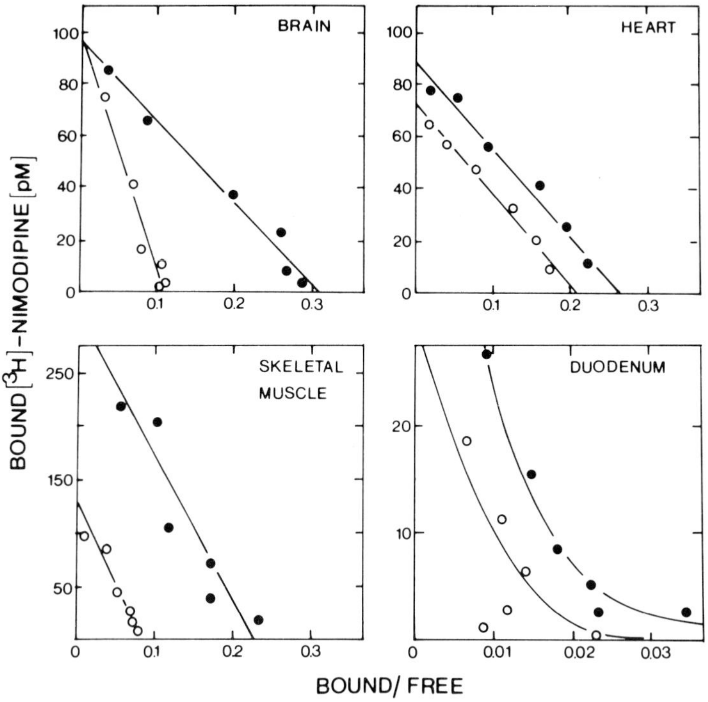

FIG. 8. Tissue-specific regulation of [^{3}H]nimodipine saturation-isotherms by *d-cis*-diltiazem. Saturation isotherms in the presence of 10 μM *d-cis*-diltiazem (●) and absence of *d-cis*-diltiazem (○). The data are shown as Hofstee plots without taking into account the racemic nature of the radioligand. Mean B_{max} and K_D values for brain and skeletal muscle membranes are given in Table IV. For guinea pig heart membranes in the absence of *d-cis*-diltiazem the B_{max} is 333 ± 13 fmol/mg of protein and the K_D of [^{3}H]nimodipine is 0.26 ± 0.003 nM. *d-cis*-Diltiazem (10 μM) increases the B_{max} to 423 ± 22 fmol/mg of protein without altering the K_D. [^{3}H]Nimodipine high affinity binding to duodenum membranes has a K_D of 0.2–0.5 nM and a B_{max} of 50–80 fmol/mg of protein; the low affinity site has a K_D of 2 to 5 nM and a B_{max} of 170–230 fmol/mg of protein.

branes are a poor source of [^{3}H]nimodipine-labeled calcium channels, and the saturation isotherms are not simple monophasic curves.

Comments. The stimulatory action of *d-cis*-diltiazem on 1,4-dihydropyridine binding is temperature dependent and optimal at 37° (vide infra).

Interactions between Class II and Class III Calcium Antagonists. In guinea pig brain membranes [^{3}H]nimodipine labeling of calcium channels is stereoselectively inhibited by (−)verapamil and (+)verapamil, but is stimulated by *d-cis*-diltiazem. When binding inhibition experiments with (−) and (+)verapamil are performed in the presence of a maximally stimulating dose of *d-cis*-diltiazem (10 μM) the latter induces characteristic alterations of the binding-inhibition curves.[21] For (+)verapamil an interconvension of the (+)verapamil drug receptor site from high to low affinity is observed. For (−)verapamil a decrease in affinity of the high affinity drug receptor site is noted (see Table IV). Although all aspects of the class II and class III drug interactions are not yet resolved, the results are reproducible and underline that without the use of optically pure phenylalkylamines spurious, misleading results are obtained. This point is emphasized by Fig. 9 in which the inhibition curves of (−) and (+)verapamil can be algebraically added to give the inhibition curve found experimentally with (±)verapamil. In guinea pig skeletal muscle microsomal mem-

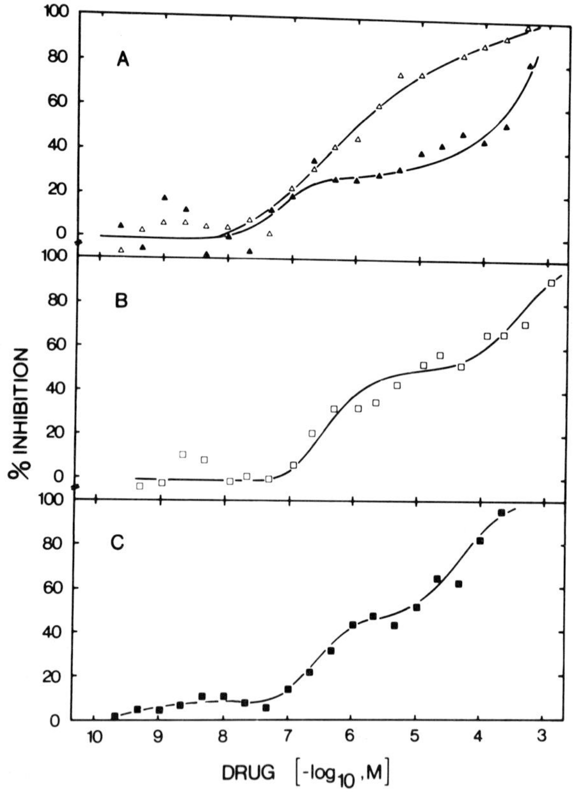

FIG. 9. Inhibition of [^{3}H]nimodipine binding by (−)verapamil, (+)verapamil, and (±)verapamil in guinea pig brain membranes. (A) (−)Verapamil (▲) and (+)verapamil (△) exhibit complex biphasic inhibition profiles, indicative of the heterogeneity of the receptor sites for these class II calcium antagonists. The experiment is performed with 0.069 mg of guinea pig brain membrane protein and 0.88 n*M* [^{3}H]nimodipine. (B) Algebraic addition of the (−) and (+)verapamil inhibition curves shown in A. (C) Inhibition profile of (±)verapamil. The experiment is performed as in A. The [^{3}H]nimodipine concentration was 0.78 n*M*.

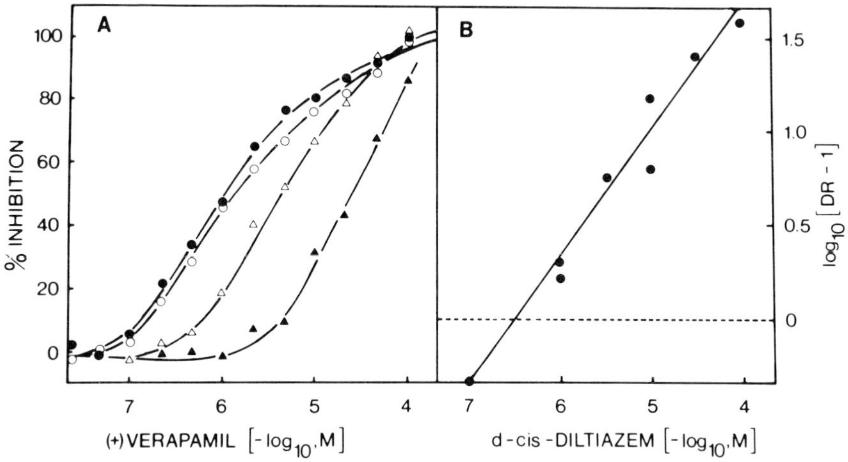

FIG. 10. Shift of the (+)verapamil inhibition curve by *d-cis*-diltiazem. Skeletal muscle microsomal membranes (at 0.04 mg of protein/ml) are incubated with 3.6 n*M* [^{3}H]PN 200-110. (+)Verapamil inhibition curves in the absence of *d-cis*-diltiazem (●), in the presence of 0.1 *M d-cis*-diltiazem (○), in the presence of 1 μM *d-cis*-diltiazem (△), and in the presence of 10 μM *d-cis*-diltiazem (▲) are shown. As (+)verapamil inhibits only 80% of [^{3}H]PN 200-110 binding to calcium channels at a saturating concentration of 100 μM, the inhibition curves are normalized (maximum inhibition by (+)verapamil is defined as 100%). The inhibition curves are then analyzed in Hill plots to obtain IC_{50} values. The IC_{50} values at the respective *d-cis*-diltiazem concentrations are used to compute dose-ratios which are plotted according to Ref. 25. In the example shown in A the IC_{50} values, Hill slopes, and regression coefficients were as follows. Zero added *d-cis*-diltiazem: 1.46 μM, 0.81, 0.97; 0.1 μM *d-cis*-diltiazem: 2.08 μM, 0.77, 0.97; 1 μM *d-cis*-diltiazem: 4.43 μM, 1.04, 0.99; and 10 μM added *d-cis*-diltiazem: 10.8 μM, 1.16, 1.00. The Schild plot in B covers a 1000-fold range of *d-cis*-diltiazem concentrations from eight experiments. The correlation coefficient is 0.98, the slope 0.68, and the *x* axis intercept at 300 n*M* *d-cis*-diltiazem. The low slope of the Schild plot illustrates that the interaction of *d-cis*-diltiazem and (+)verapamil is not one of simple competition.

branes (+)verapamil and (+)D-600 exhibit low Hill slopes for binding-inhibition and (−)verapamil causes only a partial inhibition. When calcium channels are labeled with [^{3}H]nifedipine, [^{3}H]nimodipine, or [^{3}H]PN 200-110 it can be demonstrated that the (+)verapamil binding-inhibition curve is shifted to the right by increasing concentrations of *d-cis*-diltiazem, and that the tendency is for the binding-inhibition curves to become steeper. When a family of (+)verapamil inhibition curves is constructed with increasing concentrations of *d-cis*-diltiazem and analyzed according to Arnulakshana and Schild[25] a linear plot with a slope of 0.68 is obtained, indicating that *d-cis*-diltiazem is not acting as a simple competitor of (+)verapamil (Fig. 10).

[25] O. Arunlakshana and H. O. Schild, *Br. J. Pharmacol.* **58,** 14 (1959).

Comment. The subclassification of calcium antagonists at the molecular level, performed with representative drugs, has led to the conclusion that three main drug receptor binding sites can be differentiated within the channel.[16,21,22] The selectivity of 1,4-dihydropyridines for their receptor site appears to be high, however, the selectivity of many class II and class III drugs is clearly not as high. The introduction of ligands specific for class II and class III drug receptors (vide infra) will eventually help to determine the selectivity ratios. The number of compounds active in calcium channel regulation (blockade or activation) is large and it should not be overlooked that temperature, ionic conditions, labeled drug, and tissue (e.g., brain, heart, skeletal muscle) are critical factors in the evaluation of a given drug.

Differential Labeling of Calcium Channels in Brain, Skeletal Muscle, and Heart with [^{3}H]Nimodipine, [^{3}H]Nitrendipine, [^{3}H]Nifedipine, and [^{3}H]PN 200-110

Saturation Experiments with Skeletal Muscle Microsomes. In skeletal muscle microsomes [^{3}H]PN 200-110 is a ligand with a remarkably good signal-to-noise ratio, attributable in part to the high specific density of binding sites labeled by [^{3}H]PN 200-110 (which allows assays to be performed at very low protein concentrations) and the negligible filter binding of [^{3}H]PN 200-110.

The 3-nitrophenyl 1,4-dihydropyridine derivatives [^{3}H]nimodipine and [^{3}H]nitrendipine also exhibit excellent signal-to-noise ratios for skeletal muscle microsomes, but [^{3}H]nifedipine has a worse signal-to-noise ratio attributable to higher nonsaturable binding, the lower K_D, and the smaller density of binding sites labeled by this 2-nitrophenyl 1,4-dihydropyridine derivative.[26] Table VI shows the B_{max} and K_D values in skeletal muscle microsomes for the four radiolabeled 1,4-dihydropyridines. Figure 11 shows saturation isotherms with low specific activity (ring labeled) racemic [^{3}H]PN 200-110 performed at different skeletal muscle microsomal membrane concentrations. At higher membrane concentrations failure to take into account the racemic nature of [^{3}H]PN 200-110 leads to errors in the calculation of the equilibrium-binding parameters.[27] As expected the ^{3}H-labeled (+)enantiomer of PN 200-110 has a bindability of ≥80% (not illustrated).

[26] D. R. Ferry, A. Goll, and H. Glossmann, *Naunyn-Schmiedeberg's Arch. Pharmacol.* **323,** 276 (1983).

[27] S. A. Builder and I. H. Segel, *Anal. Biochem.* **85,** 413 (1978).

TABLE V

TISSUE AND [^{3}H]1,4-DIHYDROPYRIDINE SPECIFIC LABELING OF CALCIUM CHANNELS AT 37°[a]

Tissue	Protein μg/assay	*b*	Specifically bound radioligand[b]							
			[^{3}H]Nifedipine		[^{3}H]Nimodipine		[^{3}H]Nitrendipine		[^{3}H]PN 200-110	
			Control	Diltiazem	Control	Diltiazem	Control	Diltiazem	Control	Diltiazem
Skeletal muscle	9.31	a	3391 ± 447	7890 ± 200	10593 ± 692	22506 ± 139	9713 ± 800	23190 ± 560	1078 ± 107	1513 ± 70
		b	1868 ± 246	4346 ± 111	3427 ± 224	7281 ± 45	6562 ± 540	15666 ± 378	20314 ± 2016	28511 ± 1319
Heart	15.60	a	1668 ± 207	1733 ± 51	4313 ± 164	4656 ± 125	2466 ± 151	2812 ± 245	107 ± 34	126 ± 6
		b	599 ± 68	592 ± 17	833 ± 32	900 ± 24	995 ± 61	1134 ± 99	1204 ± 383	1418 ± 68
Brain	83.67	a	3684 ± 411	4594 ± 346	7129 ± 314	10178 ± 379	5625 ± 339	6364 ± 458	251 ± 19	343 ± 36
		b	226 ± 25	282 ± 21	257 ± 11	366 ± 14	423 ± 25	478 ± 34	526 ± 40	719 ± 75

[a] Guinea pig tissues were incubated with [^{3}H]nifedipine (4.87 n*M*), [^{3}H]nimodipine (0.97 n*M*), [^{3}H]nitrendipine (3.67 n*M*), and [^{3}H]PN 200-110 (2.6 Ci/mmol, 1.16 n*M*) in the presence and absence of *d-cis*-diltiazem (10 μ*M*). Average dpm/filter of triplicates with standard error of the means are given (with blanks subtracted) and the corresponding fmol/mg of protein.

[b] Column a, dpm per filter; column b, fmol/mg of protein.

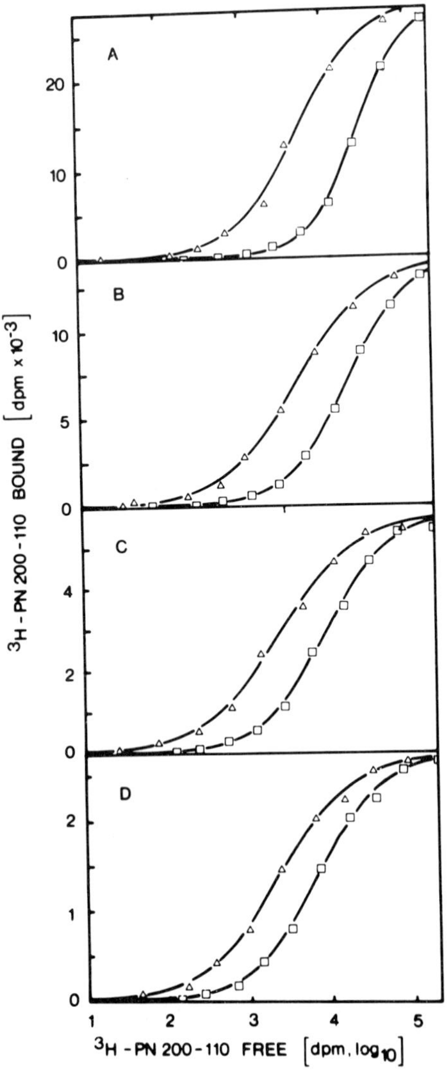

FIG. 11. Saturation isotherms of [^{3}H]PN 200-110 in skeletal muscle membranes. A to D show saturation isotherms of specific [^{3}H]PN 200-110 (specific activity: 3.31 Ci/mmol) binding to skeletal muscle microsomes at four membrane protein concentrations (A = 0.160, B = 0.077, C = 0.038, and D = 0.022 mg/ml). Each saturation isotherm was calculated with the assumption that 100% of the added radioactivity was bindable (squares) or that only 45% of the added radioactivity was bindable (triangles). The data for specific binding are plotted as "bound" dpm as a function of log "free" dpm. Lines are best fits derived from nonlinear curve fitting. The respective calculated apparent K_D values and the slope factors (± the asymptotic standard deviation) are given below. Note the high dependency of the apparent K_D on the receptor density for the 100% bindability calculation.

Differential Labeling of Calcium Channels in Brain, Heart, and Skeletal Muscle. Guinea pig membrane fractions, prepared as in methods, are incubated with [^{3}H]nifedipine, [^{3}H]nimodipine, [^{3}H]nitrendipine, and [^{3}H]PN 200-110. Typical results of parallel experiments are exemplified in Tables V and VI. Although in terms of dpm collected/filter the binding of [^{3}H]PN 200-110 is relatively low it serves to illustrate that in guinea pig heart membranes [^{3}H]PN 200-110 and the nitrophenyl [^{3}H]1,4-dihydropyridines label very similar densities of binding sites.

Labeling of the Calcium Channel with [^{125}I]Iodipine

For displacement and time-course studies the iodinated ligand is diluted in buffer B. For ~20 p*M* ligand the incubation time is 1–2 hr at 25°. The assay tubes may be counted for total radioactivity during the incubation. In a typical assay with skeletal muscle membranes, employing 10 μg of membrane protein in 0.25 ml of assay volume, when 10,000 cpm/tube is employed 4200 cpm is bound with a blank of 400 cpm. For saturation studies the ethanol solvent is removed by lyophilization in the dark; the ligand is taken up in buffer B and the assay volume is reduced to 0.1 ml. For the skeletal muscle microsomal fraction the B_{max} is 3.9 pmol/mg of protein and the K_D 390 p*M* (Table VII). [^{125}I]Iodipine is especially suited for ultrarapid autoradiographic visualization of 1,4-dihydropyridine binding sites in brain sections (vide infra).

Temperature Dependence of the d-cis-Diltiazem Effect

Labeling of calcium channels by the 1,4-dihydropyridines nimodipine and nitrendipine and the effects of *d-cis*-diltiazem, e.g., in skeletal muscle microsomes are temperature dependent (Fig. 12). The figure also illustrates that the methodology used for 1,4-dihydropyridines can be also applied to label the calcium channel with [^{3}H]*d-cis*-diltiazem (vide infra).

Protein (mg/ml)	B_{max} (fmol/mg of protein)	100% bindability		45% bindability	
		K_D (n*M*)	Slope-factor	K_D (n*M*)	Slope-factor
0.160	24,220	3.87 ± 0.18	1.48 ± 0.08	0.695 ± 0.05	1.12 ± 0.09
0.077	25,050	2.64 ± 0.08	1.19 ± 0.03	0.650 ± 0.02	0.94 ± 0.02
0.038	21,030	1.29 ± 0.07	1.10 ± 0.05	0.360 ± 0.03	0.92 ± 0.06
0.022	17,090	0.92 ± 0.03	1.08 ± 0.04	0.306 ± 0.01	0.95 ± 0.05

TABLE VI
SKELETAL MUSCLE MICROSOMES EQUILIBRIUM BINDING CONSTANTS[a]

	[^{3}H]Nifedipine		[^{3}H]Nitrendipine		[^{3}H]Nimodipine		[^{3}H]PN 200-110	
	Control	Diltiazem	Control	Diltiazem	Control	Diltiazem	Control	Diltiazem
Dissociation constant (n*M*)	4.9 ± 1.3	5.8 ± 1.6	3.6 ± 0.4	2.6 ± 0.4	3.6 ± 1.0	2.2 ± 0.3	1.4 ± 0.5	1.5 ± 0.7
B_{max} (pmol/mg of protein)	3.9 ± 0.5	8.6 ± 0.8	7.0 ± 0.8	14.7 ± 1.6	8.0 ± 0.4	14.3 ± 1.8	20.6 ± 1.4	25.4 ± 1.5

[a] The table shows the B_{max} and K_D values for the different [^{3}H]1,4-dihydropyridines and the skeletal muscle microsomal preparation. Saturation experiments were performed at 37° with and without 10 μM *d-cis*-diltiazem. From Ref. 26 with permission.

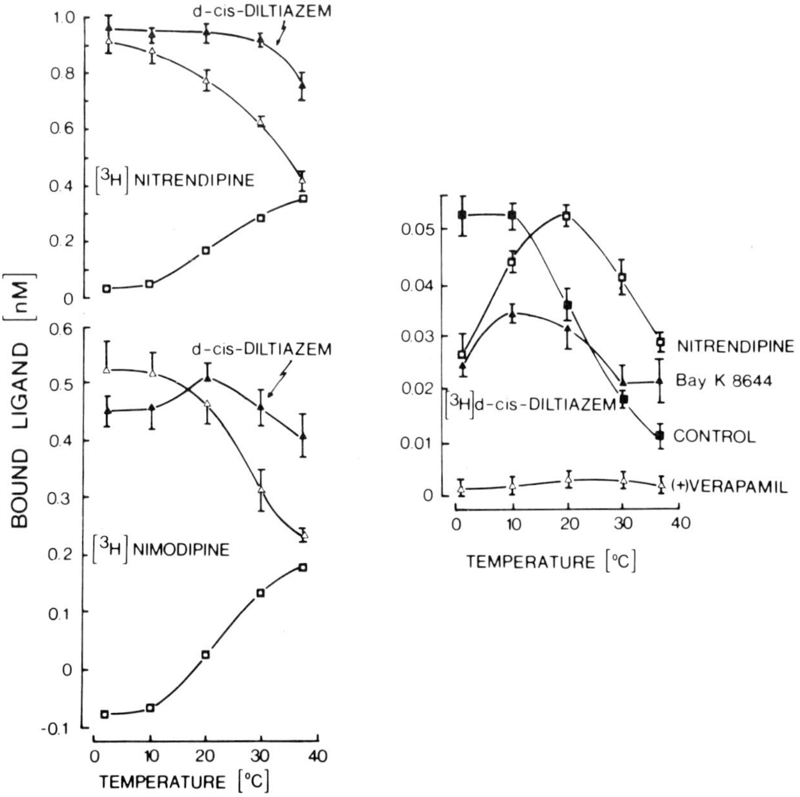

FIG. 12. Temperature-dependent binding of [^{3}H]nitrendipine (4.43 nM) and [^{3}H]nimodipine (1.56 nM) to guinea pig skeletal muscle microsomes (0.16 mg of protein/ml). Also illustrated is the temperature-dependent labeling of calcium channels with [^{3}H]*d-cis*-diltiazem (3.12 nM). For this ligand nonspecific binding is defined by either 10 μM *d-cis*-diltiazem or 10 μM (±)verapamil. [^{3}H]*d-cis*-Diltiazem labels 11 ± 1.3 pmol/mg of protein of binding sites ($n = 6$) in skeletal muscle microsomes ($K_D = 39 \pm 5$ nM) at 2° but only 2.9 ± 0.6 pmol/mg of protein ($n = 3$) at 30° with a K_D of 37 ± 9 nM. Each point in the graph is the mean of three experiments ± SEM. In the case where [^{3}H]nitrendipine and [^{3}H]nimodipine labeled the calcium channel, *d-cis*-diltiazem (10 μM) (closed symbols) was also present during the incubation; in the case where [^{3}H]*d-cis*-diltiazem labeled the channel, (±)nitrendipine (0.1 μM), (+)verapamil (1 μM), and the calcium channel agonist (see Ref. 5) Bay K 8644 [methyl-1,4-dihydro-2,6-dimethyl-3-nitro-4-(2-trifluoromethylphenyl)pyridine-5-carboxylate] (1 μM) were present during the incubation as indicated. The incubation times for the various temperatures were as follows: 2° (4 hr), 10° (3 hr), 20° (2 hr), 30° (1 hr), 37° (30 min). For the 1,4-dihydropyridine labeling experiments of calcium channels the calculated difference between the concentrations of ligand specifically bound in the absence and presence of *d-cis*-diltiazem is plotted as □. Note that the *d-cis*-diltiazem effect on 1,4-dihydropyridine labeling is temperature dependent and that the 1,4-dihydropyridines can either inhibit or stimulate the binding of [^{3}H]*d-cis*-diltiazem, depending on temperature and structure of the 1,4-dihydropyridine. From Ref. 10 with permission.

TABLE VII

BINDING CONSTANTS OF [^{125}I]IODIPINE FOR SKELETAL MUSCLE MICROSOMES[a]

Equilibrium binding and kinetic parameters	
B_{max} (fmol/mg of protein)	3900 ± 1100
Dissociation constant (p*M*)	390 ± 100
Association rate constant (nM^{-1} min^{-1})	0.100 ± 0.03
Dissociation rate constant (min^{-1})	0.06 ± 0.01

	Binding inhibition data		
Drug	IC_{50} (n*M*)	Slope factor	Max. inhibition
(+)205-033	2 ± 0.05	0.85 ± 0.12	100
(−)205-034	198 ± 40	0.88 ± 0.12	100
(−)Bay e 6927	3 ± 0.04	1.20 ± 0.16	100
(+)Bay e 6927	420 ± 67	0.92 ± 0.12	100
(+)Nicardipine	7 ± 0.8	1.24 ± 0.18	100
(−)Nicardipine	42 ± 3	1.0 ± 0.06	100
Bay K 8644	37 ± 5	0.85 ± 0.08	100
CGP 28392	1250 ± 180	0.91 ± 0.12	100
Bay M 5579	1860 ± 300	1.13 ± 0.16	100
Vo 2605	320 ± 35	0.96 ± 0.10	100
(−)Methoxyverapamil	21 ± 8	0.62 ± 0.13	70
(+)Methoxyverapamil	190 ± 50	0.50 ± 0.06	100
(−)Verapamil	98 ± 22	0.80 ± 0.12	80
(+)Verapamil	54 ± 12	0.66 ± 0.10	100
d-cis-Diltiazem	960 ± 310[b]	0.9 ± 0.27	88[c]
l-cis-Diltiazem	3000 ± 1600	1.0[d]	44
La^{3+}	110,000 ± 20,000	1.48 ± 0.38	100

[a] Average data from 3–4 experiments ± SD performed at 25°. Drug competition studies were done with 6–9 different concentrations of unlabeled drugs, each in duplicate with 20–30 p*M* [^{125}I]iodipine. The respective IC_{50} values (±asymptotic SD) and slope factors were determined by computer fitting. In the case of *d-cis*-diltiazem the experiments were performed at 37° and the EC_{50} values for stimulation were determined as above. The maximum average inhibition (%) at the highest drug concentration employed is also given. In the case of *d-cis*-diltiazem the maximal stimulation (%) is presented.

[b] In the case of *d-cis*-diltiazem: EC_{50} for stimulation.

[c] In the case of *d-cis*-diltiazem: percentage stimulation.

[d] Fitted with a slope factor fixed to 1.0. From Ref. 6 with permission.

Labeling of the Calcium Channel with [^{3}H]d-cis-Diltiazem

The assay is performed with buffer B adjusted to pH 8.2 at 2°. At this temperature the maximum number of sites is labeled in the skeletal muscle microsomal preparation.[10] Unlabeled *d-cis*-diltiazem (10 μ*M*) is used as a blank definition and the incubation time is extended to 4 hr. For equilibrium saturation studies the specific activity of the ^{3}H-labeled ligand is lowered by addition of unlabeled *d-cis*-diltiazem (hot plus cold). Care should be taken to ensure that overlapping points are obtained for the hot only case and the hot plus cold case.

Labeling of the Calcium Channel with (±) [^{3}H]Verapamil or (−)-[^{3}H]Desmethoxyverapamil

The assay is performed with buffer B adjusted to pH 8.2 at 2°. Figure 13 shows that at this temperature and pH the maximum number of sites in the skeletal muscle microsomal preparation is labeled. (±)Verapamil (10 μM) is used as a blank definition and the incubation time is extended to 4 hr with 0.02 to 0.1 mg of skeletal muscle microsomal protein. For equilibrium saturation studies the specific activity of the ^{3}H-labeled ligand is lowered by addition of unlabeled (±)verapamil. Note that the ligand is racemic but that the IC_{50} values of (+) and (−)verapamil are similar under the above conditions (Table VIII). (−)-[^{3}H]desmethoxyverapamil assays are performed at 25° for 1 hr in buffer B. The nonspecific binding definition is 1 μM (−) desmethoxyverapamil. The dissociation constant is 2.2 nM and the B_{max} 18 pmol/mg of protein.

Comment. [^{3}H]*d-cis*-diltiazem and [^{3}H]verapamil high affinity binding as well as 1,4-dihydropyridine labeling in skeletal muscle microsomes is sensitive to sulfhydryl reagents such as *N*-ethylmaleimide or *p*-chloromercuriphenylsulfonic acid. Divalent cations ($Mn^{2+} > Ca^{2+} > Mg^{2+}$) in the range 10 μM to 10 mM are inhibitory for [^{3}H]*d-cis*-diltiazem and [^{3}H]verapamil[28] binding. Care should be taken to assure that oxidation of essential sulfhydryl groups is prevented and that the ionic composition of the assay with [^{3}H]*d-cis*-diltiazem and [^{3}H]verapamil is appropriate, e.g., for the solubilized channel (vide infra) it may be necessary to dialyze. Buffer E, despite its high concentration of $MgCl_2$, is used for filtration (and precipitation) of [^{3}H]*d-cis*-diltiazem, [^{3}H]verapamil and (−)-[^{3}H]desmethoxyverapamil assays.

This section illustrates the following:

1. In order to label the "true maximum density" of calcium channels with high affinity for 1,4-dihydropyridines, a series of labeled ligands must be carefully examined.

2. In skeletal muscle microsomes and at 37°, a rank order of the density of binding sites exists, which is [^{3}H]nifedipine < [^{3}H]nitrendipine = [^{3}H]nimodipine < [^{3}H]PN 200-110.

3. In skeletal muscle microsomes the increase in density of binding sites at 37°C, induced by *d-cis*-diltiazem, is, as a percentage, smallest for [^{3}H]PN 200-110. From all the ligands so far examined [^{3}H]PN 200-110 labels the highest specific density of binding sites at this temperature and [^{125}I]iodipine the lowest.

[28] A. Goll, D. R. Ferry, and H. Glossmann, *Eur. J. Biochem.* **141,** 177 (1984).

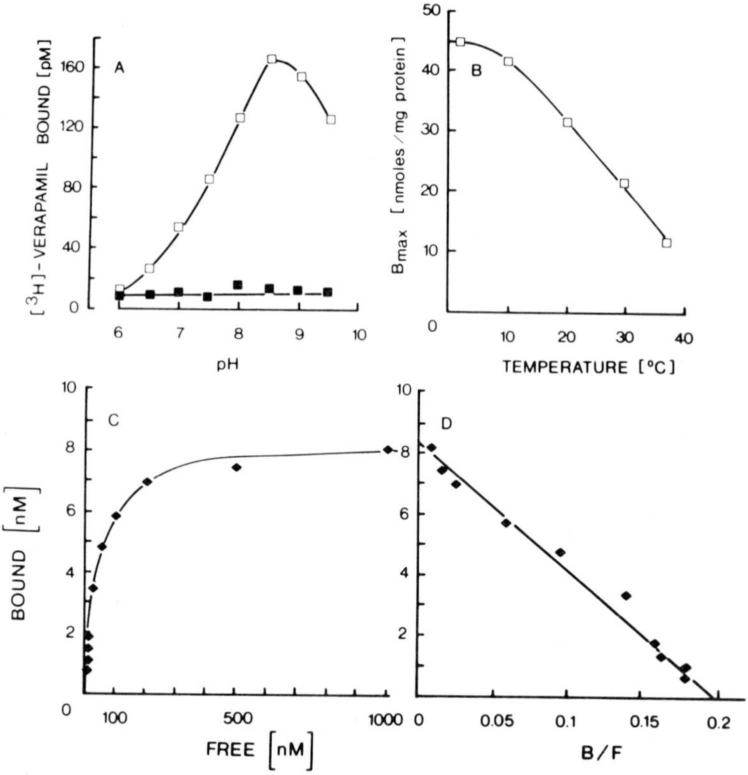

FIG. 13. (A) Dependence of [^{3}H]verapamil binding on pH. [^{3}H]Verapamil (5.89 nM) was incubated with 0.07 mg of membrane protein in the presence of 50 mM 1,3-bis-tris(hydroxymethyl) methylaminopropane buffer, adjusted with HCl to the desired pH at 2°. Open symbols are data for specific binding, closed symbols are data for nonspecific binding. (B) Saturation analysis of [^{3}H]verapamil at different temperatures. Saturation analysis was performed at different temperatures (as indicated) with 0.08 mg of membrane protein. The incubation times were 4–6 hr (2°), 4–5 hr (10°), 3–4 hr (20°), 1–2 hr (30°), 0.5–1 hr (37°). The B_{max} values, derived from Scatchard analysis, are plotted. (C) Saturation isotherm of [^{3}H]verapamil binding at 2°. The range of free ligand concentrations was 3–1000 nM, the concentration of membrane protein was 0.184 mg/ml. (D) Hofstee transformation of the data in C. The K_D was 43.0 nM, the B_{max} 8.2 nM (= 44.600 fmol/mg of protein). From Ref. 28 with permission.

4. Low specific activity radioligands (as is exemplified for ring-labeled [^{3}H]PN 200-110) can be of considerable utility in tissues with extremely high specific densities of binding sites.

5. The temperature dependence of calcium channel labeling by tritiated 1,4-dihydropyridines (as well as by tritiated verapamil and *d-cis*-diltiazem) must be investigated. At 2° and for the guinea pig skeletal muscle microsomal membranes, the 1,4-dihydropyridines nimodipine and nitrendipine, and [^{3}H]*d-cis*-diltiazem and [^{3}H]verapamil label their re-

TABLE VIII
BINDING DATA FOR [^{3}H]VERAPAMIL-LABELED CALCIUM CHANNELS[a]

Equilibrium binding data	
B_{max} (fmol/mg of protein)	37.100 ± 1.400
Dissociation constant (nM)	45.0 ± 4.2
Association rate constant (μM^{-1} min^{-1})	0.72 ± 0.14
Dissociation rate constant (min^{-1})	0.034 ± 0.002

Drug	Binding inhibition data: IC_{50} (nM)	Max. inhibition
(−)6927	3.8	94
(+)6927	102.0	92
(+)Nicardipine	12.0	97
(−)Nicardipine	90.0	95
(+)205-033	3.9	95
(−)205-034	76.0	95
Nifedipine	4.0	90
M 5579	4200.0	63
Vo 2605	430.0	88
Bay K 8644	67.0	89
CGP 28392	700.0	84
(+)D-600	67.0	100
(−)D-600	15.0	100
(+)Verapamil	30.0	100
(−)Verapamil	77.0	100
Tiapamil	221.0	100
d-cis-Diltiazem	140.0	95
l-cis-Diltiazem	14000.0	95
KB-944	700.0	85

[a] Equilibrium binding, kinetic, and binding inhibition constants for [^{3}H]verapamil-labeled calcium channels in skeletal muscle microsomal membranes of the guinea pig. All experiments were performed at 2°. Mean apparent rate constants (±SEM) were computed by fitting the experimental data as in Fig. 2. Dissociation induced by 0.3 μM (+) or (−)verapamil yields a dissociation rate constant of 0.011 min^{-1}. Binding inhibition data are means from 2 to 5 individual experiments, for which the SEM is <15%. Average IC_{50} values (at which 50% of the specifically bound radioligand is displaced) and the maximal inhibition (%) observed at the highest drug concentration (in parentheses) are given. From Ref. 28 with permission.

spective drug receptor sites maximally. In the case of [^{3}H]nimodipine and nitrendipine, addition of *d-cis*-diltiazem (10 μM) is necessary at temperatures >20° to convert the channel into a high affinity state for these radioligands.

6. The stoichiometry of the 1,4-dihydropyridine : verapamil : *d-cis*-diltiazem drug receptor sites in guinea pig skeletal muscle microsomes appears to be 2 : 4 : 1.

Autoradiographic Localization of 1,4-Dihydropyridine Binding Sites in Brain

Binding characteristics of [^{3}H]nimodipine, [^{3}H]PN 200-110, and [^{125}I]iodipine to 8 μm brain sections are essentially identical to those observed with the particulate membranes. Thus [^{3}H]PN 200-110 binds to tissue sections with a K_D of 0.6 nM in a saturable manner, and calcium channel binding of [^{3}H]PN 200-110 is inhibited by class I calcium antagonistic drugs with the appropriate potencies and stereoselectivity. Specifically bound [^{3}H]nimodipine, [^{3}H]PN 200-110, and [^{125}I]iodipine binding has a highly specific anatomical localization, which is the same for all three calcium channel labels. One of the regions with the highest density of labeling is the molecular layer of the dentate gyrus. Figure 14 shows autoradiographs for [^{3}H]nimodipine and [^{125}I]iodipine binding to sections of the guinea pig brain. The optically pure (−)-[^{3}H]desmethoxyverapamil, which binds to hippocampus membranes from the guinea pig with a K_D of 1.6 nM at 25°, labels essentially the same structures.[29]

Solubilization of Calcium Channels

Both CHAPS and digitonin can be used to solubilize calcium channels from skeletal muscle microsomal membranes.[9] Note, that the *d-cis*-diltiazem stimulation is retained in the solubilized material (see Table IV).

General Properties of Solubilized Calcium Channels. These are in many aspects similar to those of the membrane-bound channel. Thus, binding of [^{3}H]nimodipine is of high affinity, saturable, reversible, and stereoselectively inhibited by the chiral 1,4-dihydropyridines. The stability of the solubilized binding sites, kept at 0–4°, is excellent. The half-life (B_{max} decrease) is ≥60 hr. In contrast to the membrane-bound skeletal muscle calcium channel, inhibition of 1,4-dihydropyridine binding by EDTA or CDTA can be easily demonstrated (see Table IV for details). The optically pure (+)-[^{3}H]PN 200-110 at 25° labels more sites (B_{max} = 20 pmol/mg solubilized protein) with higher affinity (K_D = 0.7 nM) than [^{3}H]nimodipine and is especially suited to monitor 1,4-dihydropyridine binding during purification.

[29] D. R. Ferry, A. Goll, C. Gadow, and H. Glossmann, *Naunyn-Schmiedeberg's Arch. Pharmacol.* **327,** 183 (1984).

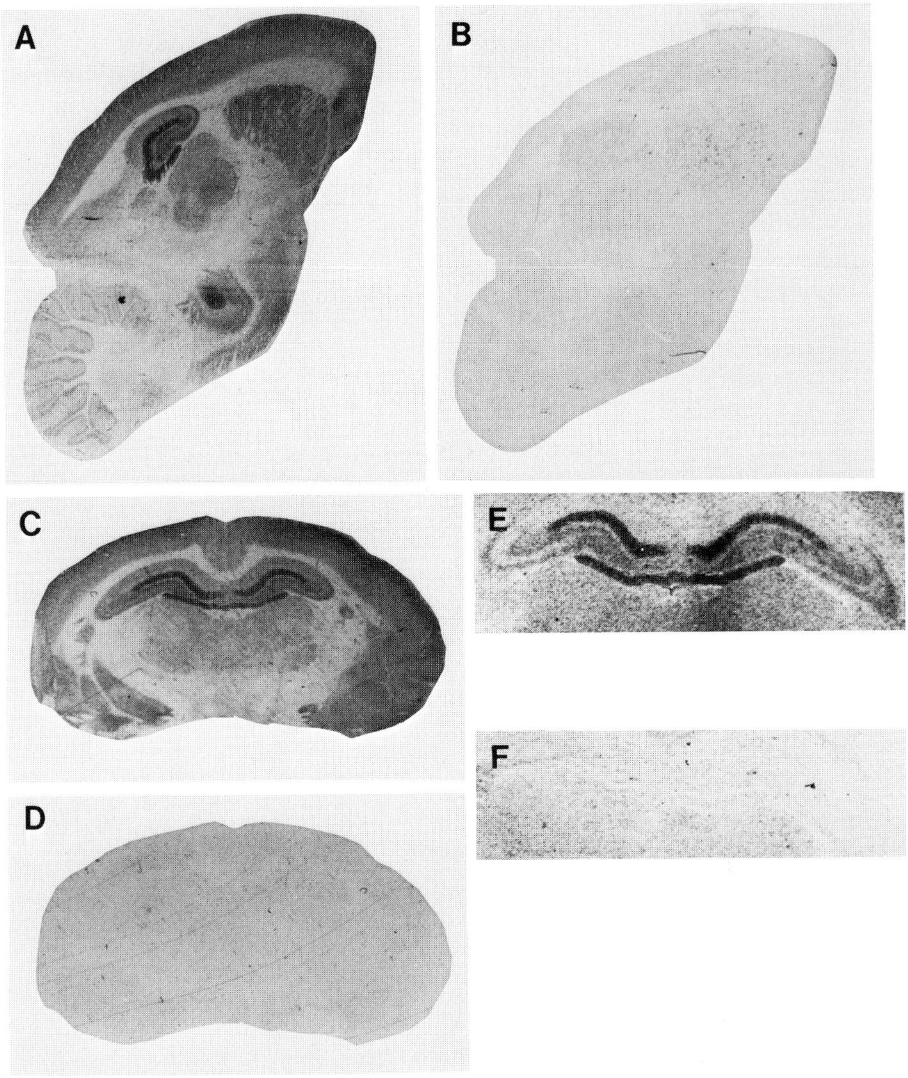

FIG. 14. Autoradiographic localization of [^{3}H]nimodipine and [^{125}I]iodipine binding in guinea pig brain. (A) Total [^{3}H]nimodipine binding to a sagittal section; (B) corresponding blank; (C) total [^{3}H]nimodipine binding to a coronal section; (D) corresponding blank; (E) total [^{125}I]iodipine binding (coronal section of hippocampus formation and dentate gyrus); (F) corresponding blank.

Sucrose Density Centrifugation of Solubilized Calcium Channels. One milliliter of the solubilized material is laid over 11.5 ml 5–20% w/w sucrose gradients in buffer D. The gradients are spun at 170,000 *g* in a Beckman SW41 rotor for 14–16 hr at 4° and then dropped out into 25–28 fractions from the bottom. Aliquots (100 μl) are taken for [^{3}H]nimodipine labeling of calcium channels, and 50-μl aliquots for protein determination and ^{14}C-marker protein counting. A typical experiment is shown in Fig. 15. The $s_{20,w}$ value of the solubilized calcium channel is 12.9 ± 0.03 S ($n = 3$) when assayed in the presence or absence of *d-cis*-diltiazem. *d-cis*-diltiazem (10 μ*M*) stimulates [^3H]nimodipine binding in all fractions where specific binding is found by 1.84 ± 0.09-fold in sucrose gradient fractions and 2.1 ± 0.13-fold in the starting material. The overall recovery of [^3H]nimodipine binding is 27–35%, with purification factors of 7–10 in the peak binding fractions and B_{max} values in the range of 25 pmol/mg of protein in the presence of 10 μ*M* *d-cis*-diltiazem.

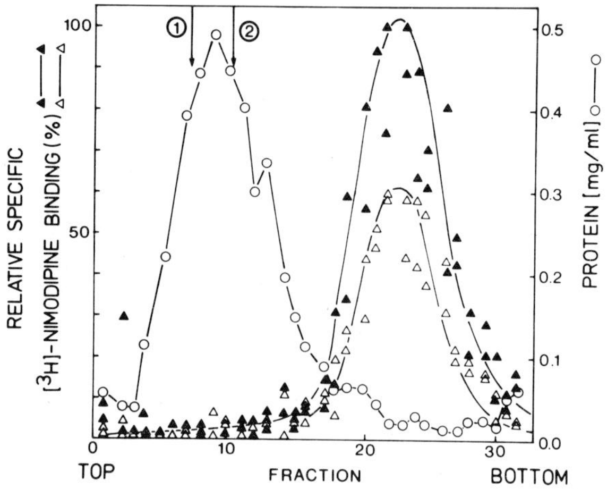

FIG. 15. Sucrose density centrifugation of solubilized calcium channels. The figure shows pooled data from two gradients: 11.5 ml of a 5–20% linear sucrose gradient was overlayed with 0.65 or 1.0 ml of solubilized skeletal muscle corresponding to 0.61 and 0.94 mg of protein, respectively. ^{14}C-labeled ovalbumin (15,000 dpm, arrow 1) or rabbit γ-globulin (10,000 dpm, arrow 2) were included in the overlays to serve as internal markers. Gradients were spun at 170,000 *g* in an SW41 rotor for 16 hr. Aliquots were assayed for specific [^3H]nimodipine (2.94 n*M*) binding in the absence (△) and presence (▲) of 10 μ*M* *d-cis*-diltiazem. Peak binding in the presence of 10 μ*M* *d-cis*-diltiazem was set to 100% and for the 1.0 ml overlay corresponds to 88,300 cpm of specifically bound [^3H]nimodipine/ml, and for the 0.65 ml overlay to 47,200 cpm/ml. The relative binding data for both overlays were then pooled and computer-fitted to the equation describing a normal distribution (using the error sums of squares principle) to determine the peak position. The peak positions of specific [^3H]nimodipine binding (± asymptotic standard deviation) in the absence of *d-cis*-diltiazem were fraction 21.98 ± 0.14; and in the presence of 10 μ*M* *d-cis*-diltiazem: 21.83 ± 0.14. The ^{14}C-labeled ovalbumin peak was at fraction 7.36 ± 0.06 and the ^{14}C-labeled γ-globulin peak at fraction 10.1 ± 0.14. The purification factor for the specific binding of [^3H]nimodipine in the presence of 10 μ*M* *d-cis*-diltiazem in the 1.0 ml overlay was 9.4-fold (corresponding to 22,800 fmol/mg of protein) and for the 0.65 ml overlay was 7.7-fold (corresponding to 18,475 fmol/mg of protein). The overall recovery of [^3H]nimodipine binding sites in this experiment was 27% measured in the presence and 35% measured in the absence of *d-cis*-diltiazem, respectively. From Ref. 9 with permission.

Chromatography of Solubilized Calcium Channels on Lectin Affinity Gels. Concanavalin A, lentil-lectin, and wheat germ agglutinin-Sepharose adsorb 1,4-dihydropyridine binding sites.[9] Adsorption and desorption conditions (with the corresponding sugars) can be tested in batch experiments. The purification factors for [^{3}H]nimodipine-labeled binding sites vary between 17- and 40-fold with recoveries between 12 and 50% when digitonin-solubilized material is run over concanavalin A-Sepharose.[9]

Summary and Conclusions

1. Calcium channels can be labeled with chiral ([^{3}H]PN 200-110, [^{3}H]nimodipine, [^{3}H]nitrendipine, [^{125}I]iodipine, [^{3}H]azidopine[30]) or symmetrical ([^{3}H]nifedipine) 1,4-dihydropyridines, [^{3}H]*d-cis*-diltiazem, [^{3}H]verapamil, and (−)-[^{3}H]desmethoxyverapamil.

2. High affinity [^{3}H]1,4-dihydropyridine binding requires certain divalent cations, e.g., Ca^{2+}, Mg^{2+}, Ba^{2+}, Sr^{2+}, or Mn^{2+}. Monovalent or trivalent cations are ineffective in this respect; instead La^{3+} (the inorganic calcium channel blocker) is inhibitory, whereas Co^{2+}, Ni^{2+}, Cd^{2+}, and Zn^{2+} can, at low doses, induce channels into high affinity for [^{3}H]1,4-dihydropyridines and at high doses they cause inhibition. The Me^{2+} requirement can be easily demonstrated for brain and heart membranes (where EDTA and CDTA inhibit [^{3}H]nimodipine binding by 95% and 60%, respectively) and solubilized skeletal muscle membranes. Divalent cations are, on the other hand, inhibitory for [^{3}H]*d-cis*-diltiazem, (−)-[^{3}H]desmethoxyverapamil, and [^{3}H]verapamil binding.

3. For chiral 1,4-dihydropyridines and at 37° only ~50% of the radioactivity added is bindable with high affinity. Calculation of the free, bindable radioligand is essential to avoid apparent dependence of the K_D on receptor concentration, spurious equilibrium binding isotherms, and wrong calculation of the association rate constants. The availability of high specific activity labeled enantiomers [as is (+)PN 200-110] circumvents this problem.

4. The optically pure enantiomers of class II and class III calcium antagonists are an absolute requirement to fully evaluate the allosteric regulation of 1,4-dihydropyridine binding.

5. Tissue heterogeneity of calcium channels is revealed by the effects of the class III calcium antagonist *d-cis*-diltiazem, the equilibrium binding dissociation constant of a given 1,4-dihydropyridine, differential effects of chelators, as well as subtype-selective drugs.

6. Skeletal muscle t-tubules are the richest known particulate source of calcium channels. A simple differential centrifugation procedure allows the preparation of a crude membrane fraction with approximately 20

[30] D. R. Ferry, M. Rombusch, A. Golland, and H. Glossmann, *FEBS Lett.* **169,** 112 (1984).

pmol/mg of protein of [^{3}H]PN 200-110 binding sites. This preparation is also useful to evaluate the binding properties of labeled verapamil (approximately 40 pmol/mg of protein binding sites) and of *d-cis*-diltiazem (approximately 10 pmol/mg of protein binding sites).

7. The temperature, at which the channels are labeled, is a critical factor as well as the radioligand itself.

8. Skeletal muscle calcium channels can be solubilized by detergents with the retention of most properties of the membrane-bound channel and can be purified by sucrose density centrifugation and lectin affinity chromatography.

9. Autoradiographic localization of 1,4-dihydropyridine binding sites in the brain exemplifies that these ligands can be employed to visualize calcium channel distribution in the CNS.

The establishment of reliable [^{3}H]1,4-dihydropyridine binding assays has facilitated the beginning of a molecular approach to calcium channels. Ligands which label other drug receptor sites within the calcium channel are complementing and extending this approach.

Acknowledgments

The authors wish to thank their colleagues at the pharmaceutical companies, especially Drs. Hoffmeister, Meyer, and Wehinger (Bayer AG), Drs. Karobath, Palacios, Tobler, Hauser, and Lindenmann (Sandoz Ltd), Drs. Bahrmann and Satzinger (Goedecke AG), and Drs. Unger, Lietz, Hollmann, and Traut (Knoll AG). C. Auriga, C. Gadow, A. Goll, B. Habermann, M. Rombusch, I. Seidel, T. Linn, A. Rücker, and W. Nürnberger shared our enthusiasm in the laboratory. We would like to pay tribute to Prof. A. Fleckenstein, the scientific father of the concept of calcium antagonism. Our work is supported by Deutsche Forschungsgemeinschaft, Bonn-Bad Godesberg.

[43] Assessment of Effects of Vasopressin, Angtiotensin II, and Glucagon on Ca^{2+} Fluxes and Phosphorylase Activity in Liver

By P. F. BLACKMORE and J. H. EXTON

Introductory Comments

Vasopressin, angiotensin II, and α_1-adrenergic angonists exert their effects on liver and certain other tissues by raising the Ca^{2+} concentration in the cytosol.[1,2] On the other hand, glucagon exerts its effects on liver by

[1] J. H. Exton, *Mol. Cell. Endocrinol.* **23,** 233 (1981).

[2] J. R. Williamson, R. H. Cooper, and J. B. Hoek, *Biochim. Biophys. Acta* **639,** 243 (1981).

ISBN 0-12-182009-2

raising cAMP, although it can increase cytosolic Ca^{2+} at high concentrations. The alterations in liver cell Ca^{2+} induced by these agents can be monitored by measuring Ca^{2+} fluxes across the plasma membrane, changes in the Ca^{2+} content of intracellular organelles or changes in cytosolic Ca^{2+}. Until recently, the latter have been very difficult to measure and changes in the activities of Ca^{2+}-sensitive intracellular enzymes have been utilized instead. The intracellular enzyme routinely used to monitor cytosolic Ca^{2+} is phosphorylase *b* kinase which phosphorylates and activates phosphorylase. Phosphorylase *b* kinase is sensitive to Ca^{2+} in the range normally encountered in the cytosol and produces a stable change in phosphorylase activity which can be assayed in cell homogenates providing certain precautions are taken (see below). Phosphorylase *b* kinase is also subject to activation by phosphorylation by cAMP-dependent protein kinase. Through this mechanism glucagon and β_2-adrenergic agonists activate phosphorylase in liver. Thus, in utilizing phosphorylase as an index of cytosolic Ca^{2+} in liver cells, it is crucial to exclude possible changes in cAMP and vice versa.

Experimental Systems—Perfused Liver and Isolated Hepatocytes

The two experimental systems currently employed to assess the effects of hormones on hepatic Ca^{2+} fluxes and phosphorylase are the isolated perfused liver preparation and suspensions of isolated liver parenchymal cells (hepatocytes). The rat is the predominant species used, although mouse, rabbit, and guinea pig have been employed.

The technique of rat liver perfusion has been described in detail previously.[3] For studies of hormone effects on Ca^{2+} fluxes and glucose release, the perfused rat liver is best used in a "flow through" mode with nonrecirculation of medium. This enables continuous assessment of Ca^{2+} release or uptake by the liver through measurements of the Ca^{2+} concentration in the influent and effluent media. Phosphorylase *a* levels can not be measured on a continuous basis since it is not feasible to take multiple biopsies from the perfused liver because of leakage and altered perfusate flow. However, glucose output can be monitored if glycogen-rich livers from fed rats are utilized. In this regard, it is important to note that glucose release is a relatively late index of phosphorylase activation, i.e., it follows a phosphorylase *a* increase by about 20–30 sec. Another important point is that glucose release may not correlate with glycogen breakdown if glycolysis is simultaneously changed.

The isolated hepatocyte preparation is generally preferable to the perfused liver for studies of hormone effects on Ca^{2+} fluxes and phosphory-

[3] J. H. Exton, this series, Vol. 37 [23].

lase. This is because it enables a larger number of variables to be studied simultaneously and greatly facilitates dose–response, additivity, and time course studies. For meaningful exploration of hormonal effects on Ca^{2+} fluxes and phosphorylase activity in isolated cells, it is essential to prepare the cells with minimal perturbation of hormone receptors and preservation of normal plasma membrane ion permeability and pumping characteristics. Hormone receptors are presumed to be intact if their affinities for agonists and antagonists resemble those of the intact liver, and membrane ion channels and pumps are considered to be normal if the ion content of the cells is within the physiological range.

The procedure for isolation of liver cells routinely used in our laboratory is based on that of Berry and Friend[4] as modified by Zahlten and Stratman.[5] Cells can be prepared from 50 to 400 g rats, but 180 to 220 g rats are routinely used. These are anesthetized with 10 mg sodium pentobarbital intraperitoneally and their livers perfused *in situ* at 37° as described in detail elsewhere.[3] The perfusion medium consists of Ca^{2+}-free Krebs–Henseleit bicarbonate buffer pH 7.4 containing 11 m*M* glucose, 5 m*M* Na pyruvate, 5 m*M* Na L-glutamate, and washed human erythrocytes (10% hematocrit). The medium is continuously gassed with $O_2 : CO_2$ (95 : 5) and the flow rate is 14 ml/min. After ~50 ml of medium is allowed to flow through the liver without recirculation, the perfusion circuit is changed to a recirculating one containing approx. 70 ml of medium. Approximately 30 mg of collagenase (Worthington Type I) is then injected into the system and perfusion is continued for ~30 min with return to the circuit of fluid leaking from the liver surface. The approximate nature of the amount of collagenase injected and of the time of perfusion is due to the fact that collagenase preparations vary in activity per milligram. An appropriately digested liver should leak from the surface profusely and be mushy to feel.

The digested liver is removed into a Petri dish containing 10 ml of the modified buffer described above containing 1% gelatin (Difco), but no erythrocytes. It is cut into small pieces with scissors, mashed gently with a spatula, and placed into a 250-ml plastic Erlenmeyer flask containing 50 ml of the same buffer. The flask is then shaken at 60 cycles/min for 10 min at 37° with continuous gassing with $O_2 : CO_2$ (95 : 5).

The flask contents (dispersed liver cells, erythrocytes, undigested liver fragments) are poured through Nylon mesh (Hanes Hosiery) into 50-ml plastic conical centrifuge tube and centrifuged at 50 *g* for 2 min (including braking). The supernatant fluids including the erythrocyte layer are

[4] M. N. Berry and D. S. Friend, *J. Cell Biol.* **43**, 506 (1969).
[5] R. N. Zahlten and F. W. Stratman, *Arch. Biochem. Biophys.* **163**, 600 (1974).

withdrawn and the sedimented cells are washed (three times) with 30 ml of oxygenated, normal Krebs–Henseleit buffer containing 1.5% gelatin (Difco). The washed cells are resuspended in the gelatin-containing Krebs–Henseleit buffer to give a final concentration of ~40 mg wet weight of cells per ml. They are then incubated for 10 min at 37° with continuous gassing with $O_2 : CO_2$ (95 : 5) prior to use in experiments. This procedure yields cells more than 95% of which exclude Trypan blue, and which have the following concentrations of intracellular constituents: 170–190 m*M* K^+, 8–12 m*M* Na^+, and 5–6 m*M* ATP.

Measurement of Hormone Effects on Perfusate Ca^{2+} in the Perfused Rat Liver Preparation

One approach to study the ability of glucagon, vasopressin, angiotensin II, and α_1-adrenergic agonists to mobilize liver Ca^{2+} utilizes the perfused rat liver preparation and measures changes in perfusate Ca^{2+} by atomic absorption spectroscopy.[6,7] Livers of rats (200–250 g) are perfused[3] at a flow rate of 7 ml/min with Krebs–Henseleit bicarbonate buffer pH 7.4 containing 3% (w/v) bovine serum albumin (Fraction V, Pentex) and washed expired human erythrocytes at 25% hematocrit. The perfusion medium is not recirculated and is continuously gassed with $O_2 : CO_2$ (95 : 5). Its Ca^{2+} concentration may be varied from the micromolar range to the millimolar range, depending on what is being studied. Samples of perfusate are collected sequentially for fixed times using a fraction collector. When the perfusate Ca^{2+} is low, e.g., 50 μM, at least 0.2 ml of perfusate per fraction needs to be collected in order to measure total Ca^{2+} by atomic absorption spectroscopy. The samples of perfusate are centrifuged at ~1500 *g* for 10 min to sediment erythrocytes and an aliquot of the supernatant is added to an equal volume of 0.6 *M* $HClO_4$ containing 0.1% (w/v) $LaCl_3 \cdot 6H_2O$ and centrifuged at ~1500 *g* for 10 min to remove denatured protein. When perfusate Ca^{2+} is in the more physiological range, the samples have to be diluted ~10-fold with the perchlorate–lanthanum diluent.

The level of Ca^{2+} in the supernatant fluid is determined using an atomic absorption spectrophotometer (Perkin-Elmer model 603). Industrial grade acetylene is the fuel and air is used as the oxidant. Calcium standards ranging up to 150 μM are prepared in $HClO_4/LaCl_3$ diluent. Hormones and agents to be tested are prepared at a 100-fold higher concentration than that required in the perfusate. The agents are then infused

[6] P. F. Blackmore, J.-P. Dehaye, and J. H. Exton, *J. Biol. Chem.* **254,** 6945 (1979).

[7] N. G. Morgan, E. A. Shuman, J. H. Exton, and P. F. Blackmore, *J. Biol. Chem.* **257,** 13907 (1982).

into the tubing leading to the portal cannula as close as possible to the liver using an infusion/withdrawal pump (Harvard Apparatus) at a flow rate 100-fold less than the perfusate flow rate. To determine the perfusate sample first exposed to hormone, a trace amount of radioactively labeled hormone may be included with the hormone being infused. With vasopressin, angiotensin II, α-adrenergic agonists, and glucagon as stimuli, an increase in perfusate Ca^{2+} is observed, although with glucagon less Ca^{2+} is mobilized than observed with the other hormones.[6] The changes in perfusate Ca^{2+} can be correlated with decreases in intracellular Ca^{2+} pools, in particular mitochondria and a fraction enriched in glucose-6-phosphatase.[6] Upon removal of the Ca^{2+} mobilizing stimulus, Ca^{2+} influx into the liver can be seen by a decrease in perfusate Ca^{2+}, especially when the perfusate Ca^{2+} concentration is in the mM range.[7]

Measurements of Hormone Effects on Total Hepatocyte Ca^{2+} Using Atomic Absorption Spectroscopy

Another approach to analyzing hormone effects on liver Ca^{2+} is to measure the changes in total Ca^{2+} in hepatocytes using atomic absorption spectroscopy. Isolated hepatocyte suspensions are equilibrated at 37° in Krebs–Henseleit bicarbonate buffer pH 7.4 with continuous gassing ($O_2 : CO_2/95 : 5$) for at least 10 min after isolation. This allows reaccumulation of Ca^{2+} by the cells since the perfusion with collagenase is carried out with perfusate containing ~50 μM Ca^{2+}. This equilibration period also allows intracellular K^+ and Na^+ to be restored to normal levels.

For time course experiments, 10 ml of hepatocyte suspensions is incubated with constant shaking (70–90 cycles/min) in 250 ml polypropylene or polycarbonate Erlenmeyer flasks with continuous gassing with $O_2 : CO_2$ (95 : 5) at a flow rate of 3–4 liters/min. For dose–response experiments, 2 to 3 ml of hepatocyte suspensions is incubated in 25 ml polycarbonate or polypropylene Erlenmeyer flasks. Aliquots of hepatocyte suspensions ranging between 0.5 and 1.0 ml are removed at appropriate times and layered on 10 ml of an ice-cold solution of 150 mM NaCl, 10% (w/v) sucrose, and 0.5 mM EGTA (pH 7.4) contained in 12-ml conical pyrex test tubes. The tubes are then centrifuged briefly (20 sec at full speed ~3000 rpm) to sediment the cells in either an IEC HN-SII centrifuge (Damon/International Equipment Division) with a 958 rotor or an IEC size 2 model SBV centrifuge fitted with a 240 rotor and a mechanical foot brake.

After centrifugation, the supernatant solution is removed by aspiration and the tubes are inverted and allowed to drain for ~5 min. The insides of the tubes are then rinsed with distilled water and allowed to drain for a further 5 min after which the insides of the tubes are wiped dry with facial tissues. The cell pellets are then dispersed into 0.5 ml of added water using a vortex mixer, and 0.5 ml of 0.6 *M* $HClO_4$ containing 0.1% (w/v) $LaCl_3 \cdot 6H_2O$ is added to each sample to precipitate protein. The solutions are then centrifuged at 2000 *g* for 5 min and the Ca^{2+} in the supernatant fluid is determined by atomic absorption spectroscopy as described in the preceding section. Typically, basal Ca^{2+} readings are 0.03 to 0.04 absorbance units when the wet weight of the cell suspensions is ~40 mg/ml. The contamination of the cells with extracellular Ca^{2+} rarely exceeds 3–5% of cell Ca^{2+} thus making correction for extracellular Ca^{2+} unnecessary.

Since the effects of many hormones on net cell Ca^{2+} content are relatively small, duplicate samples should be taken and incubations carried out in duplicate or triplicate. For separation of cells and medium 7% (w/v) bovine serum albumin (Pentex) can be used instead of sucrose, and 1% (w/v) $LaCl_3 \cdot 6H_2O$ can be used instead of 0.5 m*M* EGTA to displace extracellular bound Ca^{2+}.[8] The same results are obtained with either method; however sucrose is more economical and more effectively minimizes the mixing of the incubation medium with the cell pellet.

Measurement of Hormone Effects on Cytosolic Free Ca^{2+} ($[Ca^{2+}]_i$) in Hepatocytes

One of the most important parameters to measure in analyzing the effects of hormones on liver cell Ca^{2+} is cytosolic free Ca^{2+} ($[Ca^{2+}]_i$]. However, measurement of $[Ca^{2+}]_i$ has proved to be most difficult. A null point titration method has been developed for hepatocytes[9] and has proved very successful, but continuous measurements cannot be made with this technique. A better method involves the use of a fluorescent analog of EGTA (Quin-2).[10] This compound shows a 5-fold increase in fluorescence when Ca^{2+} is bound. Furthermore it can be introduced into cells as the membrane-permeant tetraacetoxymethyl ester (Quin-2/AM) which is then hydrolyzed by intracellular esterases to yield the membrane-impermeant free acid. The Quin-2 thus accumulates in the cytosol and permits the measurement of $[Ca^{2+}]_i$ on a continuous basis. The use of Quin-2 to monitor hormone effects on $[Ca^{2+}]_i$ in hepatocytes[11] is described below.

Following a preincubation period of 10–15 min, hepatocytes (50–60

[8] P. F. Blackmore, F. T. Brumley, J. L. Marks, and J. H. Exton, *J. Biol. Chem.* **253,** 4851 (1978).

[9] E. Murphy, K. Coll, T. L. Rich, and J. R. Williamson, *J. Biol. Chem.* **255,** 6600 (1980).

[10] R. Y. Tsien, *Biochemistry* **18,** 2396 (1980).

[11] R. Charest, P. F. Blackmore, B. Berthon, and J. H. Exton, *J. Biol. Chem.* **258,** 8769 (1983).

mg/ml) prepared as described above are loaded with Quin-2 by incubating them (5 ml in 25-ml polypropylene Erlenmeyer flasks) in the presence of 0.1 m*M* Quin-2/AM (Lancaster Synthesis, Ltd, Morecambe, Lancashire, England LA3 3DY) for 15 min at 37°. The stock solution of Quin-2/AM (50 m*M*) is in Me_2SO, hence control cells are treated with an equivalent amount of Me_2SO. The cells are then sedimented (50 *g* for 1 min) and resuspended to ~40 mg/ml in Krebs–Henseleit bicarbonate buffer containing 20 m*M* HEPES (pH 7.4), 5 mg/ml gelatin, and 0.5–2.0 m*M* Ca^{2+}. After incubation for 15 min with continuous gassing ($O_2 : CO_2$, 95 : 5) 3 ml of the cell suspension is transferred to a 12 × 50-mm borosilicate glass test tube containing a triangular shaped magnetic stirring bar. The test tube containing the cells is then placed in the cell holder of a spectrofluorometer (Varian Model SF-330) fitted with a magnetic stirrer and a thermostatically controlled cell housing. The cells are continuously stirred and oxygenated by the infusion of $O_2 : CO_2$ (95 : 5) through polyethylene tubes (Intramedic PE50, Clay Adams). Infusion of hormones and agents is made through another piece of PE50 tubing ~40 cm in length into the hepatocyte suspension, thus eliminating the opening of the cell housing.

Excitation and emission wavelengths are 340 and 500 nm, respectively, with 10 nm slit widths. The output from the spectrofluorometer (mV) may be recorded graphically directly. However, for quantitation of data it is better to channel the signal into an Analog/Digital converter (Hewlett Packard, mVs, sampled twice each second) and store the data in a computer (Hewlett Packard 3356 Lab Automation System). Routinely, data are collected for 30 sec to 1 min before hormone addition, then for various periods of time up to 30 min after hormone addition. The results are expressed as change in fluorescence (mVs) from time zero which is the time of hormone injection. The advantage of storing the data in raw data computer files is that many experiments can be averaged and, more importantly, the changes in fluorescence in control (non-Quin-2 loaded) cells can be subtracted from those in the Quin-2 loaded cells. Changes in control cell fluorescence are generally due to alterations in reduced pyridine nucleotides which are detectable at the wavelengths used to monitor Quin-2 changes. The changes in pyridine nucleotide fluorescence induced by hormones in control cells are smaller and slower than the Ca^{2+}-dependent changes in fluorescence seen in the Quin-2 loaded cells.[11] However, the changes in the control cells should always be measured and subtracted from those in the Quin-2 loaded cells to provide reliable data on Ca^{2+} changes.

To quantitate the fluorescence measurements in terms of probable $[Ca^{2+}]_i$ values, the following procedures are employed. After the fluorescence measurements are made, the cells are lysed with Triton X-100 (10 mg/ml) and the fluorescence at high Ca^{2+} (1.0 mM Ca^{2+}, F_{max}) and low Ca^{2+} (2.0 mM EGTA or 1.0 mM Mn^{2+}, F_{min}) is determined. The $[Ca^{2+}]_i$ is then calculated using the equation

$$[Ca^{2+}]_i = K_D(F - F_{min})/(F_{max} - F)$$

The K_D value used in the equation is 115 mM, which was determined under ionic conditions which closely mimic the cytosol. Since it is not certain that this K_D is the same as that for Quin-2 in the cytosol of cells, and since the values for F_{max} and F_{min} are determined under different conditions (i.e., in Triton X-100 extracts), than the values for F, the calculated values for $[Ca^{2+}]_i$ are potentially open to error. However, relative and directional changes are highly reliable.

Measurement of Hormone Effects on Phosphorylase in Hepatocytes

As described in the first section, hormones can activate phosphorylase in liver by two major mechanisms, one involving cAMP-dependent activation of phosphorylase *b* kinase and the other involving Ca^{2+} stimulation of phosphorylase *b* kinase. Thus changes in phosphorylase *a* have been used extensively to analyze hormone actions in the liver. The isolated hepatocyte preparation is preferable to the perfused liver to study hormone effects on phosphorylase for several reasons given above. In particular, it permits multiple samples to be taken from a single preparation, which is of great advantage in time course and dose-dependence studies. Since hormone effects on liver phosphorylase are very rapid, appearing within 2–10 sec, and since phosphorylase can become activated or inactivated during the handling or homogenization of cell samples, two important features that need to be considered in studying phosphorylase in hepatocytes are the rapid fixation of the samples and their careful processing prior to assay. As described below, samples are best fixed by immersion in liquid N_2 and activation artifacts are best avoided by thawing and homogenizing samples at 0° in the presence of EDTA (to inhibit protein kinase action) and NaF (to inhibit phosphoprotein phosphatase action).

Following a preliminary incubation of 10–15 min, suspensions of hepatocytes are exposed to hormones at various concentrations, and aliquots (0.5 ml) are withdrawn at various times for rapid addition to polypropylene test tubes immersed in liquid N_2. This procedure instantaneously freezes the samples. For very rapid times of hormone exposure, e.g., 1–5 sec, 0.5 ml of cell suspension is added to the hormone contained in polypropylene tubes at 37°. The tubes were then rapidly frozen at the required times thereafter. Frozen samples can be stored at −70° for many months without change in phosphorylase activity.

For homogenization of samples, the following buffer is used: 138 mM Na β-glycerolphosphate, 138 mM NaF, 28 mM Na_2EDTA, 550 mM sucrose, and 28 mM cysteine HCl. The pH of the buffer is adjusted to 6.0 and the cysteine HCl is added just before use. Prior to homogenization 1.3 ml of the buffer is added to the frozen cell suspension (0.5 ml) and the mixture allowed to thaw at ice temperature. The mixture is then homogenized for 10–15 sec at 0° (Ultra Turrax or Tissumizer, Tekmar).

For assay,[12] 50 μl of the homogenate is added to 50 μl of assay mixture of the following composition: 100 mM glucose 1-phosphate, 2% repurified oyster glycogen,[13] 1 mM caffeine, 0.5 μCi/ml [U-^{14}C]glucose 1-phosphate. After incubation at 30° for 15 min, 50 μl of the mixture is spotted onto a numbered 2 × 2 cm piece of filter paper (Whatman 3 MM) which is then placed into a beaker containing 66% ethanol stirred magnetically. The papers from an assay are washed 3 times for 20 min with 66% ethanol. To avoid disintegration, the papers are separated from the magnetic stirring bar by a stainless-steel mesh. The papers are then briefly washed with acetone and dried with a hair dryer. The dry papers are added to 10 ml of toluene-based scintillation fluid and counted in a scintillation spectrometer. Following counting, the papers can be removed from the scintillation fluid and this can be reused up to 10 times without an appreciable increase in background radioactivity.

The preceding method assays phosphorylase *a* only, since caffeine is present to inhibit the small stimulatory effect of AMP in liver phosphorylase *b*.[12] Total phosphorylase can be assayed using dimethoxyethane.[14] However, since rapidly acting hormones do not alter total phosphorylase levels in this tissue, assays of phosphorylase *a* alone are generally a valid measure of the activation state of the enzyme.

[12] W. Stalmans and H.-G. Hers, *Eur. J. Biochem.* **54,** 341 (1975).
[13] R. A. Anderson and D. J. Graves, *Biochemistry* **12,** 1895 (1973).
[14] R. J. Uhing, A. M. Janski, and D. J. Graves, *J. Biol. Chem.* **254,** 3166 (1979).

[44] Pertussis Toxin: A Tool for Studying the Regulation of Adenylate Cyclase

By RONALD D. SEKURA

Regulation of adenylate cyclase is mediated, in part, through the action of two specific GTP binding proteins which couple the interaction of hormones at specific receptors to modulation of catalytic activity.

ISBN 0-12-182009-2

These proteins, termed N_s and N_i have been purified[1-3] and function respectively in expression of stimulatory and inhibitory ligands. The susceptibility of N_i and N_s to specific bacterial toxins, which block normal function of the coupling proteins by covalent modification, has led to a clearer understanding of the role played by these proteins in regulation of adenylate cyclase.[4-7] With both proteins, toxin catalyzed modification is achieved by transfer of the ADP-ribosyl moiety of NAD to the 39 K subunit of N_i or the 42 K subunit of N_s. Covalent modification of N_s is mediated by the action of cholera toxin and is discussed in detail elsewhere.[8] Specific modification of N_i is achieved by treatment of purified preparations of protein or cell membrane with pertussis toxin.[9-11] Techniques for the preparation of pertussis toxin and its use in probing the function of N_i are discussed below.

Growth of Bordetella pertussis

Since *B. pertussis* is a pathogenic organism appropriate precaution should be taken to minimize the potential for laboratory acquired infection. Transfer of cultures should be done in a laminar flow hood and care should be taken to minimize the generation of aerosols. Use of sealed centrifuge rotors and not overfilling centrifuge tubes minimizes the potential for generation of infectious aerosols.

Strains. In the described preparation, *B. pertussis* strain 165 is used for production of pertussis toxin. The use of strain 65 is recommended since good growth and relatively high production of toxin is consistently achieved.[12] Cultures of *B. pertussis* strain 165 can be obtained from the

[1] J. K. Northup, P. C. Sternweis, M. D. Smigel, L. S. Schteifer, E. M. Rose, and A. G. Gilman, *Proc. Natl. Acad. Sci. U.S.A.* **77,** 6516 (1980).

[2] E. Hanski and A. G. Gilman, *J. Cyclic Nucleotide Res.* **8,** 323 (1982).

[3] J. Codina, N. D. Hildebrandt, R. D. Sekura, M. Birnbaumer, J. Brejan, C. Manclark, R. Iyengar, and L. Birnbaumer, *J. Biol. Chem.* **259,** 5871 (1984).

[4] G. L. Johnson, H. R. Kaslow, and H. R. Bourne, *J. Biol. Chem.* **253,** 7120 (1978).

[5] D. M. Gill and R. Mersen, *Proc. Natl. Acad. Sci. U.S.A.* **75,** 3050 (1978).

[6] T. Katada and M. Ui, *Proc. Natl. Acad. Sci. U.S.A.* **79,** 7059 (1982).

[7] J. D. Hildebrandt, R. D. Sekura, J. Codina, R. Iyengar, C. R. Manclark, and L. Birnbaumer, *Nature (London)* **302,** 706 (1983).

[8] J. Codina, W. Rosenthal, J. D. Hildebrandt, L. Birnbaumer, and R. D. Sekura, this volume [38].

[9] T. Katada and M. Ui, *J. Biol. Chem.* **257,** 7210 (1982).

[10] D. L. Burns, E. L. Hewlett, J. Moss, and M. Vaughan, *J. Biol. Chem.* **258,** 1435 (1983).

[11] J. Codina, J. Hildebrandt, R. Iyengar, L. Birnbaumer, R. D. Sekura, and C. R. Manclark, *Proc. Natl. Acad. Sci. U.S.A.* **80,** 4276 (1983).

[12] R. D. Sekura, F. Fish, C. R. Manclark, B. Meade, and Y. Zhang, *J. Biol. Chem.* **258,** 14647 (1983).

author or from the culture collection at the National Center for Drugs and Biologics, Food and Drug Administration, Bethesda, MD 20205. Stock cultures are stored *in vacuo* after lyophilization with skimmed milk.

Media. Petri plates containing Bordet-Gengou agar (Difco Base) supplemented with 15% defibrinated sheep blood are used for preparation of stock cultures and to start growth from lyophilized cultures. For growth of the organism in liquid culture, modified Stainer-Scholte media[13] is used. Each liter of media contains 10.72 g sodium L-glutamate, 2.5 g NaCl, 0.5 g KH_2PO_4, 0.2 g KCl, 0.1 g $MgCl_2 \cdot 6H_2O$, 0.02 g $CaCl_2$, 0.24 g L-proline, and 0.24 g Tris. The pH is adjusted to 7.6 and the solution is sterilized by autoclaving. At the time cultures are innoculated, a filter sterilized supplement (0.22 μm) is added at a rate of 1 ml/100 ml of culture medium. The supplement contains, for each 50 ml, 0.2 g L-cystine, 0.05 g $FeSO_4 \cdot 7H_2O$, 0.1 g ascorbic acid, 0.02 g niacin, and 0.5 g glutathione. Solution of cystine is facilitated by dissolving the amino acid in 0.5 ml concentrated HCl prior to the addition of other reagents.

Culture. In general each large scale growth of the organism is initiated by opening a tube of lyophilized bacteria. The sealed ampule is opened aseptically and the contents is suspended in about 0.2 ml of sterile distilled water. The suspension is used to innoculate about six Bordet-Genagou agar plates. After 3 days at 37° the growth from each plate is split and used to innoculate two fresh plates which are incubated overnight at 37°. The growth from these plates can be used to innoculate additional plates, or can be used to innoculate seed cultures. Growth in Stainer Scholte media (200 ml media in a 500-ml flask) is initiated by suspending the growth from two Brodet-Gengou plates in a small amount of medium and innoculating the flask with this suspension. Starter cultures are incubated overnight at 37° on a gyrorotary shaker set at about 150 rpm. For production of pertussis toxin the bacteria are grown in 2.8-liter baffled Fernbach flasks containing 1.3 liters of Stainer Scholte medium. These flasks are innoculated with starter culture to an initial A_{650} of between 0.05 and 0.1. The bacteria are grown at 37° on a gyrorotary shaker set at 120 rpm. Culture is complete after about 40 to 60 hr when the bacteria have achieved an A_{650} of about 2.7–3.0.

Assay of Pertussis Toxin

Most techniques for the assay of pertussis toxin are relatively difficult and require several days before the results are available. The hemagglutination activity of the toxin, first described by Irons and MacHennan,[14]

[13] E. L. Hewlett and J. Wolff, *J. Bacteriol.* **127,** 890 (1976).

[14] L. I. Irons and A. P. MacHennan, *Biochim. Biophys. Acta* **580,** 175 (1979).

however, provides a simple and rapid assay and is well suited for following the toxin during purification. However, the results obtained using this assay must be viewed with caution since *B. pertussis* also produces filamentous hemagglutinin (FHA) which is a much more potent agglutinating agent. Since FHA is produced in low concentration in shake culture, and since FHA and pertussis toxin are resolved during chromatography on glycoprotein affinity columns[15] this is generally not a problem in monitoring the purification described below. Final characterization of toxin preparations should include monitoring a pertussis toxin activity such as histamine sensitization,[16] leukocytosis promotion,[17,18] determination by an immunologic technique,[15] or by one of the newer *in vitro* methods.[19]

Hemagglutination Assay. The hemagglutinating activity in preparations containing pertussis toxin is monitored by following conventional assay techniques. Fractions containing toxin are serially diluted in microtiter plates with V-shaped wells. After dilution each well contains 50 μl of solution; 50 μl of a suspension containing 0.7% fresh goose red blood cells (in phosphate-buffered saline) is then added and the plates are agitated for 60 sec on a rotary shaker (such as the Dynatech Micro Shaker II). Pertussis toxin mediated hemagglutination is somewhat unique in that at higher toxin concentrations, discrete clumps of agglutinated red blood cells are seen at the bottom of the wells; at lower toxin concentrations the hemagglutination appears more conventional. One unit of hemagglutination activity is defined by the maximal dilution at which hemagglutination is observed. The hemagglutination units per milliliter of the original sample are calculated from this value.

Purification of Pertussis Toxin

Several methods have been described for the purification of pertussis toxin.[14,17,20,21] The method described here is relatively simple and produces toxin in high yields.

Pertussis toxin is prepared from the culture supernatant of bacteria grown on Stainer–Scholte media. A supernatant solution free from most bacteria is prepared by centrifugation at 6000 *g* and 4° for 30 min. After

[15] Y. Sato, J. L. Cowell, H. Sato, D. G. Burstyn, and C. R. Manclark, *Infect. Immun.* **41,** 313 (1983).

[16] J. J. Munoz and R. K. Bergman, *Bacteriol. Rev.* **32,** 103 (1968).

[17] S. I. Morse and J. H. Morse, *J. Exp. Med.* **143,** 1483 (1976).

[18] H. Aria and Y. Sato, *Biochim. Biophys. Acta* **444,** 765 (1976).

[19] M. Endoh and Y. Nakase, *Microbiol. Immunol.* **26,** 689 (1982).

[20] H. Arai and J. J. Munoz, *Infect. Immun.* **31,** 495 (1981).

[21] M. Yajima, K. Hosoda, Y. Kanbayashi, T. Nakamura, K. Nogimori, Y. Mizushimo, Y. Nakari, and M. Ui, *J. Biochem.* (*Tokyo*) **83,** 295 (1978).

removing the bacteria the supernatant is adjusted to pH 6.0 using HCl. This step is important and should be performed as soon as possible since pertussis toxin is somewhat unstable at the alkaline pH (generally greater than pH 8.0) of the original culture supernatant. The subsequent steps of the purification include the use of two affinity matrixes and a desalting step to obtain preparations of toxin which are apparently homogeneous. Although the procedure below is given for 20 liters of culture supernatant, the technique can be successfully applied to smaller volumes of culture supernatant by scaling down the quantities of AffiGel Blue (Bio-Rad 50–100 mesh) and fetuin agarose[22] used.

AffiGel Blue. Pertussis toxin (in 20 liters of culture supernatant) is adsorbed to AffiGel Blue (200 ml) in a batch wise fashion at 4°. After at least 24 hr of stirring the resin is allowed to settle and the supernatant solution is removed. The resin is collected and transferred to a column (6 × 7 cm) which is eluted with the following solutions at 25°: (1) 400 ml, 0.25 *M* sodium phosphate, pH 6.0; (2) 400 ml, 0.05 *M* Tris–HCl, pH 7.4; (3) 400 ml, pH 7.4 buffer containing 0.05 *M* Tris–HCl and 0.75 *M* $MgCl_2$. The column is monitored by following hemagglutination activity (see Fig. 1A) which is observed to elute with buffer 3. Fractions containing more than 10,000 hemagglutination units per ml are pooled and carried through subsequent purification steps.

Chromatography on Fetuin Agarose (Fig. 1B). The pooled fractions from the AffiGel Blue column are diluted with an equal volume of distilled water and applied to a fetuin agarose column (2.5 × 6 cm) which is eluted at 25° with the following buffers: (1) 40 ml pH 7.4 buffer containing 0.05 *M* Tris–HCl; (2) 40 ml pH 7.4 buffer containing 0.05 *M* Tris–HCl and 1 *M* NaCl, and (3) 40 ml pH 7.4 buffer containing 0.05 *M* Tris–HCl and 4.0 *M* $MgCl_2$. The elution of pertussis toxin by buffer 3 is monitored by following the hemagglutination activity. Fractions containing more than 50,000 hemagglutination units are pooled and desalted as follows.

Chromatography on Sephadex G-25. A sample of the pooled toxin (not more than 20 ml for each column run) is applied to a column (2.5 × 35 cm) of Sephadex G-25, fine, equilibrated with pH 7.4 buffer containing 0.05 *M* Tris–HCl and 0.5 *M* sodium chloride. The elution of protein is monitored by following the A_{280} of the column effluent, if hemagglutination activity is monitored it is seen to parallel the protein peak although some trailing of hemagglutination activity is observed. Fractions having an A_{280} of more than 0.05 are pooled and pertussis toxin is precipitated

[22] The fetuin affinity resin was prepared using fetuin prepared by the Spiro method (Gibco) and cyanogen bromide activated Sepharose 4B (Pharmacia). Fetuin (200 mg) was coupled to 25 g of resin according to the manufacturer's recommended procedure.

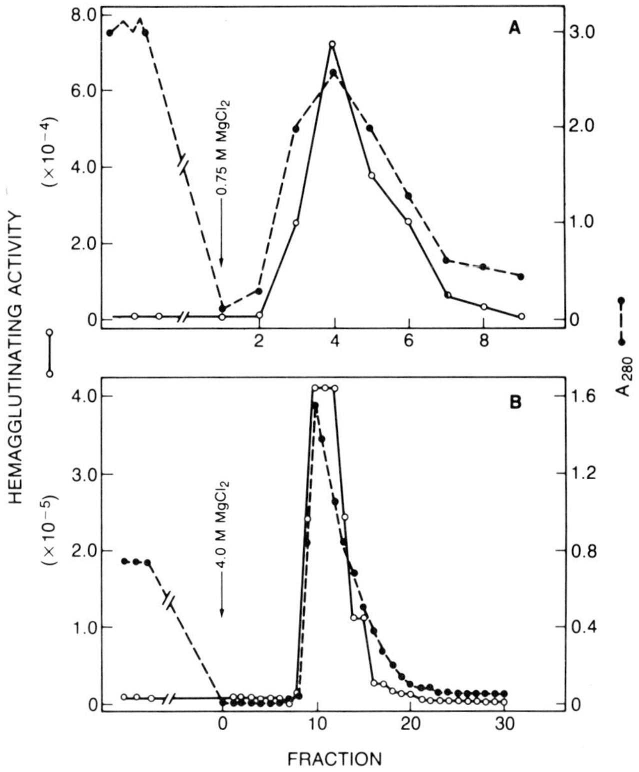

FIG. 1. Purification of pertussis toxin. (A) After absorption of toxin from the culture supernatant onto an AffiGel Blue, the resin was packed into a column and washed as described in the text. Pertussis toxin was eluted from the column with the $MgCl_2$ wash. (B) Pooled fractions from AffiGel Blue were applied to the fetuin affinity resin. The column was washed as indicated in the text and elution of pertussis toxin was accomplished with 4.0 *M* $MgCl_2$.

from solution at 4° by the addition of solid ammonium sulfate (0.65 g/ml). After standing overnight at 4° the precipitate is resuspended in an ammonium sulfate solution (0.65 g/ml) to a toxin concentration of about 2 mg/ml. Pertussis toxin prepared by this procedure and subsequently stored at 4° has been observed to be stable for more than 1 year.

A Probe for Adenylate Cyclase Function

In using pertussis toxin as a probe to examine the role of inhibitory pathways for adenylate cyclase regulation the investigator may want to employ experimental protocols in which either intact cells (i.e., tissue culture cells, tissues from organ culture, fresh tissues) or preparations of cell membranes are under investigation. The techniques for treating toxin vary. Examples are given below in which pertussis toxin is used to treat intact adipocytes and preparations of cell membranes. These methods should be readily applicable to other systems although times of preincubation and concentration of toxin might vary. The choice of appropriate methods for measuring the effect of pertussis toxin on intact cells will depend on the system under investigation but may involve measurement of changes in cellular levels of cAMP or other more specific physiologic effects. Since pertussis toxin blocks the regulation of cyclase by inhibitory pathways, care must be taken to ensure that experiments are designed in a fashion to maximize this action. Thus, in some systems it may be necessary to first activate adenylate cyclase with appropriate ligands and then examine the capacity of pertussis toxin to interfere with down regulation by other agents.

While the N_i in partially purified preparations of cell membranes is accessible for ADP-ribosylation by pertussis toxin (see below) this method is used primarily in studies where radiolabeling of the N_i regulatory component is desired. To obtain membrane preparations suitable for measuring changes in adenylate cyclase resulting from pertussis toxin treatment, we generally pretreat cells with toxin and then prepare membranes.

Treatment of Adipocytes with Pertussis Toxin. Isolated adipocytes are prepared from the epididimal fat pools of Sprague–Dawley rats as described.[23] Fat cells are suspended in buffer and treated with pertussis toxin or other agents as desired. However, prior to using pertussis toxin prepared by the techniques described above it is essential to remove excess ammonium sulfate. Even modest concentrations of ammonium sulfate (i.e., <10 m*M*) totally inhibit the action of pertussis toxin in fat cells. Since preparations of pertussis toxin have a tendency to aggregate and form a precipitate on dialysis, toxin is first dissolved in 0.1 *M* sodium phosphate buffer, pH 6.0, to a concentration of about 0.1 mg protein per ml, and then dialyzed against the same buffer for 2–3 hr. This treatment maintains the toxin in a soluble form and removes ammonium sulfate. The effect of pertussis toxin on fat cells, which can be readily monitored by following the release of glycerol,[24] is maximal after about 3 hr treatment. Figure 2 shows that pertussis toxin is effective in the fat cells over a broad concentration range, with toxin at 20 ng/ml being the ED_{50}.

ADP-ribosylation of N_i in Membrane Preparations. Cell membranes are prepared from isolated rat adipocytes as described.[25] From cells

[23] S. W. Cushman, *J. Cell Biol.* **43,** 326 (1970).
[24] J. K. Pinter, J. A. Hayashi, and J. A. Watson, *Arch. Biochem. Biophys.* **121,** 404 (1967).
[25] S. W. Cushman and L. J. Wardzala, *J. Biol. Chem.* **255,** 4758 (1980).

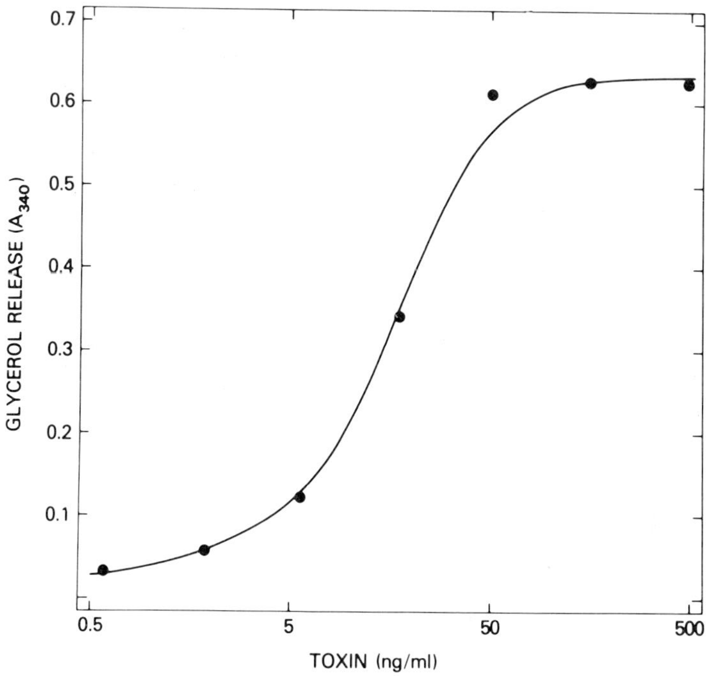

FIG. 2. Pertussis toxin stimulated glycerol release. Rat adipocytes were incubated with the indicated concentration of pertussis toxin for 180 min. At this time an aliquot of buffer was removed and the extent of glycerol release was determined in an NAD coupled assay, glycerol release is reported as the change in A_{340}.

grown in tissue culture or from small samples of tissue, crude membrane preparations can be readily prepared. The cells are first suspended in a hypotonic pH 7.4 buffer containing 50 m*M* Tris–HCl, 5 m*M* $MgCl_2$, and 0.5 m*M* EGTA. After homogenizing with a Dounce homogenizer cellular debris is removed by centrifugation at 500 *g* for 5 min. Membranes are pelleted by centrifuging at 10,000 *g* for 15 min. The membrane pellet is resuspended in the same buffer and the suspension is stored in liquid N_2.

For ADP-ribosylation of membrane preparations the ammonium sulfate suspension of pertussis toxin is dialyzed against 50 m*M* Tris–HCl buffer pH 7.9. This dialysis sometimes results in the formation of precipitate, but this does not appear to interfere with reactivity of the toxin. After dialysis pertussis toxin is preincubated with 100 m*M* dithiothreitol for at least 1 hr. For ADP-ribosylation, toxin (about 5 μg) is added to preparations of cell membranes (about 10 μg) in an Eppendorf centrifuge tube. The reaction, run in a final volume of 50 μl, is initiated by the tion of a solution containing 5.0 μmol Tris–HCl, pH 7.9, 1.5 μmol dithiothreitol, and 1 μCi [^{32}P]NAD (6 μCi/nmol). After 10 min incubation at 37°, membranes are pelleted by centrifugation and the labeled proteins are examined by SDS gel electrophoresis conducted according to the method of Laemmli.[26]

[26] J. R. Laemmli, *Nature (London), New Biol.* **227,** 680 (1970).

[45] ADP-Ribosylation of Membrane Components by Pertussis and Cholera Toxin

By FERNANDO A. P. RIBEIRO-NETO, RAFAEL MATTERA, JOHN D. HILDEBRANDT, JUAN CODINA, JAMES B. FIELD, LUTZ BIRNBAUMER, and RONALD D. SEKURA

Pertussis and cholera toxin are important tools to investigate functional and structural aspects of the stimulatory (N_s) and inhibitory (N_i) regulatory components of adenylyl cyclase.[1–3] Cholera toxin acts on N_s by ADP-ribosylating its α_s subunit. It uses NAD^+ as a cosubstrate.[4] ADP-ribosylation of N_s alters its properties so that its GTP hydrolyzing capacity is inhibited[5] and the action of GTP is potentiated.[6] In intact cells this leads to increases in cyclic AMP levels. Pertussis toxin acts on N_i by ADP-ribosylating its α_i subunit.[2,7] Like cholera toxin it uses NAD^+ as a cosubstrate. ADP-ribosylation of N_i also alters its properties, however, unlike what happens with N_s, the action of GTP on N_i is inhibited.[3] Concomitantly, the coupling of hormone receptors that affect N_i is interrupted.[8,9]

Both pertussis and cholera toxin are hexameric multisubunit molecules. Cholera toxin is composed of one A and 5 B subunits; pertussis toxin is formed of one S_1, one S_2, one S_3, two S_4, and one S_5 subunits.

[1] D. M. Gill and R. Meren, *Proc. Natl. Acad. Sci. U.S.A.* **75,** 3050 (1978).
[2] T. Katada and M. Ui, *Proc. Natl. Acad. Sci. U.S.A.* **79,** 3129 (1982).
[3] J. D. Hildebrandt, R. D. Sekura, J. Codina, R. Iyengar, C. R. Manclark, and L. Birnbaumer, *Nature (London)* **302,** 706 (1983).
[4] D. M. Gill, *Proc. Natl. Acad. Sci. U.S.A.* **72,** 2064 (1975).
[5] D. Cassel and Z. Selinger, *Proc. Natl. Acad. Sci. U.S.A.* **74,** 3307 (1977).
[6] M. C. Lin, A. F. Welton, and M. F. Behrman, *J. Cyclic Nucleotide Res.* **4,** 159 (1978).
[7] T. Katada and M. Ui, *J. Biol. Chem.* **257,** 7210 (1982).
[8] H. Kurose, T. Katada, T. Amano, and M. Ui, *J. Biol. Chem.* **258,** 4870 (1983).
[9] J. A. Hsia, J. Moss, E. L. Hewlett, and M. Vaughan, *J. Biol. Chem.* **259,** 1086 (1984).

ISBN 0-12-182009-2

The A subunit of cholera toxin and the S_1 subunit of pertussis toxin are ADP-ribosyltransferases; the remainder of each toxin is a five-membered oligomer whose function it is to attach the toxin to the cell surface, which is a prerequisite for the "injection" of the S_1 or A subunit respectively into the cell. Once inside the cell, both ADP-ribosyltransferases need to undergo an "activation" process that depends on reduced glutathione, and can be mimicked in the test tube with dithiothreitol (DTT) (for reviews, see Refs. 10–12).

By using $[^{32}P]NAD^+$ and determining the transfer of its $[^{32}P]$ADP-ribose moiety to membrane components, it is possible to obtain information on both structural and functional aspects of N_s and N_i. The structural aspects that can be studied relate to the types and size of the ADP-ribosylated molecules as determined by autoradiography of the membrane components separated by sodium dodecyl sulfate–polyacrylamide gel electrophoresis (SDS–PAGE). In this manner, it has been established that active cholera toxin A subunit ADP-ribosylates peptides that vary in size between 42,000 and 55,000 daltons (α_s subunits of N_s,[13-17]), and that activated S_1 subunit of pertussis toxin ADP-ribosylates a single polypeptide of 40,000–41,000 daltons (α_i subunits of N_i,[2,18]).

The functional aspect that can be studied by $[^{32}P]$ADP-ribosylation of membrane components relates to the fact that both cholera and pertussis toxin show preference for a given conformation or state of N_s and N_i. Thus, cholera toxin is very effective in ADP-ribosylating N_s in the presence of Mg and GTP.[10] By inference it is thought that cholera toxin has preference for the stimulated form of N_s. Pertussis toxin, on the other hand, has preference for a form of N_i that predominates in the absence of Mg (see below). By inference, this is a form that is not "activated" with respect to affecting the catalytic unit of the adenylyl cyclase system. This form, however, is not the inactive form of N_i, for its ADP-ribosylation by pertussis toxin is strongly stimulated by guanine nucleotides, either GTP or GDP provided Mg is avoided (see below). Under these conditions, the

[10] D. M. Gill, *Adv. Cyclic Nucleotide Res.* **8,** 85 (1977).

[11] P. H. Fishman, *J. Membr. Biol.* **69,** 85 (1982).

[12] M. Tamura, K. Nogimori, M. Yajima, K. Hse, and M. Ui, *J. Biol. Chem.* **258,** 6756 (1983).

[13] H. R. Kaslow, G. L. Johnson, V. M. Brothers, and H. R. Bourne, *J. Biol. Chem.* **255,** 3736 (1980).

[14] M. R. Kaslow, D. Cox, U. E. Groppi, and H. R. Bourne, *Mol. Pharmacol.* **19,** 406 (1981).

[15] E. Lai, O. M. Rosen, and C. S. Rubin, *J. Biol. Chem.* **256,** 12866 (1981).

[16] D. M. F. Cooper, R. Jagus, R. L. Somers, and M. Rodbell, *Biochem. Biophys. Res. Commun.* **101,** 1179 (1981).

[17] C. G. Malbon and M. L. Greenberg, *J. Clin. Invest.* **69,** 414 (1982).

[18] J. Codina, J. D. Hildebrandt, R. D. Sekura, M. Birnbaumer, J. Bryan, C. R. Manclark, R. Iygengar, and L. Birnbaumer, *J. Biol. Chem.* (in press).

receptors that couple to N_i are in the low as opposed to the high affinity form. It appears, therefore, that pertussis toxin prefers the conformation of N_i that exists when receptors uncouple. Thus, toxin-mediated [^{32}P]ADP-ribosylation is a tool that can be used to explore changes in both structural and functional properties of N_s and N_i.

Unless validated otherwise, [^{32}P]ADP-ribosylation is not a means to estimate absolute levels of N_s and N_i proteins in membranes.

We present here a set of protocols, as developed in our laboratory, that can be used to study simultaneously and comparatively the susceptibility of N_s and N_i to be ADP-ribosylated by cholera and pertussis toxin.

Stock Solutions

The following stock solutions are made up: 2.4 *M* potassium phosphate, [P_i(K)], pH 7.0, 300 m*M* ADP-ribose (Sigma), 120 m*M* thymidine (Sigma), 5 m*M* GTP, 50 m*M* ATP, 500 m*M* $MgCl_2$, 50 m*M* EDTA, 1% bovine serum albumin (BSA, Armour, Cohn Fraction V, filtered through 0.45-μm Millipore filters), 50 m*M* DTT (Sigma), 1 *M* Tris–HCl, pH 7.5, 10 m*M* NAD^+ (Sigma), 1 mg/ml DNase I (Sigma), 20% trichloroacetic acid, 6 mg/ml cholera toxin (Sigma, Schwartz-Mann, Cal-Biochem), ammonium sulfate suspension of pertussis toxin (see [7]), [^{32}P]NAD^+ (NEN, >300 Ci/mmol).

Activation of Toxins

Pertussis Toxin. The ammonium sulfate suspension is dialyzed exhaustively against 25 m*M* Tris–HCl, 50 m*M* NaCl, 1 m*M* EDTA, pH 7.5 and brought to a concentration of about 0.6 mg protein per ml. Ten microliters of this solution is then incubated in a capped microcentrifuge tube with 10 μl 50 m*M* DTT at 32–37° for 30 min. This mixture is diluted to 120 μg pertussis toxin/ml with 25 m*M* Tris–HCl, 0.075% BSA, pH 7.5 (*activated pertussis toxin*). This solution can be fractionated into small aliquots and kept frozen at −70° for up to 2 months until used.

Cholera Toxin. Ten microliters of cholera toxin solution (6 mg/ml in water) is incubated with 10 μl of 50 m*M* DTT in a capped microcentrifuge tube for 30 min at 32–37°. This mixture is then diluted to 1.2 mg toxin/ml with 75 m*M* Tris–HCl, 0.075% BSA, pH 7.5 (*activated cholera toxin*). This solution can be fractionated into small aliquots and kept frozen at −70° (1 to 2 months) until used.

Preparation of Membranes to be ADP-Ribosylated

Membranes to be ADP-ribosylated are diluted with 0.12 mg/ml DNase I in 75 m*M* Tris–HCl pH 7.5, to give 1–5 mg membrane protein per ml.

Preparation of [^{32}P]NAD^+ Solution

A solution of 60 μM NAD^+ containing per 10 μl between 2 to 20 $\times 10^6$ cpm of [^{32}P]NAD^+ is made in water.

Incubation Mixtures

These are mixtures which are prepared fresh every day using the stock solutions listed above; they provide the necessary ionic milieu to carry out the ADP-ribosylation with either cholera toxin (CT Mixture) or pertussis toxin (PT Mixture).

CT Mixture I. This mixture is used to ADP-ribosylate membranes with cholera toxin in the absence of ADP-ribose. It consists of 600 m*M* P_i(K), pH 7.0, 20 m*M* thymidine, 2 m*M* ATP, 0.2 m*M* GTP, 20 m*M* $MgCl_2$, and 2 m*M* EDTA.

CT Mixture II. This mixture is used to ADP-ribosylate with cholera toxin in the presence of ADP-ribose. It is identical to CT Mixture I with an additional 100 m*M* ADP-ribose.

PT Mixture. This mixture is used to ADP-ribosylate optimally with pertussis toxin. It is the same as CT Mixture I but omitting $MgCl_2$ and potassium phosphate.

Typically, CT and PT Mixtures are prepared by mixing the stock solutions of P_i(K) (2.4 *M*), ADP-ribose (300 m*M*), thymidine (120 m*M*), $MgCl_2$ (500 m*M*), EDTA (50 m*M*), ATP (50 m*M*), and GTP (5 m*M*) in the ratio of 8 : 8 : 4 : 1 : 1 : 1 : 1.

A typical ADP-ribosylation is carried out by mixing, in 10 $\times$ 75-mm glass test tubes, the following in sequence: 30 μl of toxin mixture, 10 μl of membrane suspension, and 10 μl [^{32}P]NAD^+ solution, with which the reaction is started. The final volume is 60 μl and final concentrations of all ingredients (when present) are NAD^+ 10 μM; P_i(K), 300 m*M*; ADP-ribose, 50 m*M*; thymidine, 10 m*M*; ATP, 1 m*M*; GTP, 0.1 m*M*; EDTA, 1 m*M*; $MgCl_2$, 10 m*M*; Tris–HCl, 25 m*M*; DNase I, 0.02 mg/ml; BSA, 7.5 μg/assay; DTT, 1.66 m*M*; cholera toxin, 100 μg/ml; and pertussis toxin, 10 μg/ml. Incubations are at 32° for 30–45 min. They are stopped by addition of 1.0 ml of ice cold 20% trichloroacetic acid. After up to 20 min on ice the mixtures are centrifuged in a clinical centrifuge (~2000 rpm) for 20 min in the cold. The supernatants are carefully aspirated and discarded. The pellets are washed once with 1.5 ml of ethyl ether and collected by centrifugation at the same speed for 30 min. The ether is discarded, the test tubes are brought to room temperature, and the remaining ether is allowed to evaporate. Membranes ADP-ribosylated in this manner are then ready for analysis by SDS–PAGE.

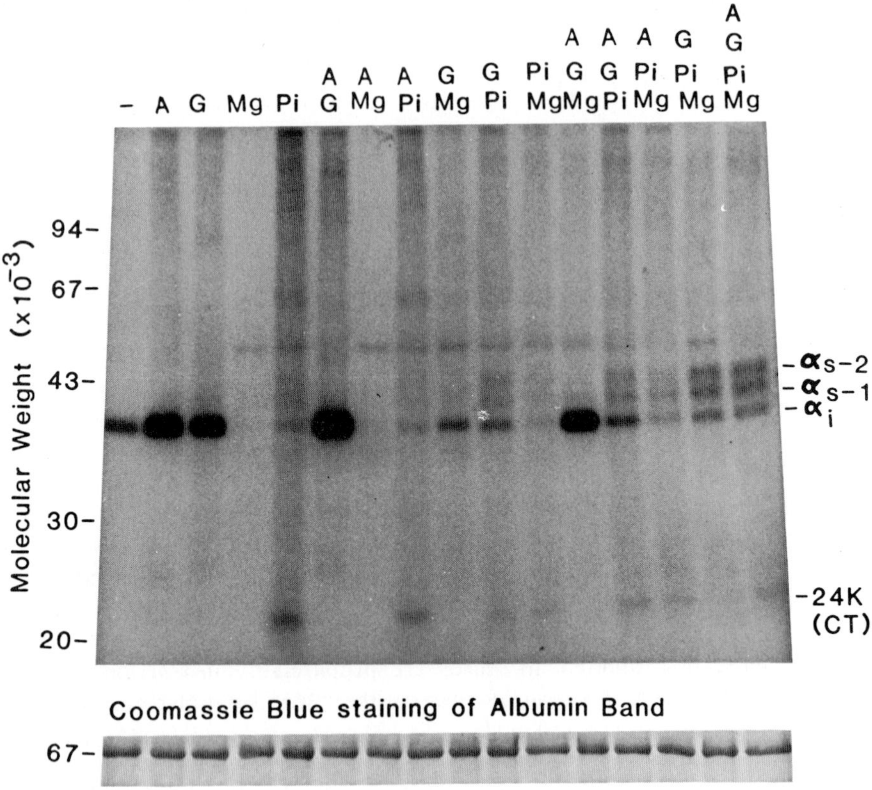

FIG. 1. ADP-ribosylation of thyroid membranes by a combination of cholera and pertussis toxin and the effect of adding or omitting the ingredients from CT Mix I. The abbreviations are A, 1 m*M* ATP; G, 0.1 m*M* GTP; P_i, 300 m*M* inorganic potassium phosphate; and Mg, 10 m*M* $MgCl_2$.

SDS–PAGE Analysis

Samples are dissolved with 50 μl of Laemmli's sample buffer[19] without heating, and 20-μl aliquots of each tube are subjected to SDS–PAGE in 10 or 12% linear polyacrylamide gels[17] of 0.75 mm thickness and using 4-mm-wide wells. Thus, each well receives 40% of the ADP-ribosylated sample containing, in addition to the membrane protein, 3 μg of BSA. The remainder of the sample is frozen and used if a repeat of the SDS–PAGE is necessary. After electrophoresis, the bottom part of the gel containing [^{32}P]NAD^+ is cut off and the remainder of the gel is stained with Coomas-

[19] U. Laemmli, *Nature* (*London*) **227,** 680 (1970).

sie blue. The staining intensity of the BSA band (M_r = 67,000) will provide an indication as to the reproducibility of the processing of the sample (i.e., trichloroacetic acid precipitation, ether wash and transfer onto the top of the gel). Gels are destained, and then dried and subjected to autoradiography.

Comments

Typically, three types of incubation are carried out (each in duplicate) on a given membrane sample. All three are performed with both the cholera toxin and the pertussis toxin being present. To this end each incubation contains 10 μl of a 1 : 1 mixture of activated cholera toxin and of activated pertussis toxin. A Type I incubation contains 30 μl of CT Mix I, a Type II incubation contains 30 μl of CT Mix II, and a Type III incubation contains 30 μl of PT Mix. Generally, cholera toxin does not ADP-ribosylate in the absence of Mg and phosphate, which are the conditions used for ADP-ribosylation by pertussis toxin. On the other hand, under the conditions used for ADP-ribosylation by cholera toxin, pertussis toxin does ADP-ribosylate the membranes, albeit in a much less effective manner. Addition of ADP-ribose has two effects on labeling with cholera toxin: it reduces background labeling (depends on membrane sample) and it alters the relative intensity with which various forms of the α subunit of N_s are labeled. The reasons for these differences are not clear.

Comparison of the labeling obtained under the three conditions described above provides information as to whether or not one or the other of the ADP-ribosylated polypeptides is present, absent, or altered in its susceptibility to be ADP-ribosylated by one or the other toxin.

Figures 1–3 illustrate the effects of additives to the ADP-ribosylation reactions on labeling of N_s and N_i components.

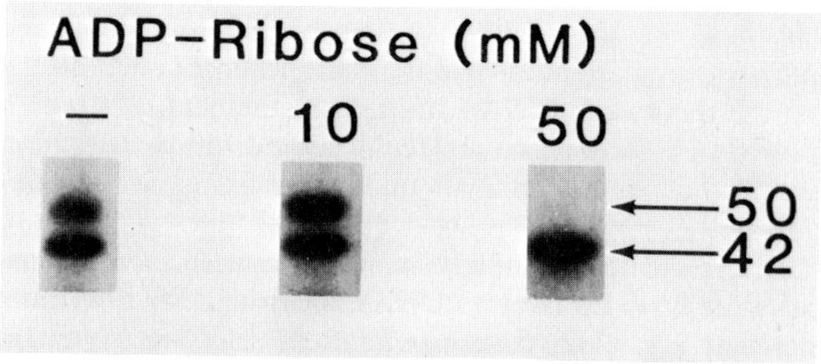

FIG. 2. The effect of ADP-ribose on the labeling of human erythrocyte membranes with cholera toxin.

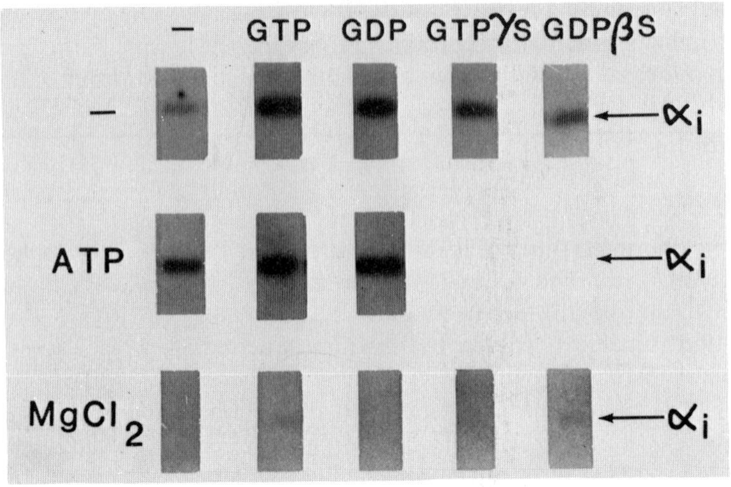

FIG. 3. The effect of ATP and GTP on the labeling of pure N_i, by pertussis toxin.

Acknowledgments

Supported in part by research grants from the U.S. Public Health Service AM-19318 and HD-09581 to L.B., AM-26088 to J.B.F., and Center Grants HD-07495 and AM-27685.

[46] Measurement of mRNA Concentration and mRNA Half-Life as a Function of Hormonal Treatment

By JOHN R. RODGERS, MARK L. JOHNSON, and JEFFREY M. ROSEN

In this chapter we will discuss the use of molecular hybridization techniques to measure mRNA concentrations in cells and tissues. Kinetic analysis of these measurements may then be employed to estimate the rate of mRNA degradation under different hormonal conditions. We describe first the theory of mRNA half-life measurement, and then several methods of RNA isolation and molecular hybridization that are particularly useful to the analysis of small amounts of tissue.

Several methods can be used to quantitate the concentration of a specific mRNA. We have routinely applied two methods to the quantitation of casein mRNA levels: (1) cDNA-excess liquid hybridization; and (2) the dot blot assay. Each approach offers selective advantages and disadvantages. The first is the method of choice for the determination of extremely low basal amounts of mRNA (i.e., on the order of a few mole-

ISBN 0-12-182009-2

cules per cell) and measures absolute amounts of mRNA, but requires larger quantities of cells and RNA than may be readily available. Also, this method is cumbersome, especially when a number of different samples are to be analyzed. The second approach is excellent for determining relative changes in mRNA levels (i.e., between different hormonally induced states versus control) and is a rapid procedure for analyzing large numbers of samples, but suffers from a lack of sensitivity at low levels of mRNA. These methods will be discussed with a few specific examples from our studies on the hormonal regulation of casein mRNA levels in the mammary gland.

Theoretical Analysis of mRNA Degradation

Messenger RNA is characteristically unstable. The half-lives of mRNAs may vary with cellular conditions; this means that mRNA turnover can determine the final extent of gene expression.[1] The list of mRNAs known to be regulated at least in part by turnover control is short but growing, and includes hen ovalbumin,[2] *Xenopus* vitellogenin,[3] rat casein,[4,5] dihydrofolate reductase,[6] adenovirus EIA and B gene products,[7] and tubulin.[8,8a] An instructive example is that of myosin, the effective half-life of which is dependent on the growth rates of differentiating myoblasts.[9] We describe in this section approaches that we and others have followed in studying the kinetics of mRNA decay.

Two approaches allow the estimation of mRNA half-lives: approach to steady-state analysis and label-chase (pulse-chase) analysis. Approach to steady-state analysis demands acceptance of a given kinetic model, but can avoid the experimental manipulations that may trouble pulse-chase experiments. It can yield estimates of synthesis rates as well as degradation rate constants. In contrast, pulse-chase experiments often require fewer assumptions and yield direct estimates of half-lives.

[1] F. Kafatos, *Acta Endocrinol.* (*Copenhagen*), *Suppl.* **168,** 319 (1972).
[2] R. F. Cox, *Biochemistry* **16,** 3433 (1977).
[3] M. L. Brock and D. J. Shapiro, *Cell* **34,** 207 (1983).
[4] W. A. Guyette, R. J. Matusik, and J. M. Rosen, *Cell* **17,** 1013 (1979).
[5] J. R. Rodgers and J. M. Rosen, in preparation.
[6] E. J. Leys and R. E. Kellems, *Mol. Cell. Biol.* **1,** 961 (1981).
[7] A. Babich and J. R. Nevins, *Cell* **26,** 371 (1981).
[8] D. W. Cleveland and J. C. Havercroft, *J. Cell Biol.* **97,** 919 (1983).
[8a] To this list should be added histone mRNAs (R. M. Alterman, S. Ganguly, D. H. Schulze, W. F. Marzluff, C. L. Schildkraut, and A. I. Skoultchi, *Mol. Cell. Biol.* **4,** 123, 1984); lactate dehydrogenase mRNA (R. A. Jungmann, D. C. Kelley, M. F. Miles, and D. M. Milkowski, *J. Biol. Chem.* **258,** 5312, 1983); and chicken thymidine kinase mRNA (G. F. Merrill, S. D. Hauschka, and S. L. McKnight, *Mol. Cell. Biol.* **4,** 1777, 1984).
[9] R. M. Medford, H. T. Nguyen, and B. Nadal-Ginard, *J. Biol. Chem.* **258,** 11063 (1983).

Steady-State Analysis

We assume that in the presence of a given concentration of hormone, the rate of change in the cellular concentration of a given mRNA is described by

$$dC/dt = k_s - \lambda C \tag{1}$$

where k_s is the rate of synthesis and λ is a first-order rate constant of degradation. If a tissue or cell system is maintained under uniform conditions long enough, a steady-state will result. Setting Eq. (1) equal to zero and solving for C we obtain

$$C_{ss} = C_{\text{steady-state}} = k_s/\lambda \tag{2}$$

C_{ss} is determined by liquid hybridization, dot blot analysis, etc. (see Methods).

Estimation of λ from C_{ss} and k_s

Equation (2) may be solved for λ if k_s also can be measured independently. From λ the half-life, τ, is given by

$$\tau = \ln 2/\lambda$$

Note that this estimation requires a single half-life component for the mRNA in question: if significant nuclear turnover occurs this approach will underestimate the cytoplasmic half-life.

Approach to Steady-State Kinetics

Frequently it is difficult to measure k_s. If two hormonal conditions define two sets of rate constants, and if a change in hormone concentration results in a rapid change of the rate constants from their old to their new values, then the characteristic time for the system to move one-half of the distance between the old and the new steady-state concentrations is equal to τ. Assay of the mRNA levels during the approach to the new steady-state allows the estimation of k_s (new) and λ (new). This is true for both induction and deinduction, and this constitutes a crude but rapid method of half-life estimation.

If the rate constants reach their new values rapidly, Eq. (1) integrates to yield

$$C = (k_s/\lambda) - [(k_s/\lambda) - C_0]e^{-\lambda t} \tag{3}$$

Substituting from Eq. (2), rearranging and taking logarithms, we obtain

$$\ln(C - C_{ss}) = \ln(C_0 - C_{ss}) - \lambda t \tag{4}$$

(For example, see Nikol *et al.*[10]) *Note that it is necessary to know* C_{ss} but it is not necessary to know C_0. It is best to collect data from the entire induction curve, but one need not worry about assaying basal levels that might fall below the sensitivity of the assay. An estimate of the basal value C_0 is obtained from the y intercept of expression (4). Note also that once λ has been estimated, k_s may be obtained from Eq. (2).

In some cases the value of C_{ss} is unknown or is not known reliably. In the case of induction of globin mRNA by DMSO, a deinduction occurs very rapidly even in the continuous presence of the inducing agent.[11] Thus it is clear that a true steady-state is never reached. We have used a formulation of Eq. (1) which is useful in the absence of reliable estimates of C_{ss}.[4] We estimate values of dC/dt by measuring the change in concentration between successive time points.

$$dC/dt \sim \Delta C/\Delta t = k_s - \lambda C \tag{5}$$

A plot of $\Delta C/\Delta t$ versus C (not time!) gives a straight line with y intercept k_s and slope $-\lambda$. The value of C plotted for each pair of successive values can be the mid-C value, or if a weighted value is desired, the geometric mean of the two values $(C_1C_2)^{1/2}$.

These methods are applicable only when hormonal manipulation results in rapid changes in the rate constants k_s and λ. It may be necessary to measure the half-life within a steady-state condition.[12] Continuous labeling experiments can be subjected to similar mathematical treatments. The value of λ is the same as previously, but now the zero-order rate constant is k^*, the rate of incorporation of radioactivity into the species. If the specific activity of the precursor is known (see Measurement of the Specific Activity of Nucleotide Precursors), k_s can be obtained from k^*. Likewise, if the specific activity of precursors is not constant during the labeling, a correction can be applied to the data before interpretation.[3,13]

Pulse-Chase Kinetics

If the value of k_s in Eq. (3) can be driven to zero, then Eq. (4) becomes

$$\ln C = \ln C_0 - \lambda t \tag{6}$$

A plot of $\ln C$ versus time gives a straight line with slope $-\lambda$.

The method by which k_s is driven to zero is a matter of judicious choice. In the chick oviduct, removal of progesterone after secondary

[10] J. M. Nikol, K.-L. Lee, and F. T. Kenney, *J. Biol. Chem.* **253,** 4009 (1978).

[11] K. Lowenhaupt and J. B. Lingrel, *Proc. Natl. Acad. Sci. U.S.A.* **76,** 5173 (1979).

[12] M. C. Wilson and J. E. Darnell, *J. Mol. Biol.* **148,** 231 (1981).

[13] M. L. Brock and D. J. Shapiro, *J. Biol. Chem.* **258,** 5449 (1983).

stimulation results in a rapid suspension of ovalbumin gene transcription. Cox[2] measured both translational and hybridization activity of ovalbumin mRNA after hormone withdrawal to obtain estimates of τ. Unfortunately, this method requires changing hormone states, so that the value of τ is not obviously that of either induced or withdrawn states.

Pulse-chase experiments allow the determination of τ under uniform hormone conditions. Label is allowed to accumulate and a chase is effected either by blocking new synthesis with an inhibitor such as actinomycin D, or by diluting the labeled nucleotide pools with excess cold nucleosides. Note that actinomycin D has been suggested both to increase[14,15] and retard[16,17] RNA decay. Similarly, the use of excess cold nucleosides may be toxic to the cells and/or alter RNA metabolism.[9] A brief treatment with glucosamine may effect a more rapid chase.[4,18] However, note that this compound is also toxic at high concentrations and during lengthy chase protocols. As the chase is rarely effective immediately label will continue to accumulate, especially into stable species; a correct value for λ can be obtained from the linear portion of the decay curve.

Ideally, the value of C is determined from a known quantity of cells and represents a concentration per cell. However if the half-life is long with respect to the half-time of cell doubling, dilution will occur and affect the value of τ. Such dilution may be physiologically important,[9] but obscures the actual measurement. In some cases the value of C can be expressed as "per dish" equivalent, which corrects for cell growth. Otherwise, if the cell doubling time, T_g, is known and if the cells are in exponential growth the correction

$$C = C'e^{+gt} \tag{7}$$

will correct for dilution ($g = \ln 2/T_g$, note the sign). Otherwise the correction can be applied after obtaining λ' from uncorrected data by

$$\lambda = \lambda' - g \tag{8}$$

where λ' is the apparent degradation rate constant. Alternatively,

$$\tau = \tau' T_g/(T_g - \tau') \tag{9}$$

In our experiments in explant cultures the effect of cell doubling is negligible. More problematic is variability in RNA yield from small

[14] W. Murphy and G. Attardi, *Proc. Natl. Acad. Sci. U.S.A.* **70,** 115 (1973).

[15] J. R. Greenberg, *Nature (London)* **240,** 102 (1972).

[16] R. L. Cavalieri, E. A. Havell, J. Vilcek, and S. Pestka, *Proc. Natl. Acad. Sci. U.S.A.* **74,** 4415 (1977).

[17] Y. Endo, H. Tominaga, and Y. Natori, *Biochim. Biophys. Acta* **240,** 215 (1971).

[18] A. Spradling, H. Hui, and S. Penman, *Cell* **4,** 131 (1975).

amounts of tissue. We solved this problem by normalizing recovered labeled RNA to the amount of 18 S rRNA in the extract. Some investigators prelabel for 24 hr with [^{14}C]uridine, then pulse-label with [^{3}H]uridine before chasing, using the ratio of ^{3}H/^{14}C to correct for recovery.[19] In order to measure the decay of rat casein mRNAs we routinely label explants for 24 hr with [^{3}H]uridine and chase for up to 80 hr. Under these conditions the percent hybridization to 18 S rDNA filters is constant during the chase.

If the specific activity of total RNA is known C can be expressed as a specific activity by multiplying ppm times total specific activity, and is then corrected for cell-doubling, if necessary, by Eqs. (7) or (8). In some experiments the specific activity may not be known precisely, e.g., if carrier RNA is used to precipitate small quantities of labeled RNA. C can also be expressed as the ratio C/C_{rib} or simply as C (ppm). We correct for recovery variability by determining the ratio of labeled RNA retained on casein DNA filters, to radioactivity retained on 18 S rDNA filters.

When using this formulation of C the effect of rRNA decay often can be ignored, but we find it necessary to correct for the 30–50 hr half-life of rRNA in mammary explant cultures. If the fraction of total radioactivity hybridizing to 18 S rRNA filters is constant, the half-life of 18 S rRNA may be estimated from the decay of the specific activity of total RNA. The value of λ is obtained after first transforming the C' values:

$$C = C' e^{-\lambda_{\text{rib}} t} \tag{10}$$

(note the sign of λ_{rib} is negative), or by applying a correction to λ':

$$\lambda = \lambda' + \lambda_{\text{rib}} - g \tag{11}$$

or to τ'

$$\tau = \tau' \tau_{\text{rib}} T_g / (\tau' T_g + \tau_{\text{rib}} T_g - \tau' \tau_{\text{rib}}) \tag{12}$$

It is often convenient to label for several hours or more before commencing the chase. If the mRNA of interest has a short half-life relative to the labeling time, its relative signal will diminish during the label procedure. This will not affect the value of τ to be obtained, but may affect the reliability of the measurements. On the other hand, if the value of τ does not adopt a single value (as is generally assumed), then labeling for a long time will obscure the presence of short-lived components. For example, we estimate the half-life of the external transcribed spacer (ETS) of rat ribosomal RNA to be approximately 1 hr. However, after a long labeling time, the bulk of radioactivity hybridizing to an ETS DNA filter has a half-life of 10–20 hr. To measure the half-life of rapidly decaying RNA species it is necessary to label for a very short time.

[19] R. H. Singer and S. Penman, *J. Mol. Biol.* **78,** 321 (1973).

Confidence Estimates of mRNA Half-Life

The value of τ or λ should be reported with an estimate of its error. This can be obtained from linear regression analysis of the data, with ln C as the dependent variable, using statistical parameters readily obtained from most hand-held scientific calculators. The first-order variance of the regression line is

$$s^2 = [\sigma_y^2(n - 1)(1 - r^2)]/(n - 2) \tag{13}$$

where σ_y^2 is the variance of the measurement of ln C, and r is the correlation coefficient. The standard deviation in the estimate of λ is

$$\sigma_\lambda = (\sigma_y/\sigma_x)[(1 - r^2)/(n - 2)]^{1/2} \tag{14}$$

σ_x^2 is the variance in the measurement of time. A t test is used if the estimates of λ from two hormonal conditions are to be compared:

$$t = (\lambda_1 - \lambda_2) \Big/ \left\{ s' \left[\frac{1}{(n_1 - 1)\sigma_{x_1}^2} + \frac{1}{(n_2 - 1)\sigma_{x_2}^2} \right]^{1/2} \right\} \tag{15}$$

where

$$(s')^2 = [(n_1 - 2)s_1^2 + (n_2 - 2)s_2^2]/(n_1 + n_2 - 4) \tag{16}$$

with $df = n_1 + n_2 - 4$. Note that a variance ratio test should be applied if the first-order variances s_1^2 and s_2^2 are not equal.

If Eqs. (8) or (10) have been used to obtain λ, then appropriate estimates of σ_λ^2 should be obtained by summing the variances $\sigma_\lambda'^2$, $\sigma_{\lambda_{rib}}^2$, and σ_g^2.

Methods

Isolation of RNA

Isolation of Purified RNA by Phenol Extraction

Homogenize the tissue in 5 tissue volumes (v/w) of Solution I (100 mM NaCl, 10 mM Na_2 EDTA, pH 8.0, 0.5% SDS) and 5 v/w of redistilled phenol saturated with Solution I. Retain the first aqueous supernatant and interface, and reextract the phenolic phase with 2.5 v/w Solution I. Combine the two aqueous fractions and extract with 5 v/w each of saturated phenol and ChIsoA (chloroform–isoamyl alcohol, 24 : 1 v/v). Retain the aqueous phase only and reextract with 7.5 v/w of ChIsoA. Add 1/50 volume 5 M NaCl and 2 volumes ethanol and precipitate for 30 min at −70°.

Dissolve the ethanol pellet in 5 v/w Solution I containing 20 μg/ml predigested Proteinase K and incubate 30 min at 37°. Extract with 2.5 v/w each of saturated phenol and ChIsoA, then with 5 v/w ChIsoA alone. Precipitate with added NaCl and ethanol.

Dissolve the deproteinized nucleic acid pellet in 3 v/w distilled H_2O (previously treated with 25 μl/100 ml diethyl pyrocarbonate to inactive RNases) and precipitate high-molecular-weight RNA with 9 v/w of 4 *M* NaOAc, pH 6.0, overnight at 4°. Wash the RNA pellet with ice cold 3 *M* NaOAc, pH 6.0, and dissolve in 3 v/w distilled H_2O. Make the solution 0.2 *M* with respect to NaCl and precipitate with 6 v/w ethanol, overnight at −20°. Dissolve the final RNA pellet (free of tRNA and DNA) in 2 v/w H_2O and determine the RNA concentration spectrophotometrically using 1 A_{260} unit = 40 μg/ml.

Cytoplasmic Dot Method[20,20a]

Stock Solutions

PBS: 50 m*M* Na phosphate pH 7.4, 0.9% NaCL
TE Buffer: 10 m*M* Tris–HCl pH 7.4, 1 m*M* EDTA
20% Nonidet NP-40 in TE Buffer
20 × SSC: 3 *M* NaCl, 0.3 *M* Na citrate pH 7.0
20% Formaldehyde

The cell pellet (corresponding to one 60 m*M* culture dish) is harvested into a 1.5-ml Eppendorf tube, washed with PBS, and pelleted at 1500 *g* for 5 min. The PBS supernatant is discarded and the cell pellet is resuspended in 180 μl of TE buffer. Ten microliters of 20% NP-40 is added and the cells incubated on ice for 5 min with occasional vortexing. Nuclei are pelleted at 10,000 *g* for 5 min and the supernatant is transferred to a new 1.5-ml Eppendorf tube. The nuclear pellet is frozen in liquid N_2 and stored until the DNA content is determined by a diphenylamine assay.[21] To the supernatant is added 120 μl of 20 × SSC and 160 μl of 20% formaldehyde and the sample incubated at 68° for 10–15 min. Aliquots (1–30 μl) of supernatant are diluted to 200 μl with 15 × SSC and spotted onto nitrocellulose.

[20] B. A. White and F. C. Bancroft, *J. Biol. Chem.* **257,** 8569 (1982).

[20a] An alternative to the cytoplasmic dot method for selectively immobilizing mRNA on nitrocellulose using cell lysates has been developed by J. Bresser, J. Doering, and D. Gillespie, *DNA* **2,** 243 (1983). The Quick-Blot method has the advantage of reducing the coimmobilization of proteins on nitrocellulose, thus permitting the application of larger amounts of cell extract and increasing the sensitivity of detection of low abundance mRNAs. In our laboratory, however, this technique has been less reliable than the cytoplasmic dot method, especially when using tissue extracts.

[21] K. Burton, this series, Vol. 12, Part B, p. 163.

DNA Labeling Techniques

In this section we will discuss two methods for the synthesis of ^{32}P-labeled single-stranded cDNA probes (^{32}P-sscDNA). These probes are used in conjunction with the dot blot assay (see Dot Blot Assays). In the case of the first method, by substitution of ^{3}H-labeled nucleotides, probes can be made which are also suitable for cDNA excess liquid hybridization (see cDNA Excess Liquid Hybridization Assays).

Synthesis of ^{32}P-Labeled Single-Stranded cDNA Using Oligo(dT) Priming on cDNA Inserts Cloned in Bacteriophage M13mp7

In this method a single-stranded probe complementary to mRNA will be synthesized using the procedure of Ricca *et al.*[22] This procedure uses oligo(dT) priming and, therefore, is suitable for use only with cDNA inserts containing the poly(A) tail (plus strand of the insert) cloned into the *Pst*I site of bacteriophage M13mp7.

Stock Solutions

Oligo(dT): 500 μg/ml
5× Salts: 0.75 *M* NaCl, 0.1 *M* Tris–HCl, pH 7.4, 0.1 *M* $MgCl_2$, 0.5 m*M* EDTA
Cold dNTPs: 20 m*M* dGTP, 20 m*M* TTP, 20 μ*M* dATP, 20 μ*M* dCTP
Stop buffer: 60 m*M* EDTA, 6% SDS, 1 mg/ml sheared salmon sperm DNA
Alkaline gradient buffer: 0.1 *M* NaOH, 0.9 *M* NaCl, 2 m*M* EDTA
5 and 20% sucrose (w/v) in alkaline gradient buffer

Reaction. Combine 3 μg of M13mp7 DNA containing the plus strand insert with 4 μl of the oligo(dT) solution, 4 μl of 10× salts and water to make 20 μl final volume. Place the reaction mix in a siliconized 100 μl volumetric capillary tube, seal both ends, and denature in boiling water for 3 min. Quick cool in ice water for 5 min. To this solution add 6 μl of cold dNTPs, 50 μCi each of [α-^{32}P]dATP and dCTP (3000 Ci/mmol, Amersham), 1 μl of 50 m*M* dithiothreitol (DTT), and 10 units of Klenow DNA polymerase I (New England Biolabs). Incubate for 3 hr at 13°. The reaction is stopped by the addition of 50 μl of stop buffer and heating at 68° for 2 min. ^{32}P-sscDNA is then separated from unincorporated [^{32}P]dNTPs by chromatography of the reaction mixture on a small Econo-column (Bio-Rad) containing BioGel P-60 (4–4.5 ml bed volume). The ^{32}P-sscDNA is collected in the first 3 ml of eluate. DNA is precipitated by the

[22] G. A. Ricca, J. M. Taylor, and J. E. Kalinyak, *Proc. Natl. Acad. Sci. U.S.A.* **79,** 724 (1982).

addition of yeast tRNA at 50 μg/ml, 0.25 *M* NaCl, and 2 volumes of ethanol at −80° for at least 15 min. The DNA is pelleted and reconstituted in 200 μl of the alkaline gradient buffer. This solution is layered on top of a 12 ml 5 to 20% linear sucrose gradient in the same buffer. The gradients are centrifuged in an SW41 rotor at 35,000 rpm overnight. Gradients are fractionated by pipetting 300 μl fractions from the top and stored on ice. The ^{32}P-sscDNA usually bands between fractions 6 and 12 with the M13mp7 DNA template pelleting. The pooled fractions are then neutralized by the addition of 1.0 *M* HCl and 1.0 *M* Tris–HCl pH 7.4 to 0.1 *M* final concentrations each. The ^{32}P-sscDNA is ethanol precipitated as before and the resulting pellet is dissolved in 400 μl of 10 m*M* Tris–HCl, 1 m*M* EDTA, pH 7.4. The probe is now ready for use.

Generally 50–100 × 10^6 cpm of ^{32}P-labeled probe can be synthesized by this method with a specific activity greater than 1 × 10^9 cpm/μg.

M13 External Primer Second Strand Synthesis

In this method an external primer specific for the M13 portion of the cloning vector is used to synthesize a ^{32}P-labeled second strand. Synthesis is terminated before the insert is copied; therefore, the unlabeled minus strand insert must be used to permit hybridization to mRNA. This procedure is a modification of the procedure described by Hu and Messing.[23] Since the M13 vector rather than the cloned insert is copied using this method, this protocol is suitable for generating a probe to any portion of the mRNA, unlike the first method which requires the presence of the poly(A) tail.

Stock Solutions

Hybridization buffer: 6 m*M* each: Tris pH 7.5, NaCl, $MgCl_2$, and dithiothreitol

dNTP mix: 0.5 m*M* each dGTP, dCTP, dTTP

M13 external primer (New England Biolabs): 6.6 ng/μl in 10 m*M* Tris–HCl, pH 7.4, 1 m*M* EDTA

Reaction. Boil 1.0 μl of primer for 2 min, quick cool in ice water, then mix with 1.1 μl of hybridization buffer and 1.0 μg of M13 DNA containing the minus strand insert. Mix and adjust the final volume to 8 μl with water. The reaction is heated at 68° for 10 min and then allowed to cool slowly to room temperature (approximately 1 hr). Add 20 μCi of [α-^{32}P]dATP (3000 Ci/mmol, Amersham), 10 units of Klenow DNA polymerase I (1 μl), and 1 μl of the dNTP mix, and incubate the reaction at 15°

[23] N.-T. Hu and J. Messing, *Gene* **17,** 271 (1982).

for 60 min. The reaction is stopped by the addition of 1 μl of 0.5 *M* EDTA and heating at 68° for 2 min. Separation of unincorporated [^{32}P]dATP from ^{32}P-labeled DNA is accomplished by chromatography on a Sephadex G-50 column (0.8 × 15 cm), the ^{32}P-labeled DNA eluting in the void volume. The M13 DNA can either be precipitated as described in the section on Synthesis of ^{32}P-Labeled Single-Stranded cDNA or used directly. It is important not to denature the newly synthesized ^{32}P-labeled DNA second strand from the unlabeled M13 template because the synthesized second strand contains the radioactive tag in this method.

Nick-Translation

Several procedures are in general use for the nick translation of DNA with ^{32}P-labeled nucleotides. Here we describe a procedure which has been adapted from Mackey *et al.*[24] and provides excellent reproducibility with regard to the amount and the specific activity of the probe generated. In our experience, probes with a specific activity between 150 and 250 × 10^6 cpm/μg give the best results in the dot blot assay (see Dot Blot Assays).

Stock Solutions

10× salts: 0.5 *M* Tris–HCl pH 7.6, 75 m*M* $MgCl_2$, 500 μg/ml bovine serum albumin

The 10× salts solution should be heated at 68° for 15 min to inactivate DNase

dNTP mix: 1.0 m*M* dGTP and dTTP, 10 μ*M* dATP and dCTP

Stop buffer: 60 m*M* EDTA, 6% SDS, 1 mg/ml sheared salmon sperm DNA

0.15 *M* 2-mercaptoethanol

Reaction. Combine 1.5 μl of the 10× salts, 1.0 μl of dNTP mix, 1.0 μl of 0.15 *M* 2-mercaptoethanol, 5.0 μl each of [α-^{32}P]dATP and [α-^{32}P]dCTP (50 μCi each, 3000 Ci/mmol, Amersham), 1–2 μl of DNA to be nick translated (10–500 ng), and sufficient DNA polymerase I to give 15–25 units/μg (generally 1 μl). We have observed that keeping the reaction volume at 15 μl is important in order to provide high enough concentrations of the labeled substrates. Also, when using fresh isotope it may be necessary to increase the unlabeled dATP and dCTP concentrations to 20 μ*M* in order to keep the specific activity of the probe in the desirable range.

The reaction is incubated at 13° for 2 hr and is terminated by the addition of an equal volume of stop buffer and heating at 68° for 5 min. Chromatography on BioGel P-60 and precipitation of ^{32}P-labeled DNA is

[24] J. K. Mackey, K. H. Brackmann, and M. Green, *Biochemistry* **16,** 4478 (1977).

performed as described in Synthesis of ^{32}P-Labeled Single-Stranded cDNA. The ^{32}P-labeled nick translated DNA probe is reconstituted in 400 μl of 10 m*M* Tris, 1.0 m*M* EDTA, pH 7.4 and can be stored either at −20° as a frozen solution or an ethanol pellet. It should be used within 48 hr of preparation to ensure the best results.

RNA Labeling Techniques

^{32}P-5′-End-Labeled mRNA[25,26]

This method is useful if the mRNA of interest is at least 5% of total mRNA and is based on the protocol of Maisel.[26] For rarer mRNAs, use the method in the section below.

Dissolve 5 μg poly(A)-containing RNA, selected from total RNA on oligo(dT)-cellulose in 20 μl of 100 m*M* Tris–HCl, pH 9.5, and heat for 15 min at 90°. This introduces 5′-OH nicks at a rate of about 0.2/kb/min). Quench the reaction on ice.

Add in a final volume of 50 μl (1) 5 μl of (500 m*M* Tris, pH 7.4, 100 m*M* $MgCl_2$, 1 m*M* Na_2EDTA), (2) 4 μl of 50 m*M* DTT, (3) 100 μCi of [γ-^{32}P]ATP at 2000–5000 Ci/mmol, (4) 30 units of T4 polynucleotide kinase, and (5) H_2O to 50 μl.

Incubate for 60 min at 37°. About 50% of the label should be incorporated into TCA-precipitable material.

Stop the reaction with 5 μl of 2% SDS, 50 μg *E. coli* RNA, and 5 μg predigested Proteinase K. Dilute to 100 μl and incubate 15 min at 37°.

Separate free counts and small RNA fragments from hybridizable ^{32}P-labeled RNA tracer by chromatography over Sephadex G-150. The column is developed in 100 m*M* NaCl, 50 m*M* Tris–HCl, pH 7.4, 5 m*M* Na_2EDTA, 0.1% SDS. The excluded peak should contain 80% of TCA-precipitable radioactivity. Add carrier RNA, NaCl to 0.2 *M*, and precipate with ethanol.

Hybridize to cloned DNA-containing filters as described in DNA-Filter Assay of Radioactively Labeled RNA, treat with RNase, Proteinase K, and then wash. Elute specifically hybridized radioactively by heating the filters for 10 min at 68° in 100 μl of 25 m*M* Tris–HCl, pH 7.4, 10 m*M* Na_2EDTA, containing 25 μg *E. coli* carrier RNA.

DNA-Directed Synthesis of Internally Labeled Tracer

Assemble in a total volume of 50 μl (1) 100 μCi of [α-^{32}P]UTP, (2) 10 μl of (250 m*M* Tris–HCl, pH 7.4, 250 m*M* KCl, 50 m*M* $MgCl_2$, 5 m*M*

[25] H. Biessmann, E. A. Craig, and B. J. McCarthy, *Nucleic Acids Res.* **7,** 981 (1981).
[26] N. Maisel, *Cell* **9,** 431 (1976).

spermidine, 0.25 mM Na_2 EDTA, and 1 mM each of CTP, GTP, and ATP), (3) 5 μl of 50 mM DTT, (4) 5–10 μg recombinant plasmid DNA, (5) 1 unit/mg of *E. coli* RNA polymerase holoenzyme, and (6) H_2O to 50 μl.

Incubate at 25° for 2 hr.

Terminate and process as described above for end-labeled tracer except that radioactively labeled RNA should be hybridized to (−) strand containing recombinant M13 phage DNA bound to nitrocellulose filters. Only (+) strand mRNA sequences will hybridize specifically.

In Vivo Labeling for Pulse-Chase

Cells in culture typically incorporate label at a high rate, but we find it necessary in some cases to stimulate labeling of mammary explants with L-glucosamine–HCl[18] or with pyrazofurin.[27] We pretreat explant cultures for 1 hr with 20 mM glucosamine in order to achieve a 2-fold enhancement of labeling. Glucosamine is toxic and cannot be used in cultures for extended times. Glucosamine combines with UTP to form UDP-glucosamine, depleting endogenous pools of unlabeled UTP. Pyrazofurin inhibits the enzyme orotidylate decarboxylase and thus all *de novo* synthesis of uridylate.[27] It has been used with success in cultured cells and also enhances labeling in mammary explants about 2-fold.

We routinely chase with a 1-hr exposure to 20 mM glucosamine and 5 mM each of uridine and cytidine. The media is then changed and replaced with 5 mM each uridine and cytidine. Higher concentrations of nucleosides have been reported to be toxic in some systems.[9,18] For cells in culture concentrations of glucosamine higher than 5 mM should be avoided.[18]

Molecular Hybridization Techniques

cDNA Excess Liquid Hybridization Assays

cDNA excess liquid hybridization is an extremely sensitive method for the absolute quantitation of low levels of RNA. Unfortunately, the method is time consuming and requires substantially larger quantities of RNA than is required by the "dot" blot assay.

Stock Solutions

4× hybridization buffer: 2.4 M NaCl, 40 mM HEPES pH 7.0, 8 mM Na_2EDTA

[26a] A more powerful method involves cloning the DNA sequence of interest immediately downstream from a phage sp6 promoter and transcribing *in vitro* using purified sp6 RNA polymerase, as described by M. R. Green, T. Maniatis, and D. A. Melton, *Cell* **32,** 681 (1983).

[27] E. Cadman and C. Benz, *Biochim. Biophys. Acta* **609,** 372 (1980).

S_1 nuclease buffer: 0.2 *M* Na acetate, pH 4.5, 0.4 *M* NaCl, 2.5 m*M* $ZnCl_2$

20% trichloroacetic acid (TCA) w/v

Reaction. The hybridization reactions are carried out in a 10–15 μl volume in tapered reaction vials (2 ml, Regis Chemical Co.). A range of RNA/cDNA ratios of between 3,000 and 50,000 : 1 is generally used, the amount of ^{3}H-labeled sscDNA (see Synthesis of ^{32}P-Labeled Single-Stranded cDNA) kept constant at approximately 0.4 ng (3000–5000 cpm). Sample RNA is first pipetted into the reaction vial and lyophilized to dryness. A 10-μl aliquot containing the hybridization buffer (1× final), ^{3}H-sscDNA, and hen oviduct carrier RNA (200 μg/ml) is added to the lyophilized RNA. The vials are capped, boiled for 2 min, and quick cooled in ice water. Duplicate vials are removed and frozen rapidly in ethanol–dry ice and stored at −20° until S_1 nuclease treatment. These represent the t_0 samples (see below). The remaining vials are incubated at 68° for 48 hr. It is important that the entire vial be submerged in the water bath to keep the reaction volume from condensing on the top or upper walls of the vial. After incubation, the vials can be frozen and stored at −20° until S_1 treatment. Duplicate reaction vials are set aside for direct TCA precipitation, i.e., no S_1 nuclease treatment. These will be used as the total input radioactivity controls (see below).

The extent of hybridization is determined by S_1 nuclease digestion. 500 μl of S_1 buffer containing 1500 units of S_1 nuclease are added to each vial and incubated at 37° for 2 hr. An equal volume of 20% TCA is added. TCA precipitable material is collected by filtration through nitrocellulose filters (Millipore Corp., HAWP, 0.45 μm). The filters are dissolved with Piersolve (Pierce Chemical Company) and then counted in Aquasol (New England Nuclear).

S_1 nuclease resistant counts are expressed as a percentage of total input counts less the background (t_0 samples) and plotted against the log equivalent R_0t determined as described by Britten.[28] From the slope of the hybridization curve compared to the slope of the standard curve it is possible to calculate the absolute number of molecules of mRNA.

Dot Blot Assays[29]

The so-called dot blot assay in which either RNA or a cytoplasmic extract containing RNA is spotted onto nitrocellulose paper has become a widely used procedure.[20,29] In this section we discuss two methods of dot blot hybridization that we have employed successfully. There are several potential problems of which one must be aware and control if meaningful

[28] R. J. Britten, *in* "Symposium on RNA in Development" (E. W. Hanley, ed.), p. 187. Univ. of Utah Press, Salt Lake City, 1969.

[29] P. Thomas, this series, Vol. 100, Part B, p. 255.

results are to be obtained from the dot blot assay. Among these are the problems of nonspecific hybridization, sensitivity, stringency, and standardization so that data from separate filters may be compared directly. Many of these problems have been discussed extensively by P. Thomas in a recent article in Methods in Enzymology.[29] The dot blot assay is useful for rapid assay of a large number of samples with both a visual and quantitative signal. The limitations of the method are that the efficiency of hybridization is low (~0.1%) and a maximum of about 3 μg of RNA can be applied to the nitrocellulose filter.

Preparation of the Dot Blot Filter

Stock Solutions

20 × SSC: 3 *M* NaCl, 0.3 *M* Na citrate, pH 7.4
TE buffer: 10 m*M* Tris–HCl, 1 m*M* EDTA, pH 7.4
20% Formaldehyde
20% Nonidet P-40 in TE buffer

RNA is prepared by phenol extraction according to the procedure described by Guyette *et al.*[4] in the section on Isolation of Purified RNA or cytoplasmic extracts can be used as prepared according to White and Bancroft[20] as detailed in the section on Cytoplasmic Dot Method. Nitrocellulose paper is equilibrated in water and placed on top of a sheet of Whatman 3 mm paper in the spotting Minifold apparatus (Schleicher and Schuell). One milliliter of 15 × SSC is then passed through each well (check that all wells are not leaking). RNA or cytoplasm which has been formaldehyde treated as described is diluted into 200 μl volumes with 15 × SSC so that 5 concentrations are spotted for each sample. Generally, between 50 ng and 1 μg of RNA or the cytoplasm from 1×10^4 to 2×10^5 cells is the range which we spot. In the case of mammary gland cytoplasmic RNA we have found that it becomes impossible to filter much more than the equivalent of 2×10^5 cells. A standard curve, in our case lactating RNA or cytoplasm prepared from lactating mammary tissue, is always spotted on each filter. This permits a direct comparison of the extent of hybridization between filters because the results can be expressed as a percentage of a lactating standard. After the filters have been spotted, they are baked at 68–80° for 4 hr. Filters can be stored indefinitely at room temperature if they are kept dry.

Hybridization with ^{32}P-Labeled sscDNA or External-Primer Synthesized M13 DNA

Stock Solutions

P/H buffer: 50% formamide (chelexed and filtered), 0.6 *M* NaCl, 10 m*M* HEPES pH 7.4, 1 m*M* EDTA, 0.1% SDS, 100 μg/ml hen oviduct RNA and poly(A) RNA

20 × SSC: 3.0 *M* NaCl, 0.3 *M* Na citrate, pH 7.0
10% SDS

Reaction. The filters are placed in plastic seal-a-bag and prehybridized in 5 ml P/H buffer. After prehybridization, the solution is removed and fresh P/H buffer (5 ml) containing 40–80 × 10^6 cpm of probe is placed in the bag. Hybridization is conducted at 37° for 2 days. The temperature and length of hybridization as well as formamide concentrations can be adjusted depending upon the stringency desired. For example, we have used this method to analyze casein mRNA levels in the T47D human mammary carcinoma cell line with a rat casein cDNA clone by conducting the hybridization at 22 instead of 37°.

After hybridization, the buffer is removed and the filter washed as follows: 2 × 15 min in 2 × SSC-0.1% SDS at room temperature, 30 min in 2 × SSC–0.1% SDS at 68°, 2 × 30 min in 0.2 × SSC–0.1% SDS at 68°. Again, depending on the stringency desired, the washing temperatures can be altered appropriately. The filters are air dryed and autoradiography performed. The autoradiograms can then be scanned by densitometry or the dots cut out and counted in a scintillation counter. If densitometry is used, care must be taken to ensure that the exposure of the X-ray film is in the linear response range of the film. When purified RNA is spotted, the slope of the curve of cpm bound per amount of RNA spotted is used for comparison, while with cytoplasm the cpm bound per equivalent amount of DNA spotted is used, or if the cellular DNA content is known it can be converted to equivalent number of cells. The DNA amount is determined by performing a diphenylamine[21] assay on the nuclear pellet obtained during the preparation of the cytoplasmic extracts.

Dot Blot Using Nick-Translated Probes

Stock Solutions

Prehybridization buffer: 50% formamide (chelexed and filtered), 3 × SSC, 5 × Denhardt's (50 × = 1% bovine serum albumin, 1% polyvinylpyrrolidone, 1% Ficoll), 50 m*M* Na phosphate buffer pH 7.4, 100 μg/ml sheared salmon sperm DNA

Hybridization buffer: 50% formamide (chelex and filter), 3 × SSC, 1 × Denhardt's, 10% dextran sulfate, 50 m*M* Na phosphate buffer pH 7.4, 100 μg/ml sheared salmon sperm DNA

We have used ^{32}P-labeled nick translated cDNA probes coupled with the dextran sulfate method of hybridization.[30,31] Filters are placed in a plastic seal-a-bag and incubated for 6–24 hr at 42° in a prehybridization

[30] G. M. Wahl, M. Stern, and G. R. Stark, *Proc. Natl. Acad. Sci. U.S.A.* **76,** 3683 (1979).
[31] J. C. Cohen, *Cell* **19,** 653 (1980).

buffer. After prehybridization, the nick translated probe is boiled for 7–10 min, quick cooled in ice water, and added to 5 ml of hybridization buffer. The prehybridization solution is removed from the bag and the probe mixed with hybridization solution is added. Hybridization is conducted for 12–16 hr at 42°. When using dextran sulfate it is important that the length of hybridization does not exceed 16 hr because it then becomes difficult to wash down the background.[30] After removing the filter from

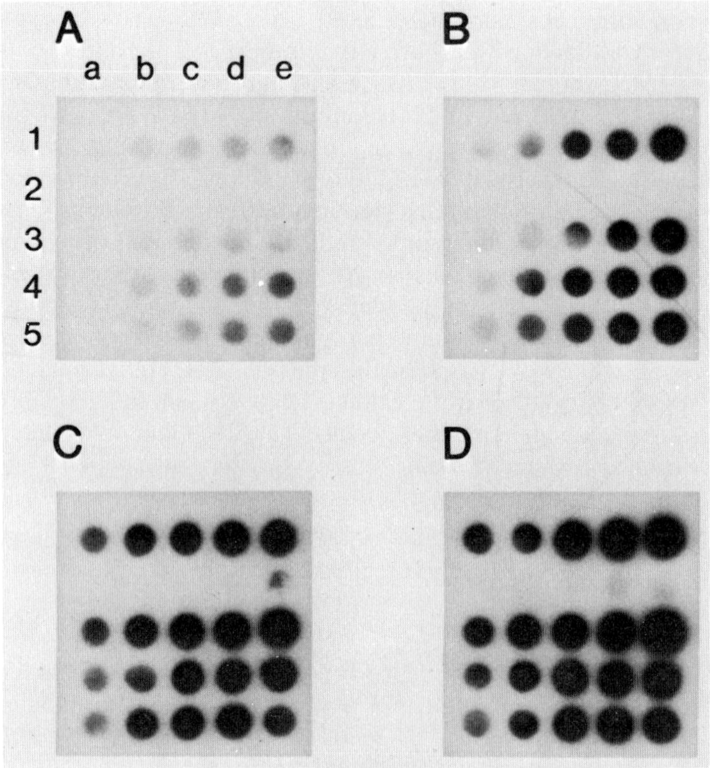

FIG. 1. Increasing amounts of cytoplasm isolated from 14-day lactating mammary gland (rows 1 and 3) or RNA isolated from 14-day lactating mammary gland (rows 4 and 5) were spotted onto four identical filters. Also, increasing amounts of cytoplasm from a mouse lymphoma cell line was spotted (row 2). In rows 3 and 5 an amount of cytoplasm corresponding to that amount spotted in row 2d was added to each sample and spotted simultaneously. (A) 10×10^6 cpm. (B) 20×10^6. (C) 40×10^6. (D) 80×10^6 cpm of a ^{32}P-nick translated β-casein cDNA clone (specific activity of 210×10^6 cpm/μg) were hybridized as described in the text. (1) LC cyto.; (2) lym cyto.; (3) LC + lym cyto.; (4) LC RNA; (5) LC RNA + lym cyto.

the bag, it is placed in the solution of 2 × SSC and carefully drawn over the sides of the washing dish several times. This greatly reduces the problems of adequate washing created by the dextran sulfate. The filter is then washed as follows: 2 × 1 hr in 2 × SSC at room temperature, 3 × 45 min in 0.1 × SSC–0.1% SDS at 50°, and finally 2 × 10 min in 0.1 × SSC. The filter is dried and processed as before.

To ensure maximum sensitivity with the dot blot assay it is essential that the assay be conducted under cDNA excess conditions. This can be demonstrated by setting up a series of identical standard curves and hybridizing with increasing amounts of probe. Also, one needs to establish the specificity of the hybridization especially when using cytoplasmic extracts. Both of these points are illustrated in Figs. 1 and 2. Increasing amounts or either lactating RNA or lactating mammary gland cytoplasm were spotted in the absence or presence of a constant amount of cytoplas-

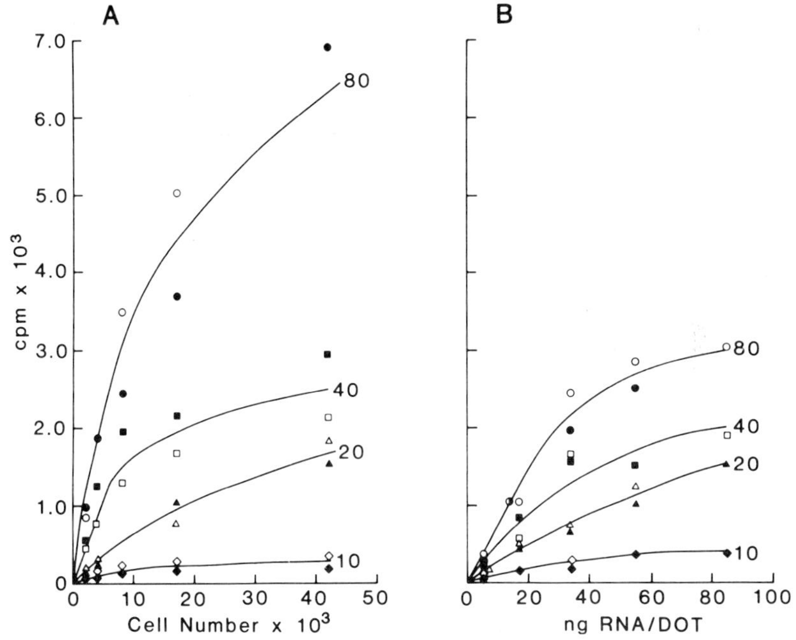

FIG. 2. Individual dots from the filters used in Fig. 1 were cut out and counted as described in the text. The data are plotted as either cpm bound versus cell number (A) for cytoplasmic RNA spotted or versus ng RNA spotted/dot for isolated RNA (B). The equivalent cell number (A) was determined from the equivalent DNA content for the amount of cytoplasm spotted and then dividing by the DNA content of a mammary gland epithelial cell (6 pg per cell). Open symbols in absence of lymphoma extract; closed symbols in presence of extract.

mic extract from a mouse lymphoma cell line which does not express casein mRNA. A curve of the lymphoma extract was also spotted. Four identical filters were hybridized with either 10, 20, 40, or 80 × 10^6 cpm of β-casein cDNA (nick translated to a specific activity of 210 × 10^6 cpm/μg DNA). As expected, the lymphoma extract gave no hybridization until the highest input of probe (only 50 cpm above background at the highest concentration, Fig. 1D) and did not significantly alter the hybridization to either lactating RNA or cytoplasmic RNA, indicating that the hybridization is indeed specific. Second, the extent of hybridization (cpm bound) increased dramatically as more probe was added, indicating that 40–80 × 10^6 cpm should be used in order to achieve maximum sensitivity at the lowest levels of spotted RNA.

One problem which is inherent in the dot blot assay is illustrated in Fig. 2. The individual dots from the autoradiograms shown in Fig. 1 were cut out and counted as described above. The resulting graph of either cpm hybridized versus equivalent cell number (2A) or versus ng RNA (2B) spotted are curvilinear. It is necessary to use the slope of the graph taken from the lower inputs because this region is more linear. Therefore, it is a good idea to use at least 5 concentrations of sample taken over a wide range to ensure usable results from this method of analysis.

DNA-Filter Assay of Radioactively Labeled RNA

Preparation of DNA Filters

Wet gridded Millipore nitrocellulose filters (type HA, 0.45 mm) in water, then in 20 × SSC.

If single-stranded recombinant phage DNA is used, dissolve 400 μg of DNA in 50 ml of 20 × SSC and go to the next step. Recombinant plasmid DNA must be nicked prior to binding to nitrocellulose, either by restriction endonuclease digestion or by base hydrolysis. Dissolve 400 μg of plasmid DNA in 100 μl 0.1 *M* NaOH and heat in a 90° water bath for 10 min.[32] Quench by diluting into 50 ml of ice-cold 20 × SSC containing 100 μl of 0.1 *M* HCl.

Filter the DNA solution slowly through the nitrocellulose. Air dry and bake at 80° for 3–6 hr.

Cut gridded squares out and label with a sharp pencil, using a number to indicate hybridization vial and dots to indicate the DNA fragment. Assemble up to 8 filters per hybridization reaction. (The efficiency of hybridization is not affected by including this many filters.)

[32] J. S. Ullman and B. J. McCarthy, *Biochim. Biophys. Acta* **294,** 396 (1973).

Prehybridize in HB (Hybridization Buffer; 50% formamide, 0.6 *M* NaCl, 50 m*M* HEPES, pH 7.4, 1 m*M* Na_2EDTA, 0.1% SDS) containing 1 × Denhardt's solution and 250 μg/ml high-molecular-weight *E. coli* RNA. Incubate 3–12 hr at 37°.

Hybridization Conditions

Aspirate prehybridization solution from filters, rinse with HB, and replace with 80 μl of HB containing 250 μg/ml *E. coli* RNA and the labeled RNA. Include a fixed quantity of ^{32}P-labeled tracer RNA (see sections on ^{32}P-5′-End-Labeled mRNA and DNA-Directed Synthesis) in each reaction.

Incubate at 37° for 16–24 hr.

Terminate the reaction by diluting with 0.5 ml of HB. Remove and count. Wash each reaction with a second wash of 0.5 ml which is also counted, yielding 95% of input counts (except when measuring rRNA levels, see below). Combine all the filters in a 50-ml plastic tube containing 25 ml of HB and wash with gentle agitation at 37° for 2 hr with one change of buffer. Rinse several times with WB (Wash Buffer, 0.6 *M* NaCl, 50 m*M* HEPES, pH 7.4, 1 m*M* Na_2EDTA) to remove SDS and formamide. Incubate for 1 hr at room temperature in WB containing 20 μg/ml RNase A (pretreated at 68° for 15 min to inactivate contaminating DNases) and 100 U/ml RNase T1. Return filters to HB and wash for another hour. Dissolve in 0.2 ml Piersolve and count in a suitable scintillant.

DNA-Filter Assay Using ^{32}P-End-Labeled RNA[25,26]

This method allows the detection of very low concentrations of RNA on the order of 10 ppm. Extracted RNA is gently hydrolyzed to introduce nicks which can be end-labeled with polynucleotide kinase and [γ-^{32}P]ATP. Labeled sequences are detected by hybridization to nitrocellulose filters containing recombinant plasmid DNA in sequence excess as described in the preceding section. Cpm (100 × 10^6) can easily be synthesized from 5 μg of RNA. With a hybridization efficiency of 30%, an mRNA present at 10 ppm (0.001% of total RNA) will yield a signal of 300 cpm above background. Thus the chief advantage of this technique over the dot assay is that the efficiency of hybridization is high, and more than 3 μg of RNA can be assayed maximally.

Procedure. Gently hydrolyze 5 μg of RNA using the procedure outlined in the section on ^{32}P-5′-End-Labeled mRNA. The labeled RNA is mixed with *E. coli* carrier RNA, deproteinized, and freed from nucleotides and short fragments by chromatography over G-175, or by repeated precipitation out of ethanolic NH_4OAc. The latter involves adding 0.5 volume of 7.5 *M* NH_4OAc, pH 7.0 and 3 volumes ethanol. Precipitate for 30 min at −70°.

Hybridize the end-labeled RNA to DNA filters as described in the previous section. An appropriate standard RNA may be similarly labeled and used to calibrate the assay.

DNA-Filter Assay of Radioactively Labeled rRNA

External transcribed spacer (ETS) and 28 S rRNA sequences are highly self-complementary and do not hybridize efficiently unless nicked to ~100–200 bases in length.[5] rRNA (18 S) hybridizes more efficiently, but also needs to be nicked to provide reliable estimates of relative 18 S rRNA labeling. Gently hydrolyze the RNA sample for a few minutes as described previously for synthesis of end-labeled tracer. Quench on ice, add *E. coli* carrier RNA, make 0.2 *M* in NaOAc, pH 5, and precipitate with ethanol. Dissolve and assay as described above with inputs in the range of 1,000–50,000 cpm. In order to account accurately for total input the RNase-release cpm should be counted as these may account for 5–10% of input.

Measurement of the Specific Activity of Nucleotide Precursors[3,9,13,33]

Recently Brock and Shapiro[3,13] have described a method of HPLC analysis suitable to the measurement of nucleotide precursor specific activities from small amounts of tissues. An attractive alternative is the method of Medford *et al.*,[9] which uses a luciferase-spectrophotometric assay. An important discussion of the problems of pool compartmentalization may be found in Khym *et al.*[33] In most cases, the use of pulse-chase techniques obviates the need for pool specific activity determinations, but for an exception to this, see Brock and Shapiro.[13]

[33] J. X. Khym, M. H. Jones, W. H. Lee, J. D. Regan, and E. Volkin, *J. Biol. Chem.* **253,** 8741 (1978).

[47] Radioactive Labeling and Turnover Studies of the Insulin Receptor Subunits

By JOSE A. HEDO and C. RONALD KAHN

Introduction

The insulin receptor is an integral membrane protein composed of two major subunits of approximate molecular weights 135,000 and 95,000. Like other cellular proteins, the concentration of the receptor is the result of a dynamic equilibrium between the processes of synthesis and degradation. The study of biosynthetic and degradative processes of the life cycle of the insulin receptor requires the use of adequate probes for measuring the receptor independently of its ability to bind insulin. In our studies, we have used as immunological probes human autoantibodies from diabetic patients with a rare form of insulin resistance and acanthosis nigricans.[1] The specificity of these autoantibodies to the insulin receptor is based on a number of studies.[1,2] Furthermore, these antibodies can immunoprecipitate quantitatively the solubilized insulin receptor.[3] Similar studies have been performed using antibodies raised in animals and monoclonal antibodies.

Our experimental approach in these studies comprises three basic steps (Fig. 1). First, intact cells are radioactively labeled on either the carbohydrate or the protein moiety of the receptor. This labeling can be accomplished by using procedures that label proteins or glycoproteins at the cell surface or by biosynthetic labeling using radioactive amino acids or sugars. Second, the receptors are identified among all the labeled cellular proteins by immunoprecipitation of the detergent-solubilized cellular extracts with anti-receptor antibodies. Third, the immunoprecipitated receptors are analyzed by sodium dodecyl sulfate–polyacrylamide gel electrophoresis and autoradiography.

[1] J. S. Flier, C. R. Kahn, J. Roth, and R. S. Bar, *Science* **190,** 63 (1975).

[2] C. R. Kahn, J. S. Flier, R. S. Bar, J. A. Archer, P. Gorden, M. M. Martin, and J. Roth, *N. Engl. J. Med.* **294,** 739 (1976).

[3] L. C. Harrison, J. S. Flier, J. Roth, F. A. Karlsson, and C. R. Kahn, *J. Clin. Endocrinol. Metab.* **48,** 59 (1979).

METHODS IN ENZYMOLOGY, VOL. 109

ISBN 0-12-182009-2

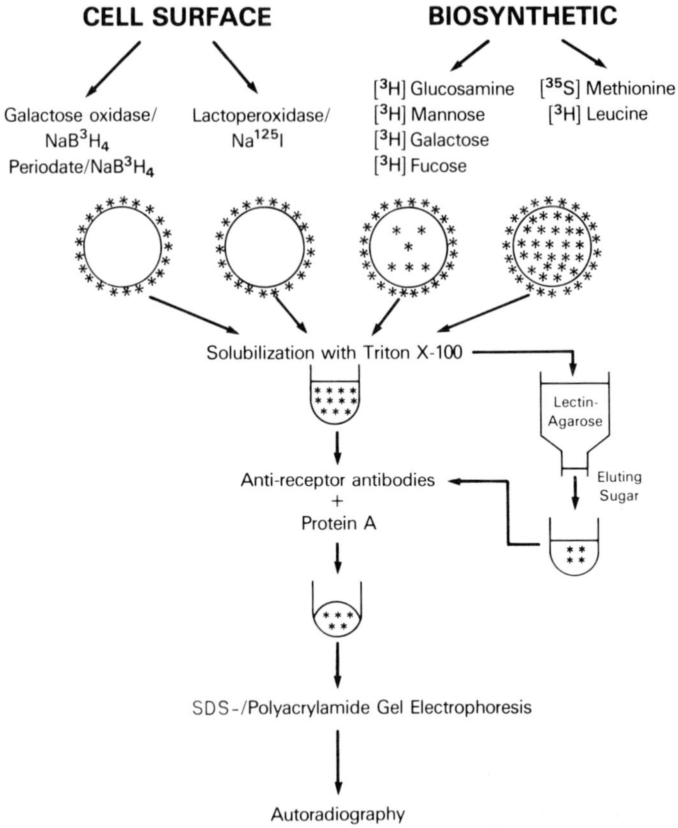

FIG. 1. Experimental procedure for the identification of the insulin receptor subunits by cell surface or biosynthetic labeling.

Cell Labeling Techniques

Surface Labeling of the Carbohydrate Moiety of the Receptor

Theory. The insulin receptor can be labeled with NaB^3H_4 by two different procedures:

1. Terminal nonreducing galactose residues can be made radioactive by sequential exposure to galactose oxidase and NaB^3H_4, as described for soluble glycoproteins by Morell and Ashwell[4] and adapted for cell surface labeling by Gahmberg and Hakomori.[5] A 6-aldehyde intermediate is enzy-

[4] A. G. Morell and G. Ashwell, this series, Vol. 28, p. 209.

[5] C. G. Gahmberg and S. Hakomori, *J. Biol. Chem.* **248**, 4311 (1973).

matically produced and subsequently reconverted to galactose containing a tritium on carbon 6. Since very often carbohydrate chains of glycoproteins are terminated by the disaccharide sialylgalactose, removal of the sialic acid by prior treatment with neuraminidase often increases the number of exposed galactose residues susceptible to labeling.

2. Sialic acid residues can be labeled by sequential treatment with metaperiodate and NaB^3H_4 under controlled conditions as described by Van Lenten and Ashwell[6] for soluble glycoproteins and modified for cell surface labeling by Gahmberg and Andersson.[7] Mild oxidation with metaperiodate selectively cleaves the two distal exocyclic carbons of sialic acid and forms an aldehyde. Subsequent reduction with NaB^3H_4 introduces a tritium atom on carbon 7.

These procedures for labeling are described below for cells in suspension (IM-9 lymphocytes), but can be used equally well with cells attached to tissue culture plates.

Labeling of Galactose Residues

Materials

IM-9 lymphocytes are used at a stationary phase of growth. A total of 2×10^8 cells are used.
Galactose oxidase (200 units/ml, EC 1.1.3.9, Dactylium dendroides, Type V, Sigma)
Neuraminidase (1 unit/ml, EC 3.2.1.18, *Vibrio cholerae,* Calbiochem)
Tritiated sodium borohydride (64 Ci/mmol, New England Nuclear) dissolved in 0.01 *N* NaOH at 1 mCi/50 μl
Dulbecco's phosphate buffered saline (containing Ca^{2+} and Mg^{2+}) pH 7.4 (DPBS)

Procedure

1. Sediment the cells by centrifugation at 600 *g* for 5 min and wash three times with DPBS in order to remove completely the culture medium. Resuspend the cells in 1 ml of DPBS.
2. Add 50 μl of galactose oxidase. If a higher degree of labeling is desired (see Theory and Comments), add also 250 μl of neuraminidase.
3. Incubate at 37° for 30 min in a waterbath with mild agitation.
4. Wash the cells with DPBS twice.

[6] L. Van Lenten and G. Ashwell, this series, Vol. 28, p. 209.

[7] C. G. Gahmberg and L. C. Andersson, *J. Biol. Chem.* **252,** 5888 (1977).

5. Suspend the cells in 0.5 ml of DPBS. Add 50 μl of the NaB^3H_4 solution (1 mCi) and incubate for 30 min at room temperature with mild agitation.

6. Wash the cells three times with DPBS.

Labeling of Sialic Acid Residues

Materials

Cells, as described above.
Sodium metaperiodate (100 mM)
Tritiated sodium borohydride (64 Ci/mmol) dissolved in 0.01 N NaOH at 1 mCi/50 μl
Glycerol 0.1 M in Dulbecco's phosphate buffered saline (DPBS), pH 7.4

Procedure

1. Wash the cells three times with DPBS to remove culture medium.
2. Suspend the cells in 1 ml of DPBS and place the container on ice at 0° and in the dark. After a few minutes, add 10 μl of sodium metaperiodate 100 mM (final 1 mM).
3. After 5 min, quench the reaction by adding 0.4 ml of 0.1 M glycerol in DPBS.
4. Wash the cells three times with DPBS at room temperature.
5. Suspend the cells in 0.5 ml of DPBS. Add 50 μl of the NaB^3H_4 solution (1 mCi) and incubate for 30 min.
6. Wash the cells three times with DPBS.

Comments. NaB^3H_4 should be dissolved in 0.01 M NaOH just prior to use. It is possible to store the solution at −70° in a nitrogen atmosphere but even under these conditions the borohydride solution is unstable.

Other neuraminidases such as that from *Clostridium perfringens* can be used with equally satisfactory results provided that they are free of protease activity.

Incorporation of radioactivity into trichloroacetic and precipitable cellular proteins ranges between 1 and 5% of the total radioactivity added.

The pattern of labeling of the insulin receptor subunits differs with each of the above described techniques (Fig. 2).[8] Galactose oxidase alone labels preferentially the M_r = 135,000 subunit of the receptor, although faint labeling of the M_r = 95,000 subunit can also be detected. Treatment with neuraminidase and galactose oxidase increases the amount of label in the M_r = 135,000 subunit and reveals clearly the M_r = 95,000 subunit. The

[8] J. A. Hedo, M. Kasuga, E. Van Obberghen, J. Roth, and C. R. Kahn, *Proc. Natl. Acad. Sci. U.S.A.* **78,** 4791 (1981).

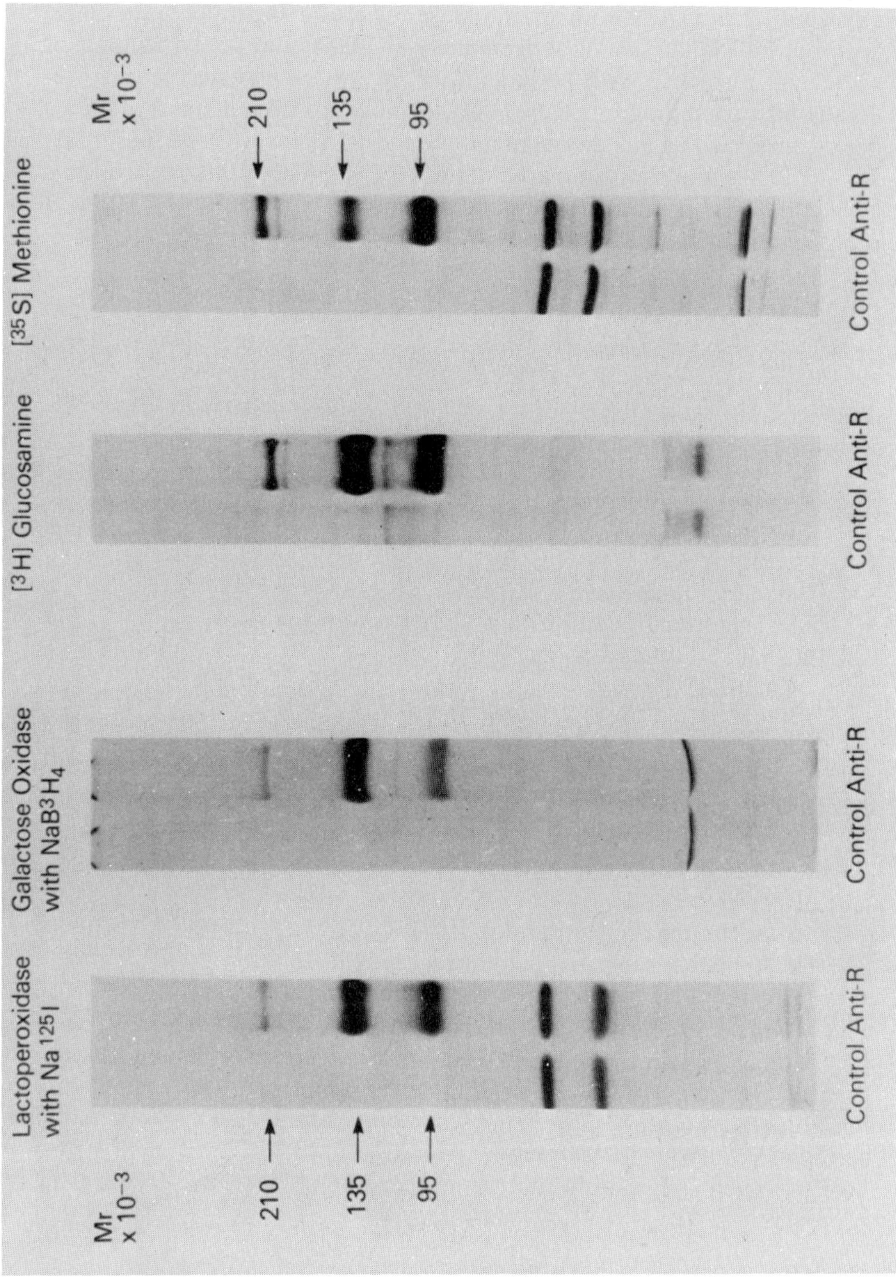

FIG. 2. Autoradiographs of insulin receptors labeled using a variety of techniques. Left, cell surface labeling; right, biosynthetic labeling.

basis for this phenomenon has been mentioned under Theory. As could be expected, treatment with neuraminidase induces a slight decrease in the apparent molecular size of both subunits (<10,000) due to the removal of sialic acid residues. With periodate, on the other hand, labeling is predominantly on the M_r = 95,000 subunit. These different patterns have been interpreted as a higher content of exposed sialic acid residues in the M_r = 95,000 subunit as compared to the M_r = 135,000. Both methods also label a minor M_r = 210,000 component believed to be a fully glycosylated proreceptor for both subunits.

Surface Labeling of the Protein Moiety of the Receptor

Theory. Lactoperoxidase catalyzes the iodination of accessible tyrosine and histidine residues of membrane proteins in the presence of H_2O_2 and NaI.[9] H_2O_2 is generated by a glucose oxidase–glucose system.

Materials

Cells (IM-9 lymphocytes), as described above
Lactoperoxidase (from bovine milk, EC 1.11.1.7, Boehringer Mannheim)
Glucose oxidase (from *Aspergillus niger,* EC 1.1.3.4, Type V, Sigma)
$Na^{125}I$ (carrier-free, 17 Ci/mg, ICN Radioisotope Division)
Phosphate-buffered saline (PBS) pH 7.4 with and without 20 m*M* glucose

Procedure

1. Sediment the cells by centrifugation at 600 *g* for 5 min. Wash the cells three times with PBS containing 20 m*M* glucose.
2. Resuspend the cells in 100 ml of PBS containing 20 m*M* glucose.
3. Add 2 mg of lactoperoxidase (400 μl of a 5 mg/ml suspension) and 2 mCi of $Na^{125}I$.
4. Start the reaction by adding 100 units of glucose oxidase and incubate for 30 min at room temperature with mild agitation.
5. Wash the cells four times with PBS.

Comments. Incorporation of radioactivity into trichloroacetic acid-precipitable protein ranges from 2 to 5% of the total added.

This procedure labels clearly both major subunits of the insulin receptor, although the M_r = 135,000 subunit incorporates twice as much radioactivity as the M_r = 95,000 (Fig. 2).[10] An M_r = 210,000 component is also labeled which presumably represents the fully glycosylated form of the proreceptor.

[9] M. Morrison and G. R. Schonbaum, *Annu. Rev. Biochem.* **45,** 861 (1976).
[10] M. Kasuga, J. A. Hedo, K. M. Yamada, and C. R. Kahn, *J. Biol. Chem.* **257,** 10392 (1982).

Biosynthetic Labeling of the Insulin Receptor

Theory. Either the carbohydrate or the protein moiety of the insulin receptor can be labeled by incubating the cells in the presence of radioactive monosaccharides or amino acids.[8,10–13] Two types of labeling procedures may be followed depending on the duration of the labeling period. In continuous labeling experiments, the cells are incubated with the radioactive precursors for a period of several hours. Since the insulin receptor is such a minor component of the cell, the incubation is usually performed at a cell density of 5- to 10-fold higher than that achieved at stationary phase of growth. In pulse-labeling experiments, the duration of the incubation with the radioactive precursor is shortened to a period of several minutes. Because of the low content of insulin receptor, in this case, the cell density is about 20-fold higher than in stationary phase of growth and the concentration of the radioactive substrate is at least 5-fold higher than in the continuous labeling type of experiments.

Biosynthetic Labeling with Tritiated Monosaccharides

Procedure. For continuous labeling, IM-9 lymphocytes (or other cell types, as desired) at stationary phase of growth ($1–2 \times 10^6$ cells/ml) are collected by centrifugation at 600 *g* and resuspended in normal culture medium, RPMI 1640, 10% fetal calf serum, 25 m*M* HEPES at a cell density of $0.5–1 \times 10^7$ cells/ml. Either D-[1,6-^{3}H]glucosamine hydrochloride (39 Ci/mmol), D-[2-^{3}H]mannose (14 Ci/mmol), D-[1-^{3}H]galactose (14 Ci/mmol), or L-[5,6-^{3}H]fucose (56 Ci/mmol) (New England Nuclear) is added to the medium at a concentration of 50–100 μCi/ml and the cells are incubated at 37° for 12–14 hr.

For pulse labeling, the cells are preincubated for 15 min at 37° in glucose-free RPMI 1640 medium containing 10% dialyzed fetal calf serum and 25 m*M* HEPES. After the preincubation, the cells are resuspended in fresh glucose-free medium at a cell density of $2–4 \times 10^7$ cells/ml. The tritiated monosaccharide is added to the medium at a concentration of 0.5 mCi/ml and the cells are incubated at 37° for 15 min. The pulse is terminated by quickly washing the cells three times in phosphate-buffered saline.

Comments. It is extremely important to add the tritiated sugars in sterile physiological saline rather than in an ethanol-containing solution as

[11] E. Van Obberghen, M. Kasuga, A. Le Cam, J. A. Hedo, A. Itin, and L. C. Harrison, *Proc. Natl. Acad. Sci. U.S.A.* **78,** 1052 (1981).

[12] M. Kasuga, C. R. Kahn, J. A. Hedo, E. Van Obberghen, and K. M. Yamada, *Proc. Natl. Acad. Sci. U.S.A.* **78,** 6917 (1981).

[13] J. A. Hedo, C. R. Kahn, M. Hayashi, K. M. Yamada, and M. Kasuga, *J. Biol. Chem.* **258,** 10020 (1983).

they are often provided by the suppliers. The presence of even small amounts of ethanol in the medium markedly inhibits the incorporation of the sugars.

Depending on the tritiated sugar and the length of the labeling, incorporation of radioactivity into trichloroacetic acid-precipitable material ranges from 2 to 12% of the total added.

[^{3}H]Glucosamine and [^{3}H]galactose label the insulin receptor to a higher specific activity than [^{3}H]mannose or [^{3}H]fucose.[8] All four sugars label major receptor subunits $M_r = 135,000$ and $M_r = 95,000$ as well as the $M_r = 210,000$ component of the receptor (Fig. 2).[13] However, the $M_r = 190,000$ high mannose precursor of the receptor can be labeled only with [^{3}H]mannose or [^{3}H]glucosamine. The ratio of radioactivity in the $M_r = 135,000$ subunit versus that in the $M_r = 95,000$ subunit is around 2 : 1 for all sugars except for [^{3}H]galactose where the ratio is about 1 : 1.

Biosynthetic Labeling of the Insulin Receptor with Radioactive Amino Acids

Procedure. For continuous labeling, IM-9 lymphocytes are incubated at a density of $0.5–1 \times 10^7$ cells/ml in methionine-free RPMI 1640 medium containing 10% fetal calf serum, 25 m*M* HEPES, and 50–100 μCi/ml [^{35}S]methionine (>1000 Ci/mmol, New England Nuclear). The incubation can be maintained at 37° for up to 8 hr. Longer incubation periods induce a progressive decrease in the viability of the cells due to the lack of methionine. Alternatively, the IM-9 cells can also be continuously labeled in complete RPMI 1640 medium in the presence of 100 μCi/ml of L-[3,4,5-^{3}H(N)]-leucine (122 Ci/mmol) for 12–14 hr.

For pulse-labeling, IM-9 lymphocytes are preincubated for 30 min at 37° in methionine-free RPMI 1640 medium containing 10% dialyzed fetal calf serum and 25 m*M* HEPES. Then the cells are resuspended in fresh methionine-free medium at a cell density of $2–4 \times 10^7$ cells/ml and L-[^{35}S]methionine is added at a concentration of 0.5 mCi/ml and the incubation is maintained at 37° for 15 min. The pulse is terminated by quickly washing the cells three times in phosphate-buffered saline.

Comments. Incorporation of ^{35}S into trichloroacetic acid-precipitable cellular proteins ranges from 10 to 30% of the total radioactivity.

Measurement of Turnover Rates of the Insulin Receptor

All the labeling methods of the insulin receptor described above appear to have little or no deleterious effect on cellular metabolism and growth. Cell viability of IM-9 lymphocytes after cell surface or biosynthetic labeling is usually greater than 90% and remains at this level for a period of time of at least 12–20 hr after labeling. Therefore, the cells can be returned to normal culture medium for turnover studies. The usual conditions for culture are RPMI 1640 medium with 10% fetal calf serum and 25 m*M* HEPES, at a cell density of 2×10^6 cells/ml. In order to measure the turnover rates of the insulin receptor, aliquots of cells are solubilized at intervals and the receptor identified by immunoprecipitation and electrophoresis as described below. A minimum of four time points are required to obtain a reliable estimate of the degradation rate. The number of cells required per time point is about $1–2 \times 10^8$. The half-life of the receptor subunits appears to be between 7 and 12 hr.[12] Ideally, the experimental period should be longer than one half-life. However, in practice, an observation period of 8–9 hr is sufficient to determine the half-time with reasonable confidence since the decay rate appears to be monoexponential.[12]

The degradation rate of the receptor can be examined after either cell surface or biosynthetic labeling of the receptor. It is obvious that cell surface labeling allows measurement of the degradation rate of the surface exposed receptors whereas biosynthetic labeling measures the degradation rate of both surface exposed and intracellular receptors. However, in practice, in IM-9 lymphocytes both methods give similar results,[12] probably due to the very small size of the intracellular pool of receptors in this cell type.[10] Likewise, estimates of the turnover rate of the insulin receptor by labeling either the protein or the carbohydrate moiety also yield very similar results.

When biosynthetic labeling is used for measuring the degradation rate of the receptor, a continuous labeling protocol is preferred to a short pulse label because it allows a higher degree of labeling. After labeling, the culture medium is supplemented with unlabeled sugar or amino acid to dilute, as much as possible, the unincorporated radioactive precursor taken up by the cells. Methionine or leucine is added during the chase at a concentration of 2 mg/ml, while sugars such as mannose, galactose, or fucose are added at 2 m*M*. Glucosamine, at millimolar concentrations, is toxic for the cells and, therefore, if this sugar has been used for labeling, a high concentration of glucose (15 m*M*) should be used instead of glucosamine in the chase. In theory, one of the potential problems in measuring the degradation rate of the receptor after biosynthetic labeling is reutilization of the radioactive substrate after receptor degradation. However, the fact that both cell surface and biosynthetic methods give very similar results suggests that under these experimental conditions, reutilization is not a significant factor.

Measurements of the biosynthetic rate of the receptor pose more difficulties than estimating the degradation rate. In particular, measurement of the absolute rate of synthesis of the receptor would require precise quantitation of the specific activity of the free intracellular pool of the radioactive precursor. Relative rates of synthesis, however, can be obtained by measuring the amount of radioactivity incorporated into the insulin receptor relative to the amount of precursor incorporated into total cell proteins. In this case, the cells are continuously labeled for a 6- to 8-hr period and aliquots are analyzed at intervals during the labeling period. The identification of the precursor forms of the receptor and the study of processing into the mature receptor subunits is achieved by using the short pulse labeling protocol followed by a chase period of up to 8 hr.

Cell Solubilization and Lectin Chromatography

In order to identify the insulin receptors among the number of cellular proteins labeled with any of the above described procedures, the cells are first solubilized in a nonionic detergent Triton X-100. Routinely, the cells are solubilized by continuously stirring for 30 min at 4° 2–4 × 10^8 cells/ml in 0.15 *M* NaCl, 50 m*M* HEPES, pH 7.6, 1% Triton X-100 (v/v). Undissolved material is sedimented by ultracentrifugation at 200,000 *g* for 45 min. The supernatants can be used directly for immunoprecipitation except when biosynthetic labeling with radioactive amino acids has been used. In this case, the background of nonspecifically precipitated bands in the gels is so intense that it impedes identification of the receptor subunits (Fig. 3). This problem is independent of the radioactive amino acid used and occurs with a variety of cell types. In our experience, repeated preimmunoprecipitation with control nonimmune serum reduces somewhat the intensity of the background, but not to a level adequate for direct quantitation of the receptor subunits. Preparation of plasma membrane fractions before solubilization circumvents this problem.[14] However, the most convenient and simple procedure to eliminate the problem is partial purification of the insulin receptors on immobilized lectins.

The insulin receptor can be purified in a number of different lectin-agarose columns (see the table).[15] Wheat-germ agglutinin is the most efficient lectin in terms of receptor purification (20- to 30-fold) and recovery (70–95%) as measured by binding of ^{125}I-labeled insulin to the eluates.[15] However, not all lectins bind the same receptor components (Fig. 4). The major receptor subunits M_r = 135,000 and M_r = 95,000, as well as the M_r = 210,000 component are all absorbed and eluted from lectin columns with high affinity for mannose (lentil lectin), galactose (ricin I), or sialic

[14] I. A. Simpson, J. A. Hedo, and S. W. Cushman, *Diabetes* **33,** 13 (1984).

[15] J. A. Hedo, L. C. Harrison, and J. Roth, *Biochemistry* **20,** 3385 (1981).

AFFINITY CHROMATOGRAPHY OF INSULIN RECEPTORS ON IMMOBILIZED LECTIN COLUMNS

Lectin	Eluting sugar
Concanavalin A *Lens culinaris* agglutinin (lentil) *Pisum sativum* agglutinin (pea)	Methyl α-D-mannopyranoside Methyl α-D-glucopyranoside
Ricinus communis agglutinin I (ricin I) *Ricinus communis* agglutinin II (ricin II)	Lactose Methyl β-D-galactopyranoside
Wheat germ agglutinin	*N*-Acetyl-D-glucosamine

acid residues (wheat germ agglutinin). The M_r = 190,000 proreceptor, however, is only retained by lectins with affinity for mannose residues (lentil lectin). This is due to the fact that this proreceptor form contains exclusively carbohydrate chains of the high mannose type[13] and therefore is only bound efficiently by lectins with affinity for mannose residues (lentil lectin, pea lectin, or concanavalin A). On the other hand, the other receptor components contain carbohydrate chains of the complex type and are also bound by lectins with affinity for galactose and sialic acid residues. Any aglycosylated forms of receptor will be lost in this procedure.

Affinity Chromatography of Solubilized Insulin Receptors on Immobilized Lectins

Materials

IM-9 lymphocytes or other cells which have been labeled as described above and solubilized in Triton X-100. Although we have used samples in which the Triton concentration of the sample varies between 0.025% and 1% without any apparent change in the efficiency of the procedure, theoretically the specificity of the procedure is increased as the concentration of detergent approaches the critical micellar concentration.

Lectin immobilized on agarose (Sepharose 4B). The amount of coupled lectin may vary from 1 to 5 mg/ml of settled gel.

Buffer I: 0.15 *M* NaCl, 50 m*M* HEPES, 0.1% Triton X-100, 0.01% sodium dodecyl sulfate, pH 7.6

Buffer II: 0.15 *M* NaCl, 50 m*M* HEPES, 0.1% Triton X-100, 0.3 *M* eluting sugars (see the table), pH 7.6

Buffer III: 0.15 *M* NaCl, 50 m*M* HEPES, 0.1% Triton X-100 , pH 7.6

Small plastic columns (15 × 0.9 cm)

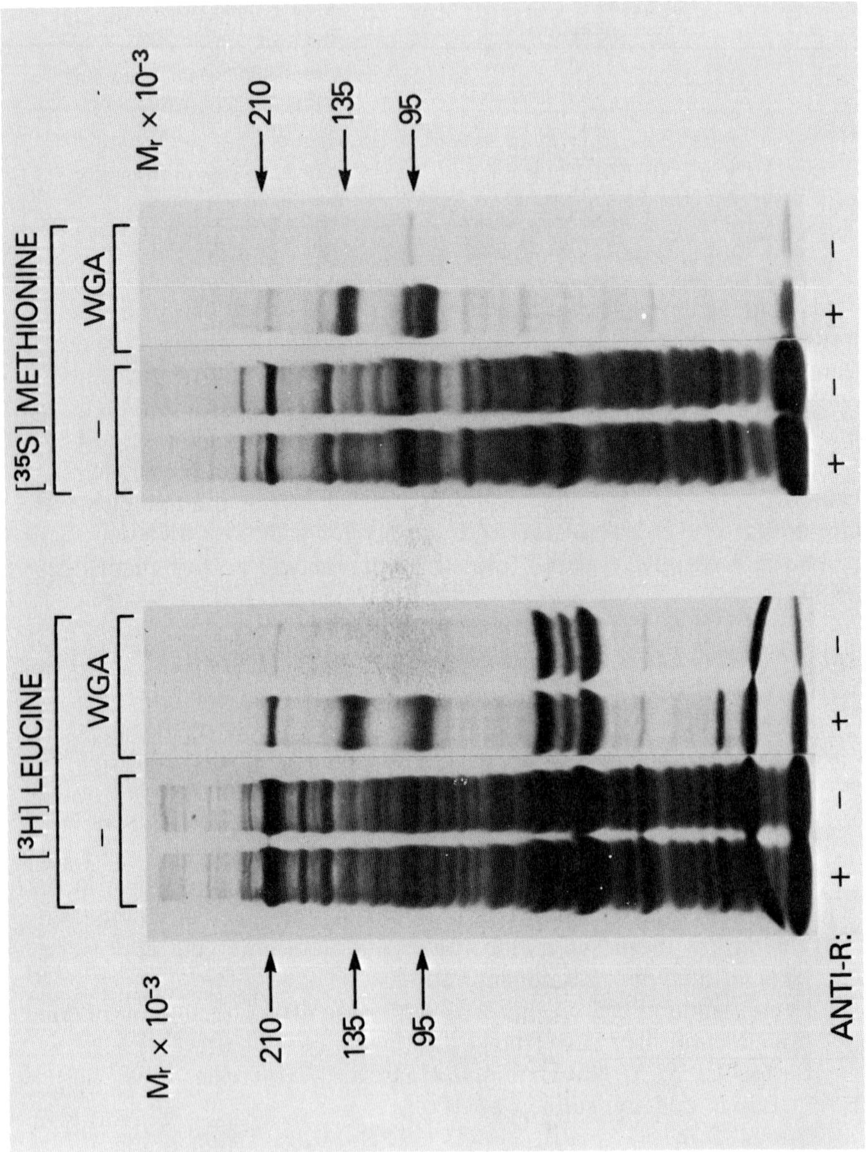

[^{3}H] LEUCINE
[^{35}S] METHIONINE
–
WGA
$M_r \times 10^{-3}$
210
135
95
ANTI-R: + – + – + – + –

Procedure

1. Prepare the column with the appropriate amount of lectin-agarose. This amount will depend on both the capacity of each particular batch of lectin-agarose and the concentration of glycoproteins in the sample to be chromatographed. As an index, 2 ml of settled gel (1 mg/ml of coupled lectin) is used to purify the solubilized insulin receptors from 2 to 4×10^8 IM-9 lymphocytes.
2. Wash the column with 50 bed volumes of buffer I.
3. Wash with 10 bed volumes of buffer II.
4. Wash with 100 bed volumes of buffer III.
5. Apply the sample and recycle it through the column at least 3 times.
6. Wash with 50–100 volumes of buffer III.
7. Elute with buffer II (at least 2 bed volumes). Collect 1 ml fractions.
8. Measure radioactivity in each eluate and pool the fractions with the highest number of counts for immunoprecipitation.

Comments. Steps 1–4 can be performed at room temperature. In order to prevent receptor degradation, steps 5–7 are performed at 4°.

For chromatography on lentil lectin, pea lectin, or concanavalin A, buffer III should contain 1 m*M* $CaCl_2$ and 1 m*M* $MnCl_2$.

The columns can be reused for periods of up to 1 year without apparent loss of activity. They should be stored at 4° in buffer II containing 0.01% sodium azide.

Immobilized lectins are commercially available from a number of different suppliers (Vector Labs, Miles Labs, Pharmacia, P.L. Biochemicals, etc.) and these seem to vary somewhat in their efficacy. Alternatively, the free lectins can be coupled to Sepharose 4B (Pharmacia) using the cyanogen bromide procedure[16] or to polyacrylic hydrazide-Sepharose (Miles) using the glutaraldehyde[17] procedure.

[16] P. Cuatrecasas and I. Parikh, this series, Vol. 34, p. 653.

[17] R. Lotan, G. Beattie, W. Hubbel, and G. L. Nicolson, *Biochemistry* **16,** 1787 (1977).

FIG. 3. Biosynthetic labeling of insulin receptors with radioactive amino acids: immunoprecipitation of cellular extracts before and after chromatography on immobilized wheat germ agglutinin (WGA). Left: Human IM-9 lymphocytes were labeled with [^{3}H]leucine for 8 hr and after solubilization in Triton X-100, the insulin receptors were immunoprecipitated before (−) and after (WGA) chromatography on wheat germ agglutinin-agarose. The immunoprecipitates were analyzed by SDS–polyacrylamide gel electrophoresis and autoradiography. Immunoprecipitation with anti-receptor serum (anti-R) is indicated by the + sign at the foot of each lane; the − sign corresponds to immunoprecipitation with control nonimmune serum. Right: Rat hepatoma Fao cells were labeled with [^{3}S]methionine for 8 hr and processed as described above for the IM-9 lymphocytes.

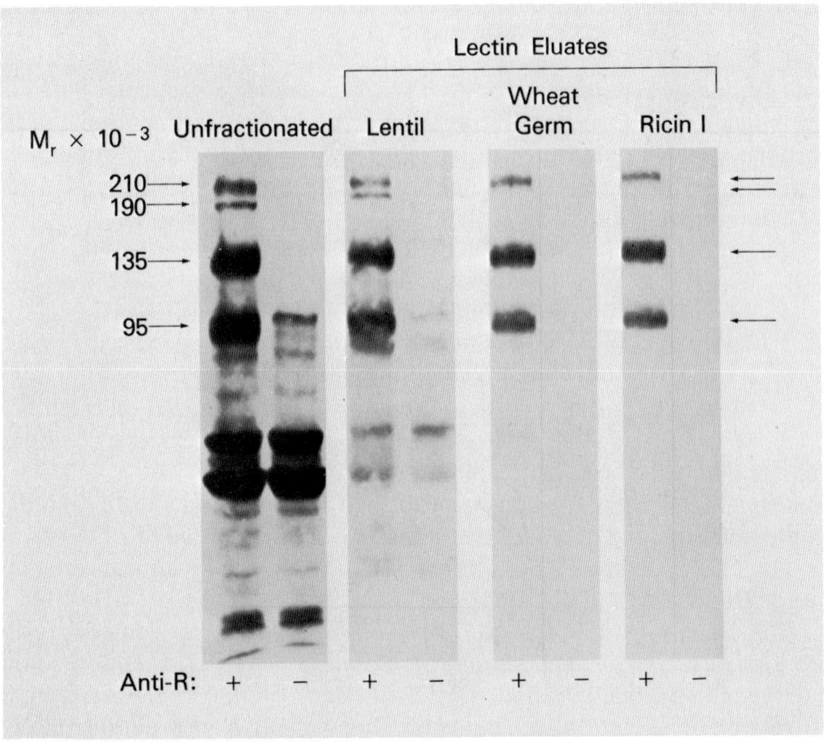

FIG. 4. Chromatography of [^{3}H]glucosamine-labeled receptors on different immobilized lectins. IM-9 lymphocytes were biosynthetically labeled with [^{3}H]glucosamine for 12 hr and after solubilization in Triton X-100, the extracts were chromatographed on lentil-agarose, wheat germ agarose, or ricin I-agarose. The insulin receptors in the eluates and in the unfractionated extract were immunoprecipitated and analyzed by SDS–polyacrylamide gel electrophoresis and autoradiography. Immunoprecipitation with anti-receptor serum (anti-R) is indicated by the + sign at the foot of each lane; the − sign corresponds to immunoprecipitation with control nonimmune serum.

Immunoprecipitation of Insulin Receptors

Sera containing autoantibodies to the insulin receptor from patients with the Type B syndrome of insulin resistance and acanthosis nigricans (or other anti-receptor serum) are used to immunoprecipitate the solubilized insulin receptors. The most extensively used serum (from patient B-2) is maximally effective at a dilution of 1 : 400. Other sera require lower dilutions (1 : 100 or less) to achieve maximal immunoprecipitation of insulin receptors. With serum B-2, there is no difference between the results obtained with whole serum or with purified IgG preparations. However, with other sera that have to be used at lower dilutions to completely immunoprecipitate the receptor, an IgG preparation purified on Protein A-Sepharose is used instead of whole serum in order to reduce nonspecific immunoprecipitation.

To adsorb the immune complexes, we use a suspension of formaldehyde-fixed *Staphylococcus aureus* cells. The *S. aureus* are treated prior to use with sodium dodecyl sulfate and 2-mercaptoethanol to remove any material that might be released during the preparation of the samples for electrophoresis.

Preparation of S. aureus Cells

Materials

10% (w/v) suspension of formaldehyde-fixed *S. aureus* cells (Pansorbin, Calbiochem)
3% sodium dodecyl sulfate and 1.5 *M* 2-mercaptoethanol in 0.1 *M* NaCl, 20 m*M* Tris–HCl buffer, pH 7.2
Phosphate-buffered saline, pH 7.4
0.15 *M* NaCl, 50 m*M* HEPES, pH 7.6

Procedure

1. Sediment the cells at 10,000 *g* for 10 min and resuspend them in buffer containing SDS and 2-mercaptoethanol at the original concentration.
2. Incubate the suspension in a boiling waterbath for 30 min.
3. Sediment the cells, discard the supernatant, and repeat the boiling once.
4. Wash the cells *at least* six times with phosphate buffered saline.
5. Resuspend the cells at 20% (w/v) in 0.15 *M* NaCl, 50 m*M* HEPES, pH 7.6 and store them at 4° after addition of 0.02% sodium azide.

Immunoprecipitation Procedure

Samples (0.5 ml) containing the labeled insulin receptors are first preincubated with 200 μl of *S. aureus* suspension at 4°. After 1 hr, the samples are centrifuged at 10,000 *g* for 5 min. The pellets are discarded and the supernatants incubated with the appropriate dilution of anti-receptor serum. Parallel incubations with an equivalent (in terms of IgG concentration) amount of nonimmune normal serum are also processed simultaneously in order to obtain control, nonspecific immunoprecipitates. After a minimum of 4 hr at 4°, 100 μl of *S. aureus* suspension is added and the incubation is maintained for 2 additional hours. The *S.*

aureus cells are then collected by centrifugation at 10,000 *g* for 5 min and washed twice with 0.15 *M* NaCl, 50 m*M* HEPES, pH 7.6, 1 m*M* EDTA, 0.1% Triton X-100, and 0.1% SDS. After washing, the adsorbed immune complexes are dissociated by boiling for 5 min in 2% SDS, 0.1 *M* dithiothreitol, 0.01% bromophenol blue, and 10 m*M* sodium phosphate buffer, pH 7.0.

Electrophoretic analysis of the immunoprecipitates in 0.1% SDS according to Laemmli[18] in 7.5% slab gels with an acrylamide to bisacrylamide ratio of 37.5 : 1.

Autoradiography and Quantitation of Results

Autoradiographic detection of the labeled receptor subunits in polyacrylamide gels is performed by exposing Kodak X-Omat AT film to the dried gel at −70° for a variable period of time. Gels containing ^{125}I-labeled bands are autoradiographed using intensifying screens (Dupont Lightning-Plus). The film is placed in close contact between the dried gel and the screen. On the other hand, bands labeled with ^{3}H or ^{35}S are best detected using fluorography. This procedure consists of impregnating the gel with a suitable fluor compound so that some of the energy from the radioisotope decay process is converted to light which exposes the film. We use a commercially prepared fluorographic solution (En^{3}Hance, New England Nuclear) according to the manufacturer's instructions.

The exposure time can range from a few hours to several months on the type and the amount of the isotope. The adequate time can be determined only by trial and error. Even the total number of radioactive counts loaded per lane in the gels may be misleading sometimes since the radioactivity in the receptor subunits may represent an extremely variable percentage of the total depending on the number and intensity of nonspecifically immunoprecipitated bands. The signal-to-noise ratio is the most critical factor and even a small amount of radioactivity such as 100 dpm above background can be enough for the detection of a band provided that the background of the gel is appropriately low. On average, to visualize the insulin receptor subunits, gels containing ^{125}I or ^{35}S require exposure times from 3 to 7 days and gels containing ^{3}H from 6 to 20 days.

Quantitative estimates of the radioactivity present in the receptor subunits are obtained by excising the labeled bands from the gels. The bands are located by superimposing autoradiographs on the original gels. ^{125}I activity in gel slices is measured directly in a gamma counter but radioactivity in the form of ^{3}H or ^{35}S is first eluted from the slices in 3% (v/v)

[18] U. K. Laemmli, *Nature (London)* **227,** 680 (1975).

Protosol in Econofluor (New England Nuclear) overnight at 37° and measured in a scintillation counter. Alternatively, the intensity of the bands can be quantitated by densitometric scanning of the autoradiograph. Although in most instances this method gives comparable results to those obtained by counting radioactivity, in our experience the latter is more accurate and reliable, especially when the intensity of the bands escapes the range of linear response of the film. Preflashing of the film before fluorography has been reported[19] to correct the nonlinear relationship between radioactivity of the sample and the film image. In any case, however, densitometric quantitation of the results should be interpreted with caution and compared when possible to direct measurements of the radioactivity in the excised bands.

[19] R. A. Laskey and A. D Mills, *Eur. J. Biochem.* **56,** 335 (1975).

[48] Phosphorylation of the Insulin Receptor in Cultured Hepatoma Cells and a Solubilized System

By MASATO KASUGA, MORRIS F. WHITE, and C. RONALD KAHN

Introduction

Phosphorylation of proteins is established as a basic regulatory mechanism in hormone action. Several enzyme systems, regulated by hormones through phosphorylation or dephosphorylation, have been described recently in this series.[1] However, beyond our knowledge of glycogen metabolism, the molecular mechanisms linking hormone binding at the cell surface to the phosphorylation and activity of specific intracellular enzymes are unknown. Accumulating evidence suggests that phosphorylation of cytoplasmic proteins is involved in insulin action. Recently, we have demonstrated that insulin binding stimulates phosphorylation of the β-subunit of the insulin receptor.[2-7] This phos-

[1] J. D. Corbin and J. G. Hardman, this series, Vol. 99, p. 3.

[2] M. Kasuga, F. A. Karlsson, and C. R. Kahn, *Science* **215,** 185 (1982).

[3] M. Kasuga, Y. Zick, D. L. Blithe, F. A. Karlsson, H. U. Haring, and C. R. Kahn, *J. Biol. Chem.* **257,** 9891 (1982).

[4] M. Kasuga, Y. Zick, D. L. Blithe, M. Crettaz, and C. R. Kahn, *Nature (London)* **298,** 667 (1982).,

[5] M. Kasuga, Y. Fujita-Yamaguchi, D. L. Blithe, and C. R. Kahn, *Proc. Natl. Acad. Sci. U.S.A.* **80,** 2137 (1983).

METHODS IN ENZYMOLOGY, VOL. 109

ISBN 0-12-182009-2

phorylation reaction, first documented with intact cells, is retained by the insulin receptor after solubilization and partial purification from a variety of sources.

In cultured cells labeled with $^{32}PO_4^{3-}$, insulin stimulates phosphorylation of serine, threonine, and tyrosine residues in the β-subunit.[3] On the other hand, phosphorylation is detected almost exclusively on tyrosine residues following incubation of solubilized receptors with [γ-^{32}P]ATP and insulin.[4] This insulin-sensitive tyrosine kinase activity is immunoprecipitated by antibodies to the insulin receptor and remains associated with the insulin receptor after 2500-fold purification by sequential affinity chromatography on wheat germ agglutinin agarose and insulin Sepharose.[5,6] These data suggest that the insulin receptor is a tyrosine-specific protein kinase. Insulin binding stimulates both phosphorylation of the receptor and phosphorylation of other substrates. Although the biological role of this reaction is uncertain, it could provide a molecular link between extracellular insulin binding and intracellular insulin responses.

In this chapter, we describe the methods which have been used successfully in our laboratory to study insulin receptor autophosphorylation in cultured cells and detergent solubilized receptor systems. The hepatoma cell line, Fao, is used in each case as an example, but our experiences show that these methods can be used with other cell systems such as IM-9 lymphocytes,[2] rat adipocytes,[6] and rat hepatocytes.[8]

Materials

Equipment

- Tissue culture facilities
- Reverse-phase high-peformance liquid chromatography system (Waters Associates)
- Microfuge B; L8-55 ultracentrifuge; 70.1Ti ultracentrifuge rotor; polycarbonate ultracentrifuge bottle assembly, #339574. (Beckman)
- Model 100 slab gel stand; Lucite comb and spacer sets with 13 teeth; 1.2-cm-thick glass sets, 15.9 × 14 cm. (Aquebogue Machine and Repair Shop, Box 205, Main Road, Aquebogue, L.I., N.Y., 11937)
- Electrophoresis power supply, EC-400 (E-C Apparatus Corporation)

---

[6] M. Kasuga, Y. Fujita-Yamaguchi, D. L. Blithe, M. F. White, and C. R. Kahn, *J. Biol. Chem.* **258,** 10973 (1983).

[7] H. U. Haring, M. Kasuga, and C. R. Kahn, *Biochem. Biophys. Res. Commun.* **108,** 1538 (1982).

[8] Y. Zick, M. Kasuga, C. R. Kahn, and J. Roth, *J. Biol. Chem.* **258,** 75 (1982).

Chromatography columns K9/15, 0.9 × 15 cm, with buffer reservoirs (Pharmacia)

Supplies

[γ-^{32}P]ATP (1000–3000 Ci/mmol); carrier-free [^{32}P]H_3PO_4; Triton X-100 (New England Nuclear)
HEPES; TRIZMA BASE; 2-mercaptoethanol; aprotinin; phenylmethylsulfonyl fluoride; *N*-acetylglucosamine; bovine serum albumin (Sigma)
Trifluoroacetic acid; trichloroacetic acid; polyethylene glycol (PEG 6000); glycerol; HPLC grade acetonitrile; sodium vanadate (Fisher Scientific)
N,*N*′-Diacetylchitobiose (I. J. Goldstein, Dept. of Biological Chemistry, The University of Michigan, Ann Arbor, MI 48105)
L-1-Tosylamido-2-phenylethyl chloromethyl ketone (TPCK) treated trypsin (Worthington)
Porcine insulin (Elanco)
Arcylamide; *N*,*N*′-methylene bisacrylamide (BIS); ammonium persulfate; dithiothreitol; *N*,*N*,*N*′,*N*′-tetramethylethylenediamine (TEMED); high-molecular-weight PAGE standards; sodium dodecyl sulfate; Bio-Rad protein assay concentrate (Bio-Rad)
Pansorbin, #507858 (Calbiochem)
Wheat germ agglutinin-agarose (Vector Laboratories or Miles Laboratories)
Microfuge tubes, 1.7 ml (Denville Scientific, Inc.)
Tissue culture supplies (Nunc or Corning)
X-Omat AR film, 8″ × 10″; developer, #190 0984; fixer, #146 0984 (Kodak)
Cronex lightening plus intensifying screen, #165 1454. (Du Pont)
μBondapak C_{18} reverse-phase HPLC column (Waters)
3a70b Scintillation mixture (Research Products International Corp.)

Solutions

Solution A: Calcium- and magnesium-free phosphate-buffered saline: 137 m*M* NaCl, 2.7 m*M* KCl, 8 m*M* $Na_2HPO_4 \cdot 7H_2O$, 1.5 m*M* KH_2PO_4
Solution B: Phosphate-free Krebs–Ringer bicarbonate: 119 m*M* NaCl, 5 m*M* KCl, 1.3 m*M* $CaCl_2$, 1.2 m*M* $MgSO_4$, 25 m*M* $NaHCO_3$
Solution C: Solubilization buffer: 50 m*M* HEPES, pH 7.4, 1% Triton X-100
Solution D: Wheat germ agglutinin agarose wash buffer: 50 m*M* HEPES, pH 7.4, and 0.1% Triton X-100

Solution E: Wheat germ agglutinin agarose wash buffer used to inhibit dephosphorylation during the chromatographic procedures: 50 m*M* HEPES, pH 7.4, 0.1% Triton X-100, 10 m*M* sodium pyrophosphate, 10 m*M* sodium fluoride, and 4 m*M* EDTA

Pansorbin suspension: Insoluble protein A from *Staphylococcal aureus* is prepared for use as follows: 0.5 g of Pansorbin is suspended in 50 ml of Solution A supplemented with 3% SDS and 10% 2-mercaptoethanol, boiled for 20 min, and sedimented by centrifugation at 10,000 *g* for 15 min. This wash of the Pansorbin is repeated twice with 50 ml of Solution A supplemented with 3% SDS only. The Pansorbin is then washed 4 times by resuspension and centrifugation with 25 m*M* HEPES and finally suspended (10% w/v) in 25 m*M* HEPES containing 10 m*M* sodium azide.

Wheat germ agglutinin agarose: Wheat germ agglutinin agarose is washed before each use with 0.1% SDS, about 50 ml of wash for 2 ml of agarose. Then the agarose is washed with 200 ml of water and equilibrated with the appropriate buffer.

PAGE Buffers

Stacking gel buffer is prepared by dissolving 6.05 g of TRIZMA base and diluting 10 ml of 0.2 m*M* EDTA and 10 ml of 10% SDS in 500 ml of water; adjust the pH to 6.7 with phosphoric acid.

Resolving gel buffer is prepared by dissolving 45.4 g of TRIZMA base and diluting 10 ml each of 0.2 *M* EDTA and 10% SDS with 500 ml of water; adjust the pH of this solution to 8.9 with HCl.

Acrylamide/BIS (30%/0.8%, w/v) solution is prepared by dissolving 30 g of acrylamide and 0.8 g of BIS in 60 ml of water. Adjust the final volume to 100 ml with water.

Electrophoresis running buffer is prepared as a 4× solution and diluted when needed. Dissolve TRIZMA base (96.9 g), glycine (456 g), EDTA (10.7 g), and SDS (16 g) in 4 liters of water.

Laemmli sample buffer is prepared as a 5× concentrated solution as follows: in a total final volume of 10 ml, mix 1 ml of 0.1% bromphenol blue, 0.5 ml of 1 *M* sodium phosphate (pH 7.0), 5 ml of glycerol, 1 g of SDS, and water. This mixture solidifies at cool temperatures so it should be warmed before dilution. Add 15 mg of DTT to 1 ml of 1× sample buffer or 75 mg to 1 ml of the 5× solution.

Cell Culture

The hepatoma cell line Fao was obtained from Dr. M. C. Weiss (Guf-sur-Yvette, France). The Fao cell line is a fourth generation clonal line of well differentiated rat hepatoma cells derived from the Reuber H-35 hepatoma.[9] It has many insulin receptors and is very sensitive to insulin as measured by stimulation of glycogen synthase and tyrosine aminotransferase activities.[10] Monolayer cultures of Fao cells were grown to confluence in 150 cm^2 plastic dishes (Nunc) or plastic bottles (Corning) containing 50 ml of Coon's modified[11] Ham's F12 medium[12] obtained as a special formulation from GIBCO. This medium is supplemented with 5% fetal bovine serum (Hyclone), 100 units/ml penicillin (GIBCO), and 100 μg/ml streptomycin.

Polyacrylamide Gel Electrophoresis and Autoradiography

Phosphoproteins are separated by polyacrylamide gel electrophoresis (PAGE) according to the method of Laemmli.[13] For separation of the insulin receptor under reducing conditions, 7.5% resolving gels are prepared by diluting 7.5 ml of acrylamide/BIS (30%/0.8%) solution with resolving gel buffer (15 ml), glycerol (3 ml), water (4.2 ml), TEMED (0.02 ml), and 10% ammonium persulfate (0.3 ml). This solution is poured into the mold to a height of 8.5 cm. The 4% stacking gel is prepared by diluting 2 ml of the same acrylamide/BIS solution with stacking gel buffer (7.5 ml), water (5.3 ml), TEMED (0.01 ml), and 10% ammonium persulfate (0.15 ml). This solution is immediately layered on top of the unpolymerized resolving gel and the Lucite comb is inserted. This slab gel polymerizes and is ready to use within 40 min. Samples (80 μl) are applied to the gel and allowed to enter the stacking gel by electrophoresis at a constant current of 7 mA. Then the current is increased to 30 mA until the tracking dye reaches the bottom of the gel. The slab gel is stained for 5 min with 0.25% Coomassie blue dissolved in 50% trichloroacetic acid and destained by incubation for 12 hr in 7% acetic acid and 20% methanol at 37°. Autoradiograms are obtained by exposing overnight the dried gels to Kodak X-Omat AR film in a cassette.

Phosphorylation of the Insulin Receptor in Cultured Cells

Labeling Intact Cultured Cells with Radioactive Phosphate

General methods for labeling cells with $^{32}PO_4^{3-}$ have been presented in Volume 99 of this series.[1] In this section we describe specific procedures used in our laboratory to study phosphorylation of the insulin receptor in

[9] J. Deschatrette, E. E. Moore, M. Dubois, D. Cassio, and M. C. Weiss, *Somatic Cell Genet.* **5,** 697 (1979).
[10] M. Crettaz and C. R. Kahn, *Endocrinology* **113,** 1201 (1983).
[11] H. E. Coon and M. C. Weiss, *Proc. Natl. Acad. Sci. U.S.A.* **62,** 852 (1969).
[12] R. G. Ham, *Proc. Natl. Acad. Sci. U.S.A.* **53,** 288 (1965).
[13] U. K. Laemmli, *Nature (London)* **227,** 680 (1970).

intact hepatoma cells (Fao). Confluent monolayer cultures (20×10^6 cells/10 cm dish) are washed twice with a solution containing 50 m*M* HEPES, pH 7.4, and 0.1% bovine serum albumin. The cells are then incubated for 2 hr at 37° in a humidified atmosphere composed of 5% CO_2/95% air with 5 ml of Solution B containing $^{32}PO_4^{3-}$ (0.125 mCi/ml) and 0.1% dialyzed bovine serum albumin. About 50 to 60% of the $^{32}PO_4^{3-}$ is accumulated by the cells after 1 hr and the steady-state occurs after about 2 hr. The amount of $^{32}PO_4^{3-}$ precipitated by 5% trichloroacetic acid increases linearly during the first 4 hr of incubation and reaches a maximum of about 6 to 8% of the radioactivity added to the incubation medium after 4 hr. The cells also can be incubated overnight at 37° in Ham's F-12 medium containing 0.125 mCi/ml $^{32}PO_4^{3-}$, 1 m*M* phosphate, and 5% dialyzed fetal bovine serum. The observed phosphorylation of the insulin receptor is identical with each method; however, prolonged labeling may lead to higher background in the autoradiograms used to visualize the proteins.[14] Insulin, or other additions are added at the various concentrations and the incubation is continued for the desired time interval which may range between a few seconds to several hours. In radioactively labeled Fao cells stimulated with 100 n*M* insulin, phosphorylation is usually maximal after 20–30 min.

Solubilization of the Labeled Cells

The phosphorylation reaction is stopped by placing the culture dish on ice, removing the radioactive incubation mixture, and washing the cell monolayers twice with ice-cold portions of Solution B containing 10 m*M* sodium pyrophosphate, 100 m*M* sodium fluoride, and 4 m*M* EDTA. The monolayer is immediately solubilized at 4° with 5 ml of Solution C containing 10 m*M* sodium pyrophosphate, 100 m*M* sodium fluoride, 4 m*M* EDTA, aprotinin (1000 trypsin inhibitor units/ml), and 2 m*M* phenylmethylsulfonyl fluoride (200 m*M* stock solution can be prepared in ethanol). After a 10 min interval, the cells are removed from the culture dish and the insoluble material is sedimented by ultracentrifugation for 60 min at 200,000 *g*, that is, 50,000 rpm in a Beckman 70.1Ti rotor.

Wheat Germ Agglutinin-Affinity Chromatography

Wheat germ agglutinin coupled to agarose (about 1.0 ml packed volume per dish) is prepared for chromatography as described above and equilibrated with Solution E. Each supernatant from the centrifugation

[14] J. C. Garrison, this series, Vol. 99, p. 20.

described above is applied onto an affinity column. The sample is collected and reapplied onto the column two more times and then the agarose is washed with 50 ml of Solution E. The glycoproteins bound to the immobilized wheat germ agglutinin are eluted with 5 ml of Solution E supplemented with 300 m*M* *N*-acetylglucosamine. Five, 1 ml fractions are usually collected and the two fractions containing the most radioactivity are combined and used for immunoprecipitation. Only 1–2% of the trichloroacetic acid precipitable $^{32}PO_4^{3-}$ elutes from the wheat germ affinity column. This chromatographic step yields a 20-fold purification of the insulin receptor and an 80 to 100% recovery as determined by ^{125}I-labeled insulin binding.[15]

Immunoprecipitation of the Phosphorylated Insulin Receptor

The eluate obtained from the wheat germ affinity column is incubated with pooled normal human serum or serum containing antibodies which react with the insulin receptor. The concentration of serum depends on its ability to immunoprecipitate the insulin receptor. The most potent anti-insulin receptor serum used in our laboratory is a human serum, designated B-2, which was obtained from a patient with an autoimmune form of insulin-resistant diabetes. This antiserum immunoprecipitates more than 90% of labeled receptor at 1 : 200 dilution. In a typical experiment, 3 to 4 μl of B-2 serum or 6 to 8 μl of control serum are added to 600–800 μl of eluate and incubated at 4° for at least 3 hr in microfuge tubes. More control serum is used because the amount of immunoglobulin in this preparation is typically half of that in the B-2 serum. A 10% (w/v) suspension of Pansorbin (0.2 ml) prepared as described above is added to the immunoprecipitation reaction mixture and the incubation is continued for at least 1 hr at 4°. After this incubation, the Pansorbin is sedimented by centrifugation (10,000 rpm, 500 *g*) of the microfuge tubes in a Microfuge B which is equilibrated at 4° in a cold room. The pellets are washed twice at 4° by resuspension and centrifugation with 1 ml of Solution C supplemented with 0.1% SDS, and once with Solution D. After this wash procedure, the pellets are suspended in 80 μl of Laemmli buffer and heated in a boiling water bath for 5 min to dissociate the immunoprecipitated protein from the Pansorbin. The Pansorbin is sedimented by centrifugation and the supernatant is removed. About 75 μl of each supernatant is applied to the SDS-polyacrylamide slab gel and the phosphoproteins are separated by electrophoresis and detected by autoradiography as described above. An example of the results obtained with this procedure is shown in Fig. 1.

[15] J.-A. Hedo, L. C. Harrison, and J. Roth, *Biochemistry* **20,** 3385 (1981).

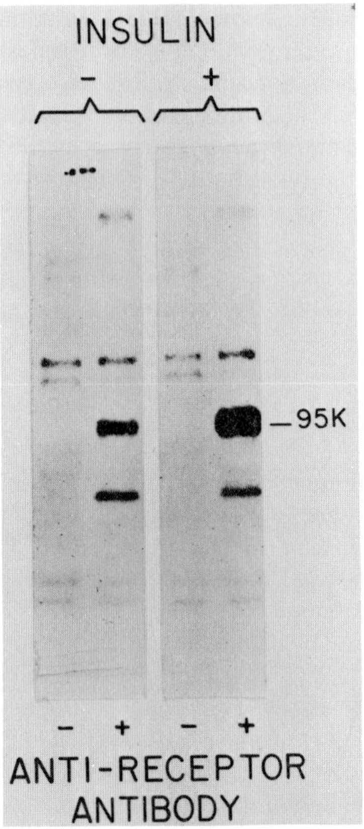

FIG. 1. The effect of insulin on the labeling of the insulin receptor in Fao cells by [^{32}P]orthophosphate. Fao cells were labeled with $^{32}PO_4^{3-}$ for 2 hr. Then 100 n*M* insulin (+) or an equal volume insulin-free buffer (−) was added to the cells for 15 min. The cells were washed, solubilized, and the insulin receptor was purified by wheat germ agglutinin chromatography. The eluate was divided into equal portions and the insulin receptor was immunoprecipitated with control human serum (−) or human serum containing anti-insulin receptor antibodies (+). This autoradiograph was obtained after 4 days of exposure.

In Vitro Phosphorylation of the Solubilized Insulin Receptor

In this section we describe a method for studying the phosphorylation of the insulin receptor kinase which is solubilized from the rat hepatoma cell, Fao. The same method should be useful with most other cell types.

Solubilization and Purification of the Insulin Receptor

Sixty confluent 150-mm culture dishes of hepatoma cells are washed at 22° with Solution A (20 ml). The cells are solubilized by adding to each dish at 22° Solution C (1.5 ml) supplemented with aprotinin (1000 trypsin inhibitor units/ml) and 2 m*M* PMSF. The cells are scraped from the dishes and the suspensions are pooled in 50-ml plastic conical test tubes which are cooled on ice. The suspension is mixed by vortexing and 10 ml portions are transferred to Beckman ultracentrifugation tubes. The detergent insoluble material is sedimented by centrifugation at 50,000 rpm in a Beckman 70.1Ti rotor for 40 min. The supernatants from each tube are combined (about 200 mg of protein in 150 ml total volume) and applied onto a wheat germ agglutinin-agarose column (6 ml packed volume) which was previously washed and equilibrated with Solution D as described above. The sample is collected and reapplied to the column two more times to obtain maximum binding of the glycoproteins. After sample application, the column is washed with 200 to 300 ml of Solution D. Finally, the bound glycoproteins are eluted (1 to 3 mg of protein) with 15 ml of Solution D supplemented with 300 m*M* *N*-acetylglucosamine, or 10 mg/ml *N*,*N'*-diacetylchitobiose, or a mixture of 300 m*M* *N*-acetylglucosamine and 5 mg/ml, *N*,*N'*-diacetylchitobiose. The *N*,*N'*-diacetylchitobiose has a high affinity for wheat germ agglutinin, a property which provides a very efficient elution of the glycoproteins.[16] An elution profile is shown in Fig. 2. The protein-containing fractions identified with the Bio-Rad protein assay standardized against bovine serum albumin are pooled (about 10 ml), separated into portions containing 24 μg of protein (typically 100 to 250 μl), and stored at $-70°$ for 3 months without apparent loss of autophosphorylation activity. This material is sufficient for about 250 to 500 assays.

An Insulin Dose–Response Curve for Phosphorylation of the Solubilized Insulin Receptor

One of the first experiments that should be tried after obtaining a solubilized insulin receptor preparation is to show that insulin stimulates the phosphorylation of the β-subunit in a dose-dependent manner. Solubilized insulin receptor (3 to 4 μg) is diluted to 50 μl having a final concentration of 50 m*M* HEPES, pH 7.4, 5 m*M* Mn^{2+}, and 0.1% Triton X-100. Then to each reaction mixture, insulin is added at the following concentrations: 0 *M*, 10^{-11}, 10^{-10}, 10^{-9}, 10^{-8}, and 10^{-7} *M*. These 6 solutions are incubated at 22° for 1 hr so that hormone binding reaches a maximum level. The phosphorylation reaction is initiated by adding 5 μl of 100 μM [γ-^{32}P]ATP (10,000–54,000 cpm/pmol). The mixtures are vortexed and the incubation is continued for the desired time interval; 10 min is often used because the phosphorylation reaction reaches steady state during this

[16] I. J. Goldstein and C. E. Hayes, *Adv. Carbohydr. Chem.* **5**, 128 (1978).

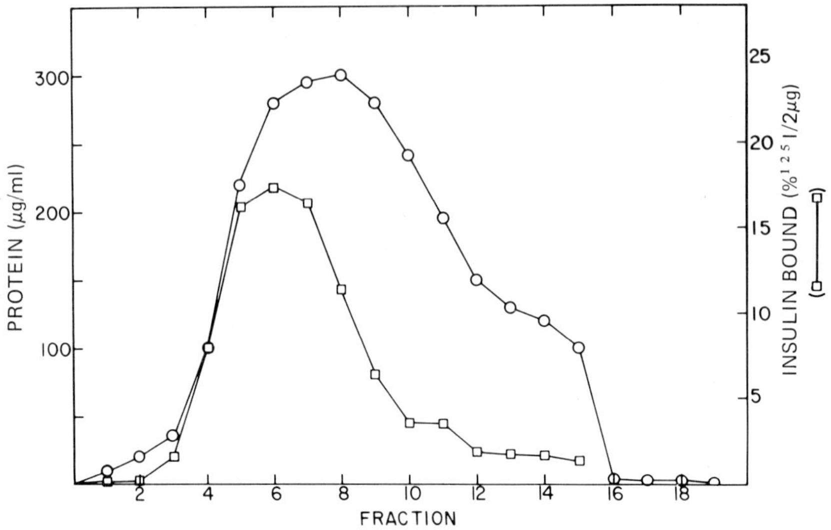

FIG. 2. An elution profile of protein and insulin binding activity obtained from a wheat germ agglutinin affinity column. Triton X-100 soluble proteins from 60 confluent dishes of Fao cells were prepared and applied onto the affinity column as described in the text. The glycoprotein fraction was eluted from the agarose with a Solution D containing both *N*-acetylglucosamine (300 m*M*) and *N,N'*-diacetylchitobiose (5 mg/ml). The elution profile of protein and specific tracer insulin binding is shown on the figure. Insulin binding was measured exactly as described previously.[17]

time interval.[17] To measure initial velocities of phosphorylation, 30 sec to 1 min time intervals must be used. The reaction is terminated precisely by adding 10 μl of 5-fold concentrated Laemmli sample buffer supplemented with 75 mg of dithiothreitol per ml. This mixture is heated immediately for 3 min in a boiling water bath. The proteins are separated by SDS–PAGE and identified by autoradiography (Fig. 3, left).

The radioactivity located in gel fragments by autoradiography is quantified by scintillation counting in 5 ml of 3a70b scintillation mixture (Research Products International Corp.). The background radiation is estimated by measuring the radioactivity in a gel fragment of the same size shown by autoradiography to be free of discrete bands of phosphoprotein. The insulin dose–response curve analyzed in this way is shown in Fig. 3, right.

A phosphoprotein of $M_r = 95{,}000$ and some other minor proteins are detected in the wheat germ agglutinin purified insulin receptor preparation (Fig. 3, left). The notion that the $M_r = 95{,}000$ phosphoprotein is the β-

[17] M. F. White, H. U. Haring, M. Kasuga, and C. R. Kahn, *J. Biol. Chem.* (in press).

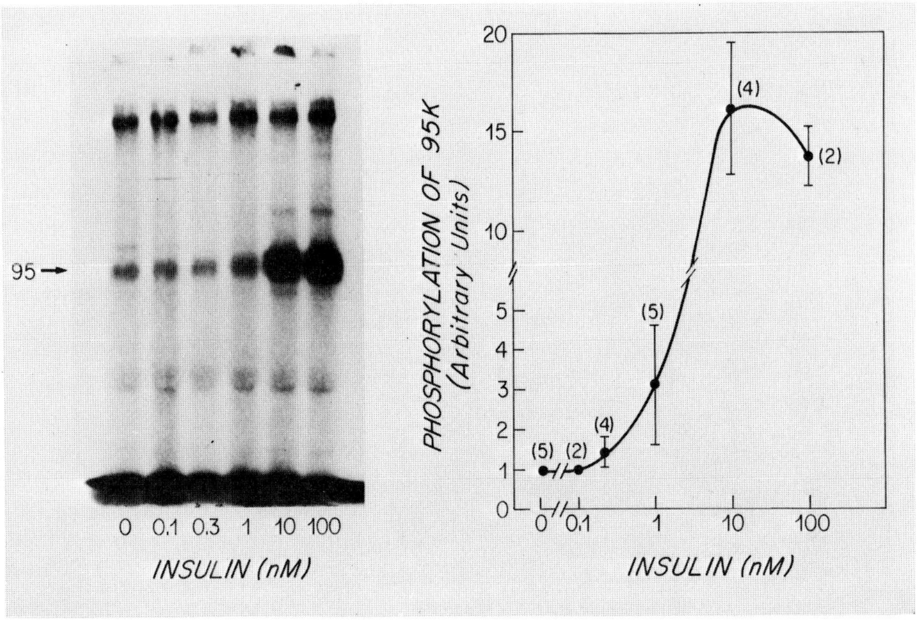

FIG. 3. An insulin dose–response curve of insulin receptor autophosphorylation. Insulin receptor (4 μg) purified from Fao cells by wheat germ agglutinin affinity chromatography was incubated for 1 hr at room temperature with the concentrations of insulin indicated on the figure. After this time interval, the phosphorylation reaction was initiated by adding [γ-^{32}P]ATP (15 μM, 10 μCi) to a 50 μl reaction for 10 min. The reaction was stopped by adding Laemmli sample buffer containing dithiothreitol (100 mM) and heating this mixture in a boiling water bath for 2 min. The phosphoproteins were separated by PAGE in a 7.5% gel; the autoradiogram is shown on the left. Right, the M_r = 95,000 phosphoprotein identified by autoradiography was cut from the gel and the radioactivity was quantitated by scintillation counting. The background radioactivity was measured in a region of the gel which did not contain discrete bands of phosphoprotein; this background was subtracted from the values shown here. The number of individual measurements obtained at each insulin concentration is shown on the figure.

subunit of the insulin receptor is confirmed by immunoprecipitation of the receptor from the *in vitro* reaction with anti-insulin receptor antibodies. In this case, the phosphorylation reaction is stopped by cooling the mixture to 4° and adding 1 ml of Solution E supplemented with 1 mM sodium vanadate and lacking sodium pyrophosphate. Anti-insulin receptor antibody is added to this mixture at a dilution of 1 : 300 and incubated at 4° for 2 hr as previously described. The immune complex is precipitated from the solution by incubation with 200 μl of 10% Pansorbin for at least 1 hr. The immunoadsorbed protein is solubilized in Laemmli sample buffer containing DTT and applied on a gel.

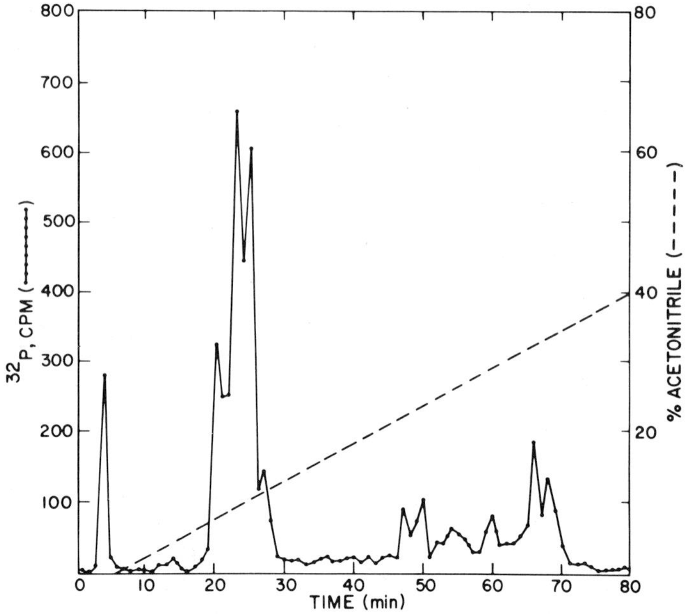

FIG. 4. The wheat germ purified insulin receptor (4 μg) was incubated in a reaction mixture (50 μl) containing 100 m*M* insulin and 10 m*M* Mn^{2+} for 1 hr. The phosphorylation reaction was initiated by adding 50 μ*M* [γ-^{32}P]ATP. After a 5 min incubation, the reaction was stopped by adding Laemmli sample buffer and heating the mixture for 2 min in a boiling water bath. The β-subunit was separated by PAGE and located by autoradiography. This gel fragment was excised from the fixed and dried gel and incubated with trypsin for 24 hr. The eluted peptides were separated by HPLC. Fractions were collected at 1 min intervals. The chromatogram obtained by measuring the radioactivity in each fraction is represented by the solid line. The linear gradient of acetonitrile containing 0.05% trifluoroacetic acid is indicated by the dashed line.

Tryptic Peptide Mapping by High-Performance Liquid Chromatography

Many phosphorproteins have multiple phosphorylation sites. These sites can be catalyzed by several different kinases. The importance of autophosphorylation of the insulin receptor and interactions between the insulin receptor and other kinases will require a detailed knowledge of the available phosphorylation sites and their relation to biological activity.[18] To obtain a tryptic peptide map of the phosphorylated β-subunit, fixed, stained, destained, and dried polyacrylamide gel fragments containing phosphoprotein located by autoradiography are washed for 12 hr at 37°

[18] H. Juhl and T. R. Soderling, this series, Vol. 99, p. 37.

with 20 ml of 10% methanol. The adsorbent paper is removed from the gel fragment. The gel is dried at 70° for 60 min and rehydrated in 2 ml of 50 m*M* NH_4HCO_3 containing 100 μg of TPCK treated trypsin. This mixture is incubated for 24 or 48 hr at 37°, the gel fragment is removed, and the supernatant is clarified by centrifugation. The supernatant is lyophilized and the residue is dissolved in 25–50 μl of 0.1% trifluoroacetic acid. The phosphopeptides can be separated using a Waters high performance liquid chromatography system equipped with a μBondapak C_{18} reversed-phase column. Phosphopeptides applied to the column were eluted at a flow rate of 1 ml/min with a linear gradient obtained by mixing water containing 0.05% trifluoroacetic acid and acetonitrile also containing 0.05% trifluoroacetic acid. Radioactivity in 0.4 ml to 1.0 ml fractions eluted from the reversed-phase column is measured in 5 ml of scintillation mixture.

As shown in Fig. 4, trypsin digestion of the phosphorylated β-subunit suggests that at least 4 peptides containing sites of ^{32}P incorporation exist in this molecule. These phosphopeptides elute with retention times between 20 to 35 min under the conditions described. The pattern of phosphopeptide elution is similar after 24 and 48 hr of incubation with trypsin suggesting that maximum digestion is attained after 24 hr.[17] These major phosphopeptides have been reproducibly identified during our studies of the receptor, whereas the fragments having longer retention times are not reproducibly detected, and are usually minor. This method should prove useful in the future to obtain more structural details of the receptor and relate phosphorylation to biological activity.

Section VII

Antibodies in Hormone Action

[49] Development of Monoclonal Antibodies against Parathyroid Hormone: Genetic Control of the Immune Response to Human PTH

By SAMUEL R. NUSSBAUM, C. SHIRLEY LIN, JOHN T. POTTS, JR., ALAN S. ROSENTHAL, and MICHAEL ROSENBLATT

Introduction

In normal physiology, parathyroid hormone (PTH) serves an important homeostatic role in the regulation of extracellular calcium levels through its effects on kidney, bone, and gut. Extracellular calcium levels are maintained within the narrow limits required for optimal functioning of nerve, muscle, enzymes, and hormones. Analysis of the relation of structure to hormonal function for parathyroid hormone has provided important insights into the physiological roles served by certain regions of the hormone molecule and the initial events and mechanisms by which hormonal messages are translated into physiological actions.

Parathyroid hormone is initially biosynthesized in a precursor form termed pre-proPTH, a single-chain protein of 115 amino acids. At the NH_2-terminus of pre-proPTH, there is a 25-amino acid region representative of a class of precursor-specific sequences found in secreted proteins, enzymes, and peptide hormones known as leader, or signal, sequences.

The principal secreted form of the hormone contains 84 amino acids. PTH's critical role in the physiology of calcium homeostasis results from its multiple actions on kidney and bone. Expression of hormone action is linked to a number of biochemical events at the surface and within target-tissue cells. PTH interacts with hormone-specific receptors on the plasma membrane of target-tissue cells.

This interaction initiates a cascade of intracellular events: the generation of cAMP, phosphorylation of specific intracellular proteins by PTH-activated kinases, intracellular entry of calcium, and activation of intracellular enzymes.

The NH_2-terminal one-third of this molecule (positions 1–34) contains all the structural requirements necessary for full biological activity in multiple assay systems, in terms of calcium and phosphate metabolism. This region has been synthesized and found to be nearly equipotent, on a molar basis, to native PTH *in vitro* and *in vivo*.[1] Also, binding of bPTH-(1–

[1] G. W. Tregear, J. van Rietschoten, E. Greene, H. T. Keutmann, H. D. Niall, B. Reit, J. A. Parsons, and J. T. Potts, Jr., *Endocrinology* **93,** 1349 (1973).

ISBN 0-12-182009-2

34) to renal membrane receptors is equal, on a molar basis, to binding of bPTH-(1–84).[2,3] No physiological function has yet been established for the remaining major (COOH-terminal) portion of PTH (positions 35–84).

A most intriguing posttranslational cleavage of PTH is an extracellular proteolytic modification that results from efficient and selective uptake of circulating PTH by cells in liver and perhaps kidney and bone. This selective uptake is followed by cleavage, which converts the hormone to a biologically active amino terminal fragment, and a larger fragment consisting of the middle and carboxyl region. Thus, the peripheral metabolism of PTH may represent final activation occuring outside of the parathyroid gland. Alternatively, the peripheral metabolism may represent an important route of destruction and metabolic clearance of the hormone, functioning with changes in rate of secretion and receptor availability to control expression of biological activity of the hormone *in vivo*. Figure 1 schematically summarizes our current knowledge of PTH biosynthesis and secretion.

Metabolism of intact PTH-(1–84) after secretion by hepatic Kupffer cells, and hormonal fragments secreted by the parathyroid glands, particularly in hypercalcemia caused by parathyroid gland hyperplasia or adenoma, contribute significantly to the concentration of the multiple circulating immunoreactive forms of the hormone. The principal fragments detected in the circulation and present in greater than 10 times the concentration of intact hormone appear to be approximately 60% as large as the intact hormone and consist of the middle and carboxyl-terminal portions of the hormone sequence. There is a general agreement by all laboratories that amino-terminal fragments, if detected at all, are present in trace quantities. Therefore, the principal circulating forms of PTH are the intact hormone containing the biologically active amino terminal portion of the hormone, and fragments devoid of the amino terminus and devoid of biological activity.

Several critical questions of parathyroid gland secretory activity that relate to the clinical differential diagnosis of hypercalcemic individuals and to physiologic studies of parathyroid hormone secretion and metabolic clearance can best be approached by using an antiserum which recognizes only the biologically active intact, or circulating amino-terminal portion of the hormone present in circulation. Until recently, antisera to PTH that contain antibodies directed against the biologically important portion of the hormone raised by traditional immunization programs in guinea pigs, rabbits, and goats have not been sufficiently sensitive to

[2] R. A. Nissenson and C. D. Arnaud, *J. Biol. Chem.* **254,** 1469 (1979).

[3] G. V. Segre, M. Rosenblatt, B. L. Reiner, J. E. Mahaffey, and J. T. Potts, Jr., *J. Biol. Chem.* **254,** 6980 (1979).

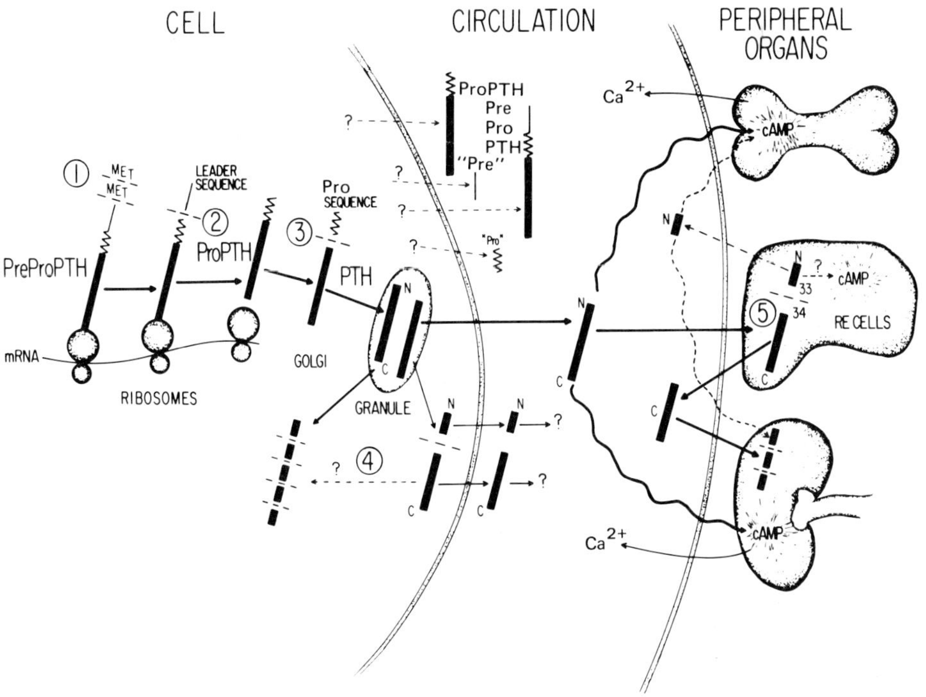

FIG. 1. Summary of PTH biosynthesis, secretion and metabolism. From J. F. Habenor, M. Rosenblatt, and J. T. Potts, Jr., *Physiol. Rev.* **64,** 985 (1984). Reproduced with permission.

detect circulating levels of intact parathyroid hormone in normal individuals with normal serum calcium concentrations. In individuals with hyperparathyroidism, sensitivities of these amino terminal antisera are occasionally sufficient to detect circulating levels in blood. A chicken antisera raised against human PTH-(1–34) by Segre[4] is sufficiently sensitive to measure picograms of circulating intact PTH and can detect hormone in normocalcemic individuals; however, this antiserum may not distinguish hypercalcemia of malignancy from hyperparathyroidism in all patients.

Most clinically applied radioimmunoassays contain antibodies directed against the middle or carboxyl-terminal portions of the hormone. These radioimmunoassays detect biologically inactive fragments of the hormone present in approximately 10-fold higher concentration than intact hormone in normals as well as in hyperparathyroidism. Such assays, at best, are only an index of parathyroid hormone secretion. In patients

[4] G. V. Segre, *in* "Clinical Disorders of Bone and Mineral Metabolism" (B. Frame and J. T. Potts, eds.), pp. 14–17. Excerpta Medica, Amsterdam, 1983.

with renal insufficiency, where there is an even greater accumulation of circulating inactive carboxyl-terminal fragments because of impaired renal clearance, interpretation of assays based on detection of carboxyl fragments can be problematic or even misleading with regard to the presence or absence of excessive secretion of biologically active intact hormone.

We therefore embarked upon a program to develop monoclonal antibodies to the biologically active amino terminal region of PTH. Using the BALB/c mouse for immunization, fully biologically active synthetic human PTH-(1–34) and bovine PTH-(1–84) as immunogens, monoclonal antibody methods and a solid-phase screening assay in which PTH-(1–34) was adhered to polyvinylchloride plates in a manner that preserved immunoreactivity, we generated 17 monoclonal antibodies against the amino-terminal portion of parathyroid hormone.

Isotypic analysis of these monoclonal antibodies was performed using affinity purified goat anti-mouse immunoglobulins specific for IgG heavy chains, γ_1, γ_{2a}, γ_{2b}, γ_3; α(IgA); and μ(IgM). All antibodies were IgM as evidenced by 40 times greater than background radioactivity when 25,000 cpm of ^{125}I-labeled goat anti-mouse IgM was used as second antibody in a solid-phase radioimmunoassay. All incubations with iodinated second antibodies to other heavy chain classes of immunoglobulins demonstrated background radioactivity.

Extensive synthetic work in our laboratory for multiple biologic studies of structure–activity relationships of PTH, as well as analog design, has led to the synthesis of many peptide analogues and fragments from 7 to 34 amino acids in length.

Study of the antibody recognition site (region specificity) by two of these monoclonal antibodies, $10A_7$, and $6B_1$, was und aken with synthetic peptides.

Twenty-five micrograms of each of bPTH-(1–27)-NH_2, bPTH-(25–34)-NH_2, ACTH (1–24), and glucagon was incubated overnight at 4° with a 1/20,000 dilution of ascitic fluid $10A_7$ or $6B_1$ in 1% bovine serum albumin 0.05 *M* phosphate-buffered saline, pH 7.4 following dissolution of the peptides in 0.1 *M* acetic acid. These solutions of antibody and hormone were then incubated in triplicate on PVC wells that had previously been coated with hPTH-(1–34), (followed by 10% horse serum in phosphate-buffered saline to coat any unoccupied sites on the polyvinylchloride that may have been available for nonspecific adherence of antibody to PTH, or of iodinated second antibody). Following a 3 hr incubation at 23°, the plates were washed copiously to remove any antibody that was not specifically bound to PTH. ^{125}I-labeled second antibody (25,000 cpm), either goat anti-mouse Fab or goat anti-mouse IgM, was incubated overnight at

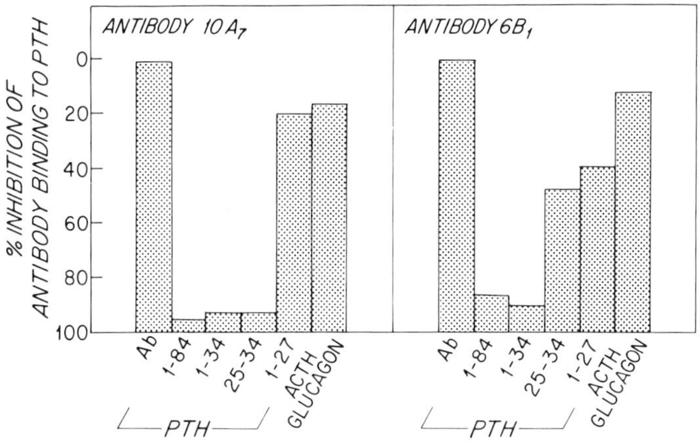

FIG. 2. Inhibition of antibody binding by PTH fragments. The binding site of antibody $10A_7$ is satisfied by the 10 amino acid peptide PTH-(25–34) whereas for antibody $6B_1$, neither the 25–34 nor the 1–27 fragments completely inhibit binding of second antibody. The bar farthest to the right in each panel represents control peptides ACTH and glucagon. From S. R. Nussbaum, M. Rosenblatt, M. Mudgett-Hunter, and J. T. Potts, Jr., *in* "Monoclonal Antibodies in Endocrine Research" (R. Fellows and G. Eisenbarth, eds.), p. 181. Raven, New York, 1981. Reproduced with permission.

4°. The microtiter wells were washed with distilled H_2O, cut, and counted. Inhibition of antibody binding to the PTH coated wells reflected antibody recognition and binding of unlabeled hormone fragments. The results of these region-specific assays are shown in Fig. 2. A fragment of PTH as short as 10 amino acids, PTH-(25–34), can completely inhibit antibody $10A_7$ binding to PTH coated on the wells; therefore, the antibody recognition site for antibody $10A_7$ resides within the sequence 25–34. In contrast, for antibody $6B_1$, although PTH-(1–34) causes complete inhibition of binding, the antigenic determinants encompass either amino acid sequences throughout the 1–34 region or a conformation present in 1–34 and not its fragments, as neither the 1–27 nor 25–34 fragment can entirely inhibit antibody binding.

Although the monoclonal antibodies generated against hPTH-(1–34) were highly specific, they had binding affinities of less than 1×10^{-8} *M* and therefore were not satisfactory for clinical application as the basis of a PTH immunoassay. The repeated generation of low affinity IgM antibodies, despite an immunization program which included repeated boosts, suggested that the BALB/c mouse used for immunization by us, as well as by many other investigators who raise monoclonal antibodies, may have been a genetic nonresponder to parathyroid hormone. Immune responsiveness to many soluble synthetic and natural protein antigens has been found to be genetically controlled.[5] These immune response (*Ir*) genes are linked to the major histocompatibility complex (MHC) of the mouse, the H-2 complex. We therefore undertook studies of the immunoresponsivity of inbred mouse strains to human amino-terminal parathyroid hormone, hPTH-(1–34), by T cell proliferation assay and correlated T cell *in vitro* responsiveness with *in vivo* serum titer in mice immunized by multiple subcutaneous injections of parathyroid hormone. We found both T lymphocyte and antibody response to PTH were genetically controlled. We initially studied routes of administration, the effects of conjugation of PTH to thyroglobulin, and PTH dosages for immunization.

In Vitro and *in Vivo* Immunologic Studies

Peptides

Human PTH-(1–34) was synthesized by solid-phase method using a resin support of 1% divinybenzene cross-linked with polystyrene. The tert-BOC group was used for α-amino group protection. The peptide was cleaved from the resin with anhydrous HF at 0° and purified by HPLC using a 0.1% TFA, 20–50% acetonitrile buffer gradient system. The hPTH-(–34) peptide was homogeneous by HPLC and isoelectric focusing and conformed to theoretical amino acid composition by amino acid analysis and Edman sequence determination.

Animals

Multiple strains of mice [B10.A, B10.BR, $B10.D_2$, B10.A(4R), B10.A(5R), DBA, BALB/c, CBA, C57B6, BALB/c–C57B6 F_1 hybrids, and C_3H.Hej] aged 6–12 weeks were obtained from Jackson Labs (Bar Harbor, Maine) for immunization with synthetic hPTH-(1–34) and for study of T lymphocyte responsiveness.

Immunization Program

For the initial immunization studies, hPTH-(1–34) was conjugated to bovine thyroglobulin using water-soluble carbodiimide (Schwarz/Mann) in a molar ratio of 40 : 1 PTH:thyroglobulin. The efficiency of coupling was monitored with ^{125}I-labeled PTH and averaged 30%. The CDI and the unreacted PTH were separated from the PTH–thyroglobulin conjugate by P_4 polyacrylamide chromatography. Only the PTH–TG conjugate was used for immunization. For high dose immunization, 100 μg of the PTH–TG conjugate, and for low dose immunization, 10 μg of the PTH-TG

[5] M. E. Dorf, *Springer Semin. Immunopathol.* **1,** 171 (1978).

conjugate were administered subcutaneously in 8–10 sites, intraperitoneally, or by heel pad to groups of 9 mice of each strain. Primary immunization was followed by boosts at days 21, 28, 35, and 42. All immunizations were administered in emulsified 1 : 1 v : v complete adjuvant, H37Ra (Difco) and PTH–TG conjugate. For the second series of studies on immune responsiveness of inbred mouse strains against PTH, all animals received 40 μg of unconjugated hPTH-(1–34) administered subcutaneously in complete adjuvant at days 1, 21, 28, 35, and 42.

Determination of Antibody Titers in Serum

All animals were bled from the retroorbital venous sinus at day 42 and peripheral titers were determined using a double antibody solution phase radioimmunoassay in which hPTH-(1–34) was iodinated with chloramine-T and a rabbit anti-mouse second antibody was used for phase separation.

T Lymphocyte Proliferation Assay

Immunization. hPTH-(1–34), 1 mg/ml in PBS, was emulsified in an equal volume of complete Freund adjuvant (0.5 mg/ml killed *Mycobacterium tuberculosis,* H37Ra, Difco Laboratories, Inc., Detroit, MI). Five mice of each strain were injected with 0.5 ml of the emulsion in each of the hind foot pads.

Cell Collection and Separation. Inguinal, periaortic, and popliteal lymph nodes were removed 11 days after immunization. The nodes were removed aseptically and single cell suspensions were obtained by pressing the excised lymph nodes between two glass slides. Following two washings, the lymph node cell suspension was passed through a nylon wool adherence column (nylon wool, Fenwell Laboratories, Morton Grove, IL) to deplete antigen presenting cells. The enriched T cell populations obtained were suspended in RPMI 1940 (Grand Island Biological Co., Grand Island, NY) supplemented with fresh L-glutamine (0.03 mg/ml), penicillin (100 U/mg), gentamicin (10 μg/ml), 2-mercaptoethanol (5×10^{-5} *M*), and 10% heat-inactivated fetal calf serum (Grand Island Biological Co., Grand Island, NY).

Cell Culture and Assay of DNA Synthesis. Aliquots of 0.2 ml of 1.5×10^5 T cell suspension were pipetted into flat bottomed well microtiter plates (Flow Laboratories). Spleen cells from normal, nonimmunized animals were irradiated (2000 rads) and added to the cell cultures as the source of antigen presenting macrophages. Ten microliters of antigen at appropriate dilutions was added to wells in triplicate. The microtiter plates were incubated for 96 hr at 37° in a humidified atmosphere of 5% CO_2 and 95% air. Sixteen to 24 hr before harvesting, 1 μCi of tritiated thymidine (6.7 Ci/mmol, New England Nuclear Co., Boston, MA) was added to each well. Cells were harvested onto glass fiber filter paper by the use of a microharvestor. Tritiated thymidine incorporation was then determined by liquid scintillation spectrometry and the results were reported as counts per minute above control.

Results

In Vivo Immunization with PTH–Thyroglobulin (PTH–TG) Conjugate

For each inbred strain selected: B10.BR, B10.A, C57B6, CBA, and C57B6–BALB/c hybrids, 6 groups of 3 animals were immunized intraperitoneally, subcutaneously or by heel pad injection with either high or low doses of PTH–TG conjugate. All failed to produce a peripheral titer against parathyroid hormone as measured in a solution-phase RIA utilizing ^{125}I-bPTH-(1–84) as tracer, and polyclonal antisera as positive controls, whereas the majority of animals responded with peripheral titer directed against thyroglobulin (Table I). We excluded the possibility that conjugation of PTH to TG may have conformationally altered PTH to make it immunologically unrecognizable, as the PTH–TG conjugate reacted with a panel of parathyroid hormone antisera. Using these antisera, immunoassayable PTH was equivalent to the amount of peptide in the prepared conjugate.

In Vivo Immunization with PTH-(1–34)

Having been unsuccessful in demonstrating any peripheral antibody titer against hPTH-(1–34) when this antigen was coupled to thyroglobulin, we chose unconjugated hPTH-(1–34) for immunization of BALB/c and B10.BR mice utilizing the identical immunization schedule and routes for

TABLE I
IMMUNIZATION WITH PTH–THYROGLOBULIN CONJUGATE

Mouse strain	Peripheral titer against PTH conjugated (1 : 100 B/F) to thyroglobulin	Peripheral titer against thyroglobulin (1 : 1000 B/F, mean $N = 18$)
B10.BR	<0.06	0.35
B10.A	<0.06	0.26
C57B6	<0.06	0.34
C57B6/BALB/c F_1	<0.06	0.42
CBA	<0.06	0.48
BALB/c	<0.06	Not studied

TABLE II
SUMMARY OF IMMUNIZATION PROGRAM IN INBRED MOUSE STRAINS

Mouse strain	Number of mice immunized	Number responding to PTH with peripheral titer	B/F at 1 : 100 dilution of serum ± SEM
B10.BR	16	12	0.74 ± 0.35
DBA	8	6	0.20 ± 0.04
B10.D2	3	0	<0.10
B10.A	6	4	0.25 ± 0.16
CBA	3	3	0.47 ± 0.19
BALB/c	15	2	<0.10
C57B6	9	0	<0.10
B10.A(5R)	6	0	<0.10
B10.A(4R)	5	2	0.14 ± 0.03
C_3H.Hej	5	5	0.734 ± 0.20

immunization previously used for study of the PTH–TG conjugate. Although mice immunized with PTH–TG conjugate, in doses as high as 100 μg, failed to develop peripheral titers against PTH, initial immunization with 40 μg unconjugated parathyroid hormone, followed by 5 μg boosts, especially in mice in which the antigen was administered subcutaneously at 5–10 sites per immunization, resulted repeatedly in peripheral titers against parathyroid hormone. Intraperitoneal immunization resulted in lower peripheral titer. Having completed studies on antigen dosages, conjugation to thyroglobulin, and routes of administration, we embarked upon a program to correlate *in vitro* T lymphocyte responsiveness with *in vivo* immunization of inbred mouse strains.

Table II summarizes our studies of peripheral titers against hPTH-(1–34) in inbred mouse strains, initially immunized with 40 μg of unconjugated hPTH-(1–34) followed by 5 μg boosts at days 21, 28, 35, and 42.

T Lymphocyte Proliferative Responsiveness to hPTH-(1–34) in Inbred Mouse Strains

Three different inbred mouse strains were first tested for their responses to hPTH-(1–34). The results (Fig. 3) indicate that the immune response to hPTH-(1–34) is genetically controlled. Inbred mouse strains respond to hPTH-(1–34) differently. The BALB/c mouse which usually is the strain of choice for the source of antibody producing spleen cells for monoclonal antibody studies was found to be a low responder to hPTH-(1–34). Another inbred strain, C57B6 mice were also found to be low responders. CBA animals were found to be high responders.

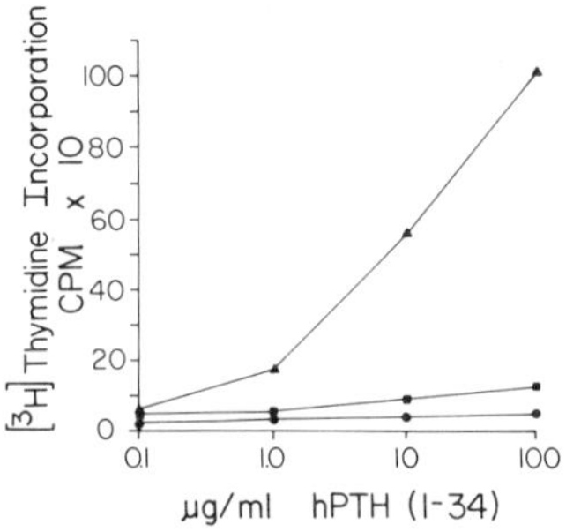

FIG. 3. T lymphocyte proliferative responsiveness to hPTH (1–34). Mice were immunized with 50 μg hPTH-(1–34) emulsified in complete Freund adjuvant in hind foot pads. Eleven days after immunization, T cells from draining lymph nodes were collected and cultured in flat-bottom well microtiter plates at 3×10^5 cells/well with the same number of irradiated spleen cells from normal mice of appropriate strain. Antigens at different dilutions were added to triplicate wells. The cultures were incubated at 37°, 5% CO_2 in air for 4 days. [^{3}H]Thymidine was added 10–24 hr prior to harvesting. Δcpm was calculated by subtracting medium control from antigen stimulated cultures. Each experimental point represents the average of triplicate cultures. Mouse strains are represented as follows: (▲) CBA/J; (●) B6; and (■) BALB/c.

Responsiveness to hPTH-(1–34) in H-2 Congeneic Mice

B10, B10.D2, and B10.BR mice are *H-2* congeneic mice. They share the same background genes except for the genes at the *H-2* locus. Strain B10 (H-2^b) is identical with the B6 strain, the B10.D2 strain is identical with B10 mice except for a *H-2*^{d} haplotype. Similarly, B10.BR mice are identical with B10 mice except that they have an *H-2*^{k} haplotype. When these three mouse strains were tested, it was found that the *Ir* genes controlling the immune response to hPTH-(1–34) were linked to the *H-2* complex. Results in Table III indicate that both B10 and B10.D2 mice minimally respond to hPTH-(1–34) while B10.BR mice respond very well to hPTH-(1–34). Since B10.D2 mice share *H-2*^{d} with BALB/c mice and B10.BR mice share H-2^k with CBA mice, we conclude from the data shown in Fig. 3 and Table III that *Ir* genes controlling the immune response to hPTH-(1–34) are located in the *H-2* locus. T lymphocyte responsiveness to hPTH-(1–34) is macrophage dependent, since T lymphocytes from immune B10.BR mice respond to hPTH only in the presence of macrophages.

Genetic Mapping of Immune Response Genes Controlling the Immune Responsiveness to hPTH-(1–34)

In order to further map the *Ir* genes controlling immune responsiveness to hPTH-(1–34), several intra-*H-2* recombinant mouse strains were

TABLE III
RESPONSIVENESS TO hPTH-(1–34) IN B10 CONGENEIC STRAINS[a]

Strain	Macrophage added	[³H]Thymidine (cpm) incorporation in the presence of: hPTH-(1–34) 1 μg/ml	hPTH-(1–34) 10 μg/ml	hPTH-(1–34) 100 μg/ml	PPD 25 μg/ml	cpm medium
B10	–	359	0	0	44,621	480
	+	374	689	377	56,451	2,264
B10.D2	–	0	635	0	41,360	948
	+	289	2,981	5,089	120,569	4,955
B10.BR	–	134	265	437	2,099	376
	+	3,222	23,698	31,329	82,840	979

[a] The immunization of the mice and the T cell proliferation assay are described in Fig. 3. T cells were cultured with (+) or without (–) macrophages.

studied. Data in Table IV show that B10 and B10.D2 mice were low responders while B10.BR mice were good responders to hPTH-(1–34). The fact that B10.A mice were also good responders maps the *Ir* genes to the left of *IE* subregion. Since B10.A(4R) mice are low responders, the *Ir* gene for hPTH-(1–34) is not located in the *IA* subregion. Since B10.A (5R) mice were shown to be nonresponders while B10.A (2R) mice were good responders, the immune response to hPTH-(1–34) is controlled by complementary *IA*^{k} and *IE*^{k} genes.

TABLE IV
RESPONSIVENESS TO hPTH-(1–34) IN B10 INTRA-*H-2* RECOMBINANT STRAINS[a]

Strain	K	I: A	I: B	I: J	I: E	I: C	S	D	Thymidine (cpm) incorporated: hPTH-(1–34) 100 μg/ml	PPD 25 μg/ml	cpm medium
B10	b	b	b	b	b	b	b	b	1,435	56,481	1,385
B10.BR	k	k	k	k	k	k	k	k	29,107	67,940	3,950
B10.D2	d	d	d	d	d	d	d	d	19,082	98,655	2,633
B10.A	k	k	k	k	k	k	k	k	41,079	95,486	2,693
B10.A (2R)	k	k	k	k	k	k	k	k	34,180	87,945	4,001
B10.A (4R)	k	k	k	k	k	k	k	k	831	109,713	10,167
B10.A (5R)	b	b	b	b	b	b	b	b	0	80,879	1,349

[a] The immunization of the mice and the T cell proliferation assay are described in Fig. 3.

Comments

A critical concern in the development of monoclonal antibodies for use as reagents for immunoassay of peptide hormones has been the achievement of requisite affinities for measurement of circulating concentrations of peptides in serum.

Overall, despite great enthusiasm for hybridoma technology and the large numbers of laboratories devoted to the development of monoclonal antibodies against peptide hormone antigens since the initial reports of successful cell fusion by Köhler and Milstein,[6] virtually no monoclonal antibody has replaced a multivalent conventionally raised antisera for clinical immunoassay of peptide hormones in man.

Monoclonal antibodies have successfully been raised against peptide and glycopeptide hormones. However, these monoclonal antibodies have generally been used for alternative application other than direct measurement of peptides in blood, such as immunocytochemistry, affinity purification, and study of antigenic determinants and recognition sites by the monoclonal antibodies. Nevertheless, monoclonal antibodies raised against human growth hormone and prolactin, which circulate in nanogram quantities in man as opposed to the picogram quantities of parathyroid hormone, have compared favorably with polyclonal antisera.[7,8]

In order to attain this goal of increased affinity, several investigators have utilized novel immunization programs. Stahli determined that the frequency of successful cell fusion correlated with the increased frequency of blast cells in the spleen, measured by cell size analysis, after antigenic stimulation.[9] These investigators were able to increase the frequency of cell fusion by an immunization schedule that included daily intravenous administration of high doses of antigen for 4 days prior to fusion. Luben has successfully utilized *in vitro* immunization to generate monoclonal antibodies against rat hypothalamic growth hormone releasing hormone using picomoles of partially purified rat GRF.[10] Van Wyk *et al.*[11] were able to increase the fusion frequency and generate a monoclonal antibody against somatomedin-C (IGF-1) with a program of *in vivo*

[6] G. Köhler and C. Milstein, *Nature (London)* **256,** 271 (1975).

[7] M. C. Stuart, P. A. Underwood, and L. Boscato, *J. Clin. Endocrinol. Metab.* **54,** 881 (1982).

[8] P. G. Bundesen, R. G. Drake, K. Kelly, I. G. Worsley, H. G. Freisen, and A. H. Sehon, *J. Clin. Endocrinol. Metab.* **51,** 1472 (1980).

[9] C. Stähli, T. Staehelin, V. Miggiano, J. Schmidt, and P. Haring, *J. Immunol. Methods* **32,** 297 (1980).

[10] R. A. Luben, P. Brazeau, P. Bohler, and R. Guillemin, *Science* **218,** 887 (1982).

[11] J. J. Van Wyk, L. E. Underwood, G. Y. Gillespie, M. E. Svoboda, and W. E. Russel, *Proc. 65th Annu. Meet., Endocr. Soc.* Abstract 64 (1983).

followed by 5 days of *in vitro* immunization. Previously, *in vivo* immunization alone against somatomedin-C led to a very low incidence of successful hybridomas.

The *H-2* complex of the mouse, consisting of genes controlling multiple functions, can be divided into *K, I, S,* and *D* regions.[12,13] *I* region genes control the immune response and also code for Ia surface molecules on macrophages and B cells. These *Ir* genes appear to operate at the level of the macrophage by a process of determinant selection.[14] Antibodies against these Ia surface markers on macrophages block *in vitro* antigen-induced lymphocyte proliferation. The *I* region can be further divided into five subregions: *A, B, J, E,* and *C*. Immunoresponse genes have been identified in *IA* and *IE* subregions. The immune response to insulin[15] is controlled by *Ir* genes in the *IA* region. Our immunologic studies indicate that *Ir* genes controlling the immune responsiveness to human parathyroid hormone-(1–34) are located in *IA* and *IE* subregions.

Immunological studies using small peptides of known sequences as antigens have made significant contributions toward the understanding of the mechanisms of immune recognition and cell–cell interaction. For example, studies comparing the immune response pattern in inbred strains of guinea pigs to insulin molecules of various species have resulted in the proposal for the function of macrophages and the mechanism of Ir gene function.

Information on immune responsiveness may have great relevance for development of monoclonal antibodies to soluble peptide antigens. Without background immunological studies, Meo recently found that BALB/c–C57B6 F_1 hybrids gave a 5- to 10-fold higher serum titer to β-endorphin than the BALB/c parent strain. Fusion only of the hybrid strain allowed successful generation of monoclonal antibodies against β-endorphin.[16]

Our studies demonstrate the importance of the choice of mouse strain for immunization in the development of monoclonal antibodies against soluble peptide antigens. The choice of nonresponder mouse strains may explain why many laboratories have been unsuccessful in utilizing hybridoma technology to raise high affinity monoclonal antibodies. Although we have correlated T lymphocyte responsiveness to PTH-(1–34) with *in*

[12] H. O. McDevitt and B. Benacerraf, *Adv. Immunol.* **11,** 31 (1969).

[13] J. Klein, "The Major Histocompatibility Complex of the Mouse." Springer-Verlag, New York, 1979.

[14] A. S. Rosenthal, M. A. Barcinski, and J. T. Blake, *Nature (London)* **267,** 156 (1977).

[15] C. S. Lin, A. S. Rosenthal, H. C. Passmore, and T. H. Hansen, *Proc. Natl. Acad. Sci. U.S.A.* **78,** 6406 (1981).

[16] T. Meo, C. Gramsch, R. Inan, V. Höllt, E. Weber, A. Herz, and G. Reithmüller, *Proc. Natl. Acad. Sci. U.S.A.* **80,** 4084 (1983).

TABLE V
SUMMARY OF IMMUNE RESPONSE TO PTH-(1–34)

Strain	*H-2* haplotype	*IA*	*IE*	T cell response	Antibody response
C3H/HeJ	k	k	k	+	+
CBA/J	k	k	k	+	+
B10.BR	k	k	k	+	+
B10.A	a	k	k	+	+
B10.A (4R)	h4	k	b	−	−
B10.A (5R)	i5	b	k	−	−
BALB/c	d	d	d	±	−

vivo immunization (summarized in Table V), a logical and practical approach to raising monoclonal antibodies would be immunization of immunologically diverse inbred mouse strains, selecting only those strains with high peripheral titer for fusion.

Acknowledgment

We thank B. F. Reiner for expert technical assistance.

[50] Monoclonal Antibodies to Gonadotropin Subunits

By PAUL H. EHRLICH, WILLIAM R. MOYLE, and ROBERT E. CANFIELD

Introduction

The production of monoclonal antibodies to peptide hormones, with their unifocal binding sites, can provide tools for understanding hormone structure and function and have been applied to conduct studies that were not possible with polyclonal antisera. This has been especially true for the gonadotropins, follicle-stimulating hormone, luteinizing hormone, and chorionic gonadotropin (FSH, LH, and CG),[1] which are homologous glycoprotein hormones composed of two different subunits designated α and β. Thyroid-stimulating hormone (TSH) is also considered to be in this class of glycoprotein hormones.

Within a particular species the α subunits of all four of these hormones have identical structures, so antisera raised against the native form of one hormone frequently exhibit a high degree of cross reactivity with the

[1] J. G. Pierce, *Endocrinology* **89,** 1331 (1971).

ISBN 0-12-182009-2

others. The potential for immunochemical distinction between hormones therefore resides in epitopes on the subunits. Since the isolated subunits have different structural conformations than when they are combined to form the native hormone, monoclonal antibodies that are produced by immunization with the whole hormone, but are selected for reaction with epitopes on the different β subunits, offer the best opportunity to achieve sensitivity combined with immunochemical specificity.

Many monoclonal antibodies to hCG and to other glycoprotein hormones have been described with examples cited in Refs. 2–5. Clones secreting antibodies with interesting properties have been described including an antibody that binds the hCGβ subunit with only 0.23% cross-reactivity with the intact hormone.[2] Most monoclonal antibodies have been characterized with regard to subunit specificity and cross-reactivity with other gonadotropins. In addition systematic attempts to map several different regions have been reported for hCG[6,7] and hTSH.[8]

Spleen cell-myeloma cell fusion, screening of hybridoma supernatants, cloning, and immunological mapping are now standard in many laboratories.[9,10] We have chosen to focus here on techniques that have been important for the study of monoclonal antibodies to gonadotropin hormones in our laboratories. This presentation is not intended to be a comprehensive review of literature in this field; rather it is intended to detail methodology that has been developed as a consequence of our interest in hCG.

Hybridoma Production

The immunization protocol for hCG subunits has been described.[6,11,12] On the basis of a limited sample, it seems that immunization with intact hormone rather than the subunits results in a higher anti-hormone titer.

[2] M. B. Khazaeli, B. G. England, R. C. Dieterle, G. D. Nordblom, G. A. Kabza, and W. H. Beierualtes, *Endocrinology* **109,** 1290 (1981).

[3] R. Kofler, P. Berger, and G. Wick, *Am. J. Reprod. Immunol.* **2,** 212 (1982).

[4] L. Wang, N. Rahamim, N. Harpaz, C. S. Hexter, and M. Inbar, *Hybridoma* **1,** 293 (1982).

[5] M. M. Federici, R. Fraser, C. Lundquist, C. Fraser, and J. C. Lankford, *Hybridoma* **1,** 222 (1982).

[6] W. R. Moyle, P. H. Ehrlich, and R. E. Canfield, *Proc. Natl. Acad. Sci. U.S.A.* **79,** 2245 (1982).

[7] C. Stahli, V. Miggiano, J. Stocker, T. Staehelin, P. Haring, and B. Takacs, this series, Vol. 92, p. 242.

[8] M. Soos and K. Siddle, *J. Immunol. Methods* **51,** 57 (1982).

[9] G. Kohler and C. Milstein, *Nature* (*London*) **256,** 495 (1975).

[10] H. Zola and D. Brooks, *in* "Monoclonal Hybridoma Antibodies: Techniques and Applications" (J. G. R. Hurrell, ed.), p. 1. CRC Press, Boca Raton, Florida, 1982.

[11] P. H. Ehrlich, W. R. Moyle, Z. A. Moustafa, and R. E. Canfield, *J. Immunol.* **128,** 2709 (1982).

[12] J. R. Wands and V. R. Zurawski, *Gastroenterology* **80,** 225 (1981).

The methods of hybridoma production including fusion procedure, cell culture methods, cloning, and antibody production have been reported.[6,10] We have used both P3-NS1/1-Ag 4-1 and P3-X-63/Ag8/653 myeloma lines for fusion. Since the 653 line does not produce myeloma light or heavy chains and the number of antibody-producing hybrids obtained is very high, this is now the recommended line.

Methods to Screen for Desired Clones

Principle. Two general methods are employed in the initial detection of hybridoma colonies that are producing antibodies: liquid phase assays and assays in which either antigen or one antibody is bound to a solid phase. While both assays can be used to screen for anti-gonadotropin monoclonal antibodies, solid phase assays are seldom used for screening by us because their extreme sensitivity and the possibility of bivalent antibody binding to two solid phase-bound antigen molecules results in inclusion, as positives, of cell supernatants containing very low affinity antibodies of limited usefulness. This problem has been encountered in our laboratory and apparently by others.[13] For this reason, a liquid phase double antibody radioimmunoassay is employed that will rarely detect hybridoma colonies producing an antibody with an equilibrium binding constant of less than $10^7\ M^{-1}$. A simple equilibrium calculation will illustrate this point. If 100 μl of cell supernatant containing the average amount of antibody (about 10 μg/ml IgG) is employed in an assay of 300 μl total volume containing tracer amounts of radiolabeled gonadotropin, and a cell line is considered to be producing antibody if the bound/free ratio of tracer is larger than 0.1, then the equilibrium binding constant must be at least

$$K = (\text{Ab–Ag})/[(\text{Ab})(\text{Ag})] = (\text{bound/free})(1/\text{Ab}_0) = 4.5 \times 10^6\ M^{-1} \quad (1)$$

where (Ab–Ag) is the concentration of the antibody–antigen complex, Ag is the concentration of free antigen, Ab is the concentration of free antibody, and Ab_0 is the total concentration of antibody (which is essentially equal to Ab in this case because only trace concentrations of antigen are present).

The buffer for the screening assay is relatively concentrated so that the effect of the cell culture medium (which is normally buffered in the incubator by carbon dioxide) on the assay pH is minimized.

[13] C. Stahli, T. Staehelin, V. Miggiano, J. Schmidt, and P. Haring, *J. Immunol. Methods* **32,** 297 (1980).

Reagents

Buffer: 0.3 *M* potassium phosphate, pH 7.5

Radiolabeled [^{125}I]hCG (or other gonadotropin) in 1% IgG-free horse serum or 1 mg/ml bovine serum albumin

Normal mouse serum

Anti-mouse IgG or anti-mouse Fab or anti-mouse Ig light chain

Procedure. Cell supernatant (100 μl) is added to 100 μl radiolabeled antigen and 100 μl buffer. The mixture is incubated for 1 hr at 37° and then 18 hr at 4° (the 18 hr incubation can be eliminated but the assay will be less sensitive). Previously titered normal mouse serum and second antibody are then added in the appropriate amounts. Anti-mouse IgG can be used as the second antibody since the immunization protocol described below results in essentially all clones being of that isotype. However, anti-mouse Fab or anti-mouse Ig light chain may be used in order to be certain to detect other isotypes.

Properties of the Monoclonal Antibodies; Subclass and Affinity

In the following text we refer to monoclonal antibodies with the prefix A or B, depending on whether they bind to the α or β subunit, respectively. The subclass of the great majority of monoclonal antibodies directed against hCG that have been examined in our laboratory is IgG_1. We have two IgG_{2b} clones. The affinity of the antibodies varies from a clone with an equilibrium binding constant of 3×10^7 M^{-1} to one with a binding constant of 1.5×10^{11} M^{-1} (determined by Scatchard analysis of double antibody radioimmunoassay data). The cross-reactivity of the antibodies with hLH varies from 100% (including some anti-hCGβ clones) to less than 0.5%. The cross-reactivity with hCG subunits (relative to intact hormone) varies from less than 0.1 to about 1200%. The characteristics of several monoclonal antibodies directed against hCG are detailed in Table I.

Effect of Monoclonal Antibodies on Hormone–Receptor Interaction

Monoclonal antibodies to gonadotropins have been useful in analyzing the orientation of the hormone after it binds to cell surface receptors. One can prepare a rough map of the orientation of hCG in the receptor complex by comparing the relative hCG binding sites of the antibodies with the ability of the antibody to bind the hormone–receptor complex.[6] Antibodies that do not bind to hormone–receptor complexes should also inhibit the biological activity of the hormone, while antibodies that bind to the complex should not inhibit the biological activity. Due to the size of

TABLE I
CHARACTERISTICS OF SEVERAL MONOCLONAL ANTI-hCG ANTIBODIES[a]

Antibody	K_{hCG} (M^{-1})	Cross-reactivity (%) relative to hCG, for: hCG subunit	hLH	hTSH	Antibody can bind simultaneously with antibodies
B101	7×10^8	9	2	<0.1	B102, B103, B105, A109
B102	3×10^7	200	11	NT[b]	B101, B105, B107, A102, A103, A109
B103	2×10^8	50	90	12	B101, B107, A102, A103, A109
B105	1.5×10^{11}	100	100	7	B101, B102, B107, A102, A103, A109
B107	4×10^{10}	0.1	<0.5	<0.1	B102, B103, B105
A102	2×10^8	25	100[c]	NT	B102, B103, B105, A103
A103	2×10^8	150	NT	NT	B102, B103, B105, A102, A109
A109	3×10^7	1200	NT	NT	B101, B102, B103, B105, A103

[a] From Ehrlich *et al.*[13a]
[b] Not tested.
[c] Cross-reactivities of anti-hCGα have not generally been tested since the α subunit is identical in hCG, hLH, and hTSH. The one case in which these assays have been performed, A102, the cross-reactivity was indeed 100%. However, clones could theoretically exist that differentiate between the α subunit bound to different β subunits.

the antibodies, it is usually not possible to obtain precise estimates of the areas of the hormone which are present in the receptor complex.

Antibodies that bind to hormone in the receptor complex have the potential to be used as tools not only to monitor the binding of unlabeled hormone to the receptor complex but also to monitor changes in hormone shape and orientation which might occur after binding. Further, if the residues that comprise the binding sites of the antibodies on hCG can be identified, it might then be possible to define the portions of the hormone which interact with the receptor.

Two major approaches are available to screen for antibodies which inhibit binding of hormone to receptor. In one, the antibody is evaluated for its ability to inhibit the binding of radiolabeled hCG to Leydig cell membrane receptors, whereas in the other the antibody is employed to block the biological response to hCG.[6]

[13a] P. H. Ehrlich, Z. A. Moustafa, A. Krichevsky, and R. Mesa-Tejada, *in* "Monoclonal Antibodies in Cancer" (S. Sell and R. Reisfeld, eds). Humana Press, Clifton, N.J. (in press).

Inhibition of Binding of Radiolabeled hCG: Reagents

Radiolabeled hCG (iodinated to a specific activity of approximately 20–50 μCi/μg)
Adult male rats (60 days or older)
Solution A: 0.25 *M* sucrose–0.001 *M* EDTA
Solution B: 0.9% NaCl–0.15 *M* HEPES–Cl buffer (pH 7.4)
Bovine serum albumin (BSA)
Human γ-globulin (hIgG)

Inhibition of Binding of Radiolabeled hCG: Procedure. After the rats are sacrificed, the testes are dissected, the capsule removed, and the testes homogenized in 5 volumes solution A : solution B (1 : 1)/g using a Teflon or glass apparatus. Filter the homogenate through glass wool and collect the sediment at 1000 *g* for 10 min. After carefully decanting the supernatant (taking care not to dislodge the pellet) resuspend the pellet in solution B containing 1 mg BSA and 1 mg hIgG/ml. As a general rule, a pair of rat testes is sufficient for 80 incubation tubes. Therefore, the pellet should be resuspended in 8 ml or less of solution B. If larger numbers of incubation tubes are needed, more rats should be utilized.

Dilute radiolabeled hCG into solution B containing BSA and hIgG to a final concentration of 20,000–50,000 cpm/10 μl (i.e., approximately 1 n*M*). Mix 10 μl of the radiolabel with 10 μl of antibody solution in a 12 × 75-mm polystyrene tube. After sufficient time to permit the antibody to bind the radiolabeled hCG (usually 1 hr at 37° is ample), add 100 μl of resuspended testes homogenate. Incubate the mixture for 1 hr at 37°, add 4 ml of solution B containing 1 mg BSA/ml, and centrifuge at 1000 *g* for 10 min. Carefully aspirate the supernatant and measure the radioactivity remaining in the pellet. Control tubes contain 0.1, 0.3, 1, 3, 10, 30, 100, and 1000 ng of unlabeled hCG in place of the antibody. A graph of radiolabel bound *versus* log of the unlabeled hCG concentration appears sigmoidal with a half-maximal value at approximately 3–10 ng/ml (i.e., between 0.3 and 1 ng). Antibodies which bind to epitopes located on or near the receptor binding site will inhibit binding of hCG to the testes preparations. Those antibodies with binding sites on hCG remote from the receptor recognition site usually do not inhibit binding of the hormone.

Inhibition of hCG Induced Steroidogenesis: Reagents

Immature male rats (approximately 3 weeks old)
Krebs–Ringer–HEPES buffer prepared as follows
Stock solutions

6.1% $CaCl_2$	19.1% $MgSO_4$
5.75% KCl	0.9% NaCl
10.55% KH_2PO_4	3.58% HEPES–Cl (pH 7.4)

Prepare a stock salt solution by mixing 3 ml $CaCl_2$, 4 ml KCl, 1 ml KH_2PO_4, 1 ml $MgSO_4$, and 36 ml water. Mix 9 ml of salt solution with 100 ml NaCl and 21 ml HEPES–Cl.

Gelatin buffer. Heat 2 g of gelatin (Difco) in 100 ml water until the gelatin is dissolved. Cool and mix with 30 ml of 0.2 *M* K_2HPO_4, 20 ml of 0.2 *M* KH_2PO_4, 1 g NaN_3, 8.2 g NaCl, and sufficient water to make 1 liter.

Charcoal slurry. Mix 1.25 g charcoal (Norit EXW), 0.125 g dextran T-70, and 400 ml of gelatin buffer.

Procedure to Evaluate Inhibition of hCG Induced Steroidogenesis: Cell Preparation. Rat testes are dissected so that they are free of the capsule. Incubate the testes in Krebs–Ringers–HEPES buffer containing 1 mg collagenase/ml (CLS grade, Worthington) in a capped plastic 50-ml centrifuge tube (laid on its side parallel to the direction of shaking) in a shaking water bath incubator (approximately 80 oscillations/minute) at 37°. A convenient volume of buffer is 7 ml, which is sufficient for up to 10 testes from the immature rats. After approximately 15 min, the testes tubules will begin to dissociate. Dilute the solution to approximately 45 ml with NaCl solution and mechanically disrupt the testes by swirling them at the end of a pasteur pipet. When the tubules are dispersed, let them sediment to the bottom of the tube (usually this requires letting the tube stand on the bench for about 1 min) and decant the supernate through cheese cloth into a second 50-ml plastic centrifuge tube. Collect the interstitial cells by centrifugation at 300 *g* for 10 min. Resuspend the cells in Krebs–Ringers–HEPES buffer containing 1 mg/ml of BSA and count them in a hemocytometer. Prepare a final dilution containing 150,000 Leydig cells per 100 μl of the same buffer.

Incubation. Prepare hCG standards containing 0, 0.3, 1.0, 3.0, 10, and 100 ng of hormone per ml of Krebs–Ringer–HEPES buffer containing 1 mg BSA/ml. (It is important to include BSA in this solution to prevent the hormone from adhering to the walls of the tube.) Add 10 μl of each standard to 12 × 75-mm plastic test tubes (in duplicate or triplicate). Add 10 μl of antibody, mix, and incubate for 1 hr at 37°. Add 100 μl of the cell suspension and incubate for 2 hr at either 34 or 37°. Analyze the testosterone produced by radioimmunoassay of the entire contents of the incubation flask.

Testosterone Measurement (Long Procedure). Heat the incubation flasks to 69° for 1 min to kill the Leydig cells and prevent further testosterone synthesis. Add 0.2 pmol of tritiated testosterone (in 5 μl of gelatin buffer) and sufficient antibody to testosterone to bind 0.06 pmol of the steroid. Incubate 1 hr at 37° and 30 min in an ice slurry. Add 400 μl of charcoal slurry, mix for 10 min at 0.4°, and sediment at 1000 *g* for 10 min. Decant the supernatant into scintillation cocktail and analyze in a scintillation counter. Testosterone standards can be analyzed if desired.

Testosterone Measurement (Short Procedure). Add tritiated testosterone and testosterone antiserum directly to the cell suspension. The final concentration should be 0.2 pmol testosterone and 0.06 pmol testosterone antibody equivalent per 150,000 cells/100 μl incubation medium. (Be certain that the radiolabel and testosterone antibody are not diluted with gelatin buffer prior to adding them to the cells.) Incubate the cells with hCG and/or antibody to hCG as described above and terminate steroid synthesis after 2 hr by immersing the incubation tubes in an ice slurry. After 30 min, add the charcoal slurry, wait 10 min, centrifuge at 1000 *g* (10 min), decant the supernatant into scintillation cocktail, and analyze in a scintillation counter.

Data Analysis. Prepare graphs of tritiated testosterone remaining in the supernatant versus log of the unlabeled hCG concentration. A series of parallel sigmoidal curves should be obtained as a function of antibody concentration. The separation of the curves will depend on antibody titer (i.e., affinity and concentration) and on the ability of the antibody to bind hCG and prevent it from binding to or activating the receptor. Antibodies which do not bind to or alter the biologically active site on hCG will not cause a rightward displacement of the sigmoidal curves. Those which interfere with hCG binding will cause the curves to be displaced to less sensitive hCG values in proportion to the titer of the antibody. Indeed, if the assumptions are correct, it should be possible to calculate the affinity of such antibodies for hCG based on the following reasoning.

Assume that the total hCG added to the incubation mixture can be expressed as the sum of that remaining free, that binding to the cellular receptors, and that binding to the antibody. This can be written as

$$H_0 = H + b + a \tag{2}$$

where H_0 is the hormone added to the incubation flask, H is the free hormone, b is the hormone bound to Leydig cell receptors, and a is the hormone bound to the antibody. At any arbitrary response (i.e., constant value of testosterone production) the values of the free and bound hormone concentrations are constant.[14] Thus, the difference in H_0 needed to induce a given response is equal to the amount of hormone bound to the antibody. Binding of hCG to the antibody can be expressed as

$$a = K_A[\mathrm{H}]A_0/(K_A[\mathrm{H}] + 1) \tag{3}$$

[14] W. R. Moyle, M. Netburn, A. E. Cosgrove, J. Krieger, and O. P. Bahl, *Am. J. Physiol.* **238,** E293 (1980).

where K_A is the affinity of the antibody for hCG and A_0 is the total concentration of antibody. Since H and b are constant at any given response and since K_A is constant, a graph of H_0 versus A_0 should be a straight line having a slope of $K_A[H]/(K_A[H] + 1)$. The value of K_A should be readily calculated from this slope.

This treatment of the data has been useful in estimating the ability of monoclonal antibodies to bind to sites on hCG that appear necessary for biological activity.[6] The results for five of our cell lines are illustrated in Fig. 1 for the case with H_0 at half-maximal response. For example, antibodies like A102, B101, and B107 block the ability of hCG to induce Leydig cell steroidogenesis with about the same potency as they bind hCG. Graphs of H_0 versus A_0 for these antibodies have very steep slopes. Antibodies such as A103, B102, B103, and B105 fail to block hCG induced steroidogenesis even though they bind to the hormone with an affinity equal or greater than A102, B101, or B107. Graphs of H_0 versus A_0 for antibodies A103, B102, B103, and B105 have very low slopes which are orders of magnitude lower than predicted based on their affinity for hCG.

Critique of These Methods. The methods used to examine the effect of antibodies on hCG are limited by the assumptions on which they are based.

1. It is assumed that the antibodies block the response only by binding to hCG at the receptor binding site. They are assumed not to alter the shape of the gonadotropin. In addition, antibodies are large molecules in comparison with hCG (i.e., molecular weight of 150,000 versus 37,000). Thus, although an antibody might block hCG binding and biological activity, it is conceivable that its effect might occur through steric hindrance on a part of hCG close to the receptor interaction site, not necessarily the actual binding site. The resolution may be improved using Fab fragments of the antibodies. However, even these are large and likely to limit the degree of certainty which can be used to identify the biological site, i.e., that region which acts at the receptor. Identification of sites outside of the receptor complex can be made with greater confidence since radiolabeled monoclonal antibodies can be used to detect the exposed epitopes of hormone–receptor complexes. It would be impossible to interpret data on the relative binding sites of antibodies if they altered the conformation of hCG. However, it has been determined that several monoclonal antibodies that we have employed in our studies do not alter the conformation of hCG (see below: section on Synergistic Interaction of Monoclonal Antibodies to hCG).

2. Antibodies are bivalent. Thus, particular antibodies might augment binding of hCG to receptor complexes by forming a "circular" complex

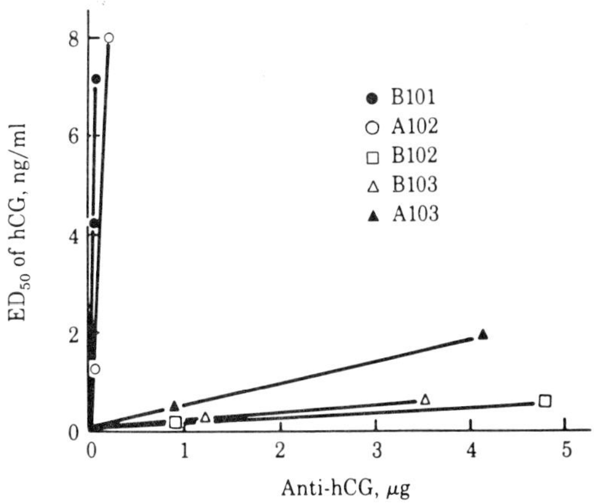

FIG. 1. Influence of monoclonal antibodies to hCG on Leydig cell steroidogenesis. Dose–response curves to hCG were generated in the presence of several concentrations of antibodies. The concentration of hCG that was needed to produce half-maximal stimulation of steroidogenesis is shown here as a function of antibody effectiveness in blocking the hCG response. These results are from Moyle *et al.*[6]

containing two receptors, two molecules of hCG, and one of antibody (see below: Synergistic Interaction of Monoclonal Antibodies to hCG). In this case, the monoclonal antibody might appear to potentiate binding of hCG rather than inhibit it. This phenomenon would only be expected for antibodies that bind to hCG receptor complexes and could be avoided by the use of Fab fragments. It should be noted here that some hybridoma antibodies are monovalent due to the presence of myeloma light chain. This may sometimes complicate interpretation of the data.

3. The binding of hCG to an antibody may not reach equilibrium before hCG begins to induce a biological response. (The approach described above is true only when binding of hCG to the antibody reaches equilibrium before the mixture is added to the cells.) If it takes a while for binding of hCG to the antibody to reach equilibrium, a larger than expected fraction of hCG will be available for binding to the receptor at the beginning of the incubation phase. Since binding of hCG to the receptor and induction of a response becomes rapidly irreversible, it is possible for one to underestimate the apparent affinity of the antibody for hCG and miss inhibitory antibodies. In preliminary experiments, we have noted that this is particularly true for antibodies that have a very high affinity for hCG (i.e., B107). Unless the B107 is incubated with hCG for a time before the hCG is incubated with receptor, estimation of the binding constant from inhibition of hCG-induced testosterone synthesis provides values

which are an order of magnitude lower than that obtained when the binding constant is measured directly.

4. It is assumed that the antibody does not bind to the receptor or the membranes. Addition of hIgG to the incubation medium is included to minimize binding of the antibodies to Fc receptors on the cell surface.[14]

5. It is assumed that the antibodies interact only with hCG and have no effect on the ability of the receptor to bind hormone or on the ability of the cell to make testosterone. An influence of antibody on the ability of the cell to make testosterone could be detected by the inability of excess hCG to overcome the inhibitory effect of the antibodies on steroidogenesis.

6. This experimental approach assumes that the incubation conditions do not interfere with the ability of the antibody to bind to hCG. This can be tested by using a precipitating antibody to pull hCG–monoclonal antibody complexes out of solution.[6]

Determination of Relative Orientation of Epitopes

A form of immunological map of the protein antigen can be derived from determinations of the relative binding sites of different monoclonal antibodies in the presence of each other or of other macromolecules that interact with the antigen. This type of evaluation is of particular importance in the design of immunoradiometric assays which employ two antibodies for measuring gonadotropins. Variations of different antibodies can then be used to design assays to measure the whole hormone or either of the free subunits.

We performed this type of study with hCG, comparing the binding of five monoclonal antibodies and the Leydig cell receptor for hCG.[6,11] To determine if two monoclonal antibodies bind to epitopes near to each other on the hCG molecule, one purified monoclonal antibody is adsorbed to a plastic microtiter plate, the plate is incubated with hCG to saturate the binding sites, and a purified radiolabeled second monoclonal antibody is allowed to react with the hCG–antibody complex adsorbed to the solid. If radioactivity above background is detected on the solid phase, one can infer that the two antibodies can bind simultaneously to the hormone.

The procedure for this assay is as follows: polyvinyl chloride microtiter plates are incubated for 4 hr at 37° with 50 μl containing 70 μg/ml of monoclonal antibody (this concentration of antibody has not been optimized). The antibody solution is removed and the wells incubated at room temperature with 150 mM NaCl, 1 mg/ml bovine serum albumin (BSA-saline). This solution is replaced with 20 μg/ml hCG in BSA-saline for 2 hr. The hCG is removed and the plate washed with BSA-saline. Radiola-

TABLE II
SANDWICH ASSAYS WITH PAIRS OF MONOCLONAL ANTIBODIES DIRECTED AGAINST hCG[a]

Tracer antibody	Antibody adsorbed to the plastic surface (cpm, %)				
	A102	A103	B101	B102	B103
A102	−110	556[b]	−171	1,981[b]	2,174[b]
A103	502[b]	−157	−239	916[b]	1,975[b]
B101	−215	−167	−181	6,005[b]	10,854[b]
B102	2,847[b]	483[c]	3,646[b]	15	−27
B103	1,255[b]	985[b]	869[b]	−145	−124

[a] Values are cpm over the control, i.e., when hCG was omitted. From Moyle *et al.*[6]
[b] $p < 0.01$ that a value this much greater than that of the control (no hCG) could have arisen by chance.
[c] $p < 0.05$ that a value this much greater than that of the control could have arisen by chance. Other values were not significantly greater than the controls.

beled second antibody (50,000 to 100,000 cpm in BSA-saline) is then added and removed after 2 hr at room temperature. The wells are washed with BSA-saline and counted. Typical results with this mapping technique are shown in Table II. There is a clear-cut distinction between the CPM bound if the two antibodies can bind simultaneously vs those that inhibit each other's binding.

One drawback with this experimental design is that both antibodies must be pure (or, in the case of the adsorbed antibody, in a concentration much higher than is normally present in cell culture supernatant). Therefore, a cell line must be cloned and expanded before its antibody can be examined. In lieu of this a screening test has been developed to determine as little as 2 weeks after fusion, if the antibodies from two cell colonies can bind simultaneously. This method takes advantage of a phenomenon that occurs frequently with anti-hCG antibodies: synergistic binding of two monoclonal antibodies (see below: Synergistic Interactions of Monoclonal Antibodies to hCG).[11] When one antibody is adsorbed to the solid phase and one antibody is in the liquid phase, all pairs of monoclonal antibodies to hCG examined so far that can bind antigen simultaneously are synergistic in their binding. Therefore, if radiolabeled hCG is mixed with supernatant from one hybridoma cell culture and placed in a microtiter well that has solid-absorbed antibody from another hybridoma cell culture, the number of counts bound to the solid will increase compared to a control mixture of radiolabeled hCG and cell culture medium if the two antibodies can bind simultaneously. The counts bound to the solid will decrease if the two antibodies cannot bind simultaneously. It is best

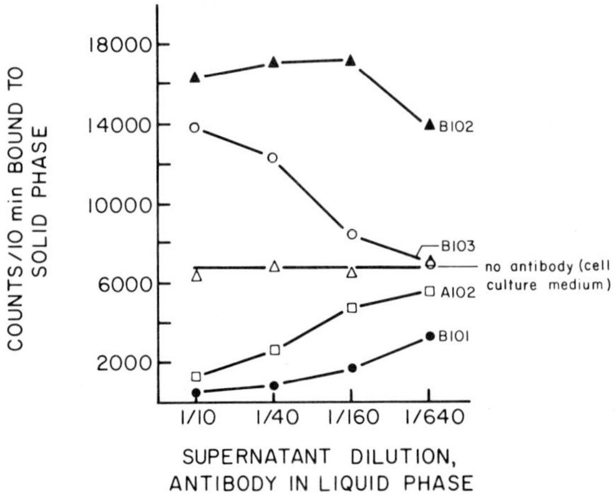

FIG. 2. Results of mapping monoclonal antibodies by determining if synergistic binding or inhibition of binding occurs with different pairs of monoclonal antibodies. Radiolabeled hCG is incubated simultaneously with the liquid phase antibody above the solid phase bound antibody (in this case, antibody B101). The control experiments are performed with cell culture medium that does not appreciably affect the assay. The cell culture supernatant from antibody producing hybridoma colonies either increases or decreases the number of counts bound to the microtiter well. An increase in counts bound indicates that the two antibodies can bind to hCG simultaneously. A decrease in counts bound indicates that the two antibodies inhibit each other in binding to the hormone.

to use several dilutions of the cell supernatant because the optimal dilution will vary with the amount and affinity of the antibody. This is illustrated in Fig. 2. The results indicate the same spectrum for B101 of inhibiting and noninhibiting antibodies as previously reported.[6] In order to avoid using a concentrated antibody solution for adsorption to the microtiter well, the well is first incubated with an anti-mouse Ig antiserum (in the case of Fig. 2, this antiserum was goat anti-mouse Fab).

This procedure works well for the great majority of hybridoma antibodies. However, for antibodies with low affinities ($K < 5 \times 10^7\ M^{-1}$) and those with very high affinities ($K > 10^{11}\ M^{-1}$) the method does not appear to work well, because in the former case not enough radiolabeled antigen is bound to detect inhibition and in the latter case the affinity is so high that essentially all the antibody binding sites contain antigen. Some improvement in the generality of the procedure is noticed when a concentrated solution of the first antibody is bound directly to the microtiter plate (without the anti-antibody). However, this is useful only when screening for monoclonal antibodies that are synergistic or inhibitory to an antibody from a cell line that has already been expanded.

The procedure is as follows: 50 μl of an appropriate dilution of goat anti-mouse Fab antiserum (for our antiserum a 100-fold dilution in phosphate-buffered saline was used) is incubated 18 hr or more at 4° in each well of a 96-well polyvinyl chloride microtiter plate. This solution is removed, the plate is washed with distilled water, and incubated for 2 hr at room temperature with 10% IgG-free horse serum. The plate is then washed again and each well incubated for 3 hr at room temperature with undiluted cell supernatant of the first hybridoma colony to be examined. (Of course, if a concentrated solution of the antibody is to be bound directly to the solid phase, the anti-mouse immunoglobulin antiserum is not used. The concentrated antibody solution is incubated 18 hr at 4° in the microtiter wells. After removing this solution and washing the plate with distilled water, the wells are incubated 2 hr with 10% horse serum. The procedure is then the same as with the antiserum approach.) After washing, each well is then incubated 18 hr at room temperature with a mixture of varying dilutions of cell supernatant from a second hybridoma colony and [^{125}I]hCG (approximately 20,000 cpm). The liquid is aspirated, the plate washed four times with distilled water, and then each well is counted.

In the experiments in which an anti-mouse antibody is adsorbed to the solid, radiolabeled antigen must be used (and not radiolabeled hybridoma antibody), because the tracer amounts of radiolabeled antibody would bind to the solid directly through the anti-mouse Fab antiserum either to unfilled sites or by exchange with bound first antibody. This is apparently not a problem with the radiolabeled antigen because the majority of sites on the anti-mouse antibodies on the solid phase are probably saturated with the first hybridoma antibody and the few remaining sites do not affect the assay.

An immunochemical mapping arrangement of this type indicating the relative positions of the epitopes of eight monoclonal antibodies, a rabbit antiserum to hCGβ (SB6), and the Leydig cell receptor is shown in Fig. 3. An additional rabbit antiserum to the unique COOH-terminal β subunit sequence (R529)[15] is also described below but not identified in Fig. 3. This is an expanded version of a figure published earlier.[6] The relative positions of the monoclonal antibodies were inferred by sandwich assays (using either the radiolabeled second antibody or the synergistic method). The relative orientation of the epitopes for the antisera, compared to the monoclonal antibodies, was determined by a double antibody radioimmunoassay with a species specific second antibody to the rabbit antibodies. In this case, if a monoclonal antibody bound to an epitope near the antise-

[15] S. Birken, R. E. Canfield, G. Agosto, and J. Lewis, *Endocrinology* **110,** 1555 (1982).

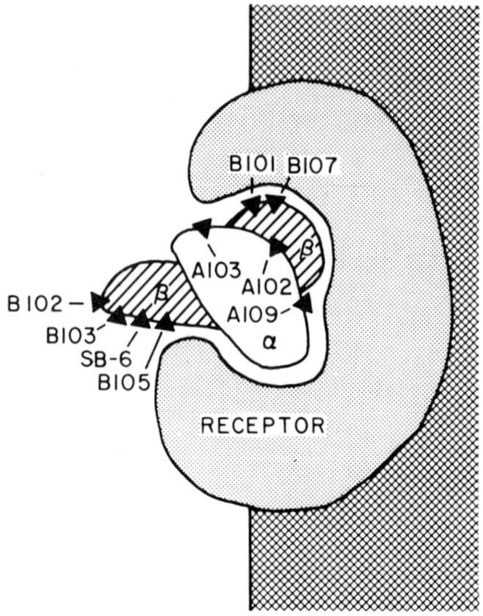

FIG. 3. Relative positions of the hCG binding sites of eight monoclonal antibodies, two rabbit anti-hCG antisera, and the rat Leydig cell hCG receptor. The locations are chosen as symbolic of the properties exhibited in the sandwich and receptor inhibition assays. The positions of monoclonal antibody A109 and of the SB-6 rabbit antiserum were determined by pairwise mapping with the monoclonal antibodies (no experiments were performed with the Leydig cell receptor). Expanded from Moyle *et al.*[6]

rum, binding of radiolabeled hCG would be inhibited by addition of the hybridoma. If the monoclonal antibody bound to an epitope distant from the majority of the antibodies in the serum, then addition of the hybridoma antibody leads to an increase in binding of radiolabeled antibody (through the synergistic effect) or no change in bound radioactivity.

Some interesting features of hCG structure emerge from these studies. First, the rabbit antiserum to the COOH-terminus of the β subunit[15] is not inhibited by any monoclonal antibody tested so far. This implies that this structure is distant from the epitopes of these antibodies and may be a separate domain of the molecule. Second, B101 and B107 inhibit many monoclonal antibodies directed against the hCGα subunit indicating that they probably bind near the interface of the two subunits. Third, antibody B103 binds hCG, hCGβ subunit, and hLH with almost the same affinity. It also cross-reacts significantly with hFSH, hTSH, and ovine LH, indicating that it reacts with a conserved region of these molecules. Fourth, B102 and the SB-6 antiserum[16] bind hCG more strongly than hLH. There-

[16] J. L. Vaitukaitis, G. D. Braunstein, and G. T. Ross, *Am. J. Obstet. Gynecol.* **113,** 751 (1972).

fore, there are at least three distinct regions of hCGβ that are immunochemically different from hLH: the B101–B107 region (the two antibodies have cross-reactivities of 2 and 0.5%, respectively), the B102-SB-6 region, and the COOH-terminal region. In addition, these studies show that at dilutions used in radioimmunoassays, the SB-6 antiserum appears to be monospecific (within the resolution of the technique utilized here). Therefore, it is probable that one antibody or a family of very similar antibodies contribute the great majority of antigen binding under these conditions.

Synergistic Interactions of Monoclonal Antibodies to hCG

Pairs of monoclonal antibodies directed against hCG are frequently synergistic in their interactions with the antigen.[11,17] An example is shown in Fig. 4. Since this cooperativity is not present with Fab monovalent fragments of the antibodies,[11] it can be concluded that the synergy is not due to a conformational change in hCG. Further support for this is derived from a computer model of the interaction of hCG with two monoclonal antibodies assuming no conformational transition that can accurately predict the experimental data.[18] The interaction is best explained by the formation of a circular complex consisting of one of each type of monoclonal antibody and two hCG molecules.[11,17–19] A diagram illustrating this complex is shown in Fig. 5. It is probable that circular complexes form with some antisera to gonadotropins since cooperative antisera directed against hCG with characteristics similar to mixtures of monoclonal antibodies have been described.[20]

As mentioned above, all pairs of monoclonal antibodies that can bind simultaneously to hCG that have been tested in an assay in which one of the antibodies is bound to a solid phase are synergistic. However, if the antigen and antibodies are all in a liquid phase, some pairs of monoclonal antibodies are not synergistic.[11,21] In order to determine if two monoclonal antibodies are cooperative in the liquid phase, the double antibody radioimmunoassay described above can be used. The amount of each antibody required to bind about one-half of radiolabeled tracer is determined. This amount of both antibodies is then added to the same assay tube. If the radiolabeled antigen bound to the mixture of antibodies exceeds the sum of radiolabeled antigen bound to the individual antibodies, then the

[17] P. H. Ehrlich and W. R. Moyle, *Science* **221,** 279 (1983).
[18] W. R. Moyle, C. Lin, R. L. Corson, and P. H. Ehrlich, *Mol. Immunol.* **20,** 439 (1983).
[19] W. R. Moyle, D. M. Anderson, and P. H. Ehrlich, *J. Immunol.* **131,** 1900 (1983).
[20] B. D. Weintraub, S. W. Rosen, J. A. McCammon, and R. L. Perlman, *Endocrinology* **92,** 1250 (1973).
[21] P. H. Ehrlich, W. R. Moyle, and Z. A. Moustafa, *J. Immunol.* **131,** 1906 (1983).

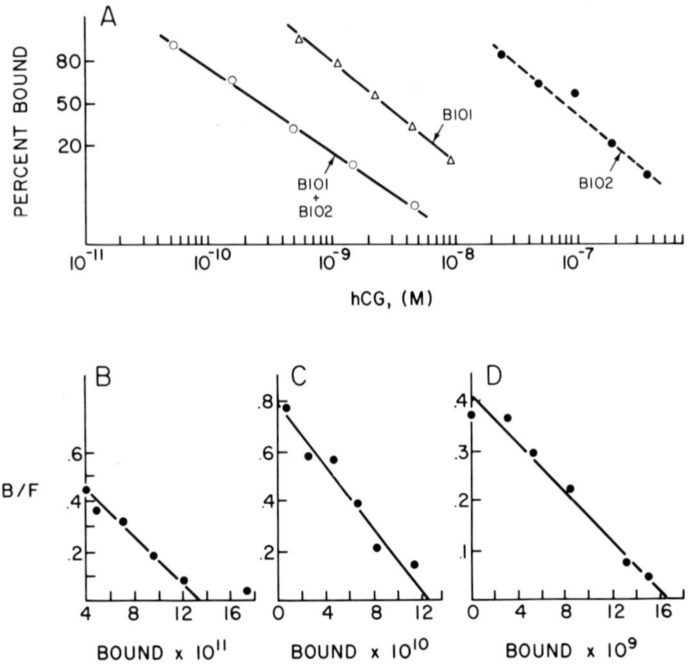

FIG. 4. (A) Radioimmunoassay for hCG by double antibody liquid-phase assay with antibodies B101, B102, and a mixture of B101 and B102. The amount of labeled hCG bound in the absence of unlabeled inhibitor was approximately the same for each antibody or antibody mixture. Rabbit anti-mouse IgG was the second antibody. The ED_{50} (effective displacement of 50% of bound radiolabeled hCG) ± SD was calculated for each antibody or mixture. The ED_{50} for the mixture of B101 and B102 is $2.44 \pm 0.74 \times 10^{-10}$ *M*, antibody B101 $2.88 \pm 0.74 \times 10^{-9}$ *M*, and antibody B102 $1.058 \pm 0.995 \times 10^{-7}$ *M*. (B, C, D) Scatchard analysis of the binding to hCG of B mixture of antibodies B101 and B102, C antibody B101, and D antibody B102. Slopes of the lines (therefore, the equilibrium binding constants) are B, 5.4×10^9, C, 5.1×10^8, and D, 1.9×10^7. The moles of each antibody can be inferred from the intercept on the abscissa of the Scatchard plots. Thus, the concentrations of B101 and B102 (from C and D, assuming bivalent antibodies) are, respectively, 6.5×10^{-10} and 8×10^{-9} *M*. The data in B indicate that the concentration of the high affinity sites is 1.3×10^{-10} *M*. The concentration of B101 and B102 added to obtain the results in B were 3.6×10^{-11} and 4.4×10^{-10} *M*, respectively. From Ehrlich *et al.*[11]

binding of the antibodies is cooperative. In this case, a radioimmunoassay based on the appropriate dilution of the mixture will appear to have a higher sensitivity[11] or a rise and then fall in radiolabeled antigen bound upon addition of increasing amounts of unlabeled antigen,[17] depending on the amount of radiolabeled antigen. The latter type of assay is termed a "cooperative immunoassay" and can be up to 1000 times more sensitive

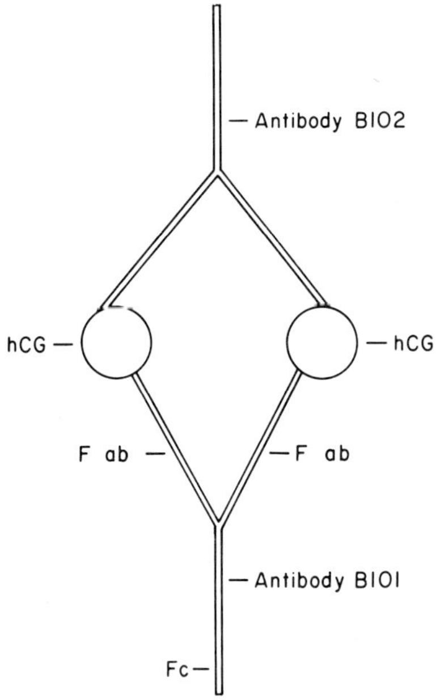

FIG. 5. Diagram showing the molecular complex that is responsible for the synergistic interactions of some pairs of monoclonal antibodies with hCG. From Ehrlich *et al.*[13a]

than an assay based on an individual monoclonal antibody.[17] The appropriate dilution of the mixture of the monoclonal antibodies is that amount of antibody that binds about 20 to 30% of the radiolabeled tracer. It should be noted that antigens other than gonadotropins, particularly molecules that are much smaller or much larger, may not have synergistic interactions with pairs of monoclonal antibodies. The useful range and slope of the inhibition curve of a cooperative immunoassay can be changed by varying the ratio of the two monoclonal antibodies.

Acknowledgments

This work was supported by the National Institutes of Health, Grants HD-15454 and CA-26636.

[51] Assay of Antibodies Directed against Cell Surface Receptors

By SIMEON I. TAYLOR, LISA H. UNDERHILL, and BERNICE MARCUS-SAMUELS

Antibodies directed against cell surface receptors for hormones and neurotransmitters play an important role in the pathogenesis of human disease (see the table).[1–16] In some disease states, the interaction between the antibody and receptor may lead to biological effects identical to those of the natural ligand. In other conditions, anti-receptor antibodies may antagonize the bioactivity of the natural ligand. In addition to their role in pathogenesis of human disease, anti-receptor antibodies play a central role as reagents in studies of receptor structure and function.[4,13,14,14a,17–19]

[1] M. Zakarjia and J. M. McKenzie, *J. Clin. Endocrinol. Metab.* **47,** 249 (1978).
[2] B. R. Smith, *Recept. and Recognition, Ser. B* 215 (1981).
[3] M. Kishihara, Y. Nakao, Y. Baba, N. Kobayashi, S. Matsukura, K. Kuma, and T. Fujita, *J. Clin. Endocrinol. Metab.* **52,** 665 (1981).
[4] C. R. Kahn, K. L. Baird, J. S. Flier, and D. B. Jarrett, *J. Clin. Invest.* **60,** 1094 (1977).
[5] J. S. Flier, R. S. Bar, M. Muggeo, C. R. Kahn, J. Roth, and P. Gorden, *J. Clin. Endocrinol. Metab.* **47,** 985 (1978).
[6] S. I. Taylor, G. Grunberger, B. Marcus-Samuels, L. H. Underhill, R. F. Dons, J. Ryan, R. F. Roddam, C. E. Rupe, and P. Gorden, *N. Engl. J. Med.* **307,** 1422 (1982).
[7] L. Tardella, L. Rossetti, R. DePirro, A. Camagna, S. Leonetti, G. Tamburrano, S. Rossi-Fanelli, M. Merli, and R. Lauro, *J. Clin. Lab. Immunol.* **12,** 159 (1983).
[8] D. B. Drachman, C. W. Angus, R. N. Adams *et al., N. Engl. J. Med.* **298,** 1116 (1978).
[9] D. B. Drachman, R. N. Adams, L. F. Josifek, and S. G. Self, *N. Engl. J. Med.* **307,** 769 (1982).
[10] J. Lindstrom, *Ciba Found. Symp.* **90,** 178 (1982).
[11] J. S. Flier, C. R. Kahn, D. B. Jarrett, and J. Roth, *J. Clin. Invest.* **58,** 1442 (1976).
[12] C. R. Kahn, J. S. Flier, R. S. Bar, J. A. Archer, P. Gorden, M. M. Martin, and J. Roth, *N. Engl. J. Med.* **294,** 739 (1976).
[13] F. A. Karlsson, E. Van Obberghen, C. Grunfeld, and C. R. Kahn, *Proc. Natl. Acad. Sci. U.S.A.* **76,** 809 (1979).
[14] C. Grunfeld, E. Van Obberghen, F. A. Karlsson, and C. R. Kahn, *J. Clin. Invest.* **66,** 1124 (1980).
[14a] S. I. Taylor and B. Marcus-Samuels, *J. Clin. Endocrinol. Metab.* **58,** 182 (1984).
[15] J. C. Venter, C. M. Fraser, and L. C. Harrison, *Science* **207,** 1361 (1980).
[16] V. Chiauzzi, S. Cigorraga, M. E. Escobar, M. A. Rivarola, and E. H. Charreau, *J. Clin. Endocrinol. Metab.* **54,** 1221 (1982).
[17] M. Kasuga, E. Van Obberghen, K. M. Yamada, and L. C. Harrison, *Diabetes* **30,** 354 (1981).
[18] E. Van Obberghen, M. Kasuga, A. Le Cam, J. A. Hedo, A. Itin, and L. C. Harrison, *Proc. Natl. Acad. Sci. U.S.A.* **78,** 1052 (1981).
[19] J. A. Hedo, M. Kasuga, E. Van Obberghen, J. Roth, and C. R. Kahn, *Proc. Natl. Acad. Sci. U.S.A.* **78,** 4791 (1981).

METHODS IN ENZYMOLOGY, VOL. 109

ISBN 0-12-182009-2

ANTI-RECEPTOR ANTIBODIES IN HUMAN DISEASE

Disease	Bioeffect of anti-receptor antibody	References
Graves' disease	Mimics thyrotropin	1–3
Hypoglycemia	Mimics insulin	4–7
Myasthenia gravis	Antagonizes acetylcholine	8–10
Type B extreme insulin resistance	Antagonizes insulin	11–14a
Asthma	Antagonizes effects of epinephrine	15
Hypergonadotropic amenorrhea	Antagonizes effects of follicle-stimulating hormone	16

Identification of anti-receptor antibody requires demonstration (1) that the antibody binds to a receptor and (2) that the same receptor binds to the ligand. These two properties can be demonstrated using several different assay techniques. No single assay method is perfect. Rather the different approaches complement one another in definitive identification of anti-receptor antibodies.

In this chapter, we will describe the three principal assay techniques which are employed to detect antibodies to the insulin receptor. Similar approaches may be applied to the detection of antibodies directed against other cell surface receptors.

Binding-Inhibition Assay

Antibodies directed against the insulin receptor were first demonstrated using an assay based on the ability of the antibodies to inhibit insulin from binding to the receptor.[11] Cells are incubated with antibody and then washed; subsequently, ^{125}I-labeled insulin binding to the cells is assayed. This method continues to be the most useful technique for analyzing clinical samples because of its simplicity, specificity, and resistance to interference from other substances which may be present in the sera of insulin-treated patients (e.g., anti-insulin antibodies, insulin, etc.). All of the reported examples of human autoantibodies to the insulin receptor inhibit insulin binding although some anti-receptor antibodies raised by immunization of laboratory animals are not detectable using this technique.[20,21]

[20] S. Jacobs, K. Chang, and P. Cuatrecasas, *Science* **200,** 1283 (1978).

[21] F. C. Kull, Jr., S. Jacobs, Y. F. Su, and P. Cuatrecasas, *Biochem. Biophys. Res. Commun.* **106,** 1019 (1982).

Reagents

IM-9 lymphoblastoid cells: Supplied by American Type Tissue Collection (Rockville, Maryland). Cultivated in RPMI-1640 medium supplemented with 10% (v : v) fetal bovine serum

Phosphate-buffered saline, pH 7.4

Binding assay buffer: NaCl, 120 mM; $MgSO_4$, 1.2 mM; KCl, 2.5 mM; Na acetate, 15 mM; glucose, 10 mM; EDTA, 1 mM; HEPES, 50 mM; bovine serum albumin, 10 mg/ml; pH 7.8

^{125}I-labeled insulin: Porcine insulin labeled with ^{125}I to a specific activity of 100–150 Ci/g.[22,23] [^{125}I]Monoiodoinsulin (receptor grade) supplied by New England Nuclear (Boston, MA) is also satisfactory

Porcine insulin. Supplied by Elanco, Inc. (Indianapolis, IN). Stock solution (1 mg/ml) in 0.01 N HCl.

Procedure

1. Sediment cultured lymphocytes (IM-9 lymphoblastoid cells) by centrifugation (200 g, 10 min, room temperature). After washing the cells once with phosphate-buffered saline, resuspend them at a density of 10^7 cells/ml in binding assay buffer.

2. Place aliquots (0–0.05 ml) of the serum to be tested in plastic culture tubes (12 × 75 mm), set up in duplicate. Adjust the total volume in the tube to 0.05 ml using binding assay buffer. A range of dilutions of serum may be employed if it is desired to titer the antiserum. A parallel set of tubes with normal serum instead of test serum should be employed as a control.

3. Add aliquots (0.45 ml) of cells suspended in binding assay buffer to the tubes containing serum samples. Incubate the cells for 2 hr at room temperature with intermittent shaking to keep the cells suspended. After addition of 2 ml of binding assay buffer, sediment the cells by centrifugation (200 g, 10 min, room temperature). Aspirate and discard the supernatant fluid, taking care not to aspirate cells.

4. Add aliquots (0.5 ml) of assay buffer containing ^{125}I-labeled insulin (~0.1 ng/ml). To estimate nonspecific binding of ^{125}I-labeled insulin, add unlabeled insulin (10 μg/ml) to one of each pair of duplicate tubes. Incubate cells for 3 hr at 15° with intermittent shaking to keep the cells suspended.

5. At the end of the incubation, layer aliquots (0.2 ml) of cells over ice cold binding assay buffer (0.15 ml) contained in microcentrifugation tubes

[22] P. Freychet, J. Roth, and D. M. Neville, Jr., *Biochem. Biophys. Res. Commun.* **43,** 400 (1971).

[23] J. Roth, this series, Vol. 37, p. 223.

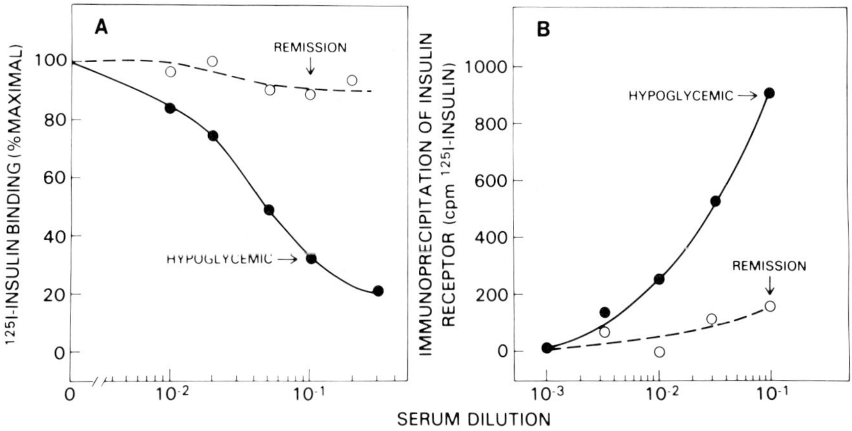

FIG. 1. Identification of anti-receptor antibody in patient's serum. Serum samples were obtained from a patient with fasting hypoglycemia resulting from autoantibodies to the insulin receptor.[6] One sample was obtained at the time she manifested hypoglycemia; the other sample was obtained at the time when the antibody titer had fallen and the clinical symptoms had remitted. The serum samples were assayed using two techniques: binding inhibition (A) and immunoprecipitation (B).

(0.4 ml capacity). After sedimenting the cells by centrifugation (~10,000 *g*, 30 sec), aspirate, and discard the supernatant fluid. Excise the tip of the tube and measure cell-associated radioactivity in a gamma counter.

6. Calculate specific ^{125}I-labeled insulin binding by subtracting nonspecific binding from total binding. Plot specific ^{125}I-labeled insulin binding as a function of the dilution of antiserum (Fig. 1A).

Comments

1. Ordinarily, we do not incubate cells in concentrations of serum exceeding 10%. So long as the concentration of serum is kept below 10%, preincubation in normal serum has little if any effect upon insulin binding. Plasma derived from heparinized blood or purified immunoglobulin fractions may be substituted for serum in these studies. However, it has been reported that heparin in sufficiently high concentrations may inhibit ^{125}I-labeled insulin binding to insulin receptors from some (but not all) tissues.[23a]

2. In the original description of this method,[11] the cells were washed three times with phosphate-buffered saline after preincubation with serum. This wash was intended to eliminate anti-insulin antibodies as well

[23a] K. Kriauciunas and C. R. Kahn, *Clin. Res.* **32,** 401A (abstract) (1984).

as insulin which might have been present in the serum sample. In most cases, our simplified washing procedure is sufficient. However, in an occasional sample, the titer of anti-insulin antibody is so high (e.g., detectable at a dilution of greater than 1 : 1000) that it is necessary to employ the original procedure for washing the cells. In addition, if the concentration of insulin in the incubation medium were to exceed 10^{-8} M, a more extensive washing procedure would be necessary.[24]

3. We have never observed a major effect of anti-receptor antiserum upon nonspecific binding of insulin to IM-9 cells. Thus, nonspecific binding need not be determined at each concentration of serum.

4. This assay method will detect only those anti-receptor antibodies which inhibit ^{125}I-labeled insulin binding to the receptor. If anti-receptor antibodies bound to the receptor without inhibiting ^{125}I-labeled insulin binding, they would not be detected in this assay. While the latter class of anti-receptor antibodies has not been definitively demonstrated in clinical samples, this type of antibody has been raised by immunizing animals with solubilized insulin receptors.[20,21]

Immunoprecipitation Assay

The major application of the immunoprecipitation technique is to detect anti-receptor antibodies which do not inhibit insulin binding. The assay as originally described[25] was based upon the ability of anti-receptor antibodies to immunoprecipitate ^{125}I-labeled insulin bound to receptor. Unfortunately, anti-insulin antibodies can interfere in this assay because of their ability to immunoprecipitate ^{125}I-labeled insulin directly. Thus, the presence of anti-insulin antibodies in the majority of insulin-treated patients greatly restricts the utility of this assay in analyzing clinical samples. In recent years, other methods have been devised for labeling the receptor either biosynthetically or by covalent modification.[17–19,26,27] These labeling methods combined with the immunoprecipitation technique have greatly expanded our understanding of receptor structure. However, because the methods are relatively cumbersome and insensitive, they have not been utilized as a routine method for the assay of anti-receptor antibodies.

[24] F. C. Kosmakos and J. Roth, *J. Biol. Chem.* **255,** 9860 (1980).

[25] L. C. Harrison, J. S. Flier, J. Roth, F. A. Karlsson, and C. R. Kahn, *J. Clin. Endocrinol. Metab.* **48,** 59 (1979).

[26] P. F. Pilch and M. Czech, *J. Biol. Chem.* **254,** 3375 (1979).

[27] M. Kasuga, C. R. Kahn, J. A. Hedo, E. Van Obberghen, and K. M. Yamada, *Proc. Natl. Acad. Sci. U.S.A.* **78,** 6917 (1981).

Reagents

Triton X-100, 20% (v : v)
Buffer A: NaCl, 150 m*M*; HEPES, 50 m*M*; pH 7.8
Buffer B: Sucrose, 250 m*M*; HEPES, 50 m*M*; pH 7.6
Buffer C: HEPES, 50 m*M*, pH 7.6; bacitracin, 100 U/ml; aprotinin, 1 TIU/ml; phenylmethylsulfonyl fluoride, 1 m*M*
^{125}I-labeled insulin: Porcine insulin labeled with ^{125}I to a specific activity of 100–150 Ci/g.[22,23] [^{125}I]Monoiodoinsulin (receptor grade) supplied by New England Nuclear (Boston, MA) is also satisfactory.
Porcine insulin: Supplied by Elanco, Inc. (Indianapolis, IN)
Sheep anti-human γ-globulin antiserum: Supplied by Miles Laboratories, (Elkhart, Indiana)
Triton extract of human placental membranes: Prepared as described below.

Preparation of Solubilized Insulin Receptors

Preparation of Placental Membranes (*A*)

1. A fresh human placenta (within 4 hr postpartum) is rinsed in ice cold buffer B. Trim away and discard large blood vessels as well as amniotic and chorionic membranes. Rinse until most of blood is washed away.

2. Homogenize the tissue at 4° (3 × 30 sec in a Waring blender or similar homogenizer) in 1.5 volumes of ice cold buffer B supplemented with bacitracin (100 U/ml), aprotinin (1 TIU/ml), and phenylmethylsulfonyl fluoride (1 m*M*).

3. Centrifuge the homogenate (600 *g* for 10 min at 4°). Save the supernatant fluid and resuspend the pellet in one-half the original volume of buffer. Repeat centrifugation step.

4. Discard the pellets and centrifuge (12,000 *g*, 30 min, 4°) the combined supernatants from step 3.

5. Discard pellet and add NaCl (5.8 mg/ml) and $MgSO_4$ (0.05 mg/ml) to the supernatant medium (final concentrations, 100 m*M* NaCl and 0.2 m*M* $MgSO_4$).

6. Sediment placental membranes by centrifugation (40,000 *g*, 40 min, 4°). Discard the supernatant and wash the pellet two times with buffer C until the supernatant fluid is colorless.

Solubilization of Placental Insulin Receptors (*B*)

1. Suspend placental membranes at a protein concentration of 3–10 mg/ml in ice cold buffer C.

2. Add 0.05 volumes of ice cold Triton X-100 (20%, v : v) to give a final concentration of 1% (v : v) detergent. Incubate for 2 hr at 4° with stirring.

3. Clarify the extract by centrifugation (200,000 *g*, 90 min, 4°).

4. Discard the pellet and the lipid layer floating on top. Save the clear red-brown infranatant which contains the solubilized insulin receptors.

5. The solubilized insulin receptors may be stored in aliquots at −70° for a period of 6 months.

Procedure

1. Distribute aliquots (0.16 ml) of buffer A containing ^{125}I-labeled insulin (0.1 ng/ml) among microcentrifugation tubes (1.5 ml capacity). Add the Triton-solubilized extract (0.0 ml) of human placental membranes containing solubilized insulin receptors to one set of tubes (series A). In a paired set of tubes (series B), substitute a solution of Triton X-100 (1%, v : v in buffer C) for the placental extract. Incubate the tubes overnight at 4°.

2. Add serum samples (0.02 ml) to paired tubes (one from series A and one from series B). To titer the antiserum, employ varying dilutions of serum diluted in normal human serum. Use normal human serum as a control to assay for "nonimmune" precipitation. Extend the incubation for an additional 6–8 hr at 4°.

3. Add an appropriate volume (vide infra) of sheep anti-human γ-globulin containing Triton X-100 (0.1%, v : v) to each tube and continue the incubation for an additional 16–18 hr at 4°.

4. Sediment immune complexes by centrifugation (10,000 *g* for 5 min at 4°). Aspirate and discard the supernatant fluid. Excise the tips of the tubes and measure the radioactivity in the immunoprecipitate using a gamma counter.

5. Plot specifically immunoprecipitated ^{125}I-labeled insulin as a function of the dilution of antiserum (Fig. 1B).

Comments

1. This assay will detect anti-receptor antibodies which bind to the receptor in such a way that the antibodies do not inhibit binding of ^{125}I-labeled insulin to the receptor. The assay can also detect anti-receptor antibodies which do inhibit ^{125}I-labeled insulin binding. However, the property of the latter class of anti-receptor antibodies to inhibit ^{125}I-labeled insulin binding does interfere with this assay. To the extent that an anti-receptor antibody inhibits ^{125}I-labeled insulin binding to the receptor, the antibody may preferentially immunoprecipitate unoccupied receptors.

2. ^{125}I-labeled insulin which is immunoprecipitated in the absence of insulin receptors (series B) represents ^{125}I-insulin which has been immunoprecipitated by anti-insulin antibodies. In the presence of insulin receptors (series A), ^{125}I-labeled insulin may be immunoprecipitated either directly by anti-insulin antibodies or indirectly by anti-receptor antibodies.

3. It is possible to modify this technique to minimize the interference by anti-insulin antibodies. This is accomplished by a two step assay. In the first step, insulin receptors are immunoprecipitated with anti-receptor antiserum plus protein A (Pansorbin; Calbiochem). In the second step, the insulin receptors remaining in the supernatant may be measured by a technique utilizing precipitation with polyethylene glycol to separate bound from free ^{125}I-labeled insulin.[28,29]

4. The concentration of sheep anti-human γ-globulin must be chosen so as to give complete immunoprecipitation of the immunoglobulin contained in 0.02 ml of human serum.

Insulin-Like Bioactivity of Anti-Receptor Antibodies

Autoantibodies to the TSH receptor were the first recognized examples of an anti-receptor antibody.[1–3] These autoantibodies were identified by virtue of their bioactivity as "long-acting thyroid stimulators." In much the same fashion, autoantibodies to the insulin receptor mimic the bioactivity of insulin both *in vitro* and *in vivo*.[4–7] Despite the biological importance of the insulin-like bioactivity of anti-receptor antibodies, the technical complexity of this bioassay has interfered with the routine use of this method to assay for anti-receptor antibodies. Two other considerations also limit the utility of this assay method. First, hormonal effects may be mimicked by antibodies directed against components other than the receptor. For example, antibodies directed against intrinsic proteins of adipocyte plasma membranes may have insulinomimetic effects upon isolated fat cells.[30] Second, some anti-receptor antibodies may fail to mimic the bioactivity of the natural ligand. For example, Roth *et al.*[31] have obtained a monoclonal antibody which binds to the insulin receptor but fails to mimic insulin action.

[28] S. I. Taylor, *in* "Regulation of Target Cell Responsiveness" (K. W. McKerns, ed.), p. 133. Plenum, New York, 1984.

[29] P. Cuatrecasas, *Proc. Natl. Acad. Sci. U.S.A.* **69,** 318 (1982).

[30] D. J. Pillion and M. P. Czech, *J. Biol. Chem.* **253,** 3761 (1978).

[31] R. A. Roth, D. J. Cassell, K. Y. Wong, B. A. Maddux, and I. D. Goldfine, *Proc. Natl. Acad. Sci. U.S.A.* **79,** 7312 (1982).

Reagents

Bovine serum albumin: Supplied by Reheis Biochemical Department (Kankakee, IL). Must be tested and shown to be free of insulin-like bioactivity

Collagenase (type I). Supplied by Worthington Biochemical Corporation (Freehold, NJ). Must be tested and shown to succeed in producing dispersed adipocytes which retain insulin responsiveness

Stock solution I: In 300 ml of water, dissolve NaCl (35.04 g), KH_2PO_4 (2.73 g), and $MgSO_4$ (1.23 g). Separately dissolve $CaCl_2$ (0.55 g) in 100 ml of water. Combine the two solutions and adjust the total volume to 500 ml

Stock solution II: Dissolve $NaHCO_3$ (4.2 g) in 500 ml of water

Stock solution III: Dissolve HEPES (37.75 g) in 500 ml of water. Adjust to pH 7.4 with NaOH

Glucose (10 mg/ml): Make up fresh daily

Krebs–Ringer bicarbonate buffer (KRBHA): Combine 100 ml of water, 20 ml each of stock solutions I–III, and 1 ml of glucose solution. Dissolve bovine serum albumin (2 g). Adjust to pH 7.4 and adjust total volume to 200 ml. Make up fresh daily

[3-^{3}H]Glucose: Specific activity, 10–20 Ci/mmol. Supplied by New England Nuclear (Boston, MA)

Econofluor. Supplied by New England Nuclear (Boston, MA)

Procedure

1. Prepare isolated adipocytes by collagenase digestion of epididymal fat pads from 3 male rats (~80–120 g) according to the method of Rodbell.[32] Resuspend cells in 50 ml of KRBHA at 37°.

2. Place aliquots (0.1 ml) of either the serum to be tested or normal human serum into 12 × 75-mm plastic tubes. To titer the serum, employ varying dilutions of the serum to be tested (diluted in normal human serum). Aliquots (0.9 ml) of suspended adipocytes are added to each tube. (Take great care to stir the cells so as to maintain a uniform cell suspension.) Incubate the cells at 37° for 15 min.

3. Separate the cells from the medium by centrifugation. Aspirate and discard the infranatant medium. Wash the cells three times with KRBHA (1 ml) at 37°. Take great care not to aspirate the cells. In practice, it may be necessary to leave a small volume of medium in order to avoid aspirating cells.

4. Resuspend the cells in KRBHA (1 ml) at 37°. Transfer the resus-

[32] M. Rodbell, *J. Biol. Chem.* **239,** 375 (1964).

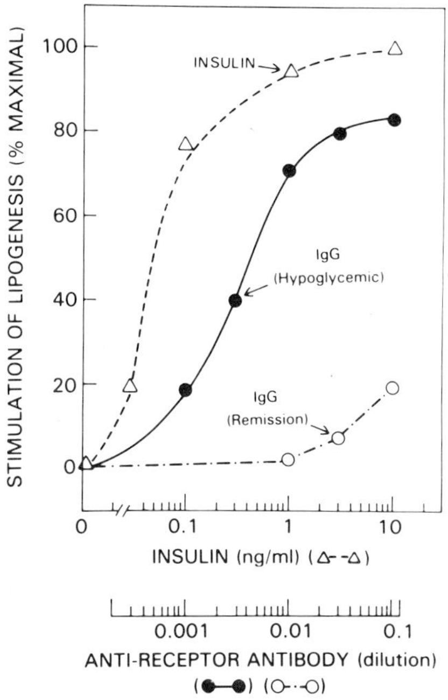

FIG. 2. Bioassay of insulin-like activity of anti-receptor antibody. Immunoglobulin G was purified from the same samples of serum that were utilized in Fig. 1. Immunoglobulin G obtained from the patient during the time she was hypoglycemic stimulated lipogenesis by isolated adipocytes. In contrast, this insulin-like bioactivity was markedly decreased at the time her clinical symptoms had remitted. For comparison, a bioassay curve for porcine insulin is also included.

pended cells to 20 ml plastic scintillation vials containing [3-^{3}H]glucose (0.1 μCi). Incubate at 37° for 2 hr with shaking.

5. Terminate the incubation by addition of scintillation fluid (Econofluor). Cap vials tightly and shake vigorously. Allow vials to incubate 4–16 hr at room temperature to complete the extraction of ^{3}H-labeled lipid into the organic phase. The "blank" is determined by addition of scintillation fluid to two vials at zero time. Quantitate the ^{3}H in the lipid using a scintillation counter. The results of a representative bioassay are shown in Fig. 2.

Comments

1. Ordinarily, we carry out a second bioassay in parallel in which adipocytes, preincubated with normal human serum, are subsequently incubated with varying concentrations of porcine insulin (0.03–10.0 ng/ml).

This control serves two functions: first, to verify that the adipocytes respond to insulin; second, to allow for quantitation of the bioactivity of anti-receptor antibody relative to the bioactivity of insulin.

2. It is important to use a scintillation fluid which is immiscible with water so that the labeled lipids will be extracted into the scintillation fluid which serves as an organic phase; the [3-^{3}H]glucose remains in the aqueous phase.[33] Scintillation fluids which accept aqueous solutions will not be satisfactory in this method.

3. We have employed the ability of insulin to stimulate lipogenesis in rat adipocytes as a bioassay for insulin-like activity. This particular bioassay has many advantages including simplicity and sensitivity. Other bioassays using either adipocytes[4] or other systems[13,14,34] have been employed successfully.

4. Some samples of serum or plasma may agglutinate adipocytes. This agglutination may be minimized by dilution of the sample. Of course, the need to dilute the sample decreases the sensitivity of the assay to detect anti-receptor antibodies.

5. A "nonspecific" effect of human immunoglobulin to stimulate lipogenesis in rat adipocytes has been described.[35–37] However, this "nonspecific" effect of immunoglobulin may be differentiated from the "specific" effect of antibodies to the insulin receptor:

a. *Difference in assay methodology*. In the original report by Kahn *et al.*,[4] a two-step assay technique was employed (similar to the one described above). Cells were first exposed to antisera. Then after the cells were washed, the *in vitro* bioassay was carried out. The wash procedure appeared to remove the majority of insulin-like factors in serum with the exception of those antibody molecules which were tightly bound to the cell surface. Using this method, Kahn *et al.*[4] did not detect insulin-like bioactivity in normal human serum or in serum from patients with a variety of diseases. The studies which have reported an insulin-like bioactivity of normal human immunoglobulin G have employed a one-step assay in which the cells were not washed prior to bioassay.

b. *Specificity of bioactivity*. In the case of the "nonspecific" bioactivity of human immunoglobulin G, the insulin-like activity is proportional to the concentration of immunoglobulin G.[35–37] In contrast, with anti-receptor antibodies, the bioactivity is not related simply to the immunoglobulin

[33] A. J. Moody, M. A. Stan, M. Stan, and J. Gliemann, *Horm. Metab. Res.* **6,** 12 (1974).

[34] Y. Le Marchand-Brustel, P. Gorden, J. Flier, C. R. Kahn, and P. Freychet, *Diabetologia* **14,** 311 (1978).

[35] M. A. Khokher, L. G. Coulston, and P. Dandona, *Diabetologia* **21,** 290 (1981).

[36] M. A. Khokher, P. Dandona, L. G. Coulston, and S. Janah, *Diabetes* **12,** 1068 (1981).

[37] M. A. Khokher and P. Dandona, *J. Clin. Endocrinol. Metab.* **56,** 393 (1983).

G concentration. Rather, the bioactivity is related (albeit in a complex way) to the concentration of anti-receptor antibody.[14] Thus, when patients with anti-receptor antibodies enter clinical remission, this is associated with a marked fall in the antireceptor antibody titer (e.g., see Fig. 1) as well as marked fall in the insulin-like activity of the immunoglobulin G fraction (e.g., see Fig. 2).

c. *Effect of enzymatic digestion of antibody molecules.* When anti-receptor antibodies are digested with pepsin, the insulin-like bioactivity is associated with the bivalent $F(ab)_2$ moiety; the Fc moiety is devoid of bioactivity.[4,38] In contrast, Khoker and Dandona[37] found the "nonspecific" bioactivity of human immunoglobulin G to be associated with the Fc rather than the $F(ab)_2$ fragment.

[38] C. R. Kahn, K. L. Baird, D. B. Jarrett, and J. S. Flier, *Proc. Natl. Acad. Sci. U.S.A.* **75,** 4209 (1978).

[52] Characterization of Antisera to Prolactin Receptors

By Paul A. Kelly, Masao Katoh, Jean Djiane, Louis-Marie Houdebine, and Isabelle Dusanter-Fourt

The action of prolactin, as is true for other peptide hormones, is mediated by a specific receptor located in membrane components of the cell. The PRL receptor has been well characterized and has been shown to respond to both a down- and an up-regulation by the hormone.[1–4] This receptor was partially purified[5] and an antibody against the partially purified receptor was obtained several years ago.[6] Antibodies to the PRL receptor have been shown to inhibit the capacity of prolactin to support the synthesis of casein and the uptake of α-isoaminobutyric acid by rabbit mammary explants in culture[7] and to attenuate the action of prolactin on

[1] J. Djiane and P. Durand, *Nature (London)* **266,** 641 (1977).
[2] P. A. Kelly, L. Ferland, and F. Labrie, *in* "Progress on Prolactin Physiology and Pathology" (C. Robyn and M. Harter, eds.), p. 59. Elsevier/North-Holland Biomedical Press, Amsterdam, 1978.
[3] J. Djiane, H. Clauser, and P. A. Kelly, *Biochem. Biophys. Res. Commun.* **90,** 1371 (1979).
[4] J. Djiane, P. Durand, and P. A. Kelly, *Endocrinology* **100,** 1348 (1980).
[5] R. P. C. Shiu and H. G. Friesen, *J. Biol. Chem.* **249,** 7902 (1974).
[6] R. P. C. Shiu and H. G. Friesen, *Biochem. J.* **157,** 619 (1976).
[7] R. P. C. Shiu and H. G. Friesen, *Science* **192,** 259 (1976).

ISBN 0-12-182009-2

mammary gland and ovary when injected into rats.[8] Recently, we have shown that anti-prolactin receptor serum is able to both inhibit and mimic the action of prolactin in rabbit mammary explants[9] and in rat liver.[10]

The present studies were designed to improve the techniques of obtaining purified prolactin receptors and to immunize several species of animals with the partially purified receptor in order to increase the yield of antisera and to better characterize the specificity of these antisera in different target organs.

Production of Anti-Receptor Antibodies

Prolactin receptors were partially purified from crude microsomal fractions of lactating mammary glands, using affinity chromatography techniques which have been described elsewhere.[5–11] Such procedures resulted in prolactin receptor preparations representing a 1000- to 5000-fold increase in purity over mammary gland homogenates.

These partially purified receptor preparations were injected at monthly intervals into male sheep, goats and guinea pigs, at a concentration of 50 μg of antigen per injection in Freund's complete adjuvant. Animals were bled at seven to 10 days after the booster immunization.[9] Antibodies began to become measurable (see below) after the second immunization and reached a maximum titer after 3 to 6 immunizations.

Prolactin Receptor Assay

For the assay of solubilized extracts of mammary gland samples at various stages of purification, ^{125}I-labeled human GH or ovine PRL were used as labeled hormones. It has previously been shown that hGH,[11a] which is also a lactogenic hormone, binds with the same specificity and affinity as labeled PRL, and does not form micelles in the presence of Triton, as is the case for the labeled prolactin molecule.[5] For the assay of

[8] H. G. Bohnet, R. P. C. Shiu, D. Grinwich, and H. G. Friesen, *Endocrinology* **102,** 1657 (1978).

[9] J. Djiane, L. M. Houdebine, and P. A. Kelly, *Proc. Natl. Acad. Sci. U.S.A.* **78,** 7445 (1981).

[10] A. A. M. Rosa, J. Djiane, L. M. Houdebine, and P. A. Kelly, *Biochem. Biophys. Res. Commun.* **106,** 243 (1982).

[11] M. Katoh, J. Djiane, G. Leblanc, and P. A. Kelly, *Mol. Cell. Endocrinol.* **34,** 191 (1984).

[11a] Abbreviations: hGH, human growth hormone; oPRL, ovine prolactin; bGH, bovine growth hormone; hCG, human chorionic gonadotropin; PEG, polyethylene glycol; IM_{50}, 50% inhibition of binding; EGF, epidermal growth factor; LHRH, luteinizing hormone releasing hormone.

nonsolubilized samples and for the characterization of antisera against the receptor, ^{125}I-labeled ovine PRL was used to eliminate the possibility of a cross-reaction of labeled hGH with any hGH antisera which might have been generated in low concentration due to presence of minute amounts of hGH in the purified receptor preparation, as a result of a possible leaching from the affinity column.[5] A modification of the technique utilizing low concentrations of chloramine-T (500 ng) and 5 μg of hGH or oPRL was employed.[12] Similarly, insulin (Schwarz/Mann, 25 U/mg), bGH (USDA bGH-I-1 for iodination), and hCG (CR121, 13,450 U/mg) were iodinated. The iodinated hormones were purified on a column of Sephadex G-75 (0.9 $\times$ 60 cm) and the tubes of radioactivity comprising the protein peak were diluted in assay buffer (25 m*M* Tris–HCl, pH 7.4, 10 m*M* $MgCl_2$, and 0.1% bovine serum albumin). Specific activities as calculated by isotope recovery for hGH, oPRL, bGH, and hCG ranged between 45 and 100 μCi/μg and that of insulin was 120 μCi/μg. The integrity of the iodinated hGH and oPRL was verified using a laboratory standard receptor preparation. Using 300 μg of standard receptor protein, greater than 30% of the iodinated hormones were specifically bound. Using excess receptor greater than 80% of the labeled hormones could be specifically bound.

For the binding assay of solubilized fractions, each tube contained approximately 100,000 cpm of [^{125}I]hGH, in the presence or absence of excess (1 μg) of unlabeled oPRL in a final volume of 0.5 ml. Tubes were incubated 14–16 hr at 20°. Solubilized binding activity was precipitated by adding 0.5 ml of 0.1% bovine γ-globulin (in 0.1 *M* phosphate buffer, pH 7.4) and 1 ml of a solution of 24% polyethylene glycol (PEG 6000) resulting in a final PEG concentration of 12%. Tubes were vortexed and centrifuged at 2500 *g* for 15 min. For the assay of particulate fractions, ^{125}I-labeled ovine PRL was used and bound and free hormones were separated by a centrifugation at 2000 *g* for 15 min.[12] Tubes were counted in a LKB gamma spectrometer with a counting efficiency of 50%.

Receptor assays were calculated using a program developed in this laboratory. Calculations were performed using a Hewlett-Packard 9845B desk-top calculator. The half-maximal inhibition (IM_{50}) values of inhibition curves were calculated by computer curve fitting on an HP9845B.

Inhibition of Binding of Prolactin by Antireceptor Serum

Sera of animals injected with partially purified receptor preparations (3 to 18% pure) were assayed for their capacity to inhibit the binding of ^{125}I-

[12] P. A. Kelly, G. Leblanc, and J. Djiane, *Endocrinology* **104,** 1631 (1979).

labeled ovine prolactin to receptors in rabbit mammary gland membranes. As shown in Fig. 1, sera from sheep, goats, and guinea pigs were capable of inhibiting the binding of prolactin to its receptors. Significant inhibition of binding was observed at a dilution of 1 : 10,000, with the sheep antibody being the most potent. The half-maximal inhibition (IM_{50}) values for the sheep, guinea pig, and goat antisera represented a dilution of 1 : 5700, 1 : 2400, and 1 : 1100, respectively. Sera of nonimmunized animals did not significantly alter the formation of the prolactin–receptor complex. That increased concentrations of PRL receptor required more antiserum is demonstrated by the linearity of the regression curve obtained between IM_{50} values and PRL receptor concentrations (data not shown).

The ability of the sheep antireceptor serum to inhibit the binding of labeled oPRL, bGH, and insulin in rabbit and rat liver and of labeled hCG in rat ovary was compared with PRL binding in mammary gland. As can be seen in Fig. 2, PRL binding was inhibited in all tissues whereas the

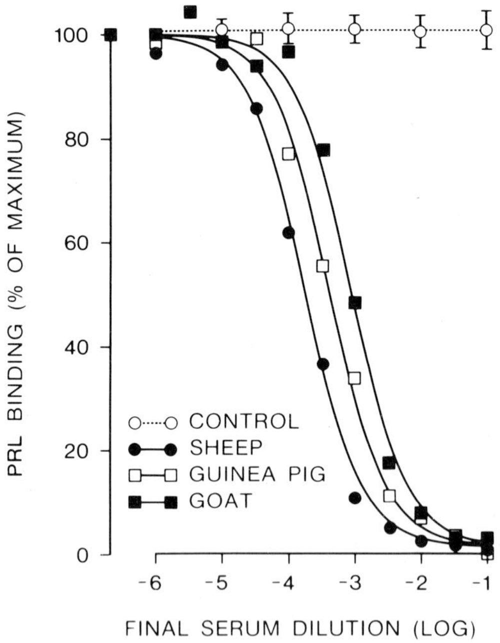

FIG. 1. Action of anti-prolactin receptor sera on the binding of ^{125}I-labeled ovine prolactin to mammary membranes. About 100,000 cpm of the hormone was incubated for 16 hr at 20° with rabbit mammary membranes (300 μg protein per incubate), in the presence of various concentrations of control sera or anti-prolactin receptor serum from a sheep, guinea pig, and goat. Binding values are expressed as a percentage of values in tubes containing no antiserum.

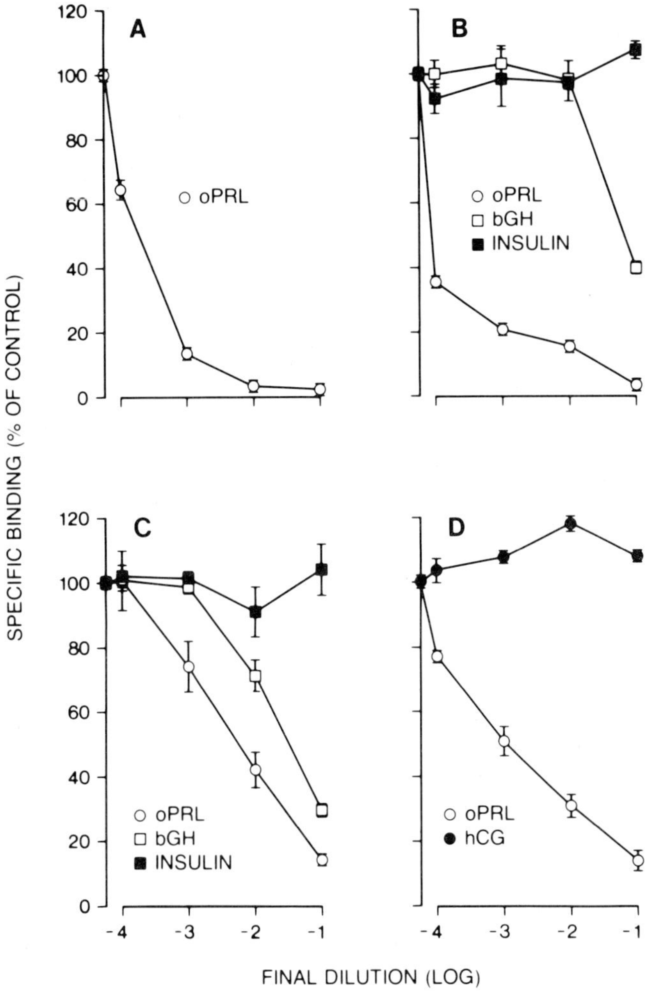

FIG. 2. Comparison of bGH and insulin binding in rabbit and rat liver and hCG binding in rat ovary compared with oPRL binding. A depicts rabbit mammary gland, B, rabbit liver, C, rat liver, and D, rat ovary. Control (100%) binding (oPRL) was 20,430 ± 240 cpm for rabbit mammary, 14,020 ± 1090 cpm for rabbit liver, 21,200 ± 1650 cpm for rat liver, and 6330 ± 400 cpm for rat ovary. Control binding for bGH and insulin was 29,460 ± 580 and 6030 ± 1830 cpm for rabbit liver, respectively and 6615 ± 190 and 4350 ± 140 cpm, respectively for rat liver. Control binding was 6330 ± 110 cpm for hCG in rat ovary. These data are the means ± SEM of 3 separate experiments.

binding of insulin and hCG was virtually unaffected. There was a slight effect on bGH binding, although the difference was over 1000-fold in rabbit liver and approximately 10-fold in rat liver.

The specificity of the sheep antiserum in 8 different rat tissues was compared with rabbit mammary tissues (data not shown). The greatest specificity was observed for rabbit mammary tissue which was considered to have the highest PRL receptor concentration. Maximal inhibition was observed at a dilution of 1 : 300. The slopes of the inhibition curves for the rat tissues were slightly less than for rabbit mammary gland, suggesting homology between the receptor proteins in the various tissues, with, however, some differences in the antigen–antibody interactions. IM_{50} values varied from tissue to tissue, although these changes cannot be attributed simply to the difference of tissue specificity since the PRL receptor concentration was not determined for each tissue.

The specificity of the sheep antiserum was further examined in rabbit ovary and adrenal and mammary tissue from a lactating pig. Figure 3 shows a similar inhibition curve for rabbit mammary tissue as has been

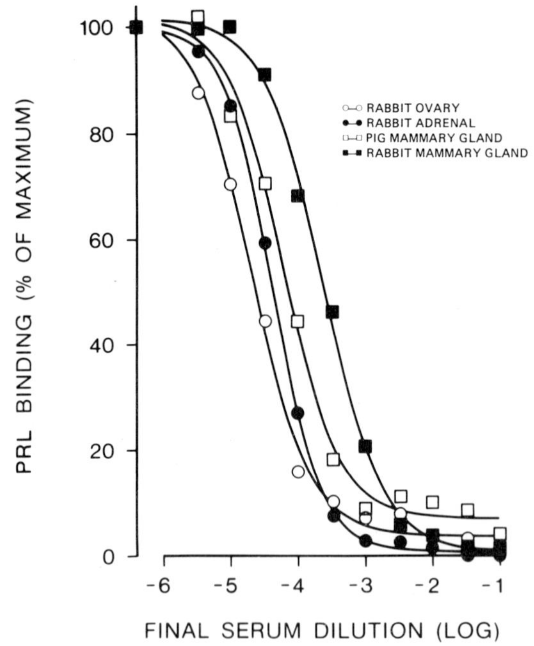

FIG. 3. Specificity of sheep anti-prolactin receptor serum on the inhibition of binding ^{125}I-labeled ovine prolactin to receptors in rabbit ovary, adrenal, and pig mammary gland compared with rabbit mammary gland. Conditions were as described in the legend to Fig. 1. Three hundred micrograms of microsomal protein per incubation was employed. Control (100%) values were 7750, 19,690, 8110, and 29,250 cpm for rabbit ovary and adrenal, pig and rabbit mammary gland, respectively.

observed previously (IM_{50} = 1 : 5200). Although rabbit ovary and adrenal and pig mammary tissue showed lower IM_{50} values (1 : 49,000, 1 : 23,000, and 1 : 16,500, respectively), these tissues usually contain lower PRL receptor concentration (30–100 fmol/mg microsomal protein) than rabbit mammary gland (200–500 fmol/mg microsomal protein).

Effect of Mono- and Bivalent Fragments of Anti-Receptor Serum

Bivalent fragments, $F(ab')_2$, of anti-prolactin receptor antibodies were prepared by pepsin cleavage, whereas the monovalent fragment Fab′ was obtained after a reduction with dithiothreitol of the $F(ab')_2$ fragment. When assayed for their ability to inhibit [^{125}I]oPRL binding to its receptor in rabbit mammary gland, whole antireceptor serum, immunoglobulin bivalent $F(ab')_2$ and monovalent Fab′ fragments had similar potencies. Half maximal inhibition was reached with 5 to 10 μg/ml for sheep antireceptor antibodies, with similar but slightly higher concentrations being effective for goat antireceptor antibodies.[13] In addition to binding to receptors, immunoglobulins as well as their $F(ab')_2$ and Fab′ fragments all induce a 50% down-regulation of prolactin receptors in mammary explants.[13]

Using a similar culture procedure, the effect of prolactin and antiprolactin receptor antibody fragments on casein and DNA synthesis was examined. As can be seen in Fig. 4, prolactin concentrations greater than 5 μg/ml had a significant effect on DNA synthesis and casein production. Maximal casein production was observed at prolactin concentrations of 100 to 1000 μg/ml, when either immunoprecipitation technique[14] or a specific radioimmunoassay[13] was used. At higher concentrations, a desensitization was observed. It was only at these higher concentrations of prolactin that an occupation of free and a down regulation of total receptors was apparent.[14]

Figure 5 demonstrates the effect of whole anti-receptor serum, and the bivalent and monovalent fragments on casein production in rabbit mammary explants. While serum was about 50–60% as effective as the optimum concentration of prolactin, the $F(ab')_2$ fragment was about equipotent, however, the Fab′ fragment was completely devoid of activity. Only bivalent $F(ab')_2$ fragments were able to stimulate DNA synthesis, as was whole anti-receptor serum, whereas, the monovalent Fab′ fragment had no effect (data not shown).[13]

It has been previously shown that either prolactin or anti-receptor serum incubated with prolactin receptors results in the production of a soluble relay which is able to stimulate the transcription of beta casein

[13] I. Dusanter-Fourt, J. Djiane, P. A. Kelly, L. M. Houdebine, and B. Teyssot, *Endocrinology* **114,** 1021 (1984).

[14] J. Djiane, L. M. Houdebine, and P. A. Kelly, *Endocrinology* **110,** 791 (1981).

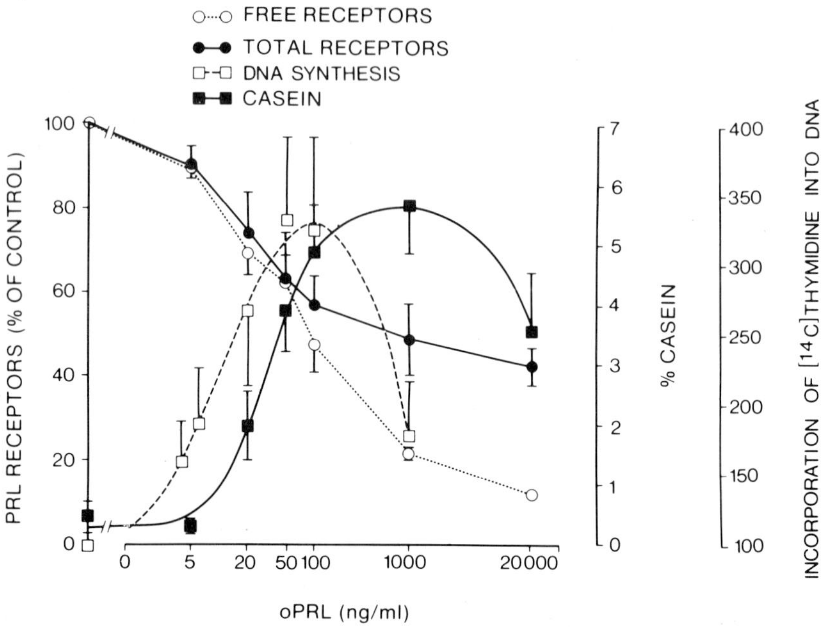

FIG. 4. Effect of increasing concentrations of oPRL on free and total PRL receptors and DNA and casein synthesis in explant cultures of rabbit mammary gland. Explants were cultured in the presence of insulin (1 μg/ml) and cortisol (500 μg/ml) for 24 hr. Results are the mean ± SEM of 7 independent cultures.

genes in isolated nuclei.[15,16] More recently, we have reported that membrane supernatants obtained with $F(ab')_2$ fragments were able to markedly stimulate the transcription of both α S_1 and β casein genes. In contrast, monovalent fragments were never able to induce any mediator capable of stimulating casein gene transcription.[13]

Relevance of Observations

Partially purified receptor preparations are of sufficient purity to make antibody production feasible. Whole antisera as well as bivalent $F(ab')_2$ and monovalent Fab′ fragments produced in sheep, goats, and guinea pigs were capable of inhibiting the binding of prolactin to rabbit mammary gland membranes. The inhibition of binding by the antireceptor was specific for PRL or lactogenic hormones. In addition, these antisera were shown to inhibit the binding of prolactin to a number of tissues containing prolactin receptors in rabbits, rats, and even in human breast cancer

[15] B. Teyssot, L. M. Houdebine, and J. Djiane, *Proc. Natl. Acad. Sci. U.S.A.* **78,** 6729 (1981).

[16] B. Teyssot, J. Djiane, P. A. Kelly, and L. M. Houdebine, *Biol. Cell.* **43,** 81 (1982).

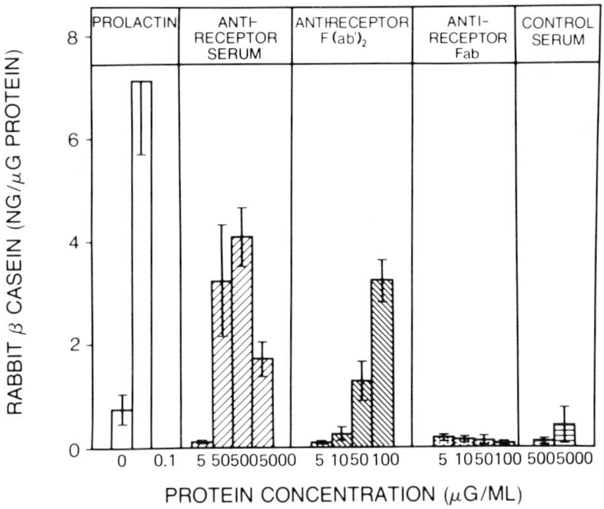

FIG. 5. Effect of prolactin, whole receptor serum, and antireceptor $F(ab')_2$ and Fab' fragments on rabbit beta casein content (measured by radioimmunoassay) in mammary glands cultured for 24 hr in the presence of insulin (1 μg/ml) and cortisol (500 μg/ml). Results are the mean ± SEM of 4 independent cultures.

biopsies (data not shown). These observations suggest some homology of the receptor molecule between species.

The anti-prolactin receptor antibody is a potent inhibitor of at least one prolactin action on the mammary cell: the initiation of casein synthesis. This fact is in good agreement with earlier observations of Shiu and Friesen who reorted inhibitory actions of anti-prolactin receptor on the mammary gland[5] and on ovary.[8] We have recently shown that at low concentrations, the antibody can also mimic prolactin actions on the mammary gland, namely the production of casein mRNA and casein synthesis.[9] Similar dose-dependent inhibitory as well as stimulatory effects of the antisera could be seen on prolactin receptor levels in rat liver cells in suspension culture.[10]

Stimulatory actions have been observed using anti-insulin receptors found in serum of some patients or obtained from purified receptors which can mimic insulin action on the uptake and oxidation of glucose.[17–19] The effects of the anti-prolactin receptor serum on the action of other tissues, such as the testis, ovary, and adrenal, is yet to be elucidated.

The studies with mono and bivalent fragments demonstrate that the whole serum as well as the $F(ab')_2$ fragments possess prolactin-like activ-

[17] J. C. Flier, C. R. Kahn, J. Roth, and R. S. Bar, *Science* **190,** 63 (1975).
[18] S. Jacobs, K. J. Chang, and P. Cuatrecasas, *Science* **200,** 1283 (1978).
[19] D. Baldwin, S. Terris, and D. F. Steiner, *J. Biol. Chem.* **255,** 4028 (1980).

ity, which is active in a dose dependent fashion, with desensitization being observed at high concentrations. It is interesting to note that anti-prolactin receptor antibodies mimic prolactin action on both casein and DNA synthesis, suggesting these two effects may involve a common mechanism, at least in their early steps.

None of the prolactin-like actions could be mediated by the monovalent Fab′ fragment, although these monovalent fragments are as efficient as bivalent fragments to inhibit prolactin binding to its receptor. Clearly, these studies indicate that the anti-receptor antibodies require their bivalency to activate casein and DNA synthesis. This suggests that the prolactin receptor antibodies contribute to cross-link or to microaggregate prolactin receptors when they mimic prolactin's actions. These models are in agreement with that proposed for insulin receptor antibodies.[20] The importance of receptor cross-linking or microaggregation is emphasized by other studies involving EGF,[21] acetylcholine[22] and LHRH.[23]

In conclusion, antisera produced from partially purified prolactin receptors are able to inhibit binding of labeled hormone to all tissues thus far tested which contain prolactin receptors. It appears that microaggregation is a key step for the prolactin like activity of prolactin receptor antibodies to be observed. It remains to ascertain if prolactin itself is a bivalent ligand, or if receptor microaggregation is induced by increased mobility of receptors in the plasma membrane, triggering the formation of chemical links between receptors or by enhancing the affinity of receptors for each other.

Acknowledgments

The authors wish to express their gratitude to the National Hormone and Pituitary Program, National Institute of Arthritis, Metabolism and Digestive Diseases for providing human growth hormone, ovine prolactin, ovine luteinizing hormone, and human chorionic gonadotropin and to the United States Department of Agriculture for iodination grade bovine growth hormone. These studies were supported in part by grants from the National Cancer Institute (Canada), Medical Research Council of Canada, the Centre National de la Recherche Scientifique, the Institut National pour la Santé et la Recherche Médicale, and the Délégation à la Recherche Scientifique et Technique.

[20] C. R. Kahn, K. L. Baird, D. B. Jarrett, and J. S. Flier, *Proc. Natl. Acad. Sci. U.S.A.* **75,** 4209 (1978).

[21] Y. Schechter, L. Hernaez, J. Schlessinger, and P. Cuatrecasas, *Nature* (*London*) **278,** 835 (1979).

[22] D. B. Drachman, C. W. Angus, R. N. Adams, J. D. Michelson, and G. J. Hoffman, *N. Engl. J. Med.* **298,** 1116 (1978).

[23] P. M. Conn, D. C. Rogers, J. M. Stewart, J. Niedel, and T. Sheffield, *Nature* (*London*) **296,** 653 (1982).

[53] Assays of Thyroid-Stimulating Antibody

By J. M. McKENZIE and M. ZAKARIJA

The thyroid-stimulating antibody (TSAb) of Graves' disease is a term generally agreed to mean the immunoglobulin G (IgG) that stimulates the thyroid gland in this syndrome and thus usually causes hyperthyroidism. There are, however, many methods for its measurement, and frequently other names have been given, resulting in confusion both in terms of what is being measured and in the understanding of the clinical relevance of resultant data. The table summarizes many of the names, methods, and acronyms. As shown, there are two major methodological approaches to the assay of TSAb: (1) some index of thyroid stimulation, usually *in vitro*, and we shall use TSAb to indicate this activity; (2) indirect recognition by assessment of the inhibition of binding of radiolabeled thyrotropin (TSH) to a preparation of its receptor, i.e., TSH-binding inhibition or TBI. To generalize, TSAb assays are laborious and logistically nonappealing but specific; TBI assays are convenient, comparatively easy to establish and maintain, but relatively nonspecific. That is, while we may accept that TSAb indeed inhibits TSH binding to the receptor, it is clear that other thyroid-directed antibodies do so also without stimulating the thyroid. It is in this summary of comparison of TSAb and TBI methods that there lies potential both for misinterpretation of data acquired by testing patients' sera by one or other basic procedure, and opportunities to understand better the mechanisms of antibody–receptor interaction.

TBI Procedures

Although the concept was illustrated first by others,[1,2] to Smith and Hall goes the credit of describing the technique that has been most widely used.[3] The details of the procedures for the radioiodination of TSH, its purification, the preparation of a suspension of receptor, and the assay itself are listed in the following sections. Various modifications of this technique have been developed and will be considered separately.

[1] S. Q. Mehdi, S. S. Nussey, C. P. Gibbons, and D. J. ElKabir, *Biochem. Soc. Trans.* **1,** 1005 (1973).

[2] S. W. Manley, J. R. Bourke, and R. W. Hawker, *J. Endocrinol.* **61,** 437 (1974).

[3] B. R. Smith and R. Hall, *Lancet* **2,** 427 (1974).

ISBN 0-12-182009-2

METHODS OF ASSAY OF ANTIBODIES TO THE TSH-RECEPTOR

Receptor-modulation procedures
- Inhibition of binding of $[^{125}I]$TSH to human thyroid membranes (particulate)[a–c]
  - guinea pig fat cell membranes (particulate)[d]
  - human thyroid membranes (solubilized)[e–g]
  - porcine thyroid membranes (particulate)[h]
  - porcine thyroid membranes (solubilized)[g]
- Inhibition of binding of LATS to human thyroid membranes (particulate)[i]

Thyroid-stimulation procedures
- Discharge in mice of ^{125}I-labeled thyroid components; increase in blood ^{125}I[j]
- ^{125}I release from mouse thyroid *in vitro*[k]
- T_4 release from mouse thyroid *in vitro*[l]
- Colloid droplet formation in human thyroid slices[m]
- Increase in cyclic AMP in human thyroid slices[m,n]
- Increase in cyclic AMP in human thyroid cells[o–q]
- Increase in cyclic AMP in functioning rat thyroid (FRT_L) cells[r]
- Stimulation of adenylate cyclase in human thyroid membranes[s–u]
- T_3 release from human thyroid slices[v]
- Free T_3 release from porcine thyroid slices[w]
- Cytochemical changes in guinea pig thyroid segments[x] or sections[y]

The assays are known by many names and resulting acronyms including long-acting thyroid stimulator, LATS[j]; LATS-protector, LATS-P[i]; human thyroid adenyl cyclase stimulator, HTACS; thyroid-stimulating immunoglobulin, TSI[a]; thyrotropin-binding-inhibiting immunoglobulin, TBII [K. Endo, K. Kasagi, J. Konishi, K. Ikekubo, T. Okuno, Y. Takeda, T. Mori, and K. Torizuka, *J. Clin. Endocrinol. Metab.* **46,** 734 (1978)]; thyrotropin-displacing immunoglobulin, TDI[b]; thyroid-stimulating antibody, TSAb[n]

[a] B. R. Smith and R. Hall, *Lancet* **2,** 427 (1974).

[b] J. O'Donnell, K. Trokoudes, J. Silverberg, V. V. Row, and R. Volpé, *J. Clin. Endocrinol. Metab.* **46,** 770 (1978).

[c] M. Borges, J. C. Ingbar, K. Endo, S. Amir, H. Uchimura, S. Nagataki, and S. H. Ingbar, *J. Clin. Endocrinol. Metab.* **54,** 552 (1982).

[d] M. Kishihara, Y. Nakao, Y. Baba, S. Matsukura, K. Kuma, and T. Fujita, *J. Clin. Endocrinol. Metab.* **49,** 706 (1979).

[e] V. B. Petersen, B. R. Smith, P. J. D. Dawes, and R. Hall, *FEBS Lett.* **83,** 63 (1977).

[f] P. Kotulla and H. Schleusener, *J. Endocrinol. Invest.* **4,** 155 (1981).

[g] G. Shewring and B. R. Smith, *Clin. Endocrinol.* **17,** 409 (1982).

[h] B. R. Smith and R. Hall, this series, Vol. 74 [27].

[i] D. D. Adams and T. H. Kennedy, *J. Clin. Endocrinol. Metab.* **27,** 173 (1967).

[j] J. M. McKenzie, *Endocrinology* **62,** 865 (1958).

[k] J. Ensor, P. Kendall-Taylor, D. S. Munro, and B. R. Smith, *J. Endocrinol.* **49,** 487 (1971).

[l] M. L. Maayan, E. M. Volpert, and F. P. Dawes, *J. Endocrinol. Invest.* **1,** 299 (1978).

[m] T. Onaya, M. Kotani, T. Yamada, and Y. Ochi, *J. Clin. Endocrinol. Metab.* **36,** 859 (1973).

[n] M. Zakarija, J. M. McKenzie, and K. Banovac, *Ann. Intern. Med.* **93,** 28 (1980).

[o] R. S. Toccafondi, S. Aterini, M. A. Medici, C. M. Rotella, A. Tanini, and R. Zonefrati, *Clin. Exp. Immunol.* **40,** 532 (1980).

Iodination and Purification of [^{125}I]bTSH

Iodination

Materials

0.1 *M* potassium phosphate, pH 7.4 (5 ml, at 4°)
0.1 *M* potassium phosphate, pH 7.4 with 0.5% bovine serum albumin (BSA) (≃100 ml)
0.1 *M* potassium phosphate, pH 7.4 with 0.02% sodium azide (100 ml)
Highly purified bTSH (provided by Dr. J. Pierce), 4.5 μg/5 μl of water stored at −70°
$Na^{125}I$ (low pH) (NEZ-033L)
Lactoperoxidase (Calbiochem 45.7 IU/mg), 100 μg/20 μl 0.2 *M* Na acetate, pH 5.6, stored at −70°
Hydrogen peroxide, 30%
Sephadex G-25 (medium) column (1.0 × 25 cm)

1. Prepare fresh phosphate buffer.
2. Sephadex G-25 column is preequilibrated with phosphate/BSA at room temperature, several hours or 1 day before use.
3. Thaw bTSH and lactoperoxidase and keep them on ice.
4. Add 313 μl of phosphate buffer to the stock of lactoperoxidase (=3 μg/10 μl).
5. Prepare 30 ng H_2O_2/10 μl of phosphate buffer: i.e., successively dilute 0.05 ml H_2O_2 (30%) with water to 1 : 100, then 1 : 10,000 and finally 1 : 100,000.

---

[p] K. Kasagi, J. Konishi, Y. Iida, K. Ikebudo, T. Mori, K. Kuma, and K. Torizuka, *J. Clin. Endocrinol. Metab.* **54,** 108 (1982).
[q] B. Rapoport, S. Filetti, N. Takai, P. Seto, and G. Halverson, *Metab., Clin. Invest.* **31,** 1159 (1982).
[r] P. Vitti, W. A. Valente, F. S. Ambesi-Impiombato, G. Fenzi, A. Pinchera, and L. D. Kohn, *J. Endocrinol. Invest.* **5,** 179 (1982).
[s] J. Orgiazzi, D. E. Williams, I. J. Chopra, and D. H. Solomon, *J. Clin. Endocrinol. Metab.* **42,** 341 (1976).
[t] K. Bech and S. N. Madsen, *Clin. Endocrinol.* (*Oxford*) **11,** 47 (1979).
[u] R. E. Kleinman, L. E. Braverman, A. G. Vagenakis, R. W. Butcher, and R. B. Clark, *J. Lab. Clin. Med.* **95,** 581 (1980).
[v] I. Takata, Y. Suzuki, K. Saida, and T. Sato, *Acta Endocrinol.* (*Copenhagen*) **94,** 46 (1980).
[w] S. Atkinson and P. Kendall-Taylor, *J. Clin. Endocrinol. Metab.* **53,** 1263 (1981).
[x] L. Bitensky, J. Alaghband-Zadeh, and J. Chayen, *Clin. Endocrinol.* (*Oxford*) **3,** 363 (1974).
[y] J. Chayen, D. M. Gilbert, W. R. Robertson, and L. Bitensky, *J. Immunoassay* **1,** 1 (1980).

6. Add appropriate volume of phosphate buffer to $Na^{125}I$ and adjust to 1.5 mCi/10 μl.

7. Incubation mixture is bTSH, 4.5 μg in 5 μl; 10 μl phosphate buffer; lactoperoxidase, 3 μg in 10 μl; $Na^{125}I$, 1.5 mCi in 10 μl; H_2O_2, 30 ng in 10 μl.

8. Incubate at room temperature for 10 min; shake the tube gently.

9. Add 0.4 ml of cold phosphate buffer.

10. Apply the radioactive mixture onto Sephadex G-25 column to separate [^{125}I]TSH from free radioiodide; with 1 ml per tube and elution rate of 30 ml per hr [^{125}I]TSH peak occurs between tubes 8 and 10 and free ^{125}I between tubes 19 and 21.

11. To measure radioactivity place 0.02 ml from each tube into 0.5 ml of distilled water and count for 0.1 min.

12. Take the highest (one or two) radioactive tubes of the first peak and aliquot for storage; store at −70° or continue with the next step of purification.

13. To calculate specific activity (SA):
   a. Add 0.02 ml iodination mixture from Step 9 to 10 ml Tris/BSA (see below); mix well.
   b. Add 0.05 ml of the above to 0.05 ml of Tris/BSA in duplicate or triplicate; this is total count.
   c. Add 0.05 ml from Step a to 1 ml Tris/BSA in duplicate or triplicate; add 3 ml 10% TCA, vortex, and leave at room temperature for 10 min.
   d. Dilute with 7 ml distilled water, centrifuge at 3000 rpm for 10 min and discard the supernatant; these are precipitated counts.
   e. Count all 4 or 6 tubes and calculate SA as follows:

$$(\text{Precipitated counts} \times 100)/\text{total count} = \%\ \text{incorporation}$$
$$(\%\ \text{incorporation} \times 1500\ \mu\text{Ci})/100 : 4.5\ \mu\text{g} = \mu\text{Ci}/\mu\text{g TSH}$$
$$\text{SA is usually } 100\text{–}140\ \mu\text{Ci}/\mu\text{g}$$

14. After free ^{125}I peak is eluted (collect total of 30 tubes), buffer is changed to phosphate/sodium azide and the column thoroughly washed.

Comments

1. Activity of lactoperoxidase is different from batch to batch.

2. Optimal pH of lactoperoxidase is acidic. With standard $Na^{125}I$ (NEZ-033H), lactoperoxidase is diluted with 0.2 *M* Na acetate pH 5.6 and 10 μl is taken. When low pH $Na^{125}I$ is used, lactoperoxidase is diluted with phosphate buffer not to iodinate bTSH excessively.

3. Tubes 7–11 should be kept on ice as they come off fraction collector.

4. Described method for SA calculation overestimates incorporation of ^{125}I into TSH because all the losses of radioactivity during iodination (adsorption to tubes, pipet tips, etc.) are not measured.

Purification of [^{125}I]bTSH

Materials

[^{125}I]bTSH 0.5 ml (see above)
Bovine crude membranes 5-ml (1 g-Eq/ml) (see below)
20 m*M* Tris–HCl, pH 7.4 (Tris), 500 ml at 4°
20 m*M* Tris–HCl, pH 7.4 with 0.5% BSA (Tris/BSA), 200–250 ml at 4°
20 m*M* Tris–HCl, pH 7.4 with 0.02% sodium azide, 200–250 ml at 4°
2 *M* NaCl in Tris–HCl, pH 7.4, 10 ml at 4°
Sephadex G-100 column (1.2 × 50 cm) at 4°

1. Sephadex G-100 column is preequilibrated with Tris/BSA at 4°.
2. Mix [^{125}I]bTSH (0.5 ml) and bovine crude membranes (5 ml) in an ultracentrifuge tube, and incubate at 37° for 40 min with shaking.
3. Cool the tube on ice and centrifuge at 12,000 *g* for 20 min at 4°.
4. Discard the supernatant.
5. Add 5 ml of cold Tris/BSA and resuspend the pellet.
6. Recentrifuge the tube under the same conditions.
7. Discard the supernatant.
8. Add 2 ml of cold 2 *M* NaCl in Tris and resuspend the pellet.
9. Incubate for 1 hr in ice with occasional shaking of the tube.
10. Centrifuge at 100,000 *g* for 1 hr at 4°.
11. Fractionate the supernatant on the Sephadex G-100 column; with 1 ml per tube and elution rate of 12 ml/hr [^{125}I]TSH peak occurs around tube 40.
12. Count tubes 20–80 for 0.1 min; before and after counting keep tubes 32–47 on ice.
13. Pool the contents of 8 to 10 tubes with highest radioactivity and store 0.5- to 1-ml aliquots at −70°.
14. After tube 80 change the buffer to Tris/sodium azide and wash the column thoroughly.

Comments

1. Aggregated or damaged [^{125}I]TSH is eluted between tubes 25 and 30 as a small peak or "hump."

2. Small free ^{125}I content is eluted between tubes 60 and 70.

Preparation of "Crude Thyroid Plasma Membranes"

1. Place frozen glands, that were trimmed of connective tissue and fat, preweighed and stored at −70°, in cold Tris.

2. When thawed, prepare slices with Stadie–Riggs microtome.

3. Keep slices in 200 ml cold Tris.

4. Mince slices in a small beaker on ice with a minimum amount of buffer (final product looks like strawberry jam).

5. Homogenize the mince gently in a Dounce homogenizer or with a Polytron unit (10 sec, setting 4); use 40 ml cold Tris for 6 g equivalent of tissue.

6. Pass the homogenate through 4 layers of gauze.

7. Centrifuge the total filtered homogenate at 600 *g* for 10 min at 0–4° and recentrifuge the supernatant at 10,000 *g* for 20 min at 0–4°; discard supernatant.

8. Resuspend the pellet, recentrifuge at 11,000 *g* for 20 min, and again discard supernatant; repeat the wash once more.

9. Transfer the pellet with buffer to a graduated cylinder and adjust the volume so that 1 ml = 1.5 g equivalents; bovine thyroid membranes, used for purification of [^{125}I]TSH, should be diluted to 1 g equivalent per ml.

10. Aliquots (0.5 ml) of human thyroid and 5-ml aliquots of bovine thyroid membranes are stored at −70°.

For each batch, membrane protein is measured except for membranes used for [^{125}I]TSH purification. In the TBI assay 20–30 μg membrane protein per tube usually results in 20–25% specific binding. Higher concentration of membranes, i.e., higher binding, decreases the sensitivity of the assay.

Assay Using IgG from Graves' Disease Patients

Total binding: 100 μl membranes (20–30 μg protein), 50 μl 0.9% NaCl–0.5% BSA, 50 μl [^{125}I]bTSH (10,000 cpm)

Nonspecific binding: 100 μl membranes, 50 μl "cold" TSH in saline–0.5% BSA (50 mU), 50 μl [^{125}I]TSH

Control binding with pooled normal human IgG: 100 μl membranes, 50 μl IgG in 0.9% NaCl (20 mg/ml), 50 μl [^{125}I]TSH

Patient's IgG treated as for control.

Buffer for the assay is 20 m*M* Tris–HCl, 0.5% BSA, pH 7.4; membranes and [^{125}I]TSH are diluted in this buffer. All procedures are on ice.

Mixture is incubated at 37° for 1 hr. The reaction is stopped with the addition of 1 ml ice-cold 20 m*M* Tris, 50 m*M* NaCl, 0.1% BSA, pH 7.4 and placing the tubes on ice. Tubes are centrifuged at 20,000 *g* for 20 min at 4°. Supernatant is aspirated and pellets counted in a gamma-sensitive spectrometer.

For routine testing of Graves' disease sera, a crude preparation of IgG is used [40% $(NH_4)_2SO_4$ precipitate, three times concentrated over original serum volume, dialyzed against saline]. It should be emphasized that this solution is not more than 70% IgG and should not be designated "IgG," implying a pure preparation. Protein concentration is adjusted by measuring OD at 280 nm and taking $Eco^{1\%}_{10\,mm} = 14.0$. (For IgG purified by DEAE-cellulose chromatography Eco = 13.5.) Concentration of samples for testing is adjusted to 20 mg/ml giving a final concentration in the assay of 5 mg/ml. It is possible to test IgG at higher concentrations changing the proportions of different assay "ingredients" and dialyzing IgG against 0.45% NaCl, but the specific control binding (normal IgG) declines appreciably.

Residual TBI Assay

One of the problems with the TBI procedure is the variable, but significant, effect of normal IgG.[4] Borges and her colleagues[5] reasoned that this TBI-like activity might be more readily dissociable from the thyroid membranes than the IgG specific for Graves' disease would be. Therefore they first incubated membranes with IgG or serum, washed the particulate material, and then added [^{125}I]TSH; the use of whole serum rather than IgG occasioned a need for more extensive washing to remove the nonspecific effect. Satisfactory results were obtained with both IgG and whole serum from Graves' disease subjects, there being a 96% frequency of inhibitory responses in testing 50 specimens of IgG, 77% in testing 35 sera, and a significant correlation between results with 27 IgG and the sera from which the IgG were obtained.

Solubilized Human Thyroid Receptor Preparations

Another development in TBI assays that appears to reduce the influence of normal IgG or serum, and even increase the sensitivity to anti-receptor antibodies, is the use of detergent-solubilized preparations of human thyroid membranes. Preliminary data by Petersen *et al.*[6] were recently amplified by a report from the same laboratory group[7] in which

[4] A. Sato, M. Zakarija, and J. M. McKenzie, *Endocr. Res. Commun.* **4,** 95 (1977).

[5] M. Borges, J. C. Ingbar, K. Endo, S. Amir, H. Uchimura, S. Nagataki, and S. H. Ingbar, *J. Clin. Endocrinol. Metab.* **54,** 552 (1982).

[6] V. B. Petersen, B. R. Smith, P. J. D. Dawes, and R. Hall, *FEBS Lett.* **83,** 63 (1977).

[7] G. Shewring and B. R. Smith, *Clin. Endocrinol.* (*Oxford*) **17,** 409 (1982).

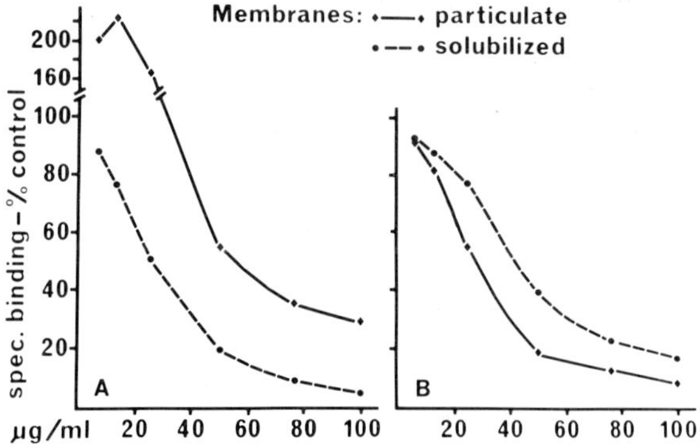

FIG. 1. Patient's IgG was tested, at the concentrations shown, in four TBI assay systems, i.e., using as receptor preparation particulate or solubilized membranes from human thyroid (A) or guinea pig fat (B), as indicated. Dilutions were prepared with normal IgG so that the total IgG concentration at each point was 100 µg/ml. The data are shown as the means of closely agreeing triplicates and the specific binding is expressed as the percentage of that obtained in the presence of buffer.

porcine thyroid membranes were solubilized with 1% Lubrol 12A9 in 50 m*M* NaCl, 10 m*M* Tris–HCl pH 7.5, 3 m*M* NaN_3, to give a satisfactory TSH–receptor solution that was stable at −70° for several months. They claimed that about 80% of patients with Graves' disease could be shown to have circulating anti-receptor antibodies when serum IgG concentrates were tested; rapid preparation of an IgG concentrate by precipitation with 15% polyethylene glycol (PEG, MW 4000) was recommended. [The advantage of PEG over the more commonly used $(NH_4)_2SO_4$ was speed and convenience; the product of the former method was about 60% IgG and of the latter about 70%, but both precipitated more than 95% of the total serum IgG.] The advantages of this technique include the use of porcine thyroids, the relative convenience of PEG as an IgG precipitant, and the overall speed, both for preparation of the test IgG and the binding inhibition process itself.

An additional aspect of the specificity of solubilized receptor for the TBI assay is illustrated in Fig. 1. The IgG under study enhanced TSH binding to particulate human thyroid membranes but this effect was not found when the TSH–receptor preparation was either solubilized human membranes or particulate or solubilized guinea pig fat cell membranes (see below for a description of the use of fat as a source of TSH receptor). Other studies[8] confirm that in this patient's IgG there exists a component that binds to the human thyroid membrane and enhances TSH binding,

[8] M. Zakarija, J. M. McKenzie, and D. S. Munro, *J. Clin. Invest.* **72,** 1352 (1983).

but this action requires both the specificity of the thyroid gland and an intact, particulate receptor complex. The precise nature of the disruption effected by detergent is not understood at present but the data clearly underline the potential for multiple components influencing the binding of TSH to its receptor.

Fat Cell Membranes as a Source for TSH-Receptor

Both TSH and TSAb have long been known to have biological actions on adipose tissue.[9,10] Subsequently the existence of a receptor for TSH in fat cell membranes was established[11] and then a receptor assay for TSAb, using guinea pig fat cell membranes, was described.[12] The particular advantage of this preparation of TSH-receptor is the absence of other thyroid antigens such as thyroglobulin and the microsomal antigen, and this fact was exploited by Endo *et al.*[13] to purify TSAb-IgG to a large degree by adsorption of IgG to, and desorption from, guinea pig fat cell membranes.

The preparation of a suspension of guinea pig fat cell membranes is similar to the process with thyroid tissue; the material also may be solubilized as was described for thyroid membranes, but there is no apparent gain of specificity in this step.[14]

Immunoprecipitation of TSAb

The title of this section is slightly misleading, but it is akin to the terminology used in recent reports in which antibodies to the TSH receptor (i.e., solubilized guinea pig fat cell membranes[15] or human thyroid membranes[16]) were precipitated by the process of adding antiserum against human IgG to the complex of (TSH-receptor + [^{125}I]TSH) + IgG from Graves' disease serum. This was attempted by Rickards *et al.*[17] who failed to obtain significant precipitation of the complex with antiserum to human IgG, but the subsequent reports by Konishi *et al.*,[15] using adipocyte-TSH-

[9] J. E. White and F. L. Engel, *J. Clin. Invest.* **37,** 1556 (1958).
[10] I. R. Hart and J. M. McKenzie, *Endocrinology* **88,** 26 (1971).
[11] C. S. Teng, B. R. Smith, J. Anderson, and R. Hall, *Biochem. Biophys. Res. Commun.* **66,** 836 (1976).
[12] M. Kishihara, Y. Nakao, Y. Baba, S. Matsukura, K. Kuma, and T. Fujita, *J. Clin. Endocrinol. Metab.* **49,** 706 (1979).
[13] K. Endo, S. M. Amir, and S. H. Ingbar, *J. Clin. Endocrinol. Metab.* **52,** 1113 (1981).
[14] M. Loeffler, M. Zakarija, and J. M. McKenzie, *J. Clin. Endocrinol. Metab.* **57,** 603 (1983).
[15] J. Konishi, Y. Iida, K. Kasagi, K. Ikekubo, K. Kuma, and K. Torizuka, *Endocrinol. Jpn.* **29,** 219 (1982).
[16] T. W. A. de Bruin and D. van der Heide, *Acta Endocrinol. (Copenhagen)* **102,** 49 (1983).
[17] C. Rickards, P. Buckland, B. R. Smith, and R. Hall, *FEBS Lett.* **127,** 17 (1981).

receptor, and de Bruin and van der Heide[16] with human TSH-receptor, clearly documented significant immunoprecipitation of antibody. Two problems with this approach were identified by both groups, viz. that precipitation of the [^{125}I]TSH-receptor complex when normal human serum IgG was added averaged about 50% of that obtained with Graves' disease IgG, and the correlation with results of a conventional TBI procedure was poor. The latter observation is of interest, indicating as it does the probability that the procedure recognizes antibodies to TSH-receptor or sites on the membrane distinct from those involved in the TBI assay, but the usefulness of the technique otherwise is not yet evident.

Thyroid Stimulation Procedures

Although most of the procedures listed in the table under this heading are probably currently in use, there are several reasons for emphasizing only two, viz. that using human thyroid slices[18] and that based on cultured human thyroid cells.[19] These combine sensitivity, specificity, and discrimination that are not met by the other procedures. The use of thyroid homogenate as a source of adenylate cyclase for direct measurement of stimulation of the enzyme by TSAb has had several proponents, particularly Bech and her colleagues[20,21]; although this is a sensitive and apparently specific procedure, discriminating powers are poor with the maximum range of response in most publications being considerably less than what is reported with whole cell systems.

A similar concern may be expressed about the use of porcine thyroid slices with the secretion of triiodothyronine (T_3), measured as the free hormone, as the end-point[22]; a 3-fold increment in release of free T_3 was maximal for the effect of TSH and, in a series of 20, most Graves' disease sera fell far short of this degree of response. In contrast, in the human thyroid systems, 10- to 20-fold increments of response are common,[18,19] facilitating clearer discrimination.

Human Thyroid Slice System[18]

Principle. TSAb stimulates the thyroid through activating adenylate cyclase; this may be assessed by the change in the concentration of cyclic AMP in the tissue.

[18] M. Zakarija, J. M. McKenzie, and K. Banovac, *Ann. Intern. Med.* **93,** 28 (1980).
[19] B. Rapoport, S. Filetti, N. Takui, P. Seto, and G. Halverson, *Metab., Clin. Exp.* **31,** 1159 (1982).
[20] K. Bech and S. N. Madsen, *Clin. Endocrinol.* (*Oxford*) **11,** 47 (1979).
[21] K. Bech, *Acta Endocrinol.* (*Copenhagen*) **103,** Suppl. 254, 11 (1983).
[22] S. Atkinson and P. Kendall-Taylor, *J. Clin. Endocrinol. Metab.* **53,** 1263 (1981).

Tissue. Relatively normal human thyroid obtained at operation (e.g., lobectomy) for a "cold" nodule; "frozen section" microscopy is advised to ensure normal thyroid architecture and absence of lymphocytic infiltration. Tissue is placed in a beaker containing 0.9% NaCl solution and immediately transported to the laboratory in crushed ice. Adventitial tissue and capsule are removed and slices of thyroid, less than 1 mm thick, are cut with a Stadie–Riggs microtome. The slices are placed in a Petri dish containing chilled Krebs–Ringer bicarbonate buffer (see below) and small portions, approximately 2 × 3 mm, are made with fine-pointed scissors.

Test Solutions. IgG is prepared from patient's sera by addition of $(NH_4)_2SO_4$ to make a 40% saturated solution ("crude IgG") or further purified by chromatography on diethylaminoethyl cellulose in 50 m*M* Tris–HCl, pH 8.0. Final dialysis of pure or crude IgG is against 0.9% NaCl. The crude IgG is reconstituted in a volume equivalent to one-third of the original serum volume; pure IgG concentration is estimated by spectrophotometry ($E^{1\%}_{10\ \mathrm{mm}}$ at 280 nm = 13.5) and the desired concentration is obtained by addition of 0.9% NaCl. Aliquots of the test solutions are pipetted into 1 × 4-cm glass vials, 0.2 ml per vial, 4–6 vials per solution; the vials are stoppered and stored at −20° until used in the assay.

Buffers. Krebs–Ringer bicarbonate (containing 1/3 recommended Ca^{2+}) is prepared fresh for each assay run; to this buffer is added glucose and bovine serum albumin, each to 0.1 or 0.2% concentration (KRB-G-A). The former buffer is used for the preparation of slices of tissue (see above) and the latter buffer containing 20 m*M* theophylline (KRB-G-A-Th) is used for the final incubation.

Assay Procedure. Vials containing test solutions are kept at room temperature to allow thawing and 0.2 ml KRB-G-A-Th added to each; they are then placed in an appropriate size test tube rack in a shaking Dubnoff incubator at 37°. Tissue slices are added to the vials, one per vial, at 30 sec intervals, being first blotted of excess fluid and weighed on a torsion balance (range of weights 3–12 mg). The vials are gassed with 95% O_2 + 5% CO_2, before and after addition of the slice, and restoppered. Incubation is for 2 hr after which the tissue is removed, one piece every 30 sec, and immediately totally homogenized in 0.5 ml ice-cold 6% trichloracetic acid (TCA) in individual 1 ml capacity Kontes all-glass homogenizers.

In each assay run, that may conveniently comprise 60–120 vials, normal human IgG is used as control (six vials) and thyrotropin, 5 mU/ml, as a semiquantitative assessment of tissue responsiveness, is assayed in an additional four or five vials.

The homogenizers are centrifuged (20 min, 4°, 3500 rpm) in a swinging bucket rotor and the supernatant solution quantitatively transferred by Pasteur pipet to 10-ml glass tubes. TCA is then removed by "washing" with 3 × 5 ml of water-saturated diethyl ether, i.e., the ether is added, the tube vortexed, the ether–water phases allowed to separate, and the ether, top phase, is removed by suction. The remaining 0.5 ml water, containing all the extracted cyclic AMP, is evaporated to dryness by a jet of nitrogen gas while the tubes are sitting in a 70° water bath.

Cyclic AMP Radioimmunoassay. Cyclic AMP is assayed using the Becton Dickinson radioimmunoassay kit; the standards are extended by addition of 0.10, 1.0, and 10 pmol from the solutions supplied. Over this range (0.05–10) the correlation coefficient for the standard curve is consistently 0.99.

For the assay, the dry residue in each tube is dissolved in distilled water. The volume added per tube is calculated on the basis of the original wet weight of tissue, so that 100 μl used in the assay is equivalent to approximately 0.65 mg (for TSH group 0.16 mg) of thyroid that contains in the control group on average 0.1–0.3 pmol of cAMP. This ensures that most results fall on the most sensitive portion of the dose–response slope and there is a clustering of data for individual test groups. Data may be expressed as pmol cyclic AMP per mg wet weight of tissue or as a percentage of the control value.

The major disadvantage of this procedure is the requirement for fresh, relatively normal tissue, directly from the operating room. This is most commonly obtained at lobectomy for a "cold" nodule but it is our experience that even this paranodular tissue is frequently macroscopically abnormal, containing small cysts and/or excessive fibrous tissue. The use of cryopreserved tissue, to alleviate part of the problem, was assessed by Knox *et al.*[23] but the thawed-out tissue was less sensitive and responsive than was the general experience with fresh thyroid.

Use of Human Thyroid Cells in Monolayer Culture. Assuming facilities are available, and the primary cultures can be established and frozen, this technique has significant advantages over the thyroid slice system, compared with which it has similar sensitivity, specificity, and discrimination. Details of the procedure are provided in reports from Rapoport's laboratory.[18,24] Briefly, 2–6 × 10^4 thyroid cells are plated per well of a Falcon microtiter plate; the buffer is Hanks' balanced salt solution, pH 7.4 (N.B., containing no NaCl) + 20 m*M* HEPES, 2 m*M* 3-isobutyl

[23] A. J. S. Knox, C. von Westarp, V. V. Row, and R. Volpé, *Cryobiology* **14,** 543 (1977).
[24] W. E. Hinds, N. Takai, B. Rapoport, S. Filetti, and D. H. Clark, *J. Clin. Endocrinol. Metab.* **52,** 1204 (1981).

methylxanthine, and 0.4% bovine serum albumin. TSH, test or control IgG is added and as the end-point cyclic AMP is measured in the medium after 4 hr incubation.

The effect of using NaCl-deficient buffer, as earlier established by Kasagi *et al.*,[25] is both to divert the increased concentration of cyclic AMP from the cells to the medium, and to increase the sensitivity of response to TSAb, at least for concentrations of IgG $\not>$ 3 mg/ml.

Use of FRTL Cells. Ambesi-Impiombato and his colleagues[26] established a line of functioning rat thyroid cells (FRTL cells) that grow well in monolayer culture, although requiring a highly hormone-enriched culture medium. These have been shown[27] to respond well to both TSH and TSAb in terms of an increase in cellular cyclic AMP over 2 hr of incubation. In our experience they are not as sensitive as human thyroid slices to TSAb nor is the response with individual samples of TSAb-IgG as great. However, recently the system was shown to be more sensitive to stimulation by either TSH or TSAb if the end-point was the uptake of radioactive iodide.[28] FRTL cells that were grown for several days in the absence of TSH were exposed to TSAb for 48 hr after which a 30 min uptake of ^{125}I was measured; stimulation of uptake was half-maximal with a 10-fold lower concentration of TSAb than was required for half-maximal stimulation of cyclic AMP accumulation.

Another use of FRTL cells developed by Kohn and his colleagues has been to assess the existence, in the serum IgG of Graves' disease patients, of growth-promoting antibodies that had been suggested by Drexhage *et al.*[29] Although these antibodies apparently interact with the TSH receptor, adenylate cyclase is not the enzymatic mediator. The FRTL cells indeed responded to IgG from Graves' disease sera by an increase in [^{3}H]thymidine uptake, reflecting cell growth.[30] Using all three assays with the cells, i.e., cyclic AMP accumulation, ^{125}I, and [^{3}H]thymidine uptakes, evidence for the presence of TSAb was obtained in 100% of 19 sera, but

[25] K. Kasagi, J. Konishi, Y. Iida, K. Ikekubo, T. Mori, K. Kuma, and K. Torizuka, *J. Clin. Endocrinol. Metab.* **54,** 108 (1982).

[26] F. S. Ambesi-Impiombato, L. A. M. Parks, and H. G. Coons, *Proc. Natl. Acad. Sci. U.S.A.* **77,** 3455 (1980).

[27] P. Vitti, W. A. Valente, F. S. Ambesi-Impiombato, G. Fenzi, A. Pinchera, and L. D. Kohn, *J. Endocrinol. Invest.* **5,** 179 (1982).

[28] C. Marcocci, W. A. Valente, A. Pinchera, S. M. Aloj, L. D. Kohn, and E. F. Grollman, *J. Endocrinol. Invest.* **6,** 463 (1983).

[29] H. A. Drexhage, G. F. Bottazzo, D. Doniach, L. Bitensky, and J. Chayen, *Lancet* **2,** 287 (1980).

[30] W. A. Valente, P. Vitti, C. M. Rotella, M. M. Vaughan, S. M. Aloj, E. F. Grollman, F. S. Ambesi-Impiombato, and L. D. Kohn, *N. Engl. J. Med.* **309,** 1028 (1983).

not all activities in all samples. Confirmation and extension of these initial data are awaited with interest.

Cytochemical Assay of TSAb. In general, cytochemical assays of hormones are exquisitely sensitive and this was found to be the case for the assay of TSAb.[31] The procedure entails incubation of a preparation of guinea pig thyroid with test material and subsequent processing of the tissue to establish the degree of increased permeability of the lysosomal membranes of individual follicular cells. This is carried out on sections adhering to microscope slides by the provision of a chromogenic substrate, leucine 2-naphthylamide, for the intralysosomal enzyme naphthylamidase and the intensity of the reaction is monitored by scanning integrating microdensitometry.

The original procedure[31] used segments of guinea pig thyroid for incubation with test material and subsequent sectioning for the cytochemistry and densitometry. This severely restricted the number of samples processed in one assay since a guinea pig thyroid could not provide realistically more than six segments. Subsequently[32] microscope sections of tissue were used for both incubation and enzyme reaction, greatly facilitating the testing on one gland of multiple samples of test IgG.

The drawbacks of this technique include the sophistication of both the instrument (i.e., the scanning microdensitometer) and the processing of the tissue, and the complexity of dose–response relationships. As shown by several[33–35] both TSH and TSAb have a bell-shaped dose–response relationship so that care has to be taken with the assay design and the objectivity of microscopy ensured by a single-blind, if not double-blind, arrangement for processing samples.

Comparison of Results of Thyroid Stimulation Type Assays with Those of TBI Procedures

As stated earlier, it now appears that a thyroid-stimulating antibody will inhibit the binding of TSH to its receptor, but the corollary is not necessarily the case, i.e., an IgG active in a TBI assay may not act as a thyroid stimulator. This was the interpretation of several series of comparisons of individual samples of IgG in which the potency in one assay

[31] L. Bitensky, J. Alaghband-Zadeh, and J. Chayen, *Clin. Endocrinol.* **3,** 363 (1974).

[32] J. Chayen, D. M. Gilbert, W. R. Robertson, and L. Bitensky, *J. Immunoassay* **1,** 1 (1980).

[33] N. Loveridge, M. Zakarija, L. Bitensky, and J. M. McKenzie, *J. Clin. Endocrinol. Metab.* **49,** 610 (1979).

[34] P. A. Ealey, N. J. Marshall, and R. P. Ekins, *J. Clin. Endocrinol. Metab.* **52,** 683 (1981).

[35] D. Neylan and P. P. A. Smyth, *Clin. Endocrinol. (Oxford)* **17,** 479 (1982).

might bear no relationship to the potency in another (i.e., comparing effects in TBI and TSAb procedures).[36–38] Confirmatory data in this regard are now appearing with monoclonal antibodies obtained from hybridomas resulting from the fusion of mouse myeloma cells with peripheral blood lymphocytes from patients with Graves' disease.[39] The interpretation, particularly from the monoclonal antibody experience, includes the probability that the TSH-receptor is a complex with multiple antigenic determinants; interaction with any one determinant might inhibit TSH binding and/or stimulate the adenylate cyclase enzyme that is in some way linked to the receptor. In fact, we have recently identified[40] in a patient's IgG three different activities: (1) an IgG that increases binding of [^{125}I]TSH to human thyroid membranes (Fig. 1); (2) another IgG that has TSAb activity; (3) an IgG, distinct from TSAb, that inhibits binding of [^{125}I]TSH and bioactivity of both TSH and TSAb.

From this complicated analysis it is clear that data obtained with one assay (e.g., TBI) may not be casually compared with data resulting from another procedure (i.e., TSAb). Much more understanding of interactions of antibodies with the receptor for TSH is required before the different assay procedures may be viewed in total perspective.

Acknowledgments

This work was supported by Grants AM04121 and 31117 from the USPHS.

[36] J. M. McKenzie, M. Zakarija, and A. Sato, *Clin. Endocrinol. Metab.* **7,** 31 (1978).

[37] A. Sugenoya, A. Kidd, V. V. Row, and R. Volpé, *J. Clin. Endocrinol. Metab.* **48,** 398 (1979).

[38] H. Bliddal, K. Bech, P. H. Petersen, K. Sirsbsek-Nielsen, and T. Friis, *Acta Endocrinol. (Copenhagen)* **101,** 35 (1982).

[39] W. A. Valente, P. Vitti, Z. Yavin, E. Yavin, C. M. Rotella, E. F. Grollman, R. S. Toccafondi, and L. D. Kohn, *Proc. Natl. Acad. Sci. U.S.A.* **79,** 6680 (1982).

[40] M. Zakarija, J. M. McKenzie, and A. Claflin, *in* "Autoimmunity and the Thyroid" (P. Walfish, R. Volpé, and J. Wall, eds.). Academic Press, New York, in press.

[54] A Monoclonal Antibody to Growth Hormone Receptors

By J. S. A. SIMPSON and H. G. FRIESEN

Introduction

The preparation of polyclonal antisera to growth hormone receptors has been previously described[1,2] and recently reviewed.[3] However, the generation of monoclonal antibodies to receptors appears to be an area of growing interest because of the advantages these reagents offer.[4] There are two main reasons why we undertook the preparation of a monoclonal antibody to the growth hormone receptor. First, attempts to purify the receptors by classical biochemical techniques[5] have proven of limited success and the use of immunoaffinity chromatography has proven of significant value in other receptor purification schemes. To this end a monospecific antibody preparation would be highly desirable and hybridoma technology is the most likely method by which this can be achieved.

Second, a growing number of reports are appearing in the literature of direct effects of growth hormone in several tissues.[6] Some of these studies undoubtedly reflect confusion of the prolactin-like and somatotropic actions of hGH[7] and a significant benefit would be derived from an estimate of the abundance of somatotropic and lactogenic receptors in tissue which is independent of the hormone preparation used in the test.

The preparation of high quality polyclonal antibodies in large quantities usually requires a series of immunizing injections of highly purified receptors in significant amounts. Growth hormone receptors are present in tissues only in small quantities. Therefore, it is not possible to prepare either sufficiently pure or sufficient quantities of receptor to produce antiserum for studies of structure and regulation of these receptors or to attempt immunohistochemical studies of their distribution.

In contrast, it is possible to prepare monoclonal antibodies in large quantities with only a few micrograms of purified antigen[8] by *in vitro*

[1] M. J. Waters and H. G. Friesen, *J. Biol. Chem.* **254,** 6826 (1979).

[2] T. Tsushima, *in* "Growth and Growth Factors" (K. Shizume, ed.), p. 171. Univ. of Tokyo Press, Tokyo, 1980.

[3] R. G. Drake and H. G. Friesen, this series, Vol. 74, p. 380.

[4] G. S. Eisenbarth and R. A. Jackson, *Endocr. Rev.* **3,** 26 (1982).

[5] M. J. Waters and H. G. Friesen, *J. Biol. Chem.* **254,** 6815 (1979).

[6] S. Eden, O. G. P. Isaksson, K. Madsen, and U. Friberg, *Endocrinology* **112,** 1127 (1983).

[7] M. J. Rowe, unpublished data (1982).

[8] R. A. Luben, P. Brazeau, P. Bohlen, and R. Guillemin, *Science* **218,** 889 (1982).

METHODS IN ENZYMOLOGY, VOL. 109

ISBN 0-12-182009-2

immunization or by immunization of larger quantities of less pure material.

There are a number of reviews of the methodology of hybridoma production[9–12] which cover the technical aspects in detail so we will concentrate our discussion on those areas most pertinent to receptors.

The screening assay used is a significant determinant of the type of antibody which will be found and consideration must be given to this technical point before antibody production can commence. There are a number of methods by which an antibody to receptor may be detected. The simplest procedures use the hormone binding characteristics of the receptor in some manner, either by using the antibody to inhibit binding of labeled hormone to the receptor[13] or using the hormone to inhibit the antibody binding to membranes.[14] These assays will detect antibodies which interact with the receptor at, or close to, the hormone binding site, but may not detect those antibodies which react at a site removed from the hormone binding site. If the binding of labeled hormone to detect or label the receptor is used, when it has been immunoprecipitated,[15] the assay will be able to detect only those antibodies which can react with the receptor without preventing hormone binding. While any particular polyclonal antiserum may react in all of these assay systems, an individual monoclonal antibody is unlikely to do so. It is also possible to use inhibition or activation of the biological action of the hormone–receptor complex as a screening assay, but this usually results in an assay which is time consuming and difficult to use with the large number of cell lines which must be screened.

It is a very significant aid for comparative testing purposes if a polyclonal antiserum is available. Thus, in this instance the simplest assay is recommended. It may be necessary to use supplementary tests to confirm initial positive results.

Immunization is an area where a number of techniques have been used and no consensus exists as to the best method to follow. The success of the *in vitro* method of Luben *et al.*[8] has increased interest in this route for

[9] R. H. Kennett, T. J. McKearn, and K. B. Bechtol, eds., "Monoclonal Antibodies." Plenum, New York, 1980.

[10] C. L. Reading, *J. Immunol. Methods* **53,** 261 (1982).

[11] G. Galfre and C. Milstein, this series, Vol. 73, p. 1.

[12] J. J. Langone and H. Van Vunakis, eds., "Methods in Enzymology," Vol. 92, Chapters 1–20. Academic Press, New York, 1983.

[13] J. S. A. Simpson, J. P. Hughes, and H. G. Friesen, *Endocrinology* **112,** 2137 (1983).

[14] W. A. Valente, Z. Yavin, E. Yavin, E. F. Grollman, M. Schneider, C. M. Rotella, R. Zonefrati, R. S. Toccafondi, and L. D. Kohn, *J. Endocrinol. Invest.* **5,** 293 (1982).

[15] F. C. Kull, Jr., S. Jacobs, Y.-F. Su, and P. Cuatrecasas, *Biochem. Biophys. Res. Commun.* **106,** 1019 (1982).

activating the lymphocyte population before fusion with the myeloma cells. It is worth considering that it has been demonstrated that actively dividing cells appear to be fused preferentially. In order to get successful activation of the lymphocyte population that will secrete an antibody of value a number of factors have to be optimized.[10,16] Attempting to control these in a culture system is undoubtedly a complex procedure in its own right and most new workers in this field will probably find they are more successful if they employ the whole animal immune system rather than a subset of the cells involved in a typical immune reaction.

We used female $CB6f_1$/j mice for the series of immunizations which was successful. This is a hybrid mouse with one BALB/c parent the other is C57BL; we also tried unsuccessfully to immunize the parent strains as well. The strain of mouse used may be important in determining how successful a course of immunization is. It is also important to bear in mind that the most common myelomas available are of BALB/c origin and if ascitic fluid is to be made the hybridomas will have either to be genetically compatible with a host mouse strain or irradiated mice will have to be used. The $CB6f_1$/j was compatible with several hybridomas which we have produced using lymphocytes from both BALB/c or $CB6f_1$/j mice and P3X20 or P3X Ag8.653 myelomas.

Materials and Methods

Iodination

Human GH (2.2 IU/mg) prepared in this laboratory was used for preparation of [^{125}I]hGH. Rat GH I–4 (NIAMDD) was used for iodination and rat GH-RP-1 was used for displacement. Iodinations were carried out by the method of Thorell and Johansson[17] with modifications described elsewhere.[18] Specific activity of the labeled hormones averaged 130 μCi/μg for [^{125}I]hGH and 145 μCi/μg for [^{125}I]rGH.

Preparation of Receptors

Liver was obtained from adult female New Zealand White rabbits and from late-pregnant (day 18) Sprague–Dawley rats. Mammary gland was removed from late-pregnant NZW rabbits. Membranes were prepared by homogenization of thawed, frozen tissue in 10 volumes of ice cold 0.2 *M* sucrose. The homogenate was then centrifuged at 15,000 *g* for 20 min. The supernatant was diluted with an equal volume of 0.025 *M* Tris–HCl buffer

[16] J. Cavagnaro and M. E. Osband, *Biotechniques* **1**(1), 30 (1983).

[17] J. I. Thorell and B. G. Johansson, *Biochim. Biophys. Acta* **251,** 363 (1972).

[18] J. P. Hughes, *Endocrinology* **105,** 414 (1979).

pH 7.6 containing 20 mM $CaCl_2$. After allowing this to flocculate for 30 min at 4° it was recentrifuged at 15,000 g for 20 min. The pellet was stripped of endogenous hormone using the method outlined by Gerasimo *et al.*,[19] washed with a minimum of 20 volumes of distilled water and lyophilized.

Solubilized receptors were prepared by treatment of this membrane preparation with 1% (v/v) Triton X-100. Purified receptor for immunization was further processed by hGH affinity chromatography and gel filtration on Sepharose 6B as described by Waters and Friesen.[5] The receptor which was prepared in this way had a binding capacity of 45.8 pmol/mg which implies that the receptor was about 1% pure. IM-9 human lymphocytes were used in the binding studies as whole cells.

Immunization Procedure

Immunization is used as a method of preparing a population of lymphocytes which when they are fused with the myeloma cells has a high probability of producing hybridomas that will secrete the desired antibody. Under certain circumstances, immunization may not be essential. For example, lymphocytes can be harvested from animals or patients which have an autoimmune disease involving antibodies of the type required.[20,21] There are also possibilities for producing monoclonal antibodies from lymphocytes derived from animals which have not been immunized but have been instead infected with certain viruses.[22] However, the procedure most likely to succeed will usually involve some form of active immunization using one of the well described adjuvants (Freund's, pertussis vaccine, alum[23]).

The antigen preparation can take many forms, whole cells,[24,25] membrane fractions,[14] more or less purified preparations,[26] or hormone–receptor complexes.[27]

[19] P. Gerasimo, J. Djiane, and P. Kelly, *Mol. Cell. Endocrinol.* **13,** 11 (1979).

[20] G. S. Eisenbarth, A. Linnenbach, R. A. Jackson, R. Scearce, and C. M. Croce, *Nature (London)* **300,** 264 (1982).

[21] W. A. Valente, P. Vitti, Z. Yavin, E. Yavin, C. M. Rotella, E. F. Grollman, R. S. Toccafondi, and L. D. Kohn, *Proc. Natl. Acad. Sci. U.S.A.* **79,** 6680 (1982).

[22] M. V. Haspel, T. Onodera, E. S. Prabhakar, M. Horita, H. Suzuki, and A. L. Notkins, *Science* **220,** 304 (1983).

[23] A. Zimmerman, A. Sutter, and E. M. Shooter, *Proc. Natl. Acad. Sci. U.S.A.* **78,** 4611 (1981).

[24] A. B. Schreiber, I. Lax, Y. Yarden, Z. Eshar, and J. Schlessinger, *Proc. Natl. Acad. Sci. U.S.A.* **78,** 7535 (1981).

[25] D. Solter and B. B. Knowles, *Proc. Natl. Acad. Sci. U.S.A.* **75,** 5565 (1978).

[26] P. Grandics, D. L. Gasser, and G. Litwack, *Endocrinology* **111,** 1731 (1982).

[27] J. L. Luborsky and H. R. Behrman, *Biochem. Biophys. Res. Commun.* **90,** 1407 (1979).

The initial injections of semipurified receptor were given subcutaneously every 7 days as an emulsion in Freund's complete adjuvant (Calbiochem-Behring Corp., La Jolla, CA), 26 μg of solubilized receptor which was approximately 1% receptor protein. One month after the start of the series a 10-fold concentrated receptor preparation was administered every 14 days by intraperitoneal injection without adjuvant. On the fifty-second day a test bleed was made from a tail vein, 7 days after an injection and tested in an assay for anti-receptor antibody. This showed the presence of a small but significant amount of antibody. Immunization was continued bi-weekly for a further month and on the eightieth day a booster injection was given; 3 days later a mouse was sacrificed for splenocyte preparation.

Fusion Method

The advent of hybridoma technology was the outcome of the current research effort into the mechanisms which control the immune system. In this area the impact of new discoveries is having a great effect with changes in the basic fusion protocols occurring more and more frequently.[28,29]

We used a modified method after Bundesen *et al.*[30] The spleen, which had been removed aseptically, was placed in a 2 × 60-mm Petri dish (Falcon, 3001 Oxnard, CA) containing 5 ml of complete medium (85% RPMI 1640, 15% fetal bovine serum, 100 IU/ml penicillin, 100 μg/ml streptomycin, and 2×10^{-3} *M* glutamine (Gibco, Grand Island, NY). A cell suspension was prepared by decapsulating the spleen with 2 × 18-gauge needles attached to 3-ml disposable syringes with the last cm of the tip bent at an angle of approximately 60°. The released cell suspension was then aspirated into a 10-ml syringe through a 22-gauge needle and ejected into a Corning 15-ml centrifuge tube. The suspension was then left to stand for 5 min at room temperature to allow larger clumps and membrane fragments to settle before it was transferred to a fresh centrifuge tube, and the cells spun down at 350 *g* for 5 min.

The supernatant was decanted from the cell pellet and the cells were resuspended in 5 ml of fresh complete medium. Viable spleen white blood cells were counted using trypan blue stain and 100×10^6 were placed into a separate centrifuge tube in a total volume of 5 ml complete medium. The P3X20 myeloma cells (secreting both MOPC $21\gamma_1$ and κ chains) to be used for fusion were washed once by centrifugation at 350 *g* for 5 min and adjusted to 5×10^6 viable cells/ml in complete medium.

[28] R. T. Taggart and I. M. Samloff, *Science* **219,** 1228 (1983).

[29] T. Block and M. Bothwell, *Nature* (*London*) **301,** 342 (1983).

[30] P. Bundesen, R. G. Drake, K. Kelly, I. G. Worsley, H. G. Friesen, and A. H. Sehon, *J. Clin. Endocrinol. Metab.* **51,** 1472 (1980).

Twenty-five × 10^6 P3X20 and 100 × 10^6 immune cells were mixed and spun at 350 *g* for 5 min at room temperature. The supernatant was decanted carefully, 2 ml of 50% (w/v) solution of polyethylene glycol (PEG MW 1540, JT Baker chemical Company, Phillipsburg, NJ) in RPMI 1640 containing 15% (v/v) dimethyl sulfoxide at 37° was added with a 5-ml glass disposable pipet, and the cells were resuspended for 30 sec with the aid of an electric pipetter (Pipet-aid, Drummond Scientific Co., PA). The PEG-cell suspension was allowed to stand for 30 sec at room temperature before adding 5 ml complete medium, drop-wise, with a fresh pipet, over a period of 90 sec with constant flicking of the tube, sufficient to ensure complete mixing of the cells with the viscous PEG solution. A further 5 ml of complete medium was added all at once and the tube mixed by inversion before being allowed to stand for 2–3 additional minutes at room temperature. The cells were then spun at 350 *g* for 5 min and resuspended in 5 ml of complete medium taking care not to break up all the clumps of cells; then 0.05 ml of the suspension was added (Tridak stepper, Bellco Glass Inc., Vineland, NJ), to each well of 4 Costar 24-well plates (Costar 3524, Cambridge, MA) containing 1 × 10^6 normal mouse spleen cells as feeders in 1 ml of HAT medium (10^{-4} *M* hypoxanthine, 4 × 10^{-7} *M* aminopterin, 1.6 × 10^{-5} *M* thymidine, and 4 × 10^{-5} *M* 2-mercaptoethanol).

Generally on day 10 or 11 following fusion 0.5 ml of the medium was removed for testing in the screening assay. Cloning was carried out using the method of limiting dilution at least twice to ensure that the antibody-producing clones were stable. Initially 3 clones were identified as positive and from one of these 4 cell lines were established which appeared to secrete identical antibodies. One of these was recloned to yield a monoclonal cell line called D6. Ascitic fluid was then produced by injecting approximately 10^7 cells into mice which had been primed with pristane (2,6,10,14-tetramethyl hexadecane, Aldrich chemical Company) 0.5 ml ip 2 weeks before the cells were given. The ascitic fluid had a high activity in the assay for antibody.

Screening Assay

Rabbit liver membranes (100 μg) were incubated with 50 μl of hybridoma supernatant for 18 hr at 22° and then 350 μl of assay buffer (0.025 *M* Tris–HCl pH 7.6, 0.01 *M* $MgCl_2$, 0.1% BSA, 0.1% NaN_3) containing 100,000 cpm of [^{125}I]hGH was added and the incubation continued for a further 24 hr. The reaction was terminated by addition of 3 ml ice-cold assay buffer followed by centrifugation at 2000 *g* for 30 min. Figure 1 shows the inhibition of binding caused by various concentrations of ascitic fluid.

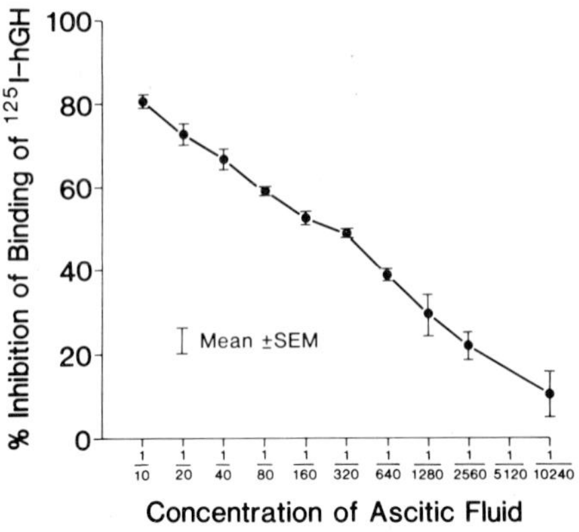

FIG. 1. The effect of anti-receptor antibody on the binding of [^{125}I]hGH to rabbit liver membranes. Antibody containing ascitic fluid diluted in Tris–$MgCl_2$ buffer (100 μl), membranes (100 μg in 100 μl), and labeled hGH (30,000–60,000 cpm in 100 μl) were incubated for 18 hr at 22°. The bound label was then determined as described in the text under screening assay. The vertical bar at each concentration shows the mean and SEM for 5–17 determinations of the percentage inhibition of the specifically bound label each of which was the mean of a triplicate.

Identification of the Immunoglobulin Class

The monoclonal antibody was precipitated from 50 ml of ascitic fluid by addition of ammonium sulfate to a concentration of 33% saturated at pH 6.6. A second fraction was precipitated by adjusting the supernatant from the first precipitation to 50% saturated with more ammonium sulfate. The first pellet contained 12% of the total antibody activity and the second 35% (Fig. 2). These two pellets were redissolved in phosphate-buffered saline (PBS) and combined in a final volume of 50 ml. They were then dialyzed twice overnight against 20 volumes of PBS. Five milliliters of this solution was then adjusted to pH 8.0 with 1 *N* NaOH and passed through a 4 ml column of protein A Sepharose 4B (Pharmacia, Piscataway, NJ) which had been equilibrated with 0.01 *M* phosphate buffer pH 8.0.[31] The column was then eluted with 35 ml of phosphate buffer, pH 8.0 followed by 0.1 *M* phosphate buffer, pH 6.0 (42 ml) and 0.1 *M* glycine/HCl, pH 3.0 (36 ml). Fractions of 3 ml were collected and the protein content measured (Fig. 3).

The protein-containing fractions eluted at each pH were pooled separately and dialyzed against Tris–HCl buffer pH 7.6 for estimation of anti-receptor activity. The unadsorbed protein (fraction I) was the only fraction containing significant activity. It contained greater than 80% of the

[31] P. L. Ey, S. J. Prowse, and C. R. Jenkin, *Immunochemistry* **15,** 429 (1978).

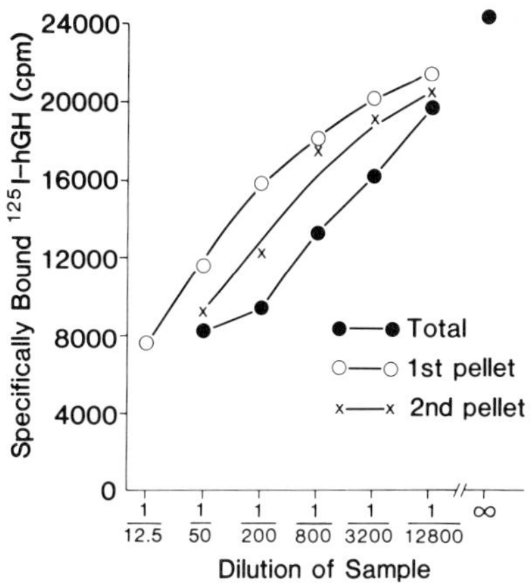

FIG. 2. This shows the inhibition of [^{125}I]hGH binding caused by solutions of the first and second pellets from the ammonium sulfate precipitation redissolved in Tris–$MgCl_2$ buffer and adjusted to the same volume as the initial sample compared with a sample from the initial ascitic fluid. The assay protocol was similar to that described for Fig. 1 except that the amount of label used was 100,000 cpm in 100 μl and the points plotted are the means of triplicate determinations.

antibody activity added to the column, indicating that the immunoglobulin class of the monoclonal antibody was not IgG. A solid phase derivative of the unadsorbed material was prepared by coating the protein onto a plastic microtiter plate; 100-μl aliquots of the solution were placed in each of

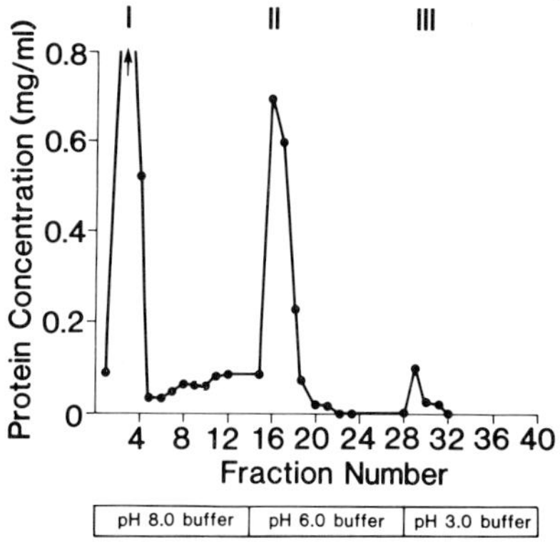

FIG. 3. The elution profile of protein of a sample of immunoglobulin derived from the ascitic fluid on a protein-A Sepharose 4B column. The column was equilibrated at pH 8.0 and loaded at the same pH with 5 ml of the mouse immunoglobulin solution. The protein eluted by each buffer solution as indicated by the horizontal bars was measured by absorption at 280 nm and referenced to a standard solution of mouse IgG.

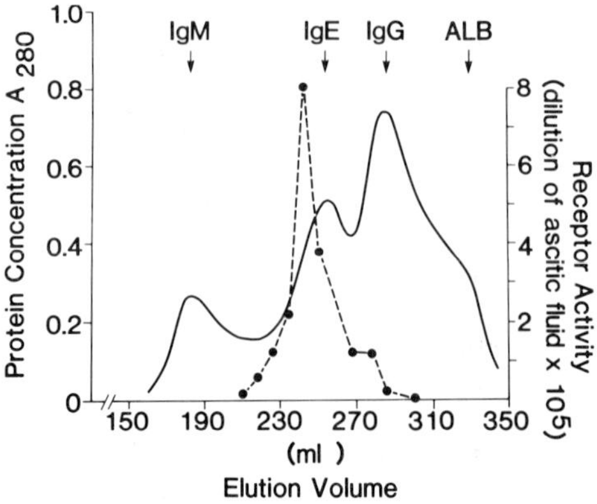

FIG. 4. The elution profile of protein (——) and anti-receptor activity (●) from an Ultrogel AcA34 column. The protein concentration was measured by absorption at 280 nm and the antibody as described in the text. The results were then referenced to the initial ascitic fluid using the data of Fig. 1.

the wells and incubated at 37° for 1 hr. The liquid was then removed and the wells washed with 0.1% BSA in saline solution 4 times to remove any unadsorbed protein. The wells were then examined using a mono-screen ID kit (Cedarlane laboratories, Hornby, Ont. Canada) for the presence of mouse immunoglobulins. This showed that there was no reactivity with antiserum against mouse α or μ heavy chains or with λ light chain but that there was a significant reaction for κ light chains. The kit does not include reagents for ε heavy chains but the pattern of reaction suggested that the monoclonal antibody was an IgE, κ.

Two tests were carried out in order to confirm this result. First, a sample of the monoclonal antibody solution was chromatographed on Ultrogel AcA 34 (Fig. 4), which is capable of partly resolving IgE and IgG due to the difference in molecular size between these two species of immunoglobulin.[32] The antibody activity was associated with a protein peak which had the same elution volume as a monoclonal IgE used to characterize the column. Second the binding of peroxidase-labeled goat anti-mouse IgE (Nordic Immunology Labs, El Toro, CA) to rabbit liver membranes pretreated with the antibody was prevented if excess ascitic fluid was included in the incubation medium but not when bovine IgG up to a concentration of 5% (w/v) was included.

The production of monoclonal antibody of the IgE class is relatively rare but not impossible.[33] It may be that the prolonged immunization protocol was in some measure responsible for this unusual finding.

[32] S. B. Lehrer and B. E. Bozelka, *Prog. Allergy Res.* **32,** 8 (1982).

[33] A. S. Tung, this series, Vol. 92, p. 47.

THE EFFECT OF ASCITIC FLUID ON OTHER RECEPTORS[a]

Species	Tissue	Receptor type	Percentage inhibition[b]
Rabbit	Liver	GH[b]	82 ± 8.6
Rabbit	Mammary gland	PRL	3.8 ± 11.6
Rat	Liver	GH[c]	0.4 ± 4.2
Rat	Liver	PRL	7.2 ± 10.1
Human (IM-9)	Lymphocyte	GH	3.0 ± 10.2

[a] Reproduced with permission from Simpson *et al.*[13]
[b] Ascitic fluid 1 : 40.
[c] Measured with [^{125}I]rGH. All other inhibition was measured with [^{125}I]hGH.

Specificity Studies

The specificity of the anti-receptor antibody for GH receptors was examined by testing the ability of the ascitic fluid to inhibit binding of [^{125}I]hGH to the prolactin receptors on rabbit mammary gland membranes and the inhibition of ^{125}I-labeled rat GH binding to rabbit liver membranes (see the table). The antibody was able to inhibit binding of the rGH to rabbit somatotropic receptors but not the binding of hGH to rabbit prolactin receptors.

Species specificity was determined by examination of the effect of the antibody on labeled rat GH binding to GH receptors in rat liver membranes and labeled hGH binding to the GH receptor on human IM-9 lymphocytes (see the table).

The antibody was incapable of preventing the binding of the hormones to either rat or human GH receptors but could inhibit binding of either tracer to rabbit GH receptors. There was also no inhibition of label binding to prolactin receptors in rabbit mammary gland. This showed clearly that the antibody was highly specific for the somatotropic receptor of rabbit tissues.

Characterization of Growth Hormone Receptors in Rabbit Liver

We have used the monoclonal antibody to investigate the binding sites for hGH in rabbit liver by affinity chromatography on a column of antibody coupled to Sepharose 4B (Pharmacia, Piscataway, NJ).[13,34] Our data are in agreement with estimates of the proportion of somatotropic and lactogenic receptors for hGH in this tissue made by Waters and Friesen,[5]

[34] J. S. A. Simpson, J. P. Hughes, and H. G. Friesen, *in* "Proceedings of the International Symposium on Hormone Receptor and Receptor Diseases" (H. Imura, ed.), Excerpta Medica, Amsterdam, 1983.

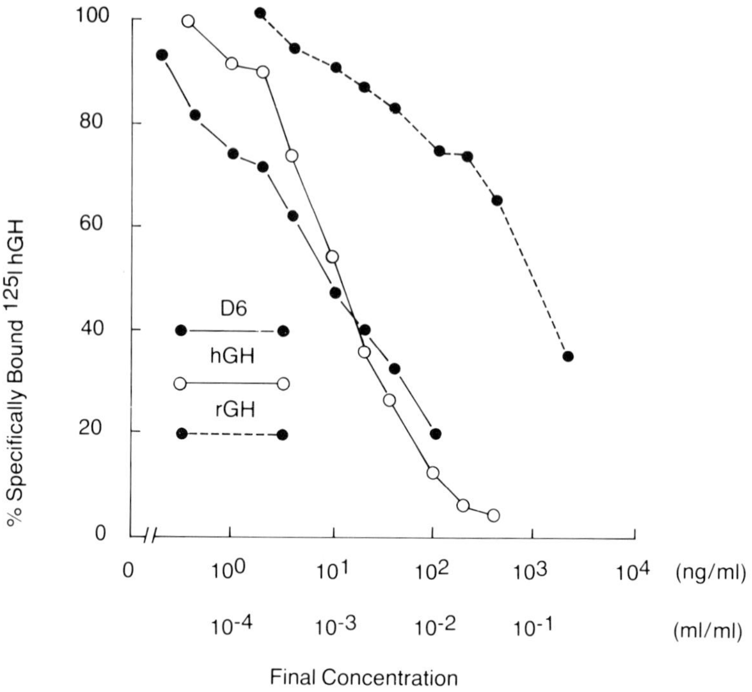

FIG. 5. Dose–response curves for hGH, rGH, and ascitic fluid (D6) at inhibiting the binding of [^{125}I]hGH to rabbit liver membranes (100 μg). The hormone concentrations are expressed as ng/ml in the final incubation mixture, whereas the concentration of the ascitic fluid (ml/ml) is given in milliliters of ascitic fluid per unit volume of the final incubation mixture. Reproduced with permission from Simpson *et al.*[34]

i.e., 20% lactogenic receptors and 80% somatotropic receptors, and show that the antibody can be used to separate the somatotropic receptor site from the lactogenic sites. The antibody is capable of inhibiting both the binding of [^{125}I]hGH (Fig. 5) and [^{125}I]rGH (Fig. 6), which argues strongly that both hormones are either interacting with a single class of receptors or that the receptors that the hormones bind to are closely immunologically related. The binding of [^{125}I]hGH to rabbit liver has been shown to be poorly inhibited by rGH[18] (Fig. 5). However, Fig. 6 shows that rat GH binds to sites on rabbit liver that have a high affinity.

The disparity in the affinity of rat GH for the receptors which are binding [^{125}I]hGH and the affinity it shows for those which bind ^{125}I-labeled rat GH (compare Figs. 5 and 6) indicates that there are likely two or more classes of receptors in the membrane preparations.

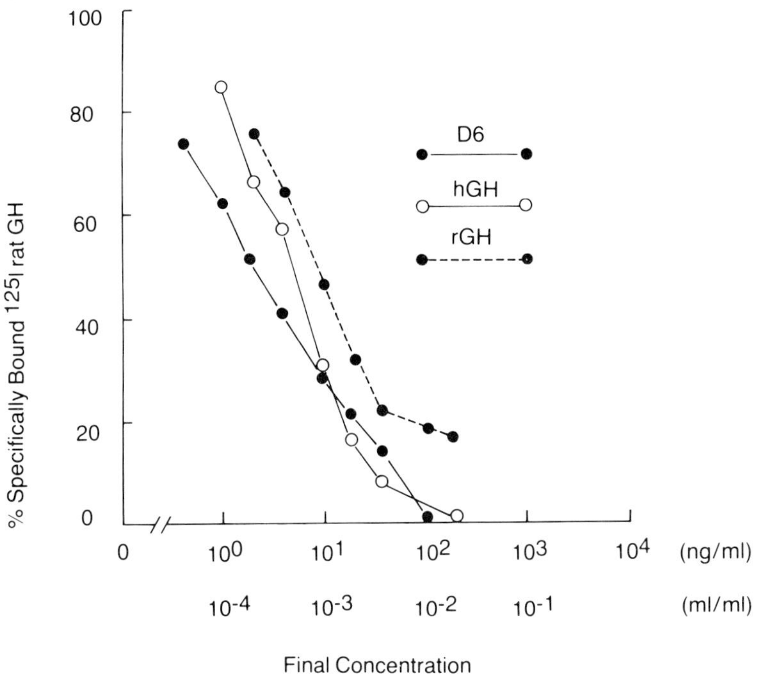

FIG. 6. Dose–response curves for hGH, rGH, and ascitic fluid (D6) at inhibiting the binding of [^{125}I]rGH to rabbit liver membranes (2 mg). The data are expressed as in the legend to Fig. 5. Reproduced with permission from Simpson *et al.*[34]

Conclusion

The production of monoclonal antibodies is significantly more time consuming than polyclonal antibody production, but this is offset by the reproducible quality of the hybridoma-derived product. Many monoclonal antibodies are also highly species specific thereby reducing their general usefulness. However, this high specificity serves to increase confidence that immunological identity reflects biochemical similarities.

In the absence of another source of antibody to a receptor, i.e., autoantibodies, hybridoma production is clearly justified by the requirement for an immunological agent for purification by affinity chromatography or immunoprecipitation of receptor for structural studies. Furthermore, these reagents may provide a useful method for isolating mRNA coding for the receptor subunits.[35] A preparation of this type would allow study of the genetic organization of receptors and of the mechanisms by which a hormone/receptor pair coevolve to control a particular function.

[35] R. A. Maurer, *J. Biol. Chem.* **255,** 854 (1980).

[55] Site Specific Monoclonal Antibodies to Insulin

By TIMOTHY P. BENDER and JOYCE A. SCHROER

Monoclonal antibodies (mAb) offer several significant advantages over immune serum antibody in the study of immune responses or as probes for structural determinants. First, mAbs have defined specificity. Immune sera are mixtures of different antibodies with different antigen binding sites. Sera which have been adsorbed and appear monospecific do not contain a single antibody specificity. Second, hybridoma cell lines can be grown as acscites tumors which produce large amounts of mAb. Immune sera are not only limited in supply but they are mixtures of antibodies which are not identical from one animal to the next or even over time from the same animal. Reproducibility of reagents is not a problem with mAbs.

The immune response to insulin is T cell dependent[1] and is controlled by genes linked to the major histocompatibility locus (H-2) in mice.[2,3] Insulin has been well characterized structurally[4] and the regions of beef and pork insulin which are required for T cell proliferation *in vitro*[5] and the development of T cell help[2,3] in responder strains of mice are known. However, little is known about the structure of antigenic sites recognized by antibodies on protein antigens. Only recently have such studies begun using mAbs in other systems.[6,7]

Since the immune response to insulin has been well studied at the cellular level, we have initated studies to produce mAbs with specificity for beef, pork, and human insulin. These reagents are being used to compare the specificity of T and B lymphocyte antigen receptors for insulin and to define the nature and extent of molecular determinants recognized by B lymphocytes. This paper will describe the approach that we have taken to produce mAbs with specificity for insulin and to characterize their epitope specificity.

[1] F. K. Jansen, U. Kiesel, and D. Brandenburg, *Ann. Immunol.* (*Paris*) **128C,** 313 (1977).
[2] K. Keck, *Nature* (*London*) **254,** 78 (1975).
[3] J. Kapp and D. S. Strayer, *J. Immunol.* **121,** 978 (1978).
[4] T. Blundell, G. Dodson, D. Hodgkin, and D. Mercola, *Adv. Protein Chem.* **26,** 279 (1972).
[5] L. J. Rosenwasser, M. A. Barcinski, R. H. Schwartz, and A. S. Rosenthal, *J. Immunol.* **123,** 471 (1979).
[6] S. J. Smith-Gill, A. C. Wilson, M. Potter, E. M. Praqer, R. J. Feldman, and C. R. Mainhart, *J. Immunol.* **128,** 314 (1982).
[7] J. A. Berzofsky, G. K. Buckenmeyer, G. Hicks, F. R. N. Gurd, R. J. Feldman, and J. G. Minna, *J. Biol. Chem.* **257,** 3189 (1982).

METHODS IN ENZYMOLOGY, VOL. 109

ISBN 0-12-182009-2

General Consideration of Insulin Structure

The insulin monomer (molecular weight ≧5750) consists of two chains which are connected by two intrachain disulfide bonds (amino acid sequence of beef insulin and several species variants are shown in Fig. 1). The A-chain has two helical regions separated by a stretch of extended polypeptide chain from A9–A12. The region consisting of bases A8–10 formed by an intrachain disulfide bond between cystines at positions A6 and A11 has been called the A-chain loop. The B-chain consists of two segments of β-pleated sheets at the amino and carboxyterminal ends and a central region of helical structure. The helical region of the B-chain rests roughly between the two helical regions of the A-chain.

Since we have been primarily interested in producing mAbs against beef insulin, a brief comparison of the structures of beef and mouse insulin is in order. There are seven amino acid differences between beef and rat insulin (each individual rat or mouse has two circulating insulins designated I and II). Mouse and rat insulins have identical sequences. However, as is indicated in Fig. 1, only five of these differences B3, A4, B30, A8, and A10, are present in both mouse insulins. Recently described computer programs[8] have been used to generate a space filling theoretical model of beef insulin based on the X-ray coordinates of insulin. Each atom is represented as a nondeformable sphere with a characteristic van der Waals radius. The amino acid substitutions between beef and mouse insulin are located on this model in Fig. 2. All five of the substitutions differing from mouse insulins are located on the same face of the molecule in the region of the A-chain loop.

Production of mAbs with Specificity for Insulin

The protocols that we have used for making somatic cell hybrids between insulin primed murine spleen or lymph node cells and myeloma cell lines are adaptions of standard fusion protocols. For detailed discussions of the principles involved in somatic cell fusion the reader is referred to articles by Galfrè and Milstein[9] and Kennett[10] in this series. This section will deal with aspects of cell fusion that specifically relate to the production of anti-insulin mAbs. It should be noted that anti-insulin-producing hybridoma cell lines are relatively difficult to make compared to other anti-protein hybridomas. Though not all cell fusion experiments have resulted in anti-insulin-producing cell lines, most yield positive clones in

[8] R. J. Feldmann, D. H. Bing, B. C. Furie, and B. Furie, *Proc. Natl. Acad. Sci. U.S.A.* **75**, 5409 (1978).

[9] G. Galfrè and C. Milstein, this series, Vol. 73 [1].

[10] R. H. Kennett, this series, Vol. 93 [28].

Insulin A Chain

	1	2	3	4	5	6	7	8	9	10	11	12	13	14	15	16	17	18	19	20	21
Beef	Gly	Ile	Val	Glu	Gln	Cys	Cys	Ala	Ser	Val	Cys	Ser	Leu	Tyr	Gln	Leu	Glu	Asn	Tyr	Cys	Asn
Sheep	—	—	—	—	—	—	—	—	Gly	—	—	—	—	—	—	—	—	—	—	—	—
Pork	—	—	—	—	—	—	—	Thr	—	Ile	—	—	—	—	—	—	—	—	—	—	—
Chicken	—	—	—	—	—	—	—	His	Asn	Thr	—	—	—	—	—	—	—	—	—	—	—
Rat	—	—	—	Asp	—	—	—	Thr	—	Ile	—	—	—	—	—	—	—	—	—	—	—
Human	—	—	—	—	—	—	—	Thr	—	Ile	—	—	—	—	—	—	—	—	—	—	—
Rabbit	—	—	—	—	—	—	—	Thr	—	Ile	—	—	—	—	—	—	—	—	—	—	—
Fish	—	—	—	—	—	—	—	His	Lys	Pro	—	Asn	Ile	Phe	Asp	—	Gln	—	—	—	—
Guinea pig	—	—	—	Asp	—	—	—	Thr	Gly	Thr	—	Thr	Arg	His	—	—	Gln	Ser	—	—	—

Insulin B Chain

		1	2	3	4	5	6	7	8	9	10	11	12	13	14	15	16	17	18	19	20	21	22	23	24	25	26	27	28	29	30
Beef		Phe	Val	Asn	Gln	His	Leu	Cys	Gly	Ser	His	Leu	Val	Glu	Ala	Leu	Tyr	Leu	Val	Cys	Gly	Glu	Arg	Gly	Phe	Phe	Tyr	Thr	Pro	Lys	Ala
Sheep		—	—	—	—	—	—	—	—	—	—	—	—	—	—	—	—	—	—	—	—	—	—	—	—	—	—	—	—	—	—
Pork		—	—	—	—	—	—	—	—	—	—	—	—	—	—	—	—	—	—	—	—	—	—	—	—	—	—	—	—	—	—
Chicken		Ala	Ala	—	—	—	—	—	—	—	—	—	—	—	—	—	—	—	—	—	—	—	—	—	—	—	—	Ser	—	—	—
Rat		—	—	Lys	—	—	—	—	—	Pro*	—	—	—	—	—	—	—	—	—	—	—	—	—	—	—	—	—	—	—	Met†	Ser
Human		—	—	—	—	—	—	—	—	—	—	—	—	—	—	—	—	—	—	—	—	—	—	—	—	—	—	—	—	—	Thr
Rabbit		—	—	—	—	—	—	—	—	—	—	—	—	—	—	—	—	—	—	—	—	—	—	—	—	—	—	—	—	—	Ser
Fish	Val	Ala	Pro	Pro	—	—	—	—	—	—	—	—	—	—	Asp	—	—	—	—	—	—	Asp	—	—	—	—	—	Asn	—	—	‡
Guinea pig		—	—	Ser	Arg	—	—	—	—	—	Asn	—	—	—	Thr	—	—	Ser	—	—	Gln	Asp	Asn	—	—	—	—	Ile	—	—	Asp

FIG. 1. Comparison of the amino acid sequences (M. O. Dayhoff, ed., "Atlas of Protein Sequence and Structure," Natl. Med. Res. Found., Washington D. C., 1972, 1976, 1978) of bovine insulin and the insulin from several other species used in this study. Identities are shown by solid lines. Insulins were obtained as follows: mouse insulin and monocomponent beef, pork, and human insulin were purchased from Novo Institute, Bagsvard, Denmark, while chicken, rat, rabbit, fish, guinea pig, desoctapeptide (DOP) pork, desoctapeptide and single component sheep insulin were all generous gifts of Dr. Ronald Chance, Eli Lilly Company, Indianapolis, IN. * Present only in rat I; † present only in rat II; ‡ deletion.

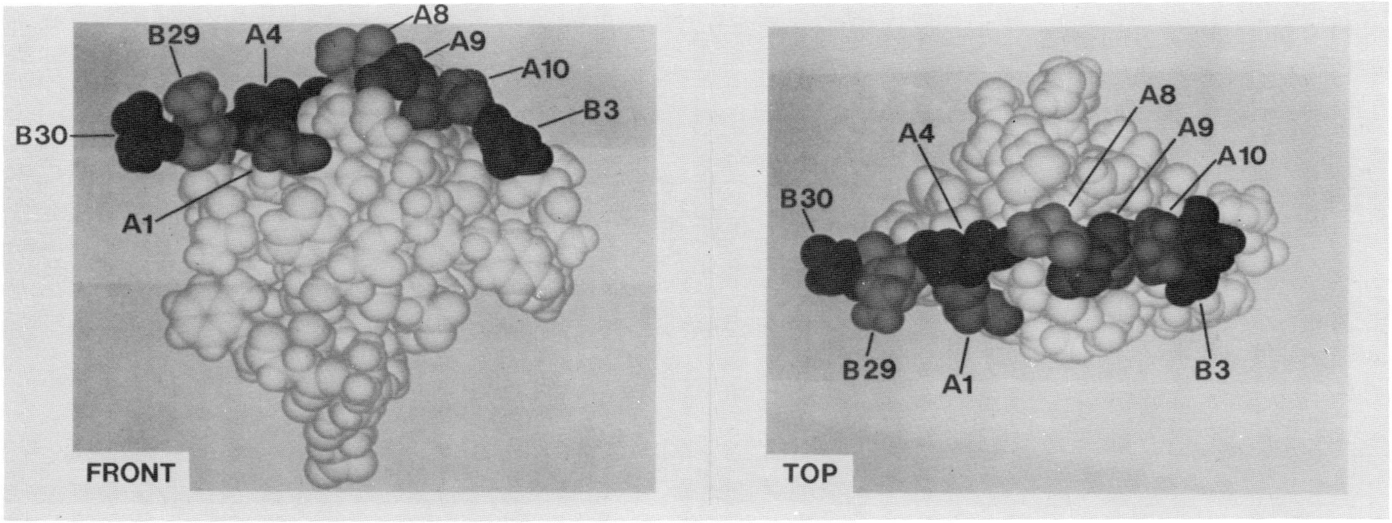

FIG. 2. Theoretical space-filling model of bovine insulin. The three-dimensional structure of beef insulin was modeled on the X-ray crystallographic coordinates of pork insulin by substituting amino acid replacements followed by rotation of side chains to minimize unfavorable van der Waals contacts. Photograph is courtesy of Dr. Richard J. Feldmann (National Institutes of Health, Bethesda, MD). The A-loop residues A8–10 as well as critical adjacent residues are shaded. The front view of insulin is arbitrary. The top view is the same as the previous panel but the molecule was rotated 90° from the front view.

about 1% of the wells seeded (about 4–10 cell lines per fusion). Occasionally up to 10% of the wells have yielded stable, anti-insulin, producing clones.

Immunizations

Since the immune response to insulin in mice is controlled by genes linked to the major histocompatibility complex (MHC) not all strains of mice make antibodies in response to insulin. Also those strains that do respond to insulin do not respond to all species of insulin. For example, BALB/c mice (H-2^d) make antibody in response to beef, pork, and human insulin but do not respond to sheep insulin. Thus, it is important to ascertain that the strain of mouse chosen for cell fusion will respond to the insulin of interest. Since most myeloma fusion partners are of BALB/c origin, hybridomas made with BALB/c immune cells do not require the use of F_1 lines or X-irradiated mice to be grown as ascites tumors. All of our mAbs with specificity for insulin were made with lymphocytes from BALB/c mice.

Lymphocytes derived from either spleens or lymph nodes have resulted in successful cell fusions. When lymph node cells were used, mice received a single dose of 50 μg of insulin in Freund's Complete Adjuvant (CFA) on day 0 in the rear footpads in a total volume of 0.1 ml and cell fusion performed 12–14 days later. Both popliteal and inguinal lymph nodes were used. When spleen cells were used mice were immunized by either of two protocols: (1) mice were primed on day 0 with 50 μg of insulin in CFA, intraperitoneally and then again with the same dose of insulin on day 21, or (2) mice were given 50 μg of insulin in CFA, intraperitoneally on days 0, 21, and 81. When spleen cells were used cell fusions were performed 3 days after the last boost. In the latter two cases, mice were given 0.1 ml of 10% glucose intraperitoneally 30 min before immunization to prevent hypoglycemia.

Fusion Partners

A number of cell lines with drug markers are presently available for use as fusion partners. The properties of the four cell lines that we have had experience with are shown in Table I. The myeloma lines 45.6TG and NS-1 have both been successfully used in the production of anti-insulin mAbs. However, they both suffer from the disadvantage of producing an indigenous light chain (NS-1) or both heavy and light chains (45.6TG). The use of these cell lines may result in hybridoma lines which produce mixed molecules with more than one heavy and/or light chain. This could result in antibodies of seriously altered specificity. The SP2/0-Ag14 cell

TABLE I
CHARACTERISTICS OF FUSION PARTNERS

Cell line	Homologous chains	Drug sensitivity	Ref.
45.6TG	γ_{2b}, κ	5 μg/ml 6-thioguanine	*a*
NS1/1.Ag4.1	Cytoplasmic κ	5 μg/ml 6-thioguanine	*b*
SP2/0-Ag14	None	20 μg/ml 8-azaguanine	*c*
X63-Ag8.653	None	20 μg/ml 8-azaguanine	*d*

[a] D. H. Marguiles, W. M. Keuhl, and M. D. Scharff, *Cell* **8,** 405 (1976).
[b] G. Köhler, S. C. Howe, and C. Milstein, *Eur. J. Immunol.* **6,** 292 (1976).
[c] M. Schulman, C. D. Wilde, and G. Köhler, *Nature (London)* **276,** 269 (1978).
[d] J. F. Kearney, A. Radbruch, B. Liesgang, and K. Rajewsky, *J. Immunol.* **123,** 1548 (1979).

line does not produce homologous heavy or light chains and is the fusion partner that we currently use. Although X63-Ag8.653 does not produce homologous heavy or light chains and has been used successfully in a number of antigen systems, we have not been successful in using it to make anti-insulin producing hybridoma cell lines.

Fusion

The two fusion protocols that we used are described in Table II. SP2/0-Ag14 is grown for 1 week with and without 8-azaguanine at 20 μg/ml in Iscoves media supplemented with 15% heat-inactivated fetal calf serum prior to cell fusion to remove possible revertants.

Assay for Production of Anti-Insulin Antibody

Two to four weeks after cell fusion the hybridoma colonies are large enough to be tested for production of anti-insulin antibody. We have used an enzyme-linked immunoabsorbent assay (ELISA) system to identify colonies making anti-insulin mAb. In our assay, insulin is allowed to adsorb to the wall of a plastic microtiter well and tissue culture supernate from a hybridoma colony is added. If a given tissue culture supernate contains anti-insulin mAb, it will bind to the insulin and is then detected using an alkaline phosphatase (AP) tagged goat anti-mouse κ chain, goat anti-mouse IgG or goat anti-mouse IgM (Cappell Laboratories, Cochranville, PA). Though we conjugate AP to these reagents by the method of Keren[11] similar reagents are available on the commercial market already conjugated with AP.

[11] D. F. Keren, *Infect. Immun.* **24,** 441 (1979).

TABLE II
COMPARISON OF FUSION PROTOCOLS

	Primary immunization	Hyperimmunized
Source of immune cells	10^8 lymph node	10^8 spleen
Fusion partner	10^7 SP2/0-Agl4	10^7 SP2/0-Agl4
Steps in protocol	Modified from Nowinski[a]	Modified from Gefter[b]
	Wash above separately in serum free media	Wash above separately in serum free media
	Pellet together at 600 *g*	Pellet together at 500 *g*
	Dropwise add 1 ml 50% PEG (MW 1000), resuspend pellet gently	Add 0.2 ml 30% PEG (MW 4000), gently agitate to break up pellet
	Add dropwise an additional 15 ml warm media over next 5 min then up to 50 ml with warm media	Centrifuge 6 min at 500 *g* resuspend in 20 ml Iscoves 20% FCS
	Pellet 600 *g* 7 min	Resuspend in 20 ml Iscoves 20% FCS plate in 2–100 mm dishes overnight at 37° in 7% CO_2
	Resuspend in 50 ml 10% FCS Iscoves containing 1× HAT and plate in microwells at 2×10^5 immune cells/well with $2-10 \times 10^4$/well normal peritoneal macrophages as feeders[c]	Add 10 ml more Iscoves 20% FCS to each plate incubate another 24 hr
		Pellet cells at 500 *g*
		Resuspend in 1× HAT in Iscoves 20% FCS in microwells at $\sim 5 \times 10^5$ immune cells/well
	Colonies appear in 5–14 days	Colonies appear in 5–14 days
	Established clones described in this paper AD2, CA4, AE9D6	Established clones described in this paper 5.7E9 and 5.4C5

[a] R. C. Nowinski, M. E. Lostrom, M. R. Tam, M. R. Stone, and W. N. Burnette, *Virology* **93**, 111 (1979).
[b] M. L. Gefter, D. H. Marguiles, and M. D. Scharff, *Somatic Cell Genet.* **3**, 231 (1977).
[c] S. F. de St. Groth and D. Scheidegger, *J. Immunol.* **35**, 1 (1980).

The assay procedure for anti-insulin antibody in hybridoma supernatants follows:

1. Add 100 μl of insulin in phosphate-buffered saline with 0.04% NaN_3 (PBS-A) at a concentration of 10 μg/ml to each well of a polyvinyl chloride round bottom microtiter plate (Dynatech Laboratories, Inc., Alexandria, VA) (coat 3 wells for each colony to be tested) and incubate for 4 hr at room temperature (RT) in a humidified chamber.

2. Wash plate 4 times with PBS-A.

3. Saturate nonspecific sites on the plastic by flooding each plate with 1% bovine serum albumin (BSA) (RIA Grade, Sigma Chemical Co., St. Louis, MO) in PBS-A for 30 min at RT.

4. Wash plate 4 times with PBS-A.

5. Transfer 30 μl of tissue culture supernates from each colony to be tested to each of the three insulin coated wells and then add 70 μl of 1% BSA in PBS-A. Incubate for 4 hr at room temperature.

6. Wash each plate 4 times with PBS-A.

7. Add 100 μl of a working dilution of each AP conjugated developing reagent (dilute AP-conjugated reagents in 0.2 *M* Tris buffer pH 8.0 containing 1% BSA, 1 m*M* $MgCl_2$, and 0.02% NaN_3). Incubate 8 hr to overnight at RT.

8. Wash each plate 4 times with PBS-A.

9. Add 100 μl of *p*-nitrophenyl phosphate (phosphatase substrate, Sigma) at a concentration of 1 mg/ml in 0.5 *M* carbonate buffer pH 9.8 containing 1 m*M* $MgCl_2$. This solution should be made up fresh just prior to use.

Assays using AP-conjugated reagents turn yellow and may be quantitated spectrophotometrically using an ELISA reader at 405 nm. However, for screening hybridoma supernates this is not necessary as visual comparison of positive wells with negative controls (fresh Iscove's + 20% FCS in place of test supernate is the best negative control) will readily differentiate positive from negative wells. It should be noted that the use of these reagents allows one to differentiate IgG from IgM mAbs. In our experience most anti-insulin mAbs bear κ light chains (35/35) and γ heavy chains (34/35). This assay has also been used to differentiate the γ subclasses of heavy chains using subclass specific antisera. When 18 anti-insulin mAbs were examined with these reagents, 14 possessed the γ_1 isotype and two each the γ_{2_a} and γ_{2_b} isotypes. To date, we have only characterized one μ heavy chain bearing anti-insulin mAb.

Growth of Hybridoma Lines as Ascites

Once hybridoma lines are identified which produce anti-insulin mAbs, they may be expanded, cloned by limiting dilution, and frozen away. These procedures have been previously discussed in this series.[9] To obtain large quantities of mAb for purification and subsequent studies hybridoma cell lines should be grown *in vivo* in mice as ascites tumors. Although these lines can be grown indefinitely in tissue culture, the yield of mAb from tissue culture supernates is only about 10–100 μg/ml while 5–20 mg/ml can be obtained from ascites fluid. Hybridoma cell lines should be expanded gradually from 96-well microtiter wells to 24-well dishes and finally 100-mm petri dishes (tissue culture grade dishes are not required for lymphocyte growth). Mice which are to be used for ascites production must be given an intraperitoneal injection of 0.5 ml Pristane 1–5 weeks prior to use. This prepares the peritoneal cavity for tumore growth. Mice are then given approximately 10^6 hybridoma cells in DMEM at total volumes of 0.1–0.5 ml. Swelling of the abdominal region will occur in 1–3 weeks signifying growth of the hybridoma cells as an ascites producing tumor. Ascites fluid containing mAb can be drawn off by puncturing the peritoneal cavity with an 18-gauge needle and allowing the ascites fluid to drain into a tube through the needle. Ascites are then pooled and stored frozen at −70°.

Purification of Anti-Insulin mAb

The standard method of coupling proteins to cyanogen bromide-activated Sepharose beads has not been useful in producing an immunoadsorbent suitable for the affinity purification of anti-insulin mAb. In coupling small proteins such as insulin to solid supports at least three points should be considered:

1. It is desirable to attach the protein by one or a very few sites to the solid support to maintain accessability for specific antibody binding.
2. The coupling sites should have little affect on the overall conformation of the protein so that most antibodies produced will bind.
3. The protein should be spaced away from the solid support surface to reduce steric hindrance.

We have adopted an earlier method which was devised to couple beef and pork insulin to sheep erythrocytes for use in plaque assays.[12]

In this protocol two hetero-bifunctional cross-linkers, 2-iminothiolane–HCl (MBI) and *m*-maleimidobenzoic acid *N*-hydroxysuccinimide ester (MBS), are used to couple beef or pork insulin to lysine-Sepharose via the lysine residue at position B29. Briefly, MBS was used to acylate amino groups of insulin via its reactive *N*-hydroxysuccinimide ester group (only one lysine is present in the insulin molecule, B29). MBI reacts with the ε-amino groups of lysine residues on the lysine-Sepharose through its imidate function. MBI-insulin and MBS-lysine-Sepharose are then combined through formation of a thioether bond by addition of the thiol group of MBI-lysine-Sepharose to the double bond of the malimide group of MBS.

[12] J. A. Schroer, J. K. Inman, J. W. Thomas, and A. S. Rosenthal, *J. Immunol.* **123,** 670 (1979).

Preparation of Beef Insulin-Sepharose Immunoadsorbent

During the preparation of beef insulin-Sepharose, the MBS-insulin and MBI-lysine-Sepharose should be prepared simultaneously to minimize the loss of reactive groups. Lysine-Sepharose (Sigma) is purchased in lyopholized form and should be swollen before use in several changes of bicine–saline buffer (0.02 *M* bicine, 0.01 *M* NaOH, 0.15 *M* NaCl, pH 8.3). The proportions given here are to prepare approximately 1.0 ml of beef insulin-Sepharose. The procedure can be scaled up to make any amount desired.

MBI-Sepharose is prepared by adding approximately 1.0 ml of swollen lysine-Sepharose to 3.0 ml of bicine–saline buffer. Four milligrams of MBI is added and the mixture is allowed to react at RT for 2.5 hr with constant mixing on a slowly rotating wheel. After 2.5 hr the reaction mix is made 0.1 *M* in dithiothreitol (DTT) by adding solid DTT and the reaction allowed to proceed another 30 min. The MBI-Sepharose is washed 5 times in bicine–saline buffer and finally resuspended in the same buffer.

MBS-beef insulin is prepared by dissolving 25 mg of insulin in 3 ml bicine buffer (0.1 *M* bicine, 0.05 *M* NaOH, pH 8.3. Thirteen milligrams of MBS is dissolved in 3.0 ml dimethylformamide and added immediately to the insulin solution, vortexing immediately. The reaction is allowed to proceed for 2 hr at RT. The MBS-insulin is then precipitated with 6.0 ml citrate-phosphate buffer (1.0 *M* citric acid, 1.0 *M* K_2HPO_4, pH 5.3), centrifuged at 12,000 *g* for 10 min, washed twice in dilute citrate-phosphate buffer (0.01 *M* citric acid, 0.01 *M* K_2HPO_4, pH 5.3), and centrifuged at 12,000 *g* for 10 min. The pellet is then dissolved in 1.5 ml bicine–saline buffer.

The washed MBS-insulin is added directly to the MBI-Sepharose and allowed to react for 2 hr at RT on a slowly rotating wheel. After 2 hr the insulin-Sepharose is washed once with bicine–saline buffer, resuspended in 3 ml of bicine–saline buffer, and made 0.01 *M* in iodoacetamide. This is allowed to react for 30 min at RT with rotation and then washed 10 times in phosphate buffer (0.078 *M* NaCl, 0.056 *M* Na_2HPO_4, 0.016 *M* KH_2PO_4, pH 7.2). The insulin-Sepharose may be stored indefinitely at 4° as a suspension in PBS-A. When a trace amount of ^{125}I-labeled beef insulin was included in the reaction, the efficiency of coupling was approximately 36%.

Affinity Purification of Anti-Insulin Antibodies

Columns for affinity purification using the beef insulin-Sepharose as an immunoadsorbent are prepared in disposable plastic syringes. A small amount of glass wool is placed in the bottom of the syringe and a slurry of insulin-Sepharose in phosphate buffer is poured over it. The column should be washed extensively before use with phosphate buffer, then 5 volumes of 0.1 *M* glycine–HCl (pH 2.2) and then PBS-A.

Ascites fluid samples containing anti-insulin mAb should be twice 40% cut with $(NH_4)SO_4$, resuspended to the original volume in phosphate buffer, and dialyzed exhaustively against phosphate buffer. Samples are applied to the top of the column and allowed to slowly pass over the immunoadsorbent at RT. The column should then be washed extensively with phosphate buffer until the OD at 280 nm of the effluant is <0.005, followed by two column volumes of 0.05 *M* carbobenzoxyglycine (CBZ-glycine; Sigma) and then PBS-A. The antibodies can then be eluted from the column with 0.1 *M* glycine–HCl pH 2.2. Fractions should be collected in an equal volume of 10-fold concentrated phosphate buffer to neutralize the glycine–HCl and assayed by optical density at 280 nm for protein. Although these columns have not been extensively characterized for capacity, a 3 ml ascites sample passed over a column made with 1 ml of insulin-Sepharose typically yields 5–20 mg of mAb.

As a cautionary note, one anti-beef insulin mAb, CA-4 (see Table III), could not be purified from this column. Presumably this is because the epitope to which CA4 binds on insulin includes the B29 lysine of beef insulin. Although this is the only anti-insulin mAb to date that could not be purified using this column, others will probably be found. As an alternative to affinity purification, mAb bearing γ-heavy chains are readily purified by ion exchange using A-25 DEAE-Sephadex (Sigma) in 0.01 *M* Tris–PO_4 buffer pH 8.2. Under these conditions mAb with γ isotype heavy chains come off the column in the "fall through." When assayed for purity by isoelectric focusing or SDS–polyacrylamide electrophoresis (PAGE) these mAbs were >95% pure.

TABLE III

CHARACTERISTICS OF ANTI-INSULIN MONOCLONAL ANTIBODIES (mAb)

Clone name	Immunogen[a]	Isotype	Specificity
AE9D6	Human	$\gamma_1\kappa$	A-chain loop
AD2-2	Beef	$\gamma_1\kappa$	A-chain loop
5.7E9	Beef	$\gamma_1\kappa$	A-chain loop
CA-4	Beef	$\gamma_{2b}\kappa$	B29, A4, A8
5.4C5	Beef	$\gamma_1\kappa$	—

[a] Species of insulin used to prime mice that were sources of immune cells for fusion.

Fine-Specificity of Anti-Insulin mAb

By correlating differences in the ability of species variants of insulin to inhibit the binding of a given mAb to its ^{125}I-labeled immunogen with differences in primary sequences of insulin we have been able to identify regions of the insulin molecule which are bound by four mAbs (Table III). To perform this analysis, nine species variants of insulin, pork, and beef proinsulin as well as desoctapeptide beef or pork insulin (insulin with the last eight residues of the B-chain removed), have been used. The assay is a modification of that reported by Desbuquois and Aurbach[13] where polyethylene glycol is used to precipitate antibody and antibody-bound ^{125}I-labeled insulin while leaving unbound ^{125}I-labeled insulin in solution. The assay is performed by first determining how much mAb is required to bind ~50% of input ^{125}I-labeled immunogen insulin. Then by adding varying amounts of cold, competing insulin while holding the amount of mAbs constant, an inhibition curve is generated by plotting the percentage ^{125}I-labeled insulin bound vs the concentration of inhibitor. Using species variants of insulin as inhibitors a series of inhibition curves are generated and the amounts of species variant insulin required for 50% inhibition (I_{50}) are compared to the amount of the cold immunogen insulin giving I_{50}. The species variant assay is performed as follows:

1. Twenty microliters of cold species variant insulin (at 10-fold increments over a million-fold concentration range) is added to a 0.5 × 5-cm glass tube.
2. Add 20 μl of antibody (predetermined dilution of ascites or purified antibody that binds ~50% of cpm ^{125}I-labeled insulin added) and vortex.
3. Add 20 μl of ^{125}I-labeled insulin (10–15 × 10^3 cpm) and vortex.
4. Allow to react at 22° for 1 hr.
5. Add 50 μl 27% polyethylene glycol (MW 6–7.5 × 10^3, J. T. Baker Chemical Co., Phillipsburg, NJ) vortex.
6. Centrifuge at 1800 *g* for 1 hr.
7. Aspirate supernate and count pellet in a gamma counter. Experimental and control tubes are done in triplicate. BSA (1%) in PBS-A is the diluent for inhibitors, antibody, and label. Controls consist of tubes which do not contain inhibitor, tubes without antibody, and tubes with high concentration of antibody added for maximum precipitation of label. The data are expressed as percentage binding vs concentration of inhibitor where 100% binding is equated to binding in the absence of inhibitor.

[13] B. Desbuquois and G. D. Aurbach, *J. Clin. Endocrinol. Metab.* **33,** 732 (1971).

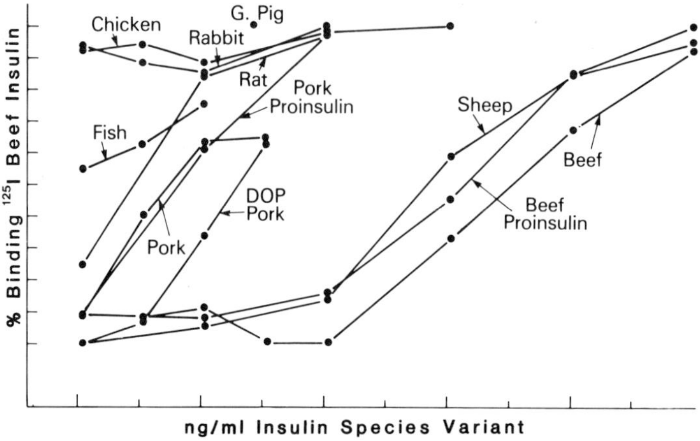

FIG. 3. Competitive binding curves by species variants of insulin between mAb AD2-2 and ^{125}I-labeled beef insulin. Each point represents the mean of triplicates.

Fine specificity curves generated using this assay for the anti-beef insulin mAb AD2-2 are shown in Fig. 3. Similar data compiled for AD2-2 and four other anti-insulin mAbs (one anti-human insulin mAb and three made against beef insulin) are shown in Table IV. The data in Table IV are expressed as the ratio of the concentration of each species variant needed to inhibit 50% of mAb binding to ^{125}I-labeled immunogen insulin compared to the concentration of cold immunogen insulin required to give the same inhibition. Numbers <5 indicate that a species of insulin is as good a competitor as the immunogen. Higher numbers indicate that many times the concentration of a species variant of insulin compared to the immunogen is needed to inhibit the binding of ^{125}I-labeled insulin. The antigenic determinants recognized by mAbs AD2-2, AE9D6, and 5.7E9 involve the A-chain loop region of insulin.[14] AD2-2 is specific for the sequence Ala-Ser-Val at positions A8–10 in beef insulin (see Fig. 1). Pork, human, and rabbit insulin which differ at positions A8 and A10 are poor inhibitors (>200). Chicken, fish, and guinea pig insulin, which differ at all three positions, are even worse inhibitors (>1000). Three- to 4-fold more rat insulin than pork, human, and rabbit insulin which has the same A-chain loop sequence but substitutions at positions A4 and B3, is required to inhibit the binding of AD2-2 to ^{125}I-labeled beef insulin. As residues A4 and B3 are in close proximity to the A-chain loop region this finding suggests that these positions may be included in the antigenic site recog-

[14] J. A. Schoer, T. P. Bender, R. J. Feldman, and K. J. Kim, *Eur. J. Immunol.* **13,** 693 (1983).

TABLE IV
50% INHIBITION RATIOS OF INSULIN SPECIES VARIANTS TO IMMUNIZING INSULIN

Insulin	Monoclonal antibodies AE9D6	AD2-2	5.7E9	CA-4	5.4C5
Beef	113[a]	1	1	1	1
Human	1	246	7	10	6
Pork	0.1	324	36	15	1
Pork proinsulin	3	527	>500[b]	>17	>26
Beef proinsulin	110	4	146	>17	>26
Rabbit	4	>3000	11	12	1
Rat	38	1200	180	23	2
Sheep	5000	7	1	1	1
Chicken	>2000	>3000	90	50	>3
Guinea pig	>5000	>1000	>112	>3	>5
Fish	>5000	>3000	101	6	>26
DOP beef	811	168[c]	2	6[c]	>26
Immunogen ng/ml giving 50% inhibition	18	645	890	10,000	9000

[a] Ratio of amount (ng/ml) of each insulin species variant divided by the amount of beef or human (AE9D6) insulin needed to inhibit the binding of a limiting dilution of each mAb.

[b] > indicates that highest concentration of species variant insulin did not give 50% inhibition.

[c] Desoctapeptide pork insulin used.

nized by this mAb. Similar reasoning shows that mAb AE9D6 is specific for Thr-Ser-Ile sequence of positions A8–10 of the A-chain of human insulin. In this case, more rat insulin than pork or human insulin is required for 50% inhibition though all have the Thr-Ser-Ile sequence. This again suggests that the substitutions in rat insulin at positions A4 and B3 may be part of this antigenic site.

mAb 5.7E9 also appears to bind the A-chain loop of beef insulin but with an order of magnitude less specificity than AD2-2. Type I specificity must be involved in this case rather than affinity as defined by Berzofsky and Schechter[15] as AD2-2 and 5.7E9 have similar affinities for insulin. The binding site of 5.7E9 on the beef insulin A-chain loop region must also differ geometrically from that of AD2-2 or AE9D6 as 5.7E9 does not bind to its determinant as it is present on proinsulin.

The antigenic determinant on beef insulin recognized by CA-4 is thought to be a conformational determinant which includes residues B29, A4, and A8. These residues lie in a line on the surface of insulin which

[15] J. A. Berzofsky and A. N. Schechter, *Mol. Immunol.* **18,** 751 (1981).

includes the A-chain loop (see Fig. 2). Residue B29 was implicated in the determinant because CA-4 does not bind to beef insulin coupled predominantly through this residue to lysine-Sepharose (see previous section). A8 appears to be involved in this determinant as 10- to 15-fold increases in concentration of human, pork, and rabbit insulin are required to achieve the same inhibition as beef insulin. Substitution of glycine for serine at position A9 in sheep insulin does not affect binding. Residue A4 or B3 appear to be involved as rat insulin is a 20-fold poorer inhibitor than beef insulin. Most likely, A4 is involved as it is located between residues B29 and A8 in the tertiary structure of insulin. However, this assignment remains tentative because desoctapeptide beef insulin, which lacks B29, is a good inhibitor of CA-4 binding to ^{125}I-labeled beef insulin.

Finally, mAb 5.4C5 represents a group of 12 mAb (out of 18 that have been studied in this manner) which bind rat insulin as well as beef insulin. We cannot localize regions of the insulin molecule bound by these mAbs although they can be differentiated by whether or not they bind beef or pork proinsulin.[14]

Dual Binding Assay

We have used a dual binding assay which examines the ability of two mAbs to bind simultaneously to insulin. These studies were initiated to look for overlapping antigenic sites on the insulin molecule. To perform this assay one mAb (mAb-1) is allowed to adsorb to the well of a polyvinyl chloride microtiter dish. After saturating nonspecific sites on the plastic with 0.1% KLH in phosphate buffer, insulin is added. Unbound insulin is washed away and a second mAb (mAb-2) which has been tagged either with ^{125}I or biotin is added. Clearly, two mAbs which can bind to insulin simultaneously must bind distinct antigenic determinants. If two mAb cannot bind simultaneously to insulin they may compete for the same determinant, overlapping antigenic determinants, or simply bind in close enough proximity on the surface of insulin that they sterically prevent each other from binding. This type of analysis has been performed in other systems[16] and discussed previously in this series.[17]

Two important controls are included for each pair of mAbs examined. First, insulin is replaced in the assay by keyhole limpet hemacyanin. Second, mAb-2 is allowed to present insulin to tagged mAb-2. This control is included to ensure that simultaneous binding of two mAbs is not due to dimerization of insulin or the possibility of repeating determinants

[16] Y. Kohno, B. I. Berkower, J. Minna, and J. A. Berzofsky, *J. Immunol.* **128,** 1742 (1982).

[17] C. Stähli, V. Miggiano, J. Stocker, T. Staehelin, P. Häring, and B. Takàcs, this series, Vol. 92 [20].

on the insulin molecule. When some mAb pairs were examined with the ELISA system we found a high background when insulin was replaced with keyhole limpet hemacyanin. In this case, an excess of insulin was included at the step where biotin-mAb-2 was added. Only cases where this decreased binding were considered to be simultaneous binding.

Two systems were used to tag mAb-2 because, with some pairs, reciprocal binding was not obtained. This is most likely the result of an alteration of the antigen binding site on mAb-2 by one of the labelling procedures. Iodination with carrier-free ^{125}I was performed by the standard chloramine T method.[18] Biotinylation was performed by the method of Jackson *et al.*[19] Each mAb pair was tested 2–5 times by both the radioimmunoassay and ELISA methods. Only radioimmunoassay values >2-fold above background were considered significant and binding of 0.2–7 ng of ^{125}I-labeled mAb (2–75 × 10^3 cpm) was detected. ELISA values of 492 nm ranged from 0.1 to 1.9 OD units above background.

Assays with ^{125}I and biotin tagged mAb-2 were run simultaneously and were treated identically as follows:

1. One hundred microliters of either an affinity purified mAb or twice 40% ammonium sulfate cut ascites fluid at a 10^{-2} dilution (approximately 10 μg/ml in phosphate buffer) was allowed to bind to a polyvinyl chloride microtiter plate (Dynatech) for 4–16 hr at room temperature.
2. Plates were washed 3 times with phosphate buffer containing 0.005% Tween 20 (PBS/TW20) and then a KLH solution (0.1% KLH in PBS/TW20) was allowed to saturate nonspecific sites for 30 min at room temperature.
3. Plates were rinsed 3 times with PBS/TW20 and 100 μl of beef insulin was added to each test well at 1 μg/ml in PBS.
4. Unbound beef insulin was washed away with 5 rinses of PBS/TW20.
5. mAb-2 was added in 100 μl/well. Biotinylated mAb-2 was used at approximately 83 ng/ml (preliminary titration of this reagent should be performed to determine an appropriate working dilution). ^{125}I-labeled mAb2 was used at 7 × 10^4–1.2 × 10^5 cpm/well (7–112 ng of mAb/well). Both were allowed to bind 8–16 hr at RT.
6. Plates were washed 5 times with PBS/TW20 and counted in a gamma counter or developed for ELISA.[19]

A summary of the results obtained by examining the five mAbs discussed above, using the dual binding assay, is shown in Table V. The

[18] F. C. Greenwood, W. M. Hunter, and J. S. Glover, *Biochem. J.* **89,** 114 (1963).

[19] S. Jackson, J. A. Sogn, and T. J. Kindt, *J. Immunol. Methods* **48,** 299 (1982).

TABLE V
SUMMARY OF DUAL BINDING RESULTS

	AE9D6[a]	AD2-2	5.7E9	CA-4	5.4C5
AE9D6	–[b]	–	–	+	+
AD2-2	–	–	–	–	+
5.7E9	–	–	–	+	+
CA-4	+	–	+	–	+
5.4C5	+	+	+	+	–

[a] Binding of ^{125}I or biotin-tagged liquid phase mAb (mAb-2) to beef insulin as presented by solid phase mAb (mAb-1).

[b] A (–) indicates no detectable binding while a (+) indicates binding of >2-fold above background by RIA or 0.1–1.9 OD at 495 nm of ELISA.

mAbs listed in the lefthand column are those immobilized on the plastic well (mAb-1) while those listed across the top were tagged with ^{125}I or biotin (mAb-2). As predicted from the fine specificity analysis using the insulin species variants AD2-2, AE9D6, and 5.7E9 compete for simultaneous binding to insulin. However, when these mAbs were studied in terms of their binding to insulin as presented by CA-4, both AD2-2 and 5.7E9 gave a positive result. AD2-2 could not bind simultaneously to insulin presented by CA-4. The fine specificity analysis suggested that either residues A4 or B3 was involved in the antigenic determinant recognized by AE9D6, AD2-2, and 5.7E9. Since CA-4 probably binds to a determinant involving residues B29, A4, and A8, the dual binding assay indicates that residue A4 is not involved in the antigenic determinant recognized by AE9D6 and 5.7E9. This result shows that A4 may be recognized by AD2-2 while residue B3 is included in the determinant bound by AE9D6 and 5.7E9.

The dual binding assay shows that the A-chain loop region may be a topographical region of insulin that includes several overlapping determinants. However, 5.4C5 binds simultaneously to insulin as it is presented by all four of the A-chain loop region binding mAb. This clearly demonstrates that the determinant bound by 5.4C5 is distinct from that recognized by the A-chain loop region binding mAbs.

Concluding Remarks

We have presented a strategy for the production, purification, and characterization of mAb made against insulin. These mAbs have been used to examine the antigenic surface of beef insulin. Clearly the area of beef insulin which includes the A-chain loop defines a topographical region of overlapping antigenic determinants on the insulin molecule. When the dual binding assay was used to examine 18 mAbs, we found that none of the mAb had an identical pattern of pluses and minuses which indicates that none of the 18 mAbs binds to identical determinants on the surface of insulin.[14] This is of particular interest in the case of mAbs AE9D6 and 5.7E9 because it indicates that even though they clearly compete for very similar antigenic determinants, they do not bind an identical determinant. Thus, we conclude that the murine B cell repertoire for beef insulin as examined by somatic cell hybridization is quite heterogeneous. Using the identical pair method of Briles and Carroll[20] for estimating repertoire size from hybridoma data to analyze our 18 mAbs to insulin, the probable number of antibodies is $\geqq 115$. Since no identical pairs of anti-insulin mAb have yet been found no upper limit can be estimated. We have only begun to explore this extensive repertoire of antibodies to what is considered to be a very small protein. Clearly mAbs of defined 3-D specificity such as those described in this paper will prove useful to immunologists and endocrinologists interested in studying the immune response to insulin[21] as well as functionally mapping the molecule.

[20] D. E. Briles and R. J. Carroll, *Mol. Immunol.* **18,** 29 (1981).

[21] T. P. Bender, J. Schroer, and J. L. Claflin, *J. Immunol.* **131,** 2882 (1983).

Section VIII

General Methods

[56] Chemical Deglycosylation of Glycoprotein Hormones

By P. MANJUNATH and M. R. SAIRAM

General Considerations

Glycoprotein hormones[1] of all species including nonmammalian that have been investigated to date possess varying amounts of carbohydrate as part of their structure. These hormones extracted from the pituitary generally contain 15–20% carbohydrate while the placental hormones contain higher amounts of up to 45% as in eCG.[1a,2] The two structural features that are unique to these hormones are (1) the presence of a quaternary structure and (2) the covalent attachment of carbohydrate units in both the α and β subunits. In this respect they are different from all other secreted hormones presently known. There has been considerable interest recently regarding the functional significance of the presence of these carbohydrate moieties. This question can be answered if suitable methods are available for the quantitative removal of carbohydrate units without affecting the structural integrity of the polypeptide moiety. At the present time, this problem has been approached in two different ways. The first approach which is enzymatic involves the use of highly purified exoglycosidases such as mannosidase, α-fucosidase, β-galactosidase, neuraminidase, β-*N*-acetylgalactosaminidase, and β-*N*-acetylglucosaminidase from various sources to treat the gonadotropins in a sequential manner with removal of the enzyme after each step. Despite the expected presence of the sugar residues at the peripheral surface of the hormone molecule these reactions are rather slow and with the exception of removal of terminal sialic acid, removal of other sugar residues is not always complete. This approach is ineffective with pituitary lutropin[3] because of the presence of sulfate on the peripheral hexosamines. This apparently renders the carbohydrate moiety of the hormone resistant to the action of mixed exoglycosidases. Several endoglycosidases are also capable of effectively removing the carbohydrate units from glycopro-

[1] Abbreviations: oLH, ovine lutropin; bLH, bovine lutropin; oFSH, ovine follitropin; hCG, human chorionic gonadotropin; eCG, equine chorionic gonadotropin; hTSH, human thyrotropin; prefix DG, deglycosylated; HF, anhydrous hydrogen fluoride; TFMS, trifluoromethanesulfonic acid.

[1a] J. G. Pierce and T. F. Parsons, *Annu. Rev. Biochem.* **50,** 465 (1981).

[2] M. R. Sairam, *in* "Hormonal Proteins and Peptides" (C. H. Li, ed.), Vol. 11, pp. 1–79. Academic Press, New York, 1983.

[3] T. F. Parsons and J. G. Pierce, *Proc. Natl. Acad. Sci. U.S.A.* **77,** 7089 (1980).

ISBN 0-12-182009-2

teins. Because of their limited availability, they have not been extensively applied to selectively degrade glycoprotein hormones.

The second approach is to use a chemical method to strip off the carbohydrate moiety from the glycoproteins. Three reagents, namely periodate,[4,5] anhydrous hydrogen fluoride (HF),[6] and trifluoromethane sulfonic acid (TFMS)[7] have been used for this purpose. Chemical deglycosylation involving periodate oxidation is invariably incomplete and also oxidizes amino acid residues such as cystine and tyrosine.[5] Chemical deglycosylation employing TFMS is quite effective but is not useful for gonadotropins (see Alternatives to the HF Method).

Chemical deglycosylation employing anhydrous hydrogen fluoride results in rapid removal of neutral sugars and acidic sugars. Sugars involved in O-glycosidic linkages are cleaved slowly but those attached in N-glycosidic linkages are resistant to HF treatment.[6] The peptide backbone is not affected. The original work-up procedure for recovery of deglycosylated product although good for most glycoproteins is unsuitable for oligomeric glycoproteins such as gonadotropins. This method has been modified (see below) by us such that it can be used for deglycosylation of gonadotropins. The modified procedure is reproducible, simple, quick, and provides an excellent alternative to the use of sequential deglycosylation using purified exoglycosidases. The following description of a method for deglycosylation is based on our own experience with the application of anhydrous HF under controlled conditions. Using this method, deglycosylated derivatives of oLH,[8,9] bLH,[10] oFSH,[11] hCG,[12] oLH subunits,[13] eCG, and hTSH[14] have been prepared. The deglycosylated hormones (DG hormones) serve as excellent probes in the study of hormone action at the molecular and cellular levels.

Methodology

HF Apparatus. The apparatus used for the HF treatment was originally designed for the deprotection of synthetic peptides after their chemi-

[4] I. I. Geschwind and C. H. Li, *Endocrinology* **63,** 449 (1958).

[5] J. Gan, H. Papkoff, and C. H. Li, *Biochim. Biophys. Acta* **170,** 189 (1968).

[6] A. J. Mort and D. T. A. Lampert, *Anal. Biochem.* **82,** 289 (1977).

[7] A. Edge, C. R. Faltynek, L. Hof, L. E. Reichert, Jr., and R. Weber, *Anal. Biochem.* **118,** 131 (1981).

[8] M. R. Sairam and P. W. Schiller, *Arch. Biochem. Biophys.* **197,** 294 (1979).

[9] P. Manjunath, M. R. Sairam, and P. W. Schiller, *Biochem. J.* **207,** 11 (1982).

[10] M. R. Sairam, *Biochim. Biophys. Acta* **717,** 149 (1982).

[11] P. Manjunath, M. R. Sairam, and J. Sairam, *Mol. Cell. Endocrinol.* **28,** 125 (1982).

[12] P. Manjunath and M. R. Sairam, *J. Biol. Chem.* **257,** 7109 (1982).

[13] M. R. Sairam, *Arch. Biochem. Biophys.* **204,** 199 (1980).

[14] M. R. Sairam, and P. Manjunath, unpublished data.

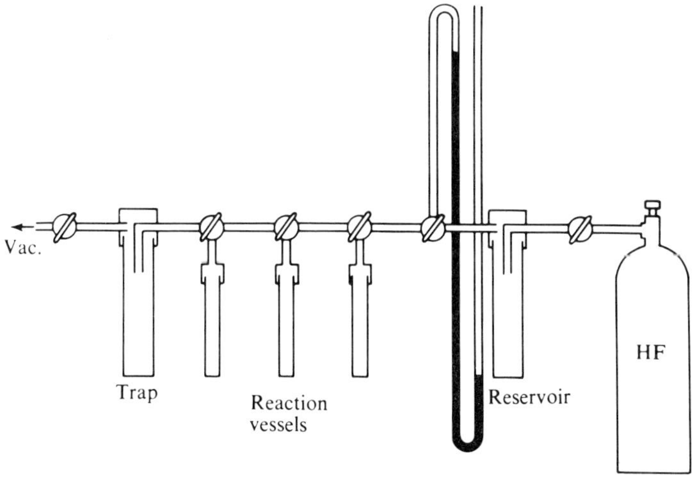

FIG. 1. Diagram of the HF apparatus[16] used for deglycosylation of the glycoprotein hormones. All materials are of Teflon and Kel-F. The reaction vessels and reservoir have stirring bars.

cal polymerization in an orderly manner. It is well known that anhydrous HF is highly effective in removing all the protecting groups used in the synthesis as well as cleaving the peptide chain from the anchoring polymer. A typical apparatus used (supplied by Protein Research Foundation, Minoh, Osaka, Japan) is constructed in Teflon-Kel-F material and is shown in Fig. 1. It consists of an HF cylinder, a reservoir, 2 or 3 reaction chambers, and a trap filled with calcium oxide to neutralize the HF when it is removed from the reaction vessel. These components are connected with Kel-F tubing and valves and a mercury manometer. The outlet from the trap is connected to a plastic water aspirator/vacuum pump. A more complete description of the apparatus can be found elsewhere.[15,16] A modified apparatus for better temperature control and more accurate timing of the reaction has been described.[16a] More recently an inexpensive microapparatus has also been designed for handling of small samples.[16b]

Precautions. Anhydrous HF like all strong chemicals causes severe burns and is hazardous material. However, it can be very easily handled in a closed system such as that indicated above in a well-ventilated hood.

[15] S. Sakakibara, *Chem. Biochem. Amino Acids, Pept. Proteins* **1,** 51 (1971).

[16] J. M. Stewart and J. D. Young, "Solid Phase Peptide Synthesis." Freeman, San Francisco, California, 1969.

[16a] A. J. Mort, *Carbohydrate Res.* **122,** 315 (1983).

[16b] M. P. Sanger and D. T. A. Lamport, *Anal. Biochem.* **128,** 66 (1983).

Such facilities including the HF apparatus can be usually found in any peptide synthetic laboratory of an Institution. It is essential that the operator wear adequate neoprene gloves and protective visors at all times when handling the apparatus.

Deglycosylation with Anhydrous HF. The freeze dried or vacuum dried hormone 10–100 mg is transferred to the Kel-F reaction vessel along with a small and thin teflon coated magnetic stirring bar and dried for about 24 hr over P_2O_5 under vacuum. The reaction vessel is then connected to the HF apparatus and the entire HF line except for the HF reservoir is evacuated. The reaction vessel is then cooled in a dry ice–ethanol bath and 10 ml of anhydrous HF is allowed to distill over from the reservoir stirring both vessels continuously. (The reservoir usually contains cobalt trifluoride as a drying agent which gives it a dark color and enables visual estimation of the liquid level.) The distilling process including the intermittent evacuation of the line (but not the reaction vessel) usually requires about 5–10 min. After distillation of HF into the reaction vessel is complete the reaction vessel is warmed up to 0° and then kept in an ice bath for the duration of the experiment. This is considered as the starting point of the reaction. The reaction vessel is kept continuously stirring during the experiment.

Recovery of the Deglycosylated Hormone. Anhydrous HF is an excellent solvent for glycoproteins. HF is very volatile and most of it can be quickly removed. However, during the reaction the HF molecules bind firmly to amino or amide groups through hydrogen bonds and ionic linkages. After completion of the reaction a very long time may therefore be required even under high vacuum to remove excess HF completely from the vessel. Even though the amounts are small the last traces of HF present can create acidic conditions upon addition of aqueous buffers, which is sufficient to promote dissociation of subunits during the recovery of the reaction products.[9] Therefore special care should be taken when oligomeric glycoprotein hormones are recovered after HF treatment: that is excess HF should be removed as soon as possible at relatively low temperature after the completion of the reaction.

We recommend the following general procedure which has given us excellent results. At the end of reaction time (generally 60 min) all visible HF is carefully removed by a water aspirator during 15–30 min. This should be done very slowly and gently to avoid excessive bubbling. When the HF evaporates the material in the vessel gets sticky and the stirring stops. To remove traces of HF the line is evacuated for an additional 2 hr using a high vacuum pump provided with adequate alkali traps. The contents of the reaction vessel which now appear sticky after solvent removal are then dissolved in 0.5 ml of ice-cold 0.2 *M* NaOH to neutralize any remaining trace of HF and the pH which is between 9 and 11 is readjusted to 7.5 by using cold 0.2 *M* HCl. As the glycoproteins are stable under such alkaline conditions there is no danger of dissociation into subunits. If the amount of NaOH added is insufficient to neutralize the residual HF the pH drops below 4 facilitating dissociation of the deglycosylated hormone into subunits. Since the ovine LH β-subunit upon deglycosylation becomes more hydrophobic it precipitates out.[13] In such instances more NaOH should be added to rise the pH to 7.5 and the solution either incubated overnight at 37° in presence of 0.02% azide or dialyzed against 0.1 *M* phosphate buffer pH 7.4 containing 0.02% azide to facilitate recombination. During this time most of the precipitated β-subunit recombines with its counter part and hormone remains in solution. The deglycosylated hormone is then purified as below.

Purification of Deglycosylated Hormone

Chromatography on Sephadex G-100 Column. The above material consisting of deglycosylated hormone is centrifuged to remove any precipitate and chromatographed on a Sephadex G-100 column equilibrated in 0.05 *M* ammonium bicarbonate at 4°. A typical pattern for hCG is shown in Fig. 2. The protein in the fractions are pooled and freeze dried and used for biochemical studies or purified further by chromatography on concanavalin A-Sepharose.

Chromatography on Concanavalin A-Sepharose. Glycoprotein hormones bind strongly to plant lectins such as concanavalin A. This prop-

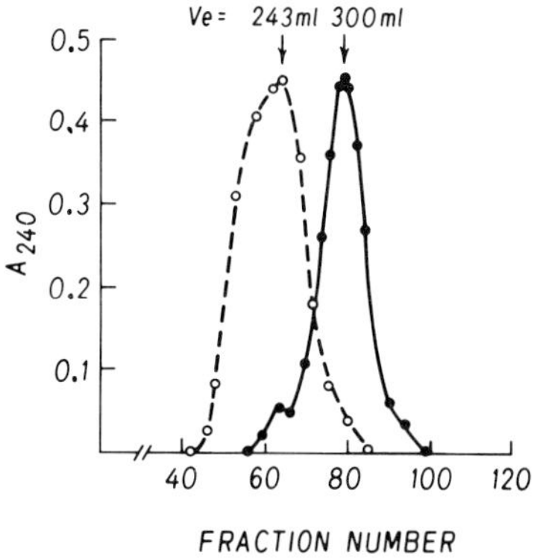

FIG. 2. Gel filtration pattern of HF treated hCG (●). Sephadex G-100 column (2.5 × 106 cm) was previously equilibrated in 0.05 *M* NH_4HCO_3. HF treated hormone (50 mg) was loaded on the column and eluted with same eluant. Fractions of 3.8 ml at a rate of 35 ml/hr were collected. Fractions 67–90 were pooled. Recovery was 26 mg by weight. Broken line indicates elution pattern of hCG on same column.

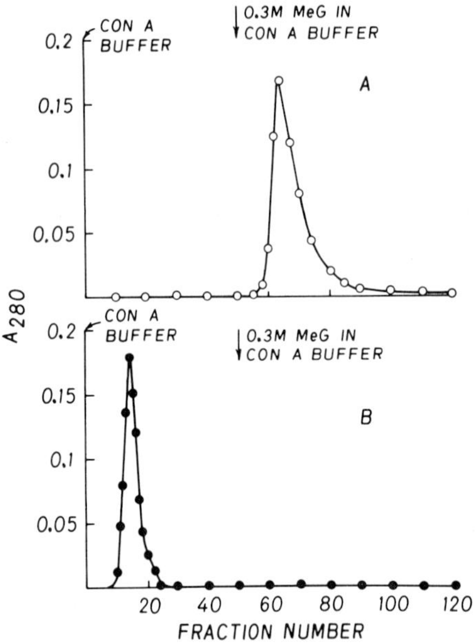

FIG. 3. Concanavalin A-Sepharose chromatography patterns of (A) hCG and (B) deglycosylated hCG. The column (1.5 × 22 cm) containing concanavalin A-Sepharose was equilibrated in the Con A buffer (25 m*M* Tris–HCl buffer, pH 7.4 containing 100 m*M* $MgCl_2$, 10 m*M* $CaCl_2$, 0.5 *M* NaCl, and 0.02% azide). The column flow was temporarily stopped for 60 min to allow for effective interaction of the protein with the matrix. Elution was then carried on as shown. Fractions of 3.8 ml at a flow rate of 45 ml/hr were collected. The proteins were recovered by ultrafiltration followed by lyophilization.

erty has been successfully used in the purification of these hormones. However, following deglycosylation by anhydrous HF as indicated above, these hormones no longer bind to the immobilized concanavalin A. Advantage may be taken of this characteristic to remove any possible unreacted hormone in the deglycosylated product. We recommend this as a routine step in the final purification of the deglycosylated gonadotropins. The deglycosylated hormone elutes in the unadsorbed fraction in contrast to the native hormone which remains strongly bound (Fig. 3).

Applications[16c]

The procedure outlined above has been successfully used for the deglycosylation of oLH, bLH, hCG, oFSH, and hTSH. For hormones such as oFSH which has tryptophan we recommend the inclusion of moisture-

[16c] Preliminary studies with deglycosylation of eCG have shown drastic losses in receptor binding activity of the hormone.

free anisole (AR grade) as a scavenger.[11,16d] In such instances, after HF removal the anisole remains as an immiscible oily layer on top of the gummy material which consists of the deglycosylated hormone. This can be carefully removed from the reaction vessel by using a Pasteur pipet and the residue of deglycosylated hormone further purified as described above. For oLH, bLH, hCG, hTSH, and eCG which do not contain tryptophan we have chosen to exclude anisole and the reaction is performed in the complete absence of any scavenger.

Comments

The chemical deglycosylation using anhydrous HF is specific to the carbohydrate moiety and permits the isolation of the modified hormone in lyophilized form in a state completely free from the native hormone.[8–14] This method of deglycosylation has distinct advantages over sequential enzymatic deglycosylation. The extent of deglycosylation by HF treatment is up to 80% as against 50–60% by enzymatic deglycosylation. In addition, the recovery of deglycosylated hormone after purification is up to 70% as against 20–25% by enzymatic deglycosylation. As far as can be ascertained no detectable change in the form of cleavage appears to have occurred in the polypeptide chains. There is no difference in amino acid composition of the hormones before and after deglycosylation in all the instances. This is also true when the isolated α and β subunits of ovine lutropin are treated with anhydrous HF.[13] Peptide bonds are stable in anhydrous HF at 0°.[15,16] Since the aqueous HF which may arise in the event of incomplete removal after the desired treatment is a weak acid, even potentially labile aspartyl bonds are unlikely to be affected. However, in susceptible hormones, such as ovine and bovine lutropin which are prone to facile dissociation into subunits under acidic conditions, even the mild acidic environment prevailing after the reaction can be responsible for the presence of subunits which were detected in our very first study.[8] However, this has now been eliminated by careful neutralization with alkali immediately after the removal of HF[9] (see Recovery of the Deglycosylated Hormones).

Anhydrous HF treatment appears to have no significant effect on quarternary structure of the gonadotropins. This is revealed by the fact that DG-hCG and DG-oLH do not show any decrease in the fluorescence of 1.8 ANS as compared to native hormones.

[16d] HF solvolysis generates fluorinated compounds, e.g., glycosyl fluorides which may alkylate aromatic amino acids, the HF acting as a Friedel-crafts catalyst. Addition of a large excess of a scavenger such as anisole protects these amino acids by competing for the alkylating compounds.

A study of the reaction conditions of HF deglycosylation of oLH revealed that deglycosylation at 0° for about 30–60 min appear to be adequate (Table I). Nearly all the accessible monosaccharide units in the hormone are effectively removed even in as short a reaction time as 30 min. Exposure to HF at higher temperature such as 23° or for longer period at 0 or 23° does not achieve complete deglycosylation but these conditions have a deleterious effect on the hormone.[9] Many of the interesting biological properties of the deglycosylated hormones, viz. antagonistic activity, are greatly reduced.

The extent of deglycosylation in hCG, oLH, bLH, and oFSH is between 70 and 80% (Table II). In ovine lutropin or follitropin all but two penultimate *N*-acetylglucosamine residues are cleaved by HF treatment.[9,11] It is also interesting to point out, that while bovine lutropin (and by inference ovine lutropin as well) is apparently stable to the action of mixed exoglycosidases,[3] anhydrous HF is very effective in removing the sugar moiety including the sulfated hexosamines.[10] In hCG the carbohydrate chains are linked both by N-glycosidic and O-glycosidic bonds. We find that some of these linkage are quite resistant to HF deglycosylation at 0° for 60 min[12] but most other sugars are cleaved almost completely.

Based on our experience with oLH optimum conditions for chemical deglycosylation of gonadotropins appear to be the exposure of the hormone to HF at 0° for 30–60 min. This results in almost 90–95% cleavage of all susceptible sugars without causing any damage to the peptide portion. The recovery of the deglycosylated hormone is high (over 80%) and the derivatives are easily soluble in aqueous buffer and are very stable in the freeze dried form. In over 5 years, we have not observed significant changes in the receptor-binding, immunological, biological, and antago-

TABLE I
CARBOHYDRATE COMPOSITION[a] OF OVINE LUTROPIN AND DEGLYCOSYLATED LUTROPIN PREPARED UNDER VARIOUS REACTION CONDITIONS[b]

Component	oLH	DGLH-1	DGLH-2	DGLH-3	DGLH-4	DGLH-5
Hexoses	7.2	0.45	0.0	0.0	0.00	0.0
Fucose	1.5	0.00	0.0	0.0	0.00	0.0
Glucosamine	5.9	3.60	3.0	3.0	1.99	1.9
Galactosamine	2.0	0.00	0.0	0.0	0.00	0.0

[a] Values are expressed as g/100 g of glycoprotein. DGLH-1, deglycosylated with HF at 0° for 30 min. DGLH-2, deglycosylated with HF at 0° for 1 hr. DGLH-3, deglycosylated with HF at 0° for 3 hr. DGLH-4, deglycosylated with HF at 23° for 1 hr. DGLH-5, deglycosylated with HF at 23° for 3 hr.

[b] Derived from Majunath *et al.*[9]

TABLE II
CARBOHYDRATE COMPOSITION OF CHEMICALLY DEGLYCOSYLATED GONADOTROPINS[a]

Sugars	hCG	oFSH	oLH	bLH
Hexoses	13	14	0	>5
Glucosamine	52	44	50	66
Galactosamine	95	0	0	<5
Sialic acid	6	0	0	0
Fucose	0	0	0	ND[b]

[a] The values given are percentage sugar remaining as compared to the content in native hormone which is taken as 100%. In each case the hormone treated with anhydrous HF for 60 min at 0° and purified subsequently on Sephadex G-100 and concanavalin A-Sepharose was used for the determinations.

[b] ND, Not determined.

nistic properties. The deglycosylated hormones prepared by treatment with HF have very interesting characteristics[8–14,17–22] and are useful in understanding the structure–function relationships and mechanism of hormone action.

The affinity of the deglycosylated hormones to their respective receptors (membrane preparations or intact cells) is enhanced after deglycosylation. All the DG hormones are generally more active than the native hormones in competing for binding sites on the receptor (Table III). Despite such increase in receptor binding ability the DG gonadotropins fail to activate adenylate cyclase and have a very low steroidogenic potency. By virtue of such differential characteristics, they act as effective and specific antagonists of the action of respective native hormones. Their immunological activity, however, remains unaffected.

Alternatives to the HF Method

The use of anhydrous HF as a convenient and effective deblocking agent in synthetic peptide chemistry has now gained wide acceptance and has helped in making the facility accessible to more and more biochemical investigators. The precautions for its use were previously stressed. For

[17] M. R. Sairam and P. Manjunath, *Int. J. Pept. Protein Res.* **19,** 315 (1982).

[18] M. R. Sairam and P. Manjunath, *Mol. Cell. Endocrinol.* **28,** 139 (1982).

[19] M. R. Sairam and P. Manjunath, *Mol. Cell. Endocrinol.* **28,** 151 (1982).

[20] M. R. Sairam and P. Manjunath, *J. Biol. Chem.* **258,** 445 (1982).

[21] M. R. Sairam and P. Manjunath, *Ann. N.Y. Acad. Sci.* **383,** 493 (1982).

[22] P. Manjunath and M. R. Sairam, *J. Biol. Chem.* **258,** 3554 (1983).

TABLE III
RECEPTOR BINDING ACTIVITY OF DEGLYCOSYLATED GONADOTROPINS[a]

	Receptor preparation (R)							
DG gonadotropins[b]	Rat testis LH-R	Rat luteal LH-R	Porcine granulosa cell LH-R	Porcine granulosa cell FSH-R	Rat granulosa cell FSH-R	Rat testis FSH-R	Bull testis FSH-R	Human testis FSH-R
DG-oLH	99 ± 7	93 ± 5	97 ± 4					
DG-bLH	85 ± 5		85 ± 5					
DG-hCG	186 ± 10	179 ± 18	200 ± 6					200–400
DG-FSH				200 ± 6	400–600	240 ± 8	252 ± 10	120–160

[a] The activity of all native hormones in the respective binding assays has been set as 100% in each instance. The data (mean ± SEM) are derived from displacement assays using ^{125}I-labeled native hormone.

[b] All DG hormones were obtained by HF treatment for 1 hr at 0° and further purified (see text).

those who feel uncomfortable with this approach or where anhydrous HF treatment facility is not available, another reagent namely trifluoromethanesulfonic acid (TFMS) may provide an alternative. Chemical deglycosylation using TFMS (available from Aldrich Company) is quite effective but is not useful for gonadotropins. When this method was applied to ovine LH,[7] over 70% of the sugars were removed but the recovered product has only 2% receptor binding activity relative to the native hormone. This should be compared with full retention of activity after the HF treatment (Table III). Our own experience with TFMS although limited until now has indicated complete dissociation of oLH into subunits even with careful recovery procedure (P. Manjunath and M. R. Sairam, unpublished results). This would obviously destroy receptor binding activity of the product[7] and negate any handling advantages the reagent may have to offer. Similar complete dissociation of hCG is apparently responsible for the destruction of receptor binding activity[23] of the hormone.

TFMS is also a corrosive, hygroscopic, and syrupy liquid which requires careful handling but it has the advantage that it can be used in glass vessels. Unlike anhydrous HF it is very acidic and nonvolatile which creates problems in recovery of the product. Neutralization of TFMS after the reaction, with aqueous pyridine or any other base, is a highly exothermic reaction. The intense heat generated coupled with the acidic environment has deleterious effects on the quaternary structure of the hormone. If preservation of the quaternary structure and receptor binding activity is not an essential objective, the TFMS method can be used. However, longer reaction times of up to 24 hr may be required for achieving deglycosylation comparable to that induced by 30–60 min of treatment using anhydrous HF.

For those glycoproteins of biological interest which are stable to strong acidic environment, TFMS may be suitable for logistic reasons. It may also be useful when only very small amounts of the glycoprotein are available for the study. But we must add the caveat that the conditions of the reaction and subsequent recovery must be carefully selected and several trials may become necessary to obtain reproducible results.

Acknowledgment

Various phases of the Research work described in this article have received support from the MRC of Canada, WHO in Geneva, and the Ford Foundation in New York.

[23] N. K. Kalyan and O. P. Bahl, *J. Biol. Chem.* **258,** 67 (1983).

[57] Disassembly and Assembly of Glycoprotein Hormones

By THOMAS F. PARSONS, THOMAS W. STRICKLAND, and JOHN G. PIERCE

Introduction

One group of polypeptide hormones, the glycoprotein hormones of the anterior pituitary and of the placenta and related tissues, exhibits features otherwise not found in the large number of polypeptides now recognized to have hormonal activities. These hormones, luteinizing hormone (LH), thyroid-stimulating hormone (TSH), follicle-stimulating hormone (FSH), and chorionic gonadotropin (e.g., hCG and PMSG, pregnant mare serum gonadotropin) consist of two distinct peptide chains, both rich in *intra* chain disulfides and both glycosylated. One chain or subunit (α) has, within a species, essentially the same amino acid sequence from hormone to hormone; the other subunits (β), while probably evolving from a single primitive gene, have different sequences. These sequences, when correctly folded, must contain the information, which upon combination with an α subunit, gives rise to a specific hormone activity. The interaction between the two subunits is solely via noncovalent forces and the formation of a hormonally active α–β dimer is accompanied by changes in conformation of both subunits. No *inter* chain disulfides are present.

The reason that the subunit nature of these hormones was not recognized in early studies on their purification and properties is because at pHs near neutrality disassembly into subunits proceeds at very slow rates. It should be noted that of the physiological conditions under which hormone activity is expressed, the hormone concentration in blood may be 10^{-10} M or lower, and for some time it was thought that the affinity between the two peptide chains was very high (perhaps 10^{12} M for an equilibrium affinity constant). Recent studies have shown, however, that the affinities between subunits are much less than previously thought (in the range of 10^{-6} M for K_d) and a slow rate of dissociation is what maintains the relative stability of the dimer at physiological pHs.[1,2] Dissociation into subunits is greatly facilitated by changing the pH to below 4 and many of the numerous methods for fractionating proteins have been applied in both analytical and preparative separations of the subunits of each of the glycoprotein hormones. Their general chemistry and biology has

[1] T. W. Strickland and D. Puett, *J. Biol. Chem.* **257,** 2954 (1982).
[2] H. Forastieri and K. C. Ingham, *J. Biol. Chem.* **257,** 7976 (1982).

ISBN 0-12-182009-2

been reviewed recently[3] as have their physiochemical properties. An earlier volume in this series contains descriptions[4,5] of the isolation of several of the glycoprotein hormones; other references to the original literature can be found in reviews.[3,6] Typical methods for isolation of individual subunits from these hormones are described herein but, in spite of overall similarity in structure between the members of this group of hormones, differences in both amino acid sequence and oligosaccharide structures do not permit a single preparative method to be used in each case, although a systematic investigation of methodology for separating subunits or for finding optimum conditions for reassembly of dissociated hormones has not been made. Depending on the species, many of the hormones are present in small amounts and it is costly both in time and materials to isolate them. In the future, high-performance liquid chromatography shows the most promise for rapid and easy separation of subunits and some initial results are described.

Detection of Dissociation into Subunits and Assembly into α–β Dimers

Several methods have been utilized to monitor the state of association of glycoprotein hormone subunits. Techniques which exploit the differences in molecular size of intact hormones and free subunits include gel filtration,[7–11] sedimentation analysis,[7,12–14] light scattering,[13,15] and electrophoresis at low temperature in the presence of sodium dodecyl sulfate (SDS).[1,16–19] Gel filtration using HPLC technology[2] appears to be a partic-

[3] J. G. Pierce and T. F. Parsons, *Annu. Rev. Biochem.* **50,** 465 (1981).

[4] A. Stockell-Hartree, this series, Vol. 37, p. 380.

[5] L. E. Reichert, Jr., this series, Vol. 37, p. 360.

[6] K. W. McKerns, ed., "Structure and Function of the Gonadotropins." Plenum, New York, 1978.

[7] P. de la Llosa and M. Jutisz, *Biochim. Biophys. Acta* **181,** 426 (1969).

[8] K. C. Ingham, S. M. Aloj, and H. Edelhoch, *Arch. Biochem. Biophys.* **163,** 589 (1974).

[9] K. C. Ingham and C. Bolotin, *Arch. Biochem. Biophys.* **191,** 134 (1978).

[10] J. G. Loeber, J. W. G. M. Nabben-Fleuren, L. H. Elvers, M. F. G. Segers, and R. M. Lequin, *Endocrinology* **103,** 2240 (1978).

[11] H. Forastieri and K. C. Ingham, *Arch. Biochem. Biophys.* **205,** 104 (1980).

[12] T. A. Bewley, M. R. Sairam, and C. H. Li, *Arch. Biochem. Biophys.* **163,** 625 (1974).

[13] R. Salesse, M. Castaing, J. C. Pernollet, and J. Garnier, *J. Mol. Biol.* **95,** 483 (1975).

[14] J. C. Pernollet, J. Garnier, J. G. Pierce, and R. Salesse, *Biochim. Biophys. Acta* **446,** 262 (1976).

[15] J. C. Pernollet and J. Garnier, *FEBS Lett.* **18,** 189 (1971).

[16] S. Schlaff, *Endocrinology* **98,** 527 (1976).

[17] K. Sorimachi and P. G. Condliffe, *Int. J. Pept. Protein Res.* **12,** 1 (1978).

[18] W. W. Chin, F. Maloof, and J. F. Habener, *J. Biol. Chem.* **256,** 3059 (1981).

[19] B. D. Weintraub, B. S. Stannard, and L. Meyers, *Endocrinology* **112,** 1331 (1983).

ularly attractive method in view of its sensitivity (<1 μg), speed of analysis (<0.5 hr), and ease of quantitation. Electrophoresis in the presence of SDS is convenient if multiple samples are to be analyzed. The SDS bound to the protein overcomes the charge heterogeneity characteristic of glycoprotein hormones and allows the hormone and subunits to migrate as discrete bands. At low temperatures and in the absence of reductant, SDS does not dissociate a glycoprotein hormone into subunits.[16] Although polyacrylamide gel electrophoresis in the absence of SDS often reveals multiple bands for a purified hormone or subunit, different banding patterns of an intact hormone versus those of free subunits also make this method useful for monitoring subunit association.[20,21]

A difference spectrum with a maximum at 287.5 nm is generated upon subunit dissociation and has been widely used to monitor the association/dissociation dynamics of glycoprotein hormones.[8,9,13,14,22,23] The spectrum results from the exposure of "buried" tyrosine residues to the aqueous environment upon dissociation. Other spectroscopic methods which have been applied to glycoprotein hormone association/dissociation include the polarization of tyrosine fluorescence[2,9,11] and circular dichroic spectroscopy.[12–14]

Since the isolated subunits are devoid of biological activity and do not bind to the hormone receptor,[24,25] a bioassay may be used to determine the amount of intact hormone. While radioligand receptor assays, which involve competition with radioactively labeled hormone for binding to a particulate receptor preparation, are most convenient,[8,10,14,26–28] the relatively less precise and more cumbersome *in vivo* bioassays have also been employed.[8,27–29] Several studies have utilized antisera which react preferentially with either an isolated subunit or the intact hormone to assess the degree of association.[30,31]

[20] J. S. Cornell and J. G. Pierce, *J. Biol. Chem.* **248,** 4327 (1973).

[21] L. C. Giudice and J. G. Pierce, *Endocrinology* **101,** 776 (1977).

[22] K. C. Ingham, B. D. Weintraub, and H. Edelhoch, *Biochemistry* **15,** 1720 (1976).

[23] J. Marchelidon, R. Salesse, J. Garnier, E. Burzawa-Gerard, and Y. Fontaine, *Nature (London)* **281,** 314 (1979).

[24] P. L. Rayford, J. F. Vaitukaitis, G. T. Ross, F. J. Morgan, and R. E. Canfield, *Endocrinology* **91,** 144 (1972).

[25] J. F. Williams, T. F. Davies, K. J. Catt, and J. G. Pierce, *Endocrinology* **106,** 1353 (1980).

[26] L. E. Reichert, G. M. Lawson, F. L. Leidenberger, and C. G. Trowbridge, *Endocrinology* **93,** 938 (1973).

[27] L. E. Reichert, C. G. Trowbridge, V. K. Bhalla, and G. M. Lawson, *J. Biol. Chem.* **249,** 6472 (1974).

[28] L. E. Reichert and R. B. Ramsey, *J. Biol. Chem.* **250,** 3034 (1975).

[29] S. M. Aloj, H. Edelhoch, K. C. Ingham, F. J. Morgan, R. E. Canfield, and G. T. Ross, *Arch. Biochem. Biophys.* **159,** 497 (1973).

[30] B. D. Weintraub, B. S. Stannard, and S. W. Rosen, *Endocrinology* **101,** 225 (1977).

[31] J. G. Pierce, G. A. Bloomfield, and T. F. Parsons, *Int. J. Pept. Protein Res.* **13,** 54 (1979).

A convenient and unique assay of association is based on the observation that several glycoprotein hormones, including hCG, hLH, bLH, and bTSH, but not their isolated subunits, enhance the fluorescence of the probe 1,8-anilinonaphthalene sulfonate (ANS).[2,8,9,22,29,32] In the presence of ANS these hormones form dimers or higher order oligomers and it is the interaction of ANS with these species which is thought to lead to the enhance of fluorescence.[33,34]

Methods Used for the Dissociation and Separation of Subunits

Dissociation into subunits is greatly facilitated by pH below 4.0 or by incubation at 37° in the presence of a denaturant such as guanidine hydrochloride or urea. Conditions such as 1 *M* propionic acid,[35] 0.04 *M* Tris-phosphate pH 7.5 + 8 *M* urea,[36] or 8 *M* guanidine hydrochloride[37] are effective even at protein concentrations as high as 10–100 mg/ml.

Although dissociation into subunits is easily accomplished, isolation of the α and β subunits is more difficult. Typical methods are described herein, and a more complete list of methods is in Tables I–III. In spite of the overall similarity in structure between the members of this group of proteins, a single preparative method for the isolation of subunits has not been reported. With most of the methods, a yield of only 25% of isolated subunits can be expected. This is presumably due to operational losses during the several steps needed to fractionate and desalt the subunits and to either the lack of complete dissociation or of recombination of the subunits during the various steps. For any of the methods listed, it is advisable to gel filter the isolated subunit over a Sephadex G-100 column in the presence of 1% ammonium bicarbonate to remove any contaminating intact hormone.

Ion-Exchange Chromatography

Numerous ion-exchange chromatographic methods have been used for the separation of α and β subunits for several of the hormones isolated from a variety of species. These procedures are listed in Table I and most involve conditions similar to those described by Morgan and Canfield[38] for

[32] K. C. Ingham, S. M. Aloj, and H. Edelhoch, *Arch. Biochem. Biophys.* **159,** 596 (1973).
[33] K. C. Ingham, H. A. Saroff, and H. Edelhoch, *Biochemistry* **14,** 4745 (1975).
[34] K. C. Ingham, H. A. Saroff, and H. Edelhoch, *Biochemistry* **14,** 4751 (1975).
[35] T.-H. Liao and J. G. Pierce, *J. Biol. Chem.* **245,** 3275 (1970).
[36] N. Swaminathan and O. P. Bahl, *Biochem. Biophys. Res. Commun.* **40,** 422 (1970).
[37] M. Ascoli, W.-K. Liu, and D. N. Ward, *J. Biol. Chem.* **252,** 5280 (1977).
[38] F. J. Morgan and R. E. Canfield, *Endocrinology* **88,** 1045 (1971).

the isolation of the subunits of human CG. Native CG is dissolved in 10 *M* urea adjusted to pH 4.5 with HCl to yield a 25 mg/ml solution and incubated for 1 hr at 40°. Glycine (0.03 *M*) is added to give a final concentration of 0.005 *M* glycine and the pH is adjusted to 7.5 with NaOH. Chromatography is performed on DEAE-Sephadex A-25 equilibrated with 0.03 *M* glycine plus 8 *M* urea, pH 7.5. The α subunit is eluted with the equilibration buffer and the β subunit is eluted by a solution of 0.2 *M* glycine which is 1 *M* with respect to NaCl, and 8 *M* with respect to urea, pH 7.5. The subunit fractions are acidified and dialyzed against 1% acetic acid and then against water and lyophilized.

TABLE I
ION-EXCHANGE METHODS USEFUL IN THE SEPARATION OF SUBUNITS OF GLYCOPROTEIN HORMONES

Hormone	Species	Conditions for dissociation	Methods used for separation of subunits	Investigators
CG	Human	0.04 *M* Tris-phosphate pH 7.5 + 8 *M* urea	DEAE-Sephadex in 0.04 *M* Tris-phosphate pH 7.5 with a linear NaCl gradient	Swaminathan and Bahl (1970)[36]
CG	Human	10 *M* urea pH 4.5 with HCL	DEAE-Sephadex in 0.03 *M* glycine pH 7.5 + 8 *M* urea with stepwise elution with glycine	Morgan and Canfield (1971)[38]
CG	Equine	10 *M* urea pH 4.5 with HCl	DEAE-Sephadex in 0.03 *M* glycine pH 7.5 + 8 *M* urea with stepwise elution with glycine	Papkoff *et al.* (1978)[a]
CG	Equine	0.04 *M* Tris-phosphate pH 7.2 + 10 *M* urea	DEAE-Sephadex in 0.04 *M* Tris-phosphate pH 7.2 with stepwise elution with NaCl	Christakos and Bahl (1979)[b]
LH	Porcine	0.025 *M* Na acetate pH 4.9 + 8 *M* urea	SE-Sephadex in 0.025 *M* Na acetate pH 4.9 with a linear Na acetate gradient	Hennen *et al.* (1971)[c]

TABLE I (*continued*)

Hormone	Species	Conditions for dissociation	Methods used for separation of subunits	Investigators
LH	Human	0.05 *M* HCl + 8 *M* urea pH 2.8	DEAE-Sephadex in 0.01 *M* Na glycinate pH 9.5 with a linear NaCl gradient	Closset *et al.* (1972)[d]
LH	Human	0.08 *M* Na acetate pH 5.0 + 8 *M* urea	CM-cellulose in 0.08 *M* Na acetate pH 5.0 + 4 *M* urea with stepwise elution with Na acetate	Stockell-Hartree (1975)[4]
TSH	Porcine	0.05 *M* HCl + 8 *M* urea pH 2.8	DEAE-Sephadex in 0.01 *M* Na glycinate pH 9.5 with a linear NaCl gradient	Closset and Hennen (1974)[e]
FSH	Human	0.04 *M* Tris-phosphate pH 7.5 + 8 *M* urea	DEAE-Sephadex in 0.04 *M* Tris-phosphate pH 7.5 with a linear NaCl gradient	Parlow and Shome (1974)[f]
FSH	Human	0.04 *M* Tris-phosphate pH 7.5 + 8 *M* urea	DEAE-Sephadex in 0.04 *M* Tris-phosphate pH 7.5 with a linear NaCl gradient	Rathnam and Saxena (1975)[g]
FSH	Human	0.007 *M* Na phosphate + 0.003 *M* Na borate pH 8.0 + 4 *M* urea	DEAE-cellulose in 0.007 *M* Na phosphate + 0.003 *M* Na borate pH 8.0 + 4 *M* urea with stepwise elution with NaCl	Shownkeen *et al.* (1976)[h]
FSH	Porcine	0.05 *M* HCl + 8 *M* urea pH 2.8	DEAE-Sephadex in 0.02 *M* Na glycinate pH 9.5 with a linear NaCl gradient	Closset and Hennen (1978)[i]
FSH	Equine	0.04 *M* Tris-phosphate pH 7.5 + 8 *M* urea	DEAE-Sephadex in 0.04 *M* Tris-phosphate pH 7.5 with a linear NaCl gradient	Rathnam *et al.* (1978)[j]

(*continued*)

TABLE I (*continued*)

Hormone	Species	Conditions for dissociation	Methods used for separation of subunits	Investigators
FSH	Bovine	1 *M* propionic acid	DEAE-cellulose in 0.04 *M* Tris–HCl pH 8.2 with stepwise elution with NaCl	Cheng (1978)[k]
FSH	Ovine	0.04 *M* Tris-phosphate pH 7.5 + 8 *M* urea	DEAE-Sephadex in 0.04 *M* Tris-phosphate + 4 *M* urea with stepwise elution with NaCl	Sairam (1979)[l]
FSH	Human	0.04 *M* Tris-phosphate pH 7.5 + 8 *M* urea	DEAE-Sephadex in 0.04 *M* Tris-phosphate pH 7.5 + 4 *M* urea with stepwise elution with NaCl	Sairam and Li (1979)[m]

[a] H. Papkoff, T. A. Bewley, and J. Ramachandran, *Biochim. Biophys. Acta* **532,** 185 (1978).
[b] S. Christakos and O. P. Bahl, *J. Biol. Chem.* **254,** 4253 (1979).
[c] G. Hennen, L. Prusik, and G. Maghuin-Rogister, *Eur. J. Biochem.* **18,** 376 (1971).
[d] J. Closset, G. Hennen, and R. M. Leguin, *FEBS Lett.* **21,** 325 (1972).
[e] J. Closset and G. Hennen, *Eur. J. Biochem.* **46,** 595 (1974).
[f] A. F. Parlow and B. Shome, *J. Clin. Endocrinol. Metab.* **39,** 195 (1974).
[g] P. Rathnam and B. B. Saxena, *J. Biol. Chem.* **250,** 6735 (1975).
[h] R. C. Shownkeen, A. Stockell-Hartree, F. Stewart, K. Mashiter, and V. C. Stevens, *J. Endocrinol.* **69,** 263 (1976).
[i] J. Closset and G. Hennen, *Eur. J. Biochem.* **86,** 105 (1978).
[j] P. Rathnam, Y. Fujiki, T. D. Landefeld, and B. B. Saxena, *J. Biol. Chem.* **253,** 5355 (1978).
[k] K.-W. Cheng, *Biochem. J.* **175,** 29 (1978).
[l] M. R. Sairam, *Arch. Biochem. Biophys.* **194,** 71 (1979).
[m] M. R. Sairam and C.-H. Li, *Int. J. Pept. Protein Res.* **13,** 394 (1979).

Countercurrent Distribution

Countercurrent distribution has been extremely useful in the preparation of large amounts of LH subunits from a variety of species. The methods are listed in Table II and most involve solvents similar to those described by Papkoff and Samy[39] for the preparation of subunits of ovine

[39] H. Papkoff and T. S. A. Samy, *Biochim. Biophys. Acta* **147,** 175 (1967).

LH. The procedure of Liao *et al.*[40] uses a stronger "carrier acid," *p*-toluenesulfonic acid, and utilizes 30 transfers instead of 10 transfers as described by Papkoff and Samy.[39] The additional transfers result in subunit preparations with less contamination by their counterparts. Distributions are carried out either in all glass apparatus with 10 ml in each phase, or in a series of glass-stoppered centrifuge tubes if lesser volumes are used. The solvent system consists of 40% (w/v) ammonium sulfate–0.15 *M p*-toluenesulfonic acid–1-propanol–ethanol (60 : 60 : 27 : 33 by volume). LH is dissolved into the upper phase to yield a 60 mg/ml solution, an equal volume of lower phase is added and the solution is stirred for an hour. After 30 transfers, the phases are broken and neutralized by the addition of 0.6 *N* ammonium hydroxide (2.5 ml/10 ml of each phase). Protein is located by absorbance at 284 nm. The α subunit strongly favors the aqueous phase and the β subunit the organic phase. The subunits are recovered by removal of the organic solvents by rotary evaporation, followed by dialysis against water and lyophilization.

Selective Precipitation

The subunits of ovine LH and bovine TSH can be isolated by NaCl precipitation as first described by Sairam and Li.[41] A slight modification, using a pH 5 buffer rather than a pH 3 buffer is described by Ascoli *et al.*[37] The hormone is dissociated by an overnight incubation in 8 *M* guanidine hydrochloride at 37°, and the subunits are rapidly desalted by gel filtration on BioGel P6 in the presence of 1% ammonium bicarbonate at 4°. After lyophilization, the subunits are dissolved in 0.1 *M* sodium acetate pH 5 to yield a concentration of 5 mg/ml or greater. Precipitation of the β subunit is achieved by addition of solid NaCl to 3 *M*. The β subunit is collected by centrifugation and the α subunit and any intact hormone remain in the supernatant. The pellet containing the β subunit is dissolved in 0.1 *M* sodium acetate pH 5 and reprecipitated by addition of NaCl. Final purification of both subunits is achieved by gel filtration on Sephadex G-100 in the presence of 1% ammonium bicarbonate and lyophilization (Table III).

Other Methods

Several other column methods have been used for the isolation of subunits and are listed in Table IV. The separation of bovine and human TSH subunits is achieved using the procedure of Liao and Pierce.[35] The

[40] T.-H. Liao, G. Hennen, S. M. Howard, B. Shome, and J. G. Pierce, *J. Biol. Chem.* **244,** 6458 (1969).

[41] M. R. Sairam and C.-H. Li, *Arch. Biochem. Biophys.* **165,** 709 (1974).

TABLE II
COUNTERCURRENT DISTRIBUTION METHODS USEFUL IN THE SEPARATION OF SUBUNITS OF GLYCOPROTEIN HORMONES

Hormone	Species	Conditions for dissociation	Methods used for separation of subunits	Investigators
LH	Ovine	Countercurrent distribution solvent	40% (w/v) aqueous $(NH_4)_2SO_4$–0.2% aqueous dichloroacetic acid–1-propanol–ethanol (60 : 60 : 27 : 33)	Papkoff and Samy (1967)[39]
LH	Bovine	Countercurrent distribution solvent	40% (w/v) aqueous $(NH_4)_2SO_4$–0.15 *M* *p*-toluenesulfonic acid–1-propanol–ethanol (60 : 60 : 27 : 33)	Liao *et al.* (1969)[40]
LH	Rat	Countercurrent distribution solvent	40% (w/v) aqueous $(NH_4)_2SO_4$–0.2% aqueous dichloroacetic acid–1-propanol–ethanol (60 : 60 : 27 : 33)	Ward *et al.* (1971)[a]
LH	Human	0.05 *M* dichloroacetic acid + 6 *M* guanidine hydrochloride	40% (w/v) aqueous $(NH_4)_2SO_4$–0.2% aqueous dichloroacetic acid–1-propanol–ethanol (60 : 60 : 27 : 33)	Shome and Parlow (1972)[b]
LH	Human	Countercurrent distribution solvent	*n*-butanol–isobutyric acid–H_2O–triethylamine (10 : 20 : 30 : 1)	Ward *et al.* (1973)[c]
LH	Ovine	Countercurrent distribution solvent	*n*-butanol–isobutyric acid–H_2O–triethylamine (10 : 20 : 30 : 1)	Liu *et al.* (1974)[d]
LH	Hamster	Countercurrent distribution solvent	*n*-butanol–isobutyric acid–H_2O–triethylamine (10 : 20 : 30 : 1)	Glenn *et al.* (1982)[e]
TSH	Human	0.05 *M* HCl	40% (w/v) aqueous $(NH_4)_2SO_4$–0.2% aqueous dichloroacetic acid–1-propanol–ethanol (60 : 60 : 27 : 33)	Sairam and Li (1973)[f]

[a] D. N. Ward, L. F. Reichert, Jr., B. A. Fitak, H. S. Nahm, C. M. Sweeney, and S. D. Neill, *Biochemistry* **10,** 1796 (1971).

TABLE III
SELECTIVE PRECIPITATION METHODS USEFUL IN THE SEPARATION OF SUBUNITS OF GLYCOPROTEIN HORMONES

Hormone	Species	Conditions for dissociation	Methods used for separation of subunits	Investigators
LH	Ovine	0.1 *M* Na acetate pH 3.0	3 *M* NaCl precipitation of β subunit out of 0.1 *M* Na acetate pH 3.0	Sairam and Li (1974)[41]
LH	Ovine	8 *M* aqueous guanidine hydrochloride	3 *M* NaCl precipitation of β subunit out of 0.1 *M* Na acetate pH 5.0	Ascoli *et al.* (1977)[37]
TSH	Bovine	8 *M* aqueous guanidine hydrochloride	3 *M* NaCl precipitation of β subunit out of 0.1 *M* Na acetate pH 5.0	T. W. Strickland and J. G. Pierce (unpublished observations)

hormone is dissociated by incubation at room temperature in 1 *M* propionic acid for 16 hr. The propionic acid is removed by lyophilization and the subunits are fractionated by Sephadex G-100 in the presence of 1% ammonium bicarbonate. Intact TSH elutes first, followed by TSH-α and then TSH-β. The subunits are recovered by lyophilization.

High-performance liquid chromatography shows the most promise for rapid and easy separation of subunits. Using a single procedure, the subunits of bovine TSH and LH can readily be isolated (Fig. 1) from amounts of hormone in the range of 1 to 100 mg. Hormone is dissolved (1–10 mg/ml) in 0.1 *M* sodium phosphate, pH 6.8, which is 1 m*M* in sodium azide and 6 *M* in guanidine hydrochloride. The solution is allowed to stand at 37° for 1 hr and then chromatographed on a Vydac 218TP10 C18 column with a linear gradient (60 min) of 0.1 *M* sodium phosphate pH 6.8 plus 1 m*M* sodium azide to 50% acetonitrile–50% 0.1 *M* sodium phosphate pH 6.8 plus 1 m*M* sodium azide, with a flow rate of 2 ml/min (Fig. 1). The

[b] B. Shome and A. F. Parlow, *Abstracts. Endocrinol. Proc. Int. Congr., 4th, 1972* Abstracts, p. 176 (1974).

[c] D. N. Ward, L. E. Reichert, Jr., W.-K. Liu, H. S. Nahm, J. Hsia, W. M. Lamkin, and N. S. Jones, *Recent Prog. Horm. Res.* **29,** 533 (1973).

[d] W.-K. Liu, K.-P. Yang, Y. Nakagawa, and D. N. Ward, *J. Biol. Chem.* **249,** 5544 (1974).

[e] S. D. Glenn, H. S. Nahm, G. S. Greenwald, and D. N. Ward, *Endocrinology* **111,** 1263 (1982).

[f] M. R. Sairam and C.-H. Li, *Biochem. Biophys. Res. Commun.* **51,** 336 (1973).

TABLE IV
OTHER METHODS USEFUL IN THE SEPARATION OF SUBUNITS OF GLYCOPROTEIN HORMONES

Hormone	Species	Conditions for dissociation	Methods used for separation of subunits	Investigators
CH	Equine	6 *M* guanidine hydrochloride	Sephadex G-75 in 0.125 *M* NH_4HCO_3	Moore and Ward (1980)[a]
LH	Bovine	0.1 *M* Na phosphate pH 6.8 + 6 *M* guanidine hydrochloride	High-performance liquid chromatography using a Vydac 218TP10 C_{18} column in 0.1 *M* Na phosphate pH 6.8 with a linear acetonitrile gradient	Parsons *et al.* (1984)[d]
TSH	Bovine	1 *M* propionic acid	Sephadex G-100 in 1% NH_4HCO_3	Liao and Pierce (1970)[35]
TSH	Human	1 *M* propionic acid	Sephadex G-100 in 1% NH_4HCO_3	Cornell and Pierce (1973)[20]
TSH	Human	1 *M* propionic acid	Pentyl-Sepharose-4B	Jacobson *et al.* (1978)[b]
TSH	Bovine	0.1 *M* Na phosphate pH 6.8 + 6 *M* guanidine hydrochloride	High-performance liquid chromatography using a Vydac 218TP10 C_{18} column in 0.1 *M* Na phosphate pH 6.8 with a linear acetonitrile gradient	Parsons *et al.* (1984)[d]
FSH	Ovine	1 *M* propionic acid	Sephadex G-100 in 0.05 *M* NH_4HCO_3	Papkoff and Ekblad (1970)[c]

[a] W. T. Moore, Jr. and D. N. Ward, *J. Biol. Chem.* **255,** 6923 (1980).
[b] G. Jacobson, P. Roos, and L. Wide, *Biochim. Biophys. Acta* **536,** 363 (1978).
[c] H. Papkoff and M. Ekblad, *Biochem. Biophys. Res. Commun.* **40,** 614 (1970).
[d] T. F. Parsons, T. W. Strickland, and J. G. Pierce, *Endocrinology* **114,** 2223 (1984).

subunit fractions are pooled, lyophilized, and desalted by Sephadex G-25 chromatography in the presence of 1% ammonium bicarbonate.

Reconstitution of α–β Dimers (Active Hormone) from Isolated Subunits

In the pH range of 5.3 to 9.5, the α and β subunits will, in a time, temperature, and concentration dependent reaction, associate to form

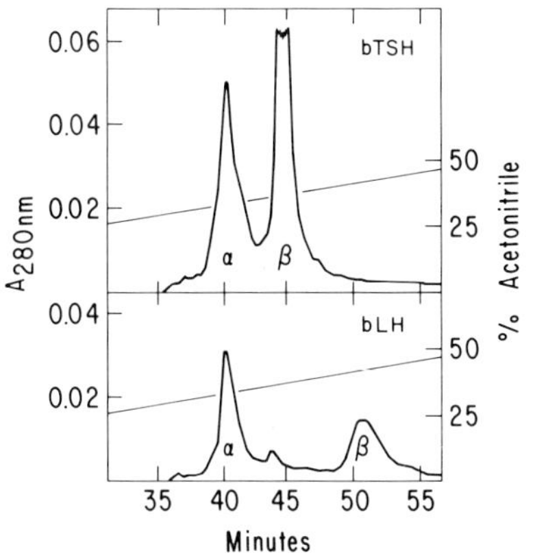

FIG. 1. Separation of bTSH and bLH subunits by high-performance liquid chromatography. Chromatography is on a Vydac 218TP10 C18 column with a linear gradient (60 min) of 0.1 *M* sodium phosphate pH 6.8 plus 1 m*M* sodium azide to 50% acetonitrile–50% 0.1 *M* sodium phosphate pH 6.8 plus 1 m*M* sodium azide, with a flow rate of 2 ml/min. The α and β subunit peaks are indicated.

active hormone.[8,9,13,21,42] Although this range of pH has been successfully employed for the recombination of many glycoprotein hormone subunits, a pH near neutrality would be the prudent choice for the reconstitution of a hormone whose pH stability has not been specifically determined in the studies cited above. The extremes of the pH range often enhance the solubility of certain β subunits. A concentrated (0.5 *M*) acetate buffer (pH 6) has been suggested by several authors to facilitate recombination by increasing the solubility of the β subunit.[7,13,37,42] No species specificity has been observed in the recombination reaction, α and β subunits from distantly related species can combine to form hybrid hormones.[22,43]

The amount of hormone formed upon incubation of subunits is dictated by the subunit concentration and the intersubunit equilibrium constant.[1,2] Table V contains intersubunit equilibrium constants as well as association rate constants and half-times of dissociation for several hormones. In Fig. 2, the equilibrium amount of intact hCG is shown as a function of hCG concentration. Reference to Fig. 2 reveals that at a total hCG concentration of 100 μg/ml (equimolar amounts of subunits), only 52% of the subunits will combine, while at 1 mg/ml 85% of the subunits is present as intact hormone. The time required to reach equilibrium is also highly dependent on the subunit concentration. For example, at a final hCG concentration of 1 mg/ml (approximately 500 μg/ml of each subunit)

[42] T. W. Strickland and D. Puett, *Endocrinology* **109,** 1933 (1981).

[43] P. Licht, S. W. Farmer, and H. Popkoff, *Gen. Comp. Endocrinol.* **35,** 289 (1978).

TABLE V
KINETIC AND EQUILIBRIUM PARAMETERS OF GLYCOPROTEIN HORMONE SUBUNIT INTERACTIONS

	hCG[a]	oLH[a]	pLH[a]	bTSH[b]
K_d (μM)[c]	0.60	0.42	0.064	—
K_a (M^{-1} min^{-1})[d]	64	180	260	30
$t_{1/2\ dissoc}$ (days)[e]	37	11	46	—

[a] At pH 6.8, 0.15 *M* NaCl, and 37°, from T. W. Strickland and D. Puett, *J. Biol. Chem.* **257,** 2954 (1982).
[b] At pH 7.0, 0.1 *M* NaCl, and 37°, from J. C. Pernollet, J. Garnier, J. G. Pierce, and R. Salesse, *Biochim. Biophys. Acta* **446,** 262 (1976).
[c] Equilibrium dissociation constant for α–β interaction.
[d] Second order rate constant for the association of α and β subunits.
[e] Time (in days) required for 50% of the hormone in a dilute solution to dissociate into subunits.

reassociation to 90% of the equilibrium amount of recombinant occurs in approximately 24 hr while at 100 μg/ml final hCG concentration, over 4 days are required for 90% of the equilibrium level of recombinant to form. These estimates of the rate of association of hCG subunits are for pH 6.8, 0.15 *M* NaCl, and 37°[1] and it should be noted that the rate of association of hCG subunits is substantially faster at lower salt concentration.[2,29] Unlike hCG, the rate of association of bTSH, bLH, and hLH subunits appears to

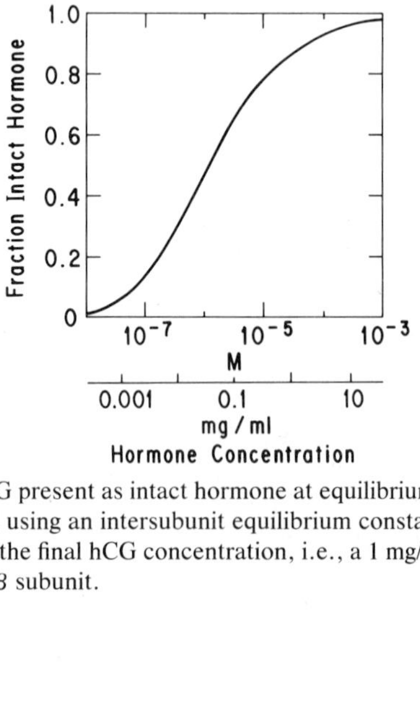

FIG. 2. Fraction of hCG present as intact hormone at equilibrium as a function of protein concentration. Calculated using an intersubunit equilibrium constant of 0.60 μM (Table V). The mg/ml scale refers to the final hCG concentration, i.e., a 1 mg/ml solution contains 0.41 mg/ml α and 0.59 mg/ml β subunit.

be unaffected by the salt concentration.[8,26,32] Consideration of the equilibrium and kinetic properties of glycoprotein hormone subunit interactions reveals that recombination should be carried out at the highest subunit concentration possible to obtain the greatest fraction of combined hormone and to minimize the incubation period.

The rate of subunit association is highly dependent on temperature with negligible recombination taking place at 0°.[8,26,29,32] The rate of association increases with temperature up to 37°,[8,26,29,32] the temperature at which most recombinations are conducted. While recombination occurs and may actually be enhanced at temperatures up to 45°,[8] temperatures higher than this favor free subunits over intact hormone.[2]

[58] Purification of Human Platelet-Derived Growth Factor

By ELAINE W. RAINES and RUSSELL ROSS

Introduction

Platelet-derived growth factor (PDGF), a 30,000 molecular weight glycoprotein released from the platelet during coagulation, has been shown to be one of the principal macromolecules in whole blood serum capable of stimulating DNA synthesis and cell growth in connective tissue cells *in vitro*.[1–3] Although the same molecule has been purified and characterized as PDGF by all four laboratories working on the platelet mitogen,[4–7] PDGF accounts for approximately 50% of the mitogenic activity found in platelets.[8,9] The other 50% is probably due to more than one growth factor, one of which is known to be EGF,[10,11] and was, therefore, not a significant enough fraction to be pursued. As a potent mitogen for fibro-

[1] R. Ross, J. Glomset, B. Kariya, and L. Harker, *Proc. Natl. Acad. Sci. U.S.A.* **71,** 1207 (1974).

[2] N. Kohler and A. Lipton, *Exp. Cell Res.* **87,** 297 (1974).

[3] B. Westermark and Å. Wasteson, *Exp. Cell Res.* **98,** 170 (1976).

[4] H. N. Antoniades, *Proc. Natl. Acad. Sci. U.S.A.* **78,** 7314 (1981).

[5] T. F. Deuel, J. S. Huang, R. T. Proffitt, J. U. Baenziger, D. Chang, and B. B. Kennedy, *J. J. Biol. Chem.* **256,** 8896 (1981).

[6] C.-H. Heldin, B. Westermark, and Å. Wasteson, *Biochem. J.* **193,** 907 (1981).

[7] E. W. Raines and R. Ross, *J. Biol. Chem.* **257,** 5154 (1982).

[8] C.-H. Heldin, B. Westermark, and Å. Wasteson, *Exp. Cell Res.* **136,** 255 (1981).

[9] E. W. Raines and R. Ross, unpublished observations (1982).

[10] Y. Oka and D. N. Orth, *J. Clin. Invest.* **72,** 249 (1983).

[11] D. F. Bowen-Pope and R. Ross, *Biochem. Biophys. Res. Commun.* **114,** 1035 (1983).

ISBN 0-12-182009-2

blasts and smooth muscle cells that is released focally at sites of injury, PDGF may play a physiologic role in wound healing and tissue repair[12,13] and a pathologic role in the formation of lesions of atherosclerosis.[14,15] Studies with highly purified PDGF will help determine the role of PDGF and other platelet mitogens in these processes and in defining the early cellular changes that occur in response to stimulation by PDGF. PDGF was originally defined by its presence in platelets, and platelets serve as the major source for purification. However, it is now known that PDGF-like molecules are also synthesized by cultured vascular endothelial cells,[16] newborn rat aortic smooth muscle cells,[17] and cells transformed by a wide variety of transforming agents.[18–20] In addition, partial amino acid sequence data for PDGF have demonstrated that one chain of PDGF is virtually homologous to the putative protein product of the transforming gene of the simian sarcoma virus.[21–23] The homology of PDGF with the viral oncogene product is the only known link of an oncogene with a growth factor. Together, these observations raise the possibilities that a PDGF analog may be involved in normal development and in neoplasia.

The low concentrations of PDGF (17.5 ng/ml) normally present in serum[24,25] and the highly cationic and hydrophobic character[26,27] of PDGF made development of purification procedures difficult. This chapter de-

[12] R. Ross and A. Vogel, *Cell* **14,** 203 (1978).

[13] C. D. Scher, R. C. Shepard, H. N. Antoniades, and C. D. Stiles, *Biochim. Biophys. Acta* **560,** 217 (1979).

[14] R. Ross and J. A. Glomset, *N. Engl. J. Med.* **295,** 369 and 420 (1976).

[15] R. Ross, *Arteriosclerosis* **1,** 293 (1981).

[16] P. E. DiCorleto and D. F. Bowen-Pope, *Proc. Natl. Acad. Sci. U.S.A.* **80,** 1919 (1983).

[17] R. Seifert, S. Schwartz, and D. Bowen-Pope, *Nature* (*London*) (in press).

[18] D. F. Bowen-Pope, A. Vogel, and R. Ross, *Proc. Natl. Acad. Sci. U.S.A.* **81,** 2396 (1984).

[19] T. F. Deuel, J. S. Huang, S. S. Huang, P. Stroobant, and M. D. Waterfield, *Science* **221,** 1348 (1983).

[20] C.-H. Heldin, B. Westermark, and Å. Wasteson, *J. Cell. Physiol.* **105,** 235 (1980).

[21] R. F. Doolittle, M. W. Hunkapiller, L. E. Hood, S. G. Devare, K. C. Robbins, S. A. Aaronson, and H. N. Antoniades, *Science* **221,** 275 (1983).

[22] A. Johnsson, C.-H. Heldin, A. Wasteson, B. Westermark, T. F. Deuel, J. S. Huang, P. H. Seeburg, A. Gray, A. Ullrich, G. Scrace, P. Stroobant, and M. D. Waterfield, *EMBO J.* **3,** 921 (1984).

[23] M. D. Waterfield, G. T. Scrace, N. Whittle, P. Stroobant, A. Johnsson, A. Wasteson, B. Westermark, C.-H. Heldin, J. S. Huang, and T. F. Deuel, *Nature* (*London*) **304,** 35 (1983).

[24] J. P. Singh, M. A. Chaikin, and C. D. Stiles, *J. Cell Biol.* **95,** 667 (1982).

[25] D. F. Bowen-Pope, T. W. Malpass, D. M. Foster, and R. Ross, *Blood* **64,** 458 (1984).

[26] C.-H. Heldin, B. Westermark, and Å. Wasteson, *Proc. Natl. Acad. Sci. U.S.A.* **76,** 3722 (1979).

[27] R. Ross, A. Vogel, P. Davies, E. Raines, B. Kariya, M. J. Rivest, C. Gustafson, and J. Glomset, *Cold Spring Harbor Conf. Cell Proliferation* **6,** 27 (1979).

scribes a method for purification of human PDGF from outdated platelet-rich plasma (PRP)[7] using commonly available laboratory reagents and yielding a mitogen purified 800,000-fold over the starting material. Methods of analysis and properties of highly purified PDGF are also discussed.

Source

Human PDGF purified from washed platelets and from outdated PRP appears to be molecularly comparable. Outdated human PRP (outdated 24 to 72 hr after collection) was used as the starting material in the method described here. PRP is approximately 2- to 3-fold enriched in PDGF over whole blood serum as determined by radioreceptor assay (see Fig. 1B and Methods of Analysis, this chapter, for more detailed description). Working with the platelet cell pellet, free of plasma proteins, would represent an even greater enrichment over serum. However, this is not possible using outdated PRP since approximately 90% of the growth-promoting activity[7] and 70 to 90% of the specific PDGF competitor[28] have already been released into the plasma when the material is received from the blood bank. Outdated PRP therefore represents an available source and allows purification of maximal amounts of PDGF.

Three pieces of evidence suggest structural conservation of the PDGF molecule and therefore the possibility that the purification method outlined here might be applicable to other species. First, a recent phylogenetic analysis of PDGF by radioreceptor assay determined that all tested specimens of clotted whole blood from phylum Chordata contain a homolog capable of competing with labeled human PDGF for specific receptor binding to mouse cells.[24] Second, human [^{125}I]PDGF binds to cells from human, mouse, rat, chicken, and fish with the same apparent affinity.[29] These two observations support significant structural homology at least in the receptor binding site. The third piece of evidence is that monospecific antisera made to human PDGF recognize a PDGF analog in a wide variety of species, including rabbit, rat, mouse, chicken, and horse.[9] In addition, preliminary observations from our laboratory, in collaboration with Dr. Hisao Kato, Kyushu University, Japan, with bovine platelets[30] and the partial purification of porcine PDGF[31] suggest that PDGF from other species fractionates in a manner similar to human PDGF.

[28] E. W. Raines, unpublished observations (1983).

[29] D. F. Bowen-Pope, R. A. Seifert, and R. Ross, *In* "Cell Proliferation: Recent Advances" (A. L. Boynton and H. L. Leffert, eds.). Academic Press, New York. In press.

[30] H. Kato and E. W. Raines, unpublished observations (1982).

[31] E. Rozengurt, P. Stroobant, M. D. Waterfield, T. F. Deuel, and M. Keehan, *Cell* **34,** 265 (1983).

PDGF-like growth factors have also been reported to be produced by cultured vascular endothelial cells,[16] a line of osteosarcoma cells,[20] cells transformed by a wide variety of transforming agents,[18,19] and by newborn rat aortic smooth muscle cells.[17] Although the absolute relationship between PDGF and the molecules produced by these cells remains to be determined, they may serve as an alternate source and the fractionation procedures described here may be useful in preparing highly purified preparations of these PDGF analogs.

Methods of Analysis

The ability of PDGF to stimulate DNA synthesis and cell growth in culture was the basis for its definition and discovery, and was used to guide its purification. [^{3}H]thymidine incorporation into DNA of cultured cells responsive to PDGF represents the most readily available method to follow its purification and define the biological activity of a purified preparation, and is discussed in detail below. Other assays to quantitate PDGF (radioreceptor assay and radioimmunoassay or ELISA) require highly purified reagents, but for the laboratory with a continuing requirement for large quantities of PDGF, one of these alternative assays may be less time consuming or more convenient.

[^{3}H]Thymidine Incorporation

We have used quiescent cultures of mouse 3T3 cells and human fibroblasts to follow the mitogenic activity during purification of PDGF by measuring [^{3}H]thymidine incorporation into trichloroacetic acid (TCA)-precipitable material. Figure 1A illustrates a typical dose–response curve for highly purified PDGF in Swiss mouse 3T3 cells (working range—0.2 to 2.0 ng/ml). Although [^{3}H]thymidine incorporation measures the biological activity of interest, there are a number of limitations of the assay that must be kept in mind with its use. First, it is not specific for PDGF. A number of different growth factors and cofactors are able to stimulate [^{3}H]thymidine incorporation. This is demonstrated in Fig. 1 where human PRP, whole blood serum (WBS), and highly purified PDGF were assayed for stimulation of [^{3}H]thymidine incorporation (Fig. 1A) and for the ability to compete for specific binding to the PDGF receptor[32] (Fig. 1B—see below and Bowen-Pope and Ross, this volume [8] for detailed discussion). In matched cultures, the amount of purified PDGF required to induce half-maximal stimulation of [^{3}H]thymidine incorporation (1.0 ng) is very comparable to the amount required to induce half-maximal competition in the radioreceptor assay (1.2 ng). However, for PRP, 7.5 μl induces half-

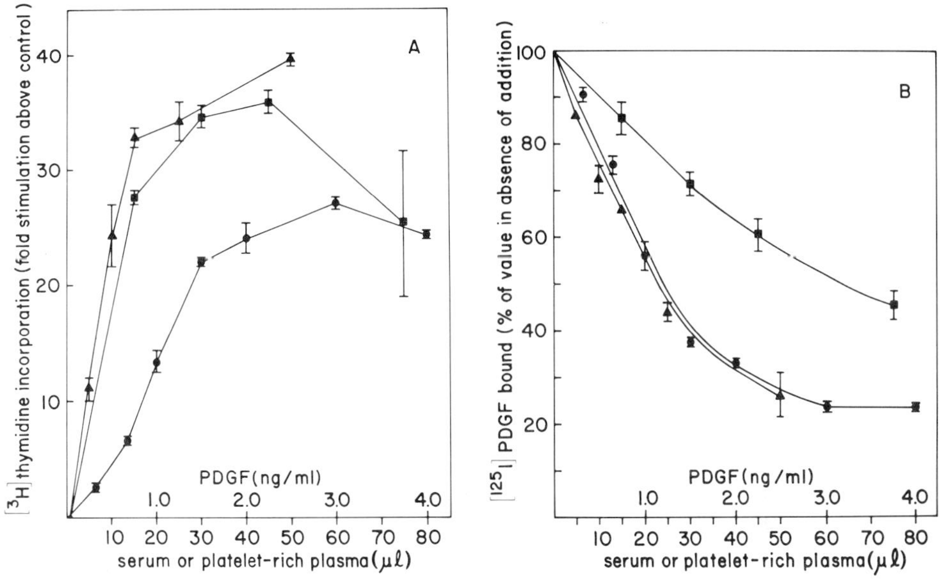

FIG. 1. Comparison between the [^{3}H]thymidine incorporation assay and the radioreceptor assay for PDGF. Defibrinogenated human platelet-rich plasma (▲), human whole blood serum (■), and highly purified human PDGF (●) were assayed on quiescent cultures of 3T3 cells prepared as described in the text for (A) stimulation of [^{3}H]thymidine incorporation into trichloroacetic acid precipitable material, and (B) competition for [^{125}I]PDGF binding in the radioreceptor assay. [^{3}H]Thymidine incorporation is plotted as the fold stimulation over control cultures which received buffer only. Binding in the radioreceptor assay was plotted without correction for nonspecific binding which was determined to be 3% (data not shown). For both assays, the mean ± standard deviation of triplicate determinations is plotted.

maximal [^{3}H]thymidine incorporation but 25.5 μl is required to induce half-maximal competition for binding to the PDGF receptor. This difference is also found with whole blood serum, where 10 μl is required to induce half-maximal [^{3}H]thymidine incorporation and 63.5 μl is required to give half-maximal competition for binding. Therefore, because PRP and WBS contain cofactors and other growth factors, the PDGF content of these preparations is overestimated by the [^{3}H]thymidine incorporation assay and, at early stages in the purification, yields and fold purification are underestimated (see Table I). A second problem is standardization from assay to assay. Serum is such a complex mixture of growth factors and cofactors, and there is considerable variation in [^{3}H]thymidine incorporation with serum lots. It is therefore important that a large number of aliquots of the reference serum be frozen or, preferably, serum and a purified PDGF preparation be aliquoted and used as a reference in all assays. Additional problems which can dramatically affect the working

TABLE I

PURIFICATION OF PLATELET-DERIVED GROWTH FACTOR FROM 4 LITERS OF OUTDATED PLATELET-RICH PLASMA[a]

Fraction	Total protein[b] (mg)	[^{3}H]Thymidine incorporation ED$_{50}$[c] (ng)	[^{3}H]Thymidine incorporation Purification factor (fold)	[^{3}H]Thymidine incorporation Yield (% of initial)	PDGF radioreceptor assay ED$_{50}$[d] (ng)	PDGF radioreceptor assay Purification (fold)	PDGF radioreceptor assay Yield (% of initial)
Defibrinogenated platelet-rich plasma	195,200	878,000	—	100.0	1,600,000	—	100.0
CM-Sephadex	60.1	560	1,568	55.0	555	2883	88.8
Sephacryl S-200	4.89	100	8,780	25.0	115	13,913	34.9
Heparin-Sepharose	0.93	20	43,900	23.8	24	66,667	31.8
Phenyl-Sepharose	0.036	1.0	878,000	18.4	1.10	1,454,545	26.8

[a] From E. W. Raines and R. Ross, *In* "Methods in Molecular and Cell Biology" (D. Barnes, D. Sirbasku, and G. Sato, eds.), p. 89. Alan R. Liss, New York, 1984.

[b] Protein was determined by absorption employing $E_{280}^{1\%} = 10.0$ for platelet-rich plasma and subsequent steps, except the final step (phenyl-Sepharose), where protein was determined by the method of Lowry *et al.*[47] using bovine serum albumin as a standard. The protein values represent the mean of 49 preparations.

[c] ED_{50} is the amount of a given fraction required to give half-maximal stimulation of thymidine incorporation in the 3T3 cell assay described in the text and illustrated in Fig. 1A. Values represent the mean of at least 5 preparations.

[d] ED_{50} is the amount of a given fraction required to give half-maximal competition for binding in the PDGF radioreceptor assay in matched cultures of 3T3 cells as described in the text and illustrated in Fig. 1B. Values represent the mean of at least 5 preparations

range of the assay are variations in cell density and the quiescence of the cells. Using the procedure described in this chapter, we have found the working range of the assay to be quite stable. Finally, because the test samples are incubated at 37° for 20 hr with the cells, chemicals present in the samples can affect the health of the cells and therefore their ability to incorporate [^{3}H]thymidine. For example, samples from the last two steps in the purification procedure, even if lyophilized, induce toxic effects when assayed for [^{3}H]thymidine incorporation unless they are first dialyzed. It is therefore important to visually examine the cells before they are harvested, and if toxicity is suspected, a control of known mitogenicity should be tested together with an additional sample of the test substance.

Materials

Quiescent cultures of mouse 3T3 cells (available American Tissue Type collection). Quiescent test cultures can be obtained by plating the cells in 10% serum and allowing them to deplete the medium (4–5 days); plating the cells in 10% serum and then changing the medium to 2% calf CMS I[32] (calf serum incubated with CM-Sephadex to remove PDGF), or plating the cells in 10% serum and then changing the media to 2% plasma-derived serum (serum prepared from anticoagulated blood, made cell free by centrifugation, and then recalcified by addition of calcium and a final centrifugation to remove the fibrin clot). Cells switched to 2% plasma-derived serum or 2% calf CMS I are best used 2 to 4 days after the media change, but unused cultures can be maintained in a healthy state by refeeding every 5 to 7 days. Stock cultures are maintained in 75-cm^2 flasks in Dulbecco's modified Eagle's medium with L-glutamine and D-glucose supplemented to give the final concentrations of the following: 10% (v/v) calf serum; 0.225% sodium bicarbonate, 100 units/ml of penicillin G sodium, and 100 μg/ml of streptomycin sulfate; and 1 m*M* sodium pyruvate. Test cultures are plated in the same medium in 2-cm^2 Costar 24-well culture dishes, 1 ml per well and 2–3 $\times$ 10^4 cells per well.

[^{3}H]Thymidine (70–80 Ci/mmol)

5% trichloroacetic acid (TCA)

0.25 *N* NaOH

Method. Test samples are lyophilized and solubilized in or directly diluted in 10 m*M* acetic acid containing 2.5 mg/ml bovine serum albumin (Pentex crystallized, Miles) and are added directly to the wells (100 μl/well) of quiescent 3T3 cells and incubated for 20 hr at 37°. At 16 to 19 hr after sample addition, the wells are examined by phase contrast micros-

[32] A. Vogel, E. W. Raines, B. Kariya, M.-J. Rivest, and R. Ross, *Proc. Natl. Acad. Sci. U.S.A.* **75**, 2810 (1978).

copy for signs of toxicity (highly vacuolized cells or a large number of floating cells) and for induction of proliferation by examining the cells for altered morphology. As shown in Fig. 2, the stimulated cells become spindly with narrow cytoplasmic extensions. This morphologic change is very reproducible for partially purified and purified PDGF and can be used for a quick screening of fractions. The altered morphology is less pronounced in the presence of high concentrations of plasma proteins (e.g., platelet-rich plasma). For measurement of [^{3}H]thymidine incorporation, the media is removed from the wells at 20 hr and replaced with 0.5 ml of fresh media per well containing 2 μCi/ml [^{3}H]thymidine and 5% (v/v) calf serum for the 2-hr labeling period (the 5% calf serum is a convenient addition to maintain the viability of the cells and has no effect on the labeling which was induced 20 hr earlier). After an additional 2-hr incubation at 37°, the cells are harvested by aspirating off the media, washing the wells twice each with 1 ml of ice-cold 5% TCA, solubilizing TCA-insoluble material in 0.25 NaOH (0.8 ml) with mixing, and counting 0.6 ml of this solution in 5 ml Aquasol in a liquid scintillation counter. Fold stimulation over control wells (100 μl of 10 mM acetic acid alone) is determined (normally 30- to 50-fold maximal stimulation) and compared to serum and purified PDGF standards.

Radioreceptor Assays

The radioreceptor assay for PDGF[33] (Fig. 1B and Bowen-Pope and Ross, this volume [8] for detailed discussion) has several advantages over the [^{3}H]thymidine incorporation assay. First, it is specific for PDGF and sensitive (0.2–2.0 ng for 3T3 cells shown in Fig. 1B, depending upon the cell type and cell density used). It is also less sensitive to toxic effects of buffers than the [^{3}H]thymidine incorporation assay. For example, samples from the last purification step, which are toxic when assayed in the [^{3}H]thymidine assay, can be assayed directly in the radioreceptor assay without dialysis. Lastly, the radioreceptor assay does not require quiescent cultures (only that the cell number be constant) and only a 4-hr incubation time, and therefore results can be obtained more readily. However, the assay requires a highly purified PDGF preparation for use as a radioligand, and the sensitivity of the assay can be affected by the presence of specific plasma binding proteins for PDGF.[25,34]

The radioreceptor assay for epidermal growth factor (EGF) can also be used to follow the purification of PDGF. Binding of PDGF to its recep-

[33] D. F. Bowen-Pope and R. Ross, *J. Biol. Chem.* **257,** 5161 (1982).

[34] E. W. Raines, D. F. Bowen-Pope, and R. Ross, *Proc. Natl. Acad. Sci. U.S.A.* **81,** 3424 (1984).

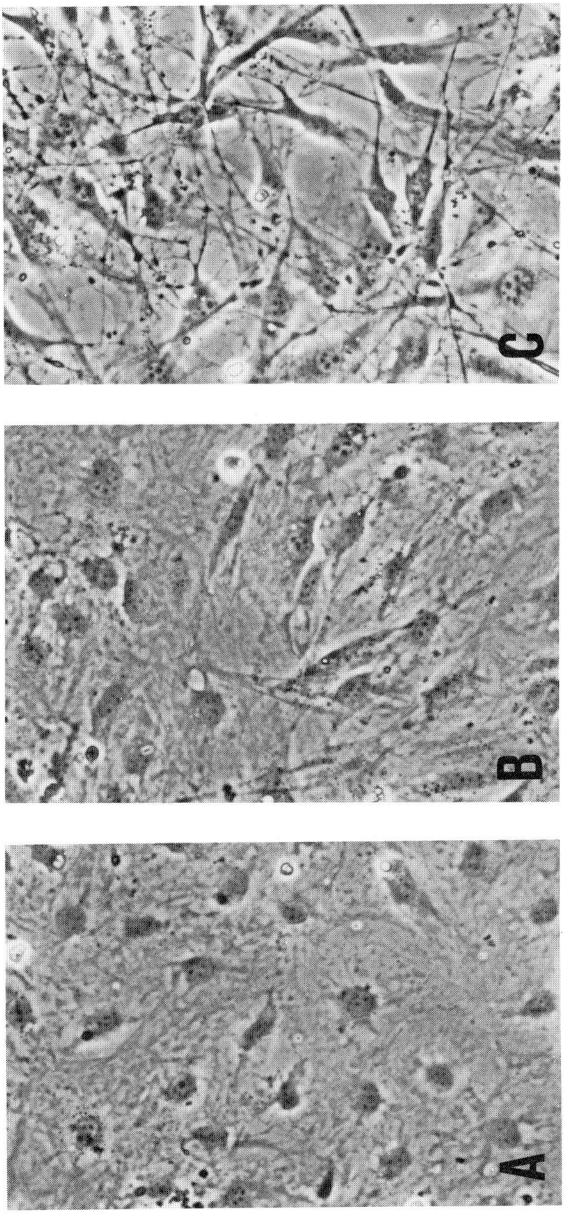

FIG. 2. Morphological change in 3T3 cells 16 hr after stimulation with purified fractions of PDGF. 3T3 cells were prepared as described in the text and (A) only buffer added to the well or (B) 0.5 ng/ml phenyl-Sepharose purified PDGF or (C) 2.0 ng/ml purified PDGF. Sixteen hours after addition of the samples, the phase contrast micrographs (×1000) were taken.

tor inhibits the subsequent binding of [^{125}I]EGF to its receptor.[35–39] This inhibition is detected at very low concentrations of PDGF (0.05–0.25 ng/ml). However, the EGF radioreceptor assay also detects EGF, fibroblast-derived growth factor[40] (a basic, heat- and acid-stable polypeptide isolated from SV40-transformed BHK cells), and potentially other, as yet undefined, growth factors. Although not specific, the EGF radioreceptor assay provides a sensitive assay for PDGF and could be readily used in a lab where [^{125}I]EGF is utilized for other studies.

Immunologic Assays

PDGF appears to be a weak antigen, but antibodies to PDGF have been raised in rabbits,[8,41,42] goat,[9] and mice.[43] Radioimmunoassays developed with these antibodies have employed antibody bound to Sepharose,[8] IgG sorb,[42] or a second antibody,[41] to precipitate antibody bound [^{125}I]PDGF, and have varied in their sensitivity from approximately 0.2 to 2 ng/ml. An ELISA developed in our laboratory[28] has been particularly useful for routine screening of column fractions. With a sensitivity of 0.2 ng/ml and rapid quantitation of samples using an ELISA plate reader, a large number of samples can be evaluated quickly with very little consumption of the purified material. An advantage of all of the immunologic assays is that cell cultures are not required. However, purified PDGF and monospecific antisera are required. Like the radioreceptor assay, the radioimmunoassays and ELISA are also affected by specific binding proteins in plasma for PDGF which will bind PDGF and prevent binding of antibody.[42]

In addition, since the PDGF antibody is a neutralizing antibody (it prevents binding of PDGF to its cell surface receptor and PDGF-induced [^{3}H]thymidine incorporation by 99%), it can be used to determine the proportion of [^{3}H]thymidine incorporation directly due to PDGF stimulation. To avoid a contribution to [^{3}H]thymidine incorporation by the antibody preparation, we have prepared the IgG fraction by sodium sulfate precipitation and DEAE-Sephacel chromatography (Pharmacia).

[35] M. Wrann, C. F. Fox, and R. Ross, *Science* **210,** 1363 (1980).

[36] C.-H. Heldin, Å. Wasteson, and B. Westermark, *J. Biol. Chem.* **257,** 4216 (1982).

[37] M. A. Shupnik, H. N. Antoniades, and A. H. Tashjiian, *Life Sci.* **30,** 347 (1982).

[38] W. Wharton, E. Leof, W. J. Pledger, and E. J. O'Keefe, *Proc. Natl. Acad. Sci. U.S.A.* **79,** 5567 (1982).

[39] D. F. Bowen-Pope, P. E. DiCorleto, and R. Ross, *J. Cell Biol.* **96,** 679 (1983).

[40] P. Dicker, P. Pohjanpelto, P. Pettican, and E. Rozengurt, *Exp. Cell Res.* **135,** 221 (1981).

[41] H. N. Antoniades and C. D. Scher, *Proc. Natl. Acad. Sci. U.S.A.* **74,** 1973 (1977).

[42] J. S. Huang, S. S. Huang, and T. F. Deuel, *J. Cell Biol.* **97,** 338 (1983).

[43] P. E. DiCorleto, E. W. Raines, and R. Ross, unpublished observations (1981).

Other Biological Assays

PDGF was purified and defined by its ability to induce [^{3}H]thymidine incorporation and cell growth in connective tissue cells which are detected at least 20 hr after addition of PDGF to the culture medium. However, it is known that cells respond very rapidly to PDGF. In addition to specific cellular binding (see Bowen-Pope and Ross[32] and this volume [8]), cellular responses such as induction of chemotaxis,[44] increased amino acid uptake,[45] or stimulation of the Na–K pump[46] could be used to monitor purification. These cellular effects, as with [^{3}H]thymidine incorporation, are not specific for PDGF, but provide alternate methods for monitoring the purification of PDGF.

Protein Determination

Protein values are determined by measuring absorbance at 280 nm with the assumption that a 1 mg/ml solution gives an absorbance of 1.0 in a 1-cm path cuvette. For highly purified PDGF (Phenyl-Sepharose eluted fraction), protein has been determined for standardization by the method of Lowry[47] using bovine serum albumin as a standard (matched with a sealed ampule standard of bovine serum albumin from Armour Chemical Co.). With highly purified PDGF this is essential for reproducible determination of protein content because protein concentration of the highly purified fraction is normally 15 to 60 μg/ml. Comparison of values determined by absorbance at 280 nm and by Lowry assay for six phenyl-Sepharose preparations gave a mean OD_{280}/Lowry ratio of 1.59 ± 0.15.[28]

Purification Procedure

General Considerations

PDGF is highly positively charged and hydrophobic[5,26,27,48] and is therefore readily adsorbed to glass and plastic surfaces,[7,32,49] particularly

[44] G. R. Grotendorst, T. Chang, H. E. J. Seppa, H. K. Kleinman, and G. R. Martin, *J. Cell. Physiol.* **113,** 261 (1982).

[45] A. J. Owen, III, R. P. Geyer, and H. N. Antoniades, *Proc. Natl. Acad. Sci. U.S.A.* **79,** 3203 (1982).

[46] S. A. Mendoza, N. M. Wigglesworth, P. Pohjanpelto, and E. Rozengurt, *J. Cell. Physiol.* **103,** 17 (1980).

[47] O. H. Lowry, N. J. Rosebrough, A. L. Farr, and R. J. Randall, *J. Biol. Chem.* **193,** 265 (1951).

[48] H. N. Antoniades, C. D. Scher, and C. D. Stiles, *Proc. Natl. Acad. Sci. U.S.A.* **76,** 1809 (1979).

[49] J. C. Smith, J. P. Singh, J. S. Lillquist, D. S. Goon, and C. D. Stiles, *Nature (London)* **296,** 154 (1982).

in the highly purified state. Therefore, exogenous protein (usually bovine serum albumin or gelatin at 2.5 mg/ml) is added to highly purified samples (heparin-Sepharose and phenyl-Sepharose fractions) prior to dialysis and in diluents for biological assays. In addition, all glass and plasticware used in the purification are siliconized (Aquasil or Surfasil, Pierce Chemical). The purification procedures, unless otherwise noted, are carried out at 4° to prevent bacterial contamination.

The procedure is described for 80 units of platelet-rich plasma (approximately 4 liters) but can be scaled up depending on facilities available. Table I summarizes the purification steps and data from each step using [^{3}H]thymidine incorporation and the PDGF radioreceptor assay.

The admixture of contaminating proteins present in partially purified fractions can dramatically affect the fractionation of PDGF. It is now known that plasma contains proteins which specifically bind PDGF[25,34,42,50] and can result in higher molecular weight complexes of PDGF and its binding protein(s) unless dissociation conditions are used. For this reason, as well as the preference for a volatile buffer to facilitate PAGE analysis, ammonium acetate is used in the last three purification steps.

Freeze-Thaw and Heat Defibrinogenation

Upon receipt from the blood bank, units of outdated platelet-rich plasma are frozen at −20° and stored until needed. They are frozen and thawed a total of 3 times before use, transferred from the bags to 1-liter beakers, and incubated at 55–57° for 8 min in a circulating water bath to precipitate fibrinogen (~4% of the plasma proteins). After heating, the plasma is transferred to 250-ml centrifuge bottles and left at 4° overnight. The precipitated protein and platelet membranes are removed by centrifugation at 27,000 *g* for 30 min at 4° and the supernatant collected by filtration over glass wool. Since the centrifuged pellet contains a significant amount of mitogenic activity that can be solubilized in high salt,[7] the pellets are combined into two bottles and incubated first with 200 ml 0.39 *M* NaCl, 0.01 *M* Tris, pH 7.4 for 1 hr, centrifuged at 27,000 *g* for 30 min at 4°, and the supernatant collected, followed by 200 ml of 1.09 *M* NaCl, 0.01 *M* Tris, pH 7.4 and the same procedure. All of the supernatants are pooled and the pH adjusted to 7.4 by the addition of 1.0 *M* Tris-base. The conductivity is determined at room temperature and adjusted with 0.01 *M* Tris, pH 7.4 to ≤13 mmho.

[50] J. S. Huang, S. S. Huang, and T. F. Deuel, *Proc. Natl. Acad. Sci. U.S.A.* **81,** 342 (1984).

Carboxymethyl-Sephadex Chromatography

The first chromatography step takes advantage of the fact that at neutral pH, PDGF is highly positively charged and readily adsorbs to cation exchangers. The supernatant from the defibrinogenated platelet-rich plasma (approximately 4 liters) is stirred overnight with 1 liter of swollen CM-Sephadex (Pharmacia, equilibrated with 0.09 *M* NaCl, 0.01 *M* Tris, pH 7.4). The supernatant gel slurry is loaded, using a large diameter tubing (0.25 cm), onto a column of CM-Sephadex (7.5 by 35 cm, approximately 1500 ml packed volume, equilibrated as above) at a flow rate of 150 ml/hr. The column is washed with 4 liters of 0.19 *M* NaCl, 0.01 *M* Tris, pH 7.4, and PDGF is eluted with 0.5 *M* NaCl as shown in Fig. 3. The mitogenic fractions (indicated by the solid bar) are pooled and concentrated using an Amicon PM-10 membrane (nominal exclusion = 10,000 daltons) to a concentration of 3–4 mg/ml and then dialyzed against 1.0 *N* acetic acid using washed Spectrapor 1 dialysis tubing (nominal exclusion = 6000–8000 daltons). The normal yield from 80 units of PRP after con-

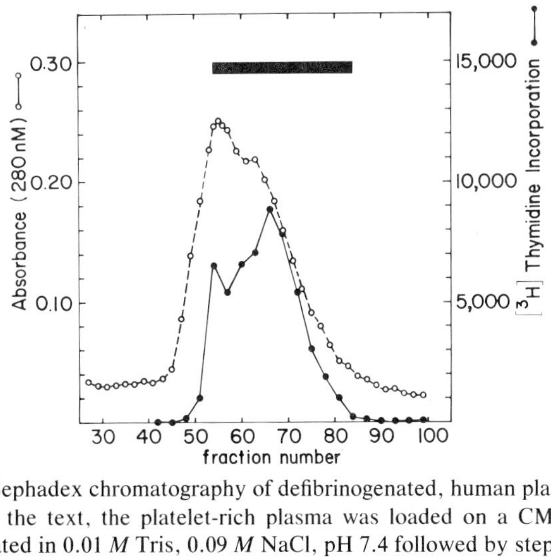

FIG. 3. CM-Sephadex chromatography of defibrinogenated, human platelet-rich plasma. As described in the text, the platelet-rich plasma was loaded on a CM-Sephadex, C-50 column equilibrated in 0.01 *M* Tris, 0.09 *M* NaCl, pH 7.4 followed by stepwise elution with 0.01 *M* Tris, 0.19 *M* NaCl, pH 7.4; and 0.01 *M* Tris, 0.5 *M* NaCl, pH 7.4. The elution of PDGF with 0.01 *M* Tris, 0.5 *M* NaCl, pH 7.4 is shown. Fractions of 25 ml were collected and protein elution monitored by absorbance at 280 n*M* (○) and mitogenic activity monitored by [^{3}H]thymidine incorporation as described in the text (●). The solid bar indicates the fractions pooled for further purification. From E. W. Raines and R. Ross, *in* "Methods in Molecular and Cell Biology" (D. Barnes, D. Sirbasku, and G. Sato, eds.), p. 89. Alan R. Liss, New York, 1984.

centration and dialysis is 60.1 mg ± 16 (n = 49). Regeneration of the column is accomplished by emptying the column, washing with 1.0 M NaCl, 0.5 N NaOH, and reequilibrating in the starting buffer.

The fraction not absorbed by CM-Sephadex contains mitogenic activity as determined by [^{3}H]thymidine incorporation (10–40% of the starting activity). This fraction represents the platelet anionic fraction previously described by Heldin *et al.*,[51] presumably epidermal growth factor, as well as other plasma growth factors and cofactors.

Sephacryl S-200 Molecular Sieving

Plasma contains specific binding proteins for PDGF[25,34,42,50] and the PDGF-binding protein complex would not be dissociated under conditions used for CM-Sephadex. Chromatography in the presence of dissociating agent (acetic acid or guanidine–HCl) has been shown to dissociate PDGF from higher molecular weight complexes.[34] Thus, the Sephacryl S-200 chromatography in 1.0 N ammonium acetate, pH 3.5, separates PDGF from its binding proteins and is a size selection step. The concentrated and dialyzed CM-Sephadex fraction is centrifuged at 27,000 g for 30 min to remove particulate material and adjusted to a pH of 3.5 at room temperature by the addition of concentrated ammonium hydroxide. This fraction is loaded (60–115 mg protein in less than 30 ml) on a Sephacryl S-200 column (Pharmacia, 5 by 92 cm, 1806 ml packed volume, equilibrated in 1.0 N acetic acid adjusted to pH 3.5 by addition of ammonium hydroxide) at a flow rate of 20 ml/hr. The elution profile (A_{280}) is shown in Fig. 4 along with the [^{3}H]thymidine incorporation data and the radioimmunoassay data for two other platelet α-granule proteins (platelet factor 4 and β-thromboglobulin). PDGF elutes with a molecular weight range of 29,000–45,000 and a peak molecular weight of 43,000, as determined by molecular weight standards (see Characteristics of Purified PDGF, this chapter, for discussion of variability in molecular weight values determined by gel filtration as compared with other molecular weight determinations). The fractions indicated by the solid bar (usually 80 to 100 ml) are combined for further purification and represent approximately 65% of the total activity recovered from the column (70% of the applied activity). The remaining 35% of the total recoverable activity contains contaminants not readily removed by subsequent steps. However, these active side fractions not included in the pool can be rechromatographed on S-200 and then further purified, or used for determination of nonspecific binding in the radioreceptor assay.

[51] C.-H. Heldin, Å. Wasteson, and B. Westermark, *Exp. Cell Res.* **109,** 429 (1977).

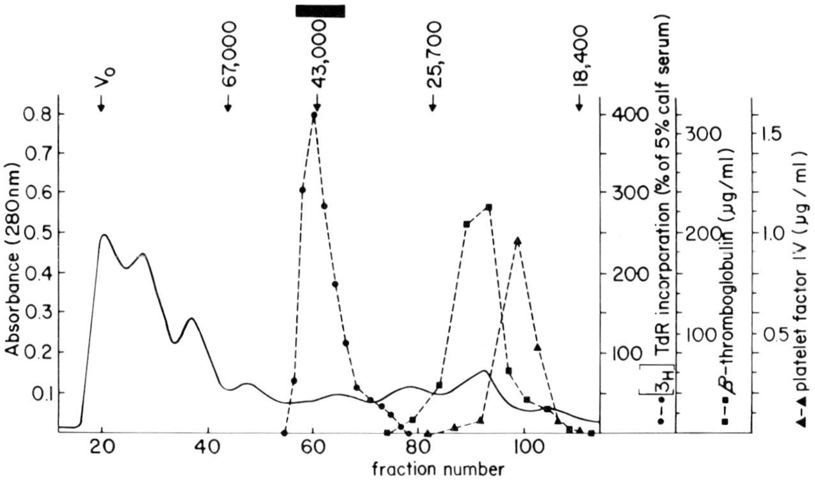

FIG. 4. Sephacryl S-200 gel filtration of CM-Sephadex purified PDGF. CM-Sephadex purified PDGF (117 mg) was prepared as described in the text and loaded on a Sephacryl S-200 column equilibrated in 1.0 *N* acetic acid pH adjusted to 3.5 with ammonium hydroxide. Fractions (7.8 ml) were collected and monitored for absorbance at 280 nm (——), for mitogenic activity by measuring [^{3}H]thymidine incorporation as described in the text (●), for β-thromboglobulin distribution by radioimmunoassay using a New England Nuclear kit (■), and for levels of platelet factor 4 by radioimmunoassay using the method of Files *et al.* [J. Files, T. Malpass, E. Yee, J. Ritchie, and L. Harker, *Blood* **58,** 607 (1981).] (▲). The molecular weight standards used to calibrate the column were: bovine serum albumin (67,000), ovalbumin (43,000), chymotrypsinogen (25,700), β-lactalbumin (18,400), and ^3H_2O (total column volume). The solid bar indicates the fractions pooled for further purification. Adapted from Raines and Ross.[7]

Heparin-Sepharose Chromatography

Gradient elution of heparin-Sepharose, a strong cation exchanger, separates PDGF from the bulk of other basic proteins selected in the initial CM-Sephadex chromatography. Pooled Sephacryl S-200 fractions (usually 500 ml, representing material from 3 to 5 CM-Sephadex preparations, depending on protein content) are adjusted to pH 7 with concentrated ammonium hydroxide and the conductivity lowered to 28.3 mmho by dilution with distilled water (both determinations at room temperature). The S-200 fractions are loaded onto a heparin-Sepharose column (Pharmacia, 1.5 by 13 cm, 23 ml of packed volume, equilibrated with ammonium acetate buffer prepared by adjusting 2.0 *N* acetic acid to pH 7 with concentrated ammonium hydroxide and then diluting with distilled water to 28.3 mmho) at a flow rate of 20 ml/hr. The maximum protein load is 30 mg or 1.5 mg/ml of gel. The column is eluted with an ammonium acetate gradient from 28.3 to 80.5 mmho (determined at 20°). The ammonium acetate for the gradient is prepared by adjusting 2.0 *N* acetic acid to pH 7 with concentrated ammonium hydroxide and then diluting with distilled water to the correct conductivity. Figure 5 shows the typical gradient elution profile with the PDGF eluting between 39 and 53 mmho being pooled for further purification (typically 120 to 140 ml). Heparin-Sepharose fractions must be dialyzed (in the presence of 2 mg/ml bovine serum albumin) before being assayed by [^{3}H]thymidine incorporation. The column is washed with 2 *M* guanidine–HCl and then reequilibrated in starting buffer.

Phenyl-Sepharose Chromatography

The hydrophobic character of PDGF is utilized for the final purification step. The heparin-Sepharose pool (from one column, normally 3 to 5 CM-Sephadex columns) is diluted to a conductivity of 34.0 mmho with distilled water (at 20°) and loaded on a phenyl-Sepharose column (Pharmacia, 0.9 by 12 cm, 7.5 ml packed volume, preequilibrated at 4° in 2.0 *N*

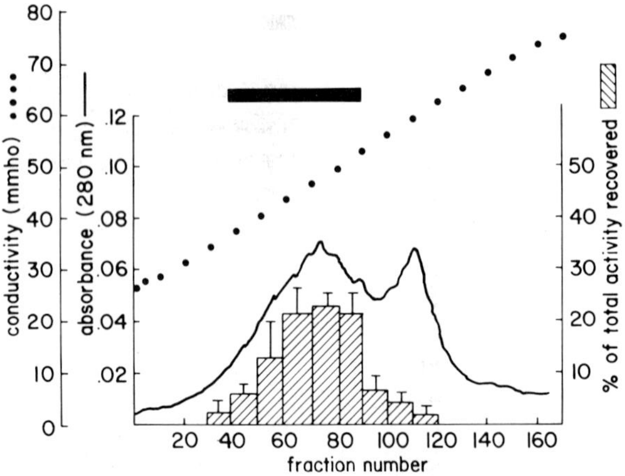

FIG. 5. Heparin-Sepharose gradient elution of Sephacryl S-200 purified PDGF. Sephacryl S-200 purified PDGF was chromatographed at 4° on heparin-Sepharose as described in the text. The linear gradient elution (ammonium acetate, pH 7, increasing from 28.3 to 80.5 mmho) of the adsorbed fraction is shown with absorbance at 280 nm indicated by the solid line (——), the conductivity measured at 20° indicated by the solid circles (●), and the percentage of total mitogenic activity recovered shown by slashed bar graphs. The mitogenic activity was determined by measuring [^{3}H]thymidine incorporation as described in the text, and the bar graphs represent the means and standard deviations for five preparations. The solid bar indicates the fractions pooled for further purification. Adapted from Raines and Ross.[7]

acetic acid adjusted to pH 7, and diluted to a conductivity of 34 mmho with distilled water) at a flow rate of 15 ml/hr and a maximal protein load of 1 mg/ml of gel. The column is washed with 30% (v/v) ethylene glycol and 10% (v/v) 1.0 *N* acetic acid adjusted to pH 7 with ammonium hydroxide. The PDGF is eluted with 50% (v/v) ethylene glycol and 10% (v/v) 1.0 *N* acetic acid adjusted to pH 7 with ammonium hydroxide as a single peak at the buffer front. One milliliter fractions are collected and absorbance at 280 nm for the peak tubes is in the range of 0.03 to 0.05. The absorbance at 280 nm of the pooled fraction is normally 0.015 to 0.025. The eluted PDGF can be tested directly in the radioreceptor assay but must be dialyzed in the presence of 2 mg/ml bovine serum albumin (or some other carrier protein to prevent adsorption to the dialysis tubing) before assay for mitogenic activity by [^{3}H]thymidine incorporation. The column is washed with 2 *M* guanidine–HCl and then reequilibrated in the starting buffer.

If it is necessary to remove the ethylene glycol or further concentrate the PDGF, the PDGF can be run on a small column of, for example, BioGel P-60 equilibrated in 0.2 *M* ammonium bicarbonate or concentrated by ultrafiltration (Centricon microconcentrators, Amicon Corp.).

Determination of Purity and Standardization

Determination of the purity of PDGF preparations is difficult because of limited quantities of purified material and the relatively large amounts required for such evaluation relative to its biological activity (10^{-11} *M*). Analysis by SDS–PAGE, with the added sensitivity of the silver stain procedure[52,53] (Fig. 6), has been the principal criteria used for evaluation of purity. A 15% separating gel with a 4.5% stacker (1.5 mm gel thickness) is used for this purpose employing the discontinuous buffer system described by Laemmli.[54] Using the silver stain procedure of Wray *et al.*,[53] highly purified PDGF run on SDS–PAGE under both nonreducing and reducing conditions at 100 ng, 500 ng, and 1 μg can be readily evaluated for other contaminating proteins. The mean percentage of stainable protein present in the 30,000 molecular weight PDGF complex (see Properties of Purified PDGF, this chapter) for six different preparations recently evaluated was 93.7% ± 0.7. Although silver-stained gels are particularly useful in assessing the purity of PDGF, we have not found them to be very quantitative (percent standard deviation of 25%) when the ratio of the

[52] R. C. Switzer, III, C. R. Merril, and S. Shifrin, *Anal. Biochem.* **98,** 231 (1979).
[53] W. Wray, T. Boulikas, V. P. Wray, and R. Hancock, *Anal. Biochem.* **118,** 197 (1981).
[54] U. K. Laemmli, *Nature (London)* **227,** 680 (1970).

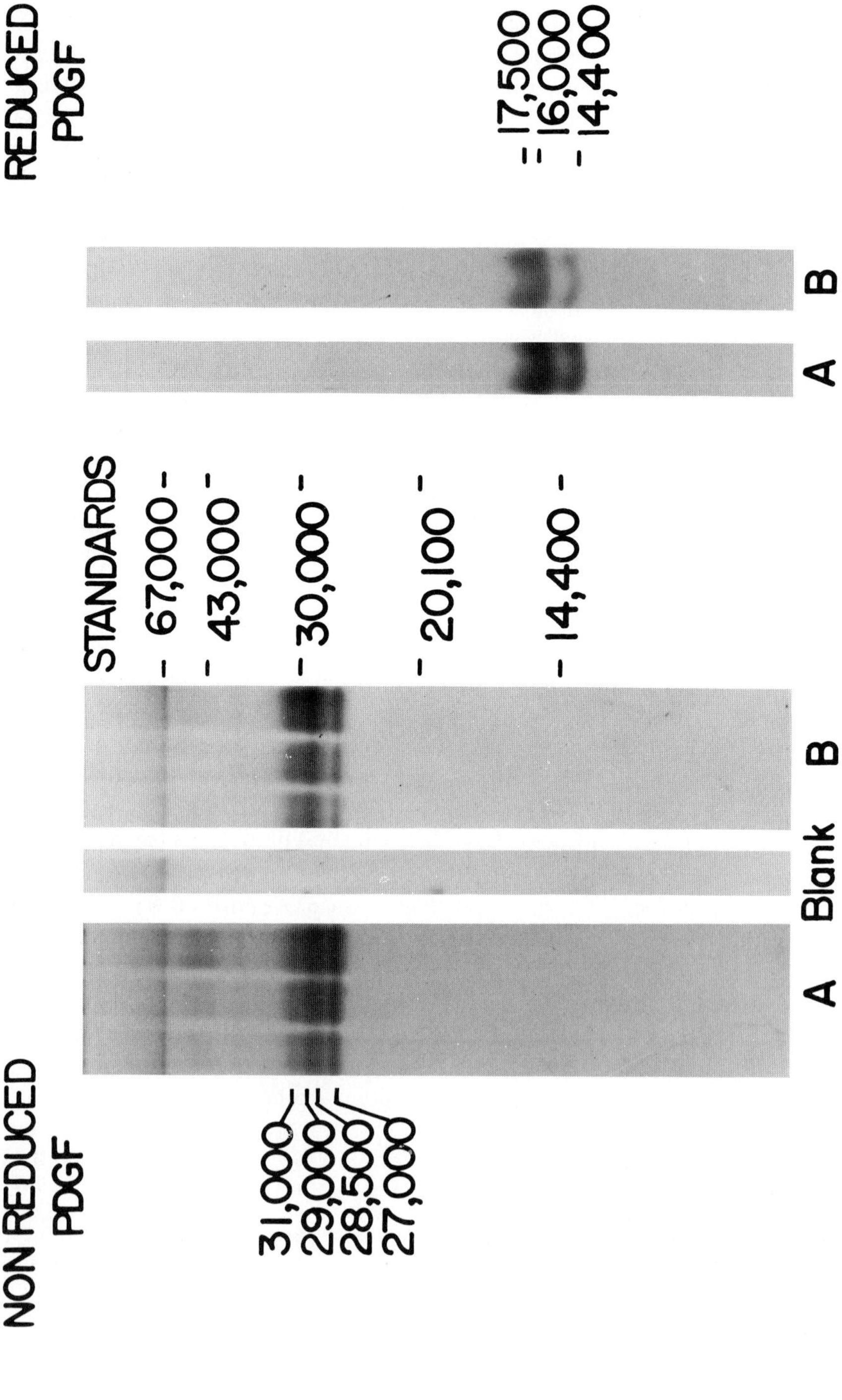

NON REDUCED PDGF
31,000
29,000
28,500
27,000
A
Blank
B
STANDARDS
67,000
43,000
30,000
20,100
14,400
A
B
REDUCED PDGF
17,500
16,000
14,400

area detected by gel scans of silver-stained gels was compared to Lowry[47] protein values for six different purified preparations. Two-dimensional electrophoresis (Fig. 7) provides additional resolution in attempting to evaluate contaminating proteins, but also requires more protein (5 μg used for gel shown in Fig. 7). It is particularly of value as applied in Fig. 7, where the gel is run under reducing conditions. Although PDGF has multiple forms and is a two-chain molecule (see Properties of Purified PDGF, this chapter) and is therefore not detected as a single spot, chances are extremely low of a contaminant meeting all three criteria: same molecular weights under nonreducing conditions, same p*I* range, and reduction to the same molecular weights.

Although a PDGF preparation may be highly purified as assessed by gel electrophoresis, there remains the possibility that not all the PDGF may be biologically active. To control for this and to establish a standard within our laboratory, two additional analyses were performed. First, a number of different preparations were evaluated by radioreceptor assay and the most potent preparation (using Lowry[47] values for determination of protein) was then adopted as the standard and aliquoted. The protein value for this preparation was then corrected for purity as determined by SDS–PAGE. In addition, preparations of [^{125}I]PDGF were evaluated by binding depletion to determine the percentage of the iodinated preparation which could specifically bind to responsive cells[32] (see Bowen-Pope and Ross, this volume [8]). These data were compared with radioreceptor assay data and silver-stained SDS gels and found to be in agreement with the above corrections. Therefore, the values used in Fig. 1 should represent the protein content of biologically active PDGF. Establishment of such a standard by laboratories studying PDGF will be important for comparison of experiments performed in different laboratories. Already the values reported for human serum levels of PDGF vary from 17.5[24,25] to 50 ng/ml.[41,42]

FIG. 6. Two preparations of PDGF varying in their proportion of four molecular weight components characterized by silver-stained SDS–PAGE. Fifteen percent SDS gels were run as described in the text and stained using the sensitive silver stain procedure of Switzer *et al.*[52] A and B represent two different preparations of phenyl-Sepharose-purified PDGF that varied in their proportion of four molecular weight components. The three lanes of the nonreduced PDGF for each preparation represent increasing amounts loaded on the gel (left to right): 220, 330, and 440 ng. The reduced gel lanes were run with a protein load of 630 ng of each preparation. The molecular weights indicated in the margins for the nonreduced and reduced PDGF were determined from a linear log plot of molecular weights of the standards indicated in the middle of the figure and their determined R_f values. The blank gel is shown to illustrate the nonspecific staining seen around M_r = 67,000 when no protein is loaded in the well. Adapted from Raines and Ross.[7]

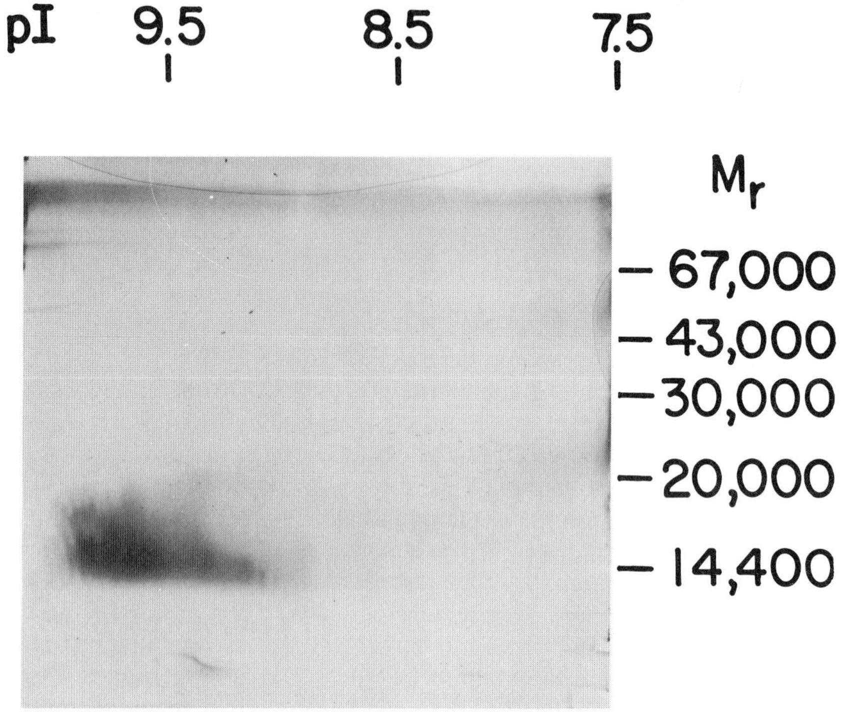

FIG. 7. Two-dimensional gel electrophoresis of highly purified PDGF using reducing conditions for the second dimension. Phenyl-Sepharose-purified PDGF was separated by isoelectric focusing (Pharmalytes, pH 8–10.5, diluted 1 to 15) in tube gels (5% acrylamide, 0.15% bisacrylamide, 8 *M* urea) using 0.01 *M* HEPES for the anode solution and 0.01 *M* ethylenediamine for the cathode. The tube gel was then frozen overnight at −70° in equilibration buffer [10% glycerol, 0.0625 *M* Tris–HCl, pH 6.8, 2.3% SDS (w/v), and 5% 2-mercaptoethanol (v/v)]. The pH gradient was determined from 0.5-cm gel slices from two identical tube gels in which cytochrome *c* was run as a reference standard. SDS–PAGE using the Laemmli discontinuous buffer system,[54] and a 15% separating gel and 4.5% stacking gel of 1.5 mm thickness was used for the second dimension. After electrophoresis, the gel was stained using the silver staining procedure of Wray *et al.*[53]

Characteristics of Purified PDGF

Multiple Forms and Subunit Structure

Highly purified PDGF prepared as described here as well as by other procedures[4,5,7] is isolated as two to four molecular weight entities ranging from 27,000 to 31,000 daltons. Figure 6 shows the silver stained 15% SDS gel of two purified preparations of PDGF under nonreducing conditions. The molecular weights of the four molecular weight entities in Fig. 6 are

31,000, 29,000, 28,500, and 27,000 as determined by the linear log plot of molecular weights of standards and their determined R_f values. The proportion and number of the different molecular weight entities varies with different preparations. By a number of different criteria, all of the multiple forms appear to be related and biologically active: (1) all forms separated by SDS gel electrophoresis or gel filtration are biologically active and the mitogenic activity closely follows the protein content or staining profile[4,5,7]; (2) the elution of cell-bound [^{125}I]PDGF and analysis by SDS–PAGE demonstrated that all forms were bound[7]; (3) peptide maps of the different species showed they were closely related[7]; and (4) amino acid analysis of two forms separated by gel filtration have similar amino acid compositions but differ in their carbohydrate content.[5] The multiple species may represent multiple gene products, varying degrees of glycosylation, or the product of proteolytic cleavage of the 31,000 molecular weight species or of a still higher molecular weight precursor.

Separation of these multiple forms appears possible only by size (gel filtration[5] or elution from SDS gels[4,7]). No separation of the multiple forms was achieved using isoelectric focusing[5] (also see Fig. 7) or gradient elution from phenyl-Sepharose[28] or elution from lectin-agaroses.[28]

Reduction of PDGF and analysis by SDS–PAGE indicates (as shown in Fig. 6) the presence of at least three chains with molecular weights of 14,400, 16,000, and 17,500.[4,6,7] An 11,000 molecular weight band[28,55] and a 20,000 molecular weight band[28] are sometimes detected on reduced SDS–PAGE. The relative amounts of the different reduced chains vary between preparations. Analysis of the reduced chains by peptide mapping[7] and separation of two different forms of nonreduced PDGF followed by reduction and SDS–PAGE[4] suggest a molecular model for native PDGF in which the reduced 14,400 chain is combined with either the 16,000 or 17,500 dalton chain. Based on separation of the reduced chains by HPLC followed by SDS–PAGE of the separated peaks, Johnsson *et al.*[55] have proposed another molecular model in which the 16,000-dalton chain is combined with a chain of molecular weight of 18,000, 15,000, 14,000, or 11,000. Partial amino acid sequence data have provided additional information about the subunit structure of PDGF.[21–23,56] Amino acid sequence analysis of PDGF and nucleotide sequence analysis of the predicted sequence of p^{28}-sis, the putative transforming gene product of simian sarcoma virus (SSV), have revealed a striking homology.[21–23] One chain of PDGF, the B chain (apparent molecular weight from SDS gel electrophoresis of 14,500 to 16,000), appears to be identical to the predicted se-

[55] A. Johnsson, C.-H. Heldin, B. Westermark, and Å. Wasteson, *Biochem. Biophys. Res. Commun.* **104,** 66 (1982).

quence of c-sis[22] (the human protooncogene) over the 109 amino acids determined. Partial sequence (75 amino acids) of the other chain, the A chain (apparent molecular weight by SDS gel electrophoresis of 18,000), demonstrates 60% homology to the B chain.[22] However, from the known sequence of c-sis, PDGF A cannot be encoded by c-sis and may be encoded by a separate locus or a locus located 5′ to the B chain, the sequence of which has not yet been determined. Analysis of sequences derived from two of the multiple forms of PDGF demonstrates that they contain approximately equal proportions of sequences from the PDGF A and B chains.[22,56] It is not known, however, whether PDGF is assembled from a single-chain PDGF precursor or whether the A and B chains are translated from separate transcripts and then assembled as homodimers or heterodimers. It is also unclear whether a homodimer of the PDGF B chains (homologous to c-sis) has PDGF-like biological activity.

Chemical and Physical Properties

As discussed above (and shown in Fig. 6), highly purified PDGF has multiple molecular weight species around 30,000 daltons as determined by SDS gel analysis. These estimates are in reasonably good agreement with sedimentation equilibrium analysis of PDGF in 1 *M* acetic acid, 1 *M* NaCl which indicated a mean molecular weight of 32,700 (determined at three concentrations).[6] Estimates of molecular weight by gel filtration have been much more variable and caused considerable confusion relative to other estimates. This is probably due to varying degrees of interaction with the gel matrices (accentuated by the highly cationic nature of PDGF, especially at acid pH), alteration in pore size of gel filtration matrices at acid pH,[28] and differing molecular shape of PDGF relative to the three-dimensional structure of the molecular weight standards used. The lowest molecular weight estimates are based on amino acid sequence data (23,000 to 24,000).[22] If confirmed, this difference from other determinations may be due to glycosylation.

As shown previously[5,26,27,48] and demonstrated in Fig. 7, PDGF is highly cationic with a peak p*I* of approximately 9.8. This property greatly affects the stability of the purified protein (see below).

The amino acid composition of PDGF is listed in Table II. As would be expected from the isoelectric point, there is a high proportion of basic amino acids and presumably many of the Glx and Asx detected are present in the molecule as glutamine and asparagine. A striking characteristic of the composition is the high proportion of half-cystine residues. For

[56] H. N. Antoniades and M. W. Hunkapiller, *Science* **220,** 963 (1983).

TABLE II
AMINO ACID COMPOSITION OF PDGF[a]

Amino acid	Residues/molecule[b]	Residues/100 residues
Lys	18.1	6.5
His	5.1	1.8
Arg	21.3	7.7
Asx	19.1	6.9
Thr[c]	19.6	7.1
Ser[c]	16.8	6.1
Glx	36.5	13.3
Pro	18.5	6.7
Gly	11.8	4.2
Ala	19.1	6.9
Half-Cys[d]	16.5	5.9
Val	29.7	10.7
Met[e]	2.6	0.9
Ile[f]	10.3	3.7
Leu[f]	17.1	6.1
Tyr	6.2	2.2
Phe	5.6	2.0
Tryp[g]	4.1	1.5

[a] Except where noted, all figures are average values of two 24-hr and one 96-hr hydrolyses in the presence of phenol.
[b] Calculated on the assumption that PDGF has a molecular weight of 31,000.
[c] Values were obtained by correction to 24-hr values, assuming 90% recovery of serine and 95% recovery of threonine.
[d] Determined as cysteic acid after performic acid oxidation and corrected for recovery by simultaneous determination of lysozyme cysteic acid to alanine ratio.
[e] Determined from 24-hr hydrolysis only.
[f] Determined from 96-hr hydrolysis.
[g] Determined after alkaline hydrolysis.

the average protein, the percent half-cystine composition is 3.4% as determined from the data of 108 separate protein families.[57] The percentage of half-cystine in PDGF is almost twice that value at 5.9%. Preliminary evidence[5] suggests that all half-cystine residues are in the disulfide form in the native PDGF. These disulfide linkages are most likely important in the extreme stability of PDGF (see below).

PDGF has also been shown to be a glycoprotein by incorporation of ^3H

[57] M. O. Dayhoff and L. T. Hunt, "Atlas of Protein Sequence and Structure," Vol. 5. Natl. Biomed. Res. Found., Washington, D.C., 1972.

after reduction with ^{3}H-labeled sodium borohydride,[58] by direct carbohydrate analysis of hydrolyzed samples using gas–liquid chromatography,[5] and by derivitization of the carbohydrates and analysis by gas–liquid chromatography and mass spectrometry.[58] Four to seven percent of PDGF was estimated to be neutral or amino sugars in the two molecular weight species of PDGF examined.[5] Sugar residues detected include mannose, galactose, glucosamine, galactosamine, fucose, and possibly sialic acid.[5,58]

Stability

PDGF is a very stable protein. Its biological activity is retained after incubation in 6 *M* guanidine–HCl, 8 *M* urea, or 2% sodium dodecyl sulfate, and over the pH range of 2–12[27] and also survives boiling for 10 min.[5,48] The extreme stability of the PDGF molecule is probably due to the maintenance of its three-dimensional structure by the large number of disulfide bonds. Chemical cleavage of the disulfide bonds by three different methods: reductive cleavage by S-sulfonation, reductive cleavage with dithiothreitol, and performic acid oxidation, in all cases reduced biological activity by 80–100%.[7] Attempts to reassociate the reduced chains did not restore biological activity.[7] In practice, complete destruction of biological activity by reduction is difficult to achieve and dependent upon complete denaturation of the PDGF molecule. 40% of biological activity remains after reduction without guanidine–HCl,[27] but even with the addition of 6 *M* guanidine–HCl, 1.4 to 10% of the mitogenic activity is retained.[7,27] Determination of the extent of reduction is therefore extremely important if this is to be used as a procedure to destroy biologically active PDGF. The only other treatment known to destroy the mitogenic activity of PDGF is incubation with trypsin.[3,27]

The strongly cationic and hydrophobic character of PDGF can result in losses in handling highly purified PDGF. A significant amount of PDGF can be lost due to adsorption after dialysis, chromatography on a gel matrix such as Sephacryl S-200 which contains some unreacted carboxyl groups, or incubation with a bare tissue culture dish.[7,32,49] These losses can be minimized by addition of exogenous protein such as bovine serum albumin or gelatin (0.25% solutions are routinely used). In cases, such as Sephacryl S-200, where ionic interaction is presumed to be the major source of loss, working near the isoelectric point can improve recovery dramatically.[28] For any manipulations of purified PDGF requiring a number of different steps without the addition of exogenous protein, trace levels of [^{125}I]PDGF are added to monitor losses.

[58] E. Bremer and E. W. Raines, unpublished observations (1982).

As isolated from phenyl-Sepharose in 50% ethylene glycol, purified PDGF appears stable for two years at 4°. Recovery of active PDGF after lyophilization of purified PDGF is extremely variable unless exogenous protein (such as BSA) is added. If exogenous protein needs to be avoided, buffer exchange or concentration is best achieved by gel filtration at a neutral pH or higher, or ultrafiltration (Centricon microconcentrators, Amicon Corp.).

Acknowledgments

We wish to acknowledge the excellent technical assistance of Billie Fortune, Janet Hansom, and Karen Polovitch; Arnie Hestness for drafting the figures; and Mary Hillman for her assistance in preparing the manuscript. The amino acid analyses were kindly provided by Dr. Ken Walsh, Roger Wade, and Lowell Ericsson. The research was supported by NIH Grant HL 18645 and by a grant from R. J. Reynolds, Inc.

[59] Isolation of Rat Somatomedin

By WILLIAM H. DAUGHADAY and IDA K. MARIZ

Starting Material

Although many liters of human serum or kilograms of human serum of Cohn Fraction IV have been the starting material for isolation of human insulin-like growth factors, it is possible to obtain sufficient rIGF I from 1 to 2 liters of rat serum for iodination and standards for RIA and for characterization with highly sensitive methods.[1] As measured in the human hIGF I/Sm C RIA, normal rat serum is from 2 to 4 times as potent as human serum. Moreover, rIGF I is only about one-third as potent as hIGF in this assay. This suggests that the true level of rIGF I in rat serum is about 600 to 900 ng/ml.

As originally shown by Peake *et al.*[2] the somatomedin bioactivity can be increased as much as 6-fold by injecting the GH secreting rat pituitary

[1] J. S. Rubin, I. Mariz, J. W. Jacobs, W. H. Daughaday, and R. A. Bradshaw, *Endocrinology* **110,** 734 (1982).

[2] G. T. Peake, I. K. Mariz, and W. H. Daughaday, *Endocrinoloy* **83,** 714 (1968).

ISBN 0-12-182009-2

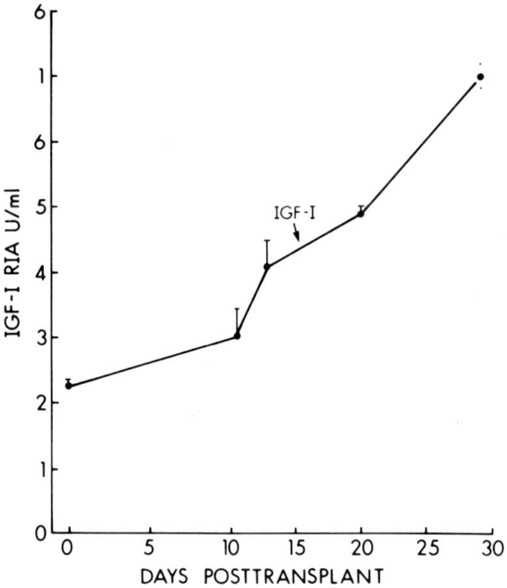

FIG. 1. Serum IGF I as determined by RIA of female Wistar Furth rats injected with MStT/W15 pituitary tumor on day 0. IGF I concentrations are compared to normal human serum which has an assigned potency of 1 unit/ml.

tumor MStT/W15 into female Wistar Furth rats. As shown in Fig. 1, there is a progressive rise in rIGF I by RIA over the first 4 weeks after transplantation but there is no increase in rIGF II by RRA. Despite the serum enrichment in rIGF I which is possible by tumor transplantation, we are not utilizing this method routinely because of the expense of the special strain of rats and their maintenance. Also, we have noted that our strain of the MStT/W15 tumor has become very aggressive. This has produced toxicity and anorexia resulting in a fall of IGF I levels to below normal rat serum.

Other workers have isolated rIGF II (MSA) from conditioned medium of Buffalo rat hepatocytes. A method which led to the isolation and characterization of a peptide very similar in amino acid sequence to hIGF II has been described by Marquardt *et al.*[3] The heterogeneity of IGF II related peptides in the conditioned medium is a problem. In addition to the rat IGF II isolated by Marquardt, there are at least six additional

[3] H. Marquardt, G. J. Todaro, L. E. Henderson, and S. Oroszland, *J. Biol. Chem.* **256,** 6859 (1981).

peptides[4] which probably represent partially processed precursor molecules.

Assay Methods

A simple and sensitive method for monitoring the purification steps is essential. In general, bioassays are too slow and time consuming and expend excessive amounts of the active components. We have relied primarily on the radioimmunoassay for hIGF I/Sm C utilizing the antibody raised against Sm C in rabbits by Furlanetto *et al.*[5] [Somatomedin C antiserum for assay of nonhuman serum may be obtained from the National Hormone and Pituitary Program, Suite 201-9, 210 W. Fayette St., Baltimore, MD 21201. Human somatomedin C for ^{125}I-iodination is obtainable from Amgen Biological, 1900 Oak Terrace Lane, Thousand Oaks, CA 91320.] When applied to native serum or the early steps in fractionation, we recommend preliminary extraction of serum with acid ethanol to reduce the binding protein content sufficiently to avoid assay artifacts.[6] In this extraction, 0.2 ml of serum or serum fraction is mixed with 0.8 ml 87.5% ethanol and 12.5% 2 *N* HCl. After centrifugation, 0.5 ml of the supernatant extract is transferred to a fresh tube and neutralized by the addition of 0.2 ml 0.855 *M* Tris base (the required molarity should be confirmed by titration when dealing with samples other than serum). The neutralized acid ethanol extract can be introduced directly into radioimmuno-, radioreceptor, and protein binding assays if all assay tubes receive the same amount of neutralized acid ethanol. A further dilution of the acid ethanol extract in assay buffer is usually required for RIAs done on rat serum.

We have used an alternative method for eliminating binding protein from rat serum prior to IGF I RIA. Serum, 0.2 ml, is gel filtered through a Sephadex G-75 40-120 mesh beads, 0.9 × 28 cm column, in 1 *M* acetic acid, 0.1 *M* NaCl. Fractions, Kd 0.45 to 0.94, are pooled and the volume recorded. Aliquots for assay are dried in a Speed Vac (Savant Instruments, Inc., Hicksville, NY).

For convenience and rapidity we have utilized a modified RIA for IGF I and a radioreceptor assay for IGF II which uses membranes from rat

[4] A. C. Moses, S. P. Nissley, P. A. Short, M. M. Rechler, and M. Podskalny, *Eur. J. Biochem.* **103,** 387 (1980).

[5] R. W. Furlanetto, L. Underwood, J. J. Van Wyk, and A. J. D'Ercole, *J. Clin. Invest.* **60,** 648 (1977).

[6] W. H. Daughaday, I. K. Mariz, and S. L. Blethen, *J. Clin. Endocrinol. Metab.* **51,** 781 (1980).

placentas.[7] These two assays are highly specific for the respective peptides. A simple radioligand assay employing [^{125}I]hIGF II (or [^{125}I]rIGF II MSA) and rat IGF binding protein will detect both peptides.

Separation of IGFs from Binding Protein

A number of methods have been utilized to separate IGF-I from its binding protein. Liberti and Miller[8] originally suggested separation of IGFs by dialysis. It is possible to dialyze rat serum in Spectra/Por No. 6 wet tubing, molecular weight cutoff 15,000 (Spectrum Medical Industries, Los Angeles, CA) against 30 volumes of 0.2 *M* acetic acid at 4°. After 48 to 72 hr nearly all the rIGF has left the dialysis bags and entered the dialyzate and the recovery of IGF I in the dialyzate is good. This method has the disadvantage of dealing with the large volumes of dialyzate.

Another method we have used is ultrafiltration with the DC Amicon hollow fiber apparatus through HIX 50 hollow fibers cartridge having a 50,000 dalton cutoff. This procedure was originally proposed by Ginsberg *et al.*[9] Initial ultrafiltration is carried out with 500 ml rat serum with 3 liters of 0.03 *M* sodium phosphate pH 7.6 at room temperature. This allows ultrafiltration of most plasma peptides and small molecular weight compounds. The final volume is concentrated to the 500 ml. The retentate is then acidified to 1 *M* acetic acid with glacial acetic acid and 3000 ml of 1 *M* acetic acid added to the reservoir. Ultrafiltration is continued until the final volume is less than 300 ml. We have encountered difficulty with stability of the Amicon Hollow fiber HIX 50 cartridge during ultrafiltration of rat serum in 1 *M* acetic acid. The serum becomes slightly viscous so that higher pressures are required to achieve an acceptable rate of ultrafiltration. At these pressures leaking of higher molecular weight proteins has resulted. It is possible that decreasing the acetic acid concentration from 1.0 to 0.2 *M* would decrease the pressure required and be easier on the 50 K cartridges.

We now extract IGF I from rat serum with SP-Sephadex C-25 (Pharmacia) as described by Svoboda *et al.*[10] Between 1 and 3 liters of rat serum are filtered through glass wool and mixed with an equal volume of 0.4 *M* acetic acid. The acidified serum is directly passed through a 35 × 5 cm column of SP-Sephadex C-25 which had been preequilibrated with 0.2

[7] W. H. Daughaday, K. A. Parker, S. Borowsky, B. Trivedi, and M. Kapadia, *Endocrinology* **110,** 575 (1982).

[8] J. P. Liberti and M. S. Miller, *J. Biol. Chem.* **255,** 1023 (1980).

[9] B. A. Ginsberg, C. R. Kahn, J. Roth, K. Megyesi, and G. Baumann, *J. Clin. Endocrinol. Metab.* **48,** 43 (1979).

[10] M. E. Svoboda, J. J. Van Wyk, D. G. Klapper, R. E. Fellows, F. E. Grissom, and R. J. Schlueter, *Biochemistry* **19,** 790 (1980).

M acetic acid + 0.075 *M* NaCl (at 300 ml/hr). After the diluted serum has passed through the column, 600 ml of 0.2 *M* acetic acid, 0.075 *M* NaCl is passed through the column followed by 600 ml of 0.2 *M* acetic acid, 0.2 *M* NaCl. The IGF I is eluted with 2.5 liter 1 *M* ammonium acetate + 1.5 *M* NaCl (pH 9.6).

The IGF I containing extracts from these three different methods are concentrated by ultrafiltration with the Amicon HIP 5 cartridge with a 5000 dalton cutoff and the addition of distilled water from the reservoir to reduce the conductivity of the ultrafiltrate to below 1 mmho. Recovery can be increased by rerunning the ultrafiltrate through the Amicon HIP 5 cartridge. This appears to be particularly important with the high salt present in the SP Sephadex method. The concentrated extract is then lyophilized. The residue is then redissolved in 15–20 ml of 0.2 *M* acetic acid.

Sephadex G-75-40 Gel Filtration

The extract in 0.2 *M* acetic acid is then added to an ascending column (2.5 × 95 cm) of Sephadex G75-40 equilibrated in 0.15 *M* acetic acid, 0.15 *M* NaCl. The same buffer is pumped through the column at 50 ml/hr. Eight milliliter fractions are collected and analyzed for IGF-I. In most runs tubes with K_d 0.66 to 0.88 contain the IGF-I and are lyophilized and dialyzed versus H_2O to reduce the salt concentration.

Narrow Range Isoelectric Focusing

The residue is dissolved in 10 ml H_2O and clarified by centrifugation before application to a 440 ml LKB isoelectric focusing column containing 5 to 50% w/v sucrose gradient and 2% ampholine pH 7–11. After the current reaches a stable minimal level for 4 to 5 hr, the column is emptied in 7 ml fractions. The fractions are dialyzed against H_2O in Spectra/Por No. 3 tubing, 3500 dalton cut-off (Spectrum Medical Industries, Inc., Los Angeles, CA). The tubes containing IGF I, usually pH 8.5 to 9.5 (Fig. 2) are pooled and lyophilized.

CM Cellulose Chromatography

The residue is taken up and dialyzed against 0.05 *M* sodium acetate pH 4.8 and then applied to a column (0.9 × 15 cm) of CM 52 (Whatman) equilibrated with the same buffer. The column is washed with 0.05 *M* sodium acetate + 0.05 *M* sodium chloride at 6 ml/hr overnight or until the OD at 230 of effluent is less than 0.02. Elution of the IGF I is accomplished with a sodium chloride gradient of 0.05 to 0.125 *M* in 0.05 *M*

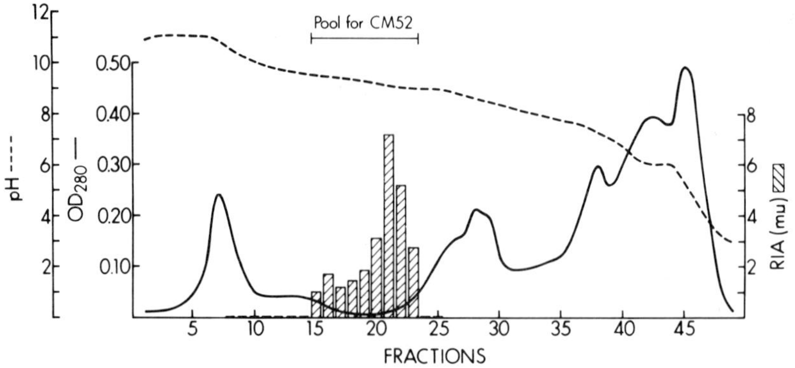

FIG. 2. Narrow pH range isoelectric focusing of partially purified rat Sm. The active material from the Sephadex G-75 column was loaded onto a 440 ml LKB isoelectric focusing column containing a 5–50% sucrose gradient and 2% Ampholine, pH 7–11. The fraction size was 7.0 ml. The Sm activity appeared at pH 9.0–9.7. Reprinted, with permission, from Rubin *et al.*[1]

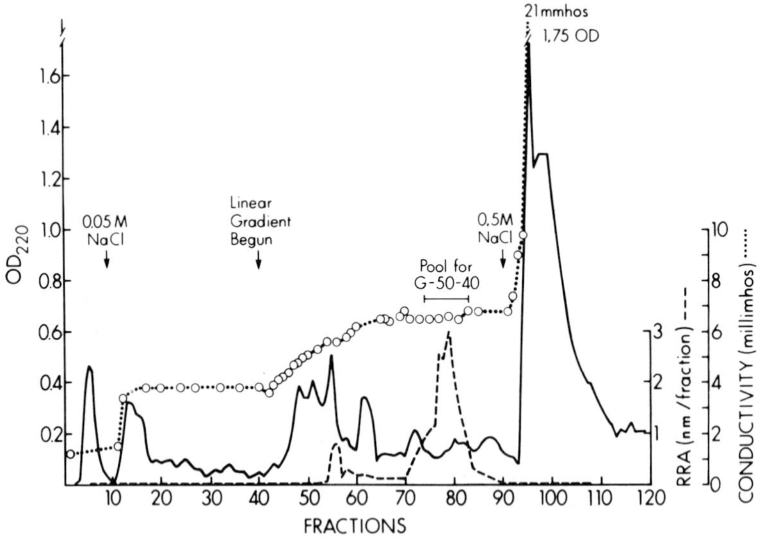

FIG. 3. CM-Cellulose chromatography of the rat Sm-containing IEF pool. After the isoelectrically focused pool was applied to the CM-52 column (0.9 × 15 cm), the column was washed extensively with 0.05 *M* sodium acetate–0.05 *M* sodium chloride, pH 4.8, before starting a shallow sodium chloride gradient (0.05–0.125 *M*). The fraction size was 3.0 ml. Reprinted, with permission, from Rubin *et al.*[1]

sodium acetate pH 4.8 (Fig. 3). Three milliliter fractions are collected and assayed for IGF I by RIA. The active fractions are pooled and lyophilized. This step achieves a 50- to 100-fold purification.

Sephadex G-50-40

The lyophilized residue is dissolved in 1 ml of 0.1 *M* acetic acid and applied to a Sephadex G-50-40 (0.9 × 160 cm) column and eluted with 0.1 *M* acetic acid with a flow rate of 6 ml/hr. The eluted fractions were assayed for IGF I by RIA.

TABLE I
AMINO ACID COMPOSITION OF RAT Sm AND RELATED MOLECULES[a]

	Rat Sm		Human IGF-s		Rat MSA[b]	
Amino acid	Obsd[c]	Integer	I[d]	II[e]	Obsd	Integer
Aspartic acid	5.7	6	5	3	4.2	4
Threonine	3.9	4	3	4	4.2	4
Serine	4.4	5	5	7	6.6	7
Glutamic acid	7.7	8	6	7	7.0	7
Proline	3.9	4	5	3	3.2	3
Glycine	5.6	6	7	5	4.7	5
Alanine	6.2	6	6	5	4.9	4
Half-cystine	ND[f]		6	6	5.8	6
Valine	3.0	3	3	4	3.1	3
Methionine	0.4	1	1	0	0.0	0
Isoleucine	1.9	2	1	1	1.1	1
Leucine	5.7	6	6	6	6.0	6
Tyrosine	1.8	2	3	3	2.8	3
Phenylalanine	3.3	3	4	4	3.7	4
Histidine	1.6	2	0	0	0.0	0
Lysine	5.9	6	3	1	1.2	1
Arginine	3.2	3	6	8	7.9	8
Tryptophan	ND		0	0	0.6	1
Total		73[g]	70	67		67
Calculated MW		8050	7649	7471		7484

[a] Reprinted, with permission, from Rubin *et al.*[1]
[b] From the data of Marquardt *et al.*[3]
[c] Mean of three 24-hr hydrolysates.
[d] From primary sequence of Rinderknecht and Humbel.
[e] From primary sequence of Rinderknecht and Humbel.
[f] ND, Not determined.
[g] Assuming the half-cystine residues are conserved in rat Sm.

TABLE II
PURIFICATION OF RAT Sm[a,b]

	Protein (mg)	Total units (mU in RRA[c])	Recovery (%)	SA (mU/mg)	Purification
Starting material: 460 ml serum	50,600[d]	81.0		0.0016	1.0
Amicon hollow fiber diafiltration					
50K neutral retentate	44,200[d]	81.1		0.0018	1.12
50K acid filtrate retentate	800[d]	86.0		0.098	61.2
Sephadex G-75-40	418[d]	96.0	100.0	0.23	144
IEF, pH 9.00–9.56	23.8[d]	18.9	19.7	0.794	496
CM-52	0.326[e]	16.3	17.0	50.0	31,200
Sephadex G-50-40					
Peak fractions	0.017[f]	8.3	8.6	500.0	312,000
Side fractions	0.026[f]	7.0	7.3	270.0	

[a] Reprinted, with permission, from Rubin *et al.*[1]
[b] To minimize losses, batches of material from different preparations were frequently combined in the latter phases of the purification scheme. Therefore, the recovery of Sm from a particular lot of starting material at a given step is not always known. Where this situation applies, the values reported in the table are estimates based on the proportion of protein and activity in a sample coming from the particular preparation being followed. These results are representative of several efforts.
[c] Activity in human placental membrane radioreceptor assay; pure human IGF-I has a specific activity of 350 mU/mg.
[d] Protein content is based on OD_{280}, using BSA as a standard.
[e] Protein content is based on OD_{220}, using BSA as a standard.
[f] Protein content is based on amino acid analysis, as described in the text.

HPLC

While we have been able to obtain homogeneous preparations of rat IGF I at the Sephadex G-50-40 stage with some preparations, purification by HPLC may be required for others. This has been successfully utilized by Svoboda *et al.*[10] for human IGF I (Sm C) and by Marquardt *et al.*[3] for rIGF II with reverse phase Bondapak C_{18} columns. In our laboratory we have successfully used a SynChropak RP-P (Syn Chrom Inc, Linden, IN) column which is a C_{18} column with a 300 Å pore size. We have employed a gradient of acetonitrile in 0.01 *M* TFA pH 2.8 for elution.

Characterization of Rat IGF I

Because of the small amounts of purified protein remaining in our final preparations, it is unreliable to depend on optical density or chemical protein determinations. We perform amino acid analysis with a Durrum D 500 Analyzer (Durrun, Sunnyvale, CA). The results of such an analysis are given in Table I.

SDS–polyacrylamide gel electrophoresis has indicated only a single peptide band when examined by a sensitive silver staining method. N terminal amino acid analysis should reveal only glycine.

Yield

The degree of purification and yield of each step in the preparation is shown in Table II. The Amicon hollow fiber initial step is shown. The overall purification is in excess of 300,000. Significant losses occurred only at the isoelectric focusing step but this step is important in separation of IGF I from IGF II. The greatest purification occurred with CM-52 chromatography step which achieved more than a 80-fold purification.

[60] Purification of Human Insulin-Like Growth Factors I and II

By PETER P. ZUMSTEIN and RENÉ E. HUMBEL

Insulin-like growth factors I and II (IGFs) are the designations for two polypeptides of approximately 7.5 kDa isolated from human serum[1] whose amino acid sequences are homologous to proinsulin.[2] *In vitro,* both factors show insulin-like metabolic effects on adipose and muscle cells and mitogenic activity on various cells, especially chondroblasts and fibroblasts. *In vivo,* IGF I promotes growth of hypophysectomized animals,[3] and its serum concentration is regulated by growth hormone. IGF I is identical to somatomedin C[4] which is thought to mediate the growth effects of growth hormone. Multiplication stimulating activity (MSA) is the rat homolog to human IGF II.[5] For a review and further references, the reader is referred to references 6 and 7.

Originally, IGF I and II were isolated from an acid ethanol extract of a plasma fraction (precipitate B).[1] Although most of the presently available IGF has been prepared by this method, the yield is unsatisfactory and the protocol very time consuming. We have therefore devised two improved methods by applying modern peptide separation techniques. The methods described here yield IGF of high purity and with high recovery within approximately 25 workdays.

Methods of Analysis

Originally, Rinderknecht and Humbel used a bioassay to monitor for IGF at various stages of purification. This bioassay measures the *n*on*s*uppressible (by anti-insulin antibodies) *i*nsulin-*l*ike *a*ctivity (NSILA) of IGF.[8] A recent modification[9] allows the handling of a larger number of

[1] E. Rinderknecht and R. E. Humbel, *Proc. Natl. Acad. Sci. U.S.A.* **73,** 2365 (1976).
[2] E. Rinderknecht and R. E. Humbel, *FEBS Lett.* **89,** 283 (1978).
[3] E. Schoenle, J. Zapf, R. E. Humbel, and R. E. Froesch, *Nature (London)* **296,** 252 (1982).
[4] D. E. Klapper, M. E. Svoboda, and J. J. Van Wyk, *Endocrinology* **112,** 2215 (1983).
[5] H. Marquardt and G. J. Todaro, *J. Biol. Chem.* **256,** 6859 (1981).
[6] J. Zapf, E. R. Froesch, and R. E. Humbel, *Curr. Top. Cell. Regul.* **19,** 257 (1981).
[7] R. E. Humbel, P. Nissley, and J. J. Van Wyk, *in* "Hormonal Peptides and Proteins" (C. H. Li, ed.), Vol. 12. Academic Press, New York, 1984 (in press).
[8] E. R. Froesch, H. Bürgi, E. B. Ramseier, P. Bally, and A. Labhart, *J. Clin. Invest.* **42,** 1816 (1963).
[9] J. Zapf, E. Schoenle, G. Jagars, I. Sand, J. Grunwald, and E. R. Froesch, *J. Clin. Invest.* **63,** 1077 (1979).

METHODS IN ENZYMOLOGY, VOL. 109

ISBN 0-12-182009-2

samples. Since insulin is used as standard, these bioassays can be performed even if no purified IGF is available. They are still used occasionally to complement radioimmunoassays.

Specific radioimmunoassays for IGF I and IGF II are the preferred routine method of analysis.[10] Briefly, IGF is iodinated by the chloramine-T method to a specific activity of 50–80 μCi/μg. Approximately 30,000 cpm in 0.1 ml is mixed with 0.1 ml of appropriately diluted sample of purified IGF as standards (0.2–30 ng), followed by 0.2 ml of polyclonal anti-IGF antiserum. All solutions are made in 0.1 *M* sodium phosphate pH 7.4 containing 1 mg/ml of bovine serum albumin (RIA buffer). After incubation overnight at 4°, 0.5 ml of 25% polyethylene glycol 6000 (Fluka) together with 0.1 ml of bovine γ-globulin (Bio-Science Products AG) are added (5 mg/ml), the mixture centrifuged immediately at 1600 *g* for 30 min, and the radioactivity of the pellet counted in a gamma counter.

Proteins are determined by the Bio-rad protein assay using bovine serum albumin as standard, or by the measurement of absorbance at 220 nm.

Purification of IGF from Human Serum

Serum is prepared from outdated blood or from plasma obtained from a bloodbank. After addition of 25 ml of 0.5 *M* $CaCl_2$ per liter of plasma, it is allowed to clot overnight at room temperature in centrifugation tubes and centrifuged at ~1500 *g*. Batches of 8 to 10 blood bags can be conveniently handled and yield 1200–1700 ml of serum. It is made 0.5 *M* by addition of glacial acetic acid and brought to pH 2.7 by addition of 6 *N* HCl.

Gel Filtration on Sephadex G-75

Acidified serum (1.3–1.5 liters) is applied to a Pharmacia sectional column KS 370 with 4 segments (37 × 15 cm, column volume 64 liters) packed with Sephadex G-75 in 0.5 *M* acetic acid. The ascending flow is at 4.4 liters/hr and the absorbance at 280 nm is continuously monitored. Fractions of 2.2 liters/30 min are collected by using a LKB Minirac 1700 fraction collector equipped with a groundplate to each hole of which tubings are attached leading to bottles. Aliquots of 0.5 ml are lyophilized, reconstituted with 1 ml of 0.1 *M* Na-phosphate pH 7.5 containing 1 mg/ml of bovine serum albumin, and analyzed by radioimmunoassay. Fractions containing the peak activity are pooled separately from the ascending and descending part, concentrated by flash evaporation to 200–300 ml, and

[10] J. Zapf, H. Walter, and E. R. Froesch, *J. Clin. Invest.* **68,** 1321 (1981).

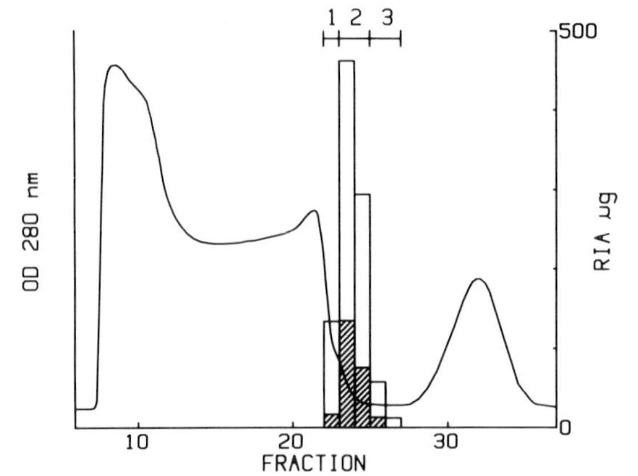

FIG. 1. Gel filtration of 1.5 liter of acidified serum on Sephadex G-75 in 0.5 *M* acetic acid. (——) Absorbance at 280 nm. Hatched bars indicate μg of IGF I per fraction, open bars μg of IGF II per fraction. IGF binding protein elutes in front of the IGF peak (not shown for clarity, as it competes with IGF in the RIA, especially in the RIA II, but yielding nonparallel displacement curves).

lyophilized. The overall yield is 90–170 μg of IGF I and approximately 450–750 μg of IGF II per liter of serum. Not all fractions give upon serial dilutions parallel displacement curves in the RIA. A typical elution curve is shown in Fig. 1. It demonstrates that IGF is separated from the bulk of serum proteins. IGF binding protein elutes as a broad zone in front of the IGF peak. Pools 1 and 2 (Fig. 1) give a lipid containing white powder whereas the descending part yields a brownish thick film in the vacuum bulb. Material from pool 2 is further used. Material from pool 1 from 4 to 6 preparations is extracted with 400 ml of 0.5 *M* acetic acid and cleared through a glass microfiber filter (Whatman, GF/C) and rechromatographed on the same column. Rechromatography of material from the pool 3 is not recommended because of notoriously low recoveries.

Acid Ethanol Extraction

Material from Sephadex G-75 is suspended in 0.01 *N* HCl (1 ml/30 mg protein) and vigorously stirred for 30 min at room temperature. Four volumes of a mixture containing 12.5 vol% HCl 2 *N* and 87.5 vol% ethanol is slowly added, stirred for 15 min at room temperature, and centrifuged for 20 min at 3000 *g*. The clear supernate is decanted. The gelatinous pellet is vigorously resuspended in 0.01 *N* HCl (1 ml/120 mg of initial protein) and extracted once more by 4 vol of the acid ethanol mixture. The final pellet is discarded. The combined supernates are brought to pH

8.3 by addition of 6 and 1 *N* NaOH. At pH 4.3–4.5, a heavy precipitate begins to form (neutral precipitate). After centrifugation, the clear supernate is immersed in an ice bath. Ice-cold acetone is added, v/v, under continuous stirring. The formation of the acetone precipitate is complete after 16 hr. After centrifugation, the precipitate is dried under a nitrogen stream and solubilized in 20 ml of 1 *M* acetic acid. The acetone supernate contains lipids and sometimes traces of IGF, especially of IGF II. The overall yield cannot be determined accurately due to nonparallel displacement curves in the RIA. As judged from the overall yield of the following gelfiltration, the yield is 70–95%. The acetone precipitate contains 9–28% of the initial protein. An acid ethanol extraction as first step followed by Sephadex G-75 is not to be recommended.

Gel Filtration on Sephadex G-50

The acetone precipitate (100 to 300 mg of protein) is dissolved in 20 ml of 1 *M* acetic acid and centrifuged for 30 min at 9000 *g*. The clear supernate is applied on a column (3.6 × 100 cm) packed with Sephadex G-50 in 0.5 *M* acetic acid. Elution is with 0.5 *M* acetic acid at 60 ml/hr. Aliquots (10–100 μl) of the 15 ml fractions are lyophilized, reconstituted with 1 ml of RIA buffer, and analyzed by RIA. The fractions containing essentially all of IGF I are pooled and lyophilized. Figure 2 represents an example. Typically, IGF I elutes as a relatively sharp peak, whereas IGF II covers a broad zone which is resolved sometimes into 2–3 peaks. Iodinated [^{125}I]IGF II added prior to chromatography coelutes with IGF II RIA activity. Insulin-like activity measured in the presence of anti-insulin anti-

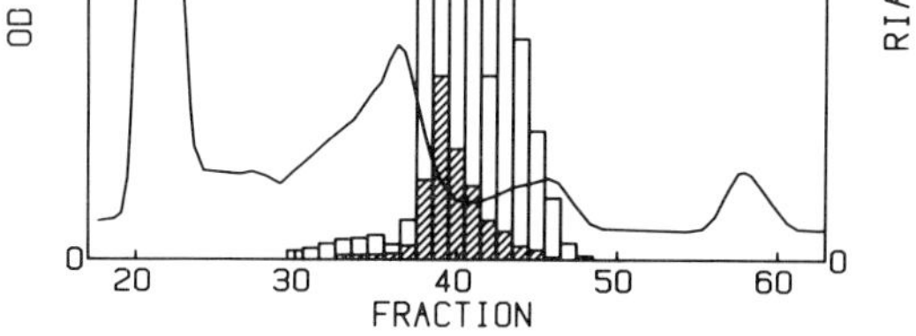

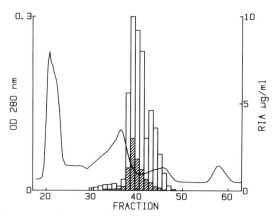

FIG. 2. Gel filtration on Sephadex G-50 of acid ethanol-extracted material 2 from Sephadex G-75 (150 mg). Hatched bars, RIA I; open bars, RIA II; (——) absorbance at 280 nm.

bodies parallels the RIA activity. The reason for this behavior is not clear. IGF is known to behave anomalously on Sephadex or BioGel in acetic acid.[1]

Chromatofocusing: Separation of IGF I and II

On a glass-sinter funnel, polybuffer exchanger (PBE 94, Pharmacia) is washed with distilled water and equilibrated with several portions of starting buffer—25 m*M* ethanolamine acetic acid pH 9.4—until the effluent reaches pH 9.4; 800 ml is usually sufficient for 20 ml of gel. All solutions and the gel are carefully degassed before use. The gel is filled into a 0.9 × 32 cm column. The lyophilized powder (up to 250 mg) from Sephadex G-50 is weighed, dissolved in 25 m*M* ethanolamine acetic acid pH 9.4, 1 ml/10 mg, and cleared by centrifugation. The pellet is extracted once more and the two supernates applied on the column at a flow rate of 13.8 ml/hr. The gradient from pH 9 to 6 is developed with Polybuffer 96 (Pharmacia) diluted 10 times with distilled water containing 0.02% Na-azide, adjusted with acetic acid to pH 5.9. Approximately 200 ml is needed to reach the final pH. Fractions of 4.6 ml are collected at 13.8 ml/hr. Thereafter, the column is rinsed with 0.1 *N* HCl to release residual proteins. Aliquots can be analyzed directly by RIA after appropriate dilution. Polybuffer diluted 4 times or more does not interfere in the RIA. The overall recovery is usually at least 80%.

IGF I predominantly elutes at pH 8.25. Subforms with p*I*s of 8.7–9, 8.5, 7.9, and 7.5 vary somewhat in their proportions in different prepara-

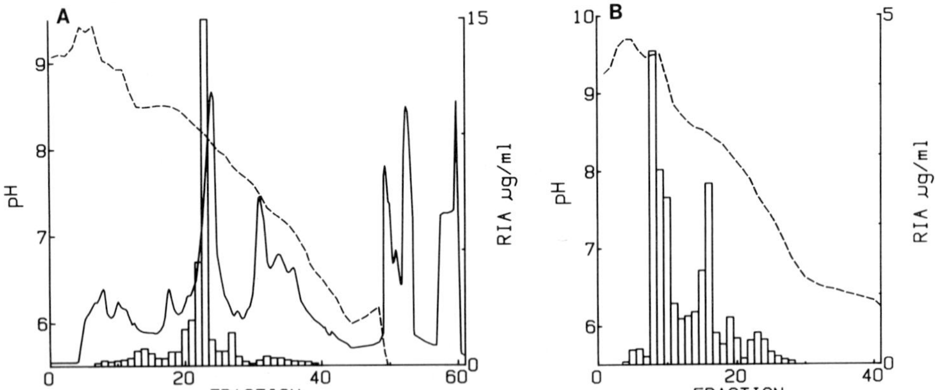

FIG. 3. Chromatofocusing of Sephadex G-50 material with a gradient from pH 9 to 6. (——) Absorbance at 280 nm; (----) pH; bars, RIA I. For clarity, the RIA II results are not shown (see Fig. 4). (A) A typical example with IGF I p*I* 8.25 as predominant form. (B) The p*I* 9 form was the main component in this preparation.

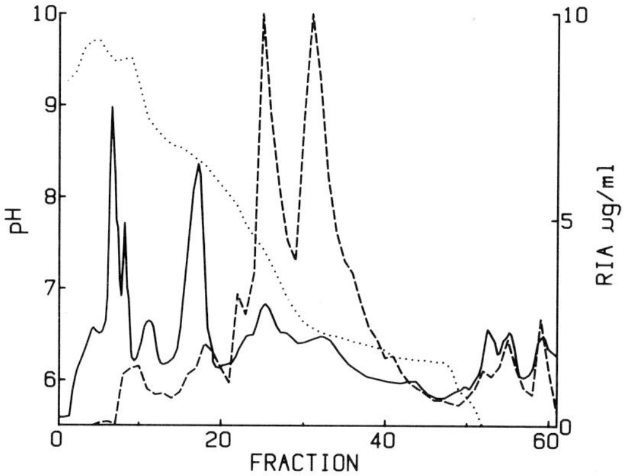

FIG. 4. Chromatofocusing of Sephadex G-50 material (110 mg) with a gradient from pH 9 to 6. This chromatogram demonstrates the main components (p*I* 6.5, p*I* 7.2) of IGF II. (——) Absorbance of 280 nm; (·····) pH; (––––) RIA II. For clarity, the RIA I results in this example (identical with that of Fig. 3B) are not shown.

tions. In some experiments, the p*I* 9 form was the predominant species. IGF I (batch I/3) purified by the original method[1] contains the p*I* 8.25 form as the main component besides p*I* 7.9 and 7.5 subforms. Two examples are shown in Fig. 3.

Heterogeneity on electrofocusing has been noted before[11,12] with main components of p*I* 8–8.5 and p*I* 9. Serial dilutions of IGF I p*I* 8.25 and p*I* 9 give parallel displacement curves in the RIA I.

The bulk of IGF II RIA activity elutes at a pH value lower than 7.8. Two main components with p*I*s of 6.5 and 7–7.5 are usually observed. However, minor forms with p*I*s of 3.7–4, 4.8–5.1, and 8 are present. Consistently, the major components p*I* 6.5 and 7.5 (Fig. 4) and subform p*I* 3.7–4 give parallel displacement curves in the RIA II. Routinely, the two major forms (IGF I, p*I* 8.25; IGF II, p*I* 6.5) are used for the following steps.

Immunoaffinity Column and Desalting

The polybuffer ampholytes are not easily separated from IGF by gel filtration. The use of a column with immobilized anti-IGF antibodies enables the separation from buffer ampholytes and a further substantial purification.

[11] M. E. Svoboda, J. J. Van Wyk, D. G. Klapper, R. E. Fellows, F. E. Grisson, and R. J. Schlueter, *Biochemistry* **19**, 790 (1980).

[12] R. M. Bala and B. Bhaumick, *Can. J. Biochem.* **57**, 1289 (1979).

Monoclonal antibody No. 43 against IGF I has the same apparent affinity of approximately 2.7×10^8 liters/mol to IGF I as well as to IGF II.[13] Antibody from 9 ml of ascites fluid was purified on a Protein-A Sepharose column. Of purified IgG 8 mg was coupled to 3 ml of CNBr-activated Sepharose. Three milliliters of gel with a capacity to bind 350 μg of IGF is equilibrated in 0.1 *M* Na-phosphate pH 7.4, 0.02% Na-azide at room temperature. Fractions from chromatofocusing containing IGF I p*I* 8.25 or IGF II p*I* 6.5 are pooled, titrated to pH 7.4 with 1 *N* NaOH or 1 *N* HCl, and applied to the column at 7.2 ml/hr.

Care is taken to rinse carefully the walls on top of the column several times. The column is then rinsed with ~30 ml of phosphate buffer at 18 ml/hr. IGF is eluted by 3 *M* guanidine in 0.25 *M* acetic acid. Fractions of 2.2 ml/10 min are collected and analyzed directly after appropriate dilution by RIA. IGF is found in fractions 2–5 after starting elution. The yield is 93–100%. Less than 3% is found in the neutral flow-through if 200 μg or less of IGF is applied. Rechromatography under identical conditions is carried out when significant amounts are found in the flow-through.

The fractions containing IGF are freed from guanidine buffer by absorption to a column (0.9 × 3 cm) of octyl Sepharose (Pharmacia) equilibrated in 0.1 *M* Na-phosphate pH 7.4 (18 ml/hr). The column is rinsed with 15 ml of phosphate buffer. IGF elutes with 50% isopropanol in 10 m*M* Na-phosphate pH 7.4 as a sharp peak within 2–3 fractions of 2.2 ml with a recovery of 95–100%.

Final Purification with Reversed-Phase High-Performance Liquid Chromatography (HPLC)

Two pumps (Altex, Model 110A) and a programmer (Model 200, Kontron, Switzerland) are used for gradient elution on C_{18} columns. Acetonitrile (HPLC grade, Baker) is used throughout. Absorbance is continuously monitored at 220 nm. For all solutions quartz distilled water is used. Two systems are used consecutively.

System I

Buffer A: 0.1% phosphoric acid (v/v) (Fluka, Switzerland)
10 m*M* Na-perchlorate (Merck)
Buffer B: 0.1% phosphoric acid
10 m*M* Na-perchlorate
60% acetonitrile

[13] U. K. Läubli, W. Baier, H. Binz, M. R. Celio, and R. E. Humbel, *FEBS Lett.* **149,** 109 (1982).

A 4.6 × 25 mm Chromosorb LC-7 column (10 μm, Brownlee, purchased from Analab) is used at room temperature at a flow of 1 ml/min. Fractions from the octyl Sepharose column are diluted by addition of 2 vol of buffer A. A thin plastic tube connected to pump A is used to load larger volumes. IGF is bound and concentrated on top of the column. The column is washed with buffer A until the absorbance declines to baseline.

Elution is perfomed by a gradient of 0–50% buffer B during 15 min followed by 50–100% buffer B during 80 min. Peak fractions are collected by hand or in 1 ml portions. IGF I p*I* 8.25 elutes at 58.5% B, IGF II p*I* 6.5 at 58% B, i.e., at the same positions as IGF I and IGF II purified by the original procedure. A typical example is shown in Fig. 5. Peaks eluting later than 80 min represent nonpeptide material as assessed by amino acid analysis. The recovery is 85–95%. Up to 500 μg of IGF can be loaded for one run (see also Fig. 9).

System II

Buffer A: 10 m*M* K-phosphate pH 7.0
50 m*M* Na-perchlorate

Buffer B: 10 m*M* K-phosphate pH 7.0
50 m*M* Na-perchlorate
60% acetonitrile

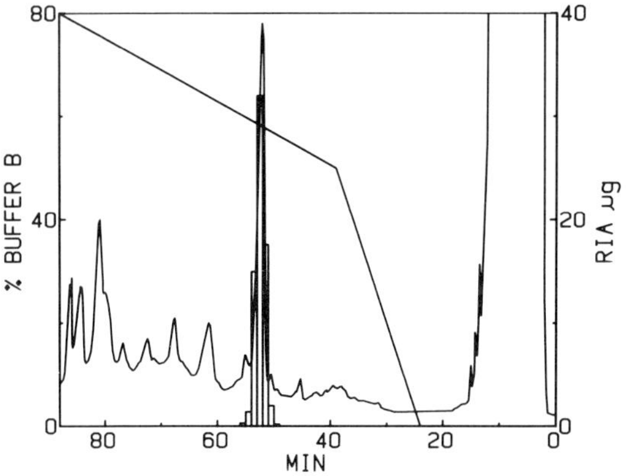

FIG. 5. Purification of 90 μg of IGF I (p*I* 8.25) after affinity chromatography on reversed-phase HPLC system I. The diluted sample is loaded via pump A on a chromosorb LC-7, 10 μm, Brownlee column, which is washed with buffer A until the absorbance declines to baseline (the absorbance peak in the flow through is not found if HPLC grade buffers of the previous desalting step on octyl Sepharose are used). IGF is eluted by the gradient of buffer B (——). Flow, 1 ml/min. Open bars, RIA I μg/fraction. Absorbance at 220 nm: 0.2 full scale.

A 4.6 × 25 mm LiChrosorb RP 18 column (10 μm, Brownlee) is used. Fractions from a HPLC run in system I are diluted by addition of 2 vol of 25 m*M* K_2HPO_4 and applied through pump A as described above. A gradient of 0–55% B in 15 min followed by 55–75% B in 45 min is formed. IGF p*I* 8.25 elutes at 59.5% B (Fig. 6) as does the major component of standard IGF I (batch I/3). IGF II p*I* 6.5 elutes significantly earlier at 58% B (Fig. 7) at the same position as standard IGF II (batch 9 SE-IV). The recovery is ~80%. If necessary, the material is easily desalted by absorption to a small precolumn filled with C_{18} material (CO: Pell ODS, Whatman) equilibrated in 0.1% trifluoroacetic acid. IGF is eluted in concentrated form by a gradient of acetonitrile (0–60%) in 0.1% trifluoroacetic acid in 15 min at high yield. The peak can be directly lyophilized or dried by flash evaporation. If stored in the dry state at 4°, IGF can be kept for years without apparent degradation. A stock solution of IGF in 1 *M* acetic acid kept at 4° is also stable for more than a year.

Upon rechromatography in system I and II or in 0.1% trifluoroacetic acid/60% acetonitrile or in 10 m*M* K-phosphate pH 7.0 without Na-perchlorate, IGF behaves as a single symmetrical peak. In all these systems, the peak position is identical to the one of IGF purified by the older method. Amino acid compositions are given in Tables I and II. They are identical to the published amino acid compositions of IGF I and II.[1,2] Glycine is the amino terminal amino acid in IGF I p*I* 8.25 as expected. In IGF II p*I* 6.5, alanine and tyrosine are the amino terminal amino acids

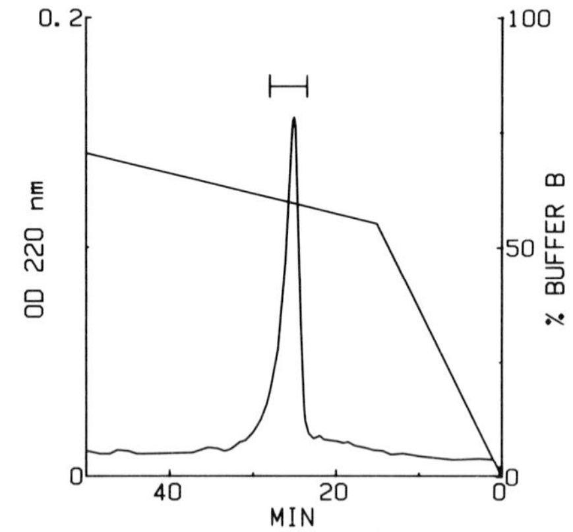

FIG. 6. Final purification of IGF I (p*I* 8.25) in HPLC system II. IGF containing fractions from HPLC system I are, after dilution, loaded on the RP 18, 10 μm, Brownlee column via pump A and IGF is eluted by the gradient of buffer B (——). Flow 1 ml/min. All the RIA activity coeluted with the absorbance peak (bar).

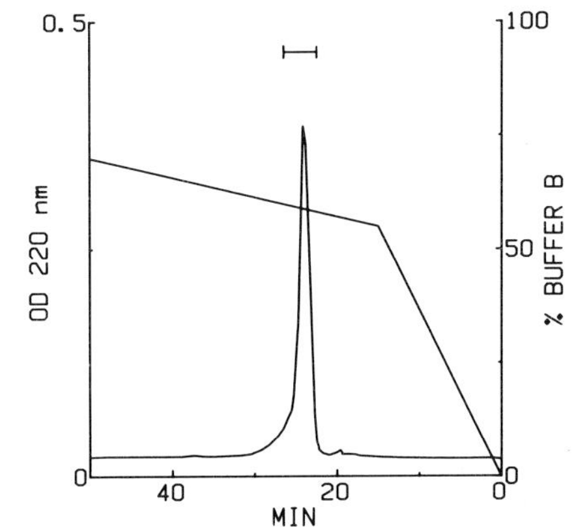

FIG. 7. Final purification of IGF II (p*I* 6.5) in HPLC system II. IGF II from HPLC system I is, after dilution, loaded on the RP 18, 10 μm, Brownlee column via pump A. IGF is eluted by the gradient of buffer B (——). Flow 1 ml/min. All the RIA II activity coeluted with the absorbance peak (bar).

TABLE I
AMINO ACID COMPOSITIONS OF IGF I

Calculated from sequence	Batch I/3	p*I* 8.25 form from Serum	p*I* 8.25 form from Extract B
Asp 5	5.3	5.2	5.1
Thr 3	2.9	3.1	3.0
Ser 5	4.7	4.8	4.9
Glu 6	7.1	6.5	6.2
Pro 5	5.0	5.1	4.9
Gly 7	6.0	7.1	6.9
Ala 6	5.7	6.4	6.1
Cys 6	n.d.	n.d.	n.d.
Val 3	2.6	2.7	2.9
Met 1	0.9	0.6	0.7
Ile 1	0.8	0.8	0.8
Leu 6	5.9	6.1	6.1
Tyr 3	1.9	2.0	2.2
Phe 4	3.5	3.7	3.9
Lys 3	3.0[a]	3.0[a]	3.0[a]
Arg 6	6.3	6.2	5.7

[a] Calculations are based on 3.0 residues per molecule assumed for Lys.

TABLE II
AMINO ACID COMPOSITION OF IGF II[a]

Calculated from sequence	Batch II/2	p*I* 6.5 form from	
		Serum	Extract B
Asp 3	3.6	3.2	3.3
Thr 4	3.8	3.9	3.9
Ser 7	6.4	6.5	6.9
Glu 7	7.9	7.4	7.1
Pro 3	3.6	3.1	2.9
Gly 5	4.9	5.2	5.3
Ala 5	4.8	5.1	5.1
Cys 6	n.d.	n.d.	n.d.
Val 4	4.1	3.7	3.8
Met 0	0	0	0
Ile 1	1.2	0.7	0.8
Leu 6	5.8	6.3	6.2
Tyr 3	2.5	2.2	2.1
Phe 4	3.8	3.6	3.7
Lys 1	2.2	1.0	1.2
Arg 8	7.9	8.3	7.5

[a] Calculations are based on 7.0 residues for Ile plus Leu.

suggesting that the two forms of IGF II[1,5] (NH_2-Ala-Tyr and NH_2-Tyr) are not separated.

The specific activities of IGF I and II as determined in the RIA and the fat cell assay are indistinguishable from conventionally purified IGF. An outline of the purification scheme is given in Table III. IGF I (23 μg) and 85 μg of IGF II are obtained from 1 liter of serum. The overall yield is thus approximately 14 and 13%.

The described method A entails the availability of an immunoaffinity column. The monoclonal antibody No. 43 which is being used in our laboratory can be supplied upon request. The advantage of being able to purify IGF from serum (as opposed to a Cohn fraction as starting material) is obvious when IGF from species other than man has to be isolated. In fact, only minor modifications are needed to adapt method A to the isolation of beef IGF.[13a]

If, however, large amounts of human IGF have to be prepared, method B starting from a Cohn fraction is the method of choice.

[13a] A. M. Honegger, P. P. Zumstein, and R. E. Humbel, in preparation.

TABLE III
PURIFICATION OF IGF FROM SERUM

Purification step	IGF I pI 8.25 Specific activity μg/mg of protein	IGF I pI 8.25 Recovery (%)	IGF II pI 6.5 Specific activity μg/mg of protein	IGF II pI 6.5 Recovery (%)
Serum	0.0027	(100)	0.009	(100)
Sephadex G-75	0.37	78	1.4	75
Sephadex G-50 after acid ethanol extraction	3.8	51	14.7	57
Chromatofocusing				
pI 8.25	11.4	23		
pI 6.5			19	24
Affinity column	540	22	600	24
HPLC system I	900	17	940	17
HPLC system II	1000	14	1000	13

Purification of IGF from Cohn Fraction "Precipitate B"

Preparation of "Extract B"

Cohn fraction "precipitate B" of human plasma,[14] obtained from the Swiss Red Cross laboratory, is treated as originally described by Bürgi *et al.*[15]: 240 g of frozen precipitate B is homogenized in 1 liter of 75% ethanol–0.18 *M* HCl and centrifuged after 15 to 30 min at room temperature. The precipitate is resuspended in 1 liter of acidic ethanol and centrifuged. The combined supernates are brought to pH 8.5 with 2 *N* NaOH. After standing for 30 min, the supernate is passed through filter paper (Schleicher and Schuell no. 588), cooled at −10° and added to 4 vol of acetone–ethanol (5 : 3, v/v). After 12 hr at −10°, the supernate is decanted and the precipitate dried under vacuum at room temperature.

The following steps to Sephadex G-50 are performed as described by Rinderknecht and Humbel[1]: 140 g of "extract B" is suspended in 1.1 liters of 0.5 *M* acetic acid, vigorously stirred for 60 min, centrifuged at 8500 *g* for 60 min, decanted, and reextracted twice with 1.1 liters 0.5 *M* acetic acid. The combined supernates contain 25 g of protein and 20–35

[14] P. Kistler and H. Nitschmann, *Vox Sang.* **7,** 414 (1962).

[15] H. Bürgi, W. A. Mueller, R. E. Humbel, A. Labhart, and E. R. Froesch, *Biochim. Biophys. Acta* **121,** 349 (1966).

mg of IGF. The specific activity of IGF I at this stage is approximately 0.8 mg/g, of IGF II approximately 0.5 mg/g. Apparently, the recovery of IGF II is much lower than that of IGF I. Uncomplete precipitation by acetone–ethanol of IGF II is probably responsible for preferential losses of IGF II (Honegger *et al.*, unpublished).

Gel Filtration on Sephadex G-75

The combined acetic acid extracts are gel-filtered on Sephadex G-75 as previously described. Peak fractions with a specific activity of at least 20 μg IGF/mg are further used. The overall yield is at least 90%.

Gel Filtration on Sephadex G-50

Five grams of G-75 material is dissolved in 1 liter of 1 *M* acetic acid and applied on a Pharmacia sectional column KS 370 (column volume 64 liters) packed with Sephadex G-50 in 1 *M* acetic acid. The ascending flow is at 7.3 liters/hr. The zone containing IGF is recycled on the same column by connecting the outlet to the inlet of the column at the appropriate time. Thereafter, fractions of 2.43 liters (20 min) are collected. The fractions containing IGF are separately concentrated by flash evaporation and lyophilized. The overall yield is at least 80%. The peak fractions with specific activities of 75–110 μg/mg account for 48–53% of the total yield and are further used. Fractions of lower specific activity are pooled and rechromatographed under same conditions.

Chromatofocusing

Material from Sephadex G-50 (100 mg) is dissolved in 2 ml of 0.01 *N* HCl and diluted with distilled water to 20 ml. Two milliliters of ethanolamine 0.1 *M*, brought to pH 9.4 with acetic acid, is slowly added and titrated back to pH 9.4 with 1 *N* NaOH. At pH 6–6.3, a precipitate forms. It is separated by centrifugation for 20 min at 3000 *g* and the clear supernate is used. Sometimes the pellet contains appreciable amounts of predominantly IGF II. The supernate containing 3.4–5.6 mg of IGF I and 1.6–2.5 mg of IGF II is applied on a PBE 94 column as previously described.

To recover the IGF from the pellet, the following procedure was used. The pellet dissolves in 3 ml of 0.01 *N* HCl. This solution contains major contaminating material which can be removed by reversed-phase HPLC. It is applied on the HPLC system I. Peak fractions containing IGF II are rechromatographed in the same system before final purification in HPLC system II is achieved. Note that the yield from this source has not been

TABLE IV
PURIFICATION OF IGF FROM "EXTRACT B"

Purification step	IGF I (p*I* 8.25) µg/mg of protein	IGF I (p*I* 8.25) Recovery (%)	IGF II (p*I* 6.5) Recovery (%)
Extract B	0.78	(100)	(100)
Sephadex G-75	n.d.	67	68
Sephadex G-50	56	35	35
Chromatofocusing			
p*I* 8.25	640	22	
p*I* 6.5			7
HPLC system I	950	17	6.5
HPLC system II	1000	13	5.5

included in the calculation of the overall yield as outlined in Table IV. Its contribution enhances the total yield to 11%.

RIA I activity elutes predominantly at pH 8.25–8.35. Subforms of p*I* 8.7–9, 7.8–8, and 7.5 are found in varying amounts (Fig. 8). The overall yield is 85–100%. IGF I of p*I* 8.25 is used for final purification by reversed phase HPLC.

RIA II activity elutes mainly at pH 6.5 and from 7.3 to 7.7. Lower amounts are found at pH below 5. In the RIA II, IGF II of p*I* 6.5 which gives parallel displacement curves is used for final purification by reversed phase HPLC.

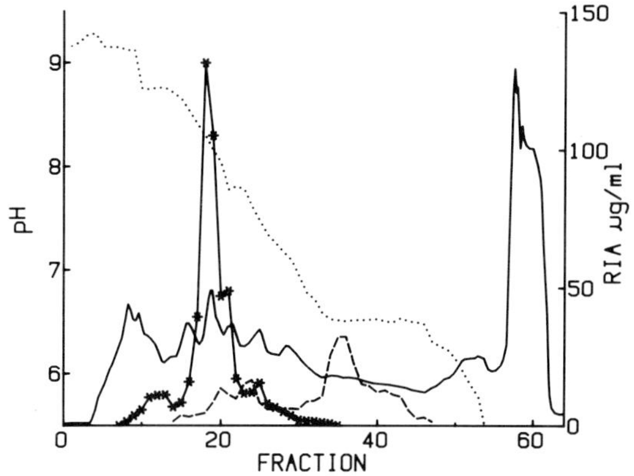

FIG. 8. Chromatofocusing of Sephadex G-50 material (100 mg) with a gradient from pH 9 to 6. The main components of IGF I (×) and IGF II (----) exhibit a p*I* of 8.25 and 6.5, respectively. (——) Absorbance at 280 nm; (····) pH.

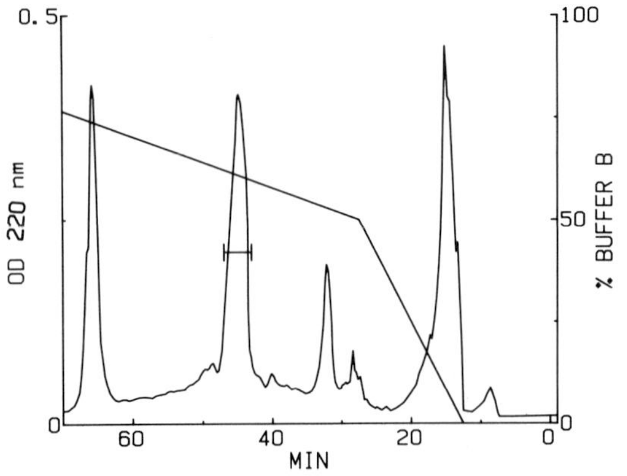

FIG. 9. Purification of 460 μg of IGF I (p*I* 8.25) from chromatofocusing on reversed-phase HPLC system I. Column Chromosorb LC-7, 10 μm, Brownlee. IGF is eluted by the gradient of buffer B (——). Flow 1 ml/min. All of the RIA I activity coeluted with the peak at approximately 59% B (bar).

Final Purification by Reversed-Phase HPLC

IGF I p*I* 8.25 and IGF II p*I* 6.5 can be obtained in homogeneous form after chromatography in system I and rechromatography in system II. Figures 9 and 10 demonstrate typical examples for IGF I (p*I* 8.25) and IGF II (p*I* 6.5). In both cases the RIA activity coelutes with the main peak. Upon rechromatography with analytical amounts, this material elutes as a sharp symmetrical peak.

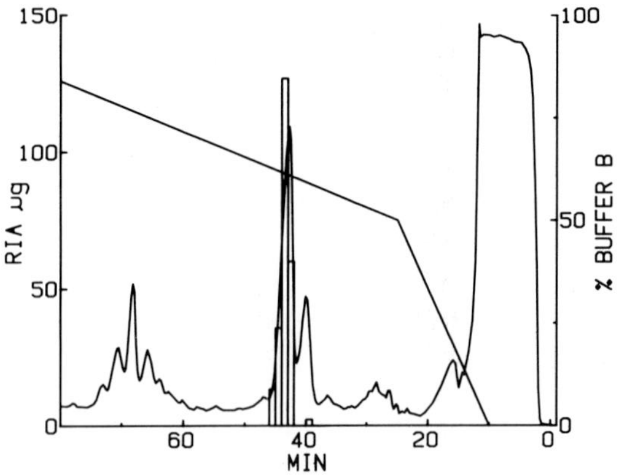

FIG. 10. Purification of 250 μg of IGF II (p*I* 6.5) from chromatofocusing on reversed-phase HPLC system I. Absorbance at 220 nm: 0.5 full scale. Open columns, RIA II μg/fraction.

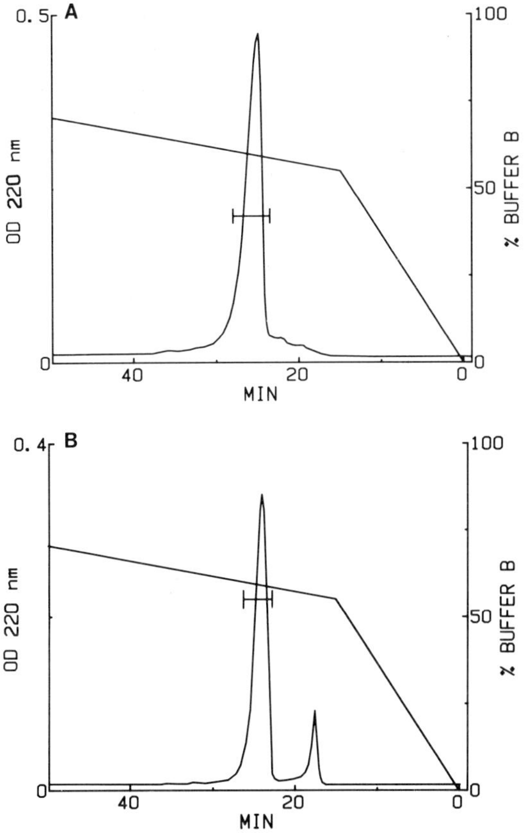

FIG. 11. Final purification on HPLC system II. (A) IGF I (p*I* 8.25). (B) IGF II (p*I* 6.5). Fractions from HPLC system I containing IGF are diluted by addition of 2 vol of 25 m*M* K_2HPO_4 and applied on the RP 18, 10 μm, Brownlee column via pump A. IGF is eluted by the gradient of buffer B (——), 0–55% B in 15 min followed by 55–75% B in 45 min. Flow 1 ml/min. RIA activity (bar) coelutes exclusively with the main absorbance peak.

Minor contaminating material is finally removed by rechromatography in system II as previously described (see Fig. 11A and B). An outline of the different steps is given in Table IV. Out of 140 g of extract B, approximately 2.5 mg of IGF I and 1 mg of IGF II are obtained. IGF prepared by this method is indistinguishable from the one prepared by method A. Amino acid compositions are given in Tables I and II.

Conclusions

The two methods described here to isolate IGF from human serum or a plasma fraction involve both a gel filtration at low pH as an early step.

This procedure dissociates and separates IGF from the plasma carrier protein to which it is bound during circulation *in vivo*. The use of separation techniques employing charge differences as an early preparative step has always been unsuccessful. This experience finds now a rational explanation by the presence of subforms differing in p*I*. The structural basis of the differing charges and their origin (different gene products, post-transcriptional or translational processing, or artifacts?) is currently being investigated.

These methods should prove useful to prepare μg or mg quantities of IGF to further study their biologic effects, for use as ligands in radioreceptor and radioimmunoassays and in affinity chromatography, and to prepare specific antibodies. For *in vivo* experiments, the higher quantities needed may be expected to be prepared in the near future by biotechnological methods.

[61] Purification of Somatomedin-C/Insulin-Like Growth Factor I

By MAJORIE E. SVOBODA and JUDSON J. VAN WYK

Introduction

Nomenclature of the Somatomedins

Somatomedin is the generic term used to describe several insulin-like peptides which contribute importantly to the mitogenic properties of serum. *In vivo*, somatomedins have been shown to restore growth to hypophysectomized animals[1] and also to play an essential role in the feedback regulation of growth hormone secretion by blocking the action of growth hormone releasing factor at the pituitary level.[2] These findings have provided strong confirmation that the growth promoting actions of somatotropin are mediated by the somatomedins.

Nomenclature of the somatomedins has been greatly clarified by the disclosure that somatomedin-C (Sm-C) and insulin-like growth factor I (IGF-I) are identical peptides[3,4] and that multiplication stimulating activ-

[1] E. Schoenle, J. Zapf, and E. R. Froesch, *Nature (London)* **296,** 252 (1982).

[2] P. Brazeau, R. Guillemin, N. Ling, J. J. Van Wyk, and R. E. Humbel, *C.R. Hebd. Seances Acad. Sci., Ser. C* **295,** 651 (1982).

[3] D. G. Klapper, M. E. Svoboda, and J. J. Van Wyk, *Endocrinology* **112,** 2215 (1983).

[4] E. Rinderknecht and R. E. Humbel, *J. Biol. Chem.* **253,** 2769 (1978).

ISBN 0-12-182009-2

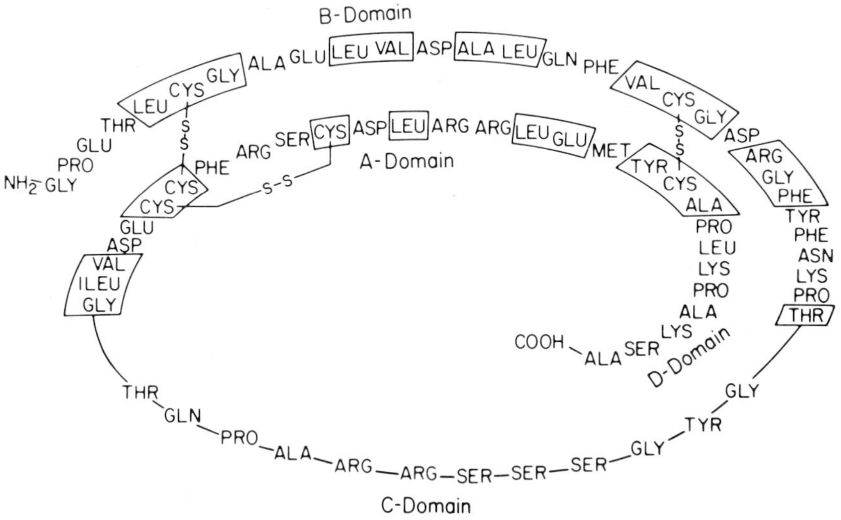

FIG. 1. Amino acid sequence of Sm-C/IGF-I. Boxes indicate amino acids identical with those in human proinsulin.

ity (MSA) differs in only 5 amino acid residues from insulin-like growth factor II (IGF-II), and thus is the rat counterpart of the human hormone.[5,6] From today's perspective, it appears that nearly all of the known biological effects of the somatomedins, insulin-like growth factors, and multiplication stimulating activities can be explained on the basis of these two major forms of somatomedin: The basic peptide, Sm-C/IGF-I (p*I* 8.4) is the more growth hormone dependent and also the more active growth promoter in *in vitro* biological systems. The neutral peptides IGF-II and MSA are more intrinsically insulin-like in their actions than Sm-C/IGF-I. IGF-II/MSA may play a particularly important role in fetal growth, since fetal fibroblasts respond to placental lactogen by secreting IGF-II/MSA rather than Sm-C/IGF-I, and serum concentrations of IGF-II/MSA are particularly high in the fetus.[7]

Chemistry of the Somatomedins

Sm-C/IGF-I is a single chain peptide of 70 amino acid residues and three disulfide bridges. The amino acid sequence is shown in Fig. 1.

[5] H. Marquardt, G. J. Todaro, L. E. Henderson, and S. Oroszlan, *J. Biol. Chem.* **256,** 6859 (1981).

[6] E. Rinderknecht and R. E. Humbel, *FEBS Lett.* **89,** 283 (1978).

[7] S. O. Adams, S. P. Nissley, S. Handwerger, and M. M. Rechler, *Nature (London)* **302,** 150 (1983).

Certain portions of its primary amino acid structure are remarkably similar to proinsulin and undoubtably these peptides share a common ancestral origin. The amino acids in those portions of Sm-C/IGF-I corresponding to the B and A chains of insulin are approximately 50% identical, whereas the connecting peptide of Sm-C/IGF-I is shorter than that of proinsulin (12 residues as contrasted with the 31 residues) and no homology is evident. In addition, Sm-C/IGF-I has an 8 residue amino acid extension beyond the carboxy terminus which is not present in proinsulin. IGF-II/MSA is similar in size and structure to Sm-C/IGF-I but 38% of the amino acids in IGF-II are different from those in similar positions in Sm-C/IGF-I, the greatest dissimilarity occurring in the C and D domains. Therefore IGF-II is clearly a different gene product than Sm-C/IGF-I.

Origin of the Somatomedins

Unlike classic hormones which can be extracted in high concentration from isolated organs of secretion, the somatomedins are produced by a large range of tissues and no tissue has been identified which contains higher concentrations than those in blood. Although MSA was purified from the conditioned media of certain lines of cells which had been cloned from cultures of rat hepatocytes, Sm-C/IGF-I and IGF-II were isolated from Cohn fraction IV of human blood plasma.[8,9]

The biogenesis of the precursors of the somatomedins and subsequent processing steps remain to be defined. It is noteworthy, however, that the somatomedins are invariably secreted by liver and other cell types as macromolecular complexes with the small active peptide attached to a binding protein. The binding proteins which are formed along with somatomedin during *in vitro* synthesis have varied in size and other characteristics but in no instance are the somatomedins secreted as the free peptide.

Physical State of Somatomedin in Blood

In serum likewise the somatomedins are associated with binding proteins within a macromolecular complex (Fig. 2). The major complex in serum has a molecular weight of about 140,000 Da and the binding protein is composed of at least two subunits. Furlanetto[10] has shown that one subunit with a molecular weight of 60K is acid stable and growth hormone dependent and a second of similar size is acid labile and less growth hormone dependent. Neither by itself can bind Sm-C/IGF-I. These sub-

[8] M. E. Svoboda, J. J. Van Wyk, D. Klapper, R. Fellows, F. Grissom, and R. J. Schlueter, *Biochemistry* **19,** 790 (1980).

[9] E. Rinderknecht and R. E. Humbel, *Proc. Natl. Acad. Sci. U.S.A.* **73,** 2365 (1976).

[10] R. W. Furlanetto, *J. Clin. Endocrinol. Metab.* **51,** 1219 (1980).

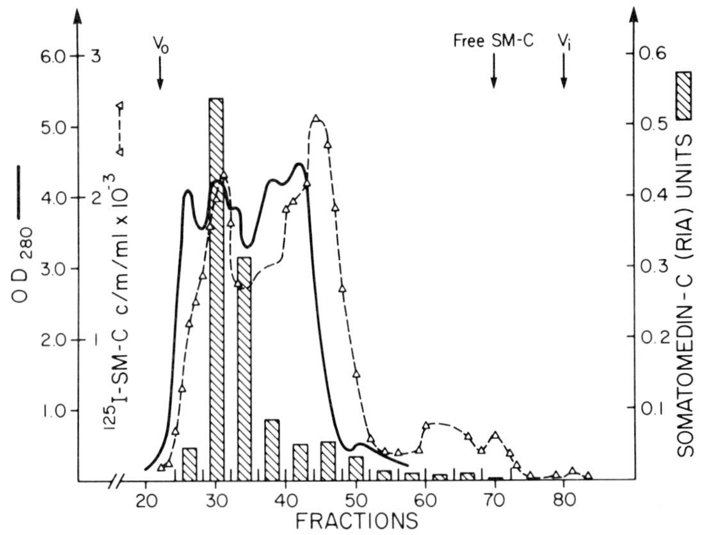

FIG. 2. Gel chromatography of 4 ml of normal fresh serum previously incubated with ^{125}I-labeled Sm-C on a 2.8 × 75 cm Sephacryl S-200 column is phosphate buffer, pH 7.4, $I = 0.1$. The majority of immunoreactive Sm-C coelutes with the immunoglobulins. Reprinted with permission from Chatelain *et al.*[29]

units, as well as the 140K complex, are virtually absent from the serum of hypopituitary subjects.[11] A third plasma binding protein is of lower molecular weight, 38K, and has a high capacity to bind radiolabeled Sm-C/IGF-I. This protein is not growth hormone dependent but its levels increase in pregnancy and may be estrogen dependent.

The serum binding proteins are clearly responsible for preventing rapid fluctuations in serum Sm-C/IGF-I concentrations. It is also because of the binding proteins that the levels of the somatomedins in plasma are higher than those of other peptide hormones which are not bound to carrier proteins. According to Zapf *et al.*,[12] the approximate mean level in adult plasma is 193 ± 58 ng/ml for Sm-C/IGF-I and 647 ± 126 ng/ml of IGF-II.

In most instances the linkage between the somatomedin peptides and the binding protein can be ruptured by exposure to acid and the binding proteins separated from the free peptide by chromatography on Sephadex G-50 in acid (Fig. 3). Acid stable forms detectable by the various assays are also present in serum. The binding proteins and acid stable forms can greatly complicate the purification of the active growth

[11] K. C. Copeland, L. E. Underwood, and J. J. Van Wyk, *J. Clin. Endocrinol. Metab.* **50,** 690 (1980).

[12] J. Zapf, H. Walter, and E. R. Froesch, *J. Clin. Invest.* **68,** 1321 (1980).

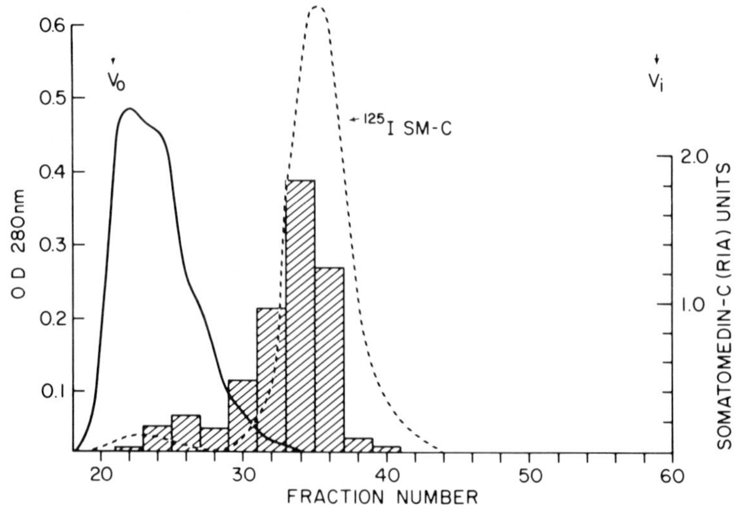

FIG. 3. Typical elution pattern of plasma chromatographed on Sephadex G-50 at pH 3.8. The large molecular weight proteins elute in the void volume along with other acid-stripped binding proteins. The somatomedin is typically recovered in the fractions which elute between 50 and 80% of the total bed volume. The addition of ^{125}I-labeled Sm-C to the acidified plasma prior to chromatography greatly facilitates identification of somatomedin-C-containing fractions.

factor since they interfere with the measurement of the somatomedins by whatever method is used.

Methods of Monitoring Purification

Methods of Assay

Bioassays. Success in isolating somatomedin depends primarily on access to a rapid, sensitive, and specific assay to monitor purification. The isolation of somatomedin-C was initially monitored by means of a dual isotope assay based on both the incorporation of $^{35}SO_4$ into proteoglycans and [^{3}H]thymidine into the DNA of cartilage fragments from hypophysectomized rats.[13] Simpler cartilage assays utilize chick embryo pelvic leaflets[14] or costal cartilage segments from young pigs[15] as the test object. The insulin-like growth factors were isolated by modifications of biological assays for insulin using the oxidation of [^{14}C]glucose to $^{14}CO_2$

[13] J. J. Van Wyk, L. E. Underwood, R. L. Hintz, D. R. Clemmons, S. J. Voina, and R. P. Weaver, *Recent Prog. Horm. Res.* **30,** 259 (1974).

[14] K. Hall, K. Takano, L. Fryklund, and H. Sievertsson, *Adv. Metab. Disord.* **8,** 32 (1975).

[15] J. L. Van den Brande and M. V. I. Du Caju, *Acta Endocrinol.* (*Copenhagen*) **75,** 233 (1974).

by adipose tissue in the presence or absence of antibodies to insulin itself.[16] Multiplication stimulating activity was isolated on the basis of its ability to stimulate DNA synthesis in secondary cultures of chick embryo fibroblasts.[17] All of these assays are cumbersome, costly, and labor intensive, and none is specific for any one of the somatomedins. Furthermore, they are influenced to various degrees by other hormones, growth factors, or inhibitors which are present in serum. Nonetheless, the isolation of the somatomedins would not have been possible without these assays.

Receptor Assays. The disclosure that somatomedin competes with insulin for binding to the insulin receptor[18] led to the quantitation of somatomedin by a radioreceptor assay for insulin in human placental membranes. Although a larger number of samples can be analyzed by the insulin radioreceptor assay than by bioassay, it consumes large amounts of sample and is more sensitive to the neutral somatomedins, IGF-II and MSA, than to Sm-C/IGF-I.

The insulin receptor assay was quickly superseded by a receptor assay for somatomedin-C[19] after it was discovered that in human placental membranes somatomedin interacts with its own receptor which is distinct from the insulin receptor. The somatomedin receptor assay is approximately 150 times more sensitive than the insulin receptor to somatomedin-C and Sm-C is 3 times more potent than IGF-II in competing with ^{125}I-labeled Sm-C for binding to the Sm-C/IGF-I receptor.

Radioimmunoassays. The most specific and sensitive method for monitoring purification is the radioimmunoassay. Specific radioimmunoassays for Sm-C/IGF-I,[20,21] IGF-II,[22] and MSA[23] have been developed. The somatomedin-C RIA, is 50 times more sensitive than the somatomedin-C receptor assay, and discriminates better than any other test between the various somatomedins. In this assay IGF-II is less than 3% as active as somatomedin-C.

[16] E. R. Froesch, H. Burgi, E. B. Ramseier, P. Bally, and A. Labhart, *J. Clin. Invest.* **42,** 1816 (1963).

[17] N. C. Dulak and H. M. Temin, *J. Cell. Physiol.* **81,** 161 (1973).

[18] R. L. Hintz, D. R. Clemmons, L. E. Underwood, and J. J. Van Wyk, *Proc. Natl. Acad. Sci. U.S.A.* **69,** 2351 (1972).

[19] R. N. Marshall, L. E. Underwood, S. J. Voina, D. B. Foushee, and J. J. Van Wyk, *J. Clin. Endocrinol. Metab.* **39,** 283 (1974).

[20] R. W. Furlanetto, L. E. Underwood, J. J. Van Wyk, and A. J. D'Ercole, *J. Clin. Invest.* **60,** 648 (1977).

[21] J. Zapf, B. Morell, H. Walter, Z. Laron, and E. R. Froesch, *Acta Endocrinol. (Copenhagen)* **95,** 505 (1980).

[22] H. Walter, R. E. Humbel, and J. Schwander, *Int. Congr. Ser.—Excerpta Med.* **481,** 248 (1979).

[23] A. C. Moses, S. P. Nissley, P. A. Short, and M. M. Rechler, *Eur. J. Biochem.* **103,** 401 (1980).

Analytical Sodium Dodecyl Sulfate–Polyacrylamide Gel Electrophoresis

SDS–polyacrylamide gel electrophoresis is a valuable technique for following the progress of purification, particularly as the preparation nears homogeneity. We routinely run vertical 1.5 mm slab gels of 15% acrylamide prepared in 0.375 *M* Tris–HCl pH 8.8 and 3% stacking gel prepared in 0.125 *M* Tris–HCl pH 6.8.

The samples are prepared in 0.01 *M* Tris, 0.1% SDS and 10% glycerol and with or without 0.1% 2-mercaptoethanol. All samples are heated at 90° for 10 min prior to application and the gels are run in a continuous electrolyte buffer system of 0.05 *M* Tris, 0.38 *M* glycine, and 10% SDS at a pH of 8.6. Low molecular weight markers at 2 concentrations, 200 ng and 1 μg/band are run along with the samples. It is advisable to prepare the samples both in sample buffer with and without 2-mercaptoethanol since degraded forms of Sm-C/IGF-I migrate with increased mobility when reduced (vide infra).

Since Sm-C stains poorly with Coomaasie blue, the gels are stained by the silver stain method of Oakley *et al.*[24] in which a band with as little as 50 ng of protein can be detected. We have found that to prevent a dark background the gel first must be thoroughly washed with deionized water of exceptionally high purity to remove the Tris and glycine. After shaking in 10% gluteraldehyde for 1 hr, it is again imperative to wash thoroughly with water until all traces of gluteraldehyde are removed. After treatment with ammoniacal silver nitrate solution (7 ml of 19.5% $AgNO_3$ solution added to a mixture of 21 ml 0.36% NaOH + 1.5 ml concentrated NH_4OH diluted to 150 ml with water), it is again necessary to wash thoroughly to remove all unbound or unreacted silver. Reduction of the silver ion to free silver takes place in a solution containing 0.01% citric acid and 0.019% formaldehyde. This reaction is allowed to continue with shaking only until the 1 μg markers are clearly visible. Development will continue more slowly after replacement of the reducing solution with water thus allowing sufficient time for development of bands containing lower concentrations of protein. When sufficient development has taken place, excessive darkening is retarded by storage in 90% methanol in a dark cold room.

We have used a number of low-molecular-weight markers and have found that both in the reduced or unreduced state Sm-C consistently migrates more rapidly than bovine lung trypsin inhibitor (BRL), 6200 Da, or aprotinin (Pierce), 6500 Da, in spite of the fact the the molecular weight of Sm-C is 7649 (Fig. 4). Although with this system a 50 ng band can be detected, it is our practice to apply approximately 1 μg to permit visualization of minor impurities.

[24] B. R. Oakley, D. R. Kirsch, and N. R. Morris, *Anal. Biochem.* **105,** 361 (1980).

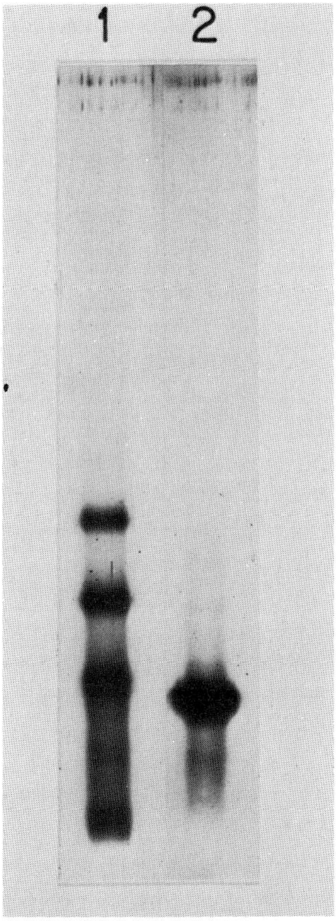

FIG. 4. Silver stained SDS–PAGE gel of reduced Sm-C. Lane 1: BRL low-molecular-weight markers, from cathode (top) to anode: β-lactoglobulin, 18,400; cytochrome *c*, 12,300; bovine trypsin inhibitor, 6200; reduced insulin, 3000. Lane 2: 1 μg reduced Sm-C. Reprinted with permission from Klapper *et al.*[3]

Purification from Cohn Fraction IV

Starting Material

The amount of somatomedin-C extractable from 1 g of rapidly frozen tissue is rarely greater than one-tenth the quantity in 1 ml of serum.[25] Thus, most investigators have purified Sm-C from blood plasma or a Cohn fraction of plasma.

[25] A. J. D'Ercole, A. D. Stiles, and L. E. Underwood, *Proc. Natl. Acad. Sci. U.S.A.* **81,** 935 (1984).

At the outset the investigator should recognize the extreme inefficiency of purifying a peptide from the myriad of other proteins and peptides present in blood. Assuming that the content of somatomedin-C in a given batch of plasma is 150 ng/ml, and the concentration of serum proteins is 75 mg/ml, it would be necessary to purify these plasma proteins 500,000-fold to obtain the pure peptide. If side fractions are reprocessed to maximize recovery, it is possible to achieve an overall recovery of approximately 5% (7% is the best we have achieved). With this degree of efficiency it would require the processing of 133 liters of plasma to obtain 1 mg of the pure peptide. This is not economically feasible since this quantity of starting material would require over 500 units of plasma. It is for this reason that most investigators have turned to Cohn fraction IV when they wish to obtain a substantial quantity of the peptide.

As shown in the table, most of the somatomedin activity in serum is segregated in Cohn fraction IV during the manufacturing processes used for the preparation of human serum albumin and other commercially useful blood products. Cohn fraction IV has no other commercial usage but is difficult to work with because it contains a high percentage of lipoproteins. Since most of our experience has been gained using this starting material,[8] this chapter will deal mostly with methods applicable to the isolation of Sm-C from Cohn IV as the starting material. Brief attention will also be given to methods of preparing microgram quantities of Sm-C/IGF-I from blood plasma.

Extraction of Cohn Fraction IV

Various types of acid extraction have been used to extract somatomedin from Cohn IV paste. After comparing extraction with phosphoric acid, formic acid, acetic acid, or acid-ethanol, we adopted extraction with 2 *M* acetic acid containing 0.075 *M* NaCl. The yield in acid ethanol extracts, for example, was only 12.5% of that achieved with 2 *M* acetic acid.

Frozen Cohn IV paste is weighed out and refrozen in an alcohol–dry ice bath. Just prior to homogenization in an industrial Waring blender, the frozen paste is pulverized with a mallet and added to the appropriate volume of extraction buffer, 2 *M* acetic acid with 0.075 *M* NaCl. One kilogram of Cohn IV paste can be homogenized in a minimum of 10 liters of extracting buffer. Yields are somewhat improved, however, if the ratio is increased to 12–15 liters of buffer/kilogram of frozen paste. The homogenate is passed through nylon mesh to remove larger particles which are then rehomogenized. After storage in the cold for 2 or more days, the homogenate is clarified by removing the precipitate with a Sharples centrifuge.

RECOVERY OF Sm-C FROM COHN FRACTIONS OF HUMAN PLASMA[a]

Cohn fractions	Percentage somatomedin-C recovered as compared to original plasma
Cohn I	5.4
Cohn II	0.01
Cohn III	8.8
Cohn IV-4	83.9
Cohn V	5.7

[a] One liter equivalent of various Cohn fractions was extracted and the somatomedin content determined by the radioreceptor assay. Reprinted with permission from J. Van Wyk *et al., Adv. Metab. Disord.* **8**, 393 (1975).

Ion Exchange Chromatography

Cation exchange chromatography is a useful next step in somatomedin-C purification as it allows most of the lipoprotein and some of the albumin to pass through the column while somatomedin-C and large amounts of other proteins are absorbed. Our laboratory uses SP Sephadex C-25 (Pharmacia) for this initial separation after first being equilibrated with extraction buffer adjusted to match the pH and conductivity of the extract. A smaller column containing the same gel is placed ahead of the main column to act as a guard column since lipids and other oily materials in the emulsion tend to clog the column matrix and cause channeling. When channeling occurs, the guard column is replaced with fresh gel. The pooled gel from the guard columns is later added to the main column for elution. We have loaded 40 liters of extract derived from 3 kg Cohn IV on a 800 ml column at 7 ml/min without exceeding the column capacity.

After loading is completed, the column is washed with starting buffer until the baseline optical density is achieved. Elution of the absorbed protein can be accomplished at the same rate with 1 *M* ammonium acetate pH 7 containing 1.5 *M* NaCl. The column eluate is monitored by its absorbance at 280 nm and by assay of the collected fractions.

Since Sm-C is most stable in acid solutions, the eluates are promptly acidified with acetic acid. The pool of active fractions is then concentrated on YM-5 filters in an Amicon TE3 ultrafiltration system. Dialysis is then carried out in the same apparatus against ≈0.5 *M* acetic acid pH 2.88 with 0.075 *M* NaCl to prepare the sample for the next step.

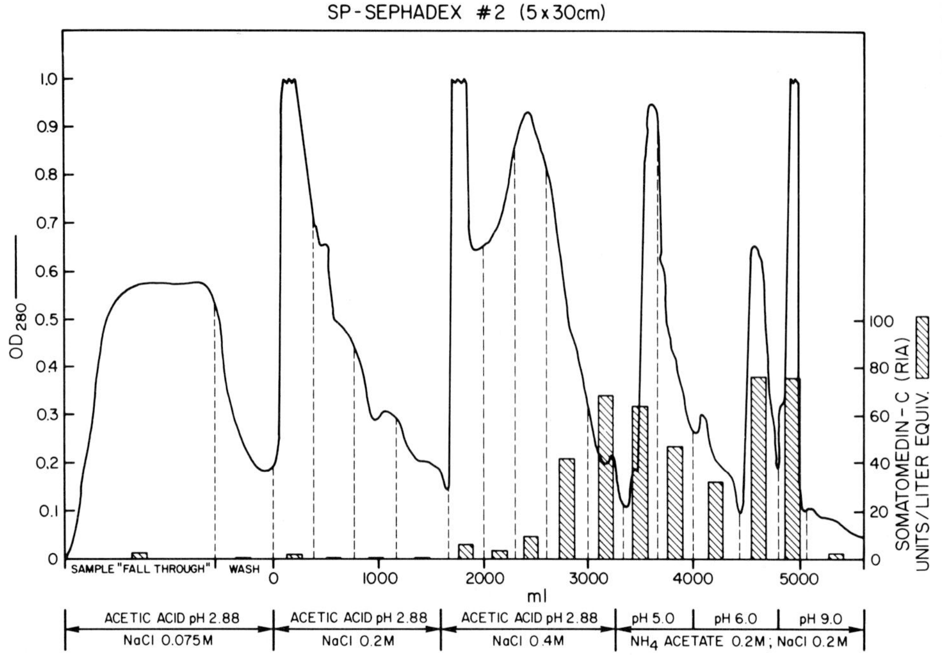

FIG. 5. Second chromatographic separation of Cohn IV extract on SP Sephadex C-25 with elution by step gradient as indicated. Reprinted with permission from Svoboda *et al.*[8]

A second SP Sephadex chromatographic step with gradient elution is carried out to remove additional acidic proteins under more optimal conditions. After the dialyzed pool is adjusted to starting buffer conditions, it is applied at 3 ml/min to a SP Sephadex C-25 column (2.5 × 30 cm), and washed with 550 ml of starting buffer. Elution is accomplished as shown in Fig. 5. We have concluded from various elution schemes that the greatest amount of inert protein can be eluted prior to the emergence of Sm-C by running a shallow salt gradient while maintaining a low pH. The somatomedin-C containing fractions are again pooled and subjected to ultrafiltration in the Amicon TE3 using the YM-5 filter. After the samples are concentrated they are dialyzed in the same apparatus against 1 *M* acetic acid and lyophilized.

Gel Filtration

After the ion exchange procedures, the Sm-C is still associated with larger or negatively charged proteins which must be removed prior to isoelectric focusing. This is accomplished by gel filtration on Sephadex G-

50 in 1 *M* acetic acid. It is preferable that this and subsequent separation processes be carried out in the cold, as somatomedin is susceptible to decomposition at higher temperatures, especially at neutral pH. The lyophilized Amicon retentate is dissolved in 1 *M* acetic acid and applied to a Sephadex G-50 column of appropriate size. When somatomedin is relatively pure, it elutes from G-50 in 1 *M* acetic acid with a K_D of about 0.33. However, when it is extremely impure, somatomedin-C is retained on the column along with many other materials and elutes at a K_D of 0.5 to 0.85.[8] Thus all of the column fractions must be assayed and only those fractions pooled which have a high somatomedin-C content. After lyophilizing the pooled fractions, the whole is again subjected to chromatography on Sephadex G-50 on a smaller column which provides better resolution.

Isoelectric Focusing

In our hands separation on the basis of isoelectric point is an essential component of any isolation scheme, since it separates somatomedin-C both from IGF-II and from a more basic peptide (p*I* 9–9.5) which crossreacts with Sm-C in all of our assay systems. These substances are sufficiently similar to Sm-C that they cannot be separated on the basis of molecular size, and we have so far been unsuccessful in separating them on the basis of hydrophobicity in reverse-phase HPLC systems.

Isoelectric focusing can be carried out by a sucrose gradient or on a flat bed tray in a matrix of Sephadex G-75 SF by the method of Radola.[26] Although our laboratory uses isoelectric focusing exclusively, some laboratories have reported good success with chromatofocusing (Pharmacia).

The general strategy is to first focus in a broad range such as pH 5–10 to remove IGF-II and the majority of contaminating proteins which are more acidic than Sm-C. The fractions focusing from pH 7.5–9.5 and then refocused in a narrower range (e.g., 7–9.5) to separate Sm-C from the more basic form of somatomedin.

Our procedure for isoelectric focusing in a flat bed tray is as follows. Approximately 35 g of pretreated Sephadex G-75 SF is added to about 750–800 ml of water, to which 25 ml of LKB Ampholine has been added. The slurry is then poured into a 20 × 40 cm flat bed tray (Brinkman Industries) and allowed to air dry to the proper consistency. The sample is dissolved in about 20–25 ml of water. If there is difficulty getting this into solution, a small volume of pH 3–10 Ampholine can be added which facilitates solution. This solution is then slurried in with the Sephadex in a 4-cm zone between the middle of the tray and the anode. After again being

[26] B. J. Radola, *Ann. N.Y. Acad. Sci.* **209,** 127 (1973).

allowed to air dry to proper consistency, the tray is placed in an isofocusing apparatus which is maintained at 4° and platinum ribbon electrodes are placed on paper strips soaked in 1 *M* phosphoric acid at the anode and 1 *M* NaOH at the cathode. Cytochrome *c* solution is streaked across the plate near the anode to serve as a marker. Preferably, isofocusing is carried out at constant power of 10–12 W, however, if this is impossible the voltage can be manually raised at periodic intervals as the current drops. Isofocusing is complete when the cytochrome *c* is at or under the cathode and when increases of voltages are not accompanied by drops in conductivity.

After isofocusing is complete, 1-cm sections of the plate are removed, placed in a container, and water added to a vol of 40 ml. Measurements of pH should be made at the running temperature. Aliquots of each supernatant are removed for assay and measurement of OD at 280 nm. An isofocusing profile at this point usually shows some somatomedin-containing fractions less than pH 8 and also greater than pH 8.7 (Fig. 6A). For this reason fractions from approximately pH 7.5–9.5 are pooled and eluted from the Sephadex. Elution is usually carried out by pouring the fractions into a chromatography column and washing the Sephadex with 2 column volumes of 1 *M* acetic acid. The eluate and the column wash then are lyophilized and ready for narrow range isofocusing.

Narrow range isofocusing is carried out in the same manner, using Ampholines from 7–9.5. The isofocusing profile of the narrow range isofocusing usually shows two distinct populations of somatomedin activity, one from approximately 8.0–8.7 and the other, which we call "Very Basic," from 8.7–9.5 (Fig. 6B). Removal of Ampholines after isoelectric focusing is accomplished by chromatography on Sephadex G-50 in 1 *M* acetic acids. In our experience the last traces of contaminating Ampholines, which are difficult to remove by Sephadex chromatography, are easily removed by reverse-phase HPLC where they elute in the void volume.

High-Performance Liquid Chromatography

The main disadvantage of HPLC is the small capacity of even the semipreparative column. This is one of the tradeoffs for the speed and great increase in resolution over conventional chromatography. It is thus advantageous to obtain a preparation by the preceding steps with as high a specific activity as possible. All HPCL was carried out using Waters Associates equipment and columns; 2 Model 6000A pumps, a Model 660 solvent programmer, and a Model U6K injector. The initial two reverse-phase HPLC steps consist of chromatography on semipreparative columns (7.8 mm × 39 cm), on an octadecylsilane (C_{18}) column using an

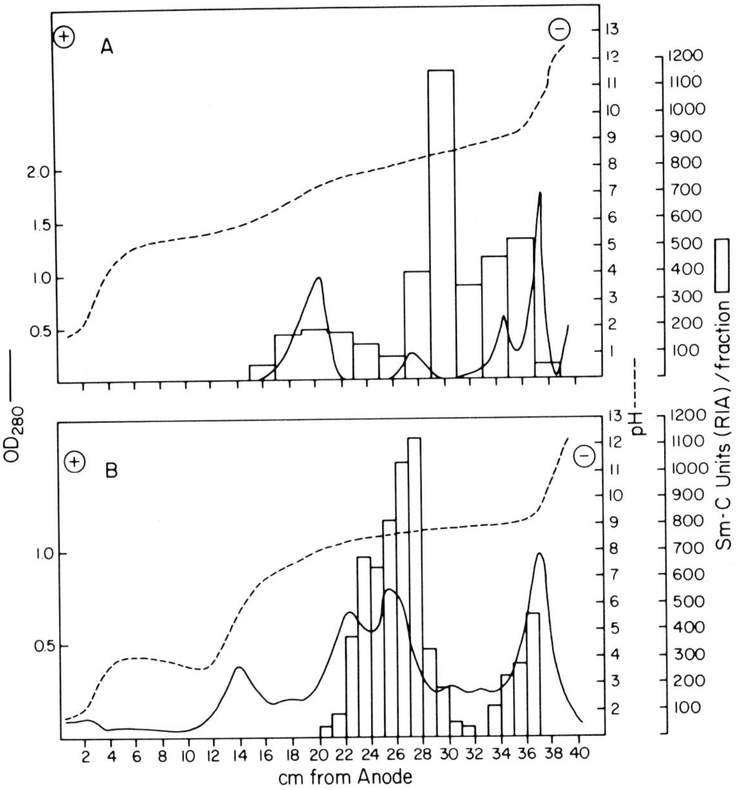

FIG. 6. Preparative thin-layer isofocusing profiles. (A) Broad range isoelectrofocusing (3–10) of the Sm-C-containing fractions derived from Sephadex G-50. Note the wide neutral band corresponding to the p*I* of IGF-II. (B) Reisofocusing of fractions 28–32 of the broad range isofocusing on a narrow range gradient of Ampholines between pH 7.0 and 9.5. Note the narrow more cathodic band termed "Very Basic." Reprinted from Svoboda *et al.*[8]

elution system of aqueous 0.1% trifluoroacetic acid/acetonitrile followed by an alkyl phenyl column with 0.01 *M* KH_2PO_4 in water/acetonitrile. In both cases a linear elution gradient is employed from 15 to 31% acetonitrile for 10 min. The elution then is continued isocratically at 31% until the bulk of the protein has been eluted. The gradient is then continued to 60% acetonitrile in 10–15 min. It is generally advantageous to use the C_{18} column as an initial step (Fig. 7) as the active fractions from the octadecylsilane HPLC after pooling and evaporation can be directly applied to the alkyl phenyl column. After pooling and evaporating the active fractions from the alkyl phenyl HPLC, desalting of the phosphate containing fractions can be accomplished by returning to a C_{18} column. The sample is

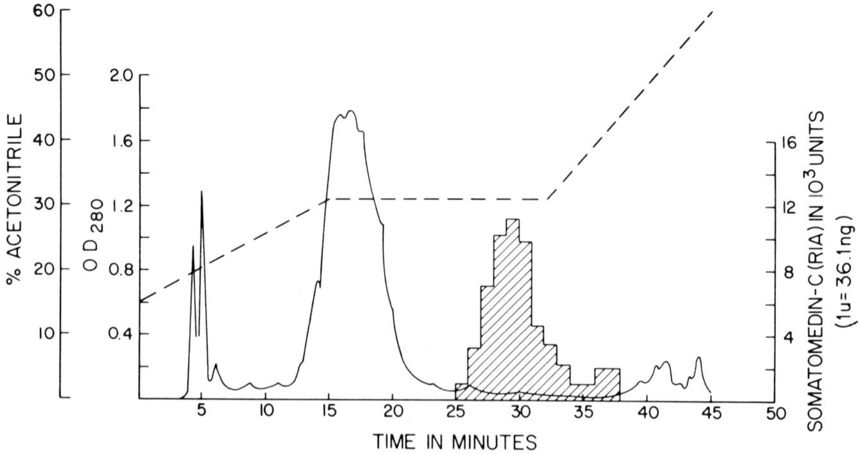

FIG. 7. HPLC of a crude Sm-C preparation (obtained from human Cohn IV extract which had been semipurified by a series of ion exchange and gel filtration chromatographic procedures) on a semipreparative μBondapak C_{18} (7.8 mm × 30 cm) column. A linear gradient was run from 15 to 31% acetonitrile in 0.1% TFA in water for 15 min followed by isocratic elution at 31% acetonitrile for 17 min, then continuation of linear gradient elution from 31 to 60% acetonitrile for 13 min at a flow rate of 3 ml/min. One minute fractions were collected. (——, OD_{280} nm; ------, percentage acetonitrile; hatched bars, units of Sm-C by RIA.)

dissolved in 0.1% trifluoroacetic acid and injected onto the column and eluted with the 0.1% trifluoroacetic acid until all of the salt has been eluted. The somatomedin is then removed from the column with 35% acetonitrile in 0.1% TFA/H_2O. This has the advantages of being a faster desalting method than with G-50 chromatography and the active material is contained in a much smaller volume.

The purity of the somatomedin preparation at this point is to a major extent dependent on the nature of the starting material. The high resolution of SDS–PAGE coupled with the great sensitivity afforded by the silver method of staining are invaluable at this stage for monitoring the purity of fractions under a protein peak. This was illustrated in the study depicted in Fig. 8 where the immunoreactivity coincided with a sharp absorbance peak. SDS–PAGE studies across the major, apparently homogeneous peak revealed the presence of two contaminants along with Sm-C, whereas the Sm-C in the minor peak migrated on the gel as a single band. When the impure pool was rechromatographed on an analytical C_{18} column using the same solvent system, Sm-C and the contaminating proteins were resolved.

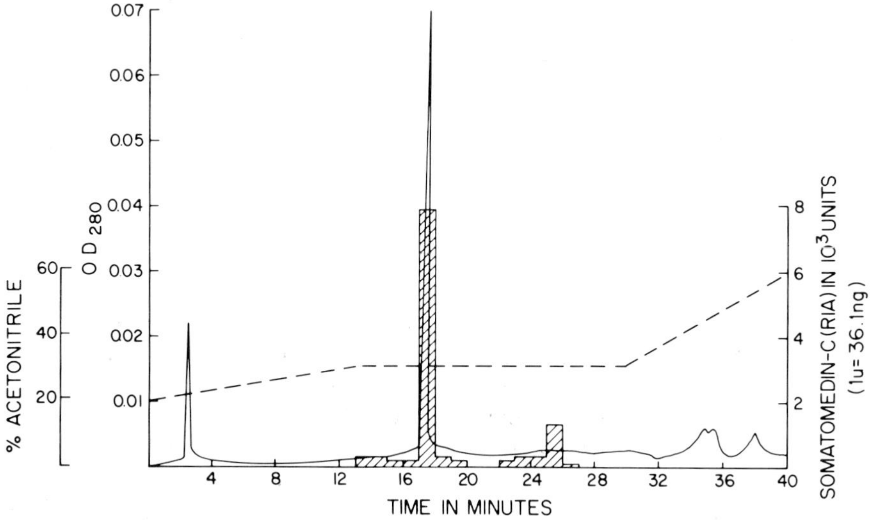

FIG. 8. The active fractions of a Sm-C preparation after 2 HPLC steps was applied to a semipreparative C_{18} column and eluted at 5 ml/min using a linear gradient program from 20 to 31% acetonitrile in 0.1% trifluoroacetic acid in water for 13 min, isocratic elution at 31% acetonitrile for 15 min, and linear gradient of 31 to 60% acetonitrile for 10 min. One minute fractions were collected. (——, OD_{280} nm; -------, percentage acetonitrile; hatched bars, units of Sm-C by RIA.) Reprinted with permission from "Handbook of HPLC for the Separation of Amino Acids, Peptides and Proteins" (W. S. Hancock, ed.), Vol. II, p. 439. CRC Press, Boca Raton, Florida, 1984.

Degradation of SM-C/IGF-I to Two Chain Somatomedin

SDS–PAGE has revealed that fragmentation of the Sm-C molecule can occur during purification. We have documented on several occasions the conversion of basic Sm-C to an acidic peptide with a p*I* of 4.8–5.0.[27]

These acid forms behave similar to Sm-C in terms of immunoreactivity, elution patterns on HPLC, molecular size based on gel filtration, and mobility on SDS–PAGE of the unreduced form. The reduced form migrates as a smaller molecule similar to the separated chains of insulin. These findings indicate that the acidic form is composed of 2 chains connected by disulfide bonds which upon reduction give rise to 2 smaller peptides. This was confirmed by sequence analysis which revealed that the peptide bond Arg^{37}-Ala^{38} had been cleaved, thus creating a 2 chain structure similar to insulin.

[27] M. E. Svoboda, J. J. Van Wyk, and D. G. Klapper, *Proc. Am. Pept. Sympo., 8th, 1983* Abstract (1983).

Although the conditions for this degradation have not been thoroughly investigated, it has been found that it is more likely to occur under neutral and basic conditions and at room and higher temperatures. For this reason, Sm-C preparations are best stored at −10° in 0.5 to 1.0 *M* acetic acid or lyophilized at −70°.

Purification of Somatomedin-C from Blood Plasma

Initial Separation Procedures

Although logistic and economic considerations make it impractical to isolate significant quantities of somatomedin from native plasma, the investigator who wishes to isolate submilligram quantities for the purpose of radiolabeling should consider starting with a pool of serum from patients with high somatomedin levels such as those with acromegaly. Since most of the somatomedin in serum circulates in a complex of approximately 140,000 Da, several strategems can be used to separate out the large-molecular-weight somatomedin containing proteins from the smaller proteins and then dissociate the small active peptide from the carrier proteins with acid. Such an approach was used by Ginsberg who dialyzed all of the miscellaneous unbound small-molecular-weight peptides away from the large proteins at pH 8.6 in a hollow fiber ultrafiltration apparatus. Then, under acid dissociating conditions, the somatomedins were passed through the membrane and recovered in the dialyzate.[28] Although we have attempted to use a similar strategem with Cohn fraction IV, the high concentration of lipoproteins has made this approach impractical.

Our current approach to purifying small quantities of Sm-C from whole plasma is to separate the small-molecular-weight peptide from binding proteins by chromatographing acidified plasma on very large (7–10 liters) Sephadex G-50 columns. The plasma sample is initially dialyzed in Seprapore 3 dialysis tubing (54 mm width) against a glycine hydrochloride buffer at pH 3.8. We have obtained better recoveries at this pH than at lower pHs.[29] A sample volume equivalent to 2–3% of the bed volume is applied and the column eluted at pH 3.8 as shown in Fig. 3. The fractions containing somatomedin elute at approximately 50–80% of the bed volume.[12] If available, ^{125}I-labeled somatomedin-C is used as a guide for pooling fractions. To obtain a significant amount of somatomedin, repeated identical aliquots are passed over the column in similar fashion,

[28] B. H. Ginsberg, C. R. Kahn, J. Roth, K. Megyesi, and G. Baumann, *J. Clin. Endocrinol. Metab.* **48,** 43 (1979).

[29] P. G. Chatelain, J. J. Van Wyk, K. C. Copeland, S. L. Blethen, and L. E. Underwood, *J. Clin. Endocrinol. Metab.* **56,** 376 (1983).

and the active eluates pooled. These are then concentrated in an Amicon TCE-3 apparatus using a YM5 filter and then dialyzed in the same apparatus against 0.5 *M* acetic acid. The preparation can then be lyophilized to concentrate it for the next step, which is usually isoelectric focusing. As indicated below, however, we have now modified the procedure to directly neutralize the eluates from the Sephadex G-50 column and without any concentration pass these large volumes directly over an affinity column prepared from a monoclonal antibody to somatomedin-C. The eluate from the affinity column is then subjected to isoelectric focusing as described below.

Isoelectric Focusing

Isoelectric focusing of plasma fractions is carried out on a smaller version of the flat bed tray described above. For some purposes, only a broad range isoelectric focusing is required to remove the IGF-II, but for more refined work broad cuts are made after broad range isoelectric focusing and the procedure is then repeated in a narrower gradient from 7.5 to 9.5. The fractions with activity between pH 8.0 and 8.6 are pooled, eluted, and lyophilized. They are then chromatographed on a Sephadex G-50 column to remove most of the ampholytes. It is impossible to fully remove the ampholytes by gel filtration but the residual ampholytes are readily removed by a single passage over a C_{18} high-pressure liquid chromatography column. At this stage, the somatomedins should be of a sufficiently high quality to permit iodination for use in radioligand assays or for other purposes.

Affinity Chromatography

Although for many years we have purified our radioiodinated Sm-C on an affinity column prepared from a polyclonal rabbit antiserum, the capacity of such columns was too limited to use this approach for large scale production. Recently, two other laboratories[30,31] and our own[32] have been successful in preparing monoclonal antibodies to somatomedin-C. Amplification of our monoclonal antibodies in ascites fluid followed by salt fractionation has now given us access to a large supply of nearly pure antibody. After coupling 1 to 2 mg of monoclonal antibody per ml of

[30] R. C. Baxter, S. Axiak, and R. L. Raison, *J. Clin. Endocrinol. Metab.* **54,** 474 (1982).

[31] U. K. Laubli, W. Baier, H. Binz, M. R. Celio, and R. E. Humbel, *FEBS Lett.* **149,** 109 (1982).

[32] J. J. Van Wyk, W. E. Russell, L. E. Underwood, M. E. Svoboda, G. Y. Gillespie, W. J. Pledger, E. Y. Adashi, and S. D. Balk, *in* "Human Growth Hormone" (S. Raiti, ed.). Plenum, New York, 1984 (in press).

Sepharose 4B we have been able to prepare affinity columns which are highly specific and which possess a very high capacity. These columns cannot be used to extract somatomedin-C from native plasma due to competition by serum binding proteins. If, however, the binding proteins are first removed by chromatography on Sephadex G-50 in acid, the recovery of somatomedin has been nearly quantitative. With the aid of affinity chromatography Chernausek was able to obtain 44 μg of iodination grade Sm-C from 500 ml of acromegalic plasma.[33] In his procedure the eluate from the first affinity column was subjected to chromatofocusing and the fractions with isoelectric points corresponding to somatomedin-C were then separated from contaminating ampholytes by a second passage over the affinity column. After HPLC as described in preceding sections, his preparation was essentially homogeneous, as judged by PAGE and silver staining.

Baxter[30] and Laubli[31] have also reported the successful use of anti-Sm-C/IGF-I affinity columns in their purification procedures. Because in nature somatomedin-C is always encountered in very low concentrations along with myriads of other proteins, the power of affinity chromatography to extract the somatomedin-C from these crude mixtures is indeed impressive and certain to render nonaffinity purification methods obsolete.

Acknowledgments

Judson J. Van Wyk is a recipient of a USPHS Research Career Award 5 KO6 AM14115. This work was supported by USPHS Research Grant AM01022.

[33] S. D. Chernausek, P. G. Chatelain, M. E. Svoboda, and J. J. Van Wyk, *Clin. Res.* **31,** 835A (1983).

[62] 5′-*p*-Fluorosulfonylbenzoyl Adenosine as a Probe of ATP-Binding Sites in Hormone Receptor-Associated Kinases

By SUSAN A. BUHROW and JAMES V. STAROS

One of the first measurable biochemical events which occurs upon binding of epidermal growth factor (EGF) to its plasma membrane receptor is the stimulation of a tyrosyl specific protein kinase which catalyzes

ISBN 0-12-182009-2

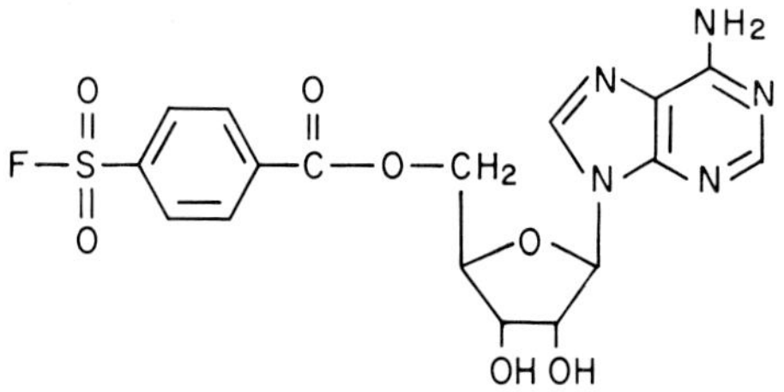

FIG. 1. The structure of 5′-*p*-FSO_2BzAdo.

the phosphorylation of a number of substrates in the cell.[1–9] In an effort to assess the physical relationship between the EGF receptor and the EGF-stimulable protein kinase, we affinity labeled the kinase site with 5′-*p*-fluorosulfonylbenzoyl adenosine (5′-*p*-FSO_2BzAdo, Fig. 1) and showed that the protein kinase site and the EGF binding site are parts of the same polypeptide chain.[10,11] Based on our evidence and that of other workers, we have proposed a model in which binding of EGF to the extracytoplasmic receptor domain of the EGF receptor/kinase allosterically stimulates the kinase on the cytoplasmic face of the membrane.[10–12] Recently, an insulin-stimulable, tyrosyl specific protein kinase activity has been reported,[13–15] and this activity appears to reside in the β-subunit of the insulin receptor.[16,17] Further, a tyrosyl specific protein kinase has been shown to be stimulated by the binding of platelet-derived growth factor to its receptor.[18] Thus, it appears that receptor-coupled protein kinases may play a role in the transmembrane signaling of receptor occupancy; identifi-

[1] G. Carpenter, L. King, Jr., and S. Cohen, *Nature (London)* **276,** 409 (1978).
[2] G. Carpenter, L. King, Jr., and S. Cohen, *J. Biol. Chem.* **254,** 4884 (1979).
[3] L. King, Jr., G. Carpenter, and S. Cohen, *Biochemistry* **19,** 1524 (1980).
[4] S. Cohen, G. Carpenter, and L. King, Jr., *J. Biol. Chem.* **255,** 4834 (1980).
[5] H. Ushiro and S. Cohen, *J. Biol. Chem.* **255,** 8363 (1980).
[6] J. A. Fernandez-Pol, *J. Biol. Chem.* **256,** 9742 (1981).
[7] T. Hunter and J. A. Cooper, *Cell* **24,** 741 (1981).
[8] R. A. Rubin, E. J. O'Keefe, and H. S. Earp, *Proc. Natl. Acad. Sci. U.S.A.* **79,** 776 (1982).
[9] R. A. Fava and S. Cohen, *J. Biol. Chem.* **259,** 2636 (1984).
[10] S. A. Buhrow, S. Cohen, and J. V. Staros, *J. Biol. Chem.* **257,** 4019 (1982).
[11] S. A. Buhrow, S. Cohen, D. L. Garbers, and J. V. Staros, *J. Biol. Chem.* **258,** 7824 (1983).
[12] J. V. Staros, S. A. Buhrow, and S. Cohen, *Biophys. J.* **41,** 198a (1983).
[13] M. Kasuga, F. A. Karlsson, and C. R. Kahn, *Science* **215,** 185 (1982).
[14] L. M. Petruzzelli, S. Ganguly, C. J. Smith, M. H. Cobb, C. S. Rubin, and O. M. Rosen, *Proc. Natl. Acad. Sci. U.S.A.* **79,** 6792 (1982).
[15] J. Avruch, R. A. Nemenoff, P. J. Blackshear, M. W. Pierce, and R. Osathanondh, *J. Biol. Chem.* **257,** 15162 (1982).
[16] M. A. Shia and P. F. Pilch, *Biochemistry* **22,** 717 (1983).
[17] R. A. Roth and D. J. Cassel, *Science* **219,** 299 (1983).
[18] B. Ek and C.-H. Heldin, *J. Biol. Chem.* **257,** 10486 (1982).

cation and characterization of these receptor-associated kinases may help elucidate the mechanisms of action of the corresponding hormones and growth factors.

5′-*p*-FSO_2BzAdo, a reagent first synthesized and utilized by Roberta Colman and co-workers,[19,20] has proven to be particularly useful in our studies of the EGF-stimulable protein kinase. The reagent stereochemically resembles ATP,[20] but it is not a substrate for enzymes which mediate phosphoryl transfer, making it stable to the phosphatases inevitably present in crude preparations. Reaction with nucleophilic groups in a target protein results in covalent incorporation of the 5′-*p*-sulfonylbenzoyl adenosine (5′-*p*-SO_2BzAdo) moiety into the protein. Nearly quantitative modification of the nucleotide binding site under study can be achieved with 5′-*p*-FSO_2BzAdo, so that inhibition of kinase activity can be correlated with the extent of modification. Although for the most part, 5′-*p*-FSO_2BzAdo has been applied to the characterization of catalytic and regulatory nucleotide binding sites of purified enzymes,[21] a number of applications have appeared in which 5′-*p*-FSO_2BzAdo[10,11,22,23] as well as the 5′-guanosine analog[24] have been successfully applied as affinity reagents in heterogeneous systems. Thus, it appears that these reagents will have utility in the identification and characterization of various receptor-associated nucleotide binding sites.

Materials

EGF[24a] and membrane vesicles from A-431 cells[25] were prepared by published procedures. 5′-*p*-FSO_2BzAdo was prepared according to Wyatt and Colman[26] by the condensation of adenosine (Sigma) and *p*-fluorosulfonylbenzoyl chloride (Sigma). Its purity was assessed by chromatographic and spectroscopic techniques as suggested by the authors.[19,26] *p*-Fluorosulfonylbenzoic acid was synthesized from *p*-chlorosulfonyl-

[19] P. K. Pal, W. J. Wechter, and R. F. Colman, *J. Biol. Chem.* **250,** 8140 (1975).
[20] R. F. Colman, P. K. Pal, and J. L. Wyatt, this series, Vol. 46, p. 240.
[21] R. F. Colman, *Annu. Rev. Biochem.* **52,** 67 (1983).
[22] J. S. Bennett, R. F. Colman, and R. W. Colman, *J. Biol. Chem.* **253,** 7346 (1978).
[23] W. R. Figures, S. Niewiarowski, T. A. Morinelli, R. F. Colman, and R. W. Colman, *J. Biol. Chem.* **256,** 7789 (1981).
[24] L. E. Limbird, S. A. Buhrow, J. L. Speck, and J. V. Staros, *J. Biol. Chem.* **258,** 10289 (1983).
[24a] C. R. Savage, Jr. and S. Cohen, *J. Biol. Chem.* **247,** 7609 (1972).
[25] S. Cohen, H. Ushiro, C. Stoscheck, and M. Chinkers, *J. Biol. Chem.* **257,** 1523 (1982).
[26] J. L. Wyatt and R. F. Colman, *Biochemistry* **16,** 1333 (1977).

benzoic acid (Eastman Chemicals) by the method of DeCat *et al.*[27] (Both 5′-*p*-FSO_2BzAdo and *p*-fluorosulfonylbenzoic acid are now listed as commercially available.[27a]) ENHANCE and 5′-*p*-FSO_2Bz[^{14}C]Ado (47 mCi/mmol) were purchased from New England Nuclear. [γ-^{32}P]ATP was purchased from ICN. Water for these experiments was deionized, then glass distilled.

Modification of Membrane Vesicles by 5′-p-FSO_2BzAdo

Due to its hydrophobicity, 5′-*p*-FSO_2BzAdo must be dissolved in an organic solvent prior to its addition to the aqueous labeling mixture.[20] Suitable solvents include ethanol and *N*,*N*′-dimethylformamide (DMF), but it is important to determine that the carrier solvent does not affect the activity of the enzyme under examination. 5′-*p*-FSO_2BzAdo is soluble at higher concentrations in DMF than in ethanol, but is less stable in DMF and therefore must be dissolved immediately prior to use.[20] We find that a stock solution of 5′-*p*-FSO_2BzAdo in absolute or 95% ethanol at a concentration of 0.4 m*M* is stable for at least several months when stored at −60° under N_2.

For the labeling of the EGF-stimulable protein kinase, an aliquot of a stock solution of 5′-*p*-FSO_2BzAdo or 5′-*p*-FSO_2Bz[^{14}C]Ado in 95% ethanol was evaporated to dryness under a stream of dry N_2. DMF was added so that the final concentration of DMF in the labeling mix would be 4%, a concentration which does not affect protein kinase activity in this system.[10] A suspension of A431 membrane vesicles (21 mg/ml of protein, final concentration) containing 1 m*M* $MnCl_2$, 15 m*M* HEPES, pH 7.4, with or without 0.10 mg/ml (final concentration) EGF was added to the 5′-*p*-FSO_2BzAdo solution. For protection experiments, the vesicles were preincubated 10–20 min at room temperature with the appropriate nucleoside or nucleoside derivative prior to modification. Although the EGF-stimulable protein kinase activity does decay somewhat during incubation at room temperature,[10] all labeling experiments were performed at that temperature due to favorable kinetics of modification.

It is important that the buffer conditions during enzyme modification be compatible with the chemistry of 5′-*p*-FSO_2BzAdo. Nucleophilic buffers such as Tris or glycine or sulfhydryl-stabilizing reagents such as dithiothreitol or 2-mercaptoethanol may accelerate the rate of hydrolysis of the fluorosulfonyl moiety of 5′-*p*-FSO_2BzAdo or may themselves react with it.[20]

[27] A. DeCat, R. Van Poucke, and M. Verbrugghe, *J. Org. Chem.* **30,** 1498 (1965).
[27a] "Chem Sources—U.S.A." Directories Publishing Co., Ormond Beach, FL., 1984.

Termination of Vesicle Modification

Aliquots of vesicles which were examined for total incorporation of 5′-*p*-FSO_2Bz[^{14}C]Ado were precipitated onto Whatman 3MM filter papers with cold trichloroacetic acid as previously described for the phosphorylation assay.[2] Although the ester linkage of free 5′-*p*-FSO_2BzAdo is susceptible to both acid- and alkaline-catalyzed hydrolysis,[20] we find that the covalently bound 5′-*p*-SO_2BzAdo has enhanced stability to acid hydrolysis and is largely unaffected by the conditions employed here. Following precipitation, the filters were dried and then were treated with 0.5 ml 1 *N* NaOH for 10 min to solubilize the radioactivity. Scintillation fluid (10 ml) was added and the solution was neutralized with acetic acid prior to counting. A representative time course of incorporation of 5′-*p*-FSO_2Bz-[^{14}C]Ado (0.04 m*M*) into protein is shown in Fig. 2.

Vesicles to be analyzed by gel electrophoresis for 5′-*p*-SO_2Bz-[^{14}C]Ado-labeled proteins were washed several times with 15 m*M* HEPES, pH 7.4, resuspended with heating in 2% sodium dodecyl sulfate, 15 m*M* HEPES, pH 7.4 to a final protein concentration of 17 mg/ml and then diluted with 1/3 vol of gel dissolution buffer[10] and heated 3 min at 100°. The samples were frozen at −60° until needed.

Vesicles which were to be analyzed for inhibition of kinase activity were diluted 30-fold into phosphorylation assay buffer at 0°, effectively terminating modification of the vesicles.

Modification of vesicles which were to be used for further purification of 5′-*p*-SO_2Bz[^{14}C]Ado-labeled receptor was terminated by solubilization of the vesicles in Triton X-100, glycerol, and HEPES, pH 7.4 to a final concentration of 5%, 10%, and 20 m*M*, respectively, and a final protein concentration of 6–7 mg/ml.[25]

Analysis of Proteins Modified by 5′-p-FSO_2BzAdo

Identification of the proteins modified by 5′-*p*-FSO_2BzAdo requires separation of the proteins by some means and detection of those proteins containing covalently bound 5′-*p*-SO_2BzAdo. Membrane vesicles were modified using various concentrations of 5′-*p*-FSO_2Bz[^{14}C]Ado as described above and the reactions were terminated by the addition of gel electrophoresis sample buffer.[10] Proteins were separated by sodium dodecyl sulfate–polyacrylamide gel electrophoresis[10] by the method of Laemmli.[28] The gels were fixed and stained[10] and photographed while

[28] U. K. Laemmli, *Nature (London)* **227,** 680 (1970).

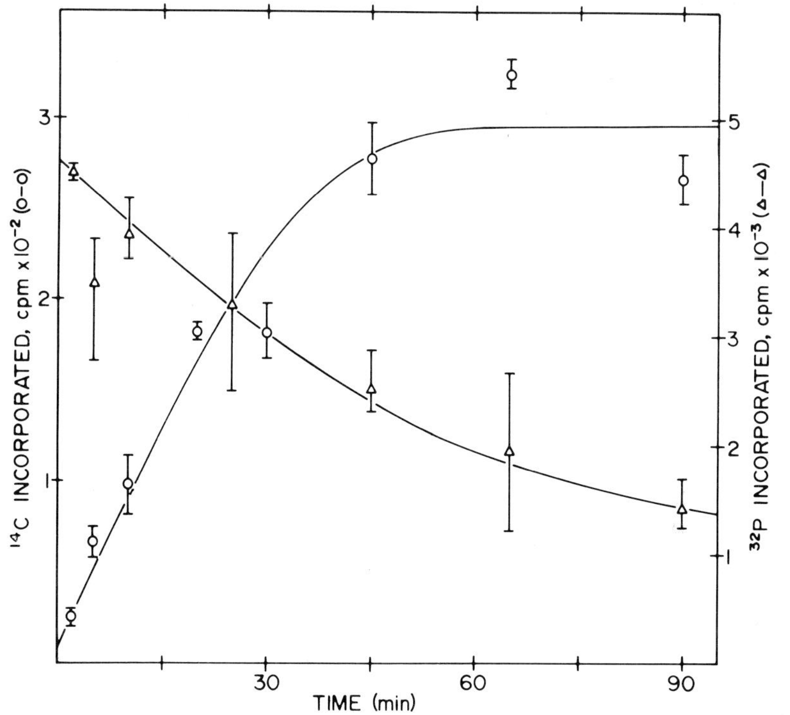

FIG. 2. Correlation of labeling of A431 vesicles by 5′-p-$FSO_2Bz[^{14}C]Ado$ with inhibition of EGF-stimulated kinase activity. A431 membrane vesicles (21 mg/ml of protein) were incubated at room temperature in the presence of 5′-p-$FSO_2Bz[^{14}C]Ado$ or unlabeled 5′-p-FSO_2BzAdo as described in the text. At the indicated times, aliquots were removed from each mixture. Phosphorylation assays were performed on vesicles modified with unlabeled 5′-p-FSO_2BzAdo as described. Aliquots which were modified with 5′-p-$FSO_2Bz[^{14}C]Ado$ were precipitated onto filters in cold 10% trichloroacetic acid and the radioactivity was determined. Symbols indicate the mean, and error bars, the range of values obtained. (Reproduced from *The Journal of Biological Chemistry* with permission.[10])

wet. The stained gels were treated with ENHANCE according to the directions supplied by the manufacturer, dried, and subjected to autoradiography for 15–25 days on Kodak SB-5 film at −60°. Identification of the labeled bands was accomplished by overlaying the autoradiograph onto the dried stained gel. On this basis, we selected a concentration of 5′-p-$FSO_2Bz[^{14}C]Ado$ which had minimal nonspecific labeling of proteins but which still gave good inhibition of the EGF-stimulable protein kinase (Fig. 3, lane 1). We used this concentration (0.04 m*M*) in subsequent modification experiments.

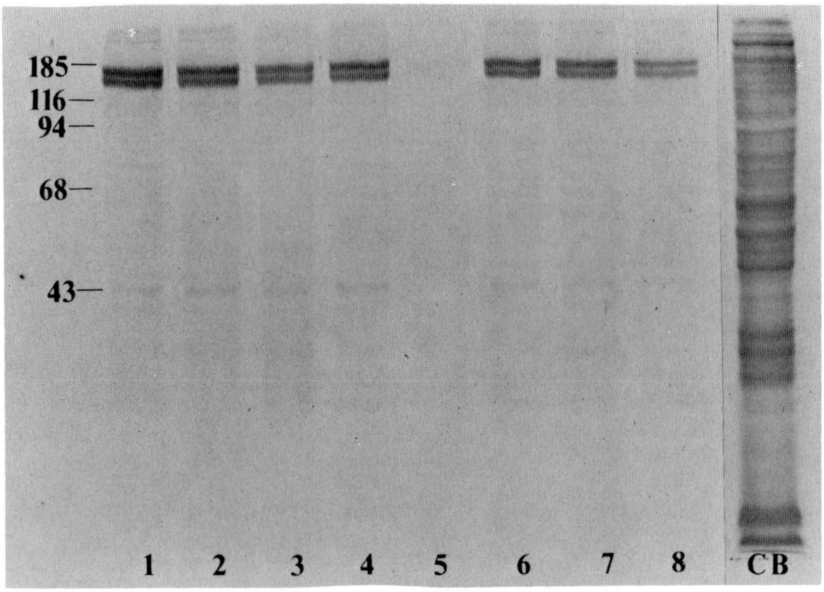

FIG. 3. Effect of preincubation with various nucleoside and nucleoside derivatives on affinity labeling of the EGF receptor/kinase by 0.04 m*M* 5′-*p*-FSO_2Bz[^{14}C]Ado. Membrane vesicles (21 mg/ml of protein) prepared from A431 cells were incubated for 20 min at room temperature with 2 m*M* of the nucleoside or nucleoside derivative indicated below prior to affinity labeling of the vesicles. Affinity labeling was carried out for 90 min at room temperature in the presence of 0.1 mg/ml EGF as described in the text. The vesicles were washed with 15 m*M* HEPES, pH 7.4, solubilized, and subjected to gel electrophoresis and autoradiography. The lanes on the autoradiograph are numbered. The lane labeled CB is a photograph of lane 8 of the Coomassie blue-stained gel from which the autoradiograph was made. The Coomassie blue staining patterns of lanes 1–8 were indistinguishable. Lane 1, no nucleoside; lane 2, adenosine; lane 3, AMP; lane 4, ADP; lane 5, AMP-PNP; lane 6, NAD; lane 7, GTP; lane 8, ATP. The numbers to the left of lane 1 are the $M_r \times 10^{-3}$ of molecular weight standards.

Specificity of Modification of Proteins by 5′-p-FSO_2BzAdo

For a chemical modification reagent to be classified as an affinity label, one must be able to demonstrate that ligands which act as natural substrates of the enzyme under examination can protect against that modification.[29] We examined the ability of a variety of nucleosides and nucleoside derivatives to protect against the 5′-*p*-FSO_2Bz[^{14}C]Ado labeling of proteins in A431 vesicles (Fig. 3, lane 1). A431 membrane vesicles (21 mg/ml of protein) were incubated for 20 min in the absence of nucleoside (lane

[29] F. Wold, this series, Vol. 46, p. 3.

1) or with 2 mM adenosine (lane 2), AMP (lane 3), ADP (lane 4), adenyl-5′-yl-imidodiphosphate (AMP-PNP), a hydrolysis-resistant ATP analog[30] (lane 5), NAD (lane 6), GTP (lane 7), or ATP (lane 8) in the presence of 0.1 mg/ml EGF, 15 mM HEPES, pH 7.4, 1 mM $MnCl_2$ at room temperature. The vesicles were then added to a solution of 5′-p-FSO$_2$Bz[^{14}C]Ado in DMF to a final reagent concentration of 0.04 mM to effect labeling as described above.

The labeled proteins were separated by gel electrophoresis and subjected to autoradiography as described above. As can be observed from the autoradiograph (Fig. 3), only AMP-PNP and ATP afforded any significant protection against the labeling of the proteins of M_r = 170,000 and 150,000. These proteins have the same M_r as the EGF receptor and an enzymatically active proteolytic degradation product of the receptor.[25,31,32] The protection data suggest that modification of these proteins occurs at one or more ATP binding sites.

Characterization of the Extent of Kinase Modification by 5′-p-FSO$_2$BzAdo

In order to establish that the protein of M_r = 170,000 which we have modified with 5′-p-FSO$_2$BzAdo is the EGF-stimulable protein kinase, it is necessary to demonstrate a change in the activity of that enzyme correlated with labeling. The EGF-stimulable protein kinase phosphorylates a number of membrane-associated proteins, most notably the EGF receptor itself,[3,4] as well as synthetic peptide substrates resembling the site of phosphorylation in the transforming protein of Rous sarcoma virus.[33,34] We quantitated the extent of modification of the kinase by 5′-p-FSO$_2$BzAdo by performing phosphorylation assays using membrane vesicles alone to examine the effect of modification on phosphorylation of endogenous substrates[10] and by using a synthetic decapeptide, Asp-Ala-Glu-Tyr-Ala-Ala-Arg-Arg-Arg-Gly ($D_{10}G_1$), to examine the effect of modification on phosphorylation of an exogenous tyrosine-containing substrate.[11]

Aliquots (2 μl; 40–70 μg of protein) of membrane vesicles which had been incubated with the appropriate concentration of 5′-p-FSO$_2$BzAdo or DMF (as a control) were added to 58 μl of a phosphorylation mix containing 15 μM [γ-^{32}P]ATP (1 μCi), 15 mM HEPES, pH 7.4, 1 mM $MnCl_2$. The

[30] R. G. Yount, D. Babcock, W. Ballantyne, and D. Ojala, *Biochemistry* **10,** 2484 (1971).

[31] R. E. Gates and L. E. King, Jr., *Mol. Cell. Endocrinol.* **27,** 263 (1982).

[32] D. Cassel and L. Glaser, *J. Biol. Chem.* **257,** 9845 (1982).

[33] C. Erneux, S. Cohen, and D. L. Garbers, *J. Biol. Chem.* **258,** 4137 (1983).

[34] L. J. Pike, H. Marquardt, G. J. Todaro, B. Gallis, J. E. Casnellie, P. Bornstein, and E. G. Krebs, *J. Biol. Chem.* **257,** 14628 (1982).

suspension was incubated for 2 or 10 min and the reaction was terminated by precipitation of a 50 μl aliquot onto a 2 × 2 cm Whatman 3MM filter paper which was immediately dropped into cold 10% trichloroacetic acid containing 1 m*M* sodium pyrophosphate. The filters were washed sequentially in the trichloroacetic acid–sodium pyrophosphate solution, ethanol, and ether, then dried and counted as previously described.[2] An example of the effect of modification of A431 membrane vesicles by 0.04 m*M* 5′-*p*-FSO_2BzAdo on endogenous kinase activity is shown in Fig. 2. At this concentration, 5′-*p*-FSO_2BzAdo gives half maximal inhibition of endogenous kinase activity after 45–50 min. The rate of kinase inactivation as measured by this assay is dependent on the concentration of 5′-*p*-FSO_2BzAdo used, up to a limiting concentration.

Phosphorylation of the synthetic peptide substrate was a modification of published procedures.[33,34] Phosphorylation was carried out in the presence of Zn^{2+} or VO_4^{3-} to inhibit endogenous phosphatase activity.[35–37] Membrane vesicles (2 μl, 0.06 mg of protein) which had been incubated in the absence of EGF with or without 0.04 m*M* 5′-*p*-FSO_2BzAdo, were added to 48 μl of phosphorylation mix containing 5 m*M* $D_{10}G_1$, 10 μM $Zn(CH_3CO_2)_2$, 15 m*M* HEPES, pH 7.4, 2 m*M* $MnCl_2$, 50 μM [γ-^{32}P]ATP (1 μCi), 0.2% Nonidet P-40 with or without 0.30 μM EGF.[34] The phosphorylated peptide was separated from the trichloroacetic acid insoluble material and the ^{32}P incorporated was measured according to Erneux *et al.*[33] The table shows the results of such an experiment, demonstrating that 5′-*p*-FSO_2BzAdo inhibits the A431 vesicle-catalyzed phosphorylation of an exogenous peptide which acts as a substrate for tyrosyl specific kinases. Similar results were obtained when VO_4^{3-} was employed as a phosphatase inhibitor.[11] These results verify that one of the proteins which is modified by 5′-*p*-FSO_2BzAdo (Fig. 3, lane 1), is the EGF-stimulable protein kinase.

Identification of EGF-Stimulable Protein Kinase as the EGF Receptor

To ascertain the identity of the proteins of M_r = 170,000 and 150,000 which are modified by 5′-*p*-FSO_2BzAdo, we utilized an affinity chromatography step in which the EGF receptor and the kinase activity are known to copurify.[3,4,25] A431 membrane vesicles which had been modified with 0.04 m*M* 5′-*p*-FSO_2Bz[^{14}C]Ado were solubilized for affinity chroma-

[35] G. Swarup, S. Cohen, and D. L. Garbers, *Biochim. Biophys. Res. Commun.* **107,** 1104 (1982).

[36] G. Swarup, K. V. Speeg, Jr., S. Cohen, and D. L. Garbers, *J. Biol. Chem.* **257,** 7298 (1982).

[37] D. L. Brautigan, P. Borstein, and B. Gallis, *J. Biol. Chem.* **256,** 6519 (1981).

INHIBITION OF A431 VESICLE-CATALYZED PHOSPHORYLATION OF $D_{10}G_1$ BY 0.04 m*M* 5′-*p*-FSO_2BzAdo IN THE PRESENCE OF $Zn(CH_3CO_2)_2$

Time of preincubation[b] (min)	^{32}P incorporated into peptide[a]: 4% DMF			0.04 m*M* 5′-*p*-FSO_2BzAdo in 4% DMF		
	−EGF (cpm)	+EGF (cpm)	Stimulation	−EGF (cpm)	+EGF (cpm)	Stimulation
15	3174	18856	5.9×	1284 (60)[c]	5507 (71)	4.3×
30	2583	13421	5.2×	683 (74)	2573 (81)	3.8×
60	2116	14781	6.8×	317 (85)	856 (94)	2.7×
120	1615	10607	6.6×	243 (85)	284 (97)	1.2×

[a] The phosphorylation of $D_{10}G_1$ by A431 membrane vesicles in the presence of $Zn(CH_3CO_2)_2$ was measured as described in the text.

[b] A431 membrane vesicles were preincubated with DMF or 0.04 m*M* 5′-*p*-FSO_2BzAdo in DMF at room temperature for the indicated time prior to the assay for kinase activity.

[c] Numbers in parentheses represent the percentage inhibition of kinase activity compared with control values at that time of incubation.

tography as above. EGF receptor was purified by chromatography on EGF-agarose as previously described.[4,25] The eluate was dissolved for gel electrophoresis as described above. Examination of the Coomassie blue-stained gel reveals an enrichment of the receptor proteins of M_r = 170,000 and 150,000, as well as the presence of small amounts of lower molecular weight proteins. Since we know that the kinase activity is purified by this technique[3,4,25] and we know from the inhibition data above that the kinase is one of the proteins modified by 5′-*p*-FSO_2BzAdo, any 5′-*p*-FSO_2Bz[^{14}C]Ado-labeled protein which is purified by this affinity chromatographic step may be the kinase. The autoradiograph of the gel shows that only one protein carries any detectable label, and that is the receptor protein of M_r = 170,000 or 150,000.[11] In a similar way, immunoprecipitation of the receptor with anti-receptor antiserum[38] under conditions which precipitate the kinase activity brings down 5′-*p*-FSO_2Bz[^{14}C]Ado-labeled proteins which appear to be the same as the M_r = 150,000 and 170,000 proteins purified by EGF affinity chromatography.[39] These results demonstrate the coidentity of the EGF receptor and EGF-stimulable protein kinase.

[38] C. M. Stoscheck and G. Carpenter, *Arch. Biochem. Biophys.* **227,** 457 (1983).

[39] A. M. Soderquist and G. Carpenter, personal communication.

Conclusions

Affinity labeling of nucleoside binding enzymes such as kinases can provide a method of identification of these proteins in heterogeneous mixtures. We have used the affinity reagent 5′-*p*-FSO_2BzAdo to identify the protein kinase which is stimulated following the addition of EGF to its target cells. We find that the receptor protein of M_r = 170,000 or 150,000 is modified by 5′-*p*-FSO_2BzAdo at an ATP binding site[10] and that this modification is correlated with a decrease in kinase activity toward both endogenous substrates[10] and an exogenous peptide substrate.[11] When the EGF receptor is purified from membrane preparations labeled with the kinase affinity label under conditions which result in the copurification of the EGF binding activity and the EGF-stimulable protein kinase activity, all of the detectable kinase affinity label is in the receptor protein. Thus, we conclude the EGF receptor and the EGF-stimulable protein kinase reside in the same polypeptide chain. Consistent with this conclusion, we[10–12] have proposed a model in which the EGF receptor/kinase is a transmembrane protein in the plasma membrane of target cells, the receptor site exposed to the extracytoplasmic milieu and the kinase site accessible to the cytoplasm (Fig. 4). Binding of EGF to the receptor allosteri-

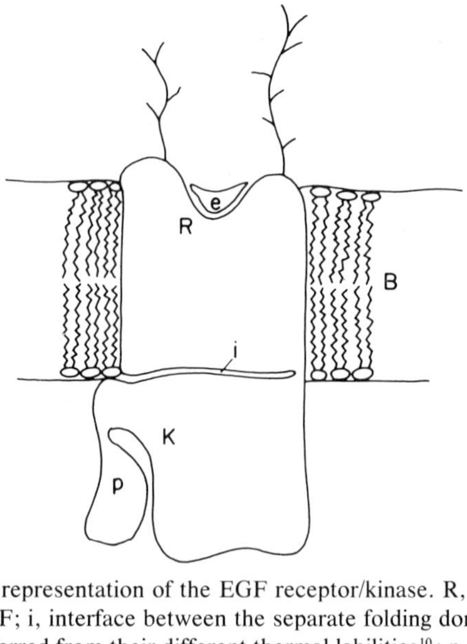

FIG. 4. Schematic representation of the EGF receptor/kinase. R, Receptor domain; K, kinase domain; e, EGF; i, interface between the separate folding domains for the receptor and for the kinase, inferred from their different thermal labilities[10]; p, fragment removed by the action of an endogenous Ca^{2+}-dependent protease,[31,32] resulting in a reduction of M_r from 170,000 to 150,000; B, lipid bilayer. See Conclusion for a discussion of this model.

cally activates the kinase, thus providing a transmembrane signal of hormone binding. Recent results from other laboratories[13–18] suggest that such a model for transmembrane signaling may be applicable for other hormone and growth factor systems as well.

Acknowledgment

Support from the National Institutes of Health (R01 AM25489 and T32 AM07061) is gratefully aknowledged.

[63] A Radioimmunoassay for Cyclic GMP with Femtomole Sensitivity Using Tritiated Label and Acetylated Ligands

By PAULINE R. M. DOBSON and PHILIP G. STRANGE

Principle

A number of assay systems exist for the measurement of cyclic GMP in biological samples, most based on the principle of saturation analysis. These include a competitive binding assay relying on either lobster muscle[1] or silk moth pupae[2] as the source of binding protein. However, these assays somewhat lack sensitivity and specificity, and the binding agents are considerably more difficult to come by than those for the comparable cyclic AMP assays.[3,4] On the other hand, the radioimmunoassay of cyclic GMP based upon an antibody raised against a succinyl derivative of cyclic GMP conjugated to a polypeptide or protein[5] showed a marked increase in sensitivity and specificity. Succinylcyclic GMP-tyrosine [^{125}I]iodomethyl ester is used as label in this system. Later the sensitivity of this assay was further increased by succinylating the standards and samples[6] to resemble the succinyl derivative used as antigen. However, this method suffered

[1] F. Murad, V. Manganiello, and M. Vaughan, *Proc. Natl. Acad. Sci. U.S.A.* **68,** 736 (1971).
[2] A. M. Fallon and G. R. Wyatt, *Anal. Biochem.* **63,** 614 (1975).
[3] B. L. Brown, R. P. Ekins, and W. Tampion, *Biochem. J.* **120,** 8p (1970).
[4] A. G. Gilman, *Proc. Natl. Acad. Sci. U.S.A.* **67,** 305 (1970).
[5] A. L. Steiner, C. W. Parker, and D. M. Kipnis, *J. Biol. Chem.* **247,** 1106 (1972).
[6] H. L. Cailla, M. S. Racine-Weisbuch, and M. A. Delaage, *Anal. Biochem.* **56,** 394 (1973).

ISBN 0-12-182009-2

from the disadvantage that one of the reaction products, succinate, interfered with the assay. Subsequently, it was demonstrated that acetylation of standards and samples[7,8] could be substituted for succinylation and did not interfere with the antibody–antigen reaction. Hitherto, radioimmunoassays using [^{3}H]cyclic GMP have shown relatively low sensitivity probably due to the lower affinity of the antibody for this unsubstituted nucleotide and the lower specific activity of the label.

Here we describe an economical assay using [^{3}H]cyclic GMP which will be of use to those laboratories either restricted from using γ isotopes or who do not possess gamma-counting facilities. In this assay, label, standards, and samples are acetylated, and interacted with an antibody raised against the succinyl derivative of cyclic GMP prepared in the usual way.[5] Using this assay procedure 20 fmol of cyclic GMP are detectable. This assay for cyclic GMP has been reported in abstract form.[9] The optimal assay conditions were calculated on theoretical grounds.[10–12] The use of acetylated ligands in this way might also be applied to the assay of cyclic AMP.[13]

Preparation of Cell Extracts for Determination of Cyclic Nucleotides

1. The cyclic nucleotides are extracted with ice-cold 65% ethanol (see notes below).
2. These samples are transferred to −20° overnight.
3. The extracts are drawn off into test tubes. The remaining precipitates are washed with ice-cold 65% ethanol and the washes added to the appropriate test tubes.
4. The extracts are centrifuged (700 *g*/15 min). The clear supernatants are transferred from the precipitates to fresh tubes.
5. The extracts are evaporated either in a stream of compressed air or nitrogen, or in a vacuum oven.
6. The dried extracts are dissolved in deionised water (usually 0.5–1 ml/10^5 cells).

[7] J. F. Harper and G. Brooker, *J. Cyclic Nucleotide Res.* **1,** 207 (1975).
[8] E. K. Frandsen and G. Krishna, *Life Sci.* **18,** 529 (1976).
[9] P. R. M. Dobson and P. G. Strange, *Adv. Cyclic Nucleotide Res.* **14,** 286 (1980).
[10] R. P. Ekins and B. Newman, *Steroid Assay Protein Binding, Karolinska Symp. Res. Methods Reprod. Endocrinol., 2nd,* p. 11 (1970).
[11] B. L. Brown, R. P. Ekins, and J. D. M. Albano, *Adv. Cyclic Nucleotide Res.* **2,** 25 (1972).
[12] R. P. Ekins, *Br. Med. Bull.* **30,** 3 (1974).
[13] T. Skomedal, B. Grynne, J. B. Osnes, A. E. Sjetnan, and I. Oye, *Acta Pharmacol. Toxicol.* **46,** 200 (1980).

Notes

Cell Monolayers

The serum-free medium is drawn off at the end of the experimental incubation (this medium can be stored for hormone assay). Immediately ice-cold 65% ethanol (1 ml) is added to the monolayers. The culture dish is transferred to −20°.

Cell Suspensions

Ice-cold absolute ethanol is added directly to the cell suspensions contained in serum-free medium in the proportion ethanol : medium = 2 : 1. Samples are transferred to −20° as before.

Protein Estimation in Cell Monolayers

The joint precipitates from each wash are dissolved in 0.1 *N* NaOH and estimated for protein content.

Assay Materials

[8-^{3}H]Guanosine 3′,5′-cyclic phosphate ([^{3}H]cyclic GMP), specific activity 16–21 Ci/mmol, was obtained from Amersham International plc, Bucks., England. Guanosine 3′,5′-cyclic monophosphate and bovine serum albumin (BSA), RIA grade, were purchased from Sigma Chemical Co., Poole, Dorset, England. GSX Charcoal was purchased from Norit Clydesdale, Glasgow, Scotland.

Assay Reagents

Buffers

A. 75 m*M* acetate buffer, pH 4.8, containing 12 m*M* theophylline and 1.65% BSA
B. 50 m*M* acetate buffer, pH 4.8, containing 8 m*M* theophylline and 3% BSA

[^{3}H]Cyclic GMP

Label is diluted with 50% ethanol to 10 μCi/ml (0.5 nmol/ml) and stored in aliquots of 200 μl at −20°. Just prior to use, an aliquot is taken to dryness and dissolved in 1 ml water.

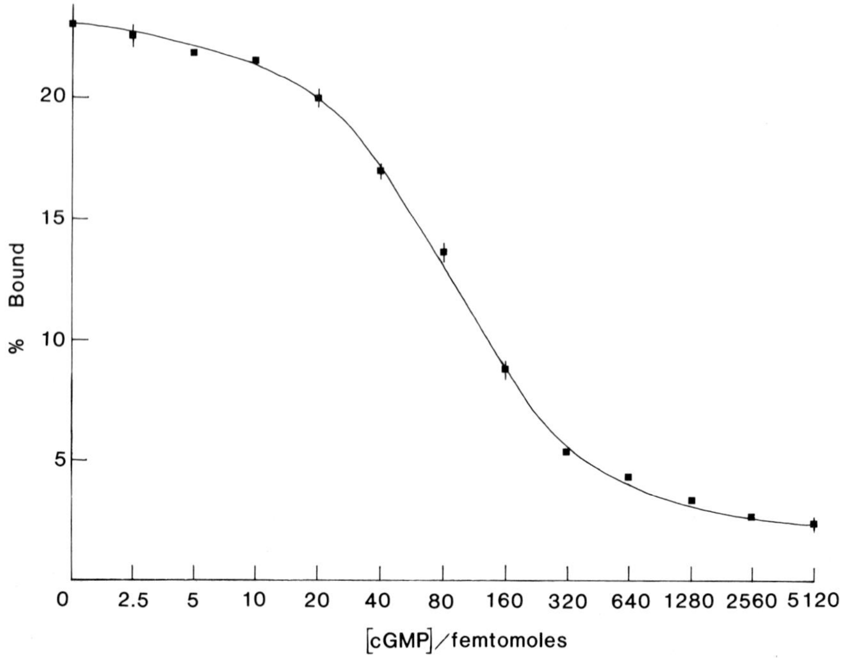

FIG. 1. Standard response curve for the radioimmunoassay of cyclic GMP using acetylated [^{3}H]cyclic GMP (80–100 fmol per tube) and acetylated unlabeled cyclic GMP. These were incubated with antibody (binding site concentration approximately 40 fmol/tube) in a volume of 150 μl for 16 hr at 4°. A charcoal suspension was used for separation of free and bound moieties.

Anti-Cyclic GMP Serum

Antisera are raised against succinylated cyclic GMP coupled to albumin according to the method of Steiner *et al.*[5] and characterized for cross-reactivity.[14] The antisera are aliquoted and stored at −80°. When required an aliquot is diluted appropriately with buffer A to give 20–25% binding of the added [^{3}H]cyclic GMP in the absence of unlabeled cyclic GMP.

Cyclic GMP Standards

A stock standard of 5 m*M* cyclic GMP is prepared in water and aliquots of 100 μl are stored at −20°. When required, one in two serial dilutions of standard are prepared in water over the range 200 fmol/ml to

[14] G. Brooker, J. F. Harper, W. L. Terasaki, and R. D. Moylan, *Adv. Cyclic Nucleotide Res.* **10,** 1 (1979).

100 pmol/ml (i.e., 10 fmol/50 μl to 5 pmol/50 μl). A zero standard (water only) is included.

Acetylation Reagents

Triethylamine
Acetic anhydride

Charcoal Suspension

A suspension of charcoal is prepared by adding GSX charcoal to buffer B to a final concentration of 20%. The suspension is kept stirred on ice for 10 min before use.

Methods

Samples, cyclic GMP standard, and [^{3}H]cyclic GMP prepared as above are acetylated by the addition of 20 μl triethylamine and 10 μl acetic anhydride to 1 ml solution. (N.B. The solutions should be mixed thoroughly after each addition[8,14].) [^{3}H]Cyclic GMP is then diluted in buffer A to 1.5–2 pmol/1 ml.

The reagents are added to test tubes (6 $\times$ 1 cm) as shown in the following tabulation:

	Totals	Blank (Vol/μl)	Standard (Vol/μl)	Sample (Vol/μl)
[^{3}H]Cyclic GMP	50	50	50	50
Standard	—	—	50	—
Sample	—	—	—	50
Antibody	—	—	50	50
Water	—	50	—	—
Buffer	200	50	—	—

After a suitable incubation period to establish equilibrium, 100 μl charcoal suspension is added to each tube excluding totals. The assay tubes are centrifuged (700 *g*/15 min) and an aliquot of the supernatant is transferred to scintillation fluid and counted.

The standard values are plotted as percentage bound vs concentration of ligand (see Fig. 1).

Acknowledgment

We would like to express our thanks to Dr. B. L. Brown for many useful discussions relating to this work. We are grateful to the Medical Research Council for financial assistance.

Author Index

Numbers in parentheses are footnote reference numbers and indicate that an author's work is referred to although the name is not cited in the text.

A

B

C

D

E

F

G

I

J

K

L

M

N

Q

R

S

Y

Z

Subject Index

A

B

C

D

E

F

H

- Histamine, effect on intracellular cAMP, in primary glial cell cultures, 342

I

N

O

P

R

S

T

U

V

W

Y